Dr Lindsay Sawyer
Department of Biochemistry
University of Edinburgh
George Square
Edinburgh EH8 9XD UK
Tel. 031-667-1011 ext. 2363

D1486000

COMPREHENSIVE
MEDICINAL CHEMISTRY

IN 6 VOLUMES

EDITORIAL BOARD

Chairman

C. Hansch
Pomona College
Claremont, CA, USA

Joint Executive Editors

P. G. Sammes
Brunel University of West London
Uxbridge, UK

J. B. Taylor
Rhône-Poulenc Ltd
Dagenham, UK

J. C. Emmett
Dialog in Science
Kensworth, UK

P. D. Kennewell
Roussel Laboratories Ltd
Swindon, UK

C. A. Ramsden
Rhône-Poulenc Ltd
Dagenham, UK

INTERNATIONAL ADVISORY BOARD

A. Albert
Australian National University
Canberra, Australia

N. Anand
Central Drug Research Institute
Lucknow, India

E. H. Ariëns
University of Nijmegen
The Netherlands

A. Burger
University of Virginia
Charlottesville, VA, USA

K. Folkers
University of Texas
Austin, TX, USA

G. H. Hitchings
Burroughs Wellcome
North Carolina, USA

Huang Liang
Institute of Materia Medica
Beijing, People's Republic of China

P. Janssen
Janssen Pharmaceutica, Beerse, Belgium

T. Kametani
Hoshi University, Tokyo, Japan

E. Mutschler
Universität Frankfurt, FRG

W. Nagata
Shionogi & Co., Hyogo, Japan

P. S. Portoghese
University of Minnesota
Minneapolis, MN, USA

P. Pratesi
Università di Milano, Italy

R. Schwyzer
Eidgenössische Technische Hochschule
Zürich, Switzerland

C. G. Wermuth
Université Louis Pasteur
Strasbourg, France

COMPREHENSIVE
MEDICINAL CHEMISTRY

The Rational Design, Mechanistic Study & Therapeutic
Applications of Chemical Compounds

Chairman of the Editorial Board
CORWIN HANSCH
Pomona College, Claremont, CA, USA

Joint Executive Editors
PETER G. SAMMES
Brunel University of West London, Uxbridge, UK

JOHN B. TAYLOR
Rhône-Poulenc Ltd, Dagenham, UK

Volume 4
QUANTITATIVE DRUG DESIGN

Volume Editor
CHRISTOPHER A. RAMSDEN
Rhône-Poulenc Ltd, Dagenham, UK

PERGAMON PRESS

Member of the Maxwell Macmillan Pergamon Publishing Corporation
OXFORD • NEW YORK • BEIJING • FRANKFURT
SÃO PAULO • SYDNEY • TOKYO • TORONTO

U.K.	Pergamon Press plc, Headington Hill Hall, Oxford OX3 0BW, England
U.S.A.	Pergamon Press, Inc., Maxwell House, Fairview Park, Elmsford, New York 10523, U.S.A.
PEOPLE'S REPUBLIC OF CHINA	Pergamon Press, Room 4037, Qianmen Hotel, Beijing, People's Republic of China
FEDERAL REPUBLIC OF GERMANY	Pergamon Press GmbH, Hammerweg 6, D-6242 Kronberg, Federal Republic of Germany
BRAZIL	Pergamon Editora Ltda, Rua Eça de Queiros, 346, CEP 04011, Paraiso, São Paulo, Brazil
AUSTRALIA	Pergamon Press Australia Pty Ltd., P.O. Box 544, Potts Point, N.S.W. 2011, Australia
JAPAN	Pergamon Press, 5th Floor, Matsuoka Central Building, 1-7-1 Nishishinjuku, Shinjuku-ku, Tokyo 160, Japan
CANADA	Pergamon Press Canada Ltd., Suite No. 241, 253 College Street, Toronto, Ontario, Canada M5T 1R5

Copyright © 1990 Pergamon Press plc

All rights reserved. No part of this publication may be reproduced, stored in any retrieval system or transmitted in any form or by any means: electronic, electrostatic, magnetic tape, mechanical, photocopying, recording or otherwise, without permission in writing from the publishers.

First edition 1990

Library of Congress Cataloging in Publication Data

Comprehensive medicinal chemistry: the rational design, mechanistic study & therapeutic applications of chemical compounds/ chairman of the editorial board, Corwin Hansch; joint executive editors, Peter G. Sammes, John B. Taylor. — 1st ed.
p. cm.
Includes index.
1. Pharmaceutical chemistry. I. Hansch, Corwin. II. Sammes, P. G. (Peter George) III. Taylor, J. B. (John Bodenham), 1939– .
[DNLM: 1. Chemistry, Pharmaceutical. QV 744 C737]
RS402.C65
615'.19—dc20
DNLM/DLC   89–16329

British Library Cataloguing in Publication Data

Hansch, Corwin
Comprehensive medicinal chemistry
1. Pharmaceutics
I. Title
615'.19

ISBN 0–08–037060–8 (Vol. 4)
ISBN 0–08–032530–0 (set)

Printed in Great Britain by BPCC Hazell Books Ltd

Contents

Introduction to Drug Design and Molecular Modelling

Quantitative Description of Physicochemical Properties of Drug Molecules

Preface

Medicinal chemistry is a subject which has seen enormous growth in the past decade. Traditionally accepted as a branch of organic chemistry, and the near exclusive province of the organic chemist, the subject has reached an enormous level of complexity today. The science now employs the most sophisticated developments in technology and instrumentation, including powerful molecular graphics systems with 'drug design' software, all aspects of high resolution spectroscopy, and the use of robots. Moreover, the medicinal chemist (very much a new breed of organic chemist) works in very close collaboration and mutual understanding with a number of other specialists, notably the molecular biologist, the genetic engineer, and the biopharmacist, as well as traditional partners in biology.

Current books on medicinal chemistry inevitably reflect traditional attitudes and approaches to the field and cover unevenly, if at all, much of modern thinking in the field. In addition, such works are largely based on a classical organic structure and therapeutic grouping of biologically active molecules. The aim of *Comprehensive Medicinal Chemistry* is to present the subject, the modern role of which is the understanding of structure–activity relationships and drug design from the mechanistic viewpoint, as a field in its own right, integrating with its central chemistry all the necessary ancillary disciplines.

To ensure that a broad coverage is obtained at an authoritative level, more than 250 authors and editors from 15 countries have been enlisted. The contributions have been organized into five major themes. Thus Volume 1 covers general principles, Volume 2 deals with enzymes and other molecular targets, Volume 3 describes membranes and receptors, Volume 4 covers quantitative drug design, and Volume 5 discusses biopharmaceutics. As well as a cumulative subject index, Volume 6 contains a unique drug compendium containing information on over 5500 compounds currently on the market. All six volumes are being published simultaneously, to provide a work that covers all major topics of interest.

Because of the mechanistic approach adopted, Volumes 1–5 do not discuss those drugs whose modes of action are unknown, although they will be included in the compendium in Volume 6. The mechanisms of action of such agents remain a future challenge for the medicinal chemist.

We should like to acknowledge the way in which the staff at the publisher, particularly Dr Colin Drayton (who initially proposed the project), Dr Helen McPherson and their editorial team, have supported the editors and authors in their endeavour to produce a work of reference that is both complete and up-to-date.

Comprehensive Medicinal Chemistry is a milestone in the literature of the subject in terms of coverage, clarity and a sustained high level of presentation. We are confident it will appeal to academic and industrial researchers in chemistry, biology, medicine and pharmacy, as well as teachers of the subject at all levels.

CORWIN HANSCH
Claremont, USA

PETER G. SAMMES
Uxbridge, UK

JOHN B. TAYLOR
Dagenham, UK

Contributors to Volume 4

Professor P. R. Andrews
School of Science and Technology, Bond University, Private Bag 10, Gold Coast Mail Centre, Queensland 4217, Australia

Dr J. M. Blaney
E. I. duPont de Nemours & Co., Inc., Medical Products Department, duPont Experimental Station, Building 400/4257, Wilmington, DE 19898-0353, USA

Dr K. Bowden
Department of Chemistry and Biological Chemistry, University of Essex, Wivenhoe Park, Colchester CO4 3SQ, UK

Dr S. K. Burt
Department 47E AP9, Abbott Laboratories, Abbott Park, North Chicago, IL 60064, USA

Dr E. N. Bush
Department of Medicinal Chemistry, Pharmaceutical Products Division, Abbott Laboratories, Abbott Park, North Chicago, IL 60064, USA

Dr F. Choplin
Centre de Recherches des Carrières, Rhône-Poulenc Recherches, 85 Avenue des Frères-Perret, BP 62, F-69192 Saint-Fons Cedex, France

Dr P. N. Craig
4931 Mariners Drive, Shady Side, MD 20764, USA

Professor G. M. Crippen
College of Pharmacy, University of Michigan, Ann Arbor, MI 48109, USA

Dr J. C. Dearden
School of Pharmacy, Liverpool Polytechnic, Byrom Street, Liverpool L3 3AF, UK

Dr W. J. Dunn, III
Department of Medicinal Chemistry and Pharmacognosy, College of Pharmacy, The University of Illinois at Chicago, PO Box 6998, Chicago, IL 60680, USA

Professor T. Fujita
Department of Agricultural Chemistry, Kyoto University, Kyoto 606, Japan

Mr D. J. Gans
Central Research Division, Pfizer Inc., Eastern Point Road, Groton, CT 06340, USA

Dr A. K. Ghose
Nucleic Acid Research Institute, ICN Plaza, 3300 Hyland Avenue, Costa Mesa, CA 92626, USA

Professor C. Hansch
Seaver Chemistry Laboratory, Pomona College, Claremont, CA 91711, USA

Dr T. E. Klein
Computer Graphics Laboratory, Department of Pharmaceutical Chemistry, School of Pharmacy, University of California, San Francisco, CA 94143–0446, USA

Professor P. A. Kollman
Department of Pharmaceutical Chemistry, School of Pharmacy, University of California, San Francisco, CA 94143-0446, USA

Professor H. Kubinyi
BASF AG Hauptlaboratorium, D-6700 Ludwigshafen/Rhein, FRG

Dr J. J. Kyncl
Department of Medicinal Chemistry, Pharmaceutical Products Division, Abbott Laboratories, Abbott Park, North Chicago, IL 60064, USA

Professor R. Langridge
Computer Graphics Laboratory, Department of Pharmaceutical Chemistry, School of Pharmacy, University of California, San Francisco, CA 94143–0446, USA

Dr A. J. Leo
Medicinal Chemistry Project, Seaver Chemistry Laboratory, Pomona College, Claremont, CA 91711, USA

Dr G. H. Loew
SRI International, Menlo Park, CA 94025, USA

Professor G. R. Marshall
Department of Pharmacology, Washington University School of Medicine, 660 South Euclid Avenue, St Louis, MO 63110, USA

Dr Y. C. Martin
Department of Medicinal Chemistry, Pharmaceutical Products Division, Abbott Laboratories, Abbott Park, North Chicago, IL 60064, USA

Professor J. A. McCammon
Department of Chemistry, University of Houston, 4800 Calhoun Road, Houston, TX 77004, USA

Dr J. W. McFarland
Central Research Division, Pfizer Inc., Eastern Point Road, Groton, CT 06340, USA

Dr C. B. Naylor
Department of Pharmacology, Washington University School of Medicine, 660 South Euclid Avenue, St Louis, MO 63110, USA

Dr M. A. Pleiss
Genentech Inc., 460 Point San Bruno Boulevard, South San Francisco, CA 94080, USA

Dr G. L. Seibel
Department of Pharmaceutical Chemistry, School of Pharmacy, University of California, San Francisco, CA 94143-0446, USA

Professor C. Silipo
Dipartimento di Chimica Farmaceutica e Tossicologica, Facolta di Farmacia, Via Domenico Montesano 49, I-80131 Napoli, Italy

Professor C. J. Suckling
Department of Pure and Applied Chemistry, University of Strathclyde, Thomas Graham Building, 295 Cathedral Street, Glasgow G1 1XL, UK

Mr P. J. Taylor
7 Radnormere Drive, Cheadle Hulme, Cheadle, Cheshire SK8 5JX, UK

Dr M. Tintelnot
Niederdorfstrasse 15, 8787 Zeitlofs/Detter, FRG

Dr M. S. Tute
Pfizer Central Research, Sandwich, Kent CT13 9NJ, UK

Dr S. H. Unger
Syntex Research, Computer Aided Drug Design R6-223, 3401 Hill Avenue, Palo Alto, CA 94304, USA

Professor A. Vittoria
Dipartimento di Chimica Farmaceutica e Tossicologica, Facolta di Farmacia, Via Domenico

Montesano 49, I-80131 Napoli, Italy

Dr D. Weininger
Daylight Chemical Information Systems Inc., 250 West First Street 344, Claremont, CA 91711, USA

Dr J. L. Weininger
809 Karenwald Lane, Schenectady, NY 12309, USA

Dr S. Wold
Research Chemometrics, Umeå University, S-901 Umeå, Sweden

Contents of All Volumes

Volume 4 Quantitative Drug Design

Volume 5 Biopharmaceutics

17.1

History and Objectives of Quantitative Drug Design

MICHAEL S. TUTE

Pfizer Central Research, Sandwich, UK

17.1.1   THEN AND NOW: THE CHANGING IMAGE OF A MEDICINAL CHEMIST

My first impression of chemical research was derived from visits to the cinema, when I was about 10 years old. In so-called horror movies, men in white coats would mix bubbling liquids with fiendish glee. They would then administer the resulting potion either to some poor caged animal, another human or even to themselves with astonishing effect. In more serious films, and I remember particularly a documentary on the Curies and the discovery of radium, the scientist was seen as a totally dedicated worker surrounded by a jumble of apparatus and dependent almost entirely on experiment. The results of these experiments seemed rarely predictable, but were always exciting, and persuaded me to request a chemistry set for my 11th birthday present.

With my chemistry set, I was soon able to demonstrate the physiological effect of chlorine gas upon my father, who suffered from asthma. I noticed that the response seemed quite disproportionate to the dose! An interest in structure–activity relationships was fostered and I was sufficiently encouraged to pursue a career in chemistry, graduating in 1956 as an organic chemist with a particular interest in synthesis. I entered the pharmaceutical industry in 1961, with the assigned task of synthesizing novel antiviral compounds.

We made hundreds of benzimidazole derivatives! Just as it had appeared to me in the old films, empiricism reigned supreme. We had our hunches about the likelihood of activity arising from various synthetic targets, but these were based on only vague notions of the importance of this or that functional group or physical property. Indeed, the target was generally chosen on the basis of expediency or perhaps because one wished to do some novel chemistry. To many of us there seemed no better way, for there had been many good drugs introduced in the 1950s by the exploitation of new synthetic methods and fermentation techniques alone.

By the mid 1960s, attitudes began to change. Good drugs must be replaced by better ones, which often seemed to result from a small change in structure of the original or 'lead' compound. If this change could be predicted from a study of the structure–activity relationships among derivatives and analogues, we could reap rich rewards. The advances in biochemistry and pharmacodynamics on the 'biological' side and advances in structural and mechanistic chemistry on the 'physical' side provided the knowledge base which was to be the foundation of a new quantitative approach to drug design. Mathematics was the catalyst, the apparatus a computer and the pioneering work was done by an organic chemist, Corwin Hansch, beginning in 1962. The structure–activity relationship was set in quantitative terms by applying numbers to describe not only activity but also structure and, most importantly, property.

The medicinal chemist is no longer an empiricist. Although still to be found in a white coat, pouring liquids or gathering crystals, he now has an additional habitat. The rational design and mechanistic study of medicinal compounds may now be performed in front of a computer display screen.

In 1969, I was able to apply QSAR principles to some of my own research into antiviral compounds, and became enraptured with the possibilities of explaining and predicting drug activity by the manipulation of numbers. A pilgrimage to meet Hansch in California followed, after which I wrote a review of QSAR techniques.[1] It was particularly gratifying to me to learn, in 1986, that this review had been cited 780 times! In 1983, I wrote another review,[2] entitled 'QSAR and Drug Design'. This sparked off some initially heated correspondence with a former research director of ICI Pharmaceuticals, Dr. F. L. Rose, who objected to my opening paragraph. The offending text read: 'The modern medicinal chemist differs from his counterpart of 20 years ago. His task, to discover new drugs, depends much less on chance and more on design. Drug design requires a certain knowledge of biological systems and how they are modulated, and an appreciation of the physico-chemical properties of molecules. Given this knowledge, suggestions for drug design may be founded on a biochemical rationale, or on the testing of bioisosteres, or perhaps on a QSAR . . . '. Rose pointed out, quite fairly, that chemists of 20 years ago had indeed considered physical properties and were well aware of concepts such as bioisosterism, by which one considers that two chemically dissimilar groups may have equivalent biological effects at receptor sites. These concepts were considered, argued Rose, but not accorded 'fancy names'. Rose was right, of course, in his assertion. But what has changed most dramatically since 1963 is that we not only have fancy names (bioisostere, QSAR, isolipophilic) but we have *numbers* with which to give these concepts some quantitative expression. Some 400 years ago, the Italian astronomer and physicist Galileo Galilei suggested that to introduce order into the universe man must pay attention to the quantitative aspects of his surroundings and discover the mathematical relationships that exist between them. He also suggested that from these relations, once discovered, certain consequences could be deduced and then verified by experiment. Galileo is generally regarded as the father of modern science, and indeed to many scientists a discipline only becomes 'respectable' if its principles can be expressed mathematically. We shall trace the exciting path by which respectability has come to the discipline of medicinal chemistry.

17.1.2   THE DESIRE FOR NUMERICAL RELATIONSHIPS: FROM GALILEO TO HANSCH

17.1.2.1   In the Beginning: Faith, Hope and Solubility

The numbers that we shall be considering in dealing with QSAR are frequently called parameters, another fancy term for a numerical descriptor of something, be it a count of atoms or a measure of property, such as solubility or partition coefficient. Parameters have been used extensively in the development of chemistry. Repeated application of the parameter value eight allowed the Russian chemist Mendeleev to develop the periodic system of the elements in about 1870. Arranging elements in order of atomic weight, it was found that similar properties belonged to each eighth member.

The 'rule of eight' worked well for the early part of the table of elements and was therefore used to predict the properties of later elements, and in this it was largely successful. The real value of the relationship, however, was diagnostic. It told us that atoms were not simply hard spheres, but that they possessed structure.

The correlation of Mendeleev (an early QSAR!) shows many of the features of our modern, biological QSARs. It was highly significant when derived and applied to early members of the table of elements and had both diagnostic and predictive value. It shows another feature: when used to predict properties of the heavier elements, it failed! As the elements become heavier, the recurrence of chemical properties begins after the 18th instead of the eighth member and, later on still, after the 32nd member. Nevertheless, this very failure told us, later on and in conjunction with other reasoning, something more about the grand design of molecular structure. The 'outlier' in any correlation can often afford the most valuable insight into the structure–activity relationship. We continue to learn by exploring the limits of correlations. Radium, when discovered by the Curies in 1898, exhibited some properties that had been predicted (a similarity to barium), but also others that no chemist had ever imagined.

Mathematics has been essential for developing our concepts of atomic structure and reactivity, from Mendeleev to the Rutherford–Bohr model of the atom in 1913, and thence to modern quantum mechanics which can only be understood by the most gifted mathematicians and is quite bewildering to most organic chemists.

But what about molecular structure, and biological 'reactivity'? At about the same time that Mendeleev was constructing his table, Richardson[3] showed that the narcotic effect of primary alcohols varied in proportion to their molecular weight, and Crum-Brown and Fraser observed that a series of quaternized strychnine derivatives could be prepared which, to a varying degree dependent on the quaternary substituent, possessed activity similar to curare in paralyzing muscle. In 1868 Crum-Brown and Fraser[4] put forward the suggestion that physiological activity depends on 'constitution', framing it in the mathematical terms of equation (1).

$$\phi = f(C) \tag{1}$$

This equation expressed an act of faith. What was 'constitution' in those days? What we now write as a structural formula was but dimly perceived in 1868 and constitution went little further than a composition of elements. This was particularly true of organic compounds: it was only three years since, in 1865, Kekulé had had his vision of the structure of benzene, and indeed it was not until 1929 that belief in a symmetrical benzene ring was confirmed by X-ray studies.

We had to wait until 1893 for Richet[5] to give credence to Crum-Brown and Fraser, by showing that the toxicities of some simple organic compounds such as ethers, alcohols and ketones were inversely related to solubility in water.

Further attempts to correlate a biological activity to molecular property occurred at the turn of the century. Both Meyer[6] and Overton[7] looked to oil/water partition coefficients to correlate and explain the potencies of narcotic substances. Overton in particular made precise histological and physiological observations on the response of plant and animal species to a large number of organic compounds. Experimenting on tadpoles, Overton found a systematic increase in narcotic potency with increasing chain length in groups of related compounds and reasoned that only one physical property, the oil/water partition coefficient, correlated with this increase. From these and other experiments Overton proposed that narcosis was due solely to physical changes brought about by solution of the narcotic in the lipid constituent of cells. In order to understand the action of an anaesthetic in man, he concluded that knowledge would be required of the ability of the compound to penetrate certain tissue cells, the fat content of the animal and the partition coefficient between water and the various lipids in question.

Overton also experimented with alkaloids and his comment on the difference in toxic effect of morphine to humans and to tadpoles is of interest (a saturated aqueous solution of morphine has little effect on tadpoles). He considered that there were differences in the structure of proteins in the two organisms and that these proteins would form salt-like complexes with morphine. He then suggested that variation in toxicity would result from differences in solubility of these complexes.[8]

Overton's work provided the impetus for other investigators. In 1904 Traube[9] found a linear relation between narcosis and surface tension, and in 1912 Seidell[10] measured both the solubility and partition coefficient of thymol in the hope that it might assist him to understand how thymol exerts its action against hookworm. Partition coefficients were not easy to measure in those days: thymol was equilibrated between castor oil and water, and estimated in water by treating the solution with bromine and titrating the resulting hydrobromic acid, after first removing excess of free bromine by potassium iodide and thiosulfate. The thymol in oil had to be separated by a steam distillation step! It is little wonder that few partition coefficients were measured until laboratories became equipped with UV spectrometers.

We had to wait until 1939 for the next advance. Ferguson,[11] working in the ICI laboratories, formulated a concept linking narcotic activity, partition coefficient and thermodynamics. Another Meyer had paved the way for this in 1935 by stating that identical (isonarcotic) effects are produced when molar concentrations of narcotics in the cell lipids are identical.[12] The partition coefficient determines the equilibrium between external phase (the water in which a tadpole is immersed, for example) and biophase (the nerve tissue of the tadpole). Ferguson reasoned that, when in a state of equilibrium, simple thermodynamic principles could be applied to drug activities, and so the important parameter to consider for the correlation of narcotic activities was the relative saturation of the substance in the applied phase. This has become known as Ferguson's principle.

Thermodynamic arguments have now been applied to many biological systems where an equilibrium is believed to exist and relationships have been obtained with solubility, partition coefficient, surface tension and parachor.[13] These relationships have, with one notable exception, been of limited use, being restricted not only to certain types of biological activity such as narcotic, anaesthetic or depressant effects but also to certain simplified test systems. In the whole animal, and where equilibria are less relevant than kinetics, such single parameter correlations have generally not been obtained. The exception referred to was the discovery of halothane as an anaesthetic, where application of Ferguson's principle suggested high enough potency to justify detailed screening and eventual development.[14]

17.1.2.2 First Correlations: Shape and Ionization

Ferguson's principle has been described fully by Albert[15] in his excellent book 'Selective Toxicity'. In 1939 Albert himself began a quantitative study of the antibacterial activity of aminoacridines, which was to reveal the importance of considering the concentration of active species, in this case the cation, of any drug capable of acid–base or other prototropic equilibrium.

Formulae (1)–(4) show some of over 100 acridines studied by Albert. Proflavine (**1**) and acriflavine (**2**) were introduced in 1913 and used in the first world war for treatment of sepsis in wounds and burns. The preliminary results of Albert showed that acridines become progressively more active as they become more highly ionized.[16]

Working at a pH of 7.3, compounds (**1**)–(**3**) are all fully ionized and equally active. In contrast, 2-aminoacridine (**4**) with a pK_a of 5.8 is less than 10% ionized and tenfold less active in a test for minimum bacteriostatic activity. The unsurprising conclusion, especially as acriflavine (**2**) can only exist as a cation, was that the cation is the active species.

But Albert went a lot further in his painstaking work and made some crucial observations. One might expect that by lowering the pH of the test medium, activity would increase because more of the cation is present. To a certain extent this is true: lowering the pH from 8 to 6 increases the activity of (**4**). However, decreasing the pH still further, to about 5, caused a decrease in activity when the percentage of cation would be rising!

Albert now changed his biological activity parameter. Rather than minimum inhibitory concentration (MIC) of *total* drug, he used the concentration of *cation*. With all acridines he now observed approximately equal MIC(cation) for any given pH. Moreover, as the pH was reduced there was a fall in potency. These results were initially interpreted to mean that acridinium cations were in competition with protons for some anionic receptor site.

As always, the truth is more complex. We now believe that acridinium cations are in competition with protonated nucleic acid bases for the phosphoric acid groups of bacterial DNA. Lerman[17] suggested in 1961 that acridinium ions insert themselves between adjacent layers of base pairs in the double helix, and this process of intercalation has subsequently been confirmed as the cause of bacteriostasis by aminoacridines and many other planar basic molecules.

Just as Overton and Meyer had found that isonarcotic effects were produced by isomolar concentrations of substances in certain cell lipids, now we had Albert showing isobacteriostatic effects from isomolar concentrations of cations. These were important discoveries, both emphasizing the need to consider thermodynamic concepts.

Albert[18] now turned his attention to other series of heteroaromatic bases, including pyridines, quinolines, benzoquinolines and phenanthridines. Simple aminopyridines and aminoquinolines were not active, despite strong basicity, but *within* each other series he found, as with the acridines, that potency was proportional to the molar concentration of the cation. Comparing strong bases *between* series, it was only necessary to draw the structures to see that neither 4-aminopyridine (**5**) nor 4-aminoquinoline (**6**) have such a large planar area as the active aminoacridines. Reduction of 9-aminoacridine to the tetrahydro derivative (**7**) gave an inactive compound, by removing planarity. As annelation of quinoline to benzoquinoline gave sufficient planar area for activity, Albert suggested that addition of styryl groups to a quinoline would also give activity. This prediction was successful, in giving the potent derivative (**8**). Clearly, size (as here expressed in terms of surface area) and shape (the area had to be flat, as it turned out, for intercalation into DNA) had to be added to solubility, partition coefficient and ionization as parameters for correlation with biological activity.

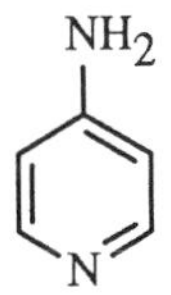

(5) 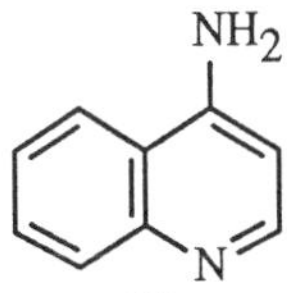   (6) 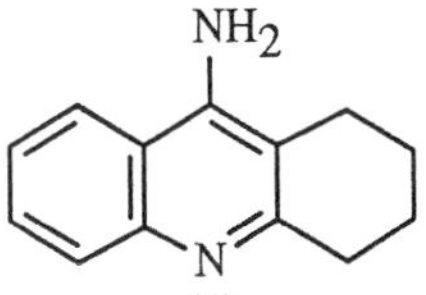   (7)

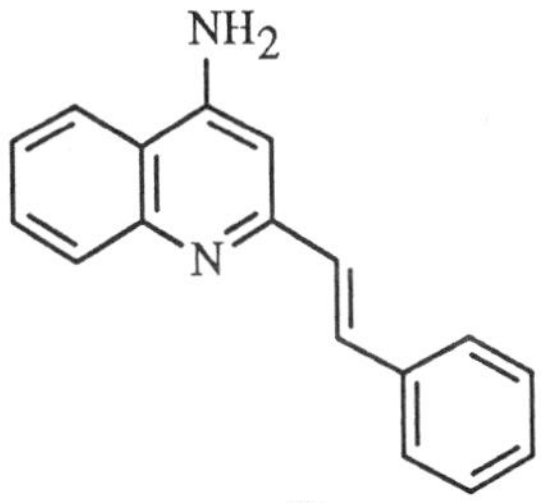

(8)

While Albert was examining ionization of bases, others were exploring ionization of weak acids. In 1935, Prontosil (**9**) was introduced as an antibacterial agent, inactive *in vitro* but effective *in vivo* through cleavage in host tissue to an active metabolite, the weak acid sulfanilamide (**10**; R = H).

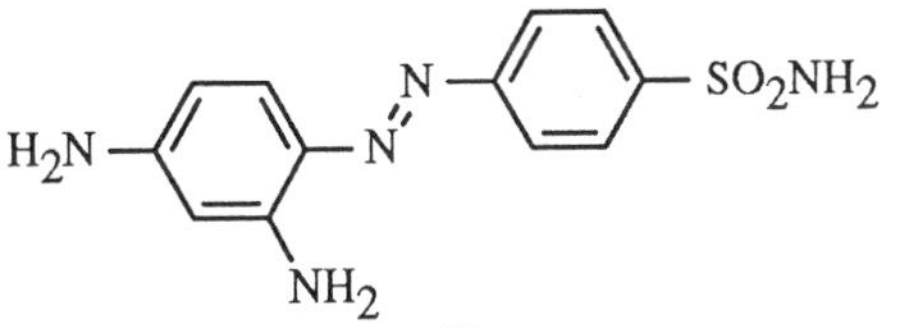

(9) 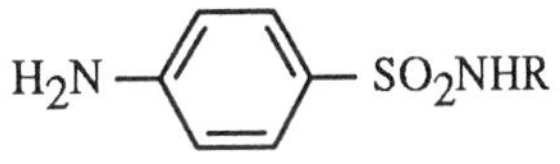   (10)

In 1938, the more strongly acidic derivative sulfapyridine (**10**; R = 2-pyridyl) was shown to be even more effective. Thousands of derivatives and analogues have since been prepared both in the search for further therapeutic advantages and in order to define the structure–activity relationship.

In 1942, Bell and Roblin[19] published results of a quantitative study of the *in vitro* activity of 46 sulfonamides, mostly of type (**10**). The potency parameter used was log(1/MIC) to inhibit *E. coli*. Plotting potency at pH 7 against pK_a of the sulfonamide, a biphasic and virtually bilinear curve resulted (Figure 1).

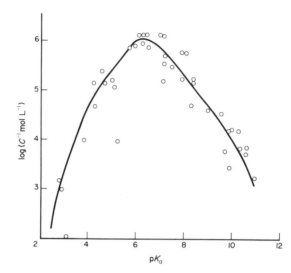

Figure 1   Effect on bacteriostatic activity of variation in the pK_a of sulfonamides (reproduced from 'Selective Toxicity', 7th edn., 1985, by permission of Chapman and Hall)

Bell and Roblin had been able to choose from active derivatives (**10**) in which group R varied widely, including both electron-withdrawing and electron-releasing functions, to provide a pK_a range of 3 to 11. They interpreted the curve as providing a rationalization for the enhanced potency of heterocyclic sulfonamides that have a pK_a near 7 (the optimum of the curve) and also to explain the potency of some sulfonamides that cannot ionize at all as acids, such as sulfaguanidine (**10**; R = NHC(NH$_2$)=NH) and dapsone (**11**). Over the range of pK_a from 11 to 7, a decrease in pK_a reflects an increasing proportion of anion, and the anion is presumably more active than the neutral species so potency increases over this range. However, from a pK_a of 5 to 3, the sulfonamide is essentially totally ionized and no significant increase in concentration of anion would be occurring. A decrease in activity over this range must reflect the group R withdrawing electrons away from the centre of negative charge, the —SO$_2$N— group of the anion. The crucial property of the sulfonamides, argued Bell and Roblin, was not ionization *per se* but *the proportion of charge existing on the oxygen atoms of the —SO$_2$N—group*. It should be noted that even in unionized sulfonamides these oxygen atoms would bear some negative charge. This would be especially true of such substances as dapsone (**11**), where the electron-releasing amino groups would transfer charge towards the oxygen atoms.

$$H_2N-\!\!\!\langle\ \rangle\!\!\!-SO_2-\!\!\!\langle\ \rangle\!\!\!-NH_2$$
(**11**)

The theory of Bell and Roblin provided a major stimulus to all subsequent investigations of the mode of action of sulfonamides. It also provided a stimulus to QSAR, in that it indicated the value of considering the potential of a substituent to influence charge distribution.

17.1.2.3   The Influence of Physical Organic Chemistry: Hammett and Taft

Through the late 1930s and 1940s, while some medicinal chemists were noting the influence of ionization on drug activity, physical chemists were developing parameters to quantify the electronic effects of substituents on chemical reactions and ionic equilibria. In particular, the effects of substituents (X) on the reactivity of a side chain (R) in aromatic compounds of the type (12) were noted. It was found that the effects of substituents in many reaction series involving *meta-* and *para*-substituted benzene derivatives could be correlated with the acid strengths of the corresponding benzoic acids.[20]

(12)

If the assumption is made that the potential for influencing a reaction is independent of the reaction conditions, then one can proceed by defining a scale of such potentials that would be useful in predicting the relative rates of chemical reactions. In 1937, Hammett[21] had shown that the chemical reactivity of *meta-* and *para*-substituted benzene derivatives could be correlated by equation (2), where K_H is the rate constant for the parent (unsubstituted) molecule and K_X is the rate constant for the derivative. The substituent constant σ_X refers to the electronic effect (potential) of the substituent relative to hydrogen and is a parameter applicable to many different types of reaction — characterized by different values of ρ — whose relative rates depend on the degree of electron release or withdrawal by that substituent.

$$\log(K_X/K_H) = \rho\sigma_X \qquad (2)$$

For the derivation of a scale of σ_m and of σ_p 'constants', the ionization of *meta-* and *para*-substituted benzoic acids at 25 °C was defined as the standard reaction for which ρ was set to unity. Dropping the subscript notation, we can write equation (2) as equation (3) or, still more generally, as equation (4).

$$\log K(\text{derivative}) = \rho\sigma + \log K(\text{parent}) \qquad (3)$$

$$\log K = k_1\sigma + k_2 \qquad (4)$$

Plotting $\log K$ against σ for a series of derivatives of benzene (for example, *meta-* and *para*-substituted phenols) undergoing a chemical reaction (for example, acylation of the phenol), we would expect a straight line of slope k_1 (equivalent to ρ) and intercept k_2 (equivalent to $\log K$ for the unsubstituted compound).

Because ionization constants were used to define σ, and ionization constants can be related to free energies through the familiar equation (5), then equation (4) or any of its many variations defines a linear free energy relationship (LFER).

$$\Delta G = -RT\ln K \qquad (5)$$

Hammett himself distinguished σ values for *meta* and *para* substituents as being different because of the much greater resonance contribution of a substituent in the *para* position and somewhat greater inductive contribution of a substituent in the *meta* position. A further modification was introduced by defining special σ values for substituents where the substituent and reactive centre were in extreme resonance situations: thus, a special σ^- constant would be used for a nitro substituent *para* to a phenol, to account for extraordinary stabilization of the phenolate anion. Similarly, a special σ^+ constant was derived for substituents capable of direct resonance interaction with an electron-deficient site, such as a carbonyl group.

In view of the complications arising both for *ortho* substituents in a benzene ring and for aliphatic systems, where both steric and polar effects may coexist, little further progress was made until 1952. Between 1952 and 1956, Taft devised a procedure for separating polar, steric and resonance effects based on a LFER analysis of the rates of base- and acid-catalyzed hydrolysis of aliphatic esters.[22] Besides introducing the first steric parameter E_s and a polar parameter σ^* for use in aliphatic systems, Taft showed how the original Hammett σ values might be factored into inductive and resonance components and then recombined in a linear equation, an idea that was to strongly influence the later development of QSAR by the Hansch school, and to be elaborated in 1968 as the $\mathscr{F}$ (field) and $\mathscr{R}$ (resonance) constants of Swain and Lupton.[23]

Despite a largely empirical development, Hammett and Taft equations have been used as the basis of practically all correlation analysis that has been done in the field of organic reaction analysis. The alternative strategy of a molecular orbital calculation is by comparison so complicated that it has been nowhere near as useful to the organic or medicinal chemist seeking a diagnostic tool with which to probe reaction mechanism.

17.1.2.4 The Biologist's Hammett Equation: Hansch and QSAR

In a review on the Hammett equation, Jaffe[20] in 1953 said that he had not been able to find any series of compounds whose *biological* action had followed such a relationship, though he did not consider such an occurrence inconceivable.

The task of setting up a 'biological Hammett equation' was approached simultaneously by three groups in the early 1960s. Hansen[24] in 1962 tried to explain data on the inhibition of growth of bacterial cultures by three sets of compounds: substituted benzoic acids (!), phenyl-substituted penicillin G derivatives and chloramphenicol analogues. Hansen postulated that the growth inhibitory reaction was bimolecular between compound and some critical enzyme, E. He represented the destruction of the enzyme as proceeding at a rate K by inhibitor I. If [I] is the concentration of inhibitor around the enzyme, then by kinetic theory we can write equation (6). Measuring the concentration [I] such that growth is inhibited by a standard amount, equivalent to making dE/dt constant, one may expect a linear plot of $\log[I]$ against $\log K$, for from equation (6) we can write equation (7).

$$dE/dt = KE[I] \qquad (6)$$

$$\log K + \log[I] = \log(dE/dt) - \log E$$
$$= a \text{ constant} \qquad (7)$$

Now, substituting for $\log K$ from the Hammett equation (4), we derive equation (8), leading to the expectation of a straight line plot between $\log[I]$ and σ.

$$-\log[I] = k_1\sigma + k_2 \qquad (8)$$

In fact, this approach gave very disappointing results. In retrospect, we can see that there are a number of reasons for this. Nevertheless, it marked an important stage in the development of QSAR theory, and provided a foundation for later thinking.

Zahradnik,[25] also in 1962, tried to develop a biological Hammett equation from more fundamental principles. He abandoned σ, but generated his own constant β in analogous fashion to Hammett. Thus, by a construction analogous to that of equation (2), Zahradnik postulated equation (9), where α characterizes the sensitivity of the biological system (analogous to the Hammett ρ) and β characterizes the biological potential of the 'substituent' R in a homologous series of compounds RX. It was assumed that β (for each R) would be independent of the functional group X.

$$\log(\gamma_i/\gamma_{Et}) = \alpha\beta \qquad (9)$$

The left-hand side of the equation is an expression of relative biological activity: γ_i is the molar concentration of the ith member of the series that will produce a defined biological end-point, such as an MIC, and γ_{Et} is the corresponding concentration for the ethyl homologue. Whereas Hammett's standard reaction was ionization of benzoic acids for which $\rho = 1$, Zahradnik chose his standard *biological* reaction as the toxicity of alcohols (X = OH) to mice. He measured LD_{50} values ($\gamma = LD_{50}$) to derive β constants for 25 different alkyl groups. The 'sensitivity' was evaluated for 39 different biological systems as the slope of equation (9). The fit was reasonably good, though in this case retrospective analysis indicates a fortuitous choice of a biological system in which activity was largely controlled by partitioning of the compound from an aqueous to a lipid phase.

The efforts of Hansen and Zahradnik failed on two counts. They concentrated on single parameter equations (β of Zahradnik, σ of Hansen) and, following too closely the Hammett recipe, looked for the simplicity of a linear relationship.

Also in 1962, Hansch[26] published the first of many brilliant contributions which were to establish our modern multiparameter approach to QSAR. Hansch and his colleagues, notably the botanist Muir, had been trying for some 15 years to develop the SAR of plant growth regulators, including indoleacetic and phenoxyacetic acids. Through testing scores of the latter compounds, it had become apparent that relative lipophilic character of the substituents was an important determinant of activity. Needing a suitable descriptor of relative lipophilicity, Hansch, Maloney and Fujita began to make measurements of partition coefficients in the octanol–water system. The precedent for using partition coefficients as a hydrophobic scale had of course been set by Meyer and Overton some 50 years before.

The scale of hydrophobic substituent constants was set up for benzene derivatives in an analogous fashion to a Hammett equation.[27] In equation (10), π is the substituent constant for group X replacing hydrogen, H. The measured quantities are partition coefficients of derivative P_X and parent P_H. Octanol–water was chosen as the standard system largely for practical reasons, but the choice was later justified on many other grounds.[28,61]

$$\log(P_X/P_H) = \pi \tag{10}$$

One missing link was now provided by Fujita, who was a visiting professor with the Hansch group. Fujita[27] suggested that because π and $\log P$ were linear free energy related variables, just like σ, the approach used by Taft of combination of two variables into a LFER might be followed. Thus, if biological response was due in part to hydrophobicity and in part to some electronic factor, one might expect an equation such as (11) to hold. In equation (11), biological reponse is also cast in LFER formalism by using the logarithmic term, where C is the molar concentration of compound producing a standard response in a constant time interval. This was the same argument as had been put forward by Hansen that same year.

$$\log(1/C) = k_1\pi + k_2\sigma + k_3 \tag{11}$$

So in 1962 the Hansch group began an analysis of many sets of biological data, to see if their empirical 'biological' LFER would hold. In fact, in many cases it gave good correlation; in others it still left a good deal to be desired. The linear relationship of equation (11) seemed poor in explaining variance in the data when a considerable spread in $\log P$ values was to be found.

The second missing link was provided intuitively by Hansch[29] himself. Molecules which are highly hydrophilic will not readily partition from water into the lipid of a membrane. If the receptor is within or beyond that membrane, such a molecule will have a low probability of reaching it in the time interval under study. Conversely, molecules which are highly hydrophobic will readily partition into the first of a series of lipid membranes but will be held there and thus slowed down in any journey to a remote site of action, such a journey being termed the 'random walk'.

The idea was born that for any particular receptor, some optimum value of $\log P$ (or π) would be found to correspond to the maximum probability of reaching the receptor in a given time. The simplest way of expressing such an idea mathematically would be to postulate that $\log(1/C)$ was parabolically dependent on $\log P$. Thus, the extremely useful general Hansch equation (12) was proposed. The applicability of this equation and variants, in particular equation (13) framed entirely in substituent constant terms and equation (14) which includes the Taft steric parameter, was convincingly demonstrated[29] by 1965.

$$\log(1/C) = k_1(\log P) - k_2(\log P)^2 + k_3\sigma + k_4 \tag{12}$$

$$\log(1/C) = k_1\pi - k_2\pi^2 + k_3\sigma + k_4 \tag{13}$$

$$\log(1/C) = k_1\pi + k_2\sigma + k_3E_s + k_4 \tag{14}$$

With equation (12), which now bears his name, Hansch initiated the full-scale development of QSAR which was to spawn, in the pharmaceutical industry, departments of computational and theoretical medicinal chemistry. We shall now trace the expansion in theory and practice of QSAR that has taken place since 1965.

17.1.3 THE OBJECTIVES OF QSAR

In this section, we shall address the question: what can be achieved by correlation analysis? The answer to this very much depends on the quality and quantity of data analyzed. It also depends on the type of dependent variable used, whether physicochemical (such as σ or π), structural (such as a frequency of occurrence of a substituent) or topological (such as a connectivity index).

A QSAR can be expressed in its most general form by equation (15). The overall objective is to find parameters from experiment or theory that, when substituted into one of the many forms of the equation along with biological activity for a series of molecules, give a statistically significant correlation. If a good model is found it may be used to predict other molecules having greater activity in the defined biological system. This predictive element is undoubtedly the most exciting aspect of QSAR but has not proved to be its most useful feature.

$$\text{Biological activity} = f(\text{physicochemical and/or structural parameters}) \tag{15}$$

17.1.3.1   Diagnosis of Mechanism

One of the great attractions of Hansch's LFER model is that the parameters used have physicochemical meaning. The chemist can therefore interpret a correlation directly in terms of a mechanism or alternatively use a QSAR to test a mechanistic hypothesis.

To exemplify the diagnostic value of casting a QSAR in LFER terms, consider equation (16). This equation was generated from data on a set of 51 substances including alcohols, ethers and amides tested as narcotics on tadpoles.[30] Here, C is the molar concentration for narcosis and only one term, $\log P$ (octanol–water), was required for excellent correlation.

$$\log(1/C) = 0.94\log P + 0.87 \tag{16}$$

$$r = 0.97, \quad n = 51$$

The equation supports the Meyer–Overton theory of narcosis for it is consistent with lipophilicity determining the accumulation of molecules in the lipid biophase and accumulation as being the only prerequisite for activity. No other parameter is required, despite wide variation in structure within the data set.

We have already discussed the work of Bell and Roblin on bacteriostatic sulfonamides. They found that weakly acidic sulfonamides became more potent as they became more ionized. Using a set of sulfonanilides (13), Seydel[31] derived equation (17) in which σ^- gives a better correlation than the original Hammett parameter, σ. This reveals information about the mechanism of action and strongly supports the Bell and Roblin theory. Despite wide variation in substituents no hydrophobic or steric parameter was needed, this being diagnostic of the nature of 'substituent space' at the receptor.

$$H_2N \text{—} \langle \quad \rangle \text{—} SO_2NH \text{—} \langle \quad \rangle \text{—} X$$

$$(13)$$

$$\log(1/\text{MIC}) = 1.05\sigma^- - 1.28 \tag{17}$$

$$r = 0.97, \quad n = 17$$

These two examples illustrate how a QSAR can be used to encapsulate an idea about mechanism, bringing order and insight from a mass of experimental data culled from the literature. There are now hundreds of published examples of QSARs giving insight into mechanism,[32] supporting or suggesting theories as to mode of action at a receptor level or delineating necessary properties for drug transport to the site of action.

17.1.3.2   Prediction of Activity

For the prediction of activity of an 'unknown' from a QSAR, any form of parametrization (physicochemical or structural) may be used. As with prediction in other fields of study (forecasting the weather or the product and yield of a chemical reaction) the reliability of such prediction depends not so much on the quality of the mathematical relationship but more on the degree of similarity between members of the 'training set' and the test case.

It is now well recognized that apparently linear relationships, for example between activity and lipophilicity as expressed by equation (16), do not 'go on forever'; that is, they do not hold true for all values of $\log P$. A compound so lipophilic as to be insoluble would most certainly not fit equation (16). The rule has often been expressed somewhat cynically by organic chemists as 'methyl, ethyl, propyl, futile'! Hansch[33] has discussed the problem of prediction in the context of LFERs and has considered two kinds of prediction: (1) *interpolative* prediction within spanned substituent spaces (SSS) and (2) *extrapolative* prediction outside SSS. The meaning of SSS can be illustrated by reference again to equation (16). If the 'training set' used to obtain this relation comprised compounds with $\log P$ values ranging from -2 to $+2$, then space between these limits is SSS and outside these limits is un-SSS. Successful interpolative prediction is much more likely than extrapolative prediction, and even then will depend on having a good spread of values for each parameter in SSS. Despite the limitations a large number of experimentally verified examples of predictions have now been reported,[34] using both classical Hansch-type equations and other techniques.

17.1.3.3 Classification

It is frequently the case at the beginning of an investigation into a new therapeutic area or when making an exhaustive literature survey that precise biological data are not available. Rather, it may only be possible to group compounds into semiquantitative categories such as 'highly active', 'active' and 'inactive' or even into qualitative categories such as agonist and antagonist.

It is also more than likely in the initial stages that chemists will not focus their attention just on a set of monosubstituted benzene derivatives for analysis! The compounds they wish to include may not all be simple derivatives of a common parent, but rather a heterogeneous collection of structures. In such cases one may not be able to attempt a correlation using substituent constants such as σ or π, but merely use the presence or absence of certain defined molecular features whose physicochemical influence remains to be established.

When heterogeneous sets and imprecise or qualitative data are involved then pattern recognition or discriminant analysis techniques are more appropriate than Hansch analysis. Useful information can still be extracted from such data, the mathematical relationships which are derived being used for classification. Thus, discriminant analysis is a technique which allows the exploration of the significance of correlation between a crude activity assignment (membership of a group) and either continuous (σ, π, *etc.*) or discontinuous indicator variables having the value of 1 or 0 according to the presence or absence of certain user-defined molecular features. The inaugural study of discriminant analysis in medicinal chemistry was by Martin[35] in 1974. It has since been much used to classify compounds as to their potential for eliciting animal toxicity, based on analysis of very heterogeneous data sets culled from toxicity databases.

17.1.3.4 Optimization

In the early stages of a medicinal project, discriminant analysis may well be used to discern the most likely chemical class for further study. Later, that particular class will be subjected to an in-depth QSAR investigation. All members will have activity, but which particular derivative should be chosen for extensive clinical evaluation?

The compound will be selected according to clinical requirements. For topical application, potency may be the property which needs to be optimized. For oral administration, however, it may well be that it is more important to optimize transport properties of the drug series provided that an acceptable level of activity can be maintained. For compounds that have a multiplicity of pharmacological effects, some of which will be undesirable ones, a degree of selectivity of action will be required. For semisynthetic antibiotics, antibacterial or antifungal agents, a particular spectrum of microorganisms will need to be killed at achievable tissue levels and this may be the criterion.

Just as yields can be optimized by careful adjustment of reaction conditions, so can biological activity, selectivity or spectrum of activity be optimized by adjustment of physicochemical property. The balance of steric, electronic and hydrophobic properties required may be predicted by a QSAR on the appropriate biological activity parameter.

The pyranenamines (14) have been found to possess antiallergic properties in a variety of animal models. Right at the beginning of an investigation into these compounds Cramer[36] applied the most simple of QSAR techniques to a series of 19 derivatives. Rather than a computer he used a piece of graph paper! Plots of activity against π and σ revealed that activity might well be increased by using more hydrophilic substituents but maintaining the σ values of these substituents near to zero. Being suitably encouraged by a few probes of this type many more derivatives were then made, until a set of 98 was available. Many of these derivatives were very carefully chosen so as to explore the differing consequences of *ortho*, *meta* and *para* substitution. This phase of the investigation required considerable synthetic effort.

(14)

The exercise paid off handsomely. A QSAR constructed for the set of 98 allowed earlier qualitative conclusions to be refined. The concept of a near to zero Hammett σ was firmed up and an optimum hydrophilicity was defined. Overall, the equation led to the prediction that a derivative of (14) with —NHCOCH(OH)CH$_2$OH groups at the 3 and 5 positions should be highly active. This compound was synthesized, and found to be highly potent, much as predicted. It was in fact 1000 times more active than the most potent member of the original set of 19. Without the benefit of QSAR such a strange derivative would most surely never have been synthesized.

17.1.3.5   Refinement of Synthetic Targets

We have traced how the relation between anaesthetic activity and lipophilicity had been studied by Hansch who took the step of using log P (octanol–water) to define lipophilicity and enable proper quantitative study.

Much has been gained over the last 20 years through use of this standard and through construction of a database of QSAR equations. One lesson is that the ideal or optimum log P value for a congener is very largely a function of the biological system. This is very clear for anaesthetics acting in mammals. Consider the parabolic equation (18) relating C, the molar ED$_{50}$ for ethers producing anaesthesia in mice, to log P.[37] This and many other equations show that the penetration of neutral compounds into the CNS is most favourable for the compounds having a log P of about 2. If we study log P values of gaseous anaesthetics that have been used over the years, we find that the weakly active ethyl ether has a log P of only 0.98, the much more active chloroform has log P of 1.97 and the modern choice, halothane, has log P equal to 2.3! The rule is not confined to gaseous compounds; the most potent barbiturates, used as sedatives and hypnotics, have log P values near 2 which give them efficient entry to the CNS.

$$\log(1/C) = 1.04\log P - 0.22(\log P)^2 + 2.16 \tag{18}$$

$$n = 26, \quad r = 0.97, \quad \log P_0 = 2.35$$

A rule has been formulated from these QSAR studies: to get a compound into the CNS, design it such that log P is near 2. To keep it out of the CNS, and so avoid possible unwanted CNS side effects such as drowsiness, design so that log P is not near 2. Bear in mind the danger of extrapolation and that this rule was derived from sets of neutral compounds of quite low molecular weight, so there will certainly be exceptions. Nevertheless, it can be a useful 'rule of thumb'.

Hansch[37] has related how this rule was made use of in design of the drug sulmazole (15) by the Thomae company. A potent cardiotonic agent, the 2,4-dimethoxy analogue of sulmazole had first been selected for clinical trial. During the trial, some patients suffered a rather bizarre side effect. This was described as 'seeing bright visions' and was clearly caused by CNS involvement. Now log P for the analogue was found to be 2.59, very close to the magic optimum for CNS involvement! Sulmazole was therefore synthesized, replacing 4-OMe by 4-S(O)Me, a group of similar size but much lower hydrophobicity. The log P of sulmazole was found to be 1.17 and the unwanted CNS effect was removed.

(15)

17.1.3.6 Reduction and Replacement of Animals

In recent years, there has been some public concern about the use of animals in medicinal research. There exists a particularly vexing dilemma with respect to use of animals in drug safety testing: on the one hand the need to avoid disasters such as occurred with thalidomide has resulted in a demand for more exhaustive toxicity testing, involving more animals. On the other hand, understandable concern for animal welfare has resulted in the desire to keep animal usage to the absolute minimum.

Use of animals in drug, pesticide and particularly in cosmetics safety testing has been vigorously attacked by some pressure groups who call for a total ban on animal experiments. In addition, groups such as FRAME (Fund for the Replacement of Animals in Medical Experiments) take the stance that this would be desirable but is impractical. FRAME has encouraged research into alternatives, advocating the refinement of animal experiments, reduction in their number and, where suitable, replacement of an animal test by an *in vitro* assay or a mathematical model.

It is good economic sense as well as humanitarian to reduce animal usage. This has in fact been happening as a result of the various pressures and the slow but sure development of alternatives. Although in the UK animal use has been falling steadily since 1976, there is still cause for concern. The latest Home Office figures show 3.3 million experiments were carried out on live animals (90% rodents) in 1985.

Use of a QSAR can be an alternative to further animal experiments. The QSAR can show when further synthesis would yield compounds of only equivalent or lower activity, thus suggesting cessation of synthesis and of course any associated testing in animals. Much effort has gone into defining models for classifying structures as to their acute toxic (*e.g.* LD_{50}) or teratogenic potential.[38] Although these models are at present crude and are certainly no substitute for a proper safety evaluation of any novel compound,[39] they can still be used to set priorities for synthesis and testing. As we learn more about the mechanisms involved in toxicity and include more relevant parameters in the equations the models will improve, and it is to be expected that QSAR will make a greater contribution to the reduction in use of animals.[40] This is a laudable objective of QSAR.

17.1.4 BEYOND HANSCH ANALYSIS: FROM EUPHORIA TO REALISM

The Hansch LFER model was applied in the late 1960s to data sets from the fields of drug, pesticide and agrochemical research. Many previously confusing observations were put in order and rationalized by applying some one or other of the forms of the equation. The twin strategies of combining variables and looking for parabolic dependence in the hydrophobic term paid handsome dividends. The American Chemical Society organized a satellite symposium on QSAR in 1970 as part of their annual meeting, and international meetings on QSAR have been held regularly ever since.

Published work emphasizes success, and the publications of the late 1960s and early 1970s led to a certain euphoria. There was nothing to stop us explaining all biological activity in terms of steric, hydrophobic and electronic effects. And from there, it would be but a short step to design drugs by computer: no more 'random screening'! However, at the meetings on QSAR we were starting to talk about some failures of the Hansch equation, to question its basic assumptions and to define some of its limitations. Worse, in the real world those series of compounds that most interested us were rarely congeneric (our active compounds were not always substituted benzenes)! We could not see how to apply LFERs to such series. Even when a series did comprise 'related' compounds, it was often impossible to assign a physicochemical constant such as σ or π to an uncommon substituent. Hansch analysis was only the beginning: much remained to be done.

17.1.4.1 Models and Variables

17.1.4.1.1 Statistical models

In 1964, Free and Wilson[41] developed a method for deriving *de novo* biological substituent constants. These were to be used when physicochemical constants were unavailable. In any series of compounds which can be considered as derivatives of some common parent structure, one may assume that the effect of a substituent on biological activity (BA) is additive and independent of the presence or absence of other substituents at other positions. The Free–Wilson model can be expressed by equation (19), where A_{ij} is the activity contribution of substituent i at position j and S_{ij} takes the value of 1 or 0 according to the presence or absence of i at j. The constant k represents the overall average activity of the series. Solution of the set of simultaneous equations (19), one for each compound, in a least-squares manner gives the 'best fit' parameter values for all A_{ij}, and k.

$$\text{BA} = \Sigma A_{ij} \cdot S_{ij} + k \tag{19}$$

This model depends on the additivity assumption. It has been useful for testing this assumption. The great disadvantage is that the A_{ij} values relate only to the particular biological activity being considered in the test, and their magnitude depends on how BA is measured. Thus, a *de novo* parameter is local to the test, unlike a σ or π parameter, which is universal. The A_{ij} values give no direct information as to mechanism and can only give interpolative predictions (any novel substituent would have no A_{ij} value). A significant improvement to the Free–Wilson method, introduced by Fujita and Ban[42] in 1971, was to cast the BA term as $\log(1/C)$. This meant that derived A_{ij} values could be compared with other LFER parameters, and that k became the calculated activity of the (perhaps hypothetical) parent structure of the series.

The greatest shortcoming of the Free–Wilson method is that it provides no immediate QSAR insight, but it does have one great virtue, which was to be incorporated into later models. A *de novo* constant generated in this way combines information on *all* properties of the substituent, known or unknown, and this feature can be used to generate physicochemical properties explicitly related to mechanism. So, in the 1970s, QSARs were often generated by combining Hansch and Free–Wilson (as the Fujita–Ban modification) approaches. This was done by the use of physicochemical variables, where appropriate, plus so-called 'indicator' variables. An indicator variable I takes the value 1 or 0 according to the presence or absence of a substituent or feature and is analogous to S_{ij} of equation (19).

A good example of this 'mixed' approach is found in the work of Hansch on the hydrolysis of esters by papain.[43] The compounds to be analyzed included two subsets, the amides (16) and sulfonamides (17). An indicator variable I was assigned a value of 0 for amides, 1 for sulfonamides, and subsets combined to derive the QSAR of equation (20). In equation (20), K_m is the Michaelis–Menten constant so the biological activity is cast in LFER terms and represents binding of the ligand. Positive coefficients in the physicochemical variables show binding to be increased by increase in molar refraction (MR) and by electron withdrawal. We also find that if $I = 1$ (sulfonamides), binding is reduced by nearly two log units (the *de novo* parameter is -1.92). Most interestingly, an X-ray analysis showed that it was possible for the amide or sulfonamide moiety to occupy a hydrophobic cleft in the enzyme, so the *de novo* parameter was rationalized as representing the effect of a difference in hydrophobicity between amide and sulfonamide.

$$RC_6H_4OCOCH_2NHCOPh \qquad\qquad RC_6H_4OCOCH_2NHSO_2Me$$
$$\textbf{(16)} \qquad\qquad\qquad\qquad \textbf{(17)}$$

$$\log(1/K_m) = 0.57\text{MR} + 0.56\sigma - 1.92I + 3.74 \tag{20}$$

$$n = 20, \quad r = 0.99$$

Until 1974, QSARs were generated only from series of active compounds, usually with activity being precisely measured on some continuous scale. In the real world, sooner or later inactive congeners or analogues are found and if not included in consideration of the SAR then of course information is lost. Often, too, biologists are only able to say that compound A is 'better than' compound B, but cannot measure the difference in the quantitative terms necessary for a Hansch analysis. In 1974, Martin[35] introduced the statistical technique of discriminant analysis (DA) as another QSAR-generating method to overcome some of these limitations.

The DA technique has proved to be the most general one available for dealing with crude or qualitative biological data (including inactives) and relating it to 'mixed' (physicochemical, indicator or other) variables. In DA, classification functions are derived and may be used to predict whether a compound should be classified as a member of a designated 'group' on the basis of the 'discriminating power' of the assigned variables. The groups can be assigned on either a crude quantitative basis ($+ + +$, $+ +$, $+$, 0) or a qualitative assessment (agonist, antagonist, no effect).

Because data analyzed by DA are so crude and variables mixed, the results have often been difficult to interpret. Without adequate spread of data, the usual significance tests are inapplicable, it is difficult to derive mechanistic information and predictions (in the form of probabilities of group membership) are uncertain.

The 1970s were great years for the mathematicians among the QSAR fraternity. DA was only the first of many statistical techniques introduced in this period with the objective of analyzing large and diverse data sets. These methods are usually referred to under the umbrella name of pattern recognition techniques and have been found most useful for qualitative assessment, classification and preprocessing of information prior to a Hansch analysis on a selected subset of compounds. Common features of such methods are:[44]

(1) The set of data analyzed, known as the 'training set', must be large and is typically in excess of 100 structures. This is necessary in order to attain some statistical significance.

(2) A trial set of N descriptors is chosen. These may include physicochemical, structural (presence or absence of a feature) or topological parameters.

(3) Each structure is represented by a point in N-dimensional space (M structures, N descriptors) with the position of each point being defined by the values of the descriptors. This process can only be visualized up to $N = 3$ (*e.g.* x, y, z axes of a three-dimensional graph) but in these techniques N is always larger, typically $N = 5$–10.

(4) If the trial set of descriptors has been correctly chosen, compounds with a characteristic biological activity will cluster together in one part of N-dimensional space.

(5) If no clustering occurs, different sets of N descriptors are tested until clustering does occur. These sets may be chosen automatically by computer or by user intervention.

To carry out pattern recognition and produce meaningful clustering (classification) a variety of statistical methods have been introduced and incorporated into computer programs. The goal of a statistical technique known as factor analysis or principal component analysis is to reduce the number of descriptors N to find a new set of mathematically derived descriptors (factors or principal components) which will represent the original M molecules almost as well. If just two principal components can be derived then these can be plotted in two-dimensional space (a plane) and the M molecules projected onto this plane so as to visualize the clusters (Figure 2).

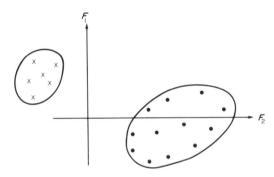

Figure 2 M molecules separated into two clusters (active and inactive) represented in two dimensions by principal components F_1 and F_2 of N descriptors in N-dimensional space

The program ADAPT (automatic data analysis using pattern recognition techniques) was introduced by Stuper and Jurs[45,46] in 1976, and in 1977 SIMCA was introduced by Wold and co-workers[47] to provide a mathematical model of each cluster and quantify the similarities between molecules in each cluster. These methods have been used extensively. Other methods of so-called 'multivariate' data analysis more recently introduced include MASCA by Mager[48] in 1980 and CASE by Klopman[49] in 1984. Such programs have given impressive results in terms of data classification, but suffer from two enormous difficulties. The first is that of comprehension of the technique—especially for the non-mathematically minded. The second is that the derived mathematical functions often cannot be translated into a communicable physicochemical model.

17.1.4.1.2 *Physical models*

The Hansch model was originally based on intuition, leading to the expectation of an optimum $\log P$ value. This expectation was justified mathematically by Penniston and co-workers[50] in 1969. They used a kinetic model to describe the movement of molecules through a series of aqueous compartments separated by lipid barriers. Dearden[51] extended this analysis and in 1976 focussed on the time dependence of both biological activity and of optimum $\log P$ value.

Building on an earlier probability argument of McFarland,[52] a very significant contribution was made by Kubinyi[53] who in 1977 derived the 'bilinear' model of equation (21), in which β represents the ratio of volumes of the lipid and aqueous compartments. This model still allowed for an optimum $\log P$, but provided linear ascending and descending parts to the curve, with different slopes. Retrospective studies have shown that data previously fitted to a Hansch parabola in $\log P$ or in π are more often than not better described by a bilinear curve.

$$\log(1/C) = k_1 \log P - k_2 \log(\beta P + 1) + k_3 \qquad (21)$$

Using both thermodynamic and kinetic arguments, a variety of physical models have subsequently been advanced, most notably by Martin who in 1978 proposed complex equations to deal with ionizable compounds in such a way as to discover whether the ionized, unionized or both fractions were active at a receptor. The QSAR practitioner nowadays has a bewildering variety of 'model based' equations from which to choose.[54]

17.1.4.1.3 Variables

Assuming high quality biological data, the 'goodness of fit' of an equation will depend on two factors: (1) the ability of the model to describe reality and (2) the relevance of the parameters (independent variables of the equation) to the physicochemical processes being modelled.

There are three sources of QSAR parameters: measurement, calculation and extrapolation from a database. There are three categories of parameter required for the classical LFER-based approach. These are (i) hydrophobic, (ii) electronic and (iii) steric. Significant improvements have been made in our ability to measure and to calculate relevant descriptors for all of these effects, since the first equations of Hansch in 1964–5.

(i) Hydrophobicity

The octanol–water system was rapidly established as the most practical experimental model system from which to derive $\log P$ or π values, even though difficulties occurred in the accurate measurement of ionizable compounds, molecules with extreme values of $\log P (< -2$ or $> 3)$ and those with poor UV absorption.[55] Some partition coefficients were measured in alternative lipid phases when compounds were not very soluble in octanol, thus steroids were measured in ether–water. The expected linear relationships[56] were found between $\log P$ values in two different systems, provided that the hydrogen-bonding characteristics of the solute molecules did not vary.[57] The thermodynamics of partitioning was studied, and there was much controversy about the nature of the so-called 'hydrophobic bond'.[58]

To overcome some of the difficulties and indeed the tedium of 'shake-flask' partition coefficient measurements, reverse phase TLC was used to derive R_M, related to the measured R_F through equation (22). That R_M could be used in place of $\log P$ or π in biological correlation was first shown by Boyce and Milborrow[59] in 1965. The method was reviewed by Tomlinson[60] in 1975. Also in 1975, McCall[61] was the first to use an HPLC-derived hydrophobicity parameter, HPLC being found extremely useful for compounds of very high $\log P$. Since it is not concentration but retention distance or retention time that is measured in chromatographic techniques, samples do not have to be either pure, plentiful or particularly stable.

$$R_M = \log(1/R_F - 1) \qquad (22)$$

The predictive use of QSAR can only be realized if parameters can be reliably calculated. Since it is overall lipophilicity ($\log P$ rather than π) that determines drug transport, much work has been devoted to methods for calculating $\log P$. The early work of the Hansch school showed $\log P$ to be not only an additive but also a constitutive property. For instance, it was not reliable to calculate $\log P$ for 4-nitrophenol by taking π values derived from substituted benzenes and applying equation (23) because resonance interaction would make π values atypical. In 1967, Bird and Marshall[62] measured $\log P$ for many penicillin derivatives and found that differences between them did not correspond to those expected on the basis of π values. It was now clear that not only electronic but also steric and intramolecular hydrogen-bonding (as in 2-nitrophenol) effects would cause simple additivity to be seriously in error.

$$\log P(\text{4-nitrophenol}) = \log P(\text{benzene}) + \pi(NO_2) + \pi(OH) \tag{23}$$

A great leap forward was taken in 1973 when Rekker[63] proposed a more general method to calculate log P. This was based on assigning 'fragmental constants' f to a variety of structural pieces, the calculated log P then being simply the sum of the f values appropriate to the molecule in question plus any interaction factors F that might be necessary to correct for intramolecular electronic, steric or hydrogen-bonding interactions between fragments. Equation (24) expresses the fragment system.

$$\log P = \sum_{i=1}^{m} f_i + \sum_{i=1}^{n} F_i \tag{24}$$

Using a large database of log P values, Rekker derived both fragment and interaction constants statistically. In 1977 he published a book[64] on his method and by 1979 had further refined his set of f values by using a database of over 1000 log P measurements.[65] Rekker further proposed that all his interaction factors F could be treated as multiples of what he called the 'magic number' (0.28), although the selection of multiples was sometimes uncertain as also was the choice of exactly what constitutes a fragment. Despite such difficulties, the method was generally accepted as being the most practical way to calculate log P. This work also greatly added to our understanding of hydrophobicity.

By 1979, Hansch and Leo had also accepted the wisdom of a fragmentation scheme,[66] and in collaboration with Chou and Jurs of Pennsylvania State University developed the first version of a computer program for calculation of log P from structural formula, CLOGP. To develop this program, care was taken to define very clearly exactly what constitutes a 'fragment' and, instead of the statistical approach adopted by Rekker, use was made of a small basis set of carefully validated log P measurements to derive fundamental fragment values. The current version of the program, CLOGP-3, is now the 'industry standard' and widely used.

(ii) Electronic parameters

Several experimentally derived electronic parameters can be used in correlation analysis, because they are all related. Most used have been pK_a or ΔpK_a, but for substances not undergoing ionization then chemical shift values from NMR spectroscopy have been useful. Kutter[67] in 1972 used redox potentials in a QSAR for nitrothiazoles, as evidence that they interfere with trichomonads 'by virtue of their reducibility'. By 1979, several groups had established correlations between inhibition of DNA synthesis by nitroheterocycles and measured redox potentials.[68, 69]

Encouraged by Taft's introduction of the polar substituent constant σ^* in 1956, an enormous variety of Hammett-like parameters have been derived from model reactions and used in correlation analysis. Franke[70] in 1984 listed 27! Of particular significance were the introduction[71] of parameters relevant to free-radical reactions; the factorization of σ into its inductive and resonance components ($\mathscr{F}$ and $\mathscr{R}$) by Swain and Lupton[23] in 1968; modified $\mathscr{F}$ and $\mathscr{R}$ values 'weighted' according to position in an aromatic ring by Norrington[72] in 1975; and charge-transfer constants proposed by Hetnarski and O'Brien[73] in 1975 and alternatively by Livingstone and co-workers[74] in 1978.

Just as hydrophobic fragments have largely replaced substituent constants for calculation of hydrophobic property, so has there been a trend towards calculating 'whole molecule' electronic indices. This was a necessary development, since Hammett parameters reflect an intramolecular influence of one part of a molecule on another part, and for characterizing drug–receptor interaction then an intermolecular parameter should be more relevant. Tute[75] in 1970 and McFarland[76] in 1971 explored dipole moment and polarizability, but these indices have not been much used. A host of electronic properties can be calculated by quantum mechanical techniques, and though attempts to relate some of these to biological activity date back to the 1930s, there was little progress until computational power became more available in the late 1960s. The Pullmans, Cammarata and Kier were the pioneers in this endeavour, focussing on atomic charge, orbital energy (especially HOMO and LUMO) and 'superdelocalizability' indices. Correlation of biological data using MO indices has not often been useful, probably because the indices relate to an unrealistic state, that of an isolated molecule in a vacuum.

With the still greater computer power available in the 1970s, a very exciting application of MO theory was made by Weinstein and co-workers.[77] In 1973, they were able to calculate the electrostatic potential around a molecule and so create a contour map where the contours represent lines of equal energy of interaction with a unit positive charge. These maps can be thought of as magnetic fields and are now believed relevant to molecular recognition and the early events in drug–receptor association. Values of the potential at strategic points in space (near a basic nitrogen atom and above an aromatic ring) were used successfully by Solmajer[78] in 1984 as QSAR parameters, relevant to the interaction of neurotransmitters with their receptors.

(iii) Steric parameters

There has been much uncertainty as to the proper treatment of steric effects. The problem is acute in flexible compounds, where both intramolecular effects (on conformation and other properties, *e.g.* log *P*) and intermolecular effects (bulk tolerance to the shape of the receptor cavity) must be entangled. Discounting the imprecise measurement of the volume of water displaced by a space-filling CPK model (!), there is no suitable experimental parameter.

The Taft E_s constant was first used[79] in QSAR in 1965, followed by Hancock's modified parameter E_s^c, which incorporated a correction for hyperconjugation.[80] In 1969, Charton[81] defined a constant v which has been used subsequently in peptide QSAR. These parameters are all related to the van der Waals radius of any symmetrical substituent or to the minimum width of unsymmetrical ones. To overcome the problem of asymmetry of substituents, Verloop[82] introduced in 1976 the STERIMOL set of five parameters (calculated by a computer program) to represent length and 'width' in four different perpendicular directions. The STERIMOL parameters have been much used and have the advantage that appropriate values can be calculated for a novel substituent, provided that structure coordinates are available from an X-ray or from a model 'built' using computer graphic techniques.

The most chameleon-like parameter in QSAR is undoubtedly molar refractivity, MR. Following the trend set by Hammett, MR was proposed by Pauling and Pressman[83] in 1945 as an electronic parameter for the correlation of dispersion forces involved in the binding of haptens to antibodies. The proposal was based on MR being proportional to the polarizability of a molecule. Now, MR can be measured by making use of the Lorentz–Lorenz equation (25) where *n* is the refractive index, MW the molecular weight and *d* the density.

$$MR = \frac{(n^2 - 1)}{(n^2 + 1)} \cdot \frac{MW}{d} \tag{25}$$

Since *n* does not vary much for most organic compounds, and MW/*d* is, of course, volume, then MR can also be considered as a crude steric parameter, characterizing bulk (but not shape) of a molecule or substituent. In the late 1960s, QSAR practitioners called attention to the fact that MR could be highly correlated with π, particularly in series rich in alkyl substituents. In such series, correlation of activity with MR may reflect merely a hydrophobic effect. Recognition of the collinearity problem has resulted in medicinal chemists paying much more attention to series design, for example by selecting deliberately a set of substituents for subsequent analysis such that their chosen parameters are not collinear (see Section 17.1.4.2 and Chapter 21.2).

In 1974, Franks[84] put forward some intriguing suggestions about the nature of hydrophobicity. He thought that there may in some instances be incomplete desolvation between ligand and receptor, in contrast to the usual complete squeezing out of all water molecules. In 1976, after finding positive coefficients with an MR term in several equations developed to describe ligand–enzyme binding, Hansch recalled Franks' suggestion and speculated that MR might be reflecting an unusual kind of hydrophobicity.[85] This idea seems now to have been abandoned, for lack of any physical evidence.

Many more enzyme–ligand QSARs were generated in the 1980s, and some of these concerned systems where the X-ray crystallographic structure of the receptor was known. By 1986, Hansch and Klein[86] were able to use computer graphics to examine the nature of these ligand binding sites and compare the pictures with the QSAR. The pictures and the QSAR told the same story. Hansch commented 'To actually see the beautiful . . . region . . . demanded by equation . . . after so many years wondering what real meaning our QSAR parameters had at the molecular level, was to experience an enormous sigh of relief'.

17.1.4.2 Strategy and Statistics

Over the past 25 years, many statistical methods familiar in other fields have been applied to the problem of correlating structure with activity. The choice of most appropriate method depends partly on the objectives: this may be classification, diagnosis of mechanism or prediction of the most fruitful direction for synthesis. The choice also depends on the quality and quantity of data available.

The technique that has been most used in QSAR is linear multiple regression, which employs the least-squares method to find the equation of 'best fit' of biological activity with a given combination of parameters. The limitations and some common pitfalls of multiple regression analysis were pointed out by Tute[1] in 1971. There must be a sufficient number of compounds included in the analysis to enable statistical significance to be reached, despite inevitable errors in measurement. A rule of thumb was evolved, that at least five data points (compounds) should be included for every parameter in the equation. The parameters themselves should be well 'spread', and one must avoid the particular danger of including one extreme value of any parameter, as a 'best line' will inevitably pass through the extreme point, which will be determinative for the whole analysis. Any error in this point would seriously influence the result and could lead to disastrous misinterpretation. Some of the parameters used are themselves interrelated, and this constitutes a major pitfall to the technique. Dependence of π on σ was noted by Fujita[27] in 1964, and was investigated by Wulfert[87] in 1969 by the measurement of $\log P$ for a number of substituted heterocycles. Wulfert compared experimental $\log P$ with values calculated by addition of $\log P$ for the nucleus and tabulated π values for the substituents. He showed that $\log P$ for the derivatives was correlated with a linear combination of π and σ. Kutter and Hansch[88] noted in 1969 that in some series π increases as the size of substituent increases, so π and steric constants can be related.

In 1971, Craig[89] recommended the use of plots, *e.g.* of σ against π (now called a Craig plot), in order to recognize problems of collinearity or occurrence of extreme values, and most of all to ensure an adequate representation of 'substituent space'. Since then, many workers have carefully considered the strategy of a QSAR analysis, addressing such questions as how to select substituents so as to (1) minimize collinearity, (2) maximize variance and (3) map substituent space with the smallest number of synthetically feasible probes.

The tendency of QSAR practitioners in the early 1970s was to choose parameters they believed to be relevant to physical models, perhaps σ, π and E_s, and test them in regression analysis, singly and in all combinations. Regression of one independent variable against another would be used to test for covariance and Craig plots would ensure adequate spread of data. But the proliferation of variables, in particular through introduction of molecular orbital indices and of connectivity indices, led many in the late 1970s to adopt the strategy of letting the computer 'search' for the best combination of variables from a database of perhaps 30 or 40 tested. Topliss[90] in 1972 warned practitioners of the danger of this procedure: the standard statistical tests on a linear regression equation take into account only the variables contained in the equation and do not consider the much larger number from which they may have been selected. Topliss[91] further examined this problem in 1979, by simulating QSAR studies. He employed random numbers as variables, and showed that there is indeed a high risk of arriving at an unrealistically high assessment of the statistical significance of a correlation when too many variables are tested irrespective of the number finally appearing in the regression equation.

Large numbers of variables and huge data sets clearly bring their own problems. A quite different problem occurs when one has few results to analyze and synthesis of new compounds is difficult. Topliss (again!) proposed a logical procedure to follow when given only *one* result. Topliss[92] put forward his 'decision tree' in 1972, and in so doing formalized a procedure often adopted by medicinal chemists, of introducing substituents into a 'lead compound' that, in their experience, were likely to give more active derivatives. Thus, the first step might be to introduce a chlorine substituent into an aromatic ring. Dependent on whether activity decreases, remains the same or increases, so the next step is to make the methoxy, methyl or else a dichloro derivative. The sequence employed in the decision tree can vary somewhat, but always assumes that lipophilicity, electron push/pull from the substituent and size are important determinants of activity. Martin and Dunn[93] have shown that by following a Topliss scheme optimum activity could have been reached rapidly in many series of compounds. However, the scheme does nothing to avoid the collinearity problem, and must be used with great care after making the first four or five derivatives. It can be misleading.[36]

In 1974, Darvas[94] proposed a graphical method which is based on the sequential simplex technique of optimization. This was to be used when results on at least three compounds were available, in order to choose a profitable fourth for synthesis. Cramer,[36] in developing his series of antiallergenic pyranenamines in 1979, commented that strict adherance to the Topliss decision tree would not have resulted in superior compounds, but for directing synthesis of early analogues he found a 'relief map' representation of the dependence of potency on π and σ of great value. There may now be a computer on every desk, but we should still not despise graph-paper technology!

17.1.4.3 Quantum Mechanics and Molecular Orbital Theory

Arising from the work of Einstein, a prediction was made in the early 1920s that electrons have associated with them waves of a frequency dependent on their mass and their velocity. This led Schrödinger in 1926 to postulate his famous wave equation and quantum mechanics was born.

Quantum mechanics is an attempt to describe, in mathematical terms, the nature of matter at the atomic and molecular level. Because of its very complexity it is limited to very simple systems, and for practical use the approximation known as molecular orbital (MO) theory has been developed. Even MO theory is intensely mathematical and until the advent of the computer was both too complex and too remote for most chemists to give it a second thought. In 1953, Ingold in his renowned textbook 'Structure and Mechanism in Organic Chemistry' dismissed the topic in a mere 11 pages, devoted to a discussion of the hydrogen molecule.[95] This was hardly changed by the second edition, in 1969! To most organic chemists quantum chemistry was irrelevant to their needs, even by 1970 when the subject in the hands of a few theoreticians with access to computers had actually progressed quite far. MO theory in 1970 was able to account satisfactorily for many aspects of chemical structure and reactivity. Indeed, some reactions which had long been a puzzle to traditional curly-arrow pushers could be rationalized in terms of MO theory at a very simple semiempirical level which considered only valence electrons and only 'frontier' orbitals.[96]

The simplest approximation considers only the π electron system, and was founded by Hückel in 1932. This theory ignores any interaction between π and σ electrons, so can be used with confidence only for relatively planar, aromatic or unsaturated systems. In the 1950s, the simple Hückel treatment was superseded by less approximate methods which could deal with all valence electrons, but still had to ignore most of the intramolecular orbital interactions. These methods were further developed in the 1960s, with EHT (extended Hückel theory) being introduced by Hoffman in 1963, CNDO (complete neglect of differential overlap) in 1965, INDO (intermediate neglect of differential overlap) in 1967, PCILO (perturbation configuration interaction using localized orbitals) in 1968 and MINDO (modified INDO) in 1969. In 1977, Dewar introduced MNDO (modified neglect of differential overlap). All of these methods are used today, programs being routinely available. There is still much debate about which is the better semiempirical method for a particular application, or whether an *ab initio* (from the beginning) calculation should be performed, despite the much heavier demands on computer time.

With the much greater computer power available by 1970, some theoreticians did indeed turn away from the semiempirical methods to the supposedly more reliable *ab initio* technique. We seem now to have arrived at the position where each chemist has their own method in which they believe, and which is fast enough for their purposes, but has little trust in or experience of anyone else's method.

Biochemical applications of MO theory were pioneered by the Pullmans in the 1960s, with calculations on nucleic acid bases, peptides and sugars. They applied the semiempirical methods to study reactivity, hydrogen bonding and conformation. Other workers in the biosciences followed the Pullmans' lead, and the first international symposium[97] on 'MO Studies in Chemical Pharmacology' was held in 1969. In 1971, Kier[98] reviewed progress in his book, 'MO Theory in Drug Research'. Much of the work in the 1970s focussed on predicting minimum energy conformations for agonist molecules such as acetylcholine (**18**) or histamine monocation (**19**). Sometimes, such conformations were at variance with those deduced from NMR studies in solution, so the validity and relevance to drug action of a theoretical (gas phase) conformation was most vigorously debated.

$$MeCO_2CH_2CH_2\overset{+}{N}Me_3$$

(**18**)

$$H-N\diagdown N\diagup\overset{+}{CH_2CH_2NH_3}$$

(**19**)

Calculations on histamine illustrate some of the accomplishments of MO theory. In the 1970s, Kier did pioneering work on the conformation of agonists, using the EHT method. Kier[99] developed thereby his concept of 'remote recognition of preferred conformation'. According to this, a low energy (thermodynamically preferred) conformation of the agonist is 'recognized' by the receptor and an appreciable interaction energy develops. This evokes an effect (agonism) long before equilibrium of the interacting pair is reached. For the histamine monocation, Kier found two conformations (*gauche* and *trans* across the ethane bond) to be of approximately equal preference and assumed the *gauche* form to be responsible for H2 receptor activation and the *trans* form for H1 receptor activation.[100] These two receptors for histamine had recently been characterized (see Chapter 12.5). This conclusion was later challenged by Ganellin,[101] whose results indicated that receptor-bound conformations were of more relevance than 'remote' conformations and that dynamic equilibria must also be taken into account.

The conformational equilibria of histamine cation were again considered by Farnell, Richards and Ganellin[102] in 1974. They pointed out that in calculating equilibrium properties one should take entropy into account (MO calculations being only of enthalpy) in order to derive free energies. They showed how to calculate the entropy using statistical mechanics, but pointed to a further problem: the need to estimate the differential effects of solvation on the various conformers! The EHT results for histamine cation do correspond to NMR experiments in solution, so perhaps (by a fortuitous cancelling of errors?) the EHT method does reproduce solvent preferences. This is not so with CNDO, PCILO or *ab initio* calculations, which have all predicted histamine monocation to be predominantly *gauche* and the dication to be predominantly *trans*. Abraham and Birch[103] showed in 1975, however, that by including a counterion in the CNDO calculation the solution results could be reproduced. From this time 'supermolecule' calculations, *i.e.* including a counterion and an inner shell of water molecules, became common. By 1980, calculations were being made not only of conformation but also of drug–receptor interaction energy using X-ray crystallographic data and the tools of computer graphics, distance geometry, molecular mechanics and very sophisticated molecular orbital methods. Good accounts have been given in a book by Richards[104] and a review by Kollman[105] (see also Chapter 18.1).

17.1.4.4 Model Interaction Calculations and Molecular Mechanics

The development of quantitative methods has benefited greatly from the criticism applied by one worker to the results of another. Thus it was that Kier, having been criticized by Gannellin for not considering the dynamic aspects of the histamine/histamine receptor interaction, devised together with Holtje the method of 'receptor mapping by model interaction calculations'. This method involves the calculation of interaction energies (electrostatic, polarization, dispersion and repulsion) between a series of active molecules and a group of model compounds simulating a likely receptor or enzyme active-site moiety. These moieties are parts of amino acid side chains. Using the monopole polarizabilities method of Claverie and Rein[106] for the numerical calculations, Kier and Holtje[107] in 1975 studied a series of acetylcholine (**18**) analogues, in which the onium group was successively replaced by *t*-butyl, isopropyl, ethyl and methyl groups. It had long been assumed that interaction of the onium head with acetylcholinesterase involved some anionic receptor site. This was modelled as an acetate ion (*i.e.* part of glutamic or aspartic acid). Surprisingly, it was found that a benzene ring (part of phenylalanine) provided a much better model to fit activity values in this series. Perhaps the assumption of a Coulombic interaction is wrong, and the interaction of acetylcholine with its receptor involves an ion-induced dipole. There have been no X-ray studies on acetylcholinesterase, so this hypothesis remains an intriguing possibility.

After five years and a series of such studies, Kier went on to develop the molecular connectivity approach, discussed in the next section. The computationally intense Claverie and Rein based method was superseded by molecular mechanics calculations, reviewed and exemplified in a paper by Blaney and co-workers in 1982 (see Chapter 18.2). Using molecular mechanics calculated interaction energies and a simple empirical method to estimate solvation energy differences the experimentally observed free energies of association of a number of thyroxine derivatives to prealbumin were put into the correct order of magnitude. These calculations were used to predict, successfully, the binding of further derivatives.[108]

Molecular mechanics (MM) was initiated by the work of Hendrickson[109] in 1961, who calculated relative energies of cycloalkanes by considering arrays of atoms connected by bonds as hard spheres connected by springs, then applying Newtonian mechanics. The first application to peptides was by Ramachandran[110] in 1963 and the method has grown rapidly since then. All MM methods assume that the energy of a molecule (or array of molecules) can be described by the sum of energies of various mechanical and electrical contributions. These contributions are rapidly calculated, given input geometry and a set of equilibrium values and force constants describing simple structures (the force-field). MM was originally used for model building and conformational analysis of isolated and strictly gas phase molecules. Although this remains its principal application, appropriate force-field parametrization, such as inclusion of a specific hydrogen-bond function and use of a distance-dependent dielectric constant for the electrical contribution, has since allowed for realistic attempts to model intermolecular interactions including both geometry and enthalpy of some model drug–receptor interactions.

MM and computer graphics technology have grown up together and complement one another. Complex molecular interactions can now be investigated by 'building' models on a suitable graphics device and optimizing the geometry with the most appropriate MM or MO tool. A relative energy is derived for each molecular association, and this can be entered into a regression analysis against biological data.[111]

17.1.4.5 Topology and Connectivity

The topological methods developed in the 1970s, the minimal topological difference (MTD) method of Simon[112] and the molecular connectivity (MC) method of Kier and Hall[113] can sometimes be applied with only paper and pencil and a modest pocket calculator.

The MTD method (originally called minimum steric difference, MSD) was initiated in 1973. The structures of all molecules of a set with some variation in activity are drawn (in two dimensions) and the skeletal structures (omitting hydrogens) overlapped in some way, as far as possible matching atom for atom, to give a topological network with all atoms as vertices. I think all medicinal chemists must do this from time to time in speculating as to what is revealed about the shape of the receptor from the shape of active molecules (which must fit) and less active analogues which may be too small to utilize all possible binding sites or, conversely, of such a shape that exact fit is not possible.

The chemist can speculate on a complementary shape of the receptor and trace this out on the network of atoms which represents the overlapping set. The MTD value for each molecule is simply the number of non-overlapping atoms when the molecule is compared to that part of the network which is hypothesized to represent the receptor cavity. Different cavity assumptions give different sets of MTD values which in turn can be entered into regression analysis against biological activity. The best regression yields information on the topology of the binding site.

This simplistic approach has yielded some interesting results, but too much cannot be expected from it. It has only proved applicable where steric fit is of paramount importance: moreover, MTD is of course a *de novo* parameter applicable only to the study from which it was derived. For those with access to molecular mechanics and computer graphic devices, MTD analysis was rapidly outmoded. It was superseded in 1980 by the philosophically similar molecular shape analysis of Hopfinger.[114]

Ever since the introduction of 'Hansch analysis', most medicinal chemists have preferred to develop a physical model for drug action by using essentially property-based parameters reflecting changes in hydrophobic, steric or electronic effect. However, property-based parameters for QSAR can be criticized. It is not always easy to measure or calculate such parameters, in particular ones relevant to steric effects. One rarely knows which properties are likely to be relevant. Furthermore, property parameters can be redundant: that is, two or more very different structures in a data set may have the same $\log P$ value or the same pK_a, and after all we are interested in designing drug *structures*. The chemist must have a structure as his target for synthesis, not a property!

The topological parameters known as molecular connectivity indexes were developed by Kier and Hall with the foregoing criticisms much in mind. In 1975 they set out to provide a set of non-redundant and fundamental descriptors of structure.

The derivation of a topological index (or set of indices) for a molecule begins with a structural drawing, sufficient to show all connections but usually omitting hydrogens. What number(s) can unambiguously represent the structure? If the universe consisted solely of straight chain hydrocarbons, the number six could be chosen unambiguously to represent hexane. But molecules branch, have rings, multiple bonding and a variety of atom types. Herein lies the problem but also the excitement of developing suitable indices for use in QSAR.

A topological index capable of characterizing the 'branchedness' of molecules was put forward by Wiener in 1947. He based his index on the number of bonds (edges of the graph) traversed in going by the shortest route from one atom (vertex of the graph) to another atom across the structure.[115] This number is summed for all pairs of atoms and turns out to be larger for a linear than for a branched isomer. It is a measure of molecular size and shape and thus was found to correlate with some physical properties in homologous series.

An improved index for characterizing shape was devised by Randic in 1975, and he is credited with the idea of assigning each atom (vertex) a 'degree of connectivity' as simply the number of other atoms to which it is attached. Each edge (connection between atoms, irrespective of whether singly or multiply bonded) is given a value which is the reciprocal square root of the product of the degrees of the two vertices. The Randic index is then the sum of these edge values over the whole molecule. This index gave even better correlations with many physical properties.[115]

Kier and Hall identified several problems with the Randic index. In particular, there was no differentiation between carbon and heteroatoms or between single and multiple bonds. Such a limitation would inevitably mean that the index was of very limited use in correlating biological phenomena. In seeking a remedy, Kier and Hall perceived a relation between the input for a simple Hückel-type MO calculation and the assignment of 'degree' to a vertex (atom). The Hückel input is in terms only of number of valence electrons on each atom and atom connections! Could a 'degree' for each vertex be constructed so as to reflect the available bonding electrons and thus include both atomic identity and potential for multiple bonding?

Thinking now in terms of electronic structure rather than pure graph theory, Kier and Hall perceived two ways of assigning an electronic 'degree' to each vertex (atom). These are defined for first row elements as connectivity delta (δ) and valence delta (δ^v) by equations (26) and (27), where σ is the number of electrons involved in sigma bonds, Z^v is the total number of valence electrons and h is the number of attached hydrogen atoms. It turns out of course that connectivity delta is equal to the Randic degree, but valence delta is new. The first-order (path between two connected atoms) index based on δ^v was called first-order valence chi ($^1\chi^v$) and has been the most widely used of all the indices. Its calculation is exemplified in Scheme 1. This development established a connection between topological indices (previously derived using pure graph theory) and the electronic structure of molecules.

$$\delta = \sigma - h \tag{26}$$

$$\delta^v = Z^v - h \tag{27}$$

$$^1\chi^v = \Sigma(\delta^v_i \cdot \delta^v_j)^{-0.5}$$

$$= 2(1/3.1)^{0.5} + (1/3.4)^{0.5} + (1/4.6)^{0.5} + (1/4.3)^{0.5}$$

$$= 1.9362$$

Scheme 1 Calculation of first-order valence chi

Indices were derived from both δ and δ^v and using all possible 'subgraphs' (*i.e.* chains, clusters and rings) of connected atoms. Each index was calculated (in the way pioneered by Randic) as the sum of reciprocal square roots of the product of delta values along the path for all paths. Some very good correlations emerged, not only with physical property but now also with biological activity. Kier and Hall rushed into print with their first book, 'Molecular Connectivity in Chemistry and Drug Design'. Unfortunately, this book[113] presented the reader with rather a heavy dose of graph theory and did not develop the electronic arguments. Also the correlations were just too good, leading to the widespread feeling that there just had to be something wrong with the statistics. A reviewer of the book recommended it 'to all who admire radical departure, courageous innovation, and unbridled controversy . . . as a very exciting and exotic approach to theoretical chemistry and drug design'. During the next decade, Kier and Hall extended the connectivity concept to elements higher than the first row of the periodic table, and found relationships between particular indices and such properties as electronegativity, molecular volume, molar refraction and surface area as well as greatly extending the number of biological correlations. A second book,[116] 'Molecular Connectivity in Structure–Activity Analysis', followed in 1986.

Despite the many studies of Kier and Hall, and others, MC has not been much used in drug design by researchers in industry. The main problem lies in the difficulty of interpreting the meaning of a correlation in such indices. In addition, by its very nature MC cannot distinguish between *cis–trans* or stereoisomers, or address conformational problems. It can never give the insight into biological phenomena that can sometimes be gleaned from MO or MM techniques. Nevertheless, there are advantages in ease of computation and it has proved helpful for the interpolative prediction of physical properties and the correlation of certain biological responses that are insensitive to conformational features.

17.1.4.6 Shape and Conformation

The conformation of a molecule refers to the relative positions of its constituent atoms, and shape refers to the distribution of molecular 'bulk' according to the conformation. The role of shape in determining the biological activity of a compound is implicit in the stereoselectivity of biological receptors: it has long been known that for a great many substrate–enzyme or drug–receptor interactions an exact fit of one partner with the other is necessary to evoke a maximal response, or indeed any response at all.[117] There are indeed many examples of activity differences between optical isomers.[118]

The importance of shape to activity is easy to imagine, but has been much less easy to quantify for QSAR. Lehmann has discussed the particular problems of developing quantitative stereostructure activity relations (QSSAR), pointing out that the use of indicator variables to distinguish isomers has not been very helpful. In 1977, Lehmann introduced the term 'eutomer' for the more potent of a pair of enantiomers and 'distomer' for the less potent. He then coined the term 'eudismic ratio' for the ratio of their activities and went on to explore relationships between the eudismic ratio and potency of the eutomer in a study of chiral discrimination at various receptors.[119]

Some oft-quoted analogies have helped our imagery of the role of shape in drug action, the most famous being that of lock-and-key suggested by Fischer[120] as long ago as 1894! A better analogy might be that of glove-and-hand which conveys the difference in optical isomers as right and left hand, only one of which will fit perfectly. A particularly apt analogy was given by Cornforth in a discourse on the stereospecificity of enzymes: he likened the substrate to a bullet placed in a rifle, an analogy which conveys not only the importance of exact fit but the enormous velocity of an enzymic reaction, once triggered by the correct conditions.

Classical lock-and-key imagery is still useful, especially for rigid molecules. For flexible molecules, and especially for proteins, then a lock-and-key concept will often be an oversimplification. One has to consider conformational (shape) changes taking place in one or both partners as association of drug and receptor takes place. Thus, Koshland[121] in 1968 discussed the evidence for an 'induced fit' between enzymes and substrates, a concept easily extended to drugs and receptors.

For 'flexible' drugs, Kier had placed great emphasis on determining the preferred conformation of the drug molecules (see Section 17.1.4.3) and developed hypotheses as to the shape of receptor sites by measuring distances across agonist molecules in their calculated low energy conformations. However, in 1975 Burgen[122] argued that a drug–receptor complex will in any case be formed, provided only that the process is not sterically hindered and that there is an overall decrease in energy, implying a cooperative process in which the drug molecule (and possibly receptor too) changes its conformation. This mechanism would surely be necessary to explain the binding of peptide hormones to their receptors, association taking place at various sites along the peptide chain in sequential fashion (the so-called 'zipper' mechanism). By 1979, it had become generally accepted that a minor conformer could be responsible for activity.[123] Moreover, thermodynamic measurements[124] made in 1979 on the binding of agonist, partial agonists and antagonists to the beta-adrenergic receptor led to a generalization of relationships between binding, conformational change and activity which strongly supported Burgen's viewpoint. For the beta-receptor, agonist binding was found to occur with a large negative enthalpy change (it was said to be enthalpy driven) and it was considered that this provided compensation for conformational changes induced in the receptor, with corresponding loss of entropy. In contrast to agonist binding, antagonists bound with increase of entropy (they usually carry large hydrophobic substituents) and were said to be entropy driven. Partial agonists bound with moderate changes in both entropy and enthalpy. From this time, it was realized that for a drug to bind to its receptor effectively it was only necessary that the bound conformation be *achievable* though highest affinity would still be realized with a binding conformation whose energy was equivalent to that of its solution minimum.

Such considerations, now that computer graphics technology was available and methods for determining the relative energies of conformers were much improved, gave rise to a variety of automated techniques for predicting the bound conformations of drug molecules at an unknown receptor. Thus, in 1979 Marshall[125] introduced a classification method, the 'active analogue approach', to determine the three-dimensional correspondence between 'pharmacophoric groups' among a set of active compounds. A philosophically similar method was introduced by Wise and Cramer[126] of Smith, Kline and French research which they called DYLOMMS (dynamic lattice-oriented molecular modelling-system). Molecules were represented as both steric and electrostatic energy maps, computed as Boltzmann-weighed sums of the maps of each molecule's contributing conformers.

For receptors of known geometry, for example dihydrofolate reductase (DHFR) from X-ray crystallography, Kuntz in 1982 introduced a fitting algorithm to determine complementarity of shape between the rigid receptor and flexible drug molecules. This algorithm was refined in 1986 to include energy minimization[127] of the crudely fitted drug molecules on the receptor, using the AMBER force-field (see Chapter 18.2).

A general method for comparing the shapes of molecules was developed in 1980 by Hopfinger, and called molecular shape analysis (MSA).[114] First using molecular mechanics to determine accessible conformations, meaningful criteria were then proposed for superimposing members of a set of active analogues. Common volumes of overlap were then computed and used as *de novo* parameters in regression analysis, either alone or in combination with π, σ, *etc.* This technique (a three-dimensional refinement of Simon's MSD method) effectively married conformational analysis with classical QSAR.

17.1.4.7 Computer Graphics: Visualization of Structure and Property

The prototype electronic molecular modelling system was developed at MIT in 1965, and consisted of a calligraphic (line-drawing) cathode ray tube controlled by a small computer. It was used to display proteins and to study their folding.[128] Within five years at least another 20 systems were in use, principally as laboratory tools for the X-ray crystallographer.

During the 1970s, interactive systems were developed specifically to aid the medicinal chemist in drug design, incorporating the latest technology. Using colour, line-drawing or space-filling representations of structure could be compared and superimposed. This proved to be an invaluable aid to conformational analysis and to the development and appreciation of models of drug–receptor interaction. In addition to structure, property was also usefully illustrated: graphs and maps were drawn to highlight structure–energy relationships; contours were drawn around structures to illustrate van der Waals surfaces, solvent-accessible surfaces and electrostatic potentials.[129]

Colour and dot surface representation[130] on stereo pictures allowed, after 1980, the use of molecular graphics as a tool to check on the meaning of QSARs generated in the previous decade.[86] Correlation equations had been derived for series of ligands interacting with enzymes, such as DHFR, carbonic anhydrase, trypsin and papain, whose three-dimensional structures were known from X-ray work. The QSARs had been based on kinetic parameters for reactions occurring in solution, and correlations with π, MR and σ had indicated features—such as hydrophobic pockets of restricted size—to exist in the receptor site. Molecular graphics was now used to 'build' models of some of the ligands included in the QSAR and these were 'docked' into the receptor constructed from X-ray coordinates. By colour-coding solvent accessible surfaces of the enzyme according to property (polar, hydrophobic or charged) it became possible to *see* the very interactions suggested by the QSAR. Moreover, by this technique new binding sites could be postulated. In 1982, a particular triumph followed the docking of trimethoprim (**20**; TMP, R = Me) into an X-ray-based structure of *E. coli* DHFR. The presence of an arginine residue was noted in a hydrophobic region of the enzyme and some few angstroms distant from the TMP molecule. Graphics modelling suggested that lengthening group R from Me to $-(CH_2)_5CO_2H$ would best allow formation of a salt bridge between ligand and enzyme, an ionic association which would be especially strong by virtue of the hydrophobic environment. The suggested compound was made and found to bind with 50 times the affinity of TMP.[131] As a drug design tool, molecular graphics had by now been fully justified.

(**20**)

In 1985, Goodford[132] introduced a very powerful quantitative technique for probing the environment of a protein, in order to identify putative binding sites for potential drugs. After assigning appropriate charges to atoms of the protein (defined by X-ray crystallography), a particular area or indeed perhaps all of the structure was literally surveyed on a grid of equally spaced points. The interaction energy of the protein with a 'probe' group was computed at each point on the grid, using molecular mechanics principles and with a dielectric constant dependent on the position of the probe within the protein. Using an oxygen anion as probe for the DHFR protein, the binding site with arginine was readily detected. This technique was made available to users of computer graphic systems, who could then plot values of the energy at each point, to visualize plausible binding locations for various probes such as carboxylate anion, amine cation, methyl or water (see also Chapters 20.1, 20.2 and 20.3).

17.1.4.8 Protein Crystallography and Protein Engineering

The first X-ray diffraction photograph was displayed by Laue in 1912. Immediately after the interruption of the first world war, X-ray pictures of simple organic structures such as naphthalene were compared with those of diamond and graphite, and by 1928 chemists had satisfied themselves that the benzene ring was hexagonal and flat. In 1927, Astbury and Bernal and his students, among them Hodgkin and Perutz, began a study of biologically significant molecules including sterols, amino acids, and proteins. Astbury coined the name of 'molecular biology' for his own work on natural fibres.[133]

Over the period from 1930 to 1980 techniques and equipment for crystallography steadily improved, particularly with the introduction of automatic diffractometers and through data processing by computer. It became clear that X-ray crystallography is the only technique that can give accurate three-dimensional structural information on proteins and, even more exciting, on some complexes formed between proteins and small molecule ligands. Enzymes were studied in stable complexes with inhibitors, though not of course with their substrates, for such complexes were short lived and underwent chemical change during the course of data collection.

In the 1980s, the introduction of synchrotron radiation (SR) made a huge impact on protein crystal structure analysis.[134] The high intensity and high resolution available from an SR source allowed even some enzyme substrate complexes to be examined in the crystal itself. Studies using conventional or SR techniques afforded wonderful insight into the mode of action and even the selectivity of some drugs at the receptor level, providing information of a type impossible to obtain by classical QSAR methods. For example, in searching for drugs to inhibit the angiotensin converting enzyme (ACE), a proteolytic enzyme which plays a key role in regulating blood pressure, several groups have studied the binding of ligands (as potential drugs) to the active site of a related protease, thermolysin, which can be readily crystallized. Two such inhibitory ligands are β-phenylpropionyl-L-phenylalanine (**21**) and carbobenzyloxy-L-phenylalanine (**22**). They differ only in respect of oxygen in (**22**) replacing the methylene group of (**21**), so were thought of as probable 'bioisosteres', likely to bind similarly to a receptor. In 1977, the X-ray studies of Matthews[135] showed totally *different* binding modes for each, yet both at the active site of thermolysin. In (**22**), the carboxyl group bonds directly to the catalytically important zinc atom of the enzyme, whereas in (**21**) it is the amide carbonyl that bonds to zinc and the carboxyl forms a salt bridge to an arginine residue. The two aromatic rings of each ligand occupy two hydrophobic pockets, but the rings of (**22**) do not occupy the same pockets as the corresponding rings of (**21**)! Such surprises have from time to time both disturbed and stimulated chemists in their quest for inhibitors of enzymes, based on analogies between active sites of related enzymes within families.

Interestingly, in 1982 the solution conformation of these compounds was studied using NMR. This showed that the lowest energy conformer of (**22**) is that which is bound by thermolysin in the crystal, but in contrast it is a higher energy conformer (by about 0.6 kcal mol^{-1}) of (**21**) which binds.[136] This must be one of the first experimental confirmations of the hypothesis that an enzyme does not always bind the lowest energy conformer of an inhibitor.

In the last 10 years, NMR techniques have provided not only information on the preferred conformation of small ligands in solution but also of some proteins. These techniques give information on proton–proton distances. From a set of distance constraints, some measured by NMR and some calculated from models, it has become possible to derive three-dimensional structures that are consistent with the given set of distances. The 'distance-geometry' method has become a powerful technique in its own right, and in favourable cases can give accuracy comparable to an X-ray determination. It can of course be used in cases when it has not proved possible to obtain crystals suitable for X-ray.[105]

In 1985 X-ray studies by Matthews,[137] in which he compared structures of the complexes between both mammalian and bacterial DHFRs with 10 different ligands, finally provided an explanation for the extraordinary selectivity of TMP in binding to the bacterial enzyme. Such selectivity had previously been quantified by separate QSARs developed for TMP derivatives.[138] These QSARs could now be rationalized and explained at the atomic level.

Another technique developed in the 1980s was that of protein (or genetic) engineering. This technique removed a further obstacle to X-ray crystallography, the difficulty of obtaining a sufficient quantity of material. Following sequence analysis of a protein (by now possible on minute amounts) the gene encoding it may be identified, then transferred to another organism where it can be 'expressed', that is, persuaded to produce relatively large quantities of the desired protein. For the study of enzyme structure–activity relationships, it became possible to systematically mutate amino acid residues in the active site of an enzyme and then to study the resultant changes in substrate binding and catalysis. In a brilliant series of experiments beginning in 1982 and involving genetic engineering, enzyme kinetic measurements and X-ray crystallography, Fersht and his colleagues have examined in atomic detail the workings of the enzyme, tyrosyl-tRNA synthetase.[139] Fersht was able to dissect out the contributions of various residues to binding of substrates and transition state as reaction progressed, and by so doing has furthered our understanding of the role of hydrogen bonds in both recognition and binding between enzymes and their substrates. By plotting rate of reaction against equilibrium constants for the wild-type and the mutant enzymes, Fersht made, in 1986, the first application of QSAR to engineered proteins.[140]

17.1.5 CONCEPTS AND PROSPECTS

In the development of a QSAR, efforts are made to differentiate between and quantify the contributions of drug structure or property to drug activity. If such a differentiation can be accomplished, a novel drug structure can then be proposed and by a process of integration its properties and activity can be predicted.

Over the years we have learned of the severe limitations to predicting the new 'wonder drug' on the basis of a single QSAR. These limitations result from the fact that a drug structure has to be optimized for so many facets of its action: solubility, stability, absorption, resistance to metabolism, transport to a site, transport away from another site; in addition to the obvious one of being just the right size, shape and polarity to perturb its target; and to be safe to both patient and environment! A QSAR can often be developed that encompasses a prediction for some of these features, but never simultaneously for all.

The real value of QSAR development has been in the order which it brings to the structure–activity data, the understanding of the relationships between physical property and activity, and the concepts that have thereby been developed and are now being used for rational drug design. The idea of 'dissecting' drug activity into physical contributions (hydrophobic, electronic, steric) has been a central theme of the Hansch approach, and has taught us in particular just how important is the hydrophobic contribution. The overall hydrophobicity of a molecule is almost always a crucial determinant of drug transport (see Chapter 19.2). No medicinal chemist nowadays is ignorant of this concept, though he may be quite unaware of its origins. 'Hydrophobic bonding' is also a frequent contributor to the association between drug and receptor in the pharmacodynamic phase of drug action, as has been made quite evident from QSARs derived for many series of enzyme substrates and inhibitors in which terms in π appear with large coefficients.

In the 1980s, the hydrophobicity of proteins was considered very carefully and advances made on two converging fronts. Firstly, to develop QSARs for peptides, hydrophobic parameters were derived for amino acid side chains. Fauchere and Pliska[141] derived parameters in 1983, by partitioning *N*-acetyl amino acid amides between octanol and water. They found that these hydrophobicities were tightly correlated with the degree of solvent exposure of the side chains in globular proteins. Secondly, the work of Eisenberg[142] showed that protein secondary structure (folding) and indeed the stability of a protein in water is a function of the accessibility of each atom to solvent (the solvent accessible surface area) and also a function of an 'atomic solvation parameter', which is in effect a measure of the hydrophobic potential of the atom. Since the work of Chothia[143] in 1974, the relation between accessible surface area of amino acid side chains and the gain of 'hydrophobic free energy', estimated at 25 cal mol^{-1} (1 cal = 4.184 J) for each square angstrom removed from contact with water, had been widely used in considerations of hydrophobic bonding. We are now able to estimate solvation energies of particular conformations of a protein,[142, 144] and it should soon be possible to estimate the hydrophobicity of any given conformation of a molecule,[123] by scaling hydrophobic potential according to calculated solvent accessible area. This will no doubt be incorporated into some future version of the CLOGP program!

An extremely satisfying feature of QSAR has been the way in which investigations from very different angles have converged to present a coherent picture of the way drugs are transported to, then recognize and finally bind to their macromolecular receptors. Consider for example some investigations of the thermodynamics of drug binding. In 1986, Contreras and co-workers[145] found by experiment that in binding to adrenergic receptors antagonists are entropy driven, consistent with a strong hydrophobic interaction between receptor and antagonist. In contrast, three steps may underlie the binding of agonists: first, an entropy driven (hydrophobic) association similar to that for antagonists; second, an agonist-induced conformational change in the receptor with a decrease in both entropy and enthalpy; and third, the formation of a ternary complex (rarely considered by the medicinal chemist), also with a decrease in both entropy and enthalpy. Overall, agonist binding is normally strongly enthalpy driven. A QSAR approach to dissect out the contributions to free energy of binding for various functional groups has been taken by Andrews.[146] In 1984, he collected from the literature binding constants for some 200 tightly binding molecules and converted binding constants to free energy values. Through regression analysis he then provided an estimate of the *average* strengths of 'bonds' associated with 10 common functional groups, these being well represented by the set (see Chapter 18.8). Based on the work of Page,[147] the equation was forced to have an intercept value of 14 kcal mol^{-1} to represent the overall rotational and translational entropy loss on binding the molecule to its receptor. Conformational flexibility was taken into account by simple summation of the number of freely rotatable bonds in each structure. The average intrinsic binding energies so derived compare well with estimates derived by other methods, for example those given by Page and Jencks[148] in 1971 or, for hydrogen-bonding groups, the measurements of Fersht[139] on site specific mutants of tyrosyl-tRNA synthetase.

As we approach 2000, we can look forward with confidence to rational drug design based on a range of QSAR-based techniques. Many pharmaceutical companies now employ individuals or groups whose responsibility it is to understand and use the relevant technique and associated computer technology for a particular problem. A wealth of information is now available from databases, such as the Cambridge Crystallographic Database[149] which holds three-dimensional coordinates for over 50 000 structures; the Brookhaven Protein Database[150] which holds over 200 protein and some nucleic acid structures; and the Pomona College Medicinal Chemistry Project which holds log P, pK_a and other parameters of direct relevance to classical QSAR studies. By using the information present in such databases, together with appropriate techniques that are described in some detail in ensuing chapters, wholly rational drug design is now a practical concept.

ACKNOWLEDGEMENT

I would like to express my gratitude to many friends and colleagues who have provided valuable comments and advice in the preparation of this chapter, but especially to D. Bawden, J. Brown, S. F. Campbell, J. C. Dearden, L. B. Kier and M. Snarey.

17.1.6 REFERENCES

1. M. S. Tute, in 'Advances in Drug Research', ed. N. J. Harper and A. B. Simmonds, Academic Press, London, 1971, vol. 6, p. 1.
2. M. S. Tute, *Chem. Ind. (London)*, 1983, 10.
3. B. J. Richardson, *Medical Times and Gazette*, 1869, **2**, 703.
4. A. Crum-Brown and T. R. Fraser, *Trans.—R. Soc. Edinburgh*, 1868, **25**, 151.
5. C. Richet, *C.R. Seances Soc. Biol. Ses Fil.*, 1893, **9**(5), 775.
6. H. Meyer, *Arch. Exp. Pathol. Pharmakol.*, 1899, **42**, 109.
7. E. Overton, 'Studien Uber die Narkose', Fischer, Jena, 1901.

8. R. L. Lipnick, *Trends Pharmacol. Sci.*, 1986, **7**, 161.
9. J. Traube, *Arch. Physiol.*, 1904, **105**, 541.
10. A. Seidell, *Am. Chem. J.*, 1912, **47**, 508.
11. J. Ferguson, *Proc. R. Soc. London, Ser. B*, 1939, **127**, 387.
12. K. H. Meyer and H. Hemmi, *Biochem. Z.*, 1935, **277**, 39.
13. P. Ahmad, C. A. Fyfe and A. Mellors, *Biochem. Pharmacol.*, 1975, **24**, 1103.
14. A. Spinks, *Chem. Ind. (London)*, 1977, 475.
15. A. Albert, 'Selective Toxicity', 4th edn., Methuen, London, 1968, p. 436.
16. A. Albert, S. D. Rubbo, R. J. Goldacre, M. E. Davey and J. D. Stone, *Br. J. Exp. Pathol.*, 1945, **26**, 160.
17. L. Lerman, *J. Mol. Biol.*, 1961, **3**, 18.
18. A. Albert, S. D. Rubbo and M. I. Burvill, *Br. J. Exp. Pathol.*, 1949, **30**, 159.
19. P. H. Bell and R. O. Roblin, Jr., *J. Am. Chem. Soc.*, 1942, **64**, 2905.
20. H. H. Jaffé, *Chem. Rev.*, 1953, **53**, 191.
21. L. P. Hammett, 'Physical Organic Chemistry', McGraw-Hill, New York, 1940.
22. R. W. Taft, Jr., in 'Steric Effects in Organic Chemistry', ed. M. S. Newman, Wiley, New York, 1956, p. 556.
23. C. G. Swain and E. C. Lupton, Jr., *J. Am. Chem. Soc.*, 1968, **90**, 4328.
24. O. R. Hansen, *Acta Chem. Scand.*, 1962, **16**, 1593.
25. R. Zahradnik, *Arch. Int. Pharmacodyn. Ther.*, 1962, **135**, 311.
26. C. Hansch, P. P. Maloney, T. Fujita and R. M. Muir, *Nature (London)*, 1962, **194**, 178.
27. T. Fujita, J. Iwasa and C. Hansch, *J. Am. Chem. Soc.*, 1964, **86**, 5175.
28. R. Nelson-Smith, C. Hansch and M. M. Ames, *J. Pharm. Sci.*, 1975, **64**, 599.
29. C. Hansch, *Acc. Chem. Res.*, 1969, **2**, 232.
30. A. Leo, C. Hansch and C. Church, *J. Med. Chem.*, 1969, **12**, 766.
31. C. Hansch, in 'Drug Design', ed. E. J. Ariens, Academic Press, New York, 1971, vol. 1, p. 271.
32. C. Hansch, *Drug Dev. Res.*, 1981, **1**, 267.
33. C. Hansch, in 'Biological Activity and Chemical Structure', ed. J. A. Keverling Buisman, Elsevier, Amsterdam, 1977, p. 47.
34. R. Franke, 'Theoretical Drug Design Methods', Elsevier, Amsterdam, 1984, p. 282.
35. Y. C. Martin, J. B. Holland, C. H. Jarboe and N. Plotnikoff, *J. Med. Chem.*, 1974, **17**, 409.
36. R. D. Cramer, III, K. M. Snader, C. R. Willis, L. W. Chakrin, J. Thomas and B. M. Sutton, *J. Med. Chem.*, 1979, **22**, 714.
37. C. Hansch, *Drug Inf. J.*, 1984, **18**, 115.
38. K. Enslein, *Pharmacol. Rev.*, 1984, **36**, 131S.
39. R. F. Rekker, *Trends Pharmacol. Sci.*, 1980, **2**, 383.
40. M. S. Tute, in 'Animals and Alternatives in Toxicity Testing', ed. M. Balls, R. J. Riddell and A. N. Worden, Academic Press, London, 1983, p. 137.
41. S. M. Free, Jr. and J. W. Wilson, *J. Med. Chem.*, 1964, **7**, 395.
42. T. Fujita and T. Ban, *J. Med. Chem.*, 1971, **14**, 148.
43. C. Hansch and D. F. Calef, *J. Org. Chem.*, 1976, **41**, 1240.
44. 'Structure–Activity Relationships in Toxicology and Ecotoxicology: an Assessment', ECETOC Monograph No. 8, European Chemical Industry Ecology and Toxicology Centre, Brussels, 1986, p. 14.
45. A. J. Stuper and P. C. Jurs, *J. Chem. Inf. Comput. Sci.*, 1976, **16**, 99.
46. A. J. Stuper, W. E. Brügger and P. C. Jurs, 'Computer-Assisted Studies of Chemical Structure and Biological Function', Wiley, New York, 1979.
47. S. Wold and M. Sjostrom, *ACS Symp. Ser.*, 1977, **52**, 243.
48. P. P. Mager, in 'Drug Design', ed. E. J. Ariens, Academic Press, New York, 1980, vol. 9, p. 188.
49. M. R. Frierson, G. Klopman and H. S. Rosenkranz, *Environ. Mutagen.*, 1986, **8**, 283.
50. J. T. Penniston, L. Beckett, D. L. Bentley and C. Hansch, *Mol. Pharmacol.*, 1969, **5**, 333.
51. J. C. Dearden, *Environ. Health Perspect.*, 1985, **61**, 203.
52. J. W. McFarland, *J. Med. Chem.*, 1970, **13**, 1192.
53. H. Kubinyi, *J. Med. Chem.*, 1977, **20**, 625.
54. Y. C. Martin, 'Quantitative Drug Design', Dekker, New York, 1978.
55. A. Leo, C. Hansch and D. Elkins, *Chem. Rev.*, 1971, **71**, 525.
56. R. Collander, *Acta Chem. Scand.*, 1951, **5**, 774.
57. A. Leo and C. Hansch, *J. Org. Chem.*, 1971, **36**, 1539.
58. R. D. Cramer, III, *J. Am. Chem. Soc.*, 1977, **99**, 5408.
59. C. B. C. Boyce and B. W. Milborrow, *Nature (London)*, 1965, **208**, 537.
60. E. Tomlinson, *J. Chromatogr.*, 1975, **113**, 1.
61. J. M. McCall, *J. Med. Chem.*, 1975, **18**, 549.
62. A. E. Bird and A. C. Marshall, *Biochem. Pharmacol.*, 1967, **16**, 2275.
63. G. G. Nys and R. F. Rekker, *Chim. Ther.*, 1973, **8**, 521.
64. R. F. Rekker, 'The Hydrophobic Fragmental Constant', Elsevier, Amsterdam, 1977.
65. R. F. Rekker and H. M. de Kort, *Eur. J. Med. Chem.—Chim. Ther.*, 1979, **14**, 479.
66. C. Hansch and A. J. Leo, 'Substituent Constants for Correlation Analysis', Wiley, New York, 1979.
67. E. Kutter, H. Machleidt, W. Reuter, R. Sauter and A. Wildfeuer, in 'Biological Correlations—the Hansch Approach', ed. R. F. Gould, American Chemical Society, Washington, DC, 1972.
68. P. L. Olive, *Br. J. Cancer*, 1979, **40**, 89.
69. D. I. Edwards, *J. Antimicrob. Chemother.*, 1979, **5**, 499.
70. R. Franke, 'Theoretical Drug Design Methods', Elsevier, Amsterdam, 1984, p. 100.
71. T. Otsu, T. Ito, Y. Fujii and M. Imoto, *Bull. Chem. Soc. Jpn.*, 1968, **41**, 204.
72. F. E. Norrington, R. M. Hyde, S. G. Williams and R. Wooton, *J. Med. Chem.*, 1975, **18**, 604.
73. B. Hetnarski and R. D. O'Brien, *J. Med. Chem.*, 1975, **18**, 29.
74. D. J. Livingstone, R. M. Hyde and R. Foster, *Eur. J. Med. Chem.—Chim. Ther.*, 1979, **14**, 393.
75. M. S. Tute, *J. Med. Chem.*, 1970, **13**, 48.

76. J. W. McFarland, *Prog. Drug Res.*, 1971, **15**, 123.
77. H. Weinstein, S. Maayani, S. Srebrenik, S. Cohen and M. Sokolovsky, *Mol. Pharmacol.*, 1973, **9**, 820.
78. T. Solmajer, M. Hodoscek and D. Hadzi, *Quant. Struct.–Act. Relat.*, 1984, **3**, 51.
79. C. Hansch, E. W. Deutsch and R. N. Smith, *J. Am. Chem. Soc.*, 1965, **87**, 2738.
80. C. Hansch and E. J. Lien, *Biochem. Pharmacol.*, 1968, **17**, 709.
81. M. Charton, *J. Am. Chem. Soc.*, 1969, **91**, 615.
82. A. Verloop, W. Hoogenstraaten and J. Tipker, in 'Drug Design', ed. E. J. Ariens, Academic Press, New York, 1976, vol. 7, p. 165.
83. L. Pauling and D. Pressman, *J. Am. Chem. Soc.*, 1945, **67**, 1003.
84. F. Franks, in 'Water: A Comprehensive Treatise', ed. F. Franks, Plenum Press, New York, 1975, vol. 4.
85. C. Hansch and D. F. Calef, *J. Org. Chem.*, 1976, **41**, 1240.
86. C. Hansch and T. E. Klein, *Acc. Chem. Res.*, 1986, **19**, 392.
87. E. Wulfert, P. Bolla and J. Mathieu, *Chim. Ther.*, 1969, **4**, 257.
88. E. Kutter and C. Hansch, *J. Med. Chem.*, 1969, **12**, 647.
89. P. N. Craig, *J. Med. Chem.*, 1971, **14**, 680.
90. J. G. Topliss and R. J. Costello, *J. Med. Chem.*, 1972, **15**, 1066.
91. J. G. Topliss and R. P. Edwards, *J. Med. Chem.*, 1979, **22**, 1238.
92. J. G. Topliss, *J. Med. Chem.*, 1972, **15**, 1006.
93. Y. C. Martin and W. J. Dunn, III, *J. Med. Chem.*, 1973, **16**, 578.
94. F. Darvas, *J. Med. Chem.*, 1974, **17**, 799.
95. G. G. Hall, *Chem. Soc. Rev.*, 1973, **2**, 21.
96. I. Fleming, 'Frontier Orbitals and Organic Chemical Reactions', Wiley, London, 1976.
97. L. B. Kier (ed.), 'Molecular Orbital Studies in Chemical Pharmacology', Springer-Verlag, New York, 1970.
98. L. B. Kier, 'Molecular Orbital Theory in Drug Research', Academic Press, New York, 1971.
99. L. B. Kier and H.-D. Höltje, *J. Theor. Biol.*, 1975, **49**, 401.
100. L. B. Kier, *J. Med. Chem.*, 1968, **11**, 441.
101. C. R. Ganellin, in 'Quantitative Approaches to Drug Design', ed. J. C. Dearden, Elsevier, Amsterdam, 1983, p. 239.
102. L. Farnell, W. G. Richards and C. R. Ganellin, *J. Theor. Biol.*, 1974, **43**, 389.
103. R. J. Abraham and D. Birch, *Mol. Pharmacol.*, 1975, **11**, 663.
104. W. G. Richards, 'Quantum Pharmacology', 2nd edn., Butterworths, London, 1983.
105. P. Kollman, *Acc. Chem. Res.*, 1985, **18**, 105.
106. P. Claverie and R. Rein, *Int. J. Quantum Chem.*, 1969, **3**, 537.
107. H. D. Holtje and L. B. Kier, *J. Pharm. Sci.*, 1975, **64**, 418.
108. P. Kollman and J. Blaney, in 'Topics in Molecular Pharmacology', ed. A. S. V. Burgen, G. C. K. Roberts and M. S. Tute, Elsevier, Amsterdam, 1986, vol. 3, p. 285.
109. J. B. Hendrickson, *J. Am. Chem. Soc.*, 1961, **83**, 4537.
110. G. N. Ramachandran, C. Ramakrishnan and V. Sasisekharan, *J. Mol. Biol.*, 1963, **7**, 95.
111. A. C. T. North, *Chem. Ind. (London)*, 1982, 221.
112. R. Franke, 'Theoretical Drug Design Methods', Elsevier, Amsterdam, 1984, p. 134.
113. L. B. Kier and L. H. Hall, 'Molecular Connectivity in Chemistry and Drug Research', Academic Press, New York, 1976.
114. M. Mabilia, R. A. Pearlstein and A. J. Hopfinger in 'Topics in Molecular Pharmacology', ed. A. S. V. Burgen, G. C. K. Roberts and M. S. Tute, Elsevier, Amsterdam, 1986, vol. 3, p. 157.
115. D. H. Rouvray, *Sci. Am.*, 1986, **255**(3), 36.
116. L. B. Kier and L. H. Hall, 'Molecular Connectivity in Structure–Activity Analysis', Research Studies Press, Letchworth, 1986.
117. Z. Simon, *Angew. Chem., Int. Ed. Engl.*, 1974, **13**, 719.
118. E. J. Ariens, *Trends Pharmacol. Sci.*, 1986, **7**, 200.
119. P. A. Lehmann, *Trends Pharmacol. Sci.*, 1986, **7**, 281.
120. E. Fischer, *Ber. Dtsch. Chem. Ges.*, 1894, **27**, 2985.
121. D. E. Koshland, Jr. and K. E. Neet, *Annu. Rev. Biochem.*, 1968, **37**, 359.
122. A. S. V. Burgen, G. C. K. Roberts and J. Feeney, *Nature (London)*, 1975, **253**, 753.
123. R. H. Davies, B. Sheard and P. J. Taylor, *J. Pharm. Sci.*, 1979, **68**, 396.
124. G. A. Weiland, K. P. Minneman and P. B. Molinoff, *Nature (London)*, 1979, **281**, 114.
125. G. R. Marshall, in 'Medicinal Chemistry VI: Proceedings of the 6th International Symposium on Medicinal Chemistry, Brighton, UK', ed. M. A. Simkins, Cotswold Press, Oxford, 1979, p. 225.
126. M. Wise, R. D. Cramer, D. Smith and I. Exman, in 'Quantitative Approaches to Drug Design', ed. J. C. Dearden, Elsevier, Amsterdam, 1983, p. 145.
127. R. L. DesJarlais, R. P. Sheridan, J. Scott Dixon, I. D. Kuntz and R. Venkataraghavan, *J. Med. Chem.*, 1986, **29**, 2149.
128. C. Levinthal, *Sci. Am.*, 1966, **214** (6), 42.
129. K. Prout, in 'Topics in Molecular Pharmacology', ed. A. S. V. Burgen, G. C. K. Roberts and M. S. Tute, Elsevier, Amsterdam, 1986, vol. 3, p. 1.
130. R. Langridge, T. E. Ferrin, I. D. Kuntz and M. L. Connolly, *Science (Washington, D.C.)*, 1981, **211**, 661.
131. L. F. Kuyper, B. Roth, D. P. Baccanari, R. Ferone, C. R. Beddell, J. N. Champness, D. K. Stammers, J. G. Dann, F. E. A. Norrington, D. J. Baker and P. J. Goodford, *J. Med. Chem.*, 1982, **25**, 1120.
132. P. J. Goodford, *J. Med. Chem.*, 1985, **28**, 849.
133. M. M. Julian, *Chem. Br.*, 1986, **22**, 729.
134. C. D. Garner and J. R. Helliwell, *Chem. Br.*, 1986, **22**, 835.
135. W. R. Kester and B. W. Matthews, *Biochemistry*, 1977, **16**, 2506.
136. T.-L. Shieh and S. R. Byrn, *J. Med. Chem.*, 1982, **25**, 403.
137. D. A. Matthews, J. T. Bolin, J. M. Burridge, D. J. Filman, K. W. Volz and J. Kraut, *J. Biol. Chem.*, 1985, **260**, 392.
138. J. M. Blaney, C. Hansch, C. Silipo and A. Vittoria, *Chem. Rev.*, 1984, **84**, 333.
139. A. R. Fersht, J. P. Shi, J. Knill-Jones, D. M. Lowe, A. J. Wilkinson, D. M. Blow, P. Brick, P. Carter, M. M. Y. Waye and G. Winter, *Nature (London)*, 1985, **314**, 235.

140. A. R. Fersht, R. J. Leatherbarrow and T. N. C. Wells, *Nature (London)*, 1986, **322**, 284.
141. J. L. Fauchere and V. Pliska, *Eur. J. Med. Chem.—Chim. Ther.*, 1983, **18**, 369.
142. D. Eisenberg and A. D. MacLachlan, *Nature (London)*, 1986, **319**, 199.
143. C. Chothia, *Nature (London)*, 1974, **248**, 338.
144. C. Frommel, *J. Theor. Biol.*, 1984, **111**, 247.
145. M. L. Contreras, B. B. Wolfe and P. B. Molinoff, *J. Pharmacol. Exp. Ther.*, 1986, **237**, 165.
146. P. R. Andrews, D. J. Craik and J. L. Martin, *J. Med. Chem.*, 1984, **27**, 1648.
147. M. I. Page, *Angew. Chem., Int. Ed. Engl.*, 1977, **16**, 449.
148. M. I. Page and W. P. Jencks, *Proc. Natl. Acad. Sci. U.S.A.*, 1971, 1678.
149. F. H. Allen, O. Kennard, W. D. S. Motherwell, W. G. Town and D. G. Watson, *J. Chem. Soc.*, 1973, **13**, 119.
150. F. C. Bernstein, T. F. Koetzle, G. J. B. Williams, E. F. Meyer, M. D. Brice, J. R. Rodgers, O. Kennard, T. Shimanouchi and M. Tasumi, *J. Mol. Biol.*, 1977, **112**, 535.

17.2

Computers and the Medicinal Chemist

FRANÇOIS CHOPLIN

Rhône-Poulenc Recherches, Saint-Fons, France

17.2.1 OVERVIEW

17.2.1.1 Needs of the Medicinal Chemist

The goal of the medicinal chemist is to design and synthesize new active molecules. Trial and error screening, which used to be considered as the normal procedure, is becoming very costly and, at the same time, less efficient. In the early 1960s, one could expect to discover a marketable compound out of 2000–3000 tested molecules, whereas this ratio is now close to 1 in 10 000 and biological testing expenses have increased dramatically. Therefore, only molecules with a good chance of activity should be prepared and tested. To this end, chemists can no longer rely only on their own experience and knowledge. Fast access to accurate and reliable information on the candidate or similar molecules and the possibility of designing and testing accurate mathematical models of chemical structures and interactions, in order to simulate the behavior and physicochemical properties of a candidate molecule and the interactions between a drug and a receptor, are necessary.

The ideal system to enable medicinal chemists to achieve their goal should be able to:

(i) obtain the stable conformations for a given molecule, and provide their energies, along with a full description of the geometries;

(ii) calculate atomic charges and interactions, electrostatic potentials, frontier orbitals, and indexes related to reactivity;

(iii) describe interactions between molecules, especially between a drug and a receptor, and compute their energies;

(iv) superimpose and compare geometric and electronic molecular models;

(v) calculate physicochemical properties of a molecule, such as heat of formation, pK_a's for different protons, partition coefficient, dipole moment, spectroscopic coupling constants, *etc.*;

(vi) display molecular models in several different ways (*e.g.* ball and stick, ORTEP diagrams, CPK space-filling); special features should be available for complex molecules like proteins and crystals (*e.g.* cylinder and ribbon representations for proteins, crystal construction from a unit cell);

(vii) find and display quantitative and qualitative relationships between representations of molecules and biological activities;

(viii) simulate the course of a chemical reaction to study its feasibility and consider the possible byproducts; and

(ix) make as easy as possible exchange of data between different systems.

Many other problems encountered by the medicinal chemist can be summarized as how to quickly handle and retrieve relevant information.[1] For example:

(i) retrieve inhouse and external information (properties, biological data, spectroscopic data, syntheses) for a given molecule, and display structures and data simultaneously;

(ii) obtain biological data for a substructural set of molecules;

(iii) access chemical reactions, either by a structural query or a reaction type.

This list is not comprehensive but reflects the problems encountered everyday by the medicinal chemist. Our aim is to show that computerized techniques are increasingly available, and in many cases indispensable, to help the medicinal chemist to solve these problems. However, in the past few years, the situation has become complex and rather confusing for the non-computer-oriented chemist, and we will try to clarify it. We will study the main techniques, discuss briefly their possibilities, and focus our attention on how to implement these techniques for successful use. We will also try to perform a cost effectiveness analysis.

17.2.1.2 Computerized Techniques

The computerized techniques and tools available today can be separated into two main groups. (i) Design techniques which allow the chemist to build models and use them to simulate and visualize molecular properties and behaviour. They can be considered as tools to increase the creativity of the chemist.[2] (ii) Information storage and retrieval techniques to retrieve, handle and compare chemical information in many different ways. Although they certainly lead to increased creativity, they are primarily intended to improve the productivity of the end user.

This set of techniques, taken as a whole, is often referred to as 'computational chemistry'. Figure 1 shows how these techniques are involved at each stage of the development of a new drug.

Most of these techniques are now well established, and correspond to theoretical and applied work carried out during the past 20 years in universities, government agencies and companies.

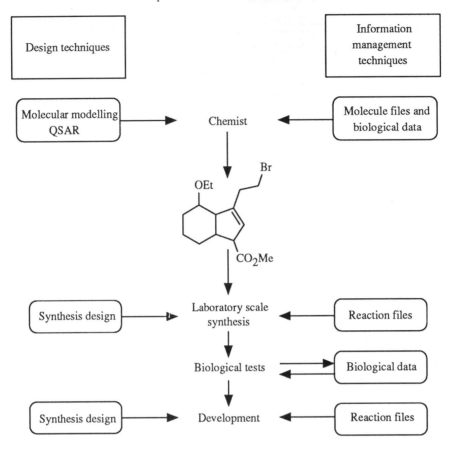

Figure 1 Drug development and computerized techniques

Theoretical developments are still underway,[3] and are included in new or existing programs as soon as their reliability has been assessed. These techniques rest on a few fundamental methods, which are described in other chapters of this work. Molecular mechanics (MM)[4,5] is especially useful for calculating geometry and strain energy of molecules with up to 100 atoms with a good degree of accuracy and reliability. Molecular dynamics (MD)[6] is used to study internal molecular motion, and it appears to be a promising tool for helping to elucidate the structure of medium-sized proteins.[7] Quantum chemistry (QM) is applicable through a large number of more or less sophisticated and reliable methods: *ab initio* at the STO-3G level, MINDO or MINDO/3[8] (which seem to be a good compromise between reliability and cost), CNDO, INDO, PCILO and the simple EHT method.[9] They are well suited to the calculation of properties of isolated molecules, such as electronic structure, ionization energies, molecular geometry, rotational barriers, binding energies, *etc.*

Determination of structure–activity relationships has always been a tool of the medicinal chemist. The goal is ultimately to design active molecules *a priori*, but, more reasonably (at least now), to find rapidly the optimal molecule in a set of analogues. There are two main methods for handling this problem. The first one requires minimal knowledge of the mechanism of action at the molecular level: it uses molecular-modelling tools to design active and inactive molecules, and simulate the drug–receptor interaction, as in the active analog approach.[10] The second method is only comparative, and has been used for many years under the name QSAR (for qualitative or quantitative structure activity relationship):[11-14] *a priori* knowledge of the mechanism is not mandatory for establishing such a relationship, but experimental activity data obtained for similar molecules are necessary. A complete evaluation of these methods applied to the prediction of toxicology and ecotoxicology end points has recently been published.[15]

To assist the chemist when synthesizing the selected drug, design tools are available, either to explore all the possible pathways leading to a target molecule,[16] or to simulate the behaviour of a reaction, given the starting products and the conditions.[17] These techniques are referred to as computer-assisted synthesis design.

In the field of information storage and retrieval, the efforts devoted to the application of graph theory in chemistry have led to very useful algorithms, such as the MORGAN algorithm,[18] or the DARC system,[19] which were designed to solve the problem of unambiguous molecular naming for computer retrieval.

During the period 1965–1980, parallel developments have been made to write more powerful and user friendly software, based on data structures and algorithms applicable to representations of chemical problems and structures. Especially outstanding is the elaboration of several computer-assisted synthesis design programs,[16] which were among the first to make extensive use of computer graphics, complex data structures and powerful algorithms and heuristics to handle the difficult (and not yet solved!) problem of simulating human reasoning in the synthesis of an organic molecule.[20]

This evolution, combined with the increasing availability of cheap and powerful hardware, has led to a completely new situation at the beginning of the 1980s: computer software is now available for treating most of the problems presented in Section 17.2.1.1, and it meets high quality and user friendliness standards. Programs are in many cases developed, distributed and maintained by independent companies. Although the fundamentals underlying these products are in many cases five to 20 years old, software has dramatically improved in quantity, quality and reliability, and the situation, which is evolving very quickly, is at the same time very favourable for the user, but also more and more complex.

17.2.1.3 Basic Requirements for a Successful Use of Computers

We have seen the problems encountered by the medicinal chemist, and some techniques which can help to solve them. Purchase of any available software and/or crude application of these techniques is not at all sufficient for a successful treatment of the needs. By successful, we mean that chemists will consider the computer as an indispensable tool to solve their everyday problems and that their creativity and productivity will be enhanced. This requires the fulfillment of four conditions (as shown in Figure 2): (a) powerful and user friendly software; (b) databases available for both internal and external data; (c) hardware, fast networks and graphics terminals; and (d) dedicated staff for software implementation, maintenance, problem solving and user training. These points are developed in the subsequent sections.

17.2.2 SOFTWARE

Before 1980 the only way to benefit from computational chemistry was to develop inhouse programs. There was practically no professional chemical software commercially available, and the Quantum Chemistry Program Exchange was the main source of programs written by research groups in universities. They were research tools, not designed and interfaced to be used on a large

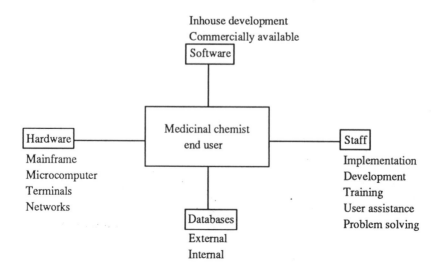

Figure 2 Requirements for successful use of computers

scale by inexperienced chemists, and most of them required extensive modifications to be used routinely.

The situation is now different, since there are many available packages. Therefore, we will study the criteria to be considered when acquiring commercial software, and will review the programs available in each field.

17.2.2.1 Criteria for Selecting Software

The items to be checked before selecting a package are summarized below. This selection is a difficult process, since it requires time and thorough comparison of programs. The major error to avoid is waiting indefinitely until a better system becomes available. A better system will always be available, but in the future. Moreover, it should be emphasized that the experience gained with a system is as important as the system itself.

17.2.2.1.1 User friendliness

All systems claim they are user friendly but there are no clear-cut standards: it simply indicates that the system has all the necessary features to make the user's task easy. The important points are reviewed below.

(i) Response time

Response time depends on the computer load but mainly on the design of the program itself and on the choice of efficient data structures. It is a very important feature for the end user, who will resent long response times on a routine basis, since it breaks the continuity of his own thinking. This point is critical for data input and database searching.

(ii) Error friendliness

Typing errors or logical errors are inevitable and a user friendly software must allow the user not only to correct an error but also to avoid it by preventing incorrect typing or commands manipulation. Many packages in chemical information management do not provide easy to use and error-checking data input modules.

(iii) Natural language

There are two kinds of user–machine interfaces: menu driven and command language driven. Many programs provide both options (if not, they should), which address novice or expert users respectively. Menu-driven interfaces can be more or less sophisticated, *i.e.* they use alphanumerical or graphical menus, including self-speaking icons. These menus are generally considered user friendly but time consuming and tedious for the expert, who prefers to employ command language. The latter has to be clear, make use of self-speaking and if possible unique mnemonics. For example, P should not mean PURGE, PRINT and PROCESS, at different points of the program: it should have the same meaning everywhere.

In chemistry, natural language is based on structural diagrams, therefore graphics is essential in all software requiring molecular input and output. Chemical names can be displayed for information along with the formula, but no longer used as an input system. This is especially true for the bench chemist as end user. However, structural input by people experienced with linear notations (such as WLN and SMILES) is generally faster than graphical input.

(iv) Online help

More and more systems provide online instant help which allows the user to get information about the current stage of the program. This module has to give clear, meaningful and as terse as possible messages in order to be read by the user.

(v) General design of the program

Although more subtle, this feature corresponds to the visual aspects of the program: design and organization of the displays, selection of colours, data being displayed, error and prompt messages which have to be clear and relevant.

17.2.2.1.2 Development tools for the database and the system supervisor

These tools are extremely important since they guarantee the possibility of adapting a system to present and future needs of the inhouse information system. Moreover, they allow the supervisor, and possibly the end user, to adapt the program to their specific requirements (such as designing customized menus and procedures). They are especially necessary for handling the following problems.

(i) Management of data security and confidentiality

Such management is a difficult task which can be solved at different levels (mainframe operating system, database management system, network, *etc.*) and involves many nontechnical issues. Systems should provide tools to control access to specific databases and, inside the database, to given datatypes and records (*e.g.* a given molecule or reaction). Very few packages provide this last functionality.

(ii) Interfaces with other systems

Interfacing two systems can be realized at two levels: an external one, in which both programs can exchange information, and an internal level, which leads to integration, *i.e.* both programs can access directly information managed by the other. The first level needs an external file to exchange data, with complete information about data structure. The basic information contained in this file is a molecular connection table and *xyz* atom coordinates. By means of this file (which can contain one or several molecules and data, and possibly reactions), data can be exchanged between systems. This involves conversion in many cases, since connection table formats differ from MDL to DARC and CAS. Wisswesser Linear Notation (WLN),[21] which was used by many companies before the advent of graphics software, can be converted in connection table format (DARC, MACCS, CAS) through a procedure involving programs such as WLNCT from Molecular Design Ltd (MDL), or DARING from Fraser Williams.[22] In this case, conversion is seldom complete and leads to 5–20% of rejected WLN-coded structures.

Conversion between DARC, MDL and CAS formats often requires inhouse programming, and is in general easily performed. DARC and LAYOUT programs (MDL) calculate automatically display coordinates from the connection table.

It should be of great interest to the user that all the packages adopt the same format for the external connection table file, in order to make exchanges and conversion procedures very easy. This is far from being the case, and even for systems developed by the same company, external file formats can differ, as is the case between REACCS and CHEMBASE, both from MDL.

This type of interface is also useful in molecular modelling to access data from external databases, or to load molecules from a chemical information database.

The second type of interface requires a subroutines tool box (generally a set of Fortran subroutines), which can be used to perform operations such as: retrieval of a given connection table from its internal registry number, and storage in an external file; display of a molecule with a given internal registry number; read/write a list of descriptors (such as registry numbers); retrieve and print data for a given datatype or record; and store data in a database from an external file.

(iii) Customization

Due to the great variety of needs of the users, the system manager must have tools to adapt the program to the users requirements; these tools can be employed to define customized forms and sheets to display and print data, to define users' profiles, to set up options for plotting and printing, to supervise the utilization of the system (logfile), and to define formats for data exchanges. Modules like the Customization Module of MDL, and RDS2 from DARC, are dedicated to these tasks.

17.2.2.1.3 Range of usable computers and terminals

All mainframe programs have a version available for VAX computers (which is an international standard in computational chemistry), and to a much lesser extent an IBM version. Microcomputer software is made for the IBM-PC, and in a few cases for McIntosh. Corporate rules can impose the use of a given computer type (generally IBM), and all packages in chemical information management should provide versions for VAX (under VMS), and for IBM (the choice between the MVS/TSO and VM/CMS operating systems is not so obvious, but MVS has clear advantages to share files).

17.2.2.1.4 Quality of documentation

This is a critical point for many programs and it is often difficult to obtain user- instead of system-oriented documentation. Both are necessary but only the last one is generally available. A good documentation should provide detailed descriptions of the operations for the most frequent queries, have a comprehensive index and give significant examples for typical searches. Short user guides should also be provided, which is the case only recently for MACCS and REACCS. On the other hand, the supervisor documentation is more system oriented and it must be comprehensive and accurate.

17.2.2.1.5 Maintenance and updates

Most companies provide maintenance (at an annual cost of *ca.* 10–20% of the licence price). This price generally includes the delivery of major updates (one per year at MDL and DARC), but does not include entirely new functionalities. Efficient maintenance is an important consideration when selecting a package.

17.2.2.1.6 Company profile

In computational chemistry, programs become obsolete very quickly. A company acquiring a system has to spend much money on training and on acquiring experience. This is why we have not considered the initial price as a key item in the choice, since it can vary, and represents only a small part of the money spent on the project. For programs of such complexity, quality and assets of the vendor company are also very important. Successful companies are those which write their own software (and not only sell it), and where the staff includes many chemists. Therefore it is important to consider the reputation of the company, the number of installations over the world (have a look at the list of customers, if available), and the availability of the source code in case of failure of the company. Table 1 gives the names and addresses of some of the main companies.

17.2.2.2 Available Software

The exploding availability of professional quality software in almost every field of interest to the medicinal chemist makes the task of setting up a comprehensive list in each field unrealistic. We have sorted systems according to the fields and have only considered packages which are publicly available and advertised: in general, programs which are only published or reported have not been included. It is also the case for many microcomputer-based packages for which we have limited information, and which are often restricted to small molecules.[52]

Besides the commercial advertisements in scientific journals, demonstrations and exhibitions at conferences and meetings, there are now numerous ways to be aware of software developments. In the microcomputers series, special chapters appear in the *Journal of the American Chemical Society*,[23] the *Journal of Chemical Information and Computer Science* and in the *Journal of Chemical Education*.[24] This latter series also provides interesting examples of the use of classical programs. General papers can also be found in *Chemistry in Britain*, *Chemical Week*, *Chemistry and Industry* and *Chemical Engineering News*.

Table 1 Software and Database Companies

Name or abbreviation	Address
ACS SOFT	American Chemical Society Software Distribution Office, Department 12, 1155 Sixteenth Street NW, Washington, DC 20036, USA
BBN	BBN Software Products Corporation, 10 Fawcett Street, Cambridge, MA 02238, USA
BIODESIGN	Biodesign Inc., 199 St. Los Robles Avenue, Suite 660, Pasadena, CA 91101, USA
BIOSYM	Biosym Technologies Inc., 9605 Scranton Road, Suite 101, San Diego, CA 92121, USA
BROOKHAVEN NL	Brookhaven National Laboratory, Chemistry Department, Upton, NY 11973, USA
CAMBRIDGE XRAY	Cambridge Crystallographic Data Centre, University Chemical Laboratory, Lensfield Road, Cambridge CB2 1EW, UK
CAS	Chemical Abstracts Services, PO Box 3012, Columbus, OH 43210, USA
CHEMDATA	Chemdata, 17 quai Joseph Gillet, 69004 Lyon, France
CHEMICAL DESIGN	Chemical Design, Unit 12, 7 West Way, Oxford OX2 0JB, UK
CISI	CISI Petrole Service, 53 Avenue Gabriel Peri, 92503 Rueil Malmaison, France
FRASER WILLIAMS	Fraser Williams (Scientific Systems) Ltd, London House, London Road South, Poynton, Cheshire SK12 1YP, UK
HAMPDEN	Hampden Data Services Ltd, Hampden Cottage, Abingdon Road, Clifton Hampden, Abingdon, Oxfordshire OX14 3EG, UK
HDI	Health Designs Inc., 183 East Main Street, Rochester, NY 14604, USA
INFODATA	Infodata Systems Inc., 5205 Leesburg Pike, Falls Church, VA 22041, USA
INFORM. DIM.	Information Dimensions Inc., 655 Metro Place, Dublin, OH 43017, USA
ISI	Institute for Scientific Information, 3501 Market Street, Philadelphia, PA 19104, USA
MDL	Molecular Design Ltd, 2132 Farallon Drive, San Leandro, CA 94577, USA
ORAC Ltd.	175 Woodhouse Lane, Leeds LS2 3AR, UK
ORACLE	Oracle Corporation, 20 Davis Drive, Belmont, CA 94002, USA
PERGABASE	Pergamon Group, 8000 Westpark Drive, Suite 400, McLean, VA 22102, USA
POLYGEN	Polygen Corporation, 200 Fifth Avenue, Waltham, MA 02154, USA
POMONA	Pomona College Medicinal Chemistry Laboratory, Seaver Chemistry Laboratory, Claremont, CA 91711, USA
QCPE	Quantum Chemistry Program Exchange, Department of Chemistry, Indiana University, Bloomington, IN 47405, USA
RELATIONAL TECHN.	Relational Technology Inc., 1080 Marina Village Parkway, Almeda, CA 94510, USA
SAS	SAS Institute, PO Box 8000, Cary, NC 27511, USA
SOFTWARE HOUSE	Software House, 1000 Massachusetts Avenue, Cambridge, MA 02238, USA
SUMITOMO	Sumitomo Chemical, Sumitomo Building, 5–15, Kitahama, Higashi-ku, Osaka 541, Japan
TDS	Technical Databases Service Inc., 10 Columbus Circle, New York, NY 10019, USA
TELESYSTEMES	Telesystemes-Questel, 83–85 Boulevard Vincent Auriol, 75013 Paris, France
TRIPOS	Tripos Associate Inc., 6548 Clayton Road, Saint Louis, MO 63117, USA

17.2.2.2.1 *Molecular modelling*

Molecular modelling is a very general term which covers many methods and functions. To try to simplify the problem we have defined five basic modules which can (and should) be encountered in a molecular-modelling system. These modules and the associated functionalities are shown in Figure 3. The software list (Table 2) compares systems with respect to these five items.

(i) *Input and sketch builder*

This module allows easy input of the problem and builds an approximate 3-D sketch of the molecule. Alphanumerical keyboard input of the xyz atom coordinates is the simplest way. Many programs are interfaced to read xyz coordinates from crystallographic data banks (Protein Data Bank, Cambridge Crystallographic Database), or from a standard file containing atom coordinates and a connection table. More user friendly input is graphic, by means of a mouse (less frequently a light pen) and a graphics display.

The sketch builder works often as a computerized Dreiding model which allows hydrogen substitutions by other atoms. Special features take care of ring closures, especially for strained systems. Most of the available programs use this type of input. The pseudo-3-D model obtained at this stage needs refining, and dihedral angles relaxation has not yet been performed.

Protein building needs special functionalities to assemble amino acids and peptides.

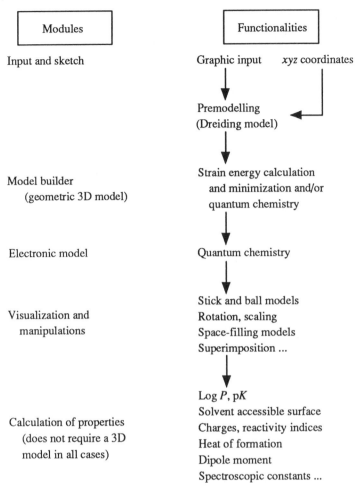

Figure 3 Modules and functionalities of molecular-modelling systems

(ii) 3-D model builder

Model refinement is necessary, especially for strained systems. It starts from the crude coordinates calculated by the INPUT module, and uses either molecular mechanics (MM) or quantum mechanics (QM) to obtain a refined model. When using MM, the software should provide the choice between several force fields and let the user select among different minimizers (steepest descent, pattern search, Monte Carlo search, Raphson Newton procedure and simplex). The model builder also includes a time-consuming conformational analysis module, which calculates a global minimum of an energy surface by relaxing several dihedral angles simultaneously.

(iii) Electronic model

Previous steps provide the geometry of the molecule. Information about electron and charge distribution and molecular orbitals can only be provided by a quantum mechanics (QM) calculation. Most systems have interfaces to standard QM programs, such as CNINDO, EHT, GAUSS8X, MINDO/3, MOPAC (available from QCPE). In most cases this can be an input interface, allowing the user to start a QM calculation from an MM-generated geometry. More interesting is the CHEM-QM module available with CHEM-X, which allows the user not only to transfer files from MM to QM, but also to explore graphically the QM calculations output.

(iv) Visualization and manipulations

This module, where user friendliness is of the utmost importance, allows the user to visualize and manipulate the molecular model as conveniently as possible. This includes the following facilities,

Table 2 Molecular Modelling Software

Name	Supplier[a] and/or author	Main features
ACACS & BIOCES	Sumitomo Chemical Co.	Graphics input and manipulation module, handles proteins and polymers. Integrates a chemical database management system. Interfaces with QCPE programs. QSAR analysis module. Runs on NEC under ACOS, with raster displays
ALCHEMY	Tripos[51]	Molecular mechanics and display program. Powerful display options with real-time rotation. Functionalities similar to those of SYBYL. Runs on MS-DOS personal computer, with math coprocessor recommended.
AMBER	Professor Kollman, University of California, Department of Pharmaceutical Chemistry, San Francisco, CA 94143, USA	Molecular mechanics and dynamics program
BIOGRAF-II	BioDesign	Graphics input and powerful manipulation module. Building module for proteins, nucleic acids and polymers. Model builder with several force fields and minimizers. Calculations of properties. Runs on VAX and IBM, ES PS3XX terminals
CHARMM	Polygen.	Batch mode program for molecular mechanics and dynamics calculations, with possible constraints definition. Analysis module. Interface with CHEMNOTE and HYDRA. Runs on VAX
CHEM-X	Chemical Design	Graphic input and manipulation module, special features for proteins and crystals. Very user friendly. CHEM-QM interface with QM-QCPE programs (for input and output). Few property calculations. Specific model builder, but interface with QCPE programs. Runs on VAX; ES PS3XX and raster terminals
CHEMLAB-II	Molecular Design Ltd (MDL)	Graphic input and manipulation module. Many modules for property calculations, especially for proteins and polymers. MM2 and MNDO as model builders. Interfaces with QCPE programs, MDL databases and software. Runs on VAX + graphic or nongraphic terminal
CHEMNOTE	Polygen.	PC-based system, graphic input and sketch builder, sequence generator for proteins. Front end to CHARMM and HYDRA
DISCOVER	Biosym.	Molecular mechanics and dynamics program. Interfaced with graphics input and manipulation module INSIGHT. Special hardware architecture for fast calculations (VAX + ST100).
FRODO	Professor F. Quiocho, Rice University, Department of Biochemistry, PO Box 1892, Houston, TX 77251, USA	Visualization program for crystallography. Runs with ES PS3XX
HYDRA	Polygen.	Graphics input and manipulation module for CHARMM. Interface with CHEMNOTE. Runs on VAX and several high quality graphics displays
MACROMODEL	Professor W. C. Still, Columbia University, Department of Chemistry, New York, NY 10027, USA	Graphic input and powerful manipulation module, very user friendly. Several force fields and minimizers. Interfaces with QCPE programs. Performs molecular dynamics calculations. Runs on VAX + graphic terminal

MANOSK	Professor J. P. Mornon, Université de Paris VI, Laboratoire de Mineralogie Cristallographie, Tour 16, 4 Place Jussieu, 75230 Paris Cedex 05, France	Visualization and model manipulation program. Interface with Cambridge data bank and PDB. Runs on VAX, NORSKDATA, ES PS3XX display
MIDAS	Professor Langridge, University of California, Department of Pharmaceutical Chemistry, San Francisco, CA 94143, USA	Display and manipulation package for proteins. Runs on IRIS + ES PS3XX display
MM2	Quantum Chemistry Program Exchange (QCPE). Author: Professor N. Allinger	Manual input of CT and coordinates. No manipulation module. Works only as a model builder. Calculates dipole moments and heat of formation. Runs on VAX, IBM, UNIVAC, PC and other systems in batch mode
MMP2	Molecular Design Ltd (MDL). Author: Professor N. Allinger	Analogous to MM2, with a more recent force field and QM calculations for aromatic and conjugated molecules
MOGLI	Evans and Sutherland	Visualization and model manipulation program. Runs on VAX and ES PS3XX display. Interface with AMBER, MM2
MOLEC. GRAPHICS	ACS Software	Molecular graphics, IBM-PC software for visualization and model manipulation
SYBYL	Tripos	Complete molecular modelling graphics package for small molecules and proteins. Runs on VAX, EX PS3XX display

ᵃ See Table 1 for explanation of abbreviations and for address.

among other possibilities: ball and stick models; space-filling models with colour and shading; Newman projections; study of the bulkiness of substituents by allowing free rotations around specified bonds; complete rotation and positioning of the model, for example easy selection of a molecular plane for better display; superposition of molecular models, either simple or with intermolecular distance optimization; ribbon and cylinder representations for proteins; atom and bond sets definition and visualization; molecular orbital representations; and electrostatic potential display.

(v) Calculation of properties

This module is certainly the least developed in many available systems, and it is of the utmost interest to the chemist. It requires a specific mathematical model for each property, and developments in chemical simulation systems which include such property calculations, using heuristic and empirical models have occurred. As an example, the CAMEO system,[17] a chemical synthesis simulation program, includes empirical calculations of heats of formation, and pK_a values in DMSO.[25] Other fast methods have been developed to calculate binding affinities,[26] atomic charges,[27] and, in some cases, incorporated in molecular modelling programs.

Other properties of interest include: dipole moment, octanol/water partition coefficients,[28] solvent accessible surfaces, thermodynamic properties. CHEMLAB-II offers modules for proteins and polymers in solution.

17.2.2.2.2 Structure–activity relationships (QSAR)

Performing quantitative QSARs requires regression analysis routines, which can be run on any microcomputer or are included in many statistical packages on mainframe computers (such as SAS, BMDP, *etc.*). The TOPKAT system, available from HDI, is a complete microcomputer (IBM-PC) based package designed to predict toxicity endpoints from molecular structure, such as rat oral LD_{50}, mutagenicity (Ames test), carcinogenicity, teratogenicity, rabbit skin and eye irritation.[29] The MEDCHEM program, a modelling and information management package (from Pomona College), includes algorithms to calculate $\log P$ values. Care must be taken in handling these statistical tools since improper use can lead to spurious results. Recent papers review typical errors and give guidelines to avoid them.[30-31]

Qualitative SARs are more difficult to perform and, due to the rather large amount of necessary data (typically 100–200 molecules), they require software connected to chemical and biological databases, and access to multivariate analysis and pattern recognition packages. To our knowledge, only one complete program is available for this type of technique: ADAPT, developed by Jurs *et al.*[12] and marketed by MDL, performs all the operations, *i.e.* vector representation of molecules (after possible extraction of interesting compounds from a MACCS database), statistical analysis by various methods, data input and management.

17.2.2.2.3 Synthesis design

Computer-assisted synthesis design can be used when synthesizing a molecule at the laboratory or development stage. The SOS program,[32] developed by Barone and marketed by CISI, runs on IBM-PC and Macintosh; it is restricted to the design of new reaction types from a few fundamental mechanisms. Highly complex systems, such as SECS,[33] and LHASA,[34] have been installed by a few companies with mixed results. The CASP experience (a computer-assisted synthesis design joint venture between several Swiss and German chemical companies)[35] uses the SECS program, and relies on a large library of chemical transforms. Its access is restricted to members of the CASP team. LHASA is mainly accessible online *via* the LHASA-UK project, managed at the University of Leeds, and supported largely by British companies: direct use of this program by the chemist seems difficult due to its complexity. Work is underway by the participating companies to improve the transforms database.

The above systems build a synthetic tree starting from the target molecule. On the other hand, the CAMEO program[25] works in the synthetic direction as a reaction simulation program: it is available for inhouse installation (from Professor Jorgensen), and runs on a VAX with standard graphics terminals. At least 10 companies are using it, and it can also be accessed online from LHASA-UK.

The CHIRON program has a more restricted goal,[36] and is of special interest for analyzing the synthesis of chiral compounds. Given a target molecule, it scans a library of chiral synthons and starting materials and compares them to the target. It has powerful features to manipulate and display chiral compounds and is available for inhouse installation on a VAX machine.

17.2.2.2.4 Chemical information packages

The goal of a chemical information system can be easily defined. (a) Retrieve a molecule from a file which contains from several thousand to several million compounds. This operation must take into account stereochemistry, tautomerism, aromaticity, charges and isotopes. (b) Retrieve a set of molecules having a given substructure. This type of query is particularly useful for studying structure–activity relationships. These goals also apply to reaction databases but searches are aimed at finding reactions from queries dealing with starting materials, products, catalysts and solvents. Moreover, knowledge and storage of the reaction centres (*i.e.* the atoms and bonds involved in the reaction) allow very specific searches, such as, for example, finding reactions to reduce selectively a ketone to an alcohol without modifying another sensitive functional group in the same molecule. The REACCS and ORAC[37] systems store the reacting centres and the latest REACCS version performs an atom mapping between both sides of the reaction.

All the systems described in Table 3 are state of the art programs. They use connection tables as internal molecular representation (except MEDCHEM, which uses the SMILES notation), and make full use of graphics. MACCS and REACCS are especially outstanding for their user friendly features and the large number of installations all over the world. DARC is identical to the online system to access CAS, and provides very good response times due to specific data structures. MEDCHEM manages THOR databases (THesaurus ORiented) of chemical structures and data, and appears to be very efficient in both response times and disk space.

CHEMBASE[38] and PSIDOM are IBM-PC based software and represent a new trend, not fully explored, of chemical database management systems at the laboratory level. This type of software is very user friendly, has no problem of response time (with a relatively small database), and is very efficient for decentralized input. The problems encountered are associated with the databases and workforce organization, which are discussed in Section 17.2.5.

17.2.2.2.5 Associated data management

All chemical information packages have an integrated system to handle alphanumerical data associated with molecules and reactions (*e.g.* references, physicochemical data, company numbers). These systems are generally not intended to treat large quantities of complex data. There is therefore a need for other programs, which can manipulate data such as biological and analytical results, but there is also a need for treatment of more general information, such as reports, laboratory notebooks, *etc*. Table 4 gives some indication about the most frequently used packages. The interfacing tools of such programs are to be studied thoroughly to integrate them into a chemical information system.

17.2.2.3 Integrated Systems

In order to reduce time and effort in accessing data, there is a definite need to provide an integrated information system for the end user. In the worst conditions data of interest are scattered in several databases, managed by different systems installed on several nonconnected machines. The minimal requirement for integration is the possibility of transferring lists of documents or molecular identifiers from one system to another, *e.g.* internal registry numbers from the file of molecules to the biological database. This can imply conversion of identifiers if the databases are not organized in the same way. However, better integration can be solved in different ways, as presented by Hagadone,[39] and these can be summarized as follows.

(i) All the data are managed by the same system. Thus, Sandoz uses MACCS and DATACCS to store and retrieve chemical structures, biological and spectroscopic data, inventory, preclinical results and observations. Their database contains *ca.* 28 000 molecules with 256 000 associated data.[40] The main argument against such a solution is that it does not provide optimal management for all types of data. However, user friendliness is impressive. Rhône-Poulenc has started a similar

Table 3 Chemical Information Systems

System	Supplier[a]	Main features
CHEMBASE	MDL	Molecules, reactions and associated data management system. Runs on IBM-PC and compatibles. Can be interfaced with other systems *via* MOLFILE, SDFILE, RXNFILE. Very user friendly. Can handle databases of *ca.* 5000 molecules or reactions
CHEMSMART	ISI	Chemical information database management system. Runs on IBM-PC
DARC	Telesystemes-Darc	Molcules, reactions and associated data management system. Runs on VAX, IBM (MVS/TSO, VM/CMS). Good response times but some features are poor. Can handle very large databases. Very powerful integration capabilities with DARC Communication Module, customized design *via* RDS2. VAX version integrated with ORACLE. System 1032, RDB, BASIS
MACCS	MDL	Molecules and associated data management system. Runs on VAX, IBM (MVS, CMS), PRIME, FUJITSU. Very user friendly. Interface *via* external file (MOLFILE). Subroutines tool box: MACCSLIB. Development and customized design *via* DATACCS and the customization module. Integration between INQUIRE, SYSTEM 1032, ORACLE and MACCS-II
MEDCHEM	Pomona College	Integrated information and modelling system (QSAR). Runs on VAX. Uses the SMILES notation for chemical databases. Efficient in response times and disk space
ORAC	Orac Ltd	Reactions and associated data management system. Runs on VAX
PSIDOM	Hampden	Chemical information database management system. Runs on IBM-PC
REACCS	MDL	Runs on VAX, IBM (MVS, CMS), PRIME, FUJITSU. Reactions and associated data management system. Very user friendly.
SYNLIB	SK&F	Reactions and associated data management system. Runs on VAX

[a] See Table 1 for explanation of abbreviations and for addresses.

Table 4 Database Management Systems

Name	Supplier[a]	Main features
BASIS	Information Dimensions	General database management system, text oriented, thesaurus. Runs on IBM, VAX; microBASIS version on PC and CD-ROM
INGRES	Relational Technology (RT)	Relational database management system. Runs on VAX, IBM, other systems
INQUIRE	Infodata Systems	General and text database management system. Runs on IBM
ORACLE	Oracle Corporation	Relational database management system. Runs on VAX, IBM, other systems
QUESTEL	Telesystemes	General and text database management system. Runs on VAX, IBM, BULL; microQUESTEL on PC
RS/1	BBN Software	Data analysis and statistics. Tables management. Runs on VAX
SAS	SAS Institute	Data analysis and statistics. Tables management. Runs on VAX, IBM, others
STAIRS	IBM	Text database management system. Runs on IBM
SYSTEM 1032	Software House	General database management system. Runs on VAX
TEXTO	CHEMDATA	General and text database management system. Runs on VAX, IBM and other systems. MicroTEXTO on PC

[a] See Table 1 for an explanation of abbreviations and for addresses.

smaller system with DARC, gathering molecules along with spectroscopic, toxicology and safety data. Figure 4 shows several displays of data associated with a molecule. Display A corresponds to the main menu, where a star (*) in a box indicates that data are available in the corresponding file. These data (safety, pK_a values) are accessed simply by pointing to the corresponding boxes and are shown by displays B and C.

(ii) Data are managed by different systems, and can be visualized (but not searched) simultaneously with molecules. This represents the minimal requirement from the user's standpoint, and MACSS-DATACCS, MACCS-II and DARC provide easy to use and efficient tools to perform this operation.

(iii) Data are managed by different systems; visualization and cross-searches are possible from inside each other. Such interfaces are running with MACCS-II, ORACLE and INQUIRE, and have been implemented on VAX for DARC and ORACLE, SYSTEM 1032, BASIS, INGRES, NOMAD. This solution provides optimal flexibility, better management of data and can be run from specially designed procedures adapted to the user's needs and experience.

17.2.3 DATABASES

Data of interest to the medicinal chemist include molecules, associated data (physicochemical properties, biological and preclinical results) and reactions. Organization of these data in databases means faster access and in many cases is the only way to retrieve and handle them. In this section, we consider the databases available to the medicinal chemist, either *via* a network or for inhouse installation under a system such as those described above. We will also consider the problems involved in the management of internal databases.

17.2.3.1 Available Databases

Databases of interest to the medicinal chemist can be accessed *via* a network or made available for inhouse installation. In the first category many databases are of infrequent use and are accessed by an information specialist. Others are extremely useful but need experience to be efficiently searched at the lowest cost. This is the case for patent documentation (Derwent, CAS, *etc.*), reaction documentation *via* the CRDS system, which cannot be handled by untrained specialists, or even in many cases for Chemical Abstracts (CAS).

Databases for inhouse installation are becoming very popular due to the availability of user friendly management systems, and they make data accessible by an internal, and possibly integrated, information system.

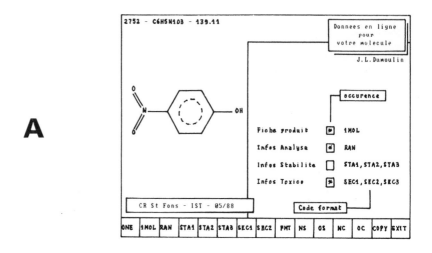

A

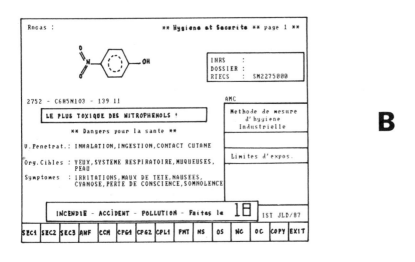

B

C

Figure 4   Integrated chemical information system with DARC

17.2.3.1.1 *Online databases*

Of the utmost interest to the medicinal chemist are Chemical Abstracts (CAS), accessible through several hosts and networks (STN, DIALOG, QUESTEL, DATASTAR). They can be searched by data (references, keywords, registry numbers), or by structure and substructure *via* CAS ONLINE or DARC systems, but these searches, except for the most simple ones (author, registry numbers, molecular formula), require practice and can be very costly. Depending on the annual volume of searches and the type of query, corporate prices can go from $60 per hour up to $150 per hour. CAS has no organized biological database connected to the structures file but does access the largest molecule file in the world (>8 000 000). At Rhône-Poulenc, training is provided to end users so that they can perform simple queries by themselves. Other similar or more advanced schemes have been reported.[41]

Other databases of interest include INDEX CHEMICUS, a file of chemical structures (*ca.* 2 000 000) connected to biological and analytical data—parts of this database are also available for inhouse installation; BIOSIS (*BIOSciences Information Service*), which contains more than 4 500 000 references in biological sciences, and publishes BIOCAS; LOGP available from TDS; NCI Drug Information System;[42] CASREACT, a file of chemical reactions (*ca.* 30 000), accessible in graphics mode, and set up by Chemical Abstracts.

Not much can be said about the future use of compact laser disks (CD-ROM) to distribute databases. The potential use is enormous and requires a rather low investment by the end user. However, editors seem reluctant to invest in a system the future of which is heavily dependent on fast and not yet settled technological evolution.

17.2.3.1.2 *Inhouse-installed databases*

Such databases are provided either by the software vendors (especially for reaction databases) as a means to boost their software sales, or by independent database producers. Reaction and molecule databases are distributed in a format compatible with major commercial packages. They are more and more favoured by the users since their cost represents only a small fraction of the time and money needed for the inhouse development of a similar database, and they are available more rapidly.

Table 5 gives the most important commercial chemical databases. Many efforts are oriented towards the development of reaction databases, and CASREACT, an online database, has been made available very recently. A major effort has been undertaken by the Institute for Scientific Information (ISI) to supply the CCR file in REACCS format, and 60 000 reactions (CCR 1986–1987) are now available. Work is underway to provide specialized databases, such as CHIRAS, dedicated to chiral reactions, by MDL.

Special attention should be given to databases which contain CAS registry numbers associated with molecules, such as CHEMQUEST, since this feature allows easy integration into an information system.

There are few databases containing physicochemical and biological data. Subsets of Index Chemicus partly fill this gap. MDL will also market a pK database, and Pomona College distributes a log P database *via* the MEDCHEM system.

17.2.3.2 Development of Internal Databases

Information management programs are like directories, *i.e.* they make access to information easier, provided that information is available. However, developing an internal database represents a major undertaking and requires manpower and time. This is necessary for the company private data, due to the increasing demand for fast access to integrated inhouse information, combined with the fact that its efficient management avoids duplication of effort and saves money. Even with the best program, this represents a major task, where long term involvement, continuity and manpower are the keywords. To explain this in more detail, we review the steps involved in developing a database.

In all cases, a database needs a unique supervisor, whose profile can be clearly defined as a documentation and computer-oriented chemist, and who has good contacts with the people involved in the development and the use of the database.

Table 5 Chemical Information Databases (for Inhouse Installation)

Name	Supplier[a]	Main features
CAMBRIDGE	Cambridge Crystallographic Data Centre	X-ray data for ca. 40000 structures. Accessible via a specific system
CCR	Institute for Scientific Information (ISI)	Current Chemical Reactions, year 1986–87. Contains ca. 60000 reactions. Accessible via REACCS
CHEMQUEST	Set up by Fraser Williams and available from Pergabase. Ex Fine Chemical Directory	Contains 60000 laboratory reagents, with 200000 suppliers references. Access via DARC, REACCS, MACCS
CLF	Molecular Design Ltd (MDL)	Current literature file. Reactions file, covers years 1982–86, ca. 21000 reactions. Access via REACCS
FCD	Fine Chemical Directory	See CHEMQUEST
INDEX CHEMICUS	Institute for Scientific Information (ISI)	File of ca. 2000000 molecules and intermediates, with spectroscopic data and references. Subsets can be installed inhouse. Access via DARC, MACCS, CHEMBASE
JSM	Molecular Design Ltd (MDL)	*Journal of Synthetic Methods.* Reactions file, covers years 1982–87, ca. 23000 reactions. Access via REACCS
LOGP	Pomona College	Octanol/water partition coefficients: 23000 measurements for more than 5000 molecules. Access via MEDCHEM
ORAC	Orac Ltd	Reactions file (ca. 45000). Access via ORAC
ORGANIC SYNTHESES	Molecular Design Ltd (MDL)	*Organic Syntheses.* Contains ca. 4800 reactions. Access via REACCS
PDB	Brookhaven National Laboratory	Protein Data Bank, contains ca. 300 structures of proteins
PK	Molecular Design Ltd (MDL)	File of pKs (ca. 10000), in several solvents and at several temperatures for ca. 3700 molecules. Access via MACCS, REACCS, CHEMBASE
SYNLIB	SK&F	Reactions file, ca. 15000 reactions. Access via SYNLIB
THEILHEIMER	Molecular Design Ltd (MDL)	Theilheimer reactions file, vols. 1–35. Contains ca. 42000 reactions and covers years 1948–80. Access via REACCS, ORAC

[a] See Table 1 for explanation of abbreviations and for addresses.

17.2.3.2.1 Database definition

First it is necessary to define items such as data to be included, relations between data, numbers of occurrences, types of data (numerical, real or integer, alphabetical, *etc.*) and ranges of data (to perform automatic checking at the input stage). In some cases, this operation can be performed by the chemist personally (*e.g.* CHEMBASE or TEXTO). However, large database management systems such as MACCS, REACCS, DARC, *etc.* require specialists to perform this task. Even when laboratory data are managed by a small system (like CHEMBASE), and data definition is made by the chemist, an information specialist should be consulted about the database specifications, to avoid any later problem when transfer of data from the laboratory level to the department database is required, or when further processing of the data is necessary (*e.g.* statistical analysis). Depending on the thoroughness of the preliminary analysis, the database complexity and the program itself, this step can take a few hours or a few days. Especially important are the functionalities of the system to modify an already existing database, since, as an example, implementation of a REACCS database at Kodak involved 14 changes after the initial design.[43]

17.2.3.2.2 Data input and checking

In the case of molecules and reactions input must be manual. At present no device or program exists to automatically interpret a handmade or printed drawing of a chemical structure and convert it into a connection table. Input with systems like MACCS or DARC on a mainframe corresponds to typical speeds of 10–30 molecules per hour by an experienced user, depending on the complexity and preliminary sorting of the drawings. This also includes the time necessary to input some information associated with the molecule, *e.g.* internal registry numbers, molecular formula (useful for automatic checking of the drawing), chemist's name and department.

This work can be performed either by a special team of trained people or by the chemists themselves in the laboratory. The first solution generally leads to a consistent input and better respect of chemical conventions but requires well-defined, explicit and unbiased formulae written by the chemists. This is an important point since, for example, phosphorus ylides entered in the ionic (P^+-C^-) or covalent form $(P=C)$ will not be recognized as identical. In the second solution the time of bench chemists is required. Programs like CHEMBASE seem to offer a good solution to decentralized input. Our experience has been limited to the first solution and we are considering a CHEMBASE approach of the second.

Entering reactions is more complex since more data have to be entered (reactants, products, solvents, catalysts, conditions, references, *etc.*), and reaction centres must be recognized and checked (automatic recognition does not work in more than 50% of the cases).[44] An interesting experience of the development of a reaction database using REACCS at Kodak has been described recently.[43] It corresponds to a centralized input by a team of technicians and chemists from carefully designed forms provided to the bench chemists. It seems that two to six reactions per hour represents a good input speed. As stated above, no mainframe chemical software currently provides powerful and user friendly tools for data input (other than chemical structures), such as a full screen editor.

From the above figures, it can be seen that a database of *ca.* 100 000 molecules needs at least four man years to be completed. Experience at Glaxo indicates that the conversion of a file of 120 000 compounds, plus thorough checking of the molecules and the associated data, requires the work of six people for one year (two full time and the remainder being done by part time chemists).[45]

17.2.3.2.3 Database maintenance and update

Addition of new data and the correction of mistakes corresponds to the input stage. Other tasks such as transferring the database to another more powerful system may be necessary.

Conversion of internal representations of molecules is necessary to move from one system into another, and this is now a standard procedure. For example, converting WLN into DARC requires no more than three hours (computer elapsed time) for 35 000 molecules, on an IBM 4381 machine.

17.2.3.2.4 Training

Until the advent of graphics-oriented packages, searching chemical information databases was restricted to information specialists, or at least to well-trained chemists. This is no longer the case, but, even with graphics systems, efficient searching by end users needs training and experience. Thus, a search over a large molecule or reaction file can be misleading if *a priori* indications about the possible aromatic or tautomer character of a bond is not given. Data searches need the knowledge of datatypes, field names and specifications, and in many cases keywords.

Training is therefore a must, and has to be made by the database or system supervisor. Our own and similar experiences indicate that for systems like DARC, MACCS or REACCS, minimal end user training requires at least half a day. Later, half-day sessions are needed to refresh knowledge, answer specific questions and improve the efficiency of searches. We have oriented our training on graphics searches with DARC and REACCS, and only limited data queries have been presented. Training must also be adapted to the most currently used databases (Chemquest, file of company molecules and reactions, Theilheimer).

We consider that many commercial manuals are inadequate for the end user. We have observed about 20–30 question types corresponding to the most frequent user's demands and have rewritten small booklets for end users, with entries corresponding to the most frequent searches, and NOT to the functionalities of the system. A comprehensive index is added, which points to entries such as: 'what to do when the system breaks down, how to save results, how to send data to another system, *etc.*', and we have tried to reproduce the most typical mistakes made by users. They have found this approach quite useful and able to meet most of their needs. More advanced training is provided when necessary or upon request.

17.2.3.2.5 Data security

Security management is essential for internal data. This is a specific task of the database supervisor.

17.2.4 HARDWARE AND NETWORKS

Hardware corresponds to computers (mainframes and/or microcomputers), interactive terminals and printing/plotting devices. Networks represent a special and very important aspect, concerned with both hardware and software. Devices associated with new technologies are more and more frequently encountered, such as CD-ROM readers and laser scanners for alphanumerical data input. Our aim is to provide general information in this highly specialized and fast-evolving field.

17.2.4.1 Computers

In molecular modelling the most widely encountered machine is the VAX (from Digital Equipment Corporation). This machine has gained worldwide acceptance due to many features of special value for scientific work. These include a user friendly operating system VMS, a good FORTRAN compiler, versatility, ascending compatibility, an easy to use assembler and the largest library of scientific packages in the world. Most of the molecular-modelling systems run only on VAX machines (microVAX, VAX 11/7XX, VAX 8XXX). However, many calculations require more computational power than can be provided by a VAX, which can, however, be connected to an array processor. More and more often, access to a very powerful machine is required, such as a CRAY X/MP, CRAY 2, or CDC ETA computer (as is the case for *ab initio* QM calculations, or molecular dynamics on proteins). Dupont, at the present time, and, since very recently, Scripps Clinic are the only chemical companies to have their own CRAY X/MP for modelling purposes.

For chemical information systems the choice of a computer is generally larger and many packages run on VAX, IBM (and IBM-like machines) or PRIME machines. For IBM, both MVS/TSO and VM/CMS operating systems can be used.

Generally IBM proves to have larger disk spaces and faster input/output operations (I/O) on disks. But a system like VM/CMS makes tedious the development of databases with multiple simultaneous input. Another problem with IBM is the preferential use of synchronous telecommunications, which is incompatible with almost all the graphic terminals commercially available, and the need for ASCII/EBCDIC conversion of internal codes, which always creates problems with graphics.

We have seen that response time is a key point for a user friendly program. To avoid bad response times due to computer overload there is a tendency to use dedicated computers. This is the case with molecular modelling, which needs batch runs of several hours (if not days), with few I/O interruptions, making difficult the simultaneous use of interactive information systems, for which fast response times are a must. Stand alone workstations (SUN, APPOLLO, SILICON GRAPHICS), running with a UNIX operating system, are now providing impressive graphics possibilities and computational power. As in the automobile and aerospace industry, they represent an obvious tool for implementing molecular-modelling software.

An interesting development which is currently underway is the use of very fast specially designed chips (transputers) in molecular modelling. According to Chemical Design, MITIE workstations built from these transputers are machines up to 100 times the power of a VAX 8600.[46] However, these machines are useful only if software developments take into account parallel processing.

17.2.4.2 Terminals

There is a definite need for the chemist to have access to internal, and possibly external, information and resources through a unique terminal, which must have graphics I/O possibilities. It is also of the utmost importance that any chemist can have easy access to a terminal, possibly a personal terminal (one for two to three chemists seems a reasonable compromise). Most of the graphics terminals work in asynchronous mode, and have the following general characteristics: graphic input with a mouse or less preferably a light pen; graphics display, with a minimal definition of *ca.* 600*400 pixels; and compatibility with graphics standards such as TEKTRONIX.

Such a terminal is quite sufficient to access chemical databases and handle simple molecular structures. Molecular modelling requires more advanced features: colour graphics; real-time local animation of display; hardware shading, clipping, perspective handling, zoom, hidden surface removal; and raster display definition of 1000*800 pixels (or more), or vector generator display.

A high definition raster display allowing real-time animation of complex images (with shading) is not yet available but announcements have been made by several companies (Pixar, E&S, Silicon Graphics). The most frequently encountered terminals in this field are PS 3XX from Evans and Sutherland.

There is an interesting solution to the problem of providing a multipurpose terminal for the chemist, which consists of a microcomputer equipped with a graphics card (EGA, HERCULES), a mouse and an emulator (*e.g.* EMUTEK, TGRAF, SMARTERM, PCPLOT, EM), which handles data transmission and makes the computer behave as an asynchronous graphics terminal. This gives more possibilities to the user, who can employ the PC as a stand-alone microcomputer to run a word processor, manipulate and display data, and use systems like CHEMBASE to manage his personal documentation. Caution should be taken in non-English-speaking countries, since many emulators are US made and do not take national keyboards into account. Such a system is also useful for accessing external databases: by means of a modem and a standard telecommunications program (such as the emulator), transmitted information is captured on the hard disk, and can be inspected and printed in nonconnected mode, which reduces the session price heavily. It also facilitates the creation of a personal bibliographic database by transferring the data from the hard disk into a file on the microcomputer itself or on a mainframe.

17.2.4.3 Printers and Plotters

Even with interactive graphics terminals there is always a need to manipulate paper copies. Besides a photograph, the simplest way is a hard copy on any matrix or inkjet printer. This is slow (at least 1.5–2 minutes per copy) and quality is average. Plotters are slow, give good quality, but are noninteractive and must be run on batch mode.

Laser printers are faster (not in graphics mode), quality is perfect (colour is announced) and prices are declining steadily (they start at *ca.* $3000), but copy price remains relatively high. The main drawback is the absence of an international standard of page description since three standards are presently competing. There are strong indications that the POSTSCRIPT language will be the winner.

17.2.4.4 Networks

The most frequently encountered internal networks use data transmission *via* an ETHERNET wiring. This classical solution can be used to transmit data at very high speeds, which is a key point in allowing future transmission of images and full text.

The PABX system uses a telephone switchboard and the internal telephone network. There is no need for extra wiring, and any phone plug can be used to connect a terminal. Voice and data can be transmitted simultaneously on the same line in asynchronous digital mode. Although this solution is presently not adapted for data requiring high transfer speeds (images, molecular modelling), it is quite flexible and much higher speeds will be available soon.

In many cases data access from the terminal is asynchronous. The main advantage is the fact that almost all terminals are adapted to this standard. The main problem remains with data security and error checking.

In an IBM environment one can have access to an asynchronous network (*via* a 7171 or 3708 adapter), or more likely to a synchronous SNA network, which requires IBM terminals and use of the GDDM graphics language for program development. The advantages lie in full data-error checking, high speeds of transmission, and easy and efficient control of the network security by the system manager.

More recently IBM has introduced the TOKEN RING network, but few installations have been made in scientific environments. The most outstanding is that installed at Carnegie-Mellon University to run the ANDREW system.[47]

17.2.5 ORGANIZATION AND STAFF

Before starting a computational chemistry group in a company it is necessary to define the tasks to be performed. Chemical information management systems are nowadays available as state of the art packages; there is no doubt that in most cases they represent an obvious choice and few companies are developing and even maintaining inhouse systems. However, work is still required to integrate systems. At the present time one cannot expect this task to be performed by vendors, due to the competition among them. Moreover, the complexity of molecular-modelling methods and the increasing number and complexity of commercial packages require special staff to manage them and to assist the chemist in making efficient use of these techniques. We will consider in this chapter the tasks to be done and staff to perform them.

17.2.5.1 Tasks to be Performed

The tasks involved in computational chemistry are oriented towards providing the user with a set of efficient computerized tools and a satisfactory solution of his problem. They can be separated into user- and system-oriented tasks.

17.2.5.1.1 *User-oriented tasks*

(i) Training

This is an absolute must if one wants to have efficient use at the end user's level. For molecular modelling training depends heavily on the background of the user. Chemists with advanced knowledge in quantum chemistry can have access to QCPE programs and run and interpret by themselves semi-empirical calculations. Any chemist should be trained to grasp the fundamental aspects of MM and QM and to use most of the manipulation and data input functionalities of a molecular-modelling system. Choice of an MM force field or a QM method adapted to a given problem, conformational analysis, QM calculations (especially at the *ab initio* level) are left to the full time specialist. It must be emphasized that the great variety and complexity of tools involved in computational chemistry require full time specialists to be properly utilized. Recent progress in computer graphics and software user friendliness have led in some cases to a belief that end users should be given free access to these tools and that they would benefit more from these techniques without the intervention of an intermediary. This is absolutely wrong. The situation is totally analogous to that encountered in analytical chemistry: end users can perform routine ^{1}H NMR, IR and chromatographic analysis, but cannot deal with complex problems which are left to the specialist, due to the large amount of time and specific skills required.

With training, a user is able to define a problem, discuss the problem with a specialist, understand the required operations, recognize difficulties, and study and visualize the results of a problem (such as conformational analysis, molecular orbital schemes, *etc.*). From our experience, training could require at least one month over a longer period of time.

Database definition and search training has been presented in Section 17.2.3.2.

If the laboratory terminal is a specially equipped microcomputer, a half-day course on the local operating system, graphics terminal, emulator and network is also necessary.

In synthesis design, systems like LHASA require training and from the experience of several users are difficult to master in order to be able to obtain significant and worthy results. Use of the CAMEO program (synthesis simulation) appears to be much easier.

(ii) Problem solving

The major task of the computational chemist is to provide help to the medicinal chemist and, in many cases, to solve specific problems. In molecular modelling, this is required at least in the following cases: conformational analysis of complex molecules; *ab initio* calculations (even at the STO 3G level); molecular dynamics; molecular mechanics on systems with many local minima; conformations of proteins; and docking studies.

17.2.5.1.2 System-oriented tasks

(i) System implementation and maintenance

Many packages are sold with maintenance contracts and major releases are delivered about once a year. This implies installation and thorough study of the new functionalities before retraining the users.

(ii) Interfacing and integrating systems

Implementing an integrated information system, including molecular modelling and QSAR programs, requires a great deal of work in writing programs, exchanging and converting data between different systems and developing and writing customized procedures and menus. These tasks require knowledge of three types of languages: scientific (*e.g.* Fortran), operating system (*e.g.* REXX2 for VM/CMS on IBM) and database management system (*e.g.* SQL with Oracle).

(iii) Developing databases

The development of databases is discussed in Section 17.2.3.2.

(iv) Scientific programming

In molecular modelling or QSAR there is often a need to modify or adapt force fields or write a program to calculate properties according to recently published or internally developed algorithms. This can give an edge to a company, with respect to standard functionalities offered by available programs, and represents a strong argument for inhouse developments.

(v) Software tests

The large amount of available software makes necessary the testing of programs, emulators, *etc.* This represents an increasing part of the work of the computational chemist.

17.2.5.2 Organization

Computational chemistry tasks are at the interface of three well-defined activities: (a) scientific and technical information management; (b) computer science (providing many of the necessary tools); and (c) chemistry and biology. This analysis of the situation has led many companies to create teams dedicated to computational chemistry, and which share the following features.

(i) A multidisciplinary group, with PhD level specialists in quantum chemistry, organic chemistry, physical chemistry, polymer science, crystallography, analytical chemistry, biology, molecular mechanics and dynamics, and at least a strong background or experience in computer science.

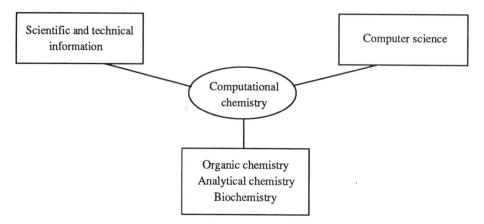

Figure 5 Interaction of computational chemistry and other activities

(ii) The chemical information management group is often connected to the molecular-modelling group in order to share similar techniques and tools and avoid duplication of effort.

(iii) In order to increase contact between the computational chemistry group and the chemists, it is necessary to have computer-oriented chemists in the laboratory, acting as 'relays' between both groups. This can be done by having voluntary bench chemists working for a given period of time (at least a few months) on a molecular-modelling problem. The computational chemistry group should act as a permanent training system to make chemists aware of the new possibilities offered by these techniques.

(iv) There are no standard figures regarding the number of people involved in computational chemistry. From our experience and contact with many such groups figures can vary as follows (in a research center of *ca.* 1000 persons): molecular modelling—four to eight PhDs, one to two MScs or technicians; chemical information—one to two MScs for reaction databases, one to two MScs for molecule databases, one to two MScs for biological databases; people involved in development are often not the same as those involved in maintenance and training, which requires three to five MScs or technicians; total—10 to 25 persons. This figure includes all the people involved in molecular modelling, management of chemical information systems and computer resources, software maintenance and training.

These figures are frequently met in many American or European companies which have been involved in computational chemistry for a long time. Smaller companies or those with less or no experience find it easier to start with a workstation and software. Hiring specialists in both chemistry and computer science proves difficult, although the situation has improved lately.

17.2.6 COST EFFECTIVENESS ANALYSIS

Cost effectiveness analysis is always difficult in a research environment. Both productivity and creativity improvements are to be considered; the latter is very important as a long term effect. In information management, productivity increases are quickly observed, and improvements in creativity are also obvious due to the new possibilities offered to the chemist for manipulating a large amount of information efficiently and rapidly. Molecular design techniques cannot be so easily interpreted in terms of enhanced productivity, but are a key factor in enlarging the chemist's creativity by providing the tools to simulate phenomena at the molecular level with speed, reliability and accuracy. This creativity enhancement is an absolute necessity for discovering effective new drugs in a highly competitive market, and these tools, when properly utilized by a team of specialists, represent an inevitable trend. This is clear from the figures for the number of worldwide installations: MDL indicates it has more than 200 mainframe installations for information management systems. In molecular modelling, a survey performed by the University of North Carolina and Research Institute of Scripps Clinic[48] gives 256 installations, among them close to 100 in chemical and pharmaceutical industries, and these figures are somewhat conservative.

The figures about costs are fluctuating rapidly, especially for software and hardware. A complete molecular-modelling workstation (with a microVAX computer and a raster display) starts at

$70 000–90 000, including software. Information management systems need to be accessed by many chemists and the hardware price must include laboratory wiring. The computer itself costs at least $50 000. A complete package costs $150 000–250 000 (including molecular structures, reactions and data management systems), depending on the size of the host and the number of users. Any graphics terminal or specially equipped microcomputer is worth $3000–8000.

Computational chemistry is expected to lead to more efficient information management and to a reduction of the number of molecules to be prepared by providing a better understanding of fundamental phenomena at the molecular level and by allowing the user to simulate experiments before actually performing them. If creativity improvements cannot be easily represented by figures, productivity increases are generally simpler to evaluate and some examples of observed or estimated savings are as follows.

Duplication of effort is reduced by access to relevant and reliable information. Thus, access to the CHEMQUEST file avoids synthesizing molecules which are commercially available. The price of preparing a molecule for testing (*ca.* $2500–3500) is similar to the annual lease price of CHEMQUEST (*ca.* $3500 per year), which can also be used to manage laboratory inventories efficiently.

Our estimation is that easy access to information can save at least 5% of the chemists time. This is partly due to elimination of duplication of effort, but also to the fact that some types of information would have been impossible to find without modern systems. This is especially true for substructure and reaction searches. This figure applied to a department with 100 chemists leads to a $500 000 saving over one year.

Drug design techniques (molecular modelling, QSAR, *etc.*) are intended to reduce the number of molecules prepared. With an annual output of 5000 molecules, a 5% improvement in research productivity leads to $500 000 per year savings.

Molecular modelling on polymers at Celanese[49] led to efficiency and effectiveness increases with estimated savings of $600 000 per year.

Dupont acquired a CRAY-1S in 1986 for modelling purposes (proteins, *ab initio* calculations, polymers and material modelling). Within two years they exchanged it for a much more powerful CRAY-X/MP/24, worth $66 million.[50] They consider the impact of modelling very important in reducing the number of unnecessary experiments and in performing carefully designed ones. Very recently, Scripps Clinic has also acquired a CRAY X-MP for modelling purposes.

These figures are somewhat conservative and many other advantages cannot be easily quantified, but are fully felt when chemists have integrated these techniques in their working habits. Especially important is the fact that these systems allow the chemist to elaborate and test models at the molecular level and provide a better understanding of microscopic phenomena.

ACKNOWLEDGEMENTS

This chapter is based on published reports, discussions with specialists belonging to universities and pharmaceutical companies and documentation provided by software, hardware and database vendors. I want to thank the following people for open and helpful discussions: Dr. Watthey, Ciba-Geigy USA; Professor W. C. Still, Columbia University; Dr. Peter Gund, Dr. G. Smith and Dr. H. Woodruff, Merck Sharp & Dohme; Dr. D. Pensak, DuPont; D. Zumbro, Carnegie-Mellon Institute; Dr. G. Moreau, Roussel-Uclaf; Dr. J. Howe, Dr. G. M. Maggiora, Upjohn; T. Hagadone, Upjohn, for providing a preprint of his paper on integrated information systems; Professor W. L. Jorgensen, Purdue University; Dr. H. Fruhbeis and Dr. Strecker, Hoechst; Dr. Gowal, Firmenich; Dr. T. Thielemans, Janssen Belgium; Dr. E. Surcouf, RP Santé; Dr. A. Tomonaga, Kureha; Dr. Iwatsubo, Mitsui PetroChemicals; Dr. Kazumi Yuki and H. Yamachika, Sumitomo Chemicals; and Dr. U. Hegi, Sandoz (Basel).

17.2.7 REFERENCES

1. H. D. Brown, *J. Chem. Inf. Comput. Sci.*, 1985, **25**, 218.
2. H. Fruhbeis, R. Klein and H. Wallmeier, *Angew. Chem., Int. Ed. Engl.*, 1987, **26**, 403.
3. U. C. Singh, F. K. Brown, P. A. Bash and P. A. Kollman, *J. Am. Chem. Soc.*, 1987, **109**, 1607.
4. N. L. Allinger, *Adv. Phys. Org. Chem.*, 1976, **13**, 1.
5. T. Clark, 'A Handbook of Computational Chemistry', Wiley, New York, 1985, p. 12.
6. M. Karplus and J. A. McCammon, *Sci. Am.*, 1986, **254**, 30.
7. J. Kuriyan, G. A. Petsko, R. M. Levy and M. Karplus, *J. Mol. Biol.*, 1986, **190**, 227.

8. D. F. V. Lewis, *Chem. Rev.*, 1986, **86**, 1111.
9. Ref. 5, p. 233.
10. G. R. Marshall, *Annu. Rev. Pharmacol. Toxicol.*, 1987, **27**, 193.
11. C. Hansch and T. Fujita, *J. Am. Chem. Soc.*, 1964, **86**, 1616.
12. P. C. Jurs, T. R. Stouch, M. Czerwinski and J. N. Narvaez, *J. Chem. Inf. Comput. Sci.*, 1985, **25**, 296.
13. P. Broto, G. Moreau and C. Vandycke, *Eur. J. Med. Chem.—Chim. Ther.*, 1984, **19**, 61, 79.
14. S. Wold, P. Geladi, K. Esbensen and J. Öhman, *J. Chemom.*, 1987, **1**, 41.
15. L. Turner, F. Choplin, P. Dugard, J. Hermens, R. Jaeck, H. Marsmann and D. Roberts, *Toxicol. in Vitro*, 1987, **1**, 143.
16. F. Choplin, *Actual. Chim. Ther.*, 1987, **14**, 179.
17. M. G. Bures and W. L. Jorgensen, *J. Org. Chem.*, 1988, **53**, 2504.
18. H. L. Morgan, *J. Chem. Doc.*, 1965, **5**, 107; P. Bawden, J. T. Catlow, T. K. Devon, J. M. Dalton, M. F. Lynch and P. Willett, *J. Chem. Inf. Comput. Sci.*, 1981, **21**, 83.
19. J. E. Dubois and H. Viellard, *Bull. Soc. Chim. Fr.*, 1968, 900.
20. E. J. Corey and W. T. Wipke, *Science (Washington, D.C.)*, 1969, **166**, 178.
21. M. F. Lynch, J. M. Harrison, W. G. Town and J. E. Ash (eds.), 'Computer Handling of Chemical Structure Information', McDonald, London, 1971, p. 24.
22. G. W. Adamson, J. M. Bird, G. Palmer and W. A. Warr, *J. Chem. Inf. Comput. Sci.*, 1985, **25**, 90.
23. E. L. Clennan, *J. Am. Chem. Soc.*, 1987, **109**, 2229.
24. J. W. Moore, *J. Chem. Educ.*, 1987, **64**, 29.
25. A. J. Gushurst and W. L. Jorgensen, *J. Org. Chem.*, 1986, **52**, 3513.
26. F. Wong Chung and J. A. McCammon, *J. Am. Chem. Soc.*, 1986, **108**, 3830.
27. J. Gasteiger and H. Saller, *Angew. Chem.*, 1985, **97**, 699; J. Gasteiger, M. G. Hutchings, B. Christoph, L. Gann, C. Hiller, P. Low, M. Marsili, H. Saller and K. Yuki, *Top. Curr. Chem.*, 1987, **137**, 19.
28. C. Hansch and A. J. Leo, 'Substituent Constants for Correlation Analysis in Chemistry and Biology', Wiley, New York, 1979.
29. K. Enslein, *Pharmacol. Rev.*, 1984, **36**, 131S.
30. S. Wold and W. J. Dunn, III, *J. Chem. Inf. Comput. Sci.*, 1983, **23**, 6.
31. P. F. Tiley, *Chem. Br.*, 1985, **21**, 162.
32. P. Azario, R. Barone and M. Chanon, *J. Org. Chem.*, 1988, **53**, 720.
33. W. T. Wipke, H. Braun, G. Smith, F. Choplin and W. Sieber, *ACS Symp. Ser.*, 1977, **61**, 97.
34. E. J. Corey, A. K. Long and S. D. Rubinstein, *Science (Washington, D.C.)*, 1985, **228**, 408.
35. P. Gund, E. J. J. Grabowski, D. R. Hoff, G. Smith, J. D. Andose, J. B. Rhodes and W. T. Wipke, *J. Chem. Inf. Comput. Sci.*, 1980, **20**, 88; W. Sieber, *191st ACS National Meeting, New York*, 1986, abstract ORGN 88; see also ref. 39, p. 361.
36. S. Hanessian, L. Foret, L. Trepanier, S. Leger, F. Major and A. Glamyan, *191st ACS National Meeting, New York*, 1986, abstract ORGN 87.
37. A. P. Johnson and A. P. Cook, 'Modern Approaches to Chemical Reaction Indexing', Gower, Aldershot, 1986, p. 184.
38. J. C. Marshall, *J. Chem. Inf. Comput. Sci.*, 1987, **27**, 47.
39. T. R. Hagadone, in 'Chemical Structures: the International Language of Chemistry', ed. W. A. Warr, Springer Verlag, Berlin, 1988, p. 23.
40. S. Barcza, H. W. Mah, M. H. Myers and S. S. Wahrman, *J. Chem. Inf. Comput. Sci.*, 1986, **26**, 198.
41. R. E. Buntrock and A. K. Valicenti, *J. Chem. Inf. Comput. Sci.*, 1985, **25**, 415; W. A. Warr and A. R. Haygarth Jackson, *J. Chem. Inf. Comput. Sci.*, 1988, **28**, 68.
42. G. W. A. Milne and J. A. Miller, *J. Chem. Inf. Comput. Sci.*, 1986, **26**, 154 and subsequent papers.
43. S. E. French, *Chemtech*, 1987, **17**, 106.
44. J. J. McGregor and P. Willett, *J. Chem. Inf. Comput. Sci.*, 1981, **21**, 137.
45. L. Tewnion, 'Scientific Information Systems Seminar', March 1986, organized by Fraser Williams, Prestbury, Cheshire, UK.
46. *Chemical Design News*, summer, 1987.
47. J. H. Morris, M. Satyanarayanan, M. H. Conner, J. H. Howard, D. S. H. Rosenthal and F. D. Smith, *Assoc. Comput. Mach. Commun.*, 1986, **29**, 184.
48. H. Thorvaldsdottir, 'List of Known Molecular Computer Graphics Installations', University of North Carolina, Chapel Hill, NC, 1987.
49. G. V. Nelson, MDL users' meeting, Paris, 1985.
50. D. A. Dixon, 'Computational Chemistry on Cray Supercomputers', Cray Symposium, Lausanne, 1986.
51. G. R. Newkome, *J. Am. Chem. Soc.*, 1988, **110**, 325.
52. D. E. Meyer, W. A. Warr and R. A. Love (eds.), 'Chemical Structure Software for Personal Computers', American Chemical Society, Washington, 1988.

17.3

Chemical Structures and Computers

DAVID WEININGER and JOSEPH L. WEININGER

Daylight Chemical Information Systems Inc., Irvine, CA, USA

17.3.1 INTRODUCTION

In any chemical endeavor one must name the substance under investigation. Whether a chemical compound is synthesized, or its properties are measured, it is essential to have a consistent, unambiguous and reproducible description of it. Historically, from the beginning of chemistry as a rudimentary science to the start of the computer age, chemical nomenclature evolved slowly, in a rather haphazard manner. On a time-scale spanning centuries, the field of chemical information has advanced more in the last two decades than in all its previous history. Advances in computer technology have increased both the speed and capacity of information storage and processing to such an extent that it is now possible to deal with chemical information problems which until recently had been considered intractable.

This chapter emphasizes the application of computers to chemical nomenclature and the advances which thereby are brought about for chemical information in general. Starting with the traditional approach of naming chemicals and their structures, the contribution of computers is related to the use of line notation in chemical nomenclature, because line notation provides the key to making the peculiarities of chemical structures amenable to computer processing. In its most recent development SMILES (*S*implified *M*olecular *I*nput *L*ine *E*ntry *S*ystem) is a simplified linear chemical notation language that is specifically designed for computer use by chemists. A SMILES-oriented chemical information system is being developed which provides solutions for a variety of problems such as generation of unique notation, structural depiction, optimal data retrieval, substructure recognition and a variety of modeling problems. It is noteworthy that its initial impact is in the field of medicinal chemistry.

17.3.2 TRADITIONAL APPROACHES TO CHEMICAL NOMENCLATURE

Wherever chemicals are used, they must be named. Naming and describing substances is, therefore, an activity that dates back millennia to the origins of chemistry itself. Its development into a systematic, although initially primitive, science occurred roughly 200 years ago when elements and compounds were being discovered which required systematic description. The development of chemical notations during this period is described by Wiswesser,[1] who mentions the contributions of many historically renowned chemists, such as Lavoisier, Berzelius, Cannizzaro, Kekulé and others. Atom-to-atom connections began to be recognized in the structure of chemical compounds. Structural features, such as open chains and fused ring systems were noted, so that more than 100 years ago the concept of commonly recognized, rational formulas evolved. This foreshadowed the introduction of linear notation to represent chemical structure. Historically, chemicals and their structures were named in a progression from common names to attempts at systematization of these names. Before serious consideration had been given to computer processing of chemical information, Chemical Abstracts evolved Collective Index Names which attempted to keep nomenclature up to date with ever proliferating and increasingly more complex structures appearing in the literature of organic and biological chemistry. Today, the description of chemical structure in terms of linear notation is necessary for efficient computer usage, which is the subject of this chapter.

Chemical structure was chosen as the key to identification, *i.e.* naming of chemicals, because it provides the best basis for collecting, storing and retrieving chemical information. Fugmann has pointed out that the special status of chemical information in the general field of information theory derives from the use of structural formulas.[2] They are used universally, without language barriers, and also clearly define a chemical in a pictorial manner that helps in the search for structural relationships between similar substances.

Listings based on a dictionary approach, such as the Chemical Abstracts Collective Index, are at the mercy of the inherent ambiguity of different names for one single structure as well as the time-consuming effort to find a name in a dictionary setting. It is not surprising, therefore, that one of the first major attempts at computerizing chemical information took place at Chemical Abstracts[3] with the introduction of registration numbers for chemical structures. At Chemical Abstracts Service (CAS) the registration process was developed to establish a method of machine searching for chemical structures. Once this was possible, every chemical structure could be given a unique identifying label which is the CAS registry number. Development of this system and also its extension to substructure searches was initiated in the 1960s. But at that time computer technology was insufficient in terms of speed and capacity to process all the necessary information. Meanwhile, in the following two decades, hardware developments have been so spectacular that computer technology has now caught up with chemical information in general so that all extant chemical knowledge, which contains information for approximately 8 000 000 discrete compounds, can now be encompassed in a single, efficiently processed database.

17.3.3 COMPUTER APPROACHES TO CHEMICAL NOMENCLATURE

The application of computer science to chemical information systems highlights the paradox that on the one hand most of the existing chemical literature is unsuitable for direct computer processing because chemical nomenclature contains many imprecise and ambiguous descriptions of chemical compounds and their properties; on the other hand, as has been pointed out previously, chemical structure is uniquely suitable as the key to identify chemical compounds. Computers can process linear strings of data with relative ease. Thus, applying linear notation to chemical structure is the logical approach to a state-of-the-art chemical information system. This has been most recently and most efficiently accomplished in programs utilizing Weininger's computer interactive chemical notation language, SMILES.[4] Originating at the Medicinal Chemistry Program, Pomona College,[5] SMILES programs are now being applied extensively in pharmaceutical and chemical processes. The language itself and some of its applications are discussed in the present chapter.

The earliest computer programs for chemical information did not use linear notation for nomenclature. Pioneering work at CAS started in the early 1960s, when structure description was combined with the registration process.[3] In this application of graph theory to the depiction of chemical structure, a chemical formula is viewed as a graph with individual nodes representing atoms, and edges representing bonds. The list of node symbols and the description of the structure as a graph comprise a 'connection table'. In one such early table,[6] non-hydrogen nodes of a structure were listed according to an exact set of rules. By themselves, these rules accomplished only a partial ordering of nodes, so that a hierarchy of rules had to be imposed. Connection tables were constructed by applying the rules in steps of successive partial ordering. The rules concern the direction of bonds, from an initially arbitrarily chosen focal node, and also deal with ring closures, node values, line values and structural modifications, while the connection tables represent the unique graphs (structures). Over the years the CAS registry system developed into the CAS ONLINE system which now performs substructure as well as full-structure searches.[7]

Search queries are normally initiated with structure diagrams in the CAS ONLINE and similar systems. With such input systems, the user specifies the desired structure (or substructure) on a graphics terminal screen which is then interpreted by the host computer as a structural query. Graphics input is done by use of light pens, mouse/digitizers, or touch sensitive screens. The primary advantages of graphics input systems are that the input process is conceptually similar to sketching a structure and is visually stimulating. Queries generated by graphics input are generally self-documenting (what you see is what you get). Disadvantages include the requirement for specialized terminal hardware, a nearly complete lack of standardization for such systems, and that specification of search queries can only be easily formulated for literal connections of atoms.

After the structure has been introduced as a diagram, the CAS ONLINE system first checks specific structural features and compares them with others stored in files. This is followed by an atom-by-atom search to verify the precise match of the search query with the file substance. The last step in this matching procedure is too slow for practical computer processing. As a result the system has been refined in terms of 'search screens' which characterize individual substances or substructures and are matched with query inputs. In its present version, CAS ONLINE still uses screen searches and atom-by-atom search procedures, but after the initial graphical input the searches are performed automatically. Manually specified screen searches can also be made.

DARC is a French computer system[8] which evolved chronologically in parallel with CAS systems, but was designed to access CAS programs, facilitating substructure searches in databases, structure–activity correlations and molecular design. These are precisely the activities which currently have highest priority in the field of chemical information. The basic principle of the DARC system is a topological approach dealing with molecular fragments. Emphasis is placed on the 'environment' of a substructure. This may consist of complete or partial ring systems, chains, functional groups and so on. The user input involves primarily structural diagrams depicting all atoms and bonds. Separate commands on the diagrams deal with atom values, bond values and atom connectivities. Search commands lead to iterative and topological screen searches and the possible use of alphanumeric as well as graphic checks.

The CAS ONLINE and DARC systems as well as most of the other important chemical computer languages and systems conceptually are related to, and, in fact, are derived from, graph theory. Outstanding descriptions of the history, basic concepts, and the applications of graph theory in chemistry are given by Read[9] and Balaban.[10] Both present as their starting point the analogy between graphs and structural formulas of molecules. This has already been mentioned in the discussion of the CAS registry system above and will be discussed in detail in the description of the SMILES chemical notation language (Section 17.3.5). Read's linear notation is a 'closed' code, *i.e.* the rules for generating the code are fixed and provide structural formulas for any existing or hypothetical molecule as long as a graph-like structure can be drawn. Retrieval of the structure from the code is simple; it is based on graph theory and does not require chemical intuition. The reader is referred to Read's paper[9] for details of the rules, definition of terms and examples of coding chemical structures and substructures. Twelve desirable properties of a chemical code (notational language) are listed. They include the requirements of a code as a linear string of symbols, devising a unique notation for a given structure, retrievable by a clearly defined process. Furthermore, both coding and decoding processes should be simple, with familiar symbols, easily comprehended by chemists, yet not requiring chemical intuition. Balaban's review goes into greater detail concerning the importance of graph theory for chemistry.[10] He deals with the definition, enumeration and coding of constitutional, steric and valence isomers. Among chemical applications of graph theory, the discussion of nomenclature is particularly pertinent to the subject of this chapter. Some of the systems mentioned in the present work, which apply graph theory to chemical nomenclature, are given by Balaban in a historic setting. The proposal of the first linear notation system by Gordon and co-workers[11] in 1948 was almost immediately followed by Dyson's system[12] (recommended by IUPAC), and by the *Wiswesser Line Notation* (WLN).[13]

A chemical nomenclature system must have a notation which systematically names every compound. From a mathematical viewpoint each such system is an algorithm which has an input of a structure or its equivalent, such as a connection table, and has as an output the unequivocal designation of the molecule. This designation need not be a name although in the most advanced systems the code consists of a string of linear, alphanumeric symbols. Over a period of approximately 30 years the WLN system has been most popular and widely used and is still in use today. It has a total of 40 symbols, the 26 letters of the alphabet, 10 numerals, three punctuation marks, (&, -, /), and a blank space. These symbols and their sequences characterize various molecular features, such as functional groups, ring structures, *etc.* The sequence of the alphanumeric string determines the starting point in the structure; alternate starting points or alternate paths are governed by the position of the symbol in the alphabet. Since compounds are indexed alphabetically, this leads to grouping of related compounds for efficient but not rigorous retrieval of associated information.

Most symbols have their normal chemical meaning. Exceptions are some new symbols, such as Q for —OH, the hydroxyl moiety, V for a connective carbonyl group, —C(=O)—, and Z for an amino group, —NH_2. Hydrogen atoms are not expressed and numerals signify the number of carbon atoms in an unbranched alkyl chain or sequence. There are 47 rules which are applied hierarchically to obtain unique structural descriptions. Nineteen of these pertain to aliphatic compounds (five for unbranched chains, eight for branched chains and six for organic salts and addition compounds). The remaining 28 rules pertain to cyclic structures (10 for compounds with benzene rings, the remaining 18 deal with a variety of structural considerations for non-benzene cyclics, concerning the possibilities, among others, of their branching, substituents, fusions or bridges). Nineteen tentative additional rules are applied to inorganic formulas and another 13 tentative rules deal with compounds containing charges, or isotopes or separate parts, *i.e.* unusual bonding, such as chelates.

WLN rules are described comprehensively in a textbook by Smith.[14] The scope of this chapter does not permit further description of the rules, except for these examples: Rule 1 states: 'Cite all chains of structural units symbol by symbol as connected'. Thus, acetone, $CH_3C(=O)CH_3$, is written as '1V1' (1 denotes the alkyl groups with one carbon, and V the carbonyl group mentioned above). *trans*-Dichloroethylene, ClCH=CHCl, is written as 'G1U1G -T' (where G is the symbol for Cl, 1 as above, U the double bond for *cis* or *trans* structures, and -T (space-hyphen-T) gives the *trans* configuration).

17.3.4 SMILES—SIMPLIFIED LINEAR NOTATION, A COMBINED APPROACH TO CHEMICAL NOTATION

An effective modern chemical nomenclature must provide an interface between humans and the computers which are their primary information management tool. The most important characteristics for such a nomenclature are: (i) the graph of a chemical structure must be described uniquely; (ii) the nomenclature must be quickly and easily learned; (iii) the nomenclature must be able to be efficiently interpreted by a computer; (iv) a unique name in the nomenclature must be able to be quickly machine-generated for any possible structure; (v) the nomenclature should be able to be expanded to specify general substructures; (vi) there must be a provision for quick display and verification of any structure and (vii) ideally, there should be no dependence on specific software or hardware.

Currently available methods fail to meet one or more of these objectives. Conventional (*i.e.* IUPAC) nomenclature is fundamentally incompatible with machine interpretation and processing (and only marginally acceptable for humans due to the expertise required). The combination of graphics input with registration numbers is much more suitable for computer processing of limited sets of known structures, but is dependent on arbitrarily assigned identifiers and thus fails to provide a general solution for chemical identification.

WLN meets the essential requirements of a deterministic chemical nomenclature and its efficacy has been proved over the years by its durability. However, the generation of a unique WLN is not easily accomplished by either a human or a computer. There are too many rules for effective use by a non-expert human. Computers are good at dealing with hierarchical systems of rules, but the nature of the WLN rules precludes efficient machine generation of WLNs. This can be attributed to the non-deterministic nature of rules of the form 'Among all possible paths choose the one which . . .'. Even the interpretation of WLN is a complex task; machine interpretation of the ideal nomenclature would be computationally trivial.

The phenomenally rapid advance in computer technology in recent years has provided both the need and the means for a more efficient chemical nomenclature. Specifically, it has become possible to use a new approach which combines a coherent chemical notation with algorithms designed around modern computer capabilities. The separate human-oriented and machine-oriented capabilities are actually combined into a new, simple and compact line-entry system, SMILES. Appearing concurrently with the accelerated rise in computer technology, SMILES is designed to be truly computer interactive, *i.e.* to take advantage of the greater speed and capacity of modern computers. The simplicity of its use is based on algorithms that recode the user's input.

The SMILES approach to chemical nomenclature separates description of a unique chemical structure (by the chemist–human) from the generation of the unique structural description (by the computer). The latter requires rules and hierarchies which are inherently difficult for the chemist. Therefore, this task is relegated to computer algorithms.

The primary objective of SMILES is to provide a notation system which is natural for the user and efficient for the computer. Unlike other chemical notation systems, brevity of notation and economy of alphabet are not primary objectives. Many of the pitfalls in other line notations can be attributed to overuse of symbols and hierarchical rules based on the length of the final notation.

Once the basic nomenclature problem is solved, other objectives are achieved rather easily. The machine-generated unique SMILES can be treated as a computer address where other data are stored (leading to constant-time data retrieval). A graphical depiction of any SMILES can be generated for structure recognition by the user. The same algorithms that code unique SMILES for complete structures can be applied to molecular fragments to provide the basis for efficient predictive models. The characteristics that make it easy for a human to use allow very fast machine interpretation which makes screening large sets of structures feasible. Although brevity was not a primary objective, SMILES is compact enough to allow all the SMILES for known structures to be stored in memory at once.

17.3.5 SMILES—SIMPLIFIED LINEAR NOTATION

SMILES denotes a molecular structure as a graph which is essentially the two-dimensional picture chemists draw to describe a molecule. Only four basic rules are necessary to accommodate graphs for the subset of almost all organic structures in normal valence states. These four rules encompass atoms, including aromatic carbons, bonds, branching and ring closures. Hydrogen atoms may be omitted (hydrogen-suppressed graphs) or included (hydrogen-complete graphs). Aromatic structures are specified directly in preference to the Kekulé form.

17.3.5.1 SMILES Specifications

17.3.5.1.1 *Atoms*

Atoms are represented by their atomic symbols; this is the only required use of letters in SMILES. Each atom, except hydrogen, is specified independently by its atomic symbol enclosed in square brackets, [   ]. Elements in the 'organic' subset B, C, N, O, P, S, F, Cl, Br and I may be written without brackets if the number of attached hydrogens conforms to the lowest normal valence consistent with explicit bonds. Atoms in aromatic rings are specified by lower case letters, *e.g.* normal carbon is represented by the letter C, aromatic carbon by c. Since hydrogens are implied, the following atomic symbols are valid SMILES notations for the indicated compounds

C	methane (CH_4)	P	phosphine (PH_3)
N	ammonia (NH_3)	S	hydrogen sulfide (H_2S)
O	water (H_2O)	Cl	hydrogen chloride (HCl)

but compounds with elements other than those specified above must be described with brackets, *e.g.* [Au] elemental gold.

Attached hydrogen and formal charges are always specified inside brackets. If unspecified, they are assumed to be zero for an atom inside the bracket. Examples are

[H+]	proton
[OH−]	hydroxyl anion
[OH3+]	hydronium cation

The SMILES program also recognizes constructions of the form [Fe + + +] as synonymous with [Fe + 3].

17.3.5.1.2 Bonds

Single, double and triple bonds have the symbols −, = and #, respectively. Single bonds may be and usually are omitted. Examples are

CC	ethane (CH_3CH_3)	O=C=O	carbon dioxide (CO_2)
C=C	ethylene ($CH_2{=}CH_2$)	C # N	hydrogen cyanide (HCN)
CCO	ethanol (CH_3CH_2OH)	[H][H]	molecular hydrogen (H_2)
C=O	formaldehyde (CH_2O)		

For linear structures, SMILES notation corresponds completely to conventional notation except that hydrogens can be omitted. For example, 6-hydroxy-1,4-hexadiene can be represented by three equally valid SMILES.

Structure	*Valid SMILES*
$CH_2{=}CH{-}CH_2{-}CH{=}CH{-}CH_2{-}OH$	(i) C=C—C—C=C—C—O
	(ii) C=CCC=CCO
	(iii) OCC=CCC=C

17.3.5.1.3 Branches

Branches are specified by enclosures in parentheses. Examples are

	Structure	*SMILES*
Triethylamine	$CH_3{-}CH_2{-}N(CH_2{-}CH_3){-}CH_2{-}CH_3$	CCN(CC)CC
Isobutyric acid	$CH_3{-}CH(CH_3){-}C({=}O){-}OH$	CC(C)C(=O)O

Branches can be nested or stacked, as shown for 3-propyl-4-isopropyl-1-heptene, $CH_2{=}CH{-}CH(CH_2CH_2CH_3){-}CH(CH(CH_3)_2){-}CH_2{-}CH_2{-}CH_3$, which is represented in SMILES notation as C=CC(CCC)C(C(C)C)CCC.

17.3.5.1.4 Cyclic structures

Cyclic structures are represented by breaking one single (or aromatic) bond in each ring. The bonds are numbered in any order, designating ring-opening (or ring-closure) bonds by a digit immediately following the atomic symbol at each ring closure. This leaves a connected non-cyclic graph which is written like a linear structure. Cyclohexane is a typical example (Figure 1).

There are many different but equally valid descriptions for the same structure, *e.g.* two of the SMILES notations for 1-methyl-3-bromo-cyclohexene are shown in Figure 2.

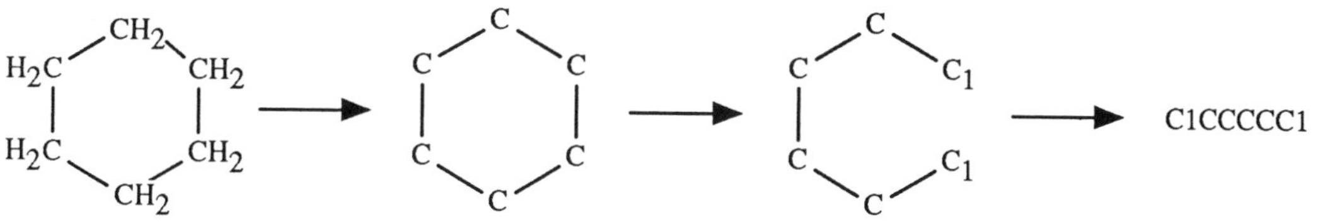

Figure 1 Generation of SMILES for cyclohexane

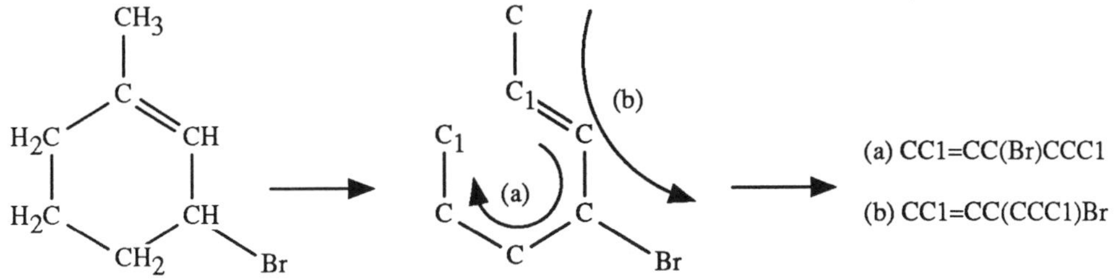

Figure 2 Alternate paths for SMILES of the same structure

SMILES does not have a preferred entry on input; although (a) in Figure 2 may be simplest, other notations are just as valid. The single digits denoting ring closures can be reused if necessary. This is illustrated in Figure 3 by the structure of cubane in which the atoms have more than one ring closure.

17.3.5.1.5 Disconnected structures

Disconnected compounds are written as individual structures separated by a . (period). The order in which ions or ligands are listed is arbitrary. There is no implied pairing of one charge with another, nor is it necessary to have a net zero charge, as for example in sodium phenoxide (Figure 4).

17.3.5.1.6 Aromaticity

Aromatic structures are distinguished by writing the atoms in the aromatic ring in lower case letters, for example as shown for benzoic acid in Figure 5.

17.3.5.2 Basic SMILES

As simple as the SMILES rules are, an even simpler four-rule subset suffices for the vast majority of organic compounds. Although SMILES allows direct specification of charges, attached hydrogens and aromaticity, often they are not required. This subset uses only the symbols H, C, N, O, P, S, F, Cl, Br, I, (,) and digits, with the following four rules: (i) atoms are represented by atomic symbols; (ii) double and triple bonds are represented by = and #, respectively; (iii) branching is indicated by parentheses; and (iv) ring closures are indicated by digits appended to symbols.

17.3.5.3 SMILES Generation Example

With the above simple rules almost all organic structures can be described in line notation. An example of a more complex structure is that of morphine in Figure 6. It contains five rings, of which one is aromatic. The breaking of the rings are shown and also the designation of ring closures by means of digits attached to the symbols of the ring atoms. The aromatic carbon atoms at ring closures 1 and 5 are in small case letters.

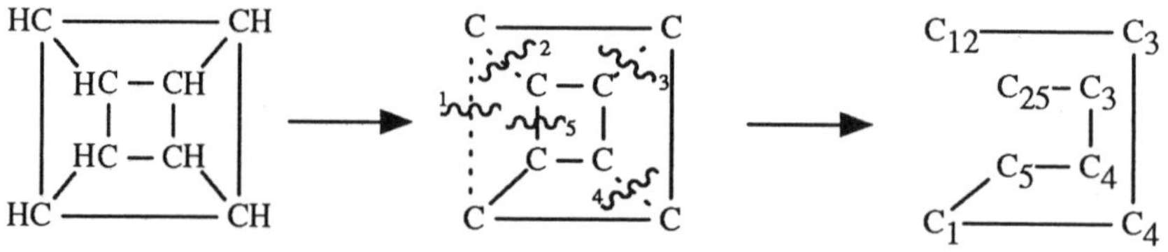

Figure 3   Generation of SMILES for cubane: C12C3C4C1C5C4C3C25

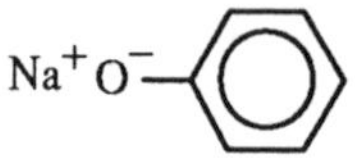

Figure 4   SMILES for sodium phenoxide

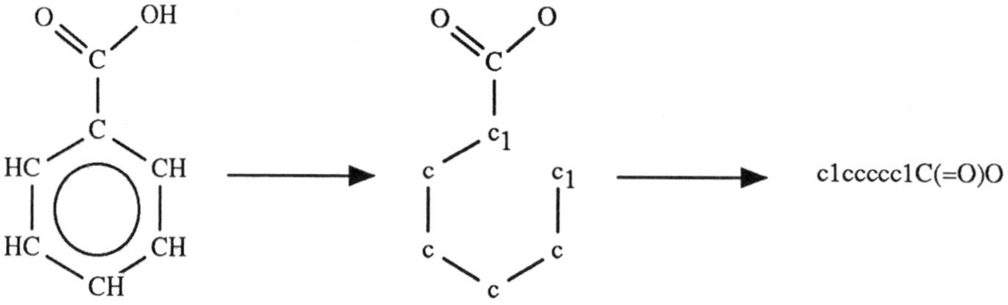

Figure 5   SMILES for benzoic acid

17.3.5.4   SMILES Notation Conventions

The above rules are refined by stipulating some conventions for writing SMILES notations of certain classes of compounds.[4] These involve bond specification, hydrogen specification in certain nitrogen ring compounds, and additional factors involving aromaticity.

17.3.5.4.1   Hydrogen specification

Hydrogen atoms do not normally need to be specified when writing SMILES. Except for special purposes, the SMILES system treats hydrogen attachment as a property of non-hydrogen atoms.

The SMILES convention regarding hydrogen attachment assumes that hydrogen makes up the remainder of an atom's lowest normal valence, consistent with explicit bond specifications. The normal valences of B, C, N, O and the halogens are 3, 4, 3, 2 and 1, respectively. 'Lowest normal valence' refers to 3 or 5 for phosphorus; 2, 4 or 6 for aliphatic sulfur. There is no distinction between 'organic' and 'inorganic' SMILES nomenclature. This ensures, for example, that OS(=O)(=O)O is interpreted as H_2SO_4 (sulfuric acid) while S is interpreted as H_2S (hydrogen sulfide). Aromatic sulfur donating a lone pair of electrons is assigned a formal valence of 3 or 5.

The number of attached hydrogens may be explicitly specified by supplying a hydrogen count inside brackets or by putting H or [H] atoms in the SMILES. Even when specified as explicit atoms, the SMILES system removes all hydrogen atoms and just retains the attached hydrogen count.

There are few exceptions to the hydrogen suppression convention. The most obvious are specifications for the proton, [H+], and molecular hydrogen, [H][H]. There are also some applications that require general specification of more than one bond to hydrogen, such as in crystallographic databases. The rule used in the SMILES system is to eliminate all hydrogen atoms except in the following three cases: (i) hydrogens connected to other hydrogens; (ii) hydrogens connected to zero or more than one other atom; and (iii) in isomeric SMILES, isotopic hydrogen specifications, *e.g.* [2H].

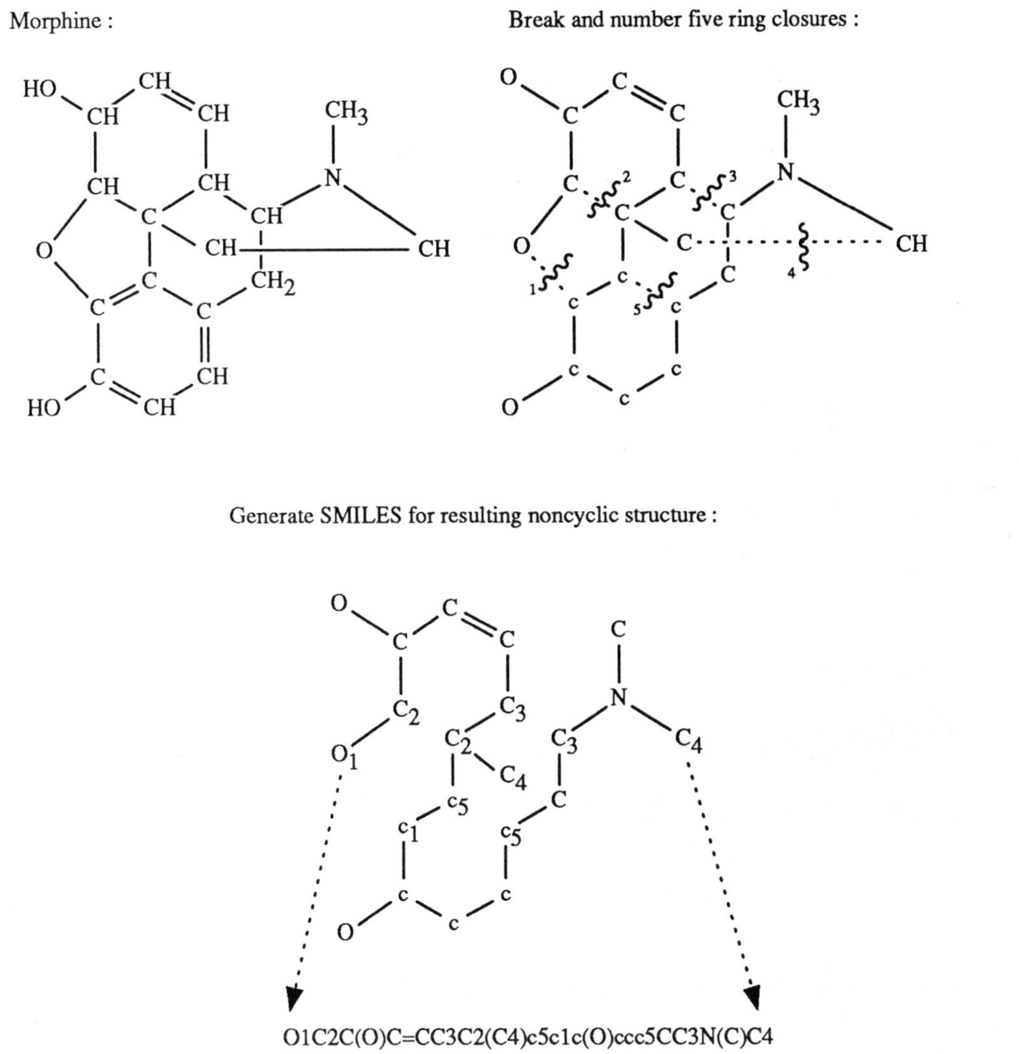

Morphine :   Break and number five ring closures :

Generate SMILES for resulting noncyclic structure :

O1C2C(O)C=CC3C2(C4)c5c1c(O)ccc5CC3N(C)C4

Figure 6   Evolution of SMILES for morphine

17.3.5.4.2   Bonds

Ionic bonds and complexes are not specified directly. They are written as separate parts of a disconnected structure. There are, however, a few types of bonds that do not fit easily into the above categories. Occasionally bonds have a hybrid character with both covalent and ionic characteristics. They can be denoted in either way, but for database applications it is necessary to choose one description and adhere to it.

Delocalized bonds, as in the nitrogen moiety, are also best written as covalent bonds to both uncharged oxygens, as in nitromethane, CN(=O)=O. This is a matter of convention because nitromethane could also be written as a charge-separated structure, C[N+](=O)[O–]. The advantage of the neutral form is that it preserves correct symmetry. When symmetry is not an issue, charge-separated structures are preferred if they avoid representing atoms in unusual valence states. For instance, diazomethane is written as C=[N+]=[N–] in preference to C=N=[N].

17.3.5.4.3   Ring compounds containing nitrogen

Two types of aromatic nitrogen must be distinguished in SMILES notation. Examples are pyridine and pyrrole (Figure 7).

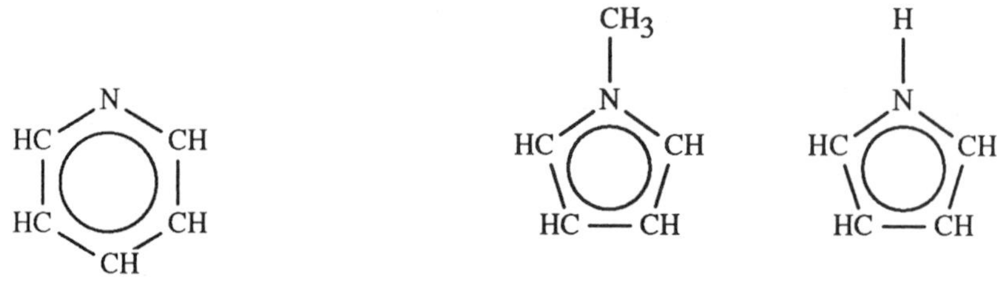

n1ccccc1   Cn1cccc1   [nH]1cccc1

Pyridine   Methyl- and 1*H*-pyrrole

Figure 7   Aromatic nitrogen compounds

Pyridine can be written as n1ccccc1. SMILES correctly deduces that no hydrogens are attached to the nitrogen in pyridine because two aromatic bonds satisfy the lowest normal nitrogen valence. In 1H-pyrrole the nitrogen is both aromatic and trivalent, with one attached hydrogen atom. Consequently, it could be written as an aliphatic structure N1C=CC=C1. However, the Pomona College MedChem algorithms will correctly detect its aromatic nature in the form of [nH]1cccc1 or Hn1cccc1. The hydrogen appearing outside brackets in SMILES is used to determine the hydrogen count for the atom to which it is bonded.

17.3.5.4.4   *Tautomers*

Tautomeric structures are explicitly specified in SMILES. The choice of one or all tautomeric forms of a structure is left to the user and will depend on the application. For databases and indexing purposes some authors prefer the enol over the keto form. For example, for the structure shown in Figure 8 the enol form, Oc1nccccc1 (2-pyridinol), is preferred over O=c1[nH]cccc1 (2-pyridone). This is a matter of convention. If a formal representation must be chosen for modeling, then the more 'stable' form is preferred. Alternatively, it may be best to model all tautomeric forms.

17.3.5.4.5   *Aromaticity*

SMILES algorithms detect aromatic compounds and ions. The system will accept either aromatic or non-aromatic inputs, check for aromaticity and convert the structure accordingly. This is accomplished with an extended version of Hückel's rule to identify aromatic molecules and ions.[15] To qualify as aromatic all atoms in the ring must be sp^2 hybridized and the number of available 'excess' π electrons must satisfy Hückel's $4n + 2$ criterion. For example, benzene is written c1ccccc1, but an entry of C1=CC=CC=C1 (cyclohexatriene)—the Kekulé form—leads to detection of aromaticity and undergoes an internal structural conversion to aromatic representation. Entries of c1ccc1 and c1ccccccc1 will produce the correct antiaromatic structures for cyclobutadiene and cyclooctatetraene, C1=CC=C1 and C1=CC=CC=CC=C1, respectively. In such cases the SMILES system searches for a structure with sp^2 hybridization, the implied hybridization, the implied hydrogen count and the specified formal charge, if any. However, some inputs may not only be incorrect but also impossible to convert, such as cyclopentadiene. Here, c1ccccc1 cannot be converted to C1=CCC=C1 because the central carbon atom is sp^3 with two attached hydrogens. The SMILES system will flag this as an 'impossible' structure.

SMILES deals with the aromaticity of charged structures as with neutral ones, except that in the charged ring structures the number of 'excess' π electrons is changed according to the in-ring charge.

Further details on SMILES nomenclature, available in ref. 4, are beyond the scope of this chapter. In summary, SMILES represents the best compromise between the human and machine aspects of chemical notation. It is easily understood by the chemist because it has a small number of simple rules. Computationally, SMILES is interpreted in a very fast, compact manner, thereby satisfying the machine objectives of time and space savings. It is based on a computer approach to language parsing so that the machine part follows algorithms consistent with rigorous hierarchical nomenclature rules. This results in a great improvement in the efficiency of information processing as compared to conventional methods.

In computer terms, SMILES notation represents a tree that can be interpreted in a single pass. The increase in efficiency derives from the language syntax. For example, the bonding arrangements

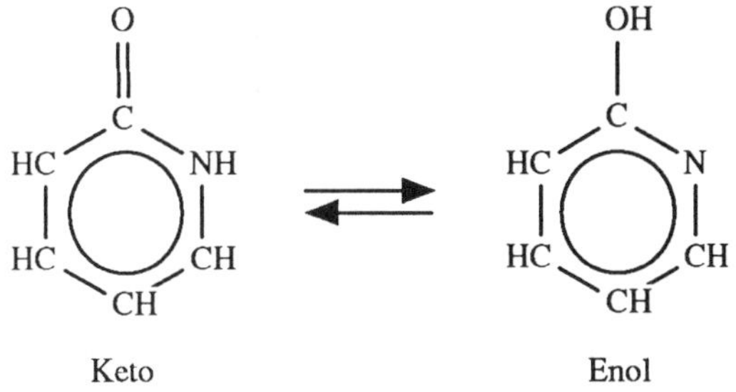

Figure 8 Tautomer example

are implied by the position of the atoms without having to be defined specifically. Thus, where previously 1000 to 2000 characters may have been needed to describe a compound such as morphine (Figure 6), SMILES stores the same information in 40 characters. It can also be shown that the necessary computer-programming time is reduced 100-fold over conventional procedures, using connection table formats, such as those used in CAS conversion tables.

Originally SMILES was developed to provide a human–machine language interface. Beyond this objective it has been valuable for the implementation of applications including data storage, structural displays, modeling new structures and substructure searches. An example is the use of SMILES in computation of partition coefficients and molecular refractivity in model compounds. These two properties are widely used in biochemical research. The logarithm of the partition coefficient is the hydrophobic parameter in Hammett methodology and has been applied to Quantitative Structure–Activity Relationships (QSAR).[16] Based on structural considerations, fragments of a modeled compound, designated and processed in SMILES nomenclature, estimate log P accurately (see Chapter 18.7).

17.3.6 GRAPHICAL DEPICTION OF CHEMICAL STRUCTURES

Computer graphics of chemical structures and formulas are most important for the interaction between chemist and computer, particularly in the fields of organic synthesis and substructure search. What appears to be well within the scope of modern technology actually is unexpectedly difficult because different conventional methods of presenting structures pictorially do not always follow the same rules. Ambiguities and exceptions are often encountered. Much has been published on this subject,[17,18] generally with reference to computer interactive graphics, *i.e.* by electronic input with a stylus on a tablet, or with light pens on graphical cathode ray tubes (CRT).

Initial guidelines at CAS concerned manual processing of graphical data. This type of formatting of two-dimensional graphs was improved and converted to computer programming, standardizing structural representation wherever possible and establishing files for fragments as well as for complete structures. Preferred formats for representation of acyclic structures, ring systems and stereoisomers were also established.[17]

Interactive graphics are introduced either by a light pen with integrated amplifier, or by a stylus on an electronic tablet. In one case a program functions either by keyboard action in a 'text' mode or in a pictorial 'normal' mode with the use of the light pen.[18]

The importance of connection tables for storing the graphics information is circumvented in some cases by storing essential information for creating the diagram as well as the connection tables. For complex structures in large databases this is impractical, so that the problem of displaying structural diagrams from connection tables has to be addressed. The procedure developed by Shelley[19] involves an initial perception of structural features as a data tree. This is followed by generating atoms, ring structures and their connecting bonds with graph invariant codes. They are used to minimize atom crowding and bond overlap, and for orienting ring systems and generating system coordinates independent of assigned sequence numbers.

Howe and Hagadone discuss the use of light pens in interactive graphics for two-dimensional substructure searches in terms of their code, search execution and system hardware. Many references are cited in their publication.[20]

A method of deriving three-dimensional structures from connection tables is described by Wenger and Smith.[21] The compound adamantane is given as an example in a perspective line drawing. Such a structure with highly fused ring systems may require specification of the normal stereochemistry at ring junctures with little conformational freedom because of distance constraints. Much more structural freedom is demonstrated for less-strained compounds, containing acyclic chains or macrocyclic rings.

17.3.6.1 Graphical Depiction by DEPICT Program

All the above methods of interactive graphics require file entries and computer codes. In the Pomona MedChem program DEPICT no file entries or coding is involved.[22] There is no explicit graphical information required to produce a drawing; therefore, graphical entry tools are not used. The structural representation is obtained directly from the input of SMILES nomenclature. Any structure for which SMILES notation can be written is represented in a drawing, which is easily recognized by the user. Structures in connection tables are also converted to graphics since the connection tables are converted by SMILES notation. There are three basic functions in the DEPICT representation of chemical structures: (i) the location of each atom is computed to give the best description of the structure in terms of simplicity, clarity and aesthetics; (ii) the structure is drawn with bonds, atomic symbols, charges and hydrogen attachments, and (iii) DEPICT catalogs the structures in a graphics library which, in turn, produces a correct terminal or plotter image.

DEPICT uses atomic symbols, hydrogen counts and bond representation in their conventional aspect. It searches for a picture which spreads out the atoms. Linear chemical structures are so represented; branched structures are spread out as symmetrically as possible without resulting in undue crowding. For example, 4-propylheptane is shown in Figure 9.

When a structure becomes crowded, DEPICT will give up symmetry to keep atoms apart (Figure 10).

DEPICT attempts to draw all rings as perfect polygons, as shown for progesterone (Figure 11).

As an aid to visual recognition in complex structures, the atomic symbols for ring atoms are drawn slightly smaller than non-ring atoms. For aromatic structures, circles are drawn inside the rings. Otherwise, they are treated identically to alicyclic rings. Fused aromatic rings are drawn as symmetrically as possible across fusion bonds, *e.g.* the DEPICT drawing for coronene (Figure 12).

Figure 9   4-Propylheptane

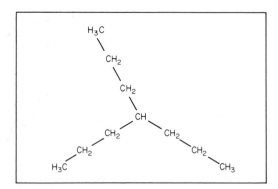

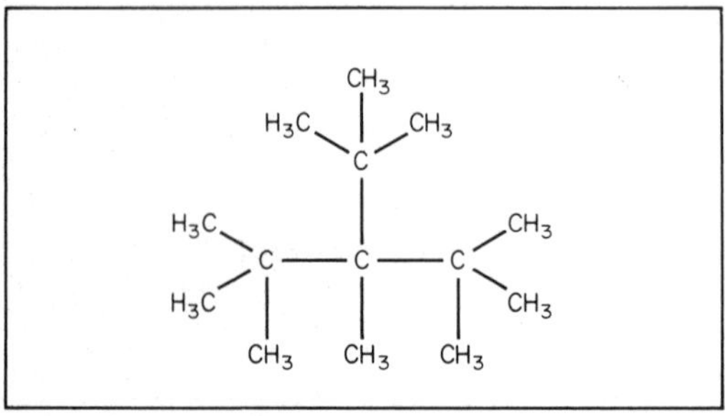

Figure 10   1,1,2,3,3-Pentamethyl-2-isobutylpentane

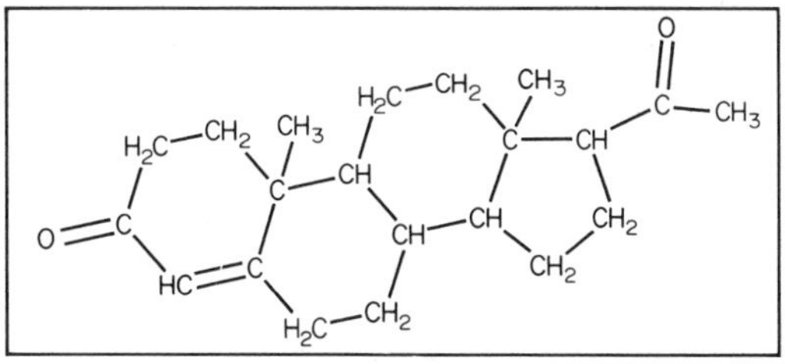

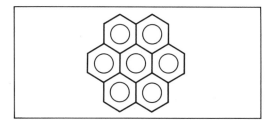

Figure 11 Progesterone

Figure 12 Coronene

Heteroatoms are always shown, as are atoms with multiple bonds external to the aromatic ring. If there is a charge or an unusual hydrogen count, the aromatic carbon is also shown. Spiro rings are drawn as two polygons with a line of symmetry through the spiro junction and both ring centers.

Multicyclic rings are difficult to display. Geometric polygonal representations are retained as much as possible while stretching one bond per multicyclic ring and reorienting atoms in the 'loose' ends away from other atoms in the structure. In Figure 13 the square and pentagonal shape of four- and five-membered rings in mirex can be seen.

DEPICT favors short over long stretching of bonds, but will stretch a multicyclic bond as far as necessary. Crossing of bonds is not explicitly prohibited. It is avoided as much as possible by spreading the atoms in the graph, but multicyclic structures may require bond crossing for clarity. This is one of the difficulties of drawing their graphs. Dieldrin illustrates the advantage of using smaller symbols for in-ring atoms and allowing bond crossing (Figure 14).

Disconnected structures are represented as two separate entities in the same display. A simple example is sodium phenoxide which can be entered in SMILES notation as [Na +].[O −]c1ccccc1 (Figure 15).

DEPICT produces good structural graphs for the vast majority of chemicals, but a few structures do not have satisfactory pictures. This may occur for a condensed multicyclic ring system. In the case of cubane (Figure 16), without drawing curved lines, the requirement of 'as many perfect polygons as possible' forces two bonds to fall directly on top of each other (compare the SMILES notation of cubane in Section 17.3.5.1.4).

However, in general, DEPICT produces excellent structural representations for organic chemicals, directly from SMILES inputs.

17.3.7 CANGEN — GENERATION OF UNIQUE SMILES

Examples in Section 17.3.5 show that a given chemical structure can be described by several equally valid SMILES notations. From a variety of such valid forms one must emerge that is 'unique' in order to serve as the identifier of the structure for databases and other chemical computer applications. This is accomplished by a method called CANGEN which combines two separate algorithms, CANON and GENES. The first stage, CANON labels a molecular structure with canonical molecular graph node labels. The structure is treated as a graph with nodes (atoms) and edges (bonds). Each atom is given a numerical label. In the second stage, GENES generates the unique SMILES notation in the form of a tree (molecular graph). This is accomplished regardless of which of the possible linear descriptions was initially introduced by the user. CANGEN rigorously defines two choices: firstly with which atom to start and secondly which direction to turn at each

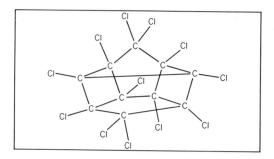

Figure 13   Mirex

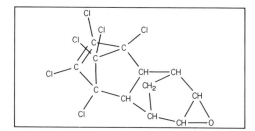

Figure 14   Dieldrin

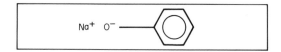

Figure 15   Sodium phenoxide

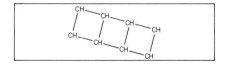

Figure 16   Cubane

branch. It selects a unique starting atom and refers to the canonical order when a branching decision must be made.

17.3.7.1   Theoretical Background

In general, graph theory has become important in applications to chemical information because it provides the basis for codification of nomenclature in chemical computer programs.[10] The classification and ordering of nodes in a graph is here applied to chemical structure notation. With an initial set of node properties and a given connectivity for a two-dimensional, non-directed graph (with N nodes and E edges), each node is assigned a rank. In CANGEN this ranking will completely discriminate each node environment with respect to all initial node properties. The classification algorithm must also recognize constitutionally symmetric nodes, *i.e.* nodes that are equivalent in all respects. This step as well as the generation of unique node order, breaking all ties, graph construction and identification are all essential parts of the CANGEN process.

The combinatorial and the sum method are two different methods which characterize and connect nodes and their environments. The combinatorial process is suitable for analyses of small graphs (simple chemical structures) but becomes too cumbersome for more complex ones because of the need to characterize each node environment completely. Simple, exhaustive solutions which have orders of max(N, E)! become impractical as N increases beyond 15. Partial characterization is therefore often attempted and is adequate for symmetry perception. Such algorithms use a general approach of breadth-first optimization of a tree.[23,24] All nodes are characterized successively deeper into the total graph until the combined characterization is adequate. This reduces the base of the algorithmic order of N or E to the number of edges in the shortest path between the most distant nodes. However, these algorithms do not avoid the problem of factorial order for the general case.

The sum method achieves greater efficiency by the limiting use of a combined description of connected nodes, while ignoring all path specific topological information. A sum vector S is modified iteratively by summing over the S elements of neighboring nodes only one edge away. On iteration, the difference in the developed sums of two nodes indicates non-equivalence. This intrinsically low order process still encounters difficulties. The original Morgan algorithm[3] uses only the local degree (number of nearest neighbors) in S and relies on subsequent combinatorial identification of other zero-order node properties, P (for example, the atomic numbers). It is possible to improve the basic sum method by incorporating node properties in a two-dimensional matrix. Yet, eventually combinations of nodes must be produced and compared. Thus, there is an inherent ambiguity in sums with respect to addenda of rational numbers. Hence, identical S values do not necessarily assure symmetry of nodes for properties P in a general case. This is the reason for eventually invoking combinatorial procedures. It will be shown below, however, that an unambiguous sums method can dispense with any subsequent combinatorial procedure. Furthermore, by eliminating the ambiguity, all zero-order node properties may be expressed in a single vector S. Tracking individual properties as 'extended sums' becomes unnecessary and redundant.

Unique SMILES notation is obtained by taking the squares of sums and processing them in parallel with the sums of node properties during the process of classifying graph nodes and searching for symmetry. This SMILES method of 'unambiguous sums' will be illustrated below with examples of labeling, ranking and unique ordering of structural notations. It guarantees canonicalization over the originally specified graph theoretical invariant properties.

17.3.7.2 CANON — Canonicalization of Molecular Graphs by Method of Unambiguous Sums

17.3.7.2.1 Initial graph invariant order

Graph theoretical invariants are properties which are independent of the way in which a graph is ordered. Examples are the atomic invariants of Table 1.

A unique linear combination of these invariants represents their initial vector in the CANGEN algorithms. This set of six variables is sufficient for the purpose of obtaining unique notation for simple SMILES, but it is not necessarily a 'complete' set. No 'perfect' set of invariants is known which will distinguish all possible graph asymmetries. However, for any given set of structures, a set of invariants can be devised to provide the necessary discrimination. The list used by CANGEN is designed for simple molecular graphs. Other graph properties may be added as needed, for instance the invariant used by isomeric SMILES contains isotopic mass, bond directionality and local chirality. Conversely, one or more invariants may be eliminated in less rigorous operations than CANGEN conversion of SMILES notation.

Table 1 Atomic Invariants

1. Number of connections
2. Number of non-H valence bonds
3. Atomic number
4. Sign of charge
5. Absolute charge
6. Number of attached hydrogens

Table 2 Invariants for Pentane

Atom type	Individual invariant						Combined invariant
Methyl carbon	1	1	6	0	0	3	10106003
Methylene carbon	2	2	6	0	0	2	20206002

The set of invariants in Table 1 have the indicated priorities (1 is first, 6 has last priority). This set conforms to the fundamental assumption, made throughout the CANGEN process, that 1:1 mapping represents any set of invariants equally well. As an example, Table 2 gives the two invariant sets for the methyl and methylene carbons of the pentane molecule, CCCCC.

17.3.7.2.2 Rank equivalence

Although the different values in an invariant set must be ordered by their priority, there is nothing intrinsically meaningful in their specific values. To avoid numerical overflow of the computer system these values are replaced by small numbers so that the rank of each invariant retains the desired properties. In the above example the initial invariants are

$$10106003\text{-}20206002\text{-}20206002\text{-}20206002\text{-}10106003$$

Their ranks are

$$1\text{-}2\text{-}2\text{-}2\text{-}1$$

giving a new invariant set which is just as usable in the CANGEN process, and more suitable for machine processing than the original set.

17.3.7.2.3 Extended connectivity

While there are only two types of carbon atoms in pentane (methyl and methylene), there are three carbon symmetry classes. Morgan[3] and Behrson[26] view sets of invariants in terms of the sums of the atoms' invariants one away, two away, *etc.* The test of symmetry classes is whether or not the 'extended connectivity sums' are different. For pentane summing the neighbors one away reveals the three symmetry classes: the first ranking, from the initial invariant is

$$1\text{-}2\text{-}2\text{-}2\text{-}1$$

Summing the ranks of the nearest neighbors, *i.e.* one away results in a new rank for the set which shows the symmetry classes

$$1\text{-}2\text{-}3\text{-}2\text{-}1.$$

The present CANGEN technique refines this process in order to avoid storing an indeterminate number of such sets. Each value is iteratively replaced by the sum of its neighbors' ranks plus another term. This latter term is the product of a large number (100 is chosen for convenience) times the invariants' own rank, as demonstrated in Table 3.

Since only one vector is used in this method, the speed of in-place computation is greatly increased. The factor 100 is added to one of the terms in the iteration process in order to maintain rank stability. When the rank vector is not changed by an iteration, there is no need to continue because once a summing iteration fails to differentiate the equivalence of a node pair, all subsequent iterations will also fail.[20] The sum method is then complete and an invariant partitioning has been developed.

17.3.7.2.4 Unambiguous sums by sums-of-squares

One pitfall of the extended sums method, as pointed out in Section 17.3.7.1, is the inherent ambiguity of the sum of integers which can lead to false symmetry perceptions. At this point earlier methods[3, 25, 26] revert to higher order, combinatorial algorithms which are very slow. Instead, the CANGEN process develops and keeps in parallel two vectors for each invariant: a straight sum and a sum of squares. Numeric overflow that might be expected from a repeated operation of squaring squares does not occur because on each iteration the two invariant vectors are replaced by their ranks, with ties being broken by the sum of squares vectors.

Details of the application of extended connectivity and the method of summing by sums of squares can be found in a publication devoted to the subject of CANGEN.[27] As one example, the search for a unique SMILES notation for the compound 6-amino-2-ethyl-5-aminomethyl-1-hexanol with the molecular structure shown may be considered (Figure 17).

Table 3 Computation of Extended Sums for Pentane

Original rank	*1*	*2*	*2*	*2*	*1*
Iteration 1	100 + 2	200 + 3	200 + 4	200 + 3	100 + 2
New values	102	203	204	203	102
New ranks	1	2	3	2	1
Iteration 2	100 + 2	200 + 4	300 + 4	200 + 4	100 + 2
New values	102	204	304	204	102
New ranks	1	2	3	2	1

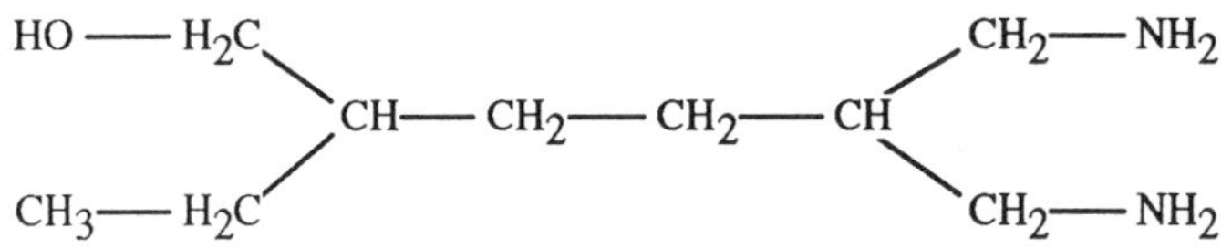

Figure 17 CANON example: OCC(CC)CCC(CN)CN

In this case there exists the potential ambiguity with respect to the connected atoms in the linkages N + N and C + O which was referred to in the beginning of this section. An initial arbitrary SMILES for this structure, OCC(CC)CCC(CN)CN, has an original ranking of the invariants of

3-4-5-(4-1)-4-4-5-(4-2)-4-2

The procedure of ranking in parallel successive iterations of straight sums and square sums results in this final ranking

3-6-10-(4-1)-8-7-9-(5-2)-5-2

17.3.7.2.5 *Breaking ties*

If there are no constitutionally symmetric node classes in the graph, the problem of ordering nodes is not difficult. Difficulties arise when ordering more symmetric graphs, such as that of cubane with the notation C12C3C4C1C5C4C3C25 (see Section 17.3.5). This molecule has eight identical carbon atoms. Iteration shows that all eight nodes are identical, so the starting point must be arbitrary. Once a starting point is chosen, however, then the remaining seven nodes are no longer identical; three atoms are one away, three atoms are two away and one atom is three away. The algorithm proceeds by doubling all ranks and reducing the value of the first (lowest-valued) atom, which is tied, by one. The set is then treated as a new invariant set and the previous algorithm for generating an invariant partitioning is repeated. For cubane the final ranking is

1-3-5-7-8-6-4-2

Table 4 briefly summarizes the canonicalization of chemical structure with the following eight listed steps of the CANON algorithm.

Table 4 CANON Algorithm

1. Establish initial invariants. Go to step 3.
2. Sum ranks and ranks squared.
3. Sort extended sums.
4. Rank sorted sums.
5. If not invariant partitioning, go to step 2.
6. On first pass, save partitioning as symmetry classes.
7. If highest rank is smaller than number of nodes, break ties, go to step 2.
8. . . . else done.

17.3.7.3   GENES—Generation of Unique SMILES

With symmetrical classes established, the structure is treated as a tree and a SMILES string is generated that corresponds to a depth first search (DFS) of that tree. As noted above, the only required decisions are where to start, *i.e.* at which node of the tree, and which branch to follow at each branching point.

17.3.7.3.1   *Initial node selection*

The lowest canonically numbered atom is chosen as the starting point of the SMILES notation. This atom becomes the root of a tree for a subsequent DFS. For the example of 6-amino-2-ethyl-5-aminomethyl-1-hexanol the final ranking has the terminal carbon of the ethyl group as the starting point (root) of the graph (see Section 17.3.7.2.4). As a rule this selection implies that a terminal atom is chosen, if one exists. This is desirable for efficiency because a pair of parentheses are eliminated and also for aesthetics.

If the chemical structure consists of separate entities, such as ions or ligands, it is considered a disconnected compound, denoted by a period (.) as the disconnection symbol. Repeated selection of starting atoms, using the same criterion of the lowest remaining canonical label, ultimately produces a disconnected SMILES (a forest).

17.3.7.3.2   *Branching decisions*

Branching decisions could be as simple as the selection of a starting atom because the algorithm directs branching towards the lowest-labeled atom at the fork in the branch. For example, acetone has the combined invariants

$$10106003\text{-}30406000\text{-}(10208000)\text{-}10106003$$

giving a rank equivalent of 1-3-(2)-1. Starting with one methyl group, the direction of branching at the central carbon atom will be towards the second methyl group which has a lower rank than the oxygen of the carbonyl. Consequently, the unique SMILES for acetone is CC(C)O and not CC(O)C.

In cyclic structures, at branches with multiple bonds, it would not be suitable to select a multiple bond for a SMILES ring closure. This is avoided by establishing a rule that directs branching in a ring towards the multiple bond over the single bond. The following two rules apply: (i) branch to a double or triple bond in the ring, if they exist, or (ii) branch to the lower canonically numbered atom. Rule (ii) is the same as the one that applies to linear structures (*cf.* CC(C)O and CCC(CO)CCC(CN)CN above).

17.3.7.3.3   *Two-pass method: treatment of cyclic and multicyclic structures*

There are several algorithms available that are based on the depth first search (DFS). A 'two-pass' method is chosen because it produces unique SMILES for complex multicyclic structures. It starts with a simple DFS, appending nodes (atomic) and edge (bond) symbols to the output SMILES as the search progresses. Each time a branch is taken, a left parenthesis is added to the output string; each time a dead end is reached, a right parenthesis is added. The first pass terminates when all nodes have been reached. If the structure is linear, the SMILES is complete and unique; there is no need for a second pass.

For cyclic structures, however, the search will encounter a node that has already been visited. At this point the ring closure nodes are known (the last node and the already visited one) so that the SMILES ring closure indicators (digits) can now be appended to the node symbols in preparation for the second DFS pass. This second DFS enables the two-pass method to cope with problems specific to multicyclic systems. They concern the search around a ring where an errant left parenthesis may be left dangling, sorting the digits on nodes with multiple ring closures and ordering them in 'opening' order since their assignment is not determined by the 'closing' order. All three of these problems are solved by introducing a second DFS in the two-pass method (for examples see ref. 27).

Summing up, the CANGEN process consists of a two-stage algorithm. The first stage involves canonicalization of structure, whereby the molecular structure is treated as a graph with nodes (atoms) and edges (bonds). Each atom is canonically ordered and labeled numerically. In the second stage starting with the lowest-labeled atom, a tree (molecular graph) is constructed which is the unique SMILES structure regardless of which of the various valid original linear SMILES notations was introduced by the user. The establishment of unique SMILES by this process provides the key to solving the basic problem of chemical nomenclature, namely that one single chemical compound may have many different names. In addition, one single structure of a compound may be depicted in many different ways. By obtaining a unique notation for structure and name, chemical nomenclature is amenable to many applications for databases where SMILES can be extended to any number of synonyms, identifiers and structures, such as common names; Collective Index names, IUPAC names, CAS numbers and others.

17.3.8   APPLICATION OF SMILES

It has been noted in Section 17.3.3 that chemical information presents the problem, unusual in computer science, that the most natural identifier—structure—can only be described by a series of imprecise and overlapping synonyms. They may refer to overall structure, isomers, congeners, conformations, mixtures, *etc.* No matter how sophisticated a dictionary is constructed, one can never include all possible forms of structures for which information is available. The SMILES system solves this chemical information problem with the implementation of an extremely efficient database, THOR (*Th*esaurus *O*riented *R*etrieval), which has many applications. In this section, only those aspects of THOR are discussed which are relevant to the subject of this chapter.

17.3.8.1   THOR Chemical Database

THOR is organized like a thesaurus, in which all data associated with a single chemical structure are stored on a 'page'. This is similar to a conventional thesaurus, where words with related meaning are grouped around key topics. Chemical names or designations are subject to ambiguous interpretation, but in THOR the graph of each chemical structure is the key topic with which data are associated. Efficient organization of the database, surpassing previous chemical information systems, is made possible by the use of unique SMILES notation as the 'key words'. The basis of the successful storage of chemical information is the combination of thesaurus construction with unique SMILES as the topic and identifier (or keyword) which may have an unlimited amount of data associated with it.

All data are stored by structure. Given the unique SMILES, THOR can look up related data without searching. This technique is essential to THOR's efficiency. Since the structure designated by the SMILES string of characters can be treated as a fixed address in memory or disk, THOR always looks up structures; it never searches for them. Not only does this speed up computer retrieval of data, but even more importantly, the retrieval speed is constant with respect to the size of the database so that data are retrieved as quickly from large as from small databases.

Other identifiers (name, CAS number, *etc.*) can be treated as data and can also be retrieved with equal efficiency. THOR can thus provide extremely fast cross-referencing between any number of identifiers.

As an example, the THOR page for 1,2-dichloroethylene (SMILES notation: ClC=CCl) is shown below in Figure 18. It may include identifiers which have a meaning related to structure, such as identifiers for *cis* or *trans* isomers, racemic and constant-boiling mixtures or any other physical or chemical property. In THOR, as in chemistry, there is no limit to the number of ways in which one can identify data.

Modern software methodology has advanced greatly in recent years. It is no longer necessary to use mainframes and limited memory, in which programs are sorted, selected and reformatted in large files of database systems. Although THOR can operate under conventional management systems, its speed and flexibility is derived from methods based on lexical analysis and language parsing. Thesaurus orientation with SMILES structure identifiers eliminates the need to reserve files for further information. Instead, a structure is presented as a string of linear characters. When new data are introduced, the vocabulary is defined and new datatypes are added, all within the same thesaurus page (in THOR, a database is a forest of datatrees in a language where the vocabulary is applied by the user). More concisely, THOR is not specifically concerned with files and records; it stores all information in a simple language by context and not by address.

SMILES	ClC=CCl			1
	LOCAL NAME	1,2-Dichloroethylene		2
	MOLFORM	C2H2CL2		3
	SUBSET	CIS		4
		LOCAL NAME	cis-1,2-Dichloroethylene	5
		WLN	G1U1G -C	6
			LOCAL NAME cis-1,2-Dichloroethylene	7
			LOGPSTAR 1.86	8
			LOGP 1.86 SOLVENT Octanol REFERENCE Chan, T. & Hansch, C., Pomona College Unpublished Results SELECTED *	9
		CAS NUMBER	156-59-2	10
	SUBSET	TRANS		11
		LOCAL NAME	trans-1,2-Dichloroethylene	12
		WLN	G1U1G -T	13
			LOGPSTAR 2.09	14
			LOGP 2.09 SOLVENT Octanol REFERENCE Chan, T. & Hansch, C., Pomona College Unpublished Results SELECTED *	15
			LOGP 2.08 SOLVENT Undecane REFERENCE Barbari, T. & King, C., Envir. Sci. Technol., 16, 624 (1982)	16
		CAS NUMBER 156-60-5		17

Figure 18   Partial display of THOR tree rooted at ClC = CCl

All datatypes may be user-defined; one can store any data that can be named or change the name of a datatype. There is no space penalty for defining more datatypes because space is not reserved for storage. Except for the actual stored data, the only necessary information about new types of data is its definition. A page can be looked up by means of any identifier; even ambiguous ones are acceptable; short or long strings of characters can be added alike. There are no minimum records or field length.

17.3.8.2 Datatypes, Identifiers and Datatrees

Datatypes and datatrees are two fundamental THOR concepts. Datatype names are short descriptive words or phrases that are used to name data. Different sets of THOR datatypes may be defined for each THOR system. A given datatype name is used whenever data are entered or requested, so that there cannot be any confusion about the meaning of the data. Simple datatypes contain one or more items of information about a chemical compound. Examples that contain just one datum each are shown in Table 5.

Usually several related types of data have to be stored together. For example, for physical properties the source of measurements, its conditions, and other information may have to be recorded in the database. For density one may define a datatype with four fields, containing density in $g\,cm^{-3}$, the temperature of measurement, a citation reference and a field for comments. Each field has a name, *e.g.* in Table 6 the first field is named by the datatype itself; others are given names that describe their content.

All fields of a data item are stored and retrieved together and the length of the field, *i.e.* the data content, may vary from null (zero characters) to an assigned limit, typically 64 kb. Sometimes more than one datum is available for the same datatype. For example, there may be two pK_a values for a compound with two labile protons. In this case a field is divided into two subfields. The data are displayed separately, but the multiple subfields must not be confused with multiple data, *e.g.* 10 different measurements of the partition coefficient P are represented by 10 different data items of datatype log P.

Identifiers used such as SMILES,[4] WLN,[13, 14] CAS NUMBER[3] and any ordinary name, are datatypes which specifically identify structures. They differ from normal data in three ways. (i) Data can be associated with an identifier. Thus, the boiling point of 1,2-dichloroethylene may be known as that of CAS number 156-59-2 (the CIS or Z isomer). By entering this item under the CAS number on the ClC=CCl (SMILES) page, one extends the meaning of the entered item to the structure itself. Any non-identifier datum can be associated with an identifier. (ii) Identifier contents are automatically cross-referenced. Once a CAS number is entered on a page, that page may be retrieved by specification of the CAS number. Only the first field of the datatype is cross-referenced. (iii) Identifiers are standardized to make them easier to find. In SMILES it is the unique form obtained in the CANGEN algorithms[27] which serves as the key to a THOR page (datatree). CAS numbers have the check digit verified and hyphens inserted correctly. NAME and WLN notations are shifted to upper case; NAME also has blanks and non-alphanumerics removed. These standardizations greatly improve the efficiency of cross-referencing.

In addition to the above identifiers, there are SUBSET datatypes which are like identifiers, except that they are not cross-referenced, allowing other identifiers to be linked with them. For example, for 1,2-dichloroethylene the SUBSET datatype allows various *cis* and *trans* identifiers to be grouped together. Other common subsets are ORGANIC SALT, INORGANIC SALT, COMPLEXES, IONS, *etc.* This is best illustrated by the example of a THOR page, *i.e.* a datatree in Figure 18.

Table 5 Datatypes

Datatype name	Example data
REMARK	Verified 10 Oct. 1986 by DW
MOLFORM	C6H6O2
PKAWATER	3.21
LOGPSTAR	−1.23

Table 6 Multiple Fields

Data tags	Example data content
Density	0.8765
Temperature	25
Reference	LePetit, G., Pharmazie 32, 289 (1977).
Comment	Difficult measurement

Each thesaurus page is represented in THOR as a datatree. This construction is particularly well suited for chemical data and their relationships, for which THOR was designed. The root of each datatree is the key linear notation of the unique SMILES for the structure. Simple data may be attached to the SMILES root, *i.e.* data apply directly to the structure identified by SMILES. Other identifiers (branches) may also be associated to the SMILES on the thesaurus page. Data may then be associated to the identifiers, always maintaining the direct relation between the data and the structure represented by the identifier. If a page is divided into subsets, the SUBSETS are attached to the SMILES; other identifiers can be attached to the SUBSETS.

Figure 18 shows part of the overall 1,2-dichloroethylene page. It displays the datatree by writing individual data items in offset boxes to show the tree relationship. The SMILES (root) box extends all the way down the page, indicating that all entries are associated with it. Only two data are directly associated with SMILES: the LOCAL NAME 1,2-dichloroethylene and the MOLFORM C2H2Cl2. The rest of the tree is divided between CIS and TRANS SUBSETs. Each of these subsets has a LOCAL NAME and two identifiers (WLN and CAS NUMBER). The original partition coefficient data were organized by WLN; therefore, all LOGP and LOGPSTAR data are associated with WLN. They are always placed under the known identifier. CAS numbers are not associated with any data.

It is often useful to refer to specific data items by number. For this purpose, THOR provides the sequential item number along the margin of the datatree displays and tables. The item number is not a fixed property of the data items; it merely locates the order of the items on the tree which is changed when other data are inserted (except for SMILES which is always item 1).

Data items are always associated with the last occurring identifier; otherwise the order of their appearance is not important. This determines whether moving an item on a page changes the meaning of the tree. For example, in Figure 18, item 16 is a measurement of log *P* (undecane/water). Its position in the tree is associated with the WLN 'G1U1G -T' in the SUBSET 'TRANS'. Moving it up one position (to item 15), changes nothing, but moving it above item 9 would associate it with the WLN and SUBSET for the CIS isomer.

In conclusion, the THOR system is uniquely suitable for storage of chemical information because of the following three aspects: (i) storage of all information by structure provides an unambiguous key for retrieval; (ii) thesaurus orientation allows accurate documentation of the source and chemical identity for all compiled data; and (iii) the use of a unique SMILES as a key permits extremely efficient retrievals (retrieval time is independent of the number of structures in the database).

17.3.9 PRESENT STATUS AND FUTURE OF CHEMICAL INFORMATION SYSTEMS

We are reaching a turning point in the history of chemical information. The fundamental chemical information problems are: 'What are the properties of a given structure?'; 'What information is relevant for the prediction of the properties of a hypothesized structure?' and 'What is known about compounds which are similar in structure or behavior?'. Historically, these have proven to be difficult questions to answer definitively because our traditional tools—journals, handbooks and word-of-mouth—are quite primitive.

While the amount of information storage and processing power available with modern computers has been growing exponentially, the amount of actual information known about chemicals is increasing at a much more modest rate. We have already reached the point where computers are able to store and efficiently access all extant data for all known discrete structures so that the ability to solve the above chemical information problems is within reach.

Several problems must be overcome to achieve an 'ideal' chemical information system: (i) access to structures of known compounds; (ii) accurate and meaningful storage of measured data; (iii) acquisition of relevant measured data; (iv) interfacing relevant data with modeling and statistical tools; and (v) effective interfacing of the information system with chemists.

All of the above capabilities require accurate, efficient and understandable chemical nomenclature. The SMILES system appears to meet the needed requirements.

The first step towards a truly complete chemical information system must be to provide efficient access to the structures of known compounds for which there are measured data. Using SMILES, all known structures can be stored in the memory of current computers. In-memory storage is important because it decreases access time 100 000-fold over disk storage, allowing all stored structures to be scanned in seconds rather than days. At present most computers used for scientific research can be configured with 0.5 Gb of memory; enough to store the SMILES of 16 000 000 structures. This appears to solve the problem of structure access, since that is about twice the number of discrete structures currently known. The newer personal computers are designed for 1–4 Gb of memory, so it is certain that this capability will soon be available to everyone.

A difficulty with current chemical information systems is that they are inherently inaccurate or cumbersome. This is the result of the nature of chemical information which requires that all measured data somehow be associated with the chemical entity which was measured. Unfortunately, any chemical entity can be broken down into further subclasses, each of which may have information associated with it. Without a formal nomenclature for structure, information for different subclasses must either be combined (inaccurate) or assigned new arbitrary keys (cumbersome). To illustrate this, consider the problem of storing information for 2-pentene ($CCC=CC$). Valid data might reasonably be generated for the *cis* and *trans* forms, racemic and constant boiling mixtures, radiolabeled for various formulations, and even for specific conformations. When information is requested we must be certain that all relevant data is presented and that the actual form which was measured is presented as well.

The thesaurus-oriented approach used in THOR appears to solve the problem of documenting chemical information. A side benefit is that operating on unique SMILES as the key, THOR access operations run at the maximum speed theoretically possible for a given disk (a single disk access).

Although the sheer amount of information is no longer a problem, gaining access to relevant information certainly is. There is a recent trend in chemistry (as in the rest of modern society) towards specialization in information generation and distribution. Activity data comes from one source, crystal structures from another, spectroscopic and physical property data from yet another. We need to combine information from different sources, and the only way to do so is to associate information with a unified key. Currently existing systems key information on an arbitrary identifier, such as a CAS Registration Number. When structural information is stored, it is done as a form of chemical data rather than as an identifier. While this approach was historically useful in making chemical databases function with computational methods, it imposes a severe limitation: information is not available unless the arbitrary identifier is known. Combining information from different sources is thus difficult or impossible.

Again, the idea of linking information to structure as a fundamental identifier solves the problem simply. Since data can be stored as a forest of datatrees rooted at each structure, the 'difficult' problem of merging disparate databases becomes the computationally trivial one of merging common-rooted trees. In practice, single datatrees or whole THOR databases can be merged freely.

The final problem to be solved is the development of an effective and natural interface between chemists, the chemical information system and other computational tools (*e.g.* predictive models, statistics, graphics and expert systems). Probably the most important aspect of these interfaces will be how well they allow complex questions to be asked while maintaining clarity for human operators. Linear notations such as SMILES and similar languages will undoubtedly play a crucial role in such systems because humans are remarkably adept at using language for communication. In the last decade there has been great progress in the use of language-oriented software methods which are suitable for such interfaces (lexical analyzers, parsers, finite automatons).

The technological capability to solve the fundamental problems of chemical information is now available. When all relevant chemical information can be accessed and applied to daily problems, there will be a profound change in the way such information is used by chemists. This potential will certainly be realized within the coming decade.

ACKNOWLEDGEMENTS

Early development of the SMILES system was initiated at the Environmental Protection Research Laboratory, USEPA, Duluth, Minnesota by one of the authors (DW) and was continued at the Medicinal Chemistry Project, Pomona College. It is now under development at Daylight Chemical Information Systems, Inc., Irvine, California. The authors wish to thank Gilman Veith for his early support of the SMILES concept, Albert Leo and Corwin Hansch for their continuing support, and Arthur Weininger and Steven Burns for computer programming assistance.

17.3.10 REFERENCES

1. W. J. Wiswesser, *J. Chem. Inf. Comput. Sci.*, 1985, **25**, 258.
2. R. Fugmann, *J. Chem. Inf. Comput. Sci.*, 1985, **25**, 174.
3. H. L. Morgan, *J. Chem. Doc.*, 1965, **5**, 107.

4. D. Weininger, *J. Chem. Inf. Comput. Sci.*, 1988, **28**, 31.
5. J. L. Weininger, D. Weininger and A. Weininger, *Chem. Des. Autom. News*, 1987, **1** (8), 2; 1988, **2** (3), 1.
6. D. J. Gluck, *J. Chem. Doc.*, 1965, **5**, 43.
7. P. G. Dittmar, N. A. Farmer, W. Fisanick, R. C. Haines and J. Mockus, *J. Chem. Inf. Comput. Sci.*, 1983, **23**, 93.
8. R. Attias, *J. Chem. Inf. Comput. Sci.*, 1983, **23**, 102.
9. R. C. Read, *J. Chem. Inf. Comput. Sci.*, 1983, **23**, 135.
10. A. T. Balaban, *J. Chem. Inf. Comput. Sci.*, 1985, **25**, 334.
11. M. Gordon, C. E. Kendall and W. H. T. Davison, Chemical Ciphering. A Universal Code as an Aid to Chemical Systematics, Royal Institute of Chemistry Monograph Report, Royal Institute of Chemistry, London, 1948.
12. G. M. Dyson, 'A New Notation and Enumeration System for Organic Compounds', 2nd edn., Longmans, London, 1949; 'Rules for IUPAC Notation for Organic Compounds', Wiley, New York, 1961.
13. W. J. Wiswesser, 'A Line-Formula Chemical Notation', Crowell, New York, 1954.
14. E. G. Smith (ed.), 'The Wiswesser Line-Formula Chemical Notation (WLN), McGraw-Hill, New York, 1968.
15. S. J. Weininger and F. R. Stermitz, 'Organic Chemistry', Academic Press, Orlando, FL, 1984.
16. A. Leo, *J. Chem. Soc., Perkin Trans. 2*, 1983, 825.
17. A. L. Goodson, *J. Chem. Inf. Comput. Sci.*, 1980, **20**, 212.
18. W. Kalbfleisch and G. Ohnacker, *J. Chem. Inf. Comput. Sci.*, 1980, **20**, 176.
19. C. A. Shelley, *J. Chem. Inf. Comput. Sci.*, 1983, **23**, 61.
20. T. R. Hagadone and W. J. Howe, *J. Chem. Inf. Comput. Sci.*, 1982, **22**, 182.
21. J. C. Wenger and D. H. Smith, *J. Chem. Inf. Comput. Sci.*, 1982, **22**, 29.
22. D. Weininger, *J. Chem. Inf. Comput. Sci.*, 1989, in press.
23. C. Jochum and J. Gasteiger, *Top. Curr. Chem.*, 1978, **74**, 93.
24. W. T. Wipke and T. M. Dyott, *J. Am. Chem. Soc.*, 1974, **96**, 4834.
25. M. Uchino, *J. Chem. Inf. Comput. Sci.*, 1980, **20**, 116.
26. M. Bersohn, *Comput. Chem.*, 1978, **2**, 113.
27. D. Weininger, *J. Chem. Inf. Comput. Sci.*, 1989, **29**, 97.

17.4

Use and Limitations of Models and Modelling in Medicinal Chemistry

COLIN J. SUCKLING

University of Strathclyde, Glasgow, UK

17.4.1 MODELS AND THE NATURE OF MEDICINAL CHEMISTRY

It is unlikely that anyone will be reading this volume from cover to cover. But in that improbable event, this chapter could come as a surprise amongst the dense discussion of the theory and practice of technique in medicinal chemistry. In this chapter, we step back from the pressure to discover new drugs to examine the nature of the scientific activity of medicinal chemistry more from the point of view of a spectator than of a participant. Hopefully this aside will be helpful not only to scientists new to the field of medicinal chemistry and wishing to gain confidence in its subtleties but also to experienced medicinal chemists to encourage them to review their own contributions. This chapter will refer to many topics in this volume for further details but will not touch upon every subject.

It is part of the scientist's training to evaluate results as objectively as possible and such evaluation must extend to the methods that are used. Chemistry is concerned with the structure and reactions of atoms and molecules and at every turn chemists use models to describe their subject. A few years ago in collaboration with other scientists in my family I analyzed what turned out to be a surprising wealth of models throughout many chemical activities.[1] Models are essential components of all experimental and theoretical chemistry, although chemists are so accustomed to their use that they may scarcely notice them. With the very familiarity lies the danger that the model will be misused, leading, perhaps, to erroneous conclusions. In the pure sciences misuse of a model will certainly be a nuisance and at worst it may lead to a loss of a scientist's reputation. But in the applied sciences, including medicinal chemistry, misuse of a model can have catastrophic consequences that could lead to death or to injury of those who have not taken part in the scientific activity. The importance of evaluating models in medicinal chemistry can therefore not be overemphasized.

Before proceeding further, it is necessary to define what a model is and to identify some of its important properties. The introductory statements in this paragraph will be expanded in the following sections. A model in its widest sense in chemistry is not simply a substitute for what it represents, which can be called the prototype of the model, it is a transformation of the prototype into another form that can be more easily or conveniently handled by theoretical or experimental work. In constructing a model, the selection of the transformations is a most important step because the transformations define the limits of validity of the model. To apply a model in a sense inconsistent with the transformations is to court the disasters noted above. Another feature of constructing and using models is that it is important to consider the points of contact of the model with the rest of the scientific or commercial situation in which it has to work. These points of contact define what can be called the environment of the model and, as will be seen in subsequent sections, consideration of the environment of a model is of special significance in medicinal chemistry.

Medicinal chemistry exists to manipulate reactions and structures to obtain compounds that will be beneficial to a sick person or will treat disease prophylactically. This simple view of the aims of medicinal chemistry conceals the great complexity of the activity. The search for a new drug begins with a stimulus to conquer a disease, a stimulus not principally initiated by chemists, although they may contribute. It is principally a demand from the public through the clinician and is commonly relayed in scientific terms to the chemist by a biologist. The biologist will provide the chemist with a number of opportunities and may offer a system related to the disease in which new compounds can be tested or may suggest a biochemical origin for the disease. Even at this stage a model has been constructed and a critical one at that because every move the chemist subsequently makes will be directed to some extent by this model, a model of the disease itself. This model will provide the means to test the compounds that the chemist's skills will be used to obtain either by synthesis or by extraction from natural sources. It may also offer a stimulus to the design of compounds that might lead to drugs.

17.4.1.1 The Need to Model

Of the many answers to the question 'Why model?', the start of a project in medicinal chemistry gives perhaps the clearest: it is self-evidently ethically unacceptable to try out potential drugs wantonly on the sick before proper screens have been passed. Since a direct test is out of the question, medicinal chemists proceed by stages to devise new compounds on the basis of the results obtained from biologists, and on the basis of their own theoretical or empirical models. Chemists must be able to study not only the action of the drug on its primary target in the body but also the mechanism by which it reaches the target, including the possibility of metabolic modification *en route*. They must evaluate the drug's toxicology, its specificity of action through pharmacology and finally its administration to the patient through pharmacy. All of the results from these studies must be correlated in terms of the molecular structure of the molecule. Synthesis or isolation of a compound is only the beginning; the medicinal chemist must be acquainted with all of this relevant science. To encompass such a complex interacting range of science is extremely demanding and challenging, and requires an understanding not only of modern synthetic organic chemistry but also of physical organic chemistry.

When people are faced with apparently impenetrable complexity they either look away for something easier or, if sufficiently determined, look harder in an effort to reduce the complexity. Apart from paramount ethical considerations, in chemistry, as in any other activity, models exist to make complexity manageable. Of all areas of applied chemistry, medicinal chemistry arguably contains more types of complexity than others and is correspondingly richer in its diversity of modelling. Because the products of medicinal chemistry will eventually be administered to a patient, it is literally vital that a medicinal chemist should be thoughtful about the models used and it is the purpose of this chapter to examine models of chemistry as they relate to the medicinal chemist's task.

What types of model are common in medicinal chemistry? Two main classes can be distinguished. Firstly, there are models used to design the structure of a potential drug or to refine the structure of a lead compound. These models lie in the realm of chemistry and relate in various ways the structure and physical properties of molecules to their biological activity. Secondly, there are the models of the disease state itself, models in the realms of biology and pharmacology. Let us look firstly at these biological models in outline so that a number of important features of the modelling process can be related directly to medicinal chemistry. We can then turn our attention to properties of chemical models and biological models in more detail.

17.4.1.2   The Scope of Models in Medicinal Chemistry

17.4.1.2.1   Biologically based models

The primary screen used to test the products of the synthesis or isolation laboratory is immensely important. It must focus upon a property that is believed to be intimately, although not necessarily directly, related to the disease that is to be treated. Its reliability must therefore be clearly understood: the statistical evaluation of imprecise biological data is a major concern of the medicinal chemist. Expense and speed of response are also important in primary screens but perhaps the most important property at this stage is reliable discrimination. A false positive result is likely to be spotted in subsequent experiments but a false negative result is an opportunity missed. The better the results from the assay, the less the risk that a strange result will escape detection because numerical results can be correlated with parameters relating to mathematical models (Section 17.4.2.2). Without satisfactory data, effective quantitative drug design or molecular modelling cannot begin.

Having selected a series of promising compounds from a primary screen, it is common to undertake further studies to take account of metabolism, either by using excised organs or intact small animals. Even the choice of animal can be important; in some situations, for example, hamsters are preferable to rats because the former possess a gall bladder and hence are a better animal model for cholesterol metabolism in humans than are rats. Similarly it is important to ask whether the source of a test enzyme in a primary screen is the most appropriate. The enzyme dihydrofolate reductase (DHFR) is an important target for antibacterial, anticancer and antimalarial drugs[2] and the effectiveness of a drug depends upon its selectivity for the parasitic DHFR compared with the human enzyme. Obviously potential inhibitors must be tested in at least two enzyme screens, a parasitic and a human enzyme screen, to provide useful data.

Toxicology and mutagenicity are also investigated and the Ames test is a well-known screen for the latter. In its simplest form, this test gives information about the intrinsic mutagenicity of a compound itself but it does not indicate whether a metabolite might be mutagenic. To include this possibility in the screen, various expedients can be employed such as the addition of extracted metabolic enzymes to the test system. The Ames test has the virtues of reproducibility and simplicity but many other biological systems, including higher plants, yeasts, *Drosophila* and mammalian cell lines, have been used to investigate mutagenicity.[3] To ignore the possibility of metabolism of a potential drug after administration could conceal serious defects. Of course this risk is well recognized and is checked for routinely; the important point for this discussion, however, is that it is always necessary to look critically at a biological screen and to ask how well it relates to the disease state itself. Anything other than a clinical trial inevitably introduces substantial changes from the ultimate test system, the patient, and medicinal chemists must constantly interrogate their biological colleagues to evaluate the relevance of their screens.

17.4.1.2.2   Chemically based models

The balancing arm to the limb of biological models in medicinal chemistry is the application of data derived from chemical measurements or calculations to the discovery of drugs. Since relatively few drugs undergo covalent bond formation with their target receptor, cytotoxic antitumour drugs being a significant exception, chemically based models have so far focused upon the physical properties of the drug molecules rather than on their chemical reactivity, although this is now changing with the advent of efficient theoretical methods to calculate electronic properties. One of the first examples of the importance of physical properties of molecules in drug action was the recognition of the relationship between the pK_a of a sulfonamide and its antibacterial action.[4] Sulfanilamide, the prototype drug (1), has a pK_a of 10.3, which means that it will scarcely be ionized at physiological pH. The activity of sulfonamides is due to their ability to interfere with the biosynthesis of dihydrofolate through competing with *p*-aminobenzoic acid, which has a pK_a of 4.9, in binding to the enzyme dihydropteroate synthetase. It is not surprising that a more highly ionized sulfonamide, such as sulfadiazine (**2**), pK_a 7.5, is more effective because it is 75% ionized at physiological pH. However, more strongly acidic analogues are less effective and the reduction in activity has been attributed to the reluctance of ionic compounds to cross biological membranes which are apolar barriers (see also Section 17.1.2.2).

$$H_2N \text{—} \langle \bigcirc \rangle \text{—} SO_2NH_2 \qquad\qquad H_2N \text{—} \langle \bigcirc \rangle \text{—} SO_2NH\text{—}\langle \text{ring} \rangle$$

(1)   (2)

Albert[5] has emphasized the importance of the distribution of a drug in the patient as a major principle for obtaining selectivity, which he labels the first principle of selective toxicity. The distribution of a drug within an organism relies chiefly upon its ability to partition between aqueous fluids, such as blood, and fatty tissues, especially biological membranes. To reach a site of action, a drug may be required to be absorbed in the intestine, dissolve in blood, reach a target cell, diffuse through the plasma membrane into the target cell and bind to its site of action. These events are controlled not only by the ionic state of a molecule but also by its overall polarity. Although the work of Hansch (Section 17.4.3.2) has greatly refined considerations of hydrophobicity, the significance of this molecular property with respect to biological activity was first recognized in the early 1900s[6] in correlations of the hypnotic action of compounds of unrelated chemical structure. Hansch's arguments have now been extended to help interpret the detailed molecular interactions in enzyme–inhibitor complexes (Section 17.4.3.2 and Chapter 20.3). It is also significant that these early physical chemical concepts led in time to the discovery of the fluorine-containing inhalant anaesthetics, including halothane.[7]

If, to a great extent, the passive transport of a drug depends upon its charge and lipophilicity, the detailed chemical structure being of secondary importance, then the potency at the site of action relates intimately to its structure. It is therefore natural to apply the best theories of chemical bonding and reactivity to investigate biological activity. The many manifestations of quantum mechanics are now a major contributor to modelling in medicinal chemistry (see Chapter 18.1). Pioneering applications of quantum mechanical methods in 1945 led to the recognition of the so-called K and L regions of the structures of carcinogenic polycyclic aromatic compounds.[8] As will be discussed later, the revolution in computer construction and hence in capability and cost caused by the microchip has enhanced the place of theoretical methods in medicinal chemical modelling (Section 17.4.4). Even so, calculations on molecules as large as enzymes are still computer intensive and other theoretical models such as molecular mechanics make the estimation of conformational properties of molecules more accessible (see Chapter 18.2). However, if calculations of electronic properties are required, quantum mechanical methods are essential.

17.4.1.2.3 *Connecting biological and chemical models*

Good biological data and precise chemical measurements or calculations are of no consequence in drug design and development unless they can be meaningfully compared with one another. This comparison is made by structure–activity relationships and quantitative structure–activity relationships (QSAR) are the focus of modern medicinal chemistry.[9] However, to make the connection between measured chemical and biological properties it is necessary to employ additional models. Typically, mathematical models including statistical treatments are used (Section 17.4.2.2). Unlike the biological and chemical models that relate respectively to the disease and the molecular structure, the form of a connecting model may express no obvious scientific relationship. The model may often be expressed in the form of an equation relating biological activity to a number of molecular parameters (Section 17.4.3), and, given enough data, impressive correlations can be obtained by regression analysis. When only a few molecular properties such as pK_a, partition coefficient or substituent constant were available, the search for correlations by means of regression analysis was not burdensome. However, the number of physical measurements possible on a drug molecule is now vast when spectroscopic methods are taken into account and it has become a real problem to decide which molecular parameter to include and which to omit. Intuition and judgement can, of course, play an inspirational role but there is a need to develop more objective and exhaustive assessments of the significant chemical inputs to a correlation. Many methods of multivariate statistical analysis have been applied to medicinal chemistry[10a] and, if many variables are found to be significant, the problem then focuses more upon the ability of the human mind to comprehend a multidimensional parameter space. Most of us can easily interpret two- or three-dimensional plots but more than three dimensions give trouble. Reliable techniques for the reduction of the dimensionality of a drug development programme are therefore most important.[10b]

17.4.1.3 Important Concepts in Modelling

The above outline of some models used by medicinal chemists illustrates some important intrinsic and general properties of models.[11] The essence of a model, as defined in the introduction, is that it is a transformation of the system that it represents; the system itself can be referred to as the prototype of the model. The transformation may be a simplification such as the use of a single enzyme or receptor assay as a primary screen to represent the key locus at which the effect must occur. Such a practical restructuring of the prototype may also be expressed in a conceptual sense. Thus biologists select from their experience of human and pathogenic biology what they consider to be the critical sequence of events that leads to disease and hence build a hypothesis to guide the search for a cure. Transformations also exist in chemical modelling. For example, a drug molecule becomes represented by a series of measured or calculated parameters which together form a model of its relevant properties. At its most detailed, the data can be accessed, manipulated and displayed by a computer graphics system, there being many additional transformations depending upon the selected display (Section 17.4.4).

A further concept of value in considering models is suggested by the importance of considering metabolism and transport of a drug. Even if a highly potent compound is found in a primary screen in the laboratory to be a useful drug, it must *in vivo* reach the site represented by the primary screen. In other words, it is essential to consider the environment in which the drug will act together with the target site. It is not only interfaces within a patient that make up the working environment of a drug. A drug must pass through several environments before entering the patient and must be equal to the demands of each. One must consider, for example, such things as the shelf life of a compound and the way in which it is administered. Significant examples of these were encountered during the development of halothane, a widely used inhalant anaesthetic.[7] It was found that the drug was subject to some decomposition on storage with exposure to light; this problem was solved by adding a little thymol and by storage in amber bottles to inhibit free radical oxidation.

Although not all of the above models are constructed by medicinal chemists themselves, they inevitably make great use of them. And it is a dictum of working with models that the user must be aware of the limitations inherent in the structure of the model to hand, otherwise the science is being built upon precarious foundations. The exquisite importance of collaboration between medicinal chemist, biologist and pharmacist is a corner-stone of drug discovery and development. What characterizes the medicinal chemist's own models is that they are principally related to chemical structures. Before examining some examples of such models, however, it will be helpful to survey the extensive use of models in a wider view of chemistry with simple examples to illustrate important features of modelling.[12]

17.4.2 THE USE OF MODELS IN CHEMISTRY

A few years ago, the first thing that came to a chemist's mind when models in chemistry were mentioned was a solid molecular model, something that conveyed an impression of the three-dimensional structure of a molecule. Partly as a result of the growth of models in medicinal chemistry that has paralleled the widespread availability of computing power, there is a much greater perception today of the importance of models in chemistry as a whole. In fact, the solid models have become much less significant and emphasis has fallen upon four types of model.[13] These are conceptual models, indicated above by a model of a disease state; mathematical models, which could for example express the kinetics of the delivery of a drug; enactive models, in which the chemist interrogates the model, commonly at the keyboard of a computer; and iconic models or look-alikes, such as computer graphics displays. Whatever model is involved, the caveat that a model is valid only within the limits defined by its transformations holds. Such is the power of modelling in chemistry and, as recent developments with computers have shown, so great is the fascination that the risk of addiction is acute. Medicinal chemistry has yet to discover the antidote to chronic infectious modelling!

17.4.2.1 Conceptual Models

The way we think about chemistry, especially organic chemistry, is a model, a conceptual model, built from the interplay of many ideas and theories. Whatever new techniques are introduced to improve the assessment of results or to stimulate the scientific imagination will be assimilated by our brains into a conceptual model developed from our formal learning as students and from later experience. Many people involved in teaching chemistry have realized that the development of chemical perception, understanding and intuition embodied in a personal conceptual model consists of becoming acquainted with a series of models that are independently constructed but that feed into each other (see Section 17.4.2.2.2). There is a hierarchy of models, which for an organic chemist, for example, will usually be grounded in some form of quantum mechanical theory of bonding. The experimental basis that can be related to concepts of bonding are properties of individual functional groups and the conceptual model is extended by considering substituent effects and interacting functional groups. A synthetic chemist would include models of reactivity related to reaction conditions and a spectroscopist would include models of energy transfer in molecules.

Whatever their specialization, chemists usually have some feel for the way in which quantum mechanics, a substantial theoretical model, relates to chemical structure, and some will add to that a connection with chemical reactivity. The advance of computing has revolutionized the conceptual model of most chemists. Whereas 15 years ago most organic chemists, for example, would consider the likely reactions and properties of a compound in terms of resonance theory, today they are as likely to look for an answer from data relating to molecular orbitals derived from a calculation, especially to understand more subtle points of chemical reactivity. Medicinal chemists today need not look simply at isolated atoms such as a positively charged nitrogen in a biologically active molecule, they can examine the whole electron distribution. They can now not only look at structural effects on physical properties such as acidity or dipole moment with the aid of qualitative concepts such as inductive and resonance effects, but also obtain quantitative estimates of the same properties at the cost of very little computer time. For the medicinal chemist, these measurements and calculations are major benefits because the properties mentioned frequently relate to biological activity.

17.4.2.2 Mathematical Modelling

17.4.2.2.1 *Empirical approaches*

In the previous section, the concept of the environment of a model was introduced; this concept is also extremely important in very familiar chemistry. Undergraduates learn early in their careers the ideal gas law formulated in mathematical terms as

$$PV = nRT \tag{1}$$

The equation, of course, empirically relates the changes in pressure, volume and temperature of a fixed mass of something called an ideal gas. Intuitively it is obvious that this model has limited applicability through the very use of the word 'ideal' and, as is well known, gases behave in ways that do not fit the equation, especially under conditions close to liquefaction. To refine the model, it was necessary to think of factors that might cause deviations from ideality, such as the conditions that pertain near the walls of the vessel. In one of the best known modifications, van der Waals introduced a term to account for intermolecular interactions (a/V^2) and a term to account for the volume of a molecule of a gas $(-b)$ to give the equation that bears his name

$$(P + a/V^2)(V - b) = nRT \tag{2}$$

In this case, the environment of the model can be considered in two ways, both of which are significant. Firstly, and more obviously, the limits of the model refers to a fixed mass of gas; in other words, a boundary condition is imposed. This is a mark of the limits of the model beyond which it is not applicable. The boundary segregates the model from its external environment, which, in the case of the van der Waals equation, is not further defined. Secondly, and more subtly, the revised model now includes reference to some internal workings of the system that are not directly measured; we can refer to these interactions as components of the internal environment. Consequently it is always important to recognize them when constructing and operating a model.

The happy place of modelling in the overall context of the scientific method is well expressed in the van der Waals equation. Just as a hypothesis, a theoretical model, is defined carefully in terms of axioms or assumptions and experiments are designed to test its predictions, so the transformations and the environment of a mathematical model can be defined and tested with equal rigour. If either fail, it is rejected and the position is reevaluated. Even if success is obtained, the results are only valid within the defined terms of the model's structure, as was emphasized above.

A conceptual input is not always necessary or available in constructing mathematical models and this situation is well illustrated by regression analysis. The limit of conceptual or intuitive support in this case is the identification of the relevant variables with which to conjure a correlation. Regression analysis[14] then produces a set of equations that relate biological activity to structural parameters. A particularly valuable property of such an analysis is that it is possible to calculate how much of the variance in biological activity can be associated with each of the structural parameters so that a significant but unsuspected contributor can be exposed.

Such models, despite their lack of intellectual satisfaction in understanding the causal relationships in the system, may nevertheless be very good representations of the system; they encompass not only the underlying scientific laws but also the idiosyncrasies of the particular prototype and they are expressed in terms of readily measurable parameters. On the other hand, since the model is so closely limited by the choice of parameters, the need to be alert for discontinuities or significant misfits is great. Experience in many fields outside medicinal chemistry has shown that with empirical models, extrapolation is very risky and even interpolation demands care.

17.4.2.2.2 *Theoretical models of structure in chemistry*

Other chapters in this volume cover the details of theoretical methods in chemistry and this section serves to indicate the opportunities offered to medicinal chemical modelling. When it is impossible to measure a property of interest for correlation with biological data, it is helpful to turn to calculations. For example, the conformation of a molecule may be critical for its interaction with its receptor. Conformational equilibria can be examined experimentally, by NMR for instance, but practical considerations such as solubility or the rate of equilibration of conformers may militate against experiment. Since conformation depends largely upon intramolecular steric interactions, made up of van der Waals interactions, torsional and angle strain, theoretical methods known as molecular mechanics or force field calculations have been constructed.[15] On the other hand, it is often the case that electronic properties are a determining factor in biological activity and then the methods of quantum mechanics become important.[16] Conformational properties can also be estimated by quantum mechanics but the calculation times are much longer than for molecular mechanics. The two methods are complementary: quantum mechanics is more general but molecular mechanics is quicker.

Theoretical and physical organic chemists have made immense efforts to devise efficient molecular and quantum mechanical models so that reliable data can be obtained with expenditure of a minimum of computer time. To achieve such efficiency, it has often been necessary to set up a mathematical description of the interactions involved and to simplify the solving of the relevant equations by including parameters derived from sound experimental data such as crystallographic bond lengths and angles or heats of formation. Alternatively, as has often been done in quantum mechanical calculations, certain orbital interactions are neglected to simplify the computation. Calculations largely free of parametrization, known as *ab initio* calculations, are possible, but in practice it is helpful if an approximate calculation using semiempirical methods is carried out before applying the more sophisticated and computer time-hungry *ab initio* method.

Whatever technique is used there are some common problems, principal of which is that of so-called local or global minima. Chemists are interested in identifying stable molecular states for correlation with biological activity and the geometries of such states can be calculated by either type of theoretical model using energy minimization techniques. These are mathematical models based upon differential calculus and are chemically ignorant. They will lead a molecule along a path down the steepest local gradient to the nearest energy minimum in which the molecule will be trapped for the purposes of the calculation. This minimum may, however, be only a local depression in the potential energy surface and not the lowest point on the whole surface, the global minimum. Well-conceived strategies, such as grid searches, need to be applied to identify global minima. Although such calculations are possible for small molecules with limited conformational freedom, they are unrealistic for large molecules such as enzymes or proteins on all but the most powerful computers.

A further problem that extends the challenge for theoretical methods is solvation. It is becoming increasingly clear that individual water molecules have functional significance in some enzyme-catalyzed reactions,[17] even when water is not a reactant. Hence hydration and hydrogen bonding become important factors for inclusion in theoretical models. Investigations of these phenomena will have increasing significance. It is fortunate for many mechanistic problems that solvation can be considered to be invariant in a series of structurally similar molecules. Relative energies of molecules in a series can often be successfully related to biological activity. This method of comparison has, of course, led to reliable correlations in traditional physical organic chemistry. Happily there are many theoretical chemists who are alert to the dual task of medicinal chemists, namely compound design and synthesis. A constructive dialogue between such scientists will encourage further progress.

17.4.2.2.3 *Enactive and iconic models in chemistry*

Despite the advances in computer technology, still the most widely used iconic models in chemistry are structural formulae. Structural formulae are a major means of communication, especially for organic chemists, and it is essential to realize what such a drawing is. Robinson[18] as early as 1917 wrote, concerning the structure of morphine, 'This formula has been adopted especially since the formulae which it is suggested should replace it cannot without hesitation be accepted as superior representations of the properties of the substance'. The emphasis upon structural formulae as representations of chemical properties, that is as models, is noteworthy. Having recognized this important feature, it is easy to appreciate the force of the common device of adding curly arrows to structural formulae to indicate pathways of reactions and chemical reactivity in general. Curly arrows convert static iconic models into enactive models, enabling the chemist to ask the most penetrating question 'What if?'. It is also necessary to realize that structural formulae and their associated embellishments are reflections of the valence bond and resonance models of bonding and reactivity. They have the great virtue that pen and paper are the only necessary equipment. For many more taxing problems, computer-based molecular orbital calculations are required and a simple visual representation of the resulting numerical data can be cumbersome. Assorted blobs and bulges, either black or white, can help but the representations impose a limit to comprehension. Certainly the enactive use of such models is compromised and the solution to the problem is through the computer.

An obvious limitation in structural formulae is the representation of three dimensions. Whilst there are many projection formulae that are useful with practice, it is still difficult to get a good feel for the shape and relative size of a molecule without resorting to solid molecular models, the archetypal iconic model in chemistry. Readers will be familiar with two major types: framework models, which are useful for investigating conformational properties and interatomic distances and bond angles, and space-filling models, which convey an impression of the overall shape and bulk of a molecule. There are enactive possibilities with both types of model in that they can be twisted about single bonds but operations of interest to medicinal chemists, such as the superimposition of two or more molecules of a series, are difficult with framework models and impossible with space-filling models. Even where comparisons can be made, framework models, from a set such as Dreiding models, contain standard geometries and are unable to take into account intramolecular inter-actions that may lead to distortions of the standard geometry. Once again, the advantage of computer-based models becomes obvious (see Section 17.4.4). Solid molecular models nevertheless remain a useful standby; unlike the more sophisticated models to be discussed later, you can get your hands on them.

17.4.2.3 Systems and Misfits

As has been alluded to already, quantum mechanics has become a major contributor to medicinal chemistry through its ability to calculate electron distributions in molecules. A major factor in the success of such techniques is that absolute values of many parameters are not required. It is simply necessary to obtain a controlled comparison of the behaviour of one molecule with another. In terms of a modelling formalism, the device is successful because there is a well-defined system of parameters and constraints within which the whole comparison is carried out. In other words, the modelling system is treated as if it is isolated, internally self-consistent, and with no perturbing external contacts. To encompass a complete product development programme typical of medicinal chemistry, what is needed is a set of such quasi-independent models that fit together so that the output from one level of model becomes the input to the next. Quasi-isolated systems are a particularly useful structure for models; they make it possible to concentrate upon one problem in a manageable way, whilst not losing sight of the overall aims of the project.

Having said that a series of quasi-isolated models should fit together, the possibility is immediately raised of a misfit. An example will be offered shortly but misfits may also arise within a model itself. For a simple example of an empirical model from classical physical organic chemistry we can consider Hammett's correlation of substituent effects upon the acidity of substituted benzoic acids and other reactions.[19] He showed how a linear free energy relationship was able to relate the influence of *meta* and *para* substituents to the pK_a of the acids but found that *ortho* substituents did not fit. The direct correlation dealt with electronic effects of substituents remote from the site of reaction and, of course, the misfit of the *ortho*-substituted compounds was due to their proximity to the reaction site, in particular to steric effects and different solvation. Similarly, alternative parameters were introduced to account for systems in which the substituents became directly conjugated to the reaction site so that so-called resonance effects could be included. The controlled comparison offered by linear free energy relationships is a major approach to the correlation of numerical data with chemical structure in medicinal chemistry and received a major impetus when Hansch[20] stressed the importance of the hydrophobicity of molecules, as will be seen later (see also Chapters 18.6 and 21.1).

However, the most serious potential misfit in a drug development programme is a clash between a model and its external environment. If the clash is only between one model and the next in the series of quasi-isolates, the damage is limited to a loss of hair in head scratching to find the scientific basis for the misfit and to repair the damage. On the other hand, if the clash is between the drug and something in its operating environment that was neglected when the programme was defined, then results could be catastrophic. The case in point that caused great public concern with the workings of the pharmaceutical industry was, of course, thalidomide. It has been argued that the scientific misfit here was a failure to accept the chirality of nature, a failure to realize that the receptors of drugs are chiral. Hence, it is possible that the two enantiomers of a compound will not have the same biological activity and in the worst situation, one isomer may have harmful effects as has been suggested for thalidomide. Experiments in rats have indicated that only the (R) isomer of thalidomide had teratogenic effects.[21] Of course there are many drugs in clinical use sold as racemic mixtures (for example β-blockers) and they have been found to be acceptable. What must be emphasized is that racemic products must be regarded as mixtures because once in the chiral natural environment, their interactions, especially with biological macromolecules, will consist of two effects, one for each enantiomer. Although the thalidomide tragedy is one of the most well-known examples of enantiomers having different biological activity, many differential effects have been demonstrated and these facts emphasize the importance of control of chirality in the synthesis of drugs.[22]

17.4.3 THE MEDICINAL CHEMISTRY ENVIRONMENT TRANSFORMED INTO MODELS

Having investigated some of the characteristics of models in a broad range of chemistry, the discussion now turns to concentrate upon models in medicinal chemistry. What distinguishes medicinal chemistry is the breadth of information that must be unified to discover, to develop and then to obtain an acceptable deposition for a new drug to bring it to the market. The quasi-isolate analysis of models is helpful to achieve this and reference has been made to a number of relevant medicinal situations above.

17.4.3.1 Models in Drug Discovery

In the course of the development of medicinal chemistry (see Volume 1), there have been many fashionable strategies for discovering new drugs. The most venerable of these is the empirical observation of a novel biological effect which could be significant as a treatment for a disease. In this class the discovery of penicillin and the antibacterial activity of sulfonamides can be included. Hopefully such chance observations will continue to occur and the medicinal chemist with a keen conceptual model both of chemistry and, more particularly, of biological possibilities will be there to profit. Recent examples of compounds with significant biological activities discovered in screens include the avermectins (3),[23a] which are antiparasitic compounds, and mevinolin (4),[24] which is the basis of drugs for the control of cholesterol biosynthesis. A screen is, of course, a model of the required biological activity of the compound, however remote it is in construction from the ultimate system in which a new drug must operate. It exhibits many features typical of models in that it retains some relevant properties of the prototype. A coarse screen will select compounds with activity at low precision or with low discrimination for the detailed biological origin of the observed effect, as may be the case for a pathogenic organism in which the biochemistry is poorly understood. On the other hand, in well-defined systems, it is possible to design fine screens with high discriminating power; the value of such a screen will be potentiated if its structure as a model has been consciously considered. Pathogenic microorganism cultures have frequently been the bases of such screens, the implied model of course being that the death of the microorganism in culture identifies a potential antibacterial compound. The coarseness of such screens with regard to the eventual production of a drug is obvious.

(3)   (4)

   As the knowledge of metabolic pathways developed, it became possible to suggest that compounds that are close analogues of natural metabolites could have medicinal properties. In a very limited sense, this concept of antimetabolites was an embryonic form of rational drug design because it modelled the potential drug on the natural metabolite. The work of Hitchings in the pyrimidine field, leading to drugs such as allopurinol (5) and eventually to trimethoprim (6) is a prime example. As understanding of biochemistry developed further, it was possible to recognize that many drugs acted as enzyme inhibitors, whence an important step in the concept of selective toxicity was taken. It was recognized that, for parasitic diseases, it was highly desirable to establish a model of metabolism that could be used to contrast the metabolism of the parasite with that of the host. In this context, a parasite may be a bacterium, a virus, a protozoan or a proliferating cancer cell. Such a model identifies enzymes specific to the parasite and these enzymes therefore became prime targets for inhibition and for drug design. Selectivity due to differences in metabolism was identified as the second principle of selective toxicity by Albert.[25] For example, the inhibition of enzymes in the biosynthesis of dihydrofolic acid has been extensively studied as a means of discovering antibacterial, anticancer and antiprotozoal drugs.[26] Further, if a difference in metabolism could not be identified, the possibility remained of exploiting substantial differences in the properties of enzymes from host and parasite; this possibility is especially strong if the parasite is a prokaryote. The classic case of species differences in enzyme properties leading to valuable chemotherapeutic agents is the

(5)   (6)   (7)   (8)

development of diaminopyrimidine antibacterial drugs such as trimethoprim. Recent extensions of the concept in a different sense include the important antiviral drug acyclovir (7) and the anti-AIDS drug AZT (azidothymidine; 8).

Although the field in which to search for a new drug was consciously chosen with a biological model in mind in the above examples, only recently have chemical models contributed to the development of drugs with commercial potential. What has been called 'rational drug design' involves a chemical model for the mechanism of a key process identified by the biological model as a possible key to a cure; in other words, the concepts from two quasi-isolated systems are brought together. For example, the basis of the action of pyridoxal phosphate, a coenzyme involved in amino acid metabolism, has been understood for many years. There have been many publications describing the design, synthesis and properties of inhibitors of pyridoxal dependent enzymes, and recent progress has led to compounds for the treatment of Parkinson's disease and to potential antibacterial agents.[27] A further stimulus here is to use the chemist's understanding of reaction mechanisms to devise new strategies for inhibiting enzymes specifically. Several categories have been described, each differing in the emphasis it places upon part of the enzyme-catalyzed reaction. Thus tight binding inhibitors have been approached with models of the transition state of the enzyme-catalyzed reaction in mind[28] often with the assistance of binding the equivalents of non-reacting parts of the substrate, a factor considered to be extremely important in enzymic catalysis.[29] Such binding can also be a source of the specificity desired in a drug. An alternative approach to the design of specific inhibitors is to invent compounds that are activated by the catalytic action of the enzyme as illustrated by the pyridoxal-based compounds mentioned above. There is also the stimulus to seek new chemistry to bring about such latent inhibition.[30]

Once this stage of sophisticated chemical thought has been reached, the way is open to the most seductive form of modelling in medicinal chemistry, namely computer-assisted molecular design. Whatever the ultimate contribution of computer-assisted molecular design to the development of new drugs, pictures of molecules on the colour graphics screen seem to impress marketing and media personnel in pharmaceutical companies as judged by the promotional uses of such images. This is not a cynical remark; an opportunity to make scientific thought more comprehensible to the commercial colleague and to the layman is to be grasped enthusiastically. People need a clear conceptual model of science appropriate to their own needs. The scientific value of computer-based models will be discussed further in Section 17.4.4.

There are also other computer-based methods of approaching the task of lead generation in drug discovery that have concentrated upon the evaluation of existing information on biological activity in contrast to the invention of new chemistry. For example, databanks have been constructed encoding the biological activities of many classes of compounds and associating specific substructures with the biological activity.[31] By searching the database, it is possible to identify a range of molecular substructures, such as a particular heterocyclic ring, that have formed part of a molecule with the biological activity of interest. It could be that a combination of such information with the chemical ideas of mechanism could be a very powerful strategy for lead generation. Other ways of focusing upon the appropriate molecular substructure have employed set theory;[32] in order to carry out analyses leading to lead generation using sets, it is necessary to embrace the problem of identification of parameters for inclusion in the description of the relevant total set of compounds. At the simplest level, sets of compounds may be established by investigating plots in parameter space using a parameter to represent the hydrophobicity of compounds (usually expressed as the octanol/water partition coefficient) and the acidity of the compound (expressed as the pK_a). If only two parameters are used, the model so constructed amounts to little more than an organization of comprehensible data. However, when more parameters are introduced, an extension of understanding and intuition is generated.

17.4.3.2 Models in Drug Development

Once a lead structure for a new drug has been established by good fortune, thorough screening or creative modelling, there remains much work to do to establish the optimal structure. At this stage, it is necessary to consider much wider aspects of the action of the drug, or, as was discussed above, its working environment. A range of compounds must be synthesized. However, it has been recognized for many years that an interaction between the synthetic chemist and the physical organic chemist is immensely important at this stage. Essentially the study becomes one of substituent effects on biological activity and it is no surprise to see that models of the type first used by Hammett have been extensively developed. Hammett's correlations refer to electronic effects, which are often the dominant effects in an organic reaction. In a drug, however, the environment includes the whole of the patient and consideration must be given to the mechanism by which the drug reaches its site of action. Hansch[20] developed extensive correlations to take account of the lipophilicity of substituents. A notion behind the endeavour was that most drugs will reach their targets by passive diffusion, it being unlikely that a specific active transport system would be available. It was necessary for molecules to traverse biological membranes which are largely composed of lipids. Hansch's parametrization used the partition of the drug between octanol and water as a model for the interaction of the drug with the lipid. With both Hammett σ and Hansch π parameters, many impressive retrospective assessments of drug optimization have been carried out (see Chapter 21.1). Although hydrophobic effects had earlier been recognized as important in drug action, Hansch was one of the first to treat the phenomenon in a way that could unite it with the electronic effects of substituents, thus creating a more coherent model. So-called Hansch analysis is conventionally carried out with the aid of regression analysis (Section 17.4.2.2) and it therefore will embody that technique's strengths and limitations. It is perhaps unfortunate that the success of retrospective analysis using Hansch's innovative ideas has led to an overemphasis upon a few molecular properties only in evaluating biological data.

There have been many other approaches to quantitative structure activity studies (QSAR), which are discussed in this volume (see Chapters 21.1–21.3). They differ in either the emphasis on the importance of a particular molecular parameter or the transformations used to relate biological activity to molecular structure. Lewi[33] has compared some of these models and has indicated how the successive inclusion of further molecular parameters can extend the applicability of Hansch analysis or how, alternatively, the use of standardized parameters can be avoided as in the Free and Wilson model. The Free and Wilson method is an additive model in which biological activity is expressed in terms of so-called 'indicators' for each substituent and substitution site in a molecule. Lewi recognizes the value in both approaches and their potential complementarity but is alert to the strategic weakness of all correlation methods: they very easily become molecular blinkers preventing the medicinal chemist from looking aside for a new series of compounds or to correlate results from very different structural types. The blinkers can, however, be removed if methods of multiparameter analysis are employed.

Many of the parameters used in such analyses can be calculated by robust theoretical models or can be estimated semiempirically as has already been mentioned. The critical thing, however, is that the model should point to the key molecular properties that relate to biological activity. The structural entities that embody these properties have often been called pharmacophores and their recognition often depends upon having a large number of synthetic analogues of the potential drug. The medicinal chemist may then be able to recognize that a property such as pK_a or dipole moment is dominant and to refine synthetic objectives accordingly. It is more likely that a combination of effects is important. An uncritical insistence upon one or two parameters is likely to lead to an error. As modellers, we would at once point out that such a strategy arbitrarily narrows the defined system and neglects careful evaluation of the full internal environment of the series of compounds under study. What is needed is a method of describing the characteristics of a molecule in a readily comprehensible way so that the molecular description relates more to the biological activity than to a conventional chemical classification. The force of this can be appreciated when it is realized that enzymes and other receptors respond not to a specific formal class of compounds, such as a peptide, but to the electric field presented by any compound that approaches. A well-known illustration is the binding of morphine, an alkaloid, to the same class of receptors as enkephalins, which are oligopeptides[34] (see Chapter 13.2). A multiparameter method is required that can analyze not only substituent effects within a series of similar compounds, as has been possible for many years, but also relates the properties of disparate types of molecules. Clearly such properties as electron distribution and conformation will be of particular significance in such an analysis.

Finally in this section, some consideration must be given to the subject of toxicology. Mutagenicity testing by the Ames method so that metabolism can be included by means of a mixture of hydroxylating enzymes has already been mentioned (Section 17.4.1.2.1). This model for a toxic effect makes no pretence to being a close parallel to an intact animal, although it does give a very valuable indication of the likelihood of mutagenicity in the animal. Taking toxicology further requires experiments in animals leading up to clinical trials but there is a wealth of evidence to indicate that the metabolism of one mammal can be very different from that of another. Useful experiments therefore require a good deal of experience on behalf of the toxicologist to choose the best animal model. As examples of extreme differences between rodents and primates, methanol is toxic to monkeys and to man but is tolerated by rats, mice and guinea-pigs.[35] It has been shown that the lack of toxicity of methanol in rodents is probably due to their ability to transform rapidly its oxidation products, formaldehyde and formate, into carbon dioxide using folate dependent enzymes. Similarly, difficulties were obtained in finding a suitable animal model for the teratogenic effects of thalidomide: in rats oral administration was marginally teratogenic but in rabbits a significant effect was found.[36] However, teratogenicity was observed in both species when the drug was administered intravenously. The model must include not only the site of action and possible metabolic enzymes but also an appropriate absorption or administration system. Quantitative consideration must be given to kinetics to evaluate an animal model in terms of its whole apparatus for transporting and transforming drugs.

17.4.4 COMPUTER-BASED MODELS

Perhaps the greatest single addition to the modelling armoury of the medicinal chemist is the development of computer methods for displaying and manipulating the three-dimensional structures of molecules, ranging in size from small drugs to macromolecular polypeptides and nucleic acids. The place of computer-based models in medicinal chemistry has already been introduced. The detailed methods for operating molecular design systems are described in Chapters 18.1–18.8, and in this section our purpose is to place the technique, for it is only another technique, in the general context of models in medicinal chemistry.

The key to the power of the computer graphics system is that in one technique it provides access to iconic, conceptual, mathematical and enactive models of medicinal chemistry; in short, all of the main classes of model. Figure 1 shows typical examples of screen displays relating to some of the following discussions.

In the iconic mode (Figure 1a and b) the terminal screen displays an image that represents the structure of the molecule under study, most usually as a stick diagram, that relates closely to a structural formula. Looking at the image, the chemist's scientific imagination is stimulated and through the image and the chemist's concept of the chemical system in question, intuitive jumps can be made that would be very difficult without the visual aid. But the image need not only be evaluated qualitatively and conceptually; the same dataset that is used to generate the image can also be used to initiate theoretical calculations relating to such properties as total energy and conformation, and electron distribution. Thus the computer provides a very important practical link between the theoretical chemist and the applied chemist in the field of drug design. Once the results have been

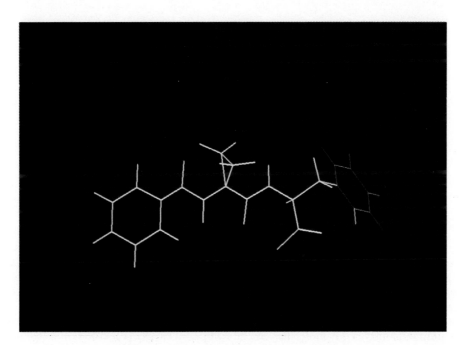

Figure 1a Stick diagram (see full caption on p. 98)

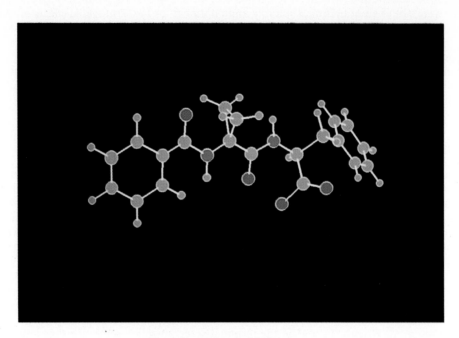

Figure 1b   Ball and stick diagram

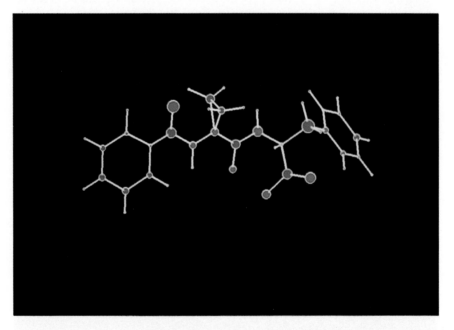

Figure 1c   Electron densities represented on ball and stick diagram

calculated, they can be displayed in ways that are helpful to the medicinal chemist; electron densities (Figure 1c) or van der Waals volumes with electron densities superimposed (Figure 1d) can, for example, be displayed. Finally, it is possible within the limits of computer power available to carry out many of these operations interactively. The ability of the medicinal chemist to ask 'What if?' of the computer raises the technique to an experimental level.

To these main modelling procedures can be added also the ability to look at two or more structures simultaneously. Thus molecules can be compared closely (Figure 1e) and ligands docked into their receptors (Figure 1f). The theoretical methods to which the computer gives access can then

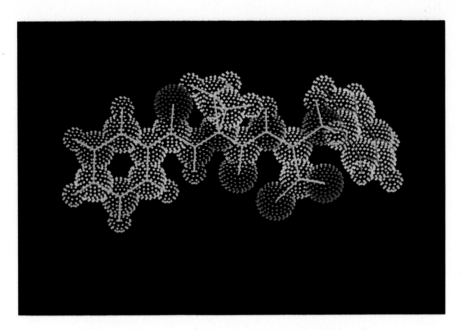

Figure 1d   Electron densities represented on van der Waals surface

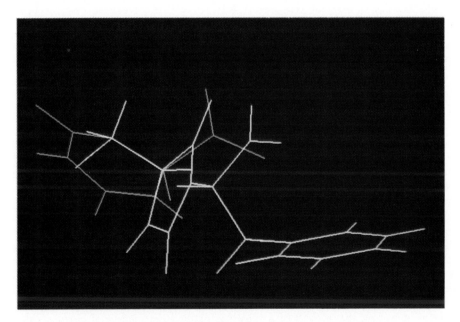

Figure 1e   Superimposition of an enzyme substrate and an inhibitor

be used to evaluate the molecular similarity or strength of interaction. There are, however, problems. As was pointed out above, many of the interesting calculations require a great deal of computer time. Since the interactive mode is so valuable to the medicinal chemist, it is very important to find ways in which important questions such as energies of interaction between enzymes and inhibitors, for example, and molecular similarity can be estimated quickly. If this can be done, the computer graphics system becomes a hands-on tool for the medicinal chemist and not the exclusive domain of the specialist. Richards has opened up a number of ways to improve the accessibility of calculated data. On the one hand, he has shown how it is possible to estimate the binding energies of inhibitors to dihydrofolate reductase by calculating the energies of interaction of small portions of an inhibitor

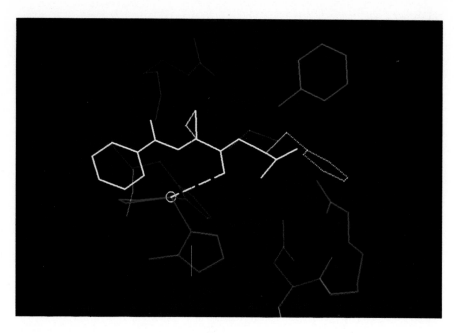

Figure 1f

Figure 1 Models of enzyme inhibitors represented by molecular graphics using the INTERCHEM programs at the University of Strathclyde.[46,52] (a) Stick diagram akin to a structural formula of an inhibitor of carboxypeptidase A. (b) Perspective ball and stick model in which the atoms are colour coded C, H = brown, N = blue, O = red. (c) Ball and stick diagram in which the colour and size of the atoms represents the sign and magnitude of the electron density on the atom as calculated by MOPAC. The partial positive charge on the apices of the cyclopropane ring that is believed to be the site of the covalent inhibitor–enzyme interaction is visible. (d) The same electron density as in (c) displayed on the van der Waals surface of the inhibitor using a palette of 256 colours with red and blue as the extremes. (e) Superimposition of substrate and inhibitor of dihydroorotate dehydrogenase[52] to suggest the possible congruence of sites of oxidation by the enzyme (see Figure 5 and Section 17.4.6). (f) The carboxypeptidase inhibitor illustrated in (a)–(d) docked at the active site of carboxypeptidase A. The amino acids are colour coded to assist in recognition: acidic = red, basic = blue, hydrophilic = magenta, aromatic = orange, aliphatic = green, substrate = white

at a time and summing the results,[37] and, on the other hand, he has proposed an *ab initio* method for calculating molecular similarity that requires only a few seconds CPU time on a VAX 11/780.[38] The development of such methods holds out much promise and scientific software houses have not been slow to seize the opportunity.

The remaining sections in this chapter examine computer models in the context of the development of models and problems. The question can well be asked whether there are any further enhancements to computer modelling that would help a medicinal chemist. Without doubt there are many. The biggest problem in medicinal chemistry is introducing information to intuition and imagination. Graphics techniques have been shown to be an important link between theoretical chemistry and practical medicinal chemistry and similar links have been forged between graphics and the devising of synthetic strategies. If the same dataset that seeded the synthesis and theoretical calculation can be used to access the literature of biological activity, including toxicology, then the germination of ideas and the growth of a scientific plan in the medicinal chemist's mind will be encouraged. But the model need not be bounded by the conventional limits of chemistry. In the future lies the possibility of including the pharmacological and toxicological properties of molecules within one and the same computer-based model.

17.4.5 THE DEVELOPMENT OF MODELS OF AN ENZYME'S ACTIVE SITE

As an example of many of the features that have been introduced into the discussion of models in the preceding section, we review here the development of models of the active site of a single enzyme, alcohol dehydrogenase. Although this enzyme does not have direct medicinal chemical applications, it has strong connections through the various studies that have used it either as a model for other enzymes or as a testing ground for models themselves. A recurrent feature is how each stage in the model's development was subsequently found to be inadequate to cope with the more stringent demands of new experimental results and how the model was modified appropriately.

The use of enzymes in organic chemistry[39] has grown at an increasing pace since the first demonstrations of the potential of enzymes to transform non-natural substrates by Prelog in the early 1960s.[40] He explored the ability of horse liver alcohol dehydrogenase (HLADH) to oxidize substituted cyclohexanols and found that alkyl substituents in certain positions, notably the 2-position, prevented the redox reaction from occurring. From a series of experiments with a wide range of substrates he built up a model of the space at the active site of HLADH and represented it in the form of a so-called diamond lattice (Figure 2a) in which the points of the lattice represented the positions at which the carbon atoms of the substrates would be positioned. Many similar studies have since been carried out not only to map the active sites of other enzymes but also to explore the topography of receptors.

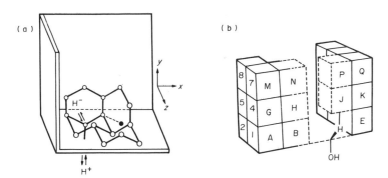

Figure 2　Models of the active site of horse liver alcohol dehydrogenase devised to summarize the substrate specificity of the enzyme. In Prelog's model (a), filled points represent forbidden positions for carbon atoms in the diamond lattice[40] but this model only has predictive value for substrates composed of six-membered alicyclic rings (reproduced from ref. 40 by permission of IUPAC). Jones's model[41] (b) is more general and indicates accessible regions by dashed lines in the cubic lattice. Sterically restricted areas are within the closed cubes. The volumes are identified by letters. Since it uses an arbitrary dissection of the active site space, this model can be used for any type of substrate

This work attracted polite interest without much suggestion of application until, some 10 years later, Jones began a systematic investigation of the potential of HLADH to transform compounds of interest in synthesis. It was obvious that most of the substrates of importance in synthetic chemistry would not be derivatives of cyclohexanones. Consequently a new model of the active site was needed. In order to accommodate as many substrates as possible, Jones[41] devised a model that did not relate specifically to any compound's geometry: he divided the space at the active site into cubes using a cubic lattice (Figure 2b). With this model it was possible to predict the substrate properties of both cyclic and acyclic compounds without restriction on ring size or requiring that molecules contain no heteroatoms in the skeleton. For most of the demands of a synthetic chemist, to whom maximum rates of reaction are less important than is maximum affinity of an inhibitor to a medicinal chemist, this model was adequate. However, it gave very little insight into the nature of the binding of substrates to the enzyme; all it said was that there will or will not be sufficient space.

As the synthetic work was being developed, two advances made it possible to investigate substrate and inhibitor interactions with HLADH in more detail. Firstly, an X-ray crystallographic study of the ternary structure of an enzyme–pseudo substrate–coenzyme complex was published[42] and secondly, computer graphics techniques were developed.

The first attempt to evaluate interactions within the HLADH ternary complex was by Dutler and Brändén,[43] who combined kinetic studies of substituted cyclohexanols as substrates with the data from the X-ray study to construct a solid framework model of the active site of HLADH. It is interesting to look at the geometry that they deduced for the reactive complex. Unfavourable interactions between substituents were detected through physical contact between substrate and enzyme or coenzyme (Figure 3). Although a quantitative correlation was not possible, they were successful in accounting for the relative reactivities of a series of methylcyclohexanols at the active site. Subsequently, a ternary complex with a genuine substrate was crystallized and the structure determined.[44] This time, the model was built in a computer. The greater precision available to the computer model together with the more appropriate ternary complex used and energy minimization

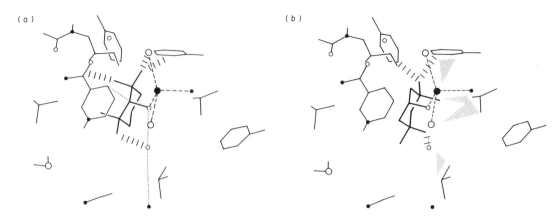

Figure 3 A drawing from a solid model of the active site of alcohol dehydrogenase constructed from data from X-ray crystallographic studies. The picture is explicit with regard to the relative positions occupied by the enzyme and substrate and has been interpreted by the addition of ladder lines to indicate unfavourable steric interactions and shaded areas to indicate favourable non-polar interactions. The substrates are methylcyclohexanols (reproduced from ref. 43 by permission of Academic Press)

calculations showed that although the position of the substrate was similar in both models, differences were evident in the positions of amino acid side chains at the active site. Those substrates (methylcyclohexanols) that reacted rapidly were found to have no contacts closer than 3 Å with the protein. From these and other results the models proposed by Prelog and Jones were refined to identify the amino acid residues that are responsible for preventing the access of some potential substrates to the active site. For example, an equatorial alkyl substituent at C-2 of cyclohexanol evidently impinges upon the space occupied by the nicotinamide ring of the coenzyme.

The connection between medicinal chemistry and HLADH through the preceding paragraphs is, perhaps, a little tenuous. However, in two respects, the studies mentioned above lead to useful exercises in medicinal chemistry. The value of HLADH in bioorganic chemistry has principally been that it has a wide substrate tolerance but a stereoselective reaction. Thus it has been possible to use HLADH as a testing ground for concepts and methodologies in both synthesis and enzyme inhibition. Two examples of the latter are of interest.

Firstly, in our work on devising new methods of selective inhibition of enzymes,[45] we used HLADH as a prototype enzyme on which to test the new principles of chemical reactivity that were of interest. We proposed that the oxidation of cyclopropane methanols would lead to a cyclopropane that could be activated towards nucleophilic addition by a suitably placed group at the enzyme's active site (Figure 4). The idea proved to be successful and the question then arose as to the identity of an enzyme nucleophile responsible for inhibition. To provide suggestions as to the detailed interaction of the inhibitor with the enzyme, we were able to make use of the molecular model defined by the second X-ray structure described above. The cyclopropane-containing substrate was docked into the active site without disturbing the normal hydrogen transfer geometry. Taking into account the expected stereochemistry of the hydrogen transfer, it was possible to suggest that a hydroxyl group of serine could act as a nucleophile to form a covalent bond with the inhibitor. This work has been valuable to establish a much wider project in which cyclopropane-containing

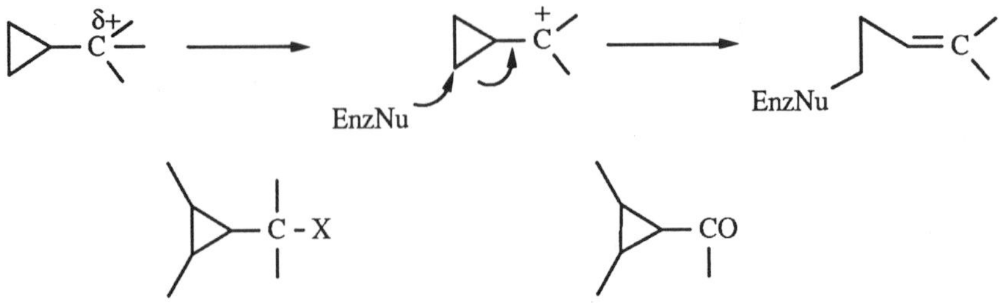

Figure 4 Activation of a cyclopropane ring to nucleophilic attack by enzyme-induced enhancement of the positive charge on the carbon atom to the ring: X = a leaving group, typically H^- in enzyme-catalyzed oxidations

compounds can be used to inhibit enzymes of chemotherapeutic importance including peptidases and dihydrofolate reductase.[30,46]

In the previous example, the computer-based model was able to interface effectively with the chemist's conceptual model of the mechanism of a reaction at a qualitative level. The second example in which HLADH relates closely to medicinal chemistry takes the investigation a stage deeper to look at a quantitative correlation between the inhibition of the enzyme by 4-alkylpyrazoles and some aromatic amides and their binding to the active site. HLADH is a zinc-containing enzyme and both classes of inhibitor are believed to function by coordination to the zinc, thereby preventing the substrate from binding productively to the zinc. In addition, the hydrophobic parts of the molecules make favourable contacts with the hydrophobic binding pocket of the enzyme. A correlation between the hydrophobicity and the inhibitory power deduced from a regression analysis was of the form

$$\log 1/K_i = 0.96 (\pm 0.25) \log P + 5.70 (\pm 0.56) \tag{3}$$

$$n = 5, r = 0.990, s = 0.207$$

Since only alkylpyrazoles were tested, there was no need to include a term relating to the electronic effects of substituents using a Hammett parameter. The series then offered a good test of the ability of a computer model to reproduce hydrophobic effects. Langridge and Hansch and their colleagues[47] made the comparison and the agreement was most gratifying. Similar experiments on dihydrofolate reductase were also successful.[48] In the alcohol dehydrogenase case, the computer model was built from a ternary complex including enzyme, coenzyme and pyrazole inhibitor. Alkyl chains were incorporated in extended conformations using standard geometries with small adjustments made to the torsion angles. The computer model then showed that the inhibitors fitted with no steric interaction between the enzyme and alkyl chain. If, therefore, the oil–water partition coefficients measured by $\log P$ or π values represent the hydrophobic effects well, then the coefficient of the regression curve should approach 1, as was found experimentally. On the other hand, if the coefficient deviates from 1, the importance of steric effects, which the computer model ruled out in this case, must be considered.

This conclusion is interesting because it supports the view that the octanol–water partition coefficient measures solubilization in a hydrophobic environment with complete desolvation of the solute from water. The same line of argument leads to the expectation that the binding of a substituent like a phenyl ring to a flat surface would be characterized by a coefficient of about 0.5 in the regression equation, indicating half desolvation.

There are a number of points of satisfaction arising from this field of research. Firstly, the concordance of models derived from distinct approaches is evident in the descriptions of both inhibitor binding and reactivity. Secondly, a molecular interpretation is offered for components of regression analysis equations dealing with hydrophobicity. To have a molecular view is a major step forward for chemists because the system is then open to scrutiny and manipulation with the aid of imagination and intuition through their conceptual models. Such opportunities are the gates to progress.

17.4.6 FROM THE RETROSPECTIVE TO THE PROSPECTIVE AND SPECULATIVE

Not surprisingly, in view of the commercial importance of medicinal chemistry, many publications from industry on structure–activity relationships are retrospective; it is often left to the more intrepid academics to stick their necks out, sometimes encouraging industrial collaborators to be similarly bold. Many of the remarks earlier in this chapter have hinted at the importance of modelling as a stimulus to creative thinking and in this final section some examples will be offered. As it happens, all of the examples relate in one way or another to the key question of molecular similarity.

A knowledgeable medicinal chemist would be able to suggest several possible analogue structures for a given biologically active molecule but would not expect to be able to decide with confidence on a quantitative rank order of activity of the analogues. A quantitative comparison of the similarity of molecules would obviously be valuable if a sufficiently quick method were available so that priorities for synthesis could be better assigned. With this in mind, Richards[49] has developed an embryonic semiempirical method to estimate the molecular similarity by replacing the complex multicentre integrals required by *ab initio* calculations of electron density with a series of spherical Gaussian functions. The overall strategy resembles that used for the calculation of binding energies of inhibitors to dihydrofolate reductase; it is assumed that molecules can be divided into a number of independent fragments. Thus from the electron densities of a series of simple molecules, it should be possible to build up a description of the electron density in large complex molecules. So far, he has obtained functions to describe alkanes and ethers which are potential bioisosteres. The similarity index so derived can be used as a criterion for judging the best superposition of two molecules and excellent agreement was obtained between the use of the semiempirical Gaussian method and the *ab initio* method, especially for the more chemically significant valence electrons. Taking a similarity index of 1 to indicate identity, the comparison between propane and dimethyl ether led to indices of 0.87 and 0.81 respectively for the *ab initio* and semiempirical methods. The importance of having a semiempirical method that gives useful numbers was hinted at earlier (Section 17.4.2.2.2) in that the speed of a semiempirical calculation (2 s on a VAX 11/780 for molecules of the size of propane and dimethyl ether) makes it reasonable to consider calculations on compounds of significance in medicinal chemistry. However, the critical step of relating such a similarity index to biological activity has yet to be taken and is awaited with interest.

Much of the preceding section related to studies of enzymes and their active sites by means of molecular graphics. In the examples given, a good basis was available in a crystallographic structure for an enzyme–inhibitor complex. What can be done when such data are not available? Firstly, it must be noted that X-ray crystallography is not the only way in which three-dimensional structural data can be obtained for proteins. For small proteins it is now possible to obtain accurate structural information in solution by means of high field NMR (typically 500 MHz) and results have begun to emerge in which the chemistry of enzyme–inhibitor interactions is considered.[50] Nevertheless, the major body of available structural evidence is crystallographically derived. In order to obtain such data, however, a good deal of background information is required, including the primary sequence of the protein. Accumulated experience concerning the expected geometries of domains of proteins is also valuable to resolve doubts. Such information for a protein of which the crystal structure is unknown can be used to generate a model for its structure.

Human renin is an aspartate protease that catalyzes the hydrolysis of angiotensinogen to angiotensin II; it is the rate-limiting enzyme of the pathway and as such is of great interest as a target for inhibitors as drugs designed to regulate blood pressure. Blundell and his colleagues[51] have constructed a computer model for this enzyme using the primary sequence derived from the gene and the sequences defined for homologous aspartic proteases, including mouse submaxillary renin, pepsins, calf chymosin, endothiapepsin and penicillopepsin. The first stage in the model building was to align the human sequence with the mouse submaxillary renin sequence and then with the other enzymes. Having defined the homology between the human enzyme and its relatives, reference was then made to the X-ray analysis results for porcine pepsin, rhizopuspepsin, penicillopepsin and endothiapepsin. The model building attempted to retain the hydrophobic residues of the core and to limit changes, both insertions and deletions, to surface regions or loops. Having achieved a satisfactory overall geometry for human renin, the positions of side chains were adjusted so that acceptable torsion angles and non-bonded interactions were obtained with endothiapepsin as the main reference system. It is interesting to note that attempts were made to carry out energy minimization operations for the active site residues but it was found that the calculations necessitated unacceptable movements of key water molecules at the active site.

Having obtained the model, detailed comparisons were made between the structure of human renin deduced in the above way with particular reference to the evolution of the enzyme. In particular, features that might relate to the specificity of human renin were identified and, bearing in mind the remarks made about substrate binding at multiple sites and ways of obtaining drug selectivity mentioned earlier in this chapter, such features could have an important bearing upon the design of specific inhibitors. If such computer-based techniques are to have value in medicinal chemistry, it is important that their results should be validated by comparison with an independent structure derived crystallographically. As far as human renin is concerned, close similarities with the prototype acid peptidases have now been characterized.[51]

Finally, let us examine the opportunities that molecular modelling can provide when an enzyme structure is not available at all. We have been interested in the inhibition of enzymes of the biosynthesis of vitamins and recently our work has been extended to study the inhibition of enzymes of pyrimidine biosynthesis.[52] The inhibition of these enzymes might reasonably be expected to lead to potential anticancer or antiprotozoal compounds. Fortuitously we found that the hydantoin (9) was a time-dependent irreversible inhibitor of dihydroorotate dehydrogenase, an enzyme that catalyzes the oxidation of dihydroorotate (10) to orotate (12). It was by no means obvious why the hydantoins should be oxidized by the enzyme since no obvious superimposition of substrate and inhibitor brought together the superficially similar cyclic amides and the carboxylic acid function, which is most probably critical in binding of substrate to the active site. A chemical reaction of the inhibitor gave the clue to what might be taking place. We found that oxidation led to the α,β-unsaturated derivative (**11**) and this compound contains a site susceptible to Michael addition of a nucleophile from the enzyme's active site. Such an oxidation also accounts for the observed importance of the benzyl group of (**9**) in that oxidation to the conjugated system of (**11**) would be favoured. With these clues, the computer was used to see whether it was possible to superimpose the oxidation site of substrate and inhibitor in such a way that the carboxylates could in each case interact with an expected positive charge at the enzyme's active site. Figure 5(a) shows the result. By superimposing the known site of oxidation of the substrate (**10**) with the corresponding benzylic site of the inhibitor (**9**), it was shown that the carboxylate groups of the two could lie such that they might form an ionic bond with the same residue on the protein. Additionally, the enantiomer of (**9**) was found to behave similarly and it could be aligned in a compatible manner (Figure 5b).

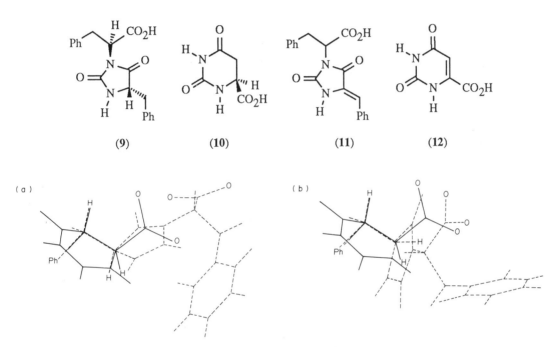

(**9**)   (**10**)   (**11**)   (**12**)

Figure 5 (a) (*R,R*)- and (b) (*S,S*)-hydantoins (**9**), inhibitors of dihydroorotate dehydrogenase, superimposed with the substrate dihydroorotic acid (**10**) to illustrate the hypothetical congruence of oxidation sites (see Figure 1e) (reproduced from ref. 52 by permission of the Royal Society of Chemistry)

The attractive features of this model are that it leads in several ways to possibilities for the creative design of inhibitors. Firstly, the potential existence of a hydrophobic pocket in which the non-reactive phenyl groups can be placed is indicated. Secondly, the location of a cationic group at the active site in relation to the hydrophobic pocket is revealed. Thirdly, and most importantly, it shows how a compound belonging to a quite different series from the substrate can reasonably react with the enzyme through a catalytic act by the enzyme. It is as if the urea portion of the hydantoin ring is an inert piece of molecular scaffolding conveniently keeping binding and reacting groups in place. If this is so, then the possibility exists to design many compounds with properties suitable for inhibition of dihyroorotate dehydrogenase. Such a molecular sidestep to a substrate surrogate[53] could lead to very favourable properties with regard to the selectivity of any resulting drug since the series of interest is no longer a close relative of an important natural intermediate on the biosynthetic pathway. Needless to say, we are following up the opportunities offered by this discovery.

17.4.7 REFERENCES

1. C. J. Suckling, K. E. Suckling and C. W. Suckling, 'Chemistry through Models', Cambridge University Press, Cambridge, 1978.
2. J. N. Champness, L. F. Kuyper and C. R. Beddell, in 'Molecular Graphics and Drug Design', ed. A. S. V. Burgen, G. C. K. Roberts and M. S. Tute, Elsevier, Amsterdam, 1986, p. 335.

3. D. E. Hathway, 'Molecular Aspects of Toxicology', Royal Society of Chemistry, London, 1984, p. 40.
4. P. H. Bell and R. O. Roblin, Jr., *J. Am. Chem. Soc.*, 1942, **64**, 2905.
5. A. Albert, 'Selective Toxicity', 6th edn., Chapman and Hall, London, 1981.
6. Ref. 3, p. 17; K. H. Meyer and H. Hemmi, *Biochem. Z.*, 1935, **277**, 39; E. Overton, 'Studien uber die Narkosen', Fischer, Jena, 1901.
7. Ref. 1, p. 291; C. W. Suckling, *Br. J. Anaesth.*, 1957, **29**, 466.
8. A. Pullman, *C.R. Seances Soc. Biol. Ses. Fil.*, 1945, **139**, 1956.
9. Y. C. Martin, in 'Quantitative Drug Design', ed. E. J. Ariëns, Academic Press, New York, 1979, vol. 8, p. 1; G. Redl, R. D. Cramer and C. E. Berkoff, *Chem. Soc. Rev.*, 1974, **3**, 273.
10. (a) P. P. Mager (ed.), 'Medicinal Chemistry', Academic Press, Orlando, FL, 1984, vol. 20; (b) A. J. Everett, in 'Topics in Medicinal Chemistry', ed. P. R. Leeming, Royal Society of Chemistry, London, 1988, p. 314.
11. Ref. 1, p. 5.
12. Ref. 1, p. 64.
13. J. S. Bruner, 'Towards a Theory of Instruction', Harvard University Press, Cambridge, MA, 1967.
14. Y. C. Martin, in ref. 9, p. 167; ref. 1, p. 48; ref. 5, p. 71; ref. 3, p. 16.
15. Ref. 1, p. 80; H. Maskill, 'The Physical Basis of Organic Chemistry', Oxford Scientific Publishers, Oxford, 1985, p. 66.
16. W. G. Richards, 'Quantum Pharmacology', 2nd edn., Butterworths, London, 1983.
17. M. J. S. Dewar and D. M. Storch, *J. Chem. Soc., Chem. Commun.*, 1985, 94.
18. R. Robinson, *J. Chem. Soc.*, 1917, 876.
19. H. Maskill, in ref. 15, p. 441; ref. 1, p. 77.
20. C. Hansch, *Acc. Chem. Res.*, 1969, **2**, 232.
21. V. G. Blaschke, H. P. Kraft, K. Fickentscher and F. Köhler, *Arzneim.-Forsch.*, 1979, **29**, 1640.
22. M. Schneider and E. H. Reimerdes, *Forum Mikrobiol.*, 1987, 65.
23. (a) M. H. Fisher, in 'Recent Advances in the Chemistry of Insect Control', ed. N. F. Janes, Royal Society of Chemistry, London, 1985, p. 53. (b) H. Mrozik, ref. 10b, p. 245.
24. A. Endo, *J. Antibiot.*, 1979, **32**, 852; B. Hesp and A. Willard, in 'Enzyme Chemistry, Impact and Applications', ed. C. J. Suckling, Chapman and Hall, London, 1984, p. 141.
25. Ref. 5, p. 103.
26. Ref. 2; H. C. S. Wood, *Chem. Ind. (London)*, 1981, 150.
27. M. G. Palfreyman, I. A. McDonald, P. Bey, C. Danzin, M. Zreika, G. A. Lyles and J. R. Fozard, *Biochem. Soc. Trans.*, 1986, **14**, 410; C. Walsh, R. Badet, E. Daub, N. Esaki and N. Galakatos, in 'Third SCI-RSC Medicinal Chemistry Symposium', ed. K. W. Lambert, RSC Special Publication No. 55, Royal Society of Chemistry, London, 1986, p. 183.
28. R. Wolfenden, *Acc. Chem. Res.*, 1972, **5**, 10.
29. M. I. Page, in 'The Chemistry of Enzyme Action', ed. M. I. Page, Elsevier, Amsterdam, 1984, p. 1.
30. C. J. Suckling, in ref. 10b, p. 128; R. B. Silverman, in ref. 10b, p. 73.
31. V. E. Golender and A. B. Rozenblit, in 'Drug Design', ed. E. J. Ariëns, Academic Press, New York, 1980, vol. 9, p. 299.
32. V. Austel and E. Kutter, in 'Drug Design', ed. E. J. Ariëns, Academic Press, New York, 1980, vol. 10, p. 1.
33. P. J. Lewi, in ref. 32, p. 308.
34. J. DiMaio, C. I. Bayly, G. Villeneuve and A. Michel, *J. Med. Chem.*, 1986, **29**, 1658.
35. Ref. 3, p. 169.
36. Ref. 3, p. 210.
37. A. F. Cuthbertson and W. G. Richards, *J. Chem. Res. (S)*, 1985, 354.
38. E. E. Hodgkin and W. G. Richards, *J. Chem. Soc., Chem. Commun.*, 1986, 1342.
39. C. J. Suckling and K. E. Suckling, *Chem. Soc. Rev.*, 1974, **3**, 387.
40. V. Prelog, *Pure Appl. Chem.*, 1964, **9**, 126.
41. J. B. Jones and I. J. Jakovac, *Can. J. Chem.*, 1982, **60**, 19.
42. H. Eklund, J.-P. Samama, L. Wallen, C. I. Brändén, A. Aakeson and T. A. Jones, *J. Mol. Biol.*, 1981, **146**, 561.
43. H. Dutler and C.-I. Brändén, *Bioorg. Chem.*, 1981, **10**, 1.
44. H. Eklund, B. V. Plapp, J.-P. Samama and C.-I. Brändén, *J. Biol. Chem.*, 1982, **257**, 14 349.
45. I. MacInnes, D. C. Nonhebel, S. T. Orszulik, C. J. Suckling and R. Wrigglesworth, *J. Chem. Soc., Perkin Trans. 1*, 1983, 2771.
46. S. K. Ner, C. J. Suckling, A. R. Bell and R. Wrigglesworth, *J. Chem. Soc., Chem. Commun.*, 1987, 480; J. Haddow, C. J. Suckling and H. C. S. Wood, *J. Chem. Soc., Chem. Commun.*, 1987, 478.
47. C. Hansch, T. Klein, J. McClarin, R. Langridge and N. W. Cornell, *J. Med. Chem.*, 1986, **29**, 615.
48. C. D. Selassie, Z.-X. Fang, R. Li, C. Hansch, T. Klein, R. Langridge and B. T. Kaufman, *J. Med. Chem.*, 1986, **29**, 621.
49. P. E. Bowen-Jenkins, D. L. Cooper and W. G. Richards, *J. Phys. Chem.*, 1985, **89**, 2195.
50. (a) J. M. Schwab, W. Li, C.-K. Ho, C. A. Townsend and G. M. Salituro, *J. Am. Chem. Soc.*, 1984, **106**, 7293; (b) S. I. Foundling, J. Cooper, F. E. Watson, A. Cleasby, L. H. Pearl, B. L. Sibanda, A. Hemmings, S. P. Wood, T. L. Blundell, M. J. Valler, C. G. Norey, J. Kay, J. Boger, B. M. Dunn, B. J. Leckie, D. M. Jones, B. Atrash, A. Hallett and M. Szelke, *Nature (London)*, 1987, **327**, 349.
51. B. L. Sibanda, T. L. Blundell, P. M. Hobart, M. Fogliano, J. S. Bindra, B. W. Dominy and J. M. Chirgwin, *FEBS Lett.*, 1984, **174**, 102.
52. I. G. Buntain, C. J. Suckling and H. C. S. Wood, *J. Chem. Soc., Perkin Trans. 1*, 1988, 3175.
53. C. J. Suckling, 'Proceedings of the XIth International Congress in Medicinal Chemistry, Budapest, 1988', Elsevier, Amsterdam, 1989, in press.

18.1

Quantum Mechanics and the Modeling of Drug Properties

GILDA H. LOEW

SRI International, Menlo Park, CA, USA

and

STANLEY K. BURT

Sandoz Research Institute, East Hanover, NJ, USA

18.1.1 REVIEW OF METHODOLOGY

In this section we present an overview of quantum mechanical methods which can be useful in rational drug design. In Section 18.1.2 we present specific examples of their use in a variety of contexts.

18.1.1.1 Introduction and Background

A drug is simply a chemical compound which, when administered to a living system, elicits a benign or adverse effect. Figure 1 indicates the possible chemical and physical processes that could lead to such effects. A commonly accepted first step in drug action is reversible drug–receptor or

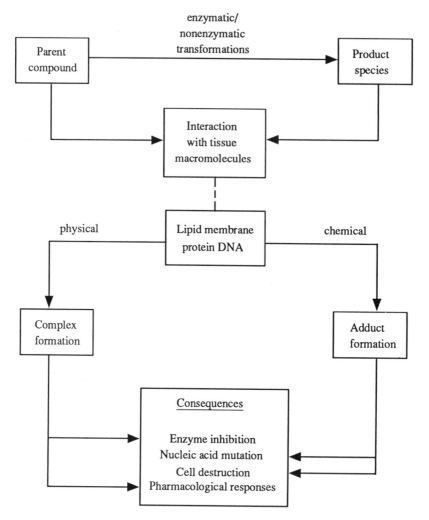

Figure 1   Physical and chemical processes involved in pharmacological or toxic response

enzyme–inhibitor complex formation. Such a 'physical' complex could be formed by a parent compound or by a chemically transformed species. The formation of a drug–receptor complex is thought to initiate a chain of activation processes ultimately leading to one or more observable pharmacological end points. In other instances, particularly for adverse or toxic effects, such as chemical carcinogens or drugs causing hepatic or renal toxicity, these effects can be initiated by covalent adduct formation or attack of chemically reactive species on tissue components.

The inherent capabilities of theoretical chemistry embodied in a hierarchy of methods, as summarized in Table 1, can be useful in the identification, selection and calculation of reliable molecular descriptors of both the physical and chemical processes involved in eliciting a pharmacological response, *i.e.* in mechanistic structure–activity studies.

As shown in Table 1 and elaborated in the following sections, the techniques used in molecular modeling are not limited to one particular method. In fact, an important first step in applying such techniques is to choose the most appropriate method for a given problem from among the diverse methods of varying degrees of complexity now available.[1] In general, empirical energy methods[2-4] are very effective for energy conformation studies, particularly for molecular systems with many rotational degrees of freedom and for characterization of intermolecular complexes. There is, however, a definite continued role for quantum mechanical methods in rational drug design. Examples of the types of molecular properties that can be calculated by these methods and how they could be useful as molecular descriptors of pharmacological activity are given in Tables 2–4. Calculation of electronic properties implicated in both physical and chemical reactions of drugs with their biological environment can only be done using quantum mechanical methods. In addition,

Table 1 Capabilities of the Hierarchy of Methods of Theoretical Chemistry

Empirical energy methods
Rapid calculation of optimized molecular conformations
Rapid characterization of energies and geometries of intermolecular drug–receptor or enzyme–inhibitor complexes

Semiempirical quantum chemical methods
Calculation of energy conformation profiles including changes in electronic properties
Calculation of explicit electronic properties of individual compounds such as chemical/biochemistry reactivity parameters
Characterization of model chemical/biochemical reactions
Model intermolecular complex formation

Ab initio quantum chemical methods
Same capabilities as semiempirical methods, but without use of any empirically derived parameters
Useful for systematic understanding of reliability of any level of calculation
Useful for compounds for which empirical parameters are not available
Require access to more computer capabilities than semiempirical or empirical energy methods

Table 2 Capabilities of Theoretical Chemistry to Calculate Molecular Descriptors of Drugs

Quantity calculated	*Potential usefulness*
(A) Electronic properties	Measures of extent of complex formation: with solvents and tissue macromolecules—DNA, proteins and membranes
1. Net atomic charges	
2. Dipole and higher moments	Modulators of relative ease and selectivity in transformation of parent compounds to specific intermediate and products
3. Ionization potentials	
4. Electron affinities	Modulators of the extent and specificity of covalent adduct formation with tissue macromolecules
5. Molecular electrostatic potentials (MEPs)	
6. Chemical reactivity properties, *e.g.* electrophilicities, nucleophilicities	
(B) Conformational energies	Possible indicators of relative affinities of family of drugs
1. Set of energy-ordered stable conformers	Identify bioactive form of drugs
2. Rotational energy profiles	Imply nature of receptors

Table 3 Capabilities of Theoretical Chemistry to Characterize Drug Receptor Complexes

What can be calculated	*How it can be useful*
Complex geometry	Identify conformational requirements of drug/inhibitor for complex formation
Complex stability	Select electronic properties of drug/inhibitor which modulate affinity

Table 4 Capabilities of Theoretical Chemistry to Characterize Chemical/Biochemistry Reactions

What can be calculated	*How it can be useful*
Reaction mechanisms Identification of reactants, intermediates, transition states and products	Allows identification of chemical reactivity parameters of parent compounds Allow identification and characterization of short-lived intermediates which could be implicated in adverse or beneficial pharmacological effects
Enthalpies and entropies of activation	Estimate of relative rates of reaction
Enthalpies and entropies of reaction	Estimate of energy requirements for reaction

calculation of energy conformational profiles and intermolecular interactions in a variety of contexts, for example (i) to include the effects of electron redistribution in response to geometry changes; (ii) when empirical energy parameters for the atom types of interest are not available; and (iii) to obtain parameters for use in molecular mechanics calculations, are best done using quantum mechanical methods. Finally, as indicated in Table 4, the thermodynamics, kinetics and mechanism of biochemical and chemical reactions can only be characterized using quantum mechanical methods. Such processes can be involved in (i) receptor activation; (ii) transformation of parent compounds to bioactive intermediates; (iii) interactions of reactive species with target tissue macromolecules; and (iv) enzyme–substrate reactions. Explicit characterization of such reactions can also lead to the identification of useful mechanistically relevant reactivity parameters which are modulators of activity for a series of drug molecules and help in the design of enzyme inhibitors as drugs.

In principle, the exact solution of the Schrödinger eigenvalue equation (equation 1), where H is the Hamiltonian operator, ψ is the wave function and E is the energy of the system, would yield a complete description of a molecular system. In practice, because it is computationally intractable for large systems of interest, approximations have to be made. The accuracy and completeness of the description obtained from a solution of the Schrödinger equation depend on three crucial factors: (i) the completeness of the Hamiltonian operator; (ii) the form of the wave function; and (iii) the method used to obtain a solution. A hierarchy of methods in the field of computational chemistry have been developed corresponding to the choice of specific levels of such approximations.

$$H\psi = E\psi \tag{1}$$

One of the most common approximations made in solving the Schrödinger equation is called the Born–Oppenheimer approximation.[5] Because the mass of nuclei are so much greater than that of electrons, the Schrödinger equation can be separated into a product of two functions: one depending only on electrons and the other only on nuclear coordinates. This approximation allows the solution of the electronic Schrödinger equation for a fixed geometry resulting in an energy-optimized electron distribution, from which electronic properties can be calculated. For additional characterization of geometries, energy gradient optimization procedures which allow nuclear coordinates to vary must be added. Thus, use of the Born–Oppenheimer approximation allows the separation of desired information into calculation of electronic properties only or, by additional optimization procedures which are more time consuming, calculating optimized geometries as well.

Focusing on calculation of electronic properties only, a hierarchy of further approximations are made. While the level of calculation chosen to solve the electronic Schrödinger equation will affect the reliability and usefulness of the electron distributions obtained, all levels of calculation can yield some useful electronic properties. The most common procedure used for solution of the electronic Schrödinger equation is an iterative one called the Hartree–Fock (HF)[6] self-consistent field (SCF)[7] method. In this approximation, each electron is assumed to move in the average field of all the other electrons. Thus, the electrons are treated independently and an SCF procedure is used to account for electron interactions. In the most common SCF procedures, which incorporate the variational principle,[8] a set of atomic orbitals (wave functions, basis sets) each of which describes the spatial distribution of a single electron is used. The total energy is then minimized by the iterative solution of the Schrödinger equation until the electron distribution or the total internal energy converge to a chosen criterion of variation between cycles. In all SCF calculations, the major computational problem is the large number of one- and two-electron integrals that must be evaluated and stored. Since the number of integrals is proportional to the fourth power of the number of electronic functions or 'basis sets' used, these computational difficulties have led to a great deal of effort to devise more effective and efficient ways to use the power of large-scale computers to solve the molecular Schrödinger equation. Two qualitatively different types of methods have emerged, called *ab initio* and semiempirical.

In *ab initio* methods,[9–11] all one- and two-electron integrals are retained and all are calculated. Improvements in these methods include more efficient algorithms for solution of integrals; use of more complete forms of the Hamiltonian function, for example including electron correlation terms and use of more accurate basis sets. In most *ab initio* methods, all electrons are explicitly included. In attempts to extend these methods to larger systems, modifications in which inner or core electrons are replaced by an 'effective core potential' have been developed.[12] *Ab initio* methods have the advantage of not requiring any parametrization and therefore can be used for all types of systems. It is also much easier to identify failings of these methods and improve them in a conceptually consistent and even-handed way.

In the second type of approach, called semiempirical, only valence electrons are explicitly included; some integrals are neglected and others are approximated by parameters derived from experiment. A hierarchy of semiempirical methods have been developed which differ mainly in the type of integrals that are neglected and in the manner in which the remaining integrals are evaluated. These methods fall into four broad classes. The first approximations made were to treat the π electrons only. The best known π-electron-only method is the simple Hückel method.[13] The Hückel method, while outdated for modern quantitative work, was useful for examining conjugate coplanar molecules. An improved π-only method, the Pariser–Parr–Pople (PPP) method,[14,15] included electron repulsion in a so-called zero differential overlap approximation and was very useful for calculating spectra of aromatic systems. The next level of approximate methods are those that include explicit consideration of all-valence electrons. Among these methods are three levels, one level in which all two-electron integrals are neglected (EHT,[16] IEHT[17]); the next level which uses the neglect of differential overlap in various approximations (ZDO) to select integrals to be retained (*e.g.* CNDO[18-21], INDO,[22] MINDO[23-26]); and a third level in which the approximation of neglect of differential overlap is improved such as MNDO[27,28] (modified NDO).

Both *ab initio* and semiempirical methods share one factor that affects the reliability of results, the choice of functions to describe the electron spatial distribution. The most common approximation is to represent each individual electron distribution in a molecule as a 'molecular orbital' (MO) comprising a linear combination of atomic orbitals (LCAOs) centered on each nucleus. This description is the so-called LCAO MO representation of electronic wave functions. Such a molecular orbital function is expressed as equation (2)

$$\phi_i = \sum_j C_{ij} x_j \qquad (2)$$

where the C_{ij}'s are the coefficients representing the contribution of the x_j atomic orbitals to the ith MO. The set of atomic orbitals chosen to represent the spatial distribution of electrons in the molecule is called the 'basis set.'[29]

One widely used basis set of atomic orbitals, particularly in semiempirical methods, are the so-called Slater orbitals,[30] which were developed by fitting analytical expressions to numerical atomic wave functions. One function is used for each atomic orbital. Slater-type orbitals (STOs) provide a very good description of atomic orbitals, but integrals involving them are difficult to evaluate computationally. For this reason, representation of atomic orbitals by Gaussian functions was devised, particularly for *ab initio* methods.[9,10] Use of several Gaussian probability functions can approximate a Slater orbital and are much easier to compute since corresponding integrals can be evaluated analytically. Gaussian basis sets are usually described as one of three types: minimal basis set, double zeta basis set and extended basis set.

In a minimal basis set there is only one basis function for each atomic orbital. The expression, STO-nG is an abbreviation for a single Slater-type orbital simulated by n Gaussians.[9] STO-3G is the most commonly used minimal basis set for *ab initio* methods. An STO-3G basis set has only as many orbitals as necessary to accommodate the electrons of the neutral atom. All STO-3G basis sets for any row in the periodic table are equivalent except for the exponents of the Gaussian function. Minimal basis sets are limited because they are unable to expand or contract to fit the molecular environment since the exponent is fixed.[31] Very often semiempirical quantum mechanical methods give more reliable results than minimum basis *ab initio* calculation. Alternatively, more extensive basis sets have been used.[32]

As the name implies, the double zeta basis set has two functions for each atomic orbital and two different exponents (zeta) assigned to each atomic orbital. The coefficients of each orbital can be varied independently during construction of the basis set. Split valence basis sets have only the valence orbitals split in this manner in contrast to the double zeta basis sets which also have two exponents for core electron orbitals. Split valence basis sets are a considerable improvement over minimum basis set and use of a 3-21G basis set is a reasonable compromise for large molecular systems generally yielding good results.

Extended basis sets contain more than two basis functions for each atomic orbital. Either triple zeta sets are used or orbitals of higher azimuthal quantum number, often called polarization functions, are added. This latter improvement means, for example, that the p orbitals are added for hydrogen atoms and d orbitals for first- and second-row heavy atoms. This addition of polarization functions allows a shift of the center of the electron distribution away from the nucleus. As the number of basis functions used to describe a single electron distribution increases, the time required for the calculation also increases.

In addition to electronic properties, quantum chemical methods can be useful in characterizing optimized geometries in a variety of contexts. For example, use of quantum mechanical methods for conformational studies is recommended for polar molecules, or in general when electron redistribution effects in response to changes in conformation can be important. They must also be used for reaction pathway studies.

To obtain such information, whatever the level of approximation used to solve the electronic Schrödinger equations, additional computations must be made. These calculations also use a hierarchy of procedures and methods. The simplest procedure is to calculate the energy and corresponding electron distributions as a function of an explicit stepwise geometry change, for example variations in bond length, bond angle or torsion angle, and identify approximate minima and maxima by this procedure. The next approximation is to perform total geometry optimizations using energy gradient procedures leading to stationary points accessible from the starting geometry.[33] The third level is the verification that a minimum rather than maximum energy conformer is indeed obtained, by calculation of force constants,[33, 34] *i.e.* second derivatives of the energy with respect to nuclear coordinates. An energy minimum will have all positive values for these force constants. With this added capability, chemical reactions can also be described since not only minima, *i.e.* reactants, intermediates and products, but transition states can be identified as stationary points (maxima) with a single negative force constant value.[33]

For studies involving conformational changes, a hierarchy of three quantum mechanical programs are recommended: (i) PCILO for relatively rapid calculations of energy as a function of varying internal rotation angles; (ii) MNDO for total geometry optimization, for comparisons of isomeric and other qualitatively different structures of the same species and for chemical reaction studies; and (iii) *ab initio* methods for the same type of studies as with MNDO, for systems which have not been reliably parametrized in the MNDO method.

18.1.1.2 Semiempirical Quantum Mechanical Methods

Despite the various approximations that are made, semiempirical methods can, by careful parametrization and calibration, provide a reliable framework for calculating molecular properties. In many instances, they are the only practical methods available for the large molecular systems frequently involved in biological processes.

Not all levels of approximation of solutions to the Schrödinger equation are appropriate to use for geometry optimizations or calculations of reaction pathways. In the choice of a particular semiempirical method, there is a balance between four considerations: the nature and size of the molecular system, the properties that one wishes to describe, the reliability of the method for those particular properties and the computer resources available. Given here is a brief review of the various types of semiempirical quantum mechanical methods and their strengths and weaknesses in calculating electronic and conformational properties, as a guide for their use.

18.1.1.2.1 *π-Electron-only methods*

The simplest and earliest description of molecular behavior was limited to π electrons only. This approximation not only ignores core electrons but also assumes complete separation of σ and π valence electron systems, an approximation justified only in aromatic systems. The simple Hückel[13] method provided early insight into the nature and symmetry classifications of π orbitals. The more developed PPP method,[14, 15] which made use of the ZDO approximation for electron repulsion was useful for calculating π electron distributions and electronic spectra.

18.1.1.2.2 *The Hückel method (EHT) and iterative extended Hückel method (IEHT)*

The first all-valence method used was the Hückel method (EHT)[16] and a refined version of it the iterative extended Hückel method (IEHT)[17] in which a one-electron Hamiltonian Schrödinger equation is solved by iteration of charge density. While use of EHT and IEHT for calculating geometries of biological systems has been reported,[35-38] this is not an appropriate use of this one-electron method. If the focus of a problem is on one-electron properties for large molecular systems, then results from an approximate one-electron Hamiltonian method such as IEHT can be used. IEHT is relatively reliable for calculating and comparing electron and spin distributions and one-electron properties calculated from them for a series of similar closed-shell or radical molecules.[39] Because of the relative ease of parametrizing the IEHT Hamiltonian, IEHT has been extended to a large number of atoms in the periodic table and was one of the first methods to provide insight into the chemistry of transition metal complexes.[39] Specifically, it is most useful for calculating net atomic charges, spin densities, dipole and higher moments, polarizabilities, electric field gradients, magnetic moments and magnetic field energies.[40-43] It has also been used extensively to evaluate chemical reactivity parameters,[44] relevant to electrostatic interactions[45] and to covalent bond formation that utilize frontier orbital indices pertaining to the highest occupied (HOMO) and lowest unoccupied (LUMO) molecular orbitals.[46] It can be used, however, only for systems with known or inferred geometry and for which total molecular energies are not relevant properties of interest.

18.1.1.2.3 *CNDO and INDO methods*

The simplest all-valence electron method which includes electron repulsion integrals is the so-called zero differential overlap (ZDO) approximation. It is the simplest level at which total state energies can be obtained and geometry optimizations can be performed. The method also provides all the electronic properties that the IEHT method does. For ground state electronic properties of closed-shell systems, complete neglect of differential overlap (CNDO) performs fairly well.[18-21] For excited electronic states, the CNDO/S method,[47] which includes certain configuration interactions, is appropriate.

The intermediate neglect of differential overlap (INDO) method,[22] which includes one-center repulsion integrals between orbitals on the same atom, is an improvement of CNDO for all properties. An ROHF/INDO CI method developed by Zerner is particularly useful for the calculations of electronic properties for molecular species with unpaired spins and a greater variety of atom types, for example transition metal atoms.[48,49] It is unique among semiempirical methods in providing an even-handed description of states of radical species with different numbers of unpaired spins and also allows calculation of electronic spectra for the widest variety of molecules. Molecular electrostatic potentials using deorthogonalized INDO wave functions are an added property that can be reasonably calculated. INDO is best used for systems with known geometries although recent efforts by Zerner have been directed toward efficient geometry optimization methods including obtaining second derivatives of energies for verification of energy minima.

18.1.1.2.4 *The MINDO/3 and MNDO methods*

These methods, developed by Dewar and his co-workers, are currently the most widely used semiempirical quantum mechanical methods for characterizing large organic molecules. The MINDO/3 method[23-26] is the next in the hierarchy of NDO methods such as CNDO and INDO in which a set of parameters are used to evaluate the one-center repulsion integrals. By contrast, the MNDO method,[27,28] in its various evolving forms, *e.g.* AMPAC, MOPAC and AM1, has a different origin. It is based on a neglect of diatomic differential overlap (NDDO) approximation and eliminates some of the errors associated with MINDO/3. Both of these methods are considerably faster than *ab initio* methods and are not as computationally demanding either in time or resources.

These methods can be used not only for electronic properties but also for conformational analysis and total geometry optimizations. They are also now coupled to capabilities which allow characterization of chemical reactions including structures of reactants, products, intermediates and transition states as well as enthalpies and entropies of reaction and activation.

MINDO/3 calculations have been extensively applied to geometry optimizations, conformational analyses and chemical reactions of many organic molecules and an extensive bibliography exists.[50] Because of the voluminous data on applications of MINDO/3, only some limitations of the method will be mentioned. MINDO/3 has a tendency to overestimate the stability of triple bonds,[51,52] exaggerate dipole moments,[50] overestimate dihedral angles between singly bonded sp^3 atoms,[50] underestimate the stability of aromatic compounds,[50] incorrectly predict some bond lengths and bond angles[50] and sometimes fails in the prediction of geometries of some transition states.[53] In general, MINDO/3 underestimates lone pair repulsion[54] but the most serious difficulty with MINDO/3 is its inability to predict hydrogen bonds, which especially limits its utility for characterizing intermolecular interactions in biological systems.[55,56]

The MNDO method, and in particular its current modified versions found in MOPAC and AM1, corrects many of the problems found with MINDO/3. In general, bond angles are predicted more accurately, lone pair repulsion is handled better, stabilities of double and triple bonds are more reliably reproduced, molecular orbital energies are ordered better and polar molecules (in general) are treated better. One of the major advantages of MNDO over MINDO/3 is the range of molecules that can be studied. It is overall the single best semiempirical quantum mechanical method to use for large organic-type molecules containing essentially first-, second- and third-row atoms, and no transition metals. If one does not exceed the large class of compounds for which this method has now been parametrized, very useful results can be obtained. Dewar has recently published a study in which a comparison of calculations by MINDO/3 and MNDO with *ab initio* results for chemical reactions was made.[57] This study shows that in most cases reported, the MINDO/3 and MNDO results were equal to or better than some large basis set *ab initio* results. They seemed to be systematically better than the minimum basis set STO-3G calculations, often the only possible *ab initio* level for modeling chemical reactions of biologically relevant systems. MNDO methods should not, however, be used for hypervalent sulfur or phosphorus or for substituent geometries on aromatic rings, and in general it overestimates ring planarities. The major problems with MNDO are the underestimation of hydrogen-bonding energies, too negative energies for four-membered rings and activation energies that are too large. These problems can all be, in general, related to a tendency to overestimate repulsion between atoms when they are approximately at their van der Waals distance apart. In an effort to overcome these difficulties, Dewar and his group have introduced a new formalism known as AM1.

AM1 is developed within the NDDO approximation, and the core repulsion function has been modified in two ways. In one strategy, attractive Gaussians were added to compensate for excessive repulsions and in the second, repulsive Gaussians were centered at smaller internuclear separation, which ultimately reduces the repulsion at larger internuclear distances. Currently AM1 is only parametrized for C, H, O and N but can easily be extended to other elements.

Applications of AM1, at this time, have only been carried out in limited studies. AM1 was found to perform well for alkenes and alkynes and, in general, performs better than MNDO for nitrogen-containing compounds. For those compounds containing both nitrogen and oxygen, AM1 appears to offer a real advantage over MNDO but is worse in calculations on some hydrocarbons. The hydrogen-bonding energies and geometries found by AM1 are still deficient. A recently developed version of MNDO called MNDO-H, however, succeeds in improving this defect.[58]

A complementary approach to the use of MNDO methods, however, would be to try to identify relevant model systems small enough to be able to use reliable levels of *ab initio* methods in reaction pathway studies.

18.1.1.2.5 The PCILO (perturbative configuration interaction using localized orbitals) method

This method is different in philosophy from the other semiempirical methods. In the other methods, the molecular orbitals, which are constructed as linear combinations of atomic orbitals (LCAO), are delocalized over the entire molecule. PCILO[59,60] used a chemical-bonding description, a linear combination of orbitals that are 90% localized to a given bond. It is also a perturbative, rather than an iterative, method. The initial ZDO approximation to the wave functions, using localized orbitals, is improved by a perturbation formalism which adds delocalization and correlation effects, usually to second order. At infinite order perturbation, PCILO, in principle, approaches an exact solution of the Hamiltonian model defined by the one- and two-electron integrals.

As PCILO is not iterative, calculations are proportional to the first rather than the fourth power of the number of electrons, and thus it can be used to rapidly calculate energy conformational profiles for large molecules such as peptides and nucleotides. However, PCILO utilizes the CNDO Hamiltonian and thus has many of the same weaknesses as CNDO such as overestimation of charge transfer. Because of bond localization, PCILO will also give different energies with bonds fixed in equivalent positions in aromatic systems. Thus it must be used with caution for aromatic systems and atoms with ambiguous hybridization. Also, this method makes no provision for verification that the energies obtained correspond to true minima, *i.e.* are stable conformers, and is most appropriately used as a first step in a rapid scan of energy conformational space to be followed by use of MNDO or *ab initio* methods to characterize minima.

18.1.2 APPLICATION OF QUANTUM MECHANICAL METHODS IN MECHANISTIC STRUCTURE–ACTIVITY STUDIES OF BIOACTIVE COMPOUNDS

There have been numerous applications of quantum chemistry to biological systems. An older review of the literature can be found in the article by Christoffersen.[38] An extensive bibliography has also been more recently published[61] and a symposium has been devoted to the role of quantum chemistry in biological systems.[62] In this section we present selected examples of the application of quantum mechanical methods to studies of pharmacologically active compounds.

18.1.2.1 Conformational Analysis

One of the most common goals of conformational analysis in structure–activity (SAR) studies is to identify the bioactive conformers of a family of drugs which bind to receptors and modulate their relative affinities. The absolute minimum energy conformer may not be the bioactive conformer since it is possible that an enhanced affinity of a higher energy conformer could more than compensate for the energy required to attain that form. In addition, geometric flexibility may be a necessary feature for agonist activity, leading to an 'induced fit' to the binding site by a drug conformer which is not a minimum energy structure. Thus the calculation of the lowest energy form is just the first step in determining a pharmacophore, *i.e.* the geometry and orientation in which a family of drugs bind to a specific receptor site. In addition, comparison of the energy conformational profiles for a series of analogs including both high and low affinity compounds must be made in order to identify candidate bioactive conformations. In such studies, the analogs with the highest affinity and selectivity in binding to the receptor of interest should be used as templates. The more rigid such a template is, the more useful it will be to define a bioactive form. Two reasonable criteria for the selection of the bioactive conformer are that the energy required to attain this form and the extent of its similarity to the highest affinity analog for a series of compounds should parallel at least their correct qualitative rank order of binding affinities.

In recent years, there has been a general trend towards using molecular mechanics methods for conformational analysis rather than quantum chemical methods. This is due to two main factors: an increase in the quality and accuracy of the empirical potential force fields and a general increase in interest in larger biomolecules, especially peptides and nucleic acids. For those force fields which have been properly parametrized and tested, molecular mechanics results can be as accurate as most quantum chemical methods. In the identification of the bioactive forms of opioid peptides leading to high affinity binding at the μ and δ receptors,[63,64] such methods were used based on ECEPP[65] parameters. However, if there are no parameter types available for particular atoms or if the electron distribution is significantly conformation dependent, as it can be in polar molecules, then quantum mechanical methods, especially semiempirical methods, are a more suitable choice.

Most conformational calculations using semiempirical quantum mechanical methods have been carried out using the PCILO or MNDO method, although a few conformational calculations with CNDO/2[66,67] and INDO[68,69] are still being reported. Small model systems can be studied by *ab initio* methods.

18.1.2.1.1 Use of the PCILO method

For large molecules or molecules with many degrees of rotational freedom, the PCILO method is an appropriate one for rapid scans of energy as a function of nested rotations about specific bonds. While it does not contain procedures that allow verification that the rotamers investigated correspond to an energy minimum, if used for a series of closely related analogs, it can yield reliable rank orders of rotational energy minima and barriers to rotation.

PCILO has been extensively used by the Pullman group to study conformational behavior of nucleic acids, nucleotides, amino acids and peptides.[70] More recently PCILO has been used in studies of nucleotide antibiotics,[71,72] histamine antagonists,[73] anesthetics,[74,75] hormones,[76] α-adrenergics,[77] β-adrenergics,[78] saccharides[79] and enzymes.[80]

A typical example of the use of the PCILO method is a study of local anesthetics of the procaine (**1**) type.[81] These compounds have five rotatable bonds and in order to reduce computational time, full incremental nested rotations were not performed. Instead local minima were found in a succession of variations of three of the angles while two were held fixed in extended values of 180°. The results of these calculations yielded a low energy structure consistent with the observed X-ray

(**1**) Procaine
(**2**) Parpanit

structure.[82] However, there were other conformers with comparable or even lower energy not found in previous theoretical studies of procaine (**1**) and parpanit (**2**).[82,83,84] These results both validate the reliability of the method and indicate that the X-ray structure need not be the bioactive form.

In a second phase of this study the conformational behavior of acetylcaine in the presence of a model receptor site was investigated. The results found that one of these three minima originally found *in vacuo* disappears. Conformational maps revealed two other effects of interaction with the anionic receptor site: the rotational barrier between the *trans–gauche* and fully extended conformation was decreased by complex formation and the ethyl side chain rotations were more hindered leading to increased barriers to their rotation. These PCILO calculations allowed the identification of a reasonable bioactive conformer.

18.1.2.1.2 Use of MNDO methods

The MNDO method is appropriate for determining total geometry optimizations for a wide variety of organic molecules. It can be effectively used to compare different geometric isomers and qualitatively different conformers as well as for barriers and minima in a scan of rotations about specific bonds. These rotational energies can be calculated as single-point or optimized energies for specified values of rotational angles and bonds.

MNDO has been used in a variety of recent conformation calculations on angiotensin-converting enzyme inhibitors,[85] 5-HT receptor agonists,[86] tautomers,[87] vitamin C radicals,[88] nicotine receptors,[89] β-carboline analogs,[90] β-adrenergics,[91] benzodiazepines[92,93] and anesthetics.[94]

18.1.2.1.3 Use of the CNDO/2 method

Conformational calculations can be used not only for determining minimum energy structures but also to determine a common pharmacophore for a family of molecules. This use is illustrated by the example of the use of CNDO/2 to determine a common receptor conformer for a series of antischistosomiasis agents.[95]

Lucanthone (**3**) is an antischistosomal drug which is probably hydroxylated *in vivo* to its active form, a 4-hydroxymethyl metabolite. Congeners of these compounds are thought to act by blocking acetylcholine-binding sites of the schistosomes. The study reported included lucanthone, related congeners and their metabolites — some with apparently dissimilar structures. When compared to acetylcholine (ACh), common structural features emerged. The protonated ethylenediamine chain of the lucanthone analogs is similar to the choline fragment of ACh and the benzene ring replaces the planar acetoxy group of ACh. Using the hypothesis that these commonalities are the significant ones, it was proposed that comparable functional groups in this series of compounds could assume a common binding mode at the receptor. To investigate this possibility, conformational calculations were carried out using CNDO/2. Calculations were done *in vacuo* and also in an aqueous environment simulated by a virtual charge method.

(**3**) Lucanthone

The results of the calculations strongly indicated a preference for an ethylenediamine chain configuration in these compounds that is similar to that of acetylcholine. Also, all molecules have a common crucial N–N distance of about 3 Å, and the distance from the center of the ring to the protonated nitrogen is also fairly similar. The results found for the solution studies were similar and, in fact, the inclusion of solvent did not alter the conformational preference but only attenuated the nonbonded interactions. The results of this study allowed the determination of a general pharmacophoric pattern for antischistosomal activity.

18.1.2.1.4 Use of ab initio methods

Another study, which illustrates the search for a common pharmacophore, but which explores both electronic and conformational properties and a variety of methods including *ab initio*, is one of muscarinic agonists.[96] Most potent muscarinic agonists have a cationic head group, an ester or ether oxygen and a terminal alkyl group or its equivalent. With the exception of the acetoxyquinuclidine structure, which is protonated, all of these compounds have a tertiary amine. In addition, stereo-isomerism is important in determining the biological activity, with the L(+) isomer (2S,3R,5S) of muscarine being the only potent compound. While there have been many models postulated for these muscarinic pharmacophores, no definite conclusions have been reached.

In an attempt to define the receptor pharmacophore, Schulman *et al.*[96] adopted the following strategy. They assumed a direct interaction of the cationic head group with an anionic receptor site in a specific orientation, namely a receptor oxygen along the threefold axis of the tetramethyl-ammonium group at a distance of 3 Å away from the nitrogen. Using a molecular electrostatic potential generated by electronic densities from STO-3G calculations, two minima were found in the vicinity of the ester and ether oxygens. Using the position of these two minima and geometric constraints, two criteria were used to find all possible common orientations of the muscarinic agonists with the receptor for all candidate conformations 3–4 kcal mol^{-1} (1 kcal = 4.18 kJ) above the global minimum. Further refinement of the model was based on known SAR data. Relaxed geometries were obtained by the MM2 molecular mechanics method and the energy of each candidate conformer was subsequently recalculated as a single point by the STO-3G *ab initio* method.

The results of this study allowed the identification of a common pharmacophore. Dihedral angles were found for low energy conformers of each active compound which satisfied the pharmacophore model. In addition, these data were able to explain the lack of activity for several of the stereo-isomeric pairs. It is interesting to note that the known X-ray structures were considerably different than those matching the pharmacophore for some compounds.

18.1.2.2 Electronic Properties

18.1.2.2.1 Net atomic charges

Charge distributions, expressed in terms of net atomic charges associated with individual atoms, are useful for making comparisons between molecules and for calculation of electrostatic inter-actions. Partial atomic charges are also used as parameters in electrostatic interaction terms in molecular mechanics programs. While the concept of atomic charge seems simple, net atomic charge is not a physical quantity and is not directly measurable.

There have been numerous theoretical constructs applied to calculations of atomic charges. The standard method used in most molecular orbital programs is the Mulliken population analysis.[97] In this method the atomic charge (q_a) is defined by equation (3)

$$q_a = Z_a - \sum_{\mu \varepsilon A} (P \cdot S) \tag{3}$$

where Z is the nuclear charge, S is the overlap matrix and P is the net atomic population of a given atomic orbital. In Mulliken's scheme, the charge is divided evenly between the two nuclei con-stituting the bond even if their electronegativities are quite dissimilar.

One of the major problems with the Mulliken population analysis is that the result is very sensitive to the basis set used. This is true for both semiempirical and *ab initio* methods. The comparisons of changes in net charges when substituents are altered are meaningful only if they are made for closely related compounds for the same basis set or method. For *ab initio* methods the calculated populations are not invariant to the basis set.[94] Larger basis sets do not necessarily provide better results and instability of atomic charges can be observed.[98] Values obtained from minimal basis sets are usually more meaningful. The IEHT, CNDO/2 and INDO methods usually give reliable estimates of charge distributions while EHT overestimates them.[99]

Quantum chemical programs are widely used to obtain partial charges for molecular mechanics programs. The source of the charges is important because different sets of net atomic charges from different methods will give different rank orders of conformational energies if electrostatic energy plays a significant role. For large biomolecules, for example peptides and polynucleotides, the determination of partial atomic charge is a major problem. One method[100] which has been used is similar to that of Cox and Williams.[101] Electrostatic potentials are calculated for smaller fragments using *ab initio* methods, usually STO-3G, and then individual atomic charges which reproduce the molecular electrostatic potential map are found by an iterative process. While this is, in principle, a reliable procedure for obtaining net charges in empirical energy programs, it is computationally time consuming and the manner in which the individual fragments are smoothed at juncture points introduces ambiguities.

Net atomic charges alone are not particularly useful for gaining insight into a biomolecular event. They are better utilized in conjunction with QSAR studies. Several examples can be found in refs. 102 and 103.

One study in which a direct correlation between formal atomic charge and biological activity was exhibited was that of a series of thiazide diuretics.[104] CNDO/2 calculations were reported for hydrochlorothiazide (**4**) and related molecules. Analogs with substituent variations in the sixth and seventh positions were included in the study. Although the study used a limited data set, the strongest correlation in a regression analysis with observed activity was found with the net atomic charge at the seventh position C_7 (equation 4). In equation (4), A is diuretic activity in the dog, π is the hydrophobic parameter of the 6-substituent, VW is the van der Waals volume of the 6-substituent and F_{C7} is the formal charge calculated for the 7-position of the benzothiadiazine ring. Some effect of the 7-substituent in lowering the energy of LUMO was also found which could affect the electron-accepting ability of the compound. A hydrophobic parameter and a van der Waals term (VW) for substituents at the sixth position contributed less significantly to observed activity.

(**4**) Hydrochlorothiazide

$$\ln A = 0.2695 \times 10^{-3}\pi + 7.62535\text{VW} + 15.7681 F_{C7} + 0.98502 \qquad (4)$$
$$n = 13, \quad r = 0.98$$

18.1.2.2.2 *Molecular electrostatic potentials (MEP)*

A more reliable indicator of electrostatic reactivity than net atomic charges is the molecular electrostatic potential (MEP) expressed as equipotential contours and positions of minimum potentials. The nuclear and electronic charge distribution of the molecule creates an electrostatic potential, which will interact with a point charge or the electron density of another molecule. In contrast to atomic charges, the electrostatic potential is a physical quantity and can be determined experimentally, for example by scattering experiments. Electrostatic fields calculated by theoretical methods have been found to agree well with such experimental results.

The electrostatic potential at any point r in space can be expressed by equation (5)

$$V(r) = \sum_A \frac{Z_A}{|R_A - r|} - \sum \frac{\rho(r')\mathrm{d}r'}{|r' - r|} \qquad (5)$$

where Z is the charge at nucleus A located at R, $\rho(r)$ is the electronic density function. The first term represents the nuclear contribution and the second term is the electronic contribution. The electron density used to calculate the potential can be obtained from either an *ab initio* or semiempirical wave function and is approximate, as is the potential. Those regions that have high nuclear contributions will yield positive MEP, corresponding to repulsive interaction energies with point positive charges, and those with higher electron contributions will yield negative MEP, corresponding to attractive energies. The drug–receptor-type recognition process should be one in which regions with opposite MEPs match.

Electrostatic potentials have been widely used to investigate numerous biological systems such as the reactive properties of nucleic acid bases,[105-107] interactions of serotonin and LSD with the 5-HT receptor,[108-110] clozapine analogs,[111] cholinesterase inhibitors,[112] carcinogens,[113] dopamine compounds[114,115] and inhibitors of enzymes such as carboxypeptidase,[116,117] dihydrofolate reductase,[118,119] serine proteases[120-121] and carbonic anhydrase.[122]

One of the classic examples of the use of MEPs was in the identification of an interaction pharmacophore for serotonin and LSD.[108-110] Serotonin (5-HT) interacts with receptor sites in both brain and periphery. LSD also interacts with different relative affinities at the same receptors, despite its apparent difference in chemical structure. In searching for molecular discriminants of relative affinity for a series of tryptamine derivatives, it was found that the electrostatic potentials have two characteristic minima above and below the indole portion of the molecules. Deviations of an 'orientation vector', defined as the vector connecting the two minima along the potential gradient, from that in 5-HT, were correlated with the relative binding affinities of other tryptamine derivatives. LSD was found to have a similar electrostatic potential map and thus predicted to interact with the 5-HT receptor in the same manner as serotonin. The C(12)—C(13) double bond of LSD produced a minimum in the potential that behaved similarly to the OH group in 5-HT. This hypothesis was supported experimentally by the suggested synthesis of an LSD-like compound that lacked the C(12)—C(13) double bond and which had, as predicted, a much lower binding affinity.

Another example of the use of MEP was in a study of β-adrenergic compounds.[123] In this study, reactivities toward electrophiles were calculated from electrostatic potentials as the optimum energy of interaction of the entire molecule with a point positive charge. The β-adrenergic activity of a series of analogs was found to decrease in the same order as this index. This result suggests that an electrophilic group of the adrenergic receptor interacts with the ring portion of the compounds.

In another study of a series of dopaminergic and ergoline compounds,[115] comparisons of electrostatic potentials also suggested that the aromatic portion of the molecules interact with a common receptor site. It is worth noting that, in this study, superposition of analogs based on simple steric overlap was not informative and the MEP was a better criterion for defining the manner in which the compounds could be superimposed at a receptor. The important conclusion is that the electrostatic potential presented to the receptor by the drug rather than its steric orientation was the determining feature in relative modes of binding of the different analogs.

In some instances the electrostatic potential map calculated for the isolated drug does not provide the most relevant description. This is particularly true for charged species in which the charge, whether positive or negative, will overwhelm the potential generated. In such cases, a plausible working hypothesis is that an initial interaction occurs between the charged group and a specific complimentary subsite of the receptor. The MEP can then be calculated for a model drug–receptor complex which is neutral. The results would then indicate the remaining regions of the drug important for receptor interactions. An example of this approach is reported in studies of opiate narcotics.[124] When the protonated form of morphine is allowed to interact with a model anionic receptor site, a very large negative electrostatic potential in the region of the phenolic ring is generated with a minimum of $-87.5 \, \text{kcal} \, \text{mol}^{-1}$. This negative region is diminished and the minimum reduced to $-46 \, \text{kcal} \, \text{mol}^{-1}$ for the isolated protonated opiate. The enhanced electron density in this region upon initial cation–anion interaction could play a significant role in activation of receptors by opiates. The results also demonstrate that electronic effects of drug–receptor interactions need not be localized only in those areas of the drug where initial receptor interactions occur.

Some molecules of biological interest are too large for quantum mechanical calculations of MEPs to be practical. In a study of a series of clozapine neuroleptic agents (**5–8**), Kollman and co-workers[111] investigated three alternative methods of calculating MEPs from atomic charges. The first method used CNDO-generated partial charges. The second used STO-3G charges. The third

(**5**) X = Cl, Y = H Clozapine
(**6**) X = H, Y = Cl
(**7**) X = Y = Cl
(**8**) X = Y = H

was a least-squares fit of atomic charges such that the potential generated was that which optimally reproduced the *ab initio* results. Since the molecules were large, the latter step was performed on fragments and the fragments were fitted together to conserve the overall total charge. Using the latter method, the electrostatic potential surfaces were found to correlate with the biological activity. The two Y = Cl compounds (**6**) and (**7**) with similar extrapyrimidal and anticholinergic activities had similar MEPs. These differed from clozapine (**5**) and the dechloro compound (**8**), which themselves had similar pharmacological profiles, no extrapyrimidal side effects and qualitatively similar MEPs. These investigators also found that the results using CNDO and STO-3G point charges were similar, although, in general, point charge models do not directly provide reliable electrostatic maps.

18.1.2.2.3 *Properties related to receptor activation*

In order for a biological response to be elicited by a drug–receptor complex, recognition must be followed by initiation of a sequence of biochemical events leading to such a response, *i.e.* ligand binding must lead to receptor activation.[125] Agonists, by definition, should cause this response, while antagonists are those compounds which are recognized by the receptor but do not activate it.

An example of the use of quantum mechanics to explore receptor activation is the investigation by Weinstein and co-workers of proton transfer between hydrogen-bonded groups as a model for activation of the histamine receptor.[126–128] Structure–activity studies on H_2 receptor agonists have shown that imidazole ring tautomerism is required for activity.[129] This inference led to the proposal of a proton-relay receptor activation mechanism, involving a tautomeric shift from N(3)—H to N(1)—H caused by a neutralization of the cationic side chain interacting with a negative region in the receptor.

In order to investigate the validity of this hypothesis, the following calculations were carried out. Geometries for the cationic and neutral N(1)—H and N(3)—H tautomers were optimized with STO-3G. MEPs were calculated using the coreless Hartree–Fock pseudopotential and LP-3G basis set to determine the regions of most electron density. The results show that there is a negative potential in the region of the unprotonated ring nitrogen and a positive potential around the rest of the molecule. The N(3) position in the N(1)—H tautomer of the cation had a higher proton affinity than the N(1) position of the N(3)—H tautomer. Thus the N(3)—H tautomer should be the preferable cationic form presumed to interact with the receptor.

(9)

(10)

(11)

(12)

More recently, another receptor activation mechanism has been proposed by the same workers,[130] as a model of 5-HT receptor activation. The interaction of 5-HT with a receptor model containing a hydrogen-bonded imidazolium–ammonium cation complex was calculated. The indole portion of 5-HT was found to stack with the imidazolium–ammonium region and the resulting alteration in the electric field facilitated proton transfer in the direction from imidazolium to ammonia.

Another example in which chemically derived parameters were able to explain modulation of receptor activation is a study of the antiinflammatory action of NSAID compounds (nonsteroidal antiinflammatory drugs).[131] This study is a good example of the combined use of quantum chemical indices and QSAR.

In this study, molecular orbital indices such as E(HOMO), E(LUMO), frontier orbital charge and HOMO and LUMO charges were calculated for simple congeners of salicylic acid and benzoic acids. A variety of regression equations were calculated using combinations of these indices and a consistent measure of biological activity. The best correlations were found using the two equations, (6) and (7) of which equation (6) contains $\Delta E[E(\text{HOMO}) - E(\text{LUMO})]$ and equation (7) contains the frontier orbital density on the phenolic hydroxy ($f_{(\text{OH})}$) and noncarboxylic acid substituents ($f_{(\text{R})}$). The correlation of biological activity with these properties was taken to imply that upon binding, a charge transfer takes place from the receptor to the acidic function of the drug and from the aromatic ring of the drug to the receptor. Similarly, good correlations with activity were found for regression equations which included dipole moments and atomic charges for those regions of the molecules considered to be involved in the interaction with the receptor.

$$pI_{50} = 0.394\Delta E + 13.97 \tag{6}$$

$$n = 16, \quad r = 0.92, \quad F = 76.0, \quad s = 0.21$$

$$pI_{50} = 1.090F_t + 2.430 \tag{7}$$

$$F_t = [(f_{(\text{OH})}) + (f_{(\text{R})})]/E_{(\text{HOMO})}$$

$$n = 16, \quad r = 0.87, \quad F = 42.0, \quad s = 0.26$$

The poor correlation between activity and ΔE alone, which did not even allow separation of active and inactive species, suggested that additional steps besides charge transfer are involved in receptor recognition. To investigate this possibility, MEP maps for several congeners were calculated. It was found that, in a manner similar to the serotonin (5-HT) compounds mentioned above, the orientation vector could be correlated with relative potency.

Using the results of these studies, the authors conclude that a two-step process is involved in receptor recognition. The proposed first step is recognition and stabilization due to electrostatic orientation, which aligns the HOMOs and LUMOs in the receptor. This alignment is proposed to increase the ease with which the second proposed step, charge transfer, can occur. This mechanistic viewpoint then allows an understanding of active and inactive compounds.

18.1.2.2.4 Use of electronic parameters for chemical reactivity

Chemicals which initiate processes leading to cancer are a particularly appropriate class to address by mechanistic structure–activity studies using quantum mechanical methods. In many cases metabolites rather than parent compounds are the active carcinogens, and these ultimate carcinogens are often electrophilic and form covalent adducts with electron rich centers of DNA bases in the initiating step thought to lead to tumor formation. Thus electronic properties which are a measure of the extent of transformation of parent compounds to ultimate carcinogens and of the ability of these active species to covalently bind to target biopolymers are among the most likely modulators of relative activity. The identification and calculation of such properties can only be accomplished by quantum mechanical methods. Examples of studies of classes of chemical carcinogens known to require transformation to active carcinogens are the halohydrocarbons,[132,133] polycyclic aromatic hydrocarbons (PAHs)[134] and amines (PAAs).[135,136] A brief description of the nature of these studies is given here.

For these classes of carcinogens, chemical reactivity properties that are reliable indicators of the relative extent of transformation of parent compounds to carcinogenic species were considered to be one important type of modulator of their relative carcinogenic activity. Studies of the reaction mechanism of these transformations allowed the selection of mechanistically relevant reactivity properties which were then calculated by quantum mechanical methods. The enzymes most frequently implicated in initial transformation of compounds to active carcinogens are the metabolizing heme proteins, called cytochrome *P*-450s. In cytochrome *P*-450-mediated oxidations, the enzyme transfers an electrophilic oxygen atom to the substrate leading to aliphatic and aromatic N- and C-hydroxylations, which are possible first steps in either formation of the active carcinogen or in detoxification pathways.

The extent of N-hydroxylation was monitored by the calculated values of the nucleophilicity of the nitrogen atom (S_N); of aromatic ring-hydroxylation and epoxidations by the nucleophilicity of the ring carbons (S_C), and by a π bond nucleophilic index $R_{AB}(\pi)$; and of aliphatic hydroxylation by the stability of radicals formed by H abstraction deduced as the first step in hydroxylation.

In addition to these potential modulators of the extent of formation of initial metabolic products, two other types of properties were calculated. One was the stability of the putative ultimate carcinogens, the identity of which was deduced from known DNA adducts. The other was both covalent and electrostatic indices of the relative electrophilicity[41] of these putative carcinogens. These properties were used as a measure of the extent of adduct formation with DNA. For example, one indicator of covalent electrophilicity used was the calculated electron density in the LUMO of the atoms which form covalent bonds to DNA, since these atoms act as electron acceptors in incipient reaction with target DNA base nucleophiles. One measure of the extent of initial electrostatic interaction with DNA prior to adduct formation used was the net calculated charge on each atom.

For each class of chemicals studied, molecular indicators of the presence or absence of carcinogenic activity were identified and calculated which could be used in screening of untested compounds. In addition, further insight into the processes, species and adducts contributing to carcinogenic activities in each class of carcinogens was obtained. For the 44 PAHs studied, the relative stability of the putative active form was the most useful indicator of relative carcinogenic activity. For the PAAs, indices related to the extent of formation of different metabolites and adduct formation were both important. For the chlorohydrocarbons, adduct formation appeared to be most important in determining relative carcinogenic activities.

18.1.2.2.5   *Characterization of reaction mechanisms and thermodynamics and kinetics of reaction*

Quantum chemical methods must be used to characterize chemical and biochemical reactions. A reaction is characterized by stationary points: minima corresponding to the reactants, intermediates and products, and saddle points between them corresponding to transition states. For all stable species, the force constants (*i.e.* a diagonalized second derivative matrix of the energy) all have positive values and any change in the geometry will cause an increase in the energy. A transition state, on the other hand, is characterized by a force constant matrix in which the force constant for one mode has a negative value. A displacement along this normal mode, which can be considered the 'reaction coordinate', will cause a decrease in energy and lead to the formation of a product or stable intermediate. Characterization of the geometries of reactants, intermediates, transition states and products constitute determination of a reaction mechanism.

Characterization of chemical reaction mechanisms by quantum chemical methods is an excellent example of symbiosis between experiment and theory. Since transition states cannot be isolated and characterized by direct experimental observations, experimental studies of reaction mechanisms are based on the determination of observable properties from which reaction mechanisms, *i.e.* transition states and reactive intermediates, are inferred. Among the experimental properties commonly used to infer mechanisms are primary and secondary kinetic isotope effects, retention or loss of configuration and product distribution. The techniques of quantum chemistry are complementary to such experimental efforts since they allow the explicit characterization of transition states and intermediates as well as reactants and products. The calculation of free energy differences between these species allows the determination of enthalpies and entropies of activation and reaction. From such explicit characterizations, a number of observable properties can be deduced such as retention of configuration, relative reaction rates for competing reactions and kinetic isotope effects on them, and an explicit mechanism can be chosen which best fits and explains experimental observations.

Semiempirical methods can be effectively used for such studies provided their ability to predict relative behavior can be assessed. Such reliability can be estimated by comparison of one of a series of similar compounds being studied with known experimental values. For example, using the known experimental value for free energies of reaction and activation for one of a series of related reactions, absolute rate constants and kinetic isotope effects can sometimes be estimated by scaling the calculated results.

The usefulness of semiempirical methods to obtain insights into enzyme mechanisms is illustrated by a series of studies of a model system for cytochrome *P*-450 aromatic,[137] aliphatic C-[138-140] and N-oxidations.[141]

The first goal of these studies was to try to distinguish between closed-shell and radical mechanisms of *P*-450 hydroxylations and epoxidations. In general, closed-shell mechanisms were found to be concerted, while radical mechanisms for hydroxylation and epoxidation both appear to proceed *via* intermediates in a two-step process. Comparisons of theoretical results with experimental observations of such reactions led to the conclusion that the radical mechanism was more consistent with experiment. Further studies of radical mechanisms allowed more detailed descriptions of their consequences. One such study of oxidation of aliphatic hydrocarbons, exemplified by model *P*-450 oxidations of propene[138] as a substrate is described below.

In this study, propene was used as a prototype asymmetric alkene substrate and five competing model *P*-450 oxidation reactions of it were investigated. These reactions were: oxygen addition to each carbon atom of the ethylene bond leading to epoxide formation and H abstraction by the oxygen radical from allyl, 2-propene and 1-propene carbon atoms leading to different hydroxylated products. All calculations were carried out using the MNDO method. Reactants, products and transition state were identified and all geometries optimized. All stationary points were characterized by means of their force constant matrix. Scaling factors, obtained from experimental studies on methane and ethylene, were used to scale the reaction rates.

α - addition

$$CH_2 = CH - Me \quad + \quad O(^3P) \quad \longrightarrow \quad \overset{\overset{\textstyle O\bullet}{\textstyle |}}{\bullet CH_2 = CH - Me} \qquad (8)$$
$$(\mathbf{13})$$

β - addition

$$CH_2 = CH - Me \quad + \quad O(^3P) \quad \longrightarrow \quad \overset{\overset{\textstyle O\bullet}{\textstyle |}}{CH_2 = \overset{\bullet}{C}H - Me} \qquad (9)$$
$$(\mathbf{14})$$

allyl abstraction

$$CH_2 = CH - Me \quad + \quad O(^3P) \quad \longrightarrow \quad CH_2 = CH - \overset{\bullet}{C}H_2 \quad + \quad \overset{\bullet}{O}H \quad (10)$$

2 - propenyl abstraction

$$CH_2 = CH - Me \quad + \quad O(^3P) \quad \longrightarrow \quad CH_2 = \overset{\bullet}{C} - Me \quad + \quad \overset{\bullet}{O}H \quad (11)$$

1 - propenyl abstraction

$$CH_2 = CH - Me \quad + \quad O(^3P) \quad \longrightarrow \quad \bullet CH = CH - Me \quad + \quad \overset{\bullet}{O}H \quad (12)$$

The results of this study found that the first step in epoxide formation is asymmetric addition of oxygen to either β or α carbon, leading to a biradical tetrahedral intermediate (equations 8 and 9). The β-oxytetrahedral intermediate (**14**) was approximately 9 kcal mol^{-1} more stable than the α-oxy intermediate (**13**).

Comparing the three H abstractions (equations 10–12), the first step in hydroxylation, allylic abstraction, was exothermic, while those for the vinylic abstractions (2-propenyl and 1-propenyl abstractions) were endothermic and the activation energy barrier was higher.

The overall conclusions from these studies were: (i) epoxide formation proceeds by oxygen addition to the β carbon (equation 9); (ii) epoxidation would occur much faster than any hydrogen abstractions leading to alcohols; and (iii) among the hydroxylation reactions, allylic carbon atoms were preferred to vinylic carbon atoms.

While these studies were done on a model system for hydrocarbon oxidation by cytochrome *P*-450, the results could be used to explain experimental observations on *in vitro* *P*-450 epoxidation of the styrenes and for suicide inactivation of *P*-450 by ethylene. This study demonstrates the utility of studying model systems by quantum mechanical methods in order to gain insight into more complex biological enzyme processes.

18.1.3 REFERENCES

1. Quantum Chemistry Program Exchange, Indiana University, Bloomington, IN, USA.
2. N. L. Allinger, *Adv. Phys. Org. Chem.*, 1976, **13**, 1.
3. S. Lifson and A. Warshel, *J. Chem. Phys.*, 1968, **49**, 5116.
4. S. J. Weiner, P. A. Kollman, D. A. Case, U. C. Singh, C. Ghio, G. Alagona, S. Profeta, Jr., and P. Weiner, *J. Am. Chem. Soc.*, 1984, **106**, 765.
5. M. Born and R. Oppenheimer, *Ann. Phys. (Leipzig)*, 1927, **84**, 457.
6. D. R. Hartree, 'The Calculation of Atomic Structures', Wiley, New York, 1957.
7. C. C. J. Roothaan, *Rev. Mod. Phys.*, 1960, **32**, 179.
8. C. C. J. Roothaan, *Rev. Mod. Phys.*, 1951, **23**, 69.
9. W. J. Hehre, R. F. Stewart and J. A. Pople, *J. Chem. Phys.*, 1969, **51**, 2657.

10. W. J. Hehre, R. Ditchfield, R. F. Stewart and J. A. Pople, *J. Chem. Phys.*, 1970, **52**, 2769.
11. R. Ditchfield, W. J. Hehre and J. A. Pople, *J. Chem. Phys.*, 1971, **54**, 724.
12. S. Topiol, R. Osman and H. Weinstein, *Ann. N. Y. Acad. Sci.*, 1981, **367**, 17.
13. E. Hückel, *Z. Phys.*, 1931, **70**, 204; 1932, **76**, 628.
14. R. Pariser and R. G. Parr, *J. Chem. Phys.*, 1953, **21**, 466, 767.
15. J. A. Pople, *Trans. Faraday Soc.*, 1953, **49**, 1375.
16. R. Hoffmann, *J. Chem. Phys.*, 1963, **39**, 1397.
17. R. Daudel and C. Sandorfy, 'Semiempirical Wave Mechanical Calculations on Polyatomic Molecules', Yale University Press, New Haven, 1971.
18. J. A. Pople, D. P. Santry and G. A. Segal, *J. Chem. Phys.*, 1965, **43**, S129.
19. J. A. Pople and G. A. Segal, *J. Chem. Phys.*, 1966, **44**, 3289.
20. J. A. Pople and M. Gordon, *J. Am. Chem. Soc.*, 1967, **89**, 4253.
21. M. S. Gordon, *J. Am. Chem. Soc.*, 1969, **91**, 3122.
22. J. A. Pople, D. L. Beveridge and P. A. Dobosh, *J. Chem. Phys.*, 1967, **47**, 2026.
23. M. J. S. Dewar and E. Haselbach, *J. Am. Chem. Soc.*, 1970, **92**, 590.
24. N. Bodor, M. J. S. Dewar and D. H. Lo, *J. Am. Chem. Soc.*, 1972, **94**, 5303.
25. M. J. S. Dewar and D. H. Lo, *J. Am. Chem. Soc.*, 1972, **94**, 5296.
26. R. C. Bingham, M. J. S. Dewar and D. H. Lo, *J. Am. Chem. Soc.*, 1975, **97**, 1285.
27. M. J. S. Dewar and W. Thiel, *J. Am. Chem. Soc.*, 1977, **99**, 4899; 4907.
28. M. J. S. Dewar and M. L. McKee, *J. Am. Chem. Soc.*, 1977, **99**, 5231.
29. J. A. Pople and D. L. Beveridge, in 'Approximate Molecular Orbital Theory', McGraw Hill, New York, 1970.
30. J. C. Slater, *Phys. Rev.*, 1930, **36**, 57.
31. T. Clark, 'A Handbook of Computational Chemistry', Wiley Interscience, New York, 1985, p. 236.
32. W. J. Hehre, L. Radom and P. V. R. Schleyer, 'Ab Initio Molecular Orbital Theory', Wiley Interscience, New York, 1986.
33. M. C. Flanigan, A. Kormornicki and J. W. McIver, Jr., in 'Modern Theoretical Chemistry', ed. G. A. Segal, Plenum Press, New York, 1977, vol. 8, p. 1.
34. E. B. Wilson, Jr., J. C. Decius and P. C. Cross, 'Molecular Vibrations: The Theory of Infrared and Raman Vibrational Spectra', Dover Publications, New York, 1980.
35. G. H. Loew and S. Chang, *Theor. Chim. Acta*, 1972, **27**, 273.
36. L. B. Kier, *J. Pharmacol. Exp. Ther.*, 1968, **164**, 75.
37. J. M. George, L. B. Kier and J. R. Hoyland, *Mol. Pharmacol.*, 1971, **7**, 328.
38. R. E. Christoffersen, in 'Quantum Mechanics of Molecular Conformations', ed. B. Pullman, Wiley, New York, 1976, p. 194.
39. M. Zerner, M. Gouterman and H. Kobayashi, *Theor. Chim. Acta*, 1966, **6**, 363.
40. G. H. Loew, C. J. Kert, L. M. Hjelmeland and R. F. Kirchner, *J. Am. Chem. Soc.*, 1977, **99**, 3534.
41. G. H. Loew and R. F. Kirchner, *J. Am. Chem. Soc.*, 1975, **97** (25), 7388.
42. R. F. Kirchner and G. H. Loew, *J. Am. Chem. Soc.*, 1977, **99**, 4639.
43. G. H. Loew and R. F. Kirchner, *Int. J. Quantum Chem., Quantum Biol., Symp.*, 1978, **5**, 403.
44. G. Klopman and R. F. Hudson, *Theor. Chim. Acta*, 1967, **8**, 165.
45. R. Rein, N. Fukuda, H. Win, G. A. Clarke and F. E. Harris, *J. Chem. Phys.*, 1966, **45**, 4743.
46. K. Fukui, *Angew. Chem., Int. Ed. Engl.*, 1982, **21**, 801.
47. J. Del Bene and H. H. Jaffé, *J. Chem. Phys.*, 1968, **48**, 1807.
48. A. D. Bacon and M. C. Zerner, *Theor. Chim. Acta*, 1979, **53**, 21.
49. M. C. Zerner, G. H. Loew, R. F. Kirchner and U. T. Muller-Westerhoff, *J. Am. Chem. Soc.*, 1980, **102**, 589.
50. D. F. V. Lewis, *Chem. Rev.*, 1986, **86**, 1111.
51. R. C. Bingham, M. J. S. Dewar and D. H. Lo, *J. Am. Chem. Soc.*, 1975, **97**, 1294, 1307.
52. M. J. S. Dewar, D. H. Lo and C. A. Ramsden, *J. Am. Chem. Soc.*, 1975, **97**, 1311.
53. P. R. Andrews and R. C. Haddon, *Aust. J. Chem.*, 1979, **32**, 1921.
54. D. B. Boyd, *J. Phys. Chem.*, 1978, **82**, 1407.
55. T. J. Zielinski, D. L. Breen and R. Rein, *J. Am. Chem. Soc.*, 1978, **100**, 6266.
56. G. Klopman, P. Andreozzi, A. J. Hopfinger, O. Kikuchi and M. J. S. Dewar, *J. Am. Chem. Soc.*, 1978, **100**, 6267.
57. M. J. S. Dewar and D. M. Storch, *J. Am. Chem. Soc.*, 1985, **107**, 3898.
58. A. Goldblum, *J. Comput. Chem.*, 1987, **8**, 835.
59. S. Diner, J. P. Malrieu and P. Claverie, *Theor. Chim. Acta*, 1969, **13**, 1.
60. J. P. Malrieu, in 'Modern Theoretical Chemistry', ed G. A. Segal, Plenum Press, New York, 1977, vol. 7, p. 69.
61. J. L. Taylor and J. C. Durant, *J. Mol. Graphics*, 1985, **3**, 158.
62. H. Weinstein, J. P. Green and R. Osman, *Ann. N.Y. Acad. Sci.*, 1981, **367**, 434.
63. G. Loew, C. Keys, B. Luke, W. Polgar and L. Toll, *Mol. Pharmacol.*, 1986, **29**, 546.
64. C. Keys, P. Payne, P. Amsterdam, L. Toll and G. Loew, *Mol. Pharmacol.*, 1988, **33**, 528.
65. F. A. Momany, R. F. McGuire, A. W. Burgess and H. A. Scheraga, *J. Phys. Chem.*, 1975, **79**, 2361.
66. M. Bohl, K. Ponsold and G. Reck, *J. Steroid Biochem.*, 1984, **21**, 373.
67. A. F. Clayton, M. M. Coombs, K. Henrick, M. McPartlin and J. Trother, *Carcinogenesis*, 1983, **4**, 1569.
68. R. Hilal and A. M. El-Aaser, *Biophys. Chem.*, 1985, **22**, 145.
69. H.-D. Höltje, G. Lambrecht, U. Moser and E. Mutschler, *Arzneim.-Forsch.*, 1983, **33**, 190.
70. B. Pullman, in 'Quantum Mechanics of Molecular Conformations, ed. B. Pullman, Wiley, New York, 1976, p. 295.
71. A. Saran, *Proc. Indian Acad. Sci., Chem. Sci.*, 1987, **99**, 119.
72. A. Saran and C. L. Chatterjee, *Int. J. Quantum Chem.*, 1984, **25**, 743.
73. A. M. Bianucci, A. Martinelli and A. Da Settimo, *Farmaco, Ed. Sci.*, 1984, **39**, 686.
74. M. Remko, I. Sekerka and P. T. Van Duijnen, *Arch. Pharm. (Weinheim, Ger.)*, 1984, **317**, 45.
75. M. Remko, V. Frecer and J. Cizmarik, *Collect. Czech. Chem. Commun.*, 1983, **48**, 533.
76. L. M. Viana and Y. Takahata, *Int. J. Quantum Chem.*, 1982, **22**, 265.
77. A. Carpy, J. M. Leger, G. Leclerc, N. Decker, B. Rouot and C. G. Wermuth, *Mol. Pharmacol.*, 1982, **21**, 400.
78. J.-M. Leger, M. Gadret and A. Carpy, *Mol. Pharmacol.*, 1980, **17**, 339.

79. I. Tvaroska and T. Kozar, *Theochem*, 1985, **24**, 141.
80. H. Van de Waterbeemd, B. Testa and J. Caldwell, *J. Pharm. Pharmacol.*, 1986, **38**, 14.
81. J. Gerhards, T.-K. Ha and X. Perlia, *Arzneim.-Forsch.*, 1986, **36**, 861.
82. E. A. H. Griffith and B. E. Robertson, *Acta Crystallogr., Sect. B* 1972, **28**, 3377.
83. B. Pullman and P. Courrière, *Theor. Chim. Acta*, 1973, **31**, 19.
84. B. Pullman and P. Courrière, *C. R. Hebd. Seances Acad. Sci., Ser D*, 1974, **278**, 1785.
85. H. Yanagisawa, S. Ishihara, A. Ando, T. Kanazaki, S. Miyamoto, H. Koike, Y. Iijima, K. Oizumi, Y. Matsushita and T. Hata, *J. Med. Chem.*, 1987, **30**, 1984.
86. L. E. Arvidson, A. Karlen, U. Norinder, L. Kenne, S. Sundell and U. Hacksell, *J. Med. Chem.*, 1988, **31**, 212.
87. D. Kocjan, M. Hodoscek and D. Hadzi, *Theochem*, 1987, **37**, 331.
88. M. Eckert-Maksic, P. Bischof and Z. B. Maksic, *Theochem*, 1986, **32**, 179.
89. C. E. Spivak, T. M. Gund, R. F. Liang and J. A. Waters, *Eur. J. Pharmacol.*, 1986, **120**, 127.
90. G. H. Loew, J. Nienow, J. A. Lawson, L. Toll and E. T. Uyeno, *Mol. Pharmacol.*, 1985, **28**, 17.
91. S. Diamant, I. Agranat, A. Goldblum, S. Cohen and D. Atlas, *Biochem. Pharmacol.*, 1985, **34**, 491.
92. T. A. Hamor and I. L. Martin, in 'X-ray Crystallography and Drug Action', ed. A. S. Horn, C. J. De Ranter, Oxford University Press, Oxford, 1984.
93. G. H. Loew, J. R. Nienow and M. Poulsen, *Mol. Pharmacol.*, 1984, **26**, 19.
94. See ref. 31, p. 28.
95. F. Peradejordi and E. L. da Silva, in 'Quantum Theory of Chemical Reactions III. Chemisorption, Catalysis, Biochemical Reactions', ed. R. Daudel, A. Pullman, L. Salem and A. Veillard, Reidel, Dordrecht, 1982, p. 135.
96. J. M. Schulman, M. L. Sabio and R. L. Disch, *J. Med. Chem.*, 1983, **26**, 817.
97. R. S. Mulliken, *J. Chem. Phys.*, 1955, **23**, 1833.
98. G. Naray-Szabo and P. R. Surjian, in 'Theoretical Chemistry of Biological Systems', ed. G. Naray-Szabo, Elsevier, Amsterdam, 1986, p. 59.
99. J. I. Fernandez Alonso in 'Quantum Mechanisms of Molecular Conformations', ed. B. Pullman, Wiley, New York, 1976, p. 118.
100. U. C. Singh and P. A. Kollman, *J. Comput. Chem.*, 1984, **5**, 129.
101. S. R. Cox and D. E. Williams, *J. Comput. Chem.*, 1981, **2**, 304.
102. J. K. Seydel (ed.), 'QSAR and Strategies in the Design of Bioactive Compounds', VCH, Weinheim, 1985.
103. D. Hadzi and B. Jerman-Blazic (eds.), 'QSAR in Drug Design and Toxicology', Elsevier, Amsterdam, 1987.
104. Y. Orita, A. Ando, S. Yamabe, T. Nakanishi, Y. Arakawa and H. Abe, *Arzneim.-Forsch/Drug Res.*, 1983, **33**, 688.
105. R. Bonaccorsi, A. Pullman, E. Scrocco and J. Tomasi, *Theor. Chim. Acta*, 1972, **24**, 51.
106. A. Pullman and B. Pullman, *Rev. Biophys.*, 1981, **14**, 289.
107. D. F. Lewis and V. S. Griffiths, *Xenobiotica*, 1987, **17**, 769.
108. H. Weinstein, R. Osman and J. P. Green, *ACS Symp. Ser.*, 1979, **112**, 161.
109. H. Weinstein, R. Osman, S. Topiol and J. P. Green, *Ann. N. Y. Acad. Sci.*, 1981, **367**, 434.
110. H. Weinstein, R. Osman, J. P. Green and S. Topiol, in 'Chemical Applications of Atomic and Molecular Electrostatic Potentials', ed. P. Politzer and D. G. Truhlar, Plenum Press, New York, 1981, p. 309.
111. H. P. Weber, T. Lybrand, U. Singh and P. Kollman, *J. Mol. Graphics*, 1986, **4**, 56.
112. A. Goldblum, *Mol. Pharmacol.*, 1983, **24**, 436.
113. P. Politzer and P. R. Laurence, *Carcinogenesis*, 1984, **5**, 845.
114. H. Van de Waterbeemd, P.-A. Carrupt and B. Testa, *J. Med. Chem.*, 1986, **29**, 600.
115. D. Kocjan, M. Hodoscek and D. Hadzi, *J. Med. Chem.*, 1986, **29**, 1418.
116. D. M. Hayes and P. A. Kollman, *J. Am. Chem. Soc.*, 1976, **98**, 7811.
117. R. Osman, H. Weinstein and S. Topiol, *Ann. N.Y. Acad. Sci.*, 1981, **367**, 356.
118. M. J. Spark, D. A. Winkler and P. R. Andrews, *Int. J. Quantum Chem., Quantum Biol. Symp.*, 1982, **9**, 321.
119. P. R. Andrews, M. Sadek, M. J. Spark and D. A. Winkler, *J. Med. Chem.*, 1986, **29**, 698.
120. P. A. Kollman and D. M. Hayes, *J. Am. Chem. Soc.*, 1981, **103**, 2955.
121. G. Naray-Szabo, *Int. J. Quantum Chem.*, 1983, **23**, 723.
122. R. P. Sheridan and L. C. Allen, *J. Am. Chem. Soc.*, 1981, **103**, 1544.
123. A. Martinelli and C. Petrongolo, *J. Phys. Chem.*, 1980, **84**, 105.
124. G. H. Loew, D. Berkowitz, H. Weinstein and S. Srebrenik, in 'Molecular and Quantum Pharmacology', ed. E. D. Bergmann and B. Pullman, Reidel, Dordrecht, 1974, p. 355.
125. H. Weinstein, J. P. Green, R. Osman and W. D. Edwards, in 'National Institute for Drug Abuse Research Monograph Series 22', ed. G. Barnett, M. Trsic and R. Willette, National Institute for Drug Abuse, Washington DC, 1978, p. 333.
126. H. Weinstein, A. P. Mazurek, R. Osman and S. Topiol, *Mol. Pharmacol.*, 1986, **29**, 28.
127. R. Osman, S. Topiol, L. Rubenstein and H. Weinstein, *Mol. Pharmacol.*, 1987, **32**, 699.
128. R. Osman, H. Weinstein, S. Topiol and L. Rubenstein, *Clin. Physiol. Biochem.*, 1985, **3**, 80.
129. S. Topiol, H. Weinstein and R. Osman, *J. Med. Chem.*, 1984, **27**, 1531.
130. S. Topiol, G. Mercier, R. Osman and H. Weinstein, *J. Comput. Chem.*, 1985, **6**, 581.
131. E. L. Mehler and J. Gerhards, *Mol. Pharmacol.*, 1987, **31**, 284.
132. G. H. Loew, E. Kurkjian and M. Rebagliati, *Chem.-Biol. Interact.*, 1983, **43**, 33.
133. G. H. Loew, M. Rebagliati and M. Poulsen, *Cancer Biochem. Biophys.*, 1984, **7**, 109.
134. G. H. Loew, J. Ferrell and M. Poulsen in 'Structure–Activity Correlation as a Predictive Tool in Toxicology', ed. L. Golberg, Hemisphere Publishing Co., New York, 1983, p. 111.
135. G. H. Loew, J. Phillips and G. Pack, *Cancer Biochem. Biophys.*, 1979, **3**, 101.
136. G. H. Loew, B. S. Sudhindra, S. K. Burt, G. R. Pack and R. MacElroy, *Int. J. Quantum Chem., Quantum Biol. Symp*, 1979, **6**, 259.
137. K. Korzekwa, W. Trager, M. Gouterman, D. Spangler and G. H. Loew, *J. Am. Chem. Soc.*, 1985, **107**, 4273.
138. A. T. Pudzianowski and G. H. Loew, *J. Mol. Catal.*, 1982, **17**, 1.
139. A. T. Pudzianowski and G. H. Loew, *Int. J. Quantum Chem.*, 1983, **23**, 1257.
140. A. T. Pudzianowski and G. H. Loew, *J. Phys. Chem.*, 1983, **87**, 1081.
141. A. Goldblum and G. H. Loew, *J. Am. Chem. Soc.*, 1985, **107**, 4265.

18.2

Molecular Mechanics and the Modeling of Drug Structures

GEORGE L. SEIBEL and PETER A. KOLLMAN

University of California, San Francisco, CA, USA

18.2.1 INTRODUCTION

Molecular mechanics is a method of calculating the potential energy of an isolated molecule or system of interacting molecules as a function of their nuclear coordinates. When the structure of a receptor is known, molecular mechanical methods can be used to probe the energetics of drug binding. In cases where the receptor structure is not known, molecular mechanics is useful in locating low energy conformations of ligands for use in pharmacophoric mapping procedures.

The molecular mechanical model considers molecules to be collections of atoms held together by classical forces. The atoms are treated as classical particles under the influence of the molecular mechanical potential or force field. The force field is a set of simple analytically differentiable functions of the nuclear coordinates that yields a potential energy for the molecule with respect to a hypothetical strain-free state. The strain-free state is one in which all bond lengths, angles and torsions are at their 'natural' or minimum energy values and nonbonded atoms are at infinite separation.

The molecular mechanical potential allows us to calculate the relative energy of different conformations of a molecule with little computational effort. Since the terms in the potential function are analytically differentiable, the gradients of potential energy with respect to coordinates, which constitute the forces on the atoms, are easily obtained. This allows one to use standard numerical optimization methods to minimize the energy of the system, resulting in the location of a local minimum or in a few cases the global minimum energy structure. The term 'molecular mechanics' is generally synonymous with such energy minimization using an analytic potential. The same forces are used in molecular dynamics through integration of Newton's equations of motion to solve for a molecular trajectory at a given temperature. This is described in detail in Chapter 18.3.

Since quantum mechanical methods also provide potential energies as a function of nuclear coordinates, it will be instructive to point out some differences in the two methods. In the most common case, quantum mechanics is concerned with the explicit calculation of the electron distribution in a fixed nuclear field. In molecular mechanics, the electron distribution is implicit in the force field in which the nuclei are allowed to move. A consequence of this difference is that bond-making and -breaking cannot be simulated by molecular mechanics unless suitable analytic representations of the event can be developed.

The most significant practical difference between quantum mechanics and molecular mechanics is the way in which their requirement for computer resources scales with the size of the problem attempted. Molecular mechanical methods scale as N^2 in the worst case, where N is the number of atoms. In large macromolecular simulations where a finite cut-off on nonbonded interactions is employed, the scaling can be close to linear. In contrast, quantum mechanical methods scale as M^3 to M^4 or higher, where M is the total number of basis functions. Typically there are 5 to 20 basis functions per atom. In addition, *ab initio* quantum mechanics can require a vast amount of temporary storage for integrals, such that space limitations are often reached before CPU limitations are felt. These factors limit quantum mechanics to relatively small systems, while molecular mechanical methods are commonly used to treat systems of thousands of atoms.

A final comparison between quantum mechanics and molecular mechanics involves the information that must be supplied to each method in order to perform a calculation. For a given geometry, quantum mechanics needs only the nuclear and net charge, quantum mechanical state multiplicity and appropriate basis functions. Molecular mechanics, on the other hand, requires that all atoms be classified into distinct types that are recognized as different by the force field. The bonding topology must be specified, equilibrium values and force constants must be supplied for all valence terms, and all atom types must have nonbonded interaction parameters specified. In addition, electrostatic information in the form of atomic partial charges or bond dipoles is generally required.

For systems within the scope of its parametrization, a molecular mechanical treatment may well give better results for certain properties than a quantum mechanical treatment, but a significant advantage of quantum mechanical methods is that they can treat a much wider range of molecules and properties.

In the following section, we will examine the various terms in typical molecular mechanical potentials, considering the physical basis for each term, the applicability of the function and sources of parameters. We will compare small molecule force fields[1, 2] exemplified by MM2,[1, 3] with typical macromolecular force fields[4-8] using that of Weiner *et al.*[7, 8] as an example. Finally, we discuss some applications of molecular mechanical methods to problems in drug design, considering first the most common case where the structure of the receptor is unknown, followed by examples of applications in which the receptor structure is known to atomic detail.

18.2.2 THE MOLECULAR MECHANICAL POTENTIAL

The molecular mechanical force field is an analytic description of the potential energy surface of a molecule. In reality, every molecule has a unique force field but to a very good approximation the force field can be broken down into components that are transferable between molecules. The force field is parametrized against experimental data for a given class of molecules, and is subsequently used to predict structural and energetic properties of related molecules.

18.2.2.1 Potential Functions and Parametrization of Force Fields

There are three general categories of data that may be used in force field parametrization: structure, energy and vibrational frequencies. Force fields were first developed in the area of vibrational analysis in order to analyze and predict vibrational spectra. The vibrational force fields were not appropriate for the calculation of structure or energy. The early molecular mechanical force fields focused on structure and energy, employing modifications of the vibrational force fields, but did not effectively reproduce spectra. In 1968, Lifson and Warshel[9] reported a 'consistent force field' (CFF) which was intended to simultaneously reproduce structure, energy and vibrational frequencies from the same set of equations. They used a least-squares optimization procedure to fit the energy function parameters to a large amount of experimental data. The idea of the CFF set the tone of force field development that succeeded it, although the least-squares fitting approach to parametrization has not been universally adopted.

The developers of a force field usually have a particular class of problems that they are interested in treating well. The potential functions will be selected with the goals of the force field in mind, and the parametrization and testing will be performed against a set of compounds representative of those the developers are interested in. This is a natural consequence of the difficulty of creating a truly general force field. A particular force field can therefore be expected to perform best when it is used for those problems for which it was designed.

Force fields in general use today tend to be focused either toward small molecules or macromolecules; both are pertinent to drug design. A small molecule force field will generally use a more elaborate potential, since computation time is not as critical. Macromolecule force fields tend to use simpler potentials that can be evaluated quickly, and place more emphasis on electrostatic interactions and hydrogen bonding. Allinger's MM2 force field[3] is the best known small molecule force field and will serve here as an example. It was originally developed for hydrocarbons but has since been parametrized for a variety of organic functionalities. As an example of a typical macromolecular force field we will consider the Weiner *et al.* force field.[7]

Typical force fields contain terms for potential energy due to bond stretch, angle bending, torsions, van der Waals interactions and Coulombic interaction (equation 1).

$$E_{total} = E_{bond} + E_{angle} + E_{torsion} + E_{vdW} + E_{Coulomb} \tag{1}$$

In addition to the terms in equation (1), a term is usually included to account for out-of-plane distortion of sp^2 centers. A hydrogen bond function may also be used, and valence cross terms that take into account, for example, the change in bond force constants as angles are deformed may be employed. We will consider each of the above components of the force field in turn.

Atoms in the molecular mechanical force field are classified into distinct types and their combinations dictate the bond, angle and dihedral types that must be parametrized. The number of possible bond types is found by taking all possible pairs of atom types. Similarly, angle and dihedral types correspond to all possible triples and quadruples of atom types. Many of these combinations will be ruled out on physical grounds, and in practice only a small subset of possible combinations will be parametrized. Considering the element carbon, a logical classification is by hybridization, giving three types of carbon: sp^1, sp^2 and sp^3. Such a limited classification will inevitably lead to compromises. In this example, carbonyl, alkene and aromatic carbon would be treated identically in some respects. Most force fields therefore provide for finer distinctions than this, attempting to strike a balance between the higher accuracy possible with more atom types and the greater difficulty in parametrization that results.

We next consider the energy required to distort bonds from their equilibrium lengths. The Morse potential is a good description of bond energy as a function of length

$$E_{bond} = D_e \{1 - \exp[-c(R - R_0)]\}^2 \tag{2}$$

where D_e is the depth of the potential energy well, R is the bond length, R_0 is the equilibrium bond length and c includes the masses of the bound atoms and the force constant of the particular bond. The Morse potential asymptotically approaches D_e as the atoms of the bond dissociate. This potential is shown in Figure 1 as a solid line. The dashed line is the simpler harmonic potential which is used in most force fields. The harmonic potential has the following form

$$E_{bond} = K_b(R - R_0)^2 \tag{3}$$

where K_b is the force constant of the particular bond. The harmonic potential is a good model near the bottom of the potential well, and is easier to implement. It is used in the Weiner *et al.* force field. In the MM2 force field the harmonic potential is augmented with a cubic term to partially correct for the true anharmonicity of the bond potential. In this case the function is

$$E_{bond} = K_b(R - R_0)^2[1 + c(R - R_0)] \tag{4}$$

where c is an adjustable parameter.

The equilibrium values for bond lengths are generally obtained from high resolution small molecule X-ray crystallography. These values are typically known to 0.001 Å. Bond lengths may also be obtained from electron diffraction and microwave spectroscopy. It should be noted that microwave spectroscopy measures moments of inertia, which are a function of nuclear coordinates, while X-ray and electron diffraction methods measure electron density. In most cases electron density is centered on the nucleus, but in the case of hydrogen the center of electron density may be displaced up to 10% toward the bound atom. This points out a difficulty in parametrization: one must obtain data that are both accurate and well defined. Force constants may be obtained from vibrational spectra. The experimental values of equilibrium lengths and force constants are only

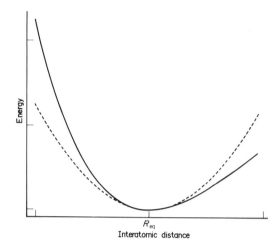

Figure 1   Morse bond potential (solid) and harmonic approximation (dashed). Note that the harmonic approximation is a very good match near the bottom of the potential well

used as starting points in the parametrization process, however. Lengths may have to be modified to account for the perturbing effect of vicinal nonbonded interactions, and while vibrational force constants have been found to reproduce geometries well, they are usually reduced 20–50% to simultaneously reproduce energies.[1] One should thus be wary of assigning too much physical significance to the individual parameters of a molecular mechanical force field. In the parametrization process, all individual parameters may be tuned to best reproduce geometries, energies and vibrational frequencies of model compounds. The actual values of individual parameters, such as the force constants mentioned above, relate to the *model*, not the physical molecule.

Angle bending terms are modeled quite effectively with a harmonic potential

$$E_{\text{angle}} = K_t(\Theta - \Theta_0)^2 \tag{5}$$

where K_t is the force constant, Θ is the bond angle and Θ_0 is the equilibrium angle. The largest bond angles found in strained organic compounds are about 15° above the strainless value, thus anharmonicity does not seem to be a problem here.[1] Excessively small angles, such as in three- or four-membered rings, are a different situation and a common approach, used in MM2, is to assign special atom types for those situations. The Weiner *et al.* force field is not parametrized for such small rings. Angle bending parameters are derived from the same sources as bond stretch parameters described above.

The MM2 potential includes a 'stretch–bend' term to account for the fact that as a bond angle is compressed, the force constants of the associated bonds decrease. This coupling is expressed as

$$E_{\text{sb}} = K_{\text{abc}}^{\text{sb}}[(R - R_0)_{\text{ab}} + (R - R_0)_{\text{bc}}](\Theta - \Theta_0)_{\text{abc}} \tag{6}$$

where $K_{\text{abc}}^{\text{sb}}$ is the stretch–bend constant for angle type abc. Cross terms are not used explicitly in the Weiner *et al.* potential, although the effects of such terms can be accounted for to some extent in the parametrization of the simpler potential.

Significant torsional barriers exist about ethane-like linkages, and they are not due solely to steric repulsion between substituents. From theoretical considerations we can account for only a very small barrier to rotation in ethane due to the nonbonded interaction of the vicinal hydrogens, yet the experimental barrier is 12 kJ mol^{-1}. For years the source of this barrier was not well understood. In 1979 Brunck and Weinhold[10] demonstrated through the use of a simple bond orbital MO formalism that much of the torsional 'barrier' in saturated systems was due to electron delocalization causing an energy *well* in the staggered conformation. Torsional barriers about multiple bonds are very large, obviously the result of disruption of the π system. Molecular mechanical potentials usually model the electronic component of the torsional barrier with a Fourier series such as the following

$$E_{\text{torsion}} = \sum_j \frac{V_j}{2}[1 + \cos(j\omega - \gamma)] \tag{7}$$

Here j is the periodicity of the barrier, V_j is the barrier height, ω is the dihedral angle and γ is the phase offset. The series is usually truncated at the third term. By summing onefold, twofold and threefold terms as indicated in Figure 2 for a hypothetical torsion a complicated barrier can be constructed. In practice the total torsional barrier obtained in the molecular mechanical potential will be a sum of the Fourier components and the 1–4 nonbonded component (nonbonded interactions between atoms separated by three bonds), although the relative magnitude of each will not necessarily have physical significance. Barrier heights for model compounds can be obtained from microwave spectroscopy. The twofold torsion about double bonds and the threefold torsion about single bonds are relatively easy to fit, with barriers for the Fourier portion adjusted to yield the experimental barrier when the 1–4 nonbonded terms are taken into account. Onefold and twofold terms for single bonds are not as straightforward to fit, but are adjusted to duplicate conformational preferences in model compounds.

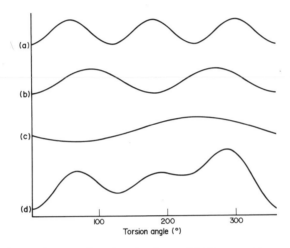

Figure 2 In this hypothetical example, a complicated torsional potential (d) is created by summing threefold (a), twofold (b) and onefold (c) Fourier components. The onefold potential in this example is phase shifted by 60° resulting in an asymmetric sum

The van der Waals parameters are very important, and are also the·most difficult to fit. At close range, the interaction potential between two nonbonded atoms is steeply repulsive, as electrons would be forced to occupy the same space in violation of the Pauli principle. The equation that best fits the repulsive potential is an exponential form, although simpler forms are often used for computational expediency. At long range, the interaction potential is attractive, due to the instantaneous dipole interactions or London dispersion forces. The dispersion potential can be calculated from second-order perturbation theory as

$$E_{disp} = -c_6 r^{-6} - c_8 r^{-8} - c_{10} r^{-10} - \cdots \tag{8}$$

The leading term is the dipole–dipole interaction and is predominant. The potential is truncated at the first term in all well-known force fields. A popular simple van der Waals potential is the Lennard-Jones equation, used in many macromolecular force fields

$$E_{vdW} = \varepsilon \left[\left(\frac{r^*}{r} \right)^{12} - 2 \left(\frac{r^*}{r} \right)^6 \right] \tag{9}$$

Here ε is the depth of the potential well at the minimum, r^* is the minimum energy internuclear separation and r is the actual internuclear distance. This potential was developed for rare gas interactions and is actually too 'hard' or steeply repulsive for organic compounds. A common modification to the Lennard-Jones potential is the use of a softer ninth power dependence for the repulsive part. A more functionally correct form is the Buckingham potential, with exponential repulsion and r^{-6} attraction terms

$$E_{vdW} = \frac{\varepsilon}{1 - (6/\alpha)} \left[\frac{6}{\alpha} \exp \left[\alpha / \left(1 - \frac{r}{r^*} \right) \right] - \left(\frac{r^*}{r} \right)^6 \right] \tag{10}$$

α is an adjustable parameter relating to the steepness of the repulsive part of the potential. MM2 uses a modification of this potential. Regardless of the functional form used, interactions are calculated over all pairs of atoms that are not bonded to each other or a common third atom. Since the van der Waals potential dies off fairly quickly, the evaluation is often restricted to atom pairs closer than some cut-off distance, typically around 7 to 9 Å. This reduces computation time dramatically in large systems.

Sources of parameters for the nonbonded potentials include molecular beam experiments, second virial coefficients and lattice spacings in small molecule crystals. Information on well depths can also be obtained from enthalpies of sublimation. In molecular mechanical potentials, one is assuming that the same nonbonded potential is appropriate for intermolecular interactions as for the intramolecular case. In practice this seems to work well. In addition, the model treats atoms as soft spheres centered on the nuclei, a reasonable approximation for large atoms but not as good for hydrogen. In the MM2 force field, hydrogens are treated as spheres displaced 10% toward their bound neighbor, which improves the model. Nearly all force fields treat the nonbonded interactions as pairwise additive, when in fact a nearby dipole may distort more polarizable atoms, altering their true potential. Nonpairwise additive potentials are an area of current research.[11]

Electrostatic interactions are of particular importance in biological systems. The physical picture one is attempting to model is that of a mobile polarizable three-dimensional cloud of electron density situated about the positive nuclei. The most common simple electrostatic models are the bond dipole approach used in MM2 and the atom-centered point charge approach used in Weiner *et al.* In the bond dipole method, a point dipole is placed at each bond center. The values of the dipoles are chosen such that they reproduce the molecular dipole moment in model compounds. The point charge method simply involves placement of point charges at each atom. Here again the goal is to reproduce the overall molecular dipole as well as quadrupole and higher moments. The two methods are essentially equivalent in capability, being dependent only on the charges or dipoles chosen. The dipole–dipole interaction may be calculated as follows

$$E_{\text{dipole}} = \frac{\mu_i \mu_j}{D r_{ij}^3}(\cos\chi - 3\cos\alpha_i\cos\alpha_j) \tag{11}$$

where μ_i and μ_j are a pair of point dipoles, D is the dielectric constant, χ is the angle between the dipoles and α_i and α_j are the angles formed between the dipoles and a vector connecting the two.[1] Interactions between point charges are calculated *via* the Coulomb potential

$$E_{\text{Coulomb}} = \frac{q_i q_j}{D r_{ij}} \tag{12}$$

where q_i and q_j are the partial charges on atoms i and j. These electrostatic models require that either bond dipoles or point charges be determined for systems of interest. The bond dipoles are generally treated as parameters for each bond type and are empirically adjusted to reproduce dipole moments in model compounds. Point charges can be obtained from quantum mechanical calculations on the molecule of interest but this presents several problems. Most serious is the fact that the magnitudes and even the signs of the charges are quite dependent on the quantum mechanical method and basis set used. The usual Mulliken population analysis represents a rather severe approximation; much of the electrostatic information available from the wavefunction is lost in this method. Significant improvement can be achieved by using methods that calculate the electrostatic potential at many points about the molecule directly from the wavefunction, then use this potential to fit point charges at the atom centers.[12] A remaining difficulty with quantum mechanical charge calculations is the large computational investment required for such calculations at the *ab initio* level, which may render them impossible for large molecules. Modern semiempirical methods[13] may prove quite useful in this regard, however. Several less demanding methods have been used for obtaining reasonable partial charges. The empirical method of Del Re[14] was originally developed for saturated organic molecules, and has been parametrized for amino acids.[15] Gasteiger's method of partial equalization of orbital electronegativity[16] is an iterative method wherein charge is transferred from atoms of lower orbital electronegativity to higher, which then reduces the difference in orbital electronegativity. The process is repeated until charge transfer falls below a threshold value. A damping factor prevents total equalization of electronegativity. The orbital electronegativity is defined in terms of ionization potential and electron affinity for a particular valence state and is thus amenable to physical measurement. Regardless of the source of electrostatic information, in the typical molecular mechanical potential charges are considered to be independent of conformation and polarization is not accounted for. This is an adequate model for hydrocarbons but is probably the largest source of error when charged or highly polar molecules are considered.

Most force fields intended for macromolecules treat hydrogen bonds differently than other nonbonded interactions. In the Weiner *et al.* force field, the 6–12 nonbonded potential is replaced by a 10–12 form for hydrogen bond pairs

$$E_{\text{Hbond}} = \frac{C_{ij}}{r_{ij}^{12}} - \frac{D_{ij}}{r_{ij}^{10}} \tag{13}$$

The parameters C_{ij} and D_{ij} are functions of the well depth and equilibrium distance of the donor–acceptor interaction. In the Weiner *et al.* force field most of the energy of a hydrogen bond interaction is electrostatic with the 10–12 function only being used to 'fine tune' the hydrogen bond length.

18.2.2.2 Minimization Methods

The previous section described how we obtain potential energies and the forces acting on the atoms as a function of nuclear coordinates. This information can be used in a variety of ways. It may be sufficient to calculate the potential energy of a single fixed conformation or to map the energy as a function of one or more internal degrees of freedom in the molecule. Usually, we wish to determine a low energy structure for the molecule at hand. This involves finding a minimum on the potential energy hypersurface. Any of a number of optimization techniques may be used, working in either Cartesian or internal coordinate space. Most common are Cartesian coordinate minimizers employing either first or second derivative methods. The simplest first derivative method is the steepest descent minimizer, which moves atoms directly down the energy gradient. It is robust and has the advantage of quickly improving bad starting geometries, but suffers from poor convergence near the minimum. A better first derivative method is the conjugate gradient minimizer. It stores information on the direction of previous moves in order to predict better movement directions. This results in somewhat better convergence properties than steepest descent, although it is less tolerant of poor starting geometry. Many programs use both minimizers, starting with steepest descent then switching to conjugate gradient when a suitable average gradient is reached.

Second derivative methods are much more efficient in locating a minimum, although they have the least tolerance for poor starting geometry. The full Newton–Raphson method requires the storage and inversion of a $3N \times 3N$ second derivative matrix, the $3N$ dimensions corresponding to the $3N$ degrees of freedom of the molecule. Because of the large storage requirement, the full Newton–Raphson method is not widely used in molecular mechanics. A modification known as the block-diagonal Newton–Raphson method is used in MM2, where only a 3×3 portion of the second derivative matrix is stored for each atom. Some information about the curvature of the energy hypersurface is lost, but the savings in storage and matrix inversion time are significant.

A molecular potential energy surface is not a harmonic surface. It is actually a very noisy surface with countless spikes and, in many cases, broad and ill-defined regions. As such it is a difficult surface to minimize, and this fact is reflected in the number of iterations required for most molecular mechanics calculations. For a complicated system, first derivative methods may take thousands of iterations to converge, while second derivative methods as commonly implemented may require hundreds of iterations. Recent advances in optimization methods have resulted in molecular mechanics energy minimizers with relatively small storage requirements that will converge in tens of cycles.[17,18] There is still a large computational requirement with these methods, involving the necessary matrix calculations, but since the force field need only be evaluated a few times a much more sophisticated potential could be used. To date these minimizers have only been used with conventional force fields.

18.2.2.3 Conformational Searching

Regardless of the method used, all minimizers will place the molecule into the first local minimum encountered on the potential energy hypersurface. There is no guarantee that this will be the global minimum. For molecules with a very small number of rotatable bonds an exhaustive search of conformational space is possible, but naïve approaches quickly become intractable as the number of rotatable bonds increases. Improvements to exhaustive search methods center on pruning the search tree to avoid searching areas that are sterically inaccessible.[19-21] For example, because the high energy conformations of the naturally occurring amino acids are well understood, one may constrain searches of peptide molecules to regions that avoid known steric contacts. In cyclic compounds, ring closure provides a powerful constraint on the number of allowable conformations. Similarly, interatomic distances obtained from the solution NMR nuclear overhauser effect (NOE) provide another source of constraints. These approaches can reduce the complexity of the search by many orders of magnitude. The number of torsional degrees of freedom that may feasibly be searched through the use of such methods is a function of the number of constraints present and of the coarseness of the search grid, but currently appears to be about 10–15.[21]

Other methods for exploring conformational space include Monte Carlo methods[22] and distance geometry approaches.[23-24] At this point it is not clear which of these methods are superior. Most likely, the strengths of each method will depend on the molecule, *e.g.* shape and number of torsional degrees of freedom. Once various starting conformations have been generated, they can be subjected to molecular mechanics minimization to move them to the nearest local energy minimum. Molecular dynamics has been used for conformational searching, but it is not very computer-time efficient in moving far from the initial geometry. It is useful for searching for local minima in a limited area of conformational space, so a combination of a method such as distance geometry to randomly sample conformation space, molecular dynamics to search local areas of this space and finally molecular mechanics to compare the relative energy of various local energy minima seems a reasonable strategy for spanning conformational space in a nonexhaustive manner. Nonetheless, one must bear in mind that no matter how efficient the conformational searching method, the ability to analyze the correct conformational behavior of a molecule depends critically on the accuracy of the calculated energies of the different conformations.

18.2.2.4   Solvent Effects

Drug–receptor interactions occur in an aqueous environment. Because of its high dielectric constant, the presence of water can have a dramatic effect on the energy and structure of an isolated molecule or system of interacting molecules. It is therefore advisable to consider the solvent environment in some way, especially when applying molecular mechanical techniques to polar molecules. The solvent environment has been treated in a number of ways within the molecular mechanical framework. The most physical model is the explicit inclusion of solvent molecules in the system. The SPC water model of Berendsen *et al.*[25] and the TIPS2, TIP3P and TIP4P models of Jorgensen *et al.*[26] are well-developed parametrizations. In order to generate meaningful solvent configurations, Monte Carlo or molecular dynamics methods are used with constant volume or constant pressure periodic boundary conditions. (see Chapter 18.3) The most serious difficulty with the explicit water model is the large number of water molecules needed, putting this technique computationally out of reach for many problems. The steadily improving performance of computers is making it a more viable option, however.

A much less costly solvent treatment is the use of a modified dielectric. The dielectric constant D is normally set to unity for a true gas phase calculation or for a calculation in which solvent is included explicitly. Increasing its value will reduce the magnitude of the Coulomb potential, as would be appropriate in a high dielectric medium. While this is an adequate approximation in some cases, it is not clear exactly what value should be used for the effective dielectric constant, although values of 2–4 have been suggested for proteins.[27] The dielectric can also be made a function of distance. Simply setting $D = r_{ij}$ in the Coulomb potential is a common approximation, although Whitlow and Teeter have found $D = 4r_{ij}$ to result in less deviation from the X-ray structure in energy minimization of a small protein in the absence of explicit solvent or crystal environment.[28] In a study of electrostatic effects in proteins, Gilson and Honig find that the distance-dependent dielectric model severely overestimates weak interactions but is reasonable for strong ones.[27] They suggest that a model which increases screening of charges near the surface of proteins would be an improvement.

In an effort to include solvent effects in molecular mechanical calculations, several groups have developed hydration shell models in which various functional groups are assigned free energies of solvation.[29-31] The intramolecular energy then includes a solvation free energy which depends on the group solvation free energies and the exposed surface area of these groups. It is not clear how well such approaches work, given that intramolecular conformational energies and empirical solvation free energies are very different quantities and their magnitudes may not be comparable.

We now briefly describe some specific studies using molecular mechanics in drug design. These studies can be divided into two classes: the first pertain to those systems where the receptor structure is unknown, and the second pertain to systems in which a detailed three-dimensional structure of a receptor is available. We wish to emphasize that molecular mechanics is used primarily to address aspects of the energetics of receptor binding, which is only one element that must be considered in the drug design process. One must also consider bioavailability, distribution, metabolism and toxicity in the search for effective therapeutic agents. Synthetic accessibility and economic factors are further practical concerns in the development of an actual drug.

18.2.3 MOLECULAR MECHANICS STUDIES ON SYSTEMS OF UNKNOWN RECEPTOR STRUCTURE

When no receptor structure is available, molecular mechanics is used in the context of pharmacophoric pattern matching. In this technique,[32] one attempts to identify and superimpose the pharmacophores of a series of active molecules. The resulting pharmacophoric pattern is then used to rationalize the biological activity of other molecules based on their ability to match the pharmacophoric pattern. The initial generation of the pharmacophoric pattern is usually based on rigid analogs. If the only available active analogs are flexible molecules, one should not make the *a priori* assumption that the binding conformation is the global minimum energy conformation of the isolated analog. There are undoubtedly cases where the binding conformation will be near the solution minimum energy conformation of a flexible ligand but this is by no means assured, given the asymmetric nature of the potential at the binding site. The ensemble of accessible conformations is the more reasonable bound on the pharmacophoric pattern in such a situation. An example of the use of molecular mechanics in the development of a pharmacophoric pattern is the work of Hagler and co-workers[33] who developed a pharmacophoric pattern for the binding of gonadotropin-releasing hormone by examining conformations of a cyclic peptide antagonist. The antagonist was subjected to 24 ps of molecular dynamics simulation, and instantaneous structures taken from the trajectory each picosecond were minimized with molecular mechanics. The resulting structures were found to fall into a common family of configurations, taken to be the putative 'binding conformation'.

In the process of pharmacophoric pattern matching, one can use molecular mechanics to ascertain the ability of a flexible analog to attain a conformation in which its pharmacophores match those of the pharmacophoric pattern. The actual quantity of interest is the overall free energy of binding in the solution phase but because the detailed atomic structure of the receptor is unknown one is forced to make a number of assumptions regarding binding energetics. We can consider the overall free energy of binding to have two components: the free energy cost ΔG_{lock} of 'locking' the flexible analog in the pharmacophorically relevant position, and the free energy ΔG_{bind} of receptor binding from the 'locked' conformation. The sequences in Scheme 1 illustrate this for a flexible analog in comparison to a rigid analog, where D^{flex} is the flexible analog, D^{rigid} is the rigid analog, D^{locked} is the flexible analog now locked in the conformation that matches the rigid pharmacophore, R is the receptor and DR is the drug–receptor complex.

$$D^{flex} \quad + \quad R \quad \xrightarrow{\Delta G_{lock}} \quad D^{locked} \quad + \quad R \quad \xrightarrow{\Delta G_{bind}} \quad DR$$

$$D^{rigid} \quad + \quad R \quad \xrightarrow{\Delta G = 0} \quad D^{rigid} \quad + \quad R \quad \xrightarrow{\Delta G_{bind}} \quad DR$$

Scheme 1

A major assumption that we must make is that if the flexible and rigid molecules have the same pharmacophores, their interaction with the receptor ΔG_{bind} will be similar. Thus one can hope to relate the binding affinity of the flexible analog to that of the rigid analog by ΔG_{lock}, the free energy cost of 'locking' the flexible molecule in the pharmacophoric conformation. This is related to P_{lock}, the classical Boltzmann probability of finding the flexible molecule in the appropriate conformation by $\Delta G_{lock} = -RT \ln P_{lock}$. The probability P_{lock} is a function of the ensemble of accessible conformations

$$P_{lock} = \frac{\exp(-E_{lock}/RT)}{\int \exp(-E(r)/RT)\,dr} \approx \frac{\exp(-E_{lock}/RT)}{\left(\sum_i \exp(-E_i/RT) \right)} \tag{14}$$

where E_{lock} is the energy of the locked conformation, R is the gas constant and T is the absolute temperature. The integral in the denominator is a partition function taken over all conformational degrees of freedom of the molecule. The integral can be crudely approximated by considering only the relevant minima E_i. A second assumption in this analysis is that the solvation free energy does not change between the flexible and locked conformation. It is clear that the necessary assumptions will be likely to hold only in cases where fairly similar analogs are being compared. A final uncertainty is that of the pharmacophoric pattern itself, which supposes a single binding mode. Thus one cannot make the above relationships too quantitative but, if the rigid molecule is a perfect template and all other energetic assumptions hold, every $5.7\,\mathrm{kJ\,mol^{-1}}$ ($-RT\ln K_{\mathrm{eq}}$) of strain energy necessary to place the flexible molecule in the pharmacophoric conformation should reduce its binding affinity to the receptor by a factor of 10 compared to an unstrained analog.

Andrea *et al.*[34] have carried out a full ensemble analysis for thyroxine analogs and have shown that ΔG_{lock}, the conformational free energy required to constrain the molecules from their ensemble of minima to the conformation *assumed* to be relevant for biological activity, could be related to binding in an *in vitro* assay with a correlation coefficient $r = 0.89$. *In vivo* activity also correlated with ΔG_{lock}. The thyroxine conformational energies were actually evaluated using semiempirical quantum mechanics, but the important principle in this application is the ensemble analysis to give free energies instead of internal energies.

Even using molecular mechanics, calculating energies at all local minima may be difficult for molecules with a large number of rotatable bonds. The following study illustrates a simplification in which ΔG_{lock} is approximated by simple differences in molecular mechanical energies.

In a recent paper, Pettersson and Liljefors used molecular mechanical methods with pharmacophoric pattern matching to rationalize the dopaminergic effect of analogs of (*R*)-apomorphine.[35] The pharmacophoric pattern consisted of four points taken from the conformationally rigid apomorphine, a potent and stereospecific dopamine receptor agonist. They constructed conformations of the other analogs which, in a least-squares sense, put the corresponding atoms as close as possible to those in apomorphine. The 12 analogs were then energy minimized with constraints on the movement of appropriate atoms in order to retain the fit to the pharmacophoric pattern. They then compared the energy difference between the global minimum of each analog and its pharmacophorically relevant conformation. They found that for the seven molecules which had low activity or were inactive, the global minimum conformation was $\geqslant 2.5\,\mathrm{kcal\,mol^{-1}}$ ($10.5\,\mathrm{kJ\,mol^{-1}}$) lower than the pharmacophorically relevant one; for the two molecules with moderate activity, the global minimum was 2–$3\,\mathrm{kcal\,mol^{-1}}$ (8.4–$12.5\,\mathrm{kJ\,mol^{-1}}$) below and for the three high activity molecules, the global minimum was $< 1\,\mathrm{kcal\,mol^{-1}}$ ($4.2\,\mathrm{kJ\,mol^{-1}}$) below the pharmacophorically relevant one.

One of the subjective decisions forced on one in such studies is what to do if, during the molecular mechanics optimization, the pharmacophoric pattern of the flexible molecule moves significantly from the rigid template. What should one consider a 'significant' movement? There is no precise answer to this question, but a movement of less than 0.3–$0.5\,\text{Å}$ would probably still allow a near optimal interaction of the pharmacophoric atoms with the corresponding receptor atoms. Such a level of movement was observed in the above paper, but we stress that, in general, one must make subjective decisions at this point on how much, if any, to *force* the molecule to adopt a given precise pharmacophoric pattern. Many molecular mechanics programs contain the necessary machinery to constrain appropriate atoms in Cartesian or internal coordinates in order to achieve a pharmacophoric pattern. Hagler and co-workers[33] call the constraining process 'template forcing' and have applied it to a number of peptide analogs to force them to mimic more rigid molecules.

In considering the appropriate balance between spatially matching the pharmacophoric pattern and seeking the lowest energy, one may ask the question how much strain energy is induced in the small molecule when it binds to a receptor? The answer to this question may influence conformational search strategies as well. If one assumes that there is a high likelihood of significant strain, then one might allow much higher energy conformations to pass filters of search strategies. Based on the data of Ponder and Richards from high resolution protein structures,[36] side chain χ angles follow closely the conformational preferences expected on the basis of isolated fragments. This implies that a significant amount of strain is not likely to be tolerated in ligand–protein interaction. One might similarly exploit small molecule crystal structures such as those found in the Cambridge Crystallographic Database.[37] By imagining the crystal as a drug–receptor complex, considering the central molecule as the drug and the surrounding molecules the receptor, a careful analysis could reveal the extent of strain and deviation from ideal torsion angles for various classes of drugs in their model 'receptor' site. It seems reasonable to expect that 15–$20\,\mathrm{kJ\,mol^{-1}}$ of strain energy is all that should be tolerated in template forcing. If more is required, and the pharmacophore is correct, it is highly unlikely that the molecule will be very active.

In summary, one can use molecular mechanics to analyze conformational energies of small molecules, not only finding all the local energy minima but those that might correspond to a 'receptor bound' conformation. The two relevant questions to ask are how accurate are the relative energies that emerge from the calculation and how does one know when the global energy minimum for the molecule has been found?

For small organic molecules in which the differential hydrogen-bond/electrostatic contribution to various conformations is not large, molecular mechanics methods typified by the MM2 approach appear to be adequate to describe relative conformational energies. However, as soon as solvent effects and competing solvent–solute and intrasolute interactions become important, molecular mechanics approaches which do not include solvent explicitly become much less reliable.

In addition to the general difficulties in simulating relative conformational free energies for polar and ionic molecules in solution, one has a second set of difficulties when treating molecules with greater than about five torsional degrees of freedom; that of enumerating all the possible local energy minima in the system and correctly evaluating their energies. A number of reasonable conformational search approaches were mentioned in the previous section.

Clearly more research is needed in the area of search strategy and correctly calculating the conformational energies of molecules in solution and these are areas of very active research. Even in the absence of receptor structure, more accurate evaluation of molecular conformational energies would be useful in understanding structure–activity relationships. However, even if one had exact conformational energies of the small molecules, it would still be difficult to make better than qualitative conclusions in the absence of receptor structure, due to the generally non-negligible differences in solvation free energy and free energies of binding between different analogs, as well as uncertainties in the positioning and identity of the groups making up the pharmacophoric pattern.

18.2.4 MOLECULAR MECHANICS STUDIES ON SYSTEMS OF KNOWN RECEPTOR STRUCTURE

Our focus in this section is on molecular mechanics studies of protein–ligand and DNA–ligand interactions where the structure of the receptor is known. Knowledge of the receptor site in atomic detail provides a fertile ground for rational drug design efforts. One must first dock the ligand of interest in the receptor site, unless its position has been solved crystallographically. This is commonly done using interactive stereographics, with the ligand positioned by hand through the aid of the chemist's intuition. Multiple orientations may be created and subjected to energy minimization, but extensive conformational searching is difficult given that the ligand–receptor system will comprise many hundreds of atoms. In addition, conformational searching is complicated by the extra translational and rotational degrees of freedom between receptor and ligand. Kuntz *et al.* have developed an algorithm that performs a directed search of the orientation space of a rigid ligand in a rigid receptor site.[38] The search is based only on steric interactions and is nonexhaustive but has been shown to be effective at locating novel binding orientations as well as reproducing the crystallographically observed orientation of known ligands. This approach was extended by DesJarlais *et al.* to consider limited flexibility in the ligand,[39] and to screen and score large numbers of rigid small molecules for their ability to be docked in the receptor site.[40]

Once a reasonable receptor–ligand geometry is obtained, the ligand and surrounding receptor atoms are energy minimized. The initial goal is to see if the relative calculated interaction energies of various analogs binding to the receptor correlate with experiment. If they do, one might use such an approach to calculate the binding energy of unknown analogs as an aid in drug design.

However, there are significant difficulties with this approach, which can be appreciated with the thermodynamic cycle shown in Scheme 2. The gas phase free energy of binding ΔG_g of drug D to receptor R to form a DR complex can be related to the association energy in solution ΔG_{aq} by the solvation free energies of drug $\Delta G(D)$, receptor $\Delta G(R)$ and complex $\Delta G(DR)$

$$\Delta G_{aq} = \Delta G_g + \Delta G(DR) - \Delta G(D) - \Delta G(R) \tag{15}$$

With molecular mechanics approaches, one calculates the internal energies rather than free energies of complexation, and often does not include solvent explicitly, so one is really calculating ΔE_g. Nonetheless, in comparing closely related analogs it is not unreasonable to assume that the *difference* in gas phase binding energies for the analogs $\Delta\Delta E_g$ would be approximately $\Delta\Delta G_g$. If one is comparing two drugs, Blaney *et al.*[41,42] argue that it is reasonable to assume the difference in the solution phase free energy of binding can be approximated by the sum of the differences

$$\Delta\Delta G_{aq} \approx \Delta\Delta E_g + \Delta\Delta G(D) \tag{16}$$

in molecular mechanical interaction energies $\Delta\Delta E_g$ and solvation free energies $\Delta\Delta G(D)$. This strategy was successfully employed in ranking the relative binding affinity of four thyroxine analogs with the protein prealbumin.[41] When such an approach was used in a predictive mode,[43] it was less successful, probably because the approximations inherent in equation (16) break down when comparing charged and neutral side chain thyroxine analogs.

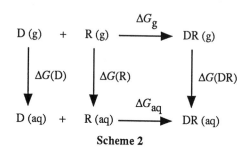

Scheme 2

Equation (16) has been used in comparing dioxin analogs binding to prealbumin[44] and trimethoprim analogs binding to DHFR[45] with reasonable success. In the latter case the binding affinities of two new analogs were successfully predicted by the model. Nonetheless, the difficulties in such an approach involve both the limited searching of conformational space and the large electrostatic effects in these systems, given that one is looking for differences in energy of a few $kJ\,mol^{-1}$ in total energies of many thousands. It is clear that using the molecular dynamics/thermodynamic cycle approach described by McCammon in Chapter 18.3 is a more accurate strategy for such systems, given that computer time is not limiting. This approach circumvents the problem of finding small differences between large energies, but still requires explicit solvent inclusion for accurate free energies and does not solve the 'local minimum problem' although local conformational sampling is accomplished by molecular dynamics.

DeTar[46,47] and Wipff *et al.*[48] have successfully used molecular mechanics methods to compare noncovalent and transition state binding of D- and L-*N*-acetyltryptophan substrates of α-chymotrypsin. Their calculations have allowed a nice rationalization of the significant stereoselectivity that is observed in catalysis but not observed in the initial binding step.

Caldwell and Kollman[49] have used the NMR NOE data of Patel[50] and molecular mechanics calculation to calculate the structure of a complex of netropsin with d(CGCGAATTCGCG)$_2$. The molecular mechanics optimized structure was compared to the independently derived crystal structure of the same complex[51] by least-squares fitting. It was found that the calculated and observed structures of the netropsin molecule in the DNA were within 1.2 Å overall, but within 0.6 Å if the terminal amidinium group was not considered. Considering that the initial model had no crystallographic input and a crude electrostatic treatment was used, this study is an indication that molecular mechanics, combined with solution NMR NOE data, can provide useful structures for analyzing and visualizing drug–DNA complexes.

A second application to drug–DNA complexes is a study of mitomycin C–DNA adducts by Rao *et al.*[52] At the time of this study, experimental evidence pointed to the O-6 atom of guanine as the favored alkylation site, with N-2 a much less favored site. Rao *et al.* noted that O-6 (major groove) and N-2 (minor groove) adducts were *both* quite stereochemically and energetically reasonable and that model building/molecular mechanics could not rationalize the then accepted observation favoring O-6 adduct formation over the N-2 site. The calculations were unable to exclude either model, reflecting the fact that only for very closely related systems can one reasonably use the relative molecular mechanics energies as a guide to determine which structure should be the lowest energy in solution. It was subsequently demonstrated[53] that, indeed, the N-2 adduct was formed with DNA, not the O-6 site previously suggested.

In summary, molecular mechanics methods can be used on a wide variety of protein–ligand or DNA–ligand complexes. These calculations suffer from a number of defects including limited search of conformational space and uncertainties in the energies. As noted above, qualitative structural insights can emerge from such studies, as can, in favorable cases, relative energies that can be qualitatively related to relative binding energies observed experimentally. However, the ability to leap well beyond the observed class of ligands and make confident predictions is not currently possible with molecular mechanics alone. Obviously, computer graphics methods and the input of reasonable structures from X-ray crystallography or model building are critical in these studies and the use of molecular dynamics to more effectively scan local variations in conformations for these macromolecule–ligand complexes is often appropriate. The use of free energy perturbation approaches in addition to molecular mechanics should allow a more effective and accurate study of relative binding affinities of a wider variety of ligands to a given macromolecule and thus be a generally useful tool in drug design.

18.2.5 REFERENCES

1. U. Burkert and N. L. Allinger, *ACS Monogr.*, 1982, **177**.
2. E. M. Engler, J. D. Andose and P. von R. Schleyer, *J. Am. Chem. Soc.*, 1973, **95**, 8005.
3. N. L. Allinger, *J. Am. Chem. Soc.*, 1977, **99**, 8127.
4. F. A. Momany, R. F. McGuire, A. W. Burgess and H. A. Scheraga, *J. Phys. Chem.*, 1975, **79**, 2361.
5. B. R. Brooks, R. E. Bruccoleri, B. D. Olafson, D. J. States, S. Swaminathan and M. Karplus, *J. Comput. Chem.*, 1983, **4**, 187.
6. W. L. Jorgensen and J. Tirado-Rives, *J. Am. Chem. Soc.*, 1988, **110**, 1657.
7. S. J. Weiner, P. A. Kollman, D. A. Case, U. C. Singh, C. Ghio, G. Alagona, S. Profeta, Jr. and P. Weiner, *J. Am. Chem. Soc.*, 1984, **106**, 765.
8. S. J. Weiner, P. A. Kollman, D. T. Nguyen and D. A. Case, *J. Comput. Chem.*, 1986, **7**, 230.
9. S. Lifson and A. Warshel, *J. Chem. Phys.*, 1968, **49**, 5116.
10. T. K. Brunck and F. Weinhold, *J. Am. Chem. Soc.*, 1979, **101**, 1700.
11. T. P. Lybrand and P. A. Kollman, *J. Chem. Phys.*, 1985, **83**, 2923.
12. U. C. Singh and P. A. Kollman, *J. Comput. Chem.*, 1984, **5**, 129.
13. M. J. S. Dewar, E. G. Zoebisch, E. F. Healy and J. J. P. Stewart, *J. Am. Chem. Soc.*, 1985, **107**, 3902.
14. G. Del Re, *J. Chem. Soc.*, 1958, 4031.
15. G. Del Re, B. Pullman and T. Yonezawa, *Biochim. Biophys. Acta*, 1963, **75**, 153.
16. J. Gasteiger and M. Marsili, *Tetrahedron*, 1980, **36**, 3219.
17. J. W. Ponder and F. M. Richards, *J. Comput. Chem.*, 1987, **8**, 1016.
18. T. Schlick and M. Overton, *J. Comput. Chem.*, 1987, **8**, 1025.
19. D. N. J. White and D. H. Kitson, *J. Mol. Graphics*, 1986, **4**, 112.
20. J. Moult and M. N. G. James, *Proteins: Struct. Funct. Genet.*, 1986, **1**, 146.
21. T. Ryhanen, F. J. Bermejo, J. Santoro and M. Rico, *Comput. Chem.*, 1987, **11**, 13.
22. G. H. Paine and H. A. Scheraga, *Biopolymers*, 1985, **24**, 1391.
23. G. M. Crippen, in 'Chemometrics Research Studies: Distance Geometry and Conformational Calculations', ed. D. Bawden, Wiley, New York, 1981, vol. 1.
24. P. K. Weiner, S. Profeta, Jr., G. Wipff, T. Havel, I. D. Kuntz, R. Langridge and P. A. Kollman, *Tetrahedron*, 1983, **39**, 1113.
25. H. J. C. Berendsen, J. P. M. Postma, W. F. van Gunsteren and J. Hermans, in 'Intermolecular Forces', ed. B. Pullman, Reidel, Dordrecht, 1981, p. 331.
26. W. L. Jorgensen, J. Chandrasekhar, J. D. Madura, R. W. Impey and M. L. Klein, *J. Chem. Phys.*, 1983, **79**, 926.
27. M. K. Gilson and B. H. Honig, *Proteins: Struct. Funct. Genet.*, 1988, **3**, 32.
28. M. Whitlow and M. M. Teeter, *J. Am. Chem. Soc.*, 1986, **108**, 7163.
29. D. Eisenberg and A. D. McLachlan, *Nature (London)*, 1986, **319**, 199.
30. Y. K. Kang, G. Nemethy and H. A. Scheraga, *J. Phys. Chem.*, 1987, **91**, 4105.
31. T. Ooi, M. Oobatake, G. Nemethy and H. A. Scheraga, *Proc. Natl. Acad. Sci. USA*, 1987, **84**, 3086.
32. C. Humblet and G. R. Marshall, *Annu. Rep. Med. Chem.*, 1980, **15**, 267.
33. R. S. Struthers, J. Rivier and A. T. Hagler, *Ann. N. Y. Acad. Sci.*, 1985, **439**, 81.
34. T. A. Andrea, S. W. Dietrich, W. J. Murray, P. A. Kollman, E. C. Jorgensen and S. Rothenberg, *J. Med. Chem.*, 1979, **22**, 221.
35. I. Pettersson and T. Liljefors, *J. Comput. Aided Mol. Des.*, 1987, **1**, 143.
36. J. W. Ponder and F. M. Richards, *J. Mol. Biol.*, 1987, **193**, 775.
37. F. H. Allen, S. Bellard, M. D. Brice, B. A. Cartwright, A. Doubleday, H. Higgs, T. Hummelink, B. G. Hummelink-Peters, O. Kennard, W. D. S. Motherwell, J. R. Rodgers and D. G. Watson, *Acta Crystallogr., Sect. B*, 1979, **35**, 2331.
38. I. D. Kuntz, J. M. Blaney, S. J. Oatley, R. Langridge and T. E. Ferrin, *J. Mol. Biol.*, 1982, **161**, 269.
39. R. L. DesJarlais, R. P. Sheridan, J. S. Dixon, I. D. Kuntz and R. Venkataraghavan, *J. Med. Chem.*, 1986, **29**, 2149.
40. R. L. DesJarlais, R. P. Sheridan, G. L. Seibel, J. S. Dixon, I. D. Kuntz and R. Venkataraghavan, *J. Med. Chem.*, 1988, **31**, 722.
41. J. M. Blaney, P. K. Weiner, A. Dearing, P. A. Kollman, E. C. Jorgensen, S. J. Oatley, J. M. Burridge and C. C. F. Blake, *J. Am. Chem. Soc.*, 1982, **104**, 6424.
42. S. J. Oatley, J. M. Blaney, R. Langridge and P. A. Kollman, *Biopolymers*, 1984, **23**, 2931.
43. P. Kollman and J. Blaney, in 'Molecular Graphics and Drug Design', ed. A. S. V. Burgen, G. C. K. Roberts and M. S. Tute, Elsevier, Amsterdam, 1986, p. 285.
44. T. Darden, J. McKinney, A. Maynard, S. Oatley and L. Pedersen, *Ann. N. Y. Acad. Sci.*, 1986, **482**, 249.
45. L. Kuyper, S. Davis and H. LeBlanc, in 'Advances in Gene Technology: Protein Engineering and Production, ICSU Short Reports', ed. K. Brew, IRL Press, Oxford, 1988, vol. 8, p. 190.
46. D. F. De Tar, *J. Am. Chem. Soc.*, 1981, **103**, 107.
47. D. F. De Tar, *Biochemistry*, 1981, **20**, 1730.

48. G. Wipff, A. Dearing, P. K. Weiner, J. M. Blaney and P. A. Kollman, *J. Am. Chem. Soc.*, 1983, **105**, 997.
49. J. Caldwell and P. Kollman, *Biopolymers*, 1986, **25**, 249.
50. D. J. Patel, *Proc. Natl. Acad. Sci. USA*, 1982, **79**, 6424.
51. M. L. Kopka, C. Yoon, D. Goodsell, P. Pjura and R. E. Dickerson, *Proc. Natl. Acad. Sci. USA*, 1985, **82**, 1376.
52. S. N. Rao, U. C. Singh and P. A. Kollman, *J. Am. Chem. Soc.*, 1986, **108**, 2058.
53. M. Tomasz, D. Chowdary, R. Lipman, S. Shimotakahara, D. Veiro, V. Walker and G. L. Verdine, *Proc. Natl. Acad. Sci. USA*, 1986, **83**, 6702.

18.3

Dynamic Simulation and its Applications in Drug Research

J. ANDREW McCAMMON

University of Houston, TX, USA

18.3.1 INTRODUCTION

The early steps in the activity of many drugs can be denoted schematically by

$$D + R \underset{}{\overset{1}{\rightleftharpoons}} DR \underset{}{\overset{2}{\rightleftharpoons}} DR' \underset{}{\overset{3}{\rightleftharpoons}} P \tag{1}$$

Here, process 1 is the diffusional encounter of drug and receptor, process 2 is a conformational change to form a stronger noncovalent complex and process 3 is one or more covalent reactions that lead to a product. The molecular motion involved in such processes must be accounted for in any detailed interpretation or accurate prediction of drug activity. This is true not only in kinetic analyses, but also in thermodynamic analyses where such properties as the flexibility of drugs and receptors enter into calculations of entropy and enthalpy changes.

During the past decade, major advances have been made both in our ability to simulate the molecular motion of complex systems (*e.g.* a drug–receptor complex in water) by the use of fast computers, and in our ability to extract critical information (*e.g.* equilibrium constants) from the results of such simulations.[1] These advances, which reflect developments in chemical theory, numerical algorithms and computer hardware, have now made it feasible to use dynamical simulations as tools for receptor-based drug design. Such applications are, however, far from routine at the present time. Each new study raises fundamental questions concerning the detailed procedures and models to be used. It is clear, however, that simulations will be used increasingly in molecular design and that the reliability and simplicity of this work will also increase with time.

Dynamic simulations can be used in a variety of ways in drug design. Two of the most highly developed approaches, the thermodynamic cycle–perturbation method and the Brownian reactive dynamics method, will be described in some detail in this chapter. The former method typically

139

makes use of molecular dynamics simulations of the Newtonian motion of a set of atoms, while the latter method makes use of Brownian dynamics simulations of molecular diffusion.

In its simplest form, the thermodynamic cycle–perturbation method is used to predict how the affinity of one molecule for another (*e.g.* the affinity of a drug for a receptor) depends on the chemical composition of the two molecules. The solvent surroundings are explicitly included in the simulation model. In other applications, the method can be used to estimate the relative free energies of activation (and thereby the rate constants) of certain kinds of chemical reactions, the relative folding stabilities of homologous proteins, or other thermodynamic differences.

The Brownian reactive dynamics method is used to predict the rate of initial diffusional encounter between reactant molecules in solution. From a design viewpoint, this rate sets the ultimate limit on the speed of enzymatic and other reactions. If the reactant molecules are such that chemical events develop very rapidly when the reactants come into contact, the net rate of reaction will be equal to the rate of diffusional encounter. Because the rates of such diffusion-controlled reactions can be calculated by Brownian dynamics methods, one can predict how changes in the reactants will change these rates.

The thermodynamic cycle–perturbation method and the Brownian reactive dynamics method usually require the availability of both experimental structural data and computational power. For calculations on complicated molecules such as enzymes, one must have available the three-dimensional structure of molecules that are closely related to the design targets. Such calculations also typically require access to supercomputers or minisupercomputers. Fortunately, developments in X-ray crystallography and other structural techniques are providing the needed structural information at an accelerating rate,[2] and the power of and access to computers continues to improve.

Although space does not allow for a full discussion here, it should be noted that the dynamical simulation techniques used in the above methods can also be used in other applications related to drug design. These include the refinement of molecular structures based on X-ray diffraction or NMR data, the building of receptor models based on homologous known structures and the prediction of possible conformations of molecules whose three-dimensional structures are not known.[1]

18.3.2 MOLECULAR DYNAMICS

In a molecular dynamics simulation, the motion of the atoms in a molecule or a group of molecules is traced out for some period of time by solving Newton's equations of motion on a fast computer. The forces on the atoms that are needed in the calculation are obtained from approximate potential energy functions of the kind described in Chapter 18.2. In principle, molecular dynamics simulations can be used to describe many of the kinds of events involved in drug–receptor interactions, including the solvation and conformational changes required for initial complex formation, and any conformational or covalent rearrangements that may occur subsequent to binding. Some of the most important of the emerging methods for drug design do not attempt to follow any of these physical processes, however. Instead, such methods make use of the thermodynamic cycle–perturbation approach, in which hypothetical processes are simulated to determine the thermodynamic effects of possible changes in the chemical structures of drug and receptor molecules. In such simulations, molecular dynamics is used simply as a powerful method for generating the samples of thermally accessible molecular configurations that are needed in calculations of entropies, enthalpies and other thermodynamic quantities. The calculations can be used, for example, to predict how changes in the chemical structure of a drug will change the equilibrium constant for binding to a receptor *if* a high-resolution structure of the original drug–receptor complex is available. Changes in the equilibrium constants that result from mutations in the receptor can also be predicted, which may be helpful in redesigning drugs in response to the development of certain types of drug resistance.

18.3.2.1 Simulation Technique

Molecular dynamics simulations involve four basic steps. First, one must define the model system. This typically includes choosing the size and composition of the system, and choosing the potential energy function that will be used to describe the interatomic interactions. Second, the system must be equilibrated at the temperature of interest. That is, the detailed distribution of atomic positions and velocities in the initial model must be adjusted to be typical of those that would be found at the desired temperature. In particular, any large interatomic forces that may be present (*e.g.* due to placing two atoms too close together in the initial model) must be relaxed; otherwise, the atoms involved would develop excessive kinetic energies during the subsequent dynamics. The third step is the simulation itself, during which the atoms are allowed to move in accordance with Newton's equations, and the final step is the analysis of the resulting trajectory. In this section, some of the key considerations involved in a simulation are described. Additional technical details can be found elsewhere.[1]

For definiteness, consider a system comprising an inhibited enzyme molecule in liquid water. The inhibitor is assumed to be a relatively small molecule. The considerations that arise in simulations of many other systems are quite similar.

The initial structure of the enzyme–inhibitor complex is typically obtained from a high-resolution X-ray diffraction analysis. Such analyses also usually provide the positions of a few dozen to a few hundred water molecules; it is useful to include these in the development of the model system, because these waters already have locations that are well adapted to the protein structure. To build the rest of the solvent surroundings, one usually immerses the X-ray structure (complex plus solvent) in a representative sample of bulk water. The latter is obtained as a single configuration (set of atomic coordinates) from a simulation of the pure liquid. Simulations of the pure liquid make use of 'periodic boundary conditions' (see below); the simulated liquid (typically 216 molecules in a cubic box of edge 16.2 Å) can therefore be replicated to fill all of the space around the complex. Bulk waters that are closer than 2.3 Å to any non-hydrogen atom of the X-ray structure are then deleted. This distance is chosen empirically to account for electrostriction and other effects that increase the local density of water near the protein surface; a slightly larger distance may be appropriate when many water molecules are present in the X-ray structure. The choice of initial structure is usually completed by determining which, if any, water molecules from the bulk should be left in the interior of the protein, and which atoms of basic groups such as histidine imidazoles should be protonated. The maximum number and quality of possible hydrogen bonds and, in the protonation case, estimates of the local electrostatic potential may be considered in such determinations.

The actual size of the system to be studied will depend on a number of factors, not the least of which are the computer resources available. To perform a simulation of the whole complex and its solvent surroundings, one usually makes use of periodic boundary conditions. Here, atoms beyond certain geometric surfaces (*e.g.* the surfaces of a cube that contains the complex and several layers of solvent molecules) are replaced by replicates of the atoms within those surfaces. The finite system is thereby extended in the same manner that a crystal is built from a single unit cell. Atoms in the finite system are allowed to interact with atoms in the surrounding replicates. The number of solvent molecules in the finite system must be chosen large enough that the central complex is not too close to any of its replicates, or 'concentration effects' will occur. Alternatives to periodic boundary conditions may require less computer time. Simply deleting all atoms beyond a certain distance from some point in the complex would leave a free surface from which waters will evaporate during a dynamic simulation. Such free surfaces are sometimes used, however, because the number of waters that escape is generally not large during the periods of typical simulations. Using a centrosymmetric potential to constrain waters at the surface is not ordinarily recommended, because such constraints can produce long-range disturbances of the water structure.[1] More elaborate surface constraints may be useful, however.

The equilibration procedure generally involves the use of both energy minimization and molecular dynamics algorithms. In each of these, the forces on the atoms in the system are computed from a potential energy function. Such functions, and their use in energy minimization, are described in Chapter 18.2. A popular algorithm for molecular dynamics is the 'leap-frog' scheme. Here, each Cartesian coordinate of each atom is moved forward in time according to the equation

$$x(t + \Delta t) = x(t) + \dot{x}(t + \Delta t/2)\Delta t \qquad (2)$$

Equation (2) shows that the *x*-coordinate of the atom at time $t + \Delta t$ can be predicted from the corresponding components of its position at time t and velocity $\dot{x}$ at time $t + (\Delta t/2)$. The velocity is in turn moved forward according to the equation

$$\dot{x}(t + \Delta t/2) = \dot{x}(t - \Delta t/2) + F_x(t)\Delta t/m \qquad (3)$$

where $F_x(t)$ is the *x*-component of the total force acting on the atom in its position at time t, and m is the mass of the atom. For these equations to be reliable, the time step Δt must be a small fraction of the shortest period of motion in the system; a typical time step is 2 fs.

In the equilibration process, it is usually sensible to allow the solvent surroundings to relax first while the complex is held fixed. Otherwise, the stresses in the system due to the unrelaxed solvation shells (*e.g.* of charged side chains) may lead to distortion of the protein conformation. In one equilibration procedure, the water molecules are first subjected to 100 steps of steepest descent energy refinement to relax the largest forces (*e.g.* overlaps with solute atoms). The solvent is then subjected to 20 ps of molecular dynamics, but with new, random velocities reassigned every ps or so from a Maxwellian distribution at the desired temperature. The total period is roughly sufficient for the formation of solvation shells, and the frequent velocity reassignments prevent local 'hot spots' from forming as the forces in the starting structure relax. At this point, the complex itself is allowed to move. The energy refinement and molecular dynamics procedures described above are repeated, but now for all the atoms of the system. The completeness of the equilibration is tested by continuing the molecular dynamics without any velocity reassignments for a few ps; a systematic rise in temperature in any region of the system means that the forces in the starting structure are still relaxing and the dynamic equilibration with velocity reassignments should be continued.

The dynamic simulation itself is simply run by continuing the molecular dynamics calculation for the desired period. Velocity reassignments are not made, but some form of weak coupling to a heat bath is usually applied to keep the average temperature of the system fixed at the desired value. In a traditional dynamic simulation (where the composition of the system is fixed), the quality of the simulation can be gauged by the similarity of the time-averaged structure of the complex to the original X-ray structure and the absence of any systematic drift in the potential energy of the system. Ideally, the root-mean-square difference of the dynamically averaged and X-ray structures will be less than about 2.5 Å. The largest differences should be at the surface of the complex, where packing effects occur in the crystal but not in the simulation.

18.3.2.2   Thermodynamic Cycle–Perturbation Method

The thermodynamic cycle–perturbation method was developed to answer questions such as the following.[3] Given the structure of a ligand–receptor complex, how will the strength of ligand binding change if one group on the ligand is replaced by another? If the complex is that of an enzyme and substrate, how will changes in the substrate affect the rates of the reaction steps that follow binding? Considering possible changes in a protein (such as might be produced by site-directed mutation), how do these changes alter the thermodynamic stability of the native conformation of the molecule? And how would changes in an enzyme or other receptor alter ligand-binding properties and subsequent reactivity? Quantitative answers to such questions can be expressed in terms of equilibrium constants or changes in free energy. Thus, the affinity of a drug D for a receptor P in the simple association reaction

$$D + P \rightleftharpoons DP \tag{4}$$

is given by

$$K = \exp(-\Delta G^\circ / RT) \tag{5}$$

where $K = [DP]/([D][P])$ is the equilibrium constant, ΔG° is the change in the standard state Gibbs free energy, R is the gas constant and T is the temperature in Kelvin. The Gibbs energy is the appropriate measure of the driving force for reactions at constant temperature and pressure. Helmholtz energy changes ΔA (strictly appropriate for constant temperature and volume) are also commonly used; ΔA is similar to the corresponding ΔG for many reactions in solution. Thermodynamic cycle–perturbation calculations usually deal with model systems in which the solutes are effectively at infinite dilution. The free energy differences that are calculated can therefore be identified with standard state differences if the usual biochemical definition of standard states is used.[4]

By focusing on *differences* in homologous processes, the thermodynamic cycle–perturbation method takes advantage of extensive cancellation of large contributions to the changes in thermodynamic functions for a single process. For example, the binding of a drug to a receptor may be accompanied by complicated changes in the conformation, solvation and other properties of both molecules. The changes that accompany the binding of two drugs that differ in a single chemical group may, however, be very similar. The thermodynamic cycle–perturbation method exploits this situation in a natural way by considering the relatively small perturbations of structure and energy associated with replacing one chemical group by another in either the drug–receptor complex (where the complicated changes have already occurred) or in the unbound drug molecule.

The general idea of the thermodynamic cycle–perturbation approach is quite simple. Suppose that one is interested in computing the relative free energy change for two different processes

$$A + B \rightarrow AB \qquad \Delta G_1 \qquad (6)$$

$$A' + B \rightarrow A'B \qquad \Delta G_2 \qquad (7)$$

The desired quantity, $\Delta\Delta G = \Delta G_2 - \Delta G_1$, can in principle be obtained from molecular dynamics simulations in which processes (6) and (7) are caused to occur sufficiently slowly in the appropriate solvent surroundings. To cause a process such as that in equation (6) to occur, an auxiliary force is applied that gradually brings molecules A and B together. Analysis of such simulations using standard methods yields ΔG_1 and ΔG_2.[1,5] Unfortunately, this direct approach is unworkable except for very simple molecules. The difficulty is that each process must be carried out slowly enough that the system remains in thermodynamic equilibrium. For molecules comprising more than a few atoms, conformation or solvation changes requiring more than a few tens of picoseconds can make it difficult to insure that representative configurations of the system develop during the simulations, which themselves typically cover only 10 to 100 ps. In the thermodynamic cycle–perturbation approach, one considers instead the nonphysical processes

$$A + B \rightarrow A' + B \qquad \Delta G_3 \qquad (8)$$

$$AB \rightarrow A'B \qquad \Delta G_4 \qquad (9)$$

Because processes (6) through (9) form a thermodynamic cycle

$$A + B \rightarrow AB$$
$$\downarrow \qquad\qquad \downarrow$$
$$A' + B \rightarrow A'B$$

the desired relative free energy change $\Delta\Delta G = \Delta G_2 - \Delta G_1 = \Delta G_4 - \Delta G_3$. Analogous relations hold for any other thermodynamic state function, *e.g.* enthalpy or entropy. The quantities ΔG_3 and ΔG_4 can again be calculated by molecular dynamics simulations. Because the changes in the processes represented by equations (8) and (9) are typically much smaller and more localized than those in equations (6) and (7), the calculations are greatly simplified.

Free energy changes for nonphysical processes such as equations (8) or (9) can be evaluated by using statistical mechanical perturbation theory.[1,5,6] For example, ΔG_4 can be obtained as

$$\Delta G_4 = -RT \ln \langle \exp(\Delta V/RT) \rangle_{AB} + \Delta G_4^m \qquad (10)$$

for small perturbations. In equation (10) $\Delta V = V_{AB} - V_{A'B}$ where V_{AB} and $V_{A'B}$ are the potential energies of the systems containing AB and A'B, respectively, and $\langle \ \rangle_{AB}$ denotes an average over the thermally accessible configurations of the system containing AB. The quantity ΔG_4^m is a contribution due to the difference in masses of AB and A'B.[7] It is usually not included in the calculation because it cancels with a similar term in the other half of the thermodynamic cycle; that is, $\Delta G_4^m = \Delta G_3^m$.

The average $\langle \exp(\Delta V/RT) \rangle_{AB}$ can be obtained from molecular dynamics simulations of AB and the surrounding solvent under isothermal–isobaric conditions in order to obtain ΔG_4. However, the results usually converge slowly except for very small perturbations. A more general way to obtain a value of ΔG is to express this quantity as

$$\Delta G = \sum_{i=1}^{n} \Delta G_i \qquad (11)$$

where

$$\Delta G_i = -RT \ln \langle \exp[(V_i - V_{i+1})/RT] \rangle_i \qquad (12)$$

In the case of ΔG_4, one has $V_1 = V_{AB}$ and $V_{n+1} = V_{A'B}$; V_i changes gradually (linearly or nonlinearly) from V_{AB} to $V_{A'B}$ when i changes from 1 to $n+1$. Other methods such as thermodynamic integration can alternatively be used to compute the free energy changes.[1,5]

The example given in equations (6)–(9) involves bimolecular (covalent or noncovalent) association. In principle, the thermodynamic cycle–perturbation approach can also be used to predict reaction rates (by calculating relative free energies of activation), to predict the folding stabilities of globular biopolymers (by calculating relative free energies of unfolding) and in other applications. The results of some early applications are mentioned below. Important theoretical and technical problems must be solved before these methods become fully reliable, however; some of these problems are indicated in Section 18.3.4.

18.3.2.3   Illustrative Applications

Several studies of molecular recognition and activity based on the thermodynamic cycle–perturbation method have been published during the past two years.[8–12] The first application of the method was to predict and rationalize the ion-binding selectivity of an organic host molecule in water.[8] Subsequent applications have included studies of the association between enzymes and inhibitors,[9–11] of the folding stability of mutant proteins[10] and of the catalytic rate of an enzyme.[12] Representative examples of some of this work are briefly described here.

In the ion-binding study, the relative free energy of binding Cl^- and Br^- to the macrotricyclic molecule SC24 (Figure 1)[13] in water was calculated.[8] The processes are described by equations (6)–(9) with $A = Cl^-$, $A' = Br^-$ and $B = SC24$. The perturbation technique was used to compute the Helmholtz free energy changes ΔA_3 and ΔA_4. All simulations were performed using the canonical ensemble (*i.e.* constant temperature, volume and number of particles) at 300 K. For the computation of ΔA_3, the system consisted of Cl^- and 214 water molecules in a cubic box of length 18.62 Å with periodic boundary conditions. The system for calculation of ΔA_4 comprised SC24, Cl^- and 191 water molecules in a cubic box as above. The initial structure for the SC24 complex with Cl^- was taken from an X-ray crystallographic study.[14] The potential energy function for the system was that from the GROMOS molecular modeling program, supplemented as described elsewhere.[9,15] In replacing Cl^- by Br^-, the essential change is an increase of the ionic radius by about 0.15 Å. The molecular dynamics simulations were performed using the program AMBER.[16] All hydrogen masses were increased to 10 u to allow a long dynamics time step, $\Delta t = 4$ fs, by slowing the librational motions of groups containing hydrogen. This mass adjustment has no effect on equilibrium properties of a classical system, but does result in a more efficient sampling of configurations.

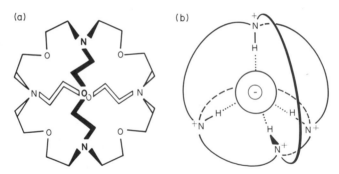

Figure 1   (a) Structure of the synthetic receptor molecule SC24.[13] (b) Schematic diagram of the complex formed between SC24 and a chloride or bromide ion (the negatively charged sphere in the center of the complex). The ability of the receptor to recognize or selectively bind one ion *versus* another in solution is quantitatively defined by the ratio of the equilibrium constants for binding the two ions. This ratio has been calculated by the thermodynamic cycle–perturbation method[8]

After extensive equilibration of each system, simulations were continued for 30 ps. Configurations were saved every 0.1 ps, and were used to calculate ΔA_3 and ΔA_4 as outlined above. The perturbations are small enough that statistically reliable values ($\Delta A_3 = 3.35 \pm 0.15$ kcal mol^{-1}, $\Delta A_4 = 7.50 \pm 0.20$ kcal mol^{-1}; where 1 cal = 4.18 J) could be obtained using equation (10) rather than the more general equations (11) and (12). The predicted relative free energy of binding ($\Delta\Delta A = 4.15$ kcal mol^{-1}) is in good accord with the experimental result that was subsequently made available (4.3 kcal mol^{-1}).[17] Examination of ΔA_3 and ΔA_4 suggests that selective binding of Cl^- to SC24 is due to the highly favorable interaction of Cl^- with the receptor, which more than compensates for the unfavorable free energy of desolvation of Cl^- *versus* Br^-. The more favorable interaction of Cl^- with the receptor relative to Br^- arises because the Br^- anion is slightly too large to be comfortably accommodated in the relatively rigid SC24 molecule.

The benzamidine–trypsin system was chosen for initial studies of biomolecular recognition using the thermodynamic cycle–perturbation method.[9,10] This choice was motivated by several factors. Benzamidine is one of the simplest inhibitors of trypsin. An X-ray structure of the enzyme–inhibitor complex is available.[18] In this, it is apparent that the benzene moiety of the inhibitor contacts several hydrophobic groups in the walls of the enzyme's specificity pocket, and that the positively charged amidinium moiety forms a salt bridge with the carboxyl group of an aspartic acid residue at the base of this pocket. Thermodynamic data are available for the binding of benzamidine with various substituent groups to trypsin.[19] Also, a variety of modified trypsins have been produced by genetic manipulations; these and other modified forms of the enzyme are obvious candidates for model studies of ligand binding.[20]

The first step of the thermodynamic calculation was to carry out a molecular dynamics simulation of benzamidine-inhibited trypsin in water. The system is a very large one, comprising 4785 water molecules plus the inhibited enzyme in a box of dimensions $49.15 \times 54.43 \times 64.28$ Å. Hydrogen atoms capable of participating in hydrogen bonds were included explicitly. The total number of atoms in the system was 16 384. Periodic boundary conditions were used; the box dimensions were chosen to be large enough that all solute atoms were separated from the closest atoms of image solutes by at least four layers of solvent. The dynamics calculations were carried out on a supercomputer using a vectorized version of the GROMOS program.[15] Starting with the X-ray structure for the complex and a bulk solvent configuration, the system was relaxed and equilibrated at 300 K during a period of 16.3 ps. A subsequent 28.8 ps simulation was carried out with the system coupled to a constant temperature bath at 300 K.

To compare the binding of differently substituted benzamidines to trypsin, or of benzamidine to differently substituted enzymes, it is necessary to supplement the simulation of the complex with simulations of the separated inhibitor or enzyme, respectively, in water. Such simulations have been carried out, as described elsewhere.[9, 10] Significant results of the analysis of these simulations include the following. The *p*-fluoro analog of benzamidine binds somewhat less strongly to trypsin than does benzamidine itself ($\Delta\Delta A \approx 0.9$ kcal mol^{-1}). Unlike the ion-binding system described above, selectivity in the present case primarily reflects solvation effects, it being less difficult to desolvate benzamidine than its *p*-fluoro derivative ($\Delta A_3 \simeq -0.8$ kcal mol^{-1}). The effects upon benzamidine binding of the mutation Gly 216 → Ala in trypsin have also been examined (Figure 2). The mutant enzyme was predicted to have the lower affinity for benzamidine ($\Delta\Delta A \approx 1.3$ kcal mol^{-1}), primarily as a result of steric crowding in the binding site due to the methyl group added in the mutation. All of the net changes in free energies of binding are in accord with experimental data.[9]

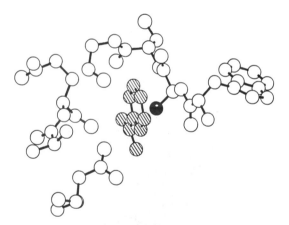

Figure 2 The inhibitor benzamidine (hatched spheres) bound in the specificity pocket of the enzyme trypsin. The methyl group (solid sphere) added in the Gly 216 → Ala 216 mutation projects toward the amidinium group. The resulting steric conflict causes the binding affinity to be reduced from that of the wild-type enzyme. The ratio of equilibrium constants for inhibitor binding has been calculated by the thermodynamic cycle–perturbation method[1, 9, 10]

Exciting applications of the thermodynamic cycle–perturbation method to other enzymes have also been reported recently. These include successful predictions of the effects of deleting a hydrogen-bonding group in an inhibitor of the enzyme thermolysin,[11] and of the change in the free energy of activation (or rate constant) for cleavage of a tripeptide substrate due to a mutation in the enzyme subtilisin.[12] Related methods have been used by Warshel and co-workers to study the catalytic effects of mutations in trypsin[21] and subtilisin.[22] The thermodynamic cycle–perturbation method has also been used to predict the relative free energies of hydration of a variety of molecules[7, 23–29] and the free energy of binding of inert gases in myoglobin.[30]

18.3.3 BROWNIAN DYNAMICS

In Brownian dynamics, the diffusional motion of one or more solute molecules in a liquid is simulated by generating trajectories that are consistent with the Smoluchowski equation. The simulation method is like molecular dynamics in that the particles of interest are moved incrementally in a way that reflects the interparticle forces. Unlike molecular dynamics, however, solvent molecules are not represented explicitly. Instead, the effects of the solvent are included implicitly by three modifications of the equation of motion. First, the forces among the particles are scaled or changed in form to reflect the average effect of the implicit solvent; *e.g.* a dielectric constant of 78 might be introduced to account for the effect of solvent water in attenuating electrostatic forces among the solute particles. In more technical terms, the potential energy function for the particle interactions is replaced by a potential of mean force. Second, frictional drag terms are added to the equation of motion to describe the viscous damping of the motion of the particles by the solvent. Finally, force terms that vary randomly in time (*i.e.* 'stochastically') are added to describe the thermal kicks that would be given to the particles by the solvent molecules if the latter were explicitly present.

Brownian dynamics can be used to simulate any diffusional phenomenon in which the motion of the particles is overdamped, *i.e.* where the frictional and stochastic forces are strong enough to randomize the velocities of the particles before they can move a significant distance. The major applications have been to determine the effects of electrostatic interactions on the rates or protein–ligand or protein–protein binding. Such calculations should be useful in redesigning certain enzymes of pharmacologic importance.

18.3.3.1 Simulation Technique

Brownian dynamics simulations involve the same four steps as molecular dynamics (*cf.* Section 18.3.2.1), but the details of each step are somewhat different.[1] In the first step, definition of the model system, one must specify what components of the molecules of interest are to be explicitly represented and what functions are to be used to describe the interactions among these components. In a typical simulation of a small ligand diffusing toward a receptor protein, the ligand and receptor are both approximated as rigid objects of appropriate dimensions and shape. The ligand may be a sphere or a small number of fused spheres, with each sphere representing a single atom or a group of atoms (*e.g.* an amino acid residue). The size and shape of larger molecules are commonly represented by assemblies of small cubes. These assemblies are easily constructed by determining which elements of a cubic lattice have their centers within the van der Waals surface of the large molecule. These assemblies of spheres and cubes are used to define the excluded volume or collision surfaces of the corresponding molecules.

Other intermolecular interactions are determined by embedding interaction centers within the molecular surfaces and specifying laws for the forces among these centers. Typical electrostatic interaction centers are the partial atomic charges located at the positions corresponding to the atomic nuclei, or sums of these charges for the atoms in a group, with the net charge located at the centroid or other appropriate position in the group. The electrostatic forces acting on these centers are usually determined by numerical solution of the Poisson–Boltzmann equation[31] (see also Section 18.3.3.3). In the ligand–receptor case, the region inside the receptor collision surface is approximated as a homogeneous medium with a low dielectric constant, and the surrounding solution as a homogeneous medium with a high dielectric constant plus additional attenuation of electrostatic interactions due to any salt ions in the solution. The Poisson–Boltzmann equation is used to determine the electrostatic potential $\phi(r)$ in the solution. The electrostatic force acting on any charge q in the ligand is computed as

$$F_{el} = -q\nabla\phi \tag{13}$$

The only other systematic forces that must generally be considered in Brownian dynamics simulations are the frictional forces associated with the motion of the molecular components through the viscous solution. For a single sphere, the frictional force F_f acting on the sphere is proportional to its velocity v relative to the solvent

$$F_f = -fv \tag{14}$$

The friction coefficient is given by Stokes' law

$$f = 6\pi\eta a \tag{15}$$

where η is the shear viscosity of the solvent and a is the hydrodynamic radius of the sphere. In practice, frictional forces enter the simulation algorithm through the diffusion coefficients of the molecules (see below). For a sphere, the diffusion coefficient is related to f by the Einstein formula

$$D = k_B T/f \tag{16}$$

Equations analogous to those above can be written for more complicated molecular shapes.

The kinds of models outlined above, together with the time propagation algorithm to be described shortly, are sufficient to carry out fairly realistic simulations of the association of many small ligands with their receptors. In such a simulation, the ligand carries out a random walk, biased by electrostatic forces, until it collides with the surface of a receptor and either binds (reacts) or is reflected, depending on whether the collision is at the active site. More detailed models can also be constructed if necessary. They can, for example, allow for internal flexibility of the ligand or receptor or both.[1] It is only when internal flexibility is included that an equilibration step is usually needed prior to the simulation itself.

The actual simulation or time propagation step in Brownian dynamics is quite straightforward. For notational simplicity, consider again the diffusion of a small, spherical ligand subject to the force $F(r)$ due to a receptor molecule. If the position of the ligand is initially $r°$, its position after a time Δt is given by the Ermak–McCammon algorithm as[32]

$$r = r° + (k_B T)^{-1} D F(r°)\Delta t + R \tag{17}$$

where D is the diffusion coefficient of the ligand and R is a vector of random numbers with the statistical properties

$$\langle R \rangle = 0 \tag{18}$$

$$\langle R_i R_j \rangle = 2D\delta_{ij}\Delta t \tag{19}$$

The labels i and j distinguish the Cartesian components of R. In generating trajectories with equation (17), the time step Δt must be short enough that $F(r) \approx F(r°)$. Similar algorithms are available for the cases of many spherical particles with arbitrary forces and hydrodynamic interactions, nonspherical particles modeled as flexible or rigid assemblies of spherical subunits, and spheres with translational and rotational motion explicitly coupled by hydrodynamic interactions.[1]

A full Brownian dynamics trajectory is computed by repeated application of equations such as the above to trace out a sequence of particle positions. The lengths of Brownian trajectories are usually longer than those in molecular dynamics simulations, primarily because the number of particles that are explicitly considered is much smaller. Also, the time steps are usually somewhat larger in the Brownian case, especially when variable time step algorithms are used.[1]

The trajectories can be analyzed to provide a variety of types of information. Perhaps of greatest interest from the standpoint of drug design is the calculation of the rates of diffusion-controlled binding processes or reactions, which is discussed in Section 18.3.3.2. Other types of information that can be obtained from Brownian dynamics simulations include molecular transport properties and, in the case of flexible molecules, information on conformational fluctuations or transitions (*e.g.* bending of DNA or helix–coil transitions of polypeptides).

18.3.3.2 Brownian Reactive Dynamics Method

The Brownian reactive dynamics method was developed to calculate the rate at which reactant molecules diffusing in solution will collide with the appropriate orientations for reaction.[33-37] If chemical events within the encounter complex unfold so quickly that the overall rate of reaction is just equal to this rate of encounter, the reaction is said to be diffusion controlled.[37-41] A wide variety of reactions exhibit diffusion control, including many reactions between ions or free radicals, and certain biological redox and enzyme-catalyzed reactions.

The rates of diffusional encounter between real molecules are influenced by a number of complicated factors.[1] Typically, only part of the molecular surface is active. Electrostatic, hydrodynamic and other interactions between the reaction partners will generally produce attractive or repulsive forces that can increase or decrease the rate of diffusional encounter. Structural fluctuations in the reactant molecules may result in a time-dependent reactivity upon contact. Effects such as these can influence reactions in a variety of interesting ways. For example, reactants that have complementary distributions of electric charge may be 'steered' into productive orientations during diffusional encounters, resulting in increased reaction rates.

In the Brownian reactive dynamics method, one carries out simulations of the diffusional motion of reactant molecules and then analyzes the trajectories to determine the mechanism and rate of reactive encounter. The model systems used in the simulations can be constructed to include virtually any feature of real molecules. Also, any appropriate algorithm can be used to simulate the motion of the reactants. Most studies to date have made use of the Ermak–McCammon algorithm, described in Section 18.3.3.1.

The calculation of a rate constant for reactive encounter is in principle straightforward in the case of a dilute solution, where it is sufficient to consider the dynamics of one reactant particle relative to a reaction partner (the 'target') that is translationally fixed at some position in space. If the target molecule is much larger than the diffusing reactant, one can also neglect the rotational motion of the target, which then is rigidly fixed. The rate constant can then be written as

$$k = k_D(b)\beta_\infty \tag{20}$$

Here, $k_D(b)$ is the steady state rate at which mobile particles first strike a spherical surface of radius b around a target particle, and β_∞ is the probability that a reactant starting at $r = b$ will react with the target rather than escape. If b is chosen large enough that the reactant pair interactions are approximately centrosymmetric for $r > b$, then $k_D(b)$ can be determined analytically. For example, $k_D(b)$ is given by the familiar Smoluchowski result[38-41]

$$k_D(b) = 4\pi b D \tag{21}$$

where D is the relative diffusion constant for the reactant pair, if the interaction force between the reactants vanishes for $r > b$.

The quantity β_∞ reflects the complicated interactions that occur at short range and is calculated from the simulated trajectories. In principle, one could carry out a number of trajectory calculations, each with a mobile particle starting at a randomly chosen point on the $r = b$ surface, and obtain β_∞ as that fraction of the trajectories that lead to reaction rather than escape. This straightforward approach has to be modified, however, because of the difficulty of determining the ultimate fate of particles that have not reacted at the end of a simulation of finite length. Given the nature of Brownian motion, even a mobile particle that has diffused a large distance away from the target could ultimately return toward the target and react. In practice, therefore, trajectories are truncated at a radius $q > b$ and the increase in the rate constant due to trajectories that would have returned and reacted is determined as an analytic correction. The formula that is used in the simplest calculations is then[33]

$$k = k_D(b)\beta[1 - (1 - \beta)\Omega]^{-1} \tag{22}$$

where β is the uncorrected probability of reaction in the presence of the truncation surface and $\Omega = k_D(b)/k_D(q)$.

Thus, to calculate the rate constant for a bimolecular diffusion-controlled reaction in dilute solution, one need only compute a number of trajectories of one reactant diffusing in the vicinity of the other, fixed reactant. Trajectories are initiated at $r = b$ and terminated upon reaction or upon reaching $r = q$. The fraction of trajectories that react is β, and this quantity yields the rate constant through equation (22). Analysis of the trajectories also provides information on the mechanistic details of the reaction, for example whether reactants tend to be steered into productive collision geometries during the diffusional encounter.

A variety of extensions of these basic ideas have also been developed. These extensions require somewhat more elaborate calculations but can lead to substantial increases in efficiency in the rate constant calculations for certain systems.[34-37]

18.3.3.3   Illustrative Applications

The Brownian reactive dynamics method has been used successfully to calculate reaction rates in two rather different systems. One system involves the diffusion-controlled reaction of the small substrate molecule superoxide (O_2^-) catalyzed by the enzyme superoxide dismutase (SOD), while the other involves the diffusion-controlled association of two proteins (cytochrome *c* and cytochrome *c* peroxidase) to form an electron transfer complex. In both systems, substantial rate enhancements result from electrostatic interactions that steer the reactants toward optimal reaction geometries during diffusional encounter.

The reaction rate of the SOD–O_2^- system is quite high and decreases with increasing ionic strength of the solvent.[42,43] Because the active sites represent a small fraction ($\sim 0.1\%$) of the surface area of the dimeric enzyme, and the electrostatic charges of the enzyme (-4) and substrate (-1) are of the same sign, it has been recognized that the approach of substrate to the active sites is likely facilitated by the anisotropic electrostatic potential field around the enzyme.[42–44] These results were qualitatively confirmed in the first Brownian dynamics simulations of simple models of the SOD–O_2^- system.[45] Current studies of SOD reactivity involve fairly realistic models of the molecular shapes and interactions.[46,47] In the calculations by Allison *et al.*,[37,45] the topography of the enzyme surface, including the important channels leading to the catalytic centers, is modeled by construction of a 1 Å resolution exclusion grid based on the X-ray structure of the enzyme.[48] Also, the electrostatic potential field is evaluated on a 1.25 Å resolution grid by using finite difference methods to solve the linearized Poisson–Boltzmann equation (Figure 3).[47,49,50] This electrostatic model reflects the detailed topography of the enzyme, the difference in dielectric constants of the enzyme and solvent, and the concentration of salt dissolved in the solvent. The calculated rate of O_2^- reaction ($k = 3.9 \times 10^9 \, \text{mol}^{-1} \, \text{s}^{-1}$ at univalent salt concentration of 0.1 M) is in reasonable agreement with experiment ($k \approx 2 \times 10^9 \, \text{mol}^{-1} \, \text{s}^{-1}$, from ref. 43), as is the variation of the rate with salt concentration (Figure 4).

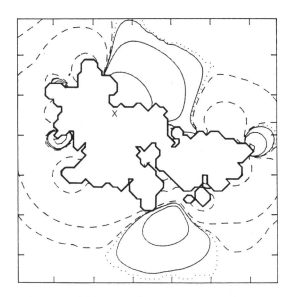

Figure 3 The electrostatic potential energy field surrounding the enzyme superoxide dismutase in an electrolyte solution (0.1 M ionic strength). Brownian reactive dynamics calculations show how the catalytic rate of the enzyme is increased by the effect of this field in steering the diffusion of substrate (O_2^-) molecules toward the active sites, one of which is indicated by an X.[1,37,45–47] The contours correspond to surfaces at which the effective potential energy of O_2^- is ± 0.006, ± 0.06 or $\pm 0.6 \, \text{kcal mol}^{-1}$; dashed and solid lines indicate positive and negative energies, respectively. The zero energy contour is represented by dotted lines. Contours with the largest magnitudes are closest to the protein, whose surface is represented by the heavy solid line. Hatch marks at the boundary of the figure are separated by 12.5 Å

The recent study by Northrup *et al.*, of cytochrome *c* association with cytochrome *c* peroxidase makes use of topographic and electrostatic models similar to those described above for the larger of the two proteins (*peroxidase*), which is treated as a stationary target.[51] Cytochrome *c* is represented as a rigid array of 34 charges, corresponding to the ionized residues. Each charge interacts with the peroxidase as an independent small test charge (*i.e.* the interior of cytochrome *c* is approximated as an extension of the solution to simplify the calculation of electrostatic forces and torques). The steric interactions with peroxidase are determined by the positions of a set of surface atoms, and the array translates and rotates independently with appropriate diffusion constants. Stringent conditions for productive association were defined in terms of the relative distance and orientation of the heme groups in the two proteins. Two reactive regions at the peroxidase surface were found to satisfy these

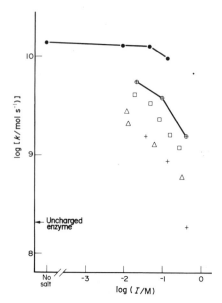

Figure 4 Dependence upon ionic strength I of the bimolecular rate constant k for O_2^- collision with the active sites of superoxide dismutase. The units of I are M and those of k are mol s^{-1}. The results of Brownian dynamics simulations are indicated by symbols connected by lines (filled circles from ref. 46; open circles from ref. 47). Experimental results are indicated by disconnected symbols (crosses for the NaCl data from ref. 42; triangles for the phosphate data from ref. 42; squares for the NaCl data from ref. 43). The ionic strengths are calculated to take proper account of the degree of phosphate ionization[47]

conditions, a primary region near Asp 34 and a secondary region near Asp 148. Good agreement with experimental data for the association rate and its dependence on salt concentration was obtained for this model. The electrostatic forces and torques were essential to account for the rate. At a univalent salt concentration of 0.1 M, and with a requirement of heme plane alignment to within 60° for reaction, the calculated rate constants in the presence and absence of the electrostatic interactions are 5.4×10^8 M^{-1}s^{-1} and 0.41×10^8 M^{-1}s^{-1}, respectively; the corresponding experimental rate constant is 4.75×10^8 M^{-1}s^{-1}.[52]

18.3.4 CONCLUDING REMARKS

The design potential of the thermodynamic cycle–perturbation and Brownian reactive dynamics methods is clear. These new methods can, in favorable cases, provide accurate predictions concerning the thermodynamic and kinetic effects of changes in molecular composition. The reliability of the predictions can be expected to depend on the system considered and the extent of the compositional changes, however. Consider, for example, a ligand and receptor that can bind in different orientations with only a small difference in free energy. Replacement of one functional group on the ligand by another could lead to a reversal of the favored orientation, yet this reversal may not be seen in the simulation if there is a sizable energy barrier to motion from the first orientation to the second. Additional theoretical work is needed to handle problems like this one, in which the systems to be compared differ by an extensive structural rearrangement.

Other limitations of the methods described here should also be mentioned. These include the implicit assumption in studies of mutant proteins that the mutations will not lead to substantial unfolding and inactivation of the protein. In principle, the thermodynamic cycle–perturbation method can be used to predict shifts in the folding–unfolding equilibrium, but practical calculations will require a better understanding of the structures of unfolded proteins than is currently available.[10] All of the types of calculations discussed here are also limited by the accuracy of the underlying potential energy functions. For example, the types of functions that are commonly used at present do not explicitly allow for fluctuations in the polarization of electronic distributions within atoms and molecules, and may therefore not be quantitatively reliable for the study of certain very strong electrostatic interactions.

Fundamental research underway in a number of laboratories will increase the power and the range of applicability of the methods described in this chapter. The development of potential energy functions is an ongoing effort, as described in Chapter 18.2. Progress in the inclusion of electronic polarizability effects has been made by groups at the University of Groningen[53] (H. Berendsen, in a personal communication) and at Caltech (W. Goddard, in a personal communication). Extensions of the current methods for solving the Poisson–Boltzmann equation will allow for the correct treatments of two diffusing reactant molecules as objects that have low dielectric constants and that exclude diffusing co- and counter-ions.[54] These methods will allow a more exact analysis of diffusion-controlled associations of proteins and of other pairs of macromolecules, such as nucleic acids and regulatory molecules. The development of more efficient methods for transforming one molecule into another and calculating the resulting free energy changes will be useful in thermodynamic cycle–perturbation calculations[55] (U.C. Singh, in a personal communication). Applications of the thermodynamic cycle–perturbation approach to calculate relative free energies of activation will be made more rigorous and accurate by new methods of accounting for changes in the location of transition states that may result from changes in molecular composition.[56]

At present, then, the dynamic simulation methodology is essentially in place for predicting the functional consequences of small changes in composition of molecules whose three-dimensional structures are known. Research efforts such as those outlined above will increase the accuracy and range of such predictions, as will the increasing power of the computers that are used in this work. Finally, early work is in progress that will allow replacement of classical dynamics by quantum dynamics in simulations of biological molecules.[57, 58] Such simulations will be useful in the design of electron transfer agents, for example.

ACKNOWLEDGEMENT

Many of the ideas and results presented here were developed in collaboration with S. A. Allison, R. J. Bacquet, W. F. Lau, T. P. Lybrand, M. H. Mazor, S. H. Northrup, B. M. Pettitt and C. F. Wong. The AMBER and GROMOS programs were generously provided by Professors P. A. Kollman and W. F. van Gunsteren, respectively. This work was supported in part by the National Science Foundation, the National Institutes of Health, the Robert A. Welch Foundation, the Texas Advanced Technology Research Program, Sterling–Winthrop Research Institute, the National Center for Supercomputing Applications, IBM, NEC and Cray Research. The author is the recipient of the G. H. Hitchings Award from the Burroughs–Wellcome Fund.

18.3.5 REFERENCES

1. J. A. McCammon and S. C. Harvey, 'Dynamics of Proteins and Nucleic Acids', Cambridge University Press, Cambridge, 1987.
2. W. G. J. Hol, *Angew. Chem., Int. Ed. Engl.*, 1986, **25**, 767.
3. B. L. Tembe and J. A. McCammon, *Comput. Chem.*, 1984, **8**, 281.
4. K. E. Van Holde, 'Physical Biochemistry', Prentice-Hall, Englewood Cliffs, NJ, 1971.
5. M. Mezei and D. L. Beveridge, *Ann. N.Y. Acad. Sci.*, 1986, **482**, 1.
6. J. P. M. Postma, H. J. C. Berendsen and J. R. Haak, *Faraday Symp. Chem. Soc.*, 1982, **17**, 55.
7. T. P. Lybrand, I. Ghosh and J. A. McCammon, *J. Am. Chem. Soc.*, 1985, **107**, 7793.
8. T. P. Lybrand, J. A. McCammon and G. Wipff, *Proc. Natl. Acad. Sci. USA*, 1986, **83**, 833.
9. C. F. Wong and J. A. McCammon, *J. Am. Chem. Soc.*, 1986, **108**, 3830.
10. C. F. Wong and J. A. McCammon, in 'Structure, Dynamics and Function of Biomolecules', ed. A. Ehrenberg, R. Rigler, A. Graslund and L. Nilsson, Springer-Verlag, Berlin, 1987, p. 51.
11. P. A. Bash, U. C. Singh, F. K. Brown, R. Langridge and P. A. Kollman, *Science (Washington, D.C.)*, 1987, **235**, 574.
12. S. N. Rao, U. C. Singh, P. A. Bash and P. A. Kollman, *Nature (London)*, 1987, **328**, 551.
13. J. M. Lehn, *Science (Washington, D.C.)*, 1985, **227**, 849.
14. B. Metz, J. M. Rosalky and R. Weiss, *J. Chem. Soc., Chem. Commun.*, 1976, **14**, 533.
15. J. Hermans, H. J. C. Berendsen, W. F. van Gunsteren and J. P. M. Postma, *Biopolymers*, 1984, **23**, 1513.
16. P. K. Weiner and P. A. Kollman, *J. Comput. Chem.*, 1981, **2**, 287.
17. E. Kaufmann, Ph.D. Thesis, Université Louis Pasteur, Strasbourg, France, 1979.
18. W. Bode and P. Schwager, *J. Mol. Biol.*, 1975, **98**, 693.
19. M. Mares-Guia, D. L. Nelson and E. Rogana, *J. Am. Chem. Soc.*, 1977, **99**, 2331.
20. C. S. Craik, C. Largman, T. Fletcher, S. Roczniak, P. J. Barr, R. Fletterick and W. J. Rutter, *Science (Washington, D.C.)*, 1985, **228**, 291.
21. A. Warshel and F. Sussman, *Proc. Natl. Acad. Sci. USA*, 1986, **83**, 3806.
22. A. Warshel, F. Sussman and J. K. Hwang, *J. Mol. Biol.*, 1988, **201**, 139.
23. W. L. Jorgensen and C. Ravimohan, *J. Chem. Phys.*, 1985, **83**, 3050.
24. C. L. Brooks, III, *J. Phys. Chem.*, 1986, **90**, 6680.
25. T. P. Straatsma, H. J. C. Berendsen and J. P. M. Postma, *J. Chem. Phys.*, 1986, **85**, 6720.
26. U. C. Singh, F. K. Brown, P. A. Bash and P. A. Kollman, *J. Am. Chem. Soc.*, 1987, **109**, 1607.

27. P. A. Bash, U. C. Singh, R. Langridge and P. A. Kollman, *Science (Washington, D.C.)*, 1987, **236**, 564.
28. C. L. Brooks, III, *J. Chem. Phys.*, 1987, **86**, 5156.
29. S. H. Fleischman and C. L. Brooks, III, *J. Chem. Phys.*, 1987, **87**, 3029.
30. J. Hermans and S. Shankar, *Isr. J. Chem.*, 1986, **27**, 225.
31. H. L. Friedman, 'A Course in Statistical Mechanics', Prentice-Hall, Englewood Cliffs, NJ, 1985.
32. D. L. Ermak and J. A. McCammon, *J. Chem. Phys.*, 1978, **69**, 1352.
33. S. H. Northrup, S. A. Allison and J. A. McCammon, *J. Chem. Phys.*, 1984, **80**, 1517.
34. S. A. Allison, N. Srinivasan, J. A. McCammon and S. H. Northrup, *J. Phys. Chem.*, 1984, **88**, 6152.
35. S. A. Allison, S. H. Northrup and J. A. McCammon, *J. Chem. Phys.*, 1985, **83**, 2894.
36. S. H. Northrup, M. S. Curvin, S. A. Allison and J. A. McCammon, *J. Chem. Phys.*, 1986, **84**, 2196.
37. J. A. McCammon, R. J. Bacquet, S. A. Allison and S. H. Northrup, *Faraday Discuss. Chem. Soc.*, 1987, **83**, 213.
38. D. F. Calef and J. M. Deutch, *Annu. Rev. Phys. Chem.*, 1983, **34**, 493.
39. O. G. Berg and P. H. von Hippel, *Annu. Rev. Biophys. Biophys. Chem.*, 1985, **14**, 131.
40. J. A. McCammon, S. H. Northrup and S. A. Allison, *J. Phys. Chem.*, 1986, **90**, 3901.
41. J. Keizer, *Chem. Rev.*, 1987, **87**, 167.
42. A. Cudd and I. Fridovich, *J. Biol. Chem.*, 1982, **257**, 11443.
43. E. Argese, P. Viglino, G. Rotilio, M. Scarpa and A. Rigo, *Biochemistry*, 1987, **26**, 3224.
44. E. D. Getzoff, J. A. Tainer, P. K. Weiner, P. A. Kollman, J. S. Richardson and D. C. Richardson, *Nature (London)*, 1983, **306**, 287.
45. S. A. Allison and J. A. McCammon, *J. Phys. Chem.*, 1985, **89**, 1072.
46. K. Sharp, R. Fine and B. Honig, *Science (Washington, D.C.)*, 1987, **236**, 1460.
47. S. A. Allison, R. J. Bacquet and J. A. McCammon, *Biopolymers*, 1988, **27**, 251.
48. J. A. Tainer, E. D. Getzoff, K. M. Beem, J. S. Richardson and D. C. Richardson, *J. Mol. Biol.*, 1982, **160**, 181.
49. J. Warwicker and H. C. Watson, *J. Mol. Biol.*, 1982, **157**, 671.
50. I. Klapper, R. Hagstrom, R. Fine, K. Sharp and B. Honig, *Proteins: Str. Funct. Gen.*, 1986, **1**, 47.
51. S. H. Northrup, J. O. Boles and J. C. L. Reynolds, *Science (Washington, D.C.)*, 1988, **241**, 67.
52. C. H. Kang, D. L. Brautigan, N. Osheroff and E. Margoliash, *J. Biol. Chem.*, 1978, **253**, 6502.
53. H. J. C. Berendsen, J. R. Grigera and T. P. Straatsma, *J. Phys. Chem.*, 1987, **91**, 6269.
54. M. E. Davis and J. A. McCammon, *J. Comput. Chem.*, 1989, in press.
55. M. Mezei, *J. Chem. Phys.*, 1987, **86**, 7084.
56. J. D. Madura, B. M. Pettitt and J. A. McCammon, *Chem. Phys. Lett.*, 1987, **140**, 83.
57. A. Kuki and P. G. Wolynes, *Science (Washington, D.C.)*, 1987, **236**, 1647.
58. C. Zheng, C. F. Wong, J. A. McCammon and P. G. Wolynes, *Nature (London)*, 1988, **334**, 726.

18.4

Three-dimensional Structure of Drugs

CARLO SILIPO and ANTONIO VITTORIA

University of Naples, Italy

18.4.1 INTRODUCTION

The study of steric requirements for interactions between ligands and corresponding biological acceptor sites is often of decisive importance in understanding the role played by structural features in promoting activity. In its most general form, drug–receptor theory requires that a ligand exerts its biological action as a consequence of binding or otherwise interacting with a specific biological acceptor site, such as a membrane protein, an enzyme, *etc.*, which may be generally termed the receptor. The concept that is the basis for modern drug–receptor theory involves the old principle that a ligand fits its receptor much as a key fits a lock. This concept, although somewhat arbitrary since a high degree of flexibility is present in biomacromolecules, indicates that the geometric arrangement of ligand functionalities, *i.e.* its three-dimensional structure, governs the principles of molecular recognition and molecular discrimination. Although stereochemistry often plays a major role in drug bioactivity, care must be taken when considering structure–activity relationships to explore whether other differences in physicochemical properties exist before one makes significant correlations with the steric properties of the structures under study.

Steric effects may arise in a number of ways. A primary factor results from repulsion between non-bonded atoms. Such repulsions may determine not only the intramolecular steric influence of substituents on molecular properties but also the specific intermolecular influence on the ligand fitting to the receptor. These influences are connected with the bulk and spatial arrangements of ligand subfragments and substituents. Moreover, a set of secondary steric effects may be involved in determining the variation of chemical reactivities; these effects may include changes in the solvation of the active site, shielding from interaction by a bulky group, variation in the electronic delocalization, *etc.*

The importance of stereoselectivity in drug action is well known. There are many examples of drugs that have activities highly dependent on their geometry and configuration. While the correlation between geometrical isomerism and biological activity has received rather little attention from the steric point of view, since vast differences in the physicochemical properties of the geometrical isomers may be involved, the differences in biological activity between optical isomers are largely dependent on their ability to react selectively with a biological system. In fact, besides the case in which marked differences in selective pharmacokinetic processes are involved, if three functions of a ligand are involved in binding to a receptor, a specific structural orientation will be necessary for ligand–receptor interaction; the stereoselectivity of optical isomers is then considered to arise for one isomer because of its ability to achieve a three-point attachment with its receptor, while the enantiomer will only be able to achieve a two-point attachment.

Another problem is that many drugs have flexible structures and their conformations may vary under different conditions. For example, the conformation of a molecule in the ground state may differ from its shape when bound to a biological receptor. In addition to three-dimensional molecular features, electronic aspects also play an important role in promoting the binding to the receptor, since a molecule also constitutes a complex three-dimensional organization of atomic nuclei and electrons controlling intrinsic and relational properties.

Following the progress of molecular modeling systems, extensively described in Volume 4, Part 20, many computerized databanks,[1-4] based on X-ray crystallography techniques, are now available to provide three-dimensional data for different types of simple structures[1] and biomacromolecules.[2] The Cambridge Crystallography Database is the most used among databanks, which are continuously fed for search and retrieval purposes.

Theoretical approaches to the study of all possible conformations of a structure are generally based on two sets of methods.[5] One involves molecular orbital calculations (*ab initio* or semiempirical) and the other involves molecular mechanics calculations. For both sets of methods many programs are available. Programs based on molecular mechanics calculations have some advantages such as simplicity and do not require great amounts of computing time.

A major advance for model building of molecules and macromolecules is real-time interactive stereo computer graphics.[6-8] This technique can serve as an independent check in the study of structure–activity relationships of biochemical reactions when the X-ray crystallographic structure of the bioreceptor is known. Although statistical correlation equations are based on biological data from reactions occurring in solution and the molecular graphics models are based on information obtained from macromolecules in the crystalline state, in the past few years many authors have found good qualitative agreement between models derived from the two different methods.

While there are many different approaches for exploring the spatial atomic arrangement and the conformational space of a drug molecule, in classical QSAR the methods for quantifying the topology of a compound in a manner comparable to the description of other physicochemical properties are still rather unsatisfactory. Only steric properties of substituents or of certain substructures can be adequately described, while much more information is required to enable the precise analysis of stereochemical effects with respect to the three-dimensional complementarity of ligands with the binding sites of biomacromolecules.

In early studies, physical organic chemists defined a number of steric parameters in order to explain steric effects of substituents on the reaction centers of organic molecules. The same type of steric effects observed in studies of the variation of physical properties and chemical reactivity with structure may be assumed to be involved in biological activity studies which, at least as a first approximation, may be treated in a similar fashion. In the last 25 years, owing to the development of drug design and the Hansch Approach,[9-11] many other parameters and methods have been developed which have the merit of trying to avoid a simple empirical correlation with given ligand properties and also trying to propose the possible geometrical features of the receptor site.

In this chapter we will analyze the ability of steric descriptors to describe structural features of molecules, their methods of calculation and their ability to treat steric bulkiness, size and shape. Finally, their intercorrelation will be considered in each corresponding section.

Steric descriptors will be classified into the following groups: (a) topological indices based upon characterization of the chemical structure by graph theory; (b) geometric descriptors resulting from the view of organic molecules as three-dimensional objects from which standard dimensions can be calculated; (c) chemical descriptors derived from steric influence upon a standard reaction; (d) physical descriptors which result when an organic molecule is considered as a three-dimensional object with size-determined physical properties; and (e) differential descriptors which result when an organic molecule is considered as a three-dimensional object which varies from a reference structure (overlapping method).

Selected QSAR results will be discussed to illustrate the significance of steric effects; through these examples the usefulness and the limits of each group of descriptors will be considered.

18.4.2 TOPOLOGICAL DESCRIPTORS

The three-dimensional structure of a molecule depends upon its topology, *i.e.* on the position of the individual atoms and the bonded connections between them. It is possible to derive from the hydrogen-suppressed molecular formula numerical steric descriptors encoding within them information relating to the number of atoms and their structural environment. These descriptors are termed topological indices and their formulation is based upon the characterization of chemical structure by graph theory.[12-14] In fact, all structural formulas of covalently bonded compounds have vertices symbolizing atoms and edges representing covalent bonds; therefore, they are graphs which can be termed molecular (or constitutional) graphs.

Although several topological indices have been proposed, only a few of them have been widely used in QSAR to take into account steric effects. According to Balaban *et al.*[15] the most convenient classification of these indices is based on their logical derivation rather than according to their chronological development. Following this procedure, topological indices can be essentially subdivided into four classes: (i) topological indices based on the adjacency matrix; (ii) topological indices based on the distance matrix; (iii) centric topological indices; and (iv) indices based on information theory.

18.4.2.1 Indices Based on the Adjacency Matrix

The first class of indices is based on the consideration that the whole set of connections between adjacent pairs of atoms may be represented in a matrix form, termed the adjacency matrix. The entries a_{ij} of the matrix equal one if vertices i and j are adjacent and zero otherwise.

It has been shown[16] that if the molecular graph represents the carbon skeleton of a conjugated hydrocarbon the eigenvalues of the adjacency matrix are similar to those of the Hückel matrix; there is a formal equivalence between the graph-theoretical treatment and the quantum chemical one in the Hückel MO method. The formulations of the adjacency matrices for two structural isomers (2,3-dimethylbutane and *n*-hexane) are given below (Figure 1). Note that to formulate the matrices the graphs may be numbered in any order.

The simplest topological index is given by the sum of all matrix entries. Since the adjacency matrix is symmetric, this index is redundant and can be reduced to the sum of the upper triangular

$$A(G_1) = \begin{array}{c|cccccc} & 1 & 2 & 3 & 4 & 5 & 6 \\ \hline 1 & 0 & 1 & 0 & 0 & 0 & 0 \\ 2 & 1 & 0 & 1 & 0 & 1 & 0 \\ 3 & 0 & 1 & 0 & 1 & 0 & 1 \\ 4 & 0 & 0 & 1 & 0 & 0 & 0 \\ 5 & 0 & 1 & 0 & 0 & 0 & 0 \\ 6 & 0 & 0 & 1 & 0 & 0 & 0 \end{array}$$

$$A(G_2) = \begin{array}{c|cccccc} & 1 & 2 & 3 & 4 & 5 & 6 \\ \hline 1 & 0 & 1 & 0 & 0 & 0 & 0 \\ 2 & 1 & 0 & 1 & 0 & 0 & 0 \\ 3 & 0 & 1 & 0 & 1 & 0 & 0 \\ 4 & 0 & 0 & 1 & 0 & 1 & 0 \\ 5 & 0 & 0 & 0 & 1 & 0 & 1 \\ 6 & 0 & 0 & 0 & 0 & 1 & 0 \end{array}$$

Figure 1 Formulation of adjacency matrices for two illustrative graphs: 2,3-dimethylbutane (G_1) and n-hexane (G_2)

submatrix; the total adjacency index A is given by equation (1).

$$A = (1/2) \sum_{i,j=1}^{n} a_{ij} \tag{1}$$

For both the examples in Figure 1 one obtains a value equal to five; this index can only distinguish between structures having different numbers of cycles. For example, it is easy to show that for cyclohexane the total adjacency index is equal to six.

Gutman *et al.*[17] were among the first to be interested in the quantitative characterization of the degree of branching. This structural feature has been taken into account by the so-called vertex degree (δ_i), which is equal to the number of σ bonds involving each atom i, excluding bonds to hydrogen atoms (Figure 2). Note that the vertex degree δ_i for an atom i is also equal to the sum of the entries in row i in the adjacency matrix. With this assumption, two topological indices M_1 and M_2 (Zagreb group indices) were derived.[17]

$$M_1 = \sum_{i=1}^{n} \delta_i^2 \tag{2}$$

$$M_2 = \Sigma \delta_i \cdot \delta_j \tag{3}$$

For the two examples of Figure 2 one obtains, respectively

$$M_1(G_1) = 4 \cdot 1^2 + 2 \cdot 3^2 = 22 \qquad M_1(G_2) = 2 \cdot 1^2 + 4 \cdot 2^2 = 18$$

$$M_2(G_1) = 4(1 \cdot 3) + 1(3 \cdot 3) = 21 \qquad M_2(G_2) = 2(1 \cdot 2) + 3(2 \cdot 2) = 16$$

Simultaneously, Randic[18,19] introduced a connectivity index (χ_R), similar to the M_2 index.

$$\chi_R = \Sigma(\delta_i \cdot \delta_j)^{-1/2} \tag{4}$$

Kier and Hall[20,21] recognized in the Randic procedure the possibility for the development of a more general method to describe organic molecular structures in terms of topological indices. Assuming $\chi_R = {}^1\chi$, where the prefix 1 indicates that the ${}^1\chi$ index is for one-edge dissection of the structural formula, Kier and Hall adopted a general scheme to calculate higher-order dissection of the molecular skeleton.

$$^h\chi = \Sigma(\delta_i \cdot \delta_j \ldots \ldots \delta_{h+1})^{-1/2} \tag{5}$$

For example, to calculate the ${}^2\chi$ index of the G_1 graph of Figure 2 the two-edge (three contiguous atoms) dissection is as shown in Figure 3.

Actually, in addition to the ${}^1\chi$ (edge) and ${}^h\chi$ (path) higher-order connectivity indices, Kier and Hall defined three other indices derived from dissections of molecular skeletons into the commonly recurring features illustrated in Figure 4. From equation (5) one calculates that the connectivity index of order zero, which corresponds to the dissection of the molecular formula into fragments of atoms, is equal to ${}^0\chi = \Sigma(\delta_i)^{-1/2}$. In general, the connectivity indices of Kier and Hall may be identified as ${}^h\chi_m$, where the prefix h indicates that the index is for an h-bond dissection of the molecule, while the subscript m denotes the type of fragment dissection.

Figure 2 Assignment of vertex degree for graphs G_1 and G_2

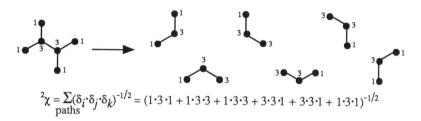

$$^2\chi = \sum_{\text{paths}}(\delta_i\cdot\delta_j\cdot\delta_k)^{-1/2} = (1\cdot3\cdot1 + 1\cdot3\cdot3 + 1\cdot3\cdot3 + 3\cdot3\cdot1 + 3\cdot3\cdot1 + 1\cdot3\cdot1)^{-1/2}$$

Figure 3 Two-edge dissection of graph G_1

cluster	path-cluster	chain
$^3\chi_c$	$^4\chi_{pc}$	$^5\chi_{ch}$

Figure 4 Commonly occurring features of molecular skeletons

A further extension of this approach has been made by Kier and Hall to include molecules containing double or triple bonds as well as heteroatoms. In the first case, the procedure takes into account the valence of each atom and therefore the vertex degree (δ_i) is replaced by the atom connectivity (δ_i^v)

$$\delta_i^v = Z_i^v - H_i \tag{6}$$

where Z_i^v and H_i are, respectively, the number of valence electrons of the atom i and the number of hydrogen atoms suppressed. For example, while for each carbon atom of benzene the δ_i value is equal to two, one obtains $\delta_i^v = 4 - 1 = 3$; the calculation of the corresponding $^h\chi_m^v$ (valence chi) proceeds as before using equation (5). The heteroatom (O, N, F) connectivity calculation is also based upon equation (6), taking into account the adjacent bonded atoms plus all π and lone pair electrons. For example, the δ^v value for oxygen in the alcohol RCH_2OH is computed from the count of two adjacent atoms (carbon and hydrogen), plus four electrons in two lone pair orbitals, minus one hydrogen ($\delta^v = 6 - 1 = 5$). For higher row atoms (P, S, Cl, Br and I), care must also be taken of the non-valence electrons. Kier and Hall proposed equation (7) to take account of both valence and non-valence electrons

$$\delta_i^v = (Z_i^v - H_i)/(Z_i - Z_i^v - 1) \tag{7}$$

where Z_i is the number of all electrons of the atom i. δ^v values for C, N and O in various hybrid states and for higher level atoms are shown in Tables 1 and 2, respectively.

As Kier pointed out,[21] the choice of the indices to be considered in a correlation analysis is very important. For example, one should not employ a type of index characterized by constant or zero values for more than a few of the compounds under study. Moreover, the valence indices should be considered when heteroatoms are varied in a list of compounds. Similar considerations hold true in cases where unsaturation or aromatic systems are prominent in the dataset. The connectivity indices $^h\chi_m$ and $^h\chi_m^v$ have been extensively used in QSAR[20, 21] on series of biologically active molecules (enzyme inhibition, anesthetic activity, *etc.*); however, interpretation of the results is often not obvious. In fact, although Kier and Hall pointed out the topological information linked to connectivity indices, these may be highly correlated to other molecular physicochemical properties such as molar refractivity, polarizability and/or lipophilicity.

Table 1 Valence δ^v Values for C, N, O and F Atoms in Hydrogen-depleted Graphs

Atom	Hybrid state	Adjacent atoms δ	Valence electrons	π electrons	Lone pair electrons	δ^v
C	sp^3	1	4	0	0	1
		2	4	0	0	2
		3	4	0	0	3
		4	4	0	0	4
	sp^2	1	4	1	0	2
		2	4	1	0	3
		3	4	1	0	4
	sp	1	4	2	0	3
		2	4	2	0	4
N	sp^3	1	5	0	2	3
		2	5	0	2	4
		3	5	0	2	5
	sp^2	1	5	1	2	4
		2	5	1	2	5
	sp	1	5	2	2	5
O	sp^3	1	6	0	4	5
		2	6	0	4	6
	sp^2	1	6	1	4	6
F[a]		1	7			7

[a] Hybrid state not specified.

Table 2 Valence δ^v Values for P, S, Cl, Br and I Atoms

Atom	$Z^v - H_i$	$Z - Z^v - 1$	δ^v
P	3	9	0.33
	4	9	0.44
	5	9	0.56
S	5	9	0.56
	6	9	0.67
Cl	7	9	0.78
Br	7	27	0.26
I	7	47	0.16

It is worth pointing out that Platt[22, 23] was the first to use a topological index in correlations with certain molecular properties. The Platt index (F) is defined by

$$F = \sum_{j=1}^{A} e_j \tag{8}$$

where e_j is the degree of the edge j, *i.e.* the number of edges adjacent to edge j. Thus, for the graph G_1 of Figure 1 one obtains the F value shown in Figure 5.

The Gordon–Scantlebury index,[24] which is defined in equation (9), is related to the F value.

$$N_2 = \Sigma(P_2)_i \tag{9}$$

In this equation P_2 is the path of length 2, *i.e.* the triatomic link C—C—C which can be superimposed on the hydrogen-suppressed molecular graph. Thus, for G_1 of Figure 1 it is possible to calculate the N_2 value shown in Figure 6. A comparison between F and N_2 shows that $N_2 = F/2$.

Other interesting topological indices based on the adjacency matrix have been introduced. For example, Gutman and Randic[25] proposed a procedure for the comparability (comparability code) of chemical structures, Randic[26, 27] proposed the smallest binary notation (SBN), while Lovasz and Pelikan[28] discussed the largest eigenvalue of the characteristic polynomial of the molecular graph as a topological index. It is worth pointing out that controversial arguments[29, 30] indicate that the use

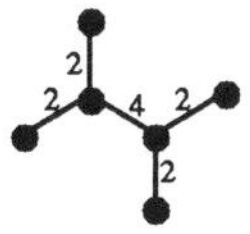

$$F = 2 + 2 + 4 + 2 + 2 = 12$$

Figure 5 F value for graph G_1

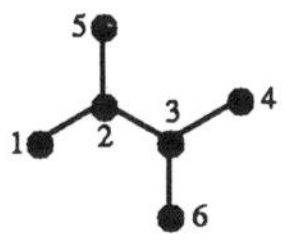

$$N_2 = \Sigma(P_2)_i = (125) + (123) + (523) + (234) + (236) + (436) = 6$$

Figure 6 N_2 value for graph G_1

of topological indices based on adjacency matrices does not lead to a unique description of the topology of a molecule.

18.4.2.2 Indices Based on the Distance Matrix

Molecular graphs can also be characterized by the so-called distance matrix; a second set of topological indices is based on this matrix. The entry d_{ij} in the distance matrix D indicates the number of edges in the shortest path between vertices i and j. That is, while in the adjacency matrix the entries are equal to one or zero (adjacent vertices or otherwise), in the distance matrix entries can be also equal to two, three, *etc.*, corresponding to the shortest path between second neighbors (two edges), third neighbors (three edges), *etc.* Thus, referring to the graphs of Figure 1, one obtains

$$
D(G_1) = \begin{array}{c|cccccc}
 & 1 & 2 & 3 & 4 & 5 & 6 \\
\hline
1 & 0 & 1 & 2 & 3 & 2 & 3 \\
2 & 1 & 0 & 1 & 2 & 1 & 2 \\
3 & 2 & 1 & 0 & 1 & 2 & 1 \\
4 & 3 & 2 & 1 & 0 & 3 & 2 \\
5 & 2 & 1 & 2 & 3 & 0 & 3 \\
6 & 3 & 2 & 1 & 2 & 3 & 0 \\
\end{array}
\qquad
D(G_2) = \begin{array}{c|cccccc}
 & 1 & 2 & 3 & 4 & 5 & 6 \\
\hline
1 & 0 & 1 & 2 & 3 & 4 & 5 \\
2 & 1 & 0 & 1 & 2 & 3 & 4 \\
3 & 2 & 1 & 0 & 1 & 2 & 3 \\
4 & 3 & 2 & 1 & 0 & 1 & 2 \\
5 & 4 & 3 & 2 & 1 & 0 & 1 \\
6 & 5 & 4 & 3 & 2 & 1 & 0 \\
\end{array}
$$

Wiener[31-34] proposed in 1947 the first topological index based on the distance matrix. He defined for saturated hydrocarbons the path number w as the sum of the number of bonds between all pairs of vertices; this index, which reflects the branching of the molecule, may be calculated from the sum of the off-diagonal elements of the distance matrix.

$$w = \Sigma d_{ij} \tag{10}$$

Thus, referring to the graphs of Figure 1, one obtains $w(G_1) = 29$, $w(G_2) = 35$. Altenburg[35, 36] modified the Wiener index w by expressing it as a polynomial

$$P_A(G, a) = \Sigma a_i g_i \tag{11}$$

where g_i is the frequency number of pairs of vertices separated by d_i bonds and a_i has been introduced only for enumeration in polynomial form. Thus, the expression for 2,3-dimethylbutane is

$$P_A(G_1, a) = 5 \cdot a_1 + 6 \cdot a_2 + 4 \cdot a_3 = 5 \cdot 1 + 6 \cdot 2 + 4 \cdot 3 = 29$$

Initially, the Altenburg expression was devised only for acyclic graphs; however, Hasoya[37] extended its application to cyclic graphs.

In addition to the w index, Wiener also proposed the polarity number p, which is the number of pairs of vertices separated by three edges; the polarity number is equal to the sum of entries of length three in the off-diagonal matrix. For graph 1, $p(G_1) = 4$ and for graph 2, $p(G_2) = 3$. The w and p indices have been employed in order to correlate boiling points and other properties of acyclic saturated hydrocarbons. Note that the Gordon–Scantlebury N_2 index can be similarly expressed as the half sum of all entries of length two in the distance matrix, *i.e.* $N_2 = (1/2)\Sigma d_{2,i}$.

Recently, Kier[38] developed the $^2\kappa$ index which is based on the count of two-bond fragments. The method considers two reference structures chosen among non-cyclic molecules. Considering, as an example, graphs corresponding to five atoms, one obtains the two reference structures of Figure 7. The first graph (Figure 7a) for normal pentane has the minimum number of two-bond fragments for five atoms; Kier introduced the term $^2P_{min}$ as the count of the number of two-bond paths, which for any non-cyclic molecule is equal to equation (12), where A is the total number of vertices in the graph. Of course, the number can be obtained by the half sum of entries of length two in the distance matrix. A second reference structure can be identified and quantified among all non-cyclic isomeric series; that is, graph (b) for neopentane in Figure 7. This graph has the maximum number of two-bond links ($^2P_{max}$) possible for five atoms and corresponds to the graph structure known as a star, in which all points but one are connected to that one point. The value of $^2P_{max}$ can be calculated for any isomeric series through equation (13).

$$^2P_{min} = A - 2 \tag{12}$$

$$^2P_{max} = \frac{(A - 1)(A - 2)}{2} \tag{13}$$

For pentane and neopentane of Figure 7, one obtains the values $^2P_{min} = 3$ and $^2P_{max} = 6$, respectively. Obviously when $A > 5$ the star structures are not real molecules because of the tetravalent nature of carbon. All non-cyclic molecules within an isomeric series have ^2P_i values at or within the bounds of the minimum and maximum value of P, *i.e.*

$$^2P_{min} \leqslant {}^2P_i \leqslant {}^2P_{max} \tag{14}$$

After appropriate considerations and refinements, Kier derived the final equation

$$^2\kappa = 2\frac{^2P_{max} \cdot {}^2P_{min}}{(^2P_i)^2} = \frac{(A - 1)(A - 2)^2}{(^2P_i)^2} \tag{15}$$

According to Kier the $^2\kappa$ index describes the molecular shape in relationship to the star and linear graph and is normalized to the number of atoms. The method, which less rigorously may also be classified among the indices based on information theory, has been extended with a more pragmatic approach to cyclic structures.

As a first-order approximation, Kier has assumed that the points in the hydrogen-depleted graph represent any atom, while the connections represent a bond of any multiplicity. Subsequently,[39] the molecular shape index $^2\kappa$ has been extended to take into account the size differences among heteroatoms and carbon atoms in different valence states; this index was expressed as

$$^2\kappa_\alpha = \frac{(A + \alpha - 1)(A + \alpha - 2)}{(^2P_i + \alpha)^2} \tag{16}$$

Since an appropriate attribute of an atom which influences shape is the covalent radius r, the ratio of radii relative to the sp^3 carbon atom was computed and the modifier α was derived as

$$\alpha_X = \frac{r_X}{r_{Csp^3}} - 1 \tag{17}$$

In the case of more than one heteroatom or carbon of another valence state, the α modifier is the sum of all $[(r_X/r_{Csp^3}) - 1]$ values.

The concept of a shape index developed from the molecular graph previously described, has been expanded[40] to include two additional indices derived from counts of one-path (^1P) and three-path

(a)   (b)

Figure 7   Linear (a) and star (b) graphs for pentane

(^3P) fragments which led to shape indices $^1\kappa$ and $^3\kappa$, respectively.

$$^1\kappa = 2\frac{^1P_{max} \cdot {}^1P_{min}}{(^1P_i)^2} = \frac{A(A-1)^2}{(^1P_i)^2} \tag{18}$$

$$^3\kappa = \frac{^3P_{max} \cdot {}^3P_{min}}{(^3P_i)^2} \tag{19}$$

Equation (19) has two solutions in terms of the number of atoms in the molecule, depending upon whether A is even or odd.

$$^3\kappa = \frac{(A-3)(A-2)^2}{(^3P_i)^2} \qquad \text{for } A \text{ even} \tag{20a}$$

$$^3\kappa = \frac{(A-1)(A-3)^2}{(^3P_i)^2} \qquad \text{for } A \text{ odd} \tag{20b}$$

Note that the non-cyclic isomers all have the same values of $^1\kappa$, which is equivalent to the atom count, A. Monocyclic molecules have a lower value and bicyclic structures have a still lower value. The numerical values of $^3\kappa$ from equation (19) are smaller than the corresponding $^1\kappa$ and $^2\kappa$ values for the same number of atoms; to have $^3\kappa$ values on an approximately equal scale, the expression was multiplied by four. According to Kier, the $^1\kappa$ index describes relative molecular size, while the $^3\kappa$ index encodes information about a specific kind of branching since it varies in numerical value when branching occurs in the middle or at the ends of a long chain fragment.

As already mentioned for $^2\kappa$, Kier proposed that equations for $^1\kappa$ and $^3\kappa$ can be modified to account for the different shape contribution of non-sp^3 carbon atoms. This was accomplished by replacing A with $A+\alpha$ and using $^1P_i + \alpha$ or $^3P_i + \alpha$ in equations (18), (20a) and (20b). The indices calculated using the α modifier were designated $^1\kappa_\alpha$ and $^3\kappa_\alpha$ respectively.

By analogy with the vertex degree δ_i proposed by Gutman *et al.*, Bonchev *et al.*[41] developed the so-called distance rank which is based on the distance sum (V_{Di}) of the vertex i defined by the sum of all entries (d_{ij}) of the ith row in the distance matrix.

$$V_{Di} = \Sigma d_{ij} \tag{21}$$

For simple graphs such as those of Figure 1, one obtains by summation the index values of 58 (G_1) and 70 (G_2), respectively (Figure 8).

Similarly to the aforementioned Randic's connectivity index (χ_R), Balaban[42,43] developed the average distance sum connectivity index

$$J = \frac{q}{\mu + 1} \cdot \Sigma \bar{V}_{Di} \cdot \bar{V}_{Dj} \tag{22}$$

where the indices i and j are for adjacent vertices, q is the number of edges, μ represents the number of cycles (*i.e.* $n = 0$ for linear graphs) and $\bar{V}_{Di}$ is the average distance sum.

$$\bar{V}_{Di} = V_{Di}/q$$

Note that for unsaturated systems fractional distances are the entries in the distance matrix. For example, if the bond order between vertices i and j is b, the entry $1/b$ is used in the row/column i/j.

Smolenskii[44] formalized in graph-theoretical language the properties of hydrocarbons by means of an additivity function of the contributions of different subgraphs of a given molecular graph. He used two orders of approximation obtaining indices which are somewhat reminiscent of the Gordon–Scantlebury index (N_2) and the Wiener polarity number (p). Atomic and molecular

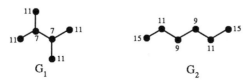

Figure 8 Distance ranks for graphs G_1 and G_2

distance code and index have been introduced by Randic[45] in order to characterize molecular structures and with the aim of studying molecular data.

Although the Hosoya's Z index[37,46,47] has been associated[13] with the adjacency matrix, according to Balaban[15] it can be classified among the distance matrix indices due to the procedure used to calculate $p(G, K)$.

$$Z = \sum_K p(G, K) \tag{23}$$

where $p(G, K)$ represents the number of ways in which K edges from all bonds of a graph G may be chosen so that no two of them are adjacent. For linear graphs only, Z can be also defined as the sum of the absolute values of coefficients in the characteristic polynomial

$$P_H(G, x) = \Sigma(-1)^K \cdot p(G, K)x^{N-2K} \tag{24}$$

where N is the number of atoms. Since by definition $p(G, 0) = 1$ and $p(G, 1)$ is equal to the number of edges in the graph, to calculate $p(G, 2)$ of the graph G_1 in Figure 1, one obtains Figure 9; that is, $p(G_1, 2) = 4$; of course, $p(G_1, 3) = 0$. Hence, $Z = 1 + 5 + 4 = 10$, while the characteristic polynomial is

$$P_H(G_1, x) = \sum_{K=0}^{2} (-1)^K \cdot p(G_1, x)x^{6-2K} = x^6 - 5x^4 + x^2$$

18.4.2.3   Centric Topological Indices

Balaban has proposed a set of five topological indices, classified as centric topological indices[48] on the basis of sequences of numbers obtained by 'pruning' an acyclic graph toward its center. As shown in graph theory, by pruning stepwise all end-points (vertices of degree one) of an acyclic graph one is left finally with one vertex (center) or two adjacent vertices (bicenter) joined by an edge. Balaban developed a centric index B, a normalized centric index C, a normalized quadratic index Q and two binormalized indices, C' and Q' respectively, defined, after algebraic manipulation, as

$$B = \sum_{i=1}^{N} d_i^2 \tag{25}$$

$$C = (1/2)(B - 2N + U) \tag{26}$$

$$Q = 3V_4 + V_3 \tag{27}$$

$$C' = \frac{B - 2N + U}{(N-2)^2 - 2 + U} \tag{28}$$

$$Q' = \frac{3V_4 + V_3}{2(N-2)(N-3)} \tag{29}$$

where d_i is the number of vertices deleted at each step, N is the number of all vertices, $U = [1 - (-1)^N]$ and V_3 and V_4 are the number of vertices of degree three and four respectively. Note that B, which is a function of the branching of the alkane, is similar to the Zagreb index M_1, while the centric index C is related to the Gordon–Scantlebury index N_2. Values of all these topological indices for graphs G_1 and G_2 are obtained as shown in Figure 10. Although for cyclic graphs no simple topological center can yet be found, Bonchev *et al.*[41] derived topological indices for any graphs on the basis of a few criteria which allow one to find a so-called polycenter.

18.4.2.4   Indices Based on Information Theory

According to Balaban the fourth class of index which can be defined on any structural basis, topological or non-topological, is founded on information theory.[49,50] This theory allows the

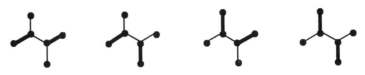

Figure 9 $p(G, 2)$ for graph G_1

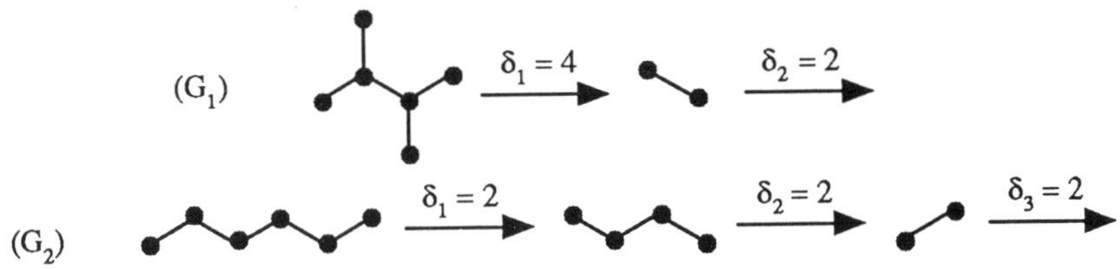

$$B(G_1) = 4^2 + 2^2 = 20; \quad C(G_1) = (20 - 12)/2 = 4; \quad Q(G_1) = 0 + 2 = 2;$$
$$C'(G_1) = (20 - 12)/(16 - 2) = 0.571; \quad Q'(G_1) = 2/(2 \cdot 4 \cdot 3) = 0.08$$

and

$$B(G_2) = 2^2 + 2^2 + 2^2 = 12; \quad C(G_2) = (12 - 12)/2 = 0; \quad Q(G_2) = 0; \quad C'(G_2) = 0; \quad Q'(G_2) = 0$$

Figure 10 Centric topological indices for graphs G_1 and G_2

introduction of structural indices termed information content of any structured system. For such a system having N elements (atoms) distributed into classes of equivalence N_1, N_2, $\cdots\cdots N_K$ a probability distribution $P(p_1, p_2, p_3, \cdots p_K)$ is constructed. According to the Shannon[49] equation, two interrelated values may be derived; the first of these is the information content per atom defined as

$$i = -\Sigma p_i \log p_i \tag{30}$$

where p_i is the probability of randomly selecting an atom from the whole. The information content in the entire molecule with N atoms is given by iN. Bruillon[51] modified this expression to give the redundancy of the molecule

$$R = 1 - \frac{i}{\log N} \tag{31}$$

Recently, Kier[52] proposed a shape index $^0\kappa$ which was made equal to either of these symmetry/redundancy indices; thus this index, which follows chronologically the development of $^1\kappa$, $^2\kappa$ and $^3\kappa$, is derived from zero-order fragments, *i.e.* from the atoms of a molecular graph treated in isolation. Since one attribute of an atom which should influence the shape of a molecule is the topological uniqueness of that atom within a molecule, the collective effect of topological uniqueness is the symmetry of the molecule. Kier discussed a few examples to illustrate the utility of $^0\kappa$ along with other κ values to encode attributes of molecular shape influencing physical and biological properties.

Although several other information indices have been devised[53–59] the most useful appear to be those developed by Bonchev and Trinajstic.[60] These indices have been defined as information content and mean information content on polynomial coefficients $(I_{pc}, \bar{I}_{pc})$, on realized distances in the graphs $(I^w_D, \bar{I}^w_D)$ and on the distribution of the distances $(I^E_D, \bar{I}^E_D)$.

$$I_{pc} = Z \log_2 Z - \sum_K p(G, K) \log_2 p(G, K) \tag{32}$$

$$\bar{I}_{pc} = I_{pc}/K \tag{33}$$

$$I^E = \frac{N(N-1)}{2} \cdot \log_2 \frac{N(N-1)}{2} \cdot g_i \log_2 g_i \tag{34}$$

$$\bar{I}^E_D = I^E_D \left/ \frac{N(N-1)}{2} \right. \tag{35}$$

$$I^w_D = w \log_2 w - g_i d_{ij} \log_2 d_{ij} \tag{36}$$

$$\bar{I}^w_D = I^w_D/w \tag{37}$$

It is clear that these indices are largely defined from a combination of the parameters used to obtain the Wiener w index, the Hosoya Z value, and the Randic connectivity index χ; concerning the last three indices, Bonchev and Trinajstic also defined their mean values as

$$\bar{Z} = Z/K \tag{38}$$

$$\bar{w} = \frac{2w}{N(N - 1)} \tag{39}$$

$$\bar{\chi} = \frac{\chi}{N(N - 1)} \tag{40}$$

18.4.2.5 Other Topological Methods

Austel and co-workers[61] described a steric descriptor S_b, derived from information on molecule branching, which can also be considered as a topological index. This constant was derived on the assumption that the steric effect of a group is mainly due to branching. Austel *et al.*, in fact, have found that only non-hydrogen atoms in α, β and γ positions, as well as branching from the α and β positions, give approximately equal contributions to substituent steric effects. Proceeding along this line it was possible to calculate a steric descriptor for every substituent by counting the number of non-hydrogen atoms in α, β and γ positions; examples are given in Table 3. Corrections have to be considered for ring structures and elements from the third, fourth and fifth row of the periodic table. The S_b constant has been derived from a previous observation of Charton[62] that the steric effect of a substituent is mainly due to branching.

A general feature of all topological indices is their degeneracy, *i.e.* the possibility that two or even more different graphs have similar values of a certain index. Besides the connectivity indices χ and J, the indices based on information theory are characterized by the lowest degeneracy.

It is worth pointing out that most of the topological indices are strongly intercorrelated,[15] hence it may be meaningless to develop a multiparametric equation involving more than one of them. Several efforts have been made to estimate their physical meaning; since systematic studies[15, 63, 64] showed that molar refractivity (MR) and van der Waals volume (V_w) are highly correlated with most of the topological indices, these may be considered useful descriptors for expressing steric features of molecules and substituents. However, log P and π hydrophobic parameters depend to a certain extent on molar volume; therefore, the collinearity between each pair of all of these vectors (topological indices, MR, V_w, π, log P) may be very high, depending on the nature of the structure. For example, log P and π are collinear with connectivity indices only for non-polar atoms or molecules; including negative π values breaks up the collinearity.

Among the systems for structure storage and retrieval that are in widespread use, the Description, Acquisition, Retrieval Computer (DARC) system of Dubois[65-67] is worth mentioning. This system is more complex than the simple topological indices and can describe conformations of molecules. It is not based on the definition of the whole structure but rather on an environment conceived as concentric modules that Dubois defined as Perturbation of an Environment, which is Limited, Concentric and Ordered (PELCO) around a focus (FO). The method considers a series of structures around the common core (FO) and various directions of development (substitution position). As in the Free–Wilson method the approach determines additive increments, but the increments are associated with sites and their bonded atom types rather than with whole groups. Proceeding in a progressive and ordered manner along each development direction it is possible to generate the sites. If the site contributions are not additive then, in order to explain a set of experimental data, interaction terms between sites may be considered until the residual error of the analysis approaches the experimental error of the data. At the end of the analysis, it is possible to identify a topological map of activity increments with the corresponding localization of important sites and site occupants; moreover, favorable or unfavorable interactions between them may be identified. The Institute for

Table 3 Examples of S_b Values of Some Substituents

Substituent	S_b	Substituent	S_b
H	0	NMe$_2$	3
Me	1	Pr^n	3
OMe	2	Bu^n	3

Scientific Information[68] has the database for the method, which has found only limited application in QSARs.

18.4.3   GEOMETRIC DESCRIPTORS

Geometric descriptors represent a class of theoretically derived parameters based on van der Waals radius (r_w). It is well known that this radius may be defined in terms of the distance l at which the repulsion between the electron densities of two approximating atoms balances the attraction forces between them. From this point of view, the van der Waals radius is a function of the electron density distribution around each atom. Bondi[69, 70] provided a qualitative understanding of the nature of the van der Waals radius by correlating it with the de Broglie wavelength of the outermost valence electrons. The critical separation distance between two unbonded atoms is given by the sum of the van der Waals radii of the two atoms concerned. Therefore the van der Waals radius of an atom is always greater than the corresponding covalent radius (r_c) and it may be considered invariable irrespective of the chemical combination of the atom and of its unbonded neighbors as well as of the phase state in which it is found. From this point of view, the stereochemistry of a molecule should include both geometric and electronic aspects and both aspects should be taken into consideration when a ligand interacts with a biological acceptor site.

18.4.3.1   van der Waals Radii and Related Dimensions

Bondi[69] showed that non-metallic-element r_w values obtained by selecting from X-ray diffraction data those which could be reconciled with crystal density at 0 K and other physical properties (gas kinetic collision cross-section, critical density and liquid state properties) are more reliable than are the frequently quoted van der Waals radii based on Pauling's[71] approximation $r_w = r_i$, where r_i is the ionic radius. Following results achieved by Bondi, Allinger[72] proposed a new set of radii for the purpose of carrying out force field calculations; these values are normally larger than the corresponding Bondi radii. In addition to these two sets of values, Bartell[73] proposed another set of radii, subsequently described and extended by Glidewell,[74, 75] which are intermediate between covalent radii and van der Waals radii. However, it is worth noting that all proposed sets of values are linearly interrelated and, as has been shown by Charton,[76, 77] they fit a general equation of the type

$$r = a_0 + a_1 r^* \tag{41}$$

where r is one of the aforementioned radii and r^* is the fundamental radius, which represents in any situation the size of the atom. Since all of the different radii r are linearly correlated to r^*, a knowledge of r^* is not necessary. On the basis of this assumption, Bondi's radii are generally used as representative of the most complete and reliable set of values. The available data for the most common elements of organic compounds are assembled in Table 4. Note that a relationship similar to equation (41) may be used to calculate from the r_w value of each atom the radius R_a which allows one to represent the envelope of a molecule at contact distance with a biomacromolecule, as constructed using the classical ball-and-stick and space-filling representations.

The problem currently faced in correlation analysis is the parameterization of steric effects of large substituent groups. For simple substituents such as F, Cl, Br and I, the r_w values can easily account for their steric hindrance. More difficult is the measure of steric influence of more complex groups. In this case the van der Waals radius may be used as a starting point to provide extensive lists of uniformly defined steric constants. An approach to this problem has been proposed by Charton,[76-80] who derived equations for calculating values of van der Waals radii for different sets of substituents. For symmetric top substituents of the type CX_3, Charton assumed that the axis of the group is the extension of the $C-G_i$ bond, where G_i is the skeletal atom to which the group is attached. With this assumption, Charton defined the minimum ($r_{v, mn}$) and the maximum ($r_{v, mx}$) van der Waals radii perpendicular to the group axis; these values can be thought of as minimum and maximum widths of the group (Figure 11).

In addition to $r_{v, mn}$ and $r_{v, mx}$ values, Charton defined the van der Waals radius parallel to the group axis ($r_{v, ax}$) which can be thought of as the length of the group. These quantities may be calculated by means of simple geometry and trigonometry using the van der Waals radius r_w the bond length $C-X$ and the $X-C-G_i$ bond angle.

Table 4 Bondi van der Waals Radii

X	r_w	X	$r_w{}^a$
H	1.20	S	1.80
C (sp^3)	1.70	Se	1.90
C (sp^2)	1.77	Br	1.85
C (sp)	1.78	Cl	1.75
N (NH_3)	1.55	I	1.98
N $(N{=}N)$	1.60	P	1.80
O (ether)	1.52	B	(2.13)
O (carbonyl parallel arrangement)	1.35	Si	2.10
O (carbonyl vertical arrangement)	1.63	Mg	(1.73)
		Hg	1.70

^{a} Values in parentheses are uncertain.

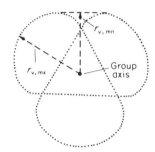

Figure 11 Bottom view of CX_3 group

Charton[78] has shown that the E_s Taft parameter (see Section 18.4.4) parallels group radii and that one can use either $r_{v,mx}$ or $r_{v,mn}$ to estimate the steric action of symmetric top-type functions on neighboring atoms. Taking advantage of these considerations, Kutter and Hansch[81] proposed the use of the average (r_{av}) of the two values given by Charton.

Although the available sets of r_v values could be considered a valuable and general scale of steric effects, Charton suggested[77] a set of steric parameters, designated v values, from the equation

$$v_X = r_{v,X} - r_{v,H} = r_{v,X} - 1.20 \qquad (42)$$

where $r_{v,X}$ is the radius of a substituent X and $r_{v,H}$ is the hydrogen atom radius. Thus, in order to characterize most substituents, v_{mx}, v_{mn} and v_{ax} have to be calculated. Of course, for monoatomic substituents the three values are coincident; for cylindrical substituents, such as —C≡N, —C≡C—Y, *etc.*, $v_{mx} = v_{mn}$, while for all other groups the three values are generally different.

Charton classified substituent groups into various categories according to the dependence of their steric effects on conformation: (i) no conformational dependence (NCD), (ii) minimum conformational dependence (MCD), (iii) intermediate degree of conformational dependence (ICD) and (iv) strong and/or large conformational dependence (SCD, LCD). However, since steric effects involve repulsions between non-bonded atoms which raise the energy of the system, a group of atoms tends to assume that conformation which minimizes these repulsions and in general the steric effect of a group should be taken as equal to the corresponding v_{mn} value. Nevertheless, the dependence of steric effects of ICD and LCD substituents on conformation makes it difficult to determine v_{mn} as well as v_{mx} and v_{ax}. Similar considerations arise with groups which have a skeleton of two or more atoms other than hydrogen. With reference to this problem, Charton[77, 80, 82–84] defined the effective steric parameter v_{eff} which is strictly related to the Taft E_s parameter (see Section 18.4.4).

Using molecular models, Bowden and Young[85] defined the steric substituent constant R, based on known bond distances and van der Waals radii. R values, in fact, are calculated as the distance in angstroms from the G_i atom to which the substituent is bonded to the periphery of the van der Waals radius of the substituent relative to the hydrogen radius. This constant, therefore, is a measure of the length of a substituent and it has been successfully applied in a QSAR of a series of amino-ester hydrochlorides active as antagonists of acetylcholine and histamine at the postganglionic receptors.[85]

18.4.3.2 STERIMOL Parameters

Although any single parameter scale is unlikely to be of general use, a possible solution to this problem has been proposed by Verloop *et al.*[86-91] who developed a set of several parameters for the same substituent. These parameters were evaluated by measuring the dimensions of substituents in a restricted number of directions with the aid of a computer program called STERIMOL, which simulates three-dimensional model building of molecules or molecular groups, using the well-known Corey–Pauling–Koltun (CPK) atomic models. Covalent and van der Waals radii, as well as bond angles used for determining the bond vectors of the various types of atoms, have been taken from the CPK catalog of the Ealing Corporation issued in the USA. For flexible substituents the minimum energy conformations were considered.

In the original derivation[86] of the STERIMOL approach, five directions were chosen in order to describe the shape of the substituent. The substituent attachment point on the skeletal (*i.e.* benzene) G_i atom is placed on the origin of the Cartesian coordinates (x, y, z), assuming that the bond axis is coincident with the x direction. Thus, the length parameter L is determined as the x coordinate of the tangential plane to the substituent van der Waals radii perpendicular to the x axis. This procedure is shown in Figure 12(a), where the projection of a substituent is shown in the plane formed by the x axis and the smallest distance from this axis to the surface of the substituent. Depending on the geometry of receptor surface the role played by L may be crucial or immaterial. Since the widths of the substituent in various directions perpendicular to the x axis may be relevant to the interaction with the receptor, Verloop *et al.* defined four width parameters, B_1, B_2, B_3 and B_4, determined by rotation of the substituents around the x axis. First, the minimum width parameter B_1 was determined as the smallest distance to the x axis of the substituent tangential planes perpendicular to the z coordinate. The additional parameters, B_2, B_3 and B_4, were derived in such a way that they represent the width parameters in four rectangular directions, *i.e.* a box is built to fit around the substituent and B_1 to B_4 represent the distances to the x axis of the box walls (Figure 12b). The corresponding values B_1 to B_4 are in ascending order; in most cases the largest width parameter B_4 is almost equal to the maximum width of the substituent.

However, the best STERIMOL parameter (or parameters) to be used in a QSAR study should be selected by regression procedures, which require the use of large sets of biological data in order to be statistically meaningful. Moreover, since for a specific substituent, such as the OMe group (Figure 13) it is possible to consider more than one plane corresponding to the minimum distance (B_1) to the x axis with consequent possible different values of B_2, B_3 and B_4, Verloop proposed a further development[88,89] of the STERIMOL approach. In order to avoid too many degrees of freedom, B_1 and L were maintained as constants, while the parameters B_2 and B_3 were omitted. This omission was also justified by the consideration that the two parameters are highly intercorrelated and that they hardly ever contributed significantly to any regression equations. Although the choice of the B_1 direction may be made in such a way that the resulting B_4 parameter is as close as possible to the maximum substituent width, a new maximum width parameter B_5 was introduced which replaced the B_4 constant. In comparison with the original STERIMOL constants, the characteristics (Figure 13) of the new parameters are essentially: (i) a different starting point is chosen for the

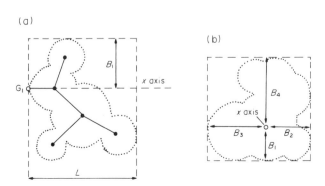

Figure 12 Projections of a substituent along (a) and perpendicular to (b) the x axis

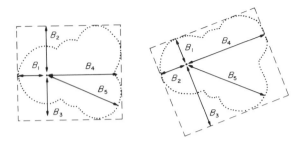

Figure 13   Different possibilities for the evaluation of the parameter B_1 for the OMe group

measurement of width; (ii) the length is maintained; and (iii) the new B_5 parameter has no directional relationships with the B_1 constant.

In Table 11 are reported about 100 of the over 1000 well-characterized substituent values determined by Verloop *et al.*[90]

A peculiar advantage of STERIMOL parameters is their directional nature; the ratios L/B_1 and B_5/B_1 are indications of the deviations of a substituent from a spherical shape. The correct interpretation of the regression equations obtained by their use might provide a better understanding of steric requirements in ligand–receptor interactions. It is of interest that the B_1 parameter shows high intercorrelation[89, 91, 92] with Charton's v_{mn} and v_{eff} constants as well as with Taft's E_s parameter. Although similar considerations hold for L and B_5 with v_{ax} and v_{mx}, respectively, the difference in importance between the STERIMOL parameters and the minimum parameters v_{eff} and E_s will only be evident if the maximum width parameter B_5 and/or the length L play a role in the QSAR under study.

18.4.4   CHEMICAL DESCRIPTORS

Although steric effects play a major role in controlling rates of reactions, the first qualitative attempts at understanding the dependence of organic compound reactivity upon steric factors were published in the late 1800s.[93-96] However, it was the seminal contribution of Meyer[97] which clearly recognized the influence of the steric hindrance of *ortho* substituents on the reactivity of benzoic acids in esterification reactions. This type of reaction, along with ester hydrolysis, being the most intensively studied and probably best understood reaction subject to both acid and base catalysis, has been elected as the most satisfactory standard reaction from which to derive substituent constants related to steric factors.

18.4.4.1   Taft's Steric Constant and Related Parameters

After the aforementioned early works, no important progress was made for about 50 years and only in the late 1930s were several studies[98] completed on quantitative estimations of steric effects. This work, however, met with only limited success because it lacked generality and the parameters obtained were largely electronic rather than steric in nature. Nevertheless, the stimulation of these investigations led to the highly successful separation of steric effects from polar and resonance effects developed by Taft[99, 100] following an early proposition of Ingold.[101] By limiting the type of reaction to esterifications and ester hydrolyses, correlations of substituent effects in aliphatic and *ortho*-substituted substrates became possible. Taft assumed that equation (43) would accomplish this. In equation (43), k_x is the rate constant in a given reaction series, while k_0 refers to the rate constant for the corresponding methyl substrate; f is a proportionality constant and it is equal to one for the appropriate standard reaction series.

$$\log(k_x/k_0) = fA \tag{43}$$

From a consideration of the nature of the transition states of acid- and base-catalyzed reactions of esters, Taft set up equation (44) to define a constant σ^*, correlating substituent polar effects (see Volume 4, Chapter 18.5).

$$\sigma^* = (1/2.48)(\log(k_x/k_0)_B - \log(k_x/k_0)_A) \tag{44}$$

In equation (44), the subscripts A and B denote acid and base hydrolyses, respectively, while the factor $1/2.48$ attempts to set the σ^* values on the same scale as the Hammett values although based on methyl as standard. This value has been derived on the assumption that the susceptibility to substituent polar effects on the same reaction of the aliphatic and *ortho*-substituted benzene substrates would be equal. It is noteworthy that Taft showed that various reactions and sets of physical data may be accommodated by various proportionality factors in equation (44).

The critical observation which led to the formulation of equation (44) was that the acid-catalyzed hydrolysis of aromatic esters is quite insensitive to electronic substituent effects, while for alkaline hydrolyses the substituent electronic characteristics greatly affect the rate of reaction.

As discussed in Volume 4, Chapter 18.5, beside this observation the basic assumptions of the approach to substituent effects in aliphatic substrates of the type XCH_2CO_2R were the following: (a) the relative free energy of activation of the transition states of acid- and base-catalyzed reactions can be treated as a sum of the independent polar, steric and resonance effects and (b) steric effects are similar and resonance effects are either absent or canceled out in the difference of the relative rate constants, $[\log(k_x/k_0)_B - \log(k_x/k_0)_A]$. While the validity of the first assumption is still questioned, and it is necessary to start to work on this problem, the second assumption requires that in the two transition states of Figure 14 the corresponding reacting carbon atoms are close to tetrahedral and have similar steric environments about them. The reliability of this assumption is supported by the small size and remoteness of the two protons by which the two transition states differ. The highly polarizing nature of the localized negative charge developed in one reaction in contrast to the dispersed positive charge developed in the other justifies the markedly greater sensitivity of the base-catalyzed reaction.

From the mode of deriving the σ^* parameter, the procedure is equivalent to assuming that acid-catalyzed hydrolysis, as well as esterification, is insensitive to polar effects, *i.e.* rearranging equation (44)

$$\log(k_x/k_0)_B - 2.48\sigma^* = \log(k_x/k_0)_A = E_s \qquad (45)$$

Equation (45) requires, as assumed by Taft, that $E_s = \log(k_x/k_0)_A$ contains no resonance contribution. This assumption has subsequently met some criticism since there is no certainty that the E_s parameter does not contain electronic components. By definition, $E_s = \log(k_x/k_0)_A = 0$ when X = H; that is, when the group attached to the ester function is Me. To obtain the value for hydrogen on this scale one must consider the acid hydrolysis of formate; this value is found to be equal to 1.24. Unger and Hansch[98-102] have subtracted the value of 1.24 from the values established by Taft to refer the E_s scale to hydrogen. These rescaled values are listed in Table 11.

MacPhee, Panaye and Dubois[103] have examined critically the basic assumptions used in obtaining the Taft E_s scale. Taft calculated average E_s constants from the corresponding $\log(k_x/k_0)_A$ values obtained using four closely related acid catalyzed reactions as the reference system: the hydrolysis of ethyl esters in 60% or 70% v/v acetone at 25 °C and esterification of carboxylic acids in methanol or ethanol at 25 °C. Moreover, Taft used results of Loening, Garrett and Newman[104] corresponding to esterification in methanol at 40 °C without taking into account the variation of τ in equation (46), related to changes in steric effects at different temperatures

$$\log(k_x/k_0)_A = \tau E_s \qquad (46)$$

MacPhee *et al.*[103] evaluated a new set of E_s' values choosing as a single reference reaction the acid-catalyzed esterification of carboxylic acids in methanol at 40 °C; under these conditions the τ term of equation (46) is equal to unity. Generally the E_s' values differ only slightly from the corresponding Taft values but in certain cases, such as halogen-containing groups (—$CHBr_2$ and —$CHCl_2$), they differ significantly.

In a subsequent study De Tar[105] also made some criticism of equation (44). His results, in fact, led to the conclusion that σ^* values of alkyl groups are an artifact since they represent a residual steric

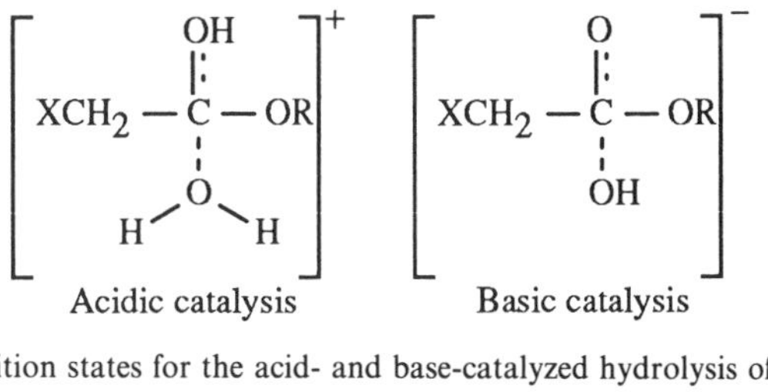

Acidic catalysis   Basic catalysis

Figure 14 Transition states for the acid- and base-catalyzed hydrolysis of aliphatic esters

effect plus an error component. Moreover, he pointed out that the E_s' values extend the Taft E_s set but are based on a slightly different scale. In addition, except for values which correct erroneous experimental data, correlations with the E_s' set are comparable to correlations with the E_s set. On the basis of these considerations, De Tar suggested and implemented a procedure to derive new E_s values or reevaluate older ones. The method is based on a stepwise procedure in which, first, a primary set of E_s values is calculated according to Taft, with the $\log k_x$ and $\log k_0$ values obtained by using the acid-catalyzed hydrolysis of RCO_2Et in 70% acetone as the reference reaction. The second step was approached by selecting four other reactions: acid-catalyzed esterification of RCO_2H with methanol, base-catalyzed hydrolysis of RCO_2Et in 70% acetone or 85% ethanol and base-catalyzed hydrolysis of RCO_2Me in 40% dioxane. Each of these reactions was used to obtain secondary E_s values. This was accomplished by computing the best values of a and b for equation (47), setting E_s equal to the primary values

$$E_s = a + b\log k_x \tag{47}$$

then each E_s secondary value was computed from its corresponding $\log k$. The last step of the procedure was to average the E_s primary values and the several secondary values for each group and to reset the origin to $E_s = 0.0$ for Me. However, the values so obtained do not represent a serious break with the Taft constants and they are compatible with the remaining E_s Taft set.

It is important to point out that steric substituent E_s parameters in no way parallel the corresponding polar σ^* constants but are in agreement with the relative van der Waals radii of the substituents. Kutter and Hansch,[81] extending an idea of Charton,[78] have developed equation (48) to formulate E_s values from the average radius (see Section 18.4.3) of symmetric top functions such as H, Br, CF_3, *etc.*

$$E_s = -1.839r_{av} + 3.484 \tag{48}$$

$$n = 6, \quad r = 0.996, \quad s = 0.132$$

This equation may be used to estimate E_s values not only of other symmetric top groups but also for unsymmetric substituents whose steric parameters cannot be easily obtained by Taft's method since the corresponding compounds are unstable under the conditions of acid hydrolysis. For example, since for substituents such as NR_2, OR and SR it was assumed that they could rotate with respect to the reaction center so that only the first atom would exert significant effects, the van der Waals radii for N, O and S, respectively, were used in equation (48) to formulate the corresponding E_s values. That is, in addition to modeling intramolecular interactions in which the effect of a neighboring atom or a group of atoms on a function within the molecule is critical for drug action, the use of E_s constants is associated with the idea that the substituent of a compound can rotate with respect to the surface of a macromolecule in such a way that steric hindrance is minimized (proximity effect). On the basis of these considerations, for substituents such as NO_2 and Ph two E_s values have been calculated by substituting the half width or thickness of the groups for r_{av} in equation (6). In fact, since steric parameters represent not only intramolecular proximity effects but also intermolecular steric factors, one must generally try both values to discover which yields the best fit in a correlation equation.

As already pointed out, Taft made use of average values of $\log(k_x/k_0)_A$ to evaluate the E_s parameters; since the dependence of $\log(k_x/k_0)_A$ on X is not independent of medium, Charton criticized this assumption and derived[77, 82, 106] a new scale of steric effective values (v_{eff}) based on a two-step procedure. Starting from the consideration that the E_s parameter varies in parallel with the group radius[78, 107] Charton correlated those $\log(k_x)_A$ values which were available under the same experimental conditions to the corresponding $v_{mn, x}$ values (see Section 18.4.3) with the equation

$$\log(k_x)_A = sv_{mn, x} + h \tag{49}$$

The s and h values obtained from the equation (49) were then used to calculate v_{eff} values of other substituents according to equation (50).

$$v_{eff, x} = \frac{\log(k_x)_A - h}{s} \tag{50}$$

Of course, $v_{eff} = v_{mn}$ for the substituents used in formulating equation (49).

A different procedure was required for the determination of v_{eff} values of substituents such as OR, SR, NR^1R^2, *etc.* since no suitable reference groups were available to make possible the use of the method described above. However, following a modified treatment,[77] Charton determined the appropriate s and h values to be used in equation (50). The Charton method results in a common scale for all groups making possible the inclusion of a wide range of substituent types in the same dataset.

Hansch and Leo[102] showed that for 104 different substituents the relationship between Taft's E_s and Charton's v_{eff} is

$$E_s = -2.062v_{eff} - 0.194 \tag{51}$$

$$n = 104, \quad r = 0.978, \quad s = 0.250$$

However, several substituents, such as CN, SMe, Cl, Br, I, NO_2, *etc.*, are rather poorly fitted by equation (51); this observation brings out a lack of understanding of steric effects since, as Hansch and Leo pointed out, any features that the outliers have in common are not obvious.

A number of authors have concluded that there are several cases where the E_s parameters are not correctly defined, *i.e.* they may contain not only steric but also electronic contributions, either direct or hyperconjugative. To separate the hyperconjugation effect of α hydrogens on the center of reaction, Hancock *et al.*[108−113] have proposed the parameter E_s^c (equation 52), assuming that the hyperconjugation effect is proportional to the number of α hydrogen atoms, N_H.

$$E_s^c = E_s - h(N_H - 3) = E_s + 0.306(N_H - 3) \tag{52}$$

In equation (52), h is a reaction constant for hyperconjugation, taken as -0.306 from quantum calculations.[114] In equation (52) using the original Taft E_s scale, by definition $E_{s(Me)}^c = 0$, while using the Unger and Hansch rescaled set this value comes out as -1.24.

In a number of cases the use of E_s^c rather than E_s led to an improvement in statistical correlation. However, it is not clear whether this finding is evidence of hyperconjugation rather than of some conformational factors. Several authors[80, 103, 115] have expressed doubts concerning the improvement of correlations using E_s^c.

Palm,[116, 117] considering that there may be a contribution not only of hyperconjugative effects of α hydrogen atoms but also a further correction for C–C hyperconjugation, proposed the modified steric parameter E_s^0

$$E_s^0 = E_s + 0.33(N_H - 3) + 0.13N_c \tag{53}$$

where N_c is the number of α C—C bonds. Several E_s^c and E_s^0 values are summarized in Table 11 (Section 18.4.10).

Fujita *et al.*[118, 119] have carried out analyses to examine whether the steric effect of substituents of the type $CR^1R^2R^3$ can be expressed in terms of the steric effect of component substituents R^1, R^2 and R^3 using E_s^c parameters. They formulated equation (54) for a set of primary, secondary and tertiary alkyl groups.

$$E_{s(CR^1R^2R^3)}^c = 3.429E_{s(R^1)}^c + 1.978E_{s(R^2)}^c + 0.649E_{s(R^3)}^c - 2.104 \tag{54}$$

$$n = 24, \quad r = 0.992, \quad s = 0.191$$

The substituents R^1, R^2 and R^3 were classified according to the relative magnitude of the corresponding E_s^c values, so that $E_{s(R^1)}^c > E_{s(R^2)}^c > E_{s(R^3)}^c$. Although the results were excellent (98.4% of the variance of the data is elucidated by equation 54), the approach is compromised by the necessity of assuming preferred conformations for the bulkier groups in order to obtain useful results.

Subsequently, MacPhee, Panaye and Dubois[120] reexamined this type of correlation, showing that E_s and E_s^c of $CR^1R^2R^3$ groups can be factored into linear combinations of subfragments as in equation (55) and (56), respectively.

$$E_{s(CR^1R^2R^3)} = \Sigma[a_i'E_{s(R^i)} + b_i'\Delta N_i] + a_0 \tag{55}$$

$$E_{s(CR^1R^2R^3)}^c = \Sigma[a_i'E_{s(R^i)}^c + b_i'\Delta N_i] + a_0' \tag{56}$$

In these equations the contribution of the $\Delta N = (N_H - 3)$ term for each component group is not fixed as in equation (52) but made adjustable.

An approach to quantify the electronic component in E_s was developed by Unger and Hansch.[98] These authors obtained equation (57) for the observed E_s values of substituents of the type CH_2X.

$$E_{s(CH_2X)} = -1.07\mathscr{F}_X - 1.05\mathscr{R}_X - 1.84 \tag{57}$$

$$n = 50, \quad r = 0.612, \quad s = 0.408$$

In this equation $\mathscr{F}_X$ and $\mathscr{R}_X$ are, respectively, the polar and resonance constants of Swain and Lupton (see Volume 4, Chapter 18.5). The correlation coefficient of equation (57) is not expected to be good although the correlation is statistically significant, since the steric factor has been omitted as an independent variable.

As Unger and Hansch pointed out, the geometry implied by the correlation is that the E_s vector lies approximately 52° off the plane defined by $\mathscr{F}$ and $\mathscr{R}$ and about midway between them. On the hypothesis that equation (57) quantifies the electronic contribution, steric parameters free of electronic components may be obtained through equation (58).

$$E_s^{\text{true}} = E_{s(CH_2X)} - f(\text{electronic effect of X})$$

$$= E_{s(CH_2X)} + 1.07\mathscr{F}_X + 1.05\mathscr{R}_X + 1.84 \tag{58}$$

The value 1.84 must be added to this equation in order to make the E_s^{true} value of the Me group, i.e. X = H in CH_2X ($E_{s(Me)} = -1.24$; $\mathscr{F}_H = \mathscr{R}_H = 0$), equal to -1.24.

Proceeding along the same lines, Unger and Hansch developed a similar equation for substituents of the type CH_2CH_2X, where the variation in E_s is essentially due to perturbation by the field-inductive component of X.

A simple relationship between the E_s parameter and the topology of a compound has been observed by Newman,[104, 121, 122] who postulated that the number of atoms in the sixth position from the carbonyl oxygen in esters (1) determines the steric effects which limit their reactivity. This idea has received a quantitative testing[123] through the formulation of equation (59), where N'_C and N'_H are the number of carbons and hydrogens in the sixth position.

$$E_s = -0.347N'_C - 0.075N_H + 0.119 \tag{59}$$

(1) Atoms in the sixth position which are supposed to be related to ester reactivity

Recently, Kier[124, 125] showed that three graph-based indices of molecular shape (see Section 18.4.2) are well correlated with the Taft steric parameter according to equation (60).

$$-E_s = -0.40\,^0\kappa + 0.78\,^1\kappa_\alpha - 0.34\,^3\kappa_\alpha - 0.63 \tag{60}$$

$$n = 46, \quad r = 0.961, \quad s = 0.280$$

This equation was based on aliphatic, heteroatomic and aryl groups. By limiting the series to aliphatic substituents in order to avoid complications of incidental non-steric effects, Kier generated a second equation.

$$-E_s = -0.36\,^0\kappa + 0.78\,^1\kappa_\alpha - 0.37\,^3\kappa_\alpha - 0.70 \tag{61}$$

$$n = 25, \quad r = 0.976, \quad s = 0.270$$

The two equations are very similar since the slopes, intercepts and statistical parameters are virtually identical. This observation led Kier to speculate that the entire set of Taft values may have about the same state of steric purity as does the set of alkyl substituents. On the basis of these considerations, Kier concluded that it is possible to use equation (60) to derive a new steric descriptor based entirely on structure and which is capable of predicting both the existing Taft values and the relative values for other groups. In fact, reducing the contributions of each term to whole numbers, by dividing through each coefficient by 0.35–0.40 and dropping the constant, Kier obtained in rounded values equation (62).

$$\Xi = 2\,^1\kappa_\alpha - \,^0\kappa - \,^3\kappa_\alpha \tag{62}$$

Kier adopted the Greek letter Ξ to represent this structure-based steric index, which is a summation of three attributes which can be easily calculated for all types of substituents. According to Kier, the Ξ parameter, which is somewhat related to the radii of the atoms embodied in the substituent, is a measure of the directed spatial influence of a group operating through the attached atom in the group; moreover, it is independent of electrical and solvent effects.

18.4.4.2 Descriptors Based on Quaternization Reactions

It is important to point out that despite some minor criticism[108, 116, 126] the E_s parameters and their corrected values are widely and successfully used in QSAR, while *ortho* steric parameters derived from aromatic compounds by Taft,[100] Farthing and Nam,[127] Kindler[128] and Hussey and Diefendorfer[129] have been shown[107] to be largely dependent on electronic effects.

Berg *et al.*[130] proposed a new scale for *ortho* steric effects based upon results of quaternization of substituted pyridines. In order to evaluate quantitatively the steric contribution of each 2-substituent to the ΔG^* value for the *N*-methylation by methyl iodide, Berg *et al.* assumed that electronic and steric effects can be separated. Since steric effects exist for all of the 2-substituted pyridine derivatives they can be estimated through the deviation from the Brønsted plot derived for a set of 3- and 4-substituted pyridines.

$$\log(k_X/k_H) = 0.35 pK_a - 1.73 \tag{63}$$

$$n = 13, \quad r = 0.967, \quad s = 0.735$$

The parameters of this equation are the basicity pK_a values, which include polar as well as resonance and field effects, and the relative rate constants $\log(k_X/k_H)$ for *N*-methylation of substituted pyridines by methyl iodide in acetonitrile at 30 °C. All of the 2-substituted derivatives are overpredicted by equation (63) and fall below the Brønsted line, indicating a reactivity lower than that expected from pure electronic effects; thus, steric effects can be calculated according to equation (64).

$$S^o = \log(k_X/k_H) - (0.35 pK_a - 1.73) \tag{64}$$

This procedure was repeated for the same substituent in different solvents. However, the comparison of individual S_i^o values and the mean value $\bar{S}^o$ for each substituent showed that they do not differ by more than 0.10. The results, which show that there is a practically insignificant variation of steric parameters S^o for *N*-methylation of pyridines in polar aprotic solvents, are in agreement with Charton's[131] and Abraham's[132] findings on the small solvent dependence of the relative rate constants of quaternization reactions. The values of parameter S^o for several substituents are summarized in Table 11 (Section 18.4.10); the value $S^o = 0$ corresponds to the hydrogen atom.

To test the validity of the scale of S^o *ortho*-steric parameters, Berg *et al.* showed that the complete lack of correlation between S^o and each of the well-known electronic substituent constants indicates that there is no significant electronic contribution to S^o. Moreover, they showed that the correlation between S^o and each of the parameters E_s, E_s^c, E_s^o and v_{eff} is satisfactory for substituents with the same symmetry but is rather poor between groups of different symmetry since conformational preferences of a different nature exist in aliphatic and heteroatomic molecules. On the basis of these considerations, Berg *et al.* concluded that S^o should be the parameter of choice to analyze steric effects of heteroaromatic molecules, which form a majority in QSAR studies.

Other quantitative measures of steric effects, not based on aliphatic acid-catalyzed reactions of esters or acids, have been developed and are adequately reviewed elsewhere.[98, 133]

18.4.5 PHYSICAL DESCRIPTORS

There is a long tradition of study in the determination of empirical parameters to take into account steric effects through physical descriptors which result when an organic molecule is considered as a three-dimensional object with size-determined physical properties. Such parameters include van der Waals volume V_w,[69] the related Exner's molar volume[134] and Traube's rule volume,[135] molar refractivity MR,[136] parachor Pr[137] and molecular weight M.

18.4.5.1 van der Waals Volume

Before considering methods for estimating the volume of specified portions of a given molecule, it is necessary to recall that a molecule may be seen as a set of vibrating nuclei surrounded by clouds of electrons. Thus, representing each atom of a molecule as a sphere, centered at the equilibrium position of the atomic nucleus with a radius equal to the van der Waals radius r_w (see Section 18.4.3), the van der Waals surface area may be artificially defined as the surface of the intersection of all the van der Waals spheres in the molecule, while the van der Waals volume is represented by the volume contained by such a surface. Since r_w is the distance at which the attractive forces between two non-bonded atoms are balanced by repulsive forces, the van der Waals volume (V_w) may be regarded as the volume impenetrable for other molecules. It must be emphasized that the atomic coordinates of a molecule change rapidly with the phase state in which it is found and the value of the van der Waals radius varies with the method by which it is determined. Bondi[69] discussed this point and proposed a method to estimate V_w from covalent bond distances and van der Waals radii. Two approximations have been made by Bondi; the contact distances have not been corrected to 0 K and all atoms have been considered as spheres and sphere segments, although it is well known that several are more nearly pear-shaped. However, the r_w values used by Bondi were obtained from a careful comparison of physical properties and are quite reliable.

The method developed by Bondi, which gives V_w in $cm^3\ mol^{-1}$, is summarized in Figure 15 for a diatomic molecule. In Figure 15, r_{w1} and r_{w2} are the van der Waals radii of atoms 1 and 2, respectively, l is the covalent bond distance, h_1 and h_2 are the heights of sphere segments, m is an auxiliary parameter and $N_A = 6.02 \times 10^{23}$ molecules mol^{-1}. Using this procedure, Bondi has determined effective radii and the corresponding volumes for several atoms and functional groups; a table of such values is given in Section 18.4.3. The Bondi volume may be obtained by summing the appropriate tabulated group contributions. However, the Bondi method, as well as the procedures developed by other authors,[138, 139] neglects the overlaps which are possible whenever three or more spheres intersect. Accurate calculation of the van der Waals volume is quite complex and alternative non-computational[140] and computational procedures are currently being developed.[141-143]

It should be pointed out that a variety of data have been successfully correlated by using volumes obtained by different approximation techniques for each set; this is not surprising since the success of the correlations indicates that the self-consistent relative V_w values obtained within a series of compounds is of more importance than the absolute values. However, one must be cautious in interpreting and comparing the slope and the intercept of such correlation equations.

Purcell and Testa[144, 145a] developed a method by using the parameter V_i, defined as the character variable whose values were estimated according to Bondi. In this method the molecule is considered as having regions in space called sectors which may be filled or unfilled by atoms or groups; the character variable V_i is defined by the volume of the atom or group in a specified sector i. In addition, Purcell and Testa defined the so-called locator variables either as the geometric distance between certain atoms or as Cartesian coordinates defining the sectors. Although the steric meaning of locator variables is rather obscure, they were used in correlating the affinity constant of carbonic anhydrase for a set of sulfonamides. Nevertheless, these results should be considered with caution. In fact, Hansch et al.[145b] showed later that electronic and hydrophobic parameters are of overwhelming importance in correlating the binding of sulfonamide inhibitors to carbonic anhydrase.

There has been some criticism[146, 147] as to whether the van der Waals volume is a true steric parameter. Molar volume and those quantities which are a linear function of it are scalar quantities while the steric effects observed in organic reactions, as well as in the interactions with acceptor sites in biological systems, should be rationalized in terms of vector quantities.

18.4.5.2 Molar Refractivity (MR)

Leo et al.[140] showed that the van der Waals volume V_w is correlated with the octanol/water partition coefficient (log P) for apolar compounds. Similar conclusions have been obtained by

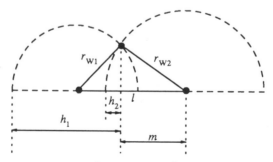

$$V_1^1 = \pi h_1^2 [r_{w1} - (h_1/3)]; \quad V_2 = (4\pi/3)(r_{w2})^3; \quad \Delta V_{2-1} = \pi h_2^2 [r_{w2} - (h_2/3)]; \quad V_w = N_A(V_1^1 + V_2 - \Delta V_{2-1})$$

Figure 15 Bondi's method of calculation for a diatomic molecule

Moriguchi *et al.*,[148] who showed that the van der Waals volume could be a parameter related to solute–solvent interactions. In addition, Charton[146] showed that V_w values are generally highly collinear with the corresponding molar refractivities, which may represent polarizability factors rather than steric effects. This observation is not surprising since the molar refractivity MR is related to the molecular volume by the Lorenz–Lorentz equation.

$$\text{MR} = \frac{n^2 - 1}{n^2 + 2} \frac{M}{d} (\text{cm}^3 \text{mol}^{-1}) \tag{65}$$

In this equation n is the refractive index of a liquid, normally measured at 20 °C, Na D line, d is the density at 20 °C and M/d is the molar volume. Since the range in the refractive index is small, MR in essence is an adjusted molar volume which contains an electronic contribution. In fact, since the molar polarization of a substance is given by equation (66), when the Maxwell relation $n^2 = \varepsilon$ is satisfied, the molar refractivity equals the molar polarization.

$$P_M = \frac{\varepsilon - 1}{\varepsilon - 2} \frac{M}{d} \quad \text{where } \varepsilon = \text{dielectric constant} \tag{66}$$

Moreover, as has been highlighted by Dunn,[149a] since the polarizability α of a molecule is related to P_M and MR through equation (67), α may also be viewed as a measure of volume. Cammarata *et al.*[150] have shown that the polarizability α is related to the lipophilicity parameter $\log P$.

$$P_M = \frac{4\pi N_A}{3} \alpha \tag{67}$$

On the other hand, the London cohesive energy ε_L for two interacting species is given by

$$\varepsilon_L = \frac{3}{2} \frac{\alpha_1 \alpha_2}{r^6} \frac{I_1 I_2}{I_1 + I_2} \tag{68}$$

where α_i and I_i refer to the polarizabilities and ionization potentials of each species, respectively, and r is the distance between them.

Equation (68) shows that the relationships between MR and ε_L is not simple since a few approximations must be made. Although r can be thought of as the center–center distance between a ligand and its binding site, its value cannot be approximated; moreover, measuring the ionization potentials of the interacting species is not an easy task since the true nature of the binding site is largely unknown. By assuming their values to be constant, Pauling and Pressman[151] found in a study of a series of benzoic and arsonic acid haptens that the hapten inhibition constant for antigen–antibody reaction was related to MR. In a subsequent study[152] it was shown that, although dispersion forces can contribute to ligand–macromolecule interactions, steric effects correlated using E_s are the primary factors in determining the value of the inhibition constant.

On the basis of these considerations, the role of MR for a ligand may be viewed as an ambivalent one. It may be that MR represents dispersion forces aiding the binding of a ligand to the receptor site, as Agin *et al.*[153] suggest. In such a case one would expect a positive coefficient for the MR term. Alternatively, since MR is to a large degree a measure of volume, it may measure a ligand's ability to distort the conformation of the receptor in such a way as to preclude union with the proper substrate. Since the conformational change is detrimental, a negative coefficient should result for the MR term in this case. Negative slopes with MR have also been assumed to reflect steric hindrance of one kind or another. Of course, some combination of these effects may be involved.[154-156]

Since MR is an additive–constitutive property of molecules, fragment values have been calculated for many groups of atoms. MR values based on Vogel's[154, 155] measurements are given in Table 11 (Section 18.4.10) for a set of selected substituents.

To test the hypothesis that MR may be related to steric hindrance, Dunn[149] carried out a regression analysis of MR as a function of the Hancock steric parameter E_s^c.

$$\text{MR} = -9.04 E_s^c + 12.75 \tag{69}$$

$$n = 142, \quad r = 0.67, \quad s = 12.8$$

The correlation coefficient of equation (69) is rather low. Considering for example *n*-alkyl groups, the values of E_s^c are practically constant for $n = 2, 3, 4, \cdots$ in C_nH_{2n-1}, while the corresponding MR values regularly increase for each CH_2 addition. On the other hand, MR is the same for all members of a set of C_nH_{2n-1} isomeric groups, whereas the E_s^c values can show considerable variation. However, the correlation is statistically significant at the 99.99% level; thus, equation (69) suggests that a negative coefficient with MR might indicate that steric effects are involved in the correlation under study. In addition to the interrelationship between MR and steric parameters, the molar refractivity of a substituent may be related to the corresponding lipophilicity as expressed by π. In fact, although for a proper set of substituents MR and π are orthogonal, both parameters depend to a certain extent on molar volume, thus the collinearity between these vectors may be high, depending on the choice of substituents. The problem of selecting a set of substituents that would be independent with respect to these two parameters is illustrated in Table 5, where the MR values have been scaled by 0.1 which makes the apolar functions essentially equiscalar with π. From an inspection of the two series of substituents, it can be easily seen that MR and π are almost completely collinear ($r = 0.93$) in set A; either of these two vectors would give almost the same result in a correlation equation. Selecting data from set B, the two vectors are practically orthogonal ($r = 0.08$) so that their relative importance may be resolved.

18.4.5.3   The Parachor

Another parameter which relates principally to molecular volume is the parachor Pr. This parameter was defined by Sugden[157] as

$$Pr = \gamma^{1/4} \frac{M}{(d - d')} \tag{70}$$

where d and d' are the densities of a liquid and its vapor, respectively, and γ is the surface tension in dyn cm^{-2} at the same temperature (1 dyn = 10^{-5} N). When the vapor density is negligibly small in comparison with that of the liquid, the relationship reduces to

$$Pr = \gamma^{1/4} \frac{M}{d} \tag{71}$$

It is interesting to note that Sugden, while stressing the additive nature of the atomic parachor, assessed structural constants for varying degrees of unsaturation and ring structures. Moreover, he recognized a constitutive effect in the difference between oxygen in alcohols and ester groups. Although various modifications have been necessary in Sugden's values summarized in Table 6, agreements within 1 or 2%, have been found between experimental and calculated parachor values. As methods of obtaining compounds of the highest purity have advanced and the techniques of the measurement of surface tension have improved, this precision has reached the limit of 0.1%. Mumford and Phillipp's reevaluations[158] of the atomic parachors, as well as those of Vogel,[159] which are the most widely used, are listed in Table 6.

The determination of atomic parachors and related structural constants has been discussed by Quayle[137] and McGowan.[160-162] Particularly, McGowan has pointed out the relationship of

Table 5   Two Sets of Substituents for which the Vectors π and MR are Collinear (Set A) and Orthogonal (Set B)

| | Set A | | | Set B | |
Substituent	π	0.1 MR	Substituent	π	0.1 MR
NH_2	−1.23	0.54	$CONH_2$	−1.49	0.98
CN	−0.57	0.63	NHCOMe	−0.97	1.49
NO_2	−0.28	0.74	OH	−0.61	0.28
OMe	−0.02	0.79	CN	−0.57	0.63
Br	0.86	0.89	NHMe	−0.47	1.03
Et	1.12	1.39	NO_2	−0.28	0.74
$CHMe_2$	1.53	1.50	F	0.14	0.09
CMe_3	1.98	1.96	Me	0.56	0.56
Cyclohexyl	2.51	2.67	CH_2Br	0.79	1.34
			CF_3	0.88	0.50
			OCF_3	1.04	0.79
			$CHMe_2$	1.53	1.50

Table 6 Atomic and Structural Parachor Values

Unit	Values assigned by		
	Sugden	Mumford and Phillips	Vogel
C	4.8	9.2	8.6
H	17.1	15.4	15.7
O	20.0	20.0	19.8
O_2(esters)	60.0	60.0	54.8
N	12.5	17.5	—
S	48.2	50.0	49.1
F	25.7	25.5	—
Cl	54.3	55.0	55.2
Br	68.0	69.0	68.8
I	91.0	90.0	90.3
Double bond	23.2	19.0	19.9
Triple bond	46.6	38.0	40.6
Five-membered ring	8.5	3.0	—
Six-membered ring	6.1	0.8	—

physical toxicity and narcosis to parachor and suggested a correction to this parameter to account for special hydrogen-bonding effects.

Since $\gamma^{1/4}$ does not vary greatly for organic compounds, parachor is, as already mentioned, a corrected molar volume term. Hansch and Leo[155] showed the great similarity between MR and Pr parameters by deriving equation (72).

$$MR = 1.30Pr - 2.76 \tag{72}$$

$$n = 75, \quad r = 0.966, \quad s = 0.288$$

The two parameters are almost perfectly collinear; they both give the same quality of correlation. Since parachor, like MR, may be considered an additive–constitutive property of organic structures, Pr substituent group values may be estimated from Vogel's values which are summarized in Table 6.

18.4.5.4 Molecular Weight and Steric Density

Each of the parameters discussed above depends largely on the molecular weight M of the compound. The dominance of M in giving apparent additivity in a parameter such as MR has been commented on by Exner.[163] Although this is true for individual substituents, Hansch *et al.*[136] have shown that for a large set of substituents the correlation between MR and M is rather low ($r = 0.76$); only 58% of the overall variation in MR is accounted for by M. This observation is not surprising since MR is an approximation of volume, whereas M is related to the substituent mass. Although the density of the substituents cannot be considered constant, Kitahara *et al.*[164] used $M^{1/3}$ as an approximation to molecular radius in diffusion studies.

Dash and Behera[165, 166] defined instead a steric density parameter (SD) combining molecular weights and van der Waals volumes according to equation (73).

$$SD = (M/V_w)_X - (M/V_w)_H = (M/V_w)_X - 0.29 \tag{73}$$

However, since they derived significant correlations with the pK_a values of various sets of X-substituted organic acids, the role of the SD parameter is of difficult delineation.

In conclusion, although all of the parameters discussed in this section are to a certain extent related to dispersion effects and to interactions involved in partition processes, they have been widely used in QSAR studies as a measure of the bulk interaction of a substituent with a particular section of space in or on the corresponding acceptor site.

18.4.6 DIFFERENTIAL DESCRIPTORS

Overlapping methods originate from structure–activity relationships where specific interactions or recognition processes are of overwhelming importance. These methods, which are a sort of receptor mapping, involve a process of deducing the structure and the physicochemical nature of the acceptor site by studying the properties of ligand structural characteristics which interact to various degrees with the receptor. Some parts of any one ligand may not interact appreciably with the receptor, but those parts of the ligands which do interact will provide indirectly the receptor structure. The ligands are then considered in superimposition and receptor groups are identified in the space around the ligands.

18.4.6.1 Similarity Indices

Several interesting ideas concerning the overlapping methods have been developed but only a few have found application in QSAR. One of the first methods was developed by Amoore *et al.*,[167] who proposed a similarity index Δ for the study of the stereochemical factors influencing the sense of smell. In this procedure, the shape of each pair of molecules (one considered as a reference) was compared by measuring the superimposition of their front, top and right molecular silhouettes, *i.e.* the projections of space-filling models in three mutually orthogonal planes. The superimposition procedure was achieved by respective superimposition of weight centers and collinearity of the main axis. Starting from the weight center toward the periphery of the superimposed silhouettes, 36 radii were traced with $10°$ angular spacing. By measuring absolute differences between corresponding radii in silhouette pairs, three Δ values for the three types of projection were obtained; from the average of these values Amoore defined the parameter $\bar{\Delta}$, which may be normalized as

$$\bar{\Delta}_N = 1/(1 + \bar{\Delta}) \tag{74}$$

The parameter $\bar{\Delta}$ is, therefore, directly related to the shape difference of molecular contours and may be considered as a kind of mean value of three-dimensional molecular differences.

A more sophisticated procedure to compare the shape of two molecules was proposed by Allinger.[168] This approach is more complicated and has not found application in QSAR.

18.4.6.2 Minimal Steric Difference (MSD)

A well-known method has been developed by Simon *et al.*,[169–171] who defined the minimal steric difference (MSD). This method is based on the assumption that ligand–site acceptor interaction is a linearly decreasing function of the steric misfit of the ligand and the site acceptor cavity, *i.e.* the activity is a function of the sum of the non-overlapping volumes of the ligand and the cavity. The term 'minimal' steric difference originates from the consideration that if the ligand has several low energy conformations it will adopt the one which fits best into the acceptor site cavity.

There are basically two requirements for the MSD approach to be applicable: (i) one must have an idea of the site acceptor cavity and (ii) a method to estimate the non-overlapping volume is necessary.

A reliable approximation for the shape of the cavity is the natural effector molecule or the most active structure in the set of compounds under study; this molecule is termed the standard (S). Then, the planar structural formulas of the other molecules are superimposed on the standard, neglecting small differences between bond lengths and angles, and counting the non-superimposable atoms. Although large simplifications are made, *e.g.* phenyl and cyclohexyl are considered equivalent, stereochemistry is taken into account, *i.e.* an *R* configuration is different from an *S* configuration. The number of non-superimposable atoms (hydrogen atoms are neglected) gives the MSD value for the considered structure. Since atoms of different rows have different van der Waals volumes a weighting factor is necessary; for second period atoms (C, Me, N, O, OH, *etc.*) the weighting factor one is used, for third period elements (S, SH, Cl, *etc.*) this factor is 1.5 and for higher period elements (Br, I, *etc.*) a factor two is taken. Following this procedure, it is also possible to calculate the MSD value when different row atoms or groups are superimposed.

For the structures reported in Figure 16 the MSD values are: $MSD(S, S_1) = 0$, $MSD(S, S_2) = 4$, $MSD(S, S_3) = 2$, $MSD(S, S_4) = 2$ and $MSD(S, S_5) = 3$. Acetylcholine is used as the standard (S) and the MSD value for structure S_1, which is completely superimposable upon acetylcholine, is equal to zero while the other MSD values are easily evaluated by the superimposition procedure.

It is worth pointing out that one can try different molecules as standards and select by trial and error that which yields the highest correlation coefficient for equation (75), in which A_i is the biological activity and α may be a function of the type $\alpha = \alpha_0 + \alpha_1 \pi_i + \alpha_2 \sigma_i + etc.$, if other structural

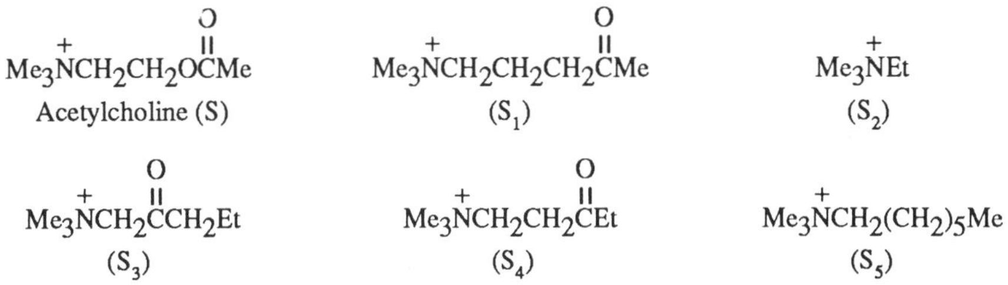

$$\overset{O}{\underset{\parallel}{Me_3NCH_2CH_2OCMe}}$$

Acetylcholine (S)

$$\overset{O}{\underset{\parallel}{Me_3NCH_2CH_2CH_2CMe}}$$

(S_1)

$$Me_3\overset{+}{N}Et$$

(S_2)

$$\overset{O}{\underset{\parallel}{Me_3\overset{+}{N}CH_2CCH_2Et}}$$

(S_3)

$$\overset{O}{\underset{\parallel}{Me_3\overset{+}{N}CH_2CH_2CEt}}$$

(S_4)

$$Me_3\overset{+}{N}CH_2(CH_2)_5Me$$

(S_5)

Figure 16 Trimethylammonium derivatives active on the cholinergic receptor

parameters are considered.

$$A_i = \alpha - \beta(MSD)_i \tag{75}$$

Despite these approximations, the MSD method has been successfully applied in many cases, such as the cholinergic activity of trimethylammonium derivatives,[171] the α-chymotrypsin-catalyzed hydrolysis of methyl esters of *N*-acetylated L-amino acids,[169] the inhibition of dihydrofolate reductase by 5-alkyl-2,4-diaminopyrimidines,[172] *etc.*

Note that, by their nature, MSD values must be calculated for each new congeneric set, so that no generally applicable MSD substituent parameters are available.

18.4.6.3 Monte Carlo Version of MSD (MCD)

An improvement of the MSD method is the MCD procedure proposed by Motoc *et al.*[172-175] In this method the minimal non-overlapping volume of the superimposed pair of structures is evaluated by a mathematical procedure, the Monte Carlo method.[176] The two structures (one is the standard, S) are described by Cartesian coordinates and the van der Waals radii of the component atoms. Starting, for example, from the lowest energy conformations, some atoms and bonds of the two structures are superimposed in order to fix the common coordinate system. The two molecular van der Waals envelopes are defined and a parallelapiped of volume V containing the envelopes is constructed. The precision of the calculation is determined by the quality of the random number of points (N_T) uniformly distributed in the volume V by a Monte Carlo technique. Since each point may belong to none, to one or to both the envelopes, if N_i is the number of points falling only on one of the two envelopes, the MCD value is evaluated according to equation (76).

$$MCD = V\frac{N_i}{N_T} \tag{76}$$

Although the MCD method is more accurate geometrically than the MSD procedure, it has not been widely used.

18.4.6.4 Minimal Topological Difference (MTD)

Several successful QSARs[171] have been obtained by a modified version of the MSD procedure. This is the minimal topological difference (MTD) method, developed by Simon *et al.* in 1976.[177-179] In this method, after defining the orientation of the standard molecule, the N compounds are superimposed one upon another, taking care that the atoms of each structure are superimposed as much as possible upon the atoms (hydrogen atoms are again neglected) occupying similar positions in the reference molecule. The superimposition of the N molecules yields a topological network, the hypermolecule which reflects the stereochemistry of ligands bound to the acceptor site. The hypermolecule, whose M vertices (j) correspond to the approximate positions of atoms in the superimposed structures and the edges to the chemical bonds between atoms, is used as the standard to calculate the MTD values. In fact, each structure is described by a vector $x_i = [x_{ij}]$, where x_{ij} is taken to be one if the vertex j is occupied by the ligand i and zero if it is not occupied. For example, the superimposition procedure will yield the hypermolecule (2) with $M = 9$ vertices for the cholinergic trimethylammonium derivatives summarized in Figure 16. Since the $Me_3\overset{+}{N}CH_2$— fragment is common to all the structures, and presumably in the interaction with the receptor it will occupy the same vertices occupied by the natural effector acetylcholine, it is used as a guide for the superimposition. Note that acetylcholine occupies vertices $j = 1, 2, 3, 4, 7$ and then the corresponding vector is $x_i = (111100100)$.

(**2**) The hypermolecule for the cholinergic derivatives of Figure 16

At this point, an initial receptor map (S^0) can be derived by the following procedure. The vertices j corresponding to the most active compound, *i.e.* acetylcholine, in the congeneric set under study, are assigned to the acceptor site cavity; these vertices are characterized by the parameter $\varepsilon_j = -1$ (beneficial effect) while the vertices j corresponding to the walls or to the exterior of the receptor are characterized by ε_j values of $+1$ (detrimental effect) or 0 (irrelevant effect), respectively. Note that the assignments $\varepsilon_j = +1$ or 0 are largely subjective and the sequence of -1, $+1$ or 0 values of the ε_j values represents the initial receptor map. Each ligand i will be characterized by a MTD$_i$ value given by

$$\text{MTD}_i = s + \sum_{j=1}^{M} \varepsilon_j x_{ij} \tag{77}$$

where s is the total number of cavity vertices.

It is worth pointing out that if the number of $\varepsilon_j = -1$ vertices is considered as the standard (S) and if there are no $\varepsilon_j = 0$ vertices the MTD$_i$ and MSD$_i$ parameters are equal.

Once the MTD$_i$ values for the first receptor map S^0 have been calculated an optimization procedure may be applied in order to perform the vertex assignments by trial and error. In fact, from the first set of the MTD$_i$ values, the ligand activities may be calculated by using a correlation equation similar to equation (75) by substituting MSD$_i$ for MTD$_i$. The mean square difference (Y) of experimental (A_i) and calculated ($\hat{A}_i$) activities (equation 78) must then be minimized. This is achieved by changing the ε_j assignments in the initial receptor map (S^0) one by one. For each monosubstituted map a new sum (Y) is evaluated and compared to the initial value. If all the new Y values are larger than the initial one no optimization is possible. In the case where one or more monosubstitutions decrease the Y value (ΔY negative), the monosubstituted map for which the lowest Y value is obtained must be selected. This map (S^1) will represent the new starting map and the procedure will be stopped when all the new ΔY values will be non-negative on a new monosubstitution. The last map is termed 'optimal' and denoted by $S^*(\varepsilon_1^*, \varepsilon_2^*, \cdots \varepsilon_M^*)$.

$$Y = \sum_{i=1}^{N} (A_i - \hat{A}_i)^2 \tag{78}$$

The MTD approach, which may be considered as a variant of the Free–Wilson method[175] has a few defects. These include the fact that the optimization procedure is performed by minimizing equation (78) in which the activity, given by equation (79), contains the term α which may be a function of hydrophobic, electronic, *etc.* factors, *i.e.* the optimal MTD$_i$ values may code globally both steric and other physicochemical factors which may control the biological response.

$$\hat{A}_i = \alpha - \beta(\text{MTD})_i \tag{79}$$

18.4.6.5 Molecular Shape Analysis (MSA)

Another approach to overlapping methods has been initiated and developed by Hopfinger[180–184] using molecular shape analysis (MSA). In this approach the congeners in a dataset are first examined by using molecular mechanics to determine the most stable conformers. After this point the procedure is somewhat similar to that of Simon *et al.*, since a reference compound must be selected against which the shape of all other congenérs can be compared. The total common overlap volume of the reference structure (S) and each of the congeners i in the dataset is defined as V_0 (i, S) which is calculated by a rather complex process[183] using highly efficient computer programs. Besides the parameter V_0, the arbitrary functions $S_0 = V_0^{2/3}$ and $L_0 = V_0^{1/3}$ were introduced as alternative molecular shape descriptors. If the shape of the reference structure is a good approximation for the acceptor site cavity, V_0 should give the part of the cavity volume occupied by the considered ligand while S_0 should be an approximation to the contact surface area of the ligand with the receptor. It is worth noting that in formulating the parabolic QSAR (equation 80) for the triazines (**3**) Hopfinger stated 'The optimum values of S_0 and, correspondingly, V_0 have no physical meaning. The S_0 represent the relative numerical scales which reflect the need to have an analogue adapt the $\theta = 310°$ conformer state such that the 3 and/or 4 substituents possess size and/or conformational freedom so that specific spaces are occupied.'

(**3**) General structure of X-phenyltriazines inhibiting dihydrofolate reductase

From the molecular mechanics calculations Hopfinger surmised that the most favorable angle (θ) between the phenyl ring and the triazine structure is $310°$. From a study[185] of about 250 congeners of structure (**3**), he selected 27 compounds as representatives of seven distinct classes and derived equation (80).

$$\log(1/C) = -15.66 - 1.384S_0 - 0.02127S_0^2 + 0.434\,\Sigma\pi_{3,4} - 0.574D_4 - 0.294D_4^2 \qquad (80)$$

$$n = 27, \quad r = 0.953, \quad s = 0.44$$

In this equation C is the molar concentration producing a 50% inhibition of dihydrofolate reductase, $\Sigma\pi_{3,4}$ is the sum of hydrophobicities of the substituents in positions 3 and 4 of the phenyl ring and D_4 is a parameter related to the length of the substituent in position 4 which is assumed to measure a hypothetical intermolecular steric interaction involving certain 4-X substituents directly along the vector defined by the —C_6H_4— bond. The correlations with V_0 or L_0 as parameters, instead of S_0, proved less successful.

As already pointed out for the other overlapping methods, the MSA values must be calculated for each new dataset so that no generally applicable substituent parameters are available.

Perhaps the most detailed and versatile procedure for mapping receptor sites is the distance geometry method developed by Crippen[186–191] and extensively treated in Volume 4, Chapter 22.4. The computerized treatment consists of a heuretic approach which combines information concerning the conformational possibilities of the ligand with the corresponding observed receptor binding affinities.

18.4.7 APPLICATIONS OF STERIC PARAMETERS IN QSARs

Since its initiation,[9] the use of QSARs has grown rapidly so that thousands of equations have been published, and they have been widely used in the study and design of bioactive compounds. In order to explore the scope and the limits of the most widely used steric parameters, selected correlations between certain biological effects and some combinations of substituent descriptors are considered in this section.

18.4.7.1 Topological Descriptors

The significance and the choice of the most commonly occurring topological descriptors have been considered in Section 18.4.2. Most of them have been applied to correlations with a number of physical properties of organic compounds such as alkanes, alkenes, alkynes and arenes. Connectivity indices developed by Kier and Hall[20,21] have found a rather wide use in QSAR. For example, Murray *et al.*[192] have analyzed a set of 2-substituted benzimidazoles for their *in vitro* ability to displace the dihydrosafrole metabolite from its complex with microsomal cytochrome P-450. Quantitative structure–activity analyses showed that a series of good correlations may be obtained for linear regressions between initial rate of displacement ($\log D$) and molecular connectivity indices. The best equation appeared to be equation (81), involving the parameter $^1\chi^v$ in which 90% of the data variance was accounted for.

$$\log D = 0.562\,{}^1\chi^v - 4.083 \tag{81}$$

$$n = 21, \quad r = 0.948, \quad s = 0.179$$

In this analysis the benzimidazole nucleus remained constant and therefore ${}^1\chi^v$ values were estimated for the 2-substituents only. Since all path-type connectivity indices yielded similar or poorer correlations, Murray *et al.* concluded that adequate topological information is contained within indices of lower order. Although equation (81) supports the assertion that steric parameters are of primary importance in the displacement process, Murray *et al.* pointed out that considerable overlap exists between many of the physicochemical descriptors (other χ indices, $\log P$ and MR) studied in developing their QSAR.

Kier *et al.* applied the molecular connectivity technique to several QSAR examples.[20,21] A study[193] which presents an opportunity to explore the meaning and the limits of connectivity indices is represented in equation (82)

$$\log PC = -\frac{21.41}{{}^1\chi} + 6.31 \tag{82}$$

$$n = 49, \quad r = 0.975, \quad s = 0.20$$

In this equation, PC represents the phenol coefficient of a set of 49 substituted phenols tested for their bactericidal action against *S. aureus*; PC is the concentration which is effective in killing bacteria after 10 min exposure, expressed relative to unsubstituted phenol.

In this analysis, several different χ variables were considered, including different orders and both connectivity and valence types. Moreover, equations of different type and order were derived. The best correlation required only the simple first-order connectivity index in a hyperbolic form. Although this type of equation does not allow a simple physicochemical interpretation, it does allow interpolative prediction. Since the coefficient associated with the $1/{}^1\chi$ term is negative, equation (82) indicates that structural features such as chain branching or adjacency of atoms influence activity in a parallel fashion to the magnitude of ${}^1\chi$. In studies on the same set of compounds, Tute[194] and other researchers[195,196] showed that bacteriostatic effectiveness of phenols increases as the lipophilic character is increased. Parabolic or bilinear dependences are observed if a wider range of lipophilicity is considered. In fact, although there is no doubt that several mechanisms may be involved in the inactivation of living cells, any general hypothesis regarding the mechanism of action of phenolic bacteriostasis must include the effects of hydrophobic–lipophilic interactions.

It is worth pointing out that a large number of simple organic molecules exert their biological effects by so-called non-specific interactions which can be accounted for by $\log P$ and π parameters. It has been proposed that such interactions are also related to various connectivity indices. Dunn *et al.*[149b] compared the results obtained by correlating $\log P$ with non-specific protein binding with those of protein binding and various connectivity indices. The results of this study indicate that care should be used in applying ${}^1\chi^v$ and ${}^1\chi$ connectivity indices to such problems if the substrates are not closely related in a structural sense, *e.g.* homologous series. It might be argued that correlations with other indices might be attempted but such an application of regression methods is unsound since, if enough independent variables are attempted, the probability is high that one significant correlation will be found.

These considerations stress the confusing problem of collinearity[197] among vectors associated with topological descriptors and other physicochemical parameters.

Subsequently, Kier[38] reconsidered the data on which equation (82) is based and showed that the log PC values are also well predicted by two different non-linear equations in which the independent variable is the shape index ${}^2\kappa$.

The ${}^2\kappa$ index has been used, in combination with ${}^3\kappa$, in a study[40] of the inhibition of the microsomal *p*-hydroxylation of aniline by alcohols. In fact, the hypothesis of Cohen and Mannering[198] that the steric requirements of the alcohols are an important factor in the inhibitory potency, prompted Kier to formulate equation (83), where I_{50} is the concentration needed to produce a 50% inhibition of the enzyme. Although equation (83) supports the hypothesis that the shape of the alcohols is a decisive factor in determining the inhibitory potency, since ${}^2\kappa$ and ${}^3\kappa$ indices imply that elongated molecules and those that are branched at the ends are more potent as inhibitors, the equation statistical parameters are rather ambiguous in spite of their quality which appears highly significant. In fact, the intercorrelation between ${}^2\kappa$ and ${}^3\kappa$ ($r = 0.73$) is not as low as one would like and the equation embodies one data point (methanol) which gives two log units of variance in the biological data. In formulating a correlation equation care must be taken not only with regard to the range of the data space (variance problem [197]) but also with regard to the spanning of the data in the range.

$$\log(1/I_{50}) = -\frac{2.236}{{}^2\kappa} - \frac{2.872}{{}^3\kappa} + 1.215 \tag{83}$$

$$n = 19, \quad r = 0.975, \quad s = 0.188$$

18.4.7.2 Geometric Descriptors

The geometric descriptors derived as a measure of the van der Waals radius appear to be workable parameters for describing intermolecular interactions. However, substituents are three-dimensional objects so that their extension in a particular direction, being a function of their conformational options, may become critical with respect to the nature of the intermolecular interactions. These considerations led Verloop *et al.*[86] to develop the STERIMOL parameters which give a measure of the dimensions of non-spherical substituents in a restricted number of directions in such a way as to delineate a more refined view of the shape of the substituents. An instructive example of the use of these parameters is the quantitative treatment of the role of different intermolecular forces in hapten–antibody interactions. In fact, Verloop *et al.*[86] reexamined a QSAR previously developed by Kutter and Hansch (see Section 18.4.7.3) and derived equation (84).

$$\log K_{\text{rel}} = -0.41B_1^o + 0.95L^p - 0.07(L^p)^2 - 0.43 \tag{84}$$

$$n = 36, \quad r = 0.961, \quad s = 0.282$$

This equation is based on a study by Pressman *et al.*,[199] who prepared antibodies by coupling diazotized *p*-aminobenzoic acid to serum proteins and injecting these modified proteins into rabbits. After isolation, the antibodies were mixed with antigen to form a precipitate. X-Substituted benzoate ions tend to prevent the union of antigen and antibody so that the corresponding affinity constant (K_{rel}) can be determined. The observation that equation (84) does not contain any parameter for *meta* X-substituents suggests that these substituents are of no importance in the interaction of hapten and antibody while, as is shown by the negative slope of the minimum width parameter B_1^o, *ortho* substitution gives poor inhibitors. Large groups in the *para* position, instead, result in effective inhibitors. However, the appearance of the squared L^p term suggests that for *para* substituents a value of $L^p = 6.8$ would lead to an optimum interaction.

In a review[119] on the application of STERIMOL parameters, Fujita and Iwamura report many interesting examples of their use. Among others, there is an analysis of the data of Bruce and Zwar[200] on a set of ring mono-, di- and tri-substituted diphenylureas (4), prepared and tested on the basis of an earlier observation[201] that *N,N'*-diphenylurea is a compound identified as a cell-division factor of coconut milk which exerts the same activity as the N^6-substituted adenines in the cytokinin tests.

(4) Substituted diphenylureas active in the cytokinin tests

By considering position-specific substituent steric and hydrophobic effects, equation (85) was derived.

$$\log(1/C) = 0.90\sigma + 1.04\pi_m - 0.85L^o - 0.27L^p + 5.00 \tag{85}$$

$$n = 39, \quad s = 0.38, \quad r = 0.91$$

In this equation, C represents the minimum molar concentration which gives a detectable response in the tobacco pith assay, σ is a measure of electronic substituent effects and π_m is the well-known hydrophobic constant for *meta* substituents. The negative slope of the STERIMOL L^o parameter for *ortho* substituents, much larger than that of L^p for *para* substituents, shows that there is a stricter steric demand at the *ortho* position than that at the *para* position. Moreover, the lack of a steric parameter for the *meta* substituents indicates that steric effects are insignificant at this position, whereas the hydrophobicity of the substituents enhances activity. These considerations rationalize and provide a physicochemical interpretation for the recognized general order of potency among isomers: *meta* > *para* > *ortho*.

In a study on cholinergic ligand interactions by QSAR, Pratesi *et al.*[202] have taken into consideration a set of X-benzyltrimethylammonium derivatives (5) as a starting point to approach the problem. Since it is not always clear which aspect of steric effects might be important, because steric substituent interactions are not easily translated from one compound or reaction type to another, in order to avoid overlooking any significant possibilities, Pratesi *et al.* considered a number of the available steric parameters. The best QSAR, subsequently extended to several related structures,[203] was obtained in terms of the maximum width parameter B_4. However, similar results can be derived by considering the more recently proposed parameter B_5.

$$pD_2 = 1.30\pi_X - 0.41B_5 + 5.68 \qquad (86)$$

$$n = 10, \quad r = 0.948, \quad s = 0.186$$

(5) X-Benzyltrimethylammonium derivatives active on muscarinic receptor

Equation (86) supports the hypothesis that the muscarinic affinity of the compounds, expressed as pD_2 determined on isolated rat jejunum, is modulated by the anchoring of the cationic end to an electron-rich binding site on the receptor and by the interaction of the X-substituent with a lipophilic pocket of limited size arranged at a suitable distance. In fact, the high positive value of the coefficient of the hydrophobic parameter strongly suggests the projection of the X-substituent into a hydrophobic pocket of the receptor. However, the negative coefficient associated with Verloop's steric parameter B_5 indicates that large groups interact to inhibit binding when X reaches a certain size.

The examples discussed show that STERIMOL parameters can be successfully used in combination with other physicochemical parameters to provide information about three-dimensional aspects of the interactions between ligands and biomolecules.

18.4.7.3 Chemical Descriptors

As already pointed out in Section 18.4.4, the first parameters for steric substituent effects in structure–activity studies originated in physical organic chemistry. A major advance was made when Taft proposed the use of the E_s parameter for defining steric constants. Although this parameter was originally derived to characterize steric effects in aliphatic reactions, it can also be applied to take into account not only intramolecular but also intermolecular interactions between aromatic substituents and the corresponding acceptor sites. In fact E_s values, as well as other related parameters, are highly intercorrelated with the van der Waals radii so that they may represent the minimum width of substituents in much the same sense as the STERIMOL B_1 parameter.

As previously mentioned (compare equation 84), Kutter and Hansch[204] gave an instructive example of the use of the E_s parameter for intermolecular steric effects on hapten–antibody reactions. In fact, from the same set of data for which equation (84) was derived, Kutter and Hansch obtained

$$\log K_{rel} = 0.86E_{s2} + 0.08E_{s3} - 0.45E_{s4} - 0.70 \qquad (87)$$

$$n = 27, \quad r = 0.934, \quad s = 0.177$$

This equation contains fewer data points than equation (84) since E_s values for nine substituents were unknown. K_{rel} in equation (87) is a measure of the affinity of benzoate hapten and antibody, but represents also a measure of the hapten's ability to prevent union of antibody with antigen. E_{s2}, E_{s3} and E_{s4} refer to E_s values for substituents on position 2, 3 and 4 of the phenyl ring ($XC_6H_4CO_2^-$), respectively. Since the larger the substituent the more negative is its E_s value, the positive coefficient with E_{s2} term indicates that *ortho* substituents lower the affinity as a result of two possible types of interaction: intramolecularly, twisting the carboxylate group out of the plane of the phenyl ring, and/or intermolecularly, preventing with their bulkiness the binding of the hapten to the antibody. However, the effect seems to be largely intermolecular, since a study of arsonic acids having the rather symmetrical group $—AsO_3H^-$ yield a QSAR with a very small coefficient for E_{s2}. The very small coefficient for E_{s3} shows that *meta* substituents have essentially no effect, while the negative coefficient for E_{s4} indicates that large groups in the *para* position are effective in preventing union of antibody and antigen. The comparison between equation (84) and (87) shows that, while it is not surprising that the same general picture is observed from the use of E_{s2} in equation (87) and B_1^o of equation (84) for *ortho* substituents, in the case of *para* substituents it is not clear if the substitution of E_{s4} with the linear and squared L^p terms will be entirely satisfactory. A possible explanation is that for the set of *para* substituents under study multicollinearity problems among L^p, $(L^p)^2$ and E_{s4} vectors are involved. The most interesting aspect of both correlation equations is that they show that hydrophobic properties of the benzoate ion play no role in preventing antibody–antigen union which, instead, seems to be governed by the negative charge on the carboxylate and the shape of the apolar portion of the benzoates.

Another instructive example of the application of E_s parameters is represented by a correlation analysis[205, 206] of the structure–activity relationships of 9-(X-phenyl)guanine derivatives (6) inhibiting xanthine oxidase. Equation (88) has been formulated for 30 derivatives having substituents in the 2-, 3- and 4-positions of the 9-phenyl moiety. C in this expression is the molar concentration of inhibitor causing 50% inhibition of the enzyme while $MR_{3,4}$ is the combined molar refractivity of substituents in the 3- and 4-positions. The signal result of equation (88) is that the X-substituents do not appear to locate themselves in hydrophobic space. Rather large substituents in the *meta* and *para* positions are well fitted by equation (88), indicating the great flexibility in the enzyme space around these positions, although 4-substituents appear to find themselves in more constricted circumstances. Actually, a good correlation was not possible without the use of E_{s2} and E_{s4} and the positive coefficients for these terms mean that large substituents have a deleterious effect on inhibitory power. Although E_s can be used to model intermolecular effects of the whole substituent in interactions with macromolecules, in the case of equation (88) it is possible to interpret the importance of E_s for the 2- and 4-substituents to signal the presence of obstacles in 2- and 4-space which hinder large groups from entering these enzymic spaces. These obstacles are only partially successful since the positive coefficient for $MR_{3,4}$ indicates that large groups in the *para* position do favor inhibitory action in spite of certain enzymic obstacles in 4-space which hinder binding. In conclusion, while molar refractivity may be assumed in this instance to be a measure of the dispersion forces between the substituent and the enzyme, the positive coefficients for the two E_s terms indicate that bulky groups in the 2- and 4-positions do not make good inhibitors; this may be considered to be a proximity effect which is related to the enzymic region near the *ortho* and *para* positions.

$$\log(1/C) = 0.20MR_{3,4} + 1.26E_{s2} + 0.43E_{s4} + 4.33 \qquad (88)$$

$$n = 30, \quad r = 0.924, \quad s = 0.228$$

(6) 9-(X-Phenyl)guanine derivatives inhibiting xanthine oxidase

A further example of the application of E_s steric constants in combination with other physicochemical parameters is shown in equation (89), which represents the QSAR[207] of a set of 2-alkylammoniumalkylthio-4-aryl-3H-1,5-benzodiazepine derivatives (7) active against *S. aureus*. C in $\log(1/C)$ of equation (89) is the minimum inhibitory concentration. The importance of hydrophobicity on the activity is underlined by the presence of the $\Sigma\pi_X$ term which shows the normal increase in antibacterial activity as lipophilicity increases. It is well known, in fact, that there are many QSARs for the inhibition of growth of microorganisms by organic compounds in which the dependence of antibacterial action on lipophilic character very closely parallels the dependence shown in equation (89). To formulate this equation, the σ^* values of R^1, R^2 and R^3 have been used in the composite sense, $\Sigma\sigma^*$; the negative coefficient for this term indicates that electron release promotes activity. Looking at the structure of equation (89), it is evident that steric factors, set by E_{s2} for *ortho* X-substituents on the phenyl moiety, play an important role in limiting the antibacterial activity. Although E_{s2} can be used to model the intermolecular effects of the whole substituent in interactions with macromolecules, in this instance it is possible to interpret the importance of E_s for 2-functions to signal the presence of obstacles which hinder free rotation of the phenyl ring with respect to the plane of the basic skeleton.

$$\log(1/C) = 0.71\Sigma\pi_X - 0.79\Sigma\sigma^* + 0.71E_s + 3.74 \qquad (89)$$

$$n = 38, \quad r = 0.929, \quad s = 0.211$$

(7) 2-Alkylammoniumalkylthio-4-aryl-3H-1,5-benzodiazepines active against *S. aureus*

Concerning the corrected steric E_s^c constant, which was first introduced by Hansch and Lien[208] into biological structure–activity studies, Fujita and Iwamura[119] reviewed a number of interesting applications to the development of pesticides. Particularly instructive was the systematic modification of structure (8), where R was varied from simple alkyls to very congested groups. During the stepwise approach, equation (90) was derived; in this equation, C is the molar concentration required for 50% inhibition against shoot elongation of the seedlings of bulrush, *Scirpus juncoides*, after 12 days. Equation (90) shows that there is an optimum in lipophilicity ($\pi_o = 3.3$) within the set of derivatives. Moreover, the steric bulkiness of the R substituents is to be augmented for higher activity. In fact, since the activity determination of these compounds requires several days of exposure on the soil, they might be partly degraded during the test period. The negative coefficient for the E_s^c term (by definition $E_{s(Me)}^c = 0$) suggests that the steric bulk of R substituents has the role of protecting the compounds against degradation.

$$\log(1/C) = -0.15\pi^2 + 0.98\pi - 0.35E_s^c + 2.88 \qquad (90)$$

$$n = 41, \quad r = 0.933, \quad s = 0.267$$

(8) *N*-(1-Methyl-1-phenethyl)acylamides active as inhibitors against
shoot elongation of the seedlings of bulrush, *Scirpus juncoides*

The Charton's v_{eff} steric parameter has been shown (see Section 18.4.4) to be rather closely related to Taft's E_s parameter. However, there are more values available of the v_{eff} constants than of E_s. As a consequence of the high collinearity between E_s and v_{eff} vectors, if v_{eff} is used in the equations so far discussed in place of E_s, similar results are obtained.

18.4.7.4   Physical Descriptors

As already discussed in Section 18.4.5, physical descriptors represent a type of steric effect which is volume dependent. For this reason, there has been some criticism[146, 147] as to whether these descriptors are really steric parameters. In fact steric effects observed in simple organic reactions are generally directionally dependent for non-symmetric substituents. However, the several successful QSAR analyses developed by using physical descriptors along with other physicochemical parameters seem to warrant their use for volume-dependent steric effects.

A classical example of the application of van der Waals volume (V_w) to QSAR has been performed by Seydel *et al.*[209] in the study of a series of 2-substituted isonicotinic acid hydrazides (9) tested against *Mycobacterium tuberculosis*. According to the hypothesis of Krüger-Thiemer,[210] isonicotinic acid (INA) is responsible for the inhibitory activity of isonicotinic acid hydrazide (INH). Following this hypothesis, INH can freely permeate into the bacterial cell through the cell wall and membrane.

Once inside the cell, an enzyme oxidizes the hydrazide group to yield isonicotinic acid which is almost completely ionized at the intracellular pH and cannot diffuse out of the cell. As a consequence of this process, the internal part of the bacterial cell accumulates INA which, instead of the natural metabolite nicotinic acid, is quaternized and incorporated into an NAD analog; thus the NAD analog disturbs the normal metabolism and leads to cell death. The rate-limiting step (*i.e.* the biological response) of this sequence seems to be the quaternization reaction.

(**9**) 2-Substituted isonicotinic acid hydrazides active against *Mycobacterium tubercolosis*

Bearing this in mind, Seydel *et al.* developed equations (91) and (92). In equation (91) the biological activity data (MIC) were obtained from serial dilution tests with *Mycobacterium tubercolosis*, while k_{rel} of equation (92) is the reaction rate constant for the quaternization of the same set of 2-substituted pyridines with methyl iodide. In both equations the electronic effects of substituents were taken into account by the experimental pK_a values. The coefficients for the van der Waals volume V_w in both the equations are very small, but one has to take into consideration that the V_w values are about 10 times higher than the pK_a values. A comparison of the two equations indicates that the rate constants show a similar dependence on the steric and electronic effects of the substituents as the antibacterial activities of the corresponding 2-substituted hydrazides. The lower coefficient in V_w of equation (92) probably reflects the smaller size of the attacking electrophile (methyl iodide) in the model reaction.

$$\log(1/\text{MIC}) = 0.26 pK_a - 0.06 V_w - 1.77 \tag{91}$$

$$n = 15, \quad r = 0.89, \quad s = 0.45$$

$$\log k_{rel} = 0.29 pK_a - 0.02 V_w - 1.94 \tag{92}$$

$$n = 15, \quad r = 0.931, \quad s = 0.32$$

In structuring the data in a quantitative context, a working hypothesis is that substituent perturbations correlated by π, or any other hydrophobic parameter, are relative to interactions with apolar residues of the acceptor sites and those correlated by molar refractivity (MR) are relative to interactions with polar residues. A better correlation with MR rather than π means that hydrophobic partitioning with its attendant desolvation is not the main driving force in the interaction of ligand with the receptor. However, MR is also a measure of substituent bulk which may be responsible for causing conformational changes in the receptor. Several QSARs show the possible applications of molar refractivity, whose values are generally scaled by 0.1. An instructive example is given by Hansch *et al.*[211] in a study on the inhibition of dihydrofolate reductase by the triazines (**10**).

(**10**) X-Phenyltriazines inhibiting dihydrofolate reductase

In equation (93), K_i represents the inhibition constant on dihydrofolate reductase highly purified from the methotrexate-resistant strain of *Lactobacillus casei* bacteria, while C in equation (94) is the inhibition concentration of a smaller number of compounds tested on methotrexate-resistant *L. casei* cell culture. The indicator variable I in both equations is assigned the value of one for substituents of the type —$CH_2ZC_6H_4Y$ and —$ZCH_2C_6H_4Y$ ($Z = O$ or NH) and zero for all other substituents. These bridged phenyl rings increased activity over what one would expect from the other physicochemical parameters. In both cases the activity is highly dependent on the hydrophobicity of the 3-X-substituents. The most striking differences between the two correlations are represented by the σ and MR parameters. While the lack of σ in equation (94) was attributable to noise in the data, an interpretation of the difference in the MR_Y term was brought out by the manner in which *N*-phenyltriazines bearing Y groups within 3-substituents of the type —$CH_2ZC_6H_4Y$ or

—$ZCH_2C_6H_4Y$ were treated in the formulation of equation (93) and (94). In both cases it was found that Y groups were apparently not interacting with the hydrophobic region in which 3-substituents fall. Since for the isolated enzyme the contribution for Y was set to zero and no additional term was needed to account for the interactions of the Y groups, it is safe to advance the hypothesis that these groups appear to extend out of the hydrophobic pocket and into the aqueous phase. Therefore they have essentially no effect on inhibition. It is worth pointing out that, after this study, it was shown by X-ray crystallography and computer graphics[212] that Y does not contact the enzyme.

$$\log(1/K_i) = 0.53\pi_X - 0.64\log(\beta 10^{\pi_x} + 1) + 1.49I + 0.70\sigma + 2.93 \tag{93}$$

$$n = 44, \quad r = 0.953, \quad s = 0.319, \quad \log\beta = -3.66, \quad \pi_0 = 4.31$$

$$\log(1/C) = 0.80\pi_X - 1.06\log(\beta 10^{\pi_x} + 1) - 0.94MR_Y + 0.80I + 4.37 \tag{94}$$

$$n = 34, \quad r = 0.929, \quad s = 0.371, \quad \log\beta = -2.45, \quad \pi_0 = 2.94$$

In the case of *L. casei* cell culture these groups have a pronounced effect on inhibitory power. To account for this effect the additional term MR_Y was utilized, *i.e.* 3-substituents were parameterized into an inner π-3 term and outer MR_Y term. The conclusion of this parameterization is that in the case of *L. casei* cell culture the negative MR_Y term might be interpreted to mean that Y groups encounter some kind of steric hindrance, probably from an adjacent macromolecule or a different conformation of the enzyme in the cell. Since this steric effect holds regardless of whether Y is hydrophobic or hydrophilic, it must be related to the bulk of Y groups. According to Hansch, the larger coefficient with the indicator variable I in equation (93) might signal a possible difference in conformation between isolated enzyme and enzyme *in situ*.

18.4.7.5 Differential Descriptors

The examples of QSAR applications so far discussed are relative to steric descriptors whose tabulations are easily available for many different substituents. When dealing with overlapping methods it is not possible to make recourse to any kind of tabulation since the values of differential descriptors are calculated anew for each set of molecules under study, different values being assigned to the same substructure when it is considered in different sets of molecules. A similar consideration holds in the case where a certain set of molecules is considered in different types of biological responses. Therefore, in reporting an example of such methods, it seems advisable to discuss how differential descriptors can be derived. Several QSARs have been established by using differential descriptors and many of them are adequately described elsewhere.[171,174] A particularly instructive example, in which Simon *et al.*[171] defined a leading example of the MTD method, is a study of a set of trimethylammonium derivatives with cholinergic action. A few of these compounds have already been considered in Section 18.4.6 and are listed in Table 7 with their biological activity. The activity is expressed as $\log(1/ED_{50})$, where ED_{50} is the dose which produces a 50% reduction in the amplitude of clam heart contraction. In the first step, one must superimpose all compounds in order to obtain the topological network, termed hypermolecule. In the case of the compounds of Table 7, the superimposition procedure gives the same hypermolecule (2) described in Section 18.4.6. In fact, all of the considered compounds are embodied in such a hypermolecule. Since the $Me_3\overset{+}{N}CH_2$— fragment is common to all the structures and it is assumed to interact with the same receptor site, it is not considered in the assignment of the MTD_i values for the 12 compounds under study. Each of the 12 structures will occupy certain vertices j of the hypermolecule, and this will be denoted by $x_{ij} = 1$ if vertex j is occupied by molecule i and $x_{ij} = 0$ if it is not occupied. Since the vertices in the hypermolecule are 9, each i molecule is described by a vector $x_i = x_{i1}, x_{i2}, \ldots x_{i9}$.

Since several low energy conformations are possible, there is a set of such vectors, one for each conformation. This has not been considered in the example described here. Each vertex j may belong either to the space of the receptor cavity ($\varepsilon_j = -1$), to the walls ($\varepsilon_j = 1$) or to the sterically irrelevant space ($\varepsilon_j = 0$). The MTD_i values are calculated by equation (77) of Section 18.4.6. Note that if there are molecules which can assume several conformations, the vector $x_{i1}, x_{i2}, \ldots x_{i9}$ which gives the lowest MTD_i value should be considered. To start the optimization procedure one must assume an initial map, S^0. Since acetylcholine is the most active compound, a good start is that in which the vertices occupied by this structure are considered beneficial ($\varepsilon_j = -1$), and the other ones as detrimental ($\varepsilon_j = 1$). The assignment of ε_j values is shown in Table 8. From these values, the evaluation of each MTD_i value is a simple task. For acetylcholine, for example, one will obtain

$$MTD_i = 5 + \Sigma\varepsilon_i x_{ij} = 5 - 5 = 0$$

Table 7 x_{ij} Assignments and MTD$_i$ Values in the Optimization Procedure

i	Compound	$Log(1/ED_{50})$	Vertices for which $x_{ij}=1$	$S^0=MSD$	S^1	S^2	$S^3=S^*$
1	$Me_3\overset{+}{N}CH_2$—O—C(=O)—Me	3.00	1–4, 7	0	0	0	0
2	$Me_3\overset{+}{N}CH_2$—CH$_2$—C(=O)—Me	1.92	1–4, 7	0	0	0	0
3	$Me_3\overset{+}{N}CH_2$——Me	1.30	1–4	1	1	1	1
4	$Me_3\overset{+}{N}CH_2$—O—Me	1.18	1–4	1	1	1	1
5	$Me_3\overset{+}{N}CH_2$—C(=O)—CH—Me	0.79	1–4, 8	2	2	2	1
6	$Me_3\overset{+}{N}CH_2$——Me	0.63	1–3	2	2	2	2
7	$Me_3\overset{+}{N}CH_2$—Me	0.48	1, 2	3	2	2	2
8	$Me_3\overset{+}{N}CH_2$——Me	0.46	1–5	2	2	2	2
9	$Me_3\overset{+}{N}CH_2$—C(=O)—Me	0.20	1–4, 9	2	2	2	2
10	$Me_3\overset{+}{N}CH_2$—Me	−1.15	1	4	3	3	3
11	$Me_3\overset{+}{N}Me$	−1.30	—	5	4	3	3
12	$Me_3\overset{+}{N}CH_2$————Me	−1.30	1–6	3	3	3	3

As shown in Table 7 the MTD$_i$ values for the map S^0 are equal to the values that one may obtain applying the simple MSD procedure. From the evaluated MTD$_i$ values and corresponding activity values one may derive equation (95) of Table 9.

The next step consists of evaluating the mean square difference Y between experimental and calculated activities (see Section 18.4.6.4). Then, by changing the ε_j assignment in the initial receptor map (S^0) one by one, new receptor maps are obtained. For each receptor map a new Y sum is evaluated and the receptor map for which the lowest Y sum is obtained will be selected. This map, which is achieved with ε_j values listed in Table 8 under S^1, will constitute the new initial map of comparison. Note that the map S^1 from the first cycle is obtained by the substitution $\varepsilon_3' = 0$ for $\varepsilon_3^0 = -1$. In the second cycle (S^2) the most negative Y value is achieved by substituting $\varepsilon_1' = 0$ for $\varepsilon_1 = -1$, while the third cycle (S^3) indicates the substitution $\varepsilon_8' = 0$ for $\varepsilon_8 = 1$. The resulting map S^3 gives in the fourth cycle only positive Y variations so that the optimization procedure is completed; thus $S^3 = S^*$ is the optimal map. The equations obtained in each cycle are listed in Table 9.

The optimal map (S^*) allows a few observations of the acetylcholine receptor site. In fact, besides the receptor anionic region for the interaction of the common $Me_3\overset{+}{N}CH_2$— fragment, the $\varepsilon_5 = \varepsilon_6 = \varepsilon_9 = \varepsilon_9 = 1$ assignments suggest that acetylcholine is enclosed in a cavity so that the hypermolecule vertices 5, 6 and 9 protrude into the receptor walls. Moreover, the $\varepsilon_1 = \varepsilon_3 = 0$ assignments support the hypothesis that the corresponding vertices are located at the boundary between wall and cavity receptor and are ineffective in promoting the biological response. Finally,

Table 8 ε_j Assignment in the Optimization Procedure

Vertex j	S^0	S^1	ε_j S^2	$S^3 = S*$
1	−1	−1	0	0
2	−1	−1	−1	−1
3	−1	0	0	0
4	−1	−1	−1	−1
5	1	1	1	1
6	1	1	1	1
7	−1	−1	−1	−1
8	1	1	1	0
9	1	1	1	1

Table 9 Correlation Equations Developed in the Optimization Procedure

Equation	$Log(1/ED_{50}) = a + b\ MTD$				
	a	b	n	r	S^i
(95)	2.18	−0.78	12	0.918	S^0
(96)	2.43	−1.04	12	0.953	S^1
(97)	2.59	−1.18	12	0.958	S^2
(98)	2.46	−1.17	12	0.960	$S^3 = S*$

only vertices 2, 4 and 7 are positively interacting in the receptor cavity. These results parallel those obtained by Pratesi *et al.*[203] in terms of physicochemical parameters.

18.4.8 STEREOISOMERISM AND THE APPLICATION OF STERIC DESCRIPTORS

Although for certain types of drugs stereoisomerism is unimportant, in most cases receptors display sharp discrimination toward the stereostructure of those ligands they recognize. Most of the methods described in Section 18.4.6 consider to a certain extent this problem. Alternative methods have been proposed[213-217] but, at present, they suffer from some limitations which reduce their general use and reliability.

Bearing in mind Pfeiffer's observation[218] that for several common chiral drugs an increase in their potency is associated with an increase in their enantiomeric potency ratio, Lehmann[219] has developed a method which should allow stereoselectivity to be analyzed quantitatively. The method, termed eudismic analysis, may be applied to any congeneric set of isomeric pairs in which both isomers show detectable activities. Lehmann has defined the eudismic index (EI) as in equation (99a), where pI_{Eu} and pI_{Dis} represent the activity (pA_2, pD_2, pI_{C50}, *etc.*) of the eutomer and distomer (see Section 18.4.11 for definitions), respectively. By plotting the eudismic index against the pI_{Eu} values, generally linear regressions (equation 99b) are obtained; the slope *a*, termed the eudismic affinity quotient, may be assumed to be a measure of stereoselectivity and may be useful in interpreting receptor binding data at the submolecular level.[220] The implications for rational drug design and receptor stereoselectivity have been analyzed by Lehmann,[219] who suggested that Quantitative Stereo-Structure Activity Relationships (QSSAR) should be made an integral part of QSAR.

$$EI = pI_{Eu} - pI_{Dis} \tag{99a}$$

$$EI \left(= apI_{Eu} + b \right. \tag{99b}$$

Of the various types of stereoisomeric drugs, chiral structures are probably the most intriguing and important to study. Only a few examples[213-217] are reported in the literature concerning attempts to correlate series of pairs of enantiomers by using typical QSAR parameters. Lien *et al.*[217] have developed a correlation analysis on the auxin activity of a set of optically active aryloxypropionic acids and structurally related derivatives of the type described in Figure 17. Note that, with

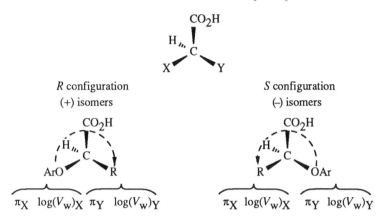

Figure 17 Parametrization of optically active aryloxypropionic acids active as plant growth regulators

a few exceptions, the (+) isomers belong to the R configuration and all the (−) isomers belong to the S series. After trying many different combinations of various parameters, equation (100) was developed.

$$\log(1/C) = -0.53\pi_X^2 + 2.51\pi_X - 2.62\log(V_w)_X - 0.36\pi_Y^2 + 2.44\pi_Y - 6.89\log(V_w)_Y \qquad (100)$$
$$+ 1.51\sigma_{Ar} + 14.61$$
$$n = 47, \quad r = 0.902, \quad s = 0.484$$

In this equation, C is the concentration of the substance that reduces growth to 50% of that of control in the flax root test. As shown in Figure 17, the physicochemical properties of the two chiral substituents on the optically active center were treated as 'independent variables', *i.e.* X was assumed to be equal to the substituent OAr if the isomer belongs to the R series and was set equal to the substituent R for the compounds in the S configuration. The substituent Y was treated in parallel fashion.

Equation (100) shows that by using parabolic relationships for the hydrophobic π constants and $\log V_w$ terms for both X and Y substituents plus the Hammett σ constant for the aryloxy ring, a correlation coefficient of 0.902 is obtained for 47 of the 56 initially examined compounds. Since the van der Waals volume (in logarithmic form in this analysis) may be used as a first approximation of the bulk steric effect, the negative dependence on $\log(V_w)_X$ and $\log(V_w)_Y$ suggests that highly bulky groups at either X or Y are incompatible with high inhibitory activity. The lower absolute value of $\log(V_w)_X$ compared with that of $\log(V_w)_Y$ may indicate that the bulk tolerance at X is greater than at Y. This finding is in agreement with the general observation that R isomers, which have the bulkier aryloxy group at the X position, are usually more active than S isomers. However, equation (100) suffers from some limitations which, as Lien *et al.* stated, may 'be due partly to the inherent nature from the method of least squares and partly to some subtle drug–receptor interactions not adequately represented by the parameters used'. For example, the slopes of the linear π terms are unusually high and the collinearities of π_X with $\log(V_w)_X$ and π_Y with $\log(V_w)_Y$ are $r_X = 0.93$ and $r_Y = 0.68$, respectively.

Another instructive example has been developed by Hansch *et al.*[156, 206, 221] in the study of chymotrypsin–ligand interactions in a large set of N-acyl esters of α-amino acids which contain a chiral center. In their approach, Hansch *et al.* have used the Hein–Nieman concept[222] of binding, illustrated in Figure 18 where the well-known double mechanism of hydrolysis catalyzed by chymotrypsin is also reported ($-CH_2OH$ is the nucleophilic moiety of serine 195). In this figure it is illustrated how an acylated ester of an L-amino acid binds to the enzyme with substituents R^1, R^2 and R^3 falling in ρ_1, ρ_2 and ρ_3 enzymic space. The α hydrogen falls below the plane of the page into the corresponding ρ_H space.

To approach an understanding of the interactions of R^1, R^2 and R^3, the enzymic parameters K_m and K_i [$K_{m(app)} = (k_{-1} + k_2)/k_1$] were correlated with physicochemical parameters related to R^1, R^2 and R^3 substituents as reported in equations (101–103b; Table 10). The subscripted MR values refer to the molar refractivities of the corresponding groups R^1, R^2 and R^3. The most important general conclusion, confirmed by computer graphic studies,[223, 224] was that ligands appear to bind in polar spaces in relatively few ways, as indicated in Figure 19. The positive coefficients of the MR terms of equation (101) show that binding is proportional to the molar refractivity of the groups

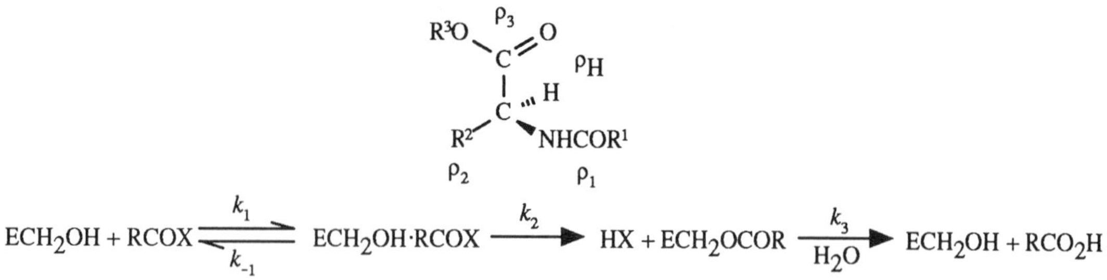

Figure 18   Hein–Nieman concept of binding and double mechanism of hydrolysis catalyzed by chymotrypsin

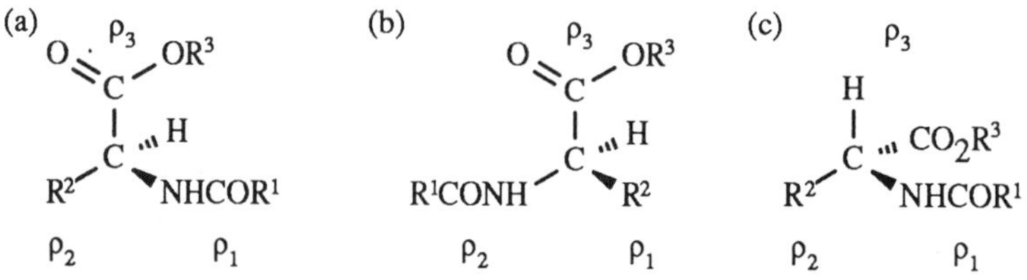

Figure 19   Substrate and inhibitor binding modes to chymotrypsin: (a) L-ester as substrate, (b) D-ester as substrate and (c) D-ester as inhibitor

falling in ρ_1, ρ_2 and ρ_3 space; interaction in ρ_2 space is stronger than interaction in ρ_1 space and binding in ρ_3 space is weakest of all. The linear dependence of binding on MR_1, MR_2 and MR_3 over a large range of bulk groups suggests a large amount of flexibility in chymotrypsin.

There are certain exceptions; for example, the poor interaction of valine residues ($-0.63I$-1 where I-1 = 1 for $R^2 = CHMe_2$). However, this linearity holds until the cross product of MR terms becomes too large. At this point, there is a gradual decrease dependent on the total bulk of the three groups. The negative coefficient associated with the cross product term $MR_1 \cdot MR_2 \cdot MR_3$ indicates that combinations of substituents with large MR values reduces binding and that binding in any one of the three spaces affects binding in the other two. This effect, however, appears to be more than additive since groups large enough to firmly establish the validity of this term have not been studied; almost as good results, in fact, can be obtained using $(\Sigma MR)^2$ or the bilinear model.

The positive sign of σ_3^*, the Taft inductive constant for R^3, indicates that binding to chymotrypsin is strengthened by electron withdrawal by R^3; this point was supported through an adequate variation of R^3 groups.[221]

Equation (102) constitutes the QSAR of a smaller set of D-esters acting as substrates. In this equation, MR_2 has the smaller coefficient and MR_1 the larger, which is the opposite to equation (101) and suggests that binding is occurring as depicted in (b) of Figure 19. Esters appear to be hydrolyzed at such a rate that even with the poorer fit of this wrong way binding, the D-esters function as substrates. Attempts to find significant cross-product terms for equation (102) similar to equation (101) were blocked by the high collinearity between MR_1, MR_2, MR_3 and $MR_1 \cdot MR_2 \cdot MR_3$ which produced a singular matrix. The indicator variable I-2 takes the value of one when $R^3 = C_6H_4$-4-NO_2. The only other R^3 substituent is Me, hence on the basis of equation (101), one assumes that I-2 is composed of electronic (σ_3^*) and molar refractivity (MR_3) components which cannot be resolved.

The D-esters acting as inhibitors led to the QSAR of equation (103a) and to the alternative (103b). Qualitatively, the coefficients with MR_1 and MR_2 of these equations parallel those of equation (101), although those of equation (103a) are somewhat larger than the corresponding ones of equation (101). The addition of the MR_3 term in equation (103a) did not improve the correlation. Anyway, in this case MR_3 behaves like an indicator variable since, out of the case where $R^3 = Me$ and Et, there is only one other in which $R^3 = CHMe_2$. The fact that the MR_3 occurs only in the cross-product term of equation (103a) or with a negative coefficient in equation (103b) points out that it has a negative influence on binding. Evidently, substituent R^3 would be forced into ρ_H space in this mode of binding; it is logical to expect ρ_H space to be limited so that a negative coefficient with MR_3 is not unexpected, being likely related to a negative bulk effect.

Table 10 Chymotrypsin: Ligand Binding ($1/K_m$ or $1/K_i$)

Equation	Intercept	MR_1	MR_2	MR_3	$MR_1 \cdot MR_2 \cdot MR_3$	σ_3^*	I-1	I-2	n	r	s	Ligands
(101)	-1.64	0.77	1.13	0.56	-0.06	1.35	-0.56	—	84	0.977	0.333	L-esters as substrates ($1/K_m$)
(102)	2.76	1.38	0.47	—	—	—	—	1.83	15	0.993	0.267	D-esters as substrates ($1/K_m$)
(103a)	-2.71	1.07	1.42	—	-0.16	—	—	—	12	0.988	0.207	D-esters as inhibitors ($1/K_i$)
(103b)	-1.20	0.73	1.18	-1.06	—	—	—	—	12	0.976	0.292	D-esters as inhibitors ($1/K_i$)

As reported in Figure 18, chymotrypsin reacts with esters to first form the $ECH_2OH \cdot RCOX$ complex which is followed by acylation of the enzyme. The acylated enzyme then reacts with water to yield the regenerated enzyme and an acid. As expected, k_2 and k_3 were also found to be strongly dependent on molar refractivity as well as the steric and electronic character of the substituents R^1, R^2 and R^3 of the basic skeleton of Figure 18.

The acylation step is correlated by equation (104) for a set of 18 L-esters.

$$\log k_2 = -0.52 MR_1 + 1.10 MR_2 - 1.56 I\text{-}1 + 0.42 \tag{104}$$

$$n = 18, \quad r = 0.971, \quad s = 0.399$$

As in equation (101), it was found in equation (104) that there was a slope of about 1.1 for MR_2 and a negative coefficient with I-1 for cases where $R^2 = CHMe_2$. No term appears for R^3; one would expect an electronic and possibly a steric effect with R^3; however, in this instance there is so little variation in R^3 that these effects cannot be assessed properly. The coefficient with MR_1 is opposite in sign to that in equation (101), showing that large substituents retard the trans-esterification step. This could be due to steric inhibition of the formation of the tetrahedral intermediate, or it could be due to the breaking of van der Waals bonding of R^1, possibly reinforced by water matrix of solvation around R^1 and the enzyme surface. The latter possibility seems more likely when one considers the following QSAR for k_3 for the deacylation step

$$\log k_3 = 0.75 MR_2 - 1.48 I\text{-}1 - 1.79 I\text{-}3 - 0.31 \tag{105}$$

$$n = 33, \quad r = 0.977, \quad s = 0.289$$

Equation (105) has been derived from data on deacylation of 33 D- and L-esters. In this expression, I-3 takes the value of one for D-isomers and zero for L-isomers. The negative coefficient with this term shows that D-isomers are more difficult to deacylate than L-isomers. I-1 has the usual meaning. R^3 is not present in the deacylation step so no term is needed for this group. The same occurs for R^1; however, as previously mentioned, this is not unexpected since the $0.77 MR_1$ term in equation (101) becomes $-0.52 MR_1$ in the acylation step (equation 104). The obvious conclusion from equations (104) and (105) is that R^1 has been removed from ρ_1 space in the acylation step and in the deacylation step it is for practical purposes out of contact with the enzyme. The positive coefficient with MR_2 seems surprising, since one might expect that in this step the desorption of R^2 from the enzyme would hinder the process. However, since MR_2 plays such an important role in both acylation and deacylation steps, its main function may be to produce and maintain a conformational change in the enzyme essential for these two reactions.

In conclusion, equations (101)–(105) do not support the view that chymotrypsin has a classical-type hydrophobic pocket. This is especially true for ρ_1 and ρ_3 space. In fact, both regions have been studied with sets of congeners for which π and MR are reasonably orthogonal and in every instance MR is far superior in the correlation equations. This view, as already mentioned, has been confirmed through comparative studies *via* computer graphics.[223, 224] The types of ligand binding in ρ_2 space have not been as extensively varied as those binding in ρ_1 and ρ_3 space; even so, MR_2 consistently gives better results than π_2. However, the role in ρ_2 space must be associated with maintaining a proper conformation and possibly a hydrophobic partitioning effect. To elucidate the latter, it would be necessary to study an appropriately designed set of R^2 groups for which π_2 and MR_2 are orthogonal.

18.4.9 CONCLUDING REMARKS

As exemplified above, a study of possible steric effects in biological systems is of great importance in delineating the intra- and inter-molecular interactions involving a ligand and its acceptor site. However, while differential descriptors are able to take into account conformational and configurational problems associated with drug action with a certain degree of reliability, most of the parameters discussed in this chapter are generally able to describe only substituent steric effects of a parent structure. In fact, if it is possible to separate out hydrophobic, lipophilic and polar interactions as well as electronic effects with some degree of certainty by the use of the well-known physicochemical parameters, steric requirements of substituent interactions can be analyzed on a more rational basis. The confusing problem of collinearity among vectors associated with parameters has already been stressed; if in a set of congeners collinearity between two parameters is very high, it becomes difficult to evaluate the quality of a structure–activity correlation, in that it is impossible to make a firm statement on the actual nature of interactions. Nevertheless, owing to the multidimensional character of substituent space, it is not an easy matter to select sets of substituents constituting orthogonal vectors. In conclusion, in structuring the data to formulate a correlation analysis, one should select equally weighted hydrophobic, polar, electronic and steric parameters which at present can be readily expressed in numerical terms. These parameters by no means contain all of the information pertinent on the role of substituents in affecting a biological response. However, they help in outlining the type of interactions which one can expect between ligands and macromolecules and the means for differentiating between them. No-one can observe what is really occurring at the acceptor site. In this respect, the procedure developed by Hansch *et al.*[224, 225] for enzyme–ligand intermediate complexes is of particular interest. The procedure, in fact, combines QSAR methodology with X-ray crystallography as well as with interactive computer graphics; three-dimensional images of the interacting partners may be visualized allowing a more reliable interpretation of correlation equations.

ACKNOWLEDGEMENT

The authors wish to thank Dr A. Verloop for helpful suggestions.

18.4.10 COMPILATION OF SELECTED DATA

Table 11 Steric Parameters for Selected Substituents

Substituent	$L^{86,90}$	$B_1^{86,90}$	$B_5^{86,90}$	$N_H 3$	$E_s^{98,102,136}$	$E_s^{c\,98,102,136}$	$E_s^{o\,98,102,136}$	v_{eff}^{77}	v_{mn}^{77}	v_{mx}^{77}	$MR^{98,102,136}$	$S^{o\,130}$
Br	3.82	1.95	1.95	—	−1.16	—	—	0.65	0.65	0.65	8.9	−0.82
CBr₃	4.09	2.86	3.75	−3	−3.67	—	−4.66	1.56	1.56	2.48	28.8	—
CCl₃	3.89	2.64	3.46	−3	−3.3	—	−4.29	1.38	1.38	2.22	20.1	—
CI₃	4.23	1.60	1.60	—	—	—	—	1.79	1.79	2.80	44.1	—
CF₃	3.30	1.99	2.61	−3	−2.4	—	−3.39	0.90	0.90	1.53	5.0	−0.89
CN	4.23	1.60	1.60	—	−0.51	—	—	0.40	0.40	0.40	6.3	—
NC	4.23	1.60	1.60	—	—	—	—	—	0.40	0.40	6.3	—
CO₂⁻	3.53	1.60	2.66	—	—	—	—	—	—	—	6.1	—
CHO	3.53	1.60	2.36	—	—	—	—	—	0.50	1.39	6.9	−2.36
CO₂H	3.91	1.60	2.66	—	—	—	—	—	0.50	1.39	6.9	—
CH₂Br	4.09	1.52	3.75	−1	−1.51	—	−1.84	0.64	—	—	13.4	—
CH₂Cl	3.89	1.52	3.46	−1	−1.48	—	−1.81	0.60	—	—	10.5	—
CH₂I	4.36	1.52	4.15	−1	−1.61	—	—	0.67	—	—	18.6	—
CONH₂	4.06	1.50	3.07	—	—	—	—	—	0.50	1.39	9.8	—
CH=NOH (Z)	3.94	1.60	4.07	—	—	—	—	—	—	—	10.3	—
CH=NOH (E)	5.05	1.60	2.88	—	—	—	—	—	—	—	10.3	—
Me	2.87	1.52	2.04	0	−1.24	−1.24	(−1.24)	0.52	0.52	1.03	5.7	−0.73
CH₂OH	3.97	1.52	2.70	−1	−1.21[a], −1.21[b]	—	−1.63, −1.64	0.53	—	—	7.2, 7.2	−0.67
CH₂NH₂	4.02	1.52	3.05	—	—	—	—	0.54	—	—	9.1	—
C≡CH	4.66	1.60	1.60	—	—	—	—	0.58	0.58	0.58	9.6	—
CH₂CN	3.99	1.52	4.12	−1	−2.18[a], −2.38[b]	—	2.65	0.89	—	—	10.1	—
CH=CH₂	4.29	1.60	3.09	—	—	—	—	—	—	—	10.1	−1.48
COMe	4.06	1.60	3.13	—	—	—	—	—	0.50	1.39	11.0	−2.28
CO₂Me	4.73	1.64	3.36	—	—	—	—	—	—	—	11.2	−1.04
CH₂CO₂H	4.74	1.52	3.78	—	—	—	—	—	—	—	12.9	—
CH₂CONH₂	4.58	1.52	4.37	—	—	—	—	—	—	—	11.9	—
Et	4.11	1.52	3.17	−1	−1.31	−1.62	−1.51	0.56	—	—	14.4	−1.08
CH₂CH₂OH	4.79	1.52	3.38	—	—	—	—	0.77	—	—	10.3	−0.86
C≡CCF₃	5.90	1.99	2.61	—	—	—	—	0.58	0.58	0.58	11.8	—
C≡CMe	5.47	1.60	2.04	—	—	—	—	0.58	0.58	0.58	14.1	—
CH₂CH=CH₂	5.11	1.52	3.78	—	—	—	—	0.69	—	—	14.1	—
Cyclopropyl	4.14	1.55	3.24	—	—	—	—	1.06	—	—	14.5	—
CH₂COMe	4.54	1.52	4.39	−1	−1.99	—	2.32	—	—	—	13.5	—
CO₂Et	5.95	1.64	4.41	—	—	—	—	—	0.50	1.39	15.1	−1.25
CH₂CH₂CO₂H	5.97	1.52	3.31	−1	−2.21	−2.52	−2.32	—	—	—	17.5	—
Pr	4.92	1.52	3.49	−1	−1.6	−1.91	−1.80	0.68	—	—	16.5	—
CHMe₂	4.11	1.90	3.17	−2	−1.71	−2.32	−2.09	0.76	—	—	15.0	−1.44

Substituent												
Cyclobutyl	4.77	1.77	3.82	-2	-1.3	-1.91	—	0.51	—	—	17.9	—
Bu	6.17	1.52	4.54	-1	-1.63	-1.94	-1.83	0.68	—	—	19.6	—
CH(Me)Et	4.92	1.90	3.49	-2	-2.37	-2.98	-2.77	1.02	1.24	1.97	19.6	-3.94
CMe_3	4.11	2.60	3.17	-3	-2.78	-3.70	-3.38	1.24	1.06	—	19.6	-2.46
2-Thienyl	6.53	1.64	3.37	—	—	—	—	—	—	2.13	24.0	—
CEt_3	4.92	2.94	4.18	-3	-5.04	-5.96	—	1.06	—	—	33.5	-2.35
2-Pyridyl	6.28	1.71	3.11	—	—	—	—	—	—	—	23.0	—
Cyclopentyl	4.90	1.90	4.09	-2	-1.75	-2.36	-1.84	0.71	—	—	22.0	—
C_5H_{11}	6.97	1.52	4.94	-1	-1.64	-1.95	—	0.68	—	—	24.3	—
$(CH_2)_3NMe_2$	6.88	1.52	5.49	—	—	—	—	—	—	—	28.0	—
Ph	6.28	1.71	3.11	—	-1.01[a] / -3.79[b]	—	—	—	0.57	2.15	25.4	-1.82
Cyclohexyl	6.17	1.91	3.49	-2	-2.03	-2.64	-2.67	0.87	—	—	26.7	—
$(CH_2)_3\overset{+}{N}Me_3$	6.88	1.52	5.49	-1	-2.59	-2.9	—	—	—	—	—	—
COPh	5.81	1.60	5.98	-1	—	—	—	—	—	—	30.3	—
CH_2Ph	4.62	1.52	6.02	-1	-1.61[a] / -1.62[b]	—	—	0.70	—	—	30.0	-1.16
1-Naphthyl	6.28	1.71	5.50	—	—	—	—	1.33	—	—	41.6	-0.75
Adamantyl	6.17	3.16	3.49	—	-0.97	—	—	0.55	—	—	40.6	-0.54
Cl	3.52	1.80	1.80	—	-0.46	—	—	0.27	0.55	0.55	6.0	—
F	2.65	1.35	1.35	—	—	—	—	1.73	0.27	0.27	0.9	—
$GeBr_3$	4.67	3.07	4.17	—	—	—	—	1.53	1.73	2.81	36.4	—
$GeCl_3$	4.47	2.85	3.87	—	—	—	—	1.05	1.53	2.51	25.8	—
GeF_3	3.88	2.20	3.03	—	—	—	—	0	1.05	1.84	6.9	—
H	2.06	1.00	1.00	—	(0)	—	—	—	0	0	1.0	—
I	4.23	2.15	2.15	—	-1.4	—	—	0.78	0.78	0.78	13.9	—
NO_2	3.44	1.70	2.44	—	-1.01[a] / -2.52[b]	—	—	—	0.35	1.39	7.4	—
NH_2	2.78	1.35	1.97	—	-0.61	—	—	0.35	—	—	5.4	-0.93
NHOH	3.87	1.35	2.63	—	—	—	—	—	—	—	7.2	—
$\overset{+}{N}H_3$	2.78	1.49	1.97	—	—	—	—	0.49	0.49	0.97	—	—
$NHNH_2$	3.47	1.35	2.97	—	—	—	—	—	—	—	8.4	—
$NHCONH_2$	5.06	1.35	3.61	—	—	—	—	—	—	—	13.7	—
NHMe	3.53	1.35	3.08	—	—	—	—	0.39	0.39	—	10.3	—
NHCOMe	5.09	1.35	3.61	—	—	—	—	—	—	—	14.9	-1.93
NHEt	4.83	1.35	3.42	—	—	—	—	0.59	—	—	15.0	—
NMe_2	3.53	1.35	3.08	—	—	—	—	0.43	—	—	15.5	-2.32
$NHCO_2Et$	7.25	1.35	3.92	—	—	—	—	1.22	—	—	21.2	—
$\overset{+}{N}Me_3$	4.02	2.57	3.11	—	-2.84	—	—	0.70	1.22	1.91	—	—
NHBu	6.88	1.35	4.87	—	—	—	—	—	—	—	24.3	—
NHPh	4.53	1.35	5.95	—	—	—	—	—	—	—	30.0	—
NHCOPh	8.40	1.70	3.97	—	—	—	—	0.32	—	—	34.6	—
OH	2.74	1.35	1.93	—	-0.55	—	—	—	—	—	2.8	—
OCF_3	4.57	1.35	3.61	—	—	—	—	0.36	—	—	7.9	—
OMe	3.98	1.35	3.07	—	-0.55	—	—	0.48	—	—	7.9	-1.28
OEt	4.80	1.35	3.36	—	—	—	—	—	—	—	12.5	-1.36

Table 11 (*Contd.*)

Substituent	$L^{86,90}$	$B_1^{86,90}$	$B_5^{86,90}$	N_H^{-3}	$E_s^{98,102,136}$	$E_s^{c\,98,102,136}$	$E_s^{o\,98,102,136}$	ν_{eff}^{77}	ν_{mn}^{77}	ν_{mx}^{77}	$MR^{98,102,136}$	$S^{o\,130}$
OCHMe$_2$	4.80	1.35	4.10	—	—	—	—	0.75	—	—	17.1	—
OPr	6.05	1.35	4.42	—	—	—	—	0.56	—	—	17.1	—
OBu	6.86	1.35	4.79	—	—	—	—	0.58	—	—	21.7	—
SO$_2$(F)	3.33	2.01	2.70	—	—	—	—	—	—	—	8.6	—
SF$_5$	4.65	2.47	2.92	—	-2.91	—	—	1.37	—	—	9.9	—
SO$_3^-$	3.33	2.03	2.70	—	—	—	—	1.03	1.03	1.73	—	—
SH	3.47	1.70	2.33	—	-1.07	—	—	0.60	—	—	9.2	—
SO$_2$NH$_2$	4.02	2.04	3.05	—	—	—	—	—	—	—	12.3	—
SMe	4.30	1.70	3.26	—	-1.07	—	—	0.64	—	—	13.8	—
SEt	5.16	1.70	3.97	—	—	—	—	0.94	—	—	18.4	—
SiBr$_3$	4.55	3.03	4.09	—	—	—	—	1.69	1.69	2.73	32.8	—
SiCl$_3$	4.35	2.80	3.79	—	—	—	—	1.50	1.50	2.44	23.8	—
SiMe$_3$	4.09	2.76	3.48	—	-3.36	—	—	1.40	1.40	2.28	25.0	—

[a] and [b] These values have been calculated by substituting the half-width or -thickness of the group for r_{av} in equation (48).

18.4.11 DEFINITIONS OF SOME STEREOCHEMICAL TERMS

This section gives definitions for selected terms currently used in the study of steric requirements involved in the interactions between ligands and biological acceptor sites.

Anticlinal	Anticlinal refers to a conformation in which the reference groups in the molecule are located at a torsion angle of $120° \pm 30°$.
Antiperiplanar	Antiperiplanar refers to a conformation in which the reference groups in the molecule are located by a torsion angle of $180° \pm 30°$.
Asymmetric carbon atom	An asymmetric carbon atom is one that bears four different substituents. It is located at a center of chirality.
Atropisomerism	See Conformational isomerism.
Chirality	A property characterizing three-dimensional forms which are not superimposable upon their mirror image.
Configuration	The configuration specifies the relative spatial arrangements of bonds and atoms in a molecule (of a given constitution) without regard to the multiplicity of spatial arrangements that may arise by rotation about single bonds.
Conformation	The conformations of a molecule denote the different spatial arrangements of the atoms in a molecule of a given constitution and configuration that may arise by rotation about single bonds.
Conformational isomerism	Conformational isomerism (atropisomerism) is an isomerism that results from rotation about a single bond. If the resulting isomers are separable and stable at room temperature they are called atropisomers.
Conformers	The term conformer is used to designate one of two or more conformations that are in such rapid equilibrium as to not represent discrete isolable substances under ordinary conditions.
Constitution	The constitution of a compound specifies: (i) the atoms in a molecule which are bound to one another and (ii) the types of bonds used for direct interatomic bonding.
Dextrorotatory enantiomer	Dextrorotatory describes the enantiomer which rotates a beam of plane-polarized light clockwise and is designated (*dextro*)-, *d*- or (+)-.
Diastereoisomers	Stereoisomers which are not enantiomers are diastereoisomers.
Distomer	See Eutomer.
Enantiomers	Enantiomers are structures which are related as an object is related to its mirror image and are not superimposable upon one another.
Epimers	Epimers are diastereoisomers which differ in the configuration of any one, and only one, of the chiral centers.
Eutomer	If in a specific biological action one of a pair of enantiomers is more active than its stereoisomer then it is called the eutomer. The less active or inactive enatiomer is called the distomer.
Eudismic ratio	The ratio of the activities (affinities, potencies *etc.*) of a pair of enantiomers in a specific biological action is termed the eudismic ratio. It is a measure of the degree of stereoselectivity in the biological action.
Geometrical isomerism	Geometrical isomerism occurs when a molecule incorporates a certain rigid element (*e.g.* double bond) to which a pair of substituents may be attached more than one way in space.
Isomers	Compounds that have the same formula but different chemical structures are called isomers.
Isosteres	Isostere is a name given to atoms or groups of atoms that by reason of similar size and electronic structure have similar physical properties.

Levorotatory enantiomer	Levorotatory describes the enantiomer which rotates a beam of plane-polarized light anticlockwise and is designated (*levo*)-, *l*- or (−)-.
Meso	*Meso* indicates that a molecule with an even number of chiral centers possesses either a center of symmetry (i) or a plane of symmetry (σ). *Meso* isomers are optically inactive.
Peri	*Peri* substituents are those located on carbon atoms separated by another carbon atom such as those on the 1,8 and 4,5 positions of naphthalene.
Positional isomers	Positional isomers are structures which have the same formula, carbon skeleton and functional groups but have the functional groups located at different positions along the carbon skeleton.
Preferred conformation	The preferred conformation of a molecule is one in which its overall energy is the minimum possible and it will be the most common species in a sample of the compound.
Racemate	A $1:1$ mixture of dextrorotatory and levorotatory enantiomers is described as a racemate or racemic compound. The net optical rotation of a racemate is zero.
Stereoisomers	Stereoisomers are either enantiomers or diastereoisomers.
Stereoselectivity	If one of two or more stereoisomers which can be formed or destroyed in a reaction (including drug–receptor complex formation) is formed or destroyed at a different rate than the others then this behavior is called stereoselectivity.
Stereospecificity	A stereospecific process is one in which stereoisomeric starting materials yield products that are stereoisomers of each other.
Synclinal	Synclinal refers to a conformation in which the reference groups in the molecule are located at a torsion angle of $60° \pm 30°$. This conformation is also called *gauche*.
Synperiplanar	Synperiplanar refers to a conformation in which the reference groups in the molecule are located at a torsion angle of $0° \pm 30°$.

18.4.12 REFERENCES

1. O. Kennard, D. Watson, F. Allen, W. Motherwell, W. Town and J. Rodgers, *Chem. Br.*, 1975, 213.
2. F. C. Bernstein, T. F. Koetzle, G. J. B. Williams, E. F. Meyer, Jr., M. D. Brice, J. R. Rodgers, O. Kennard, T. Shimanouchi and M. Tasumi, *J. Mol. Biol.*, 1977, **112**, 535.
3. J. P. Tollenaere, H. Moereels and L. A. Raymaekers, 'Atlas of the Three-Dimensional Structure of Drugs', Elsevier, Amsterdam, 1979.
4. A. S. Horn and C. J. De Ranter (eds.), 'X-ray Crystallography and Drug Actions', Clarendon Press, Oxford, 1984.
5. E. Osawa and H. Musso, *Top Stereochem.*, 1982, **13**, 117.
6. R. Langridge, *Fed. Proc., Fed. Am. Soc. Exp. Biol.*, 1974, **33**, 2332.
7. R. Langridge, T. E. Ferrin, I. D. Kuntz and M. L. Connolly, *Science (Washington, D.C.)*, 1981, **211**, 661.
8. P. A. Bash, N. Pattabiram, C. Huang, T. E. Ferrin and R. Langridge, *Science (Washington, D.C.)*, 1983, **222**, 1325.
9. C. Hansch, P. P. Maloney, T. Fujita and R. M. Muir, *Nature (London)*, 1962, **194**, 178.
10. C. Hansch, in 'Structure–Activity Relationships', C. J. Cavallito, Pergamon Press, Oxford, 1973, p. 75.
11. The journal *Quantitative Structure–Activity Relationships* (published by Verlag Chemie since 1982) is devoted to the subject.
12. A. T. Balaban, (ed.), 'Chemical Applications of Graph Theory', Academic Press, London, 1967.
13. N. Trinajstic, 'Chemical Graph Theory', CRC Press, Boca Raton, FL, 1983.
14. A. T. Balaban, *J. Chem. Inf. Comput. Sci.*, 1985, **25**, 334.
15. A. T. Balaban, I. Motoc, D. Bonchev and O. Mekenyan, *Top. Curr. Chem.*, 1983, **114**, 21.
16. A. Graovac, I. Gutman and N. Trinajstic, *Lect. Notes Chem.*, 1977, **4**.
17. I. Gutman, B. Ruščić, N. Trinajstić and C. F. Wilcox, Jr., *J. Chem. Phys.*, 1975, **62**, 3399.
18. M. Randić, *J. Am. Chem. Soc.*, 1975, **97**, 6609.
19. M. Randić, *Int. J. Quantum Chem., Quantum Biol. Symp.*, 1978, **5**, 245.
20. L. B. Kier and L. H. Hall, 'Molecular Connectivity in Chemistry and Drug Research', Academic Press, New York, 1976.
21. L. B. Kier and L. H. Hall, 'Molecular Connectivity in Structure–Activity Analysis', Research Studies Press, Letchworth, UK, 1986.
22. J. R. Platt, *J. Chem. Phys.*, 1947, **15**, 419.
23. J. R. Platt, *J. Phys. Chem.*, 1952, **56**, 328.
24. M. Gordon and G. R. Scantlebury, *Trans. Faraday Soc.*, 1964, **60**, 604.
25. I. Gutman and M. Randić, *Chem. Phys. Lett.*, 1977, **47**, 15.
26. M. Randić, *J. Chem. Phys.*, 1974, **60**, 3920.
27. M. Randić, *Chem. Phys. Lett.*, 1976, **42**, 283.

28. L. Lovasz and J. Pelikan, *Period. Math. Hung.*, 1973, **3**, 175.
29. A. T. Balaban and F. Harary, *J. Chem. Doc.*, 1971, **11**, 258.
30. M. Randić, N. Trinajstic and T. Zivkovic, *J. Chem. Soc., Faraday Trans. 2*, 1976, 244.
31. H. Wiener, *J. Am. Chem. Soc.*, 1947, **69**, 17.
32. H. Wiener, *J. Am. Chem. Soc.*, 1947, **69**, 2636.
33. H. Wiener, *J. Chem. Phys.*, 1947, **15**, 766.
34. H. Wiener, *J. Phys. Colloid Chem.*, 1948, **52**, 425.
35. K. Altenburg, *Kolloid-Z.*, 1961, **178**, 112.
36. K. Altenburg, *Brennst.-Chem.*, 1966, **47**, 100, 331 (*Chem. Abstr.*, 1966, **65**, 3714f).
37. H. Hosoya, *Bull. Chem. Soc. Jpn.*, 1971, **44**, 2332.
38. L. B. Kier, *Quant. Struct.–Act. Relat.*, 1985, **4**, 109.
39. L. B. Kier, *Quant. Struct.–Act. Relat.*, 1986, **5**, 7.
40. L. B. Kier, *Quant. Struct.–Act. Relat.*, 1986, **5**, 1.
41. D. Bonchev, A. T. Balaban and O. Mekenyan, *J. Chem. Inf. Comput. Sci.*, 1980, **20**, 106.
42. A. T. Balaban, *Chem. Phys. Lett.*, 1982, **89**, 399.
43. A. T. Balaban, *MATCH*, 1984, **16**, 163.
44. E. A. Smolenskii, *Zh. Fiz. Khim.*, 1964, **38**, 1288.
45. M. Randić, *MATCH*, 1979, **7**, 5.
46. H. Hosoya, *Theor. Chim. Acta*, 1972, **25**, 215.
47. H. Hosoya, *J. Chem. Doc.*, 1972, **12**, 181.
48. A. T. Balaban, *Theor. Chim. Acta*, 1979, **53**, 355.
49. C. E. Shannon and W. Weaver, 'The Mathematical Theory of Communication', University of Illinois Press, Urbana, IL, 1949.
50. D. Bonchev, 'Theoretic Information Indices for Characterization of Chemical Structures,' Research Studies Press, Chichester, 1983.
51. L. Brillouin, 'Science and Information Theory', 2nd edn.,.Academic Press, New York, 1962.
52. L. B. Kier, *Quant. Struct.–Act. Relat.*, 1987, **6**, 8.
53. N. Rashevsky, *Bull. Math. Biophys.*, 1955, **30**, 229.
54. E. Trucco, *Bull. Math. Biophys.*, 1956, **18**, 129.
55. A. Mowshowitz, *Bull. Math. Biophys.*, 1968, **30**, 533.
56. S. H. Bertz, *J. Am. Chem. Soc.*, 1981, **103**, 3599.
57. D. Bonchev, O. Mekenyan and N. Trinajstic, *J. Comput. Chem.*, 1981, **2**, 127.
58. W. T. Yee, K. Sakamoto and Y. J. Ihaya, *Denki Tsushin Daigaku Gakuho*, 1976, **27**, 53 (*Chem. Abstr.*, 1977, **87**, 151 563h).
59. R. E. Merrifield and H. E. Simmons, *Proc. Natl. Acad. Sci. U.S.A.*, 1981, **78**, 1329.
60. D. Bonchev and N. Trinajstic, *J. Chem. Phys.*, 1977, **67**, 4517.
61. V. Austel, E. Kutter and W. Kalbfleisch, *Arzneim.-Forsch.*, 1979, **29**, 585.
62. M. Charton, *J. Org. Chem.*, 1978, **43**, 3995.
63. A. T. Balaban, *Lect. Notes Chem.*, 1980, **15**, 22.
64. E. J. Kupchik, *Quant. Struct.–Act. Relat.*, 1988, **7**, 57.
65. J.-E. Dubois, in 'Computer Representation and Manipulation of Chemical Information', ed. W. T. Wipke, S. R. Heller, R. J. Feldmann and E. Hyde, Wiley, New York, 1974, p. 239.
66. J.-E. Dubois, D. Laurent, P. Bost, S. Chambaud and C. Mercier, *Eur. J. Med. Chem.—Chim. Ther.*, 1976, **11**, 225.
67. C. Mercier and J.-E. Dubois, *Eur. J. Med. Chem.—Chim. Ther.*, 1979, **14**, 415.
68. Institute for Scientific Information, Philadelphia (databases Questel/Darc and Index Chemicus Online).
69. A. Bondi, *J. Phys. Chem.*, 1964, **68**, 441.
70. A. Bondi, *J. Phys. Chem.*, 1966, **70**, 3006.
71. L. Pauling, 'The Nature of the Chemical Bond', Cornell University Press, Ithaca, NY, 1960.
72. N. L. Allinger, *Adv. Phys. Org. Chem.*, 1976, **13**, 1.
73. L. S. Bartell, *J. Chem. Phys.*, 1960, **32**, 827.
74. C. Glidewell, *Inorg. Chim. Acta*, 1975, **12**, 219.
75. C. Glidewell, *Inorg. Chim. Acta*, 1976, **20**, 113.
76. M. Charton, *J. Am. Chem. Soc.*, 1979, **101**, 7356.
77. M. Charton, *Top. Curr. Chem.*, 1983, **114**, 57.
78. M. Charton, *J. Am. Chem. Soc.*, 1969, **91**, 615.
79. M. Charton, *Prog. Phys. Org. Chem.*, 1971, **8**, 235.
80. M. Charton, *J. Am. Chem. Soc.*, 1975, **97**, 1552.
81. E. Kutter and C. Hansch, *J. Med. Chem.*, 1969, **12**, 647.
82. M. Charton, *J. Org. Chem.*, 1976, **41**, 2217.
83. M. Charton, *J. Org. Chem.*, 1977, **42**, 3531.
84. M. Charton and B. I. Charton, *J. Org. Chem.*, 1978, **43**, 1161.
85. K. Bowden and R. C. Young, *J. Med. Chem.*, 1970, **13**, 225.
86. A. Verloop, W. Hoogenstraaten and J. Tipker, in 'Drug Design', ed. E. J. Ariens, Academic Press, New York, 1976, vol. VII, p. 165.
87. A. Verloop and J. Tipker, *Pestic Sci.*, 1976, **7**, 379.
88. A. Verloop, in 'IUPAC Pesticide Chemistry', ed. J. Miyamoto, Pergamon Press, Oxford, 1983, vol. I, p. 339.
89. A. Verloop and J. Tipker, in 'Pharmaco Chemistry Library. QSAR in Drug Design and Toxicology', ed. D. Hadži and B. Jerman-Blăzič, Elsevier, Amsterdam, 1987, vol. 10, p. 97.
90. A. Verloop, personal communication.
91. A. Verloop and J. Tipker, in 'Pharmaco Chemistry Library. Biological Activity and Chemical Structure', ed. J. A. Keverling Buisman, Elsevier, Amsterdam, 1977, vol. 2, p. 63.
92. M. Charton, *ACS Symp. Ser.*, 1984, **255**, 247.
93. N. Menschutkin, *Justus Liebigs Ann. Chem.*, 1879, **195**, 334.
94. F. Kehrmann, *Chem. Ber.*, 1888, **21**, 3315.

95. A. Haller, *C. R. Hebd. Seances Acad. Sci.*, 1889, **109**, 112.
96. C. A. Bischoff, *Chem. Ber.*, 1890, **23**, 623.
97. V. Meyer, *Chem. Ber.*, 1894, **27**, 510; 1895, **28**, 1254.
98. S. H. Unger and C. Hansch, *Prog. Phys. Org. Chem.*, 1976, **12**, 91.
99. R. W. Taft, Jr., *J. Am. Chem. Soc.*, 1952, **74**, 3120.
100. R. W. Taft, Jr., in 'Steric Effects in Organic Chemistry', ed. M. S. Newman, Wiley, New York, 1956, p. 556.
101. C. K. Ingold, *J. Chem. Soc.*, 1930, 1032.
102. C. Hansch and A. Leo, in 'Substituent Constants for Correlation Analysis in Chemistry and Biology', Wiley, New York, 1979, p. 9.
103. J. A. MacPhee, A. Panaye and J.-E. Dubois, *Tetrahedron*, 1978, **34**, 3553; 1980, **36**, 759.
104. K. L. Loening, A. B. Garrett and M. S. Newman, *J. Am. Chem. Soc.*, 1952, **74**, 3929.
105. D. F. De Tar, *J. Org. Chem.*, 1980, **45**, 5166.
106. M. Charton, in 'Design of Biopharmaceutical Properties Through Prodrugs and Analogs', ed. E. B. Roche, American Pharmaceutical Association, Washington, DC, 1977, p. 228.
107. M. Charton, *Prog. Phys. Org. Chem.*, 1971, **8**, 235.
108. C. K. Hancock, E. A. Meyers and B. J. Yager, *J. Am. Chem. Soc.*, 1961, **83**, 4211.
109. C. K. Hancock and C. P. Falls, *J. Am. Chem. Soc.*, 1961, **83**, 4214.
110. C. K. Hancock, B. J. Yager, C. P. Falls and J. O. Schreck, *J. Am. Chem. Soc.*, 1963, **85**, 1297.
111. O. Rosado-Lojo, C. K. Hancock and A. Danti, *J. Org. Chem.*, 1966, **31**, 1899.
112. B. J. Yager and C. K. Hancock, *J. Org. Chem.*, 1965, **30**, 1174.
113. J. P. Idoux, P. T. R. Hwang and C. K. Hancock, *J. Org. Chem.*, 1973, **38**, 4239.
114. M. M. Kreevoy and R. W. Taft, Jr., *J. Am. Chem. Soc.*, 1955, **77**, 5590.
115. J. Shorter, in 'Advances in Linear Free-Energy Relationships', ed. N. B. Chapman and J. Shorter, Plenum Press, New York, 1972, chap. 2, p. 71.
116. V. A. Palm, 'Fundamentals of the Quantitative Theory of Organic Reaction', Khimiya, Leningrad, 1967, chap. 10.
117. I. V. Talik and V. A. Palm, *Reakts. Sposobn. Org. Soedin.*, 1971, **8** (2), 445.
118. T. Fujita, C. Takayama and M. Nakajima, *J. Org. Chem.*, 1973, **38**, 1623.
119. T. Fujita and H. Iwamura, *Top. Curr. Chem.*, 1983, **114**, 119.
120. J.-A. MacPhee, A. Panaye and J.-E. Dubois, *J. Org. Chem.*, 1980, **45**, 1164.
121. M. S. Newman, in 'Steric Effects in Organic Chemistry', ed. M. S. Newman, Wiley, New York, 1956, chap. 4, p. 201.
122. M. S. Newman, N. Gill and D. W. Thomson, *J. Am. Chem. Soc.*, 1967, **89**, 2059.
123. K. Bowden and K. R. H. Wooldridge, *Biochem. Pharmacol.*, 1973, **22**, 1015.
124. L. B. Kier, *Acta Pharm. Jugosl.*, 1986, **36**, 171.
125. L. B. Kier, *Quant. Struct.–Act. Relat.*, 1987, **6**, 117.
126. J.-E. Dubois, J. A. MacPhee and A. Panaye, *Tetrahedron Lett.*, 1978, 4099.
127. A. C. Farthing and B. Nam, 'Steric Effects in Conjugated Systems', Academic Press, New York, 1958.
128. K. Kindler, *Justus Liebigs Ann. Chem.*, 1928, **464**, 278.
129. W. W. Hussey and A. J. Diefenderfer, *J. Am. Chem. Soc.*, 1967, **89**, 5359.
130. U. Berg, R. Gallo, G. Klatte and J. Metzger, *J. Chem. Soc., Perkin Trans. 2*, 1980, 1350.
131. M. Charton, *J. Am. Chem. Soc.*, 1969, **91**, 6649.
132. M. H. Abraham, *Prog. Phys. Org. Chem.*, 1974, **11**, 1.
133. R. Gallo, *Prog. Phys. Org. Chem.*, 1983, **14**, 115.
134. O. Exner, *Collect. Czech. Chem. Commun.*, 1967, **32**, 1.
135. H. Høiland, *Acta Chem. Scand.*, 1973, **27**, 2687.
136. C. Hansch, A. Leo, S. H. Unger, K. H. Kim, D. Nikaitani and E. J. Lien, *J. Med. Chem.*, 1973, **16**, 1207.
137. O. R. Quayle, *Chem. Rev.*, 1953, **53**, 439.
138. J. T. Edward, *Chem. Ind. (London)*, 1956, 774.
139. A. E. Luzkii, *Russ. J. Phys. Chem.*, 1954, **28**, 204.
140. A. Leo, C. Hansch and P. Y. C. Jow, *J. Med. Chem.*, 1976, **19**, 611.
141. R. S. Pearlman, in 'Physical Chemical Properties of Drugs', ed. S. H. Yalkowsky, A. A. Sinkula and S. C. Valvani, Dekker, New York, 1980, p. 321.
142. I. Motoc, O. Dragmoir-Filimonescu and R. Vîlcearm, *MATCH*, 1981, **11**, 185.
143. T. R. Stouch and P. C. Jurs, *J. Chem. Inf. Comput. Sci.*, 1986, **26**, 4.
144. W. P. Purcell and B. Testa, in 'Biological Activity and Chemical Structure', ed. J. A. Keverling Buisman Elsevier, Amsterdam, 1977, p. 269.
145. (a) B. Testa and W. P. Purcell, *Eur. J. Med. Chem.—Chim. Ther.*, 1978, **13**, 509; (b) C. Hansch, J. McClarin, T. Klein and R. Langridge, *Mol. Pharmacol.*, 1985, **27**, 493.
146. M. Charton and B. I. Charton, *J. Org. Chem.*, 1979, **44**, 2284.
147. M. Charton, *Top. Curr. Chem.*, 1983, **114**, 107.
148. I. Moriguchi, Y. Kanada and K. Komatsu, *Chem. Pharm. Bull.*, 1976, **24**, 1799.
149. (a) W. J. Dunn, III, *Eur. J. Med. Chem.—Chim. Ther.*, 1977, **12**, 109; (b) R. L. Lopez de Compadre, C. M. Compadre, R. Castillo and W. J. Dunn, III, *Eur. J. Med. Chem.—Chim. Ther.*, 1983, **18**, 569.
150. A. Cammarata, S. J. Yau and K. S. Rogers, *J. Med. Chem.*, 1971, **14**, 1211.
151. L. Pauling and D. Pressman, *J. Am. Chem. Soc.*, 1945, **67**, 1003.
152. E. Kutter and C. Hansch, *Arch. Biochem. Biophys.*, 1969, **135**, 126.
153. D. Agin, L. Hersh and D. Holtzman, *Proc. Natl. Acad. Sci. U.S.A.*, 1965, **53**, 952.
154. A. I. Vogel, *J. Chem. Soc.*, 1948, 1833.
155. C. Hansch and A. Leo, 'Substituent Constants for Correlation Analysis in Chemistry and Biology', Wiley, New York, 1979, p. 44.
156. C. Hansch, C. Grieco, C. Silipo and A. Vittoria, *J. Med. Chem.*, 1977, **20**, 1420.
157. S. Sugden, *J. Chem. Soc.*, 1924, **125**, 1177.
158. S. A. Mumford and J. W. C. Phillips, *J. Chem. Soc.*, 1929, 2112.
159. A. I. Vogel, *J. Chem. Soc.*, 1934, 333; 1946, 133.

160. J. C. McGowan, *J. Appl. Chem.*, 1951, **1**, S120; 1952, **2**, 323, 651; 1954, **4**, 41; 1966, **16**, 103.
161. J. C. McGowan, *Recl. Trav. Chim. Pays-Bas*, 1956, **75**, 193.
162. J. C. McGowan, *Nature (London)*, 1963, **200**, 1317.
163. O. Exner, *Collect. Czech. Chem. Commun.*, 1966, **31**, 3222.
164. S. Kitahara, E. Heinz and C. Stahlmann, *Nature (London)*, 1965, **208**, 187.
165. S. C. Dash and G. B. Behera, *J. Indian Chem. Soc.*, 1980, **57**, 542.
166. S. C. Dash and G. B. Behera, *Indian J. Chem., Sect. A.*, 1980, **19A**, 541.
167. J. E. Amoore, G. Palmieri and E. Wanke, *Nature (London)*, 1967, **216**, 1084.
168. N. L. Allinger, in 'Pharmacology and the Future of Man, Proceedings of the International Congress on Pharmacology, 5th, San Francisco, 1972', ed. G. H. Acheson, Karger, New York, 1973, vol. 5, p. 57.
169. Z. Simon and Z. Szabadai, *Stud. Biophys.* 1973, **39**, 123.
170. Z. Simon, *Angew. Chem., Int. Ed. Engl.*, 1974, **13**, 719.
171. Z. Simon, A. Chiriac, S. Holban, D. Ciubotaru and G. I. Mihalas, 'Minimum Steric Difference—The MTD Method for QSAR Studies', Research Studies Press, Letchworth, UK, 1984.
172. Z. Simon, *Stud. Biophys.*, 1975, **51**, 49.
173. I. Motoc, S. Holban, R. Vancea and Z. Simon, *Stud. Biophys.*, 1977, **66**, 75.
174. I. Motoc, S. Holban, R. Vancea, *Lect. Notes Chem.*, 1980, **15**, 108.
175. I. Motoc, in 'Steric Effects in Drug Design', ed. M. Charton and I. Motoc, Springer-Verlag, Berlin, 1983, p. 93.
176. Yu. A. Shreider (ed.), 'The Monte Carlo Method; The Method of Statistical Trials', Pergamon Press, Oxford, 1966.
177. Z. Simon, A. Chiriac, I. Motoc, S. Holban, D. Ciubotariu and Z. Szabadai, *Stud. Biophys.*, 1976, **55**, 217.
178. Z. Simon, S. Holban, I. Motoc, M. Mracec, A. Chiriac, F. Kerek, D. Ciubotariu, Z. Szabadai, R. D. Pop and I. Schwartz, *Stud. Biophys.*, 1976, **59**, 181.
179. Z. Simon, I. Badilescu and T. Recovitan, *J. Theor. Biol.*, 1977, **66**, 485.
180. A. J. Hopfinger, *J. Am. Chem. Soc.*, 1980, **102**, 7196.
181. A. J. Hopfinger, *J. Med. Chem.*, 1981, **24**, 818.
182. A. J. Hopfinger, *Arch. Biochem. Biophys.*, 1981, **206**, 153.
183. C. Battershell, D. Malhotra and A. J. Hofinger, *J. Med. Chem.*, 1981, **24**, 812.
184. A. J. Hopfinger, *J. Med. Chem.*, 1983, **26**, 990.
185. C. Silipo and C. Hansch, *J. Am. Chem. Soc.*, 1975, **97**, 6849.
186. G. M. Crippen and T. F. Havel, *Acta Crystallogr., Sect. A*, 1978, **34**, 282.
187. G. M. Crippen, *J. Med. Chem.*, 1979, **22**, 988.
188. G. M. Crippen, *J. Med. Chem.*, 1980, **23**, 599.
189. A. K. Ghose and G. M. Crippen, *J. Med. Chem.*, 1982, **25**, 892.
190. G. M. Crippen, *Mol. Pharmacol.*, 1982, **22**, 11.
191. A. K. Ghose and G. M. Crippen, *J. Med. Chem.*, 1983, **26**, 996.
192. M. Murray, C. B. Marcus and C. F. Wilkinson, *Quant. Struct.–Act. Relat.*, 1985, **4**, 18.
193. L. H. Hall and L. B. Kier, *Eur. J. Med. Chem.—Chim. Ther.*, 1978, **13**, 89.
194. M. S. Tute, in 'Quantitative Structure–Activity Relationships of Drugs', ed. J. G. Topliss, Academic Press, New York, 1983, p. 23.
195. E. J. Lien, C. Hansch and S. M. Anderson, *J. Med. Chem.*, 1968, **11**, 430.
196. I. Pratesi, L. Villa, V. Ferri, C. De Micheli, E. Grana, C. Grieco, C. Silipo and A. Vittoria, *Farmaco, Ed. Sci.*, 1979, **34**, 579.
197. C. Hansch, S. H. Unger and A. B. Forsythe, *J. Med. Chem.*, 1973, **16**, 1217.
198. G. M. Cohen and G. J. Mannering, *Mol. Pharmacol.*, 1973, **9**, 383.
199. D. Pressman, S. M. Swingle, A. L. Grossberg and L. Pauling, *J. Am. Chem. Soc.*, 1944, **66**, 1731.
200. M. I. Bruce and J. A. Zwar, *Proc. R. Soc. London, Ser. B*, 1966, **165**, 245.
201. E. M. Shantz and F. C. Steward, *J. Am. Chem. Soc.*, 1955, **77**, 6351.
202. P. Pratesi, L. Villa, V. Ferri, C. De Micheli, E. Grana, C. Grieco, C. Silipo and A. Vittoria, *Farmaco, Ed. Sci.*, 1979, **34**, 657.
203. P. Pratesi, E. Grana, M. G. Santagostino Barbone, M. I. LaRotonda, C. Silipo and A. Vittoria, *Farmaco, Ed. Sci.*, 1986, **41**, 335.
204. E. Kutter and C. Hansch, *Arch. Biochem. Biophys.*, 1969, **135**, 126.
205. C. Silipo and C. Hansch, *Farmaco, Ed. Sci.*, 1975, **30**, 35.
206. C. Silipo and A. Vittoria, *Actual. Chim. Ther.*, 1980, **7th ser.**, 147.
207. C. Grieco, C. Silipo and A. Vittoria, *Farmaco Ed. Sci.*, 1977, **32**, 909.
208. C. Hansch and E. J. Lien, *Biochem. Pharmacol.*, 1968, **17**, 709.
209. J. K. Seydel, K.-J. Schaper, E. Wempe and H. P. Cordes, *J. Med. Chem.*, 1976, **19**, 483.
210. E. Kruger-Thiemer, *Ber. Borstel*, 1956, **3**, 192.
211. E. A. Coats, C. S. Genther, S. W. Dietrich, Z.-R. Guo and C. Hansch, *J. Med. Chem.*, 1981, **24**, 1422.
212. C. Hansch, B. A. Hathaway, Z.-R. Guo, C. D. Selassie, S. W. Dietrich, J. M. Blaney, R. Langridge, K. W. Volz, B. T. Kaufman, *J. Med. Chem.*, 1984, **27**, 129.
213. T. Fujita and T. Ban, *J. Med. Chem.*, 1969, **12**, 353.
214. T. Ban and T. Fujita, *J. Med. Chem.*, 1971, **14**, 148.
215. M. Yoshimoto and C. Hansch, *J. Org. Chem.*, 1976, **41**, 2269.
216. W. C. Randall, P. S. Anderson, E. L. Cresson, C. A. Hunt, T. F. Lyon, K. E. Rittle and D. C. Remy, *J. Med. Chem.*, 1979, **22**, 1222.
217. E. J. Lien, J. F. Rodrigues de Miranda and E. J. Ariëns, *Mol. Pharmacol.*, 1976, **12**, 598.
218. C. C. Pfeiffer, *Science (Washington, D.C.)*, 1956, **124**, 29.
219. F. P. A. Lehmann, *Quant. Struct.–Act. Relat.*, 1987, **6**, 57.
220. F. P. A. Lehmann, in 'Mechanism of Drug Action', ed. T. P. Singer, T. E. Mansour and R. N. Ondarza, Academic Press, New York, 1983, p. 61.
221. C. Grieco, C. Hansch, C. Silipo, R. N. Smith, A. Vittoria and K. Yamada, *Arch. Biochem. Biophys.*, 1979, **194**, 542.
222. G. E. Hein and C. Niemann, *J. Am. Chem. Soc.*, 1962, **84**, 4487, 4495.

223. C. Hansch and J. M. Blaney, in 'Drug Design: Fact or Fantasy?', ed. G. Jolles and K. R. H. Wooldridge, Academic Press, New York, 1984.
224. C. Hansch, in 'Molecular Structure and Energetics', ed. J. F. Liebman and A. Greenberg, VCH, New York, 1987, vol. 4, p. 341.
225. C. Hansch and T. E. Klein, *Acc. Chem. Res.*, 1986, **19**, 392.

18.5
Electronic Effects in Drugs

KEITH BOWDEN

University of Essex, Colchester, UK

18.5.1 INTRODUCTION

Although some early studies[1] had indicated the importance of electronic effects in controlling the activity of drugs, the Hansch analysis[2] most clearly established their importance and quantitative contribution.

Many processes have been suggested to account for the electronic or polar influences of substituents. While considerable discussion remains regarding their actual or relative importance, it is generally agreed that the major influences can be considered to be the localized effects, *i.e.* the field and/or inductive effects, and the delocalized effect, *i.e.* the resonance effect.[3]

18.5.1.1 The Field and σ Inductive Effects

The electrostatic field and σ inductive effects are usefully considered together. They are the subject of some confusion and their relative importance has excited some controversy.[4] The presence of a charged or dipolar substituent should exert an electrostatic field elsewhere in the molecule and surrounding space. Such effects should decrease fairly rapidly with distance from the charged or dipolar substituent. Bjerrum[5] appears to have been the earliest to use equations (1) (charged substituent) and (2) (dipolar substituent) to calculate the magnitude of electrostatic field effects on the ionization constants of acids.

$$\log(K/K_{\mathrm{H}}) = e^2/(2.3\,RTr\,D_{\mathrm{e}}) \tag{1}$$

$$\log(K/K_{\mathrm{H}}) = (e\mu\cos\theta)/(2.3\,RTr^2\,D_{\mathrm{e}}) \tag{2}$$

In these equations, K and K_{H} are the ionization constants of substituted and unsubstituted acids respectively, e the magnitude of the charge, μ the dipole moment, r the distance between the charges or the charge and midpoint of the dipole, D_{e} the effective dielectric constant and θ the angle between the line joining the charge and the middle of the dipole and the line joining the middle of the dipole having a charge of the same sign as the interacting charge. Bjerrum[5] in his calculations used the dielectric constant of the medium. This resulted in calculations of r values from the known ionization constants of substituted acids which were implausibly small for short chain systems. Kirkwood and Westheimer[6,7] elaborated this treatment by taking 2.0 as the dielectric constant of the molecular cavity (D_{i}) enclosed within the solvent medium of high dielectric constant (D). The value of D_{e} can then be calculated if the molecular cavities are assumed to be spherical or ellipsoidal in nature. The mathematical difficulties, in spite of more recent treatments,[8] make the systematic applications impossible. The problems are many: conformations are sometimes uncertain; few molecules may be represented plausibly as spheres or ellipsoids; embedding of the substituents will have to be assumed,[9] as will the actual centre of dipole in many substituents; furthermore, the 'microscopic' dielectric constants of molecules and solvents are required, while only the 'macroscopic' are available. Such treatments have been applied to a number of systems, particularly the 4-substituted bicyclo[2.2.2]octane-1-carboxylic acids (**1**).[10-13] In the latter system the assumptions necessary appear more reasonable. Modified treatments now give better than fair agreement for such calculations.[13]

CO$_2$H

X

(**1**)

The σ inductive effect involves propagation by successive polarization of the bonds by a dipolar or charged substituent between the substituent and the reaction site. This effect will diminish with chain length and the nature of the intervening atoms and number of paths. This can be represented in (**2**), shown below for the substituent Y on the carbon chain. This has been developed into a quantitative treatment, as shown below in equation (3), where A is a reaction constant, σ is a substituent constant and ε the transmission coefficient for the ith atom in a molecular framework. Exner[14] has offered a mathematical treatment for calculated transmission by complex networks. However, there is no general agreement as to the value of ε. Values between 0.2 and 0.7 have been suggested for different atoms and atoms of different hybridization. Bowden[15] has presented some values based on a practical analysis assuming an inductive relay.

$$\log(K/K_H) = A\sigma \Sigma \Pi \varepsilon_i \tag{3}$$

An alternative formulation[16] of the σ inductive effect has been made and can be represented as shown below (3). No experimental evidence of reactivity has yet been presented for such an induction.

$$
\underset{Y}{\overset{\delta-}{Y}} - \underset{C}{\overset{\delta+}{C}} - \underset{C}{\overset{\delta\delta+}{C}} - \underset{C}{\overset{\delta\delta\delta+}{C}} - \qquad\qquad \underset{Y}{\overset{\delta-}{Y}} - \underset{C}{\overset{\delta+}{C}} - \underset{C}{\overset{\delta\delta-}{C}} - \underset{C}{\overset{\delta\delta\delta+}{C}} -
$$

(2)   (3)

However, there is good evidence from a number of sources[4,17,18] that the electronic effect of substituents of this type is largely, if not completely, an electrostatic field effect rather than a σ inductive effect. The problem is that in many conventional molecules the major electrostatic transmissive region is occupied by the low dielectric cavity of the molecule. This then also approximates to the bond linkage system between the reaction site and substituent. The field and inductive effects are then operating in the same direction. By judicious choice of ε and D_e, comparable results from both effects can be expected. However, in specially selected and constructed model systems, the dipolar substituent can be reversed with regard to its normal 'distant' end of the dipole now being closer to the reaction site. For transmission by a field effect, the normal substituent effect should now be reversed and, for an inductive effect, no change should now be expected. In a study[17] of the four possible pseudo-substituted bromo-4-carboxy[2.2]paracyclophanes and the parent compound (4), both normal (for pseudo *meta* and *para*) and reversed (for pseudo *ortho* and *gem*) effects are observed where these would be expected on the basis of an electrostatic field effect. For the 1-(8-substituted naphthyl)propiolic acids (5), a quantitatively complete reversal has been found by comparison with suitable model compounds.[18]

(4)   (5)

The properties of substituents in the molecules under discussion are best considered in terms of an electrostatic field effect. This will be imperative for molecules with substituent interaction sites of critical geometry; but the application of simple transmissive factors may be of practical use in straightforward 'classical' situations.

18.5.1.2 π Electron Effects

The π electron system of an aromatic or unsaturated system can be modified by a substituent. If a substituent has unshared electrons in suitable orbitals, they may interact with the π electron system.

Likewise, substituents which have suitable unfilled orbitals available can also delocalize the π electron system. These two situations can be represented by the canonical structures (6) and (7). Typical substituents as X in (6) are NH_2, Cl or OH, and as Y in (7) are NO_2, CHO or CN. These effects result in a transfer of charge between the substituent and the π electron system and this is usually described as the resonance effect ($\pm R$), previously the mesomeric effect.

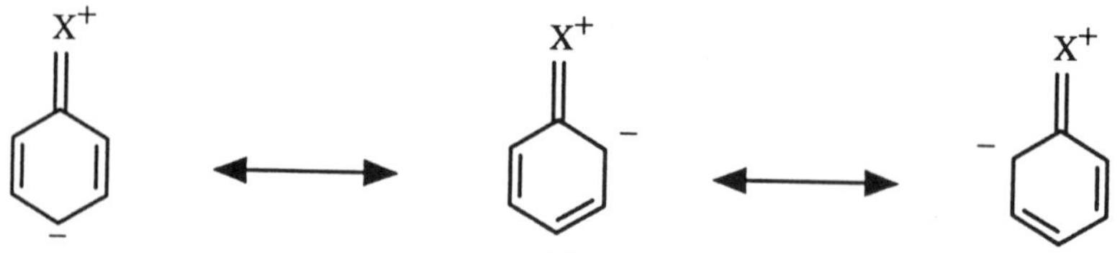

(6)

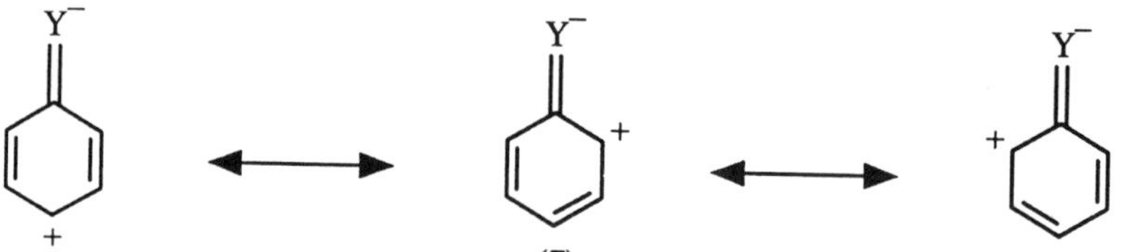

(7)

The π electron system can be affected by the substituent without any transfer of charge, a π inductive effect. This can be simply by a polarization of the π electron system by charged or dipolar substituents, just as suggested for a σ inductive effect. However, an alternative effect may arise from the repulsive interactions between the filled orbitals on the substituent and the π electron system.[3]

Studies[3,19] which review the situation indicate that the major cause of π electron disturbance in aromatic and unsaturated systems arises from the resonance effect described above. The π inductive and other secondary effects appear to be of minor importance.

18.5.2 THE HAMMETT EQUATION AND ITS EXTENSION

18.5.2.1 Introduction

Hammett in 1937[20,21] enabled the quantitative evaluation of *meta* and *para* substituents on the side chain reactivity of benzene derivatives. This relation took the forms shown below as equations (4) and (5), where k or K is the rate or equilibrium constant, respectively, for a *meta*- or *para*-substituted derivative. k_0 or K_0 refer to the constant of unsubstituted compound, or, more properly, the statistical value approximating to the latter. The substituent constant σ measures the polar effect, relative to hydrogen, of the substituent and is, in principle, independent of the nature of the reaction. The reaction constant ρ measures the susceptibility of the reaction to polar effects and depends on the nature of the reaction, including conditions such as the reaction medium and temperature. The ionization of benzoic acids in water at 25 °C was chosen as the reference system with ρ equal to 1.000. Thus, in effect, σ is simply defined by equation (6), using the pK_a values for *meta-/para*-substituted benzoic acids in water at 25 °C.

$$\log k = \rho\sigma + \log k_0 \qquad (4)$$

$$\log K = \rho\sigma + \log K_0 \qquad (5)$$

$$\sigma_X = (pK_a)_0 - pK_X \qquad (6)$$

This results in electron-withdrawing substituents being positive and electron-releasing substituents being negative. The simple Hammett equation has been applied to many thousands of reaction series with variable results. In 1953, Jaffé[22] reviewed the application of this equation to over 400 reactions with the majority being classified as 'satisfactory' to 'excellent'. Refinements enable the relation to be extended to very extensive physical organic data (see Sections 18.5.2.2–18.5.2.4). This relationship and its variants are probably the most widely employed linear free energy relation (LFER).

18.5.2.2 Duality of Substituent Constants: σ^+ and σ^-

An early refinement was the suggestion of duality of substituent constants. Marked deviations from the Hammett equation were particularly noted for *para* substituents with important resonance effects, both electron-withdrawing ($+R$) and electron-releasing ($-R$). These give rise to exalted resonance effects when such substituents engage in cross-conjugation with reaction sites of the opposite type, *i.e.* electron rich or electron deficient. This is exemplified by the structure shown in (**8**). Those *para* substituents giving exalted electron withdrawal are assigned σ^- values.[23] These are based on the ionization of *meta*-substituted phenols in water at 25 °C, giving a defining ρ value for this reaction. This ρ value is then used to derive exalted or σ^- values for *para* substituents such as NO_2, COMe, CN or CF_3. The importance of structures such as (**9**) for the ionization of *p*-nitrophenol is thus estimated.

Following the discussion above, those *para* substituents giving exalted electron release are assigned σ^+ values.[24] The reaction now used for determining such constants is the S_N1 solvolysis of substituted phenyldimethylcarbinyl (*t*-cumyl) chlorides in 90% aqueous acetone at 25 °C, using the *meta* substituents to define ρ. The exalted or σ^+ values for *para* substituents such as OMe, Me, NH_2 and Cl can be calculated and demonstrate the importance of structures such as (**10**). It should be pointed out that, in defining ρ for use in estimating σ^-, strongly $+R$ *meta* substitutents are not used and for σ^+ strongly $-R$ *meta* substituents are not used. This precludes any relayed or secondary effects.

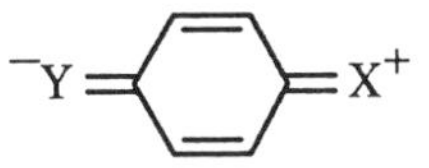

(**8**) 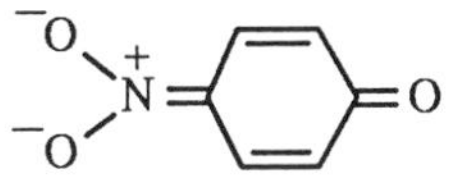   (**9**)

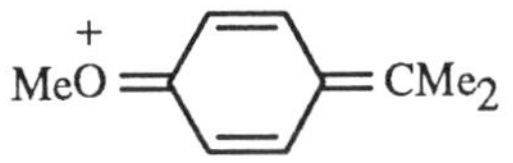

(**10**)

18.5.2.3 Normal or Unexalted Substituent Constants: σ^n and σ^0

The duality of σ values was strongly criticized by Van Bekkum *et al.*,[25] who considered cross-conjugation or through-conjugation as a continuous process. A sliding extent of such interactions was considered to be present, depending on the reaction studied. Eight 'primary' σ values of *meta* substituents, including H, were considered to be of general applicability, together with *p*-COMe and *p*-NO_2 where cross-conjugation effects can be ruled out. All other *para* substituents were excluded, as were $-R$ *meta* substituents such as OMe and NH_2 in case of relayed effects. These 'primary' values were used to calculate σ values relevant to that particular reaction for all other substituents. This analysis did indicate a sliding scale of exaltation of through-resonance. By assuming $-R$ reaction centres give 'normal' σ values for $-R$ substituents and $+R$ centres for $+R$ substituents, an averaged unexalted or 'normal' substituent constant was calculated and denoted σ^n. Large differences between σ and σ^n arise for substituents such as *p*-OMe, *p*-NH_2 and *p*-OH, where cross-conjugation between the substituent and CO_2H occur in the original reference reaction, the ionization of benzoic acids.

A closely related approach was made by Taft,[26] who selected four reaction series: the ionization of *meta/para*-substituted phenylacetic and 3-phenylpropionic acids and the alkaline hydrolysis of the corresponding ethyl phenylacetates and benzyl acetates. All these systems have substituents which are unaffected by cross-conjugation due to the insulating CH_2 groups. These reaction series were then used to generate a series of σ^0 values. Thus, *meta* and *para* substituents with $+R$ effects have σ^0

and σ values which are almost identical, whereas the substituents with $-R$ effects deviate appreciably. The σ^n and σ^0 scales are almost the same. They do appear to be more fundamental than the σ value scale. However, when the low precision of these values and the alternative treatments available (see Section 18.5.3) are considered, it seems wise to retain the σ scale in a primary role.

18.5.2.4   The Yukawa–Tsuno and Related Equations

In 1959, Yukawa and Tsuno[27] proposed a correlation procedure to enable the evaluation of $-R$ *para* substituents in reactions that are more electron demanding than the usual reference reaction (the ionization of benzoic acids). This relation can be expressed in two forms (equations 7 and 8).[27,28] In the latter, $\Delta\sigma_R^+$ equals the enhanced resonance effect ($\sigma^+ - \sigma$) and, in both, r^+ the extent of the enhanced resonance effect. The value of r^+ can vary from zero to greater than unity. If $r^+ = 0$, the correlation is with σ and if $r^+ = 1$, the correlation is with σ^+. Corresponding equations (9 and 10) employing σ^- are shown below.[29] r^- similarly measures the extent of the enhanced resonance effect and $\Delta\sigma_R^-$ equals ($\sigma^- - \sigma$). The equations have been used with σ^n or σ^0 to replace σ, which would appear more rational. A number of claims have been made that the Yukawa–Tsuno equation is a superior treatment to the simple Hammett equation using σ, σ^+ or σ^-. However, the method corresponds fairly closely to the dual parameter methods (see Section 18.5.3).

$$\log(k/k_0) = \rho[\sigma + r^+(\sigma^+ - \sigma)] \tag{7}$$

$$\log(k/k_0) = \rho(\sigma + r^+\Delta\sigma_R^+) \tag{8}$$

$$\log(k/k_0) = \rho[\sigma + r(\sigma^- - \sigma)] \tag{9}$$

$$\log(k/k_0) = \rho(\sigma + r^-\Delta\sigma_R^-) \tag{10}$$

18.5.3   THE SEPARATION OF ELECTRONIC EFFECTS

An important development was the concept that substituent electronic effects could be considered to be separable and additive, involving field or inductive, resonance and steric effects.[30] The field or inductive contribution can be considered to describe either the σ inductive and the electrostatic field effect together or the latter alone. It is described explicitly as the field effect in later treatments. Steric effects can be absent or present and, if present, can be constant within a series under examination.

18.5.3.1   Separation of Field or Inductive and Resonance Effects

Taft and co-workers[31,32] have resolved the Hammett substituent constants, σ_m and σ_p, into inductive and resonance contributions. The inductive effect, σ_I, was assumed to be equal from both the *meta* and *para* positions. The resonance effect, σ_R, operates directly from the *para* position and indirectly from the *meta* position, where it is reduced by a 'relay coefficient', α. This treatment results in equations (11) and (12). A value of α equal to 0.33 has been found to be satisfactory for Hammett σ values. The σ_I scale had been originally based on alicyclic and aliphatic reactivities.[30] An exactly analogous procedure can be applied to σ^+ and σ^- to give σ_R^+ and σ_R^-, respectively.[23] Using a relay factor, α, of 0.5, a similar dissection of σ^0 gives σ_R^0.[26]

This separation creates the possibility of a dual-substituent parameter treatment, exemplified by equation (13). A simple treatment would exist if the different σ_R scales were linearly related. Equations of type (13) could then be applied separately to *meta* and to *para* substituents. However, it is unfortunate that collinearity between the different σ_R scales is not very precise.[33] Both ^{19}F NMR[34] and IR[19,35] spectroscopic methods have been used in a direct approach to the calculation of σ_R^0.

$$\sigma_m = \sigma_I + \alpha\sigma_R \tag{11}$$

$$\sigma_p = \sigma_I + \sigma_R \tag{12}$$

$$\log(k/k_0) = \rho_I\sigma_I + \rho_R\sigma_R \tag{13}$$

This treatment has been criticized by Exner[36,37] who considered that inductive effects were more effective from the *para* than from the *meta* position in the ratio of 1.14 to 1 and who has presented revised values of σ_I and σ_R. This scale presents some major conceptual problems, whereas that of Taft has been generally accepted.

A related treatment has been offered by Dewar and Grisdale.[38] In this, σ values for substituent effects in polycyclic aromatic systems can be calculated using equation (14), in which F is a measure of the electrostatic field effect of the substituent, r the distance between the points of attachment of side chain and substituent, M a measure of the mesomeric or resonance effect of the substituent, and q a negative or positive charge produced by the substituent CH_2^- or CH_2^+, respectively, at the point of attachment of the reacting side chain. This interpretation arises from a specific interpretation of electronic effects and the values of F and M are found by application of equation (14) to the *meta* and *para* σ values. Having defined F and M using the benzene system, σ values can be calculated for various polycyclic aromatic and heterocyclic systems. These parameters cannot be considered suitable for general use in multiple regression analysis, but achieve success in the calculation of σ values for polycyclic aromatic compounds, with the obvious exception of *ortho* or similar proximity situations (see Section 18.5.4). A modified treatment,[39] with an additional mesomeric field effect, has been presented.

$$\sigma = (F/r) + Mq \qquad (14)$$

$$\sigma = f\mathscr{F} + r\mathscr{R} \qquad (15)$$

In 1968, Swain and Lupton[40] proposed a separation of electronic substituent effects into a field constant, $\mathscr{F}$, and a resonance constant, $\mathscr{R}$. *All* substituent constants are considered to be a function of these parameters as shown in equation (15), where f and r are field and resonance weighting factors. The proliferation of σ scales was strongly criticized by these authors[40] and this was demonstrated by successfully correlating other σ scales with a combination of σ_m and σ_p. Their scale was set up by making two assumptions. The first was that the effect of 4-substituents on the ionization of bicyclo[2.2.2]octane-1-carboxylic acids (1) is entirely a field effect. This enables a scale of $\mathscr{F}$ values to be set up based on the known 14 values of the latter system. The second assumption is that, for the substituent NMe_3^+, no resonance effect is possible and $\mathscr{R}$ equals 0. This enables equation (15) to be solved for the *para* σ value of the substituent and gives f as 0.56. A consistent set of 42 values can then be derived from the correlation equation of the latter values with *meta* and *para* σ values. Except for a scaling factor (see below), there does not appear to be a serious objection to the derivation of values. More recently Hansch *et al.*[41] re-examined this treatment and indicated that the Swain–Lupton procedure had not correctly scaled $\mathscr{F}$ and $\mathscr{R}$. Using a factor of 1.65 (ρ for the ionization of benzoic acids in 50% w/w ethanol–water at 25 °C), they redefined the treatment and calculated the two constants for 191 substituents on the basis of equations (16) and (17).

$$\mathscr{F} = 1.369\sigma_m - 0.373\sigma_p - 0.009 \qquad (16)$$

$$n = 14, r = 0.992, s = 0.042$$

$$\mathscr{R} = \sigma_p - 0.921\mathscr{F} \qquad (17)$$

However, the Swain–Lupton treatment has been strongly criticized.[33,42-44] The second assumption has been refuted.[42,43] The separation depends crucially on the absence of a resonance effect for NMe_3^+. Furthermore, this substituent is a monopole, whereas the vast majority of substituents studied are dipolar. It has been suggested that through-conjugation effects are very poorly correlated by the Swain–Lupton $\mathscr{R}$ and that σ_R^+ or σ_R^- scales are markedly superior when required.[42,43] Swain[45] defends the use of the field and resonance constants as being greatly preferable to the multiplicity previously suggested. An impressive survey has been presented by Swain *et al.*[46] to support the claim that these scales have universality.

18.5.3.2 Separation of Electronic and Steric Effects

As stated in the introduction to this section, electronic effects may be accompanied by steric effects. In 1952 Taft[30] demonstrated a method for the separation of polar, steric and resonance effects in aliphatic and in *ortho*-substituted aromatic systems. Ingold[47] had in 1930 suggested a method for the separation of electronic and steric effects in ester hydrolysis. Following this, Taft[30] suggested equation (18) to evaluate the inductive effect of a substituent R in the ester RCO_2R', where σ^* is an inductive substituent constant. The rate constants k and k_0 refer to reactions of RCO_2R' and $MeCO_2R'$ as the standard, respectively. The subscripts B and A refer to basic and acidic hydrolysis with the same group R', solvent medium and temperature. The assumptions underlying this relation are as follows: (i) the effects of substituents are the sum of independent contributions from inductive, resonance and steric effects; (ii) the steric and resonance effects are the same in both acidic and basic hydrolyses; and (iii) the polar effects of substituents are very much greater in the basic than in the acidic reactions.

$$\sigma^* = [\log(k/k_0)_B - \log(k/k_0)_A]/2.48 \tag{18}$$

$$\log(k/k_0)_A = E_s \tag{19}$$

The factor of 2.48 was used in an attempt to approximately place σ^* on the same scale as the Hammett σ values. The steric substituent constant, E_s, has been defined from the acidic reaction as shown in equation (19). This enabled the derivation of many σ^* and E_s values for the aliphatic system, as well as σ_o^* and E_s^o for the *ortho*-substituted benzoate system (see Section 18.5.4). These substituent parameters have enabled the application of the three related equations (20), (21) and (22) to be applied to the correlation of reactivity data,[30,48] in which the reaction constants ρ^* and δ measure the response to polar and steric effects. For σ^*, electron-withdrawing substituents are positive and electron releasing substituents are negative, as for Hammett σ values. Since their original definition, the aromatic σ_I and the aliphatic σ^* have been shown to be directly related in the manner shown in equation (23) below. The application of both equations (20) and (21), employing electronic effects, has been successful,[30,48] but the significance of σ^* values for alkyl groups is in doubt.[49,50] The independence of the σ_I scale derived from the reactivity of 4-substituted bicyclo[2.2.2]octane-1-carboxylic acids (1) and the separation of Hammett σ values must be stressed.[30-32,51]

$$\log(k/k_0) = \rho^*\sigma^* + \delta E_s \tag{20}$$

$$\log(k/k_0) = \rho^*\sigma^* \tag{21}$$

$$\log(k/k_0) = \delta E_s \tag{22}$$

$$\sigma_I(X) = 0.45\sigma^*(CH_2X) \tag{23}$$

18.5.4 *ORTHO* AND RELATED PROXIMATE SUBSTITUENT EFFECTS

18.5.4.1 Introduction

The original Hammett equation was considered not to include the effects of *ortho* substitution.[52] This can be considered to derive from the very nature of the substitution, which would be adjacent to the side chain function undergoing the reaction. Various types of proximity effects, such as steric effects, hydrogen bonding, proximity electronic effects, *etc.* can accompany the usual electronic effects discussed previously.

18.5.4.2 Separation of *Ortho* Electronic and Steric Effects

Taft[30] applied the treatment of acid- and base-catalyzed ester hydrolyses to *ortho*-substituted benzoates (see Section 18.5.3.2) to obtain a series of σ_o^* and E_s values (Me as the reference substituent). Equations of the type (20), (21) and (22) have received rather limited use in correlating *ortho* substituent effects with reactivity.[30,48] The unsubstituted members of several series tend to deviate from the correlations and ρ^* sometimes differs markedly from ρ for the same reaction. The similarity between *para* σ values and Taft's *ortho* σ_o^* values has encouraged the use of the former to represent the latter. However, the value of $\log(k_{ortho}/k_{para})$ has been taken as a measure of the *ortho* effect.[48]

Farthing and Nam[53] based a treatment on the ionization constants of *ortho*-substituted benzoic acids. Thus $\log(K_{ortho}/K_0)$ (H as the reference substituent) was defined as the sum of *ortho* electronic and steric terms, *i.e.* σ_E and σ_S, respectively. The electronic term is equated to the effect of the corresponding *para* substituent, $\log(K_{para}/K_0)$ and the steric term is defined as $\log(K_{ortho}/K_{para})$. The influence of *ortho* substituents on reactivity may be analyzed using equations (24) or (25). The original authors[53] correlated the effects of 45 substituents by this method, but it has received little interest. Charton[54] believes that the separation, as described above, is illusory and that

$\log(K_{ortho}/K_0)$ is a function of electronic effects alone.

$$\log k_{ortho} = \rho_E \sigma_E + \rho_S \sigma_S + \log k_0 \tag{24}$$

$$\log K_{ortho} = \rho_E \sigma_E + \rho_S \sigma_S + \log K_0 \tag{25}$$

Charton[54] has considered the *ortho* substituent effect to be separable into inductive, resonance and steric contributions. Equation (26) has been proposed for the correlation of *ortho* effects, where σ_I and σ_R are the inductive and resonance substituents already described (see Section 18.5.3.1) and r_v a 'bulk' steric parameter.[54] The factors α, β and ψ are essentially reaction constants, while h is the intercept. This relation has been used to analyze 269 sets of reactivity data.[54] The importance of the steric bulk term appears to be minor in most cases; but the range of substituents in terms of steric bulk is often very restricted.

$$\log k = \alpha \sigma_I + \beta \sigma_R + \psi r_v + h \tag{26}$$

A treatment of *ortho* substituent effects based on the Swain–Lupton treatment[40] has been developed by Fujita and Nishioka.[55] This treatment attempts to integrate *ortho* substituent with *meta* and *para* substituent effects in the same correlation equation (27). The σ value is the *ortho*, *meta* or *para* substituent constant and that for the *para* is used for *ortho* substituents. Further, the alternative σ^+ or σ^- values can be employed for both the *para* and *ortho* positions. The Taft steric substituent constant E_s^o (see earlier and Section 18.5.3.2) and the Swain–Lupton field constant $\mathscr{F}_{ortho}$ (see Section 18.5.3.1) are used for *ortho* substituents alone and are necessarily equal to zero for both *meta* and *para* substituents. Thus, the *ortho* effect is considered to be composed of an ordinary and proximity effect. The equation has been applied to 44 data sets[55] and is closely related to Charton's method[54] described above. It can be reformulated as a correlation based on $\mathscr{F}$, $\mathscr{R}$ and a steric term, as shown in equation (28). This relation can then be applied to *ortho* substituent effects directly. This treatment would avoid the condition set in equation (27), which assumes the equality of resonance effects for *ortho* and *para* substitution.

$$\log k = \rho \sigma + \delta E_s^o + f \mathscr{F}_{ortho} + c \tag{27}$$

$$\log k = f \mathscr{F} + r \mathscr{R} + \delta E_s^o + h \tag{28}$$

Proximate substitution not of the *ortho* type can be treated in similar ways. Thus, 1,8-naphthalene effects have been studied,[56–58] despite powerful steric effects. Examples of the reversal of normal dipolar substituent effects arising from the precise and unusual molecular geometry in systems such as (**4**)[17] and (**5**)[18] can in principle be accommodated by the use of equations similar to (26) and (28). The reaction constants can then be expected to demonstrate the reversal of the substituent effect, *i.e.* sign reversal, relative to that arising from normal substituents, of the reaction constant.

A number of cautionary and informational points may be usefully made here. Firstly, the unsubstituted member of an *ortho*-substituted or a related series often does not conform in correlations if this substituent is given a substituent constant equal to zero. Secondly, it is, in general, unwise to attempt correlations in which *ortho* and similar proximate substituent effects are treated by simple summing with other nonproximate substituent effects. One reason for this is that dependence on the solvent medium can be markedly different for different types of substitution.[59] Thirdly, it is noticeable that, for distantly transmitted effects, those of *ortho* and *para* substituents appear to be both correlated by *para* σ values *and* to have comparable reaction constants.[60a]

18.5.5 FURTHER ELECTRONIC PARAMETERS

The electronic substituent constants already considered are those based on the Hammett equation, together with its extensions and separations (see Sections 18.5.2, 18.5.3 and 18.5.4). However, it is possible to define further electronic parameters in terms of either physical properties, such as dipole moments (see Section 18.5.7), ionization constants (see Section 18.5.8), hydrogen-bonding capacity (see Section 18.5.9) or substituent constants based on types of processes unlike the classic Hammett type. A comprehensive review[60b] has recently been made of the parametrization of electronic properties, among others, used in drug design. The principal parameters are covered in this chapter in some detail, but a number of others have received brief attention.

In some biological correlations, spectroscopic parameters such as IR stretching frequencies and NMR spectroscopic shifts have been employed. However, it is generally agreed that these are mainly a function of electronic influences, which can be described by the Hammett and Taft substituent constants. While it may be permissible to use such 'parameters' in the absence of relevant substituent constants, it is more consistent to use the established parameters where they are available.

A substituent constant for charge transfer complexation has been defined from the complexation between tetracyanoethylene and aromatic hydrocarbons, aryl and arylmethyl methylcarbamates.[61] The charge transfer, C_T, is defined by the dissociation complex between tetracyanoethylene with substituted and parent molecules K_X and K_H, respectively, as shown in equation (29). It has been noted[62] that steric effects in this model process may not parallel those of a complexation under study and that charge transfer complexations can often be correlated with Hammett σ values.

An interesting development has been the use of substituent constants based on homolytic or radical reactions. However, they do not appear to have found general use in biological correlations. A parameter, E_R, has been defined using as the model process the reaction of a substituted isopropylbenzene with a styrene polymer radical, R· (**11**).[63] Equation (30) shows the relation between E_R and the chain transfer constant of the substituted and unsubstituted cumenes C_X and C_H, respectively. The value of -0.7σ was chosen to place E_R on a basis similar to that of earlier two-parameter relations. This factor apparently tends to remove the 'inductive' contribution and intensifies the resonance character of the parameter. E_R was intended to measure the ability of a substituent to facilitate radical abstraction of a benzyl hydrogen atom and was *not* found to be well correlated by Swain and Lupton's $\mathscr{F}$ and $\mathscr{R}$ constants,[64] as shown in equation (31). However, it is generally agreed that many radical reactions do obey the Hammett equation using σ or σ^+. Alternatively, a constant $\sigma\cdot$ was formulated by Hansch[65] based on substituent effects on homolytic arylation. Even σ^2 has been suggested for the correlation of radical reactions. This use may well be related to the finding that σ^2 is well correlated with E_R, in a limited series, as shown in equation (32).[66] The use of E_R values has been strongly criticized by Kieboom[67] who considers that a variant of the Yukawa–Tsuno equation (see Section 18.5.2.4) could be used. Such a relation is shown in equation (33). This equation allows for *both* exalted $+R$ or $-R$ resonance effects which are independently assessed. The use of σ^2 is subject to similar criticism and is suspect for several reasons.[66,68]

Substituent constants, σ^ϕ, for the groups attached directly to phosphorus have been defined by Kabachnik *et al.*[69,70] using the dissociation constants of $X_1X_2P(O)OH$. They do not appear to have received much attention.

$$C_T = \log(K_X/K_H) \tag{29}$$

$$E_R = \log(C_X/C_H) - 0.7\sigma \tag{30}$$

$$E_R = 0.195\mathscr{F} - 0.005\mathscr{R} + 0.056 \tag{31}$$

$$n = 14, r = 0.690, s = 0.088$$

$$E_R = 0.545\sigma^2 + 0.063 \tag{32}$$

$$n = 13, r = 0.936, s = 0.040$$

$$\log(k/k_0) = \rho(\sigma^n + r^+\Delta\sigma_R^+ + r^-\Delta\sigma_R^-) \tag{33}$$

(**11**)

18.5.6 THE USE OF REACTION CONSTANTS

18.5.6.1 Introduction

The Hammett reaction constant, ρ, has been interpreted as a measure of the susceptibility of the reaction to substituent effects. A reaction which is facilitated by reducing the electron density at the reaction centre (nucleophilic regarding the benzene derivative as the substrate) has a positive value for ρ and one facilitated by increasing the electron density at the reaction centre (electrophilic regarding the benzene derivative as the substrate) has a negative value.[21] This will be so for all reaction constants following the same conventions.

If the reaction process under study is a simple single process, the observed reaction constant can be directly related to it. With complex mechanisms, the observed value can be considered to be the sum of the values of those steps previous to the rate-determining step *plus* that of the latter step.[37]

It should be clearly noted that the Hammett and related linear relations can fail.[37,71] While the use of substituent constants like σ^+ and σ^- or dual parameter systems will often be successful when simple relations do not successfully correlate the data, curvature (*either* a concave upwards *or* downwards) may be a real result. This latter is usually considered to result from a mechanistic change. Such deviations will be particularly difficult to detect in multiparameter approaches such as the Hansch relation.

18.5.6.2 Transmission of Electronic Effects

A major factor controlling the value of the reaction constant is the distance of the substituent from the reaction centre or, as it is often expressed, the length and nature of the transmissive chain.[37] When a reaction centre is separated by a group Z, the reaction constant ρ decreases in the ratio π_Z which is referred to as the transmission factor. π_Z is related to ρ for the parent system and ρ_Z for the system employing transmission by Z as shown in equation (34). Transmission normally, but not necessarily, falls as a chain is extended. The values of π_Z do appear to roughly correspond to the product of the constituent parts. Thus π_Z for CH_2 is about 0.4 and that for $(CH_2)_2$ is 0.2. Replacement of a CH_2 group by heteroatoms gives similar results but conjugated chains transmit more effectively than saturated chains. However, the situation is much more complex than any simple analysis might indicate. As a major part of transmission is by an electrostatic field effect (see Section 18.5.1.1), the use of transmissive factors, in which the inductive route is implicit, cannot be really justified. However, it is often convenient.

$$\pi_Z = \rho_Z/\rho \tag{34}$$

Transmission in polycyclic aromatic and heterocyclic systems has received attention.[72,73] Jaffé[74] made an important contribution in an early study in which equation (35) was proposed to correlate the activity of the substituents R in a ring in which two links Y and Z can transmit the electronic effect to the reaction site X, as shown in (12). The appropriate substituent constant is used for R in respect of the link, *i.e.* for Y 3-substituents are *meta* and 4-substituents are *para*; for Z the reverse applies. This treatment has been used to assess transmission in a number of polycyclic aromatic and heterocyclic systems.[73] A number of the linear free energy relations discussed previously, the Hammett, the dual-substituent parameter and the Dewar–Grisdale equations, have been applied to a large range of polycyclic aromatic and heterocyclic systems.[72] It appears that the dual-substituent parameter approach is the most successful and provides more detailed information conerning the transmission of electronic effects.

$$\log(k/k_0) = \rho_Y\sigma + \rho_Z\sigma \tag{35}$$

(12)

In comparison to Hammett's ρ values, a distinctive interpretation of Taft's ρ_I and ρ_R values (similarly of Swain and Lupton's f and r) has yet to be well developed.[48] The ratio of ρ_R/ρ_I, equal to λ, can be related to the importance of through-conjugation and can be used to detect the inhibition of the transmission of resonance effects. Furthermore, ρ_I and ρ_R can be of opposite sign so that λ will be negative, detecting the opposing stabilization which can be afforded by inductive and resonance effects.

18.5.6.3   Temperature and Medium Effects on Reaction Constants

Reaction constants are dependent on both the temperature and the solvent medium. The isokinetic relationship indicates the dependence of ρ on T, as shown in equation (36).[37] However, in general, ρ appears to vary with $1/T$ and is found to decrease with an increase in temperature. There appear to be cases where the reverse is observed, but complex reaction mechanisms are involved.

The dependence of ρ on the solvent medium is often marked. It had originally been predicted to vary linearly with the reciprocal of the dielectric constant, D.[48] In practice, the few examples studied in detail show a more complex dependence than this. Multiple regression analysis of solvent effects on ρ, as well as ρ^*, appears to be successful in certain cases. Thus, the reaction constants for the reaction of *meta*-substituted benzoic acids with diazodiphenylmethane in aprotic solvents at 30 °C can be correlated[48] as shown in equation (37), where B is the Koppel–Palm parameter (a measure of Lewis basicity) and E_T the Dimroth–Reichardt parameter (a measure of Lewis acidity). In general, the electrostatic field model discussed earlier would predict an increase in reaction constants with a decrease in D. This would be accompanied by an increase in the dependence of the reaction constant on the medium as the transmission through the molecular cavity itself decreases.

$$\rho \propto (1 \; - \; \beta/T) \tag{36}$$

$$\rho \; = \; -0.0053B \; - \; 0.0494E_T \; + \; 4.516 \tag{37}$$

$$n \; = \; 11, \, r \; = \; 0.973, \, s \; = \; 0.092$$

18.5.6.4   Diagnostic Use

An important use of reaction constants is the information they can give to support a postulated mechanism. Both the sign and magnitude of the reaction constant can be used in this regard. These indicate the extent of charge development at the reaction site in passing from the initial to transition site, taking into account the possible transmission *via* links from the substituent to the reaction site. Further, information relating to the mechanism can be obtained by examination of the type of substituent constant required (σ, σ^0, σ^+ or σ^-) or the Yukawa–Tsuno $r^{\pm}$ values found. These will indicate the importance of any cross-conjugation effects in the reaction under study.

Although, in principle, it is possible to calculate reaction constants *a priori*,[75] they are, in practice, discussed by comparison. A few examples should suffice to demonstrate the principles.[37] Thus, the ionization of benzohydroxamic acids ($\rho \; = \; 0.98$) and phenylboronic acids ($\rho \; = \; 2.15$) can be established as involving loss of H^+ from the NH and addition of OH^-, respectively. The ethanolysis of substituted benzoyl chlorides has a positive ρ value of 1.57, which indicates a bimolecular, rather than unimolecular, pathway; *cf.* the solvolysis of *t*-cumyl chlorides (S_N1) having ρ equal to -4.54. Reaction constants can be used to distinguish between the acid-catalyzed hydrolysis of substituted isopropyl benzoates, having a ρ value of 1.99 ($A_{AL}1$), and of substituted methyl benzoates, having a ρ value of -3.5 ($A_{AC}1$). The large value of 2.1 observed for the alkaline hydrolysis of methyl (3- or 4-substituted benzoyl)benzoates indicates a pathway with rate-determining attack at the benzoyl carbonyl group, rather than the benzoate carbonyl group.[76] Transmission of the substituent effect to the latter group can be shown in model reactions to predict a ρ value of about 0.7 for direct attack. The large value of ρ, equal to 2.14, in the elimination reaction of substituted 2-phenylethyl bromides indicates a transition state with a significantly developed carbanionic character at the 2-carbon.

The Jaffé separation,[74] shown in equation (35), has been used in the elucidation of the mechanism of hydrolysis of aspirins.[77] The method enables the electronic effect to be partitioned in its influence on the carboxylate and acetate group. This indicates hydrolysis by general base catalysis, rather than general acid/specific base catalysis or a nucleophilic pathway.

Such definite interpretations are more difficult to establish in the correlation of biological activity; but they will be increasingly attempted (see Section 18.5.10).

18.5.7   DIPOLE MOMENTS

Electronic effects in drug–receptor interactions can be considered to be represented by the electric dipole moments, μ. Thus, certain noncovalent interactions between drugs and receptors arise from interactions between charges, between charge and dipole and between dipoles. Dipole–charge and dipole–dipole interactions are dependent both on the orientation of the dipoles and the distances between charges and/or partial charges. Thus, the energy of interaction between an ion and a dipole can be approximated as shown in equation (38), where e is the magnitude of the charge, μ the dipole moment (equal to the product of dipolar partial charges and their separation), r the distance between the charge and the centre of the dipole, D the dielectric constant and θ the angle between the line joining the charge and the middle of the dipole and the line between the ends of the dipole. The energy of interaction between two dipoles, μ_1 and μ_2, is shown in equation (39), where r is the distance between the dipoles and θ and ϕ the angles between the lines joining the middles of the dipoles and the lines between the ends of the dipoles. In simple studies of drug–receptor interactions it can be assumed that the receptor remains unchanged and only the relevant property of the drug is considered.

$$E = (e\mu\cos\theta)/Dr^2 \tag{38}$$

$$E = (2\mu_1\mu_2\cos\theta\cos\phi)/Dr^3 \tag{39}$$

A reasonably clear distinction can be made between the two uses of dipole moments in correlation studies. The first is the use of substituent group dipole moments to characterize the substitution. The second is the use of molecular dipole moments which are a function of *all* component dipole moments in a molecule. Group dipole moments are functions which are thermodynamically suitable for use in LFER and, as such, are vector quantities, with both additive and constitutive properties. Molecular dipole moments are similarly suitable, again being vector quantities, but are necessarily much more complex. They may be the resultant of several group dipole moments and can be very complicated in their structural definition.

In most published QSAR studies, the electronic parameters used are those derived from the Hammett and Taft treatments, *i.e.* σ, σ^*, *etc.* While it might be considered that the later treatments encompass dipole moment effects, it can be suggested that dipole–charge, dipole–dipole and related interactions are better modelled by direct use of such effects. Lien *et al.*[78] have demonstrated the complex nature of the relationships between substituent constants and group dipole moments. Previously some limited relationships of this type have been considered to exist. An analysis of 114 substituent groups for which μ and both σ_m and σ_p were available gave the results shown in equations (40) and (41), where μ is the group dipole moment.[78] The correlations can only be considered to be poor, with only about half of the variance of the data being accounted for. This must result from the significantly different type of measurement studied. The group dipole moments are essentially vector in nature, while the σ values determine the received effect at a distant site of a composite polar effect. A plot of μ *versus* σ_m or σ_p shows a large scatter with various subsets of the substituents showing utterly different slopes and intercepts. However, such subgroups have no distinct structural relationship and any subdivision will thus not be helpful. More limited series can indicate intercorrelations between μ and σ_p.[68] In physical organic studies, various physical properties have been found to be directly related to dipole moments.[78]

$$\mu = 15.99\sigma_m - 0.53 \tag{40}$$
$$n = 114, r = 0.749, s = 1.279$$

$$\mu = -3.65\sigma_p - 1.27 \tag{41}$$
$$n = 114, r = 0.706, s = 1.367$$

A collection of more than 300 aromatic group dipole moments has been published.[78] The given magnitudes of these group dipole moments of substituents are equal to those of the corresponding monosubstituted benzene. The application of dipole moments in QSAR is very much less common than that of polar substituents constants such as σ. It has been reported that μ^2 gives better correlations than μ.[79] However, this may simply arise from a narrow range of μ values being used. The justification of the use of μ^2 relates to equation (42), governing the Keesom or dipole–dipole interactions, in which the subscripts D and R refer to the drug and receptor.[79] Equation (42) relates to the average interaction for all orientations, whereas equation (42), which is closely related to equation (39), applies to the most favourable state.

$$E = (2\mu_D^2\mu_R^2)/(3r^6DkT) \tag{42}$$

18.5.8   IONIZATION CONSTANTS

18.5.8.1   Introduction

Ionization is a function of the electronic structure of an organic drug molecule. The pK_a of a drug is very important in determining its biological activity. It can influence this by affecting either the transport to or the reactivity of a drug at its site of action. Albert[80] was the first to clearly recognize the connection between ionization and biological activity, demonstrating the relationship between the antibacterial activity and the protonation of acridines. He recognized the general importance of the interaction of ionic forms of drugs with receptors having ionic groups of the opposite sign.

18.5.8.2   The Definition of pK_a

For a weak acid, HA, the ionization is described in equation (43) and the ionization constant, K_a, in equation (44). In the latter relationship, the terms in brackets are the concentration or, accurately, the thermodynamic activities of the species shown. The Henderson–Hasselbalch equation (45) can be derived from equation (44). In a similar manner, the ionization of a weak base, B, can be described by equations (46) and (47). However, the ionization of a base can also be written as in equation (48) and this allows the derivation of pK_b values. Values of pK_b can be converted to pK_a values of the conjugate acid by use of equation (49). pK_w is the ionic product of water and equals 14.0 at 25 °C. The Henderson–Hasselbalch equation allows the calculation of the percentage ionized for both weak acids, HA, and weak bases, B, as shown in equations (50) and (51), respectively.

$$HA + H_2O \rightleftharpoons H_3O^+ + A^- \qquad (43)$$

$$K_a = \frac{[H_3O^+][A^-]}{[HA]} \qquad (44)$$

$$pH = pK_a + \log[A^-]/[HA] \qquad (45)$$

$$BH^+ + H_2O \rightleftharpoons H_3O^+ + B \qquad (46)$$

$$pH = pK_a + \log[BH^+]/[B] \qquad (47)$$

$$B + H_2O \rightleftharpoons BH^+ + OH^- \qquad (48)$$

$$pK_a + pK_b = pK_w \qquad (49)$$

$$\text{For HA: } \% \text{ ionized} = \frac{100}{1 + \text{antilog}(pK_a - pH)} \qquad (50)$$

$$\text{For B: } \% \text{ ionized} = \frac{100}{1 + \text{antilog}(pH - pK_a)} \qquad (51)$$

The methods by which ionization constants can be measured have been reviewed[81] and a widely used text is available.[82]

18.5.8.3   Temperature and Solvent Effects

Equation (52) governs the variation of pK_a with temperature, T, in which $\Delta S°$ is in $J\,deg^{-1}\,mol^{-1}$.[83] Because of the pK_a and $\Delta S°$ values involved, simple carboxylic acids vary only slightly and phenols fall slightly with a temperature rise. However, the entropy change for the reaction shown in equation (46) is much less than for carboxylic acids and phenols as the number of ions and charges do not change. For the pK_a values of organic bases, equation (53) applies and has an uncertainty of ± 0.004 for $-d(pK_a)/dT$.

$$-d(pK_a)/dT = [(pK_a + 0.052\Delta S°)T] \qquad (52)$$

$$-d(pK_a)/dT = [(pK_a - 0.9)T] \qquad (53)$$

The effect of solvent on pK_a values depends on the type of ionization.[83] Thus, for acids HA ionizing according to equation (43) to give two ions from a neutral molecule, the pK_a values are very sensitive to the medium and decrease markedly with the dielectric constant of the solvent, D. For acids BH^+, ionizing according to equation (46), without any change in the number of ions, the pK_a values are insensitive, decreasing only slightly with decreasing D. Water is an obvious and a very suitable solvent for such measurements. However, problems of extremely low solubility often occur. Although 50% aqueous ethanol has been used widely, the advantages of 80% aqueous 2-methoxyethanol are, in comparison, significant. Extensive data[84] for this system have been collected and such measurements appear reliable. With care, extrapolation to aqueous solution can be made reliably from such nonaqueous systems. However, general relations for these conversions do not exist.

18.5.8.4 Ionization and Transport

Drug transport often represents a compromise between the solubility of the drug and its transport across barriers. The state of ionization of the drug will also affect solubility markedly because of the differential solubility of the ionized and unionized forms, the former usually being more soluble. Many drugs are either weak organic acids or weak organic bases. They will be partially dissociated and exist in both forms, depending on the pH of the medium.

The drugs are often bound to receptors, *etc.* in the ionized form. Organic molecules with pK_a values in the range 6 to 8 have the advantage, under physiological conditions of pH equal to about 7, of having both ionic and neutral species present.

Partition through membrane barriers is usually considered to be associated with the unionized molecules. Partition itself is strongly affected by ionization and this has been considered by Leo *et al.*[85] The use of partition coefficients obtained with buffered aqueous solutions, such that only a single species is effectively present, is strongly recommended. The use of distribution coefficients, $\log D$, has been described for the study of ionizable compounds and their biological activity.[86] When an ionizable compound is equilibrated in a two-phase system, both organic and aqueous, its concentration in the organic phase is not determined by its partition coefficient, P, alone, but also by the pH of the aqueous phase. The concentration of the compound in the organic phase is directly related to $\log D$ and is given by equations (54) and (55), where the pK_a is that of the compound and the pH that of the aqueous phase. $\log D$ and $\log P$ are identical when the drug is completely unionized. However, as $pH - pK_a$ becomes large for acids and $pK_a - pH$ for bases, the expressions reduce to equations (56) and (57). This emphasizes the care that must be taken in interpreting the dependence of biological activity on the pK_a values of a series of congeners. This is especially important as pK_a values themselves are normally a function of electronic factors which are used directly in correlation analysis.

$$\log D_{\text{acids}} = \log P + \log[1/(1 + 10^{pH - pK_a})] \tag{54}$$

$$\log D_{\text{bases}} = \log P + \log[1/(1 + 10^{pK_a - pH})] \tag{55}$$

$$\log D_{\text{acids}} = \log P + pK_a - pH \tag{56}$$

$$\log D_{\text{bases}} = \log P - pK_a + pH \tag{57}$$

18.5.8.5 Electronic and Related Effects

The modification of pK_a values by structural factors is comparatively well understood. These factors are predominantly the electronic effects already discussed (see Section 18.5.1), together with intramolecular hydrogen bonding, steric and stereochemical factors.

There are a number of excellent reviews of pK_a values. The pK_a values of about 800 compounds have been listed in order of their decreasing acidity, but without comment.[87] Very extensive compilations of the aqueous dissociation constants of organic acids[88,89] and organic bases[90] have been made. The pK_a values of heterocyclic compounds have been reviewed by Albert.[91] The texts describing the determination[82] and the prediction[83] of pK_a values have useful tables of values. The only comprehensive review of pK_a values in mixed solvents are those by Simon *et al.*[84] for 80% aqueous 2-methoxyethanol.

Electronic effects on ionization constants can be both correlated and predicted by use of the LFER discussed earlier (see Sections 18.5.2, 18.5.3 and 18.5.4).[83,92,93] Such relations would be expected as the reactions used in defining such equations are mainly ionization reactions. These reactions can be correlated and predicted very successfully as long as the relevant relationships are employed, as discussed below.

Effects other than electronic can be important in determining ionization constants.[83] Intramolecular hydrogen bonding can have a significant effect on the ionization of acids, Thus, powerful acid strengthening, arising from intramolecular hydrogen bonding, is observed in certain carboxylic acids, such as salicylic, 2,6-dihydroxybenzoic and maleic acids for pK_{a1}. The reverse can be noted for the second ionization pK_{a2} for the respective monoanionic acids. Steric effects can be important in specific cases. Both primary and secondary effects can occur. Primary steric inhibition of solvation of the ionic form can result in acid-weakening effects for carboxylic acids, such as triphenylacetic acid, with *very* bulky groups flanking the carboxylate groups. Phenols and primary amines appear less susceptible to this effect, but, in extreme cases, phenols can suffer acid weakening and amines base weakening. Secondary steric effects by bulky groups preventing coplanarity of groups with aromatic rings can cause significantly decreased resonance effects. This results in acid strengthening in compounds such as *o-t*-butylbenzoic acid and acid weakening in 3,5-dimethyl-4-nitrophenol, compared to the values expected in the absence of such a secondary steric effect. Stereochemical and conformational factors can also be very important in determining pK_a values.[93,94]

Statistical factors must be considered when a polybasic acid has a number of groups each of which has an equal probability of losing a proton. The observed pK_a will then be reduced by $\log n$, where n is the number of identical groups, compared to a closely related monobasic acid. The second proton loss requires a correction of $\log[(n - 1)/2]$ and so on. Thus, for succinic acid, pK_{a1} is then 0.3 units greater and pK_{a2} 0.3 units less than would be expected. This arises from the probability factor of having multiple sites of proton loss and gain. In correlation studies, statistical factors must be corrected for before calculations are made. In predictions, the corrections must be made after.

Tautomerism, including the formation of zwitterions, will strongly affect pK_a values. The observed pK_a can be related to the true value of the more acidic tautomer pK_{aT} and the tautomeric equilibrium K_T[95] as shown in equation (58).

$$pK_{aT} = pK_a - \log(K_T + 1) \tag{58}$$

The Hammett equation and its extension (see Section 18.5.2) can be used to correlate and predict the pK_a values of *meta* and *para*-substituted benzoic and related acids. Additivity of substituent effects is often observed.[96] Exceptions do occur, which often arise from steric interactions.[96] The treatment can be extended to phenols and anilines, in particular. The problem of *ortho* substitution is complex (see Section 18.5.4); but appropriate 'apparent' *ortho* σ values for specific acid systems have been tabulated.[83] The pK_a values of polycyclic aromatic and heterocyclic systems can be treated by use of both the Hammett and dual substituent parameter equations (see Sections 18.5.2 and 18.5.3). Thus, substituted pyridines and quinolines have been successfully correlated. Substituent constants can be derived and assigned for ring heteroatoms and for fused ring systems.

The pK_a values of substituted aliphatic carboxylic acids and bases have been correlated and predicted by the Taft equations[83] (see Section 18.5.3). Additivity of substituent effects is usually observed, but steric effects on the ionization of carboxylic acids has been quantitatively related to the Taft steric substituent constant, E_s.[97]

Many diverse acid and base systems have been analyzed in terms of the electronic effects of substituents, including benzenethiols, benzeneseleninic acids, arenearsonic acids, arenephosphonic acids, benzenesulfonamides, alcohols, hydrated aldehydes, thiols, azoles, carbon acids, tropolones, acridines, purines and pyridazines.[92,93]

18.5.9 HYDROGEN BONDING

18.5.9.1 Introduction

Hydrogen bonding is an important interaction that can change the physical properties of organic molecules. Intramolecular and intermolecular hydrogen bonds can occur in biological systems; the former generally being preferred over the latter. For intramolecular hydrogen bonding, the ring size is important in determining stability, with five- and six-membered rings being favoured. Hydrogen bonding has long been recognized as being very important in interactions between organic molecules in biological systems, especially drug actions.[98]

The more important hydrogen bond donor groups are OH and NH, with SH and CH (having strongly electron-withdrawing substituents) being capable of such interactions. The acceptor groups can be extremely varied, including PO, SO, CO, N, O, S and F. Even an electron rich π electron system can participate as an acceptor. Some groups are amphiprotic with both acceptor and donor ability, such as NH and OH. The distance between the electronegative atoms in the hydrogen bond is usually in the range 2.5–2.7 Å and heats of formation in aprotic solvents are, in general, in the range 8–24 kJ mol^{-1}.

18.5.9.2 Hydrogen-bonding Parameters

The use of hydrogen-bonding parameters to assess the importance of this factor as an independent variable appears to have been very rare and this concept seems neglected. There are some reasons for this. The hydrogen-bonding effect within a series of compounds having a common acceptor or donor group can be correlated with the electronic effects of substituents using either the parameters previously discussed (see Section 18.5.2) or other physicochemical properties, such as pK_a values. The latter are often linearly related to hydrogen bonding within limited series.[99] Hydrogen bond strengths can be directly related to the proton chemical shifts and the latter used as a parameter.[100] Such ^{1}H NMR spectroscopic and IR shifts have been related to scales of hydrogen bond complex formation as discussed below. The introduction of a hydrogen-bonding parameter was attempted by Hansch *et al.*[101] and Fujita *et al.*[99] This resulted in the introduction of an indicator variable term to designate hydrogen-bonding capacity. This was interpreted as 'extra' hydrogen bonding to that defined by the 1-octanol–water partition model.

Hansch and Leo[102] have tabulated substituent parameters for hydrogen bonding as both acceptor (H accept) or donor (H donor), designating groups as 0 or 1 for acceptor and 0 or 1 for donor ability. While such a scale will necessarily be rather insensitive and arbitrary, it does provide a method of classification and a set of indicator variables.

A more promising approach would appear to be the use of scales of hydrogen bond complex formation. Taft *et al.*[103,104] have established a hydrogen bond basicity scale based on complex formation of a base, B, with 4-fluorophenol in CCl$_4$ as shown in (13). The scale is defined as the pK_{HB} value which is $\log K_f$ for the complex in (13). A wide range of bases have been studied successfully with various OH and NH acids according to the relationship shown in equation (59).[103,104] pK_{HB} values do not correlate with aqueous pK_a values, except within a series having a common functional centre. These parameters do correlate a range of hydrogen-bonding interactions. A further analysis of 'hydrogen-bonded complexes' into hydrogen bond acceptor *and* donor scales indicates a more complex situation, which still awaits successful resolution.[105–108] Abraham *et al.*[109a] have studied the hydrogen bond complexation of 72 acids with *N*-methylpyrrolidinone in MeCCl$_3$ and suggest the use of $\log K$ for this reaction as a scale of hydrogen bond acidity. A general scale of hydrogen bond acidity *and* basicity is still awaited. The separations attempted by Swain *et al.*[107] and by Kamlet and Taft[105] are of solvent effects. Their direct application to hydrogen-bonding substituent effects in drugs cannot be recommended, even neglecting the controversy regarding their correctness.[106,107]

$$\log K = m(pK_{HB}) + c \tag{59}$$

$$F\!\!-\!\!\langle\bigcirc\rangle\!\!-\!\!OH \;+\; B \;\underset{}{\overset{K_f}{\rightleftharpoons}}\; F\!\!-\!\!\langle\bigcirc\rangle\!\!-\!\!OH\cdots B$$

(13)

Seiler,[109b] using a least-squares analysis, has derived a relationship between the difference of the logarithms of the partition coefficients in octanol/water and cyclohexane/water ($\Delta \log P$) and the sum of increments I_H each characteristic of a molecular fragment. To a first approximation I_H may be regarded as a measure of the hydrogen-bonding ability of the fragment, and a table of I_H values has been published.[109b] In an investigation of the brain penetration of H$_2$ antagonists, Young *et al.*[109c] found a significant correlation between the logarithms of the equilibrium brain/blood concentration ratios in the rat and the partition parameter $\Delta \log P$. This model suggests that brain penetration can be increased by minimizing hydrogen-bonding ability as measured by $\Delta \log P$.

18.5.10 ELECTRONIC EFFECTS IN QSAR

18.5.10.1 Introduction

Although many workers had recognized that electronic effects often influenced the activity of drugs, the situation was transformed by the advent of the Hansch and related treatments.[2] A number of important reviews of QSAR have been published.[62,68,110,114] It is only appropriate here to consider the *quantitative* role of electronic effects in such relations and to leave the relations themselves, as well as other effects, to other authors in this series. The intention here is to illustrate electronic effects either where they can be related to the modes of action, *etc.* of drugs or where they are very significant in controlling the activity of drugs.

18.5.10.2 Electronic Effects Alone and Accompanied by Partition or Steric Effects

The successful application of relationships involving *only* electronic influences, *i.e.* the Hammett and Hammett-type equations, are very much fewer than those in which electronic effects accompany partition and/or steric factors. Thus, in 1953 Jaffé[22] had found only one example and Hansch[110] less than a half-dozen distinct examples in 1969. Since that time many others have been discovered. Because of the *relative* simplicity of enzymic reactions, a number have been considered to be controlled by electronic factors alone.[111-112] However, many of these examples are based on either qualitative evidence or series which are very limited or poorly selected for the variation in their substitution.

A significant study by Fukuto *et al.*[110,115] showed that both the inhibition of fly-head cholinesterase by *para*-substituted diethyl phenyl phosphates and the toxicity of *meta*- and *para*-substituted diethyl phenyl phosphates towards house-flies could be correlated by Hammett-type equations. Equations (60) and (61) for the enzymic I_{50} and toxicity LD_{50} values show almost identical ρ values, which strongly supports their contention that the activity of the diethyl phenyl phosphates towards house-flies is a reflection of cholinesterase inhibition. Further, the ρ values found suggest that it is the step involving phenoxide loss that is involved in the activity-determining step.

$$\log(1/I_{50}) = 2.37\sigma^- + 4.38 \tag{60}$$

$$n = 6, r, = 0.98, s = 0.29$$

$$\log(1/LD_{50}) = 2.28\sigma^- - 1.28 \tag{61}$$

$$n = 14, r = 0.97, s = 0.28$$

The effects of 3- and 4-monosubstitution and 3,5-disubstitution on the phenyl leaving group of the papain-catalyzed hydrolysis of phenyl hippurates (**14**) have been studied.[116] The K_m values have been correlated as shown in equation (62) with electronic, partition and steric terms. The partition and steric terms refer to the 3- and 4-positions as subscripted. A study of the 'uncatalyzed' rate of hydrolysis of the phenyl hippurates at pH 6.0 and 8.0 gave ρ values equal to 1.91($\pm$ 0.13) and 1.66. In the enzymic hydrolysis, the binding step (K_m) shows a significantly smaller role for electronic effects than the 'uncatalyzed' hydrolysis and k_{cat} is essentially constant for all the hippurates under study. The catalysis at the enzyme site must involve a transition state close to the initial state in the structure. The 'push–pull' mechanism does not appear to require much in the way of electronic assistance in breaking the C–O linkage of the ester in the catalytic step. A later study[117] of the hydrolysis of substituted phenyl N-methanesulfenyl glycinates (**15**) catalyzed by papain yielded equation (63). The subscripts have the same significance as in the previous study. The ρ value is close to that observed for the hippurates as described above. A recent investigation[118] of the subtilisin-catalyzed hydrolysis of 31 substituted phenyl hippurates gave equation (64), with the subscripts having the same significance as before. The K_m values are shown to depend on electronic, partition and steric factors. The ρ value of about 0.4 is close to that observed for papain, ficin, actinidin and bromelain hydrolysis of similar esters. Thus, the electronic effects on the leaving group in the hydrolysis by these five enzymes are almost identical and indicate the close relationship between the pathways of these enzymes.

$$\log(1/K_m) = 0.57(\pm 0.20)\sigma + 0.61(\pm 0.29)MR_4 + 1.03(\pm 0.25)\pi_3 + 3.80(\pm 0.17) \tag{62}$$

$$n = 25, r = 0.907, s = 0.208$$

$$\log(1/K_m) = 0.55(\pm 0.20)\sigma + 0.46(\pm 0.11)MR_4 + 0.61(\pm 0.09)\pi_3 + 2.00(\pm 0.12) \tag{63}$$

$$n = 32, r = 0.945, s = 0.178$$

$$\log(1/K_m) = 0.41(\pm 0.08)\sigma + 0.20(\pm 0.08)MR_4 + 0.19(\pm 0.07)\pi_3 + 3.81(\pm 0.06) \tag{64}$$

$$n = 31, r = 0.980, s = 0.085$$

(14)   **(15)**

The mechanism of the action of dopamine β-monooxygenase has been investigated[119] using a series of *para*-substituted phenylethylamines. Both electronic and steric effects appear to control the rates of C–H bond cleavage, k_5, and of product dissociation, k_7, as shown in equations (65) and (66). The negative ρ value for the cleavage step eliminates the possibility of a carbanion intermediate and is in better accord with the formation of a radical than of a carbocation intermediate. The positive ρ for the product dissociation step appears to relate to the loss of product from an enzyme-bound Cu^{II}–alkoxide complex.

$$\log k_5 = -1.48(\pm 0.23)\sigma_p - 0.81(\pm 0.10)V_w + 2.93(\pm 0.12) \tag{65}$$

$$n = 8, r = 0.986$$

$$\log k_7 = 1.41(\pm 0.24)\sigma_p^0 + 0.24(\pm 0.08)V_w + 1.30(\pm 0.08) \tag{66}$$

$$n = 8, r = 0.938$$

A comprehensive review of structure–activity relationships of dihydrofolate reductase (DHFR) inhibitors has recently been made.[120] QSAR for human and chicken DHFR inhibitors showed the prime importance of partition and steric factors. However, the activities of the 3-substituted triazines **(16)** have a contribution from electronic factors with ρ equal to about 0.8. Very unusually, if not uniquely, in such studies, 4-substituents show *no* such contribution. It appears conceivable that the dipolar interaction between the ligand and enzyme can operate for 3-substituents, but not for 4-substituents. This could arise from the observed steric effect of the 4-substituents moving the aryl side chain on the surface of the enzyme. A similar electronic contribution is important for human and chicken DHFR inhibition by substituted benzylpyrimidines **(17)**, giving rise to a ρ value of about 0.3. However, a quite different behaviour occurs with bacterial DHFR inhibited by the same series of inhibitors. The latter activity has an electronic contribution depending on σ_R with a reaction constant of about -0.3. These results could be employed in the design of selectivity for such inhibitors.

(16)   **(17)**

Four sets of amidine inhibitors of trypsin have been studied by QSAR.[121] Studies of the inhibition of both bovine and human trypsin by 3- and 4-substituted benzamidines **(18)** showed control by both electronic and steric factors. Using σ values rather than σ^+, a ρ value of $-0.8(\pm 0.4)$ was found. A quite different series of inhibitor amidines **(19)** also indicated control by a combination of polar and steric effects. However, the ρ value for substitution in the benzoate group was found to be $-1.5(\pm 0.5)$. It is not clear why such significant dependence occurs; but it may be an effect on the hydrophobicity of the aromatic rings. Coats[122] has examined the benzamidine inhibition of thrombin, plasmin, trypsin and complement. Equation (67) shows the relation found for thrombin inhibition by 3- and 4-substituted benzamidines. This author considers that increased binding with electron donation would be expected since this would tend to stabilize the protonated amidine, which is apparently the major requirement for binding here. However, in general, there do appear to be different requirements both for 3- and for 4-substituted amidines, as well as for the different enzymes. The inhibitory activity of benzylpyridinium ions (**20**) to complement, which is a complex mixture of serum proteases, has been studied[123,124] and both electronic and partition factors have been found to be important. However, while partition terms are important for both X and Y in (**20**), electronic factors are required for *only* X, with ρ^+ equal to 1.0 (± 0.3). The original study[123] of 69 derivatives was expanded to a study[124] of 132 without significant changes in the dependence on structural terms. An indicator variable is required for the 2-SO_2F group as Y in (**20**) and equals $0.7(\pm 0.1)$. This group can irreversibly cause inhibition by sulfonation of a serine hydroxyl group. The dependence on σ^+ appears to relate to an enhancement of the anchoring of the inhibitor by the positively charged nitrogen near to the critical serine hydroxyl group.

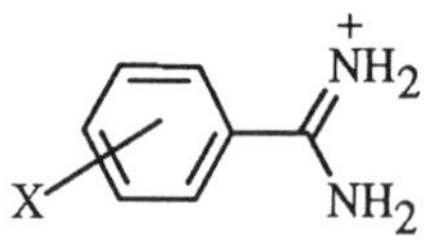

(18)

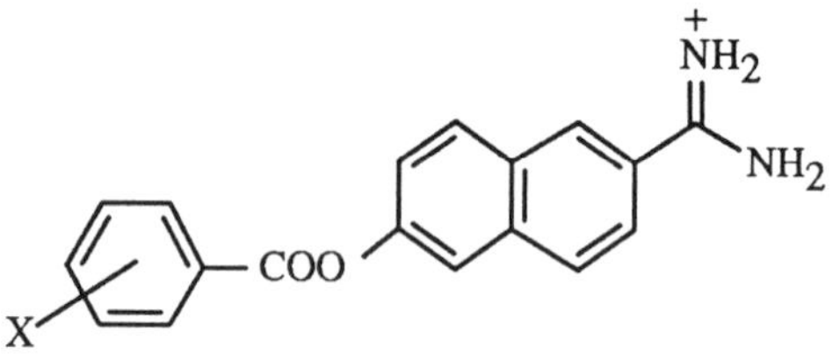

(19)

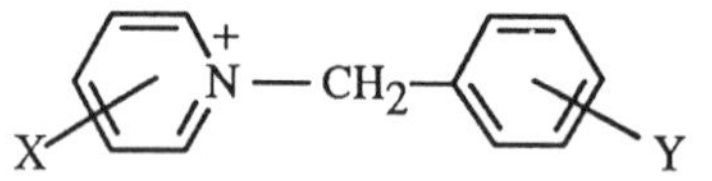

(20)

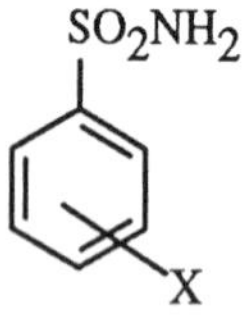

(21)

$$\log(1/K_i) = -0.82(\pm 0.31)\sigma + 0.30(\pm 0.07)\pi + 0.28(\pm 0.15) \tag{67}$$

$$n = 39, r = 0.92, s = 0.29$$

An early study[125] of inhibition of bovine carbonic anhydrase by substituted benzenesulfonamides (**21**) resulted in the correlation shown in equation (68). The role of electronic effects is considered to be dominant, compared to that for partition effects. A more recent and comprehensive study[126] of the inhibition of human carbonic anhydrase by benzenesulfonamides is shown to be governed by equation (69), where the indicator variables I_1 and I_2 refer to 3- and 2-substitution, respectively. There is considerable evidence that the anionic form of the sulfonamide is bound to the enzyme. Hansch *et al.*[126] examined the earlier results for the inhibition of bovine carbonic anhydrase by a well-selected group of benzenesulfonamides to give equation (70). The ρ value for the ionization of the sulfonamides used was found to be $0.86(\pm 0.14)$, which is very close to that observed in equation (70) (see Section 18.5.10.4).

$$pK_i = 0.886\sigma + 0.259\log P + 5.314 \tag{68}$$

$$n = 19, r = 0.923, s = 0.247$$

$$\log K = 1.55(\pm 0.38)\sigma + 0.64(\pm 0.08)\log P - 2.07(\pm 0.22)I_1 - 3.28(\pm 0.23)I_2 + 6.94(\pm 0.18) \tag{69}$$

$$n = 29, r = 0.991, s = 0.204$$

$$\log(1/K_i) = 0.80(\pm 0.22)\sigma + 0.27(\pm 0.18)\log P + 0.33(\pm 0.14) \tag{70}$$

$$n = 16, r = 0.968, s = 0.168$$

A series of phenols have been examined as reversible inhibitors of β-glucosidase.[127] The correlation of the inhibitory activity gave rise to equation (71). This is equivalent to a Brønsted coefficient of 0.26 (the ρ value for ionization of phenols is 2.23). The ρ value found in this study can be contrasted with that of 0.66 found for the binding of *meta-* and *para-*substituted phenyl glucosides to the enzyme.[128] As phenols bind much more tightly than phenoxides, the reaction constant in equation (71) suggests a hydrogen-bonding interaction and is in accord with values found for phenols with a variety of acceptors. A study[129] of the inactivation of *N*-arylhydroxamic acid *N,O*-acyltransferase (AHAT) by 7-substituted *N*-hydroxy-2-acetamidofluorenes gave a correlation with electronic and steric factors as shown in equation (72). These results support the concept that a positively charged species is involved in the inactivation of AHAT by this series of compounds. The inhibition of rat liver alcohol dehydrogenase (ADH) both *in vitro* and in isolated hepatocytes by 4-substituted pyrazoles **(22)** has been studied.[130] Correlations for the activities are shown in equations (73) and (74), having dependence on both electronic and partition terms. The σ_m values are used because the 4-substituents are considered to be '*meta*' to both nitrogens of the pyrazole. The identical ρ values for *in vivo* and *in vitro* show that both activities are governed by precisely the same electronic factors. A comparison of activities of these inhibitors against human π-ADH indicates that its active site differs from those in the other alcohol dehydrogenases. The negative ρ value found in equations (73) and (74) appears to relate to the effect of the substituents on the binding of pyrazoles to the active site zinc atom and the reaction of pyrazoles with NAD^+.

$$\log(1/K_i) = 0.57\sigma^- + 1.6 \tag{71}$$

$$n = 14, r = 0.978$$

$$\log k_i = -1.54(\pm 0.74)\sigma - 0.64(\pm 0.39)0.1MR + 0.46(\pm 0.44) \tag{72}$$

$$n = 10, r = 0.90, s = 0.29$$

$$\log(1/K_i)_{(in\ vitro)} = -1.80\sigma_m + 1.22\log P + 4.87 \tag{73}$$

$$n = 14, r, 0.985, s = 0.316$$

$$\log(1/K_i)_{(isolated\ hepatocytes)} = -1.80\sigma_m + 1.27\log P - 0.20(\log P)^2 + 4.75 \tag{74}$$

$$n = 14, r = 0.971, s = 0.320$$

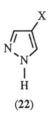

(22)

The activities of several series of monoamine oxidase (MAO) inhibitors have been examined and electronic effects shown to be important.[131] A typical relation is shown in equation (75) for the correlation of the activity of a series of substituted pargylamines **(23)** as inhibitors of rat liver mitochondrial MAO. The σ values in equation (75) refer to X in **(23)** and the E_s values are used to measure the steric effects of 4-substituents alone. A number of such inhibitor series give similar ρ values with a mean value equal to about 1.3. This is considered to be characteristic of the substrate–receptor interaction. The study of Fujita[131] has been criticized by Martin *et al.*,[132] who suggest that equation (76) is more meaningful. This relation correlates the activities of 47 propynylamines as MAO inhibitors with their pK_a values, π and an indicator variable I for *ortho* substitution. The pK_a values of the propynylamines were shown to give a good Hammett correlation for a more limited series with ρ equal to -0.82. Thus, the analysis of the effects of the pK_a values on potency indicated an effect in addition to that of changing the fraction ionized (see Section 18.5.10.5). The dependence of the activity on the electronic effects of the substituents has been related to the mechanism of action, *i.e.* the metabolism of the inhibitor. Equation (76) also can be used to derive an optimum pK_a value, which quite unusually indicates an optimum for electronic effects.

$$pI_{50} = 1.19(\pm 0.73)\sigma + 0.39(\pm 0.39)\pi + 0.76(\pm 0.43)E_s + 5.55(\pm 0.59) \qquad (75)$$

$$n = 11, r = 0.937, s = 0.271$$

$$pI_{50} = 4.38(\pm 1.38)pK_a - 0.35(\pm 0.10)pK_a{}^2 + 0.25(\pm 0.19)\pi + 1.02(\pm 0.45)I - 7.48(\pm 4.66) \qquad (76)$$

$$n = 47, r = 0.87, s = 0.58$$

A study by Seydel *et al.*[133] correlated the tuberculostatic activities of a series of 2-substituted isoniazid derivatives (**24**) with the pK_a values of the corresponding 2-substituted pyridines. The latter values themselves can be successfully related to σ_I values. Soon afterwards these activities were shown[134] to be related to electronic and steric factors according to equation (77). The indicator variable I accounts for the effects of β-methyl branching in X. Interestingly, the effects of the electronic parameters $\mathscr{F}$ and $\mathscr{R}$ are *in opposition* to each other.

$$\log(1/C) = -3.25(\pm 0.82)\mathscr{F} + 1.19(\pm 0.82)\mathscr{R} - 2.32(\pm 0.83)V_w - 0.87(\pm 0.54)I - 0.20(\pm 0.50) \qquad (77)$$

$$n = 19, r = 0.938, s = 0.325$$

(23)   (24)

The activities of a number of anticancer agents have been successfully correlated to show electronic contributions to their activity. A series of benzoquinones (**25**), related to mitomycin C, have been studied in their activity against L1210 in the mouse as a minimum effective dose (MED). Equation (78), shown below for a single-dose MED,[135] demonstrates control of the activity by electronic and partition factors. Electronic effects of the substituents, expressed in terms of $\mathscr{F}$ and $\mathscr{R}$, are found to be important in determining the activity with electron-releasing effects increasing activity. The authors considered that electronic effects would be important in electrophilic reactions of the drugs with the nucleophilic functions of biomacromolecules, for example DNA. However, no rationale was presented for the particular reaction constants observed, *i.e.* negative. It is of particular interest that the effect observed is of the same type as observed for the activity of the aniline mustards described in the next section. The antitumour activity of aniline mustards (**26**) has been evaluated by use of QSAR and electronic factors have been shown to be important.[136] Thus, the activity of the aniline mustards against Walker 256 tumour in rats in described by equation (79), where the indicator variable I is 1.0 for Y equal to Br and 0.0 for Y equal to Cl or I in (**26**). Similar correlations, in terms of electronic effects, occur for the B-16 melanoma, L1210 leukaemia and P388 leukaemia.[136,137] The mechanism of alkylation by aniline mustards has been suggested to operate by the pathway shown in equation (80). Very active nucleophiles may react directly, but less active nucleophiles will react with the aziridinium intermediate (**27**). The hydrolysis of aniline mustards and their reaction with 4-(4-nitrobenzyl)pyridine have been studied and give ρ values of about -1.4 and -1.9, respectively. A mode of biological action and mechanisms of chemical reaction having formation of an intermediate aziridinium ion, followed by intermolecular nucleophilic attack, could fit all the observed ρ values, *i.e.* the latter would be the sum of those for steps 1 and 2 in equation (80).

$$\log(1/MED) = -3.95(\pm 1.05)\mathscr{F} - 1.49(\pm 0.49)\mathscr{R} - 0.49(\pm 0.09)\,\Sigma\pi_{1,2} + 5.30(\pm 0.20) \qquad (78)$$

$$n = 35, r = 0.910, s = 0.290$$

$$\log(1/C) = -1.19(\pm 0.51)\sigma^- + 0.75(\pm 0.41)I - 1.00(\pm 0.87)\pi - 0.53(\pm 0.55)\pi^2 + 3.84(\pm 0.33) \qquad (79)$$

$$n = 14, r = 0.940, s = 0.291$$

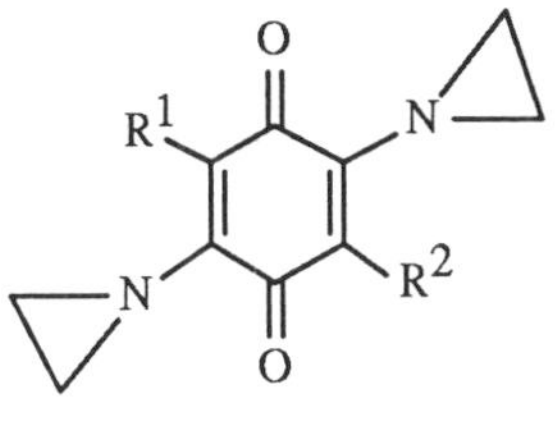

(25)

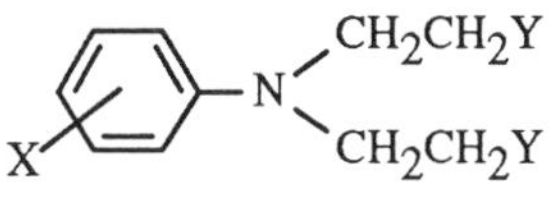

(26)

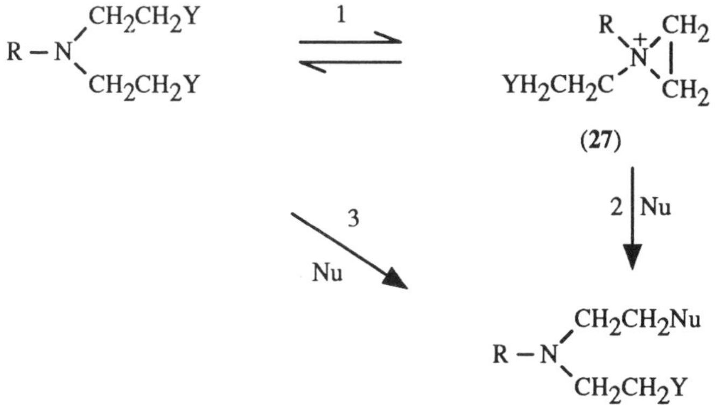

(27)

3 | Nu

2 | Nu

(80)

(28)

(29)

The antitumour activities of a very large number of 9-anilinoacridines (**28**) have been analyzed by QSAR.[138] Their activity against L1210 leukaemia in mice showed a complex dependence involving partition, steric and indicator variable terms, as well as electronic factors. The main electronic dependence found, with ρ equal to about -1.1, is that involving a summation of σ for *all* aromatic substitutents, both acridine, X and Y, and aniline, R. Electron-withdrawing substitutents in the acridine are known to stabilize 9-anilinoacridines to thiol attack, which is the main *in vivo* breakdown route for these compounds. However, the reverse is known to be the case for electron-withdrawing substitutents in the aniline ring, although such substitutents do not contribute to tighter DNA binding. It may well be that the latter interaction is the main cause of this dependence for both types of substitution.

Structure–activity relations for antimalarials have received attention and successful correlations have been developed. An early study[139] of over 100 phenanthrylethanols (**29**) showed a dominant role for partition parameters, but gave a ρ value for the sum effect of X and Y in (**29**) equal to about 0.8. A later study[140] considerably expanded the coverage of drugs to 646 diverse antimalarial 1-aryl-2-(alkylamino)ethanols. Their activity was controlled by both electronic and partition factors. Electronic factors were successfully assessed by summing σ values for *all* aryl substitutents, giving a ρ value of about 0.6. Thus, the same electronic effect arises from substitutents both in aryl groups attached directly to the alkyl chain and in aryl groups themselves attached to the aryl groups attached to the alkyl chain. Antimalarials of this type act by intercalation with DNA. The substitutents could act by moderating the ability of the arylmethanols to promote electron donation from DNA bases to the aromatic rings of the arylmethanols. This effect could be the same for both types of aryl group.

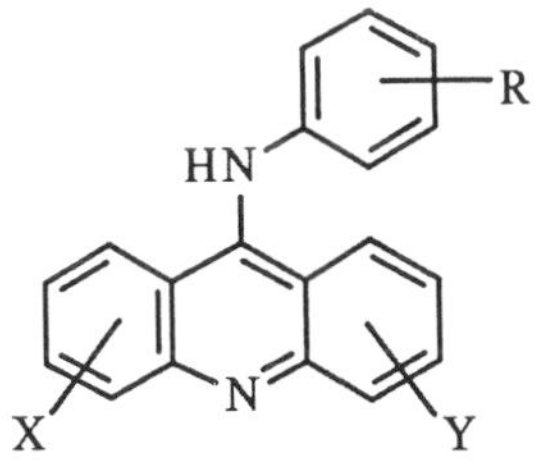

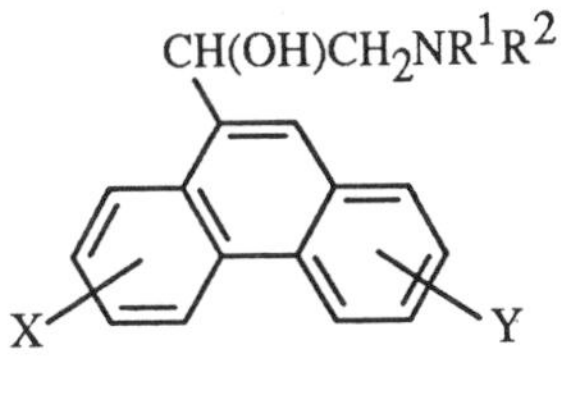

The adrenergic blocking activity of a series of β-halo-β-arylalkylamines (**30**) has been re-examined by Hansch and Unger.[141] The best simple relation found is that shown below as equation

(81). The mechanism that has been suggested for these drugs is shown in equation (82) with the formation of (**31**) being important in giving rise to the activity. The actual solvolysis reaction used in defining σ^+ (see Section 18.5.2.2) is a close model for the reverse reaction of (**31**) to (**32**). The ρ value found for the biological activity of about -1.5 agrees well with a pathway involving attachment of (**31**) to a nucleophilic centre at the adrenergic α-receptor.

$$\log(1/C) = -1.47\sigma^+ + 1.15\pi + 7.82 \tag{81}$$

$$n = 22, r = 0.94, s = 0.197$$

(30)

(82)

(31)   (32)

A survey of structure–activity relationships in antifungal agents has been made by Hansch and Lien.[142] A number of these correlations include electronic factors. In particular, the correlation equation (83) successfully describes the activity of substituted griseofulvins (**33**) against *Botyrtis allii*, especially considering that σ_p is used for the group X in (**33**). This result allows speculation on the mode of action to be an addition of essential nucleophilic groups at the activated ketovinyl double bond. This would be assisted by strongly electron-withdrawing groups as X in (**33**). The fungicidal activities of a series of alkynic sulfones (**34**) have been correlated using σ^-, as well as partition and steric parameters, as exemplified for *Cladosporium cucumerinum* by equation (84).[143] The use of σ^- for substitution in the aryl group and the ρ value found, as well as a comparative study of the addition reaction of the alkynic sulfones with the 4-nitrobenzenethiolate anion, indicate a mode of action involving a nucleophilic addition to the triple bond.

$$\log(1/C) = 2.19(\pm0.77)\sigma_x + 0.55(\pm0.17)\log P - 1.32(\pm0.61) \tag{83}$$

$$n = 22, r = 0.875, s = 0.248$$

$$pC = 1.10(\pm0.12)\sigma^- + 0.84(\pm0.14)\pi - 0.07(\pm0.04)\pi^2 + 2.10(\pm0.30)E_s + 4.17(\pm0.14) \tag{84}$$

$$n = 25, r = 0.943, s = 0.248$$

(33)   (34)

The use of σ^+ has been highlighted by Hansch[144] in a study of the reduction of substituted acetophenones using rabbit kidney reductase. A significant improvement over the use of σ results in the correlation for reduction shown in equation (85). Both the reaction constant, as well as the use of σ^+, support a pathway with reduction by hydride transfer. A similar improvement using σ^+ was shown in the correlation of the relative sweetness of 2-substituted 5-nitroanilines as shown in equation (86).

$$\log k_0 \ = \ 1.51(\pm 0.55)\sigma^+ \ + \ 1.48(\pm 0.29) \tag{85}$$

$$n \ = \ 10, r \ = \ 0.914, s \ = \ 0.39$$

$$\log RS \ = \ -1.03(\pm 0.41)\sigma^+ \ + \ 1.43(\pm 0.39)\pi \ + \ 1.58(\pm 0.30) \tag{86}$$

$$n \ = \ 9, r \ = \ 0.972, s \ = \ 0.19$$

An early study of Hansch[145] of a series of 1,3-benzodioxole (**35**), synergists for the insecticide carbaryl in flies, used the homolytic substituent constant $\sigma^{\cdot}$, together with partition terms, as shown in equation (87). Much poorer correlations resulted when using other substituent constants. It was considered that this was evidence for a mechanism in which abstraction of a hydrogen by a microsomal enzyme from the synergist could give a relatively stable free radical inhibitor of a free-radical-generating enzyme. A successful correlation of the activities of a series of substituted chloramphenicol derivatives (**36**) has been made using the radical parameter E_R, together with partition terms, as shown in equation (88). They postulated that chloramphenicol might act *via* a hydrogen transfer mechanism. The benzylic C–H bond was proposed as the point of chemical attack.[146] This was supported by the finding[147] that α-deuterochloramphenicol had only 80% of the activity of the parent compound, indicating the C–H bond is cleaved in a kinetically rate-determining step. A later study by Hansch *et al.*[148] of chloramphenicols of a more varied type was not fully in accord with the earlier results, but the general implications were confirmed. Hansch[149] has further considered the benzyl moiety in biological activities. The successful use of E_R was considered to indicate the likelihood of importance of benzylic radicals in a variety of oxidative biochemical processes.

$$\log SR5 \ = \ 1.316\sigma^{\cdot} \ - \ 0.195\pi^2 \ + \ 0.670\pi \ + \ 1.612 \tag{87}$$

$$n \ = \ 13, r \ = \ 0.929, s \ = \ 0.171$$

$$\log A \ = \ 3.07(\pm 1.2)E_R \ + \ 0.23(\pm 0.16)\pi \ + \ 0.77(\pm 0.25) \tag{88}$$

$$n \ = \ 8, r \ = \ 0.954, s \ = \ 0.140$$

(**35**)   (**36**)

18.5.10.3 Dipole Moments

A recent study[78] has reviewed the use of dipole moments in QSAR studies. Firstly, their use can be exemplified by the study of the activity of 7-substituted 1,4-benzodiazepinones with regard to both antipentylenetetrazole seizures and rotorod ataxia.[150] The resulting correlations are successful when incorporating a μ parameter and the relation for the former activity is shown in equation (89). Secondly, the use of molecular dipole moments can be illustrated in the correlations of miscellaneous anticonvulsants with regard to their antishock activity, as shown in equation (90) and a series of lactams, thiolactams, ureas and thioureas having convulsant activity in their acute lethal toxicity,[151] as shown in equation (91).

$$\log(1/C) \ = \ -0.301(\log P)^2 \ + \ 0.852\log P \ - \ 0.629\mu \ + \ 4.139 \tag{89}$$

$$n \ = \ 12, r \ = \ 0.915, s \ = \ 0.227$$

$$\log(1/C) \ = \ -0.222(\log P)^2 \ + \ 1.153\log P \ - \ 0.368\mu \ + \ 2.994 \tag{90}$$

$$n \ = \ 18, r \ = \ 0.992, s \ = \ 0.24$$

$$\log(1/C) \ = \ -0.364(\log P)^2 \ + \ 1.055\log P \ + \ 0.247\mu \ + \ 1.298 \tag{91}$$

$$n \ = \ 20, r \ = \ 0.89, s \ = \ 0.24$$

An approach using the dipole moment of the heterocyclic ring bearing a N–SCCl$_3$, group has been successful in correlating the activity of a series of fungicides in their inhibition of spore germination against the single organism *Stemphyllium sarcinaeforme*, as shown in equation (92), as well as tests using mixed organisms.[152]

$$\log(1/C) = -0.314(\log P)^2 + 2.385 \log P + 0.683\mu - 1.666 \tag{92}$$

$$n = 14, r = 0.951, s = 0.411$$

The distinct problem of the vector nature of dipole moments is important. Tute's results[153] on the inhibition of viral neuramidase by 1-(3,4-substituted phenoxymethyl)-3,4-dihydroquinolidines have been re-examined[78] and showed a somewhat better correlation using group dipole moments than Tute's division of the dipole moment into 'vertical' and 'horizontal' components along the 1,4-phenoxy axis. An interesting study by Young *et al.*[154] of the H$_2$ receptor activity of a series of cimetidine analogues resulted in the successful correlation shown in equation (93). In an attempt to improve this correlation, estimates were made of the dipole orientation by use of CNDO/2 molecular orbital calculation. The possibility of an optimum value for this orientation and the deviation of the group dipole moment to this optimum value (θ) was examined. No improvement was found using the dipole moment vector term ($\mu \cos \theta$). While a somewhat improved correlation using $\cos \theta$ alone was found, no convincing explanation of this use was offered.

$$\log K_B = 0.23\mu + 1.08 \log P + 2.74 \tag{93}$$

$$n = 7, r = 0.910, s = 0.426$$

An examination of the relationship between aliphatic dipole moments and the polar substituent constants σ^* for 214 substituent groups has been made and showed a poor correlation unless subgroups were considered.[155] The antifungal activity of 1-(3,5-dichlorophenyl)-2,5-pyrrolidine-diones (**37**) and 3-(3,5-dichlorophenyl)-2,4-oxazolidinediones (**38**) have been correlated with these aliphatic group dipole moments and steric factors, measured by $\log M$, as shown in equation (94).[155]

$$pI_{50} = 0.212\mu + 2.879 \log M - 1.252 \tag{94}$$

$$n = 28, r = 0.934, s = 0.111$$

(37)   (38)

The present position can be summarized as follows. The use of dipole moments in QSAR, while promising on the basis of theory, has yet to be sufficiently tested. At present the results appear somewhat unpromising, except where compounds of such diverse structures preclude any other electronic parametrization. The problem of the vector nature of dipole moments will require more sophisticated approaches. The sign and magnitude of the coefficients of μ in QSAR should be capable of interpretation in regard of the nature of the interaction giving rise to them. However, little or no comment has been made in published studies and the results available do not allow serious discussion.

18.5.10.4 Ionization Constants

If a correlation of drug activity with electronic effects by use of the ionization constants is to be meaningful, it is very important to exclude the simple effects of partial ionization at the pH at which the measurements have been made. An early attempt by Fujita[156] was made to modify QSAR to allow for ionization. The concentration measure of equivalent biological response, C, is multiplied by the fraction of the drug ionized, f_i, or unionized, f_u, under the pH condition of the study and results in the relations (95) and (96). K_a is the ionization constant of the drug and [H$^+$] the hydrogen ion concentration. The modified measures derived in equations (95) and (96) can then be used in correlations which are now independent of the state of partial ionization. As $f_i + f_u$ equals one, equations (95) and (96) are not independent. Hence, it is not possible to distinguish which form of the drug is the biologically active species. This may be done by other means. An alternative treatment has been suggested[86,157a] and was outlined in Section 18.5.8.4. This uses the distribution coefficient, D, which is equal to the partition coefficient, P, multiplied by the fraction of the drug unionized. Equations (56) and (57) express the conditions for highly ionized compounds. This and the treatment of Fujita[156] are closely related, although the use of $\log D$ in correlations has the advantage of a somewhat increased correlative adaptability. Both methods allow the isolation of dependence of the biological activity on the pK_a values of the drugs and, thus the electronic factors controlling ionization, separate from the effects of equilibria ionization on partition and transport. Without these corrective treatments, a number of correlations previously observed really relate to the effects of ionization on partition, rather than any real dependence on electronic effects.[1,111]

$$\log(1/Cf_i) = \log(1/C) + \log(\{K_a + [H^+]\}/K_a) \tag{95}$$

$$\log(1/Cf_u) = \log(1/C) + \log(\{K_a + [H^+]\}/[H^+]) \tag{96}$$

An analysis of the biological activity of phenols has been made by Fujita.[156,158] To exemplify the value of electronic factors, equation (97) has been shown below for the toxicity of eight nitrophenols at pH 5.5 against *E. coli*. The pK_a values of the phenols are much more varied than in an earlier study[2] and these results indicate the importance of varied substituent factors in such an analysis. The results appear to support the view that the active form of the phenol is ionic and nucleophilic centres on the receptor interact more favourably with more acidic phenols.

$$\log(1/C) + \log(K_a + [H^+]/[H^+]) = 0.339\pi + 0.659\Delta pK_a + 1.257 \tag{97}$$

$$n = 8, r = 0.997, s = 0.083$$

Bell and Roblin,[157b] in an early study, demonstrated a relationship between the bacteriostatic activity of a series of N^1-substituted sulfanilamides and their acidic dissociation constants which was parabolic. There have been some contradictory discussions of the interpretations of this and related results.[157c]

Fujita and Hansch[158] have applied the treatment described above to the bacteriostatic activity and protein binding of sulfonamides. An example of the successful correlations is shown in equation (98), which is for the minimum inhibitory concentration of a series of N^1-heterocyclic sulfanilamides against *E. coli*, considering the active species to be the neutral molecule. A marginal improvement occurs if partition factors are incorporated and an equivalent relation can be derived if the anion is considered to be the active species. An advantage of this treatment is said to be that an optimum pK_a value can be obtained for apparent activity by setting $d \log(1/C)/d \Delta pK_a$ to zero. Yoshioka *et al.*[159] determined the pK_a values of a series of substituted sulfanilamides and used a Yukawa–Tsuno type equation to correlate the substituent effects as shown in equation (99). Fujita and Hansch[158] derived a successful regression analysis for the activity of a series of *meta*-substituted sulfanilamides against *Pneumococcus*, considering either the ionic or neutral form to be the active species, as shown in equations (100) and (101). For the *para*-substituted series, the correlations are much less successful and significantly different. These are exemplified by equation (102) for the ionic form. The authors considered that the difference between the *meta*- and the more active *para*-substituted series might arise from metabolism by the bacteria of the former series.

$$\log(1/C) + \log(\{K_a + [H^+]\}/[H^+]) = 0.761\Delta pK_a - 1.995 \tag{98}$$

$$n = 17, r = 0.965, s = 0.244$$

$$\log K_a = 1.883[\sigma - 0.54(\sigma^- - \sigma)] - 8.942 \tag{99}$$

$$\log(1/C) + \log(\{K_a + [H^+]\}/K_a) = -0.676\sigma + 0.245\pi + 1.906 \tag{100}$$

$$n = 12, r = 0.930, s = 0.134$$

$$\log(1/C) + \log(\{K_a + [H^+]\}/[H^+]) = 1.204\sigma + 0.239\pi + 0.767 \tag{101}$$

$$n = 12, r = 0.982, s = 0.136$$

$$\log(1/C) + \log(\{K_a + [H^+]\}/K_a) = -1.486\sigma + 1.878 \tag{102}$$

$$n = 7, r = 0.841, s = 0.322$$

The uncoupling activity of 23 phenols covering a broad range of pK_a values has been described by equation (103).[157a] The mechanism for classical uncouplers proposes that they act as ionophores transporting protons into mitochondria and cations out. The quantitative contribution of the electronic effects (*via* pK_a values) can be directly observed in equation (103) and apparently relates to the ionic character of the compounds. These effects could also have been assessed by use of $\log D$ together with a σ term. However, the inclusion of *ortho*-substituted phenols is more directly and simply facilitated by the use of pK_a values. The bacteriostatic activity of a series of phenols against *E. coli* between pH 5.5 and 8.5 has been described by equation (104).[157a] As the pK_a term is present, the dependence on electronic factors, as distinct from the ionization itself, has been detected. The result in equation (104) is very much better than that obtained by using $\log P$ and pK_a.

$$\log(1/C) \ = \ 0.471 \log D \ - \ 0.618pK_a \ + \ 7.584 \tag{103}$$

$$n = 23, r = 0.946, s = 0.351$$

$$\log(1/C) \ = \ 0.594 \log D \ - \ 0.461pK_a \ + \ 5.231 \tag{104}$$

$$n = 31, r = 0.958, s = 0.179$$

The biological activity of 2-aryl-1,3-indanediones (**39**) has been studied in regard of both their uncoupling and antiinflammatory activities.[157a] The correlation for the former activity is shown in equation (105). In this case, the correlation shown in equation (106), using just π and σ, gives a *negative* ρ, whereas the equivalent coefficient for pK_a has a *positive* ρ. A marked contrast results when equation (107), correlating the inhibition of prostaglandin synthesis, is compared with equation (105). The reaction constants for pK_a are very dissimilar but do have the same sign. In fact, the result of increasing acidity (lowering pK_a) on the two activities is in the *opposite* direction as the pK_a values affect *both* the $\log D$ and pK_a terms. The possibility that there is a relationship between uncoupling and antiinflammatory activity does not receive any support from the significantly different dependencies on electronic factors.

(**39**)

$$\log(1/C) \ = \ 0.508 \log D \ - \ 0.308pK_a \ + \ 5.348 \tag{105}$$

$$n = 24, r = 0.967, s = 0.165$$

$$\log(1/C) \ = \ 0.504\pi \ - \ 0.347\sigma \ + \ 3.813 \tag{106}$$

$$n = 24, r = 0.968, s = 0.164$$

$$\log(1/C) \ = \ 0.382 \log D \ - \ 1.354pK_a \ + \ 9.420 \tag{107}$$

$$n = 24, r = 0.920, s = 0.228$$

18.5.10.5 Hydrogen bonding

The quantitative study of hydrogen bonding on drug activity has rarely been attempted directly. As stated in Section 18.5.9.2, such effects can be implicit in correlations employing other electronic parameters.

Fujita *et al.*[99] have reviewed the use of a hydrogen-bonding parameter in QSAR. Thus, the inhibition of acetylcholinesterase (AcChE) by *meta*-substituted phenyl *N*-methylcarbamates (**40**) has been demonstrated to depend on electronic and partition factors as in equation (108). The electron substituent terms are expressed as σ^0 and HB. The latter is essentially an indicator variable, being 1.0 for hydrogen bond acceptor groups such as NO_2, CN, CHO and NMe_2. Although it is difficult to accept that hydrogen-bonding capacity of groups can be represented by simple indicator variables, Fujita *et al.*[99] have shown that $\log P$ for octanol–water takes care of nonspecific binding of both hydrophobic and hydrogen-bonding interactions. $\log P$ for chloroform–water requires the indicator variable HB for these values of $\log P$ to be well correlated with $\log P$ for octanol–water or hexane–water. The indicator variable term in equation (108) for the inhibition of AcChE by *meta*-substituted phenyl *N*-methylcarbamates is considered likely to derive from a hydrogen bond donor group located on the enzyme close to *meta* substituents in the enzyme–inhibitor complex.

$$\log(1/K_d) = 1.55(\pm 0.26)\pi - 1.93(\pm 0.52)\sigma^0 + 1.53(\pm 0.36)\text{HB} + 2.46(\pm 0.28) \qquad (108)$$

$$n = 21, r = 0.967, s = 0.247$$

$$\text{(40)}$$

Hydrogen bonding has been implicated in determining the activity of inhalation anaesthetics. Hansch *et al.*[101] have derived the correlation equation (109) for 32 gaseous anaesthetics to relate their effective anaesthetic pressure with their partition constants for octanol–water $\log P$ and an indicator variable I. The latter is given the value of 1.0 for all compounds containing a 'polar hydrogen atom', *e.g.* $CHCl_3$, and of 0.0 for all other compounds. A later study by Yokono *et al.*[160] related the hydrogen bond strengths, in terms of proton chemical shifts, of five potent halocarbon and haloether anaesthetics to their potency in humans.

$$\log(1/p) = 1.17(\pm 0.25)\log P + 1.88(\pm 0.33)I - 2.11(\pm 0.39) \qquad (109)$$

$$n = 30, r = 0.947, s = 0.438$$

18.5.11 THE USE OF ELECTRONIC SUBSTITUENT CONSTANTS

Several reviews of electronic substituent constants have been published. Exner[161] has given a comprehensive and critical review of Hammett and Hammett-type constants. Hansch and Leo[41,102,162] have collected values of σ_m, σ_p, $\mathscr{F}$ and $\mathscr{R}$, as well as assigning hydrogen-bonding capacities to substituents. Charton[51] has published an analysis of 'electrical' effect substituent constants. More specialized substituent constants, which are considered applicable to ionization reactions in particular, have been tabulated.[83,92,93] The latter include substituent constants for *ortho* effects in ionization reactions.

Table 1 gives electronic substituent constants for 74 groups. These are a selection of those available, appearing to be those of most common interest. Values of σ_m, σ_p, $\mathscr{F}$ and $\mathscr{R}$ are given for all substituents. For consistency they are taken from the study of Hansch and Leo.[41] The values of σ_p^+, σ_p^-, and σ_p^0 are given if available. They have been selected from Exner's review[161] and are as consistent as possible, *i.e.* σ_p^+ from solvolysis reactions, σ_p^- from the ionization of phenols and anilines, and σ_p^0 from the alkaline hydrolysis of insulated systems. Similarly, the values of σ_I are derived from the study of the ionization of substituted acetic acids.[161,163] Secondary values for these four parameters are shown in brackets and are suitable for most purposes. Aromatic and aliphatic group dipole moments, μ in Debye,[78,155] are also given in Table 1. Table 2 shows the electronic substituent constants for certain aryl and heteroaryl systems. Those shown do not involve proximate substitution, being either from the literature[41,161] or derived in the same way. They result from the treatment of fused rings or heteroatom substituents in the ring itself as if they were phenyl ring substitutions.

Values of σ_R, σ_R^+, σ_R^- and σ_R^0 can be obtained directly from the listed constants by simple subtraction from the relevant *para* substituent constant of σ_I (see Section 18.5.3.1). Values of σ_R^0 have also been derived directly by spectroscopic methods.[161] The only redundancy in Table 1 could be considered to be the inclusion of both $\mathscr{F}$ and σ_I; but the above use and the need for completion requires both. Both $\mathscr{F}$ and σ_I have been considered to be applicable to both aromatic and aliphatic systems.

The additivity of electronic substituent effects appears to be reasonably good in reactivity studies.[37,92,93,96,164,165] However, in systems where two or more substituents are proximate to each

Table 1 Electronic Substituent Constants[a]

No.	Substituent[b]	σ_m	σ_p	σ_p^+	σ_p^-	σ_p^0	σ_I	$\mathscr{F}$	$\mathscr{R}$	Aromatic μ (Debye)	Aliphatic μ (Debye)
1	Br	0.39	0.23	0.15	c	0.30	0.46	0.44	-0.17	-1.57	-1.97
2	CF$_3$	0.43	0.54	c	0.68	[0.53]	0.42	0.38	0.19	-2.61	-1.94
3	CN	0.56	0.66	c	0.96	0.71	0.57	0.51	0.19	-4.08	-3.63
4	CO$_2^-$	-0.10	0.00	c	0.30	[-0.14]	-0.17	-0.15	0.13		
5	CHO	0.35	0.42	c	1.02	[0.47]	[0.25]	0.31	0.13	-3.02	-2.58
6	CO$_2$H	0.37	0.45	c	0.78	[0.44]	[0.32]	0.33	0.15	-1.30	-1.65
7	CH$_2$Cl	0.10	0.12	[-0.01]	c		0.15	0.10	0.03	-1.83	-1.93
8	CONH$_2$	0.28	0.36	c	0.62		0.27	0.24	0.14	-3.42	-3.73
9	CH$_3$	-0.07	-0.17	-0.31	c	-0.12	-0.04	-0.04	-0.13	0.36	0.0
10	CH$_2$OH	0.00	0.00	[0.01]	c		0.05	0.00	0.00	1.73	
11	C≡CH	0.21	0.23	0.18	[0.52]	[0.22]	0.35	0.19	0.05	-0.77	-0.78
12	CH$_2$CN	0.16	0.01	[0.12]	c	[0.18]	0.18	0.21	-0.18	-3.60	
13	CH=CH$_2$	0.05	-0.02			[-0.01]	0.09	0.07	-0.08	0.20	-0.40
14	COCH$_3$	0.38	0.50	c	0.83	[0.47]	0.29	0.32	0.20	-2.90	-2.77
15	CO$_2$CH$_3$	0.37	0.45	c	0.69	[0.44]	0.34	0.33	0.15	-1.92	-1.75
16	CH$_2$CH$_3$	-0.07	-0.15	-0.30	c	-0.13	-0.03	-0.05	-0.10	0.39	0.0
17	c-C$_3$H$_5$	-0.07	-0.13	-0.44	-0.08	[-0.22]	0.01	-0.06	-0.08	0.51	-0.14
18	(CH$_2$)$_2$CH$_3$	-0.07	-0.13				-0.02	-0.06	-0.08		0.08
19	CH(CH$_3$)$_2$	-0.07	-0.15	-0.28	c		-0.03	-0.05	-0.08	0.40	0.08
20	2-Thienyl	0.09	0.05	-0.33	0.19	-0.15	[0.21]	0.10	0.04	0.81	
21	(CH$_2$)$_3$CH$_3$	-0.08	-0.16		c		-0.04	-0.06	-0.11		0.08
22	C(CH$_3$)$_3$	-0.10	-0.20	[0.26]	c		-0.07	-0.07	-0.13	0.52	
23	C$_6$F$_5$	0.34	0.41	-0.21	c		0.31	0.30	0.13		
24	C$_6$H$_5$	0.06	-0.01		c	-0.16	0.12	0.08	-0.08	0.00	-0.38
25	2-Benzoxazolyl	0.30	0.33	-0.25	0.09	-0.17		0.28	0.07	-1.22	
26	2-Benzothiazolyl	0.27	0.29	-0.70	0.68	[0.27]		0.25	0.06	-0.94	
27	COC$_6$H$_5$	0.34	0.43	-0.27	0.65	[0.46]	[0.27]	0.30	0.16	-3.04	-2.90
28	CH$_2$C$_6$H$_5$	-0.08	-0.09		0.88	[-0.06]	0.03	-0.08	-0.01	0.36	-0.39
29	Ferrocenyl	-0.15	-0.18		c			-0.15	-0.04		
30	Adamantyl	-0.12	-0.13		-0.04	[-0.13]		-0.12	-0.02		
31	Cl	0.37	0.23	0.11	c	0.28	0.47	0.41	-0.15	-1.59	-1.93
32	F	0.34	0.06	-0.07	c	0.20	0.54	0.43	0.34	1.43	-1.90
33	H	0.00	0.00	0.00	0.00	0.00	0.00	0.00	0.00	0.03	
34	I	0.35	0.18	0.13	c	[0.27]	0.39	0.40	-0.19	-1.36	-1.79
35	IO$_2$	0.68	0.78	c	1.25			0.63	0.20		
36	NO$_2$	0.71	0.78	c	3.24	0.82	0.76	0.67	0.16	-4.13	-3.59
37	N$_2^+$	1.76	1.91		[0.08]	[2.18]		1.69	0.36		
38	N$_3$	0.27	0.15		c		0.42	0.30	-0.13	-1.56	-2.17
39	NH$_2$	-0.16	-0.66	[-1.36]	c	[-0.30]	[0.12]	0.02	-0.68	1.53	-1.35

40 NH$_3^+$	0.86	0.60		[0.56]		0.60	0.94	-0.27		
41 NCO	0.27	0.19		[0.34]	[0.19]	[0.36]	0.23	-0.08	-3.93	-2.81
42 NCS	0.48	0.38		[0.57]	[0.35]	[0.42]	0.29	-0.09	-2.91	
43 1-Tetrazoyl	0.52	0.50		c		[0.54]	0.50	0.02		
44 NHCONH$_2$	-0.03	-0.24		c	[0.16]	0.21	0.04	-0.28		
45 NHCSNH$_2$	0.22	0.16		c	[-0.46]	[0.29]	0.23	-0.05	-5.16	-0.16
46 NHCH$_3$	-0.30	-0.84		c		[0.18]	-0.11	-0.74	1.69	-1.01
47 N(CF$_3$)$_2$	0.40	0.53	[-0.65]	c		[0.49]	0.34	0.22		
48 NHCOCH$_3$	0.21	0.00	[-1.62]	c	[0.14]	0.26	0.28	-0.26	-3.65	-3.81
49 N(CH$_3$)$_2$	-0.15	-0.83	c	c	[-0.32]	[0.06]	0.10	-0.92	1.61	-1.26
50 N(CH$_3$)$_3^+$	0.88	0.82	[-0.15]	0.65	[0.88]	0.73	0.89	0.00		
51 N=NC$_6$H$_5$	0.32	0.39		c		0.19	0.28	0.13		
52 O$^-$	-0.47	-0.81	[-2.30]	[-0.92]	[-0.77]	[-0.16]	-0.35	-0.49		
53 OH	0.12	-0.37		c	[-0.12]	0.22	0.29	-0.64	-1.59	-1.66
54 OCF$_3$	0.38	0.35		c	-0.15	[0.55]	0.38	0.00	-2.36	-1.27
55 OCH$_3$	0.12	-0.27	-0.78	c		0.29	0.26	-0.51	-1.30	-1.81
56 OCOCH$_3$	0.39	0.31		c		[0.36]	0.41	-0.07	-1.72	-1.27
57 OCH$_2$CH$_3$	0.10	-0.24	[-0.82]	c	-0.14	0.27	0.22	-0.44	-1.38	-1.38
58 OC$_6$H$_5$	0.25	-0.03	[-0.52]	c	-0.05	0.42	0.34	-0.35	1.16	
59 PO(OCH$_3$)$_2$	0.42	0.53	[0.54]	0.80	[0.43]	[0.24]	0.37	0.19		-3.39
60 SO$_2$F	0.80	0.91	c	[1.32]		[0.75]	0.75	0.22	-4.59	
61 SF$_5$	0.61	0.68		0.77	[0.30]	0.57	0.57	0.15	-3.44	
62 SO$_3^-$	0.05	0.09		0.52	[0.06]	0.13	0.03	0.07		
63 SH	0.25	0.15		c	[0.58]	0.26	0.28	-0.11	-1.33	-1.51
64 SO$_2$NH$_2$	0.46	0.57	c	0.92		[0.44]	0.41	0.19		-4.60
65 SOCF$_3$	0.63	0.69	c			0.69	0.60	0.14		
66 SO$_2$CF$_3$	0.79	0.93	c			0.72	0.73	0.26		
67 SCF$_3$	0.40	0.50		1.49	[0.58]	0.44	0.35	0.18		
68 SCN	0.41	0.52		0.61	0.57	0.58	0.36	0.19	-2.50	-3.89
69 SOCH$_3$	0.52	0.49	c	0.60	0.57	[0.50]	0.52	0.01	-3.01	-3.88
70 SO$_2$CH$_3$	0.60	0.72	c	0.73	0.75	0.59	0.54	0.22	-3.98	-4.26
71 SCH$_3$	0.15	0.00	-0.60	1.05	0.05	0.25	0.20	-0.18	-4.75	-1.45
72 S(CH$_3$)$_2^+$	1.00	0.90		1.16	[1.06]	[0.89]	1.02	-0.04	-1.34	
73 SeCF$_3$	0.32	0.38		c		[0.42]	0.29	0.12		-2.48
74 Si(CH$_3$)$_3$	-0.04	-0.07	0.02	0.08			-0.13	-0.04		

a Values given in brackets are considered to be less certain (see text). b The arrangement is that used by Hansch *et al.*,[41] *i.e.* substituent begins with attached atom and sorted alphabetically. Within each such grouping, the order is that first, if no C or H, then alphabetically on remainder; second, if no C, then on H and alphabetically on remainder; and third, C then H then alphabetically on remainder. c Use of σ_p is considered permissible.

Table 2   Electronic Substituent Constants for Aryl and Heteroaryl Systems[a]

No. System[b]	σ	σ^+	σ^-	σ^0	$\mathscr{F}$	$\mathscr{R}$
1 H(Ph)	0.00	0.00	0.00	0.00	0.00	0.00
2 3,4-(CH$_2$)$_3$	-0.26		[-0.26]	-0.26	-0.27	-0.01
3 3,4-[(CH=CH)$_2$]	0.04	-0.14	0.08	0.08	0.03	0.01
4 3,4-(CH$_2$)$_4$	-0.48		[-0.48]		-0.49	-0.03
5 (3-Furanyl)	0.25	[-0.44]	c	0.42	0.24	0.02
6 (3-Thienyl)	0.12	[-0.49]	c	0.28	0.12	0.01
7 (3-Pyridyl)	0.65	0.57	0.67	0.72	0.65	0.05
8 (4-Pyridyl)	0.67	1.13	[1.21]	0.95	0.67	0.05

[a] See footnote to Table 1. [b] Systems referred to phenyl group (see text). [c] See footnote to Table 1.

other, serious deviations from additivity can occur.[96,164,165] These appear to derive mainly from the occurrence or enhancement of steric effects, both primary and secondary. Thus, the steric inhibition of a substituent resonance effect by the relief of 'bulk' steric effects by twisting from coplanarity or the 'buttressing' of a steric effect already present can occur.

Steric substituent constants required for the correlation of *ortho* or other proximate substitution have been collected.[161] *Ortho* substitution in the ionization of certain acid systems can be directly related by the use of special substituent constants.[83,92,93]

The use of electronic substituent constants in QSAR deserves some discussion. Most practitioners in studies of drug activity employ the values of $\mathscr{F}$ and $\mathscr{R}$ from the Swain–Lupton separation. This has the advantage of not requiring any assumption regarding the positional influence of substitution, and also of the independence of the measurements of the field and resonance susceptibilities. The advantage of using the Hammett parameters, σ_m and σ_p, lies in their comparatively simple derivation and high certainty. The σ_p^+, σ_p^- and σ_p^0 values are all less certain and less available than the ordinary σ values. In favourable situations with a suitable choice of substituents, their use can prove very enlightening in relation to the mechanistic interpretation. The possible use of the four variants of σ_R, with or without σ_I, may well be desirable, but their availability and certainty leave much to be desired. The criticism[42–44] of the $\mathscr{R}$ scale seems less realistic in relation to its use in studies of drug activity where the biological activity data are more imprecise than the data from reactivity studies. The use of group dipole moments is certainly attractive from a theoretical point of view, but requires testing and establishment. The employment of hydrogen-bonding parameters is still in its infancy, but could be greatly improved if simple, quantitative group donor and acceptor values were available.

18.5.12   REFERENCES

1. O. R. Hansen, *Acta Chem. Scand.*, 1962, **16**, 1593.
2. C. Hansch and T. Fujita, *J. Am. Chem. Soc.*, 1964, **86**, 1616.
3. A. R. Katritzky and R. D. Topsom, *J. Chem. Educ.*, 1971, **48**, 427.
4. L. M. Stock, *J. Chem. Educ.*, 1972, **49**, 400.
5. N. Bjerrum, *Z. Phys. Chem. (Leipzig)*, 1923, **106**, 219.
6. J. G. Kirkwood and F. H. Westheimer, *J. Chem. Phys.*, 1938, **6**, 506.
7. F. H. Westheimer and J. G. Kirkwood, *J. Chem. Phys.*, 1938, **6**, 513.
8. S. Ehrenson, *J. Phys. Chem.*, 1977, **81**, 1520.
9. C. Tanford, *J. Am. Chem. Soc.*, 1957, **79**, 5348.
10. J. D. Roberts and W. T. Moreland, Jr., *J. Am. Chem. Soc.*, 1953, **75**, 2167.
11. F. W. Baker, R. C. Parish and L. M. Stock, *J. Am. Chem. Soc.*, 1967, **89**, 5677.
12. C. F. Wilcox and C. Leung, *J. Am. Chem. Soc.*, 1968, **90**, 336.
13. W. H. Orttung, *J. Am. Chem. Soc.*, 1978, **100**, 4369.
14. O. Exner and P. Fiedler, *Collect. Czech. Chem. Commun.*, 1980, **45**, 1251.
15. K. Bowden, *Can. J. Chem.*, 1963, **41**, 2781.
16. J. A. Pople and M. Gordon, *J. Am. Chem. Soc.*, 1967, **89**, 4253.
17. S. Acevedo and K. Bowden, *J. Chem. Soc., Perkin Trans. 2*, 1986, 2045.
18. K. Bowden and M. Hojatti, *J. Chem. Soc., Chem. Commun.*, 1982, 273.
19. A. R. Katritzky and R. D. Topsom, *Angew. Chem., Int. Ed. Engl.*, 1970, **9**, 87.
20. L. P. Hammett, *J. Am. Chem. Soc.*, 1937, **59**, 96.
21. C. D. Johnson, 'The Hammett Equation', Cambridge University Press, London, 1973.
22. H. H. Jaffé, *Chem. Rev.*, 1953, **53**, 191.
23. S. Ehrenson, R. T. C. Brownlee and R. W. Taft, *Prog. Phys. Org. Chem.*, 1973, **10**, 1.
24. H. C. Brown and Y. Okamoto, *J. Am. Chem. Soc.*, 1958, **80**, 4979.

25. H. Van Bekkum, P. E. Verkade and B. M. Wepster, *Recl. Trav. Chim. Pays-Bas*, 1959, **78**, 815.
26. R. W. Taft, Jr., *J. Phys. Chem.*, 1960, **64**, 1805.
27. Y. Yukawa and Y. Tsuno, *Bull. Chem. Soc. Jpn.*, 1959, **32**, 971.
28. Y. Yukawa, Y. Tsuno and M. Sawada, *Bull. Chem. Soc. Jpn.*, 1966, **39**, 2274.
29. A. A. Humffray and J. J. Ryan, *J. Chem. Soc. B.*, 1969, 1138.
30. R. W. Taft, Jr., in 'Steric Effects in Organic Chemistry', ed. M. S. Newman, Wiley, New York, 1956, chap. 13, p. 556.
31. R. W. Taft, Jr., and I. C. Lewis, *J. Am. Chem. Soc.*, 1958, **80**, 2436.
32. R. W. Taft, Jr. and I. C. Lewis, *J. Am. Chem. Soc.*, 1959, **81**, 5343.
33. J. Shorter, in 'Correlation Analysis in Chemistry: Recent Advances', ed. N. B. Chapman and J. Shorter. Plenum Press, New York, 1978, chap. 4, p. 119.
34. R. W. Taft, E. Price, I. R. Fox, I. C. Lewis, K. K. Anderson and G. T. Davis, *J. Am. Chem. Soc.*, 1963, **85**, 709, 3146.
35. R. T. C. Brownlee, A. R. Katritzky and R. D. Topsom, *J. Am. Chem. Soc.*, 1966, **88**, 1413.
36. O. Exner, *Collect, Czech. Chem. Commun.*, 1966, **31**, 65.
37. O. Exner, in 'Advances in Linear Free Energy Relationships', ed. N. B. Chapman and J. Shorter, Plenum Press, New York, 1972, chap. 1, p. 1.
38. M. J. S. Dewar and P. J. Grisdale, *J. Am. Chem. Soc.*, 1962, **84**, 3539, 3548.
39. M. J. S. Dewar, R. Golden and J. M. Harris, *J. Am. Chem. Soc.*, 1971, **93**, 4187.
40. C. G. Swain and E. C. Lupton, Jr, *J. Am. Chem. Soc.*, 1968, **90**, 4328.
41. C. Hansch, A. Leo, S. H. Unger, K. H. Kim, D. Nikaitani and E. J. Lien, *J. Med. Chem.*, 1973, **16**, 1207.
42. W. F. Reynolds and R. D. Topsom, *J. Org. Chem.*, 1984, **49**, 1989.
43. A. J. Hoefnagel, W. Oosterbeek and B. M. Wepster, *J. Org. Chem.*, 1984, **49**, 1993.
44. M. Charton, *J. Org. Chem.*, 1984, **49**, 1997.
45. C. G. Swain, *J. Org. Chem.*, 1984, **49**, 2005.
46. C. G. Swain, S. H. Unger, N. R. Rosenquist and M. S. Swain, *J. Am. Chem. Soc.*, 1983, **105**, 492.
47. C. K. Ingold, *J. Chem. Soc.*, 1930, 1032.
48. J. Shorter, 'Correlation Analysis of Organic Chemistry: With Particular Reference to Multiple Regression', Wiley, Chichester, 1982.
49. M. Charton, *J. Am. Chem. Soc.*, 1975, **97**, 1552, 3691.
50. M. Charton, *J. Am. Chem. Soc.*, 1977, **99**, 5687.
51. M. Charton, *Prog. Phys. Org. Chem.*, 1981, **13**, 119.
52. L. P. Hammett, 'Physical Organic Chemistry', 2nd edn., McGraw-Hill, New York, 1970, chap. 11.
53. A. C. Farthing and B. Nam, in 'Steric Effects in Conjugated Systems', ed. G. W. Gray, Butterworths, London, 1958, p. 131.
54. M. Charton, *Prog. Phys. Org. Chem.*, 1971, **8**, 235.
55. T. Fujita and T. Nishioka, *Prog. Org. Phys. Chem.*, 1976, **12**, 49.
56. K. Bowden and D. C. Parkin, *J. Chem. Soc., Chem. Commun.*, 1968, 75.
57. K. Bowden and D. C. Parkin, *Can. J. Chem.*, 1968, **46**, 3909.
58. M. Hojo, K. Katsurakawa and Z. Yoshida, *Tetrahedron Lett.*, 1968, 1497.
59. K. Bowden and G. E. Manser, *Can. J. Chem.*, 1968, **46**, 2941.
60. (a) K. Bowden and D. C. Parkin, *Can. J. Chem.*, 1968, **46**, 3909; (b) H. Van de Waterbeemd and B. Testa, *Adv. Drug. Res.*, 1987, **16**, 85.
61. B. Hetnarski and R. D. O'Brien, *J. Med. Chem.*, 1975, **18**, 29.
62. C. Hansch, in 'Correlation Analysis in Chemistry: Recent Advances', ed. N. B. Chapman and J. Shorter, Plenum Press, New York, 1978, chap. 9, p. 397.
63. Y. Yamamoto and T. Otsu, *Chem. Ind. (London)*, 1967, 787.
64. C. Hansch and R. Kerley, *Chem. Ind. (London)*, 1969, 294.
65. C. Hansch, *J. Med. Chem.*, 1968, **11**, 920.
66. A. Cammarata and S. J. Yau, *J. Polym. Sci., Part A-1*, 1970, **8**, 1303.
67. A. P. G. Kieboom, *Tetrahedron*, 1972, **28**, 1325.
68. A. Verloop, in 'Drug Design', ed. E. J. Ariens, Academic Press, New York, 1972, vol. 3, chap. 2.
69. T. A. Mastryukova and M. I. Kabachnik, *Usp. Khim.*, 1969, **38**, 1751.
70. T. A. Mastryukova and M. I. Kabachnik, *J. Org. Chem.*, 1971, **36**, 1201.
71. J. O. Schreck, *J. Chem. Educ.*, 1971, **48**, 103.
72. M. Charton, in 'Correlation Analysis in Chemistry: Recent Advances', ed. N. B. Chapman and J. Shorter, Plenum Press, New York, 1978, chap. 5, p. 175.
73. H. H. Jaffé and H. L. Jones, *Adv. Heterocycl. Chem.*, 1964, **3**, 209.
74. H. H. Jaffé, *J. Am. Chem. Soc.*, 1954, **76**, 4261.
75. J. Hine, 'Structural Effects on Equilibria in Organic Chemistry', Wiley, New York, 1975.
76. K. Bowden and G. R. Taylor, *J. Chem. Soc. B.*, 1971, 145.
77. A. J. Kirby and A. R. Fersht, *Prog. Bioorg. Chem.*, 1971, **1**, 1.
78. E. J. Lien, Z.-R. Guo, R.-L. Li and C.-T. Su, *J. Pharm. Sci.*, 1982, **71**, 641.
79. J. W. McFarland, *Prog. Drug. Res*, 1971, **15**, 123.
80. A. Albert, 'Selective Toxicity. The Physico-chemical Basis of Therapy', 5th edn., Chapman and Hall, London, 1973.
81. R. F. Cookson, *Chem. Rev.*, 1974, **74**, 5.
82. A. Albert and E. P. Serjeant, 'The Determination of Ionization Constants', 2nd edn., Chapman and Hall, London, 1971.
83. D. D. Perrin, B. Dempsey and E. P. Serjeant, 'pK_a Prediction for Acids and Bases', Chapman and Hall, London, 1981.
84. W. Simon, G. H. Lysay, A. Mörrikofer and E. Heilbronner, 'Zusammenstellung im Losungsmittelsystem Methylcellosolve/Wasser', Juris-Verlag, Zurich, 1959, vol. 1; P. F. Sommer and W. Simon, 'Zusammenstellung im Losungsmittelsystem Methylcellosolve/Wasser', Juris-Verlag, Zurich, 1961, vol. 2; W. Simon and P. F. Sommer, 'Zusammenstellung im Losungsmittelsystem Methylcellosolve/Wasser', Juris-Verlag, Zurich, 1963, vol. 3.
85. A. Leo, C. Hansch and D. Elkins, *Chem. Rev.*, 1971, **71**, 525.
86. R. A. Scherrer and S. M. Howard, *J. Med. Chem.*, 1977, **20**, 53.
87. A. Collumeau, *Bull. Soc. Chim. Fr.*, 1968, 5087.

88. G. Kortüm, W. Vogel and K. Andrussow, 'Dissociation Constants of Organic Acids in Aqueous Solution', Butterworth, London, 1961.
89. E. P. Serjeant and B. Dempsey, 'Ionisation Constants of Organic Acids in Aqueous Solution', Pergamon Press, Oxford, 1979.
90. D. D. Perrin, 'Dissociation Constants of Organic Bases in Aqueous Solution', Butterworth, London, 1965; supplement, 1972.
91. A. Albert, in 'Physical Methods in Heterocyclic Chemistry', ed. A. R. Katritzky, Academic Press, London, 1963, vol. 1, p. 1; 1971, vol. 3, p. 1.
92. J. Clark and D. D. Perrin, *Q. Rev., Chem. Soc.*, 1964, **18**, 295.
93. G. B. Barlin and D. D. Perrin, *Q. Rev., Chem. Soc.*, 1966, **20**, 75.
94. P. F. Sommer, C. Pascual, V. P. Arya and W. Simon, *Helv. Chim. Acta*, 1963, **46**, 1734.
95. C. Pascual, D. Wegmann, U. Graf, R. Scheffold, P. F. Sommer and W. Simon, *Helv. Chim. Acta*, 1964, **47**, 213.
96. J. F. J. Dippy and S. R. C. Hughes, *Tetrahedron*, 1963, **19**, 1527.
97. K. Bowden and R. C. Young, *Can. J. Chem.*, 1969, **47**, 2775.
98. R. W. Gill, *Prog. Med. Chem.*, 1965, **4**, 39.
99. T. Fujita, T. Nishioka and M. Nakajima, *J. Med. Chem.*, 1977, **20**, 1071.
100. S. Yokono, D. D. Shieh, H. Goto and K. Arakawa, *J. Med. Chem.*, 1982, **25**, 873.
101. C. Hansch, A. Vittoria, C. Silipo and P. Y. C. Jow, *J. Med. Chem.*, 1975, **18**, 546.
102. C. Hansch and A. J. Leo, 'Substituent Constants for Correlation Analysis in Chemistry and Biology', Wiley, New York, 1979.
103. R. W. Taft, D. Gurka, L. Joris, P. von R. Schleyer and J. W. Rakshys, *J. Am. Chem. Soc.*, 1969, **91**, 4801.
104. J. Mitsky, L. Joris and R. W. Taft, *J. Am. Chem. Soc.*, 1972, **94**, 3442.
105. M. J. Kamlet, J.-L. M. Abboud, M. H. Abraham and R. W. Taft, *J. Org. Chem.*, 1983, **48**, 2877.
106. R. W. Taft, J.-L. M. Abboud and M. J. Kamlet, *J. Org. Chem.*, 1984, **49**, 2001.
107. C. G. Swain, M. S. Swain, A. L. Powell and S. Alunni, *J. Am. Chem. Soc.*, 1983, **105**, 502.
108. C. G. Swain, *J. Org. Chem.*, 1984, **49**, 2005.
109. (a) M. H. Abraham, P. P. Duce, J. J. Morris and P. J. Taylor, *J. Chem. Soc., Faraday Trans.* 1, 1987, **83**, 2867; (b) P. Seiler, *Eur. J. Med. Chem.*, 1974, **9**, 473; (c) R. C. Young, R. C. Mitchell, T. H. Brown, C. R. Ganellin, R. Griffiths, M. Jones, K. K. Rana, D. Saunders, I. R. Smith, N. E. Sore and T. J. Wilks, *J. Med. Chem.*, 1988, **31**, 656.
110. C. Hansch, *Acc. Chem. Res.*, 1969, **2**, 232.
111. A. Cammarata and K. S. Rogers, in 'Advances in Linear Free Energy Relationships', ed. N. B. Chapman and J. Shorter, Plenum Press, New York, 1972, chap. 9, p. 401.
112. J. F. Kirsch, in 'Advances in Linear Free Energy Relationships', ed. N. B. Chapman and J. Shorter, Plenum Press, New York, 1972, chap. 8, p. 369.
113. C. Hansch, in 'Structure–Activity Relationships', ed. C. F. Cavallito, Pergamon Press, London, 1973.
114. A. K. Saxena and S. Ram, *Prog. Drug Res.*, 1979, **23**, 199.
115. T. R. Fukuto, R. L. Metcalf, R. L. Jones and R. O. Myers, *J. Agric. Food. Chem.*, 1969, **17**, 923.
116. R. N. Smith, C. Hansch, K. H. Kim, B. Omiya, G. Fukumura, C. D. Selassie, P. Y. C. Jow, J. M. Blaney and R. Langridge, *Arch. Biochem. Biophys.*, 1982, **215**, 319.
117. A. Carotti, R. N. Smith, S. Wong, C. Hansch, J. M. Blaney and R. Langridge, *Arch. Biochem. Biophys.*, 1984, **229**, 112.
118. A. Carotti, C. Raguseo and C. Hansch, *Quant. Struct.–Act. Relat.*, 1985, **4**, 145.
119. S. M. Miller and J. P. Klinman, *Biochemistry*, 1985, **24**, 2114.
120. J. M. Blaney, C. Hansch, C. Silipo and A. Vittoria, *Chem. Rev.*, 1984, **84**, 333.
121. M. Recanatini, T. Klein, C.-Z. Yang, J. McClarin, R. Langridge and C. Hansch, *Mol. Pharmacol.*, 1986, **29**, 436.
122. E. A. Coats, *J. Med. Chem.*, 1973, **16**, 1102.
123. M. Yoshimoto, C. Hansch and P. Y. C. Jow, *Chem. Pharm. Bull.*, 1975, **23**, 437.
124. C. Hansch, M. Yoshimoto and M. H., Doll, *J. Med. Chem.*, 1976, **19**, 1089.
125. E. J. Lien, M. Hussain and G. L. Tong, *J. Pharm. Sci.*, 1970, **59**, 865.
126. C. Hansch, J. McClarin, T. Klein and R. Langridge, *Mol. Pharmacol.*, 1985, **27**, 493.
127. M. P. Dale, H. E. Ensley, K. Kern, K. A. R. Sastry and L. D. Byers, *Biochemistry*, 1985, **24**, 3530.
128. C. Hansch, E. W. Deutsch and R. N. Smith, *J. Am. Chem. Soc.*, 1965, **87**, 2738.
129. V. C. Marhevka, N. A. Ebner, R. D. Sehon and P. E. Hanna, *J. Med. Chem.*, 1985, **28**, 18.
130. N. W. Cornell, C. Hansch, K. H. Kim and K. Henegar, *Arch. Biochem. Biophys.*, 1983, **227**, 81.
131. T. Fujita, *J. Med. Chem.*, 1973, **16**, 923.
132. Y. C. Martin, W. B. Martin and J. D. Taylor, *J. Med. Chem.*, 1975, **18**, 883.
133. J. K. Seydel, K.-J. Schaper, E. Wempe and H. P. Cordes, *J. Med. Chem.*, 1976, **19**, 483.
134. I. Moriguchi and Y. Kanada, *Chem. Pharm. Bull.*, 1977, **25**, 926.
135. M. Yoshimoto, H. Miyazawa, H. Nakao, K. Shinkai and M. Arakawa, *J. Med. Chem.*, 1979, **22**, 491.
136. A. Panthananickal, C. Hansch, A. Leo and F. R. Quinn, *J. Med. Chem.*, 1978, **21**, 16.
137. A. Panthananickal, C. Hansch and A. Leo *J. Med. Chem.*, 1979, **22**, 1267.
138. W. A. Denny, B. F. Cain, G. J. Atwell, C. Hansch, A. Panthananickal and A. Leo, *J. Med. Chem.*, 1982, **25**, 276.
139. P. N. Craig and C. Hansch, *J. Med. Chem.*, 1973, **16**, 661.
140. K. H. Kim, C. Hansch, J. Y. Fukunaga, E. E. Steller, P. Y. C. Jow, P. N. Craig and J. Page, *J. Med. Chem.*, 1979, **22**, 366.
141. S. H. Unger and C. Hansch, *J. Med. Chem.*, 1973, **16**, 745.
142. C. Hansch and E. J. Lien, *J. Med. Chem.*, 1971, **14**, 653.
143. H. A. Selling and A. Tempel, *Pestic. Sci.*, 1976, **7**, 19.
144. C. Hansch, *J. Med. Chem.*, 1970, **13**, 964.
145. C. Hansch, *J. Med. Chem.*, 1968, **11**, 920.
146. C. Hansch, E. Kutter and A. Leo, *J. Med. Chem.*, 1969, **12**, 746.
147. E. Kutter and H. Machleidt, *J. Med. Chem.*, 1971, **14**, 931.
148. C. Hansch, K. Nakamoto, M. Gorin, P. Denisevich, E. R. Garrett, S. M. Heman-Ackah and C. H. Won, *J. Med. Chem.*, 1973, **16**, 917.
149. C. Hansch and R. Kerley, *J. Med. Chem.*, 1970, **13**, 957.

150. E. J. Lien, R. C. H. Liao and H. G. Shinouda, *J. Pharm. Sci.*, 1979, **68**, 463.
151. E. J. Lien, G. L. Tong, J. T. Chou and L. L. Lien, *J. Pharm. Sci.*, 1973, **62**, 246.
152. E. J. Lien and J. P. Li, *Acta Pharm. Jugosl.*, 1980, **30**, 15.
153. M. S. Tute, *J. Med. Chem.*, 1970, **13**, 48.
154. R. C. Young, G. J. Durant, J. C. Emmett, C. R. Ganellin, M. J. Graham, R. C. Mitchell, H. D. Prain and M. L. Roantree, *J. Med. Chem.*, 1986, **29**, 44.
155. W.-Y. Li, Z.-R. Guo and E. J. Lien, *J. Pharm. Sci.*, 1984, **73**, 553.
156. T. Fujita, *J. Med. Chem.*, 1966, **9**, 797.
157. (a) R. A. Scherrer and S. M. Howard, *ACS Symp. Ser.*, 1979, **112**, 507; (b) P. H. Bell and R. O. Roblin, Jr, *J. Am. Chem. Soc.*, 1942, **64**, 2905; (c) P. G. De Benedetti, *Adv. Drug Res.*, 1987, **16**, 227.
158. T. Fujita and C. Hansch, *J. Med. Chem.*, 1967, **10**, 991.
159. M. Yoshioka, K. Hamamoto and T. Kubota, *Bull. Chem. Soc., Jpn.*, 1962, **35**, 1723.
160. S. Yokono, D. D. Shieh, H. Goto and K. Arakawa, *J. Med. Chem.*, 1982, **25**, 873.
161. O. Exner, in 'Correlation Analysis in Chemistry: Recent Advances', ed. N. B. Chapman and J. Shorter, Plenum Press, New York, 1978, chap. 10.
162. C. Hansch, S. D. Rockwell, P. Y. C. Jow, A. Leo and E. E. Steller, *J. Med. Chem.*, 1977, **20**, 304.
163. M. Charton, *J. Org. Chem.*, 1964, **29**, 1222.
164. J. F. J. Dippy, B. D. Hawkins and B. V. Smith, *J. Chem. Soc.*, 1964, 154.
165. K. Kalfus, J. Kroupa, M. Večeřa and O. Exner, *Collect. Czech. Chem. Commun.*, 1975, **40**, 3009.

18.6

Hydrophobic Properties of Drugs

PETER J. TAYLOR

ICI Pharmaceuticals, Macclesfield, UK

18.6.1   INTRODUCTION

It is now almost a century since Overton[1] and Meyer[2] first demonstrated the existence of a relationship between the biological activity of a series of compounds and some simple physical property common to its members. In the intervening years the germ of their discovery has grown into an understanding whose ramifications extend into medicinal chemistry, agrochemical and pesticide research, environmental pollution and even, by a curious re-invasion of familiar territory, some areas basic to the science of chemistry itself. Yet its further exploitation was long delayed. It was 40 years later that Ferguson[3] at ICI applied similar principles to a rationalization of the comparative activity of gaseous anaesthetics, and 20 more were to pass before the next crucial step was formulated in the mind of Hansch.[4] This whole volume, and much of the remainder, pays tribute to what has happened since. But before we lose ourselves in self-congratulation, it is pertinent to enquire how this delay came about and what lessons, if any, there may be for our own time.

Without any doubt, one major factor was compartmentalism. The various branches of science were much more separate then than now. It has become almost trite to claim that the major advances in science take place along the borders between its disciplines, but in truth this happened in the case of what we now call Hansch analysis, combining as it did aspects of pharmacy, pharmacology, statistics and physical organic chemistry.[4] Yet there was another feature that is not so often remarked, and one with a much more direct contemporary implication. The physical and physical organic chemistry of equilibrium processes—solubility, partitioning, hydrogen bonding, *etc.*—is not a glamorous subject. It seems too simple. Even though the specialist may detect an enormous information content in an assemblage of such numbers, to synthetic chemists used to thinking in three-dimensional terms they appear structureless, with no immediate meaning that they can *visually* grasp. Fifty years ago it was the siren call of Ehrlich's lock-and-key theory[5] that deflected medicinal chemists from a physical understanding that might otherwise have been attained much earlier. Today it is the glamour of the television screen. No matter that what is on display may sometimes possess all the profundity of a five-finger exercise;[6] it is visual and therefore more comfortable and easier to assimilate. Similarly, MO theory in its resurgent phase combines the exotic appeal of a mystery religion with a new-found instinct for three-dimensional colour projection which really can give the ingénue the impression that he understands what it is all about. There are great advances and great opportunities in all this, but nevertheless a concomitant danger that medicinal chemists may forget or pay insufficient attention to hurdles the drug molecule will face if it is actually to perform the clever docking routine they have just tried out: hurdles of solubilization, penetration, distribution, metabolism and finally of its non-specific interactions in the vicinity of the active site, all of them the result of physical principles on which computer graphics has nothing to say. Such a tendency has been sharply exacerbated by the recent trend, for reasons of cost as much as of humanity, to throw the emphasis upon *in vitro* testing. All too often, chemists are disconcerted to discover that the activity they are so pleased with *in vitro* entirely fails to translate to the *in vivo* situation. Very often, a simple appreciation of basic physical principles would have spared them this disappointment; better, could have suggested in advance how they might avoid it. We are still not so far down the path of this enlightenment as we ought to be. What is more, there seems a risk that some of it may fade if the balance between a burgeoning receptor science and these more down-to-earth physical principles is not properly kept.

Hydrophobicity may be operationally defined as the tendency of organic molecules to partition away from water into some less polar medium (which in general includes their own bulk phase). In all its manifestations, it is probably the single most pervasive principle in medicinal chemistry. It is most commonly discussed in terms of the partition coefficient P, simply defined as the concentration ratio C_2/C_1 of a single species between two phases at equilibrium, where by convention phase 1 is water. However, there are other aspects and in the hope of making the whole picture comprehensible we attempt a logical sequence. Since the phenomena to be discussed have first to be defined, we start with the theoretical background to solubility and partitioning, treated in a way which is intended to help demystify the subject for the medicinal chemist. There follows a description of the structure of water and the phenomena that comprise the 'hydrophobic effect'. Applications are mixed with matters of theoretical interest in a way that is intended to emphasize interrelationships and provoke fresh thought. Much of this is relevant to the sections that follow on experimental technique. One point must be stressed. Hydrophobicity is frequently discussed almost literally in oil-and-water terms, but actual drug (and other) molecules are not like that at all. They are as much hydrophilic as they are hydrophobic. If the (hydrophobic) carbon skeleton of most drug molecules is there for structural reasons, it is their polar functionalities that dictate their character. Hence we shall play down the often arid studies that continue to labour under the misapprehension, apparently, that yet another demonstration of (say) thermodynamic rectitude in homologous series is of value to somebody.[7] In fact the great majority of fragment values (see Table 11 or Chapter 18.7) are hydrophilic and as medicinal chemists we ought to be concerned with what they can tell us about the *chemical* features of functional groups in the context of their interactions at the molecular level. Here we shall find it illuminating to compare and contrast a variety of partitioning systems; unlike most, this review is by no means confined to 1-octanol. Any solvent system is only a model for some actual biological membrane, and that aspect will be stressed. We shall also consider in some detail the problems posed by ionization, which leads to a mixture of species in one phase or both. When multiple species are present, P has to be replaced by the distribution coefficient D which takes all species into account and, unlike P which is dimensionless, depends on both concentration and pH. Further complications ensue when ion pairs can extract into the organic phase, in which case an extraction coefficient E may be defined. The interrelationships between these quantities are explored in Section 18.6.3.3.11 and some consequences for QSAR in Section 18.6.9; this is very necessary since the different behaviour of ions and neutral species towards reference solvents and actual membranes is both a complicating factor in $\log P$ estimation and a source of great confusion in QSAR studies. While QSAR applications are not our direct concern, it is inevitable that here and elsewhere there should be some overlap.

It is a measure of this subject's slow development that the 1971 review on the partition coefficient by Leo *et al.*[8] was the first ever to be devoted entirely to this topic. It has been drawn on heavily in the text as have two recent books devoted to partitioning[9] and to an account of physical properties in general.[10] Also one should mention the survey of useful physicochemical parameters by Leo and Hansch,[11] Rekker's seminal treatment of the fragmental constant,[12] and two excellent textbooks by Martin[13] and Franke[14] concerning, *inter alia*, their application to QSAR.

In any survey of this sort one has to be selective. The nagging doubt will always remain not merely that one has selected wrongly or unfairly but that one has culpably missed or—perhaps even worse—misunderstood some contribution that should have been given more prominence or treated in a different way. To anyone who feels himself unfairly traduced or neglected I can only apologize. Otherwise this is frankly an account of what I think to be important and I take responsibility for my opinions, warts and all. I have written explicitly for practising medicinal chemists who may legitimately enquire what is in this for them. I hope they will find that question answered.

18.6.2 THEORETICAL BACKGROUND TO SOLUBILITY AND PARTITIONING

We start with an outline of some fundamental thermodynamic relationships in order that the reader may appreciate the common origin of what might otherwise appear to be plausible but unrelated concepts, equations or derivations. A common nomenclature has been chosen which aims at simplicity, consistency and the use of those symbols likely to be most familiar to most readers.

18.6.2.1 Formal Thermodynamic Relationships

18.6.2.1.1 Ideal solution behaviour[15]

The chemical potential (free energy per mole) $\mu(g)$ of a perfect gas at pressure p and temperature T relates to its standard chemical potential $\mu^{\ominus}(g)$ at 1 atm (101 325 Pa) pressure ($p^{\ominus}$) by

$$\mu(g) = \mu^{\ominus}(g) + RT\ln(p/p^{\ominus}) \tag{1}$$

Since $\mu(l) = \mu(g)$ for any liquid in equilibrium with its vapour, equation (1) also describes the behaviour of the liquid phase. For any component A of a liquid mixture

$$\mu_A(l) = \mu_A^0(l) + RT\ln(p_A/p_A^0) \tag{2}$$

where zero superscripts define the pure liquid component. Raoult's law, equation (3)

$$p_A = x_A p_A^0 \tag{3}$$

expresses the empirical relation that its partial pressure is proportional to mole fraction x_A. Hence

$$\mu_A(l) = \mu_A^0(l) + RT\ln x_A \tag{4}$$

Equation (4) is one definition of an ideal solution and is closely obeyed even for quite polar liquids as x_A approaches unity; it is therefore suitable for describing solvent properties if A is taken as the major or 'solvent' component of a binary liquid mixture.

The initial free energy G_i of two liquids before mixing takes the form $\Sigma\, n_i\, \mu_i^0$ where the n_i are molar quantities. The free energy after mixing G_f sums equations of type (4) and the difference $G_f - G_i$ represents the free energy of mixing ΔG_{mix}

$$\Delta G_{mix} = nRT(x_A\ln x_A + x_B\ln x_B) \tag{5}$$

where $n = n_A + n_B$. For ideal fluids $\Delta H_{mix} = 0$, *i.e.* the heat of interaction between all particles is zero or equal; hence

$$\Delta S_{mix} = -nR(x_A\ln x_A + x_B\ln x_B) \tag{6}$$

Since any term $x_i\ln x_i$ is necessarily negative, ΔS is always positive and the mixing of fluids, gas or liquid, is always favourable in the absence of strong repulsive forces (ΔH_{mix} positive) between A and B. Hence all ideal liquids are infinitely miscible and solubility problems do not arise. They do for solids, however, since heat is required (at temperature T) to change these into liquids first. If the minor component B is a solid

$$\mu_B^0(s) = \mu_B(l) = \mu_B^0(l) + RT\ln x_B \tag{7}$$

for solid in contact with saturated solution. The difference in chemical potential between solid $\mu_B^0(s)$ and supercooled solution $\mu_B^0(l)$ relates to the free energy of melting ΔG_m, which is zero at the melting point T_m

$$\Delta G_m = \Delta H_m - T\Delta S_m = 0 \tag{8}$$

but not otherwise. By a series of transformations we obtain

$$\ln x_B = -\left(\frac{\Delta H_m}{R}\right)\left(\frac{1}{T} - \frac{1}{T_m}\right) = -\left(\frac{\Delta S_m}{RT}\right)(T_m - T) \tag{9}$$

Hence mole fraction solubility x_B will fall as its enthalpy (or entropy) of fusion rises and as the differential between T_m and the solution temperature T increases. Independently of this (at $T_m \gg T$) it should rise as T rises, as is commonly found. Equation (9) is the basis for the solubility treatment of solids to be described in Section 18.6.6.

18.6.2.1.2 *Real solution behaviour*

Real liquid mixtures do not obey Raoult's law (equation 3). If the intermolecular forces between A and B exceed those between B and B, p_B will rise with x_B more slowly than expected. If, however, B self-associates more strongly than it associates with A, p_B will rise more rapidly. As an example of the latter, *n*-butanol in alkane solvents is more volatile than as the pure liquid. However, both relations will be approximately linear when x_B is small; *i.e.* under such circumstances that B can be considered as solute to A as solvent. This tendency is expressed as Henry's law

$$p_B = x_B K^H \tag{10}$$

where K^H is the Henry's law constant. When $x_A \gg x_B$, Raoult's law remains a good approximation to solvent behaviour.

The relation between p^0 and K^H is important. Adapting equation (3) to B as solute and substituting equation (10) we have

$$\mu_B(l) = \mu_B^0(l) + RT\ln(K^H/p_B^0) + RT\ln x_B \tag{11}$$

At $x_B = 1$, the observed chemical potential $\mu_B(l)$ is no longer equal to $\mu_B^0(l)$ but to a new quantity $\mu_B^0(A)$

$$\mu_B^0(A) = \mu_B^0(l) + RT\ln(K^H/p_B^0) \tag{12}$$

which is the chemical potential of a hypothetical liquid whose composition is pure B but whose properties are those of an infinitely dilute solution of B in A. The difference between $\mu_B^0(A)$ and $\mu_B^0(l)$ is simply the transfer free energy of B from its own pure liquid to dilute solution in A; the ratio K^H/p^0 is a form of activity coefficient and is commonly written as γ^∞ for a given solute. If some unspecified solute possesses Henry's law constants K_1^H and K_2^H in solvents 1 and 2

$$\Delta G_{1\to 1} = -RT\ln(p^0/K_1^H) \tag{13}$$

$$\Delta G_{1\to 2} = -RT\ln(p^0/K_2^H) \tag{14}$$

from which we obtain $\Delta G_{1\to 2}$ (subsequently written as ΔG_{tr})

$$\Delta G_{1\to 2} = -RT\ln(K_1^H/K_2^H) \tag{15}$$

This extremely important result demonstrates that the free energy of transfer of a solute from one solvent to another may be simply expressed in terms of its Henry's law constants in those solvents. Hence the ratio

$$P_{1\to 2} = K_1^H/K_2^H = \gamma_1^\infty/\gamma_2^\infty = p_1/p_2 \tag{16}$$

is one expression for the partition coefficient P. This equation answers the question, which at first sight might appear a paradox, as to what meaning can be attached to P when the solute is miscible with one or both solvents; similarly, a hypothetical P may be derived for solvent pairs that are wholly or partially miscible. Because of the equivalence between gas and liquid chemical potentials expressed by equations (1) and (2), equations (13) and (14), or the equivalent γ^∞ values, are also ways of expressing partitioning from the gas phase, *e.g.* the gas–water partition coefficient $P_{g\to w}$ (Section 18.6.3).

Vapour pressure measurement is not a generally useful way of estimating P, especially for solids where this is commonly negligible, and a more usable empirical approach is based on solubility. For the distribution of solute between two immiscible phases at equilibrium, say water (w) and oil (o), the condition to be satisfied is equal activities, *i.e.* $a_w = a_o$. Given the definition

$$a = \gamma x \tag{17}$$

and the relation $x = C\bar{V}$, where C and x are molar and mole fraction solubilities and $\bar{V}$ is solvent molar volume, combination with equation (16) gives

$$P_{w\to o} = \gamma_w \bar{V}_w/\gamma_o \bar{V}_o \tag{18}$$

If solubility is treated as a partitioning process between solute and solution, and if by convention $\gamma = 1$ for pure solute, it may be shown that

$$S_B = 1/\gamma_B \bar{V}_A \tag{19}$$

where S_B is solute solubility while A refers to solvent. Hence

$$P = S_o/S_w \tag{20}$$

where S_o and S_w pertain to oil and water while P from here onwards stands by convention for $P_{w\to o}$ unless otherwise stated. From equation (20), factors which affect all solubilities equally vanish in the partition coefficient, an important reason for regarding P as a more objective index of hydrophobicity. As an example, anthracene (m.p. 216 °C) is fiftyfold less soluble in water than phenanthrene (m.p. 100 °C)[16] despite almost identical $\log P$ values.[11,17] This is a simple consequence of equation (9). The occasional confusion between these concepts in the (chiefly biological) literature is unfortunate.

Equation (20) will only be valid if S_o and S_w are both small enough; in practice this limits S_w to 10^{-2} mol dm^{-3} or less and S_o to ten- to hundred-fold greater. The further complications due to mutual solvent solubility have been treated by Chiou and Block.[9] If water is appreciably soluble in the organic phase but not *vice versa*, and if both γ_w and γ_o are appreciably affected, equation (21) may be derived

$$\log P = -\log S_{\text{w}} - \log \bar{V}_{\text{o}}^* - \log \gamma_{\text{o}}^* + \log(\gamma_{\text{w}}^*/\gamma_{\text{w}}) \tag{21}$$

where asterisks denote the quantities so modified. From equations (19) and (20), perturbations due to the solid state will cancel between S_{w} and γ_{o}^*. The first two terms define an 'ideal solubility line' while deviations from the expected P are due to the remainder (see Section 18.6.3.3.4).

18.6.2.1.3   Standard states

The equivalence $x = CV$ above raises the question of standard states. Medicinal chemists are most familiar with the mole ratio definition $P = C_2/C_1$ (standard state of 1 mol dm^{-3}) whereas physical chemists are liable to employ the mole fraction definition $P = x_2/x_1$. These scales interconvert as

$$P = C_2/C_1 = x_2\bar{V}_1/x_1\bar{V}_2 \tag{22}$$

where 1 and 2 are phases. At 25°C, $V_{\text{w}} = 0.018$, $V_{\text{g}} = 24.51$ and, for cyclohexane as an example, $V_{\text{o}} = 0.108$, all in dm^3 mol^{-1}. Hence if $C_{\text{o}}/C_{\text{w}} = P = 1$ at a standard state of 1 mol dm^{-3}, $P = 6$ for cyclohexane–water in mole fraction terms, *i.e.* a constant correction of $\Delta\log P = 0.78$ is required. Particular care is required for gas–liquid partitioning, where different standard states of 1 mol dm^{-3} for solution but 1 atm (101 325 Pa) for the vapour are sometimes employed; here the correction factor to 1 mol dm^{-3} for each is $\Delta\log P_{\text{g}\to\text{s}} = 1.39$ (subscript s = solvent), or in the particular case of $\Delta\log P_{\text{g}\to\text{w}}$, 3.15 if mole fraction in the aqueous phase is specified. It should be noted that any confusion concerning standard states will carry over into the associated thermodynamic quantities; however, the use of incremental quantities avoids this problem (Section 18.6.3.4).

It is a moot point which, if any, of these rival standard states is to be regarded as 'fundamental'. Thermodynamicists tend to prefer mole fraction since it cancels artefacts due to entropy of dilution, which is then the same for both phases.[18a] Ben-Naim[18b] has argued, on the contrary, that molarity is more appropriate especially when discussing, for example, conformational change since this preserves the same average density for the solvent in the vicinity of the solute. A related point concerns alkane log P values. At 1 mol dm^{-3} standard state these are virtually independent of alkane,[17] the natural consequence of an essentially identical environment. In terms of mole fraction, the same values must vary by (at least) twofold. When the two phases interpenetrate, mole fraction P may become indefinable; octanol contains around 0.2 mole fraction water at equilibrium.[19] Except where otherwise stated, all values of P or log P in this review refer to a 1 mol dm^{-3} standard state.

18.6.2.1.4   Temperature variation

The commonly assumed relation between K and T is given by the Gibbs equation

$$\ln K = -\Delta G/RT = -\Delta H/RT + \Delta S/R \tag{23}$$

where K in this context could be P or S_{B}. Hence K will become less favourable with rising temperature if the process is exothermic (ΔH negative) or more favourable if endothermic (ΔH positive). However, whether or not the overall process is favourable (exergonic; ΔG negative) or unfavourable (endergonic; ΔG positive) depends on ΔS as well as ΔH. It is common, but can be misleading (Section 18.6.3), to describe processes in which ΔG is dominated numerically by ΔH or $T\Delta S$ as enthalpy or entropy controlled respectively; misleading because this balance is itself a function of temperature. Furthermore, equation (23) assumes that ΔH and ΔS are independent of temperature, which is not always true. The rate of change of heat content with temperature is defined as the (constant pressure) heat capacity C_p which relates to H and S as

$$C_p = (\partial H/\partial T)_p = T(\partial S/\partial T)_p \tag{24}$$

If each term is separately integrated over a temperature interval ∂T, then imposing the condition from equation (24) that $\partial H = T\partial S$ we obtain the extended equation (25)

$$\ln K = -(\Delta H + T\Delta C_p)/RT + (\Delta S + \Delta C_p\ln T)/R \tag{25}$$

which collapses to equation (23) when $\Delta C_p = 0$. Equation (25) was first used for solution equilibria by Everett and Wynne-Jones;[20] it is particularly likely to be relevant for reactions which, like ionization, show a large change in heat content between initial and final (or transition) states.[21] Of course ΔH itself is a difference term; while S_B can show large variations, P as will later be seen is affected much less by temperature. As a ratio, ΔH is also insensitive to the standard state problems that bedevil ΔG and ΔS (see above).

18.6.2.2 Some Quasi-thermodynamic Approaches

Extrathermodynamic relationships are empirical attempts to explain one set of thermodynamic quantities in terms of another by equations that work and look plausible but possess no necessary validity; these dominate the discussion from Section 18.6.7 onwards. In an intermediate category come quasi-thermodynamic approaches which we shall define as attempts to explain thermodynamic non-ideality in terms of some *de novo* concept which, while plausible, is neither empirically based nor required *a priori*. Since many of these have been used to describe, explain or throw light on the nature of the 'hydrophobic effect', it is appropriate to consider them before going on to discuss how actual solvents and solutes behave.

18.6.2.2.1 *Regular solutions and the solubility parameter*

In the light of equations (5) and (6) it is possible to define excess functions G^E and S^E which denote departures from expectation for a fluid mixture (*any* value for ΔH_{mix} is H^E). Real liquid mixtures are known for which these functions (and others) are non-zero, especially when one component is water.[22] If none is too polar, however, the distribution of particles may remain macroscopically random even though microscopic interactions do occur. This leads to Hildebrand's 1929 definition of a regular solution,[23] $S^E = 0$, so that any departure of ΔG_{mix} from ideality has to be due to H^E.

Hildebrand identified the source of H^E in the cohesive energy density for the condensed phase; $c = -E/V$ if expressed in molar terms. By equating E with internal energy of vaporization ΔU_{vap} we arrive at the solubility parameter δ

$$\delta^2 = (\Delta H_{vap} - RT)/\bar{V} \tag{26}$$

as some measure of the energy required to force liquid particles apart (δ appears as δ^2 since the interaction between A and B makes the geometric mean assumption[24a]). Table 1 contains values of δ from the compilations of Barton and Abraham;[24] Barton's review[24a] presents a full account of the theory. Hildebrand supposed that any large mismatch in δ for two liquids will lead to a net repulsion that should pass through a maximum at equal mole fraction

$$H^E = (x_A \bar{V}_A + x_B \bar{V}_B)(\delta_A - \delta_B)^2 \phi_A \phi_B \tag{27}$$

where ϕ_A and ϕ_B are volume fractions. Despite Hildebrand's assumption that it should fail for immiscible liquids, since regular liquids like ideal ones are supposed to be miscible, an extension of the theory in the form of equation (28) has been used to calculate the variation of P for a given solute B between water–oil solvent systems[25]

$$\ln P = (\bar{V}_B/RT)\{(\delta_w - \delta_o)(\delta_w + \delta_o - 2\delta_B)\} \tag{28}$$

though only when a non-standard δ_w was used (Table 1). Similarly, Hafkenscheid and Tomlinson found systematic errors (by a factor of 4–5) when attempting to relate HPLC retention time to eluant

Table 1 Values of Hildebrand's Solubility Parameter δ

Solvent	δ	Solvent	δ
Water	23.4	Chloroform	9.3
Water	16.5[a]	Benzene	9.2
Formamide	19.2	Tetrachloromethane	8.6
Glycerol	16.5	Butyl acetate	8.6
Methanol	14.3	Cyclohexane	8.2
1-Butanol	11.4	Diethyl ether	7.4
Nitrobenzene	11.1	Heptane	7.4
1-Octanol	10.3	Perfluoroheptane	5.6

[a] Ref. 25.

composition; parallel problems arise in the calculation of S_w.[26] The central weakness of regular solution theory is that equation (27) is always zero or positive; it will account for repulsive deviations from ideality but not attractive ones, as Hildebrand himself fully recognized.[23] Nevertheless, these concepts have proved very valuable in the treatment of non-polar and moderately polar liquid mixtures and must be considered as milestones on the way.

18.6.2.2.2 Statistical mechanics: scaled particle theory

Statistical thermodynamics, as a sister discipline to the formal variety, has principally contributed to this subject through scaled particle theory (SPT).[27] Essentially, this considers a statistical assembly of hard spheres and splits their interactions into attractive and repulsive terms which are parametrized separately and explicitly modelled in terms of potential functions, usually of the Lennard-Jones type. A comprehensive account of its application to the 'hydrophobic effect' is given by Ben-Naim.[18] The repulsive part of the liquid–liquid interaction is approximated by a cavity term which specifies particle diameter and cavity radius; this sounds an exact approach but, in the opinion of Ben-Naim,[18c] neither cavity volume nor surface area (CSA) is defined precisely enough for real molecules. An attempt to model micelle formation correctly predicted its sign but greatly exaggerated its magnitude.[18d] The weakness of SPT once again is that, in not considering specific interactions, it is incapable of accounting for them. Nevertheless, in drawing attention to the role of cavity formation in solution processes, it identified a crucial conceptual element whose ramifications will be explored in succeeding sections.

18.6.2.2.3 Local composition theory

This approach to solution energetics originated in chemical engineering and has been pioneered in the present context by Grünbauer and his associates.[28] Like SPT it divides interaction energy into two parts, combinatorial (repulsive) and residual (attractive)

$$\ln \gamma_i = \ln \gamma_{comb,i} + \ln \gamma_{res,i} \tag{29}$$

where i is any component of a multicomponent liquid mixture. The former, which is intended to handle the entropic or size aspect, contains volume and surface area terms. The latter derives from mutual miscibility data that have allowed a large number of pairwise interaction terms A_{ij} to be calculated and compiled. Essentially this approach should allow the estimation not only of mutual liquid solubilities, which was the original intention, but also of γ_B values for the minor liquid component considered as solute. The ratio of two γ_B values in different liquids then gives P on the mole fraction scale. An enormous virtue of local composition theory (LCT) is that it allows P to be calculated for a solute distributed between mutually saturated solvent layers, since there is no limit to the number of components that can appear in the equations for each layer; in fact calculation is generally better for such systems,[28] probably because A_{ij} can be calculated more accurately. Two serious drawbacks are apparent. Firstly, these A_{ij} values are not for infinite dilution and can be seriously in error for associated liquids. Secondly, they are only obtainable for liquid solutes. Hence a more recent study[29] concentrated on the calculation of γ_{comb} and obtained γ_{res} as a difference term. For relevant $\log P$ calculation LCT is clearly limited, but there are obvious applications, for example to solubility and HPLC retention time in mixed aqueous systems, which await exploitation.

18.6.2.2.4 Cavity theories: volume vs. surface area

The crucial importance of cavity formation has been glanced at above. Among the peculiarities of water (Section 18.6.3), its abnormally high internal pressure (Table 1; equivalent to about 15 000 atm[30]) ranks high on the list. No organic liquid approaches this (typical values are six- to ten-fold less[30]). Hence, even for solutes such as the noble gases which neither attract nor repel any other molecule, $\log P$ is expected to be positive and, by equation (28), to rise monotonically with solute size. In this light, the many published relations between $\log S$ or $\log P$ for a set of non-polar solutes and their volume or some related quantity may be seen as glimpses of the obvious and will not be reviewed except where some special feature is apparent. Instead, attention will concentrate on two areas of real and recent controversy. One is whether cavity volume or cavity surface area should be used in such correlations. The other concerns just what can be meant by molecular volume or surface area at all.

For a sphere, area $\propto V^{0.67}$, and there should be little problem in deciding which is relevant. Few real molecules are spheres, however, and the exponent rises towards unity the more ellipsoidal they become. For 200 substituted aromatics, Pearlman[9] has shown that volume and area are linearly correlated over a fourfold volume range (using the algorithms SAVOL and SAREA[31]). This puts firmly in their place previous 'decisive' claims in favour of either. In fact, perfectly respectable arguments may be advanced for either view. Volume is a hallowed thermodynamic quantity on which many properties depend; in particular, cavity size must relate to the number of solvent molecules displaced. On the other hand, all interactions take place at the surface of molecules; their interiors are irrelevant. There is simply no way of deciding this issue at the present time.

The definition of both quantities is riddled with ambiguity. Given that the close packing even of spheres leaves 26% of the space as free volume, it will be realized that V need bear no constant relation to intrinsic volume V_I. In their multiparameter analysis of $\log P$ (Section 18.6.7.1), Taft and his co-workers[32] were forced to add $10 \, \mathrm{ml \, mol^{-1}}$ to all cyclic compounds to fit them to the regression. Recently, Leahy[33] has shown that use of V_I as van der Waals or Bondi volume[34] will remove this anomaly, and this suggestion has been taken up.[35] Evidently cyclic molecules pack more closely than their open-chain analogues, perhaps for entropic reasons, although V_I appears to possess no advantage as a descriptor for $\log S_w$.[33, 35] Nevertheless V_I is also valuable in that it applies equally to solids. Equally serious problems of description plague molecular surface area. Pearlman[9, 10] distinguishes three types: van der Waals, solvent accessible surface area (SASA)[36a] and contact.[36b] Superficially SASA would seem the most reliable but, as pointed out by Richards,[36b] any preference must depend critically on the choice of solvent radius, and Pearlman[9, 10] tends to favour contact for general use. The last word has not yet been said on this subject either.

Equation (30), used by Pearlman *et al.*[9, 10] to correlate S_w for 64 liquid alkyl- and halo-substituted benzenes, introduces the possibility of factoring molecular volume (or area) into fragment values along the lines of $\log P$ (Section 18.6.7.2); here the i are functional groups.

$$\log S_w = a(\text{area}) + b \Sigma c_i V_i + \text{constant} \qquad (30)$$

In the writer's view, this approach holds grave dangers. The factoring of V was initiated by Bondi himself but, because of occluded volume, the use of group contributions is an essentially unsound procedure which must now be regarded as superseded by accurate computer algorithms.[31, 34] Factoring of surface area is still more problematical since, even if this can be performed consistently, the properties of a given functional group in a series of compounds bear no necessary relation to relative surface area. Table 2 contrasts the effect of steric hindrance on the proton acceptor ability (pK_β[37]) for three classes of compound with SASA for the acceptor atom.[38] In no series is any coherent relationship evident. Ether oxygen can lose 96% of its SASA for a mere fivefold loss in acceptor ability. Ketones are little more responsive; it is indeed very difficult to hinder them at all. Even pyridines, far the most sensitive class since the single lone pair is highly directional, show no effect that could be quantified. These results help to substantiate Meot-Ner's pronouncement[39] that 'so long as there exists a *single* conformation in which the hydrogen bond can attain optimal geometry, the bond strength is not weakened by steric crowding'. Strictly this applies to enthalpy, and entropy must make some difference, but clearly not much. This principle lends much support to one implicit justification for $\log P$ additivity schemes (Volume 4, Chapter 18.7), namely that fragment values are insensitive to minor steric fluctuations. Hence we are forced to regard as largely artefactual the analysis by Franke and his co-workers[14, 40] of $\Delta G_{w \to o}$ in terms of CSA group contributions which for a given functionality appear to vary substantially along series, generally falling with chain length. Unfortunately also, through the way the equations are constructed,

Table 2 Relation between pK_β and SASA

Compound	SASA ($\mathrm{\AA}^2$)	pK_β	Compound	SASA ($\mathrm{\AA}^2$)	pK_β
MeCOMe	47.28	1.61	Pyridine	23.63	2.52
Bu^{t}COMe	35.34	1.44	2-Methyl	17.88	2.60
Bu^{t}COBu^t	21.23	1.40	2,6-Dimethyl	12.11	2.74
Tetrahydropyran	22.35	1.70	2-t-Butyl	7.17	1.88
Bu^{t}OMe	11.9	1.46	2,6-Di-t-butyl	0.96	−0.2
Bu^{t}OBu^t	0.97	1.02			

the common hydrophilic component of each polar group appears in the intercept term; $\Delta G_{w \to o}$ misleadingly appears as favourable in every case. Similarly, Pearlman's claim[9] that molecular properties are irrelevant to $\log P$ is simply absurd. Such errors are only too likely when compounds or candidate parameters are chosen with insufficient care; the literature is strewn with such examples. Others will appear in due course. The formalism of Taft *et al.*[32, 35] (Section 18.6.7.1) is a much clearer way of expressing the nature and extent of the various competing factors. The writer believes that, under this heading, most other forms of data analysis are now redundant.

18.6.2.2.5 *Solvent-dependent conformational analysis*

This procedure, due to Hopfinger and Battershell,[41] is perhaps the only related technique to escape the above strictures. It starts from the (usually MO) optimized conformation for a given molecule round which a solvation shell specific to the solvent is then constructed. Precise geometries are used so that cavity volume and surface are explicitly modelled. The interaction energies are computed and a table of solvation shell parameters for various functional groups in water and octanol is given.[41] $\log P$ then comes out as a difference term between the total interaction energies. Values for 20 assorted molecules were reproduced with a fair degree of success. The method is computationally intensive, which may explain why little more has been published. Nevertheless solvent-dependent conformational analysis (SCAP) appears to have great potential for $\log P$ calculation of individual conformers, differences between which are known to exist[42, 43] and it is rather surprising that it has not been exploited further.

18.6.2.2.6 *Molar refractivity and the parachor*

These volume-related quantities have received extensive use. The parachor P_{α}, defined by

$$P_{\alpha} = M \gamma^{1/4}/\rho \tag{31}$$

where γ is surface tension and ρ is density, was invented in 1924 by Sugden as a quantity that roughly corrects V for intermolecular forces. Its use in correlating physical properties and physical toxicity was pioneered by McGowan.[30] Equations typically take the form

$$\log Q = a P_{\alpha} + E \tag{32}$$

where Q is some property and E is a polar correction term. Equation (32) is a typical two-term equation, one term being due to volume, on the lines of equations (29) or (30). In practice P_{α} may be calculated from additive rules which involve atom and bond factors, so allowing extension to solid solutes; more recently, McGowan has defined on similar lines a characteristic volume V_x which relates closely to V_1 but does not require a computer for its calculation.[30] McGowan had great success in correlating the simpler non-specific forms of biological activity at a time when the physicochemical approach was in a state of almost total neglect. His influence in redirecting attention to basic physical principles was timely and profound.

The use of molar refractivity MR

$$\text{MR} = (M/\rho)\{(n^2 - 1)/(n^2 + 2)\} \tag{33}$$

where n is refractive index has been pioneered by Hansch, who for 75 compounds has demonstrated[11] a close relation with the parachor ($r^2 = 93\%$). Again, MR is an additive quantity. As a biological descriptor (except where it correlates with π or $\log P$),[11] MR is essentially a measure of solute bulk and implies some kind of non-specific attractive force other than simple hydrophobicity. Recent QSAR[44] and computer graphics[45] studies are relevant. In the binding of trimethoprim analogues to dihydrofolate reductase (DHFR), π appears as a substituent descriptor for the bovine liver enzyme but MR for that of *E. coli*.[44] Computer graphics studies show the substituents to line the inner face of a hydrophobic pocket in the first case but to lie along a channel pointing out of the enzyme in the second.[45] Hansch and Leo[11] discuss this situation in terms of total and partial desolvation respectively. Alternatively, where MR fits, the substrate may itself 'solvate' the receptor, displacing loosely bound water but preserving the existing pattern of interactions at the receptor surface, whereas in 'pocket filling' these totally change; *i.e.* 'surrogate solvation' when MR correlates rather than desolvation as such. In their extensive cross-comparison of volume-related quantities,

Charton and Charton[46] found evidence for some lone pair component in MR; similarly we[47] find MR to correlate mostly with V_1 but with a small β component ($r^2 = 97\%$ for 29 assorted polar liquids). Both observations emphasize the polarizability aspect. Whatever its precise significance in QSAR, MR is not a hydrophobic parameter.

18.6.2.2.7 *Molecular connectivity*

Molecular connectivity (MC) is a quantification of topology;[48] it is classified by Franke[14] as a steric parameter. As a concept based on molecular structure but which takes no account of molecular properties, it is conveniently considered here. Its use in chemistry and biology has been pioneered by Kier and Hall, whose book is the definitive text.[48] While the idea goes back a long way, its modern use begins with the invention by Randić of a branching index which handles isomers uniquely. Its use[48] is based on calculation of the connectivity index χ for a hydrogen-suppressed hydrocarbon matrix. This index has a different value for each isomer and, by an extension of the theory, can handle multiple bonds and first-row heteroatoms (arbitrary numbers are needed for halogens). Physical properties of hydrocarbons and homologous series are in general well correlated; the effects of chain branching are handled particularly well. Calculation is simple and their use has become widespread.

There has been some tendency, therefore, to use MC in contexts for which it is unsuitable, and unfortunately this has been encouraged by Kier and Hall themselves. For example, where $\log P$ will give good results in the QSAR of homologous series, then so inevitably will χ, and there is no cause for rejoicing if r is slightly better. The likely reason for this when it happens is some imprecision in the estimate of $\log P$. What else can happen is illustrated by the following cautionary tale. In Kier's most ambitious correlation between χ and $\log P$, 138 data points were fitted with $r^2 = 98\%$. Hence Lopez de Compadre *et al.*[49] were somewhat surprised to find that the binding of 42 diverse aromatics to bovine serum albumin, whose correlation with $\log P$ is excellent ($r^2 = 92\%$), is not fitted at all by χ ($r^2 = 11\%$). Inspection of the original data[48b] will soon reveal why. The 138 compounds consist of homologous and mostly long-chain series of alcohols, ethers, amines, ketones, esters and carboxylic acids, each of whose correlations with χ is of similar slope and also of similar intercept (because proton acceptor ability is nearly equal[32]); hence all these equations are readily combined. The summary equation covers five decades in $\log P$ and is dominated by χ for the alkane component. By contrast, the 42 diverse aromatics[49] vary little in size—from 7 to 16 heavy atoms—but greatly in functionality. Most compound series of interest in drug design much more closely resemble the second situation than the first; and here, inevitably, χ entirely fails to cope.

Exponents of MC are also frequently guilty of statistical sins such as the combination of closely similar indices in the same equation;[50] indeed this is difficult to avoid if they are combined at all. Curiously, MC has not been used in one context where it might have been employed to advantage. Diamond and Wright[51] report that chain branching often greatly reduces penetration into membranes, to an extent entirely out of line with expectation based on $\log P$; this is explained as being due to increased disruption of their three-dimensional structure. Except for such potential applications which make use of its real strengths, the writer is forced to regard molecular connectivity as an irrelevance which has had the unfortunate effect of diverting attention from the real work that needs doing.

18.6.3 WATER AND THE HYDROPHOBIC EFFECT

The literature on this subject is enormous and we have to be selective. Water is the subject of a comprehensive treatise,[52] a useful short monograph[53] and several valuable articles[21, 22, 54, 55] which we shall quote. The 'hydrophobic effect' is the subject of two books[18, 56] and many reviews, of which certain are specially pertinent.[57-61] Other key articles will be quoted as we go along. We shall start by discussing water structure as revealed by various physical probes and go on to describe its interactions with other substances.

18.6.3.1 The Structure of Water

Water is the universal liquid on which all life depends. Whether life *must* depend for its existence on water is an interesting question since water as a liquid is abnormal and in some ways unique.

Some of these abnormalities are widely appreciated, some not. Judged on size alone, water should freeze and boil at about $-100\,°C$ and $-80\,°C$ respectively;[55] that it does not is due to enormous cohesive forces that are generally attributed to hydrogen bonding. In fact that is not quite true: around 30% is due to dispersion (van der Waals) forces which are twice as strong in water as in liquid methane.[55] Yet, while most solids expand by about 10% on melting, ice floats on water. This is well known to be due to the diamond lattice of ice in which all possible hydrogen bonds are formed with optimum geometry (Franks[53] and Tanford[56] give good diagrams). Hence ice possesses an extraordinarily open structure relative to most solids; the coordination number of four for water in ice compares with 12 for solid argon.[53, 56]

Strange things happen on melting. Up to $4\,°C$, water contracts; its coordination number is now 4.4 (for liquid argon it falls to 10) and this continues to rise as water expands again with temperature. Hence even at $4\,°C$ the structure of water remains remarkably open and most solutes dissolve with a large net constriction in volume.[58] Two types of theory have been advanced to accommodate these facts. The continuum theory, put forward by Pople in 1951, supposes simply that some hydrogen bonds are bent or broken in random fashion. The mixture theory, originating in the 'flickering clusters' of Frank and Wen in 1957, supposes that virtually discrete regions exist, if only of very short lifetime; some part preserves the structure of the diamond lattice to give 'icebergs' that float in a sea of much more densely packed and even perhaps unbonded water molecules.[57] In recent years the 'iceberg' theory has had the better press, partly no doubt for its visual qualities but also because it can explain in simple fashion some otherwise puzzling features of the way in which water interacts with other substances.

We consider cosolvents first. Figure 1(a) displays ideal behaviour for liquid mixtures according to equations (5) and (6). Figure 1(b) exemplifies expectation for two solvents that interact; an exotherm is partially balanced by some motional constriction, but nevertheless the interaction is favourable overall. Figure 1(c), which displays no actual case, represents the type of situation that results when ethers and alcohols are mixed with water. An exotherm develops as cosolvent is added, but its maximum is quickly reached ($x_B \sim 0.15$ is typical) and, starting from the other end, addition of water to cosolvent is often endothermic. What is more, because of a large entropic term the overall interaction is at all times unfavourable. This last is the most important point of all. Water behaves to almost all adducts other than ions, even to molecules one thinks of as hydrophilic, as if to a foreign body.

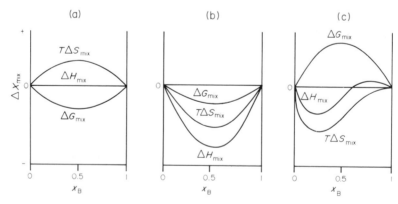

Figure 1   Excess thermodynamic functions ΔX_{mix} for hypothetical binary liquid mixtures: (a) an ideal mixture according to equation (5); (b) a mixture in which a 1:1 complex is formed; (c) typical behaviour of water A with an organic cosolvent B

The 'iceberg theory' will handle these facts in the following way. Water reacts to a non-polar solute by in a sense sealing it off; order increases in its vicinity by the formation of more (or larger) 'icebergs'. This view received a strong initial impetus from the discovery of the alkane and noble gas hydrates, though it is not now thought that these clathrate structures are actually preserved in solution (for a discussion see Neméthy[57]). This increase in order is responsible for the (slightly) exothermic dissolution of alkanes, not therefore from solute–solvent but from increased solvent–solvent bonding. However, this 'internal freezing' forces a massive drop in entropy which overrides the exotherm to give the observed net unfavourable effect. This process of structure making is the basis for the 'hydrophobic effect', which in the sense of Figure 1(c) may be regarded as entropy controlled. Polar solutes provoke less solvent reorganization so ΔS is much less negative;

this and not a larger exotherm accounts for their greater solubility—the sign of ΔH depends on the net change in bonding, which may or may not compete effectively with that of water to itself.[57,58] The hydrophobic effect 'arises not from interaction with a solute, but rather from the absence of such interaction'.[21]

Further evidence comes from other sources. Water's exceptional heat capacity is readily interpreted in terms of 'icebergs': as temperature rises these start to 'melt'.[21] Again, its increase in viscosity on adding cosolvents speaks for some gelation,[58] though otherwise its viscosity and self-diffusion rate is normal.[53] Some evidence is more neutral. Howarth[62] has demonstrated by NMR a threefold restriction of motion for alkyl groups in alkanols relative to other solvents of similar viscosity, and from this has argued that the hydrophobic effect is a function of solute more than solvent. Its variation of ΔH^0, ΔS^0 and ΔC_p^0 with temperature has been interpreted as implying about 50% each at 25 °C of four-bonded and three-bonded water molecules,[57] which is consistent with Luck's recent IR-based estimate[55] of 15% free OH. However, this does not decide between theories, all of which agree that some OH is unbonded (which is very different from alleging that wholly unbonded water is present). The effect of ions is equivocal. While ions of high charge density are structure makers as judged by their large negative ΔS, they impose their own highly directional structure which is quite incompatible with 'iceberg formation' and can lead to curious incompatibilities for molecules containing both features.[21] Ions of low charge density tend instead to be structure breakers, inducing a higher proportion of disordered water. These two types of ion salt-out and salt-in non-electrolytes respectively, and the same sequence appears in the 'Hofmeister order' for their effect on the aqueous solubility of proteins.[53]

It has recently become evident that the hydrophobic propensities of water are not quite unique. While alkanols are incapable of water's three-dimensional structure and never form more than two hydrogen bonds,[22] ethylene glycol, glycerol, formamide, hydrazine and the liquid salt $EtNH_3^+ NO_3^-$ (ethylammonium nitrate, EAN) also reject lipoidal solutes and show a strong tendency towards immiscibility with other solvents. Work on some of these has helped towards the recent reappraisal of the hydrophobic effect to be described in Section 18.6.3.4.

18.6.3.2 Hydrophobic Interactions in Aqueous Solution

Hydrophobic interaction in water 'is a partial reversal of the entropically unfavourable hydration process'.[53] Its extent will depend on size, shape, the presence and nature of polar groups and what alternatives are open to the solute. Simple inorganic ions require quite high concentrations ($> 0.1 \text{ mol dm}^{-3}$) even to begin to associate despite the stimulus of electrostatic attraction, whereas hydrophobic cations and anions can do so with $K_{ass} > 10^4$ depending on chain length.[18,58] Even ions of the same sign, e.g. R_4N^+, will self-associate if the hydrophobic portion is large enough;[53] if the shape is right, i.e. if the 'head' (polar) and 'tail' (non-polar) regions are sufficiently distinct, micelle formation can result. Since like charges repel, the charge here is counter-productive so a simple uncharged polar head might be expected to make self-association more favourable, and this is borne out: the critical micelle concentration (cmc) for non-ionic surfactants is typically a hundredfold lower than for ionic ones.[63]

The description of such associations as 'hydrophobic' is substantiated by a number of facts. For the dimerization of carboxylic acids, K_{ass} is a function of chain length in water but not in benzene.[18d] While polymethacrylate at high concentration will solubilize alkanes, polyacrylate will not; evidently the methyl groups are essential.[18e] Nucleic acid bases solubilize one another, and that these highly polar molecules dimerize not by hydrogen bonding but by dispersion forces has been demonstrated in at least one recent study.[64] Consistently, non-polar *and* polar cosolvents added to water help to solubilize non-polar solutes and even the interaction between polar solutes and non-polar cosolvents is in general only mildly endergonic.[65] Like attracts like: in the self-association of peptides in solution, $CH_2 - - - CH_2$ and $CONH - - - CONH$ contacts are favoured, whereas $CH_2 - - - CONH$ contacts are not.[66]

If enough amphiphiles self-associate, micelles are formed. These have attracted attention since they begin to bridge the gap between simple aqueous and more complex biological systems.[18] A typical micelle may contain 100 units but in fact a range of sizes is generally present.[67] Recently, the classical Hartley model of a micelle as a tight spherical pincushion has given way to the picture of a much looser structure, still with its ionic or polar head-group on the surface but with a jumbled and quite 'wet' interior (for account and good diagrams see Menger[67]). This will help explain the imprecision of the cmc; micelle formation does *not* represent a new phase—above the cmc only *most* of the material goes in to the micelle; also the existence of a premicellar stage of which not much is known.[63] The cmc falls linearly with alkyl chain length while micelle size increases.[18,63] Above the cmc, micelles will solubilize non-polar compounds, sometimes to a remarkable extent; sodium dodecylsulfonate (SDS) can absorb up to 0.9 mole fraction benzene;[67] typically K_{ass} lies in the range 10^3–10^5. Binding tends to show the Michaelis–Menten pattern and very often small molecules appear better absorbed than large ones, as if a fixed amount of space were available.[63]

Vesicles are micelles with two (or more) hydrophobic tails; these form bilayers that possess an internal solvent-filled cavity.[56] Liposomes are phospholipid vesicles and in their natural state form the core of phospholipid membranes. They have become of great interest, both as potential drug delivery systems and, within the scope of this review, through the very interesting results of partitioning studies (next section). A characteristic feature of liposomes is a transition temperature T_c which is commonly close to ambient and will always lie above T_m for the parent hydrocarbon;[56] it involves indeed a sort of 'interior melting'. Proteins lie altogether beyond the scope of this review but one or two points deserve mention. It is well established[18,58] that amino acids with polar and non-polar side chains tend to concentrate in the surface regions and interior respectively; indeed calculated $\Delta G_{g \to w}$ values for these side chains correlate well with this inside–outside balance.[59] Tanford's list of $\Delta G_{w \to alc}$ values[56] similarly ties in qualitatively with the sort of side chain π value one might expect.[68] Amphiphiles bind to proteins much as they might to a micelle.[56] Protein structure is exquisitely balanced; in general the native (folded) structures exist only in aqueous media[53] and even there are quite readily denatured, *e.g.* by high concentrations of urea. Since this will also raise the cmc, it is evident that small, large and very large molecules all obey essentially the same rules in aqueous solution.

18.6.3.3 Aspects of Partitioning from Aqueous Solution

Not all forms of partitioning are alike. In this section we discuss a number of special features, problems and peculiarities; some are confined to context while others possess a more general significance. We place particular emphasis on those which are important for the proper interpretation of $\log P$ in QSAR.

18.6.3.3.1 Gas–liquid partitioning

Since, at temperature T, mean gas velocity falls with rising molecular size,[15] increasing mass should increase the tendency to self-condensation as is well known to be the case. The same factor will oppose the (unfavourable) energy of cavity formation which also increases with molecular size, and since all ideal liquids are miscible, gas–solvent partitioning ($P_{g \to s}$) should increase with M. With one significant exception, this is found. Hine and Mookerjee[69] have compiled a list of intrinsic hydrophilicities $\log \gamma$ (defined as C_w/C_g with both standard states 1 mol dm^{-3}). Martin[13] has noted a rough inverse relation with $\log P_{oct}$ but this is superficial. Table 3 lists some selected aliphatic fragment values $\log \gamma_i$ (Hine's bond and atom contributions could do with recalculating in this form; see Section 18.6.7.2). These are contrasted with f_{oct} (see equation (66) for definition)[12] reversed in sign so as to represent the same direction of transfer. It is found that, in $\log \gamma$, *only* alkyl groups are hydrophobic; even halogens show a net attraction to water. This is particularly significant in the context of volatile anaesthesia. As Cramer[70] has pointed out, this form of biological activity is unique in that bulk and polarity both operate to increase it. In practice of course the allowed polar groups are severely limited by other effects, but this will explain why the weak proton donor property of CH in halocarbons is advantageous.[71] Gas–solvent partitioning is a useful reference point and we shall return to it later.

Table 3 Fragmental Log γ Values[69]

	CH_3	CH_2	CCl_3	Cl	Br	I	O	NO_2	$C{=}O$
$\log \gamma_i$	−0.62	−0.15	0.80	1.30	1.35	1.39	3.08	3.58	4.03
$-f_{oct}$	−0.70	−0.53	−1.79	−0.06	−0.27	−0.59	1.58	0.94	1.70

18.6.3.3.2 Association in the organic phase

This is a much feared problem in attempting to obtain authentic log P values.[8] It is much worse for non-polar organic phases than for polar ones, but since log P is generally lower in these the problem is to some extent self-limiting. Much valuable work particularly by Affsprung and his co-workers has been reviewed.[72] At low C_w the relation is linear in solute

$$P_{obs} = P + m[C_w] \tag{34}$$

where P is the authentic value (Leo *et al.*[8] give more general equations). For benzoic acid in benzene, $C_w = 10^{-2}$ mol dm^{-3} increases P sixfold; as C_w rises further higher-order terms appear, some of which result from the extraction of water into benzene. Carboxylic acids are of course prone to give dimers; $K_d > 500$ is commonly found. Other suspect solutes include phosphoric acids and primary amides. Similar effects are shown by phenol in CCl_4 as organic phase; $(P_{obs} - P)$ starts to become perceptible only at $C_w > 0.1$ mol dm^{-3} largely as a result of water extraction. For self-associating species, this last becomes important only when dimer formation has already begun; for polar species that do not self-associate, $1 < K_{ass} < 10$ typically for association with water in non-polar solvents. In neither case, therefore, does water extraction pose practical problems in log P determination given $C_o < 10^{-2}$ mol dm^{-3}. It is never going to matter in phases such as octanol for which $[H_2O] \gg [solute]$. For the extreme case of carboxylic acid dimers, C_o *ca.* 10^{-4} mol dm^{-3} is likely to be required if the solvent is non-polar, with C_w adjusted accordingly. Dimerization in water does occur ($K_d < 1$) but can be neglected. The situation may be checked along the lines of equation (34). Given a reasonably sensitive way of monitoring the solute, these problems are readily avoided; nevertheless, many published log P values have probably to be viewed with suspicion.

18.6.3.3.3 Folding of alkyl chains in aqueous solution

Early allegations of folding are now only of historical interest.[12] Nevertheless the possibility exists and we approach it *via* the cycloalkanes, which form reasonable models for the type of structure one might expect if folding occurs. Cyclohexane in the chair form fits the diamond lattice and there are no internal clashes but dispersion interactions in a C_6 chain folded this way are unlikely to overcome entropic constraints and no such effect has been demonstrated. After this ring size, internal clashes are severe all the way to cyclotetradecane, which possesses a unique and strain-free structure.[73] Hence a C_{14} chain is the shortest for which folding is likely. Nelson and de Ligny[74] showed that S_w for the linear alkanes decreases steadily with chain length till just after C_{14}, beyond which it stays constant; folding is suggested as this restricts disfavoured interaction with water. Consistently, recent partitioning work[75] finds no evidence for folding till beyond C_{12}, and hydrolytic rates for *p*-nitrophenyl esters[76] show that a C_{16} but not a C_{12} chain can fold. There are two contrary reports. Hada *et al.*[77] allege that long-chain alkanes containing a polar group in the middle may be less hydrophobic than expected since they can fold back on themselves, but inspection of their data reveals no evidence. Mayer *et al.*[78] measured log P values accurately for a range of compounds of type $Ar(CH_2)_nX$ where Ar is phenyl or pyridyl and $X = OH$ or NH_2 and alleged that their results demonstrate folding, for pyridyl, of X back on to Ar. Unfortunately, they chose to mix measured and calculated values in an illegitimate way. The results are given as $(\log P - \Sigma f)$ where the aryl f values (equation 66) are mutually inconsistent; expressed as π (equation 65) the phenomenon has disappeared by $n = 4$ and any deviations before this point are readily explained as field effects. The most curious report under this head concerns the allegation by Beezer and Hunter[79] that f_{CH_2} may alternate in value along an alkyl chain. The effect is small but appears to be definite and is not confined to P_{oct}. It is suggested that this effect goes back to the liquid phenol or alkanol; more probably, since such effects should cancel, it derives from fluctuations in the *trans–gauche* balance such as are known to occur.[73] However, there is no sign of any such phenomenon in the reported values of alkyl chains incorporated into molecules *between* polar moieties (nor should there be). We conclude that, with very few exceptions, f_{CH_2} is the constant it is always taken to be (*cf.* Figures 11 and 12).

18.6.3.3.4 Solubilization by the organic component in the aqueous phase

This is another recurrent worry on which it is difficult to find hard evidence. Some of the best so far has been provided by Chiou and Block[9] (Table 4) who, as well as determining log P for several

Table 4 Solubility Partitioning Comparisons for Very Insoluble Solutes

	$S_w(\mu g\,ml^{-1})$	$S_o(mg\,ml^{-1})$	$S_o^*(mg\,ml^{-1})$	$S_h(mg\,ml^{-1})$	$log(S_o/S_w)$	$log(S_o^*/S_w)$	$log\,P_o$	$log(S_h/S_w)$	$log\,P_h$
C_6Cl_6	5.0	3.53	2.65	5.16	5.85	5.72	5.50	6.01	5.96
DDT	5.5	41.5	31.9	30.5	6.88	6.76	6.36	6.74	6.66

very hydrophobic compounds in octanol (o) and heptane (h), measured S in the pure solvents and in those solvents saturated with one another (indicated by asterisks). The formalism they used has been given as equation (21). The discrepancy between $log\,P_o$ as measured and $log\,S_o/S_w$ is adequately explained by a small contribution due to a fall in solute solubility in octanol through added water, and a larger one ($\Delta log\,P = -0.22$ and -0.40 for C_6Cl_6 and DDT respectively) by octanol in the aqueous phase (both effects must necessarily be negative). For heptane–water where mutual solubility is slight there is almost no discrepancy. This raises the possibility of a solution to some puzzles over which there has been much agonizing. A spectacular example is the low measured $log\,P$ value (2.33)[8,11] for paracyclophane (CLOGP[80] calculates 5.79). For an HPLC system which correlates well with $log\,P_{oct}$ but is free of this solvent, Garst and Wilson[81] find $log\,P = 4.47$. Possibly this is the explanation for Leo's 'pot effect'[11] and, on a lesser scale, his 'ring clustering' factor;[11] the typical deviant molecule seems to possess a compact structure with one or more polar groups near one end. Chiou and Block[9] suggest the critical factor to be the relation between solute and cosolvent aqueous solubility. If the former is (much) less it can dissolve in the latter: DDT and octanol have S_w values of 1.8×10^{-8} and 4.5×10^{-3} mol dm^{-3} respectively. If the latter is less there will be no problems however high $log\,P$. The effect disappears as the cosolvent becomes less miscible; S_w is 3×10^{-5} mol dm^{-3} for heptane. Leo[11,80] suggests a rough-and-ready limit of $log\,P_{oct}$ *ca.* six beyond which calculation is not meaningful. It would be more accurate to say that *measurement* beyond this may be meaningless; in the writer's opinion, calculation is to be preferred for such molecules in an environmental context.

18.6.3.3.5   Ionic strength effects

No systematic study has been made. On the assumption that salts do not penetrate the organic phase one would expect the usual rules for non-electrolytes in aqueous solution to be followed: namely that ions of high charge density (structure makers: *e.g.* sulfate) will salt out, whereas ions of low charge density (structure breakers: *e.g.* perchlorate) will salt in.[82] Sodium chloride usually has a mild salting-out tendency and this is consistent with its effect on pyridine, whose $log\,P_{oct}$ value rises from 0.65 to 0.78 in the presence of 1 mol dm^{-3} NaCl.[11,17] Given that isotonic saline is 0.15 mol dm^{-3} the departure from realism in using ionic strengths close to zero for $log\,P$ determination should commonly be negligible.

18.6.3.3.6   Partitioning into structured media: liposomes

Some results in this area are of great importance. Davis and his co-workers[25,83] have studied temperature effects on ΔG_{tr} for a number of 4-alkyl- and 4-halo-phenols in L-α-dimyristoyl phosphatidylcholine (DMPC) and a number of orthodox solvent systems. In terms of ΔG_{tr} it is clear that octanol is a much better model than a hydrocarbon for the liposome[25] but, in terms of ΔH and ΔS, liposome behaviour is quite extraordinary. For the liquid–liquid systems, and especially concerning f_{CH_2} (Table 6), it is easy to appreciate from these figures what is meant by the common statement that partitioning is entropy driven. Liposomes do not behave this way at all, either below T_c or above it. Below T_c, the favourable ΔG_{tr} for whole molecules results from a very large favourable ΔS_{tr} mostly cancelled by a large unfavourable ΔH_{tr} term: the 'classical' picture but much exaggerated. Above T_c, this situation is reversed. It is also reversed on both sides of T_c for f_{CH_2}, whose unusually small ΔG_{tr} (Table 6) is composed of nearly cancelling large favourable ΔH_{tr} and unfavourable ΔS_{tr} terms. Since the temperature range is too narrow for water structure to be much affected these effects must concern the liposome itself; Davis[25] notes that 'large and compensating changes in ΔH and ΔS attributable to the change in liposome structure can mask the smaller changes in ΔH and ΔS due to the actual transfer of the solute'. Below T_c the liposome is quasi-crystalline[25,84] and

Table 5 π-Values in Liposomes for 4-Substituted Phenols

	CH_3	C_2H_5	C_3H_7	C_4H_9	F	Cl	Br	I
$T < T_c$	0.45	0.91	1.48	1.80	0.42	0.97	1.14	1.30
$T > T_c$	0.21	0.61	0.75	1.06	-0.16	0.22	0.28	0.44
π_{oct}	0.48	1.12			0.31	0.93	1.13	1.45

Table 6 Temperature Effects on Partitioning into Solvents and Liposomes (kJ mol^{-1})

Organic phase	$p\text{-}EtC_6H_4OH$			$p\text{-}ClC_6H_4OH$			f_{CH_2}		
	ΔG	ΔH	$T\Delta S$	ΔG	ΔH	$T\Delta S$	ΔG	ΔH	$T\Delta S$
C_6F_{14}	1.5	35	33				-2.5	0	2.5
C_6H_{12}	-6.7	20	27	-3.2	16	19	-3.5	-0.7	2.8
Octanol	-19.0	-8	11	-19.1	-16	3	-2.5	-0.3	2.2
DMPC ($T < T_c$)	-14.8	81	96	-14.6	19	34	-1.1	-44	-43
DMPC ($T > T_c$)	-15.9	-39	-23	-13.7	-17	-3	-1.6	-22	-20

the large positive ΔH and ΔS values are those expected for a 'mixed melting' process brought about by the solute; for any melting process note that $\Delta G_m = 0$ (equation 8). Above T_c, the liposome's own alkyl chains will possess some vibrational freedom but entry of solute promotes solidification so the sign of ΔH and ΔS reverses. Similarly, on entering the liposome, whether above or below T_c, the alkyl chain of the solute loses its comparative freedom of motion so the signs of ΔH and ΔS are those for a freezing process. The net result is a drop in π value (Table 5)[83] relative to liquid–liquid systems although, in other respects, ΔG_{tr} is not a lot different above and below T_c. An earlier study by Diamond and Katz[84] on a mixed set of solutes showed a good correlation with olive oil log P but at a slope of 0.38. Klein[60] has calculated that the complete freezing of motion in an alkyl chain would produce a reduction in the expected favourable ΔG_{tr} equivalent to $\Delta f_{CH_2} = -0.24$. These results have powerfully contributed to the new wave of thinking on the hydrophobic effect (Section 18.6.3.4).

18.6.3.3.7 *Partitioning into structured media: enzymes and proteins*

Partitioning of this sort comprises, or contributes to, a very high proportion of published QSARs. Since these are the subject of Volume 4, Part 21 only certain aspects will be considered here. In 1972, Hansch and Dunn[85] published a compilation of 147 by then known relations of linear type

$$\log(\text{RBR}) = a \log P + b \qquad (35)$$

where RBR (relative biological response) is generally expressed as $1/C$. Equation (35) is a form of the Collander equation (Section 18.6.7.1) and is that for a typical LFER. Slopes a were found that ranged from 0.16 to 1.41 and these authors detected some tendency for clustering at > 0.85 ($n = 66$) and at 0.66 ± 0.12 ($n = 71$). This became codified into the view that a *ca.* unity implies solute partitioning to the interior of a hydrophobic pocket, with maximum desolvation, while a *ca.* 0.5 entails contact along a surface, with only partial desolvation.[8,86] This idea has been widely taken up.[14] It is suggested, in particular, that binding to membranes and proteins follows $a = 1$ and $a = 0.5$ respectively,[14,86] though Franke[14] points out, with respect to the former, that the membrane will have to match octanol. However, this rationale is severely flawed. As just seen, a structured medium, as many membranes must be, will certainly lower f_{CH_2}[60] and probably other f values as well;[83,84] up to a 45% drop in a from unity could result from this cause alone. Hansch's suggestion[86] cannot be generally valid.[16] Nevertheless, in combination with that above, it may help to explain some very low slopes ($a < 0.35$) reported by these authors.[85] It may also explain one long-standing puzzle in a way that has wider implications. In attempting to characterize the buccal membrane in solvent terms, Rekker[12] noted that on one criterion, $a < 1$, it appears less lipoidal than octanol, whereas on another, that polar groups are severely discriminated against, it behaves more like a hydrocarbon.

On present evidence, the second criterion is much the sounder.[16] We take up this point in Section 18.6.8.

Two other points deserve highlighting. In a comprehensive review of pharmacokinetics, Seydel and Schaper[87] note that protein binding often involves a charged species (especially carboxylate) and therefore represents a curious blend of ionic with lipoidal character. In a parallel review to ref. 85, which deals with parabolic $\log P$ relationships, Hansch and Clayton,[88] again on pharmacokinetics, observe that a shift with time from a parabolic to a linear relation indicates that transport not receptor binding is involved. In view of the perpetual ambiguity as to the significance of a in QSAR—it could be due to either factor or both—this becomes a valuable criterion.

18.6.3.3.8 *Partitioning into structured media: chromatography*

If hydrophobic molecules can escape from water into other liquids, structured media or even the gas phase, it is equally likely that they will stick to surfaces. In reverse phase HPLC (RPLC) this is precisely what happens and the extent of retention has been widely used as another sort of hydrophobic index. Basic principles are discussed by Horvath *et al.*[89] The typical column packing consists of silica particles coated with octadecylsilane (ODS) though shorter and longer chains and other variants have been used. This hydrocarbon coating forms a 'molecular fur'[89] with which molecules associate according to their hydrophobicity, so that solute moves along the column more slowly than the eluant. Retention is defined by the capacity factor k'

$$k' = (t_R/t_0 - 1) = (v_R/v_0 - 1) \tag{36}$$

where t_R (v_R) is the retention time (volume) of the solute and t_0 (v_0) is that of the eluant itself, usually determined by using as marker some unretained solute. The use of water alone as eluant is not feasible in RPLC, except for small polar solutes, and cosolvents (*e.g.* methanol or acetonitrile) have to be added, usually to 30% or greater. For isocratic (constant eluant composition) mobile phases then, in the absence of adsorption, k' is a (relative) partition coefficient.

A large part of the voluminous literature has concerned itself with the possibility of using $\log k'$ to predict $\log P$, especially $\log P_{oct}$, and this subject has been comprehensively reviewed.[90,91] Hydrocarbon stationary phases with water as eluant *never* correlate with $\log P_{oct}$[92] and the debate devolves around which cosolvent, if any, will allow an acceptable fit to equation (37).

$$\log P = a \log k' + b \tag{37}$$

There is clear evidence,[90] some of it by default,[93] that acetonitrile, THF and isopropanol are unsuitable, and the only one so far to merit serious consideration is aqueous methanol. Ideally $a = 1$ but this is inevitably raised by cosolvents since they diminish the distinction between the phases; the resulting compression of the time-scale is a practical advantage[90,91] which may be offset by some lack of precision in $\log P$. Typical relations[90] are given by equations (38) and (39)

$$\log P_{oct} = 2.30 \log k'_{75} + 2.62 \tag{38}$$

$$\log P_{oct} = 1.63 \log k'_{50} + 1.10 \tag{39}$$

which refer to 75% and 50% aqueous methanol respectively. Here 83 solutes were examined giving $r^2 = 92\%$ (four outliers) for (38) and 96% (seven outliers) for (39). The rationale behind such correlations is that some methanol is adsorbed by the column so that partitioning is really from aqueous methanol to a stationary phase with some of the characteristics of an alcohol. Acting on this suggestion, Hafkenscheid and Tomlinson[94] measured k' as a function of eluant composition and obtained the hypothetical k'_w for pure water (strictly, for a methanol monolayer) by means of equation (40)

$$\log k' = \log k'_w - B \phi_m \tag{40}$$

where ϕ_m is volume fraction of methanol. The validity of this procedure depends on the constancy of B. In fact, there is ample evidence[81,95] for curvature, and in both directions. It is known that aqueous acetonitrile and aqueous methanol both show anomalous variations in physical properties with composition[96a] that may be related to the 'extremum behaviour'[22] already described (Figure 1). In addition, the hints of family-dependent behaviour that appear elsewhere[81,90] are amplified by Terada[91a] who shows separate intercepts b in equation (37) for proton donor, proton acceptor and inert solutes; this effect diminishes as k' approaches k_w but does not disappear. Unger[9] believes that RPLC can never reproduce $\log P_{oct}$ with sufficient precision.

In this continuing debate, several distinct issues are in danger of becoming confused. If a hydrophobicity index is required, *e.g.* for correlation with a set of biological results, then if RPLC is found to work, there can be no objection to using it. Claims are quite often made that RPLC gives better correlations than $\log P_{oct}$ and these may well be true. However, such results are not in general transposable. They have to be redetermined for the next series, and there is so far no method that will allow the chemist to estimate $\log P_{RPLC}$ ahead of synthesis which is one of the prime advantages of $\log P_{oct}$. Such correlations can only be retrospective. The claim that $\log P_{HPLC}$ can reproduce $\log P_{oct}$ also seems to be flawed at this deeper level. Such correlations are quite good enough for estimating $\log S_w$,[90] where there is no compelling reason in any case why $\log P_{oct}$ should be preferred. However, if one reason for studying $\log P_{oct}$ is to create an additive scheme that will permit extrapolation,[11,12] this must include an accurate knowledge of π values and $\Delta\pi$, *i.e.* cross-interaction terms; and here the correlations are simply not precise enough, as a plot of equation (37) for other than homologues will very soon show. What is more, there is increasing evidence that $\Delta\pi$ follows different rules. The π values of El Tayar *et al.*[95] are non-additive according to a pattern quite different from octanol or a hydrocarbon and which does not relate to either.[97] On the other hand, as seen, RPLC is probably to be preferred where octanol in the aqueous phase may interfere.[81] It will also be interesting to see whether RPLC proves to have advantages in correlating partitioning towards other structured media such as liposomes. New packing materials[91a] may eventually change the present picture, but till then it is 'horses for courses'.

Other forms of chromatography deserve brief mention. TLC has been reviewed.[98] It is prone to adsorption, will not tolerate highly aqueous eluants, and has a limited dynamic range (1.5 decades); nevertheless, Dearden *et al.*[99] found polyamide plates to give a good correlation with albumin binding. To a large extent it is superseded. Even more so is GLC since most relevant compounds are involatile; nevertheless, a group contribution scheme goes back to 1968,[100] it has been successfully used in volatile anaesthesia,[71] and there is a recent claim[101] that $\log P_{oct}$ can be reproduced by the difference between two sets of retention times.

18.6.3.3.9 *Some environmental aspects*

This is an area of rapid growth and has been reviewed.[102,103] So far, results seem useful but unspectacular, perhaps because the type of molecule that causes most concern tends to be both very hydrophobic and essentially non-discriminating in its ability to characterize the nature of the medium into which it is partitioning. We have noted in Section 18.6.3.3.4 that, for many such molecules, calculated rather than measured $\log P$ may provide a better guide to bioaccumulation. There is also the probability, however, based on the rough inverse relation between $\log S_w$ and $\log P$ as shown by Chiou and Block,[9] that the point may eventually be reached where potentially noxious compounds are not harmful any more since they cannot jump the gap between their own inert mass and their potential target; liquid paraffin, for example, is simply not absorbed, while for more polar compounds, rising melting point tends to impose its own barrier (equation 9).

Such diverse types of binding as that of acetanilides to nylon,[102] of phenoxyacetic acids to activated carbon[102] and of a mixed set of solutes to soil[104] have all been shown to depend in linear fashion on $\log P$. The study of aquatic toxicity might be said to have begun with Overton[1] and Meyer[2] and studies on bioaccumulation especially into fish are now legion.[103] In one such study, $\log P$ was shown to correlate with the ratio of clearance to uptake rates of halogenated hydrocarbons into the trout.[102] Chiou and his co-workers[105] have carried out some careful studies on partitioning into soil. They demonstrate that absorption (*i.e.* true partitioning) not adsorption of organic molecules is involved (ions tend to adsorb) and that the soil organic phase is probably polymeric and more polar than octanol, with δ *ca.* 13 (*cf.* Table 1). Finally, in the realm of human health, King and Moffat[106] report a positive correlation between $\log P$ and the statistical chance of a fatal overdose, especially from barbiturates!

18.6.3.3.10 *Solute subspecies: conformation and tautomerism*

The measured or macro partition coefficient is, in principle, the weighted mean resultant of the separate micro partition coefficients for all subspecies present.[42] In practice, hard information is extremely difficult to come by. The 'proximity correction' for two polar groups separated by one or two carbon atoms[11,12] is clearly, in the second case, a conformationally sensitive quantity, but real molecules of type XCH_2CH_2Y consist in general of a rapidly equilibrating conformational mixture and quantification of this process in water and octanol (both would be needed) for compounds large enough to be relevant does not yet exist. We are forced, therefore, to seek data on molecules of fixed geometry. The best so far come from a study by Pleiss and Grunewald[43] on the $\log P_{oct}$ values for pairs of tricyclics exemplified by (1) and (2), which in terms of the relation between X and Y are *gauche* (*exo*) and *trans* (*endo*) isomers respectively. Pleiss and Grunewald fit their results to $\Sigma f(\text{Rekker})$[12] or $\Sigma f(\text{Leo})$[11] by a small constant term C_{trans} or C_{gauche} (Rekker) or F_{trans} or F_{gauche} (Leo). The difference between the two sets averages at about $\Delta \log P = 0.2$ with the *gauche* set more lipophilic. Close inspection, however, shows that these authors have severely under-utilized their data. For most compounds, X is CH_2 or CH_2CH_2 and here $\Delta \log P$ is very slight: 0.09 ± 0.04 for $n = 8$ (Y is always NH_2 or NHR). However, for X = O, $\Delta \log P$ values of 0.35 and 0.32 result. The first set almost certainly represents a shielding effect peculiar to this structure; the second, after correcting for this, becomes a differential proximity correction $(F_{gauche} - F_{trans})$ of about 0.25. It is difficult to correct exactly for the difference between $X = CH_2$ and X = O since f_O in this context may be abnormal, but if it is not, F_{gauche} and F_{trans} values of about 0.65 and 0.40 respectively result for this particular combination (O and N). Either is less than CLOGP[80] would calculate but Rekker's fixed value[12] of 0.57 comes close to their mean. The *gauche* correction is expected to be greater since the hydration shells of X and Y must overlap more so that net hydration is reduced.[12] There is extensive evidence that, at least when X and Y are heteroatoms, *gauche* interactions tend to predominate for reasons not to do entirely with hydrogen bonding; it occurs for XCH_2CH_2Y even when $X = Y = F$.[107] So most proximity corrections may be predominantly *gauche* in nature but this whole area represents a serious gap in our understanding and one of the major remaining stumbling blocks to reliable $\log P$ calculation. It is even more serious in its effect on $P_{g \rightarrow w}$[69] and up to now is beyond the scope of the LSER approach[32] (Section 18.6.7.1). One hopes that Hopfinger's SCAP[41] (Section 18.6.2.2.5) will eventually contribute to a solution. Effects are equally possible in *cis–trans* isomerism but this has not so far been explored with discriminating substituents.

(1)   (2)

The related question of tautomerism has scarcely been considered. It is most readily approached *via* a free energy diagram. Figure 2(a) represents the case where one tautomer A predominates in both phases so that tautomeric ratio K_T is an irrelevance in interpreting $\log P$. If, however, B comes to be preferred in the organic phase, the situation of Figure 2(b) results. This reveals the fact that the switch to a hitherto minor species must necessarily increase $\log P$ beyond that otherwise expected (or reduce it, if the switch occurs in the aqueous phase). This conclusion is absolutely general and applies equally to a minor tautomer, a minor conformer or, for example, the formation in the organic phase, but not in water, of an intramolecular hydrogen bond. No use seems to have been made of this principle so far. However, some modest use is possible in a limited context of the properties of Figure 2 as a box equilibrium. The characteristics of a box equilibrium are that any three K values define the fourth and that the ratio between any pair defines that between the remainder. Figure 2(c) summarizes known information on 2-pyridone ($pK_T = 3.0$ in water[108] and $\log P_{oct}$[17] $= -0.58$) along with the value for 2-pyridinol ($\log P_{oct} = 0.86$) calculated by our algorithm for the azines.[109] This results in $pK_T = 1.56$ for 2-pyridone in octanol; the actual value is unknown but $pK_T > 1$ is expected from extrapolation of Beak's algorithm[110] for 6-chloro-2-pyridone as a function of solvent by the LSER approach.[111] This at least is good evidence that the same species predominates. Occasionally it may not; in doubtful cases a check is highly desirable.

The possibility of a switch between internal hydrogen bonding and that to solvent represents a third sort of equilibrium. Leo[112] has explored intramolecular hydrogen bonding in *ortho*-disubstituted benzenes and finds a mean correction factor F_{HB} of 0.63 (Volume 4, Chapter 18.7). With one exception, *o*-nitroaniline, all these cases involve carbonyl as acceptor, but that may merely reflect the paucity of other examples. It is tacitly assumed[112] that the intramolecular bond is formed in both phases or neither, which may not be the case. Since if formed only in the organic phase (we

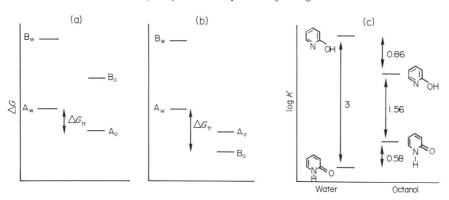

Figure 2 Free energy diagrams for subspecies in water (w) and oil (o). In (a), A is the favoured subspecies in both; in (b), B becomes that in oil; while (c) represents the actual situation for 2-pyridone (see text)

may safely discount the reverse) a rise in $\log P$ is inevitable (Figure 2b), this may account for some part of the effect described or, more probably, of any effects that seem anomalously large. Dearden[113] uses a very different $\Delta H / \Delta S$ balance in the partitioning of 2- and 4-nitrophenols despite almost identical P_{oct} ($\Delta \log P_{4 \to 2} = -0.11$) as evidence for an aqueous intramolecular hydrogen bond in the 2-isomer. This is an interesting argument though other evidence appears to be in conflict.[114] The near-zero $\Delta \log P$ in this case may derive from near cancellation between loss of weak *net* OH donor properties ($\Delta \log P$ negative) and of weak NO_2 acceptor properties ($\Delta \log P$ positive). Where the acceptor is stronger a larger effect is expected; for the corresponding hydroxy-acetophenones, $\Delta \log P_{4 \to 2}$ is 0.60. Very much more work is needed on this and all other topics under this section.

18.6.3.3.11 Solute subspecies: ionization

Ionization complicates the measurement and interpretation of $\log P$ in a number of quite distinct ways. Firstly, since ΔG_{tr} of ions into other than an aqueous medium is in general highly disfavoured, apparent P measured at a pH where the ionized form predominates will grossly underestimate the true or neutral species P value. The measured distribution ratio D that results from this cause relates to P for an acid or a base by equations (41) and (42) respectively.

$$\log D = \log P - \log[1 + \text{antilog}(\text{pH} - pK_a)] \tag{41}$$

$$\log D = \log P - \log[1 + \text{antilog}(pK_a - \text{pH})] \tag{42}$$

Figure 3, line (a) illustrates equation (42) for some base B; since BH^+ does not extract, $D < P$ at $\text{pH} < pK_a$. This is the basis of the ion correction technique for $\log P$ measurement (Section 18.6.5). Hence two drugs of similar $\log P$ may differ greatly in $\log D$ at physiological pH, with profound consequences for QSAR (see Section 18.6.9). Log P values of ionizable species are often quoted at $\text{pH} = 7.4$, sometimes unfortunately without any clear indication of this fact being given.

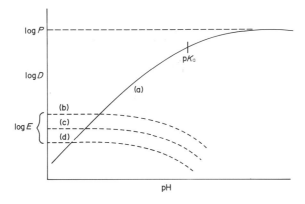

Figure 3 The relation between $\log D$, $\log P$, $\log E$, pK_a and pH according to equations (42), (43) and (53)

If the ionic subspecies extracts into the organic phase, it does so most usually as an ion pair. In that case the dimensionless form of equations (18) and (20) no longer applies. We may define an empirical extraction coefficient E which, for (say) a base, represents the total concentration of cation in the organic phase divided by its total concentration in water. Introducing the simplifying assumption that only ion pairs exist in the organic phase, we obtain

$$E = [BH^+X^-]_o/([BH^+X^-]_w + [BH^+]_w) \qquad (43)$$

We now define a (dimensionless) ion pair partition coefficient P_{ip} and an ion pair association constant K_{ip} for the aqueous phase

$$P_{ip} = [BH^+X^-]_o/[BH^+X^-]_w \qquad (44)$$

$$K_{ip} = [BH^+X^-]_w/[BH^+]_w[X^-]_w \qquad (45)$$

Combining equations (43), (44) and (45) and making the additional assumption that K_{ip} is small so that $K_{ip}[X^-] \ll 1$

$$E = P_{ip}K_{ip}[X^-] \qquad (46)$$

Hence E depends on the nature and concentration of the counterion, falling to zero if any term is zero. Equation (46) has been amply borne out in practice.[115,116] Lines (b)–(d) on Figure 3 represent hypothetical values of E which might result from variation in any of these quantities. It follows that the single ion π or f values that appear in so many compilations are, quite literally, meaningless. It is only too easy to find statements such as that ionizable compounds should be examined in both 0.1 mol dm^{-3} HCl and 0.1 mol dm^{-3} NaOH so as to 'measure log P for both species'.[12] In practice, there is internal evidence that 0.1 mol dm^{-3} Na$^+$ (for acids) and 0.1 mol dm^{-3} Cl$^-$ (for bases) is often implicitly intended, but it is extremely important that this be made explicit. In view of the presence of NaCl as the dominant physiological electrolyte at not far from 0.1 mol dm^{-3}, the writer would like to suggest that the above be adopted as defining conditions for E. A great deal of muddle might be avoided this way.

Other errors may enter if the dimensionality of E is not understood. Scherrer[117] treats the relation between P_{oct}, E_{oct} and the pK_a values of some benzoic acids in water and octanol as a box equilibrium as indicated in Figure 4. This is invalid, as inspection of Figure 3 will speedily demonstrate: any change in log E (Scherrer's log P_{A-Na^+}) as a result of varying [Na$^+$] or changing it for another cation would then affect pK_a in octanol (pK_a'), which it cannot. Scherrer's relation amounts to equation (47).

$$\log P - \log E = pK_a' - pK_a \qquad (47)$$

Defining K_{ip}' for the octanol layer on the lines of (45) and assuming K_{ip} to be negligible as in deriving (46), equation (48) may be obtained.[97]

$$\log P - \log E = pK_a' - pK_a - \log K_{ip}' - \log[Na^+]_w \qquad (48)$$

A curious feature of Scherrer's treatment, if valid, is that π for benzoate anions and for the parent acids must differ since ΔpK_a, and hence $\Delta \log P$, appears as a function of acid strength. If so there can

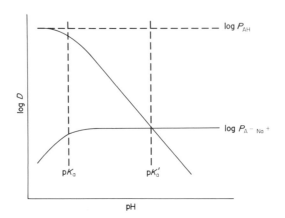

Figure 4 The relation between log D, log P, pK_a, pK_a' and pH according to ref. 117 (see text)

be no hope of transferring π or f values from neutral to charged species as one would wish. From equation (46) variation in E may plausibly result from changes in K_{ip}, rather than in P_{ip} as Scherrer's treatment would imply.

Much more sophisticated treatments than the above have been used by Biles and his co-workers[118] and by Takayama *et al.*[119] The former studied E as a function of ion pair (not counterion) concentration, analyzing their results in terms of equation (49).

$$(C_o)^{1/2}/C_w = (P_{ip}K_{ip})^{1/2} - (C_o/C_w)(K_{ip}/P_{ip})^{1/2} \qquad (49)$$

This can be solved to obtain both association constants. The latter workers studied picrate ion pair partitioning for a varied set of aliphatic amines and, by allowing for competing equilibria, obtained $K = P_{ip}/K_{ip}$ which was then compared with P for the neutral amines. Different steric factors were clearly identified. One interesting result is that each cationic proton elevates K by about as much as a methylene unit; another is the diminished slope of K relative to P as a function of size which may be the effect of an ionic atmosphere in the octanol phase. This lower slope is incorporated by Rekker[12] and Leo[11] into their empirical correction factors for amine protonation (Volume 4, Chapter 18.7).

A further potential complication concerns ampholytes (Scheme 1). At $pK_A < pH < pK_B$ where pK_A and pK_B are the titrational or macro pK_a values, an ampholyte will exist as a mixture of zwitterion and uncharged species whose ratio is determined by the relation between the micro pK_a values, pK_1–pK_4. This like Figure 2 is a box equilibrium and possesses similar properties. It generates a problem in nomenclature which has never been considered. Over what is often a considerable range in pH, P_{obs} can be a constant, so the behaviour typical of a single definite form is followed; yet it is not a single species but a mixture of two (or more) in some constant proportion that may not be known, so qualifies as a distribution ratio D. If D_{pH} is commonly used to mean some mixture of P and E of standardized (by pH) but unspecified composition (the use of P_{app} for this quantity is to be discouraged), then perhaps D_{max} would fit the present case. If Scheme 1 can be solved it may be possible to determine the composition of D_{max}; two examples follow. Log P (*sic*) $= -3.21$ for glycine[17] could be due either to the dominant zwitterion, or to the small fraction of uncharged species present. From published pK_a values, the latter is $10^{-5.29}$; $\log P = 2.08$ follows if this species is responsible for D_{max}. In fact, CLOGP[80] calculates $\log P = -1.00$ for glycine in the neutral form, so this conclusion would be absurd and the zwitterion must extract as such. Perhaps in such cases P might be termed P_z by analogy with equation (44) since P, P_{ip} and P_z are all dimensionless quantities. The other example concerns the β-adrenergic blocker sotalol where pK_A and pK_B are too close together for a true central plateau to be attained.[120] Here solution of Scheme 1 yields an estimate of 4.2% uncharged species from which $\log D_{max} = -0.79$ gives $\log P = 0.59$. This value is entirely reasonable by calculation[120] so no extraction of the zwitterion need be assumed. Figure 5 shows $\log D$ for sotalol as a function of pH. The dashed lines represent what would be calculated for a base (pK_2) or acid (pK_4) of $\log P = 0.59$, while the full line is observed and is the envelope of these. At pH $\ll pK_A$ and pH $\gg pK_B$, these lines converge. Hence, for these conditions, $\log D$ may be estimated if $\log P$ and the relevant micro pK_a can be calculated accurately enough. We have found this a useful device where $\log D$ is experimentally inaccessible.[97]

Not much work has utilized Scheme 1 but an exception is that of Streng[121] on cephalosporins; his paper refers to a useful methodology for determining micro pK values (there are several, all essentially equivalent). The complete pH profile for some β-lactams and sulfanilamides[122] shows

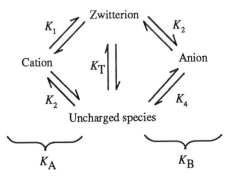

Scheme 1

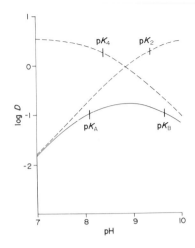

Figure 5   Relation between log D, log P, pH, and the macro- and micro-pK_a values for sotalol (see text)

that the former, which are dominated by the zwitterion, possess $D_{max} < E$ (at both ends of the pH range), whereas the second, for which the reverse holds, possess $D_{max} > E$. This is qualitatively reasonable though it must be remembered that E depends critically on the conditions chosen (D_{max} like P does not).

18.6.3.3.12   The effect of temperature

Solubility can change quite steeply with temperature but so far as this is due to the pure solute state it must cancel in P and, in any case, P as a ratio is likely to vary much less than S_o or S_w. Leo et al.[8] estimate a typical $\Delta \log P_{oct} = \pm 0.1$ for $\Delta T = 10$ K, equivalent to ΔH_{tr} *ca.* ± 16 kJ mol^{-1}. This remains true in the light of subsequent work, and represents a practical point of some importance in favour of octanol; it may be due to the increasing mutual miscibility of octanol and water as the temperature rises.[123] Elsewhere, variability is in general much greater. Usually $\Delta H_{w \to o}$ is positive (P rises with T), increasingly so as the organic phase becomes less polar.[25] For isooctane (2,2,3-trimethylpentane, TMP), Riebesehl et al.[124] find $\Delta H_{w \to o}$ for one mixed set of solutes to cover the range 0–35 kJ mol^{-1}. Equally large ΔH_{tr} but with their signs reversed are found in RPLC; *i.e.* for liquid-liquid and RPLC partitioning there is some resemblance to DMPC below and above T_c respectively[25,83] (Table 6) though the effect is not so pronounced. Except that homologues and near-homologues behave fairly regularly, little correlation with chemical structure can be discerned.

For most such systems ΔH and $T\Delta S$ change more than ΔG, and a majority of these show enthalpy–entropy compensation:[53] the tendency for ΔH and ΔS to be related in linear fashion. This has been well reviewed by Tomlinson,[125] who demonstrates its use for identifying solutes whose behaviour is out of line with the remainder. It can also show when solvent behaviour is anomalous: Anderson et al.[25] find that, on this criterion, P_{oct} falls out of line with P for various inert solvents. In fact P_{oct} in general does not show $\Delta H / \Delta S$ compensation[125] despite previous reports to the contrary; it should be noted that ΔH and ΔS contain self-compensating errors and the correct way to demonstrate compensation is through plots of ΔH or ΔS *vs.* ΔG.[126] Even for hydrocarbon solvents it is most reliably a phenomenon of homologous series, and polar solutes tend to break the pattern.[125] No clear conclusions are reached as to when, or why, it happens. We return to this subject in the following section.

18.6.3.4   A Reconsideration of the Hydrophobic Effect

Received opinion states that the hydrophobic effect is entropy controlled. By this it is meant that its main driving force is the release of structured water to water in bulk, its reverse being the endoergic freezing of water ('iceberg formation') around hydrophobic foreign bodies. In this process, enthalpy changes play almost no part. Associated with this is the puzzling business of enthalpy–entropy compensation. On one level, this is well understood: ' . . . the satisfaction of attractive forces imposes constraints (ΔH and ΔS both negative) and release from constraints confers greater freedom (ΔH and ΔS both positive) . . . '.[21] At a deeper level, it is not understood at all.

Recently, several lines of evidence have converged on a new way of thinking. One of these involves the study of hydrophobic interactions in water over a much wider temperature range than heretofore. Figure 6(a) indicates schematically what is expected according to equation (23): a constant ΔH while $T\Delta S$ changes with temperature so that ΔG alters to compensate. Figure 6(b) indicates, also schematically, what has been found by Evans and Ninham[61] for the formation of a cationic micelle. Here ΔH and $T\Delta S$ both fall steeply so that beyond $100\,°C$ the 'hydrophobic effect' shows a negative entropy and micelle formation appears as enthalpy driven; yet the actual equilibrium position is not greatly affected. This is behaviour according to equation (25) where ΔC_p is large and negative. Similar behaviour has been found for $P_{g \to w}$.[127] More evidence comes from work in solvents other than water that share its solvophobic propensities, such as EAB[127] but especially hydrazine,[61,127] whose properties have earned it the soubriquet 'non-aqueous water'.[128] Table 7 contrasts water and hydrazine as solvents towards hydrophobic solutes in a variety of contexts. Transfer free energies are generally similar but ΔH and ΔS of desolvation are both much more negative for hydrazine, which behaves near ambient temperature rather as water does at $> 100\,°C$ higher ('inhibited water'[61]). The crucial observation lies in the near-zero ΔC_p: hydrazine does not form 'icebergs'. Neither apparently does water at $170\,°C$, or much less so, yet its 'hydrophobic' properties are scarcely diminished. The conclusion seems inescapable: the $\Delta H/\Delta S$ compensation effects on which so much attention has been lavished are of little concern to the solute.

Gas–solvent partitioning provides crucial evidence. Cramer[129] plotted noble gas and alkane $P_{g \to s}$ data against solute radius and noted the apparent existence of two lines. Osinga[130] observed for $P_{g \to oct}$ that certain rigid solutes follow an extrapolation of the noble gas line, or fall in between, and

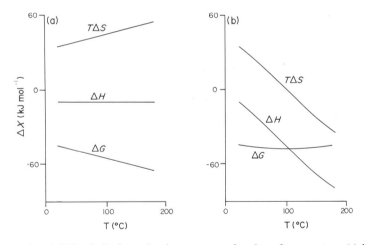

Figure 6 Thermodynamics of SDS micelle formation in water as a function of temperature: (a) hypothetical relation according to equation (23); (b) actual relation according to equation (25)

Table 7 Thermodynamics of Transfer in Water and Hydrazine (kJ mol^{-1} or J mol^{-1} K^{-1})

	Water				Hydrazine			
	ΔG	ΔH	$T\Delta S$	ΔC_p	ΔG	ΔH	$T\Delta S$	ΔC_p
Micelle formation								
SDS (35 °C)	−40	−26	14		−33	−56	−23	
$R_3NC_{14}H_{29}^+Br^-$ (40 °C)	−42	−36	6					
$R_3NC_{14}H_{29}^+Br^-$ (166 °C)	−36	−77	−41					
$R_3NC_{12}H_{25}^+Br^-$ (35 °C)					−28	−63	−35	
f_{CH_2}	−3.2				−3.0			
Transfer of argon								
$P_{g \to s}$ (25 °C)	26.2	−12.2	−38.4	178	28.0	8.5	−19.5	0
$P_{g \to s}$ (170 °C)	40.1	7.8	−32.3	111				
$P_{s \to C_6H_{12}}$	−15.6	11.2	26.8		−14.9	−9.5	5.4	

suggested the difference to lie in solute rigidity: flexible solutes incur an extra entropic penalty due to restricted rotation. Abraham[131] pointed out that the common practice of discussing ΔH and ΔS for whole molecules is vulnerable to the choice of standard state; this vitiates, for example, the claim by Wertz[132] that $\Delta S_{g \to w}$ is a constant fraction of overall molecular S^0. Concentrating on the methylene increment, for which this choice is irrelevant, Abraham[131] showed that noble gases and alkanes lie on a single line if $\Delta G_{g \to s}$ for any solvent is taken as standard. The sole exception is $\Delta G_{g \to w}$ for the alkanes, which is far more positive (unfavourable) than expected. By use of the noble gas regression line it was possible to dissect the observed small positive ΔG_{tr} into an 'expected' negative ΔG_{tr} and a large positive deviation which can be used to quantify the hydrophobic effect. Subsequent dissection in terms of ΔH and ΔS (Table 8) showed this to be almost entirely enthalpic in origin.

Table 8 Thermodynamics of Transfer for Methylene ($kJ\,mol^{-1}$)

	ΔG	ΔH	$T\Delta S$
Gas → Hexane	−3.1	−5.5	−2.4
Gas → Water			
total	0.7	−3.2	−3.9
'expected' portion	−1.6	−5.1	−3.6
'hydrophobic' portion	2.3	1.9	−0.3
Water → Hexane	−3.8	−2.3	1.5

Abraham[131] doubted Cramer's lines and avoided the decision between these and a curve by his method of analysis. Abraham used solute radius as abscissa; a replot of some of his data in terms of V_I (Figure 7) shows clearly that these lines exist.[33] Except for water, however, the distinction between them is *not* noble gases *vs.* alkanes. It enters with propane, *i.e.* with the first flexible molecule, consistently with Osinga's hypothesis.[130] Leo *et al.*[133a] studied the relation between $\log P_{oct}$ and (estimated CPK) volumes for a number of apolar solutes and found two parallel lines: the lower for noble gases and alkyl-substituted benzenes, the upper for alkanes subsequent to ethane, with methane and (later) hydrogen converging to $\log P = 0$ at zero volume. This drop in slope beyond ethane gave rise to the bond factor F_b,[80] rationalized[11] as being due to greater disorder in water which then partly cancels the entropy gain expected on desolvation; an explanation entirely inconsistent with Osinga's.[130] The difference between these parallel lines was explained on the 'reversible sweater' hypothesis.[133a] Molecules with an outer surface of electrons are surrounded mostly by water's protons, those with a protonic outer surface by its oxygen atoms; if the former

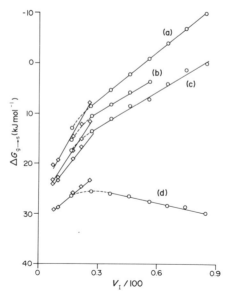

Figure 7 Free energy of transfer from the gas phase (1 atm) to solution (mole fraction) for noble gases ($\diamondsuit$) and linear alkanes ($\bigcirc$) as a function of intrinsic volume V_I: (a) to hexane, (b) to octanol, (c) to methanol and (d) to water

interaction is more favourable the observed difference could result. Figure 8 shows Leo's and some other data as a function of V_1. In fact three and probably four lines are apparent: the noble gas (a) and aromatic (c) lines are separate, and in between (c) and the linear alkane line (b) may come a fourth for the totally rigid spherical alkanes neopentane and $SiMe_4$ plus (perhaps fortuitously) CCl_4 with the cycloalkanes in an intermediate position. Again, hydrogen, methane, and ethane lie on a line of different slope.

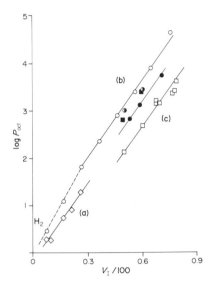

Figure 8 Log P_{oct} as a function of V_1 for (a) noble gases ($\diamond$), (b) linear alkanes ($\bigcirc$), (c) alkyl-substituted benzenes ($\square$); also for some spherical molecules ($\bullet$) and cyclic alkanes ($\circleddash$). The filled squares ($\blacksquare$) represent log P in chloroform for benzene and toluene

A self-consistent picture is the following. Partitioning from the gas phase to *any* solvent is disfavoured entropically by cavity formation; however, this is offset by dispersion interactions (enthalpy) and becomes more favourable as solute mass increases[133b] as is evidenced by rising boiling-point within related series. With increasing solvent internal pressure (δ, Table 1) the overall advantage diminishes so partitioning into water is not only disfavoured absolutely but is relatively less favoured as a function of solute size (Figure 7). This applies even to the noble gases and shows as the difference between $\Delta G_{g \to hexane}$ and 'expected' $\Delta G_{g \to w}$ in Table 8, most of which is entropic. Alkyl chains, however, which can tumble unrestricted in the gas phase, suffer a special entropic loss on solution which is shared by all solvents[130] (Figure 7) but almost cancels on partitioning;[131] Beezer and Hunter[79] may have identified a slight residuum. However, Howarth's observations,[62] while confirming the expected electrostriction of alkyl chains in water, cannot account for F_b^{11} since this has the wrong sign. The essential clue comes from Figure 9, which is similar to Figure 8 except that log P is for hexane. Here all alkanes lie on one line and this line goes off at a steep angle to that for the noble gases; it is as if the dashed line of Figure 8 were to continue. Hence alkyl chains suffer greater resistance, on a per CH_2 basis, from octanol than from water. Just possibly this relates to solvent viscosity; whatever its explanation it accounts for F_b, which may be peculiar to P_{oct} so far as solvent systems are concerned, though there is an obvious analogy in some aspects of liposome behaviour.[25,60,83,84]

The hydrophobic effect, however, is something else again. This is mostly enthalpic[131] (Table 8) and appears as the reversed slope for alkanes in Figure 7 or, which is the same thing, as the difference in slope between lines (a) and (b) on Figure 9. Its origin has not been clearly elucidated but we may speculate. One part of Leo's 'reversible sweater' hypothesis[133a] is valid: OH bonds weakly to the aromatic π-cloud. For example, the fivefold lower P_{oct} of benzene than of cyclopentane at virtually identical V_1 is composed of a thirty-fivefold greater value of $P_{g \to w}$ partly offset by a sevenfold greater value of $P_{g \to oct}$; and the black squares on Figure 8 are for benzene and toluene in $CHCl_3$/water, which now lie close to the alkane line since both solvents are almost pure proton donors.[111] The second part of the hypothesis is untenable. Water's very excess of protons oversaturates its oxygen atoms[54] to the extent that their ability to present to anything other than a strong proton donor must be seriously called in question. In fact the distribution of H and O around butane in water appears to

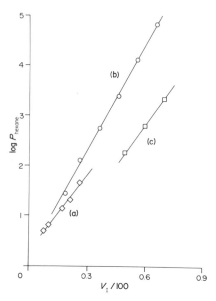

Figure 9   Log P_{hexane} as a function of V_{I} for (a) noble gases ($\diamond$), (b) linear alkanes ($\bigcirc$), and (c) alkyl-substituted benzenes ($\square$)

be entirely random[134] and what remains are dispersion interactions which are much less than for water on its own.[55] In Hildebrand's words[135] 'There is no hydrophobia between water and alkanes; there only is not enough hydrophilia . . .'. We have in this fact a plausible origin for an enthalpy-based hydrophobic effect.

Two familiar but neglected properties of water deserve re-emphasis. Water is (almost) unique in that *every atom* can take part in hydrogen bonding; on this criterion, even methanol is 67% hydrophobic. And at 55 mol dm^{-3}, the number density of water in itself is five to ten times greater than for most organic liquids. These factors must contribute greatly to the enthalpic repulsion term. Not that entropy is to be dismissed entirely: it is the major contributor to the endergonic process of cavity formation, in hydrazine as in water (Table 7), and differences in the $\Delta H/\Delta S$ balance have correctly been used to characterize certain solutes[113,125] and solvents[25] as out of line with the remainder. It has proved of great value in the study of liposomes[25,83,84] and its use in pharmacology is only just beginning; for example, Weiland *et al.*[136] present evidence that β-adrenergic agonists and antagonists may show characteristically different $\Delta H/\Delta S$ behaviour. There may be a factor common to all these contexts which will help to explain why $\Delta H/\Delta S$ compensation functions as it does. Evans and Ninham[61] point out that large coupled changes in ΔH and ΔS are characteristic of systems in which a great number of closely similar energy states are available, such that any reorganization results from a host of rather small linked processes. Living systems show a curious tendency to mimic water in this respect. Occasionally its possibilities are exploited: Klein[60] notes that the same enzyme in warm- and cold-blooded creatures often shows large and small ΔH values respectively, as if to compensate for its variable surroundings. But that very fact tends to devalue its significance: whatever minor adjustments the enzyme may undergo, its structure and function remain essentially the same.

The time has come for a change in emphasis. It is only too easy to encounter statements which appear to imply that 'entropy' does the work all on its own without the need, apparently, for any molecular mechanism. But with $\Delta H/\Delta S$ compensation reduced mostly to the status of a local artefact in 'cool bulk water',[127] attractive forces in aqueous solution can no longer be dismissed as of minor importance. The study of specific interactions in the context of partitioning will form a major theme of the sections to follow.

18.6.4   EXPERIMENTAL DETERMINATION OF SOLUBILITY

Before a drug can be absorbed it must dissolve, generally in water. Once solubility attains a certain level dissolution ceases to be rate limiting, so that interest in practice concentrates on the measurement and estimation of low aqueous solubilities[10] and, in the context of QSAR, there is not nearly so much interest in log S_{w} as in log P. Because drug dosage can vary so much it is difficult to be precise, but as a general guide, above 10 mg ml^{-1} there should be no problems, whereas less than 1 mg ml^{-1} may give cause for concern. Hence our treatment of this subject will be relatively cursory; good texts exist[137] and we confine our attention here to general principles and to the highlighting of certain problems that must not be neglected.

Solubility is measured, in principle, by shaking the well-powdered or well-dispersed solute in excess with solvent until equilibrium is attained. Solute and solvent are sealed in vials immersed in a thermostatted water bath, agitated continuously and removed at intervals for analysis of the supernatant liquid. In practice, since dissolution rate tends to parallel solubility, the attainment of equilibrium is most difficult to establish for just those compounds whose solubilities it is most important to know. Various techniques have been employed to overcome this difficulty. One is to prepare two sets of vials, one of which approaches equilibrium from a higher temperature so that solubility is initially exceeded. Another is to use a large excess of solute, since the rate of approach to equilibrium is proportional to excess solid present and independent of the solute. A third technique is to have present a small amount of a water-immiscible solvent in which the solute is easily soluble, so replacing a single slow step by two much faster ones; this can reduce equilibration time from weeks to days. Excess solute for both phases must of course be present.

Solubility measurement is complicated by two main problems: impurities and polymorphism. Both operate in different ways to exaggerate its true value. Any impurity can contribute to the amount of solid that dissolves and, if undetected by the analytical procedure, may be mistaken for the principal component. In addition, any impurity even if quite insoluble may increase the solubility of the principal component through a mixed melting effect, *i.e.* by lowering T_m; *cf.* equation (9). Still worse complications attach to polymorphism.[138] Polymorphs are identifiably distinct solid states of the same compound, each with its own heat of formation and therefore its own T_m; from this it follows that each must differ in solubility. Since the most stable polymorph is necessarily the highest melting and least soluble, it will also be the slowest to dissolve and the least bioavailable. Very often compounds exist as a mixture of polymorphs which have crystallized together, so that the least stable ones dissolve faster to give a pseudo-equilibrium state of exaggerated S_w. Re-equilibration will eventually occur but the process can be slow. Sometimes this can be turned to advantage: since interconversion is solvent mediated, use of a metastable polymorph may enhance bioavailability provided that rate of absorption *in vivo* is greater than reconversion and that this does not take place when stored in formulation.

Single crystal X-ray determination is the definitive test for polymorphism[138] but, in the absence of this, powder X-ray analysis or solid-state IR may provide 'fingerprint' evidence. In the latter, a KBr disc may itself induce polymorphism and mulls are to be preferred. Differential scanning calorimetry may detect some phase transition short of T_m but the absence of this is not definitive since, at elevated temperatures, this process can occur over a considerable range and so be missed. Polymorphism is probably a far more widespread phenomenon than the run of chemists tend to assume. It seems specially likely in compounds for which a multiplicity of hydrogen-bonding patterns may be envisaged. For an example, as well as the familiar dimers (**3**) carboxylic acids can form alternating chains (**4**) and these alternatives for a single compound have been detected.[97]

(3)   (4)

Ionization causes problems related to, but different from, those present in partitioning. The solubility of a salt Q_mX_n is governed by its solubility product K_{sp}

$$K_{sp} = \{Q^{n+}\}^m \{X^{m-}\}^n \qquad (50)$$

where braces denote activities. Hence common ions salt out, *i.e.* excess of Q or X will depress solubility, and while equation (50) is strictly valid only when K_{sp} is small, it remains qualitatively true under any circumstances. In addition, uncommon ions salt in; this comes about since γ_Q and γ_X

fall at high ionic strength so that concentrations rise to compensate (equation 17). These points have to be remembered when buffers are used in determining the solubility of an ionizable solute as a function of pH. If, for example, the solubility of a hydrochloride is measured in an acetate buffer, too high a concentration of buffer may exaggerate solubility. If the pH of (say) an acetate is adjusted using further acetate, common ion depression will result; if using HCl or another foreign anion, the salt of that anion may precipitate; and if the pH is lowered to below the pK of the counteranion, the solubility now measured will be that of the salt with the acid used to adjust the pH. Typically the salt will be much more soluble than the neutral species and these problems become unimportant if only the latter's solubility is accurately needed. The point to note is that a solubility–pH profile is meaningless unless the nature and concentration of all ionic species can be accurately defined. Also it must be remembered that the solubility of neutral species is also affected by ionic strength, if not so drastically as that of salts. One way round this problem is extrapolation to zero ionic strength; this however is unrealistic in a pharmaceutical context, and the value in saline is to be preferred.

18.6.5 EXPERIMENTAL DETERMINATION OF THE PARTITION COEFFICIENT

Despite some notable recent advances, log P measurement remains in the horse-and-buggy era. It may never get much further, since all problems related to solubility seem inherently resistant to the automated techniques that have transformed most other fields of analysis, essentially because these processes take time and will not be hurried. The main incentive for the calculation of log P (Chapter 18.7) lies in this fact. The account below attempts to provide guidance for the novice and perhaps help to others also. It is not and cannot be a fully fledged instruction manual, but it does attempt to pass on the type of practical hint that can make all the difference between success and failure. If largely based on the experiences of the writer and his colleagues, the pioneering work of the Pomona school[8, 11] and others[12, 138, 139] is hereby acknowledged. We start with the shake-flask method, primitive but still the foundation for all else; go on to consider counter-current methods with special emphasis on the filter-probe; and finish with solvent-based liquid chromatography (RPLC technique[96a] is outside the scope of this review).

18.6.5.1 The Shake-flask Method and its Derivatives

In principle, this is extremely simple. The solute is dissolved in one phase; the two phases are shaken together; then, after separation, each is analyzed for solute. In practice, certain precautions must be observed. It is *essential* to presaturate both phases; this applies equally to counter-current and HPLC techniques. This is best carried out by storing each phase in the dark at constant temperature in contact with some of the other, when the stock should be stable indefinitely. Equilibrium is usually reached within 24 h but can take a week. Without presaturation, solute equilibration takes much longer and the phase volume ratio may be affected. If the aqueous layer requires a buffer that should be incorporated at this stage. Buffers should be non-extractable; phosphate, for example, is greatly superior to acetate and citrate and, in general, mixed buffers are to be avoided. If the solute is unlikely to exceed 10^{-4} mol dm^{-3} then 10^{-2} mol dm^{-3} buffer should be acceptable, otherwise *pro rata*, but in any case pH should always be checked at the end of the experiment. The solution once prepared should be filtered to remove any suspension and then its UV spectrum recorded (if UV analysis is to be employed) before adding the second phase, which is conveniently carried out in a 10 ml test-tube with an ungreased ground-glass stopper. Cork, rubber and all organic matter must be avoided. A typical initial solute concentration is about 2 mg%, depending on the initial and likely final UV absorbance. Shaking should be vigorous: Leo *et al.*[8] recommend 100 inversions in 5 min and Rekker[12] recommends shaking for 30 min. Very low or high log P values may require more time than this. After standing for 15 min the contents need to be centrifuged for a period that is largely determined by the degree of emulsification that has to disperse; octanol takes longer than most hydrocarbons and propylene glycol dipelargonate (PGDP, 1,2-propylene dinonanoate) takes longer than octanol. Octanol will generally separate in 10 min at 3600 r.p.m. but PGDP takes longer. Nevertheless it is possible for a fine emulsion to escape visual detection and each layer to be analyzed should after separation be filtered through some material known not to adsorb the solute. The aqueous layer is generally the lower one but can be safely removed by inserting a Pasteur pipette under slight positive pressure. Each layer is then analyzed and P is the concentration ratio C_o/C_w provided that $\gamma = 1$, *i.e.* that these do not exceed about 1 mol dm^{-3} and 10^{-2} mol dm^{-3} respectively (see Sections 18.6.3.3.2, 18.6.3.3.4, 18.6.3.3.5, 18.6.3.3.11

and 18.6.3.3.12 for a discussion of the main perturbing factors that can invalidate or distort the above conclusion). It is good practice not only to replicate experiments but to approach equilibrium from both directions. Varying C will detect artefacts due to self-association (Section 18.6.3.3.2), whereas extreme values of P may be partially compensated by changing the phase volume ratio (but see later). Temperature requires reasonably precise control (Section 18.6.3.3.12).

The above describes the ideal procedure in which both phases are analyzed (BPA). This is always desirable and becomes essential if volatile material can be lost in the headspace or very involatile material can stick to the interface between phases or on the vessel walls (HPLC avoids both problems). However for $\log P$ outside the limits ± 1.5, and certainly ± 2, it becomes unreliable as it stands because most analytical methods cannot measure such ratios with sufficient accuracy. (The use of radioactive tracers may help but is very vulnerable to the radiochemical purity of the solute). In addition BPA may not be possible if the organic layer is UV opaque. Various ways exist to overcome this problem, all with their own disadvantages. Higher concentrations may be used and the more concentrated phase analyzed after dilution; in this case the condition $\gamma = 1$ may be at risk. The more concentrated phase may be extracted with a high volume ratio of the other; if this is excessive, say above 50:1, it may be difficult to ensure that the phase volume ratio is accurately known, especially if the organic phase is volatile. A useful variant on this is the multiple extraction method, since lower phase volume ratios r are then possible

$$C_f = C_i(1 + r/P)^{-n} \tag{51}$$

where C_i and C_f are solute concentrations before and after n extractions. If $(C_i - C_f)$ is small and n is large this simplifies to

$$P = nr[C_i/(C_i - C_f)] \tag{52}$$

This method is particularly valuable when $P \ll 1$. When $P \gg 1$ and the solute ionizes, an alternative method is the ion correction technique: most of the solute is trapped in the aqueous phase as cation or anion and C_o/C_w as obtained is D not P

$$D = P(1 - \alpha) + E \tag{53}$$

where α is fraction ionized at any pH, for which pK_a must first be known. A minimum of three determinations is needed at well-separated pH values for a reason apparent on inspection of Figure 3: $D = P(1 - \alpha)$ is valid only for the full line and error will be introduced if any point is influenced by ion pair extraction E. If this is absent, $\log P$ may be calculated using equations (41) or (42). If present however it is perfectly possible, with enough points, to fit D to the full sigmoidal equation (53) provided that E is not too high, which for practical purposes means $\log E \not> 1$.

An elaboration of the ion correction method involves potentiometric titration of the solute in the presence of the organic phase. There are many variants on this method, some of them adaptable for simultaneous measurement of pK_a and $\log P$, and the area has been reviewed.[141] We have successfully employed the variant due to Kaufman *et al.*[142] The method is particularly valuable when ionization produces no UV change and, for $pK_a > 8$ for bases or < 6 for acids, can enable $\log P$ values up to 5 or 6 to be determined accurately. It *must* however be checked that ion pair extraction is inappreciable, or serious error will again result.

The shake-flask method is most accurate at $P = 1$ and accuracy declines away from this even when the special elaborations above are used. The commonly claimed range of $\log P = 0 \pm 6$ is in the writer's experience excessive except in the one case noted above. A more realistic range is 0 ± 4 with an error of ± 5–10% towards the centre but approaching $\pm 100\%$ at its extremes. Even these standards demand high purity for all reagents. Outside these limits, any sort of accuracy becomes extremely difficult. Detection methods other than UV (GLC, HPLC) affect this conclusion slightly but not much. Nevertheless shake-flask remains the primary standard and no other method can be valid without calibration this way.

18.6.5.2 Counter-current and Related Methodologies

This area has been reviewed.[140a] Normal counter-current methods are complex to operate and have received little use. An ingenious variant, employing segmented flow with phase-splitters, proved rapid and flexible but of limited range, and unsuited to relatively viscous liquids such as octanol.[140a] A rapid mixing and separating device known as the AKUFVE, adapted from engineering science, suffers from a number of disadvantages such as cleaning difficulties and the need for skilled operators and large quantities of material, but has proved useful where a large amount of information, *e.g.* as a result of varying T or pH, is required on one or a very few compounds.[143] The most recent and interesting recruit to this area is Tomlinson's filter-probe,[140] another borrowing from engineering science. Essentially this samples the dominant (by volume) phase, generally the aqueous, and pumps it through a suitable, *e.g.* UV, detector; a most valuable feature of the method is that the approach to equilibrium is continuously monitored. The core of the equipment consists of a heavy metal probe, attached to the circulating stainless steel tubing, which contains a carefully designed filter that will prevent entrainment of the unwanted phase (see original literature[140] for diagrams). This is placed near the base of the reaction vessel from which it transmits the major phase to the detector and back again. As we employ the method, the solute typically at about 2 mg % is dissolved in 150–200 mL of the aqueous phase in a covered beaker and circulated until an equilibrium UV reading is attained. The organic phase is then added down the side of the vessel remote from the probe, the solution is agitated with a self-reversing stirrer, and the UV trace is monitored until there is no further drop in absorbance. If vortexing and emulsification is avoided, as is essential for octanol or the filter (Whatman No. 50) will not cope, this process can take as much as 12 h. Hydrocarbon solvents will accept much more vigorous conditions hence equilibrium is reached much faster[140] but these are correspondingly more volatile so errors can enter that way. In essence the filter-probe technique is related to shake-flask methodology at high phase volume ratio and the same equation serves for both

$$P(\text{or}\,D) = r(C_i - C_{eq})/C_{eq} \qquad (54)$$

where C_{eq} is that attained in the major phase at equilibrium. It is probably safer to use at high volume ratios without risk of serious error, and the approach to equilibrium is accurately monitored, but it is difficult to operate this as a closed system, without which loss of organic phase by evaporation represents a potential problem. In practice we find this a valuable technique especially for $\log P$ or $\log D = 0$–2 but above this errors can arise from base-line instability as C_{eq} becomes small unless both phases are analyzed, which is an elaboration that loses the method much of its convenience. Tomlinson[140] has however demonstrated its utility for unstable compounds and, as a closed system, over a wide temperature range.

18.6.5.3   Liquid–Liquid Chromatography

Since our original demonstration[144] that octanol can be made to stick to an inert support well enough to behave as a stationary phase in its own right, a number of authors, especially Unger[139] and ourselves,[109] have modified or improved the methodology. In this account we shall mostly present our own up-to-date techniques. Since the support material has the task of entraining octanol while remaining entirely self-effacing otherwise, its choice is crucial to success. While Unger[139] and at one time ourselves[109] have used a persilated C_{18} packing material, we started with,[144] and have returned to, the diatomaceous earth Hyflo-supercel (*ex* Manville Inc.: a celite diatomite filter aid). This is an irregularly shaped porous material of variable particle size which absorbs non-polar liquids like a sponge and, properly treated, appears capable of retaining octanol and (even more effectively) PGDP for considerable periods. It requires pretreatment with concentrated HCl, successive washings with which produce and then remove a bright yellow discoloration. This is followed by repeated water washings till neutral after which the material is air dried and stored. No silanation appears necessary. Columns may be slurry packed using methanol but we now prefer dry packing according to standard techniques.[96a] These columns are then coated with excess of the presaturated organic phase through a Luer lock syringe under slight finger pressure. Columns are stainless steel, of 5 mm inner diameter (i.d.), and are protected by a 10 cm guard column which is also coated. Eluant, which always consists of the presaturated aqueous phase, is passed at 6 mL min^{-1} for 20–60 min to condition the column; droplets of the organic phase are usually visible initially and during this period the column is *not* connected to the detector as these can cause problems in the UV cell. Once this is complete the column is ready for use and can be stored fully charged indefinitely in the absence of buffer. For octanol we have used a column of 2–10 cm in length at a flow rate of 0.2–4.0 mL min^{-1} which entails a pressure of 20–500 lbf in^{-2} (1 lbf in^{-2} $\simeq$ 6.9 kPa) or more. For PGDP the corresponding ranges are 5–25 cm and 0.1–3.0 mL min^{-1}, giving the same pressures. Other solvents have been employed in liquid chromatography[96a] but we have not attempted their use. Using buffers, a pH range of two to nine or higher has caused no apparent damage; it seems probable that this support is stable well outside the range 1.5 < pH < 8 to which ODS columns are limited,[140a] probably because of its thorough coating. Unger[139] recommends purging the eluant with nitrogen; we have preferred partial degassing under reduced pressure and failure to do this can be responsible for 'spiking' caused by air bubbles in the UV cell, though more recently we have not found this necessary. Before the first injection of the day, the column should be run for 15–30 min to settle down. Samples are typically injected, if not in water then in 20% methanol, and, if possible, a longer injection loop is preferred to more methanol so as to optimize column life. In between coatings, this life can range from, for example, two to three days for a 3 cm octanol column at 4 mL min^{-1} to about six weeks for a 25 cm PGDP column at 0.1 mL min^{-1}. Recoating simply repeats the initial procedure but the opportunity should be taken to inspect both ends of the column and top up the packing material if necessary. Total repacking is needed only if and when repeated recoating fails to give adequate column stability.

Since a given retention time means nothing on its own, these columns require calibration with compounds of known shake-flask log P. Unger[139] calibrates once per session. We alternate calibration with 'active' runs; the latter contain one compound only, while the former will contain three or four standards evenly spaced across a log P range of 1–1.5. It is important to alternate strong proton donor and acceptor standards so as to ensure that the hydrogen bonding characteristics of the column are what they ought to be. Non-polar standards, or those of any single class, do not provide this information. A list of suitable standards for octanol and PGDP is given in Table 9. For as long as deviations do not exceed $\Delta \log P = \pm 0.02$ from a line of slope 1.00 ± 0.02 in equation (37), the column is considered satisfactory. Unger[139] uses similar criteria. As solvent is slowly lost, the column thins to the point where these criteria are no longer satisfied; this usually occurs quite suddenly and presumably results from exposure of the packing material. The immediately preceding moment is ideal for measuring compounds of otherwise inconveniently high log P. A typical calibration run is shown as Figure 10. The HPLC specialist tends to be horrified by such a trace but, as will easily be seen, even a broad peak such as (e) can be read with quite sufficient accuracy. In fact we find a typical peak accuracy of $\pm 2–3\%$, much better than for shake-flask and ample, for example, for establishing reliable f values. Furthermore, and in sharp contrast to any other technique, this accuracy is independent of log P itself over the entire accessible range. The lack of sensitivity of any HPLC technique to impurities needs no emphasis.

Table 9 HPLC Calibration Standards for Octanol and PGDP

Octanol Compound	log P	*PGDP* Compound	log P
Acetanilide	1.16	N-Methylbenzamide	−0.05
Acetophenone	1.58	Acetanilide	0.40
Quinoline	2.03	Phenol	1.17
Anisole	2.11	Quinoline	1.62
3-Chlorophenol	2.50	N-Methylaniline	1.87
3-Bromoquinoline	3.03	4-Chlorophenol	2.13
Benzophenone	3.18	Nitrobenzene	2.16
Naphthalene	3.37	Anisole	2.41
Acridine	3.39	Phenazine	2.66
Diphenylamine	3.50	3,4-Dichlorophenol	3.00
3,5-Dichlorophenol	3.68	3-Bromoquinoline	3.04
2-Phenylquinoline	3.90	Benzophenone	3.40

In most respects, Unger's technique and ours are in close agreement. In some ways ours seems the less sensitive, for example in that we have never encountered the adsorption of proton acceptor solutes which Unger counters by adding hydrophobic amines to the eluant.[139] Unger notes that ion correction sometimes gives unreliable results. We have also encountered this in the sense that E relative to P in equation (53) is larger than expected, but only sometimes, and Hyflo-supercel seems somewhat less prone than other packing materials. Possibly some ions can be adsorbed by the column. Another problem which is rather more troublesome than by the shake-flask method is solute insolubility; this when it happens gives elongated peaks.[139]

The point has been reached where we prefer LC to all other techniques when it can be used. Useful ranges are log P = 1–4 for octanol and 0–3.5 for PGDP, but these are not limits. In principle only a stop-gap until someone can produce bonded columns with the right characteristics, in practice its

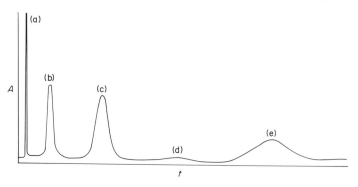

Figure 10 HPLC trace for an octanol column with Hyflo-supercel as support showing UV absorbance A as a function of retention time t: (a) pyridine 1-oxide as marker for t_0; (b) quinoline, log P 2.03; (c) 3-chlorophenol, log P 2.50; (d) unidentified impurity; (e) 3-bromopyridine, log P 3.03

limited column life has attracted more censure than it deserves.[90] On its day, for accuracy and convenience LC is in a class of its own.

18.6.6 EMPIRICAL APPROACHES TO THE ESTIMATION OF SOLUBILITY

Most attempts to estimate S_w start with log P. If a set of liquid solutes is miscible with the organic phase, then $\gamma_0^* = 1$ in equation (21) so that the whole change in P as defined by equation (20) lies in S_w not S_o (*cf.* equation 19). If solids are added to the set, their solubilities will be further reduced relative to the supercooled liquid solute, in any solvent, according to equation (9). Yalkowsky and co-workers[145] have used these principles to·derive equation (55)

$$\ln x_B = -\log P - 0.01(T_m - T) \tag{55}$$

where the coefficient of $(T_m - T)$ derives from arguments[145a] that lead to a roughly constant value of 56 J mol^{-1} for ΔS_m in equation (9). In fact, these authors find equation (56)[145a]

$$\log S_w = 0.87 - 1.05 \log P_{oct} - 0.012(T_m - 25) \tag{56}$$

at 25 °C where the constant results mostly from translation to the molar scale. (The potentiality of this equation for estimating solubility differences between polymorphs does not seem to have been explored so far). Use of P_{oct} is justified[145] by the fact that most polar organic liquids possess $8 < \delta < 12$ (Table 1). The non-unit slope of log P does however demonstrate that the log γ_0^* term of equation (21) cannot be entirely ignored, and this is confirmed by later work[145c] which shows only a moderate correlation of log x_{oct} with the T_m term alone ($n = 36$, $r^2 = 85\%$, $s = 0.32$). Chiou and Block[9] have shown that, while P for any solvent system will correlate S_w for very non-polar solutes, P_{oct} gives much better results than $P_{heptane}$ for those that can self-associate. However, inspection of the results for substituted benzenes shows that π values cannot be discussed entirely in terms of $\Delta \log S_w$ even when self-associated molecules are excluded; for example, -0.28 for π_{NO_2} results from a small reduction in aqueous solubility ($\Delta \log S_w = -0.14$) more than offset by a larger fall in octanol ($\Delta \log S_{oct} = -0.42$). Other interesting differences are found: for instance, the less negative π value of OH than of NH$_2$ results because the latter, unlike the former, does not increase S_{oct}. Hence equations such as (56) may conceal more than they reveal and ideally one should probably choose P such that the organic phase structurally resembles as closely as possible the set of solutes to be predicted. In view of this it is unsurprising that HPLC, in the form of ODS columns with aqueous methanol as eluant, has also been used as a successful predictor of log S_w;[90] such success does not really depend on the resemblance between log P_{oct} and log k' even though that was its rationale. Here[90] it was found that amphiprotic solutes (ROH, RCO$_2$H) lie on a different line from the remainder, S_w being higher than would have been expected; since the use of log $P_{heptane}$ overpredicts their solubility,[9] the source of this deviation probably lies in the properties of the eluant.

The LSER approach of Taft and Kamlet, to be described in the next section, has been used extensively to correlate solubility-related phenomena.[35] At present it is applicable mostly to simple monofunctional solutes and prediction outside this area is of doubtful value, but its usefulness for characterizing the solvent is already apparent. Not only log S_w but log S values for blood, brain and lung have been investigated and show significant differences; it appears, for example, that the affinity of brain for proton acceptors is much less than that of the other tissues.[35] Also the correlation of observed with calculated S_{blood} values shows that these reach a minimum value which may indicate the presence of hydrophobic pockets in hemoglobin; one consequence of this is a less than predicted toxicity of very non-polar substances to the golden orfe.[35] Further advances are likely both in this area and, for actual S_{w} prediction, from the development of appropriate fragment values as is now in progress.[146]

18.6.7 EMPIRICAL APPROACHES TO ESTIMATION OF THE PARTITION COEFFICIENT

18.6.7.1 Linear Free Energy Relationships

18.6.7.1.1 *Linear solvation energy relationships (LSERs)*

The LSER approach of Kamlet and Taft represents the most fundamental attempt yet made to rationalize all forms of solubility-related phenomena. These are discussed in terms of the short-range forces that must be responsible for solute–solvent interactions.[111] As in Snyder's earlier treatment of chromatographic phenomena,[96b] and as suggested by Davies,[71] three features of the solvent appear to dominate: its proton donor and acceptor properties, and its general dipolarity/polarizibility characteristics. These have been quantified as the α, β and π^* scales respectively[111] and used to correlate properties which range from spectroscopic shifts[111] through chemical reaction rates[111] to equilibrium properties such as tautomerism.[110] Recently, they have been applied to solubility and partitioning.[32,35] Here a volume term to allow for cavity formation (Section 18.6.2.2.4) is also required and equations take the general form

$$Q = Q_0 + A(\delta_A^2)\bar{V}_B + B\pi_A^*\pi_B^* + C\alpha_A\beta_B + D\beta_A\alpha_B \qquad (57)$$

where Q is some free energy related quantity relative to a standard Q_0 and A and B are solvent and solute terms respectively. Given a constant solvent as for S_{w} or a constant pair of solvents as for P_{oct} all terms in A may be subsumed under the regression constants, leading to equation (58)

$$Q = Q_0 + m\bar{V}_B/100 + s\pi_B^* + c\beta_B + d\alpha_B \qquad (58)$$

This is illustrated by equations (59) and (60) for a group of non-polar and dipolar aprotic monofunctional solutes (hence the α term is absent at this stage). Here the more positive $\bar{V}/100$ in the case of cyclohexane reflects its greater affinity for hydrocarbon fragments, whereas the reverse for β equally reflects this solvent's inability to form hydrogen bonds. In later work,[33] V_I has partly replaced $\bar{V}$,[35] and the addition of amphiprotic solutes has started to allow the importance of α to be assessed.

$$\log P_{\text{oct}} = 0.20(\pm0.07) + 2.74(\pm0.05)\bar{V}/100 - 0.92(\pm0.08)\pi^* - 3.49(\pm0.09)\beta \qquad (59)$$

$$n = 102, r^2 = 98\%, s = 0.17$$

$$\log P_{\text{cyclohexane}} = -0.01(\pm0.15) + 3.96(\pm0.13)\bar{V}/100 - 1.25(\pm0.12)\pi^* - 5.80(\pm0.16)\beta \qquad (60)$$

$$n = 25, r^2 = 98\%, s = 0.16$$

Two quite separate complicating factors exist. Since α and β were originally derived for bulk liquids,[111] whereas their use in (58) implies an isolated functional group, any resulting change in properties will lead to inaccuracies. This problem is specially acute for amphiprotic groups: the bulk properties of alkanols, for example, are no guide to that of the OH group in isolation. Work is now in progress[37,147a] to derive functional group donor and acceptor scales more suitable for use in equation (58). However, success even here is not the complete answer. All such values are based on one-to-one contact; solutes however are surrounded by solvent and, especially in water, more than one contact is possible. This is particularly significant for acceptors with varying numbers of lone pairs. An extreme example is provided by SO and SO_2, whose nearly equal fragment values result from a near-cancellation of quality against quantity (Table 11). Possibly the special correction required for pyridines[32] originates this way. This problem is only just beginning to be faced.[147b] The second problem concerns how to treat compounds containing multiple functionalities. While it is possible (but not yet certain) that α and β may be treated as additive, at least if the groups are not too close, it is by no means so obvious that this applies to π^*, which is much more of a global quantity and shows some tendency to fall along homologous series.[111] Contiguous functional groups present severe problems to any additive scheme[11,12,69] but here these are particularly acute since already present in substituted aromatics as the ring possesses its own β value.[32] These conditions are severely restrictive and indeed the authors[32] recommend CLOGP[80] in most practical contexts. Nevertheless LSER is the most chemically literate dissection of solvent properties yet devised and has enormous potentiality for characterizing membranes *via* a well-chosen solute set[35,147] (*cf.* Section 18.6.8).

18.6.7.1.2 *The Collander and related equations*

Collander[148] in 1951 found that log P values of a mixed set of solutes for various alkanol–water systems gave straight lines when plotted against one another. The Collander equation (61) was the first explicit LFER in this field and has gone on to provide the inspiration for much subsequent work: equations (29), (30), (32), (37), (40) and the basic QSAR equation (35) are all essentially its relatives.

$$\log P_{(2)} = a \log P_{(1)} + b \tag{61}$$

The simplicity of (61) is so appealing that it has been much overused. It will *not* correlate partitioning systems in general,[16] as Collander himself was entirely aware, and there are many published examples of its failure;[8,12,16,92] it even begins to fail in alkanol solvents for multi-substituted solutes.[148] Figure 11 expresses the Collander equation for a set of alkanols as solutes against carbon number as a neutral reference point; as the organic phase becomes less polar, a rises while b becomes more negative. In this set, a represents f_{CH_2} (*cf.* equation 66); more generally, it will approximate to the (relative) value of m in equation (58) for two partitioning systems. This comparison will show why (61) cannot be generally valid: its validity would involve a constant ratio between s, c and d in (58) from one system to another, these being the 'hidden' contributors to b in (61). Any appreciable change in solvent hydrogen bonding characteristics[111] will change this ratio. It is of course easy to 'prove' the validity of (61) in an inappropriate context by the use of non-discriminating solutes, and all too many such studies have been published.

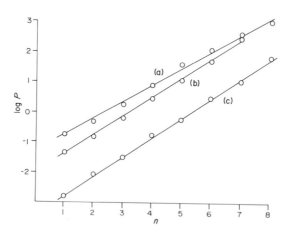

Figure 11 Log P for alkanols ROH as a function of n=number of alkyl carbon atoms: (a) octanol/water, slope 0.54; (b) CHCl$_3$/water, slope 0.62; (c) cyclohexane/water, slope 0.65

Attempts have been made to get round this problem. Acting on a suggestion by Leo *et al.*[8] that a should be unity in the absence of hydrogen bonding, Seiler[149] derived equation (62).

$$\log P_{\text{cyclohexane}} = \log P_{\text{oct}} - \Sigma I_H - 0.16 \tag{62}$$

In fact (62) is invalid since a is far from unity, especially for this solvent pair; *cf.* equations (59) and (60). Hence the derived substituent group I_H values will be log P-dependent (*cf.* Figure 11). Even without this problem they could only apply to this solvent pair or to others very similar. The frequent use of I_H in QSAR equations is unsound. There is more chance of success by treating donors, acceptors and neutrals as separate classes,[8] but even this in practice is not too successful.[12] We return to this subject in Section 18.6.8.

If homologous series RX are plotted for the same partitioning system against carbon number, Figure 12 results: parallel straight lines whose intercepts depend on X. This may be seen as a form of equation (61) whose slope and intercept terms represent the non-polar and polar contributions to partitioning respectively. A more sophisticated treatment involves calculating the former as a cavity term and obtaining the latter by difference; this is then specific for a given functional group. McGowan's use of the parachor in this way[30] has more recently been refined by Moriguchi[150] and other examples already discussed are due to Pearlman[9, 10] and Grünbauer.[28] Testa and Seiler[151] applied similar reasoning to fragment values to obtain equation (63)

$$f = 0.0534\ V + \varLambda \tag{63}$$

where $\varLambda$ is the functional group polar contribution. There is nothing the matter in principle with such approaches, but in their lack of scope for further analysis they represent a dead-end. They are neither so convenient to use as the fragment system nor do they provide the insights of LSER, and they must now be considered as overtaken by events.

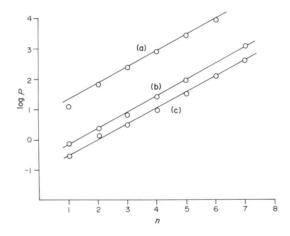

Figure 12 Octanol/water log P for homologous series as a function of n = number of carbon atoms: (a) RH, slope 0.53; (b) RCO_2H, slope 0.54, (c) RNH_2, slope 0.53

18.6.7.1.3 *Principal components analysis (PCA)*

PCA is a technique which attempts to identify the hidden variables in any system instead of assuming, as in multiple regression (MRA), that we know these already, and in many contexts has proved of great value (see Volume 4, Chapter 22.3 for a full exposition). Its most ambitious use in chemistry so far is that by Cramer.[152] Six physical properties, one of them log P, were transformed to give five new independent variables ('components') **B**, **C**, **D**, **E** and **F** (**A** is the regression constant). **B** and **C** together account for 95.7% of the variance, hence the brackets round the remainder: **B C (D E F)**. Each property (Q) equation takes the form

$$Q = \mathbf{A} + b\mathbf{B} + c\mathbf{C} + d\mathbf{D} + e\mathbf{E} + f\mathbf{F} \tag{64}$$

such that **A** and the coefficients b–f vary between properties and each compound possesses its own individual values of **B**–**F**. This procedure is then reversed by re-expressing **B**–**F** as linear functions of four of the original properties, one of them log P; where all these can be supplied for a compound outside the set, any other of its properties definable by equation (64) can now be calculated. Cramer[152] suggests that **B** correlates with bulk, and **C** is a sort of 'bulk-corrected cohesiveness', but no identification of either with a known molecular property has been possible and even V, which belongs to the set, correlates poorly with **B** alone. While interesting QSAR analyses have followed,[70] it is plain that PCA as used in this study has in no way simplified our perception of reality.

Similar comments apply to the various attempts to analyze log P. For a varied set of 51 solutes in six partitioning systems, Dunn *et al.*[153] found 77% of the variance to be accounted for by the first component t_1 and an additional 17% by t_2. The first appears to be associated with the non-polar fraction of SASA while the second, suggested to be electronic in origin, has so far eluded more explicit description. A study of interaction terms in disubstituted benzenes[154] identified a single component accounting for non-additivity in terms of a cross-product $\beta_i \theta_k$ where i and k are substituents and β and θ are coefficients specific to each. Unfortunately, while both show a trend with σ, neither can be identified with any recognized parameter. A previous analysis of a similar but smaller set[155] found t_1 to be identified with the additive part of $\log P$ while, for t_2, a number of possible σ-type relations were suggested. Wold and Dunn[156] analyzed data for 26 solutes in six solvent systems and extracted two components of which one is essentially the same for all solvents while the second shows large variations; the cause of this was not identified. Breakdown of the solute components revealed one to be associated with chain length while the other 'appears to be the result of a dipolar interaction of solute and solvent'. A similar analysis of solvent systems by Franke *et al.*[157] comes to similar conclusions: one component is associated with size while the other 'is connected with hydrogen bonding'.

All these analyses are undoubtedly valid so far as they go; the trouble is that they fail to take us very far. The vague evocation of σ values or 'polarity' above contrasts with the precise use of the former in the CLOGP algorithm[80] or the success of the LSER approach in quantifying the latter[32] in terms of values derived from other areas of chemistry.[111] Shotguns are pointless when precision rifles are available. It should also be noted that PCA is every bit as sensitive as MRA to the initial choice of candidate parameters. Any derived component can only be expressed in terms of the original variables; if these were ill chosen the analysis may be abortive. A case in point is a recent analysis of solvent properties by Chastrette *et al.*,[158] which examined 83 solvents in terms of such reasonable quantities as dipole moment and boiling point but also HOMO and LUMO and contained no term directly related to hydrogen bonding. This study succeeded, *inter alia*, in classifying TFA as dipolar aprotic, THF as an alcohol, octanol as a polar aromatic solvent and hexane as an electron pair donor. Such exercises are of academic interest only. Faced with a mass of facts but little prior idea on how to organize them, PCA undoubtedly has a valuable part to play. Mature areas of science are likely to respond rather differently: the problem may be identified or restated but it is unlikely to be solved. Nothing so far published contradicts that expectation.

18.6.7.2 Additive Treatments of Log *P*

18.6.7.2.1 *The π and fragment systems*

This subject is frankly empirical. π is a difference term

$$\log P_{RX} = \log P_{RH} + \pi_X \tag{65}$$

and as originally conceived, it was not intended that $\log P$ should be calculated this way.[11] Attempts to do so by faulty means are now history. That task was accomplished by the fragment system, introduced by Rekker[12] and since refined by the Pomona school[11,80]

$$\log P_{XYZ} = f_X + f_Y + f_Z + \Sigma F \tag{66}$$

where f values are fragment values and F stands for some factor which may be needed to allow for non-additivity. A simple relation between π and f exists[11]

$$f_X = \pi_X + f_H \tag{67}$$

and this is helpful particularly for aromatic systems where the use of π remains common and can still be convenient.[11,109]

Chapter 18.7 is entirely concerned with this topic but a few comments are pertinent here. Rekker's f system was derived by statistical means but that of the Pomona school in a stepwise manner; hence they may be described as reductionist and constructionist respectively.[11] Mayer *et al.*[159] regard them as better for large and small molecules respectively; there is certainly not a vast difference for those to which both are applicable but Rekker's methodology contains no electronic correction terms and has ceased to develop since 1979,[160] whereas the CLOGP algorithm[80] is in a constant state of evolution. By hand calculation the latter is very tedious, whereas Rekker's is easy; though, in the writer's experience,[97] its later version[160] is generally less accurate than the earlier one. This arises largely since his original value[12] of 0.235 for f_{CH} unconsciously incorporated a branching factor which the later 'normalized' value of 0.362 abolishes. The difference lies close to Leo's F_{cbr} (-0.13). At the same time there are infelicities in CLOGP which need attention. One is that multialkylated aromatics are grossly overpredicted, largely because alkyl f values are generated by a complicated procedure which involves, for a chain, subtracting the bond factor F_b (-0.12) from the sum of f_C and f_H. The latter is deduced as half log P_{H_2}, *i.e.* 0.227, while Rekker derived it as the difference between f_{CH_3} and f_{CH_2}, *i.e.* 0.17. Since F_b does not apply to the first link[11] this leads to (uncorrected) methyl f values of 0.70 (Rekker) or 0.876 (Leo). Part of the problem may lie in an inconstant value for f_H. Mayer *et al.*[159] point out that H in H_2 and aromatics is much more exposed than in an alkyl chain; the difference is compatible with f_H values of 0.23 and 0.17 respectively. However, it needs to be realized that *all* ways of analyzing log P in terms of fragment values (except the repeating unit CH_2) are in essence arbitrary, and so is further breakdown of the fragment values themselves; one must be grateful for any system that can be made to work.

Group branching (F_{gbr}) and proximity corrections represent another area of vulnerability. These certainly exist and the CLOGP algorithm is explicit on both, but on dangerously little evidence so far. In particular, both are based very largely on the behaviour of a small number of mostly amphiprotic functional groups, *i.e.* O, OH, NR_2 (R = H or alkyl) and CO_2H. Most of these are small enough to be partially shielded from octanol at a branching centre; larger less polarizable groups may not require F_{gbr}. Similarly, in molecules of type XCH_2CH_2Y, the functional groups are effectively further apart when, for example, CO or CN than when N or O, and again the corrections may be less than predicted. Much more work is needed.

Finally we consider the possibility that log P may be quantized as implied by Rekker's 'magic constant' C_M, which was proposed in the form $k_n C_M$ where k_n is an integer to rationalize observed spacings between f values of various sorts.[12] For instance, the suggested proximity corrections for XCH_2Y and XCH_2CH_2Y, which in Rekker's system are independent of substituent,[12] lie close to $3C_M$ and $2C_M$ respectively. Its rationale is that any molecule or fragment has to displace a discrete number of water molecules. The rationale for k_n is entirely *post hoc*; its magnitude cannot be predicted. The unit value is C_M 0.28 and some of the best evidence comes from the noble gases, which from Ne to Xe have log P values of 0.28, 0.28, 0.74, 0.89 and 1.28, or close to 1, 1, 3, 3 and $5C_M$.[12] Others have toyed with this idea. There is nothing inherently implausible about Rekker's rationale, though since all log P values are statistical averages, especially for conformationally flexible molecules,[42] it seems unlikely that such an effect would be observable in practice. However, these quoted regularities depend entirely on the system considered, namely log P_{oct} at a standard state of 1 mol dm^{-3}. The noble gas log P_{hexane} values are 0.71, 0.80, 1.14, 1.33 and 1.67 in sequence; a k_n range of 1–4.5 becomes 2.5–6. Similarly methane and pentane apparently displace 4 and 12 waters when entering from octanol but 5 and 15 when entering from hexane. Such discrepancies reduce C_M to an absurdity, and the situation changes yet again if mole fraction standard states are used. The attempt to subdivide C_M still further, to a unit value close to mean experimental error,[95] is not to be taken seriously.

18.6.7.2.2 *Bond and atom systems*

If a set of log P values is additive in the sense of equation (66) with $\Sigma F = 0$, then any other self-consistent method of decomposing these structures must also be additive. Hence bond and atom methodologies are feasible and have been tried.[69, 161–163] Their problems are twofold. Firstly, vastly more descriptors are required, which makes for questionable statistics. Secondly, it is far more difficult to choose these descriptors in a chemically meaningful way, so that the results tend at best to be impenetrable and at worst may conceal confusions between categories that make the analysis a nonsense. The illustrative classification system of Table 10 will make the latter point. It results in equation (68).

$$\log P = 0.05 + 0.33(C\text{-}sp^2) + 0.36(C\text{-}sp^3) + 0.01(C\!\!=\!) - 1.80(S) - 0.54(O) - 0.05(\!\!=\!\!O) \quad (68)$$

$$n = 10, r^2 = 80\%, s = 0.77, F = 2$$

According to this analysis sp^2 oxygen is more hydrophilic than the sp variety, whereas far the most hydrophilic atom is sulfur. This absurd result has been (deliberately) brought about essentially by classifying all forms of sp oxygen as the same. The result of Klopman *et al.*,[162] that NO_2 is far more hydrophobic than C or H, stems from the same sort of cause, while similarly Ghose and Crippen[163] find some forms of C to be hydrophilic and most forms of N (except in NO_2!) to be hydrophobic. All these are system artefacts. Broto *et al.*[161] have a more complex system and fare somewhat better;

nevertheless they lose information badly. For example, their most hydrophilic atom is tertiary alkyl N, but the equally hydrophilic amide group is somehow 'lost' between its constituent parts. All these systems can be made to predict log P for the reason stated, but all are chemically meaningless. In the writer's opinion, they are to be discouraged.

Table 10 Atom Classification for Selected Solutes

	$log\ P_{oct}$	$C(sp^2)$	$C(sp^3)$	$C=$	S	O	$=O$
Benzene	2.13	6	0	0	0	0	0
PhCH$_3$	2.69	6	1	0	0	0	0
PhCH$_2$CH$_3$	3.15	6	2	0	0	0	0
PhCOCH$_3$	1.58	6	1	1	0	0	1
PhCO$_2$CH$_3$	2.12	6	1	1	0	1	1
PhCH$_2$OCH$_3$	1.44	6	2	0	0	1	0
PhSOCH$_3$	0.55	6	1	0	1	0	1
PhSO$_2$CH$_3$	0.50	6	1	0	1	0	2
PhCO$_2$CH$_2$CH$_3$	2.64	6	2	1	0	1	1
Naphthalene	3.37	10	0	0	0	0	0

18.6.7.3 Non-additivity in Octanol–Water Log P

We chiefly consider that of stereoelectronic origin. The basic problem is encapsulated by equation (69)

$$\Delta \log P = \Delta f_X = \Delta \pi_X = \pi_{X(PhY)} - \pi_{X(PhH)} \tag{69}$$

i.e. the π or f value of X inserted into PhY (or other aromatic) is not that found for X in PhH. Almost always, deviations are positive. Following his early work with Hansch, Fujita[164] has proposed bidirectional Hammett-type relationships in which X and Y act on one another. A number of equations of type (70) have been derived.

$$\pi_{X(PhY)} = a\pi_{X(PhH)} + \rho_Y\sigma_X + \rho_X\sigma_Y + c \tag{70}$$

If X does not form hydrogen bonds then $\rho_X = 0$, so that only the effect of X on Y ($\rho_Y\sigma_X$) need be considered. Hence the value of ρ_Y may be determined by examining sets of PhY for constant non-bonding X. Use of this value of ρ_Y for further substituents X from which this restriction is removed allows $\rho_X\sigma_Y$ to be obtained by difference and, since σ_Y is known (σ^0 values were used throughout), ρ_X can be established. By successive approximation, 14 equations of type (70) resulted; correlations were excellent.

Leo[112] has adapted Fujita's approach for the CLOGP algorithm. He classifies fragments as inducers (electron acceptors) which possess σ values or as responders (electron donors) which have ρ values; some are both. His equation to cover electronic effects alone is (71)

$$OLP = 0.991(\pm0.017)ALP + 0.925(\pm0.074)\rho_R\sigma_I + 1.144(\pm0.334)\rho_I\sigma_R + 0.006(\pm0.042) \tag{71}$$

$$n = 213, r^2 = 98.6\%, s = 0.098, F = 39.5$$

where OLP and ALP are observed and additive log P, R = responder, I = inducer, and the $\rho_I\sigma_R$ term is to be used only for combinations that are both. Leo's treatment differs from Fujita's in that his σ values are not Hammett's but, like ρ, are best-fit values obtained by successive approximation (Volume 4, Chapter 18.7). In fact ρ and σ bear some resemblance to electron donor and acceptor σ values respectively, but this is not exact. Essentially they are empirical constants which are justified because they work.

There are two problems with the bidirectional approach. Firstly, it is statistically dubious. Comparison of equations (69) and (70) will reveal that what is being fitted is not $\Delta\pi$ but a quantity which contains the floating coefficient a, so that a large slice of the dependent variable is being used as if independent. This point is non-trivial since 95% of the variance in OLP for disubstituted benzenes is accounted for by ALP.[155] Since for Fujita's 14 equations $a = 0.932 \pm 0.048$, his statistics are in fact much less impressive than they appear. With $a = 0.991$ this objection applies much less to Leo and it is significant, therefore, that orthodox σ-values could not be made to fit. We return to this point later.

The second flaw is chemical. The bidirectional approach assumes that all interactions are positive, but this cannot be so. The nitrogen atom in 4-methoxypyridine is a stronger proton acceptor than in pyridine itself,[37,165] hence from this cause alone, $\Delta \log P$ should be negative. That it is positive ($\Delta \log P = 0.37$) can only be due to more loss of electron density from OMe than N gains.[109] Scheme 2 generalizes the situation in terms of some aromatic nucleus containing two polar moieties. Here Y and Z are proton acceptors while XH, which must belong to one or both, is a proton donor. Possible directions of electron drift are as shown. Consider proton acceptors first.

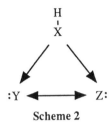

Scheme 2

If $Y = Z$, or more generally if their electronegativities are equal, there is no net electron drift so $\Delta \log P = 0$. As Z increases in electronegativity, however, so does electron drift and, since this has to traverse the σ skeleton, an increasing fraction will be lost on the way: Z gains less than Y loses. Hence $\Delta \log P$ will rise, sometimes as above, sometimes with loss of proton acceptor ability for both. Now consider the effect of adding XH. Since Y and Z both act on XH, any increase in electron pull will increase its proton donor ability and this is true even if $Y = Z$. $\log P$ rises for this reason since octanol is a better proton acceptor than bulk water[111] just as it rises as solute acceptor ability is lost since water is a net proton donor solvent (Scheme 3).[109] Figure 13 shows all available $\Delta \log P$ data for m- and p-$C_6H_4Y_2$ as a function of σ for Y. In the absence of donor groups, electronegativity has no effect. For donors however this is very pronounced, ranging from a small negative $\Delta \log P$ for $Y = OH$ to a large positive value for $Y = SO_2NH_2$. It follows that proton donor and acceptor effects *must* be separated if $\Delta \log P(\Delta \pi)$ is to be understood. So far, only one study[109] has attempted this. For example, for diazines Y-substituted α to nitrogen (but not between two), we find equation (72)

$$\Delta \pi_Y = 1.4 - 1.8\sigma_I + \sigma_I(\alpha\text{-NH}) + 0.5\sigma_I(\beta\text{-NH}) \tag{72}$$

where σ_I is that for Y while α-NH is NH adjacent to the ring (*e.g.* NHR) and β-NH is one group removed (*e.g.* CONHR). (This is not a correlation equation since the substituent set is too lopsided for valid statistics, but for 116 compounds across five such equations, $\Delta\Delta\pi = -0.03 \pm 0.12$). Equation (72) is easily recast to give the formal relation (73)

$$\Delta \pi_Y = a(\sigma_{NN} - \sigma_Y) + (b + c)\sigma_{NN}\sigma_Y \tag{73}$$

where σ_{NN} represents the (unknown) effective σ-value for the diazine ring. That is, while $\Delta\pi$ as due to proton acceptor interactions depends on their difference in electronegativity, that due to their donor properties depends on their sum (or product). This chemically essential distinction is not yet part of any algorithm. It will explain a wide variety of phenomena and can be generalized. For example, the $\Delta \log P$ values of 0.47 and 0.30 for (5) and (6) respectively are as expected, since (5) shows the extra proton donor effect, and 0.46 for (7) appears regular also, but -0.19 for (8) seems at first sight anomalous. This results from total incorporation of the nitrogen lone pair of indole into the π system so that there is nothing more to lose and what is observed for (8) is carbonyl's gain.[166] Similarly we find 0.51 for (9) but -0.13 for (10). These examples will emphasize that the behaviour of π donor (or π acceptor) heteroatoms *inside* aromatic rings is totally different from their normal behaviour as substituents and the same Hammett-type approach cannot be expected to handle both. As another example, $\Delta \log P$ for 4-nitrophenol is 0.73 in octanol but -0.37 in cyclohexane; this comes about because the sign of the donor term is reversed, since any form of increase in hydrogen bonding must now favour water (contrast Scheme 3).

Figure 13 re-emphasizes the only partial success of σ in correlating $\log P$. Others[11,112,154,155] have noted that $\Delta \log P$ is little different for *meta* and *para* substituents and deduced from this that the major effect is inductive; we[109] found σ_I to provide the best descriptor for the azines. The real explanation lies deeper.[97] With major exceptions typified by (5)–(8), lone pair electrons belong to the σ skeleton and do not interact *via* π resonance as commonly understood. They interact *via* σ resonance, which in this context means that between lone pair electrons and the LUMO of an

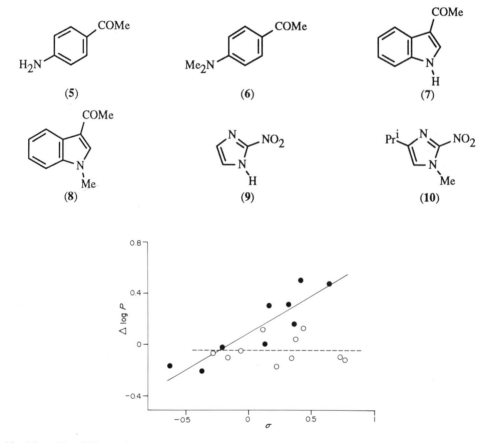

antiperiplanar electron-deficient σ-bond.[107] This is still an unfamiliar concept, partly because there is no agreed way of depicting it, but we attempt this using structures (11)–(19). The familiar electron drift of π-resonance in structure (11) contrasts with that of σ resonance in (12); when Z = O this is about the same in both directions, so that ketones and esters are roughly equal as proton acceptors (see pK_β in Table 11). If R is aryl, the C—R bond is a σ acceptor and carbonyl lone pair density can be lost that way. This is the main reason why aliphatic and aromatic f values are different and, in particular, will explain[97] the much smaller rise from -1.49 for (13) to -1.18 for (14) as compared to -0.56 for (15).[80] The other effect of course is steric—one carbonyl lone pair is partly shielded—but this will not explain the latter contrast and large differences also exist, *e.g.* for nitrile where no steric effect is possible. σ resonance will also explain the very different $\Delta \log P$ values, in terms of the second substituent, of 0.60 for pyrazine (16) in contrast to -0.05 for (17) and -0.11 for (18); as indicated, this is a simple matter of propinquity. For pyrimidine (19) the effect is less (0.43) since each nitrogen atom partially supplies the other. Until σ resonance has been parametrized, which has yet to be attempted, a fully quantitative treatment of electronic interactions in $\log P$ will not be possible. Its random intervention probably accounts for much of the scatter in Figure 13 and similarly elsewhere.[109]

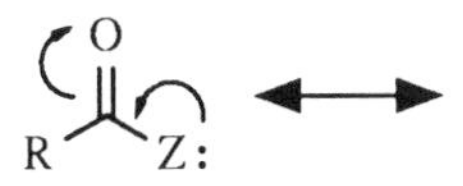

(11)    (12)

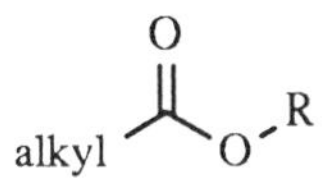

(13) 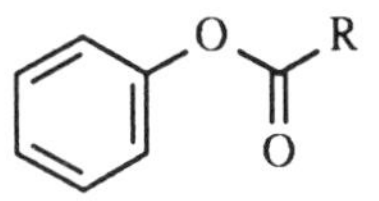   (14) 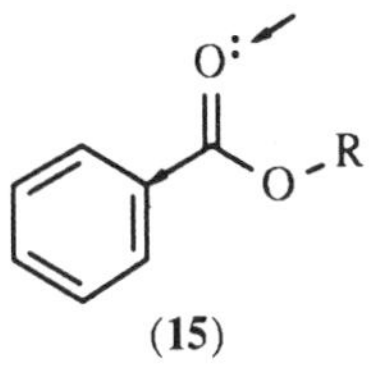   (15)

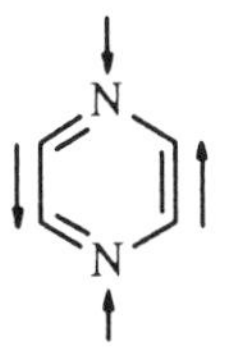

(16) 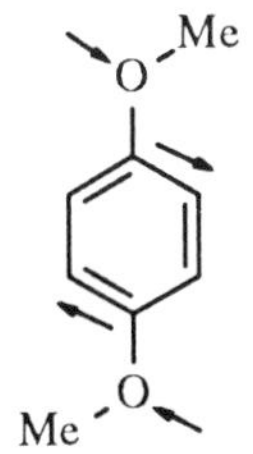   (17) 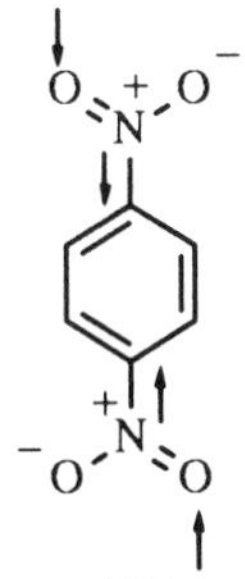   (18) 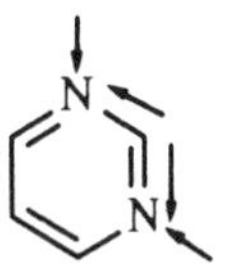   (19)

Some other topics deserve brief mention.[97] The α effect, *i.e.* lone pair repulsion between adjacent heteroatoms, always lowers $\log P$ when it occurs; this will account for the difference between compound classes (20) and (21), where the latter are in general about 0.5 lower,[167] and also that between pyrazine (16) $\log P = -0.23$ and pyridazine (22) $\log P = -0.65$ where σ resonance effects should be about equal. Other forms of lone pair repulsion are also found. The $\Delta\pi$ values of (23) and (24) are 0.01 and -0.41 respectively; CF_3 cannot form hydrogen bonds but mutual repulsion with nitrogen increases the latter's affinity for water.[109] A similar clash must occur for either planar conformation of (25) so that, again, $\log P$ is greatly lowered.[109] Finally we comment on a familiar anomaly: the π value of $CONMe_2$, -1.51 in benzene, becomes -1.27 in thiophene where planarity

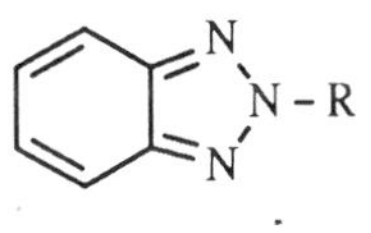

(20) 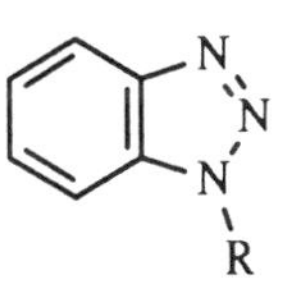   (21) 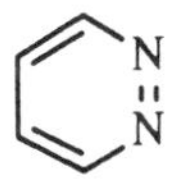   (22)

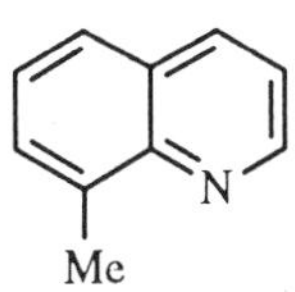

(23) 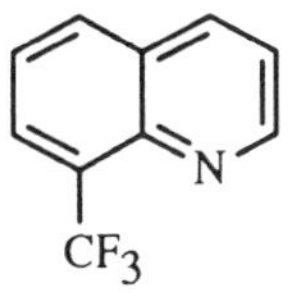   (24) 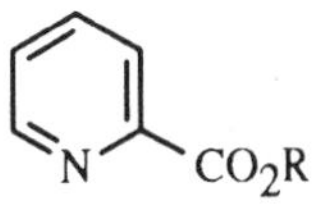   (25)

can more nearly be attained,[166] so confirming Rekker's view of the case.[12] Incidentally π and σ resonance show a similar steric dependence so that steric uncoupling of resonance[12, 168] will apply to both.

18.6.8 MEMBRANES AND THEIR MODELS: A GENERAL SURVEY OF PARTITIONING SYSTEMS

Olive oil was the first partitioning medium to be employed[1, 2] and has been followed by many others. But since its first use by Collander[148] and subsequent espousal by Hansch,[4] octanol has come to possess no serious rival. Hansch and his co-workers[4, 19, 85] have listed its advantages. These include easy availability, hydrogen bonding characteristics not too unlike water, and, in consequence, $\log P$ values in an accessible range; very non-polar solvents tend to overrespond to structural variations and suffer from other problems such as solute association. These are cogent arguments and the espousal by others[16] of hydrocarbon solvents has not in general been regarded as biochemically realistic, though Rekker[12] has suggested heptane as a model for the 'blood–brain barrier'. Nevertheless, it would be surprising were any one solvent system to match all possible membranes, and there are indeed many indications to the contrary. While Meyer and Hemmi,[169] in re-evaluating the original work of Overton[1] and Meyer,[2] concluded that oleyl alcohol would fit the data better than olive oil, Diamond and Wright[51] found the latter to correlate partitioning into the *Nitella* cell membrane much better than isobutanol, and isopropyl myristate has been proposed[170] as a suitable model for skin lipids in permeability studies; here octanol gives poor results.[171] Parallel studies on lipid bilayers suggest olive oil as the best model.[172] From work involving volatile anaesthetics, Davies *et al.*[71] deduced that tissues of less than a certain water content possess virtually no proton donor characteristics, whereas at the other extreme, Collander[173] found some botanical membranes to be unusually permeable to proton acceptors, so that a donor solvent might better model these. Rekker[12] carried out a general classification of solvent systems as hypodiscriminative (lower f_{CH_2} and higher water content than octanol) or as hyperdiscriminative (the reverse) but his attempt to use this for characterizing, for example, the buccal membrane foundered for the reasons discussed[16] (Section 18.6.3.3.7).

It has been seen in Section 18.6.7.2 that equation (61) is not in general valid and that no single adjustable parameter, *e.g.* Seiler's[149] I_H, can be used as a conversion factor, for reasons sufficiently encapsulated in the multiparameter equation (58). At least till LSER comes of age, therefore, any valid characterization must be based on actual or calculated $\log P$ values for highly discriminating solvent systems. Again in the light of equation (58), four distinct types may be suggested: apolar, amphiprotic, proton donor and proton acceptor. For the first two categories hydrocarbons and octanol pick themselves, while $CHCl_3$ is the classical donor solvent and, taking the hint from Davies *et al.*,[71] we have developed the diester PGDP to complete the quartet. Table 11 assembles some fragment values for these four systems,[47] along with the hydrogen bonding parameters[37] pK_α and pK_β. All are derived from $\log P$ for simple compounds by a constructionist methodology (Section 18.6.7.2) but with Rekker-style assumptions as to f_H and without use of F_b since we have insufficient evidence to define this (if it exists), so that the f_{oct} do not correspond precisely to any existing system.[12, 80] However, these four sets are then strictly comparable. For alkanes we assume, with Rekker[12] but *contra* Seiler,[149] that all behave alike or would do in the absence of experimental error. Some values are means and many must be considered very tentative. This series continues our earlier treatment.[29] Of course alkyl f values are crucial to all others and have to be estimated with particular care; ours are compatible with previous analyses.[12, 16] It is particularly pleasing that f_C is essentially a constant, since alkyl C is always shielded from solvent. These and some other f values also appear, relative to octanol, on Figure 14, which may be regarded as summarizing the solvent–solvent comparison in the absence of water. For alkanes, amphiprotic f values all fall further below the isolipophilic line than the proton acceptor points, consistently with the view that every term except V in equation (58) must be negative since any form of polarity will be disfavoured. The pattern elsewhere is much more interesting. Fujita *et al.*[174] present evidence that a comparison of $\log P_{CHCl_3}$ with $\log P_{oct}$ for substituted benzenes PhR shows parallel lines for R = apolar and R = proton acceptor substituents, with the latter about twice as favoured by chloroform. This is traced back to 12.4 mol dm^{-3} as the concentration of CH in $CHCl_3$ but 6.2 mol dm^{-3} as that of OH in octanol; *i.e.* it relates simply to the number density of donor groups in the solvent, with the implication that proton acceptor ability, and in principle that of donors as well, may be treated as an indicator variable.[174] This picture dissolves in the light of Figure 14(b). Not only is Δf for proton acceptors by no means constant, but for the very strong acceptors its sign reverses, *i.e.* these transfer preferentially to octanol. This is a classic strong–strong/weak–weak force distinction and will help to emphasize that all fragment values must be considered on their merits: there are no blanket rules. That is even clearer for PGDP whose overall variability is less but certainly shows nothing like a simple reversal of the trend for chloroform. In fact some subtle effects are apparent: PGDP f values tend to resemble those in octanol for weak acceptors and in $CHCl_3$ for strong amphiprotics, whereas for strong acceptors they are more negative than in either. Each set is highly individual and together they clearly form a potential base from which the task of membrane characterization can begin. As a start, we have attempted it for the liposome log P values reported by Diamond and Katz[84] (at 40 °C, *i.e.* above T_c) using the eight solutes whose log P values are calculable on the basis of Table 11 (it is important in such contexts *not* to mix measured and calculated values). For log P_{oct} and log P_{PGDP} we find $r^2 = 92\%$ and 94% respectively with the others much inferior (r^2 near 50%).[47] In both cases $a \ll 1$ in terms of equation (61), so re-emphasising the point that f_{CH_2} on its own gives no information.[16]

It is often believed, and sometimes stated, that the actual choice of partitioning system in QSAR is somewhat irrelevant. As a gloss on this we produce Table 12, where a set of imaginary biological

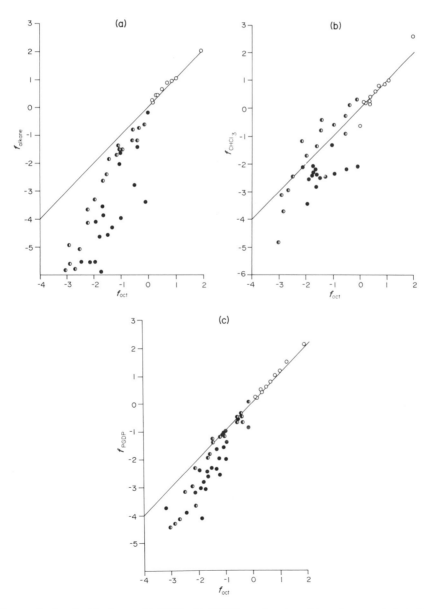

Figure 14 Fragment values for (a) alkanes, (b) $CHCl_3$, (c) PGDP, as a function of those for octanol; the isolipophilic line is drawn. Key to fragments: open circles, apolar; full circles, amphiprotic; half-filled circles, proton acceptors

Table 11 Fragment Values for Selected Solvent Systems

Fragment[a]	Alkane	Octanol	CHCl$_3$	PGDP	pK$_\alpha$	pK$_\beta$
C	0.20	0.19	0.20	0.19		
H	0.22	0.17	0.21	0.16		
CH$_3$	0.86	0.70	0.83	0.67		
CH$_2$	0.64	0.53	0.62	0.51		
CH	0.42	0.36	0.41	0.35		
C$_6$H$_5$	2.03	1.96	2.59	2.20		
a F	0.44	0.31	0.26	0.30		
a Cl	0.91	0.88	0.87	0.88		
a Br	1.05	1.03	1.02	1.07		
A CN	−2.37	−1.34	−0.82	−1.29		1.2
a CN	−0.99	−0.40	0.12	−0.54		1.0
a NO$_2$	−0.59	−0.11	0.34	−0.04		(0.7)
A NH$_2$	−3.87	−1.61	−2.47	−2.70		2.8
a NH$_2$	−2.03	−1.06	−1.33	−1.25	0.6	(1.0)
A NH A	(−3.57)	−1.96	(−2.64)	−2.95		(2.8)
A N (A) A	−3.68	−2.22		−2.75		2.8
A OH	−3.77	−1.66	(−2.21)	−2.48	1.2	1.4
a OH	−2.84	−0.50	−2.24	−1.03	2.1	(0.2)
A O A	(−2.39)	−1.55		−1.71		1.5
a O A	−0.80	−0.55	−0.30	−0.46		0.3
A CO A	−2.55	−1.75	(−1.42)	−1.81		1.6
a CO A	−1.77	−1.06	−0.63	−1.26		1.4
A CO$_2$ A	−1.82	−1.41		−1.34		1.4
a CO$_2$ A	−1.19	−0.55	−1.15	−0.54		1.2
A CO$_2$ H	−3.90	−1.08	(−2.43)	−2.12	2.0	
a CO$_2$ H	−3.42	−0.09	−2.09	−1.05	2.0	
A CONH$_2$	−5.53	−2.07	−2.24	−3.30	(0.7)	(3.0)
a CONH$_2$	−4.33	−1.32	−2.48	−2.56	(0.7)	(2.8)
A CONH A	(−5.55)	−2.45		−3.87	0.7	(3.0)
a CONH A	−4.65	−1.80	−2.46	−2.92	(0.7)	(2.8)
A CONH a	−4.59	−1.50	−2.54	−2.47	1.3	(2.5)
A CON (A) A		(−2.85)		−3.83		3.0
A NHCONH A		(−1.89)		−3.97	2.1	3.2
A NHCSNH A	(−5.86)	(−1.71)	(−2.40)	(−2.48)	2.1	2.0
A SO$_2$NH$_2$	−5.49	−1.95	−3.53	−2.60	(1.2)	(1.5)
a SO$_2$NH$_2$		−1.65	−2.83	−2.23	1.2	(1.4)
A SO A	−5.77	−2.69	−3.04	−4.10		3.0
A SO$_2$ A	−5.08	−2.52	−2.56	−2.80		1.6
a SO$_2$ A		−2.16	−1.55	−2.40		(1.4)

[a] A = alkyl, a = aryl

Table 12 Hypothetical Activity Data *vs.* Log P

Compound	log (1/C)	Alkane	Octanol	CHCl$_3$	PGDP
PhBr	2.85	3.08	2.99	3.61	3.27
PhCl	2.75	2.94	2.84	3.46	3.08
PhMe	2.66	2.87	2.69	3.41	2.89
PhCO$_2$Et	2.54	2.34	2.64	2.89	2.84
PhF	2.41	2.47	2.27	2.85	2.50
PhH	2.31	2.25	2.13	2.80	2.36
PhOMe	2.26	2.09	2.11	3.12	2.41
Me(CH$_2$)$_5$NHCSNHMe	2.17	0.94	2.31	2.36	1.41
PhNO$_2$	2.05	1.46	1.85	2.93	2.16
PhCOMe	1.88	1.16	1.58	2.79	1.61
PhCN	1.84	1.04	1.56	2.71	1.66
Ph(CH$_2$)$_2$NH$_2$	1.51	−0.56	1.41	1.36	0.52
PhNH$_2$	1.38	0.00	0.90	1.26	0.95
PhCO$_2$(CH$_2$)$_4$CONH$_2$	1.22	−2.13	1.39	1.68	0.41
O$_2$N-4-C$_6$H$_4$O(CH$_2$)$_3$SOMe	0.96	−2.53	0.93	2.00	−0.13
PhCONH$_2$	0.87	−2.30	0.64	0.11	−0.36
O$_2$N-4-C$_6$H$_4$O(CH$_2$)$_3$SO$_2$NH$_2$	0.87	−3.11	0.97	0.68	0.66

data is matched against log P for the 17 compounds listed. These linear correlation equations result for $\log(1/C)$

$$\text{Alkane:} \quad \log(1/C) = 0.310\log P + 1.694 \tag{74}$$
$$r^2 = 94.9\%, \quad s = 0.16$$

$$\text{Octanol:} \quad \log(1/C) = 0.879\log P + 0.300 \tag{75}$$
$$r^2 = 93.7\%, \quad s = 0.15$$

$$\text{CHCl}_3: \quad \log(1/C) = 0.596\log P + 0.510 \tag{76}$$
$$r^2 = 81.9\%, \quad s = 0.29$$

$$\text{PGDP:} \quad \log(1/C) = 0.559\log P + 0.985 \tag{77}$$
$$r^2 = 92.8\%, \quad s = 0.19$$

Here $\log P_{\text{alkane}}$ and $\log P_{\text{oct}}$ fit the data equally well (this result was designed) but none would be considered as giving an unacceptable fit to a QSAR correlation equation. The point of the comparison appears in Table 13, in which $\log(1/C)$ is predicted on the basis of each equation for three compounds excluded from the training set. Predictions vary enormously, and in their light a medicinal chemist might, or might not, have decided to make any of these three compounds if conceived as potential drugs. Table 13 makes a point that is crucial to medicinal chemistry. The purpose of a QSAR equation is to predict: its retrospective use for mere correlation is an academic game which helps to explain, but does not excuse, much that appears in the QSAR literature. The demonstration of some slight improvement in r, *e.g.* from the use of molecular connectivity or other form of mathematical construct, is an irrelevance unless it can be demonstrated that something predictively different will result. The converse can also happen. It is sobering that, in the present (imaginary) example, $r^2 = 79\%$ for the comparison of alkane and octanol log P values, yet these would lead to entirely different predictions just where this might matter. This point would be even clearer for multifunctional compounds or a non-linear dependence on log P. And again that figure shows the ease with which apparent resemblances, *e.g.* to HPLC retention times, may be demonstrated if the solute set is not chosen carefully enough. Similarly, one gains little insight from blunderbuss techniques such as PCA which merely succeed in identifying the features common to any partitioning system while blurring the distinctions that actually count. All these examples illustrate the pitfalls of blind statistics. There is no substitute for a scientific understanding.

Table 13 Log P Values and Predicted Activities

Compound		Alkane	Octanol	CHCl$_3$	PGDP
PhOH	$\log P$	-0.77	1.46	0.39	1.17
	$\log(1/C)$	1.45	1.58	0.72	1.64
Ph$_3$PO	$\log P$	0.18	2.83	2.95	1.60
	$\log(1/C)$	1.75	2.79	2.27	1.88
	$\log P$	-1.27	1.33	2.25	0.69
	$\log(1/C)$	1.30	1.47	1.85	1.37

The characterization of actual membranes by the use of log P for discriminating solutes is an important future task for medicinal chemistry. It has an unexpected potential pay-off. To the extent that membranes may differ in this chemical sense, the partitioning of solutes between membranes is analogous to that between solvents in the manner of Figure 14. Hence solutes may penetrate preferentially in a way that is independent of lipophilicity as such. If two such membranes contain differing receptors such that one is responsible for wanted and the other for some form of unwanted biological activity, the potential exists for biological selectivity based on differing partitioning behaviour.[97] Here we believe that the contrast between octanol as a model for the more amphiprotic regions of the membrane, and PGDP with its close resemblance to the lipid end of a typical phospholipid, may turn out to be specially relevant. The 'fluid mosaic' model of membrane structure,[175] which shows how either sort of region might present at the surface of different membranes or indeed the same one, makes this all the more likely. There is already one report[176] of the use of $\Delta \log P$ between two partitioning systems (octanol and cyclohexane) to model brain penetration. The proper choice of solvent system is more open now than at any time in the last 50 years.

18.6.9   COMPLICATIONS IN QSAR DUE TO IONIZATION

Thus far we have been concerned almost exclusively with neutral species. While the complications of ionization for log P measurement and interpretation have been noted above (Sections 18.6.3.3.11 and 18.6.5) we now need to face its much more far-reaching consequences for QSAR. Many drugs, perhaps a majority, are partly or largely ionized at physiological pH, and two sorts of problem arise. There is first what Martin[10] has described as the enigma attaching to compound sets of varying pK_a: how may these be placed on a common scale of hydrophobicity? And secondly, if only the neutral form is transported but either this or the ion may be responsible for activity, how does this affect the analysis? Finally, are there any recognizable structural principles that may influence partitioning between one membrane and another, analogously to those we have reviewed above?

There is ample evidence[87, 177] that many membranes, perhaps most, are essentially impermeable to all but formally neutral species (active transport excepted). Intestinal absorption rates of cephalosporins show virtually no contribution from the anionic species;[177b] this goes entirely against expectation based on log D in isobutanol–water[122] and demonstrates very clearly one major trap, that of expecting a solvent system that may be quite a good model for other sorts of structural variation to handle that between neutral and ion pair species (*cf.* Figure 3 and equation 53). For this reason, replacement of log P by log D as defined by equations (41) and (42) has frequently been advocated for QSAR.[142, 178] As Martin[10, 13, 179] points out, in principle this is incorrect. Consider the simplest possible case. From equation (53) with $E = 0$, replacement of (35) by (78) is equivalent to the use of equation (79), whereas the correct equation is (80)

$$\log(1/C) \ = \ a \log D \ + \ b \tag{78}$$

$$\log(1/C) \ = \ a \log P \ + \ a \log(1 \ - \ \alpha) \ + \ b \tag{79}$$

$$\log(1/C) \ - \ \log(1 \ - \ \alpha) \ = \ a \log P \ + \ b \tag{80}$$

$$\log(1/C) \ - \ \log \alpha \ = \ a \log P \ + \ b \tag{81}$$

i.e. it is *potency* not *concentration* that requires correction (there is an analogy here with equation 70). This argument requires qualification. It does not apply to simple rate processes covered by the pH partition hypothesis where only effective concentration counts so the use of log D is correct (and where apparent pK_a often appears anomalous as a result of boundary phenomena:[180] *cf.* Chapter 19.2). It *must* apply where binding is a factor. Use of log D is fortuitously acceptable in two other circumstances: where a approaches unity, and where pK_a [and hence $\log(1-\alpha)$] is substantially a constant, as for the β-blockers studied by Hellenbrecht *et al.*[181] Fujita[182] suggested that equations of type (81) and (80) are appropriate when the ionized and neutral forms, respectively, are the active species. Since α and $(1 - \alpha)$ relate as

$$\log(1 \ - \ \alpha) \ = \ \log \alpha \ \pm \ (pK_a \ - \ pH) \tag{82}$$

the presence of a receptor binding term related to pK_a will often allow a good fit to either (Fujita[182] cautions that while s may be a useful distinguishing criterion r is not, since the dependent variable has changed). In that case, any coefficient c of pK_a in (80) will become approximately $(1 - c)$ in (81). Hence Tenthorey *et al.*[183] concluded that the protection against tachycardia afforded by a set of amines of very varying log P and pK_a, where a fit to $\log(1/C)$ required only terms in log P, pointed to the cation as active species. An alternative fit[97] to equation (80) requires a pK_a term of near unit slope, which helps to substantiate this conclusion. Treated in the same way,[97] the CNS toxicity of these compounds shows $0 < c < 1$ for the pK_a term, perhaps an indication of binding as the neutral species. The only definitive way of deciding this question, but one not always possible, is to study the comparative activity of neutral and partially ionized compounds as a function of pH; this technique has recently been used to demonstrate that histamine H_2-antagonists are active as the neutral species.[184] Elsewhere, ambiguities are likely to be present: the conclusion of Hansch and Glave,[185] that ions are more effective by 10^2–10^3 than neutral compounds in the haemolysis of red blood cells, depends critically on a presumed equivalence between octanol and membrane partitioning.

Much the most comprehensive attack on this general problem has come from Martin.[10, 13, 179] This is covered by Chapter 19.1, so we confine ourselves to the mechanistic aspects that Martin[10] explicitly excludes from consideration. One of Martin's simpler equations is based on Scheme 4, according to which the concentrations C of drug in aqueous, receptor and organic loss phases are governed by partitioning processes which relate to P for the model system by exponents b and c respectively. Such a model can give rise to bilinear or parabola-like relations between $\log(1/C)$ and $\log P$, the rising and falling regions possessing sequential slopes of b and $(c-b)$. More complex relations result when the ionic species z can bind to the receptor, such that the combined effect of ion partitioning (in whatever form) and ion binding, relative to that for the neutral species, is handled for the series by the constant factor Z. Scheme 5 splits interaction with the receptor into its separate steps. These schemes as box equilibria necessarily possess equivalent algebraic consequences; for each box one equilibrium process is redundant and this is indicated by dashed lines. For simple enzyme interactions *in vitro* Scheme 5 may appear an unnecessary complication, but for the more complex *in vivo* situation where partitioning can be a sequential process and true equilibrium may not be attained, this conceptual division may prove helpful. For example, it is possibly more obvious that any change which can increase the 'leakage' of ionized species from the initial aqueous (w) to the biophase (b) will increase its degree of binding to the receptor and that this is entirely equivalent to increased ionization in the biophase relative to water so may be discussed, as a function of chemical structure, in either terms. Similarly, a very hydrophobic membrane could have the opposite overall effect (allowing for partial cancellation by tighter binding) and this in its turn is entirely equivalent to less 'leakage'.

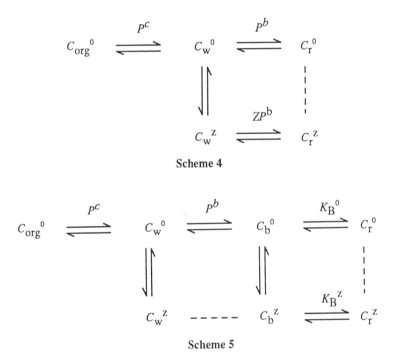

Scheme 4

Scheme 5

Ion pair partitioning as E_{QX} has been extensively investigated with results which provide clues as to the type of structural change that may encourage or discourage this process *in vivo*. From equation (83), results may be expressed either in terms of varying cations Q against a common anion X or *vice versa*. Gordon and Kutina[186] give much data of the latter sort especially by Scandinavian workers[187] and some of this relative to Cl⁻, *i.e.* as E_{QX}/E_{QCl}, is summarized in Table 14 for some contrasting solvent systems. While there is a general tendency for large polarizable anions to be (relatively) favoured by all solvents and small anions of high charge density to be disfavoured, solvent differences also appear. For instance, the amphiprotic decanol appears to favour oxygen anions, *e.g.* acetate more than the proton donor chloroform, whereas the opposite appears true for halides. Superimposed on this, however, the overall values of E_{QX} are largely a function of general solvent polarity,[187b] as is confirmed by a recent LSER study.[188] Information on cations is largely confined to amines, where again interesting solvent differences show up. Whereas in octanol for the same number of carbon atoms E_{QX} follows the order primary > secondary > tertiary,[119] this

Table 14   Log(E_{QX}/E_{QCl}) for Various Anions X in Certain
Solvents

Anion	1-Decanol	CHCl$_3$	PhNO$_2$
Picrate		6.0	
Naphthyl-2-sulfonate		3.5	
Salicylate	3.1	2.5	
Perchlorate	2.1	3.6	3.8
Iodide	2.0	3.1	2.7
Tosylate	1.4	2.4	
p-Toluate	1.3		
Bromide	0.6	1.4	1.1
Nitrate	0.5	1.5	1.8
Benzoate		0.5	
Acetate	−0.6	−2.0	
Sulfate	−2.0	−1.7	−0.9

order reverses in dichloromethane[187b] presumably because NH$^+$ can bind to the former solvent but not to the latter; in both, the neutral amines follow the octanol order above. Quaternary cations may behave like primaries[187b] or like tertiaries[119] according to solvent. Similarly, the OH group of choline gives it a powerful advantage over other quaternary cations for partitioning into amphiprotic or acceptor solvents *vis-à-vis* chloroform.[187d]

$$E_{QX} = P_{ip}K_{ip} = [QX]_o/[Q]_w[X]_w \qquad (83)$$

Since Na$^+$ and Cl$^-$ so dominate the scene *in vivo*, there is no way in which the use of foreign counterions is likely to assist ion pair partitioning in a physiological context. Nevertheless, the extraction of zwitterions might be improved by the incorporation of ions chosen in accordance with the sort of indications that appear above. For example it could be that, other things being equal, the above amine cation extraction orders are appropriate to amphiprotic and weakly donor membranes respectively; or that the use of ions of low charge density may be advantageous for membrane penetration generally. In part this amounts to setting aside Martin's assumption of constant Z (Scheme 4). Another way in which her model might be modified lies in the replacement of P^b and P^c by partition coefficients drawn from contrasting solvent systems (Section 18.6.8). In view of the increasing emphasis on peptides or peptide mimics as drugs, many of them zwitterionic or of higher charge type, speculation along these lines may carry considerable implications for drug design in the future.

18.6.10   TAILPIECE

We end with two quotations. From Menger:[189] 'Models are to be used, not believed'. And from Dewar,[190] concerning MO theory but with application far beyond its subject: 'MO theory is not a description of reality. It is only the embodiment of another molecular model, the MO model'. If some of the models we have discussed can be made to advance the cause of science, they will have done all that can reasonably be expected of them.

ACKNOWLEDGEMENTS

It is a pleasure to acknowledge the help afforded by Drs M. H. Abraham, J. Bradshaw, A. J. Leo and E. Tomlinson, by my colleagues Messrs D. G. Barratt, R. Evans, and A. R. Wait, and most especially Drs D. E. Leahy and J. J. Morris, who, however, are not to be held responsible for any views expressed.

18.6.11   REFERENCES

1. E. Overton, *Z. Phys. Chem.*, 1897, **22**, 189.
2. H. Meyer, *Arch. Exp. Pathol. Pharmakol.*, 1899, **42**, 109.
3. J. Ferguson, *Proc. R. Soc. London, Ser. B*, 1939, **127**, 387.

4. C. Hansch, *Acc. Chem. Res.*, 1969, **2**, 232.
5. P. Ehrlich, *Lancet*, 1913, 445; P. Ehrlich and J. Morgenroth, 'Studies in Immunity', Wiley, New York, 1910, pp. 24, 47.
6. 'A woman's preaching is like a dog's walking on his hinder legs. It is not done well; but you are surprised to find it done at all'. S. Johnson, quoted in J. Boswell, 'Life of Johnson', London, 1791.
7. 'There needs no ghost, my lord, come from the grave to tell us this.' Horatio, in W. Shakespeare, 'Hamlet', Act I, scene V, London, 1623.
8. A. Leo, C. Hansch and D. Elkins, *Chem. Rev.*, 1971, **71**, 525.
9. W. J. Dunn, III, J. Block and R. S. Pearlman (eds.), 'Partition Coefficient: Determination and Estimation', Pergamon Press, Oxford, 1986.
10. S. H. Yalkowsky, A. A. Sinkula and S. C. Valvani (eds.), 'Physical Chemical Properties of Drugs', Dekker, New York, 1980.
11. C. Hansch and A. J. Leo, 'Substituent Constants for Correlation Analysis in Chemistry and Biology', Wiley, New York, 1979.
12. R. F. Rekker, 'The Hydrophobic Fragmental Constant', Elsevier, Amsterdam, 1977.
13. Y. C. Martin, 'Quantitative Drug Design', Dekker, New York, 1978.
14. R. Franke, 'Theoretical Drug Design Methods', Elsevier, Amsterdam, 1984.
15. P. W. Atkins, 'Physical Chemistry', 2nd edn., Oxford University Press, Oxford, 1982.
16. S. S. Davis, T. Higuchi and J. H. Rytting, *Adv. Pharm. Sci.*, 1974, **4**, 73.
17. C. Hansch and A. J. Leo, Log P Database, Pomona College Medicinal Chemistry Project, Claremont, CA 91711, USA.
18. (a) A. Ben-Naim, 'Hydrophobic Interactions', Plenum Press, New York, 1980, p. 193; (b) p. 42; (c) p. 38; (d) p. 85; (e) p. 176.
19. R. N. Smith, C. Hansch and M. M. Ames, *J. Pharm. Sci.*, 1975, **64**, 599.
20. D. H. Everett and W. F. K. Wynne-Jones, *Trans. Faraday Soc.*, 1939, **35**, 1380.
21. D J. G. Ives and P. D. Marsden, *J. Chem. Soc.*, 1965, 649.
22. F. Franks and D. J. G. Ives, *Q. Rev., Chem. Soc.*, 1966, **20**, 1.
23. J. H. Hildebrand, *Chem. Rev.*, 1949, **44**, 37; J. H. Hildebrand and R. L. Scott, 'The Solubility of Nonelectrolytes', 3rd edn., Reinhold, New York, 1950; G. Scatchard, *Chem. Rev.*, 1931, **8**, 321.
24. (a) A. F. M. Barton, *Chem. Rev.*, 1975, **75**, 731; (b) M. H. Abraham, *Prog. Phys. Org. Chem.*, 1974, **11**, 1.
25. N. H. Anderson, S. S. Davis, M. James and I. Kojima, *J. Pharm. Sci.*, 1983, **72**, 443.
26. T. L. Hafkenscheid and E. Tomlinson, *J. Chromatogr.*, 1983, **264**, 47.
27. H. Reiss, H. L. Frisch and J. Lebowitz, *J. Chem. Phys.*, 1959, **31**, 369; H. Reiss, *Adv. Chem. Phys.*, 1965, **9**, 1; R. A. Pierotti, *Chem. Rev.*, 1976, **76**, 717.
28. H. J. M. Grünbauer, T. Bultsma and R. F. Rekker, *Eur. J. Med. Chem.—Chim. Ther.*, 1982, **17**, 411; H. J. M. Grünbauer and E. Tomlinson, *Int. J. Pharm.*, 1984, **21**, 61.
29. A. L. J. de Meere, P. J. Taylor, H. J. M. Grünbauer and E. Tomlinson, *Int. J. Pharm.*, 1986, **31**, 209.
30. J. C. McGowan, *J. Appl. Chem.*, 1954, **4**, 41; J. C. McGowan and A. Mellors, 'Molecular Volumes in Chemistry and Biology', Horwood, Chichester, 1986.
31. R. S. Pearlman, *Quantum Chem. Prog. Exch. Bull.*, 1981, **1**, 15; R. S. Pearlman, to be published (SAVOL and SAREA).
32. M. J. Kamlet, M. H. Abraham, R. M. Doherty and R. W. Taft, *J. Am. Chem. Soc.*, 1984, **106**, 464; R. W. Taft, M. H. Abraham, G. R. Famini, R. M. Doherty, J. L. M. Abboud and M. J. Kamlet, *J. Pharm. Sci.*, 1985, **74**, 807.
33. D. E. Leahy, *J. Pharm. Sci.*, 1986, **75**, 629.
34. VIKING, molecular modelling software developed by Imperial Chemical Industries, PLC, London, UK, 1985.
35. M. J. Kamlet, R. M. Doherty, J. L. M. Abboud, M. H. Abraham and R. W. Taft, CHEMTECH, 1986, **16**, 566.
36. (a) R. B. Hermann, *J. Phys. Chem.*, 1972, **76**, 2754; (b) F. M. Richards, *Annu. Rev. Biophys. Bioeng.*, 1977, **6**. 151.
37. M. H. Abraham, P. P. Duce, D. V. Prior, D. G. Barratt, J. J. Morris and P. J. Taylor, *J. Chem. Soc., Perkin Trans. 2*, 1989, in press.
38. J. J. Morris, in preparation.
39. M. Meot-Ner and L. W. Sieck, *J. Am. Chem. Soc.*, 1983, **105**, 2956.
40. R. Kühne, K. Bocek, P. Scharfenberg and R. Franke, *Eur. J. Med. Chem.—Chim. Ther.*, 1981, **16**, 7.
41. A. J. Hopfinger and R. D. Battershell, *J. Med. Chem.*, 1976, **19**, 569.
42. R. H. Davies, B. Sheard and P. J. Taylor, *J. Pharm. Sci.*, 1979, **68**, 396.
43. M. A. Pleiss and G. L. Grunewald, *J. Med. Chem.*, 1983, **26**, 1760.
44. R. L. Li, S. W. Dietrich and C. Hansch, *J. Med. Chem.*, 1981, **24**, 538.
45. R. L. Li, C. Hansch, D. Matthews, J. M. Blaney, R. Langridge, T. J. Delcamp, S. S. Susten and J. H. Freisheim, *Quant. Struct.–Act. Relat.*, 1982, **1**, 1.
46. M. Charton and B. I. Charton, *J. Org. Chem.*, 1979, **44**, 2284.
47. (a) D. E. Leahy, P. J. Taylor and A. R. Wait, *Quant. Struct.–Act. Relat.*, 1989, **8**, in press; (b) in preparation.
48. L. B. Kier and L. H. Hall, 'Molecular Connectivity in Chemistry and Drug Research', Academic Press, New York, 1976; L. B. Kier, in ref. 10, p. 277; W. J. Murray, L. H. Hall and L. B. Kier, *J. Pharm. Sci.*, 1975, **64**, 1978.
49. R. L. Lopez de Compadre, C. M. Compadre, R. Castillo and W. J. Dunn, III, *Eur. J. Med. Chem.—Chim. Ther.*, 1983, **18**, 569.
50. J. T. Edward, *Can. J. Chem.*, 1982, **60**, 480, 2573; J. G. Topliss and R. P. Edwards, *J. Med. Chem.*, 1979, **22**, 1238.
51. J. M. Diamond and E. M. Wright, *Annu. Rev. Physiol.*, 1969, **31**, 581.
52. F. Franks (ed.) 'Water: A Comprehensive Treatise', Plenum Press, New York, 7 vols., 1972–1982.
53. F. Franks, 'Water', Royal Society of Chemistry, London, 1983.
54. G. E. Walrafen, in 'Hydrogen-Bonded Solvent Systems', ed. A. K. Covington and P. Jones, Taylor and Francis, London, 1968, p. 9.
55. W. A. P. Luck, *Acta Chim. Acad. Sci. Hung.*, 1986, **121**, 119.
56. C. Tanford, 'The Hydrophobic Effect: Formation of Micelles and Biological Membranes', John Wiley, New York, 1973.
57. G. Némethy, *Angew. Chem., Int. Ed. Engl.*, 1967, **6**, 195.
58. W. P. Jencks, 'Catalysis in Chemistry and Enzymology', McGraw-Hill, New York, 1969, chap. 8.
59. R. Wolfenden, *Science (Washington, D.C.)*, 1983, **222**, 1087.
60. R. A. Klein, *Q. Rev. Biophys.*, 1982, **15**, 667.

61. D. F. Evans and B. W. Ninham, *J. Phys. Chem.*, 1986, **90**, 226.
62. O. W. Howarth, *J. Chem. Soc., Faraday Trans. 1*, 1975, 2303.
63. J. H. Fendler and E. J. Fendler, 'Catalysis in Micellar and Macromolecular Systems', Academic Press, New York, 1975.
64. Y. Yanuka, J. Zahalka and M. Donbrow, *J. Chem. Soc., Perkin Trans. 2*, 1986, 911.
65. M. Roseman and W. P. Jencks, *J. Am. Chem. Soc.*, 1975, **97**, 631.
66. G. M. Blackburn, T. H. Lilley and E. Walmsley, *J. Chem. Soc., Chem. Commun.*, 1980, 1091.
67. F. M. Menger, *Acc. Chem. Res.*, 1979, **12**, 111; F. M. Menger and D. W. Doll, *J. Am. Chem. Soc.*, 1984, **106**, 1109.
68. J.-L. Fauchère and V. Pliška, *Eur. J. Med. Chem.—Chim. Ther.*, 1983, **18**, 369.
69. J. Hine and P. K. Mookerjee, *J. Org. Chem.*, 1975, **40**, 292.
70. R. D. Cramer, III, *Quant. Struct.–Act. Relat.*, 1983, **2**, 13.
71. R. H. Davies, R. D. Bagnall and W. G. M. Jones, *Int. J. Quantum Chem., Quantum Biol. Symp.*, 1974, **1**, 201.
72. S. D. Christian, A. A. Taha and B. W. Gash, *Q. Rev., Chem. Soc.*, 1970, **24**, 20.
73. J. Dale, 'Stereochemistry and Conformational Analysis', Verlag Chemie, New York, 1978.
74. H. D. Nelson and C. L. de Ligny, *Recl. Trav. Chim. Pays-Bas*, 1968, **87**, 623.
75. A. R. Ketring, D. E. Troutner, T. J. Hoffman, D. K. Stanton, W. A. Volkert and R. A. Holmes, *Int. J. Nucl. Med. Biol.*, 1984, **11**, 113.
76. J. Xi-Hui, H. Yong-Zheng and F. Wei-Qiang, *Hua Hsueh Hsueh Pao*, 1984, **42**, 1276.
77. S. Hada, S. Neya and N. Funasaki, *J. Med. Chem.*, 1983, **26**, 686.
78. J. M. Mayer, B. Testa, H. van de Waterbeemd and A. Bornand-Crausaz, *Eur. J. Med. Chem.—Chim. Ther.*, 1982, **17**, 453.
79. A. E. Beezer and W. H. Hunter, *J. Med. Chem.*, 1983, **26**, 757.
80. CLOGP, Log P Calculation Algorithm, Pomona College Medicinal Chemistry Project, Claremont, CA 91711, USA, version 3.53.
81. J. E. Garst and W. C. Wilson, *J. Pharm. Sci.*, 1984, **73**, 1616.
82. F. A. Long and W. F. McDevit, *Chem. Rev.*, 1952, **51**, 119.
83. J. A. Rogers and S. S. Davis, *Biochim. Biophys. Acta*, 1980, **598**, 392.
84. J. M. Diamond and Y. Katz, *J. Membr. Biol.*, 1974, **17**, 101, 121.
85. C. Hansch and W. J. Dunn, III, *J. Pharm. Sci.*, 1972, **61**, 1.
86. C. Hansch, *J. Med. Chem.*, 1976, **19**, 1; C. Hansch, in 'Drug Design', ed. E. J. Ariens, Academic Press, New York, 1971, vol. 1, p. 271; C. Hansch, in 'Structure–Activity Relationships', ed. C. J. Cavallito, Pergamon Press, Oxford, 1971.
87. J. K. Seydel and K.-J. Schaper, *Pharm. Ther.*, 1982, **15**, 131.
88. C. Hansch and J. M. Clayton, *J. Pharm. Sci.*, 1973, **62**, 1.
89. C. Horváth, W. Melander and I. Molnár, *J. Chromatogr.*, 1976, **125**, 129; C. Horváth and W. Melander, *Int. Lab.*, 1978, November/December, 11.
90. E. Tomlinson and T. L. Hafkenscheid, in ref. 9 p. 101; T. L. Hafkenscheid and E. Tomlinson, *Adv. Chromatogr. (N.Y.)*, 1986, **25**, 1; T. L. Hafkenscheid and E. Tomlinson, *Int. J. Pharm.*, 1983, **17**, 1.
91. (a) H. Terada, *Quant. Struct.–Act. Relat.*, 1986, **5**, 81; (b) R. Kaliszan, *CRC Crit. Rev. Anal. Chem.*, 1986, **16**, 323.
92. T. L. Hafkenscheid and E. Tomlinson, *Int. J. Pharm.*, 1984, **19**, 349.
93. K. Valkó, *J. Liq. Chromatogr.*, 1984, **7**, 1405.
94. T. L. Hafkenscheid and E. Tomlinson, *J. Chromatogr.*, 1981, **218**, 409.
95. N. El Tayar, H. van de Waterbeemd and B. Testa, *Quant. Struct.–Act. Relat.*, 1985, **4**, 69.
96. (a) L. R. Snyder and J. J. Kirkland, 'Introduction to Modern Liquid Chromatography', 2nd edn., Wiley-Interscience, Chichester, 1979, chap. 6; (b) L. R. Snyder, *J. Chromatogr.*, 1974, **92**, 223.
97. P. J. Taylor, unpublished observations.
98. E. Tomlinson, *J. Chromatogr.*, 1975, **113**, 1; R. F. Rekker, *J. Chromatogr.*, 1984, **300**, 109.
99. J. C. Dearden, A. M. Patel and J. M. Tubby, *J. Pharm. Pharmacol.*, 1974, **26**, Suppl., p. 74.
100. I. Brown, I. L. Chapman and G. J. Nicholson, *Aust. J. Chem.*, 1968, **21**, 1125.
101. K. Valkó, O. Papp and F. Darvas, *J. Chromatogr.*, 1984, **301**, 355.
102. A. J. Leo, in 'Environmental Health Chemistry', ed. J. D. McKinney, Ann Arbor Science, Ann Arbor, MI, 1981, p. 323.
103. W. J. Birge and R. A. Cassidy, *Fundam. Appl. Toxicol.*, 1983, **3**, 359.
104. G. G. Briggs, *J. Agric. Food Chem.*, 1981, **29**, 1050.
105. C. T. Chiou, P. E. Porter and D. W. Schmedding, *Environ. Sci. Technol.*, 1983, **17**, 227.
106. L. A. King and A. C. Moffat, *Med. Sci. Law*, 1983, **23**, 193; L. A. King, *Human Toxicol.*, 1985, **4**, 273.
107. A. J. Kirby, 'The Anomeric Effect and Related Stereoelectronic Effects at Oxygen', Springer-Verlag, Berlin, 1983.
108. J. Elguero, C. Marzin, A. R. Katritzky and P. Linda, *Adv. Heterocycl. Chem., Suppl.*, 1976, **1**.
109. S. J. Lewis, M. S. Mirrlees and P. J. Taylor, *Quant. Struct.–Act. Relat.*, 1983, **2**, 1, 100.
110. P. Beak, J. Covington and J. M. White, *J. Org. Chem.*, 1980, **45**, 1347.
111. M. J. Kamlet, L. M. Abboud and R. W. Taft, *Prog. Phys. Org. Chem.*, 1981, **13**, 485; M. J. Kamlet, J. L. M. Abboud, M. H. Abraham and R. W. Taft, *J. Org. Chem.*, 1983, **48**, 2877.
112. A. J. Leo, *J. Chem. Soc., Perkin Trans. 2*, 1983, 825.
113. J. C. Dearden, *EHP, Environ, Health Perspect.*, 1985, **61**, 203.
114. M. J. Kamlet and R. W. Taft, *J. Org. Chem.*, 1982, **47**, 1734; R. Wolfenden, Y. Liang, M. Matthews and R. Williams, *J. Am. Chem. Soc.*, 1987, **109**, 463.
115. E. Tomlinson and C. M. Riley, in 'Ion-Pair Chromatography', ed. M. T. W. Hearn, Dekker, New York, 1985, p. 77; also see ref. 9, p. 96.
116. M. J. Harris, T. Higuchi and J. H. Rytting, *J. Phys. Chem.*, 1973, **77**, 2694.
117. R. A. Scherrer, *ACS Symp. Ser.*, 1984, **255**, 225.
118. G. J. Divatia and J. A. Biles, *J. Pharm. Sci.*, 1961, **50**, 916; R. L. Hull and J. A. Biles, *J. Pharm. Sci.*, 1964, **53**, 869.
119. C. Takayama, M. Akamatsu and T. Fujita, *Quant. Struct.–Act. Relat.*, 1985, **4**, 149.
120. P. J. Taylor and J. M. Cruickshank, *J. Pharm. Pharmacol.*, 1985, **37**, 143.
121. W. H. Streng, *J. Pharm. Sci.*, 1978, **67**, 666.
122. K.-J. Schaper, *J. Chem. Res., Synop.*, 1979, 357; *J. Chem. Res., Miniprint*, 1979, 4480.
123. A. Brodin, B. Sandin and B. Faijerson, *Acta Pharm. Suec.*, 1976, **13**, 331 (*Chem. Abstr.*, 1977, **86**, 78 626).
124. W. Riebesehl, E. Tomlinson and H. J. M. Grünbauer, *J. Phys. Chem.*, 1984, **88**, 4775.

125. E. Tomlinson, *Int. J. Pharm.*, 1983, **13**, 115; E. Tomlinson, in 'QSAR in Design of Bioactive Compounds', ed. M. Kuchar, Prous, Barcelona, 1984, p. 219.
126. R. R. Krug, W. G. Hunter and R. A. Grieger, *J. Phys. Chem.*, 1976, **80**, 2335, 2341.
127. D. Mirejovsky and E. M. Arnett, *J. Am. Chem. Soc.*, 1983, **105**, 1112; *cf.* E. Wilhelm, R. Battino and R. J. Wilcock, *Chem. Rev.*, 1977, **77**, 219.
128. M. Sh. Ramadan, D. F. Evans and R. Lumry, *J. Phys. Chem.*, 1983, **87**, 4538.
129. R. D. Cramer, III, *J. Am. Chem. Soc.*, 1977, **99**, 5408.
130. M. Osinga, *J. Am. Chem. Soc.*, 1979, **101**, 1621.
131. M. H. Abraham, *J. Am. Chem. Soc.*, 1979, **101**, 5477; 1980, **102**, 5910; 1982, **104**, 2085.
132. D. H. Wertz, *J. Am. Chem. Soc.*, 1980, **102**, 5316; M. H. Abraham, *J. Am. Chem. Soc.*, 1981, **103**, 6742.
133. (a) A. Leo, C. Hansch and P. Y. C. Jow, *J. Med. Chem.*, 1976, **19**, 611; (b) M. H. Abraham and R. Fuchs, *J. Chem. Soc., Perkin Trans. 2*, 1988, 523.
134. W. L. Jorgensen, *J. Chem. Phys.*, 1982, **77**, 5757.
135. J. H. Hildebrand, *Proc. Natl. Acad. Sci. USA*, 1979, **76**, 194; also see J. H. Hildebrand, *J. Phys. Chem.*, 1968, **72**, 1841; K. Miller and J. H. Hildebrand, *J. Am. Chem. Soc.*, 1968, **90**, 3001.
136. G. A. Weiland, K. P. Minneman and P. B. Molinoff, *Nature (London)*, 1979, **281**, 114.
137. A. Weissberger and B. W. Rossiter (eds.), 'Physical Methods of Chemistry', Wiley-Interscience, New York, 1971, vol. 1, part 5.
138. For a review see: S. R. Byrn, 'Solid-State Chemistry of Drugs', Academic Press, New York, 1982, chap. 4; and especially M. Takasuka, H. Nakai and M. Shiro, *J. Chem. Soc., Perkin Trans. 2*, 1982, 1061.
139. S. H. Unger, P. S. Cheung, G. H. Chiang and J. R. Cook, in ref. 9, p. 69, and references cited therein.
140. (a) E. Tomlinson, S. S. Davis, G. D. Parr, M. James, N. Farraj, J. F. M. Kinkel, D. Gaisser and H. J. Wynne, in ref. 9, p. 83; (b) E. Tomlinson, *J. Pharm. Sci.*, 1982, **71**, 602.
141. F. H. Clarke, *J. Pharm. Sci.*, 1984, **73**, 226.
142. J. J. Kaufman, N. M. Semo and W. S. Koski, *J. Med. Chem.*, 1975, **18**, 647.
143. H. Reinhardt and J. Rydberg, *Acta Chem. Scand.*, 1969, **23**, 2773; S. S. Davis, G. Elson, E. Tomlinson, G. Harrison and J. C. Dearden, *Chem. Ind. (London)*, 1976, 677.
144. M. S. Mirrlees, S. J. Moulton, C. T. Murphy and P. J. Taylor, *J. Med. Chem.*, 1976, **19**, 615.
145. (a) S. H. Yalkowsky and S. C. Valvani, *J. Pharm. Sci.*, 1980, **69**, 912; (b) S. C. Valvani and S. H. Yalkowsky, in ref. 10, p. 201; (c) S. H. Yalkowsky, S. C. Valvani and T. J. Roseman, *J. Pharm. Sci.*, 1983, **72**, 866.
146. K. Wakita, M. Yoshimoto, S. Miyamoto and H. Watanabe, *Chem. Pharm. Bull.*, 1986, **34**, 4663.
147. (a) R. W. Taft, J. L. M. Abboud, M. J. Kamlet and M. H. Abraham, *J. Solution Chem.*, 1985, **14**, 153; *cf.* B. Frange, J. L. M. Abboud, C. Benamou and L. Bellon, *J. Org. Chem.*, 1982, **47**, 4553 ; J. L. M. Abboud, K. Sraidi, G. Guiheneuf, A. Negro, M. J. Kamlet and R. W. Taft, *J. Org. Chem.* 1985, **50**, 2870; (b) R. W. Taft, W. J. Schuely, R. M. Doherty and M. J. Kamlet, *J. Org. Chem.*, 1988, **53**, 1537.
148. R. Collander, *Acta Chem. Scand.*, 1951, **5**, 774.
149. P. Seiler, *Eur. J. Med. Chem.—Chim. Ther.*, 1974, **9**, 473.
150. I. Moriguchi, *Chem. Pharm. Bull.*, 1975, **23**, 247.
151. B. Testa and P. Seiler, *Arzneim.—Forsch.*, 1981, **31**, 1053.
152. R. D. Cramer, III, *J. Am. Chem. Soc.*, 1980, **102**, 1837, 1849.
153. W. J. Dunn, III, S. Grigoras and E. Johannson, in ref. 9, p. 21.
154. W. J. Dunn, III, E. Johansson and S. Wold, *Quant. Struct.-Act. Relat.*, 1983, **2**, 156.
155. R. Franke, S. Dove and R. Kühne, *Eur. J. Med. Chem.—Chim. Ther.*, 1979, **4**, 363.
156. W. J. Dunn, III and S. Wold, *Acta Chem. Scand., Ser. B*, 1978, **32**, 536.
157. R. Franke, R. Kühne and S. Dove, in 'Quantitative Approaches to Drug Design', ed. J. C. Dearden, Elsevier, Amsterdam, 1983, p. 15.
158. M. Chastrette, M. Rajzmann, M. Chanon and K. F. Purcell, *J. Am. Chem. Soc.*, 1985, **107**, 1.
159. J. M. Mayer, H. van de Waterbeemd and B. Testa, *Eur. J. Med. Chem.—Chim. Ther.*, 1982, **17**, 17.
160. R. F. Rekker and H. M. de Kort, *Eur. J. Med. Chem.—Chim. Ther.*, 1979, **14**, 479.
161. P. Broto, G. Moreau and C. Vandycke, *Eur. J. Med. Chem.—Chim. Ther.*, 1984, **19**, 71.
162. G. Klopman, K. Namboodiri and M. Schochet, *J. Comput. Chem.*, 1985, **6**, 28.
163. A. K. Ghose and G. M. Crippen, *J. Comput. Chem.*, 1986, **7**, 565.
164. T. Fujita, *Prog. Phys. Org. Chem.*, 1983, **14**, 75; *J. Pharm. Sci.*, 1983, **72**, 285.
165. R. W. Taft, D. Gurka, L. Joris, P. von R. Schleyer and J. W. Rakshys, *J. Am. Chem. Soc.*, 1969, **91**, 4801.
166. J. Bradshaw and P. J. Taylor, in preparation.
167. F. Sparatore, C. Grieco, C. Silipo and A. Vittoria, *Farmaco, Ed. Sci.*, 1979, **34**, 11.
168. G. Van den Berg, T. Bultsma, R. F. Rekker and W. T. Nauta, *Eur. J. Med. Chem.—Chim. Ther.*, 1975, **10**, 242.
169. K. H. Meyer and H. Hemmi, *Biochem. Z.*, 1935, **277**, 39.
170. N. Barker and J. Hadgraft, *Int. J. Pharm.*, 1981, **8**, 193.
171. W. E. Jetzer, A. S. Huq, N. F. H. Ho, G. L. Flynn, N. Duraiswamy and L. Condie, Jr., *J. Pharm. Sci.*, 1986, **75**, 1098.
172. A. Walter and J. Gutknecht, *J. Membr. Biol.*, 1986, **90**, 207.
173. R. Collander, *Trans. Faraday Soc.*, 1937, **33**, 985.
174. T. Fujita, T. Nishioka and M. Nakajima, *J. Med. Chem.*, 1977, **20**, 1071.
175. S. J. Singer and G. L. Nicolson, *Science (Washington, D.C.)*, 1972, **175**, 720.
176. R. C. Young, R. C. Mitchell, T. H. Brown, C. R. Ganellin, R. Griffiths, M. Jones, K. K. Rana, D. Saunders, I. R. Smith, N. E. Sore and T. J. Wilks, *J. Med. Chem.*, 1988, **31**, 656.
177. (a) J. Gutknecht and A. Walter, *Biochim. Biophys. Acta*, 1981, **649**, 149; (b) J. L. DeYoung, H. G. H. Tan, H. E. Huber and M. A. Zoglio, *J. Pharm. Sci.*, 1978, **67**, 320.
178. R. A. Scherrer and S. M. Howard, *J. Med. Chem.*, 1977, **20**, 53.
179. Y. C. Martin, in 'Drug Design', ed. E. J. Ariëns, Academic Press, New York, 1979, vol. 8, p. 2.
180. N. F. H. Ho and W. I. Higuchi, *J. Pharm. Sci.*, 1974, **63**, 686; K.-J. Schaper, *Quant. Struct.-Act. Relat.*, 1982, **1**, 13; A. L. J. de Meere and E. Tomlinson, *Int. J. Pharm.*, 1983, **17**, 331.
181. D. Hellenbrecht, B. Lemmer, G. Wiethold and H. Grobecker, *Naunyn-Schmiedeberg's Arch. Pharmacol.*, 1973, **277**, 211; D. Hellenbrecht, K.-F. Müller and H. Grobecker, *Eur. J. Pharmacol.*, 1974, **29**, 223.

182. T. Fujita, *J. Med. Chem.*, 1966, **9**, 797; T. Fujita and C. Hansch, *J. Med. Chem.*, 1967, **10**, 991.
183. P. A. Tenthorey, A. J. Block, R. A. Ronfeld, P. D. McMaster and E. W. Byrnes, *J. Med. Chem.*, 1981, **24**, 798.
184. E. E. J. Haaksma, B. Rademaker, K. Kramer, J. Ch. Eriks, A. Bast and H. Timmerman, *J. Med. Chem.*, 1987, **30**, 208.
185. C. Hansch and W. R. Glave, *Mol. Pharmacol.*, 1971, **7**, 337.
186. J. E. Gordon and R. E. Kutina, *J. Am. Chem. Soc.*, 1977, **99**, 3903.
187. (a) K. Gustavii and G. Schill, *Acta Pharm. Suec.*, 1966, **3**, 241, 259; (b) K. Gustavii, *Acta Pharm. Suec.*, 1967, **4**, 233;
     (c) R. Modin and A. Tilly. *Acta Pharm. Suec.*, 1968, **5**, 311; (d) R. Modin and S. Back, *Acta Pharm. Suec.*, 1971, **8**, 575.
188. R. W. Taft, M. H. Abraham, R. M. Doherty and M. J. Kamlet, *J. Am. Chem. Soc.*, 1985, **107**, 3105.
189. F. M. Menger, J. M. Jerkunica and J. C. Johnston, *J. Am. Chem. Soc.*, 1978, **100**, 4676.
190. M. J. S. Dewar, *J. Am. Chem. Soc.*, 1984, **106**, 669.

18.7

Methods of Calculating Partition Coefficients

ALBERT J. LEO

Pomona College, Claremont, CA, USA

18.7.1 INTRODUCTION

The study of how solutes are distributed, or partitioned, between two immiscible liquid phases was begun in earnest in the latter part of the 19th century by Berthelot and Jungfleish.[1] They showed that the ratio of concentrations of small solutes, such as iodine and bromine, when distributed between water and either carbon disulfide or ether, remained constant even when the solvent ratios varied widely. This type of equilibrium was put on a more sound thermodynamic basis by Nernst,[2] who demonstrated that constancy of the distribution ratio was to be expected only if the molecular species in the two solvent phases was identical. The work he began was pursued by many others to determine ionization constants, self-association constants and hydrate formation.[3-7] In this chapter the currently accepted terminology will be used. The ratios are given in concentration terms, not molfraction, and non-polar phase in the numerator and water in the denominator. The distribution ratio, D, refers to the ratio of concentrations of all related species, whether neutral or ionized, monomer or polymeric, while the partition coefficient, P, refers to the ratio of the neutral monomer. The only exception to this is the distribution ratio of totally ionic organic solutes, such as quaternary ammonium salts. As will be seen later in this chapter, some useful estimations of the ratio of concentrations of 'quats' between octanol and water can be made just on the basis of structure if certain standard conditions are specified. This holds even though they are ionized in the aqueous phase and exist mostly as ion pairs in the octanol. Following earlier conventions these ratios are referred to as 'partition coefficients', but one should be aware that unless the standard conditions are met, they are more accurately termed distribution ratios, D.

It will be seen that several calculation methods are satisfactory for solutes of simple structure, but for most of those of biological interest, it is very desirable to have computer assistance. User friendly, fully computerized systems are obviously desirable, and at least one (CLOGP) is in widespread use (see Section 18.7.4).

It should be pointed out that in developing methods for calculating partition coefficients for use as a hydrophobic parameter we are not trying to eliminate the need for their measurement. Indeed, the two should complement each other.[8] And in view of the successful use of log $P_{(oct/water)}$ over the past years in a variety of structural activity studies connected with anesthesiology, pharmacokinetics, drug design, toxicology, bioaccumulation and enzyme binding,[9] it is hardly necessary to justify its importance to medicinal chemists. But its use in host–guest complexation, protein folding, environmental transport and soil binding and enzymic reactions in non-aqueous solvents may not be so familiar.[9]

18.7.2 CALCULATION FROM OTHER EQUILIBRIA

The focus of this chapter is on methods of calculation of partition coefficients directly from solute structure, but brief mention can be made of the calculation of D or P for any solvent pair from a value in another pair. Whether one begins with a structure or with a related log P, it is assumed that linear free energy relationships of the Hammett type[10] can be applied.[11] This is discussed more thoroughly in Chapter 18.6. Smith[12] was perhaps the first to suggest converting P values from one solvent system to another, but Collander[13] was the first to present these calculations in the standard linear free energy format

$$\log P_2 = a \log P_1 + b \tag{1}$$

He was able to show a good linear relationship when the polar solvent was water and the non-polar solvents, 1 and 2, were alkanols. However, the practical utility of this calculation is limited. If, for example, it is desired to calculate a solute's log $P_{(oct/water)}$ when none has been measured, it is unlikely that a value will be found in another alkanol solvent system. However, the Collander equation has prompted studies aimed at assigning physical significance to the coefficients a and b.[14, 15]

A great deal of effort has been directed to obtaining log $P_{(o/w)}$ (o = octanol, w = water) values from HPLC retention times.[16, 17] It is beyond the scope of this chapter to deal with these methods, but three points should be mentioned. The method which uses octanol as the stationary phase[16] has the advantage of requiring only one curve to relate retention time to conventional log P, but it takes a great deal of skill to overcome inherent experimental difficulties, especially with very hydrophobic solutes which require a very short column. Secondly, methods which employ an alkane-coated (usually C-18) silica stationary phase and use methanol/water mixtures as eluant, require more than one standard curve to relate retention time to log P. The structural type of the solute determines which relationship is the best to use, but this assignment is often ambiguous, and is even more so if the organic portion of the eluant is acetonitrile. Thirdly, it appears that when the aqueous fraction of the eluant drops below 50%, HPLC methods are insensitive to solutes whose log P is less than 1.5.[8] On the positive side, the latter method (C-18 HPLC) appears to be a reproducible measure of hydrophobicity of solutes whose log P is over 6.5, a range where shake-flask measurements fail,[18] but HPLC and the generator column method[19] have been evaluated at this high range chiefly with polychlorinated hydrocarbons. As might be imagined, there is great interest in any calculation methodology which will handle these very hydrophobic solutes. The fragment method described later (Sections 18.7.4–18.7.6) deals satisfactorily with the limited variety of structures measured so far, but when and if anomalies arise in this range, one must ask whether the 'fault' lies with calculation or with measurement. The highest shake-flask value in the Medchem database is for permethrin (**1**), and it compares favorably with calculation by fragments.

(1) Permethrin

Calculated = 6.81

Measured = 6.50

Calculation of partition coefficients between all possible solvent pairs is, as one might expect, a most formidable task, for it requires that the more pertinent 'solvatochromic' parameters, such as molar volume, hydrogen bonding donor–acceptor potential and group polarity be known for both solute and solvent.[20] From a practical standpoint, this field can be narrowed considerably. Most often one wants the polar phase to be water, or at least water-like. Examples of the latter might be whole blood or serum. The most frequently studied non-polar phase is 1-octanol, which has been shown to correlate well with purely physicochemical phenomena, as well as distribution into membranes, organelles, organs and whole animals.[21] The number of partition coefficients reported in the next two most popular solvents, diethyl ether and cyclohexane, are about equal, and together they amount to about one-fourth those in octanol.[22] Numerous measurements have been made of the distribution of solutes between blood and fatty tissues, such as brain, nerves and liver, but there still is a great need for extending these data by calculation. Treating these more complex biological systems as immiscible phases and calculating the partition coefficient from solvatochromic parameters looks very promising,[23] but it will not be dealt with any further here.

18.7.3 CALCULATION FROM SOLUTE STRUCTURE

18.7.3.1 Calculation by Substitution

The calculation procedures discussed from this point on will be limited to the octanol/water solvent system and will use structure or structural differences as input. Thus, whenever D or P is used the octanol/water system will be assumed unless a subscript indicates otherwise.

The first methodology for calculating $\log P$ was proposed by Fujita *et al.* in 1964.[24] Log P was considered to be an additive–constitutive property and numerically equal to the sum of the $\log P$ of the 'parent' solute plus a π term which represented the difference in $\log P$ between a particular substituent and the hydrogen atom which it replaced. Thus, π for substituent X can be defined as

$$\pi(X) = \log P(RX) - \log P(RH) \qquad (2)$$

It is obvious, then, that $\pi(H) = 0$.

For $\log P$ calculation, the desired relationship might be

$$\log P(YRX) = \log P(HRH) + \pi(Y) + \pi(X) \qquad (3)$$

For example

$$\log P_{Cl(C_6H_4)Me} = \log P_{C_6H_6} + \pi_{Cl} + \pi_{Me}$$

$$2.13 + 0.71 + 0.56 = 3.40; \text{ Measured} = 3.33$$

It is very important for newcomers to the field to notice that $\log P$ is not the sum of π values. π values must be added to a $\log P$.

The π system of Fujita *et al.* was at first applied only to substitution on aromatic rings where the hydrogen atom being replaced was definitely of a 'hydrocarbon' nature. Some investigators later attempted to apply this methodology to calculations in which the hydrogen was clearly part of a polar moiety, such as a hydroxyl or amine. The results were often, but not always, in error. But not even all 'aromatic hydrogens' could be substituted without some correction factor. π for a substituent which is capable of hydrogen bonding is greater when it replaces a hydrogen on an electron-deficient ring than it is when replacing one of benzene's hydrogens. An example would be the amino substituent on either nitrobenzene or pyridine. Thus the early π calculation methodology[24] provided eight sets of π values which served as models for almost any aromatic system, either electron rich or deficient. An extensive list of π values for commonly encountered substituents on a benzene ring appears in Table 1.

Table 1 Substituent π Constants for Benzene
or other Electron Normal Rings

In C–H–N–O–P–S order	
	π
Br	0.86
Cl	0.71
HgCl	0.05
F	0.14
SO_2F	0.16
SF_5	1.23
I	1.12
IO_2	-3.46
NO	-1.20
NO_2	-0.28
NNN	0.46
H	0.00
OH	-0.67
SH	0.39
SO_2OH	-1.86
$B(OH)_2$	-0.55
NH_2	-1.23
NHOH	-1.34
SO_2NH_2	-1.82
$NHNH_2$	-0.88
$NHSO_2NH_2$	-1.73
5-Cl-1-Tetrazolyl	-0.65
$N{=}CCl_2$	0.41
CF_3	0.88
OCF_3	1.04
SO_2CF_3	0.55
SCF_3	1.44
CN	-0.57
NCS	1.15
SCN	0.41
CO_2^-	-4.36
1-Tetrazolyl	-1.04
NHCN	-0.26
CHO	-0.65
CO_2H	-0.32
$OCH({=}O)$	-0.87
CH_2Br	0.79
CH_2Cl	0.17
CH_2I	1.50
NHCHO	-0.98
$CONH_2$	-1.49
$CH{=}NOH$	-0.38
$C({=}O)NHOH$	-1.87
$OCONH_2$	-1.05
Me	0.56
$CONHNH_2$	-1.92
$NHCONH_2$	-1.30
$NHC{=}S(NH_2)$	-1.40
OMe	-0.02
CH_2OH	-1.03
SOMe	-1.58
SO_2Me	-1.63
OSO_2Me	-0.88
SMe	0.61
SeMe	0.74
NHMe	-0.47
$NHSO_2Me$	-1.18
CF_2CF_3	1.23
$C{\equiv}CH$	0.40
$NHCOCF_3$	0.08
CH_2CN	-0.57
$CH{=}CHNO_2$-(*trans*)	0.11
$CH{=}CH_2$	0.82
$NHC{=}O(CH_2Cl)$	-0.50
HgOCOMe	-1.42
COMe	-0.55

Table 1 (*Contd.*)

In C–H–N–O–P–S order	π
SCOMe	0.10
OCOMe	−0.64
CO$_2$Me	−0.01
NHCOMe	−0.97
NHCO$_2$Me[a]	−0.37
C=O(NHMe)	−1.27
CH=NOMe	0.40
NHC=S(Me)	−0.42
CONHNHCONH$_2$	−2.63
CH=NNHC=S(NH$_2$)	−0.27
Et	1.02
CH=NNHCONHNH$_2$	−1.32
CH$_2$OMe	−0.78
OEt	0.38
SOEt[a]	−1.04
SEt	1.07
SeEt[a]	1.28
NHEt	0.08
SO$_2$Et[a]	−1.09
N(Me)$_2$	0.18
NHSO$_2$Et[a]	−0.64
CH=NNHCONHNH$_2$	−1.32
P(Me)$_2$	0.44
PO(OMe)$_2$	−1.18
C(OH)(CF$_3$)$_2$	1.28
CH=CHCN	−0.17
Cyclopropyl	1.14
COEt[a]	0.06
SCOEt[a]	0.64
CO$_2$Et	0.51
OCOEt[a]	−0.10
CH$_2$CH$_2$CO$_2$H	−0.29
NHCO$_2$Et	0.17
CONHEt[a]	−0.73
NHCOEt[a]	−0.43
CH=NOEt[a]	0.94
NHC=SEt[a]	0.12
CH(Me)$_2$	1.53
Pr	1.55
NHC=S(NHEt)	−0.71
OCH(Me)$_2$	0.85
OPr	1.05
CH$_2$OEt[a]	−0.24
SOPr[a]	−0.50
SO$_2$Pr[a]	−0.55
SPr[a]	1.61
SePr[a]	1.82
NHPr[a]	0.62
NHSO$_2$Pr[a]	−0.10
$\overset{+}{N}$(Me)$_3$	−5.96
Si(Me)$_3$	2.59
CH=C(CN)$_2$	0.05
1-Pyrryl	0.95
2-Thienyl	1.61
3-Thienyl	1.81
CH=CHCOMe	−0.06
CH=CHCO$_2$Me[a]	0.32
COPr[a]	0.53
SCOPr[a]	1.18
OCOPr[a]	0.44
CO$_2$Pr[a]	1.07
(CH$_2$)$_3$CO$_2$H[a]	0.25
CONHPr[a]	−0.19
NHCOPr[a]	0.11
NHC=OCH(Me)$_2$	−0.18
NHCO$_2$Pr[a]	0.71
CH=NOPr[a]	1.48

Table 1   *(Contd.)*

In C–H–N–O–P–S order	π
NHC=S(Pr^a)	0.66
Bu	2.13
C(Me)$_3$	1.98
OBu	1.55
CH$_2$OPr^a	0.30
N(Et)$_2$	1.18
NHBu^a	1.16
P(Et)$_2$	1.52
PO(OEt)$_2^a$	−0.10
CH$_2$Si(Me)$_3$	2.00
CH=CHCOEt^a	0.48
CH=CHCO$_2$Et	0.86
CH=NOBu	2.02
C$_5$H$_{11}^a$	2.67
CH$_2$OBu^a	0.84
Ph	1.96
N=NPh	1.69
OPh	2.08
SPh	2.32
SO$_2$Ph	0.27
OSO$_2$Ph	0.93
NHPh	1.37
NHSO$_2$Ph	0.45
2,5-di-Me-1-pyrryl	1.95
CH=CHCOPr^a	1.02
CH=CHCO$_2$Pr^a	1.40
Cyclohexyl	2.51
2-Benzthiazolyl	2.13
COPh	1.05
CO$_2$Ph	1.46
OCOPh	1.46
N=CHPh	−0.29
CH=NPh	−0.29
NHCOPh	0.49
CH$_2$Ph	2.01
CH$_2$OPh	1.66
C≡CPh	2.65
CH=NNHCOPh	0.43
CH$_2$Si(Et)$_3^a$	3.26
CH=CHPh-*(trans)*	2.68
CH=CHCOPh	0.95
Ferrocenyl	2.46
N(Ph)$_2$	3.61
P=O(Ph)$_2$	0.70

^{a} Calculated from next lower homolog.

18.7.3.2   Calculation using Molecular Orbitals

As molecular orbital (MO) calculations were refined sufficiently to make two-dimensional conformation analyses, several investigators were attracted by the possibility that the free energy of solvation could be calculated and thus partition coefficients calculated directly. Rogers and Cammarata[25] developed such an equation for aromatic solutes using a charge density term, Q_s^T, together with an induced polarization term, S_s^E. They suggested that partitioning into the aqueous phase was 'charge-controlled', while into the non-polar phase it was 'polarizability controlled'. Judging from the original set of 30 solutes, this method held forth some promise, even though the measured values used for benzene and carbazole are highly suspect. Apparently the method was not developed any further, and judging from the lack of literature references, is not widely used.

Hopfinger and Battershell[26] developed a method of 'solvent-dependent conformational analysis' (SCAP) using semi-empirical procedures which could make calculations hundreds or even thousands of times faster than MO techniques. The software component, CAMSEQ, requires essentially only a connection table input. For simple aliphatic or aromatic hydrocarbons and for monofunctional solutes, the error in the original SCAP estimations of log *P* was not much greater than with the π method of Fujita *et al.*[24] However, the following shortcomings were evident. The size of the 1-octanol molecule makes complete configurational analyses impractical, and so its solvation shell parameters must be estimated by extrapolating those from the lower alkanols. Also, SCAP must neglect the water present in the saturated octanol phase (2M), and this water surely plays an important role in the overall structure of the non-polar phase.[27] Finally, SCAP can take into account only the first hydration shell layer, which is probably sufficient for the hydrocarbon portions of a solute molecule but which may be inadequate for strongly polar groups. For whatever reasons, the SCAP method does not work as well as the purely empirical procedures (π or fragment) for solutes containing polar groups that can interact electronically or are in close proximity.

The molecular orbital approach to partition coefficient estimation has been pursued further by Klopman using MINDO/3 and Hückel-type calculations in a technique which initially was referred to as the 'charge density method'.[28] In a test set of 61 simple monofunctional solutes, results were better with the charge density methodology when compared to an early fragment methodology. No attempt was made to deal with the crucial issue of fragment interactions in solutes containing multiple functional groups.

In a later extension of this work,[29] it was discovered that charge densities play only a marginal role, which is what one would expect from the small coefficient of the π* term in the solvatochromic equations of Kamlet *et al.*[20, 23] The significant descriptors as obtained from multivariate analysis are reduced to merely the number of heteroatoms appearing in the various functionalities, such as acid/esters, nitriles, amides, *etc.* In the end, this appears to be another approach based on fragment additivity, without, however, making any attempt to account for fragment interaction.

18.7.3.3 Calculation using Atomic Contributions

A calculation method based on atomic contributions was proposed by Broto and his colleagues[30] who allowed for different bonding environments with a set of 222 'contributing substructures'. These were derived from either Monte Carlo or linear regression methods. The method lends itself to computerization, and a precision of 0.4 log units was claimed for most solutes. It is too early in its development to judge how well it will perform in the 'real world' of complex pharmaceutical or pesticide chemicals.

Another atomic contribution procedure was developed by Ghose and Crippen[31] in which the atomic classification was automated and the number reduced to only 110 types. The authors are comfortable with the fact that their method assigns a greater hydrophobicity to hydrogen atoms and correspondingly reduces that of carbon in the hydrocarbon portions of solutes. But in the case of diethyl ether, for example, one wonders what insights are gained by considering the oxygen as slightly hydrophobic ($+0.04$) while the carbons flanking it are very hydrophilic (-0.95 each). Also it appears that there is a serious weakness in the fact that the interaction of polar groups on vicinal carbons is not allowed for. They state (ref. 31, p. 575) that 'the approach of Hansch (see Sections 18.7.4–18.7.6) works very well for simple molecules and often very poorly for complex molecules'. This statement is difficult to support in view of the problems encountered with their atomic contribution method when strong electronic interactions exist. One can take hexafluoro-propan-2-ol as an example, where the Ghose–Crippen (G–C) calculation is as follows:

	F	C	C	H	H	O
Ghose–Crippen:	6(# 83) +	2(# 13)	+ 1(# 8)	+ 1(# 49)	+ 1(# 50)	+ 1(# 56)
	6(0.1172) +	2(0.6278)	− 0.9463	− 0.2232	− 0.3703	− 0.0517 = 0.367

As seen in the example of structure (**10**) (page 308), the measured value is 1.66 and the CLOGP calculation yields 1.59. A deviation of opposite sign is seen in the G–C calculation for atrazine (**14**) (page 312), which yields a value of 3.954, while the measured value is 2.75 and CLOGP-3 yields 2.82.

The most disturbing aspect of any atomic contribution approach is the difficulty in accounting for interactions at a distance. Yet Ghose and Crippen state (ref: 31, p. 575) that 'The Hansch approach is important for getting the overall hydrophobicity and does not give a good picture of its distribution, since the correction factors often have large values and do not point out the atom or group undergoing changes.' Certainly in the hexafluoropropan-2-ol example above, the hydrophilic character of the hydroxyl group is greatly reduced, because the fluorines decrease the H bond basicity of the oxygen. CLOGP calls direct attention to this with the Factor $F_{(XCCY)} = 2.70$, but the Ghose–Crippen procedure does not even take it into account, and therefore greatly underestimates this solute's hydrophobicity. Furthermore when the electronic interaction is effected through an aromatic system, as in atrazine (14), it is a *mistake* to try to 'point out' an atom on which it is supposed to reside.

For the near future at least, it appears that empirical methods of calculating $\log P$ will remain more accurate than the more fundamental approaches of molecular orbital or solvatochromic parameters. Since inter- and intra-molecular forces in solvation interactions act in a complex fashion, the 'constitutive' portion of $\log P$ calculation must be quite involved to reflect this complexity — so much so that only a few dedicated practitioners are likely to become proficient in a manual procedure. Therefore, if widespread application is desirable, it is essential that the method be reduced to a computer algorithm. As pointed out above, for the π scheme of Fujita *et al.*[24] to be viable requires an inordinate number of measured values for prospective 'parents'. It is for this reason, and not because of superior methodology, that fragment methods of calculating $\log P$ have become the most widely used. The next sections will be devoted to examining them in some detail.

18.7.4 CALCULATION BY FRAGMENT CONSTANTS

To 'construct' a solute molecule 'from scratch', that is from parts which have been previously evaluated for their hydrophobic contribution, one must decide just how large the 'parts' should be. In the π scheme, as we have seen, we want the 'parent part' to be as *large* as possible so as to contain as much of any interaction terms as possible. In a fragment scheme one might think it desirable to use the *smallest* parts, or fragments, possible, *i.e.* atoms. However, constructing a solute totally from atomic contributions[29] leaves a great many interaction factors to be evaluated, and its extension to complex solutes, *i.e.* to a drug such as a penicillin analog, has yet to be demonstrated. It appears, therefore, that a practical compromise would be to consider atomic contributions for the halogens and for carbon and hydrogen atoms in the hydrocarbon portions of the solute, but to leave intact some multi-atom polar fragments.

The 'Fragmental' approach to the calculation of $\log P$ was pioneered by Rekker and his colleagues.[32-34] From a comprehensive library of measured values, they used statistical methods to determine the average contribution of simple fragments, such as C, CH, CH_2, CH_3, OH, NH_2, *etc.* Although the atomic values for carbon and hydrogen were determined (0.15 and 0.175), the combined fragments as shown were actually used, for this eliminated the need for a branching factor. Likewise, for aromatic hydrocarbons, only combined fragments are listed; *i.e.* $C_6H_5 = 1.866$; $C_6H_4 = 1.688$; $C_6H_3 = 1.431$. Very early in their work they realized the need for assigning different values for a polar fragment depending on whether the carbon atom to which it was attached was aliphatic or aromatic. Also they found it necessary to introduce corrections if two polar fragments were separated by only one or two aliphatic carbons. Thus their formula took this form

$$\log P = \Sigma a_n f_n + \Sigma b_m F_m \tag{4}$$

where a is the number of occurrences of fragment f of type n and b is the number of occurrences of correction factor F of type m. Rekker reasoned that all constitutive factors could be directly attributable to a fundamental property of the structured water in the solvation shell in the aqueous phase. His method attempts to allow for both corrections (aliphatic *versus* aromatic and polar proximity) as a product of a 'magic constant' ($C_M = 0.28$) times a key number, k, which is simply the number of structured water molecules involved. It appears that the precision of the data presently available is inadequate to clearly support or refute this 'quantum' correction hypothesis, because, with very little to restrict one's choice of 'k', the maximum deviation in this calculation would be ± 0.14. The deviation of measured values, when taken from a variety of sources, is nearly this large. Pursuing this approach even further, van de Waterbeemd and Testa[35] proposed a hydration factor, Ω, which is just one-fourth of Rekker's C_M. The reality of Ω is even more difficult to justify, since the average precision of measurement in a single laboratory with good technique is often ± 0.03 or more. Although the 'magic constant' and the 'Ω factor' were received with some skepticism, the utility of Rekker's calculation method is widely accepted.

Rekker's method can now be applied with computer assistance.[36] By its nature the method cannot be fully computerized, because the breakdown of solute structure into fragments is left to human decision. Ideally, it should not matter how this is accomplished, as long as all the fragment values can be found in the tables furnished. This is not true in all cases, as is illustrated in Section 18.7.5.3.3 below. Another task for the human operator, which often requires a great deal of experience, is the choice of the 'key number', k, which is the multiplier of the magic constant, C_M, and is required to evaluate some of the interaction factors.

A fully computerized $\log P$ calculation should meet a minimum of three objectives: (1) it should be user friendly and capable of being operated by investigators with a minimal understanding of solvation theory; (2) it should be capable of displaying the full extent of the calculation steps needed to reach the result; and (3) it should readily accept new fragment values and new correction factors so that it can keep up with the knowledge base as it is developed. The development of the SMILES language (see Chapter 17.3) and the DEPICT algorithm for display helped achieve the first objective. The latest version of the Pomona Medchem's CLOGP shows each calculation first as a summary with error estimate, and, if requested, in one of two levels of detail. All the fragment and factor values are accessible and available for change. Adding or changing a fragment value and associated parameters concerned with electronic and steric interactions takes less than five minutes. This version is presently being used in over 60 organizations worldwide. Reports indicate that the number of regular users at each site varies between 10 and 30.

For full computerization of the $\log P$ calculation it was necessary to define both polar and non-polar fragments in such a way that any solute structure could be 'fragmented' in only one way. Even after a satisfactory fragment definition was proposed, it was not obvious how to incorporate it into an algorithm, and in the first prototype of CLOGP[37] fragment structures were 'hard wired' into the program instead of being determined by the algorithm.

To eliminate any ambiguity in how a structure is to be 'broken' into fragments, the current CLOGP program uses the following definitions. An 'isolating carbon' atom (IC) is one which is not doubly or triply bonded to a heteroatom. Isolating carbons can, however be multiply bonded to one another. An IC is an atomic fragment which, for calculation purposes at least, is always hydrophobic. Any hydrogen atom attached to an isolating carbon (ICH) is also a hydrophobic atomic fragment.* All atoms or groups of covalently bonded atoms which remain after the 'removal' of ICs and ICHs are polar fragments. Thus a polar fragment contains no ICs but each has one or more bonds to ICs, which is termed its 'environment'. Note that a carbon can be *aromatically* bonded to a heteroatom and still be 'isolating'. An example would be the carbon in position 2 of pyrimidine. Note also that in this last point, the definition currently used by CLOGP differs from the very first one proposed.[11]

The computer in our current version (CLOGP-3) does not consult a 'look-up' table to see if it can construct the structure in question from the fragments provided in advance. Instead it has the means to reduce any structure to its fragments, even structures with moieties not yet synthesized. Since we have also provided it with a definition of aromaticity (as well as defining styryl, vinyl and benzyl attachment; see below), it can quickly ascertain if every fragment found in the target structure has been evaluated in the necessary 'bonding environment'. Examples of fragments produced by this definition are

Monovalent: $-\text{Cl}$; $-\text{CN}$ Divalent:
$$-\text{O}-\overset{\displaystyle \overset{\text{O}}{\|}}{\text{C}}-\text{N}\overset{\text{H}}{\diagup}_{\diagdown}$$

Trivalent: $-\text{O}-\overset{\displaystyle \overset{\text{O}}{\|}}{\text{C}}-\text{N}\overset{\diagup}{}_{\diagdown}$ Tetravalent: $\overset{\diagdown}{}_{\diagup}\text{N}-\overset{\displaystyle \overset{\text{O}}{\|}}{\text{C}}-\text{N}\overset{\diagup}{}_{\diagdown}$

The largest 'indivisible' fragment which we have encountered so far is (**2**).

$$-\text{O}-\overset{\displaystyle \overset{\text{O}}{\|}}{\text{C}}-\overset{\displaystyle \overset{|}{}}{\text{N}}-\text{S}-\text{N}-\overset{\displaystyle \overset{\text{O}}{\|}}{\text{C}}-\text{O}-\text{N}=\overset{\displaystyle \overset{\text{S}}{\overset{|}{}}}{\text{C}}-\overset{\displaystyle \overset{\text{O}}{\|}}{\text{C}}-\text{N}\overset{\diagup}{}_{\diagdown}$$

(**2**)

---

* It can readily be seen that, after taking into consideration interaction factors in neutral solutes and electronic bond corrections for organic cations, that a methylene group in close proximity to one or more very strongly polar fragments may not appear hydrophobic. Even though it is suitable for the calculation methodology to assign this as a correction factor, for graphical representation of hydrophobic/hydrophilic areas of macromolecules by color, it is less confusing to include such hydrocarbon sections as hydrophilic.

At this point it may not be out of place to reemphasize the difference between equations (3) and (4). π values must be added to a log P, and not just to themselves, while it is proper to sum fragment values by themselves. Of course, both methods assume that allowance has been made for interaction factors. There are some occasions to use the fragment method in a manual calculation starting with a reliable log P value. One can perform such 'hybrid' calculations by subtracting f_H from the log P to get a composite fragment value for the 'parent', as long as that hydrogen is from a hydrocarbon portion of the solute. For example

$$\log P_{\text{3-indoleacetic acid}} = \log P_{\text{indole}} - f_H + \log P_{\text{acetic acid}} - f_H$$
$$= 2.14 \quad - 0.225 - 0.17 \quad - 0.225 = 1.52; \quad \text{Measured} = 1.41.$$

Two points are worthy of note here: firstly, composite fragments are never used in CLOGP, and secondly, the CLOGP methodology would add a bond factor (-0.12), since indoleacetic acid has an element of flexibility not present in either part. Thus the preferred 'hybrid' procedure would then give a result of 1.42.

18.7.4.1 Bond Environments

Following a 'constructionist' approach — that is building the foundation of the basic fragment values for aliphatic carbon and hydrogen from the simplest members of the series and polar fragments from monofunctional, unbranched analogs — it was apparent that polar fragments had their most negative value (*i.e.* were most hydrophilic) when the IC to which they were bonded was sp^3. It appears that water competes best when the charge separation in the solute is localized. A benzyl IC appears to be able to delocalize this charge somewhat, and the value of any attached fragment is raised by nearly 0.2 log units. A further increase in fragment value is seen in the series: vinyl, styryl and aromatic. CLOGP-3 keeps track of these five types of bonding environments, but arguments could be made for a lesser or greater number. As mentioned above, fragment interaction, with resulting delocalization and hydrophobicity increase, can occur through either a resonance or field effect, but it is less confusing to treat these latter effects as a separate correction type, rather than to proliferate the IC types.

The five types of ICs and the polar fragment environments are given computer designations as: 'A', 'Z', 'V', 'Y' and 'a', standing for: Aliphatic, benzyl, vinyl, styryl and aromatic, respectively. It is possible that all of the environments will eventually be independently evaluated for the most important monovalent fragments. However, this goal is certainly unlikely to be realized for the tri- and tetra-valent fragments. For example, a complete evaluation of the trivalent N,N-disubstituted amide fragment, $-C(=O)N{<}$, for the combination of five bond environment types would require 75 measured values. Fortunately a study of monovalent fragments yields a system of estimating some environments from others. Most often values can be found for the lowest (aliphatic) and the highest (aromatic) environment types. The intermediate types can be estimated as:

$$Z = A + 0.2 \quad V = (a - A)/2 + A \quad Y = 3(a - A)/4 + A$$

18.7.4.2 Fragment Classes

Evaluating the effect of bond environment was a necessary first step in characterizing polar fragments, but further classification was required before the all-important task of evaluation of correction factors could be completed. At first one might suppose that it would be sufficient to classify fragments in only two types, polar and non-polar. But as more and more fragments were carefully evaluated, it became apparent that when several were present in the same solute structure, they could interact in a number of ways that were best rationalized if the polar fragments were subdivided into several classes.

Fragments with a localized dipole but no appreciable hydrogen-bonding potential are best treated separately from other polar types. In essence this separates the halogens from those containing N, O, S and P. A difference was noted in those H-bonding fragments that contained an —OH moiety and those which did not.[11,20] Finally, it was noted that, for fragments with a formal charge, anions behaved in a much more predictable way than cations,[11] indicating that more structures are capable of delocalizing a positive charge than a negative one. Most of these observations fit in with the later development of solvatochromic relationships.[20]

As explained above, the hydrophobic 'hydrocarbon' portions of a solute structure are made up of two fundamental atomic fragments, isolating carbons (IC) and their attached hydrogens (ICH). The log P of a solute consisting solely of these fragments depends almost entirely on its size. Isolated double bonds have a slight negative effect (-0.08), but when conjugated in either chains or rings this polarity disappears. Even with no partial charge to delocalize, an extension of the resonating system has a positive effect. In a comparison with 1,2-diphenylethane, the double bond in *trans*-stilbene is seen to compensate for the loss of two hydrogen atoms.

Measured log P

$PhCH_2\text{---}CH_2Ph = 4.79$

$PhCH\text{===}CHPh = 4.81$

$PhC \equiv CPh = 4.78$

For a triple bond in a hydrocarbon, these effects are even more pronounced. In isolation, a triple bond contributes -0.5, but when conjugated between two phenyl rings, resonance appears to have compensated for the 'loss' of four hydrogens.

Halogens are the most important members of the class of non-hydrogen-bonding polar fragments. In CLOGP-3 this fragment class is designated as X. The bond between a halogen and an sp^3 IC has a very localized charge separation, the polar effect of which counteracts much of the effect of the halogen's larger size. This explains why methyl bromide is only slightly more hydrophobic than methane (1.19 *versus* 1.09), even though the bromine is almost five times the size of hydrogen.[38] When attached to an aromatic IC, however, all halogens are more hydrophobic than the hydrogens they replace. Recent evidence indicates that the $-C\equiv$ group in an alkyne exerts a field effect on H-bonding polar groups that is similar to a halogen, and for this reason we propose to treat them in CLOGP as a pseudohalogens.

Fragments which can hydrogen bond are referred to as H polar and designated as type Y in CLOGP-3. The present system requires two further subdivisions of this type: the first asks if the fragment contains the hydroxyl moiety, and the second asks for assignment to three levels of sensitivity to field effects of nearby halogens. The most sensitive, Y3, is restricted to fragments containing the substructure $-SO_2-$; the intermediate class, Y2, presently consists of the types: $-CONH-R$, $-O-R$, $-S-R$ and $-NH-R$. All others are assigned to Y1. All three classes react to the field effect of the first α-halogen in the same manner, requiring a correction factor of $+0.9$. There is no further correction for the second and third α-halogen in the case of Y1 fragments, but with Y3 it is additive (total $= +2.70$). With Y2, the second and third halogens have roughly half the effect of the first.

18.7.5 CORRECTION FACTORS

As correction factors were encountered it was seen that they were in accord with accepted solvation theory, and we labeled them accordingly. However, one should never lose sight of the fact that, by their very nature, partitioning data are so 'general' they could as well support mechanisms different from those proposed here. The following is only a brief outline of this subject from a mechanistic point of view, highlighting only those aspects that influenced the development of our method. It is covered in greater detail in Chapter 18.6.

18.7.5.1 Structural Factors: Flexibility

Collander[13] and others[20, 39] noted the positive correlation between log P and solute size. It would appear that the size contribution of an atom or group would be constant and no correction factor should be required. However, both the branching and flexibility of chains can affect solute volume if this is obtained directly from atomic fragment sums. This can most easily be seen in calculations of the simple alkanes.

The fragment value for hydrogen (one-half log P for hydrogen gas) can be used in conjunction with the measured values of methane and ethane to derive a reasonably consistent value for an sp^3 carbon in the following manner

$$\log P_{H\text{---}H} = 2f_H = 0.45; \quad f_H = 0.225$$

$$\log P_{CH_4} = 1.09 = f_C + 4f_H; \quad f_C = 1.09 - 4(0.225) = 0.19$$

$$\log P_{Me\text{---}Me} = 1.81 = 2f_C + 6f_H; \quad f_C = (1.81 - 1.35)/2 = 0.23$$

Calculation of the higher normal alkanes with these values leads to deviations that are progressively more positive. Even using the lower value for f_C (0.19), butane calculates as 3.01 and the measured value is 2.89. Either the precision of the above three measurements was much poorer than we had reason to believe or some other phenomenon is taking place in the higher alkanes. One difference between the $C_{1 \text{ or } 2}$ and the $C_{>2}$ alkanes is clear, and its effect would be in the observed direction: rotation and/or bending of the C—H and C—C bonds in hydrogen, methane or ethane could have no appreciable effect on solute size, while these motions of the C—C bonds in the longer chains would slightly reduce average molar volume and, perhaps, surface area. At any rate, CLOGP-3 treats this correction as 'chain flexibility' and calculates it as $-0.12(n-1)$, where 'n' is the number of bonds (not counting those to hydrogen) in each chain *outside* of the fragment. Of course any flexibility within the fragment is lumped in the fragment value itself. Since methane was less likely to contain any unforeseen correction factors, we gave its value for f_C some preference, and CLOGP-3 uses the value of 0.20. As the database grows, this value will be 'fine tuned' by statistical analyses.

One would expect less flexibility in aliphatic rings, and our early data supported the use of a similar factor, $-0.08(n-1)$. However, the data for the simple alicyclics are not too reliable, and the factor, $-0.09n$, appears to work as well. There is no 'flexibility factor' in aromatic ring bonds, but there is an unsaturation effect which is included in that of the aromatic carbon ($f_{CA} = 0.20$; $f_{Ca} = 0.13$).

18.7.5.2   Structural Factors: Branching

Branched structures have greater aqueous solubility than their normal isomers and the effect is greater if the branch is at the polar group, as in 2-butanol, than if in the chain, as in isobutyl alcohol. Intuitively, one expects that octanol cannot match this solvation increase, and, indeed, the log *P* factor is negative and it is greater for the 'group branch' ($F_{gbr} = -0.22$) than for the chain branch ($F_{cbr} = -0.13$). They are additive when they occur together as in *t*-butyl alcohol. Alicyclic fusions can be considered as chain branches, and an alicyclic substituent, like the hydroxyl in cyclohexanol, is considered a group branch. At present, CLOGP-3 treats branches the same, regardless of the length of the chain on which it occurs.

Again intuition might lead us to expect an increasingly larger correction as a polar group is moved toward the middle of a long alkyl chain, since this might result in a slightly increased tendency for the hydrocarbon 'arms' to overlap. Recent measurements on isomeric heptanols[41] do not support this hypothesis, but they do indicate the need for a correction scaled to chain length. For propanols, $F_{gbr} = -0.20$; for butanols, $F_{gbr} = -0.27$; for heptanols, the average $F_{gbr} = -0.47$.

Measured	*Measured*	F_{gbr}
$MeCH_2CH_2OH = 0.25;$	$MeCH(OH)Me = 0.05$	-0.20
$Me(CH_2)_3OH = 0.88;$	$MeCH(OH)CH_2Me = 0.61$	-0.27
$Me(CH_2)_6OH = 2.72;$	$MeCH(OH)(CH_2)_4Me = 2.31$	
	$MeCH_2CH(OH)(CH_2)_3Me = 2.24$	-0.47
	$(MeCH_2CH_2)_2CH(OH) = 2.22$	(average)

It is important to note that CLOGP-3 does not consider that halogens qualify for F_{gbr}. Multiple halogenation is so frequent it would dominate the evaluation of F_{gbr}. It is more convenient to allow for any halogen branching factor in the XCX and XCCX factors as discussed below.

In the two examples above, f_{cbr} and f_{gbr}, the branching center was an IC, but it can also be a polar fragment, as is the case with a tertiary amine. In this case the correction is clearly chain length dependent. CLOGP-3 has provision for reducing the hydrophobicity of alkyl chains attached to any such 'branching fragment' by incrementing the negative bond factor. So far this correction appears to be needed for only two fragments: tertiary amines, where it reduces each attached methylene by -0.08, and phosphate esters, where there is a larger reduction, -0.19. As yet the data on *N,N*-dialkylamides are insufficient to make a decision on whether or not they qualify as a branching fragment.

$$\text{Et} - \text{N} - \text{Et}$$
$$\overset{|}{\text{Et}}$$

$$f_N \quad + 6f_C \quad + 15f_H \qquad + (6-1)F_b + (6-1)\text{Frag. Br.}$$
$$-2.18 + 6(0.2) + 15(0.225) \quad + 5(-0.12) \quad + 5(-0.08) \qquad = 1.40$$

(3) \qquad\qquad\qquad\qquad\qquad\qquad\qquad\qquad\qquad Measured $= 1.45$

$$(Et_2O)_3P = O$$

$$f_{PO_4} + 6f_C \quad + 15f_H \qquad + (6-1)F_b + (6-1)\text{Frag. Br.}$$
$$-2.29 + 6(0.2) + 15(0.225) \quad + 5(-0.12) \quad + 5(-0.19) \qquad = 0.74$$

(4) \qquad\qquad\qquad\qquad\qquad\qquad\qquad\qquad\qquad Measured $= 0.80$

18.7.5.3 Polar Fragment Interaction Factors

Perhaps the choice in the terminology, 'isolating carbon', was a poor one, for if these carbons were truly isolating, polar fragments could never interact with one another. In actuality, all three types of interaction are observed: $X \leftrightarrow X$, $X \leftrightarrow Y$ and $Y \leftrightarrow Y$. The situation is even more complex, because, to obtain even moderate precision, one must consider the cases of one or two intervening ICs, and for higher precision, three. Furthermore, there is the matter of the additivity when two Y fragments on the same carbon interact with another Y or X, and the very frequent occurrence of two or three X fragments on the same IC. These factors, together with the values of the fundamental fragments, C and H, are listed in Table 2, which is taken directly from the CLOGP-3 program.

18.7.5.3.1 X versus X — aliphatic

The positive correction to $\log P$ for this interaction appears to be the result of dipole shielding and is limited to halogens on the same (geminal) or adjacent (vicinal) ICs. Adding a second halogen to an IC creates the first XCX pair, and the correction required is $+0.60$; adding a third halogen to the same IC creates two more such pairings, each of which requires another $+0.50$. If the fourth halogen is added the three additional pairings require a correction of $+0.40$ each. Thus for CCl_4 the total correction for halogen interaction is

$$\Sigma F_{XCX} = 0.6 + 2(0.5) + 3(0.4) = 2.80$$

For the vicinal halogen correction, F_{XCCX}, the bond between ICs must be single. The correction is evaluated by subtracting one from the number of halogens meeting the structural requirement and multiplying by the coefficient 0.28 (Rekker's C_M?). For example, in halothane (5), the vicinal halogen correction is 1.12, and the geminal is 2.19. (CLOGP shows five geminal halogens which create two 'first' and two 'second' pairings. The value for the 'second' pairing has been fine tuned from 0.5 to 0.495.) In trilene (6) there is no F_{XCCX}.

$$
\begin{array}{l}
f_{Cl} + f_{Br} + 3f_F \quad + 2f_C + f_H \quad + 5F_b + 5F_{XCX} + (5-1)F_{XCCX} \\
0.06 + 0.2 + 3(-0.38) + 0.4 + 0.225 - 0.6 + 2.19 \quad + 4(0.28) \quad = 2.45 \\
\qquad\qquad\qquad\qquad\qquad\qquad\qquad\qquad\qquad\qquad\qquad \text{Measured} = 2.30
\end{array}
$$

(5)

$$
\begin{array}{l}
3f^v_{Cl} + 2f_C \quad + f_H \quad + (4-1)F_b + F_{XCX} + F_= \quad \{\text{No } F(XCCX)\} \\
3(0.5) + 2(0.2) + 0.225 + 3(-0.12) + 0.6 \quad -0.09 \qquad = 2.27 \\
(v = \text{vinyl}) \qquad\qquad\qquad\qquad\qquad\qquad \text{Measured} = 2.42
\end{array}
$$

(6)

Considering the difference in size between fluorine and iodine atoms, it is surprising that no distinction need be made among the halogens in $X \leftrightarrow X$ interactions. If, as postulated above, the basis for the positive correction is dipole shielding, then some compensating factor, such as polarizability, appears to be operating.

Table 2 CLOGP-3 Model Constants[a]

HYDROGEN	0.227	FBRANCHMAX	7.000	XCCXVAL	0.280	YCY(R0, 00)	−0.320
IC-ALIPHAT	0.195	SR-FUSED	0.500	FCCXVAL	0.280	YCY(R0, 01)	−0.420
IC-AROMAT	0.130	SR-FUSED2	0.250	XCCYVAL	0.350	YCY(R0, 02)	−0.420
FUSION	0.100	SR-JOINED	0.200	FCCYVAL	0.450	YCY(R1, 00)	−0.320
BIPHENYL	0.100	SIGMA-DROP	0.500	Y1—C—X(1)	0.900	YCY(R1, 01)	−0.370
HET-FUSION	0.310	RHO-DROP	0.500	Y1—C—X(2)	1.150	YCY(R1, 02)	−0.370
CHAINBRANC	−0.130	ORTHOVAL	−0.280	Y1—C—X(3)	1.150	YCY(R2, 00)	−0.320
GROUPBRANC	−0.220	HBONDVAL	0.630	Y2—C—X(1)	0.900	YCY(R2, 00)	−0.320
DOUBLEBOND	−0.090	OCL-AROMAT	21.000	Y2—C—X(2)	1.300	YCY(R2, 00)	−0.320
TRIPLEBOND	−0.500	OCL-BENZYL	20.000	Y2—C—X(3)	1.700	YCCY(R0)	−0.260
CHAINBOND	−0.120	XCX(2)	0.600	Y3—C—X(1)	0.900	YCCY(R0)	−0.230
RINGBOND	−0.090	XCX(3)	1.590	Y3—C—X(2)	1.800	YCCY(R0)	−0.150
ZWITTERION	−2.300	XCX(4)	2.800	Y3—C—X(3)	2.700	Z-APPROX	0.200

[a] F = fluorine; X = other halogen; Y = hydrogen bonding fragment; (#) = number of preceding atom type.

18.7.5.3.2   X versus Y—aliphatic

The interaction considered here is that taking place over single bonds, and is, therefore, probably due to an inductive or field effect. In some cases fluorine requires a larger correction than the larger halogens, but not for XCY. The effect for the three H-polar types, Y1, Y2 and Y3 was discussed in Section 18.7.4.2. The corrections for XCX and XCY are additive, as seen by (7).

$$F-\underset{\underset{F}{|}}{\overset{\overset{F}{|}}{C}}-SO_2\text{—}\langle\text{phenyl}\rangle$$

CLOGP-3 = 2.61; Measured = 2.68
$F_{(XCX)} = +1.59; F_{(XCY3)} = +2.70$
Total corrections = +4.30

(7)

For the XCCY correction, CLOGP-3 currently makes fluorine a more reactive halogen, but does not distinguish between Y types. There is a constant correction of 0.45 for fluorine and 0.35 for Cl, Br or I, as seen in (8) and (9).

$ClCH_2CH_2CN$

$$f_{Cl}+f_{CN}+2f_C\quad+4f_H\quad+(3-1)F_b+F_{XCCY}$$
$$0.06-1.27+2(0.2)+4(0.225)+2(-0.12)+0.35\quad=0.20$$
$$\text{Measured}=0.18$$

(8)

FCH_2CH_2OH

$$f_F\quad+f_{OH}+2f_C\quad+4f_H\quad+(3-1)F_b+F_{FCCY}$$
$$-0.38-1.64+2(0.2)+4(0.225)+2(-0.12)+0.45\quad=-0.51$$
$$\text{Measured}=-0.67$$

(9)

It would not be surprising if data in the future will better fit a correction which is proportional to the hydrophilic level of Y, such as: $F_{FCCY}=0.28\,F_Y$ and $F_{XCCY}=0.23\,F_Y$. In (10) note that the XCCY correction appears additive even when applied six times on a branched Y fragment.

$$F-\underset{\underset{F}{|}}{\overset{\overset{F}{|}}{C}}-\underset{\underset{H}{|}}{\overset{\overset{OH}{|}}{C}}-\underset{\underset{F}{|}}{\overset{\overset{F}{|}}{C}}-F$$

$$f_{OH}+6f_F\quad+3f_C\quad+f_H\quad+F_{gbr}+(9-1)F_b+6F_{XCX}+6F_{XCCY}$$
$$-1.64+6(-0.38)+3(0.2)+0.225-0.22+8(-0.12)+3.18\quad+2.70\quad=1.59$$
$$\text{Measured}=1.66$$

(10)

18.7.5.3.3   Y versus Y—aliphatic

One widely accepted theory postulates that the hydrophobicity of a hydrocarbon solute arises from the entropy of the structured water which forms a 'sweater' around it. (However, see Section 18.6.3.1 in the previous chapter.) The negative value associated with Y fragments might well be due to their ability to form hydrogen bonds with solvent water. Thus they act more like bulk solvent, rather than inducing an 'iceberg-sweater'. The magnitude of their negative fragment value may then be a measure of the extent to which they reduce entropy by breaking up the sweater. If two Y solutes are placed near one another on a hydrophobic backbone, some overlap of this sweater-breaking ability should occur; for example some negativity is being counted twice, and so the hydrophilicity 'loss' should be proportional to what was potentially there if no interaction were taking place. Thus to evaluate the Y ↔ Y interaction correction, the CLOGP algorithm uses a set of coefficients by which the Y fragment sum is multiplied. The size of the coefficient is determined by the following factors: (1) the number of ICs separating the pair; (2) whether one or both are *in* a ring (a Y fragment *on* a ring is treated the same as *on* a chain); and (3) whether at least one of the pair has an —OH group. These coefficients are listed in Table 2.

Perhaps the best evidence to support the calculation of the YCCY correction as proportional to the Y fragment sum, rather than as a constant figure which disregards the type of fragments involved, can be seen when two ether oxygens are involved.

	Rekker's p.e.2	$0.23(f_1 + f_2)$
(i) $MeCH_2OCH_2CH_2OCH_2Me$	+0.574	+0.84
(ii) $PhOCH_2CH_2OPh$	+0.574	+0.28

In 1,2-diethoxyethane, all four bonds of the oxygens to ICs are aliphatic and the fragment values are as negative as possible (-1.81). In this case the 'proportional' and the 'constant' interaction corrections lead to very similar calculated values: CLOGP-3 = 0.95; Rekker's = 0.94. But in 1,2-diphenoxyethane, the oxygen fragments are not as hydrophilic because of the aromatic environment ($f_O^a = -0.61$). CLOGP-3 scales the interaction correction accordingly, and the calculated value for diphenoxyethane is 3.75, which compares favorably with measured, 3.81. Rekker's method adjusts the fragment values of the oxygens for the aromatic attachment, but leaves the interaction correction at the same constant value, and so the calculated value is 4.54. Rekker observed this difficulty with the phenoxyacetic acids, and included a 'combined fragment', —O—CH$_2$—C(=O)OH in his 'Table of Hydrophobic Fragmental Constants'.[33] If this phenomenon is as general as we believe it to be, then the use of combined fragments is more of a 'band-aid' than a solution, for as the fragment tables grow larger, the user will never know for sure if it is best to use simple fragments, or if he should search for a combined one.

18.7.5.4 *Ortho* Effects

A pair of fragments *ortho* to each other on an aromatic ring can require a correction factor which may be either negative or positive. CLOGP-3 allows for the possibility of this *ortho* interaction for any fragment pair entered in the *ortho* matrix provided,[9] but of those encountered so far, it appears necessary that at least one of the pair must be a Y. Furthermore, it appears that the magnitude of this negative effect depends both on the size and polar nature of the Y-group's other partner. These are shown in Table 3. This supports the mechanism that 'twisting' results in a decoupling of the Y fragment from the ring, making the fragment's environment appear more aliphatic-like. It seems, also, that as the fragments are placed in very close proximity, there may be a reversal of the normal field effect.[9,41] *o*-Methylacetanilide (**11**) shows the need for this correction factor.

$$f^a_{NHCO} + 2f_C + 6f^a_C + 10f_H + (2-1)F_b + F_{ortho}$$
$$-1.51 \quad + 2(0.2) + 6(0.13) + 10(0.225) + (-0.12) \quad -\mathbf{0.84} \quad = 0.96$$

(a = aromatic)   Measured = 0.85

(**11**)

A positive correction for an interacting *ortho* pair results when one of two Y fragments is attached to the ring through a carbonyl and the other is —OH or —NH—. It appears safe to ascribe this effect to an intramolecular hydrogen bond which reduces the solute's water affinity, as illustrated by salicylic acid (**12**).

$$f^a_{CO_2H} + f^a_{OH} + 6f^a_C + 4f_H + F_{\sigma/\rho}^* + F_{Hb}$$
$$-0.03 \quad -0.44 \quad + 6(0.13) + 4(0.225) + 0.34 \quad + \mathbf{0.63} = 2.19$$

Measured = 2.26

(**12**)   * See Section 18.7.5.5

But if the solute's affinity for octanol is reduced to the same or greater extent by an intramolecular H bond, then the correction should be zero or negative. This is apparently the case for *o*-nitrophenol, which is much more positive than the *meta* and *para* isomers in a solvent system such as CCl$_4$/water, but is slightly more negative in octanol/water. Why octanol's solvating power should be lessened by one type of intramolecular hydrogen bond but not by the other is not apparent.

Early data on the magnitude of the negative *ortho* effect seemed to fit Rekker's hypothesis that it occurred as multiples of 0.28 and our first *ortho* matrix indicated it as such.[9] Later data, especially from *N*-phenylsuccinimides, failed to support a 'quantized' effect.[8]

18.7.5.5 Aromatic Interactions: $(X \leftrightarrow Y)_a$ and $(Y \leftrightarrow Y)_a$

As previously noted, even in the absence of a heteroatom, extension of aromaticity increases log *P*. The fragment value for each carbon atom which either joins two rings (biphenyl) or fuses two

Table 3   Matrix of *Ortho* Value Corrections by *Ortho* Class (1–21); Tabulated Values are Multiples of -0.28

	1	2	3	4	5	6	7	8	9	10	11	12	13	14	15	16	17	18	19	20	21	
(1)	—	0	0	(1)	(1)	1	(1)	1	1	0	1	3	0	—	—	0	3	2	H*	0	—	—CN
(2)		—	0	(1)	(1)	1	—	1	1	0	1	3	1	—	—	0		2	—	—	1	—NO₂
(3)			—	(1)	1	—	—	(2)	(1)	—	2	(3)	1	—	—	0	(2)	(2)	—	—	—	—CF₃
(4)				—	1	—	—	(2)	(1)	—	2	3	1	(2)	—	1	2	2	0	0	—	—I
(5)					—	0	—	2	1	—	2	3	0	—	—	1	2	(1)	0	—	—	—Br
(6)						—	0	(1)	0	—	—	1	0	—	—	1	1	1	0	0	—	—Cl
(7)							—	—	0	—	—	—	—	—	0	—	—	—	0	—	—	—F
(8)								—	3	—	3	(4)	1*	—	0	H	H	H	H*	—	—	—SO₂N—(V)
(9)									—	—	4	5	1	—	—	H	H	H	H	0	—	—CO—(W)
(10)										—	—	—	2	—	—	H	H	H	H	0	—	—CHO
(11)											—	—	—	—	—	H	H	H	H	1	—	—CO₂H
(12)												—	—	—	—	1	—	2	(0)	2	2	—CONH—(X)
(13)													—	—	0	—	0	—	—	0	—	—O—(Y)
(14)														—	—	—	—	—	—	—	—	—SMe
(15)															—	0	0	—	0	3	—	—SH
(16)																—	—	—	H*	3	—	—OH
(17)																	—	—	0	—	—	—NH—(Z1)
(18)																		—	—	—	—	—NH—(Z2)
(19)																			—	—	0	—NH₂
(20)																				0	—	—Me
(21)																					—	—Phenyl

Symbol	Meaning		Symbol	Meaning
Normal	Measured		(V)	—H
(#)	Estimated		(W)	—Me, —OMe, —N(Me)₂
0	No data		(X)	—H, —NH₂
*	Zero effect		(Y)	—Me, —COMe, —CONHMe, —CON(Me)₂, —CH₂CO₂H
	5 if Y-phenyl		(Z1)	—CONH₂
H*	+0.40 to +0.50		(Z2)	—COMe
H	+0.63 (H-bond)			

(naphthalene) is increased by 0.1. If either of these carbons is also bonded to a heteroatom, as in quinoline or 2-phenylpyridine, the increase as presently programmed is 0.31 instead (see Table 2). This will be changed in the next version of CLOGP, because the data now indicate that the amount of the positive correction is dependent upon the number of electron pairs in the 'fused in' fragment.

When two or more X and/or Y fragments are attached to an aromatic ring system, their electronic interaction requires a correction factor which can be calculated by a method very similar to that used by Hammett[10] to calculate the electronic effects in other equilibria, such as acid ionization. Acceptable calculations could be made using those ordinary Hammett σ values which have a positive sign. In CLOGP they are 'fine tuned' from the actual partitioning data,[9] and appear in Table 4. In this application, 'ρ' serves a somewhat different function than it does in classic Hammett methodology. It is still a measure of susceptibility. We believe it measures the reduction in H bond acceptor strength of the Y fragment, which reduces the solvating ability of water in comparison to octanol and thus raises log P. ρ seems to have little or no relationship with substituent negativity on the Hammett σ scale. A few fragments appear to act 'bidirectionally' in that they require both a σ and ρ value. Fragments fused in aromatic rings (*e.g.* —N=) may also be assigned both ρ and σ constants. CLOGP presently treats these 'fused in' fragments slightly differently from that proposed earlier,[9] in that it requires a σ/ρ correction for multiple occurrences, such as in pyrimidine, *sym*-triazine (*e.g.* **14**) and purines. In *o*-chloroacetanilide (**13**), the positive $F_{\sigma/\rho}$ is added to any negative F_{ortho} just as it was added to a positive F_{Hb} for (**12**).

$$f^a_{Cl} + f^a_{NHCO} + f_C + 6f^a_C + 7f_H \qquad + F_b + F_{\sigma/\rho} + F_{ortho}$$
$$0.94 + (-1.51) \quad + 0.2 + 6(0.13) + 7(0.225) \quad -0.12 + 0.30 - 0.84 \quad = 1.34$$
$$\text{Measured} = 1.28$$

(**13**)

As might be expected, this electronic effect on log P is attenuated upon extension of the aromatic system through which it acts. If the rings are fused, such as in 7-hydroxyquinoline, the effect is reduced to one-half of its intrinsic value. If another aromatic ring intervenes, it is reduced to one-fourth. For joined rings, as in biphenyl, it is reduced to 0.2 of its intrinsic value.

Table 4 σ and ρ Constants of Generalized Structures

Number	σ	ρ	Generalized structure	Examples
1	1.00	1.17	=N(=O)—	Pyridine-*N*-oxide
2	0.84	0.21	—N=	Pyridine, quinoline
3	0.71	0.00	—SO$_2$F	
4	0.65	0.00	—SO$_2$X	X = alk, N(Me)$_2$
5	0.65	0.00	—CN	
6	0.60	0.00	—NO$_2$	
7	0.49	0.00	—CF$_3$	
8	0.28	0.00	Halogens	
9	0.58	0.44	—CHO	
10	0.51	0.27	—C(=O)X	X = alk, OMe, Ph, N(Me)$_2$
11	0.32	0.35	—CO$_2$H	
12	0.32	0.72	—CONHX	X = H, NH$_2$, Ph, Alk
13	0.17	0.50	—OX	X = alk, CONHMe, CON(Me)$_2$, CH$_2$CO$_2$H, PO(O-alk)$_2$, Not C$_6$H$_6$
14	0.25	0.88	—SO$_2$NHX	X = H, C$_6$H$_5$
15	0.00	0.50	—SH	X = H
16	0.00	0.30	—SX	X = alk
17	0.00	0.61	—NH$_2$	—N(Me)$_2$, —N=NN(Me)$_2$
18	0.00	1.06	—OH	
19	0.00	1.08	—NHX	X = alk, COMe, CON(Me)$_2$, CHO, C$_6$H$_5$, SO$_2$CF$_3$, CONHPh
20	0.00	0.90	—N(X)NH$_2$	Subst. hydrazine
21	1.34	0.00	Arom =NN=	Diazine
22	0.50	0.00	Arom ==NO—	Isoxazole
23	0.00	0.50	Arom —O—	Furan
24	0.00	0.40	Arom —S—	Thiophene
25	0.00	0.70	Arom —NHN==	Pyrazole
26	0.00	0.80	Arom —NH—	Pyrrole

In the classical Hammett system, several substituents may be influencing one responding group, and full additivity of σ values usually applies. For hydrophobicity, the electronic effect fades with multiple substitution, and it must be scaled down in a rather complex fashion. Currently CLOGP-3 takes the following steps. (1) The full potential σ/ρ product for each possible interaction is calculated and placed in descending value order, *after* determining if the fragment pair are on the same or separate rings. (2) Except for the σ for a few aromatic nitrogen fragments (*e.g.* in pyridine), each use of σ or ρ causes it to fade or 'age'. The first interaction at the top of the ordered list is entered at full potential because the current age of its σ and ρ components is 'zero' for each. Each use reduces the effective σ and ρ to one-half its previous value (except as noted above for never aging nitrogens), and so if each were at 'age 1' the increment to the correction would only be one-quarter as much as a 'fresh' interaction. A detailed example of such an electronic correction for atrazine (**14**), is given in Table 5.

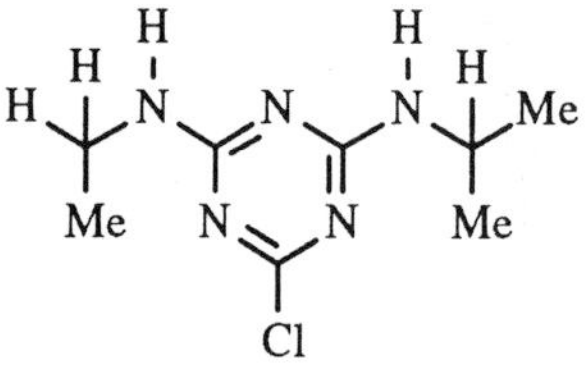

$F_{\sigma/\rho} = 17$ potential interactions; 8.3 used $= +4.032$

Calculated $= 2.82$

Measured $= 2.75$

(**14**) Atrazine

Table 4 contains no negative numbers, and, to a first approximation, one can say that no electronic interaction reduces $\log P$. However, a nitro substituent does appear to interact with another electron-withdrawing substituent, whose ρ value is zero, to lower $\log P$ by an average of -0.15 units. The other substituent may be a halogen, cyano, trifluoromethyl or another nitro, but not sulfonyl. It appears that the alkanesulfonyl substituent should be assigned a ρ value of about 0.4. At the present time the only other group to act in the way nitro does in lowering $\log P$ in an electronic interaction is the fused in fragment, oxadiazolyl: $=N-O-N=$. There are no data to support such an effect between two cyano groups or two trifluoromethyls, and the effect between either of these and a halogen is too small to consider, as is the effect between two halogens.

18.7.6   TAUTOMERS

When one of two tautomeric forms predominates, then calculations using the dominant structure are usually satisfactory. The pyridine analogs provide examples of this. For example, 2-pyridinamine (**15**) is largely in the amine form (rather than the imine) and calculated as such, there being good agreement with measurement. On the other hand, the hydroxy analog (**16**), if calculated as such, gives poor agreement.

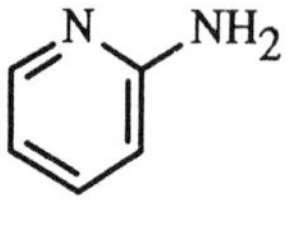

(**15**) Calculated $= 0.35$
Measured $= 0.49$

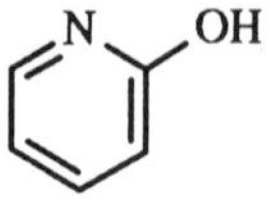

(**16a**) Calculated $= +0.89$
Measured $= -0.58$

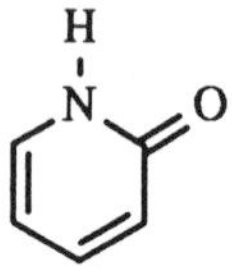

(**16b**) Calculated $= -0.57$

When the latter is calculated as the dominant tautomeric structure, 2($1H$)-pyridone, a *new fragment* is required, which can be evaluated from this measured value and that for 2($1H$)-quinolone. In a similar vein, calculation based on the structures 2-pyridinethiol (**17**), 2-hydroxypyrimidine (**18**), and pyrimidine-2-thiol (**19**) all give values much higher than observed. In each case the dominant tautomeric structure is the '-one', and new fragment values must be established from these measured partition coefficients. Since no values for other analogs have been reported in these cases, the agreement between the newly calculated and measured now is perfect.

Table 5 Atrazine: Details of Electronic Corrections

		React: list of electronically active fragments		
Fragment No.	*Ring No.*	*Type*	σ	ρ
1	1	Attached	0.280	0.000
2	1	Fused-in	0.840	0.210
3	1	Attached	0.000	1.080
4	1	Fused-in	0.840	0.210
5	1	Attached	0.000	1.080
6	1	Fused-in	0.840	0.210

		Pact: list potential electronic activity					
Fragment	*Ring*	σ	*Fragment*	*Ring*	ρ	*Distance*	*Value*
1	1	0.280	2	1	0.210	0	1.000
1	1	0.280	3	1	1.080	0	1.000
1	1	0.280	4	1	0.210	0	1.000
1	1	0.280	5	1	1.080	0	1.000
1	1	0.280	6	1	0.210	0	1.000
2	1	0.840	3	1	1.080	0	1.000
2	1	0.840	4	1	0.210	0	1.000
2	1	0.840	5	1	1.080	0	1.000
2	1	0.840	6	1	0.210	0	1.000
4	1	0.840	2	1	0.210	0	1.000
4	1	0.840	3	1	1.080	0	1.000
4	1	0.840	5	1	1.080	0	1.000
4	1	0.840	6	1	0.210	0	1.000
6	1	0.840	2	1	0.210	0	1.000
6	1	0.840	3	1	1.080	0	1.000
6	1	0.840	4	1	0.210	0	1.000
6	1	0.840	5	1	1.080	0	1.000

Elect: Details of σ–ρ corrections used
(the 17 possible interactions allocated as indicated)

	σ		ρ			
Fragment	*Drop*	*Fragment*	*Drop*	*Potential correction*	*Net correction*	
2	1.0000	3	1.0000	0.9072	0.9072	
2	1.0000	5	1.0000	0.9072	0.9072	
4	1.0000	3	0.5000	0.9072	0.4536	
4	1.0000	5	0.5000	0.9072	0.4536	
6	1.0000	3	0.2500	0.9072	0.2268	
6	1.0000	5	0.2500	0.9072	0.2268	
1	1.0000	3	0.1250	0.3024	0.0378	
1	0.5000	5	0.1250	0.3024	0.0189	
2	1.0000	4	1.0000	0.1764	0.1764	
2	1.0000	6	1.0000	0.1764	0.1764	
4	1.0000	2	1.0000	0.1764	0.1764	
4	1.0000	6	0.5000	0.1764	0.0882	
6	1.0000	2	0.5000	0.1764	0.0882	
6	1.0000	4	0.5000	0.1764	0.0882	
1	0.2500	2	0.2500	0.0588	0.0037	
1	0.1250	4	0.2500	0.0588	0.0018	
1	0.0625	6	0.2500	0.0588	0.0009	

(17)
Calculated = 1.48
Measured = −0.13

(18)
Calculated = 0.21
Measured = −1.76

(19)
Calculated = 0.56
Measured = −0.97

Solvation forces play a dominant role in tautomeric equilibria,[42] and can be a source of unexpected anomalies in calculation of $\log P$. In the case of the 4-hydroxypyridines, the -one/-ol ratio is 10^{-5} in the vapor phase but > 10 in chloroform.[43] The strong tendency for these pyridones to associate (K_a in chloroform $= 30\,000$) casts suspicion on determination of tautomeric equilibrium constants in non-polar solvents,[44] but, what is more important for $\log P$ calculation, this K_a is very dependent on the steric parameters of groups flanking the ring nitrogen.[45] None of these observations, however, provide a clear explanation of the large deviation between the calculated and measured values for this solute, as seen by (20).

$$f^a{}_{NH} + f^a{}_{C=O} + 4f^a{}_C + 4f_H + F_{\sigma/\rho}$$
$$-0.67 + (-1.53) + 0.52 + 4(0.225) + 0.80 = 0.03$$
$$\text{Measured} = -1.30$$

(20)

When each tautomeric form may be present in significant amounts, one might suppose that the calculated value for the two structures would bracket the measured value, and could serve as a measure of K_t. At the present state of our knowledge, this would appear to be a very risky procedure, because the equilibrium *between* the phases must be satisfied as well as the equilibrium *within* each phase.

Acetylacetone illustrates some of these problems:

CLOGP-3 calculates both structures as moderately hydrophilic (keto $= -0.50$; enol $= -0.30$), but the measured value is $+0.24$ (the average of two determinations: 0.34 and 0.14). The enol form is greatly stabilized by an intramolecular hydrogen bond, and there is about two-thirds of the enol present in both the liquid and vapor phases.[45] The effect of solvation forces on K_t has not been thoroughly studied, but one would expect water to solvate the keto form more efficiently while the opposite would be true for octanol. In fact, we would expect that this type of intramolecular hydrogen bond, which is presently not in the CLOGP algorithm, to add somewhere between $+0.6$ and $+0.9$ units to the above calculation: $-0.30 + 0.75 = +0.45$. From this one could reach the very tenuous conclusion that the overall, biphasic K_t would favor the enol by about $3:1$, which is very near the unsolvated value.

18.7.7 ZWITTERIONS

Octanol can accommodate a zwitterion, such as is present in α-amino acids, without too much difficulty, and so their measurement and calculation are matters of considerable interest.[46] However, it should be pointed out that as a parameter for membrane transport, $\log P$ may not serve as well for zwitterions as it does for neutral solutes, because nature has provided special methods for the passage of the former.

Calculations of reasonable accuracy can be made for the $\log P$ of a zwitterion at the pH where the net charge is zero.[47] Using the values for the neutral amino and carboxyl fragments, and also the interactions based on them, reasonable estimates are obtained if a zwitterion correction factor of -2.30 is employed. Rather surprisingly, this appears to hold for GABA (23), and for zwitterions where a sulfonic acid replaces a carboxyl. However, it appears to hold only where the difference between pK_a values of the amino and acid moieties is over 6.5. If a polar group, such as a carbonyl, is close to the protonated amine, and its pK_a drops into the range 7.5–8.0, the negative zwitterion correction may disappear completely. The absence of a zwitterion effect in ampicillin (21) is apparent when its calculated value is compared to benzylpenicillin's (22)

(21) Ampicillin

Calculated with F_{ZI} = -1.30
Calculated without F_{ZI} = $+1.00$
Measured = $+1.35$ and 0.57

(**22**) Benzylpenicillin

Calculated = 1.68
Measured = 1.83

$$O$$
$$H_2NCH_2CH_2CH_2\overset{\|}{C}OH$$

(**23**) GABA

Calculated with $F_{ZI} = -3.36$
Measured $= -3.17$

It appears that, in peptides, the zwitterion correction should be ignored for any analog larger than a dipeptide; therefore dependable calculations in this area must await reliable information on solute conformation, which may differ from crystal conformation.

18.7.8 ION PAIRS

As noted in the introduction, the true partition coefficient is the ratio of the concentrations of the same solute species in the two immiscible solvents. Ions which are completely dissociated in water must form into pairs to enter the octanol where the dissociation constant is about 10^{-3}. This set of dependent equilibria is often treated in terms of an extraction constant, E_{QX},[48] but we have proposed certain standard conditions which result in reproducible values which can be compared to those of neutral solutes. These serve as the 'target' for calculations. The calculation procedures have been described thoroughly in a previous publication[11] and will be reviewed only briefly here. They have not yet been included in a computer algorithm.

It appears that the loss of a proton from either a carboxyl or a sulfonic acid fragment results in a change in log P of -4.1, as long as there are no large hydrophobic cations to interfere. The value of -4.1 appears to be constant regardless of the structure to which the ionic fragment is attached, indicating, perhaps, that a negative charge tends to be localized. However, a phenolate ion is stabilized by resonance with a keto form, and so the difference between ionic and neutral forms is less, namely, -2.80.

On the other hand, a positive charge seems to be delocalized for a considerable distance even along a chain of sp^3 carbon atoms.[49] It is important to note that this delocalization of a positive charge, over either sp^3 or sp^2 carbons, makes them appear more hydrophilic, and thus acts in an opposite direction to delocalization in neutral solutes. For most amines, the neutral form does not contribute significantly if the aqueous phase is 0.1 N HCl. Obviously one is not concerned with interference from the neutral form in the case of quaternary ammonium compounds. We prefer to measure them at very low concentrations in the presence of 0.1 M of small anion and extrapolate to zero concentration. Octanol only registers a difference of 0.35 between Cl^- and I^-, and so one can make this correction and safely compare data from different halides.

In the simplest situation, the calculation should rationalize how an organic cation becomes more hydrophilic as the charge is allowed to delocalize to the maximum extent along alkane chains. We believe that the approach which is easiest to understand and to computerize is to impose this effect on the 'geometric' bond factor which applies to neutral solutes (see Section 18.7.5.1). The fragment value for the protonated amine or 'quat' is given the value which it would have if attached to the minimum ICs possible. Then, for analogs with more or longer chains, on top of the geometric bond factor, the method imposes a negative 'electronic bond' factor which decreases in magnitude with the square of the distance away from the central nitrogen atom. For mono- and di-amines, the normal geometric bond factor is -0.12. Because a tertiary amine is a 'branching fragment', the geometric bond factor is -0.20. We can consider a quat to be 'doubly branched' and, if a simple geometric bond factor were possible, it would be about -0.27 or -0.28.

For protonated amines, it appears that the 'electronic bond' factor should extend through four alkane carbons, and for the quats, through five. Since we deal in log terms, the factor should double for each bond as the center is approached, and be in the ratio $1:2:4:8$ for amines and $1:2:4:8:16$ for the quats. As seen in Table 6, some adjustment to the theoretical ratios is required to better fit the current data. The values of -0.07 and -0.04 were arrived at empirically.

Table 6 Negative Bond Factors for Cation Delocalization

(a) Quaternary Ammonium Bond Factors

		Geometric		Electronic		Total
6th bond and beyond,	F_{bx}	-0.27	+	None	=	-0.27
5th bond	F_{bx+5}	-0.27	+	$-(0.04)$	=	$-0.31 (0.30^a)$,
4th bond	F_{bx+4}	-0.27	+	$-2(0.04)$	=	-0.35
3rd bond	F_{bx+3}	-0.27	+	$-4(0.04)$	=	$-0.43 (0.45^a)$
2nd bond	F_{bx+2}	-0.27	+	$-8(0.04)$	=	$-0.59 (0.60^a)$
1st bond	F_{bx+1}	-0.27	+	$-16(0.04)$	=	$-0.91 (0.90^a)$

^{a} For mnemonic reasons, we have adjusted the values very slightly as indicated.

(b) Protonated Amine Bond Factors

	Mono- and di-					Tri-		
	Geometric	Electronic				Geometric	Electronic	
5th and beyond	-0.12	+ None	$= -0.12$	$= F_b$	F_{by}	$= -0.20$	+ None	$= -0.20$
4th	-0.12	+ (-0.07)	$= -0.19$	$= F_{b+4}$	F_{by+4}	$= -0.20$	+ (-0.07)	$= -0.27$
3rd	-0.12	+ $2(-0.07)$	$= -0.26$	$= F_{b+3}$	F_{by+3}	$= -0.20$	+ $2(-0.07)$	$= -0.34$
2nd	-0.12	+ $4(-0.07)$	$= -0.40$	$= F_{b+2}$	F_{by+2}	$= -0.20$	+ $4(-0.07)$	$= -0.48$
1st	-0.12	+ $8(-0.07)$	$= -0.68 (-0.78)$		F_{by+1}	$= -0.20$	+ $8(-0.07)$	$= -0.76 (-0.78)$

Currently we do not have the data to clearly define what happens when an aromatic ring is attached to such a 'charged' alkyl chain. The data for phenylpropane amine hydrochloride (**24**) indicate that the phenyl ring on the fourth bond from the nitrogen sees little effect, as would be expected.

$$\text{PhCH}_2\text{CH}_2\text{CH}_2\overset{+}{\text{N}}\text{H}_3\ \text{Cl}^-\qquad 6f_c\quad +3f_C\quad +11f_H\quad + f_{NH_3Cl}+F_{b+1}+F_{b+2}+F_{b+3}+F_{b+4}$$

$$\text{(24)}\qquad\qquad 6(0.13)+3(0.2)+11(0.225)-3.40\quad +(-0.78)+(-0.4)+(-0.26)+(-0.19)$$

$$= -1.17$$

$$\text{Measured} = -1.13^{50}$$

With a quaternary nitrogen in an aromatic ring, the fragment value is taken to allow for distribution of the charge in that ring. As shown by *N*-butylpyridinium bromide (**25**), the *N*-alkyl chain is calculated according to the above method and the values in Table 5.

$$f^{a+}_{N}\text{Br}\quad +5f_c\quad +4f_C\quad +14f_H\quad +F_{B+1}+F_{B+2}+F_{B+3}+F_{B+4}$$

$$-5.02\quad +5(0.13)+4(0.2)+14(0.225)+(-0.9)\ +(-0.6)\ +(-0.45)\ +(-0.35)$$

$$= -2.72$$

$$\text{(25)}\qquad\qquad\qquad\qquad\qquad\qquad\qquad\text{Measured} = -2.69^{22}$$

When aromaticity is extended by either fused or joined rings, it seems appropriate to allow for extra cation delocalization by multiplying the number of 'extra' aromatic carbons by the factor -0.25. Since this is not a field effect, *i.e.* falling off with the square of the distance from the charge, there is no need to determine bond distance. This is illustrated by *N*-propylquinolinium bromide (**26**).

$$f^{a+}_{N}\text{Br}\quad +9f_c\quad +4f_C\quad +14f_H\quad +4F_{xc}\quad +F_{c\bullet}+F_{c*}+F_{B+1}+F_{B+2}+F_{B+3}$$

$$-5.02\quad +9(0.13)+4(0.2)+14(0.225)+4(-0.25)+0.1\ +0.31\ +(-0.9)\ +(-0.6)\ +(-0.45)$$

$$= -2.44$$

$$\text{Measured} = -2.52^{22}$$

where F_{xc} = the extended aromaticity factor for cations; $F_{c\bullet}$ = aromatic carbon fusion factor; F_{c*} = aromatic hetero fusion factor. The latter two are the usual factors applying to neutral solutes.

18.7.9 CURRENT DEVELOPMENTS AND CONCLUSIONS

Although $\log P$ calculations are quite complex, even at our present state of knowledge, they are bound to become more so as more data are available for study. A great deal of effort was given in designing the computer program to assure sufficient flexibility to take care of future needs. At the present time, four significant shortcomings rate the highest programming priority.

In some cases, two structural isomers can have widely different $\log P$ values. For instance, the difference in measured values between fumaric and maleic acid is nearly one log unit, but at present CLOGP is not given the information to distinguish them. It is not a simple task to write a general algorithm to clearly and unambiguously identify all cases of E and Z or *exo* and *endo* pairs, but the structure input to CLOGP, a linear notation called SMILES,[51] is being extended to include all isomerism. Together with an efficient substructure searching algorithm, GENIE,[51] the problem will then be reduced to one of proper evaluation. This promises to be anything but simple, however, as the maleic/fumaric pair exemplifies. The lower first pK_a and higher second pK_a of maleic acid as compared to fumaric seems adequately explained on the basis of the structures (27) and (28) respectively.

(27) Maleate monoion	(28) Fumaric acid
H-bond stabilized	First $pK_a = 3.02$
First $pK_a = 1.94$	Second $pK_a = 4.38$
Second $pK_a = 6.23$	

Admittedly the maleic monoion provides the most favorable condition for intramolecular hydrogen bonding, but a somewhat weaker one ought to form with the neutral solute as well. As seen in Section 18.7.5.4 such carbonyl–hydroxy pairings are expected to raise $\log P$. This should make maleic acid about 0.7 higher than fumaric, but the measured values show a slightly greater difference but in the opposite direction. It is true that maleic acid has a slightly larger dipole moment (3.17 *versus* 2.45)[52] but this is an unconvincing explanation in view of larger dipole difference between *cis-* and *trans-*1,2-dichloroethylene (1.91 *versus* 0.0)[52] with a difference in $\log P$ of only 0.23, or *o-* and *p-*dichlorobenzene (2.27 *versus* 0.0)[52] with no difference in $\log P$. In the case of the dimethyl esters, the fumarate is more hydrophobic than the maleate, but the difference is less than 0.3.[53] Work is in progress to resolve this apparent anomaly.

When certain multi-atom H-polar fragments are present in a ring, their fragment values are shifted downward almost one log unit as is apparent from (29) and (30). Lactones provide the most common example of this effect. Lactams do not exhibit this effect as seen in (31), probably because two lone pairs on the hetero adjacent to the carbonyl are required. As soon as sufficient data are acquired to define the effect properly, the programming itself ought not to present any difficulties.

(29) Calculated = 0.66	(30) Calculated = 1.20	(31) Calculated = −0.56
Measured = −0.35	Measured = 1.21	Measured = −0.46

Presently CLOGP-3 measures only the topological distance between potentially interacting fragments. Considerable evidence has accumulated which indicates that some preferred conformations can bring fragments into effective interaction distance even when separated by four or more ICs. Many alkaloids, such as strychnine, seem to exhibit this behavior, with the amine function brought close to either a strongly polar group, such as a carbonyl, or else to the π cloud of a phenyl ring. Not only is the measured $\log P$ higher than expected, but the pK_a is lower than anticipated.

Both of these are undoubtedly solvation effects that may well be treated by the same algorithm. In other fused-ring structures whose conformations are rather fixed, such as some gibberellin analogs, steroids with both 17α- and 17β-substituents, and the elephantin analogs, the anomaly seems to come about through intramolecular hydrogen bonding. A somewhat similar anomaly is exemplified by adenosine. CLOGP-3 calculates both 'parts', adenine and ribose, reasonably well (deviation = 0.3 or less), but the deviation for adenosine is +1.4. We have referred to this as 'hydrophilic overlap', but it could result from some intramolecular hydrogen bonding. 'Topologically remote' hydrogen bonding may also be rather common in linear solutes, such as amino acids and peptides containing polar side chains. Allowing for this effect may be more involved than merely determining whether a five- or six-membered ring can be formed, because some intramolecular hydrogen bonding appears to be mediated by a 'bridging hydrate'.[47]

A present shortcoming of CLOGP-3, which is encountered more frequently than first supposed, involves $F_{(YCY)}$. When the correction is applied more than once over the same ring IC, it overshoots. This can be seen in calculations for a number of diazepines (**32**), for example.

A	B		Calculated	Measured
H	H	Diazepam	3.18	2.82
OH	H	Oxazepam	3.33	2.25
OH	Me	Temazepam	3.66	2.19

(**32**)

As the fragments and correction factors become securely identified and evaluated, we intend to switch our approach from 'constructionist' to 'reductionist'; that is, we will periodically make a computer account of each use of every fragment and factor when calculating the file of preferred measured values (Starlist in CLOGP), along with the deviation of each calculation. We can then apply statistical procedures to 'fine tune' both the fragment and factor values to minimize deviation and eliminate any skewing.

Future measurements will surely indicate ways that the methodology of calculating partition coefficients can be improved. However, it is important to realize that even at the present stage of development, computer calculation of log P has often prompted a remeasurement of a solute with the result that the more carefully measured value agrees well with the calculated. And it is more important, perhaps, to keep in mind that $\log P_{(o/w)}$ values are not just 'numbers' to be used as parameters in a regression equation. Becoming aware of why each value is what it is — that is what correction factors come into play in that particular structure, what solvation forces are competing in the transfer between lipid membrane and serum or cell plasma — this knowledge could give additional insights into pharmacology and drug design.

As supportive as we are to the idea of calculating $\log P_{(o/w)}$ from structure, we are totally opposed to using these values in reports for technical journals unless the calculation methodology is documented. All too frequently a table of values is referenced merely 'calculated by Rekker's (or Leo's) methods', but even someone very familiar with those methods may find it impossible to duplicate the published results. Details of each calculation would be too lengthy, but the most significant one could be chosen as an example. In this respect, calculation by computer has many distinct advantages. The format of the output is standardized, and will become familiar to those following this field. If the version used is stated, one can become aware of any recent improvements and can more easily reconcile any apparent discrepancies. There is a trap in manual calculations: if a calculated value fits into the pattern one expected to see, there is a tendency not to check quite as carefully for errors or omissions. A computer program may have flaws, but it delivers the same result each time, and can save one from falling into that trap. This observation is made from personal experience.

18.7.10 REFERENCES

1. M. Berthelot and E. Jungfleish, *Ann. Chim. Phys.*, 1872, **26**, 396.
2. W. Nernst, *Z. Phys. Chem.*, 1891, **8**, 110.

3. J. J. Banewicz, C. W. Reed and M. E. Levitch, *J. Am. Chem. Soc.*, 1957, **79**, 2693.
4. M. Davies and D. M. L. Griffiths, *J. Chem. Soc.*, 1955, 132.
5. E. N. Lassettre, *Chem. Rev.*, 1937, **20**, 259.
6. R. Van Duyne, S. A. Taylor, S. D. Christian and H. E. Affsprung, *J. Phys. Chem.*, 1967, **71**, 3427.
7. H. W. Smith and T. A. White, *J. Phys. Chem.*, 1929, **33**, 1953.
8. A. J. Leo, *J. Pharm. Sci.*, 1987, **76**, 166.
9. A. J. Leo, *J. Chem. Soc., Perkin Trans. 2*, 1983, 825.
10. L. P. Hammett, in 'Physical Organic Chemistry', 2nd edn., McGraw Hill, New York, 1970.
11. C. Hansch and A. J. Leo, in 'Substituent Constants for Correlation Analysis in Chemistry and Biology', Wiley, New York, 1979.
12. H. W. Smith, *J. Phys. Chem.*, 1921, **25**, 204, 605.
13. R. Collander, *Acta Chem. Scand.*, 1951, **5**, 774.
14. A. J. Leo and C. Hansch, *J. Org. Chem.*, 1971, **36**, 1539.
15. P. Seiler, *Eur. J. Med. Chem.—Chim. Ther.*, 1974, **9**, 473.
16. S. H. Unger, P. S. Cheung, G. H. Chiang and J. R. Cook, in 'Partition Coefficient Determination and Estimation', ed. W. J. Dunn, III, J. H. Block and R. S. Pearlman, Pergamon, Oxford, 1986, p. 69.
17. E. Tomlinson and T. L. Hafkensheid, ref. 16, p. 101.
18. R. Rapaport and S. Eisenreich, *Environ. Sci. Technol.*, 1984, **18**, 163.
19. K. Woodburn, W. J. Doucette and A. W. Andren, *Environ. Sci. Technol.*, 1984, **18**, 457.
20. M. J. Kamlet, J. -L. M. Abboud, M. H. Abraham and R. W. Taft, *J. Org. Chem.*, 1983, **48**, 2877.
21. A. J. Leo, *EHP, Environ. Health Perspect.*, 1985, **61**, 275.
22. Pomona Medchem Database, Issue # 30, Jan. 1987.
23. M. J. Kamlet, R. M. Doherty, V. Fiserova-Bergerova, P. W. Carr, M. H. Abraham and R. W. Taft, *J. Pharm. Sci.*, 1987, **76**, 13.
24. T. Fujita, J. Iwasa and C. Hansch, *J. Am. Chem. Soc.*, 1964, **86**, 5175.
25. K. S. Rogers and A. Cammarata, *Biochim. Biophys. Acta*, 1969, **193**, 22.
26. A. J. Hopfinger and R. D. Battershell, *J. Med. Chem.*, 1976, **19**, 569.
27. R. N. Smith, C. Hansch and M. A. Ames, *J. Pharm. Sci.*, 1975, **64**, 599.
28. G. Klopman and L. D. Iroff, *J. Comput. Chem.*, 1981, **2**, 157.
29. G. Klopman, K. Namboodiri and M. Schochet, *J. Comput. Chem.*, 1985, **6**, 28.
30. P. Broto, G. Moreau and C. Vandycke, *Eur. J. Med. Chem.—Chim. Ther.*, 1984, **19**, 71.
31. A. K. Ghose and G. M. Crippen, *J. Comput. Chem.*, 1986, **7**, 565.
32. G. G. Nys and R. F. Rekker, *Chim. Therap.*, 1973, **8**, 521.
33. R. F. Rekker, 'The Hydrophobic Fragmental Constant', Elsevier, New York, 1977.
34. R. F. Rekker and H. M. DeKort, *Eur. J. Med. Chem.—Chim. Ther.*, 1979, **14**, 479.
35. H. van de Waterbeemd and B. Testa, *Int. J. Pharm.*, 1983, **14**, 29.
36. H. van de Waterbeemd, 'Hydrophobicity of Organic Compounds: How to Calculate It by Personal Computers', Compudrug International, Vienna, 1986.
37. J. T. Chou and P. C. Jurs, *J. Chem. Inf. Comput. Sci.*, 1979, **19**, 172.
38. A. Immirzi and B. Perini, *Acta Crystallogr., Sect. A*, 1977, **33**, 216.
39. A. J. Leo, C. Hansch and P. Y. C. Jow, *J. Med. Chem.*, 1976, **19**, 611.
40. G. Gould and C. Hansch, 'Pomona Medchem Database', Issue # 30, Jan 1987.
41. A. Ogino, S. Matsumura and T. Fujita, *J. Med. Chem.*, 1980, **23**, 437.
42. P. Beak, F. S. Fry, Jr., J. Lee and F. Steele, *J. Am. Chem. Soc.*, 1976, **98**, 171.
43. P. Peak, J. B. Covington and J. M. Zeigler, *J. Org. Chem.*, 1978, **43**, 177.
44. J. Frank and A. R. Katritzky, *J. Chem. Soc., Perkin Trans. 2*, 1976, 1428.
45. A. H. Lowrey, C. George, P. D'Antonio and J. Karle, *J. Am. Chem. Soc.*, 1971, **93**, 6399.
46. J. L. Fauchere and V. Pliska, *Eur. J. Med. Chem.—Chim. Ther.*, 1983, **18**, 369.
47. D. Abraham and A. J. Leo, *Proteins: Struct. Funct. Genet.*, 1987, **2**, 130.
48. R. Modin and G. Schill, *Acta Pharm. Suecica*, 1967, **4**, 301 (*Chem. Abstr.*, 1968, **68**, 92786).
49. B. Pullman, P. Courriere and J. L. Coubeils, *Mol. Pharmacol.*, 1971, **7**, 397.
50. J. M. Mayer, B. Testa, H. van de Waterbeemd and A. Bornand-Crausaz, *Eur. J. Med. Chem.—Chim. Ther.*, 1982, **17**, 461.
51. D. Weininger, *J. Chem. Inf. Comput. Sci.*, 1988, **28**, 31.
52. A. L. McClellan, 'Tables of Experimental Dipole Moments', Freeman, San Francisco, 1963.
53. Pomona College Medchem Project, unpublished data.

18.8

Intermolecular Forces and Molecular Binding

PETER R. ANDREWS

Bond University, Queensland, Australia

and

MARINA TINTELNOT

Victorian College of Pharmacy Ltd, Parkville, Victoria, Australia

18.8.1 OVERVIEW

Intermolecular forces are of fundamental importance in biological processes. As mediators of all interactions between small molecules, peptides, proteins, nucleic acids and membrane lipids, they are responsible for all the regulating functions in life other than those which are simple physical processes. Intermolecular forces account for recognition, binding, catalysis and specificity. Enzyme–substrate, antibody–antigen, hormone–receptor, drug–receptor, drug–serum protein, receptor–neurotransmitter and protein–protein recognition are all examples of highly specific intermolecular interactions.

Although there are some very recent studies exploring protein–antibody interactions[1-4] or the attachment of antiviral agents to viruses[5] at the molecular level, most of the knowledge we have concerning molecular interactions comes from enzyme–inhibitor binding studies. In these cases, as distinct from drug–receptor interactions, the proteins are isolated, their structure described with high accuracy and details of their binding studied at a molecular level. The question of whether this information is fully transferable to drug receptors, which are mainly not free but membrane-bound, and whose mechanisms are not comparable to enzymatic catalysis, remains unanswered. In enzymes, the strongest interactions are with the transition state rather than the substrate, so that the characteristics of drug–receptor binding are really only strictly equivalent to the first step of enzyme binding. Nevertheless, enzymes undoubtedly provide a useful starting point in developing an understanding of drug–receptor interactions.

Even at the enzyme level, many questions remain concerning the nature of the interaction. Is the relationship between enzyme and ligand really like that between lock and key, or is a less precise match sufficient? How many of the functional groups present in a drug molecule contribute to the interaction and how much does each contribute?

Also, unlike intramolecular reactions, where no net change in the number of degrees of freedom of molecular translation or rotation occurs, intermolecular reactions lead to a remarkable change in entropy.[6-11] How does the overall interaction energy overcome this loss in translational, rotational and conformational entropy?

In this chapter, we shall first describe the forces that cause a drug to bind strongly and specifically to its target, next we shall describe methods of calculating drug–receptor interactions and then we shall look at experimental observations of drug–receptor interactions, showing examples of data from X-ray and NMR studies. Using these data, we shall deal with the strength of drug–receptor interactions and discuss the geometries of interactions and structural properties that control intermolecular forces. Finally, we shall use examples of receptor-based drug design to demonstrate the practical application of our knowledge of molecular binding, and indicate likely future developments in the modelling of drug–receptor interactions.

18.8.2 FORCES ASSOCIATED WITH DRUG–RECEPTOR INTERACTIONS

The free energy difference between the unbound ligand or receptor and the drug–receptor complex is most important for the characterization of binding. This free energy change ΔG is related to the equilibrium constant K for the drug–receptor interaction by equation (1). The observed equilibrium constant thus provides a direct measurement of ΔG. (Contributions of different partial interactions to this energy and questions of additivity and cooperativity are discussed by Jencks.[12]) Conversely, estimates of total intermolecular interaction energy due to the various bonds formed between drug and receptor can be used to estimate the likely equilibrium constant for the interaction. For the latter purpose, the free energy change for a drug–receptor interaction is usually expressed by equation (2), where ΔH and ΔS are the changes in enthalpy and entropy, respectively. Of these, the enthalpy term is often approximated by the change in internal energy ΔE, which can be calculated, at least in the gas phase, from classical or quantum chemical calculations. The entropy term, as we shall see below, is more difficult to quantify but nevertheless makes a substantial contribution to many drug–receptor interactions. Both ΔH and ΔS can also be determined experimentally from the temperature dependence of ΔG.

$$\Delta G = -2.303\,RT\log K \tag{1}$$

$$\Delta G = \Delta H - T\Delta S \tag{2}$$

The strongest intermolecular interactions are covalent bonds, which involve the sharing of one or more electron pairs between bonded atoms, and have energies ranging from 40 to 140 kcal mol^{-1} (1 kcal = 4.18 kJ). Even at the lower end of this energy range, covalent bonds are clearly irreversible under physiological conditions ($K \approx 10^{30}$ mol dm^{-3}). Their use in drug binding has therefore traditionally been limited to applications for which irreversible binding to the target receptor sites is desirable (*e.g.* antibacterials, insecticides) although, more recently, covalently bound suicide inhibitors have been developed for a number of pharmacodynamic applications. It is well known that covalent binding occurs, for example in the case of DNA-alkylating agents in cancer therapy, in the reaction of organophosphorus compounds with acetylcholinesterase, and in the action of the α-receptor antagonist phenoxybenzamine. We shall restrict discussion in this chapter to the much more pharmacologically common phenomenon of non-covalent binding.

The majority of drug–receptor interactions consist of a combination of the bond types listed in Table 1, all of which are reversible under physiological conditions. However, although there is general agreement that reversible bonds are primarily responsible for the overall strength of drug–receptor interactions, there is considerable debate on the relative importance of the various reversible bond types. Some authors adhere to the view that polar interactions, particularly ionic bonds, are the dominant forces,[13] whilst others argue that these forces are largely negated by corresponding interactions between free ions and solvent water, making hydrophobic and dispersion forces the deciding factors in the formation of the drug–receptor complex. The various reversible bonds also differ from each other in their directional character and their dependence on the distance between binding groups. The role of each of these bond types in drug–receptor interactions will therefore be considered individually.

Table 1 Major Contributions to Reversible Drug–Receptor Interactions

Interaction type	Example
Electrostatic	
ion–ion	$-\overset{+}{N}H_3\cdots{}^-O_2C-$
ion–dipole	$-\overset{+}{N}H_3\cdots O=C-$
dipole–dipole ⎫ hydrogen bond ⎭	$-N-H\cdots O=C-$
Charge redistribution	
polarization	$-\overset{+}{N}H_3\cdots\overset{\delta-}{O}=\overset{\delta+}{O}$
charge transfer	$-OH\cdots$
Non-polar (van der Waals)	
dispersion ⎫ exchange repulsion ⎭	$-Me\cdots Me-$
Entropy based	
loss of rotational/translational entropy	
hydrophobic interactions	

18.8.2.1 Electrostatic Interactions

The attractive force between opposite charges leads to three main types of non-covalent bonds, namely ion–ion, ion–dipole and dipole–dipole interactions. Of these, ionic bonds are the most important, since many of the functional groups present in drugs are predominantly anionic (*e.g.* caboxyl, sulfonamide) or cationic (*e.g.* aliphatic amino) at physiological pH. These groups interact strongly with oppositely charged groups in the receptor site; in the case of proteins, the receptor groups may be either positively (lysine, arginine) or negatively (aspartate, glutamate) charged amino acid residues, but there are also charged groups in other receptors. The phosphate residues of nucleic acids, for example, are probably responsible for the initial step in the intercalation of the positively charged aminoacridines between the base pairs of DNA. Empirical estimates of the strength of ionic interactions have been obtained from the decrease in binding energies resulting from removal of carboxyl or protonated amino groups from small molecules interacting with proteins.[7,8] These estimates are in the range 4–8 kcal mol^{-1}.

Ion–dipole and dipole–dipole interactions are weaker than ionic bonds, but are even more prevalent, since they apply to any molecule in which electronegativity differences between atoms result in bond, group or molecular dipole moments. The hydration of ions in aqueous solution is one example of ion–dipole interactions, as is the initial interaction between zinc and the carbonyl group of a labile peptide bond in the catalysis of peptide hydrolysis by zinc endopeptidases.

A quantitative estimate of the potential energy of any electrostatic interaction is given by equation (3), where q_i and q_j are two charges separated by a distance of r_{ij} Å in a medium of dielectric constant D. The constant term (331.9) is selected to give a result in kcal mol^{-1}. This equation can be applied equally to ionic interactions, where the charges q_i and q_j are integer values, or to polar interactions, in which the total energy is summed over the contributions calculated from the partial charges on all the individual atoms. These are generally derived from molecular orbital calculations as required, but compilations of calculated partial charges for common molecular fragments (*e.g.* amino acids[14]) are also available. Besides molecular orbital calculations there are also fast empirical methods for the calculation of atomic charges which are based on electronegativity equalization.[15–17] These methods are used, for example, in the molecular modelling packages SYBYL[18] and Chem-X.[19]

$$E = \frac{331.9\, q_i q_j}{D r_{ij}} \tag{3}$$

It follows from equation (3) that the strengths of ionic interactions are inversely proportional to the distance separating the two charges and generally independent of the orientation of the two ions, although it should be noted that steric effects will result in clear orientational preferences. Stricter geometric requirements apply to dipolar interactions, which may be either attractive or repulsive, depending on the orientation of the dipole moments. Also following from this equation is the crucial dependence of this type of interaction on the dielectric constant D of the surrounding medium. It is highly dependent on the microenvironment, and is handled on different levels of accuracy in different approaches to modelling (uniform, distance dependent, cavity dielectric or microdielectric).[20] For the case of ion–dipole interactions, equation (3) reduces to equation (4), where z is the valence of the ion, μ is the dipole moment in debyes, r is the distance between the ion and the centre of the dipole, and θ is the angle between the dipole and the line joining the ion to the centre of the dipole.[21] The corresponding approximate equation for the strength of a dipole–dipole interaction (in kcal mol^{-1}) is equation (5), where r is the distance between the centre of the dipoles and A is a geometric factor depending on the relative orientation of the two dipoles. The interaction between two dipoles aligned head-to-tail is twice as strong ($A = 2$) as that where the dipoles are parallel to each other.

$$E = \frac{69.1\, z\mu}{D r^2} \cos \theta \tag{4}$$

$$E = -14.4 \frac{A \mu_A \mu_B}{D r^3} \tag{5}$$

18.8.2.1.1 *Hydrogen bonds*

The interaction of two or more electronegative atoms linked by a hydrogen atom is called a hydrogen bond. In biological systems the most important hydrogen bonds are those involving the oxygen and nitrogen atoms of the carboxyl, hydroxyl, carbonyl, amino, imino and amido groups, which are responsible for maintaining the tertiary structure of proteins and nucleic acids, as well as the binding of many drugs. Other atoms occasionally involved in drug–receptor hydrogen bonding are sulfur, phosphorus, fluorine and other halogens, all of which can function as either donors or acceptors of protons. Empirical estimates of hydrogen bond strength in aqueous solution are of the order of 1–3 kcal mol^{-1},[22] while the corresponding value derived from the observed heats of sublimation of amide crystals is 5 kcal mol^{-1}.[8]

The strongest hydrogen bonds are formed between groups with the greatest electrostatic character. Thus carboxylates are better acceptors than amides, ketones or unionized carboxyls, whilst substituted ammonium ions are better donors than unsubstituted ammonium ions or trigonal donors.[23] There is thus an increased probability of charged groups being involved in the formation of multiple hydrogen bonds or multicentre hydrogen bonds.

Some hydrogen bonds, such as that in the difluoride ion, can be classified as resonance structures in which both electronegative atoms are covalently linked to the hydrogen atom. In most cases, however, only one of the electronegative atoms is covalently bound to the hydrogen, and the overall effect is best described as an electrostatic interaction with a small, but significant, charge-transfer component.[24] The approximate strength of individual hydrogen bonds can therefore be calculated from the dipole moments of the interacting groups using equation (5), or from equation (3) using partial charges. Studies comparing empirical energy calculations and experimental structures[25] on amides and carboxylic acids showed also that the hydrogen-bonding interaction in crystals is reasonably well accounted for by electrostatic and van der Waals forces, and that no explicit additional term is needed for their calculation.

18.8.2.2 Charge Redistribution Effects

The formation of a drug–receptor complex is usually accompanied by some charge redistribution, either within the drug or the receptor (polarization), or between the two molecules (charge transfer). In either case, the resulting interaction is always attractive.

18.8.2.2.1 Induced polarization

Interactions due to polarization occur when a temporary dipole is induced in a group by the field of a permanently charged group, either ion or dipole. The consequent polarization contribution to the total energy is given in kcal mol^{-1} by equation (6) for the ion-induced dipole interaction and by equation (7) for a dipole-induced interaction. In both cases, α is the molecular polarizability in Å^3, obtained from molecular refraction measurements, and A in equation (7) has the same significance as in equation (5). Although of almost universal occurrence in drug–receptor interactions, polarization terms do not usually contribute much to the total energy and are frequently omitted from semi-empirical energy calculations. One method of empirical interaction energy calculations[26] considers bond polarizabilities and will be described later (Section 18.8.3).

$$E = \frac{-166\alpha z^2}{D^2 r^4} \qquad (6)$$

$$E = \frac{-14.4\,A\alpha\mu^2}{D^2 r^6} \qquad (7)$$

18.8.2.2.2 Charge-transfer interactions

These interactions occur when an electron donor makes sufficiently close contact with an electron acceptor to allow electron transfer from a high energy occupied molecular orbital of the donor to a low energy empty molecular orbital of the acceptor. This energy has an exponential dependence on the atom–atom distance and a directional dependence related to the orientation of the overlap of the orbitals. Such a transfer is most likely to occur between large aromatic molecules and could contribute, for example, to the interaction of intercalating agents with DNA. Charge transfer may, however, also make a significant contribution to other interactions.[10]

18.8.2.3 Non-polar Interactions

While electrostatic and charge redistribution interactions generally involve polar molecules, there are also strong interactions between non-polar molecules, particularly at short intermolecular distances.

18.8.2.3.1 *Dispersion forces*

Dispersion or London forces are the universal attractive forces between atoms that hold non-polar molecules together in the liquid phase. Their occurrence is due to the fact that even non-polar molecules will, at any given instant, be likely to possess a finite dipole moment as a result of movement of electrons around the nuclei. Such fluctuating dipoles tend to induce opposite dipoles in adjacent molecules, thus resulting in a net attractive force. Although individually weak, the total contribution to binding from dispersion forces can be very significant if there is a close fit between drug and receptor. A quantitative estimate of the energy of this interaction (in $kcal\,mol^{-1}$) is given by equation (8), where I_1 and I_2 are the ionization potentials (in $kcal\,mol^{-1}$) of the interacting groups

$$E = -1.5 \frac{\alpha_1 \alpha_2}{r^6} \frac{I_1 I_2}{I_1 + I_2} \tag{8}$$

18.8.2.3.2 *Exchange repulsion*

Balancing the various attractive forces discussed above are the short range repulsive forces which result from overlap of the electron clouds of any two molecules. The potential energy associated with these interactions increases exponentially at short internuclear separations (inverse 12th-power law) and thus determines the minimum and most favourable non-bonded separation between atoms. For non-polar molecules this distance is defined by the Lennard-Jones 6–12 potential (equation 9), which gives rise to the general energy curve for non-polar interactions illustrated in Figure 1. A similar curve is obtained from the alternative Buckingham (6-exp) potential of the generalized form given by equation (10), for which a typical set of coefficients is given by Giglio.[27] The equilibrium distance d_e from Figure 1 can be determined from crystal data, and represents the sum of the van der Waals radii of the two atoms involved.

$$E = \frac{A}{r^{12}} - \frac{C}{r^6} \tag{9}$$

$$E = \frac{A\exp(-Br)}{r^d} - \frac{C}{r^6} \tag{10}$$

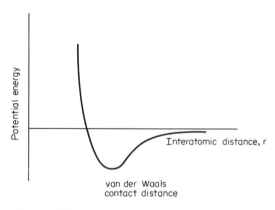

Figure 1   Distance dependence of the interaction energy

18.8.2.4   Entropic Contributions

In addition to the largely enthalpic components of the energy of interaction discussed above, there are two major entropic contributions. These are the overall entropy loss that accompanies formation of the drug–receptor complex, and the entropy gain associated with hydrophobic bonding.

18.8.2.4.1 *Entropy loss*

The formation of a drug–receptor complex is accompanied by the replacement of the three rotational and three translational degrees of freedom of the drug molecule by six vibrational degrees of freedom in the complex. The extent of this change depends greatly on the relative 'tightness' of the resulting complex. Page[7] has estimated values ranging from $10 \, \text{cal} \, \text{K}^{-1} \, \text{mol}^{-1}$ (equivalent to $3 \, \text{kcal} \, \text{mol}^{-1}$ at $310 \, \text{K}$) for a loose drug–receptor interaction to a maximum of $50 \, \text{cal} \, \text{K}^{-1} \, \text{mol}^{-1}$ ($15 \, \text{kcal} \, \text{mol}^{-1}$) for an extremely tightly bound complex.

In addition to this loss of rotational and translational freedom, there is a further entropy loss due to the conformational restriction which accompanies binding of flexible drug molecules. This is variously estimated at $3–10 \, \text{cal} \, \text{K}^{-1} \, \text{mol}^{-1}$ ($1–3 \, \text{kcal} \, \text{mol}^{-1}$) for a bond with substantial conformational degeneracy, but again clearly depends on how tight a complex is formed between the drug and the receptor. In the case of rigid analogues there is no such loss of conformational entropy, and it is this factor which provides the free energy advantage of rigid analogues relative to more flexible drugs.

18.8.2.4.2 *Hydrophobic interactions*

This contribution is particularly important in the case of interactions between non-polar molecules, and arises for the following reason. When a non-polar molecule is placed in water, stronger water–water interactions are formed around the solute molecule to compensate for the weaker interactions between solute and water. This results in an increasingly ordered arrangement of water molecules around the solute and thus a negative entropy of dissolution. The decrease in entropy is roughly proportional to the non-polar surface area of the molecule. The association of two such non-polar molecules in water reduces the total non-polar surface area exposed to the solvent, thus reducing the amount of structured water, and therefore providing a favourable entropy of association. Empirical data suggest that the free energy contribution due to hydrophobic bonding is approximately $0.7 \, \text{kcal} \, \text{mol}^{-1}$ per methylene group and $2 \, \text{kcal} \, \text{mol}^{-1}$ for a benzene ring.[28] Although individually small, the total contribution of hydrophobic bonds to drug–receptor interactions is substantial. Indeed, many authors see hydrophobic interactions as the most important driving force for non-covalent intermolecular interactions in aqueous solution.[29]

18.8.3 CALCULATIONS OF DRUG–RECEPTOR INTERACTIONS

Having looked at the different forces involved in molecular binding, we shall next discuss approaches to quantifying intermolecular interactions by means of interaction energy calculations. Inevitably, a lot of simplifying assumptions are necessary to describe relatively large systems, and the extent of these suppositions is more or less dictated by the size of the molecular components to be included. Sometimes a procedure is chosen which combines more accurate sophisticated calculations for small fragments or active sites with evaluation of larger fragments by more empirical methods. On the other hand, quantum mechanical and other more complicated calculations can be used to improve simple formulae for the expression of intermolecular interaction energies.

Following the classic text of Hirschfelder,[30] a good overview of the field of intermolecular force calculations can be gained from two books edited by Pullman.[31, 32] Most theoretical studies of non-covalent interactions start with a comparison of theoretical and experimental data on very small molecule dimer combinations or complexes, attempt a parametrization based on these small molecules, compare results of calculations on different levels of accuracy and seek a suitable method for tackling more extensive problems. Only some of the numerous approaches, which range from *ab initio* supermolecule calculations, *via* semi-empirical methods and energy decomposition studies, to empirical potential functions for intermolecular interactions, can be mentioned here. Furthermore, we restrict ourselves to the application of these methods to calculations of drug–receptor interactions. The methods themselves are dealt with in detail in Chapters 18.1 and 18.2 of this volume. A brief description of the *pro* and *contra* of quantum mechanics and molecular mechanics in drug design is given by Boyd.[33]

18.8.3.1 Molecular Orbital Calculations

Kollman[34] in particular has discussed *ab initio* supermolecule calculations on small-molecule interactions of different strengths, and shown them to be in reasonable agreement with the experimental data. Among other things, he points out that electrostatic energy is a key component in intermolecular interactions and that its directionality and relative magnitude are good predictors of the strength and directionality of non-covalent binding. He also expresses the opinion that hydrogen bonding, van der Waals and charge-transfer interactions are, to a first-order approximation, electrostatic interactions.

Morokuma,[24] on the other hand, claims that at relatively small separations (*e.g.* for hydrogen bonding), electrostatic, charge-transfer and polarization interactions are independent attractive contributions, competing with the exchange repulsion. Charge transfer in particular may be essential for the stabilization of weaker hydrogen bonds.

Kollman[35] has further extended his analyses to a broad comparison of SCF calculations, Morokuma's component analysis,[24] electrostatic potentials and simple model calculations on weak, moderate and strong complexes. He demonstrates that basic energetic, structural and spectroscopic properties of the intermolecular complex formation can be described by focussing on the features of the fragments, and concludes that, while electrostatic potentials are useful for a simple preliminary treatment, inclusion of the exchange repulsion term is necessary to account for the distance dependence of the intermolecular potential. He also suggests that difficulties in finding correct geometries in 'pure' electrostatic models may be due to the crudeness of the electrostatic models, *e.g.* a point quadrupole model.

Other, more recent, studies show that interaction approaches can be based on far less elaborate calculations than the *ab initio* method, making possible calculations even on large molecules. As an example, a comparison of *ab initio* and semi-empirical MO-IEHT results for multipole moments, hydrogen-bonding systems and ion pairs is described by Rein.[36] Here, hydrogen bonding is equally reproduced with an *ab initio* supermolecule approach on the one hand and, on the other, interaction energy calculations based on molecular multipole moments calculated from IEHT or *ab initio* wavefunctions.

The advantages of semi-empirical functions, combined with a separate exchange repulsion and dispersion energy contribution, over small basis set *ab initio* calculations are described in another study,[37] where it is demonstrated that conventional *ab initio* Gaussian programs, but not semi-empirical INDO calculations, may lead to an underestimation of exchange repulsion for the distance ranges we are concerned with in intermolecular interactions. This results in an overestimation of binding energies and intermolecular distances that are too short.

An alternative approach based on *ab initio* calculations[38] reduces the size of the problem by mimicking the binding site with a set of partial charges while treating the substrate with a full *ab initio* calculation. Model interaction calculations prove this approximation useful at distances around the van der Waals contact radii, and in cases where partial charges are involved.

18.8.3.2 Classical Potential Energy Calculations

The total energy of interaction can also be calculated reasonably accurately with the classical equations given for the individual forces associated with drug–receptor interactions in Section 18.8.2. In most applications, this takes the form of a single equation accounting for electrostatic, dispersion and repulsion terms.

More elaborate empirical potential functions describe the total energy of an interaction complex as the sum of different partial energy contributions due to atomic pair interactions, plus atom–bond and bond–bond contributions. Parameterization of the single parts of the equation, inclusion of different partial energy terms, choice of charge representation, the dielectric constant used, *etc.*, are the main differences between many of the approaches used. Parameters are taken from crystal, spectroscopic and other experimental data on standard compounds, so that the suitability of available parameter sets for more unusual chemical characteristics, which are possible in new drug compounds, is one limitation of these methods.

The value of any empirical method depends primarily on whether it is to be applied for quantitative (*e.g.* calculating binding energies) or more qualitative (*e.g.* calculating binding geometries) purposes. One quantitative approach that is used relatively frequently[36, 39] evaluates the interaction energy as the sum of three long-range contributions (electrostatic, polarization and dispersion) and the short-range repulsive contribution.[26] Normal long-range interaction formulae have to be used with modifications at close intermolecular distances.[40] The $1/r^6$ formula (see equation 8), for example, overestimates the dispersion term at short distances. Equations with which the single terms are calculated are equations (11) and (12) for the electrostatic and polarization contributions, respectively

$$E_{el} = \sum_{i=1}^{N_1} \sum_{j=1}^{N_2} \frac{q_i q_j}{R_{ij}} \tag{11}$$

$$E_{pol} = -\frac{1}{2} \sum_{k=1}^{B_1} E_k \mathbf{A}_k E_k - \frac{1}{2} \sum_{l=1}^{B_2} E_l \mathbf{A}_l E_l \tag{12}$$

where N_i and B_i are the number of atoms and bonds in molecule i, q_i is the charge at nucleus i and R_{ij} the distance between the nuclei i and j. E_k is the electric field at the centre of bond k in molecule 1, due to the monopole charges q_j of molecule 2, given by

$$E_k = \sum_{j=1}^{N_2} q_j \mathbf{IR}_{jk} R_{jk}^{-3} \tag{13}$$

with R_{jk} being the distance from nucleus j to the midpoint of bond k, and $\mathbf{IR}_{jk}$ the vector of magnitude R_{jk} pointing from the centre j to the bond midpoint, k. $\mathbf{A}_k$ is the polarizability tensor for bond k. Equation (12) accordingly describes the polarization of molecule 1 by molecule 2, and *vice versa*. For the dispersion contribution

$$E_{dis} = -\frac{1}{4} \times \frac{I_1 I_2}{I_1 + I_2} \sum_{k=1}^{B_1} \sum_{l=1}^{B_2} R_{kl}^{-6} \, Tr\,(\mathbf{T}_{kl} \mathbf{A}_k \mathbf{T}_{kl} \mathbf{A}_l) \tag{14}$$

with I_1 and I_2 given by the ionization potentials of the two molecules, R_{kl} as the distance between the midpoints of bonds k and l and the tensor $\mathbf{T}_{kl}$

$$\mathbf{T}_{kl} = 3 \frac{\mathbf{IR}_{kl} \times \mathbf{IR}_{kl}}{R_{kl}^2} - 1 \tag{15}$$

Finally, the repulsion is derived empirically from crystal energies of hydrocarbons by Kitaygorodski, considering the van der Waals radii V.

$$E_{rep} = 30\,000 \sum_{i=1}^{N_1} \sum_{j=1}^{N_2} \exp\left[-5.5 R_{ij} (V_i V_j)^{-\frac{1}{2}} \right] \tag{16}$$

We have written here the monopole version of the basic equations, but the same principles are used also with charge distributions represented by sets of multipoles. The above equations have also been expanded by inclusion of an explicit charge-transfer term,[41] and various refinements[39,40] to the short-range repulsion term.

Rein[42,43] did the pioneering studies applying these interaction energy calculations to models involving nucleic acids, whilst Höltje and Kier[44] developed their method of receptor mapping by model interaction calculations based on Claverie and Rein's monopole and multipole bond polarizability method.[45-47]

18.8.3.3 Solvent Effects

Strictly speaking, the preceding discussion relates mainly to the calculation of the total energy of interaction in the gas phase. In solution, however, the process is complicated considerably by uncertainty as to the best choice of the dielectric constant D. Commonly used values range from unity, the gas phase value, through four, for hydrophobic environments like the interior of a protein molecule, to 28, for interactions occurring near the surface of a protein, and finally to 80, the bulk-phase value for water.[48] Some authors have sought to partially accommodate these extremes by using a distance-dependent dielectric of which the simplest form is $D = r_{ij}$, which is defined as the dimensionless value equal to the distance separating the two charges r_{ij} in angstroms.[14]

The effect of this uncertainty in the choice of dielectric constant on the calculated potential energy of interaction between two ionic groups is, as illustrated in Table 2, enormous. Although the values in the table should be adjusted to compensate for the entropy changes associated with formation of the ion pair before they are compared with empirical free energy changes, it seems that the best choice of dielectric constant for drug–receptor interactions may be either $D = 4$ or the distance-dependent $D = r_{ij}$.

Table 2 Dependence of the Potential Energy (kcal mol^{-1})^a of Ion–Ion
Interactions on the Choice of Dielectric Constant D

Distance between ions r_{ij} (Å)	Dielectric constant D					
	1	*4*	r_{ij}	$6r_{ij} - 11$	*28*	*80*
3	110	28	37	16	3.9	1.4
4	83	21	21	6.4	3.0	1.0
5	66	17	13	3.5	2.4	0.8
6	55	14	9.2	2.2	2.0	0.7
7	47	12	6.8	1.5	1.7	0.6
8	41	10	5.2	1.1	1.5	0.5
9	37	9	4.1	0.9	1.3	0.45

^{a} 1 kcal = 4.18 kJ.

The situation in solution is further complicated by the direct role of solvent in electrostatic interactions. In the case of an ionic bond in aqueous solution, for example, the interaction is not simply a matter of

$$\text{drug} + \text{receptor} \rightleftharpoons \text{drug–receptor}$$

but rather

$$\text{drug}(H_2O)_i + \text{receptor}(H_2O)_j \rightleftharpoons \text{drug–receptor}(H_2O)_k + (H_2O)_{i+j-k}$$

An adequate calculation of the potential energy for this interaction must therefore include an extensive bath of water molecules around the interacting groups. Under these circumstances, because all the water molecules are explicitly included, the appropriate choice of dielectric constant is the gas phase value of unity. It is clear, however, that the scale of such calculations will almost invariably be prohibitive.

For all these reasons, calculations of drug–receptor interaction energies in solution are at best qualitative. It is impossible, for instance, to accurately predict free energies of association, even for simple organic molecules in aqueous solution, let alone interactions between drugs and their receptors. Conversely, the observed binding energy tells us very little about the three-dimensional quality of the interaction between drug and receptor. Is the match like that of hand and glove, or more like that of square peg and round hole? To answer these questions, we need some way of visually observing the interactions between drugs and their receptors.

18.8.3.4 Molecular Shape, Matching and Docking

An alternative to computational approaches to the energetics of molecular interactions is a completely different procedure, shape matching and docking, which attempts to find geometrically realistic alignments of macromolecule and ligand by evaluating their mutual orientation in terms of steric overlap. This can be done at different levels of sophistication, and was initially a tool for building up, by steric fit, a reasonable starting conformation for further energy minimization *via* molecular mechanics and computer graphics studies.[49] However, its development has made it an almost independent method. By avoiding repulsive forces through short atom–atom distances, perhaps combined with the possibility of observing electrostatic interactions using colour-coded representations of the electrostatic potentials,[50] it now provides a means of indicating inter-molecular specificity and a starting point for the design of new compounds. Even approximate binding site surfaces can be postulated from molecular surfaces of ligands with conformationally restricted structures.

The simplest method of matching complementary molecular shapes is the docking of rigid ligands to rigid receptors.[51] Extension of this method includes some degree of flexibility in the ligand molecule, which is done by docking rigid molecule fragments, screening them for possible sets of rejoinable fragments, joining the fragments and, finally, minimizing the full ligand binding.[52] In a recent paper another approach to docking, using a constrained optimization known as the ellipsoid algorithm, is described.[53] The best method for modelling depends on available information about the ligand, the nature of the interaction or the binding site.

18.8.3.5 Flexible Interactions

So far we have described interactions between two rigid partners which have to be changed systematically in their mutual orientation in order to find the energetically most preferable complex geometry. We have considered neither conformational changes in the course of the interaction nor the role played by the solvent molecules, which could saturate binding sites, alter the complex conformation, change the probable low-energy conformations of the substrate, *etc.* Including conformational flexibility, the overall energy of complexes can be decomposed into two parts. On the one hand there is the energy used for the conformational adjustment of drug or receptor (*i.e.* the change in conformational energy between the conformations of the fragments in their isolated form and those adopted in the complex), and on the other is the actual energy of interaction. Here again, a different number of terms can be included in both parts of the calculation, as in the following equations used for theoretical calculations on enzyme–substrate complexes by Scheraga,[54] which give

$$E = \sum_{i \neq j} \varepsilon_{ij} \left[\left(\frac{r_{ij}^0}{r_{ij}} \right)^{12} - 2 \left(\frac{r_{ij}^0}{r_{ij}} \right)^{M} \right] + \sum_{i \neq j} \frac{q_i q_j}{D r_{ij}} + \sum_k \frac{A_k}{2} \ (1 \pm \cos n\theta_k) \qquad (17)$$

the energy as the sum of non-bonded, hydrogen-bonding, electrostatic and torsional energy terms. ε_{ij} and r_{ij}^0 in this equation are the potential well depth and the position of the energy minimum; $M = 6$ for non-bonded energy, 10 for hydrogen-bonded energy; A_k is the barrier height for rotation around the kth bond, θ_k is the dihedral angle and n is the n-fold degeneracy of the torsional potential. The conformational energy (E_{tot}) of an enzyme–substrate complex may be expressed by the sum

$$E_{tot} = E_{sub} + E_{enz} + E_{int} \qquad (18)$$

where E_{sub} is the conformational energy of the substrate, E_{enz} of the enzyme and E_{int} for the enzyme–substrate interaction.

Considering the effect of solvent, it is often assumed from crystal data that most water molecules are located on the enzyme surface rather than at the active site and therefore do not significantly change binding characteristics. But besides the solvent shell due to non-specific hydration there is also the case of specific hydration, in which water competes for a hydrogen-bonding site in the macromolecule–ligand interaction. Inclusion of specific and non-specific hydration in empirical conformational energy computations on peptides has been described by Hodes.[55]

Returning to molecular mechanics studies of molecular interactions, a few recent examples of different types can be mentioned. Probably the most frequently used method to describe larger interaction complexes in a way that allows a certain degree of flexibility during energy refinement is the molecular mechanics software package AMBER.[56] This program evaluates total energies of a system in terms of bond stretching, angle bending, torsional and non-bonded intra- and inter-molecular interactions, as in equation (19). It minimizes this energy with respect to the positions of all atoms. Because of the minimization algorithms used, these refinements inevitably lead only to local energy minima. On the other hand, systematic conformational analysis options like SEARCH in the SYBYL program package,[18] again combined with an energy minimization procedure, try to avoid getting trapped in local energy minima.

$$E_{total} = \sum_{bonds} K_r (r - r_{eq})^2 + \sum_{angles} K_\theta (\theta - \theta_{eq})^2$$

$$+ \sum_{dihedrals} \frac{V_n}{2} [1 + \cos(n\phi - \gamma)] + \sum_{i < j} \left[\frac{A_{ij}}{R_{ij}^{12}} - \frac{B_{ij}}{R_{ij}^6} + \frac{q_i q_j}{\varepsilon R_{ij}} \right]$$

$$+ \sum_{H \ bonds} \left[\frac{C_{ij}}{R_{ij}^{12}} - \frac{D_{ij}}{R_{ij}^{10}} \right] \qquad (19)$$

Wipff *et al.*[57] studied the interaction of L- and D-*N*-acetyltryptophanamide with α-chymotrypsin in order to assess the ability of such methods to model enzyme–ligand interactions, and to understand the stereoselectivity of the catalyzed reaction. Starting with *X*-ray crystal structures and considering different models of the active site, modelling both the intermediate tetrahedral covalent enzyme–substrate complex and the initial non-covalent Michaelis complex, the AMBER calculations enabled them to explain the stereoselective activity. This was accomplished by different positioning of the enantiomers with respect to the functional groups involved in catalysis. Different stabilization of the transition states is the determining factor here, whilst the initial binding seems less important. The ability of the active site to respond to ligand binding through small conformational changes was also shown to take part in stereoselectivity. This example shows the great advantages of molecular mechanics in molecular modelling compared to the rigid picture from X-ray analyses.

The same enzyme was studied in a different interaction complex and using a different MM calculation procedure and software, by Naruto *et al.*[58] in an investigation of the mechanism of irreversible inhibition by suicide substrates. As in the previous study, the size of the problem is reduced to surrounding residues within a certain radius. In contrast to that study, a systematic conformational search first scanned all possible torsion angles around rotatable bonds and then checked the resulting van der Waals contact distances. Thus families of allowable sets of torsion angles were obtained and taken as starting points for minimizations, which were done with constraints fixing special distances between substrate and active site that are required by the chemical reaction. The observed bioactivities could be well rationalized. Estimates of van der Waals interaction energies correlate with the inactivation binding constants, and energy differences between the alkylation complexes and the suicide compounds parallel the inactivation rate constants.

Other recent studies using AMBER to model covalent and non-covalent binding of pyrrolo[1,4]benzodiazepine antitumour antibiotics to DNA segments give more information about its possibilities and limitations.[59, 60] The modelling results are in good agreement with NMR studies, but it is less certain how useful the results obtained are in designing new analogues, and what the differences in binding energy really mean quantitatively. Finally, Kollman has developed an interesting approach to describing pathways through the protein interior using basically the same computational methods.[61]

18.8.3.6   Calculating Ideal Ligand Positions

A computational approach to predicting non-covalent interactions between a molecule of known three-dimensional structure and a small probe showing the characteristics of a certain chemical group, either of the ligand or of the binding site, is Goodford's GRID program.[62, 63] It reads in the coordinates of the structure under study and systematically checks the energy of interaction with a probe group of defined chemical properties as the probe is moved around the 'grid'. Contour surfaces obtained with this procedure could be a valuable aid in drug design. The GRID program uses an extended atom representation and calculates the interaction energy between the probe at position *xyz* and the target molecule as the sum of pairwise interactions calculated in a Lennard-Jones function, an electrostatic function and a hydrogen bond function. In calculating the electrostatic energy for the protein phase, a dielectric depending on the number of neighbouring atoms within 4 Å is chosen which considers the depth of the probe in the protein phase. The hydrogen bond function included is a direction-dependent 6–8 function.

Goodford[62] has shown, in different examples, that the positions of water molecules or single functional groups of ligands in their binding to a protein could be well predicted by this method. We tried out the opposite case, taking small molecules and looking for favourable positions to locate hypothetical receptor-binding sites around them. However, this procedure gives less clear results, which seem to be predetermined by the hydrogen-bonding energy whilst the electrostatic term is always a minor influence. A careful parametrization for the respective conditions seems unavoidable.

18.8.4   EXPERIMENTAL OBSERVATIONS OF DRUG–RECEPTOR INTERACTIONS

18.8.4.1   X-Ray Crystallography

To date, X-ray crystallography is the only available method for obtaining accurate information about the three-dimensional structure of small molecules, macromolecules and their interaction complexes. Combined with computer graphics, quantum mechanics, molecular mechanics and molecular dynamics, X-ray structures of complexes are a promising starting point for rational design of new compounds.[64] Stezowski[65] gives a good synopsis of recent small molecule, drug–DNA, protein–DNA and protein–substrate interaction studies.

Although these X-ray data provide an enormous information resource, some general caveats must be considered when evaluating X-ray data,[66, 67] and despite the rapid progress in X-ray crystallography, the determination of new protein crystal structures remains a much more complex process than that for small molecules. In addition to the data for the protein itself, complete diffraction patterns for one or more isomorphous heavy atom derivatives are usually needed in order to determine the phase of scattered radiation and thus calculate an electron density map. The interpretation of the map is then an extended trial-and-error process beginning with a visual or computer search of the map for a continuous path of electron density corresponding to the peptide backbone, and proceeding through successive cycles of refinement until the difference between the observed and calculated electron densities reaches a minimum. Throughout this process a standard set of bond lengths and bond angles for the peptide chain is assumed, and extensive use made of any available sequence data. Indeed, although it is possible to identify many amino acid side-chains directly from an electron density map, the primary structure of a protein has not yet been determined from X-ray crystallography alone.

A major difference between the crystal structures of proteins and those of small molecules is the accuracy of the atomic coordinates. In small-molecule crystal structures errors are commonly less than one hundredth of an angstrom, whereas in proteins they are usually at least an order of magnitude higher. One reason for this is the relative flexibility of protein molecules, some parts of which may be able to occupy several different positions, even in the crystal. Since the structure obtained from X-ray crystallography is averaged over both time (several hours) and space (several thousands of molecules), these regions appear as blurs in the electron density map.

An idea of the accuracy of the X-ray data can be obtained from the resolution of the electron density map. In general, a resolution of 6 Å is sufficient to provide an overall picture of the macromolecular shape, but higher resolution (3–4 Å) is required to follow the path of the peptide chain. At 2 Å resolution the amino acid side-chains are visible, and atomic coordinates obtained by refining the data will generally give an average error of around 0.2 Å. At still higher resolution individual atoms are visible; the 1.2 Å structure of insulin[68] locates most atoms, including many of the hydrogens in both insulin and its bound water. The eventual limit to the resolution of X-ray analysis is determined by the degree of perfection of the crystal and its structural similarity to its heavy atom derivatives; for most proteins this limit is not much better than 2 Å.

A potential problem with the use of protein crystal structures to study drug–receptor interactions is that conditions during crystallization or in the crystal itself are far from physiological. There are, however, two major lines of evidence which suggest that there is little difference between the two.[69]

(i) The contacts between protein molecules in the crystal are so few and so tenuous that they are unlikely to substantially modify the three-dimensional structure of the protein as it crystallizes. Thus it is observed that similar proteins crystallized under widely different conditions form very similar tertiary structures (root-mean-square deviations between corresponding atoms of 1.8 Å or less).

(ii) Studies of catalytic activity show that many enzymes retain their catalytic activity in the crystal. The observation of significant differences in enzyme conformation in response to the presence or absence of substrates, inhibitors or coenzymes also tends to support rather than deny the relevance of the crystal structure to the solution environment.

A more general problem is the lack of suitable crystal structures for most likely target proteins. This has led to the increasingly common use of the crystal structures of related proteins as an aid to receptor-based drug design. Thus, for example, amino acid sequences have been used in conjunction with the crystal structures of other acid proteases to deduce three-dimensional structures of both mouse submaxillary gland renin[70, 71] and human renin, for which no crystal structures have yet been published. The observed similarities between the crystal structures of functionally related proteins, which include clear examples of both convergent (thermolysin/carboxypeptidase A) and divergent (dehydrogenases) evolution, generally support this development.

Some examples are given here of the use of high resolution X-ray crystallography to examine a protein–drug complex which is stable enough to obtain a picture of binding characteristics. A good report on the ideas and aims in crystallographic binding studies is given by Knox[72] using the example of β-lactam target enzymes. The most intensively studied example of dihydrofolate reductase complexes, where X-ray methods in fact led to the design of new DHFR inhibitors, will be discussed later, as will thyroid–hormone binding, which is described in detail in Chapter 20.3. What will be reviewed here are binding studies on hemoglobin, one of the few examples deserving the name 'drug–receptor binding study', as no chemical reaction between the two sides is involved. (Other complexes of this type analyzed so far are myoglobin and hemoglobin with CH_2Cl_2, carbonic anhydrase–sulfanilamide binding and the well-known dihydrofolate reductase–methotrexate interaction.) However, this drug-binding protein differs in its characteristics from the majority of receptors, which are probably membrane proteins and subject to interactions with neighbouring molecules.

Hemoglobin is an allosteric protein tetramer whose four chains (two α and two β subunits) are differently arranged in two alternative structures: the deoxy or T-form with low oxygen affinity and the oxy or R-structure with high oxygen affinity. The different binding characteristic of the two forms is due to additional bonds between the four subunits in the form of salt bridges between carboxylate groups and cationic amino, guanidinium or imidazolium groups. The equilibrium between these two forms is controlled by the partial pressure of oxygen and the concentration of the natural allosteric effector, 2,3-diphosphoglycerate (DPG). DPG stabilizes the deoxy form of hemoglobin, decreases the oxygen affinity and thereby enhances oxygen release.

In the case of the genetic disorder sickle-cell anaemia, mutation causes the replacement of a single glutamate by valine in each β chain. This amino acid is positioned at the outside of the protein. The single amino acid change to a neutral hydrophobic residue in hemoglobin S results in decreased solubility in the deoxygenated state, which dominates at low oxygen tension. This leads to precipitation of hemoglobin S, formation of fibrous structures and, finally, the characteristic sickle shape of erythrocytes. Approaches to the treatment of sickle-cell disease might be, for example, to antagonize the action of DPG and decrease the concentration of the insoluble deoxy form, or to inhibit the gelling of sickle-cell hemoglobin.

Based on crystal structures of deoxyhemoglobin[73] incorporating DPG, Beddell et al.[74] have designed dibenzyldialdehyde compounds which should stabilize the deoxy conformation in the same way as does DPG, and which were indeed found to lower oxygen affinity. This binding mode is supported by a study of the binding affinities of DPG and designed compounds at hemoglobin variants with changes in amino acid residues around the DPG binding site,[75] although no confirming crystal structure yet exists. Compounds were later synthesized by the same group[76] which do just the opposite; they stabilize the oxygenated form, increasing oxygen affinity, and have an antisickling effect. The structural basis of this activity is less certain.

The binding mode of several potential antisickling agents is shown in crystallographic studies of the compounds' complexes with hemoglobin. Compounds developed by Walder et al.[77] and directed specifically to the DPG binding site caused cross-linking between two β chains. Besides changing the oxygen binding properties, these compounds directly inhibit the polymerization of deoxyhemoglobin S. The work of Abraham[78] and Perutz et al.[79] reveals quite different sites of binding of antisickling agents. Depending on the available van der Waals space and, within it, on interactions of a wide range of polar (from strong hydrogen bonding of charged groups to weak forces between aromatic quadrupoles) and non-polar interactions, their stereochemistry of binding is determined individually. All the compounds studied bind to the central cavity at sites between the α chains, far from the site for DPG between the two β chain termini. Binding of the drugs induces small distortions in tertiary and quaternary structure, which have no influence on the polymerization of deoxyhemoglobin S, in a manner which has not yet been clearly explained.

The most recent approach to the design of antisickling agents is described by Sheh et al.,[80] who designed a peptide to fit and bind to the acceptor site and thus block the features that enable polymer formation. The development of their structure is based on the three-dimensional model of deoxyhemoglobin S and aims to mimic the region around the mutation site by use of a cyclic tetrapeptide. Unfortunately, their new compounds show only slight or no activity, and it is difficult to ascertain which site the tetrapeptide really binds to. However, the idea demonstrates that much can be done with well-refined three-dimensional structures where the points of interaction are known.

18.8.4.2 NMR Spectroscopy

Various data on macromolecules in solution, their structure and function can be obtained by NMR spectroscopy. Here, structural studies can be performed under conditions close to physiological, providing very useful information even if the determination of an entire large three-dimensional structure is not possible. The observation of complexes in solution with measurable kinetic and thermodynamic properties allows the evaluation of the importance of certain complex structures for catalysis. Environmental effects of special functional groups, broadening of lines in the course of substrate binding, flexibility of segments (*e.g.* the aromatic ring), dependence of the structure on pH or pK_A, binding, *etc.* can be studied by NMR techniques. In this sense, X-ray and NMR studies really complement each other to provide more detailed pictures of binding mechanisms and mutual adaptation of binding site and ligand.[81]

A good example is given by Roberts[82,83] on ligand binding to dihydrofolate reductase. He shows that lock-and-key concepts and simple structure–activity assumptions can be quite inappropriate when, for example, exchange of a single ligand atom can substantially alter the mode of binding and conformationally different enzyme–ligand complexes, built from the same compounds, exist. He describes in detail[83] what kind of information can be gathered from chemical exchange processes studied by NMR. One experimental observation, which should be remembered in connection with our attempts, described below, to find out more about the preferred environment of special functional groups, is the flipping or rotation of aromatic rings studied using changes in ^{19}F NMR chemical shifts.[84]

18.8.4.3 Strengths of Component Interactions

18.8.4.3.1 Anchor principle

The strength of the bond formed by a specific functional group in a given drug can be estimated by comparing the binding energies for pairs of compounds which differ only in the presence or absence of the specific functional group. This approach has been described by Page,[7] who refers to it as the 'anchor principle'. It has the major advantage that the difference in binding of a drug molecule with or without the particular functional group incorporates only the factors associated with that group, while excluding the loss of overall rotational and translational entropy associated with the remainder of the drug molecule (the anchor). The same principle applies to the addition or removal of a single functional group from the receptor, as achieved by single-point mutagenesis.

18.8.4.3.2 Active-site mutagenesis

Some very recent studies, mainly based on protein engineering,[85,86] give more detailed information about the strength of individual functional group interactions. They are based on the idea that it should be useful to vary either the one or the other side quite systematically to get more knowledge about the importance of single contributions to the overall interaction.

Fersht has done most extensive work in this direction on aminoacyl-tRNA synthetases.[87,88] He is able to measure approximate energy values corresponding to the binding of special groups, but he also shows that catalysis does not result solely from a few specific catalytic residues. It is also the result of contributions of small binding energies with less obviously interacting groups, meaning that even if there is a possibility of quantification of the effect of special interactions by comparative kinetic studies, the whole binding process, including conformational changes, varying solvent interactions and contribution of minor interactions, has to be kept in mind. Examples of protein engineering of hydrogen bonding[89,90] and electrostatic interactions[91] follow.

Calculations of energies involved in hydrogen bonding are normally difficult because of competition with water for hydrogen-bonding sites in aqueous solution. The study using site-specific mutagenesis at tyrosyl-tRNA synthetase[89] has some advantages compared to other investigations, in that there are a well-defined number of hydrogen bonds which are known from X-ray structures. Structural changes in the mutants can be detected quite well and, finally, contributions of single side-chain changes can be calculated reliably from observed k_{cat}/K_M values. Hydrogen bonds between the enzyme and an uncharged substrate group are proved here to provide 0.5–1.5 kcal mol^{-1} or a factor of 2.5 to 15 towards specificity, while hydrogen bonds to a charged group contribute by 3.5 to 4.5 kcal mol^{-1}, corresponding to a factor of 1000 in specificity.

Studies on hydrogen bonding and specificity by systematic modification of the ligand rather than the protein have also been done.[92] Binding data of deoxy and fluorodeoxy sugars at glycogen phosphorylase in comparison with X-ray crystallographic data resulted in similar data to that of Fersht's complementary approach. Net contributions for hydrogen bonds of the neutral–neutral and neutral–charged type are found in the same energy range as in the work mentioned previously.

The second example of designed substrate specificity by specific mutagenesis investigates the strength of electrostatic interactions.[91] Difficulties in calculating electrostatic interactions are often caused by complications in considering a realistic dielectric constant. The following study on very small structural changes avoids this problem. Mutations at two different sites in the substrate-binding cleft of subtilisin allow an estimation of electrostatic interactions from the effect of changes on k_{cat}/K_M values. Average free energies for ion pair interactions at the two different positions were determined as -1.8 and -2.3 kcal mol^{-1}, respectively. The measured electrostatic effects showed additivity.

Calculations which give a quantitative description of observed effects from such substitution experiments[93] go one step further than the previous work. Using the empirical valence-bond method and a free energy perturbation method, it was possible to reproduce changes in free energies caused by chemical modifications quite well. This may be the beginning of an era of quantitative structure–function correlation in macromolecules. This study also proves the importance of the electrostatic energy or solvation energy of the charges in the reacting system for the catalytic free energy.

In the context of Fersht's studies,[89] there is another approach that tries to express properties of biological systems by means of intermolecular force equations.[94] These equations describe an observed quantity, the ΔG value, as a function of a series of intermolecular forces. High coefficients lead to the assumption that polarizability, ionic side-chains and steric effects determine the $\Delta\Delta G$ values of ATP and tyrosine binding by tyrosyl-tRNA synthetase in the transition state.

Most of the preceding examples make the unspoken assumption that binding is a process between two more or less rigid partners. However, as will be described more explicitly later, mutual recognition and binding of two molecules can be a somewhat dynamic problem, and cannot simply be described by two isolated conformationally fixed molecules.[95] The difference between observed free energies of binding of ligands to receptors and intrinsic binding energies shows that part of the binding energies may be used in conformational changes and changes in solvent interactions.[96] Exclusion of structural perturbation in the course of a variation in substructures can seldom be confirmed with certainty, but was possible in the following case, where X-ray crystal data,[97] calculations of binding energies from kinetic data[98] and molecular dynamics calculations[99] are in fascinating agreement.

Crystallographic analyses of two thermolysin-inhibitor complexes, differing only in a single peptide or ester linkage respectively, show that the two inhibitors bind in a completely comparable manner with the exception of a single hydrogen bond. Inhibition constants for this system gave a reduction in binding energy by a factor of 840 for the replacement of the NH with an O. The NH/O difference represents 4.0 kcal mol^{-1} in binding energy, attributed to the hydrogen-bonding interaction. Calculations of the relative changes in free energy using a thermodynamic perturbation method with molecular dynamics result in a very similar free energy difference of 4.2 kcal mol^{-1}, which may prove the suitability of such methods for rational drug design.

It is also true, however, that the values obtained in this way will vary with the quality of the drug–receptor interaction, as well as the role that the individual functional group plays in binding. If the additional group is not properly aligned with the receptor site, a zero or repulsive interaction may result. Alternatively, the strength of the additional bond may be offset by a reduction in the strengths of the existing bonds. Under these circumstances the anchor principle would lead to an underestimate of the true bond strength. Similarly, it is also possible to envisage instances in which large positive or negative contributions may occur due to factors other than direct attractive or repulsive interactions between the group and the receptor. For example, the presence of the group may lead to substantial destabilization or stabilization of the biologically active conformation relative to other conformers. In the latter case, the anchor principle would lead to an overestimate of the true bond strength. All these factors, as well as the need for binding data for many pairs of compounds which differ only in the presence of a single functional group, make the anchor principle a difficult basis for determining generally applicable values for the binding energies of individual functional groups.

18.8.4.3.3   *Average group contributions*

An alternative approach has been developed by Andrews *et al.*,[100] who undertook a statistical analysis of 200 drug–receptor interactions in aqueous solution. For this purpose the total free energies of interaction ΔG, determined from experimental association constants according to equation (20), were expressed as a function of the intrinsic binding energies E_X of the individual functional groups, where $T\Delta S_{r,t}$, the average loss of overall rotational and translational entropy which accompanies drug–receptor binding, was estimated at approximately 14 kcal mol^{-1} at 310 K.

$$\Delta G = T\Delta S_{r,t} + n_{DOF}E_{DOF} + n_X E_X \qquad (20)$$

The second term in equation (20) reflects the entropy loss associated with the loss of each degree of conformational freedom on receptor binding, n_{DOF} being the number of internal degrees of conformational freedom in the drug molecule and E_{DOF} the energy equivalent of the average entropy loss on binding. This value was empirically estimated at 0.7 kcal mol^{-1}, which may be compared to the average value for the total loss of conformational freedom of 1.3 kcal mol^{-1} for a single bond in a saturated hydrocarbon at 298 K.[6] The slightly smaller number obtained empirically implies that conformational freedom is not fully lost for all the bonds in an average drug–receptor interaction.

The final term in equation (20) is the sum of the intrinsic binding energies E_X associated with each functional group X, of which there are n_X present in the drug. It should be noted that each intrinsic binding energy E_X incorporates a number of terms, including the enthalpy of interaction between the functional group and its corresponding binding site on the receptor, the enthalpy changes associated with the removal of water of hydration from the functional group and its target site, and the subsequent formation of bonds between the displaced water molecules, the corresponding entropy terms associated with the displacement and subsequent bonding of water molecules, and the low frequency vibrational entropy associated with the bonds formed between the functional group and its partner (Figure 2). It is apparent that these factors may be regarded, at least approximately, as properties of the functional group that are relatively independent of the groups to which the particular functional group is attached. Such intrinsic binding potentials may thus reasonably be used in an additive manner to provide an overall estimate of the drug–receptor interaction.

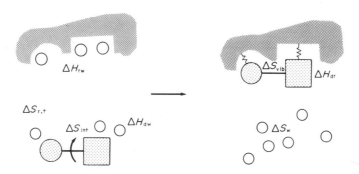

Figure 2 Binding interactions between a drug and an optimal receptor. The small circles represent water molecules, the enthalpies of hydration of the drug and receptor being ΔH_{dw} and ΔH_{rw}, respectively. The free drug has an overall rotational and translational entropy of $\Delta S_{r,t}$ and an internal entropy of ΔS_{int}. On binding, both terms are lost, but this unfavourable contribution may be compensated for by an increase in entropy due to loss of structured water on binding (ΔS_w), as well as an increase in entropy (ΔS_{vib}) due to new low frequency vibrational modes associated with the drug–receptor non-covalent bonds. While $\Delta S_{r,t}$ is essentially independent of the size of the drug (with limits), the other entropic and enthalpic terms depend on the number and nature of the functional groups present[100]

Application of equation (20) to 200 drug–receptor and enzyme–inhibitor interactions led to the average binding energies for each of 10 commonly occurring functional groups listed in Table 3. These values follow the expected trends in that charged groups lead to stronger interactions than polar groups, which in turn are stronger than non-polar groups such as sp^2 or sp^3 carbons. The magnitudes of the values for particular functional groups are also generally in accord with the ranges given above for different bond types.

Table 3 Intrinsic Binding Energies (kcal mol^{-1})[a]

Group	Energy	Range[b]
N^+	11.5	10.4–15.0
OPO_3^{2-}	10.0	7.7–10.6
CO_2^-	8.2	7.3–10.3
$C{=}O$	3.4	3.2–4.0
OH	2.5	2.5–4.0
Halogen	1.3	0.2–2.0
N	1.2	0.8–1.8
O, S	1.1	0.7–2.0
C (sp^3)	0.8	0.1–1.0
C (sp^2)	0.7	0.6–0.8
DOF[c]	−0.7	−0.7––1.0

[a] 1 kcal = 4.18 kJ. [b] Range of energies for six random 100-compound datasets. [c] Degrees of internal conformational freedom.

18.8.4.4 Geometries of Specific Interactions

So far we have considered only the type and strength of intermolecular forces and their experimental and theoretical investigation. But being interested in the construction of hypothetical receptor models, it is also important to consider their three-dimensional orientation. Should we, for example, think of interaction geometries for aromatic residues as overlapping arrangements of parallel planar molecules at van der Waals separation distances? And should we propose minimal energy orientations for the direction of lone pairs or hydrogen atoms? The selection of these geometries is obviously of substantial importance, but unfortunately the empirical function and parametrization chosen for calculations often determines the preference for one or another inter-action geometry. This is obvious, for example, in the case of calculations on phenyl interactions.[101] Depending on different energy contributions (*e.g.* dispersion or electrostatic term), either parallel (sandwich-like) or perpendicular configurations are favoured.

The same problem is also observed with Goodford's GRID program.[62] Here the choice of the maximum number of hydrogen bonds an atom can donate or accept for parametrization, drama-tically changes predicted hot-spot geometries. This means that even though the program should be applicable to any molecular interaction problem, a careful reparametrization, or at least a check, may be necessary to make sure that all geometric possibilities, which are often obvious to the eye, are included for each calculation. It is clear that additional support from experimental data is required.

Geometries of intramolecular forces have been studied extensively using X-ray crystal structures of large samples of small molecules from the Cambridge Crystallographic Database, and a broad analysis on proteins has been done from the Brookhaven Protein Databank. In particular, Taylor *et al.*,[102-105,23] Jeffrey *et al.*[106-110] and Murray-Rust *et al.*[111-115] have studied special types of interaction geometries in detail, whilst Baker *et al.*[116] reported quite extensive investigations of hydrogen bonding in globular proteins. There have also been theoretical and experimental analyses of binding patterns and specificity of stacking interactions with nucleic acids.[117] Orientational preferences in protein–ligand complexes, however, have been far less systematically analyzed. Do they show the same orientations as interactions formed within small molecules or protein molecules?

In order to gain more information in this area, we systematically studied the atomic surroundings of five different functional groups of ligands in all the complex strutures recorded in the Brookhaven Protein Databank. These (approximately 40) sets of coordinates for enzyme–ligand complexes do not provide very satisfactory material statistically, especially since certain criteria (a resolution of ≤ 2.5 Å and R factor ≤ 0.25) and variations restricted the set to 18 structures in the final interpretation (Table 4). Nevertheless, we attempted to collect information on the three-dimensional surroundings of phenyl, carboxyl, carbonyl, hydroxyl and amino groups of bound ligands in different enzyme structures by transferring the Brookhaven data to the Victorian College of Pharmacy molecular-modelling system MORPHEUS. We then focused on the ligand molecule and its surroundings, measured all distances within 3.8 Å of the ligand molecule (4.5 Å in the case of the phenyl ring), stored the fragments containing the atomic coordinates in this range around functional groups of the ligand, superimposed the fragments from different enzymes, and in this way obtained clouds showing different densities and types of atoms in particular orientations to the functional groups. We could not do a quantitative analysis of these relatively few data, but visual evaluation of ± 2 Å slices around the functional group in different planes through these clouds gave some information on favoured interaction geometries. For this purpose, each functional group was placed in the xy plane along the x axis and cuts were done in the xy, xz and yz planes, giving views of the functional group environment looking along the z axis to the planar partial structure, along the y axis for a sideways view, and from the x direction for a head-on view. Figures 3–7 show plots of these data, and the amount of available material included in each case is given in Table 5.

What conclusions can we draw from this? In the case of the phenyl ring (Figure 3a–c) we observed firstly that although the phenyl surroundings are flattened (like a hamburger bun), there are no real coplanar arrangements of groups; secondly, polar and hydrophobic amino acid residues are both equally represented by the surrounding atoms; and, finally, there seems to be a preferred interaction with oxygen atoms at the edge of the planar aromatic ring (Figure 3a). This may mean that we should desist from using coplanar arrangements in hypothetical receptor models, as has been done in the past,[118] or at least refrain from assuming that the coplanar 'receptor' groups consists of another aromatic ring system rather than a number of separate atoms or groups.

More promising for modelling is the observed carboxyl environment (Figure 4a–c). As expected, this proved to have almost totally polar surroundings in which we found two distinct types of binding, a close chelate-type interaction with the guanidino group of arginine (Figure 4a) and a lateral binding involving only one carboxyl oxygen atom and different possible amino acid residues

Table 4 Enzyme–Ligand Complexes in the Brookhaven Databank (January 1987)

Code	Enzyme	Ligand	No. of residues	No. of molecules	Solvent molecules	Resolution (Å)	R factor	Structure included in superimposition
5ADH	Alcohol dehydrogenase	ADP–ribose, 2-methyl-2,4-pentanediol	374	1	Yes	2.9	0.22	No
6ADH	Alcohol dehydrogenase	NAD, DMSO	374	2	No	2.9	0.38	No
3CPA	Carboxypeptidase A	Glycyl-L-tyrosine	307	1	No	2.0		Yes
4CPA	Carboxypeptidase A	Potato carboxypeptidase A inhibitor	307	1	No	2.5	0.196	Yes
7CAT	Catalase	NADPH	506	1	Yes	2.5	0.212	Yes
8CAT	Catalase	NADPH	506	2	Yes	2.5	0.191	Yes
2CHA	α-Chymotrypsin A	Tosyl (covalent!)	245	1	Yes	2.0		Yes
2CTS	Citrate synthase (pig)	Co*A, citrate	437	1	Yes	2.0	0.161	Yes
3CTS	Citrate synthase (chicken)	Co*A, citrate	437 (unknown)	1	Yes	2.0	0.192	Yes
4CTS	Citrate synthase	Oxaloacetate	437	2	Yes	1.7	0.182	No
3DFR	Dihydrofolate reductase (*L. casei*)	NADPH, methotrexate	162	1	Yes	2.9	0.152	Yes
4DFR	Dihydrofolate reductase (*E. coli*)	Methotrexate	159	2	Yes	1.7	0.155	Yes
2GRS	Glutathione reductase	FAD	478	1	No	2.0		Yes
1GPD	D-Glyceraldehyde-3-phosphate dehydrogenase	NAD	334	2	No	2.9		No
2YHX	Hexokinase B	o-Toluoylglucosamine	457	1	No	2.1		Yes
2MCP	Immunoglobulin	Phosphocholine	220	1	No	3.1	0.185	No
2PKA	Kallikrein A	Benzamidine	232	2	Yes	2.05	0.220	Yes
2KAI	Kallikrein A	Pancreatic trypsin inhibitor	232	1	Yes	2.5	0.224	No
3LDH	Lactate dehydrogenase (dogfish)	NAD, pyruvate	330	1	No	3.0		No
5LDH	Lactate dehydrogenase (pig)	S-Lac-NAD, citrate	334	1	No	2.7	0.196	Yes
2MDH	Malate dehydrogenase	NAD	324/325	2	No	2.5	0.324	No
1PPD	Papain	2-Mercaptoethanol (covalent)	212	1	Yes	2.0	0.145	No
3PGK	Phosphoglyceratekinase	ATP, 3-phosphoglycerate	416	1	No	2.5		No
2SNS	Staphylococcal nuclease	Deoxydiphosphothymidine	149	1	No	1.5		Yes
4TLN	Thermolysin	L-Leucylhydroxylamine	316	1	Yes	2.3	0.169	Yes
5TLN	Thermolysin	HONH-Benzylmalonyl-L-alanylglycine-p-nitroanilide	316	1	Yes	2.3	0.179	Yes
7TLN	Thermolysin	CH$_2$CO(N—OH)Leu—OCH$_3$ (covalent)	316	1	Yes	2.3	0.170	No
3PTB	β-Trypsin	Benzamidine	223	1	Yes	1.7	0.182	Yes
1TPP	β-Trypsin	p-Amidinophenylpyruvate (covalent)	223	1	Yes	1.4	0.191	Yes
3PTP	β-Trypsin	Diisopropylphosphoryl.	223	1	Yes	1.5	0.154	No
2PTC	β-Trypsin	Pancreatic trypsin inhibitor	223	1	Yes	1.9	0.187	No
1TPA	Anhydrotrypsin	Pancreatic trypsin inhibitor	223	1	Yes	1.9	0.175	No

Table 5   Analysis of Enzyme–Ligand Complexes in the Brookhaven Databank

	No. of available structures	No. of atoms within 3.8 Å	No. of atoms within 3.8 Å/structure	Superimposition	Closest distance	Dominant aminoacids
Phenyl	12	72	6.0	Plane 6 atoms	3.1/3.2 H$_2$O ~3.6	Mostly hydrophobic
Carboxyl	17	134	7.9	Plane 3 atoms	Arg 2.6/2.9 ~2.6	*No* hydrophobic
Carbonyl	23	84	3.65	Plane 4 atoms	From 2.6 ~2.8	Mostly hydrophobic
Hydroxyl	37	165	4.46	4 atoms	~2.4/2.5/2.6	2/3–1/3 hydrophilic–hydrophobic
Amino	22[a]	118	5.36	Plane 4 atoms	~2.8/2.9	1/2–1/2 hydrophilic–hydrophobic[b]

[a] Planar structures only.   [b] Including chain.

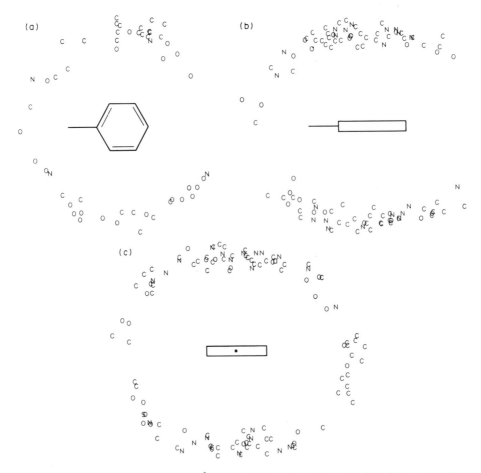

Figure 3   Phenyl environment within 4.5 Å: (a) cut in *xy* plane; (b) cut in *xz* plane; (c) cut in *yz* plane

(Figure 4a and c). These results are in good agreement with many other studies concerning the importance of carboxyl–arginine interactions,[119–123] and should be kept in mind for further receptor-modelling studies.

In the case of carbonyl groups, Taylor *et al.*[102–104,23] found for 889 small molecules from the Cambridge Crystallographic Database, containing a total of 1509 N—H ··· O=C hydrogen bonds, mean values of 120° to 130° for the C=O ··· N(H) angle, and for the R(1 or 2)—C=O ··· N(H) dihedral angle 0° or 180°, respectively. This is comparable with empirical models of hydration of small peptides.[124] In our studies, however, preferred planar orientations in the supposed lone pair direction like these could not be found (Figure 5a and b). Instead we saw a more or less random spherical orientation.

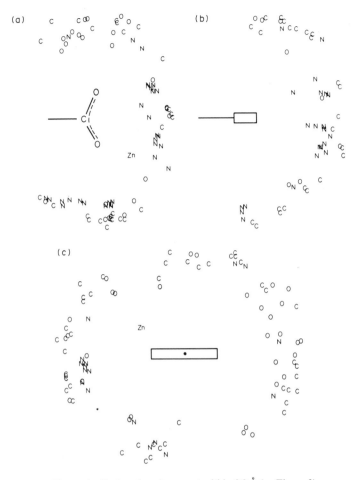

Figure 4 Carboxyl environment within 3.8 Å (as Figure 3)

Distribution in a clear hemisphere was also found for the hydroxyl group (Figure 6a–c), where on average more atoms were found within the studied range than in the carbonyl environment, but there was no spatial orientation corresponding to electron densities of non-bonded electrons. This was in accordance with the almost free rotation of the hydroxyl group and its different donor and acceptor capabilities. Again, this corresponds with analyses based on data from the Cambridge Crystallographic Database.[125]

A more directional preference can be seen in the amino environment (Figure 7a–c). Here interaction takes place preferably in a planar orientation. Oxygen-containing groups in particular, often aspartic or glutamic acid, as well as parts of the peptide backbone, interact with the amino hydrogen atoms. A second slight accumulation is assumed above and below the plane, perhaps indicating interaction with the nitrogen lone pair.

Altogether we find some similarities in the orientation of interactions formed within small molecules or between proteins and ligands, and although our questions are not yet fully answered, we have some valuable leads for receptor modelling. It would be useful to extend these studies to a wider set of data (or subsets with different substituent influence) in order to obtain further hints for a rational construction of binding site models.

18.8.5 RECEPTOR-BASED DRUG DESIGN

In this final section we shall first give examples of the successful application of receptor-based drug design to particular problems, and then discuss the reverse approach to the question of molecular interactions, that is, model studies in molecular recognition and studies on artificial enzymes. Other promising aspects which we believe will be of major importance in the future such as molecular

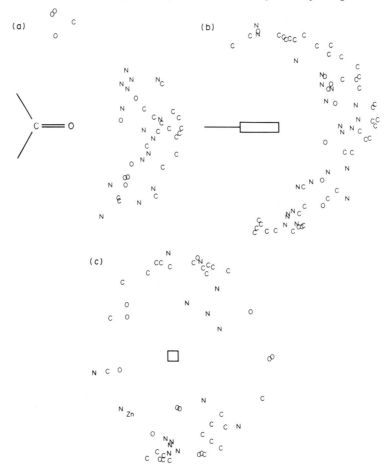

Figure 5 Carbonyl environment within 3.8 Å (as Figure 3)

dynamics will not be dealt with here, since they are covered in other chapters of this book (*e.g.* Chapter 18.3).

The most obvious starting point is the ubiquitous example of dihydrofolate reductase which, because of the well-known structures of different isoenzymes and of various complexes, is almost the standard test for new ideas.[126] X-Ray and NMR data afford the most detailed insight into its mode of action, and even stereochemical factors accounting for the species selectivity of different inhibitors have been explained in detail.[127,128] Accordingly, attempts to synthesize new active compounds benefit from advanced understanding of the enzyme's stereochemical requirements.

Kuyper *et al.*[129] used this knowledge of the methotrexate–dihydrofolate reductase interaction to make systematic changes in trimethoprim. They observed that trimethoprim (TMP) does not reach the positively charged arginine group to which methotrexate (MTX) forms an ionic linkage, and proposed that TMP should show increased activity when substituted by an appropriately placed carboxylic acid group. The realization of this idea resulted in more active compounds through a better fit to the observed binding sites. The results were supported by X-ray crystal data for two of the new complex structures, and by molecular mechanics calculations done by the same group.[130]

Another specific approach to the design of DHFR inhibitors was demonstrated by Maag *et al.*[131] They started with the structure of MTX when bound in the active site of the enzyme, constructed new compounds which kept special functional groups in comparable positions, matched them to MTX, docked them into the active site of *E. coli* DHFR, identified possible interactions with the binding site and predicted activities from computer modelling. The compounds were synthesized and tested on DHFR from different species and the measured inhibitory activities proved to correspond with predictions based on the molecular-modelling experiments.

Promising results of the design of potent inhibitors from the three-dimensional structure of another enzyme, alcohol dehydrogenase, have been presented by Freudenreich *et al.*[132] Studies

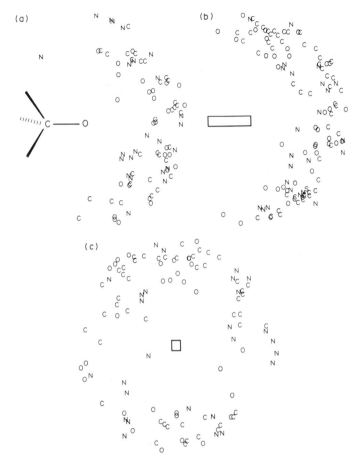

Figure 6 Hydroxyl environment within 3.8 Å (as Figure 3)

with several series of related structure allowed them to describe different features of the substrate binding site.

Design, synthesis and testing of another type of small-molecule binding is discussed by Manallack *et al.*[133] They aimed to design compounds to stabilize the insulin hexamer, thereby extending the duration of insulin action through slower dissociation to the active monomer. As can be seen from the crystal structure of long-acting dizinc insulin, each of the two zinc ions binds to three histidine nitrogen atoms stabilizing the insulin hexamer structure in this way. Between the Zn^{2+}/histidine layers there is a well-defined space in the centre of the hexamer surrounded by hydrophilic residues. The idea, based on these observations, was to design a molecule capable of fitting into this cavity between the six subunits, spanning the two groups of histidine residues, and containing functional groups able to provide binding to all six subunits. As shown by computer graphic modelling, these requirements were met in the benzene-1,4-disulfonic acid molecule. In comparison to other compounds tested, it allowed optimal nitrogen/oxygen contact distances to the six histidine residues. Sedimentation equilibrium experiments seem to confirm this theoretical mode of action.

By analogy to receptor-based drug design, the possibility of a knowledge-based prediction of receptor structures[134] should also be mentioned. The identification of similarities between ligand interactions with proteins of known three-dimensional structure and those which have to be modelled could be a further aid in drug design. While this may sound even more like 'castles in the air' than the approach from the other direction, it will undoubtedly become more realistic with increasing facilities for the computer modelling of proteins and protein engineering. One attempt in this direction, which involves calculating the three-dimensional structural changes in proteins due to amino acid substitutions has been developed by Snow *et al.*[135] Their approach, the coupled perturbation procedure based on common semi-empirical potential energy functions, combines conformational search and energy refinement for a certain region around a replacement. Of course, its use is restricted to problems where substitutions only result in local structural changes. Testing of the modelling procedure on immunoglobulin fragments gave promising results.

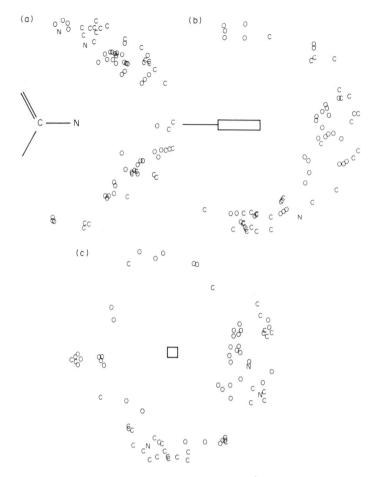

Figure 7   Amino environment within 3.8 Å (as Figure 3)

While model studies of biological recognition and binding have not yet led to the design of new compounds, we believe that they could be of immense importance for further design strategies. Some recent developments, for example, have led to the idea that it should be possible to mimic molecular recognition, binding and catalysis in biochemical systems with far smaller molecules. Features such as size, shape and functionality determining chemical processes could be realized in an artificial microenvironment, which again would enable systematic testing of the behaviour of different compounds in a system which functions in a similar way to a biochemical active site. We shall not at this stage describe different approaches to artificial structures resembling certain enzyme reaction mechanisms,[136-139] but there are several other approaches to the desgin of model receptors that could be valuable for drug design.

Lehn[140] uses the term 'supramolecular chemistry' for the study of the structure and function of those supermolecules resulting from non-covalent binding of substrates to molecular receptors. He describes macrocyclic and macropolycyclic structures of different size and chemistry suitable for molecular recognition of varous types of ligands. Crown ethers, cyclic polyamines, cyclophanes and cyclodextrins are the main classes of compounds available which present a three-dimensional structure with obvious pockets or holes able to bind to small molecules through electrostatic interactions, hydrogen bonding, van der Waals forces and so on. Suitable structures for various kinds of substrate binding have been developed.

Systematic studies on macromonocyclic polyamines are reported by Kimura *et al.*[141-143] Varying the ring size of their model, they found stable selective 1:1 complex builders for groups of compounds such as dicarboxylates or phosphates and nucleotides, and even describe a macrocyclic model receptor for histamine H_2-antagonists.[144] The ability to test interactions of new drug compounds in this way could be of enormous value.

Different models have been described by Rebek,[145,146] handling the problem of building certain functional groups in a prebuilt cavity through the construction of 'molecular clefts' instead of cyclic structures. Whilst functional groups attached to macrocycles tend to diverge and become directed away from the 'active site cavity', functional groups converge in this type of model, and better resemble the directionality of binding site groups. Different sized rigid spacer molecules such as annelated aromates substituted by 'arms' holding the binding groups, and further substituents blocking the rotation of those arms, provide a wide spectrum of possibilities.

A combination of these synthetic models and the experimental data described above could be a promising tool for the future of molecular-interaction-based design.

18.8.6 REFERENCES

1. P. M. Colman, W. G. Laver, J. N. Varghese, A. T. Baker, P. A. Tulloch, G. M. Air and R. G. Webster, *Nature (London)*, 1987, 326, 358.
2. A. G. Amit, R. A. Mariuzza, S. E. V. Phillips and R. J. Poljak, *Science (Washington, D.C.)*, 1986, **233**, 747.
3. J. A. Berzofsky, *Science (Washington, D.C.)*, 1985, **229**, 932.
4. E. D. Getzoff, H. M. Geysen, S. J. Rodda, H. Alexander, J. A. Tainer and R. A. Lerner, *Science (Washington, D.C.)*, 1987, **235**, 1191.
5. T. J. Smith, M. J. Kremer, M. Luo, G. Vriend, E. Arnold, G. Kamer, M. G. Rossmann, M. A. McKinlay, G. D. Diana and M. J. Otto, *Science (Washington, D.C.)*, 1986, **233**, 1286.
6. M. I. Page and W. P. Jencks, *Proc. Natl. Acad. Sci. USA*, 1971, **68**, 1678.
7. M. I. Page, *Angew. Chem., Int. Ed. Engl.*, 1977, **16**, 449.
8. A. Fersht, 'Enzyme Structure and Mechanism', 2nd edn., Freeman, New York, 1985.
9. M. I. Page, 'The Chemistry of Enzyme Action', Elsevier, Amsterdam, 1984.
10. P. Kollman, in 'X-Ray Crystallography and Drug Design', ed. A. S. Horn and C. J. De Ranter, Oxford University Press, Oxford, 1984, p. 63.
11. B. M. Pettitt and M. Karplus, *Top. Mol. Pharmacol.*, 1986, **3**, 75.
12. W. P. Jencks, *Proc. Natl. Acad. Sci. USA*, 1981, **78**, 4046.
13. A. Warshel, *Acc. Chem. Res.*, 1981, **14**, 284.
14. J. M. Blaney, P. K. Weiner, A. Dearing, P. A. Kollman, E. C. Jorgensen, S. J. Oatley, J. M. Burridge and C. C. F. Blake, *J. Am. Chem. Soc.*, 1982, **104**, 6424.
15. J. Gasteiger and M. Marsili, *Tetrahedron*, 1980, **36**, 3219.
16. W. J. Mortier, K. Van Genechten and J. Gasteiger, *J. Am. Chem. Soc.*, 1985, **107**, 829.
17. R. Skorczyk, *Acta Crystallogr., Sect. A*, 1976, **32**, 447.
18. TRIPOS Associates, Inc., St Louis, MO, USA.
19. Chemical Design Ltd., Oxford, UK.
20. N. K. Rogers, *Prog. Biophys. Mol. Biol.*, 1986, **48**, 37.
21. J. L. Webb, 'Enzyme and Metabolic Inhibitors', Academic Press, New York, 1963.
22. Z. Simon, 'Quantum Biochemistry and Specific Interactions', Abacus Press, Tunbridge Wells, UK, 1976.
23. R. Taylor, O. Kennard and W. Versichel, *Acta Crystallogr., Sect. B*, 1984, **40**, 280.
24. H. Umeyama and K. Morokuma, *J. Am. Chem. Soc.*, 1977, **99**, 1316.
25. P. Dauber and A. T. Hagler, *Acc. Chem. Res.*, 1980, **13**, 105.
26. P. Claverie, in 'Intermolecular Interactions: From Diatomics to Biopolymers', ed. B. Pullman, Wiley, New York, 1978, p. 69.
27. E. Giglio, *Nature (London)*, 1969, **222**, 339.
28. C. Tanford, 'The Hydrophobic Effect: Formation of Micelles and Biological Membranes', Wiley, New York, 1973.
29. W. P. Jencks, 'Catalysis in Chemistry and Enzymology', McGraw-Hill, New York, 1969.
30. J. O. Hirschfelder, *Adv. Chem. Phys.*, 1967, **12**.
31. B. Pullman (ed.), 'Intermolecular Interactions: From Diatomics to Biopolymers', Wiley, New York, 1978.
32. B. Pullman (ed.), 'Intermolecular Forces, The Jerusalem Symposia on Quantum Chemistry and Biochemistry', Reidel, Dordrecht, 1981.
33. D. B. Boyd, *Drug Inf. J.*, 1983, **17**, 121.
34. P. A. Kollman, *Acc. Chem. Res.*, 1977, **10**, 365.
35. P. A. Kollman, *J. Am. Chem. Soc.*, 1977, **99**, 4875.
36. R. Rein and M. Shibata, in 'Intermolecular Forces, Jerusalem Symposia on Quantum Chemistry and Biochemistry', ed. B. Pullman, Reidel, Dordrecht, 1981, p. 49.
37. K. Bhanuprakash, G. V. Kulkarni and A. K. Chandra, *J. Comput. Chem.*, 1986, **7**, 731.
38. A. F. Cuthbertson, C. B. Naylor and W. G. Richards, *J. Mol. Struct.*, 1984, **106**, 287.
39. N. K. Sanyal, R. P. Ojha and M. Roychoudhury, *J. Comput. Chem.*, 1986, **7**, 13, 20; N. K. Sanyal, R. P. Ojha, M. Roychoudhury and S. N. Tiwari, *J. Comput. Chem.*, 1986, **7**, 30.
40. J. Caillet and P. Claverie, *Acta Crystallogr., Sect. A*, 1975, **31**, 448.
41. N. Gresh, P. Claverie and A. Pullman, *Int. J. Quantum Chem.*, 1986, **29**, 101.
42. R. Rein and M. Pollak, *J. Chem. Phys.*, 1967, **47**, 2039.
43. M. Pollak and R. Rein, *J. Chem. Phys.*, 1967, **47**, 2045.
44. H.-D. Höltje and L. B. Kier, *J. Pharm. Sci.*, 1974, **63**, 1722.
45. P. Claverie and R. Rein, *Int. J. Quantum Chem.*, 1969, **3**, 537.
46. M.-J. Huron and P. Claverie, *Chem. Phys. Lett.*, 1969, **4**, 429.
47. J. R. Rabinowitz, T. J. Swissler and R. Rein, *Int. J. Quantum Chem., Symp.*, 1972, **6**, 353.
48. E. W. Gill, *Prog. Med. Chem.*, 1965, **4**, 39.
49. R. Langridge, T. E. Ferrin, I. D. Kuntz and M. L. Connolly, *Science (Washington, D.C.)*, 1981, **211**, 661.
50. P. K. Weiner, R. Langridge, J. M. Blaney, R. Schaefer and P. A. Kollman, *Proc. Natl. Acad. Sci. USA*, 1982, **79**, 3754.
51. I. D. Kuntz, J. M. Blaney, S. J. Oatley, R. Langridge and T. E. Ferrin, *J. Mol. Biol.*, 1982, **161**, 269.

52. R. L. Des-Jarlais, R. P. Sheridan, J. S. Dixon, I. D. Kuntz and R. Venkataraghavan, *J. Med. Chem.*, 1986, **29**, 2149.
53. M. Billeter, T. F. Havel and I. D. Kuntz, *Biopolymers*, 1987, **26**, 777.
54. M. R. Pincus and H. A. Scheraga, *Acc. Chem. Res.*, 1981, **14**, 299.
55. Z. I. Hodes, G. Nemethy and H. A. Scheraga, *Biopolymers*, 1979, **18**, 1565.
56. P. K. Weiner and P. A. Kollman, *J. Comput. Chem.*, 1981, **2**, 287.
57. G. Wipff, A. Dearing, P. K. Weiner, J. M. Blaney and P. A. Kollman, *J. Am. Chem. Soc.*, 1983, **105**, 997.
58. S. Naruto, I. Motoc, G. R. Marshall, S. B. Daniels, M. J. Sofia and J. A. Katzenellenbogen, *J. Am. Chem. Soc.*, 1985, **107**, 5262.
59. W. A. Remers, M. Mabilia and A. J. Hopfinger, *J. Med. Chem.*, 1986, **29**, 2492.
60. S. N. Rao, U. C. Singh and P. A. Kollman, *J. Med. Chem.*, 1986, **29**, 2484.
61. R. F. Tilton, Jr., U. C. Singh, S. J. Weiner, M. L. Connolly, I. D. Kuntz, Jr., P. A. Kollman, N. Max and D. A. Case, *J. Mol. Biol.*, 1986, **192**, 443.
62. P. J. Goodford, *J. Med. Chem.*, 1985, **28**, 849.
63. Molecular Discovery Ltd., Oxford, UK, 1986.
64. W. G. J. Hol, *Angew. Chem., Int. Ed. Engl.*, 1986, **25**, 767.
65. J. J. Stezowski and K. Chandrasekhar, *Annu. Rep. Med. Chem.*, 1986, **21**, 293.
66. P. G. Jones, *Chem. Rev.*, 1984, **13**, 157.
67. P. Murray-Rust, in 'Molecular Structure and Biological Activity', ed. J. F. Griffin and W. L. Duax, Elsevier, New York, 1982, p. 117.
68. N. Sakabe, K. Sakabe and K. Sasaki, in 'Structural Studies on Molecules of Biological Interest', ed. G. Dodson, J. P. Clusker and D. Sayre, Clarendon Press, Oxford, 1981, p. 509.
69. B. W. Matthews, in 'The Proteins', 3rd edn., ed. H. Neurath and R. L. Hill, Academic Press, New York, 1977, vol. III, p. 403.
70. T. Blundell, B. L. Sibanda and L. Pearl, *Nature (London)*, 1983, **304**, 273.
71. W. Carlson, E. Haber, R. Feldman and M. Karplus, in 'Peptides, Structure and Function. Proceedings of the 8th American Peptide Symposium', ed. V. J. Hruby and D. H. Rich, Pierce Chemical Co., Rockford, 1983, p. 821.
72. J. R. Knox and J. A. Kelly, in 'New Methods in Drug Research', ed. A. Makriyannis, Prous, Barcelona, 1985, vol. 1, p. 1.
73. A. Arnone, *Nature (London)*, 1972, **237**, 146.
74. C. R. Beddell, P. J. Goodford, F. E. Norrington, S. Wilkinson and R. Wootton, *Br. J. Pharmacol.*, 1976, **57**, 201.
75. C. R. Beddell, P. J. Goodford, D. K. Stammers and R. Wootton, *Br. J. Pharmacol.*, 1979, **65**, 535.
76. C. R. Beddell, P. J. Goodford, G. Kneen, R. D. White, S. Wilkinson and R. Wootton, *Br. J. Pharmacol.*, 1984, **82**, 397.
77. J. A. Walder, R. Y. Walder and A. Arnone, *J. Mol. Biol.*, 1980, **141**, 195.
78. D.J. Abraham, M. F. Perutz and S. E. V. Phillips, *Proc. Natl. Acad. Sci. USA*, 1983, **80**, 324.
79. M. F. Perutz, G. Fermi, D. J. Abraham, C. Poyart and E. Bursaux, *J. Am. Chem. Soc.*, 1986, **108**, 1064.
80. L. Sheh, M. Mokotoff and D. J. Abraham, *Int. J. Pept. Protein Res.*, 1987, **29**, 509.
81. O. Jardetzky and G. C. K. Roberts, 'NMR in Molecular Biology', Academic Press, New York, 1981.
82. G. C. K. Roberts, in 'Quantitative Approaches to Drug Design', ed. J. C. Dearden, Elsevier, Amsterdam, 1983, p. 91.
83. G. C. K. Roberts, *NATO Adv. Study Inst. Ser., Ser. A*, 1986, **107**, 73.
84. G. M. Clore, A. M. Gronenborn, R. Birdsall, J. Feeney and G. C. K. Roberts, *Biochem. J.*, 1984, **217**, 659.
85. G. Winter and A. R. Fersht, *TIBS*, 1984, **2**, 115.
86. W. G. J. Hol, *TIBTECH*, 1987, **5**, 137.
87. A. R. Fersht, *TIBS*, 1984, **9**, 145.
88. A. R. Fersht, R. J. Leatherbarrow and T. N. C. Wells, *TIBS*, 1986, **11**, 321.
89. A. R. Fersht, J.-P. Shi, J. Knill-Jones, D. M. Lowe, A. J. Wilkinson, D. M. Blow, P. Brick, P. Carter, M. M. Y. Waye and G. Winter, *Nature (London)*, 1985, **314**, 235.
90. A. R. Fersht, *TIBS*, 1987, **12**, 301.
91. J. A. Wells, D. B. Powers, R. D. Bott, T. P. Graycar and D. A. Estell, *Proc. Natl. Acad. Sci. USA*, 1987, **84**, 1219.
92. I. P. Street, C. R. Armstrong and S. G. Withers, *Biochemistry*, 1986, **25**, 6021.
93. J.-K. Hwang and A. Warshel, *Biochemistry*, 1987, **26**, 2669.
94. M. Charton, *Int. J. Pept. Protein Res.*, 1986, **28**, 201.
95. M. Karplus and J. A. McCammon, *Annu. Rev. Biochem.*, 1983, **52**, 263.
96. M. I. Page, in 'Quantitative Approaches to Drug Design', ed. J. C. Dearden, Elsevier, Amsterdam, 1983, p. 109.
97. D. E. Tronrud, H. M. Holden and B. W. Matthews, *Science (Washington, D.C.)*, 1987, **235**, 571.
98. P. A. Bartlett and C. K. Marlowe, *Science (Washington, D.C.)*, 1987, **235**, 569.
99. P. A. Bash, U. C. Singh, F. K. Brown, R. Langridge and P. A. Kollman, *Science (Washington, D.C.)*, 1987, **235**, 574.
100. P. R. Andrews, D. J. Craik and J. L. Martin, *J. Med. Chem.*, 1984, **27**, 1648.
101. J. M. Morris, *Mol. Phys.*, 1974, **28**, 1167.
102. R. Taylor, O. Kennard and W. Versichel, *J. Am. Chem. Soc.*, 1983, **105**, 5761.
103. R. Taylor and O. Kennard, *Acta Crystallogr., Sect. B*, 1983, **39**, 133.
104. R. Taylor, O. Kennard and W. Versichel, *J. Am. Chem. Soc.*, 1984, **106**, 244.
105. R. Taylor and O. Kennard, *J. Am. Chem. Soc.*, 1982, **104**, 5063.
106. C. Ceccarelli, G. A. Jeffrey and R. Taylor, *J. Mol. Struct.*, 1981, **70**, 255.
107. G. A. Jeffrey and H. Maluszynska, *Int. J. Biol. Macromol.*, 1982, **4**, 173.
108. G. A. Jeffrey and J. Mitra, *J. Am. Chem. Soc.*, 1984, **106**, 5546.
109. G. A. Jeffrey, H. Maluszynska and J. Mitra, *Int. J. Biol. Macromol.*, 1985, **7**, 336.
110. G. A. Jeffrey and H. Maluszynska, *J. Mol. Struct.*, 1986, **147**, 127.
111. R. E. Rosenfield, Jr. and P. Murray-Rust, *J. Am. Chem. Soc.*, 1982, **104**, 5427.
112. P. Murray-Rust, W. C. Stallings, C. T. Monti, R. K. Preston and J. P. Glusker, *J. Am. Chem. Soc.*, 1983, **105**, 3206.
113. N. Ramasubbu, R. Parthasarathy and P. Murray-Rust, *J. Am. Chem. Soc.*, 1986, **108**, 4308.
114. P. Murray-Rust and J. P. Glusker, *J. Am. Chem. Soc.*, 1984, **106**, 1018.
115. R. E. Rosenfield, Jr., S. M. Swanson, E. F. Meyer, Jr., H. L. Carrell and P. Murray-Rust, *J. Mol. Graphics*, 1984, **2**, 43.
116. E. N. Baker and R. E. Hubbard, *Prog. Biophys. Mol. Biol.*, 1984, **44**, 97.

117. R. Rein, in 'Intermolecular Interactions: From Diatomics to Biopolymers', ed. B. Pullman, Wiley, New York, 1978, p. 307.
118. H.-D. Höltje and M. Tintelnot, *Quant. Struct.–Act. Relat.*, 1984, **3**, 6.
119. W. Bode, J. Walter, R. Huber, H. R. Wenzel and H. Tschesche, *Eur. J. Biochem.*, 1984, **144**, 185.
120. M. Marquart, J. Walter, J. Deisenhofer, W. Bode and R. Huber, *Acta Crystallogr., Sect. B*, 1983, **39**, 480.
121. E. E. Howell, J. E. Villafranca, M. S. Warren, S. J. Oatley and J. Kraut, *Science (Washington, D.C.)*, 1986, **231**, 1123.
122. F. A. Cotton, E. E. Hazen, Jr. and M. J. Legg, *Proc. Natl. Acad. Sci. USA*, 1979, **76**, 2551.
123. J. F. Riordan, K. D. McElvany and C. L. Borders, Jr., *Science (Washington, D.C.)*, 1977, **195**, 884.
124. F. Vovelle, M. Genest, M. Ptak and B. Maigret, in 'Intermolecular Forces. The Jerusalem Symposia on Quantum Chemistry and Biochemistry', Reidel, Dordrecht, 1984, p. 299.
125. A. Vedani and J. D. Dunitz, *J. Am. Chem. Soc.*, 1985, **107**, 7653.
126. J. N. Champness, L. F. Kuyper and C. R. Beddell, *Top. Mol. Pharmacol.*, 1986, **3**, 335.
127. D. A. Matthews, J. T. Bolin, J. M. Burridge, D. J. Filman, K. W. Volz, B. T. Kaufman, C. R. Beddell, J. N. Champness, D. K. Stammers and J. Kraut, *J. Biol. Chem.*, 1985, **260**, 381.
128. D. A. Matthews, J. T. Bolin, J. M. Burridge, D. J. Filman, K. W. Volz and J. Kraut, *J. Biol. Chem.*, 1985, **260**, 392.
129. L. F. Kuyper, B. Roth, D. P. Baccanari, R. Ferone, C. R. Beddell, J. N. Champness, D. K. Stammers, J. G. Dann, F. E. Norrington, D. J. Baker and P. J. Goodford, *J. Med. Chem.*, 1985, **28**, 303.
130. B. Roth, *Fed. Proc., Fed. Am. Soc. Exp. Biol.*, 1986, **45**, 2765.
131. H. Maag, R. Locher, J. J. Daly and I. Kompis, *Helv. Chim. Acta*, 1986, **69**, 887.
132. C. Freudenreich, J.-P. Samama and J.-F. Biellmann, *J. Am. Chem. Soc.*, 1984, **106**, 3344.
133. D. T. Manallack, P. R. Andrews and E. F. Woods, *J. Med. Chem.*, 1985, **28**, 1522.
134. T. L. Blundell, B. L. Sibanda, M. J. E. Sternberg and J. M. Thornton, *Nature (London)*, 1987, **326**, 347.
135. M. E. Snow and L. M. Amzel, *Proteins: Struct. Funct. Genet.*, 1986, **1**, 267.
136. R. Breslow, *Adv. Enzymol. Relat. Areas Mol. Biol.*, 1986, **58**, 1.
137. V. T. D'Souza and M. L. Bender, *Acc. Chem. Res.*, 1987, **20**, 146.
138. C. A. Venanzi and J. D. Bunce, *Enzyme*, 1986, **36**, 79.
139. C. A. Venanzi and J. D. Bunce, *Int. J. Quantum Chem., Quantum Biol. Symp.*, 1986, **12**, 69.
140. J.-M. Lehn, *Science (Washington, D.C.)*, 1985, **227**, 849.
141. E. Kimura, A. Sakonaka, T. Yatsunami and M. Kodama, *J. Am. Chem. Soc.*, 1981, **103**, 3041.
142. E. Kimura, M. Kodama and T. Yatsunami, *J. Am. Chem. Soc.*, 1982, **104**, 3182.
143. E. Kimura, A. Sakonaka and M. Kodama, *J. Am. Chem. Soc.*, 1982, **104**, 4984.
144. E. Kimura, T. Koike and M. Kodama, *Chem. Pharm. Bull.*, 1984, **32**, 3569.
145. J. Rebek, Jr., *Science (Washington, D.C.)*, 1987, **235**, 1478.
146. J. Rebek, Jr., B. Askew, M. Killoran, D. Nemeth and F.-T. Lin, *J. Am. Chem. Soc.*, 1987, **109**, 2426.

19.1

Quantitative Description of Biological Activity

YVONNE C. MARTIN, EUGENE N. BUSH and JAROSLAV J. KYNCL
Abbott Laboratories, North Chicago, IL, USA

19.1.1 GENERAL CONSIDERATIONS

This chapter will discuss the various issues and techniques for providing the most meaningful description of biological properties of molecules as a prelude to the theoretical analysis of structure–activity relationships. Most theoretical drug design methods are based on an attempt to derive an explicit or implicit model of the physical chemistry of the interactions of the drug molecules with a molecule of the biophase. Thus it is important that the biological properties of the drug molecules be quantified in physicochemical terms such as equilibrium or rate constants.[1,2] In this chapter we will develop some of the key proposed relationships between equilibrium or rate constants and the observed biological response. A further concern is how the medicinal chemist uses the biological data to decide which molecules form a set within which structure–activity comparisons will be valid.

The choice of the biological measure is often obscure because usually the three-dimensional structure of the target biomolecules for the drugs we wish to design is not known. Even if we do know the structure, we are only now beginning to understand the relationship between the three-dimensional structure of proteins and their simple function. As a result we do not necessarily know how to measure in a physicochemically meaningful way the effect of a foreign molecule on a biological system. This chapter merely offers our experience and a summary of the state of the art.

In addition, there are strategic concerns to the selection of biological data.[3] Everyone understands that predictions from theoretical analyses are only as precise as the data on which they are based. If the data are weak, the predictions will also risk being incorrect. However, it may be that when there are only tenuous biological data, then that is just the time that a bold hypothesis may be most useful to move a project forward. Such a hypothesis might provide a framework for experimental work giving insight, even though it ultimately may prove to be partially incorrect. On the other hand, when the biological system is well understood, predictions may be expected to be more accurate. So while we can stress the importance of having *valid* biological data for doing predictive theoretical analyses, *useful* theoretical analyses may be done with whatever data are available.

Finally, if the objective of the investigation is the design of a therapeutically useful agent,[4] one must be aware of the extent to which the biological endpoint in the calculation is a good predictor of therapeutic usefulness. The predictions will be of potency in the test in which the potency was measured—extrapolation to clinical effectiveness is just an extrapolation. Thus the biological test must be chosen to quantify how closely a compound meets the objective of the study. The classic case in this regard is the *in vitro* potency of a number of penicillin analogues.[5] The measured log(potency) in the normal broth assay was shown to be independent of partition coefficient ($\log P$), but that from the assay in which serum albumin was included was negatively correlated with $\log P$. It was concluded that the potency measured in the latter assay was an artifact and merely measured the protein binding of the compound. However, if one is interested in whole animal activity of the compound, it is not irrelevant to consider protein binding. Finally, since the goals of an industrial medicinal chemistry program may change as deficiencies in current compounds or biological hypotheses are identified, so too may the biological endpoint of the theoretical analysis change.

19.1.2 STATISTICAL BACKGROUND: THE MEASUREMENT OF BIOLOGICAL PHENOMENA

This section will examine statistical methods to analyze, quantify, summarize and compare biological data.[6] The selection of data for structure–activity relationships and the pitfalls and artifacts that can affect the interpretation will be addressed.

The experimental approach to biological investigation requires that there be at least one aspect of biological function that can be measured. The measured event may be a molecularly defined biochemical process, such as enzyme activity or receptor affinity,[7-11] or it may be a complex physiological function,[4, 12, 13] such as arterial blood pressure. A response is a change in such a measured event that follows an experimental change in one or several controllable factors which are believed to affect the measured function. The factors are often varied over several levels or categories. Factors are also considered independent variables, because they can be experimentally controlled. Administration of a drug would be an example of a factor, and the doses of the drug would be the levels of the factor.

Scientists must appreciate that even in the most carefully designed biological experiment there is usually a considerable variability in the response. It is, therefore, important that statistical methods be used to summarize and compare biological information.[6]

19.1.2.1 Qualitative *Versus* Quantitative Data

Many characteristics of biological phenomena can be measured quantitatively. With the exception of those *in vitro* experiments where a true equilibrium can be reached, most biological phenomena are constantly changing. Some such dynamic responses can be measured continuously; for example, a pressure transducer can measure arterial blood pressure at any point in time, and even very rapid alterations in blood pressure produced by a drug can be quantified. Similarly, a spectrophotometer can be used to follow the rate of formation or disappearance of a compound in a biological sample. However, other responses, such as the plasma concentrations of a drug as assayed by a radioimmuno or HPLC assay, cannot be easily measured continuously. The minimum value, maximum value and/or the range of the response may be important in compound evaluation.

Either continuous or discontinuous data can be used to define such temporal relationships of the response as latency, time to peak, duration and integrated response over time, as illustrated in Figure 1. In certain studies only the magnitude of the time-independent peak response is of interest. This is the case with relatively rapidly occurring responses such as muscle contraction. On the other hand, for time-dependent responses, such as motor activity of conscious animals, one would typically measure the total response or the area under the time–response curve. In still other cases, the frequency with which a measurement changes is more important than the amplitude of the observation; for example, one might measure the rate of the systolic rise in left ventricular pressure as an index of the cardiac contractility response to the administration of a compound. An example of a case for which both components of the response, the magnitude of the peak and the total response, are of importance might be shown with analogues of gonadotropin-releasing hormone (Figure 2). In this case, the apparent potency calculated from peak response is about threefold greater than that calculated from the area under the curve. Finally, there are situations where the duration of action is the response measured, as in the length of time an animal is hyperactive after a dose of a potential stimulant. Interpretation of such biological properties of compounds for structure–activity analysis may require the use of both thermodynamic and kinetic principles (see Sections 19.1.3.3 and 19.1.3.4).

Some biological responses cannot be quantitatively measured, either because the tools for their measurement are crude or do not exist, or because the response is intrinsically all-or-none. All-or-none (quantal) events have only two possible response outcomes. An example of a quantal phenomenon is mortality or lethality, in which a treatment or disease either does or does not result in death. Still other biological events may have only a few possible outcomes: an example would be eye color. Such responses are categorical because the responses can be grouped into a few categories. Quantal and categorical responses can be described quantitatively in terms of an incidence, that is the rate at which the event occurs per unit time or per sample size. Incidences are often expressed as percentages or proportion of a sample that responded in a certain manner. For example, the dose of a drug which is lethal to 50% of a group of animals (LD_{50}) is a measure of its toxicity. Some categorical responses can logically be arranged in a certain hierarchical order and are considered rank-order events. An example would be number of offspring in a litter.

19.1.2.2 Descriptions of Variability

The measurement of a biological response in a single subject does not allow the observer to conclude that such a response is representative of all subjects. For example, some patients are much more sensitive to the sedative effects of antihistamines than others. Furthermore, the same person or laboratory animal can respond differently at different times to the identical treatment. Within-group variability of responses is variability observed within a single treatment group, while between-group variability is that observed between different groups. The term 'group' in this case refers to the different treatments, not different subjects. This distinction is especially important when a single subject receives different challenges at different times.

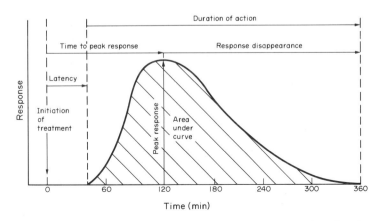

Figure 1 A general diagram of a time–response plot with definitions of the various time-dependent terms

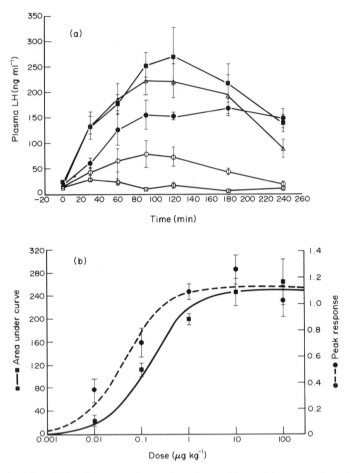

Figure 2 Plasma levels of leuteinizing hormone after various doses of leuprolide, a gonadotropin-releasing hormone analogue: (a) the time course of response to 0.01, 0.1, 1.0, 10.0 and 100 μg kg^{-1}; (b) the difference in log-dose *versus* response curves when peak *versus* integrated responses are considered

To measure the effect of a compound on a biological system, we are most interested in comparisons of responses between different treatment groups: therefore, it is essential that within-group differences be minimized and themselves quantified.

To provide estimates of response variability, the biologist makes multiple independent observations of the response to summarize the average and the variability of the response. For this purpose one selects a sample, which is a group of individual subjects, animals, tissues or test tubes that are to be treated as identically as possible. One must be careful that the only differences between the treatment groups are the treatments themselves, lest a difference be spuriously attributed to the treatment. For example, animals of different body weight may respond differently to a drug, and the researcher must, therefore, be certain that the treatment and control groups have a similar distribution of body weights. The proper selection of the sample is extremely important, because only if it is properly chosen will the sample measurements (statistics) be representative of the true values (parameters) of the whole population, and only then can the conclusions from a sample be generalized to a population.

19.1.2.2.1 *Measures of central tendency*

When data have inherent variability, some sort of average value is often used to represent the usual response. The mean ($\bar{x}$) is calculated by equation (1)

$$\bar{x} = \Sigma x/n \qquad (1)$$

where Σx is the sum of n values of response x within a treatment group. The median is the value at the middle of a sorted set of response observations. The mode is the most commonly occurring response value in the set. Of these three measures, the mean has the most practical value because it can be easily compared among groups by several statistical tests. The mean is considered accurate if it closely represents the true mean of the population from which the sample was selected.

19.1.2.2.2 Measures of dispersion

As well as the average response of members of a treatment group, we are also interested in the magnitude of its variability. Often comparisons are made between the means of two treatment groups. In order to make a statistically meaningful case for a difference between the means of two groups, one must demonstrate that there is not a high degree of overlap between the sets of data. This is done by measuring their variances. The variance(s) is the sum of squares of the differences of the observed values from the mean, dividing by the sample size minus one, *i.e.* equation (2) or equation (3). If more than two samples are to be compared, the variances are evaluated rather than the sample means, in the analysis of variance test.

$$s^2 = \Sigma (x - \bar{x})^2/(n - 1) \tag{2}$$

$$s^2 = \Sigma x^2 - (\Sigma x)^2/(n - 1) \tag{3}$$

The standard deviation (sd) describes the spread of individual data values in a sample about the sample mean. It is calculated from the variance (s) by taking the square root (equation 4).

$$sd = \sqrt{s} \tag{4}$$

The standard error (se) of the sample mean describes the distribution of normally distributed sample means about the true population mean. It is derived from the standard deviation by taking the sample size into account. The magnitude of the se decreases with increasing sample size and is inversely proportional to the precision of measurement (equation 5)

$$se = sd/\sqrt{n} \tag{5}$$

19.1.2.2.3 Distribution of data

Often one wishes to compare responses among several treatment groups. Many methods for such comparisons are based upon assumptions of the distribution of data in the population from which the sample group was selected.

The most commonly used statistical tests are the parametric tests. They are based on the assumption that the response: (i) is continuously variable; and (ii) is measured for a sample drawn from a population that adheres to the normal Gaussian distribution. Thus, before a parametric test is applied to the data, the adherence to a normal distribution must be established. Homogeneity of variance in the different treatment groups should be observed and the experiment should have been designed to have an equal interval between the several treatment levels. In Figure 3 the magnitude of a hypothetical response is dependent on the intensity of the stimulus, and the variability associated with the response is proportional to the response. Log transformation of the responses results in equal variability at each stimulus. Thus, parametric statistical tests are now appropriate.

In a normal distribution the mean, median and mode are identical; this means that the middle and most frequently occurring observations coincide with the average. Furthermore, in a normal distribution, the dispersion of the data is symmetrical above and below the mean. In a sample of $n > 30$ of normally distributed observations, 95% of the observations lie within the range mean ± 2.0 sd. With smaller samples, the distance from the mean required to encompass 95% of the observations is proportionally larger than 2.0, depending on sample size (n).

Many biological response data (such as potencies of drugs and their receptor affinities) are not normally distributed. However, often the data can be transformed using logarithms, square roots or various other functions into a normal distribution.[14] For example, Figure 4 shows frequency distributions of the potency of norepinephrine in rabbit aorta and guinea pig atria. Notice that in neither tissue does the response follow a normal distribution, but in both cases a lognormal distribution is seen. It is proper then to use only the transformed values in comparisons using parametric tests. It is also important that the transformation selected gives equal weight to all observations if they were all generated with similar precision and accuracy. For example, in Figure 4 a comparison of the transformed and untransformed data is shown.

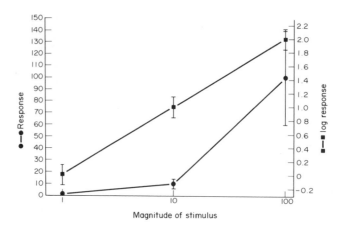

Figure 3   An example of how a transformation of data produces a better representation of that data. In this case the log transformation produces a response variable for which the error is independent of dose

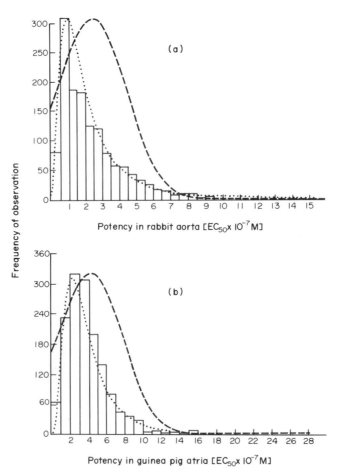

Figure 4   Frequency distributions of more than 1500 independent observations of the EC_{50} of norepinephrine in (a) rabbit aorta and (b) guinea pig atria: ——, observed value; · · ·, lognormal distribution; – – –, normal distribution

Since the normal distribution applies to sample sizes greater than 30 and most biological experiments use many fewer than 30 cases per treatment, it is often difficult to establish normality of the response. One can use a variety of nonparametric tests for data analysis if the population distribution cannot be assumed normal. Alternatively, an approximation of normality, known as the Student's *t*-distribution, can be used for small sample sizes. It is based on the assumption that the sample is drawn from a normal distribution. This distribution is also symmetrical about the mean, but the thickness of the tails, that is the relative spread of the data, is progressively larger with decreasing sample size. The t-statistic is calculated from the ratio of the sample mean ($\bar{x}$) divided by the standard error of the mean (se), as shown in (equation 6)

$$t = \bar{x}/\text{se} \tag{6}$$

A confidence limit, which establishes the expected spread of a mean response, can be established according to equation (7)

$$\text{cl} = \bar{x} \pm t_{\text{tab}}(\text{se}) \tag{7}$$

where cl represents the upper and lower confidence limits for response $\bar{x}$, and t_{tab} is a tabulated reliability coefficient available in many handbooks. If a 95% confidence interval is constructed, with repeated sampling the mean response would be expected to lie within this interval 95% of the time.

19.1.2.2.4 *Number of observations*

In order to reliably demonstrate reproducible results one must use enough subjects per group. A large sample size is associated with a small standard error; this makes it easy to detect differences between group means. However, for ethical and practical reasons, biological experimentation with animals uses the least number of animals per group that will demonstrate a difference. Although it is nearly impossible to establish the minimum sample size before a totally novel experiment is initiated, many experiments can be conducted with six or fewer subjects per group, especially when several dose levels of a compound are compared in a study.

19.1.2.3 Experimental Design and Hypothesis Testing

Careful planning of experiments can yield the proper information for the quantitative analysis of the factors that affect a response. To obtain such results each experiment should be carefully planned to meet a certain objective. Additionally, close attention must be paid to factors that could potentially influence the outcome of the experiment. These factors can be systematically varied to allow a maximum amount of information to be generated or they can be strictly controlled to one level to examine the influence of other factors on the response. For example, sexually mature male rats metabolize barbiturate sedative/hypnotic compounds more rapidly than do female or immature male rats.[15] Therefore, the sleeping interval induced by a hypnotic dose of barbiturate is shorter for mature male rats than it is for female or immature male rats. Thus, lack of attention to sex or maturational status of the test animals could confound the interpretation of the sleeping interval induced by a compound. Conversely, a limitation of the experiment to a narrowly defined combination of factors, immature male rats in the example, greatly narrows the population to which the results can be generalized.

It is common practice to test a compound at several different doses, that is to develop a dose–response relationship. For such studies it is essential that differences in sequence of treatment, age, weight, vehicle composition or other controllable factors cannot be found between different dose levels. This goal is achieved by using standard procedures for random assignment of subjects among the various test groups and careful attention to environmental conditions that could affect the experiment.

19.1.2.4 Comparative Statistical Tests

A hypothesis is a statement concerning the parameters of the population or populations: it is tested by examination of the data collected from the sample(s). The null hypothesis states that there is no difference between sample groups, that is they are constructed by repeated sampling of the same population. The alternative hypothesis is that the samples are different, that is they are from separate populations.

Statistical inference is based upon probability. It allows the experimenter to predict how often an observed difference will occur by chance, given the observed variability in the data. It cannot prove that a true difference exists, only establish its likelihood. Conversely, if a difference cannot be established, it does not mean that no true difference exists: the statistical test may not be sufficiently powerful or the sample size may be too small to reveal a difference. The commonly accepted practice in biological experimentation is to tolerate a 5% level of uncertainty.

There are many computer programs for statistical analysis of data: examples are SAS,[16] RS/1,[17] BMDP[18] and PHARM/PCS.[19] Each provides such a great variety of different tests for the statistical analysis of data that it can be difficult to decide which test is most suitable for a specific dataset. As discussed in detail in the following paragraphs, the choice of statistical method will depend on: (i) if the data conforms to a normal distribution, and (ii) the complexity of the experimental design.

The Shapiro–Wilk and the Kolmogorov–Smirnov tests can be used to test whether the sample data follows a normal distribution.[14] If the data are normally distributed, then it is appropriate to use parametric statistical tests. A statistically significant difference is more easily demonstrated with parametric tests. Thus if the data are normally distributed they are more powerful; however, if the data are not normally distributed, the use of a parametric test could lead to over-interpretation of the data.

The simplest parametric test is the Student's *t*-test discussed above, in which a sample mean or difference between two sample means is evaluated for uniqueness from zero. A variation of this test, called the paired *t*-test, is used when both conditions were evaluated in the same subject. Where more than two sample groups are compared to each other, an analysis of variance (ANOVA) may be used.

The ANOVA is considered one-way if there is only one grouping factor (*e.g.* several drug doses), or two-way if there are multiple factors (*e.g.* doses and sex). If there are multiple observed variables, such as blood pressure and cardiac output, the interaction may be established with analysis of covariance, an extension of ANOVA.

The nonparametric tests[20] must be used if the data in the population from which a sample is selected cannot be assumed to be normally distributed. Nonparametric data that can be ordered (ordinal data) can be analyzed with rank-order tests, such as the Wilcoxon signed rank test, the Kruskal–Wallace test or the Friedman test. These correspond to the parametric *t*-tests, one-way ANOVA and two-way ANOVA tests, respectively. Categorical data, where counts or frequency of incidence in one or more categories are compared, can be evaluated with the Fisher exact test (one category) or the χ^2 test (more than one category).

Correlation analysis is a measure of the extent to which the value of a response is related to the value of one or more factors. It can also be used to compare several samples of data collected from the same population to determine the degree to which the two datasets are related. Multiple and partial correlations can also be used to test the degree to which two or more observed responses to the same treatment in the same animal are related.

19.1.3 PHARMACOLOGIC BACKGROUND: THE MEASUREMENT OF THE BIOLOGICAL RESPONSE TO A MOLECULE

Observations of the consequences of the introduction of a compound, a drug, into a functioning biochemical or physiological system represents the basis of pharmacology.[4,13] The compound can affect one or several biological functions in a number of ways: it might stimulate or inhibit them in a predictable and dose-dependent manner. The interaction of the drug with a primary biochemical system to alter a physiological function is called pharmacodynamics or mechanism of drug action. Inasmuch as the molecular mechanisms are frequently hypothetical and our understanding of them is continually changing, it is only natural that the primary mechanism of the pharmacodynamic action of any compound is a constantly developing concept.

A pharmacologist in quest of determining the primary mechanism of drug pharmacodynamic action must have dual expertise: (i) a thorough understanding of the molecular mechanisms involved in a particular physiological function; and (ii) skill in the application of the most fundamental pharmacological tool, 'the dose–response relationship analysis'.

The medicinal chemist is somewhat less concerned with the complexity of the physiological function itself and pays, naturally, more attention to the molecule of the biophase with which the drug combines. It is the physical–chemical properties of these molecules that determine the nature and magnitude of the drug–tissue interaction.

19.1.3.1 Dose–Response Analysis

The biological effect of a single dose of a drug has little value for characterization of its pharmacological properties. Rather, each specific biological activity of a compound has to be measured throughout the dose range from the threshold to maximum attainable effect. Dose–response analysis is essential: to provide quantitative descriptions of the activity of a compound (median effective dose, slope and maximum attainable effect); to distinguish dose ranges at which different activities of the same compound occur; and to establish the therapeutic index, *i.e.* the ratio of the median active *versus* median toxic doses.

Depending on the purposes of the study, a variety of coordinate systems are employed in recording and/or transforming the dose–response data. Mathematical equations for the dose–response relationship were derived by analogy of enzyme–substrate interaction as developed by Michaelis and Menton (hyperbola) and Lineweaver and Burk (reciprocal plots). More frequently log dose–response plots are used. The mathematical formulas, practical aspects of their use and methods of construction are comprehensively described elsewhere.[21–24]

The response of a functioning biological system to an added compound is dependent on the concentration of the drug at the site of the primary molecular interaction (and thus the administered dose) and the time required for the response to develop and dissipate. Typically, the independent variables of dose or time are plotted on the abscissa and the observed response, the effect, on the ordinate. For examples see Figure 2.

An example of the log dose–response relationship for compounds A, B and C is shown in Figure 5(a) and (b). These *agonists* exhibit the sigmoid curve typical for log dose–response relationships: the response is zero when the dose is zero and rises to a maximum attainable effect, the *intrinsic activity* (E_{max}), of the compound in the particular biological system. Further increasing the

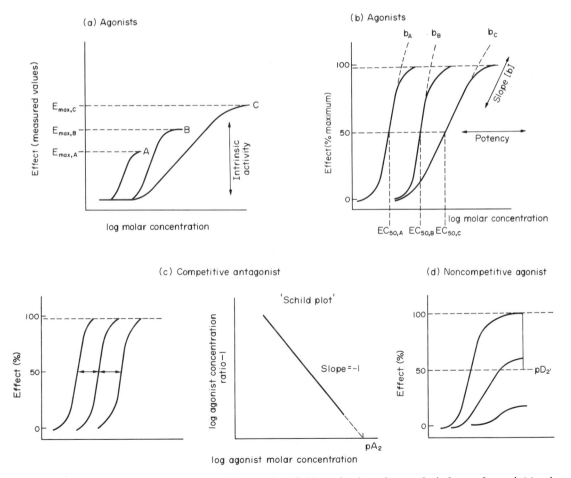

Figure 5 Hypothetical dose–response curves to illustrate the definitions of various pharmacological terms. In panels (a) and (b) the dose–response curves of three different agonists are compared. In panels (c) and (d) the dose–response curves of the agonist in the presence of different concentrations of antagonists are shown

dose (concentration) does not affect the plateau of the curve unless a new, additional mechanism is invoked. The position of the dose–response curve *vis-à-vis* the dose axis indicates the potency of the compound. The potency is usually expressed as the median effective dose ED_{50}, the dose that produces a response 50% of E_{max}. EC_{50} is used where concentration of the drug is considered rather than dose. The *slope*, *b*, of the dose–response curve provides some information regarding the mechanism of action and the binding of the drug to the receptor. Transformation of the measured values of the response into percent of that compound's maximum (Figure 5b) allows a convenient comparison of the relative potencies and slopes for a series of compounds. Parallel slopes in this transformation support the assumption that the compounds act by the same mechanism, whereas a lack of parallelism excludes that possibility.

Often more than one phase can be discerned on the dose–response curve. This indicates that the compound elicits the response by more than one mechanism. The phases may have different or equal slopes. In 'bell-shaped' dose–response curves a compound, after eliciting the response in the lower dose range, actually shuts off the response at higher doses. To properly characterize a compound in pharmacological experiments one has, therefore, to take into consideration all possible confounding phenomena and focus strictly on one mechanism of action at a time.

Certain biological responses to a compound are dependent on time (Figure 1); examples are the diuretic response (sodium excretion) or motor activity of the experimental animals. Pharmacologists often quantify such responses as the integrated area under the time–response curve. The effect is represented by the amount of activity per time unit at a certain dose of the drug. Generally, the 'peak response' provides more specific information about the mechanism of action of the compound than does the integrated response since the latter reflects additional phenomena and includes the pharmacokinetic properties of the compound. For example, the peak response in diuretics distinguishes subclasses of low-ceiling and high-ceiling agents (low and high peak excretion rates of sodium), whereas the dose–response based on the total integrated response does not provide this information.

Within certain series of compounds, the potency and duration of action might be theoretically independent phenomena and are best evaluated independently. For example, within a series of amino acid analogues of vasopressin, the inactive prodrugs administered are compounds that are converted through the activity of endogenous peptidases into active vasopressin. If one uses the area under the curve to evaluate the potencies of these compounds, the dose–response plots are usually not parallel, despite the fact that the mechanism of action of all members of the family is identical, *i.e.* *via* the *in vivo* generated vasopressin. It is, therefore, practical in such cases to measure some time-independent activity of the drug such as the peak effect as a function of dose and to evaluate the duration of action separately. In large series of compounds such as those shown in Figure 6, the individual analogues could thus be ranked by their relative potencies and their relative duration of action.[25]

The duration of pharmacological action is a very complex phenomenon that is dependent on many factors. It might or might not reflect directly the time course of the plasma levels of the compound, which is usually quantified in pharmacokinetics as a series of *half-lives* or *clearances*. A practical treatment of this problem is to calculate by analogy with pharmacokinetic calculations, first-order exponential constants of the time course of the *in vivo* response. Thus, formal onset and/or elimination constants (K; min^{-1}) are calculated as a dose-independent measure of the kinetics of the response.[25,26] Several variations of calculating the duration constants have been used, for example[25]

$$K = 2.3(\log D - \log d)/(t_D - t_d) \tag{8}$$

where D and d are a larger and a smaller dose of the given compound and t_D and t_d are the durations of the corresponding responses measured as the time in minutes from the onset of the response until its return to the base-line value. The great advantage of this representation of the drug time course to the medicinal chemist is the dose-independent nature of the duration constants. These constants can be used to calculate a dimensionless index that describes the relative duration of action (onset or elimination) of analogues for comparison within a series (Figure 6).[25]

19.1.3.2 Receptor Theory

A qualitative leap in the rational characterization of biologically active molecules for the medicinal chemist was accomplished by introducing the concept of receptors. 'Receptor theory' describes the biological response as an interaction between a molecule of the drug and a molecule of

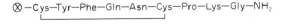

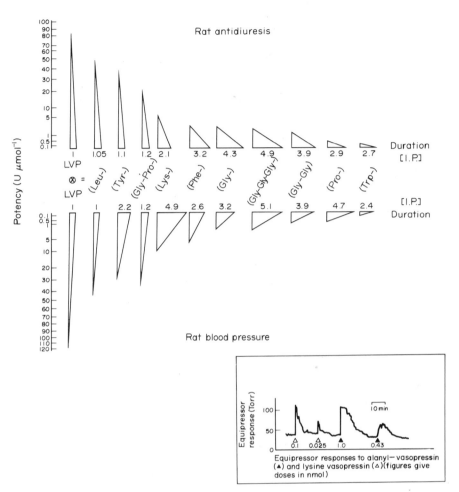

Figure 6 Rat pressor and antidiuretic potencies and indices of persistence for hormonogen analogues of lysine vasopressin (LVP) (adapted from ref. 25)

the biophase, the receptor. The interaction follows the law of mass action and the drug–receptor complex initiates the cascade of biochemical processes that control the physiological function coupled to the receptor. Medicinal chemists focus their interest primarily on the receptor and details of the formation of the drug–receptor complex.

Research in the past two decades has revealed that the receptors for the common neuro-transmitters are membrane-bound proteins.[27] For several of these the amino acid sequence is known. Thus receptors are no longer hypothetical postulates, but well-established macromolecules. They are activated by specific agonists released from nerve or endocrine cells. Once activated, a receptor elicits a change in the physiological function of the target cell such as muscle contraction or rate of glucose utilization.

Of course, the target of drug action may be a macromolecule other than a specific receptor. Ion channels, enzymes or nucleic acids are also popular targets for drug design.[27] For the purposes of this chapter, however, we shall use the term receptor in its original broad sense as the macromolecule of the biophase with which a drug must combine in order to elicit a biological response. Our principal concerns are the practical consequences of receptor theory to the quantification of the biological properties of a molecule.

Whereas the drug–receptor interaction is the primary focus of medicinal chemists for new drug design, for pharmacologists and other biologists and biochemists the receptor is of interest primarily because it is the first step in the cascade of events that lead ultimately to the biological response that is their primary focus. It is the quantitative analysis of the drug dose–response relationship which is shared equally by medicinal chemists and biologists as it inherently provides both drug-dependent and tissue/function-dependent information.[19,20]

Several physical–chemical models of receptor theory (such as occupational theory, rate theory, receptor inactivation theory and two-state models) have been advanced to explain drug–receptor interactions mathematically (for reviews see refs. 21–24). In all of them an agonist (a hormone or a receptor-specific stimulating molecule) is postulated to reversibly bind to the receptor to form a bimolecular agonist–receptor complex; this complex is the activated form of the receptor. Scheme 1 shows this diagramatically. In this context a drug could be either a mimic of a natural agonist or an antagonist that prevents the agonist action by occupying the receptor.

Although the models differ in many aspects, they all propose two basic properties of drugs:[21] (i) the capacity to bind to receptors; and (ii) the ability to stimulate receptors to produce a response. The former is usually referred to as *affinity* and the latter as *intrinsic efficacy*. Receptor theory thus provides the medicinal chemist with two biological attributes of molecules that can be derived from diverse pharmacological measurements and that are based on physical–chemical models. Reducing complex pharmacological responses to the affinity and intrinsic efficacy of compounds is, therefore, very advantageous to the analysis of structure–activity relationships and thus to modern drug design. In the following paragraphs the most frequently used terms of pharmacological nomenclature are explained using the classical equation of receptor theory as evolved in the occupational model.

The biological data obtained in drug testing do not always provide unconfounded measures of affinity and intrinsic efficacy of a compound. More often the measured phenomena are heavily affected by the special characteristics of a particular tissue or physiological function. Therefore, the analysis of the pharmacological response focuses primarily on distinguishing and separating the drug-dependent from tissue/function-dependent phenomena.

Receptor theory reduces compounds into three classes. It distinguishes: (i) *agonists*, which stimulate the receptor to initiate the physiological function; (ii) *antagonists*, which prevent the receptor from being stimulated by its natural agonist; and (iii) *partial agonists*, which possess both agonistic and antagonistic properties and thus stimulate the receptor to the level of only a partial response.

Affinity is an attribute common to agonists and antagonists. It is defined as the reciprocal of the equilibrium dissociation constant, K_D, of the drug–receptor complex. In contrast to antagonists, agonists not only possess affinity for the receptor, but activate the receptor to produce its functional response in the particular tissue—this is its *intrinsic activity*. Intrinsic activity is equal to E_{max} as shown in Figure 5(a); it describes the maximum degree of the pharmacological response attainable with the drug in the particular tissue. As seen from Figure 5, both affinity and intrinsic activity are measured in the experiment.

Since the intrinsic activity of a compound differs from tissue to tissue, agonists are best characterized by their *intrinsic efficacy*, a theoretical attribute that is independent of the biological test system. The calculation of intrinsic efficacy is based on the concept that the different agonists have a different capacity to stimulate the receptor to produce the response (the details of the

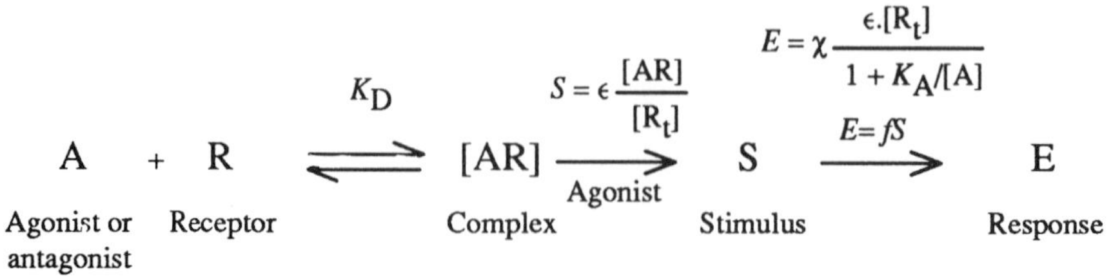

$$A \; + \; R \; \underset{}{\overset{K_D}{\rightleftharpoons}} \; [AR] \; \xrightarrow[\text{Agonist}]{} \; S \; \xrightarrow{E=fS} \; E$$

$$S = \epsilon \frac{[AR]}{[R_t]} \qquad E = \chi \frac{\epsilon.[R_t]}{1 + K_A/[A]}$$

A	+	R		[AR]		S		E
Agonist or antagonist		Receptor		Complex		Stimulus		Response

Drug dependent: K_D — equilibrium dissociation constant (K_A — agonist, K_B — antagonist)

                     $1/K_D$ — affinity

                     ϵ — intrinsic afficacy

Tissue dependent: E_{max} — intrinsic activity

                     $[R_t]$ — receptor concentration

Scheme 1 A theoretical model of drug–receptor interactions (modified according to T. P. Kenakin[21])

calculation will be discussed in Section 19.1.3.3). Only the intrinsic efficacy and affinity are strictly drug-related properties, unlike the intrinsic activity which has a strong tissue-dependent component (such as receptor concentration, R_t) that might impart confounding elements into structure–activity considerations.[21]

Caution is also necessary in the consideration of the potency of compounds, another attribute common to both agonists and antagonists. In the ideal relationship between the dose of drug and the magnitude of its response according to receptor theory, the ED_{50} corresponds to the dissociation constant for the drug–receptor complex and thus describes the compound's affinity for the receptor. However, this is not always the case since potency may not be measured in such a way that this would be true. The potency is not an independent attribute of a drug, but depends on many tissue or organ-related phenomena such as receptor concentration, drug distribution to the receptor and biotransformation. Thus, in such experiments the descriptive value of potency as a biological correlate for theoretical structure–activity analysis may be confounded.

Pharmacological antagonism arises when the agonist and the antagonist exert their effect *via* the same receptor—this is to be distinguished from physiological or therapeutic antagonism which arise from competing mechanisms of action of the agonist and antagonist. Two types of pharmacological antagonists are recognized: (i) competitive or surmountable; and (ii) noncompetitive or irreversible. They are illustrated in Figure 5(c) and 5(d), respectively. Only competitive antagonism provides useful quantitative information about the affinity of the agent for the receptor. Three conditions must be fulfilled in order for an antagonism to be competitive: (i) in the presence of antagonist, the agonist dose–response curve must move to the right along the dose axis in a parallel shift, the magnitude of which is dependent on antagonist concentration; (ii) the E_{max} of the agonist curve must not be suppressed at any antagonist concentration; and (iii) there must be a slope of unity in the 'Schild plot'.[28] In a Schild plot the logarithm of the concentration of antagonist is plotted on the abscissa against a function (S_A) of the magnitude of the shift of the agonist dose–response curve calculated from equation (9). C_0 is the concentration of agonist that produces a given response in the absence of antagonist, and the C_A is the concentration of agonist that produces the same response at the given concentration of antagonist. In a Schild plot the intercept on the abscissa is the negative logarithm of the antagonist equilibrium dissociation constant K_D, known as the pA_2 value. (Some authors distinguish the equilibrium dissociation constants of agonists and antagonists as K_A and K_B, respectively.)

$$S_A = \log[(C_A/C_0) - 1] \tag{9}$$

The pA_2 value is a good descriptor of antagonist affinity for drug design purposes. As a proof of its validity, it is usually presented together with the slope of the Schild plot from which it was derived.

Quantitative descriptions have also been developed for noncompetitive antagonists. An arbitrary measure, pD_2', is often used. It is the concentration of the antagonist that produces a 50% decline of the agonist E_{max} (Figure 5d). However, such values must be used very cautiously in drug design because the derivation does not satisfy the theoretical requirements to describe affinity. The same caution should be taken in dealing with competitive antagonists when the pA_2 values are calculated from experiments where only one concentration of antagonist was employed.

Partial agonists are agents that stimulate a receptor-mediated function to a lesser degree than the E_{max} of the natural agonist for that receptor. Typically they are described by their potency in a given tissue and E_{max} as related to the natural, 'full', receptor agonist. It is somewhat frustrating that many therapeutically useful agents are partial agonists, yet there seems to be no physical–chemical features which can be used to impart partial agonism to a molecule. The understanding of the phenomenon of partial agonism has been attempted in all main receptor theories. Occupational receptor theory interprets partial agonists by assuming a nonlinear relationship between fractional receptor occupancy by an agonist and fractional tissue response; the difference is *receptor reserve*. Receptor reserve is a drug-related variable;[21] yet the receptor reserve of a compound is characteristic of a given tissue. Since the stimulus generated by an agonist–receptor complex is proportional to not only the intrinsic efficacy of the agonist, but also the receptor density and efficiency of receptor coupling to the response mechanisms in the given tissue, a partial agonist in one tissue can act as a full agonist in another tissue.

As mentioned above, intrinsic activity is a heavily tissue-dependent attribute of an agonist. The reliance of the description of agonist activity on specific organ systems may be avoided if the intrinsic efficacy could be calculated. The *relative intrinsic efficacy* within a series of compounds *vis-à-vis* a standard agonist has a value for a medicinal chemist as it has a theoretical molecular basis and transcends species, organ type and function.

19.1.3.3 Strategy in Measurements of Affinity and Intrinsic Efficacy

There are an enormous number of biological responses that a drug can elicit which can be or have to be used in drug design. The binding of a compound to a receptor or tissue is a response that characterizes the affinity of the compound for the receptor or tissue. Cascades of biochemical events elicited by the binding of the drug can provide a multitude of measurable responses that might be quite removed from the primary specific interaction of drug and receptor and yet represent the same event. With increasing complexity, eventually the responses of a tissue (such as hormone secretion, cell contraction or frequency response) or the whole organism (such as decrease in blood pressure, increase in motor activity or sleep) are directly linked to the primary specific drug–tissue interaction and thus from the drug design standpoint might provide the same information. Therefore, in drug-response analysis it is very important for the medicinal chemist to discern the drug-dependent information provided by a given response. It is equally important for the pharmacologist to assure that the type of response selected most accurately represents the primary bimolecular drug–receptor interaction. The level of intimacy between the primary reaction and the observed response has to be considered judiciously (Table 1).

While simple drug–receptor binding in the test tube is the most intimate and least confounded response characterizing a property of the drug, it does not provide any information as to whether the drug can indeed elicit or inhibit the physiological function. Alternatively, eliciting a biological response characteristic of a certain receptor might or might not provide reliable information about the affinity of the drug. Therefore, to characterize biological properties of a molecule, a simultaneous evaluation of a multitude of responses might need to be correlated in a hierarchy of complexity at various levels of intimacy to the primary reaction. For example, in Scheme 2 is a strategy for evaluating adrenergic agonists and antagonists in a program attempting to design agents selective for the two adrenergic receptor subtypes. The primary strategy focused on the molecular mechanism of action. For this purpose we carried out parallel evaluation of compounds in radioligand binding assays and functional tests in contractile tissue *in vitro*. The secondary strategy focused on mechanisms of therapeutic reaction and addressed aspects of drug pharmacokinetics *in vivo*.

19.1.3.3.1 Radioligand binding studies

The introduction of radioligand binding technology has greatly facilitated the evaluation of the affinity of a drug for a given receptor. The semipurified receptors in fragmented or intact cellular

Table 1 Information Relevant to Molecular Design Obtained from Various Pharmacological Experiments

	Radioligand binding	Functional tests		
		In vitro	In vivo	
Degree of intimacy between binding and response	$+++$	$++$	$+$	
Biological property revealed	Affinity	Affinity Intrinsic activity Intrinsic efficacy Antagonism Potency Reversibility Receptor reserve	Affinity Intrinsic activity Intrinsic efficacy Antagonism Potency Reversibility Receptor reserve	Drug disposition Pharmacokinetics Therapeutic activity Toxicity

Confounding influences	Secondary receptors
	Specificity of ligand
	Receptor heterogeneity
		Secondary receptors
		Specificity of ligand
		Receptor heterogeneity
		
		Tissue dependent factors:
		  receptor concentration
		  intrinsic activity
			Secondary receptors
			Specificity of ligand
			Receptor heterogeneity
			
			Tissue dependent factors:
			  receptor concentration
			  intrinsic activity
			
			Drug disposition
			Whole body integration
			  of studied function	

Therapeutic relevance	$+$	$++$	$+++$	

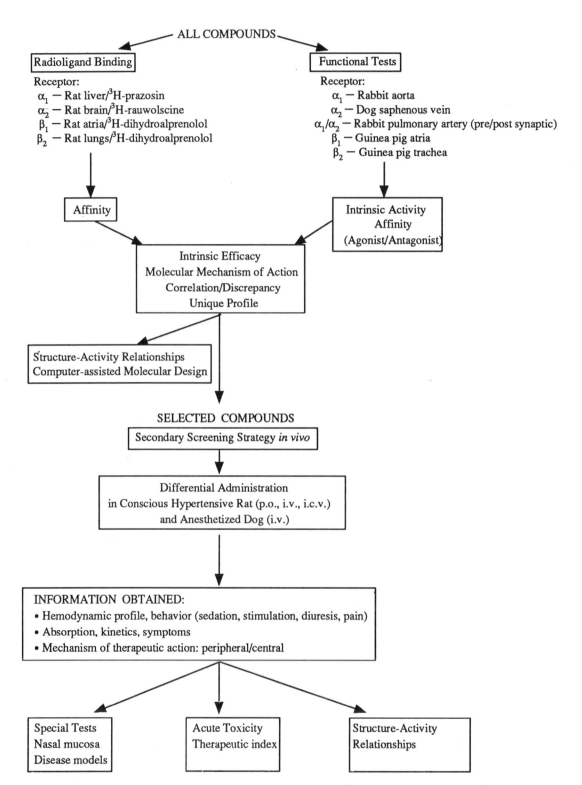

Scheme 2 Primary screening strategy *in vitro* for design of selective adrenergic agents, and secondary screening strategy *in vivo* for therapeutic relevance and profile optimization of selected compounds

membranes can be harvested from almost any tissue, and almost any ligand can be radiolabeled to test its affinity and/or be used in a displacement assay. Previously, only very few functional tests allowed one to measure affinity with such efficiency, accuracy and theoretical soundness. In the past, the biological effect of a compound was observed first and the affinity information was extracted from its quantitative responses. Radioligand techniques, on the other hand, measure the drug's affinity without any proof of its biological effects. Thus, the affinities derived from binding studies, although very valuable and pertinent to drug design, should not be considered alone without a proof of certain assumptions that assure their validity. One such assumption is that the compound is an agonist or an antagonist: this has to be demonstrated in a functional test. Another assumption to be satisfied is that the receptors involved in the functional and binding assays are the same. A final, very important assumption is that of specificity of the receptor–radioligand test compound interaction; the results could be considerably confounded if additional or nonfunctional binding sites are involved; such problems are revealed in the analysis of the binding curve.[21]

19.1.3.3.2 *Functional studies*

For functional studies one can easily evaluate affinity for antagonists by measuring their pA_2 values. The affinity evaluation of agonists in functional tests is more complicated for theoretical reasons, although techniques have been developed[29] to measure affinity of α adrenergic agonists through fractional inactivation of the receptor by alkylating agents. Therefore it seems to be most prudent to describe agonist action by both relative potency and intrinsic activity and to keep in mind the tissue dependency of these variables. Ideally, one would like to determine the tissue-independent values of intrinsic efficacy, but that might not always be practical in functional tests.

However, one can calculate relative intrinsic efficacies within a series of agonists by using affinities obtained from binding assays and intrinsic activities from functional tests. The calculation is based on the assumption that these binding constants reflect the true affinities in the isolated tissue assays. The intrinsic efficacies relative to the standard agonist are estimated from the relationship of fractional occupancy *versus* response.[21,30,31] The example in Figure 7 shows the occupancy of receptors (derived from the relationship $[A/(A + K_1)]$ where A is the concentration of agonist and K_1 the equilibrium dissociation constant from binding studies) plotted on the abscissa, and the intrinsic activity (percent of maxima from the functional test) plotted on the ordinate. Notice that A-54741 produces its half-maximal effect when less than 10% of the receptors are occupied. For norepinephrine, roughly 50% of the receptors are occupied at the EC_{50}. Thus, any structural model of the interaction of these compounds with the α_1 receptor must also explain the structural basis for this interesting observation.

The evaluation of the drug–receptor interaction in the functional *in vitro* tests requires that special attention be paid to the specificity of both the tissue and the compound. Very few compounds display an exclusive specificity for a single receptor, and very few tissues possess a homogeneous population of a single type of receptor capable of eliciting the response. Therefore, it is necessary to employ selective standard agents in nonselective tissues and selective tissues for nonselective agents. An example of this strategy is shown in Figure 8. An independent assessment of α_1 and α_2 adrenergic receptor interactions was carried out in a series of standard adrenergic agonists, agents with reported variable selectivity for these two closely related receptors.

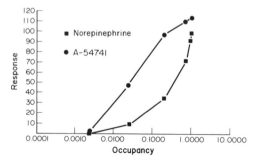

Figure 7 An example of the data analysis used to establish the relative intrinsic efficacy of two agonists. For each dose the fractional occupancy of the α_1 receptor was calculated from K_D's established from radioligand binding (adapted from ref. 30)

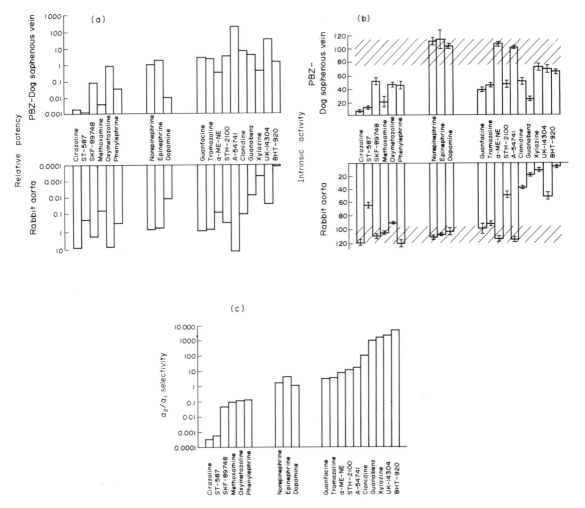

Figure 8 An analysis of the α_1- *versus* α_2-selectivity of 19 standard adrenergic agonists: (a) the relative potency $I = (EC_{50}$ compound$/EC_{50}$ NE$)$ and (b) contractile intrinsic activity E_{max}(% of NE control) at the α_2 (upper) and α_1 (lower) receptors, and (c) the selectivity based on relative potency α_2 *vs.* α_1 from (a)

For the evaluation of the α_1 interaction we selected the rabbit aorta, which is known to have a homogeneous population of α_1 receptors that mediate its contraction. Thus even nonselective compounds would act in this tissue strictly by virtue of the α_1 affinity. Dog saphenous vein was selected as the best available tissue for the α_2 interaction even though its contractile response is mediated by both α_1 and α_2 receptors. In order to use this tissue for evaluation of the α_2 component of nonselective compounds, we alkylated the α_1 receptors with phenoxybenzamine to render the tissue a homogeneous α_2 preparation.[32,33] Figure 8 illustrates the differentiating capacity of these two selective tissues for the independent evaluation of α_1 and α_2 receptor interaction of the 19 standard α adrenergic agonists. The compounds are grouped according to their reputed selectivity from left to right as α_1 selective, nonselective and α_2 selective agents, respectively. The figure shows that even the most selective agents have a component of interaction with the other α-adrenergic receptor. In addition, the figure also shows that some of the standard agents, although relatively selective, possess in these tissues intrinsic activity such as to allow them to be classified as only partial agonists. Finally, the comparisons of the intrinsic activities (E_{max}), and relative potencies put in perspective the true nature of relative specificity and selectivity of the compounds for the α_1 and α_2 receptors as pertinent properties for molecular design considerations.

19.1.3.3.3 *Correlation of binding and functional data*

While the rapid expansion of the binding technology will no doubt bring about further development in receptor theory and might satisfy certain critical assumptions in the future, at present the validity of the binding data is best assured through judicious multivariant correlations with functional data. While a strong correlation suggests that the assumptions might be correct, a lack of correlation definitely invalidates the data.

An example from our laboratory shows a correlation of the α_1 receptor affinity derived from binding assays (using rat liver receptors and ^{3}H-prazosin) *versus* that derived from contractile responses in rabbit aorta (agonist phenylephrine) for 31 competitive antagonists (Figure 9). Since a strong correlation has been established, this supports the assumption of the identity of the rat liver and rabbit aorta receptors, as well as the receptor selectivity defined by ^{3}H-prazosin and phenylephrine on one hand, and that of the utilized tissues on the other hand. Obviously, for drug design purposes the affinity values obtained from the simple radioligand binding experiments and functional tests are equivalent.

Although the correlations between binding and functionally derived affinity is relatively straightforward for the antagonists, it is much less so for the agonists. The technique[32] of measuring the α agonist affinity for receptor subtypes in functional tests through partial receptor inactivation by phenoxybenzamine is no longer practical for certain agonists since phenoxybenzamine inactivates the α_1 and α_2 receptor unevenly.[30, 34] Therefore, potency (EC_{50}) rather than affinity (K_D), derived from contractile response in the rabbit aorta, was used in the correlation with the binding-derived affinity (Figure 9). Several interesting observations were made: (i) within a group of 19 agonists (designated as full agonists, *i.e.* $E_{max} > 75\%$ of norepinephrine), there was a very good correlation between binding affinity in rat liver membranes and potency in rabbit aorta; (ii) the slope of the

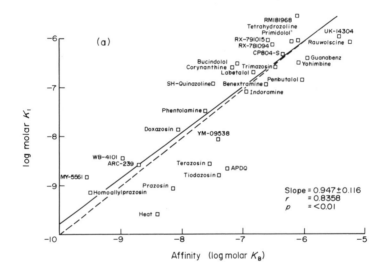

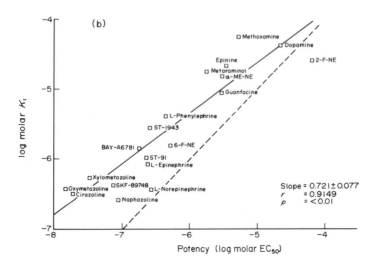

Figure 9   A demonstration of the correlation between functional and binding data for α_1 antagonists (a) and agonists (b). The solid line is the linear correlation and the broken line that for equivalent affinity and potency

correlation was significantly different from unity, which might be explained by several structure-independent mechanisms such as experimental conditions of the receptor in the two systems; (iii) the partial agonists ($E_{max} < 75\%$ of norepinephrine; not shown in Figure 9) did not display any correlation and; (iv) agents of one specific structural family (not shown in the figure), although powerful full agonists in the functional assay, where they displayed competitive relationship with prazosin, failed to displace ^{3}H-prazosin in the binding studies. These phenomena illustrate problems associated with the nondiscriminative use of the binding affinity data in the drug design of agonists.

On the other hand, the binding data are very important for the final validity and interpretation of the *in vitro* functional tests. Figure 10 is an example. The selective tissues and selective agonists were used to determine whether atriopeptin (ANF) possesses α-blocking properties. The figure shows that ANF does not interact with the α_2 receptor since it does not inhibit norepinephrine-mediated contraction in the α_2 selective tissue, nor does it inhibit the α_2 selective agonist, BHT-920. These results are in good agreement with the results of radioligand binding (not shown), since ANF does not displace [^{3}H]-rauwolscine, a standard α_2 ligand. On the other hand, the functional data opened the possibility that ANF might be an α_1 blocker since it antagonizes the effects of an α_1 agonist, and in the α_1 selective tissue displaces the α agonist in a 'competitive-like' manner ($b \sim 1$). However, since ANF does not displace [3H]-prazosin in binding studies, the above results cannot be interpreted as α_1 inhibition and an α-receptor independent mechanism must be involved.[30]

19.1.3.4 Pharmacokinetic Comparisons of Molecules

The ultimate goal of drug design is to discover a compound that is active in living intact organisms. Thus, frequently one wishes to optimize within the series the rate and extent of absorption of the compound from its site of administration, minimize biotransformation to inactive products, or delay excretion into the urine.

The classic experiment to characterize such properties of molecules is to measure blood or plasma levels of the compound and metabolites as a function of time. Many different pharmacokinetic models of the relationships between rates of transfer of compound between various compartments of different volumes have been devised and shown to fit such observed data.[35] Usually the compounds are characterized by one or more rate constants for the transfer of compound from one idealized compartment to another, and volume ratio terms that are analogous to a solvent–water partition coefficient.

In the case of a straightforward intravenous administration, one sometimes observes a simple exponential decline of the plasma level (equation 10)

$$C_t = \frac{X_0}{V} e^{-Kt} \tag{10}$$

in which C_t is the plasma level at time t in $mg\,l^{-1}$, X_0 is the dose in $mg\,kg^{-1}$, V is the relative volume of distribution of the compound in $l\,kg^{-1}$, and K is the first-order rate constant for disappearance of the compound from plasma by excretion or metabolism. The relative volume of distribution describes how many kg of tissue are needed to contain the equivalent of the drug in $1.0\,l$ of blood plasma. Thus higher values of V suggest that more of the drug is in tissues—values greater than 1.0 suggest tissue binding, whereas values less than 0.2 suggest that drug is found only in the circulating blood.

Much more elaborate pharmacokinetic models have been derived to fit plasma level *versus* time curves measured following single or multiple dose oral administration or continuous intravenous infusion and/or for disappearance curves that are not monoexponential.[35]

One common mistake in interpretation is to compare the peak plasma levels or areas under the plasma level–time curve and to take these as measures of the relative pharmacologic or antibacterial effectiveness of members of a series of compounds. Such a comparison is valid strictly only for plasma level measurements of active drug following the administration of different prodrugs of the same active compound. The comparison is not valid if the active compounds distribute at different rates to the target tissue or if the ratio of partitioning between the plasma and the target tissue is not constant across the series. Rather a full pharmacokinetic calculation should be done for each compound and for at least some of them one should try to correlate the plasma and tissue levels with the 'therapeutic' property to be optimized. With these data in hand, one can decide if the appropriate pharmacokinetic property to optimize is the area under the plasma level *versus* time curve or that under the tissue level *versus* time curve.

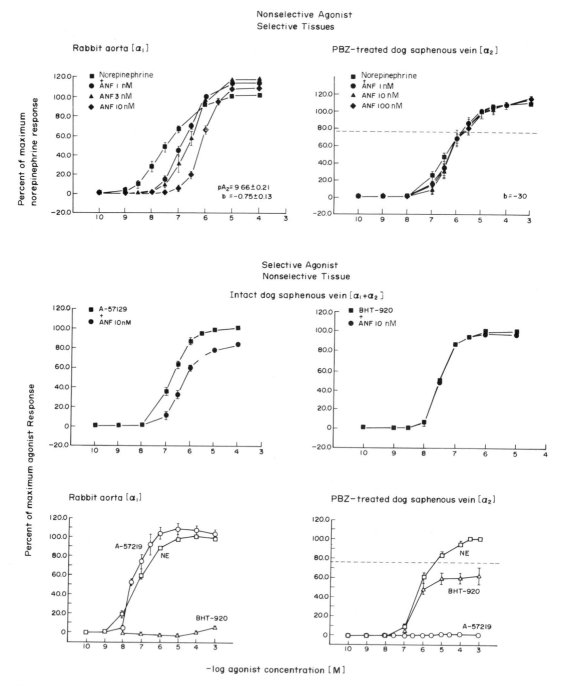

Figure 10 Plots showing an assessment of the hypothetical α-adrenergic activity of atriopeptin (adapted from ref. 42 and 43). In the top row are shown the responses of α_1 and α_2 selective tissues to norepinephrine and ANF. In the second row are shown the minimal inhibition by ANF of the effect of the α_1 agonist A-57129 and lack of effect on the α_2 agonist BHT-920. The bottom row shows the data that support the selectivity of the latter agonists

A second common mistake is to correlate raw data (such as percent absorbed) rather than data transformed by a model into more physicochemically relevant parameters (such as a rate constant).

19.1.3.5 Artifacts in Biological Response Determination

There are many factors that can affect the responses to a treatment and thus interpretation of drug effects.[7-13] Fortunately, many factors can be controlled to decrease response variability. For example, all measurement devices should be calibrated routinely. This principle applies not only to the devices that generate the response data (*i.e.* polygraphs, A/D converters, spectrophotometers), but also to the balances and pipettors used to weigh and dilute drug samples. Devices must be operated only within the limits of reasonable documented performance. Additionally, in our laboratories, a standard condition or test is incorporated into all runs of a protocol to verify the interassay reproducibility. For example, an isolated rabbit aorta's responsiveness to a tested adrenergic agent can be verified by also testing the response of (−)-norepinephrine. Careful technique, intraassay replication and the operating range for measuring devices will maximize the precision of measurement.

Several factors can interfere with a response measurement. The composition and purity of the tested samples, reagents and diluents should be known. If possible contaminants of a test compound are extremely potent, this requires a very stringent test for purity, because only a trace of the contaminant material could produce a false positive test for an inactive compound. This is especially pertinent in the assay of the less potent of a pair of stereoisomers. It is common to discover that the vehicle for a compound has biological activity itself or interferes with the assay or stability of the compound. Reproducibility of sample preparation is also important. For example, sample-to-sample differences in buffer pH, composition, strength, concentrations of contaminants or metabolites, or temperature can drastically affect a biochemical test. Especially for *in vitro* assays, there is a problem with agents that: (i) adhere to glass or plastic surfaces; (ii) are not stable because of photolytic, enzymatic or chemical sensitivity; or (iii) are not soluble. Although automatic dilutor/pipettors can increase a laboratory's productivity greatly if carryover of sample is negligible, the negligibility must be verified.

Equal attention must be paid to the properties of the test system. Biological preparations, especially *in vitro* tissues and partially purified receptor/enzyme preparations, can deteriorate and controls must be run to establish the integrity of the preparation. Microbiological growth can occur in nutrient buffers and media; prevention of this requires rigorous cleaning of reusable laboratory apparatus, proper storage of solutions, and incorporation of microbiocides where possible.

Even in *in vitro* assays one rarely uses purified receptors or enzymes; instead one uses cell fraction or membrane preparations, whole cells, organs or intact animals. Thus, there is a distinct possibility of artifactual binding to a component of the assay system.

A further complication is that there are usually at least several different receptors for natural neurotransmitters; there are at least four for norepinephrine and serotonin, three for acetylcholine, and two for dopamine.[27] Until the recent cloning of a few receptors, these different receptor subtypes could be distinguished only by artificial ligands synthesized or isolated by the medicinal chemist. For the above reasons a careful pharmacologist is constantly aware of the possibility that the assay system may contain more than one macromolecular receptor. A biphasic dose–response curve or one with less than theoretical slope may suggest two populations of receptors.

Finally, the affinity of a receptor for a ligand or an enzyme for an inhibitor may increase by several orders of magnitude upon the addition of a third substance. For example, membrane-bound receptors often are also bound to a protein that binds GTP, a circumstance that increases affinity for an agonist.[27] Similarly, the affinity of dihydrofolate reductase for methotrexate increases substantially in the presence of the cofactor NADPH.[36] Indeed a substantial part of the binding site for methotrexate is made up of close contacts with the cofactor and not the enzyme. If the concentration of a second ligand is such that the target protein is not saturated, then the binding curve for the other ligand will appear biphasic or shallow. Thus shallow or biphasic dose–response curves are hints that something is not ideal, but it takes more experimentation to understand the reason for this.

19.1.4 SELECTION OF BIOLOGICAL DATA FOR STRUCTURE–ACTIVITY ANALYSIS

19.1.4.1 Criteria for Receptor Mapping

In certain cases the objective of the laboratory and theoretical work is to discover as many atomic features of the receptor or target enzyme as is possible from structure–activity analyses of the ligands that bind to it; alternatively, one might wish to establish the active conformation of a natural substance from a structure–activity analysis of conformationally constrained analogues.[3, 37] This implies that one has at least a simple yes/no categorization of *in vitro* binding of compounds to the receptor or enzyme.

First one must decide if all the compounds may be included in the same structure–activity analysis. One must be certain that the compounds indeed displace each other from the receptor or enzyme. For receptors usually this means that one or both of the compounds is labeled and its displacement by the other can be measured. One would also expect that all compounds show the same slope of the dose–response curve. A final test compares the ability of compounds to displace the lower-affinity *versus* the higher-affinity of a pair of enantiomers. This ratio should be identical in compounds that interact with the same receptor.

However, if the radioligand binds to several receptor or enzyme subtypes with similar affinity, and each unlabeled compound binds to only one of the subtypes, it might be difficult to interpret the data. Thus apomorphine (**1**) would be a poor choice for a radioligand for D-1 receptors because although it binds to D-1 receptors it also binds to D-2 receptors.[38] On the other hand, compounds (**2**) and (**3**) are selective for D-1 and D-2 receptors, respectively.[38] As discussed above, the biologist can sometimes increase the selectivity of the assay by choosing a tissue that has only one receptor or enzyme subtype. Alternatively, selective destruction of one of the subtypes may be used.

(**1**)   (**2**)   (**3**)

It may be difficult to establish that the effect of apparent antagonists or inhibitors is indeed receptor or enzyme mediated. Certain compounds disrupt the membrane environment in which many target proteins are located. A signal that such an effect is occurring is if the compound seems to be an antagonist of too wide a variety of receptors.

Although at first blush whole animal tests may seem to offer little direct evidence of affinity of compounds for a receptor, measures of the biological response pertinent to drug design might be obtained. For example, the measurement of water and ion excretion following the intravenous administration of a diuretic may be a very direct measure of the intrinsic diuretic properties of a compound. Another example might be the increase in renal or mesenteric blood flow in adrenoceptor-blocked animals as a measure of its D-1 dopaminergic agonist properties.

There is no necessity to believe that agonists and antagonists of a particular receptor necessarily interact with the same conformation and/or binding pocket of the receptor. Thus it is not correct to include both agonists and antagonists in deriving a receptor map. (The maps for the two may be compared after derivation and combined at that time if so indicated.) This means that it will be necessary to perform a functional assay to distinguish agonists from antagonists. Again one must be alert to the possibility of artifacts that could be misleading.

If one is quite certain that an enzyme inhibitor binds to the same site on the protein as does the substrate, one may include substrates and inhibitors in the same qualitative analysis.

To date no one has established the three-dimensional structure of an activated receptor–agonist complex. Thus there are no structural guidelines to help one decide what to do about including partial agonists in a receptor map. A conservative approach is to exclude them, at least those that have less than 50% the effect of the full agonists, and to use the receptor mapping results on the full agonists to try to understand why certain compounds are partial agonists.

How can one use the K_i to decide what compounds should be considered as part of the active class? It is probably obvious that the type of biological properties that can be predicted from such an analysis will depend on the biological properties that were considered in its derivation. In other words, if only strongly bound compounds are included, the analysis will predict only strongly bound compounds and miss weaker ones; on the other hand, if everything that binds to the receptor or enzyme is included, then the analysis will predict a wider range of compounds to bind. This means that the cut-off between active and inactive should reflect the goals of the work.

In our own work we derive three-dimensional coordinates of the most potent conformationally defined compounds first.[2] Weaker compounds are modeled later and only if they will supply unique and unambiguous structural information or if it is expedient to do so. Two possible definitions of highly active compounds are: (i) those that bind at least as strongly as the natural ligand; or (ii) those that bind within 100-fold of the most strongly bound ligand. These definitions can be in error if one is comparing compounds that differ in receptor occupancy at the EC_{50}. Thus one should verify that there is a constant ratio between binding affinity and EC_{50}.

What biological assays establish a compound to be inactive? If a compound does not displace radioligand but is soluble, is chemically and metabolically stable, and does not react with any of the components of the test system, then one can probably conclude that it does not bind to the target site on the biomolecule. Often some compromise definition is needed. If the compound is not soluble, it may be mixed with serum albumin to increase its solubility. What if the compound has extraordinarily high affinity for serum albumin compared to its affinity for the target? It would appear to be inactive because not enough was available in solution to bind to the receptor or enzyme. Similar comments can be made about the use of water-miscible organic solvents such as propylene glycol or DMSO: since they dissolve the compound by solvating it, the compound has to shed the solvent shell before it can bind to the receptor or enzyme. At a minimum both the biologist and the chemist should be alert to apparent discrepancies in the data and try to devise tests to be certain that the classification of certain compounds as inactive is on firm ground. Since the above considerations often lead to the exclusion of certain compounds from consideration, the theoretical analysis is not complete until one can understand why these compounds are outliers.

19.1.4.2 Criteria for Discriminant Analysis

Sometimes one is not sure that a biological response is due to interaction with a receptor or other protein, but one still would like to know what it is that distinguishes active from inactive molecules so that other active compounds can be designed.[39]

If possible, it is sensible to consider in one's analysis only those compounds that exert the effect by the same mode of action. For example, one could imagine that it might be harder to devise a function that distinguished active from inactive β-lactam antibiotics at the same time that it distinguished active from inactive sulfonamides than it would to devise one function for each class of compound. The necessary information may be available from studies of synergism or competition with other compounds, it might be suggested by the profile of activities of the compounds, or it might be assumed from the structures of the compounds.

The ideal discriminant analysis considers compounds whose activity classes have been assigned by the biologist who tested them. Ideally also there is a clear distinction between 'active' and 'inactive' compounds.

However, sometimes the chemist must assign the compounds to groupings. As with receptor mapping, for the active class one should usually consider only those compounds whose potency falls within a certain range. Probably this should be not more than 100-fold, because it would be frustrating to predict a compound to be active and discover that it has only 1/1000 the potency of the most potent analogue. (Additionally, the omission of weakly active compounds often makes it easier to find a function that classifies the data.)

The inactive class is more difficult to define in the absence of a receptor-binding assay. Compounds might be inactive because they are metabolized or chemically decompose before they can exert their effect, because they bind to functionally irrelevant components of the assay system, or because they have multiple effects on the system. An example of the latter effect was seen in our work on the effect of compounds on the partial pressure of oxygen in the coronary sinus.[40] Certain compounds raised coronary P_{O_2} at low doses but decreased it at high doses. The explanation was found when it was discovered that at the higher doses the compounds lower blood pressure. If the dose of the compound lowered blood pressure too far, then the perfusion pressure of the coronary sinus is lowered to such an extent that the partial pressure of oxygen was lowered also. Such experiences suggest that a dose–response curve be used to verify that the inactive compounds are indeed inactive at all non-toxic doses.

19.1.4.3 Criteria for QSAR Analyses

For QSAR one usually describes each analogue as a parent molecule to which substituents have been added.[1] The change in potency as substituents are changed is correlated with the effect of these same substituents on the various types of physicochemical equilibrium constants, such as changes in the logarithm of the octanol–water partition coefficient. Alternatively, one may use a free-energy descriptor derived from theoretical calculations. Therefore, for QSAR it is most logical also to describe the biological properties of the molecule in terms of some sort of equilibrium or rate constant. Since potency is to be predicted, the relative potency of each existing compound must be supplied. The standard deviation of each potency value sets the ultimate precision that can be expected of the QSAR.

As with the receptor- or enzyme-mapping analyses, one should have information that all compounds have the same mechanism of action and that inactive compounds are truly inactive.

The nature of the statistical methods of regression or partial least squares analyses is such that the best result is obtained if one includes a large number of compounds that also show a wide range in potency; the wider the potency range the better.[1] In an ideal dataset there is an equal number of compounds tested in each potency interval. Additionally, the statistical nature of QSAR makes one aware of the advantage of testing as many compounds as possible.

These requirements may suggest that time be spent measuring potency values of relatively inactive analogues even though they are not of direct interest in the program except to avoid making more like them. This biological testing is not necessary if an analysis of the more potent analogues correctly predicts the potency of compounds that are weakly active. Similarly, one may compare the predicted potency with the best estimate of the potency of insoluble compounds to support or reject a possible QSAR equation.

For QSAR one may choose to correlate data that do not directly reflect an equilibrium or rate constant if those data are of more interest. Further, if substructural features are used as molecular descriptors then one could also include compounds with multiple modes of action in the same analysis.

In other cases one may gain important information from an analysis of the ratio of two potency measures of a compound. For example, if oral bioavailability of a set of compounds is to be optimized, one might investigate: (i) the ratio of oral to intravenous; (ii) area under the time–blood level curve; (iii) total urinary excretion; or (iv) total time-integrated response. A further example is seen in the practice of microbiologists to explain *in vivo* antibacterial activity of a compound as a function of its blood level, binding to serum proteins and *in vitro* antibacterial potency.

19.1.4.4 Criteria for Statistical Analysis of Activity Profiles of Compounds

Again, the first criterion is that the data on which the analysis is based are relevant to the objective of the synthetic program. For the most direct comparison, the various compounds should be tested under the same conditions in the variety of tests. For example, potential antibiotics would be tested under identical conditions for inhibition of a number of different species or strains of bacteria, or potential enzyme inhibitors would be tested under identical conditions against a number of enzymes. The problem is that biology might not cooperate: not all microorganisms grow in the same conditions, nor do all enzymes catalyze reactions under the same conditions. Such simple variables as pH, temperature, or the presence or absence of specific ions might drastically affect one's ability to measure one aspect of the profile of the compounds. Sometimes each biological response is measured under its optimum conditions. One must then be alert to the differences in conditions of the assays so that potential artifacts can be avoided. Again one is faced with the problem of making choices with an eye to the planned use of the structure–activity analysis.

19.1.5 REFERENCES

1. Y. C. Martin, 'Quantitative Drug Design', Dekker, New York, 1978.
2. Y. C. Martin and E. B. Danaher, in 'Receptor Pharmacology and Function', ed. M. Williams, P. Timmermans and R. Glennon, Dekker, New York, 1988, p. 137.
3. Y. C. Martin, E. Kutter and V. Austel, 'Modern Drug Research: Paths to Better and Safer Drugs', Dekker, New York, 1988, p. 161.
4. C. Hansch, P. G. Sammes and J. B. Taylor (eds.), 'Comprehensive Medicinal Chemistry', Pergamon Press, Oxford, 1989, vol. 1.
5. A. E. Bird and A. C. Marshall, *Biochem. Pharmacol.*, 1967, **16**, 2275.
6. W. W. Daniel, 'Biostatistics: A Foundation for Analysis in the Health Sciences', Wiley, New York, 1974.
7. T. P. Kenakin, *Pharmacol. Rev.*, 1984, **36**, 165.
8. H. I. Yamamura, S. J. Enna and M. J. Kuhar (eds.), 'Neurotransmitter Receptor Binding', 2nd edn., Raven Press, New York, 1985.
9. T. Chard, 'An Introduction to Radioimmunoassay and Related Techniques', 2nd edn., 'Laboratory Techniques in Biochemistry and Molecular Biology', vol. 6, part II, Elsevier, Amsterdam, 1982.
10. W. H. Evans, 'Preparation and Characterization of Mammalian Plasma Membranes', 'Laboratory Techniques in Biochemistry and Molecular Biology', vol. 7, part I, Elsevier, Amsterdam, 1980.

11. R. L. P. Adams, 'Cell Culture for Biochemists', 'Laboratory Techniques in Biochemistry and Molecular Biology', ed T. S. Work and R. H. Burdon, vol. 8, Elsevier, Amsterdam, 1980.
12. W. I. Gay, (ed.) 'Methods of Animal Experimentation', Academic Press, New York, 1965, vol. I; 1986, vol. VII.
13. A. Schwartz (ed.), 'Methods in Pharmacology', Plenum Press, New York, 1971, vol. 1; D. M. Paton (ed.), 1985, vol. 6.
14. A. A. Hancock, E. N. Bush, D. Stanisic, J. J. Kyncl and C. T. Lin, *Trends Pharmacol. Sci.*, 1988, **9**, 29.
15. G. P. Quinn, J. Axelrod and B. B. Brodie, *Biochem. Pharmacol.*, 1958, **1**, 152.
16. 'SAS User's Guide: Statistics Version, 5th Edition', SAS Institute, Cray, NC, 1985.
17. 'RS1 User's Guide', BBN Software Products, Cambridge, 1987.
18. W. J. Dixon, M. B. Brown, L. Engelman, J. W. Frane, M. A. Hill, R. I. Jennrich and J. D. Toporek, 'BMDP Statistical Software. 1983 Printing with Additions', University of California Press, Berkeley, 1983.
19. R. J. Tallarida and R. B. Murray, 'Manual of Pharmacologic Calculations with Computer Programs', 2nd edn., Springer-Verlag, New York, 1987.
20. S. Siegel, 'Nonparametric Statistics for the Behavioral Sciences', McGraw-Hill, New York, 1956.
21. T. P. Kenakin, 'Pharmacologic Analysis of Drug–Receptor Interaction', Raven Press, New York, 1987.
22. R. J. Tallarida and L. S. Jacob, 'The Dose–Response Relation in Pharmacology', Springer-Verlag, New York, 1979.
23. E. J. Ariëns, G. A. van Os, A. M. Siminis and J. M. van Rossom, 'A Molecular Approach to Pharmacology', Academic Press, New York, 1964.
24. E. M. Ross and A. G. Gilman, 'The Pharmacological Basis of Therapeutics', ed. L. S. Goodman and A. Gilman, Macmillan, New York, 1986, pp. 35–48.
25. J. Kyncl, K. Rezabek, E. Kasafirek, V. Pliska and J. Rudinger, *Eur. J. Pharmacol.*, 1974, **28**, 294.
26. V. Pliska, *Arzneim.-Forsch.*, 1966, **16**, 886.
27. C. Hansch, P. G. Sammes and J. B. Taylor (eds.), 'Comprehensive Medicinal Chemistry', Pergamon Press, Oxford, 1989, vol. 3.
28. O. Arunlakshana and H. O. Schild, *Br. J. Pharmacol.*, 1959, **14**, 48.
29. R. F. Furchgott and P. Bursztyn, *Ann. N.Y. Acad. Sci.*, 1967, **44**, 882.
30. J. J. Kyncl, J. F. DeBernardis, E. N. Bush, S. A. Buckner and H. Brondyk, *J. Cardiovasc. Pharmacol.*, 1989, **13**, 382.
31. T. P. Kenakin, *J. Pharmacol. Methods*, 1985, **13**, 281.
32. J. J. Kyncl, S. A. Buckner, E. N. Bush, J. F. DeBernardis, C. T. Lin and R. B. Warner, *Fed. Proc., Fed. Am. Soc. Exp. Biol.*, 1985, **44**, 1466.
33. P. B. M. W. M. Timmermans and P. A. Van Zwieten, *J. Med. Chem.*, 1982, **25**, 1389.
34. K. P. Minneman, *Eur. J. Pharmacol.*, 1983, **94**, 171.
35. M. Gibaldi, 'Biopharmaceutics and Clinical Pharmacokinetics', 3rd edn., Lea and Febiger, Philadelphia, 1984.
36. J. T. Bolin, D. J. Filman, D. A. Matthews, R. C. Hamlin and J. Kraut, *J. Biol. Chem.*, 1982, **257**, 13650.
37. C. Humblet and G. R. Marshall, *Drug Dev. Res.*, 1981, **1**, 409.
38. P. Seeman and H. B. Niznik, *ISI Atlas Pharmacol.*, 1988, **161**, 1988.
39. Y. C. Martin, 'Quantitative Drug Design', Dekker, New York, 1978, p. 242.
40. Y. C. Martin, in 'Strategy in Drug Research', Pharmacochemistry Library, ed. J. A. Keverling Buisman, Elsevier, Amsterdam, 1982, vol. 4, p. 269.
41. P. J. Munson and D. Rodbard, *Anal. Biochem.*, 1980, **107**, 220.
42. J. J. Kyncl, S. A. Buckner, J. F. DeBernardis, M. Winn, H. Brondyk and R. E. Dudley, in 'New Cardiovascular Drugs', ed. A. Scriabin, Raven Press, New York, 1987, p. 251.
43. J. J. Kyncl, E. N. Bush and S. A. Buckner, *J. Cardiovasc. Pharmacol.*, 1987, **10** (suppl. 4), S87.
44. J. W. Constantine, W. Lebel and R. Archer, *Eur. J. Pharmacol.*, 1982, **85**, 325.

19.2
Molecular Structure and Drug Transport

JOHN C. DEARDEN
Liverpool Polytechnic, UK

19.2.1 INTRODUCTION

It has been acknowledged from the very earliest days of the study of drug action that transport plays an important role in controlling biological response. Thus Overton[1] in 1899 interpreted the correlation between tadpole narcosis and the partition coefficient as being due to the ability of compounds to penetrate the lipoidal membranes of the tadpole.

The history of drug transport is, in effect, the history of partitioning; although it has never been proved unequivocally that drugs move through a living organism *via* a partitioning mechanism, it is universally accepted that they do, for all the evidence points in that direction. That evidence will be reviewed later in this chapter.

Drug transport, by its very nature, is a dynamic process, and so it is perhaps ironic that, in the correlation of drug action with molecular structure, many of the properties used as structural or physicochemical parameters (*e.g.* the partition coefficient (P) and the dissociation constant) are equilibrium properties. It may be this fact that has led to the relative neglect of time as an important variable in drug action. However, by definition, an equilibrium constant is the ratio of forward and reverse rate constants. Thus, in partitioning, $P = k_{wo}/k_{ow}$, where k_{wo} refers to the rate constant of partitioning from water to oil and k_{ow} is that in the opposite direction. It is, in fact, these rate constants that control drug distribution, which is a non-equilibrium process, although it may be steady state. Figure 1 illustrates this.

It is the aim of this chapter to examine each stage in drug transport and distribution from dissolution to excretion, to discuss the processes involved and to show how they are controlled by molecular structure.

19.2.2 DISSOLUTION

Most drugs are administered in the solid form, and so the first step in the drug transport process is dissolution, usually in an aqueous medium such as stomach contents. The Noyes–Whitney equation (equation 1; D = diffusion coefficient of solute; δ = thickness of stagnant diffusion layer; and C = concentration of solute in bulk solution) gives the relationship between the rate of dissolution, solubility (S) and surface area (A) of the solid. It can be seen that, provided A remains constant and sink conditions are maintained, dissolution is a zero-order process; this has been confirmed in practice.[2]

$$\text{Rate of dissolution} = \frac{DA}{\delta}(S - C) \tag{1}$$

Figure 2 shows the process of dissolution. Assuming adequate stirring in the bulk liquid, dissolution will be controlled by the actual removal of molecules from the solid surface into the unstirred boundary, and/or by diffusion across the boundary layer. Whilst these individual processes have been extensively studied,[3] this has not been done with respect to solute properties. However, some studies have been made of the variation of overall dissolution rate with molecular structure and properties. Collett and Koo[2,4] and Dearden and Patel[5] found, for series of benzoic acids/acetanilides and acetaminophen derivatives respectively, a dependence of dissolution rate upon $\log P$ or the hydrophobic substituent constant π. Equation (2), taken from Dearden and Patel,[5] shows a very high correlation between intrinsic dissolution rate D_i at 25 °C and the octanol–water partition coefficient. However, when D_i was converted to the dissolution rate constant (k) by dividing it by solubility, the correlation was much poorer (equation 3).

$$\log D_i = -0.990 \log P - 4.655 \tag{2}$$

$$n = 6, \quad r = 0.994, \quad s = 0.106$$

$$\log k = -0.181 \log P - 1.849 \tag{3}$$

$$n = 6, \quad r = 0.635, \quad s = 0.219$$

19.2.3 RELATIONSHIP BETWEEN SOLUBILITY AND THE PARTITION COEFFICIENT

For a solute partitioning between two liquid phases such as water and octanol, the dynamic equilibrium situation is described by the law of mass action (equations 4 and 5).

$$k_{wo} \times C_w = k_{ow} \times C_o \tag{4}$$

$$P = \frac{k_{wo}}{k_{ow}} = \frac{C_o}{C_w} \tag{5}$$

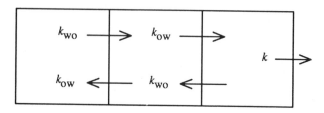

Figure 1 Simple three-compartment model of drug partitioning through an organism; k represents an irreversible excretion rate constant

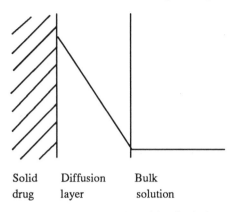

Solid Diffusion Bulk
drug layer solution

Figure 2 Schematic representation of the dissolution process

Taking equation (5) to the ultimate, the partition coefficient (P) can be defined as the ratio of solubilities (S_o/S_w) of a compound in the two phases, provided that the phases do not have an appreciable degree of mutual solubility (in which case high concentrations of solute can cause the relative compositions of the two phases to alter) and that self-association of the solute does not occur at concentrations approaching saturation. Yalkowsky *et al.*[6] have shown (equation 6), for a heterogeneous set of compounds, that this is so in practice. They also demonstrated the effect of mutual solubility of the liquid phases, and the effect of concentration. These effects are seen (Table 1) to be quite pronounced and probably account for the regression coefficients being different from unity and the intercepts being different from zero in equation (6).

$$\log P = 0.900 \log(S_o/S_w) + 0.390 \tag{6}$$

$$n = 36, \quad r = 0.985, \quad s = 0.233$$

One other factor that can cause the solubility ratio to be different from the partition coefficient is polymorphism: different crystalline forms of a compound can have different solubilities. So far as is known, however, no studies on this aspect of partition have been carried out.

Finally, since the partition coefficient is a measure of relative affinity of two liquid phases for a solute, the question must be asked as to whether control of the partition coefficient is exerted largely or exclusively by one phase, or whether both phases contribute appreciably. This question will be examined next.

19.2.3.1 Aqueous Solubility

Chiou *et al.*[7] developed a theoretical relationship showing that, for liquid solutes, the partition coefficient is related to aqueous solubility. In developing this relationship they made a number of simplifying assumptions, namely (i) that the partition coefficient is measured at low concentration, so that the mole fraction equals the molar concentration multiplied by the molar volume of each liquid phase; (ii) that the liquid solute has a low solubility in water and *vice versa*; (iii) that the organic liquid phase is only slightly soluble in water, so that the molar volume of water saturated with the organic phase may be taken as equal to the molar volume of pure water; (iv) that the solute

Table 1 Effect of Solvent Mutual Solubility and Solute Concentration on Ratio of Solubilities in Octanol and Water[6]

Solute	$\log(S_o/S_w)$ (pure phases)	$\log(S_o/S_w)$ (mutually saturated phases)	$\log(C_o/C_w)$ (dilute)
Antipyrine	−0.73	−0.50	0.26
Ethyl 4-aminobenzoate	1.86	2.21	1.96
Caffeine	−0.75	−0.42	−0.20
Theophylline	−0.57	−0.08	−0.09

forms an ideal solution in the organic phase; and (v) that the solubility of the solute is the same in pure water and in water saturated with the organic phase.

By judicious selection of solutes, concentrations and organic phase most of these assumptions can be made valid. Assumption (iv) seems perhaps rather sweeping and not easily verified, but it will be shown in Section 19.2.3.2 that it is reasonable.

With the above assumptions, Chiou et al.[7] showed that the partition coefficient is closely related to aqueous solubility (S_w) (equation 7; V_o^* = molar volume of water-saturated organic phase).

$$\log S_w = -\log P - \log V_o^* \tag{7}$$

In confirmation of this relationship, Yalkowsky and Valvani found,[8] for a large number of liquid solutes, the relationship denoted by equation (8) (where V_w = molar volume of water).

$$\log X_w = -1.08 \log P - 1.04 \tag{8}$$

$$n = 417, \quad r = 0.946, \quad s = 0.356$$

$$\log X_w = \log S_w + \log V_w$$

Hansch et al.[9] observed (equation 9) rather less agreement with theory, the regression coefficient being considerably more negative than that of equation (7). This discrepancy may have two causes: some of the simplifying assumptions made in the derivation of equation (7) might not have held for the solutes examined by Hansch et al.,[9] and the calculated $\log P$ values used in the correlation may not have been correct.

$$\log S_w = -1.339 \log P + 0.978 \tag{9}$$

$$n = 156, \quad r = 0.935, \quad s = 0.472$$

For solid solutes, the situation is complicated by the fact that the crystal lattice must be disrupted in order for dissolution to occur. This can be allowed for by incorporating into equation (7) a term involving the entropy of fusion, which yields equation (10) (ΔH_f = enthalpy of fusion and T_m = melting point, K).

$$\log S_w = -\log P - \log V_o^* - \frac{\Delta H_f(T_m - T)}{2.303 R T T_m} \tag{10}$$

Yalkowsky and Valvani[8] argued that most small solute molecules are quite rigid, and estimated the entropy of fusion to be approximately constant at 13.5 eV (1 eV = 1.6×10^{-19} J). Substituting this value into equation (10), they obtained equation (11), which predicts the aqueous solubility of a solid at 25 °C (note that melting point, MP, is in °C in this equation). They confirmed the validity of equation (11) by examining a large number of solid solutes, for which they obtained the empirical equation (12), in extremely good agreement with equation (11).

$$\log S_w = -\log P - 0.01MP + 1.05 \tag{11}$$

$$\log S_w = -1.05\log P - 0.012MP + 0.87 \tag{12}$$

$$n = 155, \quad r = 0.989, \quad s = 0.308$$

Thus both theory and practice indicate a strong correlation between the partition coefficient and aqueous solubility. Put another way, the variation of aqueous solubility (together with melting point, for solid solutes) accounts almost entirely for the variation of the partition coefficient.

19.2.3.2 Organic Phase Solubility

One might expect that a compound with a high partition coefficient would have high solubility in the organic phase. Indeed the term 'lipophilicity' implies as much. Occasionally, too, one sees the phrase 'lipid solubility' used as a synonym for the partition coefficient.[10] In fact, very few studies have been made of organic phase solubility. Yalkowsky et al.[6] point out that, if a solute forms an ideal solution with an organic liquid, then the van't Hoff equation (equation 13; X_o = mole fraction solubility and ΔS_f = entropy of fusion of solute) applies.

$$\log X_o = [\Delta S_f/2.303RT](T_m - T) \tag{13}$$

As before, it can be assumed for rigid molecules that the entropy of fusion is constant at about 13.5 eV, so that at 25 °C equation (13) reduces to equation (14); that is, organic phase solubility should be independent of the nature of the organic phase (for an ideal solution), should bear no relation to the partition coefficient, but should correlate solely with solute melting point. Yalkowsky *et al.*[6] showed (equation 15) that this was indeed so for a range of drugs of widely differing structures, despite many of these drugs presumably forming less than ideal solutions.

$$\log X_o = -0.01\text{MP} + 0.25 \tag{14}$$

$$\log X_o = -0.012\text{MP} + 0.26 \tag{15}$$

$$n = 36, \quad r = 0.92, \quad s = 0.32$$

It should be pointed out here that melting point itself is a property dependent on a number of molecular parameters, especially those relating to intermolecular forces.[11] For example, Dearden and Rahman[12] obtained equation (16) ($\mathscr{F}$ = Swain–Lupton field parameter; $\mathscr{R}$ = Swain–Lupton resonance parameter; L = STERIMOL length parameter; I = indicator variable for *p*-substitution; H_A = indicator variable for hydrogen bond acceptor ability; H_D = indicator variable for hydrogen bond donor ability; and $^3\chi^v$ = third order valence-corrected molecular connectivity) predicting the melting point (K) of substituted anilines. Because of the number of variables this correlation must be used and interpreted with caution but it does represent the first published QSAR of melting points.

$$T_m = 285 + 147\mathscr{F} + 51.6\mathscr{R} - 28.6L + 35.2I + 39.2H_A + 115H_D + 35.7\,^3\chi^v \tag{16}$$

$$n = 38, \quad r = 0.932, \quad s = 26.8$$

The pragmatic question still remains: does lipid solubility correlate with the partition coefficient? Dobbs and Williams[13] found, for a heterogeneous group of compounds, that the average solubility in a range of vegetable oils correlated negatively (equation 17) with the octanol–water partition coefficient. However, the correlation coefficient is so low as to make the correlation non-significant. Figure 3 shows the octanol solubility data of Yalkowsky *et al.*[6] plotted against the octanol–water partition coefficient. Clearly there is a lack of correlation, which is quantified in equation (18).

$$\log P = -1.2\log S_o + 6.2 \tag{17}$$

$$n = 31, \quad r = 0.40, \quad s \text{ not given}$$

$$\log X_o = 0.293\log P - 1.925 \tag{18}$$

$$n = 35, \quad r = 0.465, \quad s = 0.512$$

Thus it may be concluded that partitioning is controlled largely by solvation in the aqueous phase and not by solvation in the organic phase. The term lipophilicity should, therefore, be discarded in favour of hydrophobicity to describe the partitioning ability of a solute.

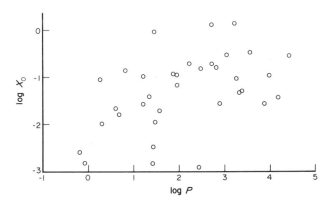

Figure 3 Octanol solubility *versus* octanol–water partition coefficient for a heterogeneous range of drugs (data from ref. 6)

19.2.4 THE TRANSFER PROCESS

It has been generally accepted for many years[14] that drug transport through lipoidal membranes is by passive diffusion; that is, the drug crosses an absorption barrier at a rate proportional to its chemical gradient across the barrier.

The so-called pH-partition hypothesis[14] put forward by Brodie and Hogben in the late 1950s postulated that for ionizable molecules only the unionized species could cross lipoidal membranes by passive diffusion; hence the flux depends on the pK_a of the drug and the pH of the aqueous phases as well as on the overall concentration gradient across the membrane. The pH-partition hypothesis has been confirmed experimentally many times; Figure 4 shows this for an *in vitro* situation.[15, 16] However, *in vivo* membranes are not homogeneous, and it is possible that ionized species can cross them *via* pores or water-filled channels. In addition, it is possible for ionized species to be transferred *via* ion pair formation,[17] although this phenomenon does not occur so readily with hydrocarbon organic phases as with octanol; there is no conclusive evidence that ion pair formation is a significant factor in transfer through lipid membranes. Broadly speaking, the pH-partition hypothesis may be assumed to hold, and it forms the basis for the vast amount of work that has been done on the transfer process.

19.2.4.1 Transfer Across Model Membranes

Much work has been done, and continues to be done, on transfer across a single membrane from aqueous to aqueous phase, for the process has relevance not only to general drug transport through an organism, but also to absorption through the skin, some sustained release devices, absorption of pesticides and packaging,[18] to name but a few. There have been several reviews of the subject.[18, 19]

There are several stages to mass transfer across a membrane (Figure 5). In the donor bulk aqueous phase, drug concentration is assumed to be uniform. (In model studies uniformity is achieved by stirring.) Adjacent to the membrane, but still within the donor aqueous phase, is a more or less stagnant boundary layer, across which the drug must move by diffusion. The drug then undergoes

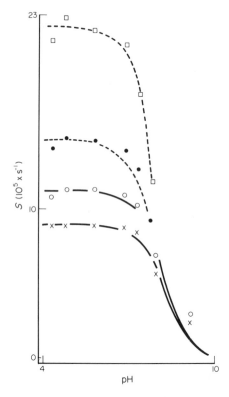

Figure 4 Dependence of transfer rate constant in the rotating diffusion cell on pH for 5-allyl-5′-isopentylbarbituric acid (□, 375 rpm; ●, 167 rpm) and ethyl *p*-hydroxybenzoate (○, 94 rpm; ×, 60 rpm), using a 0.3 M phosphate buffer–octanol–0.3 M phosphate buffer system [reproduced from ref. 16 by permission of Elsevier Science Publishers (Biomedical Division)]

interfacial transport into the oil/lipid membrane, followed by diffusion across the membrane, interfacial transport to the acceptor aqueous phase and diffusion across the second stagnant aqueous boundary layer to the bulk acceptor aqueous phase. Each one of these stages offers some resistance to drug transport, the total resistance being the sum of the individual resistances in series. It is assumed that the membrane is uniform and unstirred.

$$G = (R_{\mathrm{D}} - R_{\mathrm{A}}) \bigg/ \left(\frac{d_{\mathrm{D}} + d_{\mathrm{A}}}{D_{\mathrm{w}}} + \frac{d_{\mathrm{L}}}{PD_{\mathrm{L}}} \right) \tag{19}$$

Stehle and Higuchi[20] presented the theoretical principles of transport across a single membrane, but they ignored interfacial resistances, assuming them to be negligible compared with the diffusive resistances. They obtained equation (19) (R_{D} = concentration in bulk donor aqueous phase; R_{A} = concentration in bulk acceptor aqueous phase; d_{D} = thickness of stagnant layer in donor aqueous phase; d_{A} = thickness of stagnant layer in acceptor aqueous phase; d_{L} = thickness of membrane; D_{w} = diffusion coefficient in aqueous phase; D_{L} = diffusion coefficient in membrane; and P = membrane/aqueous phase partition coefficient) for the overall flux G across the membrane. When P is small, the second term of the denominator is much larger than the first, so that G increases with P. As the partition coefficient continues to increase, however, the second term of the denominator becomes smaller, eventually becoming negligible so that the flux is entirely diffusion controlled. This is depicted in Figure 6. Several workers[21-23] have since confirmed experimentally the theoretical relationship shown in Figure 6. Yalkowsky and Flynn[24] also proposed, from qualitative theoretical considerations, a relationship similar to that shown in Figure 6. They further postulated that at very high values of log P, the flux across the membrane would decrease because

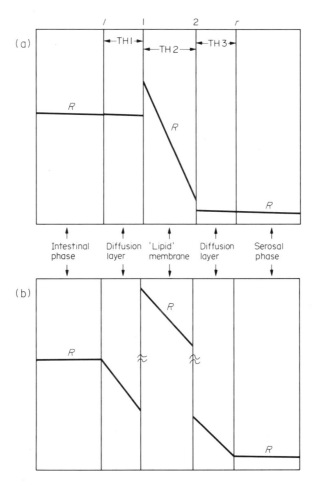

Figure 5 Schematic representation of concentration gradients in steady state transport of neutral solutes of (a) low and (b) high partition coefficient (reproduced from ref. 20 by permission of the American Pharmaceutical Association)

decreasing aqueous solubility would make it impossible to achieve the requisite solute concentrations in the aqueous donor phase. This was confirmed by an *in vivo* study on goldfish. Yalkowsky and Flynn claimed that their model thus explained the often-observed biphasic ('parabolic') dependence of biological activity on hydrophobicity, but this is only partly correct, as is discussed in Section 19.2.4.3.

In 1976 Albery *et al.*[25] devised a novel apparatus for investigating membrane transfer. Known as the rotating diffusion cell (RDC) (Figure 7), it enabled the membrane to rotate at different speeds, thus altering the thickness of the stagnant diffusion layers; this enabled the contribution of the different resistances to transfer to be determined. The Levich equation[26] (equation 20; v = kinematic viscosity) gives the thickness of the stagnant diffusion layer, d, as a function of rotation speed, w(Hz). From equation (20), the diffusion layer vanishes at infinite rotation speed, so extrapolation to that speed will enable the sum of the other resistances to transfer to be determined.

$$d = 0.643 w^{-\frac{1}{2}} v^{\frac{1}{6}} D_w^{\frac{1}{3}} \tag{20}$$

Unlike Stehle and Higuchi,[20] Albery *et al.*[25] did not assume interfacial resistance to be negligible, and derived equation (21) for the resistance to transfer. The term $2d/D_w$ is the resistance of the two stagnant aqueous diffusion layers (their thicknesses d being assumed to be equal), $2/\alpha k_{wo}$ is the sum of the two interfacial resistances and $d_L/P\alpha D_L$ is the diffusional resistance of the membrane (where α = pore area of membrane filter as fraction of total membrane area). Figure 8 shows RDC results obtained by Guy and Fleming[27] for methyl nicotinate with a variety of organic liquids in the membrane. It follows from equations (20) and (21) that the slope of each line in Figure 7 should be given by $1.286 v^{1/6} D_w^{-2/3}$, and that each intercept is equal to $2/\alpha k_{wo} + d_L/P\alpha D_L$.

$$\frac{1}{k} = \frac{2d}{D_w} + \frac{2}{\alpha k_{wo}} + \frac{d_L}{P\alpha D_L} \tag{21}$$

Using a previously determined value[25] for D_L in isopropyl myristate, Guy and Fleming[27] calculated k_{wo} for that organic phase. By assuming the same k_{wo} value for the other organic phases represented in Figure 8, they then calculated D_L for the different organic phases. It should be commented here that the assumption of a constant value for k_{wo} is not supported by the results of van de Waterbeemd,[22] who obtained quite different values of k_{wo} for different water–organic phase solvent pairs.

These workers and their associates have continued to investigate interfacial transfer using the RDC, including some recent work on the thermodynamics of transfer.[28] In that study Fleming *et al.*

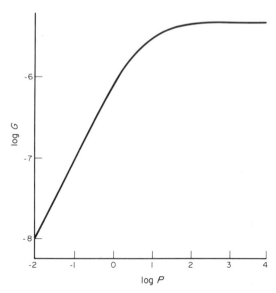

Figure 6 Calculated transport rate (G) across a membrane as a function of the partition coefficient (P) (reproduced from ref. 20 by permission of the American Pharmaceutical Association)

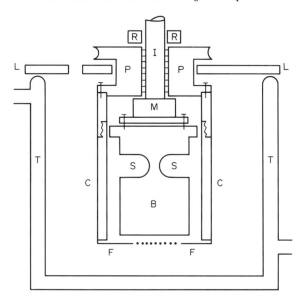

Figure 7 Cross-sectional diagram of the rotating diffusion cell. The central cylinder (C) rotates within a thermostatted jacket (T). The treated Millipore filter (F) divides the cell into two compartments. Rotating disc hydrodynamics are established in the inner compartment by the incorporation of a PTFE baffle (B), which is positioned rigidly by means of a hollow stainless steel shaft (I). The slots (S) in the baffle are necessary to achieve the correct hydrodynamic flow. Rotation of the cell is achieved *via* the pulley (P) and vertical movement of the inner cylinder is prevented by the mantle (M) and a removable locking collar (R). The rotating assembly is attached to a Perspex lid (L) which prevents excessive evaporation. The lid is drilled to facilitate periodic sampling of the solution in the outer compartment [reproduced from ref. 27 by permission of Elsevier Science Publishers (Biomedical Division)]

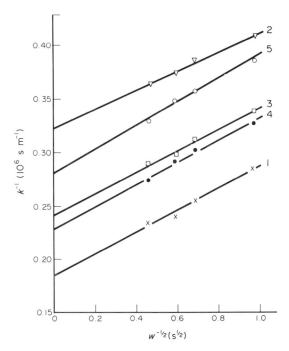

Figure 8 Plots of reciprocal permeability of methyl nicotinate as a function of rotation speed for various water–oil–water systems: 1, isopropyl myristate; 2, tetradecane; 3, linoleic acid; 4, 2% egg lecithin in isopropyl myristate; 5, 1% DL-$\beta\gamma$-dipalmitoyl-α-lecithin in isopropyl myristate [reproduced from ref. 27 by permission of Elsevier Science Publishers (Biomedical Division)]

determined, from the temperature dependence of transfer, the enthalpy and entropy of interfacial transfer for methyl and ethyl nicotinate crossing an isopropyl myristate membrane. They observed a large positive entropy of transfer in each direction, and suggested that this might arise from the passage of a solute molecule disrupting ordered structure at the isopropyl myristate–water interface.

The falling/rising drop technique has been used by Brodin and co-workers[29] to determine transfer rates between water and various organic solvents. In this technique droplets of one solvent, containing initially a known concentration of solute, are allowed to rise or fall through the other solvent and the collected droplets are then analyzed for final solute concentration. It has been shown that, with this technique, unstirred boundary layers are of negligible thickness, so that the main resistance to transfer is interfacial. Brodin found rectilinear correlations between $\log P$ and both $\log k_{wo}$ and $\log k_{ow}$. This work has recently been independently confirmed by Miller[30] who has commented that, for *water* permeation through membranes, the main barrier to transfer is interfacial. Fleming *et al.*[28] also point out that, for membrane thicknesses of $< 100 \ \mu$m (which is the case for many biological membranes, and for emulsion systems and sustained-release formulations), interfacial resistance can be a dominant factor in membrane transfer.

Recently the importance of interfacial transfer in the transfer process has been challenged by Leahy and Wait.[31] Using the RDC, these workers studied water–oil–oil as well as water–oil–water transfer. Clearly, in the water–oil–oil system there is only one interfacial transfer, so that the intercept in a plot of k^{-1} *versus* $w^{-\frac{1}{2}}$ is the membrane resistance plus one interfacial resistance, whilst for the water–oil–water system the intercept will include an additional interfacial resistance. Leahy *et al.*[31,32] found (Figure 9) that the plots for water–oil–oil and water–oil–water systems had the same intercept value, indicating negligible interfacial resistance. They suggest that the apparent importance of interfacial transfer found by other workers may have been caused by errors. Similar results have been obtained by Byron and Rathbone,[16] again using the RDC; de Haan and Jansen[33] also conclude, from transfer studies across a single interface, that no noticeable energy barrier is present at the interface.

Houk and Guy[34] have recently reported a parabolic relationship, passing through a minimum, between $\log$(resistance to membrane transfer in the RDC) and $\log P$. They claim that this supports the hypothesis that interfacial transfer is the permeation-controlling step in membrane transport. This, however, seems in contradiction of the theoretical predictions of Stehle and Higuchi[20] and Yalkowsky and Flynn,[24] who argued that if the interfacial transport alone were rate controlling, then flux ($\equiv 1$/resistance) should increase steadily as $\log P$ increased.

Clearly, much remains to be resolved in the area of interfacial transport, and further developments will no doubt be forthcoming in the not-too-distant future. Meanwhile, it is appropriate to examine more closely the diffusional resistance to phase transfer, which may turn out to be the factor of overriding importance. Flynn *et al.*[18] point out that for solutes whose molar volume is less than that

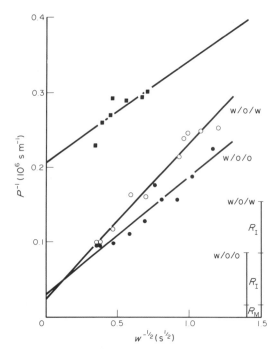

Figure 9 Plots of reciprocal permeability of methyl nicotinate as a function of rotation speed for the water–octanol–octanol (w/o/o) and water–octanol–water (w/o/w) systems: (●) w/o/o; (○) w/o/w; (■) w/o/w data from R. H. Guy and D. H. Honda, *Int. J. Pharm.*, 1984, **19**, 129 (reproduced from ref. 31 by permission of the American Pharmaceutical Association)

of the solvent, diffusivity (D) is given by equation (22), whilst for solutes with a molar volume greater than that of the solvent, it is given by equation (23) (k = a constant; η = solvent viscosity; F = frictional ratio, dependent on solute shape; and V = molar volume of solute). For water at 25 °C equations (22) and (23) reduce to equations (24) and (25) respectively. Flynn *et al.*[18] showed that most solutes fall between the limits of equations (24) and (25), assuming F = 1. They found, however, different relationships of diffusivity and partial molar volume for different classes of compound, and suggested that this is because the hydrogen bonding capabilities of polar solutes to water decrease their hydrodynamic volume.

$$ D = \frac{kT}{6\pi\eta F}\left(\frac{4\pi}{3V}\right)^{\frac{1}{3}} \tag{22} $$

$$ D = \frac{kT}{4\pi\eta}\left(\frac{4\pi}{3V}\right)^{\frac{1}{3}} \tag{23} $$

$$ D_w = \frac{3.3 \times 10^{-6}}{V^{\frac{1}{3}}} \tag{24} $$

$$ D_w = \frac{4.95 \times 10^{-6}}{V^{\frac{1}{3}} F} \tag{25} $$

The Wilke–Chang relationship, an empirical equation relating diffusivity to solute molal volume at the boiling point, indicates no such dependence on the class of solute.[35] Similarly Leahy *et al.*[32] report an empirical correlation between diffusivity and a computer calculated estimate of the van de Waals or intrinsic volume V_I (equation 26), which shows no such family dependence.

$$ D = 9.28 \times 10^{-5} V_I^{-0.58} \tag{26} $$

19.2.4.2 Transfer Across a Single Interface

The discussion of diffusivity leads us into a consideration of the extensive work that has been done on rates of drug transfer across a single aqueous–organic phase interface. The apparatus used in these studies has usually comprised a single vessel with the organic and aqueous phases sitting one above the other, both being stirred. Such an arrangement does not enable the various contributions to the overall transfer rate to be determined. It has been found, in confirmation of the early water–oil–water studies of Lippold and Schneider,[21] that k_{wo} increases initially with the partition coefficient, but eventually levels off; by contrast k_{ow} is initially constant as the partition coefficient increases, but then begins to fall. Some results of van de Waterbeemd[22] for partitioning between different solvent pairs are shown in Figure 10. The levelling off observed for both k_{wo} and k_{ow} has been interpreted, in line with theoretical predictions,[20,24] as diffusional control taking over from partition control of transfer. Furthermore, the fact that k_{wo} and k_{ow} show almost mirror image behaviour suggests that for k_{wo}, diffusion through the aqueous boundary layer is the limiting factor, whereas for k_{ow} diffusion through the organic phase boundary layer is limiting. A somewhat puzzling feature of all the work reported, with one exception,[36] is that the relationships typified in Figure 10 display horizontal plateaux; from equations (23) and (24), diffusivity decreases as solute molar volume increases, and the latter generally increases as the partition coefficient increases, particularly within an homologous series. Consequently the plateaux for both rate constant curves would be expected to have a slight negative slope. It is possible that such an effect is small enough to be hidden by experimental error.

It is also surprising, in view of the above, that the plateau levels for both aqueous and organic phase diffusional control have been found[37] to be solute independent and thus presumably molar volume independent. Again, perhaps any differences are masked by experimental error. Leahy *et al.*[32] have in fact observed a small but significant effect of solute size on permeability, perhaps because of the large number of observations in their study.

Kubinyi[38] has shown that the relationships typified in Figure 10 can be described utilizing his bilinear equation,[31] to yield equations (27) and (28); van de Waterbeemd[22] has confirmed this experimentally, for a series of sulfonamides partitioning between water and octanol (equation 29). It was also shown by van de Waterbeemd[22] that the β term of equations (27) and (28) is equal to the

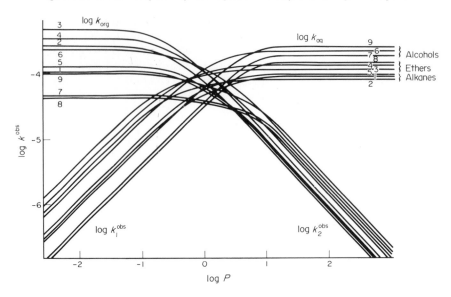

Figure 10 $\log k$ *versus* $\log P$ in various solvent systems (reproduced from ref. 22 by permission of Elsevier Science Publishers)

ratio of the plateau diffusion rate constants k_{org}/k_{aq} (see Figure 10). Since diffusivity is a function of solvent viscosity, it might be expected that β would also be related to solvent viscosity. de Haan and Jansen[40] observed correlations of k_{org} with organic phase kinematic viscosity (v_{org}), for example with alcohols (equation 30); v_{aq} was found to be almost constant, when water was saturated with nine different organic phases. Similarly, van de Waterbeemd[22] observed a good correlation between β and organic phase viscosity (equation 31), whilst de Haan *et al.*[41] obtained equation (32). From these relationships van de Waterbeemd considered β, since it is linked to transport across membranes, to be of value in membrane characterization. It has, however, not yet been applied for that purpose.

$$k_{wo} = \frac{cP}{\beta P + 1} \tag{27}$$

$$k_{ow} = \frac{c}{\beta P + 1} \tag{28}$$

$$\log k_{wo} = \log P - \log(\beta P + 1) - 3.996 \tag{29}$$

$$n = 27, \quad r = 0.993, \quad s = 0.040$$

$$\log k_{org} = -0.833 v_{org} - 3.143 \tag{30}$$

$$n = 4, \quad r = 0.955, \quad s = 0.088$$

$$\log \beta = -0.502 \log \eta_{org} + 0.156 \tag{31}$$

$$n = 9, \quad r = 0.961, \quad s = 0.130$$

$$\log \beta = -0.527 \log(v_{org}/v_{aq}) + 0.207 \tag{32}$$

$$n = 9, \quad r = 0.962, \quad s = 0.129$$

Figure 10 shows that k_{aq} varies with the nature of the organic solvent phase, although not by as much as does k_{org}. This variation is puzzling, since k_{aq} reflects the diffusional resistance of the *aqueous* unstirred boundary layer. Although the water is saturated with the organic phase, that has only a negligible effect on aqueous viscosity;[33] it has been suggested[41] that the organic solvent produces water-structuring effects at the interface and recently de Haan and Jansen[33] have shown that this effect occurs under shear conditions, *i.e.* when the phases are stirred. There is as yet, however, no independent evidence for such structuring at the interface.

The foregoing section, whilst of necessity giving but the briefest of outlines of *in vitro* phase transfer, has indicated that our understanding of the processes involved are still far from complete and there remains much careful work to be done. The words of Leo *et al.*[42] in their 1971 review of partitioning remain true today: 'the basic importance of partitioning rate studies cannot be seriously questioned, but the interpretation of the results is still subject to some ambiguity'. Nevertheless, our current knowledge is considerable and is aiding the design of drug delivery systems such as topical formulations, as well as providing insight into *in vivo* drug penetration.

Consideration of the importance of the dynamic partitioning rate constant *vis-à-vis* the equilibrium partition coefficient led Dearden and Bresnen[43] to use k_{wo} and k_{ow} in a QSAR correlation. For a series of analgesic paracetamol (acetaminophen) derivatives tested in mice, Dearden and O'Hara[44] obtained equation (33). Dearden and Bresnen obtained equation (34), of equal significance, by the use of $\log k_{ow}$; the use of $\log k_{wo}$ gave a poor correlation ($r = 0.601$), although, interestingly, the use of $\log(k_{wo}/k_{ow})$ ($= \log P$) gave a better correlation ($r = 0.931$) than did $\log P$ itself. The good correlation shown in equation (34) indicates that the compounds studied fall in the range where $\log k_{ow}$ varies rectilinearly with $\log P$ (*cf.* Figure 10).

$$\log(1/ED_{30}) = -0.133 + 1.541 \log P - 0.829 (\log P)^2 \tag{33}$$

$$n = 8, \quad r = 0.919, \quad s = 0.103$$

$$\log(1/ED_{30}) = -6.050 + 7.379 \log k_{ow} - 2.050 (\log k_{ow})^2 \tag{34}$$

$$n = 8, \quad r = 0.921, \quad s = 0.101$$

Finally, in this section, it is interesting to note a study by Collett and Koo[4] which correlated a combined dissolution and partitioning process with molecular parameters. Collett and Koo studied a series of 4-substituted benzoic acids dissolving in 0.1 N HCl and then partitioning into octanol layered above the aqueous phase. They found not only the dissolution step, but also the appearance of solute in the octanol phase to follow zero-order kinetics, suggesting that dissolution was the rate-controlling step. This could have implications for all drugs administered in a solid dosage form.

Collett and Koo were able to correlate the rate constant k_p for the appearance of solute in octanol with the hydrophobic substituent constant π (equation 35) and also with certain molecular orbital indices devised by Cammarata and Rogers (equation 36).[45] Equation (35) is not surprising, when it is remembered that dissolution rate itself correlates[2] with π; the correlation shown in equation (36) (in which Q_r^T = absolute sum of σ and π net charges on a given atom and $^1S_r^E$ = Fukui's electrophilic delocalizability index) suggests that the change in the solvation state with phase transfer is the rate-controlling factor.[45]

$$\log k_p = -1.70 \pi - 5.17 \tag{35}$$

$$n = 7, \quad r = 0.905, \quad s = 0.505$$

$$\log k_p = 1.53 \Sigma_r Q_r^T - 0.90 \Sigma_r {}^1S_r^E - 3.99 \tag{36}$$

$$n = 5, \quad r = 0.992, \quad s = 0.199$$

19.2.4.3 Mathematical Modelling of Drug Activity

The random walk of a drug molecule from site of administration to site of action and thence to excretion can, from the foregoing, be represented (to a first approximation at least) as transfer across a series of alternating aqueous and lipoidal compartments. A number of attempts to describe drug transport and distribution have utilized this alternating compartment model and much useful information has resulted therefrom.

Before examination is made of these models, it is pertinent to consider the general relationship between biological activity and hydrophobicity. Hansch *et al.*[46] showed in 1962 that the effects of a series of *m*- and *p*-substituted phenoxyacetic acids on the growth of *Avena* coleoptiles could be described by a quadratic function of hydrophobicity (equation 37; C = concentration inducing 10% growth in 24 h and σ = Hammett substituent constant); π is defined as $\log(P_{derivative}/P_{parent})$. This biphasic dependence of activity upon hydrophobicity has been observed in many systems; Hansch and Clayton[47] in 1973 reported 173 such correlations and hundreds more have been published since. In all probability the dominance of hydrophobicity in controlling activity reflects the importance of hydrophobicity in transport, although receptor binding may also involve a hydrophobic, as well as an electronic and/or steric contribution. Hence, the generalized Hansch equation may be written as equation (38), where $[E]$ represents an electronic parameter and $[S]$ a steric parameter.

$$\log(1/C) = 4.08 \pi - 2.14 \pi^2 + 2.78 \sigma + 3.26 \tag{37}$$

$$n = 20, \quad r \text{ not given}, \quad s \text{ not given}$$

$$\log(1/C) = a \log P + b (\log P)^2 + c[E] + d[S] + e \tag{38}$$

Kubinyi,[39] observing that many experimental biphasic relationships appeared to be two straight lines joined by a curvilinear portion rather than to be parabolae, devised an equation to fit his observation—the so-called bilinear equation (equation 39). This equation does in fact fit many sets of experimental data better than does the quadratic equation, although it is not so easy to use, as the constant β has to be obtained by iteration. Where the bilinear equation is really superior is in the description of biphasic relationships where the positive and negative slopes are different, or where one slope may be zero (*cf.* Figure 10).

$$\log(1/C) = a\log P - b\log(\beta P + 1) + c \tag{39}$$

At this point it is useful to inquire why biphasic quantitative structure–activity relationships are observed. Hansch and Clayton[47] list nine possible reasons, each of which is probably tenable under certain conditions (*e.g.* micelle formation, enzyme poisoning). However, the almost ubiquitous nature of the 'parabolic' QSAR dictates a single, overriding reason. This can be put simply as follows: a drug molecule with a low partition coefficient will have a low value of k_{wo} and so will partition slowly into lipid membranes, although it will partition rapidly from lipid to aqueous phase; k_{wo} will therefore be the rate-controlling step in the random walk to the site of action. Consequently, *at a given time after administration*, provided that time is not too long, only a low concentration will have reached the site of action and hence the biological response will be low. Conversely, for a drug with a high partition coefficient, k_{ow} will be the rate-controlling factor and (being low) will again result in a low concentration at the site of action at a given time after dosage. Optimal transport conditions are clearly achieved by drugs of intermediate partition coefficient, with neither k_{wo} nor k_{ow} being too low.

Yalkowsky and Flynn[24] proposed that decreasing aqueous solubility with increasing partition coefficient accounted for the oft-observed decrease in biological activity at high values of $\log P$, and demonstrated this in an *in vivo* study on goldfish. However, they used a steady state model (the goldfish were immersed in a constant concentration of drug), whereas most QSAR data are obtained following single dose administration, which brings in the important factor of time. Their model is not, therefore, applicable to most QSAR data sets, although there can be little doubt that solubility limitation is a response-controlling factor in some cases. Hansch[48] has pointed out, in opposing solubility limitation of biological activity as a general phenomenon, that the $\log P_o$ values of compounds active against Gram-positive bacteria (5.9) and against Gram-negative bacteria (4.2) would be expected to be similar if lack of aqueous solubility were the prime reason for decrease in activity as $\log P$ increased.

Franke and Oehme,[49] observing that many biphasic QSARs had different positive and negative slopes, proposed a combination of rectilinear and quadratic equations in $\log P$ to describe activity (equations 40 and 41; $\log P_x$ = limit of straight line). Their proposals were, however, based on receptor binding rather than on transport; they argued that up to the point where the hydrophobic region of the receptor was covered, rectilinearity between binding ($\equiv$ activity) and $\log P$ is observed; beyond that point, binding will start to decrease owing to steric hindrance to binding and/or to conformational changes induced in the receptor.

$$\log(1/C) = a\log P + b \quad \text{for} \quad \log P < \log P_x \tag{40}$$

$$\log(1/C) = \alpha\log P + \beta(\log P)^2 + \gamma \quad \text{for} \quad \log P > \log P_x \tag{41}$$

Hyde[50] devised an asymptotic equation (equation 42) based on receptor occupancy theory to describe those QSARs which level off but display no negative slope. A number of experimentally observed QSARs are of this form, including some to which quadratic equations have been fitted[47] on the assumption that at a sufficiently high partition coefficient, activity must eventually fall. However, the Hyde equation cannot be considered of such general applicability as, say, the Kubinyi equation,[39] since many QSARs do show a negative slope at high $\log P$ values.

$$\log C = \log(a + 10^{-\pi}) + c \tag{42}$$

All of the above equations apply to non-ionizable drugs; or rather, they contain no terms to allow or correct for ionization, but in practice have been applied to ionizable compounds, sometimes with excellent results.[51] Nevertheless, ionization causes marked changes in the partition coefficient and other molecular parameters; Leo *et al.*[42] estimate the average change in $\log P$ upon ionization to be -4.0. One way to allow for ionization is to incorporate a term such as pK_a or degree of ionization[52] in the QSAR, or to use, as did Moser *et al.*[53a] when correlating the anti-inflammatory activities of a series of phenylbutazone derivatives, and Scherrer and Howard[53b] when correlating the colonic absorption of acids, the apparent partition coefficient P' at the appropriate pH (equations 43a and b respectively).

$$\log(1/C) = -2.10 + 0.77\log P' - 0.57(\log P')^2 \tag{43a}$$

$$n = 16, \quad r = 0.798, \quad s = 0.255$$

$$\log(\%\,abs) = 0.236\log P' - 0.079(\log P')^2 + 1.503 \tag{43b}$$

$$n = 10, \quad r = 0.965, \quad s = 0.096$$

Martin and Hackbarth[54] developed, for an equilibrium model, equations to cover various situations involving ionization, such as whether the neutral or ionic species interacts with the receptor. For example, with a model comprising one aqueous compartment, one non-aqueous compartment and a receptor, and assuming that only the neutral form of the drug interacted with the receptor, they derived equation (44) (in which a, b, c, d are constants, α = degree of ionization and x = proportionality constant between amount of drug at receptor and potency). Figure 11 shows the form of QSAR predicted by this equation, whilst Figure 12 shows that predicted by an equation

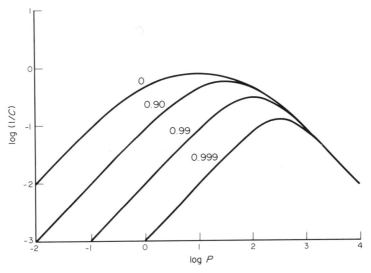

Figure 11   A plot of equation (44), with $a = b = c = 1.0$, $d = 0.01$, $x = 0.0$; fraction of drug ionized is indicated on each line (reproduced from ref. 54 by permission of the American Chemical Society)

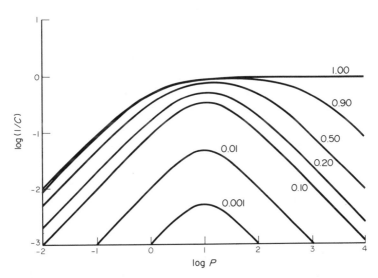

Figure 12   A plot of equation (44) modified to allow only the ionic form of a drug to interact with the receptor (reproduced from ref. 54 by permission of the American Chemical Society)

which assumes that only the ionized form of the drug interacts with the receptor. The degree of ionization clearly markedly affects the optimal value of $\log P$ ($\log P_\text{o}$) when the neutral form of the drug interacts with the receptor, but not when it is the ionic form which interacts. This type of information can thus be used to provide insight into the nature of the drug–receptor interaction. Martin[55,56] has reviewed this work extensively.

$$\log(1/C) = \log\left[1 + dP^c + \frac{1}{aP^b(1 - \alpha)}\right]^{-1} + x \qquad (44)$$

19.2.4.4 Analogue Modelling of Drug Transport and Distribution

We have seen earlier how drug movement through an organism can be represented as a series of partitioning steps through alternating aqueous and lipid compartments. A number of workers have attempted to model and explain experimentally observed QSARs by considering drug permeation through such compartments. The first such attempt was by Penniston et al.[57] who represented an organism as a series of 20 alternating aqueous and lipid compartments (Figure 13), with reversible transfer between each compartment save the last, which was thus a total sink, making the model of the non-equilibrium type; the model was used in the non-steady state mode by having an initial concentration of 'drug' (the dose) in the first compartment. The partition coefficient is defined as k_wo/k_ow and for simplicity the product of k_wo and k_ow was specified as unity. This is now known not to be realistic and the model has been criticized because of this;[58,59] however, Dearden and Townend[60] have recently shown that the assumption of $k_\text{wo} \times k_\text{ow} = \text{constant}$ does not invalidate the model.

Differential equations were written for the rate of change of concentration in each compartment with time; these were then solved using an appropriate computer program, to yield the concentrations in each compartment at different times after dosage, for selected values of k_wo and k_ow. Figure 14 shows the results obtained under one set of conditions; it will be seen that the model predicts an approximately parabolic (or rather bilinear) relationship between $\log C$ (taken to be equivalent to response) and $\log P$, as observed for many experimental QSARs. It will be shown later that this model can be developed further to yield much more information about drug transport.

McFarland[61] used a similar arrangement of alternating aqueous and lipid compartments to develop a probabilistic model of the relationship between potency and hydrophobicity, based on equation (45), where $\text{Pr}_{1,2}$ is the probability of a drug molecule with partition coefficient P moving from an aqueous to a lipid compartment. McFarland showed that this approach led to a prediction of bilinear relationships (Figure 15) with an invariant optimal $\log P$ value. Kubinyi[39] recognized that this invariance was due to the assumption of equal volumes of aqueous and lipid phases, which is, of course, not the case *in vivo*. Kubinyi's β term in his bilinear equation,[39] which allows $\log P_\text{o}$ to vary, can be considered as a phase volume ratio term, although, as has been shown earlier, it relates to other properties also.[22]

$$\text{Pr}_{1,2} = \frac{k_\text{wo}}{k_\text{wo} + k_\text{ow}} = \frac{P}{P + 1} \qquad (45)$$

Higuchi and Davis[62] considered a pseudoequilibrium model of drug distribution comprising a single aqueous phase, a series of tissue, lipid and protein phases and a receptor. To simplify matters they presented results for a model comprising the aqueous phase, a single lipoidal phase and a

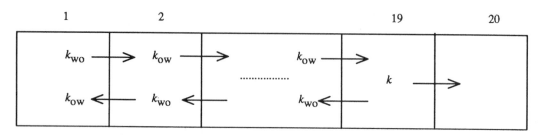

Figure 13   The compartmental model of Penniston *et al.* (ref. 57)

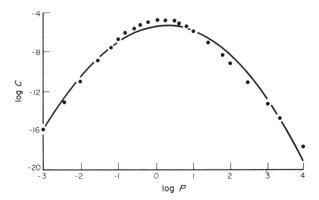

Figure 14 Penniston model prediction of variation of concentration in compartment 20 with $\log P$ at $t = 10$ (arbitrary time units) (reproduced from ref. 57 by permission of the American Society for Pharmacology and Experimental Therapeutics)

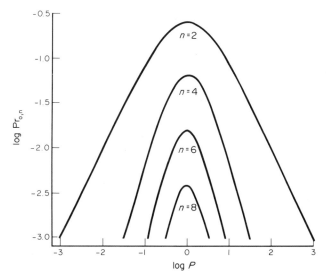

Figure 15 The probability $\mathrm{Pr_{o,n}}$ of a drug reaching a receptor region as a function of partition coefficient P and of n, the number of interfaces separating the site of drug application from the receptor region (reproduced from ref. 61 by permission of the American Chemical Society)

receptor and showed that the relationships between amount of drug at the receptor and hydrophobicity varied according to whether or not the receptor was less hydrophobic than the lipoidal phase. When hydrophobicities of lipid and receptor were equal, the amount of drug at the receptor increased and then levelled off, as hydrophobicity of the drug increased; when receptor hydrophobicity was lower than that of the lipid phase (the situation to be expected *in vivo*) then the amount of drug at the receptor first increased and then decreased, as drug hydrophobicity increased (*i.e.* the oft-observed shape of experimental QSARs). The Higuchi and Davis model also predicted that $\log P_o$ should decrease as lipid volume increased.

Dearden and Townend developed the model of Penniston *et al.*[57] further in a series of papers. They showed[63, 64] that $\log P_o$ increased with the time after dosage (Figure 16). Cooper *et al.*,[65] on the other hand, concluded from an analysis of the Penniston model that it predicted that $\log P_o$ should decrease as time after dosage increased. Dearden and Patel,[66] using a physical model consisting of 11 consecutive stirred compartments, containing alternately water and octanol, observed little or no variation of $\log P_o$ with time. However, Tichý and Roth[67] found that $\log P_o$ for a series of analgesic 1-(2-arylethyl)-4-piperidinyl-*N*-phenylpropanamides increased with testing time. They modelled these and other *in vitro* data using the Dearden and Townend modification[63] of the Penniston model.[57]

Dearden[68] reported that the model predicted $\log P_o$ decreasing with the number of partitioning steps to the 'receptor' compartment (Figure 17). This indicates different optimal hydrophobicities for

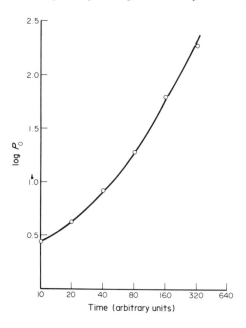

Figure 16 Variation of optimal partition coefficient (P_o) with time after dosage, in the Dearden and Townend modification of the Penniston model (reproduced from ref. 68 by permission of the National Institute of Environmental Health Sciences)

organisms of different complexity,[69] and portends difficulties in predicting, say, human response from that in an animal species. It also suggests that different routes of administration could yield different $\log P_o$ values; in support of this, Dearden and Tomlinson[70] observed $\log P_o = 1.48$ for oral administration of analgesic paracetamol (acetaminophen) derivatives to mice, whilst Dearden and O'Hara[71] found $\log P_o = 0.90$ for subcutaneous administration.

Dearden and Townend,[63,64,72] drawing attention to the importance of time as a factor in the correlation of activity with molecular structure, showed that the Penniston model predicted that the time to maximal response varied approximately parabolically with $\log P$ (Figure 18), being shortest for intermediate values of $\log P$. This is readily understood when it is recognized that at low $\log P$ values, k_{wo} is the rate-controlling factor, whereas at high $\log P$ values, k_{ow} is rate-limiting (see Figure 10). That such variation exists is hardly recognized, for most series of drug candidates are tested for activity using a fixed time protocol; probably many potential drug candidates have been discarded because of this. There is, however, some experimental evidence to support the prediction of Figure 18; the results of Kutter *et al.*[73] for the effects of morphine-like analgesics in the rabbit can be correlated with $\log P$ (measured in heptane/buffer pH 7.4) (equation 46).[67]

$$t_{max}(\text{min}) = 0.826 \log P + 0.660 (\log P)^2 + 3.457 \tag{46}$$

$$n = 11, \quad r = 0.879, \quad s = 3.103$$

The Penniston model also predicts that maximal concentration in a given compartment rises and then levels off as $\log P$ increases.[63,64] The results of Kutter *et al.*[73] can again be used to confirm this; although Kutter *et al.* fitted a parabola to their results ($r = 0.970$), the bilinear equation (equation 47) gives a slightly better fit and the slope of the second portion of the curve (-0.055) is almost zero.

$$\log(1/C) = 0.778 \log P - 0.833(\beta P + 1) + 0.775 \tag{47}$$

$$n = 11, \quad r = 0.974, \quad s = 0.298, \quad \log \beta = 1.775$$

Dearden and George[74] carried out an interesting test of the model prediction regarding maximal concentration. They determined the ability of aspirin derivatives to inhibit blood platelet aggregation in the rabbit at various times after dosage. The results are shown in Figure 19 and indicate that for a fixed time (1 h) after dosage an approximately parabolic QSAR is obtained, with very hydrophobic compounds showing no detectable activity; the maximal response, however, shows a picture approximating to the model prediction of a levelling-off of activity at high $\log P$ values. This again emphasizes the importance of time in QSAR correlations.

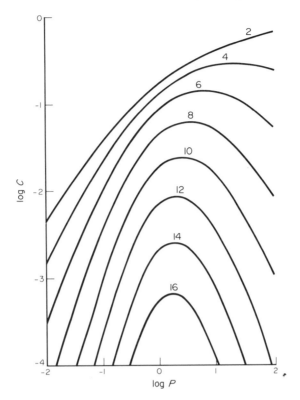

Figure 17 Variation of concentration in the nth compartment with $\log P$ at $t = 10$ (arbitrary time units), in the Dearden and Townend modification of the Penniston model (reproduced from ref. 61 by permission of the National Institute of Environmental Health Sciences)

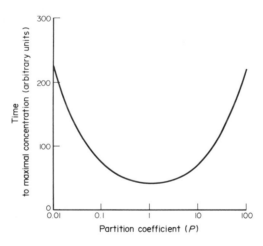

Figure 18 Time to maximal concentration in compartment 10 as a function of $\log P$, in the Dearden and Townend modification of the Penniston model (reproduced from ref. 64 by permission of the Society of Chemical Industry)

Related to the time to maximal response is duration of action. This is generally controlled by metabolism, although so far as is known no QSAR correlations between duration of action and rates of metabolism have been sought. There is in fact a dearth of duration of action data in the literature altogether. Dearden and Townend[75] have shown that, in the absence of metabolism, duration of action is predicted, like time to maximal response (and for the same reasons), to vary approximately parabolically with $\log P$. Confirmation of this again comes from the work of Kutter *et al.*[73] and Herz and Teschemacher[76] concerning analgesic effects of intravenously administered morphine-like compounds in the rabbit (equation 48). It may be noted that the model also predicts[75] a similar,

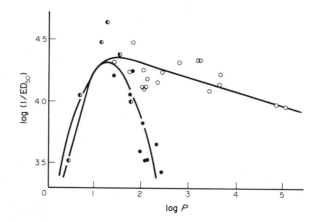

Figure 19 Variation of platelet aggregation inhibitory potency in rabbit blood of aspirin derivatives with $\log P$, one hour after dosage (●) and at time of maximal response (○)

although less pronounced, variation in time of onset of action with hydrophobicity.

$$\text{Duration(h)} = 0.053\log P + 0.149(\log P)^2 + 0.449 \tag{48}$$

$$n = 11, \quad r = 0.981, \quad s = 0.324$$

Mention has been made in Section 19.2.2 of the possible absorption rate-limiting effect of dissolution. Dearden and Townend[60] incorporated varying dissolution rates into the Penniston model and showed (Figure 20) that the usual biphasic curves were obtained, with a decrease of $\log P_o$ and of 'activity' as dissolution rate decreased.

Occasionally there is reported in the literature a double-peaked QSAR. Such a phenomenon suggests the presence of two receptor sites with different hydrophobicity requirements. Franke and Kühne[77] postulated adjacent receptors, with increasing substrate hydrophobicity eventually permitting binding with the second receptor. Dearden *et al.*,[78] believing this postulate to be too restrictive, modified the Penniston model by incorporating two side compartments, one of low and one of high hydrophobicity, and found it to predict a double-peaked curve fitting very well with observed data from a series of analgesic paracetamol (acetaminophen) derivatives.

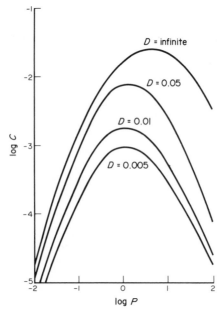

Figure 20 Effect of various dissolution rate constants D on variation of concentration in compartment 10 with $\log P$ at $t = 20$ (arbitrary time units) in the Dearden and Townend modification of the Penniston model (reproduced from ref. 60 by permission of VCH Verlagsgesellschaft mbH)

Kubinyi[79] has also modified the Penniston model, varying the number of compartments by incorporating side compartments to represent a receptor and by incorporating realistic transfer rate constants. He, like Dearden and Townend, observed biphasic correlations of drug concentration in a given compartment with log P (Figure 21), for the model depicted in Figure 22.

In very recent years Baláž and co-workers have also modified the Penniston model, examining the behaviour of both closed and open systems. They have emphasized the importance of time in drug distribution and have pointed out the effects of varying the lipid/aqueous volume ratio in the models used. Figure 23 shows their calculated curves for a four-compartment closed model[80] at various times after dosage, for a lipid/aqueous volume ratio of 0.1. Similar behaviour is observed for a 10-compartment closed system,[81] save that at very long times after dosage considerable distortion of the curves occurs.

Using a multicompartment unidirectional open model, Baláž et al.[82,83] again obtained typical biphasic curves, but with a minimum developing at longer times after dosage (Figure 24), as also observed by Dearden and Townend.[64] This is due to the more rapid loss from the nth compartment of compounds with intermediate log P values. It may be noted that although this leads to a curve that could be construed as a double-peaked QSAR (*vide ultra*), this explanation of such experimentally observed phenomena is unlikely to be realistic, since the times involved to produce curves such as 3, 4 and 5 in Figure 24 are so great.

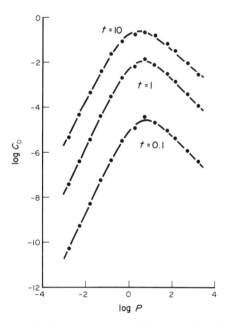

Figure 21 Dependence of drug concentration in compartment D of the model shown in Figure 22 on log P at various times t (reproduced from ref. 79 by permission of Editio Cantor)

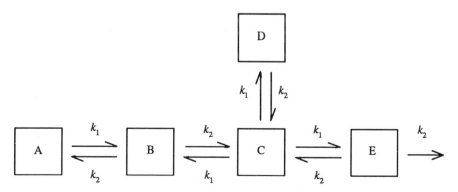

Figure 22 Five-compartment non-equilibrium model used by Kubinyi to generate Figure 21 (reproduced from ref. 79 by permission of Editio Cantor)

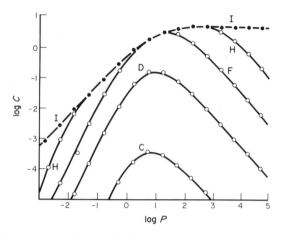

Figure 23 Concentration–hydrophobicity profiles for the fourth compartment of a four-compartment closed model at: (C) 0.01 h, (D) 0.1 h, (F) 1 h, (H) 31.62 h and (I) 1000 h or ∞ (solid circles) after dosage (reproduced from ref. 80 by permission of Editions Scientifiques Elsevier)

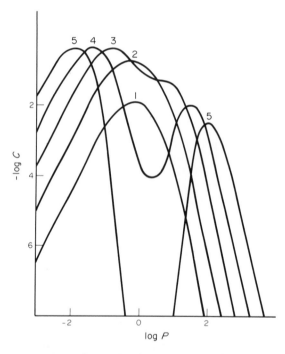

Figure 24 Concentration–hydrophobicity profiles for the fifth compartment of a unidirectional *n*-compartment model after (1) 3.2 h, (2) 10 h, (3) 32 h, (4) 100 h and (5) 320 h after dosage [reproduced from ref. 82 by permission of Elsevier Science Publishers (Biomedical Division)]

Baláž and Šturdík[84] showed that the curves produced by their models could generally be described by a modification (equation 49; C = concentration in selected compartment and A, B_i, C_i, D are constants) of Kubinyi's bilinear equation. The curves in Figure 23, for example, have been fitted using such an equation.

$$\log C = A \log P + \sum_{i=1}^{M} B_i \log(C_i P + 1) + D \qquad (49)$$

These authors also pointed out[81] that for equation (49) to be applied to *in vivo* systems, a correction term is needed, since according to the Collander equation[85] (equation 50; subscripts 1 and 2 refer to different partitioning systems), the partition coefficient of a compound in one system (P_1) may not equal its partition coefficient in another system (P_2). Hence Baláž and Šturdík obtained equation (51), and were able to show[81,84] that it described many experimental data sets well. For example, equation (52) describes the mutagenic potencies of alkylamides of 3-(5-nitro-2-furyl)acrylic acid in *S. typhimurium* TA100 rfa$^+$. The authors did not, however, compare the fit of their equation with the bilinear or the quadratic equation.

$$\log P_2 = a \log P_1 + b \tag{50}$$

$$\log C = A \log P^E + \sum_{i=1}^{M} B_i \log(C_i P^E + 1) + D \tag{51}$$

$$\log(1/C) = 3\log P^{0.638} - 5\log(0.439 P^{0.638} + 1) + 6.027 \tag{52}$$

$$n = 7, \quad r = 0.988, \quad s = 0.120$$

Aarons *et al.*[86] showed that a simple, three-compartment open model (Figure 25) gave a parabolic relationship between the maximal concentration in compartment 3 and $\log P$. The nature of k was not defined but it could presumably be either a transport or a metabolic rate constant. The authors observed that $\log P_o$ increased as $\log k$ increased. No indication was given as to whether fixed-time concentrations were also parabolic in $\log P$.

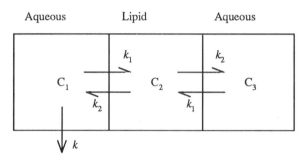

Figure 25 Three-compartment model of Aarons *et al.* (reproduced from ref. 86 by permission of the Journal of Pharmacy and Pharmacology)

Berner and Cooper[87] have applied diffusion theory to both closed and open models consisting of a number of alternating aqueous and oil phases. They obtained parabolic plots of log('receptor' concentration) *versus* $\log P$ for both finite and infinite reservoirs (doses) in the absence of a solubility limitation; the observation in respect of an infinite reservoir is in contrast to the predictions of Dearden and Townend[64] and Yalkowsky and Flynn.[24] It would be useful to have results from a physical model in order to make an assessment of the various mathematical models available.

The extensive work on compartmental modelling, particularly using the Penniston model and its modifications, has confirmed the overwhelming dominance of partitioning in drug transport, and has thrown much light on such aspects as the importance of time and lipid/aqueous volume ratio in modelling experimental drug activity. The Penniston model *per se* does not yield quantitative equations, but the work of Baláž and co-workers in fitting equations to model-produced curves and then applying those equations, or modifications of them, to fit experimental data sets, is an important step in the quantitative modelling of drug activity.

19.2.5 MODELLING OF *IN VIVO* AND *EX VIVO* TRANSPORT

19.2.5.1 Transfer Across Lipoidal Membranes

In 1954 Collander[88] showed, from experiments with *Nitella* cells, that cell penetration by non-electrolytes was a rectilinear function of the partition coefficient (equation 53), thus supporting earlier predictions[89] by Davson and Danielli. Collander was, however, working with compounds of restricted $\log P$ range, for we can predict, from Section 19.2.4, that with an infinite reservoir of drug, the penetration rate will increase and then level off as $\log P$ increases[20,24] and with a finite reservoir, penetration rate will rise and then fall again as $\log P$ increases.[64,90] This does, in fact, occur in practice as exemplified by Figure 26[91] and by equation (54), which correlates the permeability constants of some unspecified non-electrolytes into green algae with their olive oil partition coefficients.[92]

$$\log P_{en} = a \log P + b \tag{53}$$

$$\log k = -3.633 \log P - 0.757 (\log P)^2 + 0.104 \tag{54}$$

$$n = 9, \quad r = 0.988, \quad s = 0.248$$

Data of the type shown in Figure 26 do, incidentally, provide strong support for limiting aqueous solubility's *not* being the reason for a fall-off in penetration rate at high $\log P$ values, as Yalkowsky and Flynn[24] proposed. Figure 26 does, however, confirm aqueous boundary layer resistance as the limiting factor at high $\log P$ values, since increased flow rate of drug solution across the jejunal membrane reduces boundary layer thickness and thus increases permeability. Such data can readily be correlated using Kubinyi's bilinear equation (equation 39).[39] Kubinyi[79] derived equation (55) for the buccal absorption rate constants of *n*-alkanoic acids at pH 3.1–3.8.

$$\log k_{abs} = 0.339 \log P - 0.318 \log(\beta P + 1) - 1.246 \tag{55}$$

$$n = 8, \quad r = 0.995, \quad s = 0.030, \quad \log \beta = -2.450$$

Wagner and Sedman,[93] assuming a model with no aqueous diffusion layer resistance, developed equation (56) (where n and Q are constants for the system), which is a hyperbola with an asymptotic value of k_m (where k_m = rate constant for transfer of drug out of the membrane). This asymptotic value corresponds to that observed for interfacial transfer studies (*cf.* Figure 10), which other workers[22,41] have attributed to limiting transfer across the aqueous diffusion layer. Their equation fits experimental data, despite their model being at odds with most others, which accept or assume appreciable aqueous diffusion layer resistance.

$$k_{abs} = \frac{k_m P^n}{Q + P^n} \tag{56}$$

Ho and co-workers[94,95] developed a general non-linear absorption model that assumed a stagnant aqueous diffusion layer (equation 57; A/V = surface area/volume ratio of the absorption site; P_a, P_m = permeability coefficients in aqueous diffusion layer) which, like the Wagner–Sedman model, gives a plateau value. The model is difficult to handle, however, requiring the estimation of many parameters, and so its application is restricted.

$$k_{abs} = \frac{A}{V} \left(\frac{P_a}{1 + (P_a/f P_m)} \right) \tag{57}$$

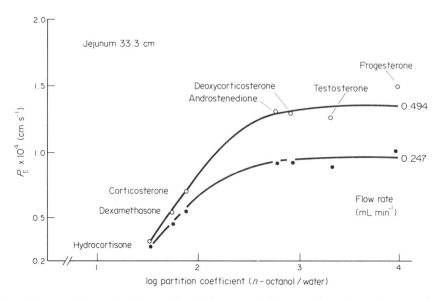

Figure 26 The influence of hydrophobicity on the effective permeability coefficient of steroids across the rat jejunal membrane (reproduced from ref. 91 by permission of Marcel Dekker Inc.)

Schaper,[96] assuming that the rate of absorption of electrolytes is related to the fraction which is unionized, derived equation (58) to predict absorption rate constants *in vivo*. Schaper points out that most *in vivo* absorption studies present data as percentage absorbed in a given time; however, such data are readily converted into rate constants, and Schaper reports the fitting of a large number of data sets by equation (58) (in which a, b, c, e, f = constants characteristic of a particular biosystem; f_u = fraction unionized; and f_i = fraction ionized) and a modification of it to account for behaviour over a wide $\log P$ range.

$$\log k = \log\left(\frac{aP^b f_u + eP^f f_i}{1 + c(aP^b f_u + eP^f f_i)}\right) \tag{58}$$

By using a term for the true rate limit (k_{aqo}) imposed by the aqueous diffusion layer at high $\log P$ values, Schaper also derived equation (59) and used it to fit various data sets. For example, for the buccal absorption of phenylacetic acids he obtained equation (60). It has to be commented here that equations like equation (59), with numerous constants, are not easy to handle and could not be used for small data sets.

$$\log k = \log\left(\frac{k_{aqo} \cdot gP^h(aP^b f_u + eP^f f_i)}{k_{aqo} \cdot gP^h + (k_{aqo} + gP^h)(aP^b f_u + eP^f f_i)}\right) \tag{59}$$

$$\log k = \log\left(\frac{0.342 \times 0.019P^{0.809}(0.033P^{0.603} f_u + 4.64 \times 10^{-4}P^{0.341} f_i)}{0.342 \times 0.019P^{0.809} + (0.342 + 0.019P^{0.809})(0.033P^{0.603} f_u + 4.64 \times 10^{-4}P^{0.341} f_i)}\right) \tag{60}$$

$$n = 131, \quad r = 0.990, \quad s = 0.072$$

de Haan and Jansen[97] have recently pointed out that whilst Schaper's equations give very good fits on the whole, there are a number of discrepancies. By assuming that k_{aqo} remains constant, they derived an equation which, they claimed, modelled observed behaviour more closely. In fact, for the buccal absorption of phenylacetic acids, Schaper's correlation ($n = 131$, $r = 0.99$; equation 60) is better than that of de Haan and Jansen ($n = 62$, $r = 0.958$), although the latter authors point out that their equation fits the data better at low $\log P$ values.

Despite the use of various sophisticated models to represent absorption across *in vivo* membranes, the great majority of the QSARs developed for membrane transport show good correlation with simple hydrophobicity terms, as Table 2 shows. The negative coefficient on the hydrophobicity term of equation (62) is puzzling, but may be due to increasing size of the molecules causing reduced penetration.

Equations (65) and (66) show that octanol is not always the best organic phase for determining partition coefficients for correlation with *in vivo* parameters. A number of workers have made this point over the years.

19.2.5.2 Modelling of Membranes with Pores

Natural membranes are not homogeneous, and quite a substantial body of evidence exists to indicate that they contain pores through which small molecules may pass by a non-partitioning process. A number of workers have devised models to incorporate the existence of such pores. Houk and Guy[34] have recently reviewed modelling of penetration through skin.

The Ho–Higuchi model mentioned above[94, 95] was modified by its authors to give equation (74) (in which P_p = permeability coefficient in pores), to account for the fact that some plots of k_{abs} *versus* $\log P$ were sigmoidal, having a minimum as well as a maximum plateau value.

$$k_{abs} = \frac{AP_a}{V}\left/\left(1 + \frac{P_a}{fP_m + P_p}\right)\right. \tag{74}$$

Plá-Delfina and Moreno[110] modified the Wagner–Sedman model[93] to take pore permeation into account and developed equation (75). They gave several examples of the fit of their equation to intestinal absorption data, of which Figure 27 is one.

$$k_{abs} = \frac{k_m P^n}{Q + P^n} + \frac{k_p Q^1}{Q^1 + P^n} \tag{75}$$

Walter and Gutknecht[111] examined the diffusion of small molecules through egg phosphatidylcholine–decane bilayer membranes. They found that the permeabilities of small

Table 2 Dependence of Membrane Transfer of Drugs on Molecular Properties

Equation	Membrane	Species	Drug	Correlation	n	r	s	Ref.
61	Skin	Mouse	Alcohols	$\log k = 0.34 \log P - 0.55$	6	0.948	—	98
62	Skin	Pig	n-Alkanoic acids	$\log k = -0.48 N^a - 1.17$	6	1.00	—	99
63	Surface	Schistosoma japonicum	Heterogeneous	$\log \mathrm{TUI} = 14.1 \log P + 57.1$	17	0.76	—	100
64	Epidermis	Human	Alcohols	$\log k = 0.544 \log P - 2.884$	8	0.979	0.150	101
65	Skin	Human	Phenylboronic acids	$\log C = 0.573 \log P - 3.749$	8	0.907	0.227	102
66	Skin	Human	Phenylboronic acids	$\log C = 0.417 \log P_{benzene} - 2.463$	7	0.954	0.148	102
67	Ocular	Rabbit	Heterogeneous	$\log k = 0.072 \log P - 0.067(\log P)^2 + 0.134$	7	0.991	0.062	103
68	Buccal	Human	Organic acids	$\log (\%\,\mathrm{abs}) = 1.29\log P_{pH6.8} - 0.154(\log P_{pH6.8})^2 + 0.66(pK_a - pH)^2 - 0.66(pK_a - pH) - 0.013$	31	0.968	0.138	104
69	Placenta	Human	Heterogeneous	$\log \mathrm{TR} = 0.35 \log P_{pH7.4} - 0.466\log(0.22 P_{pH7.4} + 1) + 0.11$	21	0.95	0.11	105
70	Cornea	Rabbit	β-Blockers	$\log k = 0.972 \log P_{pH7.65} - 0.112(\log P_{pH7.65})^2 - 2.71\log MW - 9.26\alpha^b + 0.219$	11	0.97	—	106
71	Plasma/milk	Cow	Sulfonamides	$\log M/P = 0.136\log P - 0.123(\log P)^2 + 0.191(pK_a - 7.4) - 0.330$	8	0.98	0.11	107
72	Everted intestine	Rat	Carbamates	$\log k = 0.103 \log P - 0.090(\log P)^2 - 0.833$	10	0.973	0.126	108
73	Skin	Human	Phenols	$\log \mathrm{STR}^c = -10.63\log P + 10.27(0.403 P + 1) + 10.03$	19	0.944	0.208	109

[a] N = Number of carbon atoms in chain. [b] α = Degree of ionization. [c] STR = Skin transport resistance.

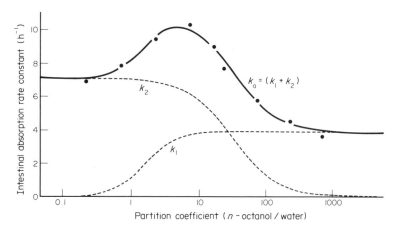

Figure 27 The variation of intestinal absorption rate constant of alkyl-substituted carbamates with partition coefficient (reproduced from ref. 110 by permission of Plenum Publishing Corporation)

molecules (molecular weight < 50) were 2- to 15-fold higher than predicted; these permeabilities could not be correlated with the partition coefficient, but were found to be inversely proportional to the molar volume. They concluded, however, that these results could *not* be explained by the existence of transient aqueous pores, but rather by the membrane's behaving more like a polymer than a liquid hydrocarbon.

Leahy *et al.*[112] have pointed out that there is evidence that small hydrophilic molecules can cross barriers such as the intestinal mucosa entrained in the flow of water; this effect is known as solvent drag and can be thought of as a convective flux. They were able, using radiolabelled flux markers, to calculate sieving coefficients for five drugs and found these coefficients to be inversely proportional to molecular volume; this observation is similar to that of Walter and Gutknecht[111] and others.[113,114]

Lien has in recent years repeatedly drawn attention to the use of molecular weight as a QSAR parameter, provided that it is not collinear with hydrophobicity (as would be the case in, for example, an homologous series). Lien and Wang[115a] showed (equations 76 and 77) that the permeability of various non-ionizable drugs across rat jejunum could be well correlated only if molecular weight was included as a parameter. The negative coefficient of the log MW term suggests that it is reflecting a steric resistance to permeation, which would be consistent with pore transfer.

$$\log k = 0.125 \log P - 4.082 \tag{76}$$

$$n = 13, \quad r = 0.679, \quad s = 0.203$$

$$\log k = 0.237 \log P - 0.657 \log MW - 2.884 \tag{77}$$

$$n = 13, \quad r = 0.933, \quad s = 0.105$$

Hansch *et al.*,[115b] in a re-examination of the data of Levin,[115c] showed that for a heterogeneous range of drugs with molecular weights up to 1400, rat brain capillary permeability coefficient PC was a function of both log P and MW (equation 78).

$$\log PC = 0.50 \log P - 1.43 \log MW - 1.84 \tag{78}$$

$$n = 23, \quad r = 0.927, \quad s = 0.461$$

An interesting observation was made recently by Fisher *et al.*,[116] who found that the rat nasal absorption of water-soluble compounds ranging in molecular weight from 190 to 70 000 was inversely related to molecular weight. From their data, equation (79) can be formulated. Similar observations have been made for skin penetration.[34]

$$\log(\% abs) = 7.62 - 1.35 \log MW \tag{79}$$

$$n = 5, \quad r = 0.995, \quad s = 0.173$$

Thus it may be said that, whilst there is overwhelming evidence that partitioning controls the major part of drug permeation across membranes, the sieving of water-soluble molecules *via* aqueous shunts or possibly other mechanisms should not be neglected.

19.2.5.3 The Blood–Brain Barrier

Much has been written about the blood–brain barrier (BBB) and the ability of substances to cross it and hence reach the CNS. The BBB is generally believed to consist of a layer of capillary endothelial cells,[117] with tight junctions inhibiting the passage of hydrophilic compounds.

Hansch *et al.*[118] deduced, for eight different sets of barbiturates in different animals, parabolic equations in $\log P$, with a mean $\log P_o$ of 2.1. Hansch[48] has also pointed out that eight other classes of hypnotics show similar behaviour, with a mean $\log P_o$ of 1.83. Hansch suggested that a $\log P$ value of about 2 represents the ideal hydrophobic character for penetration into the CNS. Hansch *et al.*[119] also found the penetration of phenylboronic acids into mouse brain to follow a similar pattern (equation 80), with a $\log P_o$ value of 2.3.

$$\log(1/C) = 2.53\log P - 0.56(\log P)^2 - 1.07 \tag{80}$$

$$n = 14, \quad r = 0.912, \quad s = 0.217$$

Lien[92] has reported a large number of correlations of CNS activity with hydrophobicity, all of which display a maximum; the $\log P_o$ values found range from < -0.13 to 4.17, but the majority (26 out of 34) lie between 1.5 and 2.7, with a mean of 2.07. Lien *et al.*[120] have similarly observed, for a variety of compounds exerting CNS depressant or stimulant activity, $\log P_o$ values ranging from 1.38 to 2.66; equation (81), involving muscle relaxant activity of cyclohexanones in mice, is an example.

$$\log(1/C) = 1.601\log P - 0.313(\log P)^2 - 0.276\mu + 1.895 \tag{81}$$

$$n = 8, \quad r = 0.92, \quad s = 0.21, \quad \log P_o = 2.56$$

T'ang and Lien[103] have also found parabolic relationships for the brain uptake of organic acids (equation 82), as have Simon-Trompler *et al.*[121] for the brain appearance of esters of lorazepam and oxazepam. Dischino *et al.*[122] observed $\log P_o \approx 1.5$ for the baboon brain extraction of a heterogeneous range of simple polar compounds such as ethers and alcohols.

$$\log(\%\,\text{uptake}) = 0.665\log P - 0.102(\log P)^2 + 0.902 \tag{82}$$

$$n = 7, \quad r = 0.985, \quad s = 0.124, \quad \log P_o = 3.26$$

In contrast to these findings, Oldendorf[123] observed a sigmoidal plot between brain clearance and $\log P_{\text{olive oil}}$, which reached a plateau at $\log P \approx 0.5$; he also showed that compounds for which an active transfer mechanism existed (*e.g.* glucose and amino acids) were cleared rapidly despite being hydrophilic.

Fenstermacher[124a] observed, for a wide range of different types of compound, a rectilinear relationship between brain uptake and $\log(P \cdot MW^{-1/2})$; such a correlation does not necessarily mean that there is no optimal value of $\log P$, since his higher molecular weight compounds were long-chain compounds such as hexanoates and octanoates. A similar correlation was observed by Levin.[115c]

Recently Young *et al.*[124b] have found, for a set of brain-penetrating drugs, three of which were histamine H_2 receptor antagonists, that the ratio of brain to blood concentrations in the rat was a function of the *difference* between the octanol–water and cyclohexane–water $\log P$ values (equation 83). The authors suggest that this indicates that BBB penetration increases as the hydrogen-bonding ability of the drugs decreases.

$$\log(C_{\text{brain}}/C_{\text{blood}}) = -0.604(\log P_{\text{oct}} - \log P_{\text{cyc}}) + 1.23 \tag{83}$$

$$n = 6, \quad r = 0.980, \quad s = 0.249$$

In general, then, the available evidence is that the BBB is crossed most readily by compounds having $\log P$ values in the range 1.5–2.7 and this fact should be of value in the design of CNS-active drugs.

19.2.6 PHARMACOKINETICS

The term 'pharmacokinetics' means literally drug transport, and as such could be said to apply to the whole of this chapter. It has, however, come to have a more specific meaning, concerning itself with such parameters as *in vivo* absorption and elimination rate, volumes of distribution within an organism, binding constants to proteins and other sites, rates of metabolism and organ clearance. Nonetheless, all these are properties related to transport, and it would be expected that the same molecular properties as discussed in previous sections would control pharmacokinetics.

The literature concerning pharmacokinetics is vast, but publications in the main deal with just one or two drugs. Only a relatively small amount of QSAR work has been done in pharmacokinetics, although there have been several reviews of the subject.[92, 101, 125–130]

The models that are used in pharmacokinetics are essentially those described in Sections 19.2.4 and 19.2.5; however, pharmacokinetics often considers a process such as absorption to be a single step, so that a three-compartment model could represent absorption and excretion, with the middle compartment representing blood, tissue or an organ. To add to the confusion, such a model is referred to in pharmacokinetics as a one-compartment model, meaning a straight-through model with no side compartments representing protein binding or other factors. The various models and their derivations have been dealt with by Wagner[131] and Rowland and Tozer,[132] and it is not appropriate to examine them further here.

19.2.6.1 Absorption

Examples of absorption across specific membranes such as the buccal membrane and the stratum corneum were discussed in Section 19.2.5. Most drugs are administered orally, so that absorption is from the stomach and/or the intestine. As a consequence, most pharmacokinetic studies of absorption have been made on these two organs. Absorption rate constant would be expected, as shown in Section 19.2.4, to be a function of hydrophobicity, with a correction for degree of ionization if appropriate; with an infinite drug reservoir, absorption rate constant would be expected to increase and then level off as $\log P$ increased, but to fall again if a single dose regimen were used.

Many correlations show a rectilinear relationship of absorption rate constant and hydrophobicity which may be taken to indicate merely a restricted $\log P$ range of the compounds examined. An example is of gastric absorption of sulfonamides in the rat[92] (equation 84; P was determined in isopentyl acetate/water). Where a wider $\log P$ range is used, parabolic or bilinear equations may be fitted. Equation (85) relates to intestinal absorption of sulfonamides in the rat[125] and equation (86) (I is an indicator variable for the presence of tertiary alkyl groups) to intestinal absorption of carbamates in the rat.[125]

$$\log k_a = 0.31 \log P - 1.159 \tag{84}$$

$$n = 17, \quad r = 0.942, \quad s = 0.122$$

$$\log k_a = 0.44 \log P - 0.09 (\log P)^2 - 0.396 \tag{85}$$

$$n = 12, \quad r = 0.89, \quad s = 0.16$$

$$\log k_a = 0.30 \log P - 0.57 \log(0.34P + 1) - 0.15I - 0.74 \tag{86}$$

$$n = 13, \quad r = 0.98, \quad s = 0.04$$

For compounds that are partly ionized under experimental conditions, some correction for ionization is necessary, for example by the use of the distribution coefficient D (apparent partition coefficient) as in equation (87) for the colonic absorption of organic acids in the rat[133] or by the use of a parameter such as pK_a, as in equation (88) for the colonic absorption of organic bases in the rat.[92] Leaving out the ionization term in equation (88) yields a much poorer correlation ($r = 0.814$).

$$\log k_a = 0.268 \log D - 0.082 (\log D)^2 + 3.96 \tag{87}$$

$$n = 10, \quad r = 0.986, \quad s = 0.095$$

$$\log(\% \, abs) = 0.869 \log P - 0.330 (\log P)^2 - 0.059 (pH - pK_a) + 0.817 \tag{88}$$

$$n = 10, \quad r = 0.910, \quad s = 0.187$$

19.2.6.2 Volume of Distribution

The volume of distribution (V) relates the amount of drug in an organism to its concentration; only rarely does it represent a true physiological volume, as it is affected by degree of ionization and by protein and tissue binding, as well as by hydrophobicity. Seydel[127] makes the point that volume of distribution must be corrected for protein binding if significant correlations are to be obtained.

Equations (89) and (90)[127] show this for a series of penicillins.

$$V = -0.05 \log P + 1.25 \tag{89}$$

$$n = 7, \quad r = 0.25, \quad s = 0.18$$

$$V_{\text{free}} = 120.4 \log P - 98.1 \tag{90}$$

$$n = 7, \quad r = 0.95, \quad s = 39.9$$

Watanabe and Kozaki[134] found a relationship between V and degree of ionization for a series of basic drugs; Seydel and Schaper[126] simplified their equation to equation (91), which they refer to as an example of a general relationship between V and P. However, if quasi-equilibrium conditions obtain, then $\log V$ would be expected to be directly proportional to $\log P$. One can envisage a situation where, say, rapid $\log P$-related metabolism caused V to increase less rapidly, or even to decrease, at high $\log P$ values, but this cannot be described as the general case. If, on the other hand, the distribution of drug is far from being at equilibrium or steady state in the organism, then one would expect, from Section 19.2.5, a parabolic type relationship, at a given time after dosage, between V and $\log P$, or between the distribution in given organs and $\log P$. Thus Lien *et al.*[135] found the prostatic fluid to plasma ratio of sulfonamides in the dog to vary parabolically with $\log P$ (equation 92).

$$\log V = \log(1.325 + 0.19 P^{0.83}) \tag{91}$$

$$n = 15, \quad r = 0.91, \quad s = 0.281$$

$$\log \text{PF/PL} = 0.097 \log P - 0.214 (\log P)^2 + 0.33 \log(pK_a - pH) - 0.57 \tag{92}$$

$$n = 16, \quad r = 0.905, \quad s = 0.211$$

It should perhaps be remembered here, however, that the equilibrium models of Higuchi and Davis[62] and Martin and Hackbarth[54] predict parabolic-type relationships under certain conditions. It remains a fact, nevertheless, that most correlations of V with hydrophobicity are rectilinear.[125, 126]

19.2.6.3 Protein Binding

Although binding to protein is not strictly a transport property, it undoubtedly affects transport; protein binding by a drug on its random walk from the site of administration to the site of action initially lowers the rate at which the drug reaches the latter, but later provides a prolonged release effect.[136]

Most protein binding is non-specific and apparently involves hydrophobic regions of the protein surface. It is, therefore, in order to regard protein binding as a type of partitioning and numerous correlations show a rectilinear relationship between $\log K$ (binding constant) and $\log P$.[125] It would be expected,[137] since protein binding does not completely desolvate the drug molecule, that the regression coefficient between $\log K$ and $\log P$ would be considerably less than unity and this is often the case,[138, 139] as is shown by equation (93) for the binding of sulfonylurea derivatives to bovine serum albumin (BSA)[138] and the binding of a heterogeneous range of organic compounds to the same protein (equation 94; C = concentration of organic compound producing a 1:1 complex with BSA).[139] Scholtan[140] has reported several correlations with regression coefficients of 1 or greater (*e.g.* equation (95) for the binding of penicillins to HSA), but this is probably because his $\log P$ values were measured in butanol/aqueous buffer. Seydel *et al.*[141] have shown that steric effects can affect even non-specific binding, as equation (96) (ΔR_m = chromatographic hydrophobic substituent constant and I_o = indicator variable for *o*-substitution) shows for the rat serum binding of 2-sulfapyridines.

$$\log K = 0.33 \log P + 0.24 pK_a + 1.48 \tag{93}$$

$$n = 15, \quad r = 0.947, \quad s = 0.090$$

$$\log(1/C) = 0.75 \log P + 2.30 \tag{94}$$

$$n = 42, \quad r = 0.960, \quad s = 0.150$$

$$\log K = 1.32 \log P + 0.37 \tag{95}$$

$$n = 8, \quad r = 0.944, \quad s = 0.140$$

$$\log K = 0.33 \Delta Rm - 0.15 pK_a - 0.44 I_o - 0.95 \tag{96}$$

$$n = 17, \quad r = 0.85, \quad s = 0.19$$

Hansch and Klein[137] have reported that even specific binding to enzymes is $\log P$-related, although electronic and steric parameters may also be involved. Equation (97)[137] (in which σ_{meta} = Hammett substituent constant for 3-substituents) shows the inhibition of human liver alcohol dehydrogenase by pyrazoles.

In general, then, non-specific protein binding may be regarded as a simple function of hydrophobicity, and as such can readily be incorporated into pharmacokinetic models.

$$\log(1/K_i) = 0.87 \log P - 2.06 \sigma_{meta} + 4.60 \qquad (97)$$

$$n = 13, \quad r = 0.977, \quad s = 0.303$$

19.2.6.4 Metabolism

Like protein binding, metabolism is not strictly a transport property but it can have a marked effect on the rate at which active drug arrives at the receptor site. *A priori*, one would expect that metabolism would be controlled primarily by electronic factors, which affect bond order and therefore group lability. However, steric effects could also play a part, by shielding vulnerable groups or by hindering enzyme binding. In addition, binding to hydrophobic regions of enzymes is controlled by substrate hydrophobicity; furthermore, it is known that organisms preferentially metabolize hydrophobic compounds in an attempt to make them more water soluble and hence enhance their elimination. Thus metabolism could be expected to correlate with any or all of these three parameters. Seydel[127] has correlated the N-dealkylation of amphetamines by equation (98). Seydel[127] also reports that N^4-acetylation of sulfonamides in humans is, surprisingly, negatively correlated with hydrophobicity (equation 99).

$$\log(\% \, \text{metab.}/(100 - \% \, \text{metab.})) = 0.927\pi - 1.106 \qquad (98)$$

$$n = 13, \quad r = 0.956, \quad s = 0.287$$

$$\log k_m = -0.42 \log P + 0.14 \text{p}K_a - 2.89 \qquad (99)$$

$$n = 11, \quad r = 0.84, \quad s = 0.34$$

In vitro enzyme-mediated metabolism has been extensively examined by Hansch and co-workers. For example, Smith *et al.*[142] correlated the hydrolysis of phenyl hippurates by papain by equation (100) ($\pi'_3 = \pi$ value of the more hydrophobic *meta*-substituent and MR_4 = molar refractivity of *para*-substituents), using the Michaelis constant K_m as a measure of hydrolysis; molar refractivity, although having the units of molar volume and, therefore, incorporating a steric component, also contains a polarizability component and is believed[137] to model binding to surfaces of intermediate polarity.

$$\log(1/K_m) = 1.03\pi'_3 + 0.57\sigma + 0.61 MR_4 + 3.80 \qquad (100)$$

$$n = 25, \quad r = 0.907, \quad s = 0.208$$

This and other examples by Hansch and co-workers demonstrate the position-specific nature of the factors affecting enzymic metabolism and can be construed as receptor mapping. In a similar vein Tong and Lien[143] have reported the dependence on hydrophobicity only (equation 101) of pork liver microsomal NADPH N-oxidation of tertiary amines.

$$\log(\text{oxidation rate}) = 0.37 \log P - 0.03(\log P)^2 - 7.06 \qquad (101)$$

$$n = 10, \quad r = 0.962, \quad s = 0.106$$

Apropos of metabolism, Seydel and Schaper[125] commented that relatively hydrophobic drugs are absorbed and reach the site of action quickly, but are generally metabolized more rapidly; they argued, however, that the longer exposure of polar drugs to metabolic inactivating factors means that only the more hydrophobic drugs can reach the receptor before being metabolized, so that the effective dose decreases as hydrophobicity increases.

19.2.6.5 Clearance and Elimination

Total clearance, or elimination, is the sum of the clearance by each eliminating organ (equation 102; CL = clearance, k = elimination rate constant and V = volume of distribution). By and large, the mechanisms controlling elimination are the same as those controlling absorption (*i.e.* hydrophobicity and degree of ionization), and so one would expect clearance to depend on those factors. This is, in fact, found to be the case[101,125] as is illustrated by the examples in Table 3.

$$CL = kV \qquad (102)$$

The negative coefficient on the ΔRm term in equation (105) is related to the fact that sulfonamides are excreted largely by glomerular filtration; that is, only the non-protein-bound fraction is filtered.[127] This being so, one would expect a negative correlation between clearance and protein binding (K), which is observed in, for example, *p*-substituted sulfapyrimidines[127] in the rat (equation 118). For other compounds, such as penicillins (equation 104), clearance increases with hydrophobicity, at least initially, as would be expected from, say, the Penniston model.[57]

$$\log CL = -1.75 \log K - 3.73 \qquad (118)$$
$$n = 8, \quad r = 0.98, \quad s = 0.12$$

Equations (113) to (115) give an interesting interspecies comparison and clearly show that excretion of sulfapyrimidines, at least, is very similar in rat, goat and man.

This brief overview, coupled with that presented in Section 19.2.6, shows that *in vivo* pharmacokinetic parameters correlate with the same properties, hydrophobicity and ionization, that control all membrane transport. The use of pharmacokinetic QSARs does, however, permit interpretation in terms of mechanisms, and can indicate which species may validly be compared with man for testing purposes as well as providing the rationale for improved drug design.

19.2.7 PRODRUGS

All drug activity is a consequence of molecular structure, and so specific requirements—$\log P$, pK_a, shape and so on—can, in theory, be designed into a molecule. However, a change in molecular architecture designed, say, to increase hydrophobicity will also affect a host of other properties, some of them perhaps adversely.[148] This is, of course, one of the main reasons why drug design is so difficult and why the success rate is still only 1 in 5000 or so.

A useful way to have one's cake and eat it, as it were, is to use the prodrug concept. A prodrug is defined as a chemical substance having superior physical and/or chemical properties, which after administration converts to an active drug substance in the body.[149]

Prodrugs may be designed to improve chemical stability, to increase absorption, to increase site specificity, to increase aqueous solubility, to reduce metabolism, to alter duration of action and to increase patient acceptance.[150] In fact, the vast majority of prodrugs are designed to increase absorption of polar drugs—that is, to increase hydrophobicity. An interesting example of this is in the use of adrenalin (epinephrine) for the topical treatment of glaucoma. Adrenalin is very polar and therefore not well absorbed; consequently high concentrations have to be used, resulting in systemic absorption causing undesirable side effects. Hussain and Truelove[151] esterified both the aromatic OH groups of adrenalin with the pivaloyl group, $-COCMe_3$, which should increase $\log P$ by about 2.3 log units. This had the effect of making the drug some 100 times more effective, and 300–400 times weaker in affecting the cardiovascular system of dogs and cats.[152] The prodrug was shown to be readily hydrolyzed back to adrenalin by enzymes present in the eye.

Other examples are quoted by Yalkowsky and Morozowich.[153] Erythromycin propionate ($\log P_{pH6} = 2.5$) is absorbed much better than is erythromycin itself ($\log P_{pH6} = -0.13$). Pivampicillin shows about 90% absorption upon oral administration in humans, whereas ampicillin shows about 33%; the ester has a $\log P$ value about 2.7 log units higher than that of the parent compound. Psicofuranine is not absorbed upon oral administration in humans, having $\log P = -1.95$; the triacetate ester ($\log P = 0.72$) is well absorbed. ($\log P = 0$ is recognized[153] as the cut-off point below which absorption is very poor.)

Yalkowsky and Morozowich[153] point out, however, that absorption does not increase indefinitely with hydrophobicity, because of competitive lumenal hydrolysis and decreased aqueous solubility. Equation (119) has been calculated from data which they quote for the absorption of lincomycin esters from rat jejunal loops.

$$\log(\% abs) = 0.220 \log P - 0.082(\log P)^2 + 1.610 \qquad (119)$$
$$n = 6, \quad r = 0.989, \quad s = 0.040$$

Table 3 Dependence of Clearance of Drugs on Molecular Properties

Equation	Type of clearance	Drug	Species	Correlation	n	r	s	Ref.
103	Biliary	Sulfathiazoles	Rat	$\log(\%\text{excr.}) = 0.860 \log P - 0.719(\log P)^2 - 0.401 \, \mathrm{p}K_a + 3.214$	9	0.937	0.242	92
104	Biliary	Penicillins	Rat	$\log(\%\text{excr.}) = 0.792 \log P - 0.132(\log P)^2 + 2.269$	9	0.931	0.083	92
105	Metabolic	Sulfonamides	Rat	$\log \mathrm{CL} = -0.57 \, \Delta Rm_u^a + 0.22 \, \mathrm{p}K_a - 2.25$	7	0.98	0.13	125
106	Renal	Sulfonamides	Rat	$\log \mathrm{CL} = 0.97 \, \Delta Rm_u + 0.23 \, \mathrm{p}K_a - 1.78$	7	0.99	0.12	125
107	Renal	Probenecid analogues	Not given	$\log \mathrm{CL} = -0.035 \log P_{\mathrm{CHCl_3}} - 0.242(\log P_{\mathrm{CHCl_3}})^2 + 0.578$	5	0.980	0.163	92
108	Renal	Probenecid analogues	Not given	$\log \mathrm{CL} = -\log(0.35 + 0.013 \, P_{\mathrm{CHCl_3}}^{1.12})$	7	0.99	0.14	125
109	Renal	Sulfapyridines	Rat	$\log \mathrm{CL} = -0.74 \, Rm + 0.22 \, \mathrm{p}K_a - 1.73$	19	0.96	0.15	127
110	Renal	Nitroimidazoles/Nitrothiazoles	Rat	$\log(\%\text{excr.} \times 10) = 0.069 \log P - 0.788(\log P)^2 + 2.469$	11	0.936	0.136	144
111	Hepatic	Barbiturates	Rat	$\log \mathrm{RHC} = 1.86 \log P - 0.25(\log P)^2 - 3.09$	32	0.89	0.31	127
112	Hepatic	Barbiturates	Rat	$\log \mathrm{CL} = 0.66 \log P_{\mathrm{pH7.4}} - 0.033(\log P_{\mathrm{pH7.4}})^2 - 0.54$	11	0.945	0.28	145
113	Total	Sulfapyrimidines	Rat	$\log k_{\mathrm{el}} = 0.441 \log k_r^b - 1.069$	6	0.90	0.13	127
114	Total	Sulfapyrimidines	Goat	$\log k_{\mathrm{el}} = -0.33 \log k_r - 0.74$	5	0.95	0.08	127
115	Total	Sulfapyrimidines	Human	$\log k_{\mathrm{el}} = -0.49 \log k_r - 1.58$	5	0.94	0.07	127
116	Total	Sulfonamides	Human	$\log t_{\frac{1}{2}} = 0.24 \log P + 0.361 \, I^c + 1.35$	7	0.90	0.15	146
117	Total	Sulfonamides	Rabbit	$\log k_{\mathrm{el}} = -0.20 \log P - 0.18 \, \mathrm{p}K_a + 0.68$	11	0.81	0.18	147

^{a} ΔRm_u is a thin-layer chromatographic hydrophobic substituent constant for the unionized molecule. ^{b} k_r is the HPLC capacity factor (a measure of hydrophobicity). ^{c} I is an indicator variable for o-substitution.

Similarly Biagi et al.[154] found that the ability of testosterone esters to hemolyze rat red cells was expressed by equation (120). Simon-Trompler et al.[121] observed that brain–blood concentration ratio integrated with respect to time (AUQ) for a series of esters of lorazepam and oxazepam in mice varied parabolically with hydrophobicity (equation 121). The same authors observed that the time lag for the compounds to appear in the brain also varied parabolically with Rm, but with the relationship (equation 122) passing through a minimum (*cf.* Figure 18).

$$\log RBR = 2.25 \log P - 0.17 (\log P)^2 - 5.47 \tag{120}$$

$$n = 14, \quad r = 0.938, \quad s = 0.199$$

$$\log AUQ = 1.35 \, Rm - 4.77 \, Rm^2 + 2.05 \tag{121}$$

$$n = 6, \quad r = 0.97, \quad s \text{ not given}$$

$$t_{lag} = 0.93 \, Rm + 28.65 \, Rm^2 + 1.76 \tag{122}$$

$$n = 7, \quad r = 0.97, \quad s \text{ not given}$$

James et al.,[155] working with testosterone esters in the rat, measured times of maximal effect; their data yield equation (123) (P_{eo} = partition coefficient in the ethyl oleate/water system), showing that at intermediate $\log P$ values t_{me} is shortest, *i.e.* duration of action is shortest (*cf.* ref. 75).

$$t_{me} = -16.886 \log P_{eo} + 2.789 (\log P_{eo})^2 + 28.142 \tag{123}$$

$$n = 6, \quad r = 0.968, \quad s = 1.659$$

Charton[156] has shown that the lability of many different classes of prodrugs can be correlated with molecular properties. For example, the hydrolysis rate constants of 5′-acyl derivatives of aracytidine in human plasma can be correlated (equation 124; v = Charton's steric constant and α = substituent polarizability) with steric and electronic parameters.

$$\log k_r = -3.50 v + 6.86 \alpha + 4.17 \tag{124}$$

$$n = 8, \quad r = 0.953, \quad s = 0.439$$

The making of a prodrug in order to increase aqueous solubility is usually done so as to enable a liquid formulation to be prepared, for example for pediatric use. However, an increase in aqueous solubility can also bring about increased absorption, when the parent drug has very poor aqueous solubility; if the prodrug is water soluble because of low melting point (see Section 19.2.3.1), rather than by incorporation of polar moieties, then the partition coefficient will probably remain high.[150,153]

It is thus clear that the transport of prodrugs follows the same pattern as that of any other type of compound. Hence there is no difficulty in principle in designing appropriate transport properties into a prodrug; whether that will yield a safe and efficacious product is beyond the scope of this chapter.

19.2.8 SUMMARY

The dominant factor governing all aspects of drug transport is hydrophobicity, modelled by $\log P$ or related parameters. It has been shown to control initial dissolution of a solid dosage form, transfer through lipid membranes, protein binding (and even metabolism to some extent), binding to receptor sites and clearance and excretion.

Hydrophobicity is closely related to aqueous solubility, and in fact partitioning is controlled largely by affinity of a solute for the aqueous phase rather than affinity for the organic phase; the latter is governed largely by the solute melting point (see equation 14).

Transport across a membrane involves a number of steps: diffusion across a stagnant aqueous boundary layer, interfacial transfer into the membrane, passage across the membrane, interfacial transfer out of the membrane and diffusion across the second stagnant aqueous layer (see Figure 5). Much work has been done on such transport, and it is generally acknowledged that the overall transfer rate constant across the membrane increases initially with hydrophobicity, but eventually reaches a maximum as diffusion across the aqueous boundary layer becomes rate limiting. Controversy exists as to whether or not the interfacial transfer step offers significant resistance to transport.

Since *in vivo* transport involves passage through more than one lipid membrane, various models have been devised to represent such passage. These models have extended our understanding of the drug transport process and have correctly predicted the effect of hydrophobicity not only on biological activity but also on time to maximal response and on duration of action.

Numerous equations have been derived to fit *in vivo* absorption data; some of these contain a large number of terms, to account for factors such as ionization. In general, however, *in vivo* absorption varies biphasically with hydrophobicity, the second leg of the relationship having zero slope if there is an infinite drug reservoir and a negative slope if a single dose of drug is given. The blood–brain barrier seems to favour passage of drugs with $\log P \approx 2$.

The various pharmacokinetic parameters such as absorption rate constant, volume of distribution and clearance are all found to be controlled predominantly by hydrophobicity, together sometimes with an electronic parameter which may model metabolic processes and/or ionization.

Finally, it is seen that the transport of prodrugs follows expected behaviour; such entities are useful because transport (or other property) can be modified by the incorporation of a labile moiety without the required properties of the parent drug being affected.

19.2.9 REFERENCES

1. E. Overton, *Vierteljahrsschr. Naturforsch. Ges. Zuerich*, 1899, **44**, 88.
2. J. H. Collett and L. Koo, *Pharm. Acta Helv.*, 1976, **51**, 27.
3. J. T. Carstensen, in 'Formulation and Preparation of Dosage Forms', ed. J. Polderman, Elsevier, Amsterdam, 1977, p. 197.
4. J. H. Collett and L. Koo, *J. Pharm. Sci.*, 1976, **65**, 753.
5. J. C. Dearden and N. C. Patel, *Drug Dev. Ind. Pharm.*, 1978, **4**, 529.
6. S. H. Yalkowsky, S. C. Valvani and T. J. Roseman, *J. Pharm. Sci.*, 1983, **72**, 866.
7. C. T. Chiou, D. W. Schmedding and M. Manes, *Environ. Sci. Technol.*, 1982, **16**, 4.
8. S. H. Yalkowsky and S. C. Valvani, *J. Pharm. Sci.*, 1980, **69**, 912.
9. C. Hansch, J. E. Quinlan and G. L. Lawrence, *J. Org. Chem.*, 1968, **33**, 347.
10. G. Lepetit, *Pharmazie*, 1980, **35**, 696.
11. M. Charton and B. I. Charton, in 'QSAR in Design of Bioactive Compounds', ed. M. Kuchař, Prous, Barcelona, 1984, p. 41.
12. J. C. Dearden and M. H. Rahman, *J. Pharm. Pharmacol.*, 1987, **39**, Suppl., 110P.
13. A. J. Dobbs and N. Williams, *Chemosphere*, 1983, **12**, 97.
14. P. A. Shore, B. B. Brodie and C. A. M. Hogben, *J. Pharmacol. Exp. Ther.*, 1957, **119**, 361.
15. A. L. J. de Meere and E. Tomlinson, *Int. J. Pharm.*, 1983, **17**, 331.
16. P. R. Byron and M. J. Rathbone, *Int. J. Pharm.*, 1986, **29**, 103.
17. P. -H. Wang and E. J. Lien, *J. Pharm. Sci.*, 1980, **69**, 662.
18. G. L. Flynn, S. H. Yalkowsky and T. J. Roseman, *J. Pharm. Sci.*, 1974, **63**, 479.
19. G. J. Hanna and R. D. Noble, *Chem. Rev.*, 1985, **85**, 583.
20. R. G. Stehle and W. I. Higuchi, *J. Pharm. Sci.*, 1972, **61**, 1922.
21. B. C. Lippold and G. F. Schneider, *Arzneim.-Forsch.*, 1975, **25**, 843.
22. H. van de Waterbeemd, in 'Quantitative Approaches to Drug Design,' ed. J. C. Dearden, Elsevier, Amsterdam, 1983, p. 183.
23. F. H. N. de Haan and A. C. A. Jansen, *Int. J. Pharm.*, 1984, **18**, 311.
24. S. H. Yalkowsky and G. L. Flynn, *J. Pharm. Sci.*, 1973, **62**, 210.
25. W. J. Albery, J. F. Burke, E. B. Leffler and J. Hadgraft, *J. Chem. Soc., Faraday Trans. 1*, 1976, **72**, 1618.
26. V. G. Levich, 'Physicochemical Hydrodynamics,' Prentice Hall, Englewood Cliffs, NJ, 1962, p. 69.
27. R. H. Guy and R. Fleming, *Int. J. Pharm.*, 1979, **3**, 143.
28. R. Fleming, R. H. Guy and J. Hadgraft, *J. Pharm. Sci.*, 1983, **72**, 142.
29. A. Brodin and A. Ågren, *Acta Pharm. Suec.*, 1971, **8**, 609; A. Brodin and M. -I. Nilsson, *Acta Pharm. Suec.*, 1973, **10**, 187; A. Brodin, *Acta Pharm. Suec.*, 1974, **11**, 141.
30. D. M. Miller, *Biochim. Biophys. Acta*, 1986, **856**, 27.
31. D. E. Leahy and A. R. Wait, *J. Pharm. Sci.*, 1986, **75**, 1157.
32. D. E. Leahy, A. L. J. de Meere, A. R. Wait, P. J. Taylor, J. A. Tomenson and E. Tomlinson, *Int. J. Pharm.*, 1989, **50**, 117.
33. F. H. N. de Haan and A. C. A. Jansen, *Int. J. Pharm.*, 1986, **29**, 177.
34. J. Houk and R. H. Guy, *Chem. Rev.*, 1988, **88**, 455.
35. R. C. Reid, J. M. Prausnitz and T. K. Sherwood, 'The Properties of Gases and Liquids', 3rd edn., McGraw-Hill, New York, 1977, p. 544.
36. J. C. Dearden and J. Williams, *J. Pharm. Pharmacol.*, 1978, **30**, Suppl., 50P.
37. H. van de Waterbeemd, P. van Bakel and A. Jansen, *J. Pharm. Sci.*, 1981, **70**, 1081.
38. H. Kubinyi, *J. Pharm. Sci.*, 1978, **67**, 262.
39. H. Kubinyi, *Arzneim.-Forsch.*, 1976, **26**, 1991.
40. F. H. N. de Haan and A. C. A. Jansen in 'Quantitative Approaches to Drug Design,' ed. J. C. Dearden, Elsevier, Amsterdam, 1983, p. 215.
41. F. H. N. de Haan, T. de Vringer, J. T. M. van de Waterbeemd and A. C. A. Jansen, *Int. J. Pharm.*, 1983, **13**, 75.
42. A. Leo, C. Hansch and D. Elkins, *Chem. Rev.*, 1971, **71**, 525.
43. J. C. Dearden and G. M. Bresnen, *J. Pharm. Pharmacol.*, 1980, **32**, Suppl., 7P.
44. J. C. Dearden and J. H. O'Hara, *J. Pharm. Pharmacol.*, 1976, **28**, Suppl., 15P.

45. A. Cammarata and K. S. Rogers, *J. Med. Chem.*, 1971, **14**, 269.
46. C. Hansch, P. P. Maloney, T. Fujita and R. M. Muir, *Nature (London)*, 1962, **194**, 178.
47. C. Hansch and J. M. Clayton, *J. Pharm. Sci.*, 1973, **62**, 1.
48. C. Hansch, *Farmaco, Ed. Sci.*, 1968, **23**, 292.
49. R. Franke and P. Oehme, *Pharmazie*, 1973, **28**, 489.
50. R. M. Hyde, *J. Med. Chem.*, 1975, **18**, 231.
51. J. C. Dearden and E. George, *J. Pharm. Pharmacol.*, 1979, **31**, Suppl., 45P.
52. T. Fujita, *J. Med. Chem.*, 1966, **9**, 797.
53. (a) P. Moser, K. Jäkel, P. Krupp, R. Menassé and A. Sallmann, *Eur. J. Med. Chem.—Chim. Ther.*, 1975, **10**, 613; (b) R. A. Scherrer and S. M. Howard, *J. Med. Chem.*, 1977, **20**, 53.
54. Y. C. Martin and J. J. Hackbarth, *J. Med. Chem.*, 1976, **19**, 1033.
55. Y. C. Martin, 'Quantitative Drug Design', Dekker, New York, 1978, p. 142.
56. Y. C. Martin, in 'Drug Design,' ed. E. J. Ariëns, Academic Press, New York, 1979, vol. 8, p.1.
57. J. T. Penniston, L. Beckett, D. L. Bentley and C. Hansch, *Mol. Pharmacol.*, 1969, **5**, 333.
58. H. Kubinyi, *Prog. Drug Res.*, 1979, **23**, 97.
59. J. Th. M. van de Waterbeemd, A. C. A. Jansen and K. W. Gerritsma, *Pharm. Weekbl.*, 1978, **113**, 1097.
60. J. C. Dearden and M. S. Townend, in 'QSAR and Strategies in the Design of Bioactive Compounds,' ed. J. K. Seydel, VCH, Weinheim, 1985, p. 328.
61. J. W. McFarland, *J. Med. Chem.*, 1970, **13**, 1192.
62. T. Higuchi and S. S. Davis, *J. Pharm. Sci.*, 1970, **59**, 1376.
63. J. C. Dearden and M. S. Townend, *Spec. Publ.-Chem. Soc.*, 1977, **29**, 135.
64. J. C. Dearden and M. S. Townend, *Pestic. Sci.*, 1979, **10**, 87.
65. E. R. Cooper, B. Berner and R. D. Bruce, *J. Pharm. Sci.*, 1981, **70**, 57.
66. J. C. Dearden and K. D. Patel, *J. Pharm. Pharmacol.*, 1978, **30**, Suppl., 51P.
67. M. Tichý and Z. Roth, in 'QSAR and Strategies in the Design of Bioactive Compounds,' ed. J. K. Seydel, VCH, Weinheim, 1985, p. 190.
68. J. C. Dearden, *EHP, Environ. Health Perspect.*, 1985, **61**, 203.
69. M. S. Tute, *Adv. Drug Res.*, 1971, **6**, 1.
70. J. C. Dearden and E. Tomlinson, *J. Pharm. Pharmacol.*, 1971, **23**, Suppl., 73S.
71. J. C. Dearden and J. H. O'Hara, *J. Pharm. Pharmacol.*, 1976, **28**, Suppl., 15P.
72. J. C. Dearden and M. S. Townend, in 'Quantitative Structure–Activity Analysis,' ed. R. Franke and P. Oehme, Akademie-Verlag, Berlin, 1978, p. 387.
73. E. Kutter, A. Herz, H. -J. Teschemacher and R. Hess, *J. Med. Chem.*, 1970, **13**, 801.
74. J. C. Dearden and E. George, *J. Pharm. Pharmacol.*, 1977, **29**, Suppl., 74P.
75. J. C. Dearden and M. S. Townend, in 'Quantitative Approaches to Drug Design,' ed. J. C. Dearden, Elsevier, Amsterdam, 1983, p. 219.
76. A. Herz and H. -J. Teschemacher, *Adv. Drug Res.*, 1971, **6**, 79.
77. R. Franke and R. Kühne, *Eur. J. Med. Chem.—Chim. Ther.*, 1978, **13**, 399.
78. J. C. Dearden, J. H. O'Hara and M. S. Townend, *J. Pharm. Pharmacol.*, 1980, **32**, Suppl., 102P.
79. H. Kubinyi, *Arzneim.-Forsch.*, 1979, **29**, 1067.
80. S. Baláž, E. Šturdík, M. Hrmová, M. Breza and T. Liptaj, *Eur. J. Med. Chem.—Chim. Ther.*, 1984, **19**, 167.
81. S. Baláž and E. Šturdík, in 'QSAR in Toxicology and Xenobiochemistry,' ed. M. Tichý, Elsevier, Amsterdam, 1985, p. 257.
82. S. Baláž, E. Šturdík and J. Augustin, *Biophys. Chem.*, 1986, **24**, 135.
83. S. Baláž, E. Šturdík and J. Augustin, *Gen. Physiol. Biophys.*, 1987, **6**, 65.
84. S. Baláž and E. Šturdík, in 'QSAR in Design of Bioactive Compounds,' ed. M. Kuchař, Prous, Barcelona, 1984, p. 289.
85. R. Collander, *Acta Chem. Scand.*, 1951, **5**, 774.
86. L. Aarons, D. Bell, R. Waigh and Q. Ye, *J. Pharm. Pharmacol.*, 1982, **34**, 746.
87. B. Berner and E. R. Cooper, *J. Pharm. Sci.*, 1984, **73**, 102.
88. R. Collander, *Physiol. Plant.*, 1954, **7**, 420.
89. H. Davson and J. F. Danielli, 'The Permeability of Natural Membranes', 2nd edn. Cambridge University Press, London, 1952.
90. L. Rodriguez, V. Zecchi and A. Tartarini, *Pharm. Acta Helv.*, 1984, **59**, 95.
91. N. F. H. Ho, H. P. Merkle and W. I. Higuchi, *Drug Dev. Ind. Pharm.*, 1983, **9**, 1111.
92. E. J. Lien, in 'Drug Design,' ed. E. J. Ariëns, Academic Press, New York, 1975, vol. 5, p. 81.
93. J. G. Wagner and A. J. Sedman, *J. Pharmacokinet. Biopharm.*, 1973, **1**, 23.
94. A. Suzuki, W. I. Higuchi and N. F. H. Ho, *J. Pharm. Sci.*, 1970, **59**, 644, 651.
95. N. F. H. Ho, W. I. Higuchi and J. Turi, *J. Pharm. Sci.*, 1972, **61**, 192.
96. K. -J. Schaper, *Quant. Struct.-Act. Relat.*, 1982, **1**, 13.
97. F. H. N. de Haan and A. C. A. Jansen, in 'QSAR and Strategies in the Design of Bioactive Compounds', ed. J. K. Seydel, VCH, Weinheim, 1985, p. 198.
98. D. Southwell and B. W. Barry, *J. Invest. Dermatol.*, 1983, **80**, 507.
99. Z. Liron and S. Cohen, *J. Pharm. Sci.*, 1984, **73**, 538.
100. E. M. Cornford, *J. Membr. Biol.* 1982, **64**, 217.
101. E. J. Lien, *Annu. Rev. Pharmacol. Toxicol.*, 1981, **21**, 31.
102. E. J. Lien and G. L. Tong, *J. Soc. Cosmet. Chem.*, 1973, **24**, 371.
103. A. T'ang and E. J. Lien, *Acta Pharm. Jugosl.*, 1982, **32**, 87.
104. E. J. Lien, R. T. Koda and G. L. Tong, *Drug Intell. Clin. Pharm.*, 1971, **5**, 38.
105. J. P. Akbaraly, J. J. Leng, G. Bozler and J. K. Seydel, in 'QSAR and Strategies in the Design of Bioactive Compounds', ed. J. K. Seydel, VCH, Weinheim, 1985, p. 313.
106. R. D. Schoenwald and H. -S. Huang, *J. Pharm. Sci.*, 1983, **72**, 1266.
107. E. J. Lien, *J. Clin. Pharmacol.*, 1979, **4**, 133.
108. J. B. Houston, D. G. Upshall and J. W. Bridges, *J. Pharmacol. Exp. Ther.*, 1974, **189**, 244; 1975, **195**, 67.

109. J. C. Dearden, unpublished information.
110. J. M. Plá-Delfina and J. Moreno, *J. Pharmacokinet. Biopharm.*, 1981, **9**, 191.
111. A. Walter and J. Gutknecht, *J. Membr. Biol.*, 1986, **90**, 207.
112. D. E. Leahy, J. Lynch, R. E. Pownall and D. C. Taylor, *J. Pharmacokin. Biopharm.*, 1989, to be published.
113. H. Ochsenfahrt and D. Winne, *Naunyn-Schmiedeberg's Arch. Pharmacol.*, 1974, **281**, 175, 197.
114. A. Karino, M. Hayashi, S. Awazu and M. Hanano, *J. Pharmacobio-Dyn.*, 1982, **5**, 670.
115. (a) E. J. Lien and P. H. Wang, *J. Pharm. Sci.*, 1980, **69**, 648; (b) C. Hansch, J. P. Björkroth and A. J. Leo, *J. Pharm. Sci.*, 1987, **76**, 663; (c) V. A. Levin, *J. Med. Chem.*, 1980, **23**, 682.
116. A. N. Fisher, K. Brown, S. S. Davis, G. D. Parr and D. A. Smith, *J. Pharm. Pharmacol.*, 1987, **39**, 357.
117. C. Crone and A. M. Thompson, in 'Capillary Permeability,' ed. C. Crone and N. Lassen, Academic Press, New York, 1970, p. 447.
118. C. Hansch, A. R. Steward, S. M. Anderson and D. Bentley, *J. Med. Chem.*, 1968, **11**, 1.
119. C. Hansch, A. R. Steward and J. Iwasa, *Mol. Pharmacol.*, 1965, **1**, 87.
120. E. J. Lien, G. L. Tong, J. T. Chou and L. L. Lien, *J. Pharm. Sci.*, 1973, **62**, 246.
121. E. Simon-Trompler, G. Maksay, I. Lukovits, J. Volford and L. Ötvös, *Arzneim.-Forsch.*, 1982, **32**, 102.
122. D. D. Dischino, M. J. Welch, M. R. Kilbourn and M. E. Raichle, *J. Nucl. Med.*, 1983, **24**, 1030.
123. W. H. Oldendorf, *Proc. Soc. Exp. Biol. Med.*, 1974, **147**, 813.
124. (a) J. D. Fenstermacher, in 'Topics in Pharmaceutical Sciences 1983', ed. D. D. Breimer and P. Speiser, Elsevier, Amsterdam, 1983, p. 143; (b) R. C. Young, R. C. Mitchell, T. H. Brown, C. R. Ganellin, R. Griffiths, M. Jones, K. K. Rana, D. Saunders, I. R. Smith, N. E. Sure and T. J. Wilks, *J. Med. Chem.*, 1988, **31**, 656.
125. J. K. Seydel and K. -J. Schaper, *Pharmacol. Ther.*, 1982, **15**, 131.
126. J. K. Seydel, in 'Strategy in Drug Research,' ed. J. A. Keverling Buisman, Elsevier, Amsterdam, 1982, p. 179.
127. J. K. Seydel, in 'Quantitative Approaches to Drug Design,' ed. J. C. Dearden, Elsevier, Amsterdam, 1983, p. 163.
128. V. Austel and E. Kutter, in 'Quantitative Structure–Activity Relationships of Drugs,' ed. J. G. Topliss, Academic Press, New York, 1983, p,. 437.
129. J. K. Seydel, *Methods Find. Exp. Clin. Pharmacol.*, 1984, **6**, 571.
130. J. M. Mayer and H. van de Waterbeemd, *EHP, Environ. Health Perspect.*, 1985, **61**, 295.
131. J. G. Wagner, 'Fundamentals of Clinical Pharmacokinetics,' Drug Intelligence Publications, Hamilton, 1975.
132. M. Rowland and T. N. Tozer, 'Clinical Pharmacokinetics: Concepts and Applications', Lea and Febiger, Philadelphia, 1980.
133. R. A. Scherrer and S. M. Howard, *J. Med. Chem.*, 1977, **20**, 53.
134. J. Watanabe and A. Kozaki, *Chem. Pharm. Bull.*, 1978, **26**, 665, 3463.
135. E. J. Lien, J. Kuwahara and R. T. Koda, *Drug Intell. Clin. Pharm.*, 1974, **8**, 470.
136. J. C. Dearden and E. Tomlinson, *J. Pharm. Pharmacol.*, 1971, **23**, Suppl., 68S.
137. C. Hansch and T. E. Klein, *Acc. Chem. Res.*, 1986, **19**, 392.
138. S. Goto, H. Yoshitomi and M. Nakase, *Chem. Pharm. Bull.*, 1978, **26**, 472.
139. F. Helmer, K. Kiehs and C. Hansch, *Biochemistry*, 1968, **7**, 2858.
140. W. Scholtan, *Arzneim.-Forsch.*, 1968, **18**, 505.
141. J. K. Seydel, D. Trettin, H. P. Cordes, O. Wassermann and M. Malyusz, *J. Med. Chem.*, 1980, **23**, 607.
142. R. N. Smith, C. Hansch, K. H. Kim, B. Omiya, G. Fukumura, C. D. Selassie, P. Y. C. Jow, J. M. Blaney and R. Langridge, *Arch. Biochem. Biophys.*, 1982, **215**, 319.
143. G. L. Tong and E. J. Lien, *J. Pharm. Sci.*, 1976, **65**, 1651.
144. G. Cantelli-Forti, M. C. Guerra, A. M. Barbaro, P. Hrelia, G. L. Biagi and P. A. Borea, *J. Med. Chem.*, 1986, **29**, 555.
145. L. H. M. Janssen, *Eur. J. Med. Chem.—Chim. Ther.*, 1987, **22**, 131.
146. J. K. Seydel and K. -J. Schaper, 'Chemische Struktur und Biologische Aktivität von Wirkstoffen; Methoden der Quantitativen Struktur-Wirkung-Analyse,' Verlag Chemie, Weinheim, 1979, p. 225.
147. M. Yamazaki, M. Aoki and A. Kamada, *Chem. Pharm. Bull.*, 1968, **16**, 707, 721.
148. C. R. Ganellin, in 'Quantitative Approaches to Drug Design', ed. J. C. Dearden, Elsevier, Amsterdam, 1983, p. 239.
149. T. Higuchi, in 'Drug Absorption,' ed. L. F. Prescott and W. S. Nimmo, MTP Press Ltd., Lancaster, UK, 1981, p. 177.
150. V. Stella, in 'Pro-drugs as Novel Drug Delivery Systems,' ed. T. Higuchi and V. Stella, American Chemical Society, Washington, D. C., 1975, p. 1.
151. A. Hussain and J. E. Truelove, *J. Pharm. Sci.*, 1976, **65**, 1510.
152. D. A. McClure, in 'Pro-drugs as Novel Drug Delivery Systems,' ed. T. Higuchi and V. Stella, American Chemical Society, Washington, D.C., 1975, p. 224.
153. S. H. Yalkowsky and W. Morozowich, in 'Drug Design,' ed. E. J. Ariëns, Academic Press, New York, 1980, vol. 9, p. 121.
154. G. L. Biagi, M. C. Guerra and A. M. Barbaro, *J. Med. Chem.*, 1970, **13**, 944.
155. K. C. James, P. J. Nicholls and M. Roberts, *J. Pharm. Pharmacol.*, 1969, **21**, 24.
156. M. Charton, *Methods Enzymol.*, 1985, **112**, 323.

20.1

Introduction to Computer Graphics and Its Use for Displaying Molecular Structures

ROBERT LANGRIDGE and TERI E. KLEIN
University of California, San Francisco, CA, USA

20.1.1 INTRODUCTION

20.1.1.1 History

The first device recognizably similar in concept and design to the modern programmable computer is the Analytical Engine of Charles Babbage. Although never completed, we are fortunate that a detailed description of the machine was written by Menabrea and still more fortunate that this report was translated and extensively, elegantly and eruditely annotated by Ada Augusta Lovelace (neé Byron).[1] This report is reproduced in full in the book by Bowden[2] and, although the high opinion of Lady Lovelace's mathematical understanding which has been formed by most of us who have read her commentary has recently been questioned by Dorothy Stein,[3] it remains a landmark in the bibliography of computer science.

The advent of the *programmable* digital computer revolutionized crystallography, and in particular protein crystallography. It is hard to realize now that when Bernal, Hodgkin and Perutz pioneered protein crystallography in the 1930s, not only was there no foreseeable method for the solution of the phase problem, but, even if there was, the staggering amount of calculation required would be an almost insurmountable barrier, although the mathematics was well understood. The first digital computer application to protein structure was made in 1951[4] using the Cambridge EDSAC.[5]

The first model of DNA was built in 1953,[6] but the first application of digital computing to DNA structure was made in 1956 with the programming of the Fourier transform of a helical molecule for an IBM 650.[7, 8] The IBM 650 became a workhorse for many of the small molecule crystallographic calculations of the late 1950s and early 1960s.[9]

20.1.1.2 Molecular Models

The first models of protein and of DNA were built on a scale of 5 cm = 1 Å. The protein electron densities were plotted in three dimensions by placing colored clips, coded according to electron density, on flexible vertical rods with the wire model constructed in the same space.[10, 11] Although the first protein model was built in this manner, it is extremely inconvenient. In 1968, Richards[12] proposed a very neat way of keeping the electron density map and the wire model in separate physical spaces, while to the observer putting them in the same visual space, using a half-silvered mirror system. This method was used for a number of years. Computer graphics gradually came into use at about the same time and has since completely taken over the fitting of electron densities in protein crystallography.

20.1.1.3 Computer Graphics

The use of the computer to draw pictures began fairly early with production of figures such as electron density maps by creative use of a printer.[4] Illustrations of molecular structures on a cathode ray tube output began in the 1950s,[13] and by 1965 the classic molecule drawing program, ORTEP (Oak Ridge Thermal Ellipsoid Plotter),[14] was available to those with suitable computers and pen plotters. Almost concurrently, the new age of *interactive* computer graphics began with the seminal work on 'Sketchpad'.[15] Shortly afterward, the first truly *three-dimensional* interactive graphics system was built.[16] Funded by the Advanced Research Projects Agency of the US Department of Defense, it allowed the user to manipulate, through a 'crystal ball' controlling three rate controllers at right angles, the orientation of a three-dimensional object on the screen. Among the first objects to be displayed (somewhat to the surprise of the Defense Department, I suspect) were proteins and DNA[17–19] (it is interesting to note that the paper describing the excellent space-filling Corey–Pauling–Koltun (CPK) models[20] appeared at about the same time).

Early attempts were made to use three-dimensional graphics for electron density fitting,[21] while working systems were first put into general use in the mid-1970s,[22, 23] and are now in almost universal use for electron density fitting in protein crystallography laboratories.

Computer graphics and other molecular modeling techniques have now proven their value as research tools for the medicinal chemist[24–30] in studying the relationship between the three-

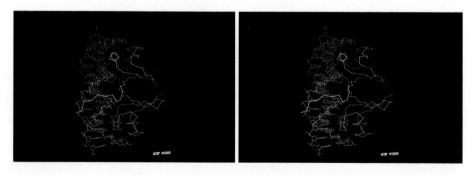

Figure 1 Stereoview of DNA and cro protein model

dimensional structure of a molecule and its biological function. Numerous computer graphics programs are available for molecular modeling applications.[21, 23,31-39] In this chapter we illustrate the basic concepts involved in using computer graphics for the display of molecular structures (Figure 1).[40-42]

20.1.2 WHY COMPUTER GRAPHICS?

20.1.2.1 The Representation of Chemical Complexity

The three-dimensional macromolecular structures determined to date are overwhelmingly the result of X-ray diffraction analysis. The final result of these analyses is an electron density map into which a molecular model is fitted, yielding a three-dimensional structure for the molecule as seen in Figure 2.

The structure of idealized DNA can be represented quite well with a simple two-dimensional illustration because idealized DNA is a highly symmetric structure with a regular repeat. Although very attractive and informative, there is very little one can do with this representation besides look at it. For the more complex asymmetric structures of proteins such a two-dimensional approach is clearly not suitable.

Representation and manipulation of such large molecular structures in three dimensions is considerably more difficult. Two main types of physical models are available. One is the space-filling kind first developed by Corey, Pauling and Koltun, as illustrated in Figure 3. These models give a reasonable sense of the space-filling properties of the structure but suffer from several major disadvantages. Firstly, the model is heavy and must be supported by scaffolding of some sort (usually steel rods). Secondly, the model is flexible and putting the atoms into reasonably uniform and reproducible positions is difficult and extremely tedious. Furthermore, if the model is not carefully protected it will gradually degrade over time as it is left on the laboratory bench and curious fingers manipulate it. Finally, and most serious of all, it is virtually impossible to measure accurate atomic coordinates from such a model.

For this purpose a 'wire' or skeletal model is required. Most chemists are familiar with the Dreiding-type models used both in research and in teaching. The models used by crystallographers in the construction of proteins and nucleic acids are somewhat more precise and sturdy. Originally designed by Kendrew for myoglobin and hemoglobin, these models also suffer from disadvantages. The first one is similar to the space-filling models in that the models are heavy. Unless a very well-designed and sturdy scaffolding system is built, the models will sag and the atomic positions will not be reproducible. The antibody represented in Figure 4 is this type of model. Furthermore, precise positioning is quite difficult. Coordinates are measured either with a plumb line dropped from the atomic positions to measure the xy coordinates in the plane of the table, with the z coordinate

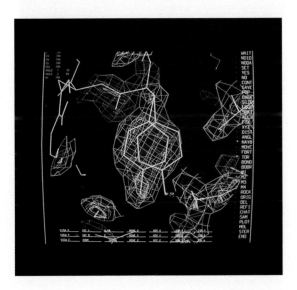

Figure 2   Electron density map with molecular model using an enhanced version of FRODO[12]

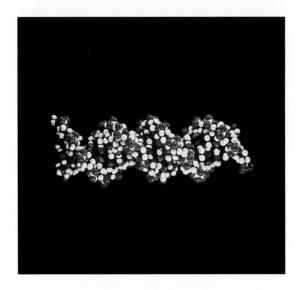

Figure 3   CPK model of DNA

Figure 4   Skeletal model of an antibody

measured vertically with a ruler (in more accurate and sophisticated systems, with the aid of a computer-assisted cathetometer). Each atom is associated with specific Cartesian coordinates and the bond lengths and bond angles are calculated. In general this will yield a set of bond lengths and bond angles which vary somewhat from standard values and there will probably be implausible interatomic contacts. Computer programs can refine these coordinates to give bond lengths and bond angles and stereochemistry which agree with generally accepted values, but at the end of the calculation the coordinates in the computer and the structure on the bench no longer agree. There is the further difficulty that if one is studying the interaction of one molecule with another, for example a small molecule such as a substrate with an enzyme active site, the positional accuracy required is considerably more precise than that needed for a general overall view of the folding of a protein or nucleic acid. Finally there is a psychological problem. It is very frustrating to spend long hours studying interactions with models which cannot be precisely positioned and once modified are very hard to return to their original positions.

Although, after refinement, the numbers in the computer represent more precisely the desired type of model, to the human eye these numbers alone are completely incomprehensible. Exact numeric coordinate data are essential if accurate studies are to be made.

We thus have an inaccurate and intractable model on the laboratory bench and accurate but uninterpretable numbers in the computer. The obvious answer is to use the computer to draw the pictures for us, preferably in three dimensions and in such a way that we may not only look at them but also manipulate them.

The establishment of national facilities such as the Princeton University Computer Graphics Laboratory in 1970 was a result of progress in both structural biology and computer science. Graphics displays, which were expensive, required a large dedicated computer to drive the black and white display. Images from the black and white display could be photographed in color through the appropriate use of filters, as seen in Figure 5. Additionally, at that time, most macromolecules were too large for the display capacity of the graphics display system. In the mid-1970s, Evans and Sutherland produced a high-performance interactive black and white three-dimensional graphics display system (Picture System 2). The introduction of the color display scope by Evans and Sutherland in 1979 had a profound effect on computer graphics and its use in displaying molecular models.

20.1.3 COMPUTER GRAPHICS DISPLAYS

There are a variety of ways for a computer to paint a picture, but two main modes should be distinguished—raster and vector.

20.1.3.1 Vector Graphics

A vector graphics machine draws lines from one point to another as directed by the program but does not have to waste time filling in the blank space corresponding to background information, giving it the ability to manipulate complex models in real-time. It is eminently suitable for drawing complex line drawings or similar representations of large molecules because matrix multiplication is only done to transform the endpoints of the lines.

20.1.3.2 Raster Displays

A standard television picture is built up of a series of interlaced lines (or raster) traced on the phosphors of the screen. The appropriate colors and shades at each point on the screen (pixel) are produced by manipulation of the electron beam. It is relatively straightforward to control the beam by a computer rather than by a TV video signal, and the majority of computer displays are now of

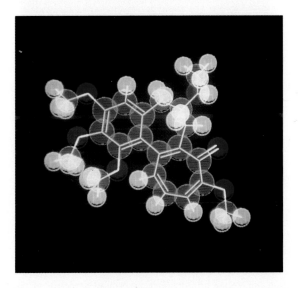

Figure 5 Color picture using filters on a black and white CRT (Princeton University Computer Graphics Laboratory, 1978)

this type. In order to create an image which is as accurate a portrayal of reality as the image on our TV screen, a raster display is the clear choice.

Raster computer displays are often used to produce a version of reality which approximates the plastic CPK space-filling models, but they are only presently becoming capable of real-time motion of molecules as complex as proteins at reasonable cost. There is a clear trade-off between the potential realism of the image created by the computer and the cost and speed of interaction. If single pictures are required for illustrations or time-lapse animation movies are to be made, and direct interaction is sacrificed, then any defined degree of realism can be attained (Figures 6 and 7). Representations of solid space-filling models of molecules have a high degree of realism.

Real-time display of and interaction with shaded raster surface displays are now becoming more available.[43] Raster-based systems can manipulate three-dimensional wireframe models as a vector display would, and can locally render static shaded images on one color display. Special VSLI chips designed for real-time manipulation of antialiased raster lines eliminate the 'jaggies' (or stair-stepping lines) that were previously seen on raster displays for wire models. These newer

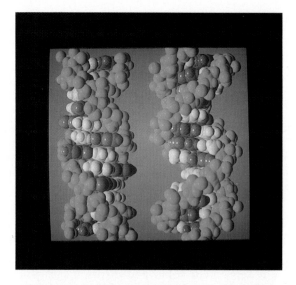

Figure 6   Surface imaging of B and Z DNA (courtesy of R. J. Feldmann, NIH)

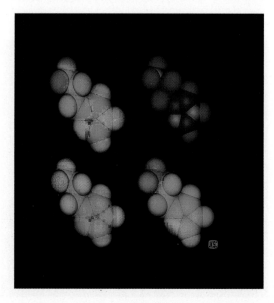

Figure 7   CPK rendering of small molecule with quantum mechanical properties mapped on to the surface (courtesy of Graphics Systems Research Group, IBM UK Scientific Centre, Winchester, UK)

raster-based graphics displays will essentially displace vector graphics displays for molecular modeling applications.

20.1.4 MOLECULAR MODELING SOFTWARE

20.1.4.1 Manipulation and Display

Efficient methods of displaying and interacting with molecular models in three dimensions in *real-time* on the computer screen are essential. In our case real-time means that, for simple manipulations, such as global rotation and translation of molecules, bond rotations and distance calculations, the response to the user's motion of the joystick or knob must appear instantaneous. More complex operations, such as the calculation of interaction energies while fitting two macromolecules together, may require too much computer time to be monitored in real-time. Therefore, as such calculations become more complex, they are usually done as background or batch calculations and the results are viewed later.

The computer graphics system we use will draw color line drawings containing up to approximately 30 000 vectors. The non-selective display of such a complex model is next to useless; there is simply too much information. Thus, an important feature of molecular modeling software is the ability to select those parts of the structure which are interesting for detailed study. 'Interesting' is, of course, a definition made interactively by the user and will change with time. Although it is possible to display an entire macromolecule, usually we are only interested in a specific region (*e.g.* the active site) on the molecule and do not wish to have to deal with the entire structure, which may contain up to several thousand atoms. Figure 8, which shows the interaction of methotrexate with dihydrofolate reductase, focuses only on the active site region of the enzyme *without* displaying the rest of the molecule.

The selection of which molecules to manipulate at any given time in the molecular modeling program is made using a menu or a switch panel, while joysticks are often used to move the molecules on the screen; the left- and right-hand sticks control translational and rotational motion, respectively, along the x, y and z axes of the screen. Bond rotations may be controlled with potentiometer knobs. A critical feature of the design of the molecular modeling system is the human interface. The interactions should appear natural; that is, when you wish to rotate a bond, *you should be given something to turn*. In response to the motion of the joysticks and bond rotation knobs, the atomic coordinates of each molecule are transformed by special-purpose hardware built into the system between each update of the display (30 times per second), providing smooth motion of the molecule(s).

It might be asked at this point how one can have a z axis control in a basically two-dimensional display. Of course, the display is indeed two dimensional on the screen, but hardware may be built into the system to transform by perspective and by depth cueing the object, so that as it travels back into the screen the object becomes both smaller and dimmer. Depth cueing, along with perspective, provides a good sense of three dimensionality which is further heightened by the real-time motion capability of the system. The color wheel pictured in Figure 9 is an example that utilizes depth cueing (and color).

Full three dimensionality is provided by a stereoviewer mounted on a vertical rod in front of the screen. A black cylinder at the top has on its right a stepping motor, whose axis is horizontal, which

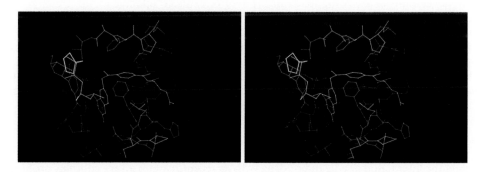

Figure 8   Stereoview of methotrexate and the active site of dihydrofolate reductase

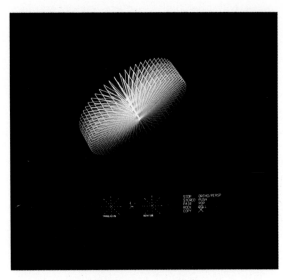

Figure 9 Color wheel

drives a thin metal cylinder with two slits cut at right angles. If the left-hand slit is open, the right-hand slit which is 180° opposite to this is obscured. The stepping motor drives this cylinder around the horizontal axis and reports the position of the cylinder to the computer, which then displays the left eye view of the object on the screen only when the left eye is open and the right eye view only when the right eye is open, resulting in a stereoscopic three-dimensional image which moves smoothly (flicker free) in real-time. More recently, stereoviewing can be obtained by placing a liquid crystal screen over the cathode ray tube (CRT) which alternates mode between left and right circular polarization while the user looks through circularly polarized glasses. Stereoviewing is extremely helpful in studying intermolecular interactions.

Using both joysticks, one is able smoothly to rotate and translate the object on the screen about any axis or combination of axes. Since one can move each molecule relative to another, it is reasonably simple to dock one molecule into another while constantly monitoring their interactions. This may not seem like a major advance if one is dealing with two small molecules, but imagine the difficulties created if one attempts to study the association of macromolecules with other macromolecules or ligands, such as a repressor protein with DNA (Figure 1). Attempts to dock physical models of this size and complexity would result in a great deal of bent wire.

The keyboard is used for input of numerical and textual information to the program. Information can also be communicated to the program using a tablet with its attached cursor (similar to a light pen). The cursor position is monitored by the computer and displayed on the graphics screen, so that a command may be picked from a menu of commands on the screen or an atom may be picked from a molecule being displayed on the screen simply by touching the atom. The main point is that the keyboard, tablet and cursor must interact in a very easy and straightforward way with objects or commands on the screen.

The most important component of the molecular modeling system is, of course, the occupant of the empty seat in front of the display. It must be emphasized that the computer graphics system is merely a tool dependent on the talent, experience, intuition and imagination of the user. Although impressive developments are being made in the area of artificial intelligence, at present computers and computer graphics systems have a rather inadequate background in medicinal chemistry. Thus, it is important to work chemistry into the molecular model by combining chemical information and properties with the structural information inherent in the molecular model, providing the user with a visual approximation of the 'chemistry' of the molecules, yet still giving the user the flexibility to use his or her own intuition in interpreting and using the information on the screen. Ideally, the computer graphics system should be completely transparent to the user, who should have the sense of interacting directly with the molecules not the computer.

20.1.4.2 Ligand–Receptor Interactions

Modeling the interaction of a ligand or drug with its receptor site is a very complex problem; exact quantitative methods for constructing and evaluating such interactions are only slowly being developed. It is generally accepted that the forces important for intermolecular association are hydrophobic, van der Waals, hydrogen bonding and electrostatic (ion-pairing) interactions. Unfortunately, the immense number of degrees of freedom in a protein–ligand complex and the lack of an adequate representation of solvent make the modeling of intermolecular association extremely difficult. Thus, it is still a major theoretical problem to determine the optimum fit and interaction energy of a ligand into a receptor site of known structure. The solution to this problem is an essential first step to the solution of the more general problem of finding the optimum ligand for a receptor site of known structure and predicting the binding affinities of ligands for this receptor. Although a simple method has been developed for solving the 'docking' problem of a rigid ligand with a rigid receptor[41] and a flexible ligand with a rigid receptor,[44] the more general problem involving a conformationally flexible ligand and receptor remains unsolved. Interactive computer-graphics-assisted modeling offers a practical qualitative method of studying such complex interactions in that it allows the user to use intuition rather than being tied to an inflexible and unforgiving algorithm.

Molecular structures are clearly not static; for example, it was known from the early 1950s that DNA could adopt a variety of forms (Figures 6 and 10); these are now multiplying rapidly and intriguingly and evidence is accumulating for significant motions in both proteins and nucleic acids. Modeling these motions requires extensive and detailed calculations; the results can be abbreviated to terse sets of parameters relating to the equations of motion but these are as hard to comprehend as atom coordinate lists. Production of a time series of frames, either to be viewed directly on the display or for filming and later viewing, is an immensely useful adjunct to real-time interactive display of the static structure.

A major difficulty in modeling intermolecular interactions is the overwhelming amount of complex structural information present within a macromolecule. For example, in trying to model the fit of an inhibitor into an enzyme active site one is faced with the difficult problems of determining which portions of the site are most likely to contact the inhibitor and to find the best fit of the inhibitor into the site in a reasonable conformation without contacting atoms of the site too closely. An example of the inhibitor sterically contacting the active site is seen in Figure 11.

20.1.4.3 Surfaces

A display which combines the standard wire molecular model with the molecular surface of the molecule(s) provides a much better feeling for the three-dimensional shape, topography and chemistry of the molecule and has proven to be extremely powerful in modeling complex intermolecular interactions. This technique is based on the definition of a molecular surface proposed by Richards[45] and developed first by Greer and Bush[46] into a computer program for the calculation of molecular surfaces but restricted to one side of a molecule. This was developed for the black and white Evans and Sutherland PS2 at the University of California, San Francisco (UCSF) by Jones[47]

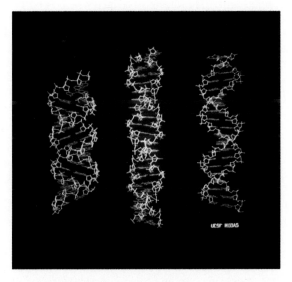

Figure 10 Three forms of DNA; A, Z and B

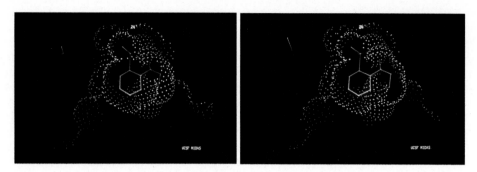

Figure 11 Stereoview of *ortho*-substituted sulfonamide inhibitor with molecular surface of carbonic anhydrase C

in 1978 and generalized by Connolly[48] when the first PS2 color system was installed at UCSF in 1979. Instead of calculating van der Waals spheres and solving the hidden surface problem, one calculates the surface which corresponds to the solvent-accessible surface. A sphere corresponding to the van der Waals radius of a water molecule (1.4 Å) traverses the surface of the molecule; a dot is placed at each point of contact of the sphere with the molecule or at the inward-facing surface of the sphere when it is simultaneously in contact with more than one atom. The resulting model resembles a transparent smoothed-over CPK model. Figure 12 shows the methotrexate–dihydrofolate reductase interaction in a similar orientation as in Figure 8, displaying the molecular surface of the enzyme active site. We find that this transparent representation of the molecular surface, combined with the usual wire model representation of the bonds, provides a significant improvement over the solid space-filling display implemented on raster graphics systems. Additionally, Connolly has developed a method which provides values for the surface area and volume based on an analytical method for calculating a molecular surface.[48, 49] A useful method for docking has been developed by Barry,[50] which extends the molecular surface by one van der Waals radius, eliminating the need for displaying a ligand surface.

Furthermore, due to the use of vector graphics hardware, the molecular surface–wire model combination can be manipulated (global rotation and translation, internal rotations) in real-time. However, as internal rotations are adjusted the molecular surface associated with those atoms breaks open, ultimately requiring recalculation of the molecular surface. For smaller molecules, a van der Waals surface[51] display using dots (Figure 13) which mimics the CPK space-filling model rather than calculating the solvent-accessible surface is useful. This method of calculating a van der Waals surface makes it possible to adjust internal torsion angles in real-time while maintaining a smoothly moving surface, as illustrated in Figures 14 and 15.

Molecular modeling is successful in combining the molecular surface for the macromolecule and the van der Waals surface for the smaller molecule. Until 1987, there has not been a method for calculating a surface to approximate the molecular surface which is both dynamic and continuous. Hence, when the user wished to change the positions of a side chain for a protein following the calculation of the molecular surface, it was necessary to either use the van der Waals surface or recalculate the molecular surface.

Most recently, a dynamic and continuous surface has been developed and implemented which can be manipulated interactively by the user on a three-dimensional graphics display.[52, 53] This surface is manipulated by moving points on the surface which control the shape and topology of the surface. Figure 16 illustrates the dynamic nature of this surface; a point has been moved on the surface but not yet saved. Figure 17 illustrates how this surface can be rendered for a more *solid-looking* representation.

By assigning a different color code to the molecular surface of hydrophilic and hydrophobic atoms, the hydrophilic and hydrophobic areas of the drug molecule and active site are immediately recognizable, as shown in Figure 12 where hydrophobic surfaces are red and hydrophilic surfaces are blue (hydrophilic = N and O, and hydrophobic = C). An extension to this type of surface characterization to include a semi-hydrophilic surface has been made where sulfur, α-carbons, carbonyl carbons, etc. have been represented in yellow.[54] Color coding the molecular surface based on the electrostatic potential at each surface point has also proven to be a valuable technique. The potential is calculated one probe sphere radius beyond the molecular surface. An electrostatic potential gradient can also be displayed. An example using this method is seen in Figure 18 for a possible trajectory for the superoxide anion nearing superoxide dismutase.[55] At present, these

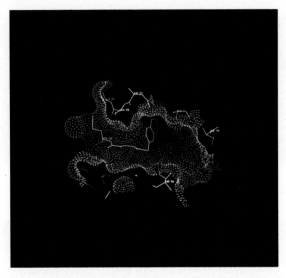

Figure 12 Methotrexate–dihydrofolate reductase complex displayed with molecular surface

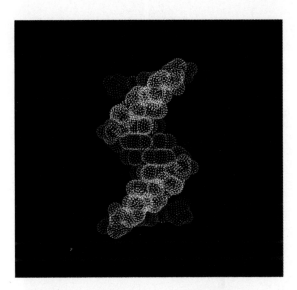

Figure 13 VDW surface of DNA, major groove

methods appear to provide the best way of combining useful *visual* chemical information with the molecular model at the interactive level. In modeling intermolecular association, we use the electrostatic and hydrophobic potential molecular surfaces to optimize the fit between chemically and topographically complementary regions of the two molecules.

20.1.5 APPLICATIONS

20.1.5.1 Qualitative Structure–Activity Relationships and Ligand Design

There are relatively few reports in the literature of attempts to explain experimental structure–activity relationships in terms of a three-dimensional structure of the ligand–macromolecular receptor complex. An early example of this approach was the study by Lowe *et al.*,[56] who used wire models to explain in a general way the structure–activity relationships of substrate binding to papain. The first real demonstration of the potential of the molecular modeling approach to drug

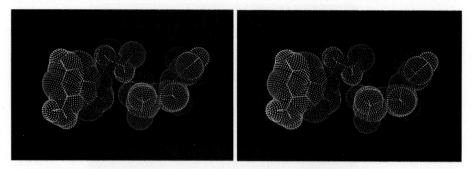

Figure 14 Stereoview of ATP with VDW surface (before rotation)

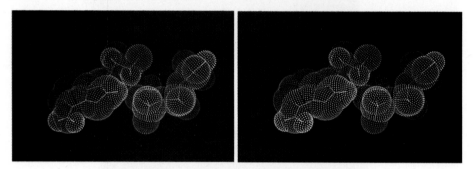

Figure 15 Stereoview of ATP with VDW surface (after rotation)

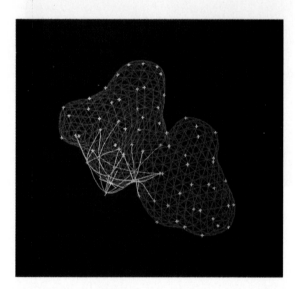

Figure 16 Dynamic and continuous surface model; original surface (red); portion of surface which has moved (green)

design was the work by Beddell *et al.*,[57] on the *de novo* design of compounds intended to bind to the 1,3-diphosphoglycerate binding site of deoxyhemoglobin. Wire models of the protein and candidate ligands were hand-built and fitted together, resulting in the successful design of new active compounds which mimicked the oxygen-displacing effect of 1,3-diphosphoglycerate *in vitro*, although the new analogs had very little structural resemblance to 1,3-diphosphoglycerate. Poe also used wire molecular models to design a highly active inhibitor of bacterial dihydrofolate reductase.[58] Dutler and Branden[59] used a similar approach to develop models of the enzyme–substrate complexes of alkylcyclohexanols with horse liver alcohol dehydrogenase, by correlating kinetic data on the reduction rates of the various alkylcyclohexanones with the three-dimensional structure of

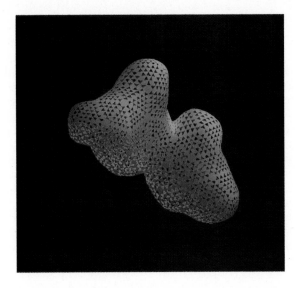

Figure 17   Rendered model based on dynamic and continuous surface

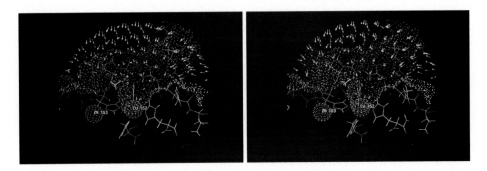

Figure 18   Anion approaching superoxide dismutase

the enzyme. Although a tedious and labor-intensive effort due to the construction of wire models, these studies nevertheless highlighted the power of molecular modeling for studying intermolecular interactions, which can only increase as computer graphics is used to replace wire models.

Computer graphics has found increasing applications in the study of structure–activity relationships. Feldman *et al.*[60] used computer graphics molecular modeling with space-filling displays to study the structure–activity relationships of trypsin–inhibitor interactions, finding a good qualitative correlation between the relative K_i values of substituted benzamidine inhibitors with their fit (steric, hydrophobic and electrostatic) into the active site of the enzyme. This technique was recently extended to the modeling of proteins whose crystal structure is undetermined, based on sequence homology with proteins of known structure, and has been used to study the active sites of the serine proteases Factor IXa, Xa and thrombin involved in blood coagulation.

A computer graphics molecular modeling study of the interaction of thyroid hormones with the plasma protein prealbumin led to a model which accounted for thyroid hormone–prealbumin structure–activity relationships and ultimately to the qualitative prediction and determination of the relative binding affinities of four previously untested thyroid hormone analogs to prealbumin.[61] In general, one can usually sort out those analogs of a series which are simply too big to fit into a binding site due to steric reasons and qualitatively rationalize or predict low activity for these analogs. Molecular surface models clearly indicated the potential significance of an empty pocket in the prealbumin hormone-binding site as being ideally positioned to accept an additional substituent on the phenolic ring of thyroid hormones. The recognition and subsequent exploitation of such empty pockets in the molecular surface of a ligand-binding site provides a route for the design of new more tightly bound ligands.

20.1.5.2 Quantitative Structure–Activity Relationships

In a structure–activity relationship study of drugs binding to a receptor, ideally one would like to have a high resolution X-ray structure for each different drug–receptor complex. Unfortunately this is rarely possible, but one can use the few such structures available along with QSAR studies to build models to direct the search for new analogs. QSAR deductions on the nature of enzyme space in which a particular substituent binds are interpreted in terms of molecular models derived from X-ray crystallography, without which one has such limited knowledge of the chemistry and topography of the binding region. Such models are admittedly speculative and are not intended to provide the definitive structure for an enzyme–inhibitor complex. The real power of this method is its ability to generate reasonable structural models based on known structures which are compatible with existing QSAR results, which can then provide experimentally testable predictions of the activity of new analogs. This combination of QSAR, high-resolution X-ray crystal structures of the receptor and (ideally) drug–receptor complexes, and high-performance computer graphics molecular modeling provides an exceptionally powerful method to study ligand–receptor interactions, as discussed in Volume 4, Chapter 20.3. These studies have begun to sort out what correlation with the various QSAR parameters, particularly π and MR, means at the atomic level in terms of ligand–receptor interactions.

20.1.5.3 Energy Refinement

Another application of molecular modeling is to rapidly generate different starting geometries for energy refinement by molecular mechanics or dynamics calculations. Although this does not solve the local minimum problem inherent in these types of calculations, it allows one to consider a large number of possible structures very quickly, in order to search as much conformational space as possible. The use of computer graphics for the initial construction and evaluation of different geometries, followed by complete geometry optimization of the molecule(s), provides one of the best methods for modeling non-covalent and covalent molecular interactions. This approach has been used to model ligand–macromolecule interactions involving the binding of thyroid hormone analogs to the plasma protein prealbumin, the non-covalent and covalent complexes of L- and D-N-acetyl-tryptophanamide with α-chymotrypsin, and the binding of 4-nitroquinoline N-oxide to DNA sequence isomers. In each case, computer graphics molecular modeling was used to build a variety of starting conformations, each of which was then energy-refined with complete geometry optimization using molecular mechanics calculations. This method resulted in a consistent structural and energetic model for ligand–macromolecule binding in each system which qualitatively reproduced the experimentally observed relative free energies of binding of the ligands.[62]

20.1.5.4 Modeling Receptors of Unknown Structure

Given the structures and receptor-binding activities of a series of ligands, one would hope to be able to construct a hypothetical three-dimensional model of the receptor site consistent with these structure–activity data. Promising methods for achieving this goal include the distance-geometry approach of Crippen,[63] the excluded volume mapping technique of Marshall,[64] both of which assume a rigid receptor, and the knowledge-based receptor mapping system of Langridge and Klein.[52] However, modeling the interaction of ligands with receptor sites is an extremely difficult problem even when starting with high-resolution X-ray structures for the receptor and ligand–receptor complexes. The many different possible modes of binding for different ligands, the large variety of conformations available to the active site and the large shifts in the position of amino acid side chains or even the main chain which may occur upon ligand binding make it difficult to generalize from one ligand to the next. The difficulty we have experienced in accurately modeling and understanding the interactions of known macromolecular structures with relatively simple ligands having few conformational degrees of freedom suggests that such modeling should be undertaken with careful experimental justification and, preferably, X-ray crystallographic backing.

The images produced by computer graphics can be so seductively beautiful that it is easy to believe that they must represent the truth. Unfortunately, stereochemically accurate and precise pictures can be drawn of structures which bear little relation to reality, and although in the course of research, such as studies of protein chain folding, unlikely structures are frequently created, extreme caution must be exercised in their use and interpretation. This is nowhere more true than in drug design, and a conscious effort must be made to develop methods with constant reference to experimental data—in particular to X-ray diffraction analyses.

20.1.5.5 Study of Protein Structures

Computer graphics can be extremely useful for generating schematic structure representations of protein molecules. Computer-generated secondary structure representations similar to the hand-drawn illustrations by Richardson,[65] allow the user to not only study and compare the structures of proteins solved by X-ray crystallography, but also provides a tool for those researchers exploring tertiary structures for proteins whose structures are unknown. Figures 19 and 20 illustrate three types of representation including the use of arrows for β-sheets as well as B-splines to generate ribbon models.[66, 67]

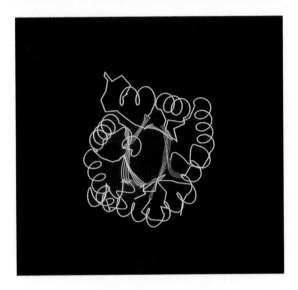

Figure 19 Chicken triose phosphate isomerase dimer (courtesy of Graphics Systems Research Group, IBM UK Scientific Centre, Winchester, UK)

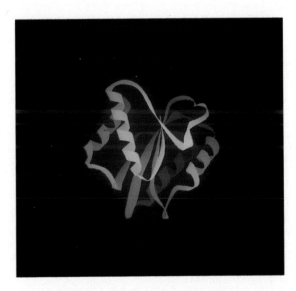

Figure 20 Ribbon model of flavodoxin

Figure 21   Semliki forest virus modeling the icosahedral symmetry of the virus and the interaction of the glycoprotein with the membrane of the virus (courtesy of R. J. Feldmann, NIH)

20.1.6 SUMMARY

Computer graphics and molecular modeling are far from reaching their full potential. Current methods rely heavily on visual qualitative molecular modeling and evaluation of the 'quality' of a particular model or the use of rather crude empirical methods which, until recently,[68] neglected the role of solvent (molecular mechanics, dynamics) to calculate the energy of a molecule or inter-molecular interaction.

Computer graphics molecular modeling has now made it possible to study and provide answers to problems which were previously impossible or difficult to deal with. An unexpected benefit is that the power of three-dimensional visualization of complex molecular models (Figure 21) very often leads one to new questions as well as answers, opening up new ideas and areas of investigation, as discussed throughout this series.

20.1.7 REFERENCES

1. L. F. Menabrea and A. A. Lovelace, 'Sketch of the Analytical Engine Invented by Charles Babbage, Esq.,' by L. F. Menabrea, of Turin, Officer of the Military Engineers. Translated and with notes by A. A. Lovelace, *Taylor's Scientific Memoirs*, 1843, **3**, 666.
2. B. V. Bowden, in 'Faster Than Thought: A Symposium on Digital Computing Machines', ed. B. V. Bowden, Pitman, London, 1953, Appendix I, p. 341.
3. D. Stein, 'Ada; A Life and a Legacy', MIT Press, Cambridge, MA, 1985.
4. J. M. Bennett and J. C. Kendrew, *Acta Crystallogr.*, 1952, **5**, 109.
5. M. V. Wilkes, D. J. Wheeler and S. Gill, 'The Preparation of Programmes for an Electronic Digital Computer, with Special Reference to the EDSAC and the use of a Library of Sub-Routines', Addison-Wesley, Cambridge, MA, 1951.
6. J. D. Watson and F. H. C. Crick, *Nature (London)*, 1953, **171**, 737.
7. R. Langridge, Ph. D. thesis, University of London, 1957.
8. R. Langridge, M. P. Barnett and A. F. Mann, *J. Mol. Biol.*, 1960, **2**, 63.
9. G. A. Jeffrey, R. Shiono and L. H. Jensen, in 'Computing Methods and the Phase Problem in X-Ray Crystal Analysis,' ed. R. Pepinsky, J. M. Robertson and J. C. Speakman, Pergamon Press, London, 1961, p. 25.
10. J. C. Kendrew, R. E. Dickerson, B. E. Strandberg, R. G. Hart, D. R. Davies, D. C. Phillips and V. C. Shore, *Nature (London)*, 1960, **185**, 422.
11. R. E. Dickerson, J. C. Kendrew and B. E. Strandberg, in 'Computing Methods and the Phase Problem', ed. R. Pepinsky, J. M. Robertson and J. C. Speakman, Pergamon Press, London, 1961, p. 236.
12. F. M. Richards, *J. Mol. Biol.*, 1968, **37**, 225.
13. W. R. Busing, in 'American Crystallographic Association Meeting, Abstracts', 1960, p. 23.
14. C. K. Johnson, *Oak Ridge Natl. Lab., [Rep.] ORNL*, 1965, 3794, (*Chem. Abstr.*, 1966, **64**, 18 545f).
15. I. E. Sutherland, Ph. D. thesis, Massachusetts Institute of Technology, Cambridge, MA, 1963.
16. R. H. Stotz and J. F. Ward, 'MIT Electronic Systems Laboratory Internal Memorandum, 9442-H-129', 1965 (also Project MAC Internal Memorandum MAC-M-217).
17. R. Langridge and A. W. MacEwan, in 'Proceedings: IBM Scientific Computing Symposium', 1965, p. 305.

18. C. Levinthal, in 'Proceedings, IBM Scientific Computing Symposium on Computer Aided Experimentation', IBM, Yorktown Heights, NY, 1965, p. 315.
19. C. Levinthal, *Sci. Am.*, 1966, **214**, 42.
20. W. L. Koltun, *Biopolymers*, 1965, **3**, 665.
21. C. Levinthal, C. D. Barry, S. A. Ward and M. Zwick, in 'Emerging Concepts in Computer Graphics', ed. S. Nievergelt, Benjamin, Reading, MA, 1968, p. 231.
22. R. Diamond, in 'Computational Crystallography', ed. D. Sayre, Oxford University Press, Oxford, 1982, p. 318.
23. T. A. Jones, in 'Computational Crystallography', ed. D. Sayre, Oxford University Press, Oxford, 1982, p. 303.
24. P. Goodford, *J. Med. Chem.*, 1984, **27**, 557.
25. A. J. Hopfinger, *J. Med. Chem.*, 1985, **28**, 1133.
26. G. Jolles and K. R. H. Wooldridge (eds.), 'Drug Design: Fact or Fantasy?', Academic Press, New York, 1984.
27. T. Klein, C. Huang, T. E. Ferrin, R. Langridge and C. Hansch, *ACS Symp. Ser.*, 1986, **306**, 147.
28. 'Computer-Aided Molecular Design. Second European Seminar, Basel, Switzerland', IBC Technical Services Ltd., 1985.
29. G. R. Marshall, J. G. Vinter and H. D. Joltje (eds.), 'Computer-Aided Molecular Design', ESCOM Science Publishers, Leiden, 1985.
30. P. Gund, T. A. Halgren and G. M. Smith, *Annu. Rep. Med. Chem.*, 1987, **22**, 269.
31. T. E. Ferrin, C. C. Huang, L. E. Jarvis and R. Langridge, *J. Mol. Graphics*, 1988, **6**, 13.
32. MMS: University of California, San Diego, La Jolla, CA 92039, USA.
33. CHEM-X: Chemical Design Ltd., Oxford, UK.
34. The software for this is available in the program SYBYL, available from TRIPOS Corp., St. Louis, MO 63117, USA.
35. MOGLI: Evans and Sutherland, Salt Lake City, UT 84108, USA.
36. INSIGHT: Biosym Technologies Inc., San Diego, CA 92121, USA.
37. BIOGRAF: Biodesign Inc., Pasadena, CA 91101, USA.
38. HYDRA: Polygen Corp., Waltham, MA 02154, USA.
39. CHEMLAB: Molecular Design Ltd., San Leandro, CA 94577, USA.
40. E. F. Meyer, Jr., in 'Drug Design', ed. E. J. Ariens, Academic Press, New York, vol. IX, p. 267.
41. J. M. Blaney, E. C. Jorgensen, M. L. Connolly, T. E. Ferrin, R. Langridge, S. J. Oatley, J. M. Burridge and C. F. Blake, *J. Med. Chem.*, 1982, **25**, 785.
42. R. J. Feldmann, in 'Computer Applications in Chemistry', ed. S. R. Heller and R. Potenzone, Jr., Elsevier, Amsterdam, 1983, p. 9.
43. 'Lab. Notes', Evans and Sutherland Molecular Science Group, 1987, vol. 1.
44. R. L. DesJarlais, R. P. Sheridan, J. S. Dixon, I. D. Kuntz and R. Venkataraghavan, *J. Med. Chem.*, 1986, **29**, 2149.
45. C. Chothia, M. Levitt and D. C. Richardson, *Proc. Natl. Acad. Sci. U.S.A.*, 1977, **74**, 4130.
46. J. Greer and B. L. Bush, *Proc. Natl. Acad. Sci. U.S.A.*, 1978, **75**, 303.
47. O. Jones, unpublished method.
48. M. L. Connolly, *J. Appl. Crystallogr.*, 1983, **16**, 548.
49. M. L. Connolly, *J. Am. Chem. Soc.*, 1985, **107**, 1118.
50. D. Barry, unpublished method.
51. P. A. Bash, N. Pattabiraman, C. Huang, T. E. Ferrin and R. Langridge, *Science (Washington, D.C.)*, 1983, **222**, 1325.
52. T. E. Klein, Ph. D. Dissertation, University of California, San Francisco, 1987.
53. T. E. Klein, C. C. Huang, E. F. Pettersen, G. S. Couch, T. E. Ferrin and R. Langridge, *J. Mol. Graphics*, 1989, in press.
54. M. Recanatini, T. Klein, C. Z. Yang, J. McClarin, R. Langridge and C. Hansch, *Mol. Pharmacol.*, 1986, **29**, 436.
55. J. M. Blaney, P. K. Weiner, A. Dearing, P. A. Kollman, E. C. Jorgensen, S. J. Oatley, J. M. Burridge and C. C. F. Blake, *J. Am. Chem. Soc.*, 1982, **104**, 6424.
56. G. Lowe and Y. Yuthavong, *Biochem. J.*, 1971, **124**, 107.
57. Beddell *et al.*, unpublished results.
58. Poe, unpublished results.
59. Dutler and Branden, unpublished results.
60. R. J. Feldmann, D. H. Bing, B. C. Furie and B. Furie, *Proc. Natl. Acad. Sci. U.S.A.*, 1978, **75**, 5409.
61. R. Langridge, unpublished results.
62. G. Wipff, A. Dearing, P. K. Weiner, J. M. Blaney and P. A. Kollman, *J. Am. Chem. Soc.*, 1983, **105**, 997; J. M. Blaney, unpublished results.
63. T. F. Havel, G. M. Crippen and I. D. Kuntz, *Biopolymers*, 1979, **18**, 73.
64. C. Humblet and G. R. Marshall, *Drug. Dev. Res.*, 1981, **1**, 409.
65. C. Chothia, M. Levitt and D. Richardson, *J. Mol. Biol.*, 1981, **145**, 215.
66. J. M. Burridge and S. J. P. Todd, *J. Mol. Graphics*, 1986, **4**, 220.
67. D. Kneller, B. Cohen and R. Langridge, unpublished method.
68. P. A. Bash, U. C. Singh, F. K. Brown, R. Langridge and P. A. Kollman, *Science (Washington, D.C.)*, 1987, **235**, 574.

20.2

Use of Molecular Graphics for Structural Analysis of Small Molecules

GARLAND R. MARSHALL and CHRISTOPHER B. NAYLOR
Washington University, St. Louis, MO, USA

20.2.1 INTRODUCTION

In this chapter, we will focus on those aspects of molecular modeling which are useful in the comparative analysis of the properties of small molecules. By historical imperative, our emphasis shall be primarily on the structure–activity problem and attempts to rationalize biological activity in the absence of three-dimensional structural information on the receptor. While most of the techniques and approaches we shall describe have broader application, the examples chosen from this area should be sufficient to illustrate their use. A number of recent reviews[1-4] of computer-aided drug design have relevant sections covering portions of this chapter.

20.2.2 DISPLAY OF MOLECULES AND ASSOCIATED PROPERTIES

20.2.2.1 Display Complexity

20.2.2.1.1 Stick models

The earliest and most simple computer-generated displays of molecules were stick models (Figure 1).[5] These drawings represented molecules as collections of monochromatic lines ('bonds') drawn between the centers of pairs of bonded atoms, and as such they closely resembled the familiar hand-held Dreiding models. Atomic positions were optionally labeled by characters denoting the atomic symbols of the atoms.

Currently, stick models are still one of the most popular modes of molecular display and are the standard mode of molecular representation in many molecular modeling systems. There are a variety of reasons for this popularity: such representations are simple to construct computationally (requiring only a knowledge of atomic coordinates and connectivities along with a simple line-drawing facility); they are easily displayed on both vector and raster graphics screens,[6-7] where they are easily transformed in real-time; and they can be turned into hardcopy on plotting devices with little difficulty.

Furthermore, such simple images are readily superimposable, enabling the user to examine two or more molecules simultaneously in the same frame of reference. However, without in-built hardware

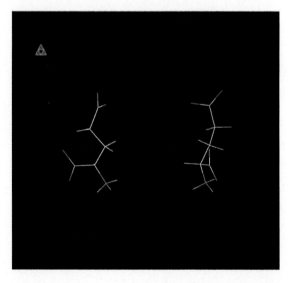

Figure 1 Orthogonal views of stick model of amino acid glutamic acid with color-coded line segments

depth-cueing techniques, the images do not appear particularly three-dimensional. The introduction of color can alleviate this deficiency considerably—each molecule may be given a different color to facilitate discrimination between several different molecules, and the atoms within any single molecule may be color-coded according to atom type. This is achieved by dividing each bond in half and coloring each end in accordance with the relevant atom type.

20.2.2.1.2 Ball-and-stick models

In the slightly more complex ball-and-stick molecular displays (Figure 2), molecules are again drawn as lines joining the centers of bonded atoms but the atoms themselves are now represented as small spheres centered on each of the atomic coordinates. By displaying the atoms as solid spheres, various more distant parts of the molecule become obscured by atoms lying towards the front of the image and thus a certain three-dimensionality is achieved. Because of the added complexity of these images they are more suited to display on raster screens where additional three-dimensionality can be obtained with shading and highlighting techniques.

As for stick model displays, the incorporation of color-coding by atom type into ball-and-stick models increases the perception of depth in the picture. This is achieved by leaving the bonds as a neutral color, while the atomic spheres at each end of a bond are colored according to the type of atom that they represent. Additionally, the atomic spheres may be cross-hatched[8] or size-coded[9] according to their z coordinates, such that those atoms furthest away from the viewer are drawn slightly smaller than those at the front.

A common criticism leveled at both stick and ball-and-stick displays is that, despite their ease of manipulation, they do not give much indication as to the size or shape of a molecule—identification of atomic positions is all very well, but chemically speaking it is the electron cloud surrounding the atomic framework (which provides the 'volume' and 'surface' of the molecule) that is of interest. Accordingly, several other molecular display representations have been developed.

20.2.2.1.3 Space-filling models

Space-filling models[10-12] represent each atom by a sphere, whose radius depends on the atom type (Figure 3). Generally the radii of the spheres are chosen to match the van der Waals radii of the atoms they represent so that bonded atoms intersect each other. Thus the final image looks very much like a photograph of a Corey–Pauling–Koltun (CPK) model. Again, coloring of the spheres by atom type and/or coding their radii according to their z coordinates enhances the three-dimensionality of the picture greatly. However, such displays are almost entirely dependent upon the use of raster display devices and can take some considerable time to draw. This effectively removes the

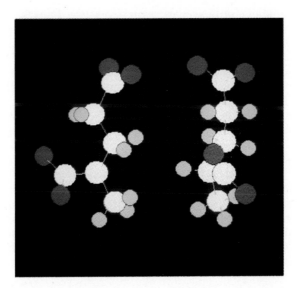

Figure 2 Ball and stick model of amino acid glutamic acid (orthogonal views)

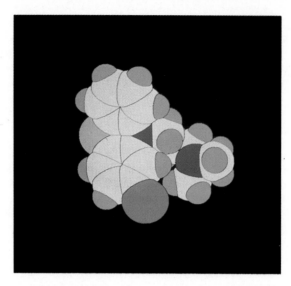

Figure 3   Space-filling model of chlorpromazine, a neuroleptic drug, with color-coded atom types

ability to transform the image in real-time, although recent advances in hardware and software are now alleviating this problem to a large extent.[13-15]

Because of the static nature of such displays, the enhancement of three-dimensionality obtained from real-time rotation of a two-dimensional image is lost. To compensate for this, sophisticated shading and highlighting techniques have been developed[16-21] that give a tremendous impression of depth. Simpler techniques that represent the atomic spheres as polygonal surfaces[8, 11,22-23] have also been employed and they produce a similar three-dimensional effect by generating a much clearer perception of the curvature of the spheres and hence of the molecular surface. Additional clarity can also be achieved by superimposing a stick diagram on to the space-filling model.

However, a significant disadvantage of all these space-filling displays is that since they are opaque they cannot be superimposed upon one another without obscuring much of the information that is of interest.

20.2.2.1.4   *Dotted surface representations*

Dotted surface representations of molecules, as developed by Langridge *et al.*,[24] represent the molecule as a stick diagram (often color-coded) surrounded by a regularly spaced array of dots that lie on the molecular surface (Figure 4). The molecular surface can be defined as the van der Waals surface,[25] the solvent-accessible surface[26-27] or other analytical surfaces.[28-30] Thus these images not only display molecules as readily identifiable stick models but also give an accurate impression of the volume that they occupy and hence of their shape and size. Furthermore, the images are transformable in real-time, superimposable and, to a certain extent, transparent: that is the front and back surfaces of a molecule may be seen at the same time. These features have resulted in the widespread use of such displays in very many different molecular modeling studies.[24, 31-35]

As in other types of display, the images may be significantly improved by the introduction of color. Thus the surface dots may be colored according to the type of atom with which they are associated.

One problem associated with dotted surface displays is that the surfaces lose their integrity as molecular conformations are altered by the twisting of flexible bonds within the molecule. However, recent algorithmic improvements in the generation of dot surfaces[36-37] have resulted in their being calculated so rapidly that a displayed molecule's conformation may be changed in real-time and its altered surface updated and redisplayed on the fly, adding greatly to the utility of such displays. In other approaches to dotted surfaces,[36] the dots can be replaced with short lines drawn outwards from the atomic centers, which gives the molecular surface a 'furry' appearance.[38]

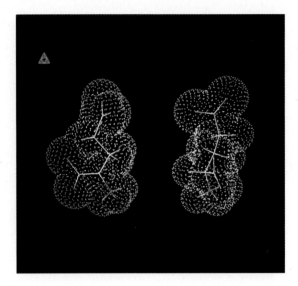

Figure 4 Volume representation of glutamic acid by dot surface colored by atom type (orthogonal views)

20.2.2.1.5 Contoured surfaces

An alternative method of displaying molecular surfaces that retains the ability to transform the display interactively has been developed by Marshall and Barry.[39] The procedure involves computing a molecular pseudoelectron density map by means of a three-dimensional grid that surrounds the molecule whose atoms are replaced by dummy Gaussian atoms. Atom types are characterized by a half-width and an integrated density, chosen so that the Gaussians have a fixed value at a distance equal to the van der Waals radius. Such density maps may be contoured in three dimensions to provide a 'chicken-wire' envelope around the molecule which corresponds to the van der Waals surface (Figure 5).

A concomitant benefit of this technique is that estimates of the molecular surface area and volume are generated as by-products of the contouring routines, whether the surface is being drawn around one or several molecules. Additionally, the generated surfaces and volumes are readily susceptible to logical operations, such as union, intersection or subtraction, enabling the rapid determination of, for example, union or difference volumes among a series of molecules.

20.2.2.2 Associating Property with Substructure

Although the relatively straightforward displays of molecular structure outlined in the previous section have already proved to be extremely useful tools in enabling medicinal chemists to visualize molecules and to compare their structural properties in three dimensions, of even greater potential utility is the display of the various chemical and physical properties of molecules in addition to their structures. Such displays allow the comparison not only of molecular shapes and structures, but also of molecular properties such as internal energy, electronic charge distribution and hydrophobic character. A number of different properties have been displayed in this manner.

20.2.2.2.1 Electron density

As stated earlier, the relative positions of the atomic nuclei within a molecule represent a skeletal framework for the molecule, the bulk of which is provided by the surrounding electrons. In terms of chemical and biological reactivity, it is the molecular surface generated by the orbiting electrons that is important rather than the positions of the centers of the atomic nuclei. To take this into account, some attempts have been made to display the shapes of molecules not as intersections of atomic van der Waals spheres (be they space-filled, dotted or contoured) but as contoured surfaces of electron

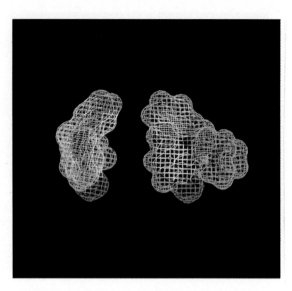

Figure 5   Contoured electron density representation of chlorpromazine

density derived from quantum mechanical calculations.[40-41] The main problem encountered in such a procedure is the exact choice of the value of the electron density to be contoured, since the electron density surrounding a molecule extends outwards to infinity.

Richards and co-workers[42-44] tried to define the 'hard edge' of a molecule by monitoring the interaction energy between the molecule and a probe water molecule as the latter was brought closer to the former along several different trajectories. They concluded that the repulsion between the two molecules became significant at a position around the molecule that corresponded roughly to the 0.001–0.01 atomic unit of charge contours. However, as they noted, this approach does not really solve the problem, since the interaction energy between the two molecules is strongly dependent on the orientation of the probe molecule as it approaches the target.

Because of the problems associated with the above technique, an alternative method of displaying the electron density around a molecule was investigated by Quarendon *et al.*[45] In this approach the surface of the molecule is predefined as the van der Waals surface, and then the value of the electron density over this surface is computed directly from a quantum mechanically derived wavefunction. The results of the computation are then color-coded by value on to a space-filling model of the molecule displayed on a raster terminal. The result is a picture of a molecule (Figure 6) that looks like a CPK model, with the molecular electron density information coded on to the surface. By comparing such images for a series of molecules, differences in their electronic properties may be discerned. However, as it turned out in this study, the electron density itself, and the related superdelocalizability,[46] which was also computed and displayed, were not as useful in highlighting differences between two different sets of similar molecules as was the electrostatic potential.

20.2.2.2.2   *Electrostatic potential*

Any distribution of electrical charge, such as the electrons and nuclei of a molecule, creates an electrical potential in the surrounding space which at any given point represents the potential of the molecule for interacting with an electrical charge at that point. This potential is a very useful property for analyzing and predicting molecular reactive behavior.[47-51] In particular, it has been shown to be a particularly useful indicator of the sites or regions of a molecule to which an approaching electrophile is initially attracted or from which it is repelled.

The electrostatic potential is a rigorously defined quantum chemical property which can be determined experimentally as well as theoretically. Computationally, there have been two approaches to the determination of electrostatic potentials. In the more rigorous and hence time-consuming approach, the electrostatic potential is calculated directly from the quantum mechanically derived electron density distribution within the molecule, whereas in the simpler approach the molecule is represented by a series of point charges and the electrostatic potential is then

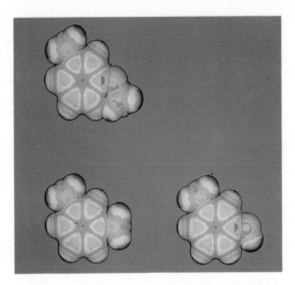

Figure 6   Electron density coded by color on CPK models of toluene derivatives; upper left, *m*-methoxy; lower left, *m*-methyl; lower right, *m*-hydroxy

calculated from a simple $q_i q_j / r$ summation procedure. There has been much debate in the literature over the choice of procedure for generating these partial charges,[52-60] with those procedures that optimize the charges to reproduce more rigorously derived potentials and dipoles around a molecule being preferred.[61-63]

Irrespective of the derivation of the electrostatic potential values around a molecule, the major obstacle to their easy use in the comparison of different molecules has been the sheer volume of information produced. The traditional means of displaying such large amounts of data has been to display the electrostatic potential around a molecule as a two-dimensional contour map on paper.[64] Such representations are of most use though for molecules that have a readily identifiable plane of symmetry in which to draw the contours. Unfortunately, most molecules of biological interest have no such symmetry and alternative display techniques have been sought. The advent of computer graphics techniques has improved the situation somewhat by allowing these two-dimensional maps to be displayed in color on the graphics screen and manipulated in real-time along with a display of the molecule itself.[65-68] Contours in three dimensions have also been displayed in this manner,[42] but the complexity of the resulting images generally limits to a minimum the number of contours and molecules that can be displayed usefully.

In a technique developed by Miller *et al.*,[69-70] the electrostatic potential contours are drawn in three dimensions over a molecular steric surface which is itself contoured in horizontal planes and is derived by positioning a test atom at various points around the molecule and then computing a steric interaction energy according to a 6-*n* potential function (Figure 7).

A similar representation has been developed by Cohen.[8, 23] It involves the display of electrostatic potential contours over the van der Waals or solvent-accessible surface of a molecule (Figure 8). The space-filling outline of the molecule itself can be included or omitted as desired, and the contours themselves can be color-coded by value.

Both these techniques attempt to give some idea of the shape of the molecules at the same time as their electrostatic profiles. However, the images are still fairly complex entities, even with only a small number of contours displayed, and this does not aid the ease with which they may be compared among sets of similar molecules.

A different approach was developed by Quarendon *et al.*[45] Here the electrostatic potential is rigorously computed on a regular grid (of 0.2 Å spacing) around the molecule, and the value of the electrostatic potential at points on the van der Waals surface of the molecule obtained from linear interpolation between the eight nearest grid points. The color-coded electrostatic potential surface is then displayed with shading and highlighting as a space-filling CPK-like model on a raster graphics screen (Figure 9). Images were produced for a series of substituted toluenes[45] and for a series of substituted nitrosamines[71] and proved capable in both cases of distinguishing between structurally very similar compounds. The technique has also been applied to a short polymer of DNA.[72]

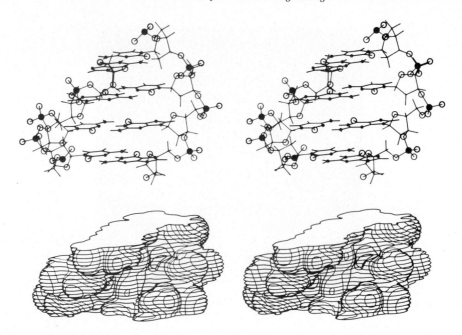

Figure 7   Three-dimensional contoured planes coded by electrostatic potential of B-DNA showing major and minor grooves in stereo. Most dense lines indicate most positive contours (reproduced from ref. 70 by permission of John Wiley & Sons, Inc.)

Hehre and co-workers have used a very similar type of display in a series of seven papers.[73-79] They produced space-filling electrostatic potential color-coded models of various dienes and were able to quantify reactivity by examination of the images. In particular, these electrostatic potential surfaces were found to be better guides than other indices, such as frontier orbital energies, in assigning regiochemistry in Diels–Alder cycloadditions and in assigning the stereochemistry of electrophilic additions to chiral allylic double bonds.

A similar type of representation has been developed by Nakamura *et al.*[80-81] In their model, the van der Waals or solvent-accessible surface of the molecule is approximated by a polyhedron, the surface color of which can represent the value of the electrostatic potential (or any other suitable property) at the position of each polygon. As in the previous example, this method is especially suited to raster display devices, but it has the advantage that different views of the molecule can be generated relatively rapidly because of the speed with which such devices can draw polygons. An additional benefit of the use of polyhedra to approximate spherical surfaces is that the boundaries between the individual polygons give a good representation of the curvature of the molecular surface which produces a three-dimensional effect without the need for any shading or highlighting.

In an interesting application of the technique,[82] the electrostatic potential of NAD^+ was displayed over the surface of the molecule in the conformation found in the crystal complex with the enzyme glyceraldehyde phosphate dehydrogenase (GAPDH; Figure 10). The atoms of the enzyme were then replaced by partial charges and the electrostatic potential due to the enzyme was then computed over the same surface. The complementarity of the two surfaces was demonstrated both graphically and numerically and serves to illustrate the importance of electrostatic interactions in molecular recognition processes.

An alternative mode for displaying molecular electrostatic potentials, which removes the essentially static nature of the previously described raster-based displays, is to employ a dotted surface representation (Figure 11), with the dots taking on an appropriate color according to the electrostatic potential value at the relevant location.[83] Such techniques were initially derived to display empirically determined potentials on the surface of proteins but have since been used widely to display the electrostatic potentials on sets of small molecules for comparative purposes.[84-85]

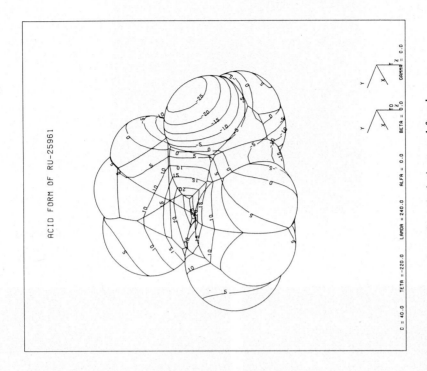

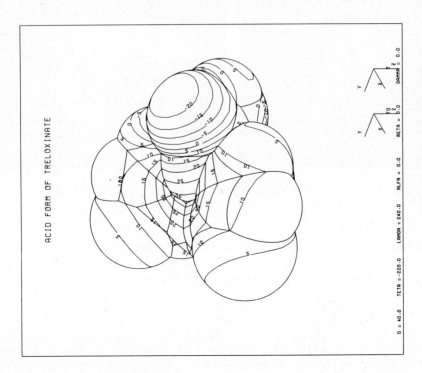

Figure 8 Electrostatic potential contours on van der Waals surface for comparison of two drugs (treloxinate on left and RU-25961 on right) (reproduced from ref. 8 by permission of the American Chemical Society)

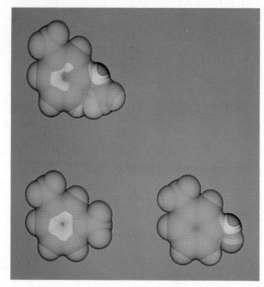

Figure 9 Electrostatic potential coded by color on CPK models of toluene derivatives; upper left, *m*-methoxy; lower left; *m*-methyl; lower right; *m*-hydroxy

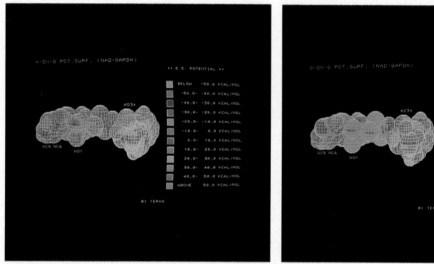
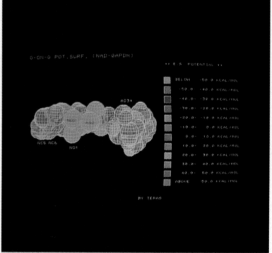

Figure 10 Electrostatic potential of guest molecule (NAD^+) displayed on guest (left) and host molecule (enzyme GAPDH) displayed on guest to illustrate complementarity (reproduced from ref. 82 by permission of Butterworth Publishers)

In another technique, Cuthbertson and Thomson[86] have displayed the electrostatic potentials of various tumor promoters on surfaces which are hybrids between dotted surfaces and solid CPK-like surfaces. Points are generated on the molecular surface in the same manner as for a dotted surface, but at each location a polygon (generally a hexagon) is drawn and colored according to the value of the electrostatic potential at the point. However, since these polygons do not tessellate exactly over spherical surfaces, the final image is neither transparent nor cohesive.

Other graphical applications of the electrostatic potential have been used by Davis *et al.*,[67] who were able to graphically align cAMP and cGMP, based on the superimposition of their respective electrostatic potential minima, and by Weinstein *et al.*,[87] who oriented 5-hydroxytryptamine and 6-hydroxytryptamine based on the alignment of an electrostatically derived 'orientation vector'.

While the electrostatic potential and its three-dimensional display have proven to be effective tools for analyzing and predicting molecular reactivity, it is important to recognize certain intrinsic limitations upon their use. Because the potential is a property of a molecule in a particular state, it cannot take into account the changes that occur within the molecule as it begins to interact with

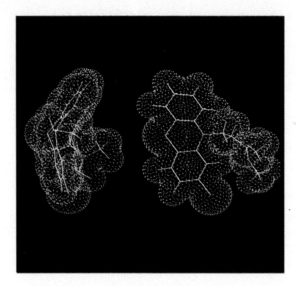

Figure 11 Use of dotted surface representation with electrostatic potential coded by dot color. Shown are orthogonal views of chlorpromazine

some approaching species. Thus it is most useful as a guide to the early stages of a reaction in which the molecule and the attacking species are not yet very close. Furthermore, the electrostatic potential cannot be applied as readily to the analysis of nucleophilic processes as to electrophilic ones, since the positive charges of nuclei, being very highly concentrated, can create strong positive potentials that may outweigh the negative contributions of the dispersed electrons but may not truly reflect a corresponding affinity for nucleophiles.

20.2.2.2.3 Electric field

The electrostatic field at a point around a molecule is simply the negative of the derivative of the potential with respect to distance at the same point. Perhaps because of this close relationship to the electrostatic potential, coupled to the added difficulties involved in the illustration of a three-dimensional vector quantity, it seems to have been used very sparingly in the molecular graphics of small molecules.[1, 88-89] However, results with macromolecular systems, notably nucleic acids[90-93] and the enzyme superoxide dismutase,[94] have shown that the electric field can be a more precise predictor of relative interaction energies and geometries than the electrostatic potential, especially for interactions between dipolar molecules.[62, 95] Indeed, Purvis and Culberson[96] have shown that an otherwise uninteresting isopotential or isodensity surface around a molecule can reveal striking features when color-coded according to the value of the gradient of the potential or density. Such findings are likely to lead to an increase in the use of the electric field as a useful molecular property for rationalizing biological activity.

20.2.2.2.4 Frontier molecular orbitals

Frontier orbital theory was originally developed to explain the differences in reactivity at each position in substituted aromatic hydrocarbons. It was based on the simple idea that the reaction should occur at the position with the largest density of electrons in the frontier orbitals: the HOMO (highest occupied molecular orbital) for electrophilic reactions, the LUMO (lowest unoccupied molecular orbital) for nucleophilic reactions and both in a radical reaction. The importance of these orbital interactions in rationalizing chemical reactions has been well documented,[97-101] and this has led to their increased use in biological systems.

Like the electrostatic potential, the first attempts to display the properties of the frontier orbitals of a molecule, as determined from quantum mechanical calculations, were simple two-dimensional contour maps of the electron density within the frontier orbital of interest.[102-104] These contour maps along with the values of the orbital energies were then compared among a series of molecules, and attempts were made to rationalize the observed differences in biological activity on the basis of the molecular orbital information.

The advent of sophisticated graphics systems soon allowed the three-dimensional display of individual frontier orbitals, superimposed upon the molecular framework.[105-107] Recently, a technique has been developed by Koide and co-workers[108] that allows the frontier orbitals to be displayed either as three-dimensional contours, dotted surfaces or as solid polyhedral surfaces (Figure 12).

The ability to display these properties in three dimensions and in color and to be able to manipulate the displays interactively gives the user a much greater feel for the similarities and differences between molecules.

20.2.2.2.5 *Van der Waals interaction energy*

In a similar procedure to that described earlier for the display of electrostatic potential, Cohen and co-workers have developed a technique whereby the steric field surrounding a molecule can be displayed on a graphics screen as a three-dimensional isopotential contour map.[23] The map is generated by calculating the van der Waals interaction energy between the molecule and a probe

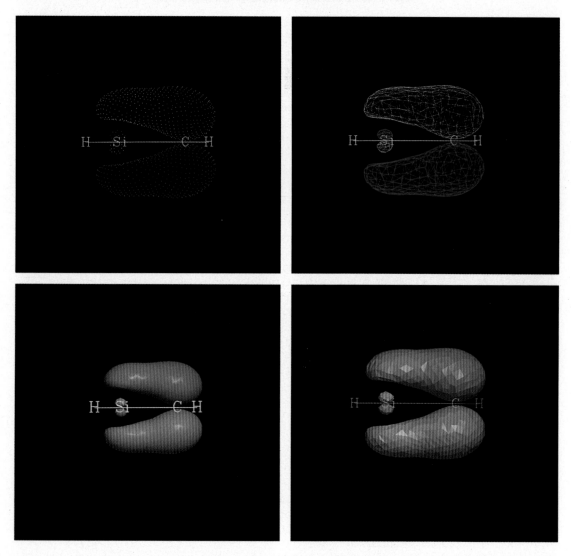

Figure 12 Four methods of displaying molecular orbital data: dots, contours, smooth shading and polyhedra (reproduced from ref. 108 by permission of Butterworth Publishers)

atom or molecule placed at varying points around the molecule of interest. This interaction energy is then contoured at specific levels to give the most stable van der Waals contour lines around the molecule, *i.e.* the contour that represents the most favorable steric position for the probe as it is moved around the target.

A similar three-dimensional contour representation of a molecule can be obtained for both the electrostatic and steric fields of a molecule within the comparative molecular field analysis (COMFA) methodology which has been developed by Cramer[109] to investigate three-dimensional quantitative structure–activity relationships (3D-QSARs). In this procedure, the molecule is surrounded by a regular lattice of points, at each point of which a van der Waals and an electrostatic interaction energy between the molecule and a probe atom are computed. Isoenergy contours can then be generated around individual molecules, displayed graphically, and they can be statistically compared throughout a series of molecules in an attempt to generate 3D-QSARs and hence to rationalize activity data.

20.2.2.2.6 *Hydrophobicity*

In situations where, either from previous QSAR work or from experimental evidence, it is known or suspected that differences in the activity of a set of molecules are due primarily to their hydrophobic rather than their electrostatic properties, it is probably of more use to compare molecular surfaces that display hydrophobicity or polarity information. Indeed, dotted molecular surfaces color-coded by hydrophobic character have been used very successfully by Hansch and co-workers to rationalize QSARs from several different systems.[110–115] For further details the reader is directed to Volume 4, Chapter 20.3.

20.2.2.2.7 *Other properties*

A number of techniques have been described for the graphical illustration of several other molecular properties. Among these, Davis *et al.*[67] have displayed, simultaneously, stick models of all the different conformers of cAMP generated by conformational analysis, with each conformer color-coded according to its total internal energy. For molecules with few enough numbers of rotatable bonds, these displays give a very clear indication of the relative energies associated with the various regions of conformational space available to a molecule.

Another group of informative displays has been developed by Weiner *et al.*,[116] who have managed to display the overwhelming amounts of output data generated by molecular dynamics simulations in several different and easily assimilable manners. In the normal course of events, the results from a dynamics simulation can be examined in two ways. The numerous coordinate sets can be listed and examined individually or they can be 'replayed' on a computer graphics screen by animation techniques. The latter method is infinitely preferable, but even so it can be difficult to extract the desired information from this procedure simply because of the amount of information being displayed. Fortunately, color graphics techniques can be employed to improve the situation.

In the first method described by Weiner *et al.*, atoms in the molecule being displayed are drawn and color-coded in accordance with the magnitude of the forces acting upon them, as defined by the dynamics force field. These images enable the viewer to identify immediately any highly strained parts of the molecule and to see how this strain is redistributed throughout the rest of the molecule as the simulation proceeds.

In the second example, atoms are color-coded according to the distance they have moved from their original positions. This enables the easy identification of those parts of the molecule that are highly flexible and mobile.

In the final example, internal coordinates of the molecule, such as dihedral angles, are color-coded according to the conformational ranges that they can adopt during the simulation. This technique was found to be particularly useful for recognizing and characterizing the sugar puckers of nucleic acids.

Of course, there is no reason why such displays should be limited to examining the output from molecular dynamics simulations—molecules undergoing normal molecular mechanical minimization are equally susceptible to these sorts of techniques, and indeed some molecular modeling systems[117–118] already possess such features.

20.2.2.3 User Interface

In addition to the types of molecular displays already described, whose three-dimensional features are enhanced by such various techniques as depth-cueing, real-time transformation, shading and highlighting, overall three-dimensional perception can be improved still further by other external methods.

Production of a realistic three-dimensional stereoimage generally requires the generation of two distinct images. Most, if not all, molecular modeling systems in common use have the ability to display two distinct images of the same display, either directly overlayed upon each other or separated into two halves of the screen. The only difference between these images is a small rotation, about a vertical axis, of one image relative to the other. Various devices are then employed to ensure that each eye of the viewer only sees one of the two images, which thus creates the illusion of true three-dimensionality from the two-dimensional screen.

One of the first devices employed to achieve this effect was a half-silvered mirror which was placed across the screen so that one image could be seen through the mirror while the other was reflected. Use of spectacles with orthogonally polarized filters ensured that the viewer's eyes each saw only one of the two images.

A later development was the vertically rotating cylinder which possessed two vertical slits through which the user viewed the screen. By synchronizing the cylinder's rotations to the refresh of the display, each eye only saw its allotted image and the three-dimensional effect was achieved.

More recently, a different technique has been developed which synchronizes the alternation of left and right images on the screen with the darkening over and clearing again of liquid crystal eye lenses which are worn by the user as a pair of spectacles. This system is much more user-friendly than previous methods, enabling the whole screen to be used for the display and removing the flickering image particularly associated with the rotating cylinder.

In addition to stereoimages, mutually orthogonal images of molecules have found widespread use in molecular modeling studies. In this type of display, the display screen is split in half vertically, one half of the screen then displays the molecule, generally as a stick model with or without an associated dotted surface, in a specified orientation, while in the other half of the screen the same molecule is displayed rotated by 90° about the vertical axis. Both halves of such displays can be manipulated interactively and in concert, affording two views of the same molecule simultaneously.

20.2.3 COMPARISON OF SMALL MOLECULES (CONGENERIC SERIES)

A congeneric series implies that the basic chemical framework of the molecule remains constant and that groups on the periphery are either modified, *e.g.* aromatic substitution, or substituted, *e.g.* tetrazole for carboxyl functional group. Implicit in this concept is the notion that the compounds bind to the receptor in a similar fashion and, therefore, the changes are localized and comparable for each position of modification. Introduction of degrees of freedom in the substituents as well as consideration of differences in properties which are conformationally dependent, such as the electric field, require conformational analysis in an effort to determine the relevant conformation for comparison.

20.2.3.1 Simple Comparisons

Comparisons can be divided into two categories: those which are independent of the orientation and position of the molecule and those which depend on a known frame of reference. When comparing molecules, it is essential that one has a common frame of reference when dealing with properties which are expressed in terms of a specific orientation and position. Simple comparisons would deal with properties independent of a reference frame. For example, the magnitude of the dipole moment is independent, but the dipole itself is a vectorial quantity dependent on the orientation and conformation of the molecule. Similarly, the bond lengths, valence angles and torsion angles, and interatomic distances are independent of orientation. The distance matrix composed of the set of interatomic distances is a convenient representation of molecular structure which is invariant to rotation and translation of the molecule but which reflects changes in internal degrees of freedom. Such a matrix when contoured has proven useful in the comparison of two known structures of proteins as an aid in identifying secondary structures and their relationships, as shown by Nishikawa and Ooi.[119] A visual comparison of the difference matrix, *i.e.* the current distance matrix subtracted from the initial matrix, has helped the study of the evolution of structural changes in molecular dynamics simulations of proteins[120] and should be equally useful in comparison of small molecules. The distance range matrix is an extension which has two values for each interatomic distance representing the upper and lower limits or range allowed for a given interatomic distance. These ranges can often be derived from experimental measurements such as nuclear Overhauser effects by NMR experiments. Crippen[121] has developed a procedure, the EMBED subroutine, which will generate conformations which conform to the constraints represented by such a distance range matrix. The use of distance range matrices in the identification of pharmacophoric patterns was initially discussed by Marshall *et al.*[122]

20.2.3.2 Fitting Procedures

In órder to compare molecules in a general way, a means of superposition or correctly orienting the molecules in the same reference frame must be available. A procedure for positioning an atom in the molecule at the center of the coordinate frame with other atoms positioned along coordinate axes can be used, or the molecules can be successively fitted to one which is used as the standard orientation. Least-squares fitting procedures for designated atoms allow selectivity in orienting the molecules with predetermined geometries in the most appropriate manner. In some cases, the use of dummy atoms allows geometric superposition of groups such as aromatic rings without requiring superposition of the atoms composing the ring. By defining the centroid of the ring and erecting a normal to the plane of the ring, the dummy atom at the end of the normal and the centroid dummy atom can be used to superimpose the ring on another ring with similar dummy atoms. This method leads to coincidence and coplanarity of the two ring systems without requiring the atoms composing the rings to be coincident. In other words, the rings can be viewed as two tori of electron density without overemphasizing the positions of the atomic nuclei. In numerous studies (see a review by Andrews *et al.*[123]) of biogenic amine ligands, this method of comparison of the aromatic ring components is essential in order to allow alignment of the nitrogens.

20.2.3.3 Constrained Minimization

In cases where one has internal degrees of freedom, besides the six associated with position and orientation, the use of constrained minimization procedures becomes a useful technique. Often the standard molecule for comparison has a fixed conformation and the molecule to be fitted has internal degrees of freedom. Several groups have published methods for dealing with this problem. In cases where one has simultaneous degrees of freedom in both the molecule to be fitted and the target, a different approach with simultaneous minimization of all variables as outlined in Section 20.2.4.2.1 is recommended (see references cited in ref. 133).

20.2.3.4 Volume Mapping

Once one has fixed the molecules in a common frame of reference, then comparison by a variety of techniques becomes feasible. As an example, difference in volume may be important in understanding the lack of activity in compounds which appear to possess all the prerequisites for activity seen in others in the series. In a congeneric series, a significant portion of the molecular structure is common to the molecules under comparison. This common volume which is shared logically should not contribute differences in activity. By subtraction of the volume shared by two molecules, one obtains a difference map in which the volume occupied by one molecule and not the other remains.[122] Correlations between the shared volume and the biological activity of a congeneric series of inhibitors of DHFR has been shown by Hopfinger.[124] Motoc[125] has emphasized the use of both overlapping volume and non-overlapping volume in QSAR studies.

20.2.3.5 Field Effects

Once the frame of reference has been established, other properties of molecules such as the electrostatic field can be compared as well. As the electrostatic properties can be sampled on a grid, differences between the values of two molecules can be calculated and a difference map contoured.

Such difference maps[66] highlight more clearly the similarities and differences between molecules. Hopfinger[126] has integrated the difference between potential fields and shown this parameter to be useful in QSAR studies.

An approach to statistically quantifying the similarity between two molecular electrostatic potential surfaces has been developed by Dean and co-workers.[127–128] Here, the previously determined molecular electrostatic potential surfaces are projected outwards on to surrounding spheres which provide a common surface of reference, and then statistical analyses are performed over the points on this common surface in an attempt to quantify the similarities or differences between the two molecules under consideration.

20.2.3.6   Conformational Clustering

In a congeneric series, the correspondence between torsional rotation variables is maintained as one compares molecules and a direct comparison of the values allowed for one molecule with those allowed for another is meaningful. Two- or three-dimensional plots of torsional variables against energy often provide considerable insight into the difference in conformational flexibility between two molecules. As more than three torsional variables become necessary to define the conformation of the molecule under consideration, then multiple plots become necessary to represent the variables. Unless unique graphical functions are included in the software, correlations between plots become difficult as each plot is a projection of a multidimensional space. One approach to this problem is to use cluster analysis programs to identify those values of the multidimensional variables which are adjacent in N-space. The clusters of conformers which result have been referred to as families. A member of a family is capable of being transformed into another conformer belonging to the same family without having to pass through an energy barrier, *i.e.* the members of a family exist within the same energy valley.

20.2.4   SEARCH FOR COMMONALITY (NON-CONGENERIC SERIES)

The primary reason for three-dimensional comparison of molecules is the rationalization of the spatial properties which either are held in common or differ between the molecules under consideration. In a congeneric series, one can often make the assumption of a common binding mode to provide a common frame of reference. In a non-congeneric series, the frame of reference on which to base the comparison is the key factor to a successful analysis. A number of tools to help establish that frame of reference have been developed which focus on a common direction for potential interaction.

20.2.4.1   Directionality

If one is comparing molecules which share interaction at a common site on a biological macromolecule, it is logical to assume that they may do so by interacting with similar sites in the receptor with optimal interaction shown by molecules with correctly oriented functional groups. If one does not have a three-dimensional model of the receptor from which to deduce potential interactions, then one can only attempt to deduce the potential interactive receptor subsites by examination of the molecules which interact with them. Systematically, one can vary the conformation of a molecule and record the relative orientation of groups postulated or shown experimentally to play a dominant role in intermolecular interactions.

20.2.4.1.1   Locus maps

One can generate a locus plot in coordinate space showing all the potential locations of one group relative to another by fixing one group in a particular orientation as a frame of reference and recording the coordinates of the other. An example would be the relative positions of the basic nitrogen to the aromatic ring in compounds such as dopamine interacting with biogenic amine receptors as shown in Figure 13. For comparison, one must choose the common fragment of each molecule and its orientation to generate a similar frame of reference so that comparing the coordinates of the atom whose locus of positions is being determined leads to a meaningful comparison.

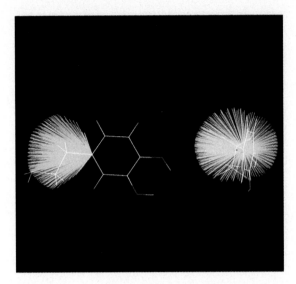

Figure 13 Orthogonal views of locus of nitrogen position for neurotransmitter dopamine derived from systematic grid search with scan of 10°

20.2.4.1.2 Vector maps

Often, one is more interested in accessing the directionality of potential interaction rather than simply looking for overlap of atoms such as the basic nitrogen. In this case, one is interested in determining both the locus of the lone pair of the nitrogen, for example, and its relation to the nitrogen as the pair of coordinates determines a vector in the coordinate space determined by the choice of molecular fragment and orientation. The resulting plot of the locus of all possible orientations of the nitrogen lone pair relative to the aromatic ring, if we continue our dopamine example, constitutes a vector map as shown in Figure 14. The combination of positional information with relative orientation offers considerable insight into potential interactions with a hypothetical receptor. The work of Andrews et al.,[123] postulating a common theme in CNS receptors based on an underlying biogenic amine pattern, can be rationalized using the vector map approach.

20.2.4.1.3 Conformational mimicry

The use of vector maps is essential to the assessment of conformational mimicry in which one attempts to determine the statistical probability that the conformation essential for activity will be preserved with a given chemical modification. Two examples will serve to illustrate this concept and its application. First, modification of amide bonds in peptide drugs to increase metabolic stability may alter the potential accessible conformations to preclude the correct orientation for receptor recognition and activation. In the general case, one has no specific information regarding which particular conformation is biologically relevant and can only assess whether the chemical modification mimics the amide bond in its conformational effects. This can be quantitatively assessed by comparison of the percentage of vectors of the vector map of the parent amide bond which can be found in a comparable vector map of the analog. A recent publication by Zabrocki et al.[129] on the use of 1,5-disubstituted tetrazole rings as surrogates for the *cis*-amide bond illustrates this application. The linear dipeptide acetyl-Ala-Ala-methylamide, with the amide bond between the two alanine residues in the *cis* conformation, and the tetrazole analog acetyl-Ala$\Psi[CN_4]$ Ala-methyl-amide (Figure 15) were modeled using the coordinates derived from diketopiperazines for the *cis*-amide bond or from the crystal structure of the cyclic tetrazole dipeptide. SEARCH, a module in SYBYL[130] which determines the sterically allowed conformations by systematically varying the torsional degrees of freedom by a specified increment, was used to generate a Ramachandran plot for each of the pairs of backbone torsional angles (Φ, Ψ) associated with each amino acid residue, as shown in Figure 16. The rigid geometry approximation was used with the set of scaled van der Waals radii shown by Iijima et al.[131] to reproduce the experimental crystal data for proteins and

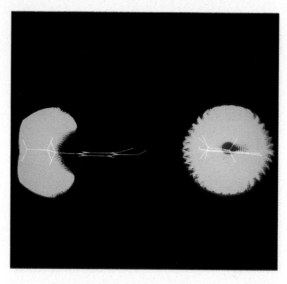

Figure 14   Orthogonal views of locus of nitrogen lone pair vector for neurotransmitter dopamine derived from systematic grid search with scan of 10°

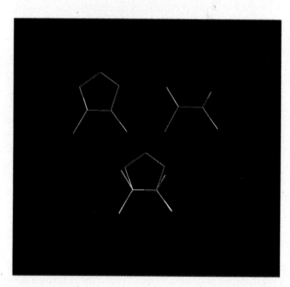

Figure 15   Comparison of crystal structure results for *cis*-amide and 1,5-disubstituted tetrazole, overlap of structures shown at bottom

peptides. When the *cis*-amide dipeptide model was calculated, an option to record the vectors associated with any atom pair was used to record the orientations of the $C(\alpha)$—$C(\beta)$ bond of Ala-1 with the methylamide fixed as a frame of reference. Using the same orientation of the methylamide in the tetrazole allowed the program to determine which vectors or orientations of the Ala-1 sidechain relative to the methylamide were common to both dipeptides. Alternatively, the acetyl group was used as the fixed frame of reference and the sidechain orientation of Ala-2 was used to monitor conformational mimicry. Since the quantitative results were essentially the same, the measurement of mimicry was shown to be independent of the chosen frame of reference. A torsional increment of 10° was used and a sidechain vector was assumed to correspond if both the C-α and C-β were within 0.2 Å of the coordinates of another vector. The percentage of orientations available to the analog which are available to the parent is referred to as the conformational mimicry index. For the tetrazole surrogate of the *cis*-amide bond, the conformational mimicry index is 88% (the number of vectors (747) common to both the tetrazole and *cis*-amide divided by the total number of vectors (849) allowed for the *cis*-amide).

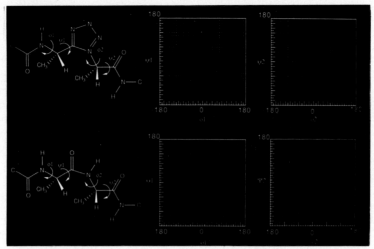

Figure 16 Comparison of Ramachandran plots for alanine dipeptide containing *cis*-amide and tetrazole surrogate

The tetrazole analog has more conformational freedom than the *cis*-amide model with 33 359 conformers allowed compared to 14 912 allowed for the *cis*-amide of the 36^4 (or 1 679 616) possible conformations. This difference may be easily visualized in plots of the vector maps for the two dipeptides shown in Figure 17. The increased number of sterically allowed conformations available to the tetrazole analogs probably reflects the increase in valence angle between the $C(\alpha)$—C═N of the tetrazole corresponding to the $C(\alpha)$—C═O angle of the *cis*-amide. This increases the number of values available to Ψ^1 for the tetrazole, as can be seen in the Ramachandran plots (Figure 16). On the other hand, the increase in steric bulk in the tetrazole where the amide hydrogen is replaced by the nitrogen is clearly reflected in the Ramachandran plot as an enlarged area around $\Phi^2 = -180°$, $\Psi^2 = 0°$ or $180°$. The relative importance of these two opposite effects determines whether the tetrazole will have more conformations available or less. In this case, the increased freedom of the torsional rotation Ψ^1 dominates and the resulting increased flexibility should be reflected in increased loss of entropy upon binding of tetrazole analogs compared with the *cis*-amide. One important aspect of this vector map visualization is the transformation of a four-dimensional problem which we had viewed as two two-dimensional projection plots, the Ramachandran plots, with loss of the correlation between the variables into a three-dimensional plot where each vector represents an allowed setting of the four variables (Φ^1, Ψ^1, Φ^2, Ψ^2).

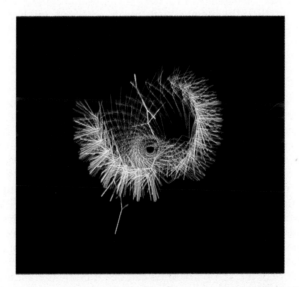

Figure 17 Comparison of vector plots for alanine dipeptide containing *cis*-amide and tetrazole surrogate

A second use of conformational mimicry is the comparison of potential transition state analogs for amide hydrolysis with the theoretical tetrahedral intermediate, a gem diol adduct for serine proteases. In this case, a direct comparison of torsional angles becomes more difficult as more degrees of freedom are introduced by the removal of the constraint associated with the planar amide bond. In addition, several candidate modifications have chemistry (*e.g.* phosphonates) in which geometrical changes in bond lengths and bond angles make direct comparison of torsional variables meaningless. In preliminary studies of the transition state inhibitors found effective against renin, an aspartic proteinase, Hodgkin and Marshall[132] have found that the P_1 sidechain and carbonyl groups of the inhibitors are geometrically situated to bind in the same subsites.

20.2.4.2 Pattern Searches

If one assumes that a common binding mode exists for two or more compounds, then one can use the computer to verify the geometric feasibility of the assumption. The question becomes whether it is possible for the two molecules to present a common geometric arrangement of the designated 'important' functional groups for recognition. There are two distinct approaches to this problem. The first, which is associated with minimization methodology, focuses on the existence issue. Is there a conformation which is energetically accessible to each of the molecules under consideration which will place the designated functional groups in a similar orientation? The second approach attempts to systematically enumerate all possible conformations and derive, thereby, all possible orientations or patterns in order to determine the set of patterns shared by the compounds under study. The latter approach, when it can be applied, can address the question of uniqueness of the common pattern directly.

20.2.4.2.1 *Minimization methods*

The combination of molecular mechanics with flexible minimization routines allows penalty functions to be assigned to force geometrical correspondence of groups, while individual molecules have their internal energy evaluated, but are invisible to the other molecules under consideration. A program MAXIMIN has been described[133] with this capability and its use illustrated on histamine antagonists by Naruto *et al.*[134]

An alternative approach uses the distance geometry paradigm in which all the constraints are combined to form the distance matrix from which energetically feasible conformations of the set of molecules are sought mathematically. Sheridan and Venkataraghavan[135] have demonstrated this approach on acetylcholine analogs which are muscarinic agonists. Either of these approaches asks the same question and suffers from the same limitations, and they differ only in computational technique. Each will suffer from the local minima problem, in that each utilizes a minimization technique and the results will be dependent on the starting geometries of the initial set of molecules. Each has the advantage that the unique constraints imposed by particular molecules enter consideration at an early stage and minimize comparison of conformations.

20.2.4.2.2 *Systematic search*

Once the question of existence of a common pattern has been affirmed, then the issue of uniqueness needs to be addressed. The active analog approach[122] utilized systematic search to generate the set of sterically allowed conformations based on a grid search of the torsional variables at a given angular increment. For each sterically allowed conformation, a set of distances between the postulated pharmacophoric groups is measured. The set of distances, each of which represents a unique pharmacophoric pattern, constitutes an orientation map (OMAP). Each point of this map is simply a submatrix of the distance matrix discussed in Section 20.2.3.1, and as such is invariant to global translation and rotation of the molecule. If the initial assumption that the same binding mode of interaction or pharmacophoric pattern is common to the set of molecules under consideration is valid, then the orientation map for each active molecule must contain the pattern encrypted in the set of distances. By logically intersecting the set of orientation maps, one can determine which patterns are common to all molecules. In other words, all potential pharmacophoric patterns consistent with the activity of the set of molecules can be found by this simple manipulation and the question of uniqueness addressed directly.

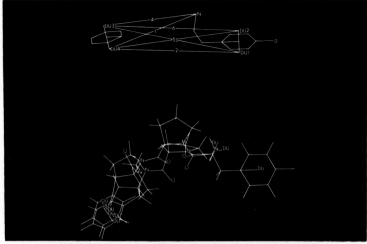

Figure 18 Overlap of two morphiceptin analogs illustrating use of normals at centroids of rings to insure coincidence and coplanarity

A good example is the work of Nelson *et al.*[136] on the receptor-bound conformation of morphiceptin. Based on structure–activity data, the tyramine portion and phenyl ring of residue three of morphiceptin, Tyr-Pro-Phe-Pro-NH$_2$, was postulated to be the pharmacophoric groups responsible for recognition and activation of the opioid m-receptor. It was assumed further that the aromatic rings bound to the receptor in the different analogs were coincident and coplanar. A series of active analogs with a variety of conformationally constrained amino acid analogs in positions two and three were analyzed and a unique conformation shown in Figure 18 was found for the two most constrained analogs which allowed overlap of the Phe and Tyr portions of the molecules. In this case, a five-dimensional OMAP with distances between the nitrogen and normals to the two aromatic rings was used in the analysis.

20.2.4.3 Beyond the Pharmacophore

One major deficiency in the approach described above is the requirement for overlap of functional groups in accord with the pharmacophore hypothesis. While it is true that molecules having functional groups which show three-dimensional correspondence can interact with similar sites, it is also true that a particular geometry associated with one site is capable of interacting with equal affinity to a variety of orientations of the same functional groups. One has only to consider the cone of equal energetic arrangements of a hydrogen bond donor and acceptor to realize the problem. Sufficient examples from crystal structures of drug–enzyme complexes and from theoretical simulation of binding have been described to realize that the pharmacophore is a limiting assumption. Clearly, one is optimizing the position of the drug ligand in an asymmetric force-field represented by the receptor subject to perturbation from solvation and entropic considerations. Less restrictive is the assumption that the receptor binding site remains relatively fixed in geometry when binding the series of compounds under study. In recent years, therefore, there has been an increasing effort to focus on the receptor groups as being the common features for a set of analogs. If we assume the receptor site points remain fixed and can augment our drug with appropriate molecular extensions which include the receptor site, *i.e.* a hydrogen bond donor correctly positioned next to an acceptor, we can then examine the set of geometrical orientations of site points to see if one is capable of binding all the ligands. In a recent study of the active site of angiotensin converting enzyme (ACE) by Mayer *et al.*,[137] the active analog approach has utilized this less restrictive assumption by incorporating the active site components as parts of each compound undergoing analysis. As an example, the sulfhydryl portion of captopril is extended to include a zinc bound at the optimal bond length and bond angle seen for zinc–sulfur complexes. The OMAP derived is based on the distances between receptor site points such as the zinc atom with the introduction of more degrees of torsional freedom to accommodate the possible positioning of the zinc relative to captopril. Analyses of over 30 different chemical classes of ACE inhibitors led to a unique arrangement of the components of the active site postulated to be responsible for binding of the inhibitors. Comparison of the receptor-bound conformation for captopril bound in the active site with that of Gly-Tyr bound in the active

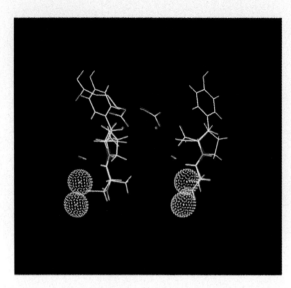

Figure 19 Captopril in bound conformation superimposed on glycyltyrosine as bound in active site of carboxypeptidase A. Pink sphere is zinc in active site of CPAse; yellow sphere is zinc as deduced for ACE. Arginine sidechain from CPAse fits positive charge site for ACE. Hydrogen bond donor represented by two-toned blue line for ACE

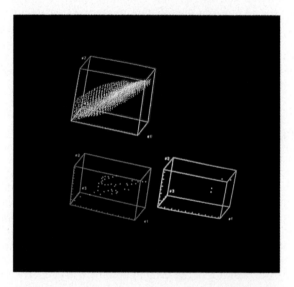

Figure 20 Orientation maps (OMAPs) used in the analysis of angiotensin converting enzyme inhibitors by Mayer *et al.*[137] Upper left plot shows OMAP of three distances with largest variation for most rigid analog. Lower left plot shows common points of several analogs. Lower right shows OMAP near end of analysis when common active site geometry has almost been determined

site of carboxypeptidase A is shown in Figure 19. The displacement of the zinc atom in ACE to a location more distant from the carboxy binding Arg seen in carboxypeptidase A is compatible with the fact that ACE cleaves dipeptides from the C terminal of peptides while carboxypeptidase A cleaves single amino acid residues.

Visualization of the OMAP as each new compound is introduced is useful in order to judge the additional information introduced (see Figure 20). Computationally, it is much more efficient to treat the set of non-congeneric compounds simultaneously (Dammkoehler *et al.*[138]), but reassuring when identical results are obtained if one uses the sequential procedure where intermediate results may be visually verified. The use of computer graphics to confirm intermediate processing of data in convenient display modes becomes increasingly more important as the individual computations and numbers of molecules under consideration increases.

20.2.5 RECEPTOR MAPPING

20.2.5.1 Overall Strategy

The ultimate goal in comparison of molecules with respect to their biological activity is insight into the receptor and its requirements for recognition and activation. Conjecture regarding the receptor is often a necessary part of rationalizing a set of structure–activity data. As an example, one can cite the now almost classical study of volume mapping by Sufrin *et al.*[139] on amino acid analogs of methionine which inhibited the enzyme methionine:adenosyl transferase, in which the data for a set of rigid amino acid inhibitors required the postulation of competition between the inactive analogs and the enzyme for a particular area of space. Summation of the volume requirements for the set of compounds when oriented on the amino acid framework yielded a minimum space from which the receptor could be excluded. In other words, one of the compounds under consideration could not bind if the receptor required occupancy of some of the space. Examination of compounds incapable of binding to the receptor, but which possess the appropriate functional groups and the ability to orient them in a complementary manner, should indicate the requirement for occupancy of novel volume when compared with the set of active analogs. Each amino acid had the necessary binding elements, but several were inactive. When the volume requirements for all the active analogs were combined, each of the inactive analogs required extra volume not required by the active analogs and shared a small common unique volume whose occupancy by the enzyme would be sufficient to rationalize their inactivity. In a similar manner, a number of volume maps for various receptors have been determined. In some cases, comparison of such maps for two receptors have allowed optimization of activity at one receptor with respect to the other. The work of Hibert *et al.*[140] using receptor mapping to increase the selectivity of a lead compound for the 5-HT$_{1A}$ receptor over the α_1-adrenoreceptor has resulted in clinical trials for the derived novel chemical class. Ortwine *et al.*[141] have used this approach to design selective adenosine receptor agonists.

While there are several feasible algorithms to deal with unions of molecular volumes, the use of pseudoelectron density functions, calibrated to reproduce van der Waals radii[39] with three-dimensional contouring to represent the surface, has allowed mathematical manipulation of the density associated with each lattice point to allow for union, intersection and subtraction of volumes. Analytical representation by Connolly[29-30] may be an alternative which would allow optimization of volume overlap, for example, by minimizing the difference in volume between two structures.

20.2.5.2 Model Receptor Sites

The basic assumption behind efforts to infer properties of the receptor from a study of structure–activity relations of drugs which bind is the idea of complementarity. It follows that the stronger the binding affinity, the more likely that the drug fits the receptor cavity and aligns those functional groups which have specific interactions in a complementary way to those of the receptor itself. Certainly, our understanding of intermolecular interactions from studies of known complexes does not dissuade us of this notion, but may make us somewhat skeptical of the naive models which often result from such efforts. Andrews *et al.*[123] have reviewed efforts of this type with regard to CNS drugs.

One of the first visualizations of a receptor model is that of Beckett and Casey[142] for the opiate receptor, published in 1954. As morphine and many other compounds active at this receptor are essentially rigid, the model did not have to address the interaction of myriad numbers of flexible naturally occurring opioid ligands such as endorphins and enkephalin which were only subsequently discovered. The model receptor had an anionic site to bind the charged nitrogen, a hydrophobic flat surface with a cleft to bind the phenyl ring and hydrophobic hydrocarbon bridge seen in morphine. Kier[143] published a number of papers attempting to define the pharmacophore based on semi-empirical molecular orbital calculations of *in vacuo* minimum energy conformations.

Humber *et al.*[144] used semi-rigid antipsychotic drugs, the so called neuroleptics, which antagonize CNS dopamine transmission and displace dopamine from its receptor, to formulate a geometrical arrangement of receptor groups to rationalize their activity. Olsen *et al.*[145] used this model to design a novel stereospecific dopamine antagonist and successfully predicted its stereochemistry.

A conceptually similar approach has been taken by Crippen[146] who uses the distance geometry paradigm to analyze his site points and drug interactions. A site model is postulated by the investigator with some initial estimates of force constants between the appropriate portion of the ligand and the site point. The binding energy for a particular binding mode can be calculated

$$E_{\text{calcd}} = cE_{\text{c}} + \Sigma x_{ti,\,tm}$$

where E_{c} is the conformational energy, c is a coefficient to be fitted, x is the interaction of a site point i with the bound ligand point m which depends on their types. The novel aspect of this approach is the use of distance geometry to generate a variety of conformers binding within the postulated site and then finding a set of force constants between the postulated site points and ligand points which will predict the affinities of the compounds in the dataset when bound in their optimal manner. With a site model of 11 attractive site points and five repulsive ones for DHFR, Ghose and Crippen[147] were able to derive force constants which fitted 62 molecules with an $r^2 = 0.90$ and to predict the activity of 33 molecules with an $r^2 = 0.71$. The compounds, however, are basically an extended congeneric series, as the essential recognition portion of the inhibitor, the pyrimidine ring, is common to all the compounds.

Linschoten *et al.*[148] extended Crippen's method by using lipophilicity to describe the binding of parts of the ligand to lipophilic areas of the receptor. Using only a nine-point model of the turkey erythrocyte β-receptor and six energy parameters, they successfully modeled 58 compounds. No statistical analysis of their fit between calculated and observed was presented, although the differences were tabulated. A review by Donne-Op den Kelder[149] of the distance geometry approach to receptor site construction has appeared.

Since we are reasonably convinced the receptor is a protein, construction of hypothetical sites from amino acid fragments and calculation of affinity for these sites should correlate with observed affinity assuming that the type of interactions and their geometry is represented by the site in some reasonable manner. An individual fragment such as an indole ring from tryptophan does a good job of simulating a flat hydrophobic surface. Holtje and Tintelnot[150] constructed a site for chloramphenicol from arginine and histidine by varying the distances of the amino acid from its postulated binding position and finding the optimal distance for correlation with observed affinity for the ribosome. Correlations for His and Arg were computed separately, but none for the combined site were presented.

20.2.5.3 3D-QSAR and Receptor Modeling

In much of the analysis presented so far, compounds have been characterized as active or inactive. In practice, most sets of data show a wide range in affinity for an isolated receptor or in potency when measuring a biological response. These differences reflect the details of binding and activation events and should prove illuminating provided they can be interpreted in a consistent manner. Traditionally, the relation between structure and quantitative biological data has been the province of quantitative structure–activity relationships (QSAR) which has begged the conformational question by focusing on congeneric series in which the three-dimensional relation of various molecular fragments could be assumed to be similar because of a common chemical framework (Hansch and Klein,[151] Franke[152]). As we have seen above, the active analog approach[122] provides a mechanism to define a set of orientation or alignment rules, and thus a basis for comparison of non-congeneric series. One should, therefore, be able to extract the importance of molecular fragments in regard to their localization in receptor space.

The first efforts to extract correlations from three-dimensional data were based on congeneric series and traditional regression analysis. Comparison of properties for a molecule with another were reduced to single numbers with the resulting loss of directionality, *e.g.* the focus on overlapping volume or electrostatic field in the work of Balaban *et al.*,[153] Hopfinger[124] or Motoc.[125] This results from the traditional use of regression analysis and its limitation on the number of variables for a given set of observations. New statistical techniques, such as partial least-squares (PLS), which are capable of dealing with large sets of variables and a limited set of observations have become available largely through the efforts of Wold *et al.*[154] Cramer and Milne[155] first attempted to compare molecules on the basis of the field that they presented to their surroundings by mapping the field on a grid. DYLOMMS, as their approach was called, has one published application[156] on gabanergic agonists, but was extremely useful in identifying the problems associated with such an approach. Cramer *et al.*[109] have revised the approach and combined it with PLS,[154] resulting in a method designated comparative molecular field analysis (COMFA). The steps associated with such an analysis are given below.

(1) Postulate a set of orientation rules (these may be deduced with the active analog approach).

(2) Align the set of molecules and establish a grid or lattice which surrounds the set in potential receptor space.

(3) Calculate for each molecule the field which a probe atom would experience at each lattice point (steric and electrostatic effects are normally kept separate for ease of interpretation).

(4) Use PLS statistics to determine a minimal set of lattice points necessary to distinguish the set of compounds according to their measured activities.

(5) Check the predictive value of the lattice model by successively eliminating observations and determine the predictive value of the newly derived model.

(6) Repeat steps (4) and (5) (and step (1), if necessary) until a lattice model is found with high predictive value.

Visualization of the coefficients of significance of the lattice points by contouring with respect to either steric or electronic effects can indicate those parts of the receptor which have special properties, and provide a basis for a hypothetical receptor composed of amino acid fragments. As one is dealing in coordinate space with an oriented set of molecules, clear signals from the analysis can lead to hypothetical site points in the appropriate frame of reference.

An instructive application of COMFA has analyzed the binding of trimethoprim analogs to DHFR, a case in which the enzyme–drug complex has a known crystal structure determined by Matthews *et al.*[157] This analysis by Naylor *et al.*[158] shows a clear steric signal consisting of two prominent contours adjacent to one of the *o*-methoxy substituents in trimethoprim. Upon examination of the X-ray structure of the ternary complex, the amino acid sidechain of residue Met-20 clearly provides the steric discrimination which the COMFA analysis suggests is essential. The electrostatic map is not nearly so well focused, which suggests that no specific electrostatic interactions are critical in explaining the differences in activity in this series. In view of the set of dominant interactions seen in the common pyrimidine ring of this series and the well-known long distance nature of electrostatic interactions, the results are not surprising. This example helps to validate the general approach; clearly, other comparisons with known systems where the structures show more structural variation should help define the capabilities and limitations of the COMFA approach. Preliminary results using COMFA where the three-dimensional structure of the receptor is not known have been encouraging. A dataset of affinity for two steroid-binding proteins has been successfully analyzed[109] when other QSAR methods were ineffective; when tested predictively against data not included in the analysis the results were quite positive, especially as different classes of steroids were represented in the test case. Initial studies of the model of ACE inhibition by Mayer *et al.*[159] gave high correlations with relatively high predictive value. The wealth of structural variation and experimental data for ACE inhibitors makes this an excellent testbed for methodological development.

20.2.6 PROBLEMS AND PROSPECTS

20.2.6.1 Current Status

Due to the prevalence of computer graphics hardware and molecular modeling software, many of the display and analysis techniques described above have become commonplace, at least within the pharmaceutical industry. With the advent of personal computers with graphics hardware capable of supporting such modeling software, the widespread application of these approaches in chemistry is obvious and numerous applications, many of which will require novel graphical displays besides those discussed above, are underway. What is clear is the need to selectively present complex data, whether experimental or theoretical, in meaningful ways in order to highlight the significant in the midst of the mundane. There is ample opportunity for progress in this arena.

20.2.6.2 Limitations

The major limitation is no longer hardware. If one can justify the expense, real-time rotation of ray-traced images is quite feasible. The problem is not to produce more complex displays, but rather to develop the conceptual framework to simplify the complexity and retain the significant. Our problem is the conceptualization of multidimensional correlated data and finding ways of visualizing the relationships between variables without losing the correlation. The problem is one common to many fields of endeavor. The opportunity to use computer graphics to explore alternatives of visualization is relatively new, however, and may lead to more useful methods.

Scientifically, the major difficulty in focusing on the properties of small molecules in isolation is the limiting relevance of such a view. Molecules are interesting because of their interactions with other molecules. In biological systems, the interacting species is often a macromolecular receptor, whose structure is unknown. From structure–activity data, one can attempt to deduce features of the receptor, but one must always keep in mind that the system is clearly underdetermined and the models derived only useful to suggest the next compound to be made and tested. Once the results are known from the assay, the model must often be rejected, or at best refined, to reflect the new information.

20.2.6.3 Prospects

The trend is quite clear as molecular graphics becomes more commonplace. Computational chemistry has developed to the point where its results are predictive in many circumstances. As molecular modeling is the natural input and output mechanism, the increased availability of molecular modeling systems will increase the number of applications dramatically and reduce the visibility of the techniques as they become accepted tools. There is a clear need for more heuristic approaches to help assist in analysis of problems; to provide guidance based on experience as well as keep track of the decision tree which has been traversed in order to respond appropriately when new data becomes available.

Graphics capability no longer needs to be a limiting factor. Even the current generation of personal computers can generate dynamic images of useful complexity. Enhanced computational power will be more tightly coupled to graphical output in the future to allow monitoring of simulations and analyses as they evolve. The human brain remains the ultimate in pattern recognition and absurdity detection. A synergistic relation between computational power and human reasoning is made possible through computer graphics. A rapid evolution in novel graphical representations of relevant chemical and conformational information is necessary as we learn to deal with the wealth of numbers available through computation.

20.2.7 REFERENCES

1. P. M. Dean, 'Molecular Foundations of Drug–Receptor Interaction', Cambridge University Press, Cambridge, 1987.
2. G. R. Marshall, *Annu. Rev. Pharmacol. Toxicol.*, 1987, **27**, 193.
3. H. Fruhbeis, R. Klein and H. Wallmeier, *Angew. Chem., Int. Ed. Engl.*, 1987, **26**, 403.
4. R. P. Sheridan and R. Venkataraghavan, *Acc. Chem. Res.*, 1987, **20**, 322.
5. C. Levinthal, *Sci. Am.*, 1966, **214**, 42.
6. J. D. Foley and A. van Dam, 'Fundamentals of Interactive Computer Graphics', Addison-Wesley, Reading, MA, 1982.
7. R. Langridge and T. E. Klein, in 'Comprehensive Medicinal Chemistry', ed. C. Hansch, Pergamon Press, Oxford, 1989, vol. 4, Chap. 20.1.
8. N. C. Cohen, *ACS Symp. Ser.*, 1979, **112**, 371.
9. D. R. Henry, *Comput. Chem.*, 1983, **7**, 119.
10. R. J. Feldmann, D. H. Bing, B. C. Furie and B. Furie, *Proc. Natl. Acad. Sci. U.S.A.*, 1978, **75**, 5409.
11. G. M. Smith and P. Gund, *J. Chem. Inf. Comput. Sci.*, 1978, **18**, 207.
12. P. K. Warme, *Comput. Biomed. Res.*, 1977, **10**, 75.
13. J. E. Pearson, *J. Mol. Graphics*, 1984, **2**, 59.
14. M. Pique, *J. Mol. Graphics*, 1984, **2**, 59.
15. J. Staudhammer, *Comput. Graphics*, 1978, **12** (3), 167.
16. N. L. Max, *Comput. Graphics*, 1979, **13** (2), 165.
17. T. K. Porter, *Comput. Graphics*, 1979, **13** (2), 234.
18. J. Brickmann, *J. Mol. Graphics*, 1983, **1**, 62.
19. K. Knowlton, *Comput. Graphics*, 1981, **15** (4), 352.
20. T. K. Porter, *Comput. Graphics*, 1978, **12** (3), 282.
21. N. L. Max, *J. Mol. Graphics*, 1984, **2**, 8.
22. D. Humbert, M. Dagnaux, N. C. Cohen, R. Fournex and F. Clemence, *Eur. J. Med. Chem.—Chim. Ther.*, 1983, **18**, 67.
23. N. C. Cohen, *Adv. Drug Res.*, 1985, **14**, 41.
24. R. Langridge, T. E. Ferrin, I. D. Kuntz and M. L. Connolly, *Science (Washington, D.C.)*, 1981, **211**, 661.
25. A. Bondi, *J. Phys. Chem.*, 1964, **68**, 441.
26. B. Lee and F. M. Richards, *J. Mol. Biol.*, 1971, **55**, 379.
27. R. B. Hermann, *J. Phys. Chem.*, 1972, **76**, 2754.
28. F. M. Richards, *Annu. Rev. Biophys. Bioeng.*, 1977, **6**, 151.
29. M. L. Connolly, *Science (Washington, D.C.)*, 1983, **221**, 709.
30. M. L. Connolly, *J. Appl. Crystallogr.*, 1983, **16**, 548.

31. B. V. Cheney and J. Kalantar, *J. Mol. Graphics*, 1986, **4**, 21.
32. J. M. Blaney, E. C. Jorgensen, M. L. Connolly, T. E. Ferrin, R. Langridge, S. J. Oatley, J. M. Burridge and C. C. F. Blake, *J. Med. Chem.*, 1982, **25**, 785.
33. V. Cody, *J. Mol. Graphics*, 1986, **4**, 69.
34. U. Christensen, S. Ishida, S.-I. Ishii, Y. Mitsui, Y. Iitaka, J. McClarin and R. Langridge, *J. Biochem. (Tokyo)*, 1985, **98**, 1263.
35. C. Hansch, R.-L. Li, J. M. Blaney and R. Langridge, *J. Med. Chem.*, 1982, **25**, 777.
36. L. H. Pearl and A. Honegger, *J. Mol. Graphics*, 1983, **1**, 9.
37. P. A. Bash, N. Pattabiraman, C. Huang, T. E. Ferrin and R. Langridge, *Science (Washington, D.C.)*, 1983, **222**, 1325.
38. R. A. Palmer, J. H. Tickle and I. J. Tickle, *J. Mol. Graphics*, 1983, **1**, 94.
39. G. R. Marshall and C. D. Barry, 'Proceedings—American Crystallographers Association, Honolulu', 1979.
40. R. F. W. Bader, W. H. Henneker and P. E. Cade, *J. Chem. Phys.*, 1967, **46**, 3341.
41. R. F. W. Bader, M. T. Carroll, J. R. Cheeseman and C. Chang, *J. Am. Chem. Soc.*, 1987, **109**, 7968.
42. W. G. Richards and L. Mangold, *Endeavour*, 1983, **7**, 2.
43. W. G. Richards and V. Sackwild, *Chem. Br.*, 1982, **18**, 635.
44. V. Sackwild, D. Phil. Thesis, University of Oxford, 1981.
45. P. Quarendon, C. B. Naylor and W. G. Richards, *J. Mol. Graphics*, 1984, **2**, 4.
46. K. Fukui, T. Yonezawa and C. Nagata, *Bull. Chem. Soc. Jpn.*, 1954, **27**, 423.
47. E. Scrocco and J. Tomasi, *Adv. Quantum Chem.*, 1978, **11**, 115.
48. R. Bonaccorsi, A. Pullman, E. Scrocco and J. Tomasi, *Chem. Phys. Lett.*, 1972, **12**, 622.
49. P. Politzer, P. R. Laurence and K. Jayasuriya, *EHP, Environ. Health Perspect.*, 1985, **61**, 191.
50. H. Weinstein, J. Rabinowitz, M. N. Liebman and R. Osman, *EHP, Environ. Health Perspect.*, 1985, **61**, 147.
51. J. J. Kaufman, H. E. Popkie and P. C. Hariharan, *ACS Symp. Ser.*, 1979, **112**, 415.
52. R. S. Mulliken, *J. Chem. Phys.*, 1955, **23**, 1833.
53. R. F. W. Bader, T. T. Nguyen-Dang and Y. Tal, *J. Chem. Phys.*, 1979, **70**, 4316.
54. S. M. Dean and W. G. Richards, *Nature (London)*, 1975, **256**, 473.
55. A. N. Barrett, *J. Mol. Graphics*, 1983, **1**, 71.
56. G. Del Re, G. Pepe, D. Laporte, C. Minichino and B. Serres, *Prog. Clin. Biol. Res.*, 1985, **172B**, 115.
57. D. M. Hayes and P. A. Kollman, *J. Am. Chem. Soc.*, 1976, **98**, 3335.
58. A. Pullman and B. Pullman, *Q. Rev. Biophys.*, 1981, **14**, 289.
59. L. E. Chirlian and M. M. Francl, *J. Comput. Chem.*, 1987, **8**, 894.
60. R. J. Abraham, L. Griffiths and P. Loftus, *J. Comput. Chem.*, 1982, **3**, 407.
61. S. R. Cox and D. E. Williams, *J. Comput. Chem.*, 1981, **2**, 304.
62. K. Zakrzewska and A. Pullman, *J. Comput. Chem.*, 1985, **6**, 265.
63. P. A. Kollman, *J. Am. Chem. Soc.*, 1978, **100**, 2974.
64. P. Politzer and D. G. Truhlar (eds.), 'Chemical Applications of Atomic and Molecular Electrostatic Potentials: Reactivity, Structure, Scattering and Energetics of Organic, Inorganic and Biological Systems', Plenum Press, New York, 1981.
65. C. A. Venanzi and J. D. Bunce, *Enzyme*, 1986, **36**, 79.
66. H.-D. Höltje and S. Marrer, *J. Comput.-Aided Mol. Design*, 1987, **1**, 23.
67. A. Davis, B. H. Warrington and J. G. Vinter, *J. Comput.-Aided Mol. Design*, 1987, **1**, 97.
68. J. Weber and M. Roch, *J. Mol. Graphics*, 1986, **4**, 145.
69. K. J. Miller, P. J. Kowalczyk, W. Segmuller and G. Walker, *J. Comput. Chem.*, 1983, **4**, 366.
70. K. J. Miller and P. J. Kowalczyk, *J. Comput. Chem.*, 1984, **5**, 89.
71. C. B. Naylor, D. Phil. Thesis, University of Oxford, 1984.
72. J. M. Burridge, P. Quarendon, C. A. Reynolds and P. J. Goodford, *J. Mol. Graphics*, 1987, **5**, 165.
73. S. D. Kahn, C. F. Pau, L. E. Overman and W. J. Hehre, *J. Am. Chem. Soc.*, 1986, **108**, 7381.
74. S. D. Kahn, C. F. Pau and W. J. Hehre, *J. Am. Chem. Soc.*, 1986, **108**, 7396.
75. S. D. Kahn and W. J. Hehre, *J. Am. Chem. Soc.*, 1986, **108**, 7399.
76. S. D. Kahn, C. F. Pau, A. R. Chamberlin and W. J. Hehre, *J. Am. Chem. Soc.*, 1987, **109**, 650.
77. S. D. Kahn and W. J. Hehre, *J. Am. Chem. Soc.*, 1987, **109**, 663.
78. S. D. Kahn and W. J. Hehre, *J. Am. Chem. Soc.*, 1987, **109**, 666.
79. A. R. Chamberlin, R. L. Mulholland, Jr., S. D. Kahn and W. J. Hehre, *J. Am. Chem. Soc.*, 1987, **109**, 672.
80. H. Nakamura, M. Kusunoki and N. Yasuoka, *J. Mol. Graphics*, 1984, **2**, 14.
81. K. Komatsu, H. Nakamura, S. Nakagawa and H. Umeyama, *Chem. Pharm. Bull.*, 1984, **32**, 3313.
82. H. Nakamura, K. Komatsu, S. Nakagawa and H. Umeyama, *J. Mol. Graphics*, 1985, **3**, 2.
83. P. K. Weiner, R. Langridge, J. M. Blaney, R. Schaefer and P. A. Kollman, *Proc. Natl. Acad. Sci. U.S.A.*, 1982, **79**, 3754.
84. H. P. Weber, T. Lybrand, U. C. Singh and P. A. Kollman, *J. Mol. Graphics*, 1986, **4**, 56.
85. Y. C. Martin and K. H. Kim, *Drug. Inf. J.*, 1984, **18**, 95.
86. A. F. Cuthbertson and C. Thomson, *J. Mol. Graphics*, 1987, **5**, 92.
87. H. Weinstein, R. Osman, S. Topiol and J. P. Green, *Ann. N. Y. Acad. Sci.*, 1981, **367**, 434.
88. J. P. Ritchie and S. M. Bachrach, *J. Comput. Chem.*, 1987, **8**, 499.
89. E. E. Hodgkin, D. Phil. Thesis, University of Oxford, 1987.
90. A. Pullman, *Prog. Clin. Biol. Res.*, 1985, **172A**, 71.
91. R. Lavery and B. Pullman, *Stud. Biophys.*, 1982, **92**, 99.
92. R. Lavery, A. Pullman and B. Pullman, *Biophys. Chem.*, 1983, **17**, 75.
93. A. Pullman, B. Pullman and R. Lavery, *J. Mol. Struct.*, 1983, **93**, 85.
94. E. D. Getzoff, J. A. Tainer, P. K. Weiner, P. A. Kollman, J. S. Richardson and D. C. Richardson, *Nature (London)*, 1983, **306**, 287.
95. P. A. Kollman, *J. Am. Chem. Soc.*, 1977, **99**, 4875.
96. G. D. Purvis and C. Culberson, *J. Mol. Graphics*, 1986, **4**, 88.
97. K. Fukui, *Bull. Chem. Soc. Jpn.*, 1966, **39**, 498.
98. K. Fukui and H. Fujimoto, *Bull. Chem. Soc. Jpn.*, 1966, **39**, 2116.

99. G. Klopman, *J. Am. Chem. Soc.*, 1968, **90**, 223.
100. L. Salem, *J. Am. Chem. Soc.*, 1968, **90**, 543.
101. K. Fukui, 'Theory of Orientation and Stereoselection', Springer-Verlag, Heidelberg, 1975.
102. B. V. Cheney, J. B. Wright, C. M. Hall, H. G. Johnson and R. E. Christoffersen, *J. Med. Chem.*, 1978, **21**, 936.
103. H. Weinstein, R. Osman and J. P. Green, *ACS Symp. Ser.*, 1979, **112**, 161.
104. R. Hilal, *J. Comput. Chem.*, 1980, **1**, 358.
105. R. F. Hout Jr, W. J. Pietro and W. J. Hehre, *J. Comput. Chem.*, 1983, **4**, 276.
106. J. G. Vinter, *Chem. Br.*, 1985, **21**, 32.
107. T. K. Dickens, C. K. Prout and M. R. Saunders, *J. Mol. Graphics*, submitted.
108. A. Koide, A. Doi and K. Kajioka, *J. Mol. Graphics*, 1986, **4**, 149.
109. R. D. Cramer, III, D. E. Patterson and J. D. Bunce, *J. Am. Chem. Soc.*, 1988, **110**, 5959.
110. M. Recanatini, T. Klein, C.-Z. Yang, J. McClarin, R. Langridge and C. Hansch, *Mol. Pharmacol.*, 1986, **29**, 436.
111. C. D. Selassie, Z.-X. Fang, R.-L. Li, C. Hansch, T. Klein, R. Langridge and B. T. Kaufman, *J. Med. Chem.*, 1986, **29**, 621.
112. C. Hansch, T. Klein, J. McClarin, R. Langridge and N. W. Cornell, *J. Med. Chem.*, 1986, **29**, 615.
113. C. Hansch, J. McClarin, T. Klein and R. Langridge, *Mol. Pharmacol.*, 1985, **27**, 493.
114. L. Morgenstern, M. Recanatini, T. E. Klein, W. Steinmetz, C.-Z. Yang, R. Langridge and C. Hansch, *J. Biol. Chem.*, 1987, **262**, 10767.
115. R. N. Smith, C. Hansch, K. H. Kim, B. Omiya, G. Fukumura, C. D. Selassie, P. Y. C. Jow, J. M. Blaney and R. Langridge, *Arch. Biochem. Biophys.*, 1982, **215**, 319.
116. P. K. Weiner, S. L. Gallion, E. Westhof and R. M. Levy, *J. Mol. Graphics*, 1986, **4**, 203.
117. K. Müller, *J. Mol. Graphics*, 1984, **2**, 58.
118. MENDYL software, Tripos Associates Inc., 1699 S. Hanley Road, St. Louis, MO 63144, U.S.A.
119. K. Nishikawa and T. Ooi, *J. Theor. Biol.*, 1974, **43**, 351.
120. A. J. Morffew, *J. Mol. Graphics*, 1983, **1**, 43.
121. G. M. Crippen, 'Distance Geometry and Conformational Calculations', Wiley, Chichester, 1981, Chemometric Research Studies vol. 1.
122. G. R. Marshall, C. D. Barry, H. E. Bosshard, R. A. Dammkoehler and D. A. Dunn, *ACS Symp. Ser.*, 1979, **112**, 205.
123. P. R. Andrews, E. J. Lloyd, J. L. Martin, S. L. Munro, M. Sadek and M. G. Wong, *Top. Mol. Pharmacol.*, 1986, **3**, 216.
124. A. J. Hopfinger, *J. Am. Chem. Soc.*, 1980, **102**, 7196.
125. I. Motoc, *Quant. Struct.–Act. Relat.*, 1984, **3**, 43.
126. A. J. Hopfinger, *J. Med. Chem.*, 1983, **26**, 990.
127. S. Namasivayam and P. M. Dean, *J. Mol. Graphics*, 1986, **4**, 46.
128. P. L. Chau and P. M. Dean, *J. Mol. Graphics*, 1987, **5**, 97.
129. J. Zabrocki, G. D. Smith, J. B. Dunbar, Jr., H. Iijima and G. R. Marshall, *J. Am. Chem. Soc.*, 1988, **110**, 5875.
130. 'SYBYL Users Guide', Tripos Associates Inc., 1699 S. Hanley Road, St. Louis, MO 63144, U.S.A.
131. H. Iijima, J. B. Dunbar, Jr. and G. R. Marshall, *Proteins: Struct. Funct. Genet.*, 1987, **2**, 330.
132. E. E. Hodgkin and G. R. Marshall, unpublished results.
133. J. Labanowski, I. Motoc, C. B. Naylor, D. Mayer and R. A. Dammkoehler, *Quant. Struct.–Act. Relat.*, 1986, **5**, 138.
134. S. Naruto, I. Motoc and G. R. Marshall, *Eur. J. Med. Chem.—Chim. Ther.*, 1985, **20**, 529.
135. R. P. Sheridan and R. Venkataraghavan, *Acc. Chem. Res.*, 1987, **20**, 322.
136. R. D. Nelson, D. I. Gottlieb, T. M. Balasubramanian and G. R. Marshall, *NIDA Res. Monogr.*, 1986, **69**, 204.
137. D. Mayer, C. B. Naylor, I. Motoc and G. R. Marshall, *J. Comput.-Aided Mol. Design*, 1987, **1**, 3.
138. R. A. Dammkoehler, S. F. Karasek, E. F. B. Shands and G. R. Marshall, *J. Comput.-Aided Mol. Design*, 1989, **3**, 3.
139. J. R. Sufrin, D. A. Dunn and G. R. Marshall, *Mol. Pharmacol.*, 1981, **19**, 307.
140. M. F. Hibert, M. W. Gittos, D. N. Middlemiss, A. K. Mir and J. R. Fozard, *J. Med. Chem.*, 1988, **31**, 1087.
141. D. F. Ortwine, A. J. Bridges, C. Humblet and B. K. Trivedi, Abstracts of Papers, American Chemical Society, 1988, no. 92.
142. A. H. Beckett and A. F. Casy, *J. Pharm. Pharmacol.*, 1954, **6**, 986.
143. L. B. Kier, in 'Fundamental Concepts in Drug–Receptor Interactions', ed. J. F. Danielli, J. F. Moran and D. J. Triggle, Academic Press, New York, 1968, p. 15.
144. L. G. Humber, F. T. Bruderlein, A. H. Philipp, M. Gotz and K. Voith, *J. Med. Chem.*, 1979, **22**, 761.
145. G. L. Olsen, H.-C. Cheung, K. D. Morgan, J. F. Blount, L. Todaro, L. Berger, A. B. Davidson and E. Boff, *J. Med. Chem.*, 1981, **24**, 1026.
146. G. M. Crippen, *Mol. Pharmacol.*, 1982, **22**, 11.
147. A. K. Ghose and G. M. Crippen, *J. Med. Chem.*, 1985, **28**, 333.
148. M. R. Linschoten, T. Bultsma, A. P. IJzerman and H. Timmerman, *J. Med. Chem.*, 1986, **29**, 278.
149. G. M. Donne-Op den Kelder, *J. Comput.-Aided Mol. Design*, 1987, **1**, 257.
150. H.-D. Höltje and M. Tintelnot, *Quant. Struct.–Act. Relat.*, 1984, **3**, 6.
151. C. Hansch and T. E. Klein, *Acc. Chem. Res.*, 1986, **19**, 392.
152. R. Franke, 'Theoretical Drug Design Methods', Elsevier, New York, 1984, Pharmacochemistry Library vol. 7.
153. A. T. Balaban, A. Chiriac, I. Motoc and Z. Simon, *Lect. Notes Chem.*, 1980, **15**, 1.
154. S. Wold, C. Albano, W. J. Dunn, III, U. Edlund, K. Esbensen, P. Geladi, S. Hellberg, E. Johansson, W. Lindberg and M. Sjöström, in 'Chemometrics: Mathematics and Statistics in Chemistry', ed. B. R. Kowalski, Reidel, Dordrecht, Netherlands, 1984, p. 17.
155. R. D. Cramer, III and M. Milne, Abstracts of Papers, American Chemical Society, 1979, no. 44 of Computers in Chemistry Section.
156. M. Wise, *Top. Mol. Pharmacol.*, 1986, **3**, 183.
157. D. A. Matthews, J. T. Bolin, J. M. Burridge, D. J. Filman, K. W. Volz and J. Kraut, *J. Biol. Chem.*, 1985, **260**, 392.
158. C. B. Naylor, J. Labanowski, I. Motoc, D. Mayer, R. A. Dammkoehler and G. R. Marshall, to be published.
159. D. Mayer, C. B. Naylor, R. D. Cramer III and G. R. Marshall, to be published.

20.3

Application of Molecular Graphics to the Analysis of Macromolecular Structures

JEFFREY M. BLANEY

E. I. duPont de Nemours & Co., Wilmington, DE, USA

and

CORWIN HANSCH

Pomona College, Claremont, CA, USA

20.3.1 INTRODUCTION

High performance computer graphics has become nearly indispensable for the study of macro-molecular structures and interactions. Mechanical model construction requires heroic patience and endurance to complete a structure which may contain several thousand atoms, while computer graphics can build and display a three-dimensional model in seconds. X-Ray crystallographers have been using computer graphics for building and interpreting electron density maps of macro-molecules for over 15 years. By 1980 only a few groups had general macromolecular-modeling capabilities, due to the high cost of the very specialized hardware and the large programming effort required. Fortunately, recent dramatic cost reductions in computer graphics hardware and improve-ments in software availability have made high performance macromolecular modeling available to far more users—especially medicinal chemists—in the last few years. We will focus on drug design related approaches and applications of macromolecular modeling in this chapter.

X-Ray crystallography and macromolecular modeling provide the most detailed possible view of drug–receptor interactions and have created a new, rational approach to drug design where the structure of a drug is designed based on its fit to the three-dimensional structure of the receptor site, rather than by analogy to other active structures or random leads.[1, 2] Active research into this approach began in the late 1970s, with widespread application beginning in the early 1980s. Biochemical targets (*e.g.* an enzyme or receptor) for drug design should now be selected with serious consideration of the likelihood of obtaining structural information on the target, since this provides the best possible situation for rational drug design.

20.3.2 SOURCES OF MACROMOLECULAR COORDINATES

20.3.2.1 X-Ray Crystallography

Over 300 X-ray crystal structures of proteins and nucleic acids have now been solved, including several ligand–macromolecule complexes; most are available in the *Brookhaven Protein Data Bank.*[3] Although relatively few structures of potential drug receptors have been solved (dihydrofolate reductase,[4] carbonic anhydrase,[5] alcohol dehydrogenase,[6] elastase,[7] phospholipase A_2,[8] hemo-globin,[9] influenza virus hemagglutinin[10] and neuraminidase,[11] rhinovirus,[12] poliovirus,[13] mengo virus,[14] β-lactamase,[15] photosynthetic reaction center,[16] purine nucleoside phosphorylase,[17] calmodulin[18] and troponin-C,[19] thymidylate synthetase,[20] various DNAs and drug–DNA com-plexes[3]), the rate of solving these structures has increased steadily during the last few years and will continue to increase due to improvements in crystallographic methods and the availability of new proteins through recombinant DNA approaches.

X-Ray crystallography requires relatively large amounts (milligrams) of pure sample. Macromol-ecules are notoriously difficult to crystallize and the preparation of high quality crystals which diffract well is still an art.[21] The entire process of sample purification, crystallization, preparation of heavy atom derivatives, X-ray diffraction data collection, construction of the initial electron density map and refinement to a final well-determined three-dimensional structure can now be completed in one to two years in favorable cases. Difference maps comparing a ligand–receptor complex with the known structure of the native receptor can be solved much more quickly; this method is commonly used for solving enzyme–inhibitor complexes, where the inhibitor is diffused into crystals of the enzyme or the enzyme–inhibitor complex is cocrystallized directly from solution. High resolution (< 2.5 Å), refined X-ray structures offer the potential of designing drugs to fit their receptor with high affinity and selectivity. Medium resolution structures (2.5–3.0 Å) may be unreliable regarding the exact placement of side chains but still provide a good starting point for qualitative modeling since the overall shape of the binding site is generally correct. The protein-folding pattern should be well determined with resolutions up to 4–5 Å.

Crystals of macromolecules usually have a very high solvent content (40–60%) and therefore mimic the solution state quite well; in fact, many enzymes retain catalytic activity when crystal-lized.[22] The X-ray crystal structure of a macromolecule is usually a good model of the biologically active conformation in solution. X-Ray crystallography provides a static, time-averaged model of a dynamic structure. Crystallographic temperature factors indicate the most mobile portions of a structure (usually surface residues) and molecular dynamics simulations[23] estimate the actual intramolecular motions, but it is extremely difficult to hit a moving target, so virtually all design efforts begin with the static, time-averaged model.

20.3.2.2 Nuclear Magnetic Resonance

Nuclear magnetic resonance (NMR) can be used to determine the three-dimensional structure of small to medium-sized ($< 15\,000$ Da) macromolecules which do not crystallize.[24] 2-D-NMR techniques have advanced tremendously and can now provide three-dimensional structural information on small proteins (up to 150 residues) and DNA in solution, using distance geometry and/or restrained molecular dynamics to build models consistent with distance constraints derived from NOE (nuclear Overhauser enhancement) and coupling constant data. 2-D-NMR has been used in several cases to solve a complete protein structure; Tendamistat, the 75-residue α-amylase inhibitor, was solved independently by 2-D-NMR[25] and X-ray crystallography,[26] resulting in very similar overall structures. However, 2-D-NMR currently provides a low resolution model which reveals the overall folding pattern of the peptide main chain but with little information about side chain locations. Although clearly not as detailed or accurate as X-ray, such a structure should still be useful for building an initial crude model of a binding site. 2-D-NMR has also been used to study DNA conformation in solution.[24]

20.3.2.3 Protein Structure Prediction Based on Sequence Homology

Many more protein sequences are available than crystal structures, and the gap will continue to grow as DNA-sequencing methods become even faster. Fortunately, protein sequences occasionally show high sequence homology with proteins whose X-ray structure is known, suggesting the possibility of modeling the unknown structure based on the crystal structure of the homologous protein. This has become a popular approach and has been applied to modeling several protein structures, including renin, a target for potential antihypertensive drugs. The approach was reviewed by Blundell *et al.*[27] Sequence homologies of 50% or more are probably required for this approach to be successful; the resulting models in several cases have been accurate to within 1–1.5 Å for the main chain atoms, although side chains are not as well located. The major difficulties are modeling insertion and deletion regions, particularly when they occur in a loop. Jones and Thirup[28] showed that it is possible to fit most secondary structure elements using fragments from other proteins of known structure; this approach should prove useful for building models for insertion and deletion regions and for homology model building in general. They raised the interesting possibility that complete three-dimensional protein structures can be constructed from a linear combination of fragments from other, unrelated protein structures.

20.3.2.4 Protein Structure Prediction Based Solely on Sequence

While there is very strong evidence that the folded structure of most tertiary proteins is dictated by their amino acid sequence, the 'protein folding' problem remains unsolved despite almost 20 years of intensive research.[29] For the majority of protein sequences with little significant homology to known structures, the problem of predicting secondary and tertiary structures accurately enough for drug design applications is still overwhelming and beyond the scope of this chapter. Few predictions of complete tertiary structures have been reported; Cohen *et al.* predicted the tertiary structure of interleukin-2.[30] X-Ray crystallography of this protein is in progress at several laboratories.

20.3.3 SPECIAL GRAPHICS REQUIREMENTS FOR MACROMOLECULAR APPLICATIONS

The complexity and size of macromolecules require sophisticated graphics software and hardware to provide a real time, interactive response along with selective display and manipulation.[31] A macromolecule-modeling system should be capable of simultaneously handling several molecules with several thousand atoms and thousands of molecular surface points in depth-cued color and time-sliced stereo, where each molecule can be individually controlled in three dimensions, while simultaneously monitoring inter/intramolecular distances and adjusting multiple contiguous or noncontiguous torsion angles—all in real time. Dials, joysticks and a mouse are used to translate and rotate molecules and rotate bonds. Several new systems have much faster processors which perform complete bump checking (checking for interatomic contacts closer than van der Waals) and even molecular mechanics energy calculations in real time; this provides excellent feedback during interactive modeling. Selective control of which molecules or portions of molecules (*e.g.* an enzyme active site) are displayed, which distances and torsions are active, *etc.* requires a powerful command language along with interactive 'picking' of atoms and bonds with a mouse or stylus. The ability to display a molecular dynamics simulation by animation (rapidly switching from one saved coordinate set to the next) can be very helpful; dynamics simulations produce a tremendous amount of data which can be difficult to interpret without graphics.

Raster graphics has recently become the dominant technology in interactive molecular modeling. Although raster displays have apparent advantages in providing beautiful 'realistic' color solid-shaded images (assuming a plastic CPK model is realistic), these images cannot be updated fast enough for real time modeling of macromolecules yet, so vector and dot images (on raster displays) still provide the best approach for interactive molecular modeling. The major performance advantage provided by raster over calligraphic displays for interactive modeling is flicker-free display. Vector/dot images have the tremendous additional advantage of providing full transparency and clipping while displaying a complex, color-coded molecular surface and bonds, which are essential for studying interactions deep inside a macromolecular binding site.[31] Time-sliced stereo, where the left and right eye views are alternately displayed every 1/30 s and viewed through a mechanical shutter or liquid crystal glasses synchronized to the display, provides a very convincing three-dimensional illusion and is extremely helpful for modeling complex interactions. A recent stereo viewing innovation is to place a liquid crystal screen over the graphics scope, allowing the user(s) to wear ordinary circularly-polarized plastic glasses.

The simultaneous development of real-time interactive color graphics[31] and Connolly's molecular surface program[32] in 1980 revolutionized macromolecular computer graphics modeling. Connolly's original program implemented Richard's definition[33] of molecular surface by rolling a probe sphere (usually 1.4 Å radius, the effective radius of a water molecule) over the surface of the macromolecule, resulting in a smooth surface which represents the surface accessible to a water molecule, including internal cavities (Figure 1a). Bash *et al.*[34] and Pearl and Honegger[35] independently developed very fast van der Waals dot surface programs which are several orders of magnitude faster than Connolly's molecular surface program, although they are not as effective at eliminating buried surfaces and produce a more complicated surface display (Figure 1b). A combination of the two surfaces provides a good compromise, where the molecular surface is calculated for the protein and the nearly instantaneous van der Waals surface is used for the ligand and any side chains which may be adjusted in the protein. Connolly also developed an analytical method for calculating molecular surface,[36] which provides nearly exact values for the area and volume[37] enclosed by a surface along with spectacular shaded raster graphics images,[38] which give a very different 'feel' for a macromolecular surface than the conventional CPK-like raster surfaces.[39] Static shaded-image raster graphics is also used for producing descriptive pictures of protein secondary structure,[40] similar to the original hand-drawn illustrations using cylinders (to represent α-helices) and arrows (for β-sheets) by Richardson.[41] Interactive vector graphics approaches for illustrating secondary structure have also been described.[42,43]

Color-coded molecular surfaces can provide qualitative or quantitative displays of hydrophobic and hydrophilic regions, neutral and charged amino acid side chains, electrostatic potential and mobility (based on X-ray crystallographic refinement or molecular dynamics simulation). Color coding by hydrophobicity and electrostatic potential are particularly useful in drug design applications, where the goal is to design a molecule which is complementary in shape, hydrophobicity and charge to a binding site. Hydrophobic color coding was originally done simply by coloring all surface points associated with carbon 'hydrophobic' (*e.g.* red) and all nitrogen and oxygen surface points 'hydrophilic' (*e.g.* blue); a more detailed approach[44] includes 'neutral' or 'semihydrophilic' surface (*e.g.* yellow) for sulfur, α-carbons of amino acids, the carbon between the imidazole nitrogens in histidine and carbonyl carbon (Figure 1). Electrostatic potential molecular surfaces[45] are usually calculated using quantum mechanically derived partial atomic charges for each atom.[46] The potential is typically calculated one probe sphere radius above the molecular surface, which should be a reasonable estimate of what an incoming ligand 'sees' as it approaches the macromolecule. The molecular surface is then color coded by the value of the electrostatic potential at each point. The electrostatic potential gradient can also be displayed graphically, where the gradient at each point

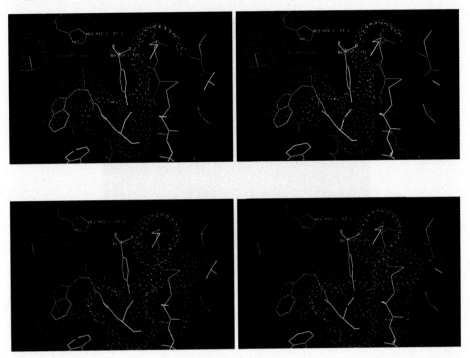

Figure 1 Molecular (a) and van der Waals (b) surfaces of chymotrypsin–tosyl inhibitor complex. The surface is color coded by hydrophobicity as described in the text: red = hydrophobic, blue = hydrophilic, neutral = yellow. The tosyl group is covalently attached to the side chain hydroxyl of Ser-195. The 'catalytic triad' of His-57, Asp-102 and Ser-195 is shown in green. (The coordinates for this and all other molecules in the following figures are from the *Brookhaven Protein Data Bank*,[3] except where otherwise noted)

on a grid above the molecular surface is displayed as a short vector. This method was used to locate the probable trajectory for superoxide anion as it approaches superoxide dismutase (Figure 2).[47]

Barry introduced the very useful 'extra radius' surface, where the surface is calculated one van der Waals radius beyond the normal surface, collapsing the surface of a binding site on to the vector model of its ligand and eliminating the need for displaying the ligand's surface.[48] This simple graphics trick makes it much easier to visualize the 'docking' of a ligand into a binding site. For example, chymotrypsin's specificity for aromatic amino acid side chains is not immediately apparent from a conventional surface of its active site (Figure 1), while the 'extra radius' surface reveals an almost perfectly planar pocket (Figure 3), which is obviously complementary to an aromatic ring.

20.3.4 LIGAND–MACROMOLECULE 'DOCKING'

The major interactions involved in drug–receptor binding are electrostatic (including hydrogen bonding), dispersion or van der Waals and hydrophobic.[49] Hydrophobic interactions usually provide the major driving force for binding, while hydrogen-bonding and electrostatic interactions primarily provide specificity and often contribute little to the free energy of binding in solution.[50] Hydrogen-bonding and electrostatic interactions are usually *not* the driving force for ligand binding, although analyses of ligand–macromolecule complexes often focus on them. Ligand–macromolecule binding requires desolvation of the free ligand and the free macromolecule and breaking ligand–water hydrogen bonds and macromolecule–water hydrogen bonds. Although many or all of these hydrogen bonds may be reformed between the ligand and macromolecule if the two are highly complementary, the net change in free energy is often close to zero or only slightly negative. Macromolecular hydrogen-bonding groups which are shielded from solvent provide the best targets, since they have little hydrogen-bonding competition from solvent and therefore greatly stabilize the ligand–macromolecule complex relative to unshielded groups. Hydrogen-bonding mismatches destabilize the ligand–macromolecule complex relative to the free ligand and macromolecule, reducing the free energy of binding. Binding affinity is therefore increased primarily by optimizing hydrophobic and van der Waals interactions, which can be achieved to a first approximation by maximizing shape complementarity between the drug and its receptor, while simultaneously

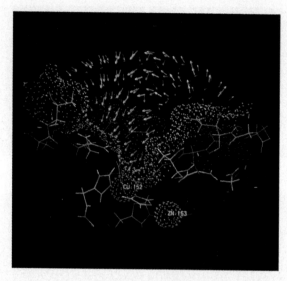

Figure 2 The deep active site channel of Cu, Zn superoxide dismutase with electrostatically color-coded molecular surface and electrostatic field vectors showing the guidance of the anionic substrate superoxide ($O_2^- \cdot$) to the catalytic site. Both the molecular surface and the field vectors are color coded by electrostatic potential: red, $< -88\ \mathrm{kJ\,mol^{-1}}$; yellow, -88 to $-29\ \mathrm{kJ\,mol^{-1}}$; green, -29 to $+29\ \mathrm{kJ\,mol^{-1}}$; cyan, $+29$ to $+88\ \mathrm{kJ\,mol^{-1}}$; blue, $> +21\ \mathrm{kJ\,mol^{-1}}$ (adapted from ref. 47)

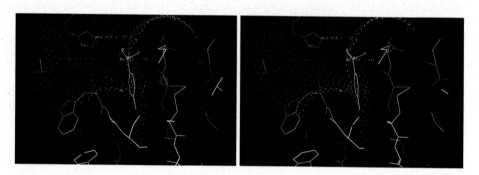

Figure 3 Extraradius surface, color coded by hydrophobicity, for chymotrypsin–tosyl inhibitor complex for comparison with Figure 1

ensuring that hydrogen bonding and electrostatic complementarity are maintained to ensure specificity. Fersht *et al.*,[51] who used site-directed mutagenesis to remove specific hydrogen-bonding groups from the enzyme in the tyrosyl tRNA synthetase–substrate complex, and Street *et al.*,[52] who removed specific hydrogen-bonding groups from the ligand in the glycogen phosphorylase–glucose complex, found that neutral hydrogen bonds contributed only 2–$6\ \mathrm{kJ\,mol^{-1}}$ to the free energy of binding and that ionic hydrogen bonds contributed up to $16\ \mathrm{kJ\,mol^{-1}}$. Bartlett and Marlowe[53] determined a possible upper limit for intrinsic hydrogen-bonding energy of $16.7\ \mathrm{kJ\,mol^{-1}}$ for an unusually favorable phosphoramidate N—H hydrogen bond to a carbonyl oxygen on the peptide backbone of thermolysin. Hydrogen bond lengths range from about $2.6\ \text{Å}$ (distance between heteroatoms) for ionic interactions to $2.8\ \text{Å}$ (strong) and $3.2\ \text{Å}$ (weak) for neutral interactions.

Several other themes occur regularly in protein structure and protein–ligand complexes and represent other stabilizing interactions. The α-helix dipole is well known and can be described as having partial charges of $+1/2$ at its N terminus and $-1/2$ at its C terminus; this is a subtle but important effect.[54] The N-terminal end of an α-helix is often involved in anion (*e.g.* phosphate) binding sites. Another common observation is the nearly perpendicular interaction of aromatic rings in proteins and protein–ligand complexes, contrary to the often assumed based stacking type of interaction.[55] These concepts are useful to keep in mind when modeling ligand–macromolecule interactions or when designing a new ligand. Excellent references are available in the texts by Creighton,[56] Cantor and Schimmel,[57] Schulz and Schirmer,[58] Fersht,[22] and in ref. 4.

'Docking' is typically done interactively with molecular surface displays (*e.g.* 'extra radius' surface) used to guide the fit, based on hydrophobic or electrostatic potential color coding. The binding site is usually treated as completely rigid initially, while the conformation of the ligand is adjusted interactively. Physically impossible models are easily built with current systems, which allow molecules and atoms to collide and pass through each other; the visual cues provided by molecular surface displays are essential for realistic modeling. Recent systems are fast enough to calculate simple molecular mechanics energies in real time while docking (future systems may use this information to provide feedback and prevent collisions or high energy conformations). High energy contacts can be highlighted with color-coded vectors.[59] While this is a great help, it provides only a crude estimate of the enthalpy of interaction—not free energy (solvation energy is often critical in determining the free energy of binding; inclusion of solvation effects is several orders of magnitude beyond what is possible with interactive modeling). Even with slow hardware it is possible to rigid-body minimize[60] the interaction energy of a typical ligand–macromolecule interaction with respect to translation and rotation of the ligand (six degrees of freedom) in about 30 s. Interactive docking typically alternates between continuous motion, possibly with real time updates of the interaction energy, and periodic cycles of simple rigid-body (plus selected torsion angle) energy minimization to clean up the initial fit. Future modeling systems (which should be available by press time) will be fast enough to perform this minimization with a completely flexible ligand (and possibly a flexible protein) in close to real time. Finally, energy minimization of the entire complex in Cartesian coordinates, where all atoms are allowed to relax (which requires large amounts of computer time and cannot be performed interactively with current systems), provides a good indication of the plausibility of the model and a rough estimate of the relative interaction enthalpy of the candidate drug. Ionic interactions and hydrogen bond energies are usually overestimated in the typical gas phase calculation (even if 'corrected' by using damped partial atomic charges or a distance dependent dielectric constant) due to the omission of solvent hydrogen-bonding competition; these effects are treated properly in the free energy perturbation theory method described below. A distance dependent dielectric constant is often used for calculations which do not explicitly include solvent; the dielectric constant is set proportional to the interatomic distance.[61]

Conventional energy minimization with this many degrees of freedom is easily trapped in local minima and can give deceptive results; energy minimization rarely produces a structure which is significantly different from the starting coordinates. Short molecular dynamics simulations[23] (10–20 ps) (see Chapter 18.3) are much better at escaping local minima and can give much lower energy structures; a good strategy is to begin with a short dynamics run and follow it with energy minimization. Such short dynamics simulations contain no meaningful information about the actual motions of the structure (up to 30 ps may be required just for thermal equilibration); they simply provide a more efficient method of energy minimization and a good indication of the stability of the model (poor models tend to fly apart very quickly).

Multiple binding modes are often possible, as illustrated by the X-ray structure of the elastase–acetyl-Ala-Pro-Ala complex in which the ligand, the product from the elastase-catalyzed hydrolysis of acetyl-Ala-Pro-Ala-*p*-nitroanilide, is bound backwards relative to the established mode of productive binding.[62] It is very difficult with interactive methods to be confident of finding the most likely binding mode candidates. Naruto *et al.*[63] used a systematic search procedure to find potential chymotrypsin tetrahedral intermediate conformers given a covalent bond linking the ligand with the site and Wodak *et al.*[64] used a similar approach to find the most favorable conformation of glutathione–cysteine in glutathionyl hemoglobin. Kuntz *et al.*[65a] developed a more general docking method for rigid ligands based on a fast sphere-matching algorithm; the approach was recently extended to flexible ligands by docking each rigid fragment of the ligand (fragments between rotatable bonds) independently.[65b] Blaney and Crippen[66a] used a distance geometry-based method to generate random sterically allowed dockings of a flexible ligand into a rigid site. Billeter *et al.*[66b] applied a new approach, the ellipsoid algorithm, to enzyme inhibitor docking with encouraging results.

The most significant advance in modeling drug–receptor interactions is the recent application of free energy perturbation theory methods to calculate surprisingly accurate (within $4 \, \text{kJ mol}^{-1}$) differences in the free energy of binding of two closely related analogs to a protein-binding site.[23, 67] This approach takes advantage of the properties of a thermodynamic cycle to simulate a physical process which is very difficult to calculate (the transfer of a drug from solution into a receptor-binding site, compared with the transfer of its analog) by an equivalent nonphysical process (the 'mutation' of a drug into its analog, performed both in solution and in the binding site) which is relatively easy to calculate. This 'mutation' is carried out by gradually changing the properties of the initial drug molecule to the properties of the final drug molecule during a molecular dynamics simulation, which is performed once in 'solution', usually in a box of several hundred water molecules with periodic boundary conditions, and again in the macromolecule. The simulation starts with 100% initial drug character and ends with 100% final drug character; intermediate steps in the simulation have nonphysical hybrid drug molecules. Molecular dynamics generates a statistical mechanical ensemble average at each point along the simulation as the properties of the initial molecule are varied. These simulations require large amounts of supercomputer time; the method is described in detail in Chapter 18.3.

20.3.5 DESIGNING LIGANDS TO FIT A SPECIFIC MACROMOLECULAR SITE

The ability to model both small organic molecules and macromolecules in the same system is essential; several of the systems currently available were originally designed for handling the regular, repeating polymeric structure of proteins and nucleic acids and deal rather poorly with the more arbitrary structures found in small organic molecules. Other systems were initially designed for modeling small molecules and do not handle macromolecular structures well. Only a few systems come close to combining the best of macromolecular and small molecule modeling, providing the ability to interactively design and build potential ligands *directly into* a macromolecular binding site.[23]

Full interactive control over the position (translation and rotation along the *x*, *y* and *z* coordinate axes) and conformation (adjustment of torsion angles) in both the macromolecule and the ligand(s) should be simultaneously available. Good torsion angle adjustment facilities are essential, since this is usually where most of the time is spent in interactive modeling. Simultaneous control of six to eight torsion angles should be possible, where the torsions may be spread over several residues or even several molecules. It should be possible to adjust both ends of a rotatable bond without having to continually reassign the bond rotation. The system should be capable of handling several molecules simultaneously to enable the comparison of different ligands in the binding site or different fits of the same ligand. Molecular surface displays (dot surfaces) should associate a set of dots with each atom, so that the dots move together with the atom as the molecule is moved or during bond rotations. Solvent accessible molecular surface calculation[32] requires at least several minutes and usually must be precalculated; the system should support reading a precalculated surface correctly into the current modeling session even when the stick model associated with the surface has been translated and rotated from its original position. The system should also include the option to calculate fast van der Waals surfaces,[34, 35] which are useful for surfacing small molecules and small portions of a macromolecule. Additional useful features include the ability to read new molecules into the current modeling session at any time (not just the beginning of the session) and to save individual molecules at any time. Since the conformation of the macromolecule is usually not changed during initial modeling, it is unnecessary to store its updated coordinates along with each saved 'docked' ligand; this eventually results in confusion due to the accumulation of multiple copies of the same macromolecule coordinate set saved in different orientations relative to the screen. It is much more convenient to always store each 'docked' ligand conformation in an orientation relative to the *initial* macromolecule coordinates (some systems provide an automatic facility for this), so that only one saved copy of the macromolecule is necessary. Finally, a facility to associate arbitrary three-dimensional graphical objects with individual molecules is very useful; such objects could be electrostatic potential maps, molecular orbital plots, electron density maps, *etc.*

The best approach is usually to design and build the developing ligand piece by piece in the binding site by combining preformed three-dimensional fragments from a library. The library may contain several hundred different ring systems, chains and functional groups, which should be conveniently selectable from within the modeling system. Small molecules can be built very rapidly in this way, and the resulting structures are usually accurate enough for initial fitting or 'docking' into the site model.

Computer graphics enables us to qualitatively visualize drug–receptor interactions and molecular mechanics can provide rough estimates of the interaction energy, which allow us to design molecules that are apparently complementary to a binding site. For close analogs this can be sufficient to both rationalize the relative activities of a series of analogs and design new analogs; in particularly fortunate cases the X-ray structure of a suitable lead compound complexed with the receptor is available and new analogs can be docked by analogy with the lead compound. An integrated approach combining recent developments in molecular modeling with over 20 years of QSAR development[68] has proven to be especially powerful for this application (described in Section 20.3.6), since the QSAR can help differentiate between possible binding modes and conformations by revealing the physical nature of the surface surrounding each substituent. We have much less experience in the *de novo* design of novel molecules (without a lead compound complexed in an X-ray structure with its receptor). The design by Beddell *et al.* of 2,3-diphosphoglycerate mimics[69] and antisickling compounds[70] based on the hemoglobin X-ray structure is still one of the best examples of this approach, despite the fact that this work was done with mechanical wire models!

All of the approaches we have described so far are analytical and oriented towards modeling known structures. Even the powerful free energy perturbation methods currently require a *known* lead compound to base subsequent modeling on. Where do the structures of novel candidate drugs come from? Actual molecular structure design is still a formidable challenge dependent on the creativity, ingenuity and experience of the medicinal chemist. Once we overcome the initial challenge of *how* to model a macromolecular-binding site with computer graphics, we are faced with the much greater challenge of *what* to put in the site and where to put it. Goodford developed a simple molecular mechanics-based approach for calculating optimal ligand atom locations in a binding site,[71] which is an important step. The method calculates the molecular mechanics interaction energy for each of a variety of probes (*e.g.* hydroxyl oxygen, carbonyl oxygen, carboxyl oxygen, amide nitrogen, amine nitrogen, *etc.*) at each point on a three-dimensional grid superimposed on the binding site. Solvation is considered in a simple way by modifying the dielectric constant based on the solvent accessibility at each grid point. The grid is then contoured by energy and the resulting contours are graphically displayed (as color-coded contour maps or dot clouds) in the binding site. The contours indicate predicted 'hot spots' where a ligand atom of a given type should prefer to bind (Figure 4). Unfortunately it is not a simple matter to connect each of these 'hot spots' together into a single molecule, but the method does provide useful visual clues for structure design. It is extremely difficult to design a three-dimensional structure which contains the majority of the predicted optimum atoms in a low energy conformation which simultaneously places these atoms near the calculated locations, and which in addition has a reasonable predicted log P, probable metabolic stability and can be synthesized in a reasonable amount of time.

Although it is tempting to speculate that all the information required for the design of an optimal ligand is present in the high resolution structure of a binding site, no computational approaches exist yet for complete *de novo* design. State of the art design techniques combine Goodford's probe method (or related approaches) with the other previously described interactive methods, where the medicinal chemist fits a variety of organic fragments in a trial and error fashion into the site, attempting to eventually combine the fragments into a complete molecule. There is no systematic method to lead to an optimum design; very different, apparently reasonable designs are found by different researchers. The docking approaches of DesJarlais *et al.*,[65b] Blaney and Crippen,[66a] or Billeter *et al.*[66b] may eventually be able to achieve this, by docking fragments from a large library and then combining the fragments into complete molecules.

20.3.6 APPLICATIONS

We have chosen the following examples to highlight the variety of structures and methods which have been used in recent studies; our list is necessarily brief and unfortunately must omit several other interesting studies.

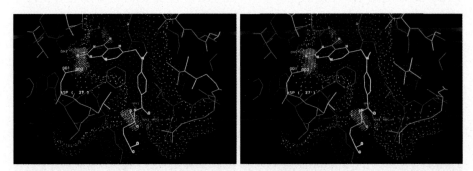

Figure 4   Probe map of *E. coli* dihydrofolate reductase–methotrexate complex. The calculated minimum energy positions for a NH_3^+ probe (red) and CO_2^- probe (cyan) closely match the experimental positions for the pteridine amino groups and the carboxyl group of methotrexate (adapted from ref. 71)

20.3.6.1 Drug–DNA Interactions

The X-ray crystal structures of several drugs bound to DNA have been solved: actinomycin D,[72] daunomycin,[73] triostin A,[74] *cis*-diaminedichloroplatinum(II),[75] and netropsin.[76] The X-ray crystal structures of several DNA-binding proteins have also been solved: catabolite activator protein (CAP),[77] lambda repressor,[78] catabolite repressor operon (CRO),[79] Trp repressor,[80] and the restriction enzyme ECO-R1 in complex with its target DNA sequence.[81] As more drug–DNA and protein–DNA structures are solved, the challenge of exploiting the information in them for the design of new sequence selective DNA-binding drugs has become increasingly important. Current DNA-binding drugs inhibit DNA replication and/or transcription and have important clinical uses as antibiotics and anticancer agents, but they also have severe side effects, which are related to their poor sequence selectivity. These compounds recognize only a few DNA base pairs at best; selective recognition of up to 17–18 base pairs may be required for a selective DNA-binding drug, based on the probability of multiple random occurrence of the same sequence in the human genome.[82]

20.3.6.1.1 Netropsin

Kopka *et al.*[76] solved the X-ray crystal structure of netropsin (**1**) complexed with CGCGAATT^{Br}CGCG at 2.2 Å resolution. Netropsin binds to the minor groove of double-helical B-DNA (Figure 5) and is selective for four or more A–T (adenine–thymine) base pairs; a single G–C (guanosine–cytosine) pair prevents binding. Netropsin's amide NH groups hydrogen bond to adenine and thymine in the minor groove, replacing the water molecules which form a 'hydration spine' along the minor groove in native DNA. While these hydrogen bonds determine the orientation and location of netropsin in the minor groove, the A–T specificity is due to the steric hindrance of the N2 amino group on guanine, which would collide with netropsin's pyrrole CH groups. This observation led to modeling[83] which showed that replacing a pyrrole by imidazole should favor G–C recognition, since the imidazole nitrogen can hydrogen bond with the guanine N2 amino group. Kopka *et al.*[83] suggested that this approach could provide synthetic 'lexitropsins' selective for any short sequence of DNA (the amide—heterocycle unit of netropsin cannot be repeated indefinitely since its repeat distance does not exactly match B-DNA).

(1)

20.3.6.1.2 Actinomycin D

The study by Lybrand *et al.*[84] of DNA–actinomycin D interactions combines molecular graphics, molecular mechanics and 2D-NMR. They built models of intercalation sites into several deoxy-

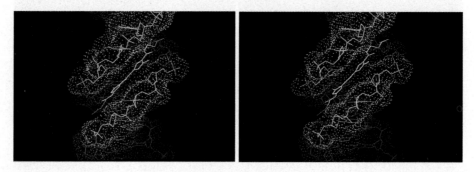

Figure 5 Netropsin (**1**; red) bound to the minor groove of DNA (blue)

hexanucleoside fragments with different sequences, followed by interactive computer graphics docking of actinomycin D into each of the intercalation sites and molecular mechanics energy minimization. Their results were consistent with actinomycin D selectivity for binding on the 3'-side of guanine, due to hydrogen bonding between threonine side chains in the cyclic pentapeptide portion of actinomycin D and guanine. Their intercalation structure, which was model built and energy minimized without using any NMR data, was compared with the 214 NOE distances in the complex observed by NMR. Good qualitative agreement between the model-built structure and the experimental NOE distances suggests that the model is reasonably close to the solution structure of the complex.

20.3.6.2 Rhinovirus: An Antiviral Drug Receptor

Smith *et al.*[12] solved the structure of human rhinovirus 14 (HRV14), one of about 100 known rhinovirus serotypes which cause the common cold, complexed with the structurally related antivirals WIN 51711 (**2a**) and WIN 52084 (**2b**) at 3 Å resolution. These compounds inhibit viral replication by preventing uncoating of the viral RNA. This is the first description of a drug–virus complex at atomic resolution. HRV14 is a member of the picornavirus family, which includes polio virus and mengo virus, whose structures have also been solved by X-ray crystallography.[13, 14] The WIN drugs bind in an extended conformation to a very deep binding site (Figure 6), with the oxazoline end of the drugs in a hydrophilic area, probably with a hydrogen bond from the oxazoline nitrogen to an asparagine side chain. The isoxazole end binds to a buried hydrophobic pocket with no apparent hydrogen bonds to the heteroatoms of the isoxazole ring. In fact, the only proposed hydrogen bonds involve the oxazoline nitrogen and possibly a weak hydrogen bond to the ether oxygen, leaving the other polar atoms buried in apparently hydrophobic areas. The (4*S*)-methyl-oxazoline of WIN 52084 is 10 times more active than the (4*R*) enantiomer, which is attributed to a hydrophobic pocket available to only the (4*S*) position. These structures should provide a good starting point for the design of new antivirals selective against different picornaviruses due to their sequence and structural homology.

(2a) R = H
(2b) R = Me

20.3.6.3 Hemoglobin: Design of Antisickling Compounds

Beddell *et al.*[69] published a remarkable paper in 1976 in which they described the design of a simple but structurally novel compound which mimicked the action of 2,3-diphosphoglycerate (DPG; **3**) on hemoglobin by selectively binding to deoxyhemoglobin and promoting oxygen release.

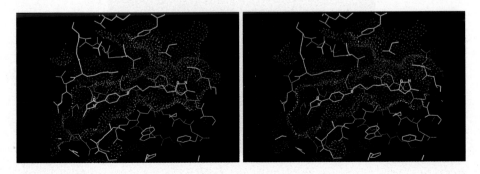

Figure 6 Human rhinovirus 14 (color coded by hydrophobicity as in Figure 1) complex with antiviral WIN 52084 (**2b**; green)

This structure was designed using Kendrew wire models of the DPG–deoxyhemoglobin complex and is the first example of the successful *de novo* design of a small molecule based on the three-dimensional structure of a specific receptor site. The carboxyl and phosphate groups of DPG bind to the site through ionic interactions with the amino termini of the β_1 and β_2 subunits and lysine and histidine side chains. Beddell *et al.* found that bibenzyl-4,4'-dicarbaldehyde (**4**) nicely spanned the site and should be able to react with the amino termini by Schiff base formation (Figure 7). They added the OCH_2CO_2H group at the 2-position to provide additional interaction with a lysine side chain and converted the formyl groups into their bisulfite adducts to increase solubility. This compound was as active as DPG and indirect NMR evidence[85] was consistent with its designed binding mode. While this work demonstrated the feasibility of receptor-based drug design, there is no therapeutic value in molecules which mimic DPG by shifting the hemoglobin oxygen dissociation curve to the right.

(3)   (4)

Sickle cell anemia is characterized by aggregation of deoxyhemoglobin into large insoluble fibers, resulting in the characteristic sickle erythrocyte cell shape. Beddell *et al.* followed their original work with the design of DPG antagonists which stabilize oxyhemoglobin,[70] thereby shifting the oxy–deoxyhemoglobin equilibrium toward oxyhemoglobin and reducing sickling. This is the first description of a drug designed *de novo* by receptor-based molecular modeling which has reached clinical trials. The DPG site is nearly collapsed in the oxy conformation, so design of a compound to bind selectively to oxyhemoglobin at this site was not possible. Beddell *et al.* observed that another potential binding site existed at the amino termini of the α_1 and α_2 subunits, with the amino groups 20.7 Å apart in the deoxy form and 12.4 Å apart in the oxy form, suggesting that it might be possible to design compounds selective for the oxy form by placing substituents which could only interact with both amino groups at the shorter distance. Beddell *et al.* designed 5-(2-formyl-3-hydroxy-phenoxy)pentanoic acid (**5**) to interact with the two amino groups by forming a Schiff base and a salt bridge; the hydroxy group *ortho* to the formyl was included to promote Schiff base formation (Figure 8). This compound is active in shifting the oxygen dissociation curve to the left and was also found to be a potent antisickling agent *in vitro*. Although attempts to crystallize the drug–hemoglobin complex have failed, the binding of this compound to its intended site has been confirmed by borohydride reduction of the Schiff base formed on drug–oxyhemoglobin binding followed by

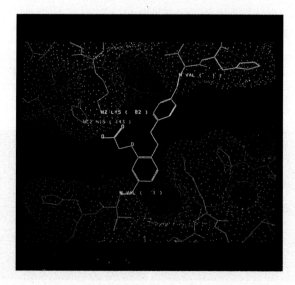

Figure 7 β_1 and β_2 subunits of human deoxyhemoglobin with model for binding of bibenzyl-4,4'-dialdehyde (4; yellow) through Schiff base formation with the amino termini of each subunit (adapted from ref. 69)

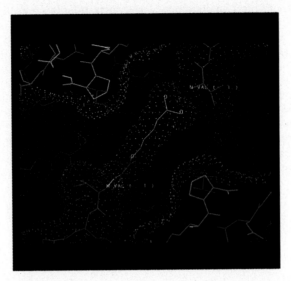

Figure 8 α_1 and α_2 subunits of human oxyhemoglobin with model for 5-(2-formyl-3-hydroxyphenoxy)pentanoic acid (**5**) forming a Schiff base with the α_1-amino terminus and a salt bridge with the α_2-amino terminus (adapted from ref. 70)

tryptic digestion and localization of the covalently bound drug to the α_1-terminal amino group;[86] additional weaker noncovalent and Schiff base binding at other sites was also found.

$$\text{HO} \quad \text{CHO} \quad -\text{O(CH}_2)_4\text{CO}_2\text{H}$$

(**5**)

20.3.6.4 Dihydrofolate Reductase: Design of Antibacterial Compounds

Dihydrofolate reductase (DHFR) has been intensively studied since the discovery of methotrexate and many other potent DHFR inhibitors in the late 1940s and 1950s. DHFR inhibitors are used as antibiotics, antimalarials and anticancer agents due to the crucial role DHFR plays in thymidylate and purine biosynthesis. The X-ray crystal structures of several different DHFRs have been solved by Kraut and coworkers:[4, 87] *E. coli, L. casei,* chicken liver and most recently a trimethoprim-resistant DHFR from *E. coli* R-plasmid R67. These crystal structures provided the first look at atomic resolution of a real drug–receptor complex and stimulated renewed interest in the structure–activity relationships and design of DHFR inhibitors.

Kuyper *et al.*[88] modeled the binding of the bacterial DHFR selective antibiotic trimethoprim (**6a**) using the X-ray structure of the *E. coli*–methotrexate complex. They assumed that the pyrimidine

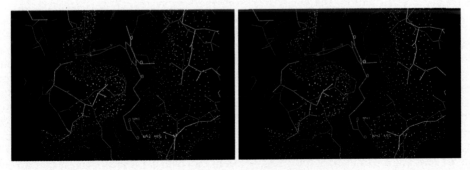

Figure 9 *E. coli*–dihydrofolate reductase complex with 3'-carboxyalkoxytrimethoprim analog (**6b**), shown interacting with Arg-57 (adapted from ref. 88)

ring of trimethoprim would bind analogously to the corresponding pteridine ring of methotrexate (their model was later shown to be qualitatively correct by X-ray crystallography of the *E. coli*–trimethoprim complex). Based on their model for trimethoprim binding, they designed new trimethoprim analogs (*e.g.* **6b**) by adding 3′-carboxyalkoxy groups targeted to interact with an arginine side chain at the entrance of the active site (Figure 9). The arginine guanidium group is almost completely buried in the active site surface and is probably poorly hydrated, suggesting that an ionic interaction between the inhibitor and this arginine might be unusually strong. The binding mode of these new trimethoprim analogs was verified crystallographically; the best compound was 55 times more active than trimethoprim *in vitro* against *E. coli*–DHFR. This work clearly demonstrated the potential for rational design of analogs with greatly improved *in vitro* activity. Unfortunately these structures showed poor *in vivo* antibacterial activity, possibly due to the highly polar carboxy group which may prevent entry into bacteria.

(6a) R = Me
(6b) R = $(CH_2)_5CO_2H$

20.3.6.5 Phospholipase-A$_2$: Design of Potential Antiinflammatory Compounds

Phospholipase-A$_2$ (PLA$_2$) catalyzes the hydrolysis of phospholipids, releasing arachidonic acid as one of the products. PLA$_2$ occupies a critical position in the arachidonic acid biochemical pathway, which leads to the leukotrienes and prostaglandins, key mediators of inflammation. PLA$_2$ inhibition should therefore produce an antiinflammatory response. Potent PLA$_2$ inhibitors have been designed by a classical approach based on the structure of the substrate by making nonhydrolyzable substrate analogs.[89] There are few other examples of potent, selective PLA$_2$ inhibitors, while there are no unequivocal examples of *in vivo* antiinflammatory activity due to PLA$_2$ inhibition. The antiinflammatory corticosteroids are believed to elicit their response by stimulating the biosynthesis of a controversial potent PLA$_2$-inhibiting protein (lipocortin);[90] lipocortin is poorly characterized and claims of antiinflammatory activity have been difficult to reproduce. The X-ray crystal structures of native bovine[8] and porcine pancreatic[91] PLA$_2$ and rattlesnake[92] (atrox) PLA$_2$ are very similar, which is not surprising due to the high sequence homology shown by the majority of (soluble) PLA$_2$s and the seven disulfides common to each structure. It is not yet clear whether the antiinflammatory relevant PLA$_2$ is cytoplasmic or membrane bound and no structural information is available for a macrophage or inflammatory cell PLA$_2$. Although pancreatic PLA$_2$ itself is not involved in inflammation, it may be a good model for the PLA$_2$ present in an inflammatory cell.

In one of the few successful reported attempts of *de novo* enzyme inhibitor design, Ripka *et al.* designed PLA$_2$ inhibitors based on the X-ray crystal structures of the pancreatic and atrox PLA$_2$s.[93] PLA$_2$ inhibitor design presents an unusual challenge due to its unusually large hydrophobic active site, with few hydrogen-bonding groups to offer potential specific points of attachment. After considering several structures, the benzylacenaphthene ring system (**7**) was selected, based on its excellent fit into the active site, limited conformational freedom and opportunities for adding a variety of different substituents to optimize activity. Hydrogen-bonding substituents were added to the *meta* position of the benzyl group to interact with His-48 in the active site (Figure 10). Several benzylacenaphthenes were designed and synthesized; they showed high *in vitro* PLA$_2$ inhibitory activity (10^{-7} M–10^{-8} M) based on release of ^{14}C-labeled arachidonate from a phospholipid substrate. These inhibitors have a well-defined and specific structure–activity relationship with respect to substitution at several positions, consistent with competitive binding to the PLA$_2$ active site, although there is no physical or kinetic evidence yet to prove this. These inhibitors show no *in vivo* antiinflammatory activity, possibly due to poor absorption and/or distribution (which might be anticipated based on their high lipophilicity), metabolism, or unsuitability of pancreatic PLA$_2$ as a model for relevant inflammatory cell PLA$_2$—recent biochemical results suggest that macrophage PLA$_2$ may be considerably different from previously described PLA$_2$.[94] Despite the lack of *in vivo* success, these results provide additional evidence that *in vitro* enzyme inhibition with novel

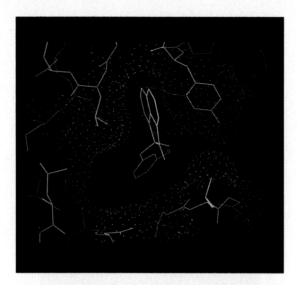

Figure 10 Bovine phospholipase-A$_2$ with model of benzylacenaphthene binding (7; R = OH). The active site molecular surface is color coded by hydrophobicity as in Figure 1

structures unrelated to the natural substrate can be achieved by a rational molecular-modeling approach based on the X-ray crystal structure of the enzyme.

(7)

20.3.6.6 Thermolysin: Comparison of Closely Related Inhibitors

Bartlett and Marlowe[53] designed a series of five phosphonamidate analogs (**8a**) of the peptide carbobenzoxy-Gly-Leu-X (X = NH$_2$, Gly, Phe, Ala, Leu), where a —PO$_2^-$—NH— group replaces the Gly-Leu peptide bond, and showed that these compounds were potent transition-state analog inhibitors of the zinc endopeptidase thermolysin. They also synthesized the corresponding phosphonate analogs, where the —NH— is replaced by —O— (**8b**). The phosphonates were uniformly 840-fold less active than the corresponding phosphonamidates, corresponding to a 16.7 kJ mol^{-1} difference in binding free energy, a remarkably large difference for such a small structural variation. X-Ray crystallography of a phosphonamidate–thermolysin complex and the corresponding phosphonate–thermolysin complex at 1.6 Å resolution[95] showed the two inhibitors bind identically to the active site and that the phosphonamidate NH hydrogen bonds to the carbonyl oxygen of Ala-113 (Figure 11). The corresponding phosphonate oxygen is in the same position, but is unable to donate a hydrogen bond. A molecular dynamics free energy perturbation simulation[67] comparing the phosphonamidate and phosphonate in solution and in the enzyme active site found a free energy difference of 17.6 ± 2.2 kJ mol^{-1}, remarkably close to the experimental value. The calculated solvation free energy difference was 14.4 kJ mol^{-1}, favoring the less hydrophilic phosphonate, which is more easily desolvated than the phosphonamidate, and the calculated difference in enzyme-binding energy was 32.0 kJ mol^{-1}, favoring the phosphonamidate due to its strong NH—Ala-113 hydrogen bond. These studies elegantly demonstrate the compensating effects of solvent–inhibitor and enzyme–inhibitor hydrogen bonding and provide the clearest example yet of a complete study of rational enzyme inhibitor design, followed by X-ray crystallographic determination of inhibitor binding and accurate free energy simulations to quantify and pinpoint the differences in activity between inhibitors.

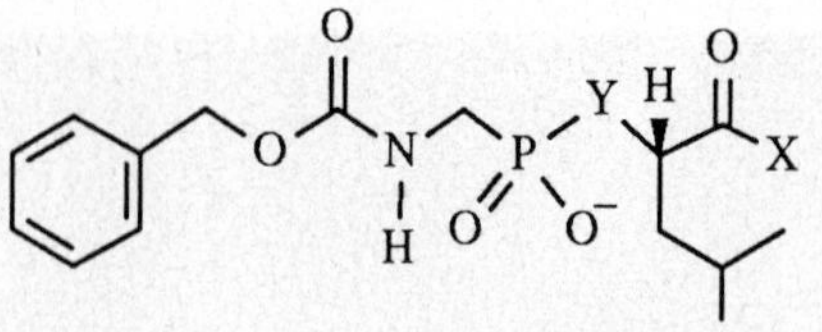

(8a) Y = NH
(8b) Y = O

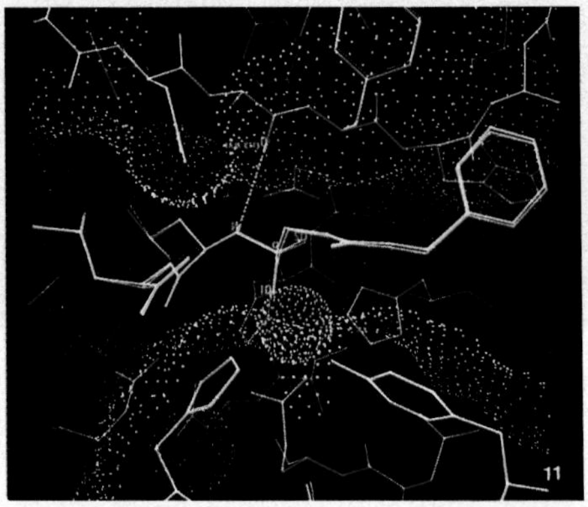

Figure 11 The complex between thermolysin (red) and phosphonamidate (**8a**; red) is contrasted with the thermolysin (blue)–phosphonate (**8b**; green) complex. The magenta and yellow lines correspond to overlapping regions of the two complexes. The blue dots represent the molecular surface of the active site, and the magenta sphere shows the catalytic Zn ion. The Ala-133 C=O · · · (**8a**) hydrogen bond is shown by the dotted yellow line (reproduced by permission of the AAAS from ref. 95)

20.3.6.7 Prealbumin: Modeling of Thyroid Hormone, Dioxin and PCB Analog Binding

Prealbumin is a transport protein for thyroid hormones which may be a good model for the nuclear thyroid hormone receptor. Blaney *et al.*[96] used a qualitative computer graphics molecular-modeling approach in their work on modeling thyroid hormone–prealbumin interactions and predicting thyroid hormone analog-binding affinities to prealbumin. They used interactive graphics with molecular surface displays to model the interaction of a variety of thyroid hormone analogs with prealbumin based on the 1.8 Å resolution X-ray crystal structure of the thyroxine–prealbumin complex. The molecular surface of the prealbumin thyroid hormone-binding site revealed six unusually well-defined pockets capable of binding substituents on thyroid hormones; Blaney *et al.* found that binding affinity correlated qualitatively with the number of pockets which were filled, presumably due to increasing surface complementarity promoting stronger van der Waals or hydrophobic interactions. L-Thyroxine (L-T_4; **9a**) and other high affinity analogs filled four pockets and a crystallographically well-defined water molecule filled a fifth pocket, leaving one empty pocket (Figure 12). This suggested that analogs which could occupy this last pocket should have increased binding affinity. The relative binding affinities of a series of α-naphthyl thyroid hormone analogs were predicted, based on modeling the number of pockets filled by each analog and their potential steric 'collisions' with the binding site. The qualitative predictions were confirmed experimentally, showing that the simple concept of surface complementarity can be extremely useful for predicting relative binding affinity and the design of new compounds to fit a receptor site.

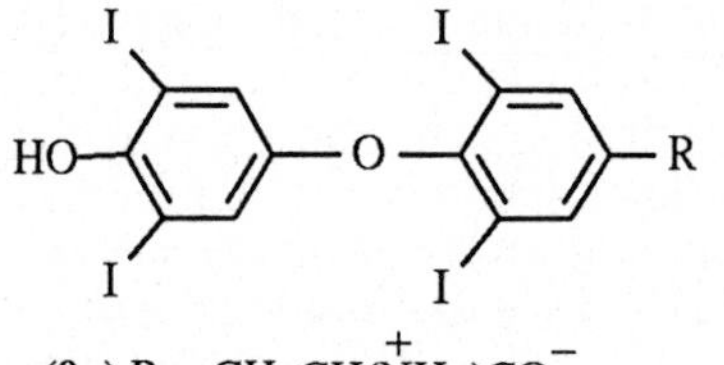

(9a) R = $CH_2CH(\overset{+}{N}H_3)CO_2^-$
(9b) R = $CH_2CH_2CH_2OH$

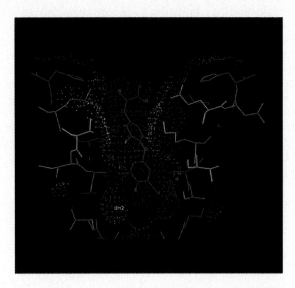

Figure 12 Prealbumin complex with L-thyroxine (**9a**). The empty binding pocket is highlighted in yellow and a bound water molecule is shown in green (adapted from ref. 96)

Blaney *et al.*[61] also used molecular mechanics to attempt to provide a more quantitative understanding of thyroid hormone–prealbumin structure–activity relationships. Calculations comparing L-T_4, D-T_4, deamino-T_4 and decarboxy-T_4 showed that a major factor in determining the free energy of binding was desolvation of the charged thyroxine side chain, which was estimated from the gas phase — aqueous phase free energies of transfer of methylamine, acetic acid and glycine (as models for the side chains of decarboxy-T_4, deamino-T_4 and T_4, respectively; the free energy perturbation theory method was not yet developed at the time of this work). Although previous thyroid hormone structure–activity research focused on the apparent need for a negatively charged carboxyl-containing side chain or a zwitterionic amino acid side chain, these results suggested that a neutral side chain might also produce high binding affinity, since a neutral side chain would be much more easily desolvated. T_4 with a neutral hydroxypropyl side chain (**9b**) was synthesized and found to have a binding affinity comparable to deamino-T_4, the most tightly bound known analog to prealbumin.[60] A neutral side chain had never been considered before, but actually produced very high binding affinity, demonstrating that molecular modeling and calculations can in fact provide new insight and suggest valuable new design ideas which otherwise might be overlooked.

McKinney *et al.* discovered that dioxin and polychlorinated biphenyl (PCB) analogs also bind to prealbumin,[97, 98] and suggested this may be related to their toxicity. They modeled the binding of several dioxin and PCB analogs to the prealbumin-binding site and predicted their relative binding affinities with good overall success. Although the dioxin analogs do not fit the site as tightly as thyroxine, they have surprisingly high binding affinity, up to 15% the affinity of L-T_4. Even more surprising was the discovery that the simple compound 2,4,6-triiodophenol (**10**) binds nearly four times better than L-T_4 and that 3,5-dichloro-4-hydroxybiphenyl (**11**) binds 850 times better than T_4! These compounds obviously cannot fill as much of the prealbumin hormone-binding site as T_4 and

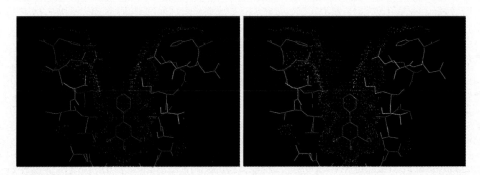

Figure 13 Prealbumin with model for binding of 3,5-dichloro-4-hydroxybiphenyl (**11**) (adapted from ref. 98)

might be expected to have low binding affinity due to the few possible interactions available to them. They probably bind deep at the bottom of the binding site, where the phenolic ring of L-T$_4$ binds, making primarily hydrophobic and van der Waals interactions. The complementarity of these *o*-halophenol structures to the binding site in this deep, solvent-shielded region is even better than L-T$_4$, so that they are almost completely buried within the site (Figure 13), while the outer portion of L-T$_4$ is well exposed to solvent. McKinney *et al.* illustrated that there are still surprises and discoveries to be made even when designing molecules based on the X-ray crystal structure of a receptor site; it is not obvious what the best candidates are and even simple molecules can be easily overlooked. Pedersen *et al.*[99] used molecular mechanics calculations with simple empirical solvation corrections to simulate the binding of the dioxin and PCB analogs to prealbumin; their results showed good qualitative agreement with the experimental relative binding affinities.

(10)   (11)

20.3.7 MOLECULAR GRAPHICS AND QSAR

One of the central problems in medicinal chemistry is that of understanding how drugs (ligands) react with their receptors in enormously complex biological systems, that is, what are the features of the drug which promote or limit the strength of its interaction with a given receptor. Unfortunately we are severely restricted in our study of this problem in that very few drug receptors have been isolated and identified.

There are a variety of ways in which one might study ligand interaction with macromolecules; however, our discussion will center on the use of two kinds of independent models.

Since 1962[100] there has been increasing interest in using physicochemical parameters and multivariate analysis to correlate structural changes in sets of congeners with changes in their ability to induce a particular biological response in a well-defined system.[101, 102] These relationships, often called quantitative structure–activity relationships (QSAR), are developed using parameters[103] from model systems. The hydrophobic parameters and steric parameters of the QSAR provide some insight on the geometry and surface characteristics of the receptor. The normal way of checking the validity of the QSAR has been to make and test new compounds.

The advent of color stereo molecular graphics, where the surfaces of macromolecules can be defined and color coded, provided a new and independent means for the analysis of structure–activity relationships. When the X-ray crystallography of the macromolecule is available, preferably with an inhibitor bound to the receptor site whose coordinates provide help in docking other ligands, comparison of QSAR and the stereo models is a powerful method for mechanistic studies. The results in hand show, for a variety of enzymes, that expectations from the QSAR for substrates as well as inhibitors are confirmed by the study of graphics models.[68] The following discussion will concentrate on examples where X-ray structures of the enzymes *and* QSAR have been compared.

20.3.8 PARAMETERS FOR CORRELATION ANALYSIS

For the present examples the hydrophobic parameters log P and π, Hammett–Taft σ constants and molar refractivity (MR) will suffice. P is the octanol–water partition coefficient of a molecule, while π is the log P of a substituent referred to the value of 0 for H. Positive values of π indicate a preference of the substituent for the hydrophobic interaction and negative values indicate hydrophilic character.

The presence of a term in log P or π in a QSAR indicates the interaction of the ligand with a hydrophobic surface of the enzyme. The occurrence of an MR term in a correlation equation (without a corresponding π term) suggests that the substituents to which it applies in a set of congeners are not contacting hydrophobic space. It is generally assumed that polar space (oxygen or nitrogen on the surface of the macromolecule) is involved. There is essentially a continuum of π

values for substituent parts of drugs indicating their relative affinity for a hydrophobic surface. There is of course a continuum of surface types in a macromolecule but there is no way of characterizing the relative hydrophobicity of small patches of surface in numerical terms. Many workers are calculating electrostatic surface potentials of macromolecular surface which yields numerical values; however, it is not clear how these relate to the hydrophobicity of the surface. In fact, when one gets beyond the rare gases and hydrocarbons, the concept of hydrophobicity becomes blurred. Still, the last 25 years have shown that the parameters $\log P$, π and MR can be employed to greatly enhance our understanding of how drugs interact with receptors.

As discussed below, in studying enzyme surfaces it is of great value to color code regions. Although one could attempt to do this continuously, we have elected, as a first approximation, to define three types of surfaces. Wherever the water molecule probe used in defining the surface contacts carbon of an amino acid side chain, the surface is colored red (hydrophobic). Oxygen and nitrogen surfaces are coded in blue. In addition, an intermediate surface is colored yellow. These surfaces are sulfur, the α carbon in amino acids, C-2 in histidine, C in C=O and C located within 3 Å of the ε-NH_3^+ in lysine or the C in the guanidine group of arginine. There is some evidence that the intermediate surface correlates better with MR than with π.[44]

The partitioning of model compounds between octanol–water has been used as a basis for defining protein surface character. Table 1 lists partition coefficients for certain model compounds.[44]

Note that replacing the CH_2 in pentane (1 in Table 1) with S yields a considerably more hydrophilic compound, but not nearly so hydrophilic as when O and NH replace CH_2. The disulfide moiety (6 in Table 1) shows almost a neutral effect, that is, dimethyl disulfide has essentially the same $\log P$ as ethane. Therefore sulfur is considered to be of intermediate character.

When carbon is placed near electron-attracting groups, its hydrophobic character is diminished,[104] as illustrated in Table 2.

Normally the addition of a CH_2 unit to a molecule increases $\log P$ by about 0.5. In the examples in Table 2 no such increase in hydrophobicity is observed. Hence, it is assumed that C-α and C-2 in histidine would not be hydrophobic or hydrophilic (like oxygen or nitrogen) and these examples are color coded yellow.

Table 3, from the work of Rekker,[105] illustrates the powerful effect of a positive charge on a nearby hydrophobic carbon.

When CH_2 moieties are added to a charged nitrogen atom the normal hydrophobic increment of 0.5 $\log P$ units is not attained until the fourth carbon is reached. For this reason the binding by such

Table 1 Octanol–Water Partition Coefficients of Selected Reference Compounds

Substance	Log P	Substance	Log P
1. Et(CH$_2$)Et	3.39	5. MeMe	1.77
2. Et(S)Et	1.95	6. MeSSMe	1.81
3. Et(O)Et	0.89	7. MeCH$_2$CH$_2$Me	2.89
4. Et(NH)Et	0.58		

Table 2 Octanol–Water Partition Coefficients Illustrating Electronic Effect of Substituents

Substance	Log P	Δ log P
1. PhCN	1.56	
PhCH$_2$CN	1.56	0.0
2. I$_2$	2.49	
ICH$_2$I	2.30	−0.19
3. PhCONH$_2$	0.65	
PhCH$_2$CONH$_2$	0.45	−0.20
4. PhNHCHO	1.15	
PhNHCOMe	1.16	0.01
5. PhCHO	1.48	
PhCOMe	1.58	0.10

Table 3 Octanol–Water Partition Coefficients of
Diphenhydramines[a]

Substance	Log P	Δ log P
1. $R\overset{+}{N}H_3$	0.12	
2. $R\overset{+}{N}H_2Me$	−0.01	−0.13
3. $R\overset{+}{N}H_2CH_2Me$	0.20	0.19
4. $R\overset{+}{N}H_2CH_2CH_2Me$	0.54	0.34
5. $R\overset{+}{N}H_2(CH_2)_3Me$	1.12	0.58

[a] $R = Ph_2CHOCH_2CH_2-$

carbon atoms is assumed not to correlate with π. Evidence for the comparable effect of a negative charge has not yet been established.

20.3.9 APPLICATIONS OF QSAR TO MOLECULAR GRAPHICS

20.3.9.1 Chymotrypsin

The hydrolases, such as chymotrypsin, have been extensively studied *via* probing with various ligands. However, as yet, rather few QSARs have been established for chymotrypsin. An early survey of chymotrypsin QSAR clearly shows this enzyme to be well suited for such analysis.[106] Equation (1) is the first to be formulated from a large number of data points.[107]

$$\log(1/K_m) = 1.09(\pm0.11)MR_2 + 0.80(\pm0.11)MR_1 + 0.52(\pm0.13)MR_3 - 0.63(\pm0.26)I_1$$
$$+ 1.26(\pm0.28)\sigma^* - 0.057(\pm0.013)MR_1 MR_2 MR_3 - 1.61(\pm0.47) \quad (1)$$
$$n = 71, \quad r = 0.979, \quad s = 0.332$$

In equation (1) K_m is the Michaelis constant for substrates of general structure (**12**)

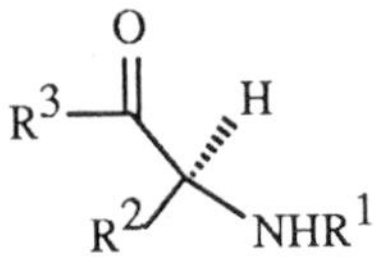

(**12**)

The subscripted MR parameters represent the molar refractivity (scaled by 0.1 to make them more equiscalar with π) of the corresponding R groups. I_1 is an indicator variable which takes the value of 1 for examples where R^2 = isopropyl. All other R^2 substituents are assigned a value of 0. Thus the negative coefficient with this term shows the deleterious effect of the branched isopropyl group on the formation of the ES (enzyme–substrate) complex.

The Taft inductive parameter σ^* applies only to R^3 which is in every instance part of an ester, R^3 = R'O. The positive coefficient in this term reveals that electron withdrawal promotes formation of the ES complex. This is to be expected with esters and is in fact found for a variety of enzymes (see Table 4).

The MR terms in equation (1) are all compromised in that MR, while being the parameter of choice, is highly collinear with π so that one cannot say that hydrophobic factors are ruled out in each of the three positions. In order to clarify this point a number of new derivatives were prepared in which only R^3 was varied (R^2 = Me; R^1 = —CO-3-pyridyl). The resulting MR_3 vector was reasonably orthogonal with respect to π_3. Rederiving equation (1) with the new data points[108] produced equation (2)

$$\log(1/K_m) = 1.13(\pm0.11)MR_2 + 0.77(\pm0.11)MR_1 + 0.47(\pm0.11)MR_3 - 0.56(\pm0.25)I_1$$
$$+ 1.35(\pm0.22)\sigma^* - 0.055(\pm0.01)MR_1 MR_2 MR_3 - 1.64(\pm0.46) \quad (2)$$
$$n = 84, \quad r = 0.977, \quad s = 0.333$$

Table 4 ρ Values from Correlation Equations of K_m For Various Hydrolases[66, 112]

ρ	Enzyme	Class	Ester	pH of assay
1. 0.56	Papain	Cysteine	II	6
2. 0.55	Papain	Cysteine	IV	6
3. 0.57	Ficin	Cysteine	II	6
4. 0.62	Ficin	Cysteine	IV	6
5. 0.74	Actinidin	Cysteine	II	6
6. 0.70	Bromelain B	Cysteine	II	6
7. 0.68	Bromelain D	Cysteine	II	6
8. 0.49	Subtilisin	Serine	II	7
9. 0.42[a]	α-Chymotrypsin	Serine	II	7
10. 0.53	Trypsin	Serine	II	6
11. 0.71	Trypsin	Serine	II	7
12. 1.66[a]	Buffer		II	8
13. 1.91	Buffer		II	6

[a] In these instances σ^- yields a better correlation than σ, and the value of ρ is for σ^-

where R′ = CO-3-pyridyl, COMe, CO-furyl-H$_4$, CO-furyl, COCH$_2$Cl, CO-4-pyridyl, CO-2-pyridyl, CO-2-thienyl, COPh, COC$_6$H$_4$-2-NH$_2$, CO-2-quinolyl, OCOCH$_2$Ph, SO$_2$Me, R^2 = Me, Pr^i, Et, CO$_2$Et, Pr, Bu, Bu^i, Ph, C$_6$H$_{13}$, CH$_2$Ph, C$_5$H$_{11}$, CH$_2$C$_6$H$_4$-4-OH, CH$_2$-c-C$_6$H$_{11}$, CH$_2$-indolyl, CH$_2$CONH$_2$, R^3 = Me, Et, Pr^i, CH$_2$CH$_2$Cl, Bu^s, CH(Me)-c-C$_6$H$_{11}$, C$_6$H$_4$-4-NO$_2$, CH(Me)Ph, Ph, C$_6$H$_4$-Me, C$_6$H$_4$-4-OMe, C$_6$H$_4$-4-Cl, C$_6$H$_4$-4-COMe, C$_6$H$_4$-3-NO$_2$, CMe$_3$, isopentyl, CH$_2$-furyl-H$_4$, CH$_2$CH$_2$OMe, CH$_2$COMe, CH$_2$OMe, CH$_2$CN, CH$_2$Ph.

At least for R^3 it is now established that it is not dependent on π and hence is assumed to interact with polar space. The fact that the parameters of equation (2) are quite close to those of equation (1) illustrates the predictive value of QSAR.

Construction of a graphics model of chymotrypsin with a typical substrate shows that when the carbonyl group of the ester is placed so that it contacts the anionic binding site with R^2 in the so-called hydrophobic hole, then R^3 falls on to very hydrophilic space confirming the result of equation (2).[109] This is illustrated in Figure 14.

Another interesting example of good agreement between the graphics of chymotrypsin and the QSAR for a set of phosphorus esters [PO(Me)(OR$_2$)(SR$_3$)] phosphorylating chymotrypsin has been reported.[109]

Returning to equations (1) and (2), the meaning of the cross-product term must be considered. The negative coefficient indicates that when the product of MR$_1$, MR$_2$ and MR$_3$ reaches a certain value, a negative contribution to $\log(1/K_m)$ occurs. This has been interpreted to mean that there is, at least within the bounds of the present data set, a cooperative negative steric effect of substituents in the three positions. This can be significant if three moderately large groups or two very large and one small group are used to derive K_m. It is not yet clear just how far this term will serve to correlate the binding of substrates. However, it does provide an interesting hypothesis suggesting further study.

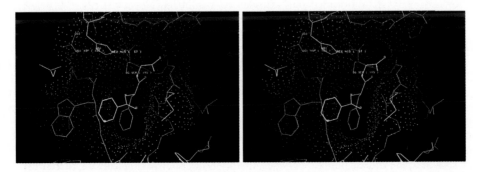

Figure 14 Chymotrypsin, color coded as in Figure 1, with model for binding of (12; R^1 = CO-3-pyridyl; R^2 = CH$_2$Ph; R^3 = OCH$_2$COMe)

A further question is: what do the coefficients with the linear MR terms signify? In the case of R^2 the positive coefficient may well represent a combined steric and hydrophobic effect. Although adding a term in π_2 does not improve the correlation as we have found in certain cases,[110] the high collinearity between MR_2 and π_2 may preclude defining two roles for R^2. Even though the hole into which R^2 binds contains patches of polar and hydrophobic space, most workers consider it as hydrophobic. Clearly substituents would have to be desolvated to fit in this site. However, there is some evidence that steric effects of R^2 may also be important.[107] The role of MR^1 *versus* π_1 is also ambivalent, but R^3 does seem to be more straightforward. MR_3 and π_3 are not collinear, binding does occur in polar space but the question remains as to what the positive coefficient with MR_3 means. One is tempted to attribute this to attractive van der Waals forces. However, as will appear in the discussion below, this seems unlikely. A more likely possibility, better documented in the case of papain, is that of a buttressing effect on the substrate helping to hold it in the most favorable position for binding. Evidence is accumulating showing that small changes in placement of reactive groups or factors which increase their time of contact can have large effects on the rate of organic reactions.[111]

Finally, one must bear in mind that QSAR such as those discussed in this review must sooner or later, as larger and larger changes are made in a set of congeners, break down. The varied surfaces and geometry of active sites mean that eventually substituents can be increased in size so that they extend beyond a hydrophobic or polar patch. However, those compounds which do fit a correlation equation well very probably have a common mechanism of action.

Probing chymotrypsin with a set of ester substrates (**13**) yields equation (3)[112]

(**13**)

$$\log(1/K_m) = 0.42(\pm 0.08)\sigma^- + 0.28(\pm 0.06)\pi'_3 + 3.87(\pm 0.05) \tag{3}$$

$$n = 28, \quad r = 0.945, \quad s = 0.081$$

where X = 4-NO_2, 4-SO_2Me, 4-SO_2NH_2, 4-CN, 4-OMe, 4-Me, 4-Cl, 4-F, 4-NH_2, 4-I, H, 3-Cl, 3-F, 3-Me, 3-NO_2, 3-SO_2Me, 3-SMe, 3-SO_2NH_2, 3-Et, 3-CN, 3-CF_3, 3-I, 3-$NHCONH_2$, 3-NHCOMe, 3,5-Cl_2, 3-Me-5-Et, 3,5-$(CF_3)_2$, 3,5-$(Me)_2$.

In equation (3) σ^- is the Hammett constant for cases where there is direct resonance interaction between the substituent and the reaction center (in this case the lone pair electrons on the phenolic O). Since σ^- yields a better correlation than σ, through resonance appears to play a role in the formation of the ES complex. The hydrophobic term π'_3 applies only to substituents in the 3- or 5-position. 4-Substituents are correlated by σ^- alone, which suggests that they do not in any way contact the enzyme in ES.

Meta substituents behave in a more complex fashion. Hydrophobic ($\pi > 0$) *meta* substituents require parametrization *via* π'_3, but hydrophilic *meta* substituents are well fitted using σ alone. This indicates that they orient themselves away from the hydrophobic binding possibility; this same propensity has been observed for hydrophilic *meta* substituents in the case of the cysteine hydrolases (see Section 20.3.9.3). In a number of examples on which equation (3) rests where both *meta* substituents are hydrophobic, both must be parametrized by π_3.

A graphics analysis, where the carbonyl group of the ester is placed in the oxyanion hole and the NHCOPh moiety is directed into the hydrophobic hole, shows that 4-substituents simply do not contact the enzyme.[112] Hence, it is not surprising that π_4 or MR_4 terms do not occur in equation (3). Figure 15 addresses the necessity for hydrophobic parametrization of both *meta* substituents. Both *meta* groups can make partial contact (resulting in the small coefficient of 0.28) with a hydrophobic portion of the surface. If the substituent is polar, a slight tilt of the phenyl ring would locate it in aqueous space.

The problem is to find a reasonable conformation to account for the shape of the QSAR and in particular the simultaneous hydrophobic contact of two *meta* substituents. The best model is shown in Figure 15 where the dihedral angle between the phenolic phenyl ring and the planar ester moiety is about 20°. Schweizer and Dunitz[113] have found that the preferred torsional angle for such esters in crystals is about 90°. To obtain support for the 20° angle, hippurates in the ES complex with papain

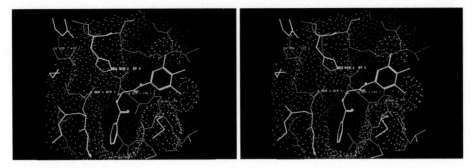

Figure 15 Chymotrypsin, color coded as in Figure 1, with model for binding of (13)

were energy minimized *via* molecular dynamics using the program AMBER.[112] The energies obtained for two possible models, one with the phenyl ring planar and the other with it perpendicular to the ester plane, were found to differ by only about 0.03%, with the planar conformation having the lower energy. Thus, the geometry of the ester in the model does not seem to be a problem.

Another likely possibility is that in the transition state to ES formation the serine hydroxyl of the enzyme is far advanced in its reaction with the carbonyl group of the ester and it is essentially the tetrahedral intermediate which is involved. This mechanism avoids the problem raised in the studies by Schweizer and Dunitz about the preferred geometry of aromatic esters. The small change in geometry from about 120° in the planar ester to about 109° in the tetrahedral intermediate leads to small differences in the placement of the moieties attached to the tetrahedral carbon.

An interesting conclusion from this study, as well as others of this type, is that the graphics model could not have been deduced without the aid of the QSAR and the QSAR could not have been understood without the graphics model.

20.3.9.2 Trypsin

Another serine hydrolase similar to chymotrypsin is trypsin. Hydrolysis of esters (13) by trypsin yields equation (4),[114] where X = 4-SO$_2$NH$_2$, 4-CONH$_2$, 4-NH$_2$, 4-CN, 4-NO$_2$, 4-NHCONH$_2$, 4-OMe, H, 4-Me, 4-Cl

$$\log(1/K_m) = 0.71(\pm 0.17)\sigma + 3.31(\pm 0.09) \tag{4}$$

$$n = 10, \quad r = 0.961, \quad s = 0.100$$

All of the 10 examples upon which equation (4) is based are *para*-substituted phenyl hippurates, and, as with chymotrypsin, no terms in MR$_4$ or π_4 are required to obtain the good fit of equation (4). Using σ^- in equation (4) in place of σ yields a poorer correlation ($r = 0.912$), which brings out the fact that the electronic effects in the ES complex transition state differ from chymotrypsin. In the case of chymotrypsin, breakage of the phenolic oxygen—carbon bond may be further advanced than in the case of trypsin, or the geometry of the phenyl ring in relation to the ester plane may be such that through resonance is not facilitated in trypsin. Figure 16 shows how 4-substituents avoid enzymic contact with trypsin.

Several correlation equations for the inhibition of trypsin by amidines have been derived and compared with graphics models.[6] Three QSARs were derived for benzamidines using different data from three different laboratories. The number of data points were 10, 14, 104 and the corresponding ρ for the σ terms are -0.75, -0.71 and -0.74. In each case other terms in addition to σ were needed to obtain a good correlation and in the case where 104 data points were involved seven terms in addition to σ were needed; nevertheless, it was possible to clearly separate the electronic effects of both *meta* and *para* substituents. Unfortunately there are few published examples of this type showing that hydrophobic, steric and electronic terms in QSAR can be treated on the assumption that they are additive. There is still a lingering feeling among many, probably based on unsatisfactory results where σ constants alone have been used to correlate sets of enzymic data, that enzymic processes are too complex to be treated in such a simplistic fashion. Of course we cannot expect to obtain the excellent correlations found for homogeneous solutions by physical organic chemists, but it is becoming clear that correlation analysis can be of great help to those studying biochemical reactions.

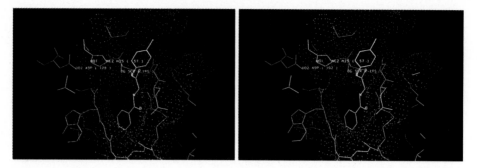

Figure 16 Trypsin, color coded as in Figure 1, with model for binding of (**13**)

Equation (5) correlates the inhibition of the case of 104 benzamidine inhibitors of trypsin[44] from the study of Labes and Hagen. The MR_4 term applies only to *para* substituents and π'_3 applies only to *meta* substituents. The *I-M* indicator variable takes the value of 1 for *meta* substituents, which appear to be slightly more active, on the average, than *para* substituents. The *I-1*, *I-2* and *I-3* indicator variables account for special structural features. The point of interest is the bilinear MR_4 portion of the equation. Initially there is a negative steric effect which increases until MR = 1.03, at which point a positive contribution occurs. A study of the graphics shows[44] that the OH of Ser-195 makes a bad contact with *para* substituents. Large substituents (MR > 1.03) appear to brush this OH aside and contact enzyme surface to assist in binding. This is a good illustration of enzyme flexibility. When making small changes in ligands, one does not usually encounter a yes or no interaction. There seems to be considerable room for mutual adjustment between ligand and enzyme surface. Equation (6) correlates[44] the inhibition of trypsin by amidines (**14**).

$$\log(1/K_i) = -0.59(\pm 0.49)MR_4 + 0.88(\pm 0.52)\log(\beta 10^{MR_4} + 1) + 0.23(\pm 0.07)\pi'_3 - 0.74(\pm 0.20)\sigma$$
$$+ 0.20(\pm 0.30)I\text{-}M + 0.65(\pm 0.22)I\text{-}1 + 0.43(\pm 0.19)I\text{-}2 + 0.51(\pm 0.15)I\text{-}3 + 1.38(\pm 0.28) \qquad (5)$$

$$n = 104, \quad r = 0.924, \quad s = 0.222, \quad \text{optimum } MR_4 = 1.03$$

$$\log(1/K_i) = -1.40(\pm 0.40)\sigma + 0.47(\pm 0.19)MR_4 + 2.59(\pm 0.24) \qquad (6)$$

$$n = 21, \quad r = 0.915, \quad s = 0.322$$

(**14**)

There is no reason to expect the σ term in equation (6) to resemble those of the benzamidines. In fact, the favorable role of electron-releasing substituents appears from a graphics analysis[44] to be associated with hydrogen bonding between the carbonyl group of the amidine and the OH of Ser-195.

A study of the graphics model reveals that X falls on the surface of a disulfide linkage. The fact that MR_4 is a far better parameter in equation (6) than π_4 indicates that such a surface is not typically hydrophobic as discussed in Section 20.3.2.[44]

20.3.9.3 Papain

Again using esters (**13**), equation (7) has been developed for comparison with other hydrolases[115]

$$\log(1/K_m) = 1.03\pi'_3(\pm 0.25) + 0.57(\pm 0.20)\sigma + 0.61(\pm 0.29)MR_4 + 3.80(\pm 0.17) \qquad (7)$$

$$n = 25, \quad r = 0.907, \quad s = 0.208$$

where X = H, 4-NH$_2$, 4-F, 4-Me, 4-OMe, 4-Cl, 4-CN, 4-SO$_2$NH$_2$, 4-CONH$_2$, 4-COMe, 4-NO$_2$, 3-NHCOMe, 3-F, 3-CONH$_2$, 3-CN, 3-NO$_2$, 3-Me, 3-SO$_2$NH$_2$, 3-Cl, 3-I, 3-CF$_3$, 3,5-(NO$_2$)$_2$, 3,5-Me$_2$, 3,5-(OMe)$_2$, 3,5-Cl$_2$, 3-Me-5-Et. [(3,5-OMe)$_2$ not used to derive equation (7)].

In equation (7), π'_3 is similar to equation (3) except that when two hydrophobic *meta* substituents are present only the more hydrophobic is parametrized. This would indicate that only one of the *meta* substituents can contact hydrophobic space. The larger coefficient with π_3 compared to equation (3) suggests much more desolvation of the substituent on hydrophobic binding. The coefficient with σ is about the same as for equations (2) and (3) (*i.e.* rather small).

In the case of papain a term in MR$_4$ is required to obtain the best correlation indicating that *para* substituents do not contact hydrophobic space.

In Figure 17 it is clear that when a hydrophobic *meta* substituent interacts with red hydrophobic space the other *meta* substituent is forced to reside in the surrounding aqueous phase. The hydrophobic *meta* substituents used in this study were all relatively small so that when they bind in the available shallow hydrophobic pocket, they would appear to be completely desolvated, hence the larger coefficient with π_3 compared to equation (3).

Although enzyme–ligand interactions are complex and poorly understood, it seems reasonable that coefficients with π convey information of value to the chemist. The interpretation of such figures will, we believe, have to be developed from empirical studies of many examples, because our understanding of the hydrophobic interaction is so rudimentary.

The MR$_4$ term in equation (7) cannot be replaced with π, implying that 4-substituents do not contact hydrophobic surface. From the graphics analysis it is clear that 4-substituents collide[68, 115, 116] with the highly polar amide moiety of Gln-142. This side chain of glutamine, on the surface of the enzyme, would appear to be easily movable so that groups as large as Ph, Bu^t or OBu can fit on this region. The function of the glutamine side chain appears to be that of a buttress helping to position the substrate for hydrolysis. Again we see evidence for the easy movement of a surface residue.

In building the papain models discussed in this article, the coordinates of the enzyme inhibitor complex benzyloxycarbonyl–L-phenylalanyl–L-alanylmethylene–papain (ZPA–papain)[3] were used to determine the placement of the substrates. Of course the models are built from the static structure obtained from X-ray crystallography. Since more and more evidence is revealing the fluctuating nature of enzymes,[117] we cannot be confident about exactly how the substrates are positioned in the enzyme at the important transition states. The shape of the hydrophobic pockets may be somewhat different than those we have pictured.

Turning now to another set of substrates (**15**), equation (8) has been formulated[116]

(15)

$$\log(1/K_m) = 0.61(\pm0.09)\pi'_3 + 0.55(\pm0.20)\sigma + 0.46(\pm0.11)MR_4 + 2.00(\pm0.12) \tag{8}$$

$$n = 32, \quad r = 0.945, \quad s = 0.178$$

where X = H, 4-Me, 4-F, 4-Cl, 4-OH, 4-OMe, 4-CHO, 4-COMe, 4-NO$_2$, 4-CMe$_3$, 4-OBu, 4-Ph, 3-Me, 3-Et, 3-CHMe$_2$, 3-CMe$_3$, 3-OMe, 3-F, 3-Cl, 3-Br, 3-I, 3-CF$_3$, 3-Ph, 3-OCH$_2$Ph, 3-OCH$_2$C$_6$H$_4$-4-Cl, 3-OCH$_2$-2-naphthyl, 3-COMe, 3-CN, 3-NO$_2$, 2,5-Cl$_2$, 3-CMe$_3$-4-NO$_2$, 3,4,5-(Cl)$_3$.

The parameters of equation (8) bear the same connotation as those of equation (7). The experimental conditions for equation (8) are slightly different (35°C *versus* 25°C) so that the intercepts cannot be compared; however, a smaller data set (13 compounds) yielded an equation with essentially the same coefficients with π_3 and σ (4-substituents were held constant),[116] but with an intercept at 25°C of 2.38.

Subtracting intercepts of equations (7) and (8) (3.80 − 2.38 = 1.42) shows that the intrinsic affinity for the ES complex of substrates (**13**) is about 30 times greater than substrates (**15**). The difference in π for NHCOPh and NHSO$_2$Me is 0.49 − (−1.18) = 1.67. The coefficient of 1 with π in equation (7) would lead one to expect a 1.67 difference in $\log(1/K_m)$ between the two parent congeners. This is in reasonable agreement with 1.42, suggesting that only the hydrophobic effect is responsible for the different affinities of the two systems.

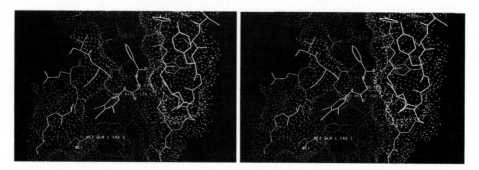

Figure 17   Papain with model for binding of (**13**). The catalytic residues Cys-25 and His-159 are highlighted in green. The active site molecular surface is color coded by hydrophobicity as in Figure 1

There is little difference between the coefficients in equations (7) and (8) except for that with π.

From a study of the graphics model of papain it appears that substituents of the size of t-butyl and smaller could be accommodated in the type of binding shown in Figure 17. In deriving equation (8) it was decided to investigate much larger groups (*e.g.* 3-Ph, 3-OCH$_2$C$_6$H$_4$Cl, 3-OCH$_2$-2'-naphthyl); these large substituents cannot possibly fit the configuration displayed in Figure 17. However, if the phenyl ring is rotated 180°, the large substituents fall on a large hydrophobic surface where only one side of the substituent is desolvated. This would explain the lower coefficient with π for the large flat groups; however, the small substituents such as Cl, Me, *etc.*, also fit equation (8) well. These have the option of binding as shown in Figure 17 or 18, but if they bind as in Figure 17, one would expect a coefficient with π of 1 and if they bind as in Figure 18 they do not fall in typically hydrophobic space (red). Hence, it is not yet completely clear why congeners in (**15**) have a different dependence on π from congeners in (**13**).

Again, as with equations (3), (7) and (8), we find that when hydrophilic *meta* substituents are present they only fit the regression equation when their π values are set $= 0$, that is, their only effect on $\log(1/K_m)$ is modeled by σ. The hydrophilic substituents seem to be repelled by the hydrophobic surface and remain in the aqueous phase by 180° rotation of the phenyl ring.

As in the case of the phenoxy esters being hydrolyzed by chymotrypsin, we encounter the problem of picturing the relationship between the phenyl ring and the ester plane. It seems most likely that in the transition state to ES for papain, the carbon of the carbonyl group is close to tetrahedral[116] with attachment of —S⁻ of the catalytic cysteine being almost complete. In fact, the models of Figures 14–17 have been built with this assumption using the recently refined[118] coordinates for papain.

20.3.9.4   Actinidin

Another cysteine hydrolase whose refined X-ray crystallographic structure is known to be very similar to that of papain is actinidin. The QSAR of equation (9) has been derived from data on congeners in (**13**).[119, 120] In equation (9) X = H, 4-F, 4-Cl, 4-Me, 4-COMe, 4-CN, 4-NO$_2$, 4-OMe, 4-NH$_2$, 4-CONH$_2$, 4-SO$_2$NH$_2$, 3-F, 3-Cl, 3-Br, 3-I, 3-Me, 3-CMe$_3$, 3-CF$_3$, 3-CN, 3-NO$_2$,

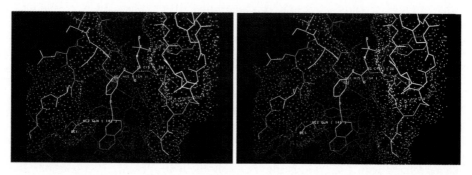

Figure 18   Papain with model for binding of (**15**), displayed as in Figure 17

3-NHCOMe, 3-CONH$_2$, 3-SO$_2$NH$_2$, 3,5-Me$_2$, 3-Me-5-Et, 3,5-(OMe)$_2$, 3,5-Cl$_2$, 3,5-(NO$_2$)$_2$, 3,4,5-Cl$_3$ [4-SO$_2$NH$_2$ and 3,5-(OMe)$_2$ not included in the derivation of equation (9)].

$$\log(1/K_m) = 0.50(\pm0.13)\pi'_3 + 0.74(\pm0.15)\sigma + 0.24(\pm0.21)MR_4 + 2.90(\pm0.12) \qquad (9)$$

$$n = 27, \quad r = 0.927, \quad s = 0.158$$

The terms have the same meaning as those in equation (7) and (8). The intercept is lower than 3.87 found for papain showing actinidin's lower affinity (about 10 times) for the phenyl hippurates. The π and σ coefficients are in good agreement with equation (8) and not equation (7), but MR_4 is of borderline importance. In actinidin the amino acid residue corresponding to Gln-142 in papain is Lys-145. Thus it is the ε-NH$_3^+$ moiety of the lysine side chain with which 4-substituents interact. The greater flexibility of this side chain probably lies behind its much smaller effect revealed by the MR_4 term. In constructing the model in Figure 19 *meta* substituents have been treated (as in Figure 18) as binding in the planar region rather than in a shallow pocket similar to the one shown in Figure 17.

Two other cysteine hydrolases (ficin and bromelain), whose X-ray crystallographic structures have unfortunately not yet been determined, have yielded QSAR[121, 122] which are similar to papain and actinidin.

An interesting generalization for ρ which comes from the study of the hydrolysis of esters (13) and (15) with various hydrolases is given in Table 4.

The mean and standard deviation of ρ for the 11 examples of enzymic hydrolysis in Table 4 is 0.6 ± 0.1. It will be interesting to see how far this relationship extends in terms of other esters and hydrolases. In all of the examples in Table 4 except trypsin, the QSAR from which ρ was taken contain other terms (besides σ) in π or MR or both. Thus the 'true' value of ρ only becomes clear when hydrophobic and steric effects are separated from electronic factors.

The electronic effect on the formation of ES (there is little variation in k_{cat}) by substituents is reduced by more than a factor of 10 in the enzymatic process compared to buffer hydrolysis.

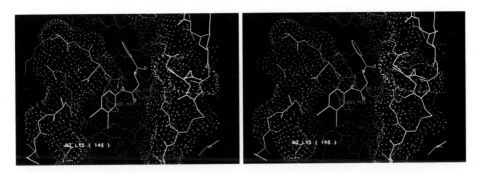

Figure 19 Actinidin with model for binding of (13). The catalytic residues Cys-25 and His-162 are highlighted in green. The active site molecular surface is color coded by hydrophobicity as in Figure 1

20.3.9.5   Carbonic Anhydrase

Carbonic anhydrase (CA), a zinc-containing enzyme of about 30 000 molecular weight which governs the hydration of CO_2 ($CO_2 + H_2O \rightarrow H_2CO_3$), is of importance to medicinal chemistry. Inhibitors of this enzyme have found use as diuretics and antiepileptic agents and may be of value in the treatment of glaucoma.

An interesting study of the binding of simple sulfonamides (16) by King and Burgen yielded the QSAR equation (10)[123]

$$\log K = 1.55(\pm0.38)\sigma + 0.64(\pm0.08)\log P - 2.07(\pm0.22)I_1 - 3.28(\pm0.23)I_2 + 6.94(\pm0.18) \qquad (10)$$

$$n = 29, \quad r = 0.991, \quad s = 0.204$$

where X = H, 4-Me, 4-Et, 4-Pr, 4-Bu, 4-C_5H_{11}, 4-CO_2Me, 4-CO_2Et, 4-CO_2Pr, 4-CO_2Bu, 4-$CO_2C_5H_{11}$, 4-$CO_2C_6H_{13}$, 4-CONHMe, 4-CONHEt, 4-CONHPr, 4-CONHBu, 4-$CONHC_5H_{11}$, 4-$CONHC_6H_{13}$, 4-$CONHC_7H_{15}$, 3-CO_2Me, 3-CO_2Et, 3-CO_2Pr, 3-CO_2Bu, 3-$CO_2C_5H_{11}$, 2-CO_2Me, 2-CO_2Et, 2-CO_2Pr, 2-CO_2Bu, 2-$CO_2C_5H_{11}$.

In this expression, K is a binding constant, and I_1 and I_2 are indicator variables. I_1 takes the value of 1 for *meta* substituents and I_2 assumes the value of 1 for *ortho* substituents. All *ortho* and *meta* substituents are of the type —CO_2R, where R is a normal alkyl group. The negative coefficients with indicator variables suggest a steric effect which makes the *ortho* substituents over 1000 times less able to bind to CA and *meta* substituents are 100 times as poor binders as *para* derivatives. However, once this steric effect is accounted for, the role of hydrophobicity of the alkyl group of 2- and 3-substituents is the same as 4-substituents (*i.e.* the single $\log P$ term suffices for all three classes of substituents).

The positive coefficient with σ brings out the fact that electron-withdrawing groups promote binding. This is in line with evidence that it is the anionic form of the SO_2NH_2 which is effective in holding the sulfonamide to the positively charged Zn atom.

Figure 20 brings to life QSAR (10). In this view the 'wire' model of $Me(CH_2)_5OCO$-4-$C_6H_4SO_2NH_2$, in yellow, is shown with the N binding to the fourth coordination site on Zn. The sulfamido group is oriented so that there is a hydrogen bond to the oxygen of Thr-199. This position allows one of the oxygens of the sulfonamido group to occupy a more distant fifth coordination site on the Zn according to Kannan *et al.*[5] In this configuration 4-substituents fall on a large slightly concave hydrophobic surface, where one would expect them to be somewhat more than 50% desolvated if only simple partitioning is involved in binding.

In Figure 21 the unfavorable interaction of *ortho* substituents with Pro-201 is displayed. After the initial bad contact with Pro, the alkyl moiety can bind to a long red (hydrophobic) surface eventually reaching the area where 4-substituents bind. *Meta* substituents make an unfavorable contact with Leu-198 and then bind in much the same area as *para* substituents. The overall agreement between

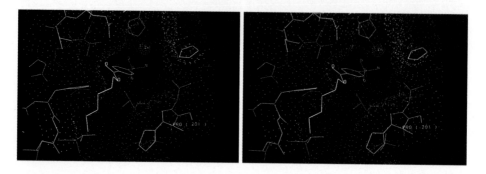

Figure 20 Carbonic anhydrase C with model for binding of (**16**; X = 4-$CO_2C_6H_{13}$). The catalytic Zn ion is shown by the green sphere and the active site molecular surface is color coded by hydrophobicity as in Figure 1

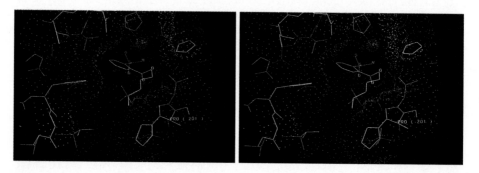

Figure 21 Carbonic anhydrase C with model for binding of (**16**; X = 2-$CO_2C_5H_{11}$), displayed as in Figure 21

the QSAR and the graphics models is gratifying. Equation (10) has been used to suggest much more potent carbonic anhydrase inhibitors.[123]

20.3.9.6 Alcohol Dehydrogenase

The enzyme alcohol dehydrogenase (ADH) catalyzes the oxidation and reduction of alcohols and aldehydes: $RCH_2OH \rightleftharpoons RCHO$. Its X-ray crystallographic structure has been established and QSAR for a number of inhibitors and substrates have been formulated[124-127] so that considerable data are available for model comparison. Although a variety of inhibitors have been found[124, 125] to yield similar QSAR, the present discussion is limited to 4-substituted pyrazole inhibitors (**17**).

X

N – NH

(**17**)

QSAR (11) is based on horse ADH and examples where X is restricted to normal alkyl groups, while QSAR (12) is based on rat ADH and a wide variety of substituents.

$$\log(1/K_i) = 0.96(\pm 0.25)\log P + 5.70(\pm 0.56) \qquad (11)$$

$$n = 5, \quad r = 0.990, \quad s = 0.207$$

$X = H, Me, Pr, C_5H_{11}, C_6H_{13}$

$$\log(1/K_i) = 1.22\log P - 1.80\sigma_{meta} + 4.87 \qquad (12)$$

$$n = 14, \quad r = 0.985, \quad s = 0.316$$

$X = C_6H_{13}, C_5H_{11}, Pr, Me, H, I, OPr, O\text{-}Pr^i, OEt, OMe, CN, NO_2, NH_2, NHCOMe.$

Equation (11) lacks the electronic term present in equation (12) because the electronic effect of alkyl groups is essentially constant. In equation (12) the negative coefficient with the σ term (σ_{meta} has been used since X is *meta* with respect to both N atoms) reveals that binding (inhibition) is increased by electron-releasing groups. This is in agreement with the fact that one of the nitrogen atoms binds to the positively charged Zn of ADH.

Equation (11) is the most reliable QSAR for ADH because of the simplicity of the substituents. Its slope of 0.96 agrees with a mean value of 0.96 found for four other QSARs of ADH inhibitors.[124] From this rather high coefficient with log P one would expect from our limited present experience to find binding in a hole or cleft with complete desolvation.

Figure 22 shows a side view of 4-pentylpyrazole bound in a deep channel surrounded on all sides by red hydrophobic surface. The side of the channel nearest the viewer has been removed for the purpose of clarity. The pyrazole N is shown binding to the critical Zn atom. Graphics analysis shows that all water of hydrophobic hydration must be removed for the pyrazoles to fit into the cavity.

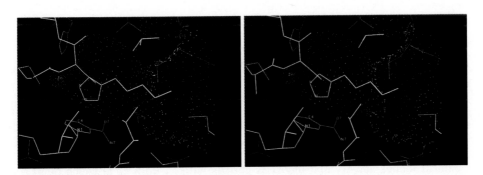

Figure 22 Alcohol dehydrogenase with model for binding of (**17**; $X = C_5H_{11}$). The catalytic Zn ion and NADH cofactor are shown in green and the active site molecular surface is color coded by hydrophobicity as in Figure 1

The channel is rather narrow so that, depending on their position, branched substituents show steric effects.[124]

A question which has long concerned biochemists is: Do enzymes in the living cell have the same conformation and behave in the same way as purified enzymes in buffered solution? Some insight on this problem can be gained from equation (13), which correlates the inhibition of oxidation of ethanol by the ADH in rat liver cells by pyrazoles.

$$\log(1/K_i) = 1.27(\pm 0.33)\log P - 0.20(\pm 0.12)(\log P)^2 - 1.80(\pm 0.87)\sigma + 4.75(\pm 0.29) \quad (13)$$

$$n = 14, \quad r = 0.971, \quad s = 0.320$$

Equation (13) has an additional term (compared to equation 12) because the two most lipophilic pyrazoles (4-hexyl, 4-pentyl) are limited in their access to ADH in the random walk process,[128] that is, they bind so tightly to other lipophilic centers in the cell that the availability for specific binding by ADH is limited. Less lipophilic compounds are not so restricted. Deleting these two compounds yields equation (14).

$$\log(1/K_i) = 1.08(\pm 0.29)\log P - 1.52(\pm 0.97)\sigma + 4.61(\pm 0.34) \quad (14)$$

$$n = 12, \quad r = 0.948, \quad s = 0.371$$

Both equations (13) and (14) have terms in σ close to that of equation (12), showing that the electronic effect of substituents is the same in isolated enzyme and enzyme *in situ*. The intercepts of all three equations are also in close agreement, showing the same intrinsic activity of ADH *in vitro* and *in vivo*.

20.3.9.7　Dihydrofolate Reductase

Inhibitors of dihydrofolate reductase (DHFR) are well established in the field of chemotherapy. The powerful DHFR inhibitor methotrexate has long been important in the treatment of cancer, and trimethoprim is a highly successful antibacterial agent. X-Ray crystallographic structures have been determined for DHFR from *E. coli*, *L. casei* and chicken liver.[4] Considerable QSAR work has been compared with graphics analyses of various DHFRs.[129] Our consideration will be restricted to triazines (**18a**) and (**18b**)

(18a)　　　　　　　　　　　　　　　　**(18b)**

In (**18a**), X represents a wide variety of substituents, including those in (**18b**), where Y and Z are also varied. Z is limited to O, NH, S and Se. Equation (15) correlates the inhibition of purified *L. casei* DHFR.[130]

$$\log(1/K_i) = 0.83(\pm 0.13)\pi'_3 - 0.91(\pm 0.19)\log(\beta 10\pi'_3 + 1) + 0.71(\pm 0.20)I + 4.60(\pm 0.13) \quad (15)$$

$$n = 38, \quad r = 0.961, \quad s = 0.244, \quad \pi_0 = 2.69(\pm 0.61)$$

X = H, 3-SO$_2$NH$_2$, 3-COMe, 3-OH, 3-CF$_3$, 3-F, 3-I, 3-NO$_2$, 3-Me, 3-Et, 3-C$_6$H$_{13}$, 3-C$_9$H$_{19}$, 3-C$_{12}$H$_{25}$, 3-OMe, 3-OEt, 3-OPr, 3-OC$_6$H$_{11}$, 3-OC$_9$H$_{19}$, 3-OC$_{12}$H$_{25}$, 3-OCH$_2$-1-adamantyl, 3-O(CH$_2$)$_2$OC$_6$H$_4$-4'-CF$_3$, 3-O(CH$_2$)$_4$OC$_6$H$_4$-3'-CF$_3$, 3-OCH$_2$Ph, 3-OCH$_2$C$_6$H$_3$-3',4'-Cl$_2$, 3-OCH$_2$C$_6$H$_4$-4'-CONH$_2$, 3-CH$_2$O-cC$_6$H$_{11}$, 3-CH$_2$NHC$_6$H$_4$-4-SO$_2$NH$_2$, 3-CH$_2$NHC$_6$H$_3$-3',5'-(CONH$_2$)$_2$, 3-CH$_2$OPh, 3-CH$_2$OC$_6$H$_4$-3'-C(Me)$_3$, 3-CH$_2$OC$_6$H$_4$-3'-NHCOMe, 3-CH$_2$OC$_6$H$_4$-3'-CN, 3-CH$_2$OC$_6$H$_4$-3'-Et, 3-CH$_2$OC$_6$H$_4$-3-Ph, 3-CH$_2$OC$_6$H$_4$-4'-C$_5$H$_{11}$, 3-CH$_2$SPh, 3-SCH$_2$Ph, 3-CH$_2$SePh (3-CN was not used to derive equation 15).

In this bilinear model, activity first increases linearly with π'_3 until π_0 at which point activity is essentially constant with a slope of $0.83 - 0.91 = -0.08$, that is, only part of large substituents contact the enzyme. As will be discussed below, this is a rather crude measure of substituent interaction since a long alkyl chain might partly extend beyond the enzyme, while a large spherical group of equivalent hydrophobicity might not extend much beyond the active site. The prime with π_3 signifies that π for Y in congeners (**18b**) is set equal to 0. It was discovered that regardless of whether π_Y is hydrophobic or hydrophilic it has little effect on K_i. Hence it was assumed that Y is not in a position to contact the enzyme and this was confirmed by graphic analysis.

The indicator variable I is assigned the value of 1 for all substituents containing the group —CH_2ZPh— and 0 for all other X. These congeners are about five times (antilog of 0.71) as active as π_3 alone would predict. This must be the result of a specific interaction of the bridged phenyl ring with the enzyme possibly causing a conformational change. The disposable parameter β is estimated by an iterative procedure of Kubinyi (see Chapter 19.2).

In Figure 23, the *L. casei* DHFR active site is shown with $X = CH_2OPh$. The model has been constructed using the coordinates for DHFR with a methotrexate bound to the active site. The nitrogen atoms in the 1-, 2-, 3- and 4-positions of the triazines were positioned over the corresponding atoms of methotrexate. Binding as indicated by the yellow wire model seems more likely since neither hydrophobic nor hydrophilic *meta* or *para* Y-substituents have much effect on K_i.

The bridged phenyl group binds in the large hydrophobic cavity where desolvation must be rather complete.

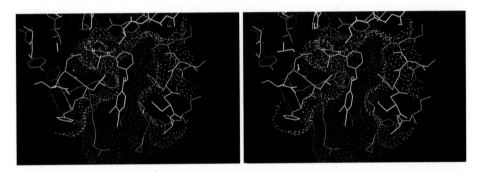

Figure 23 *L. casei* dihydrofolate reductase with model for binding of (**18**; $X = CH_2OPh$). The active site molecular surface is color coded by hydrophobicity as in Figure 1

20.3.10 COMPARISON OF *IN VITRO* AND *IN VIVO* QSAR

Using congeners (**18**) to inhibit *L. casei* cell culture provides another opportunity to compare *in vivo* activity of an enzyme with *in vitro* activity. In equation (16), C is the molar concentration of (**18**) which produces 50% inhibition of normal *L. casei* cell culture.[131]

$$\log(1/C) = 0.80(\pm 0.15)\pi_3' - 1.06(\pm 0.27)\log(\beta 10^{\pi_3'} + 1) - 0.94(\pm 0.39)MR_Y$$
$$+ 0.80(\pm 0.56)I + 4.37(\pm 0.19) \tag{16}$$
$$n = 34, \quad r = 0.929, \quad s = 0.371, \quad \pi_0 = 2.94$$

While $1/C$ in equation (16) is not directly related to $1/K_i$ in equation (15), the two parameters are similar. Except for the term in MR_Y the other parameters of equation (16) are quite close to those of equation (15). It is particularly interesting that the coefficients with I are essentially identical, showing that the highly specific effect of the bridged phenyl moiety appears to be the same *in vivo* and *in vitro*. The MR_Y term refers only to Y substituents and its negative coefficient suggests a steric effect not present when the inhibitors interact with the purified enzyme. In fact, no role at all could be discerned for Y in equation (15). The cause of the steric effect is unknown; it might be due to a change in the conformation of the enzyme or it might be due to an effect of a nearby macromolecule in the living cell. The latter explanation seems more likely since if there were a large change in the conformation of the DHFR it seems unlikely that this would occur without affecting the coefficient with I. To our knowledge this is the first instance where a specific difference in the *in vitro* and *in vivo* behavior of an enzyme has been delineated.

Still another view of *L. casei* DHFR can be seen in QSAR (17) for the 50% inhibition of *L. casei* cells highly resistant to methotrexate.

$$\log(1/C) = 0.42(\pm 0.05)\pi + 1.09(\pm 0.33)I - 0.48(\pm 0.24)MR_Y + 3.39(\pm 0.14) \tag{17}$$

$$n = 38, \quad r = 0.960, \quad s = 0.274$$

The major difference between equations (17) and (15) and (16) is in the hydrophobic parameter. With the resistant cells it is not possible to determine π_0 with confidence, but it would appear to be about six. Although the MR_Y term is less important than in equation (16), it is still significant. Equation (17) is slightly different from our previously published results[131] due to refined π constants.

Normal *L. casei* cells appear to actively transport the DHFR inhibitors, but the resistant cells have a greatly impaired transport system. Therefore it would seem that in the case of the resistant cells, the inhibitors enter *via* passive diffusion in which lipophilicity plays a major role in overshadowing the hydrophobic interaction with the DHFR.[132] It is interesting that cells which lack an active transport system show a similar QSAR to equation (17) for resistant cells. Equation (18), derived by Wooldridge,[133] correlates the inhibition of *S. aureus* cell culture by triazines. A similar equation has been derived for *E. coli* cell culture inhibition.[134] In these examples, where active transport is not involved, the initial slope with π is closer to 0.5 than to 1.

$$\log(1/C) = 0.60\pi - 1.89\log(\beta 10^\pi + 1) + 2.84 \tag{18}$$

$$n = 66, \quad r = 0.963, \quad s = 0.344, \quad \pi_0 = 5.86$$

Active transport of DHFR inhibitors is brought about by a transport protein. The inhibitor must bind to the protein and then be carried into the cell where it then reacts with DHFR. What is surprising is that equations (15) and (16) are so similar in all respects, except for MR_Y, even though reaction with two receptors is involved for equation (16). The similarity for the two different systems can be rationalized in two different ways: the transport protein may recognize the basic 2,4-diaminopyrimidine structure and not contact the rest of the inhibitor or the transport protein has the same structural features that DHFR has, and, in fact, it might even be a special form of DHFR.[132]

It should be noted that the enzyme used to obtain the data for equation (15) was obtained from the type of cells used for equation (17). Thus the difference in structure between equations (15) and (17) cannot be attributed to differences in the enzyme.

Equations (15) to (17) suggest that we can, if we are willing to test large numbers of well-designed congeners, make some inferences about the structure of enzymes in living cells. DHFR may be a more favorable case than many to study because of its active transport system, but with enough well-designed congeners one might be able to separate the QSAR of the enzymic reaction from the QSAR of passive transport. From our present state of knowledge it seems safe to say the DHFR in normal *L. casei* cells must be very similar to that of the enzyme in buffer. In other words the picture of triazine binding as seen in Figure 23 largely holds *in vivo* as well as in *in vitro*.

Equation (19) correlates the inhibition of chicken liver DHFR[130] by triazines, while equation (20) does the same for L1210 leukemia DHFR.[132]

$$\log(1/K_i) = 1.01(\pm 0.14)\pi_3' - 1.16(\pm 0.19)\log(\beta 10^{\pi_3'} + 1) + 0.86(\pm 0.57)\sigma + 6.33(\pm 0.14) \tag{19}$$

$$n = 59, \quad r = 0.906, \quad s = 0.267, \quad \pi_0 = 1.89(\pm 0.36)$$

$$\log(1/K_i) = 0.98(\pm 0.14)\pi_3' - 1.14(\pm 0.20)\log(\beta 10^{\pi_3'} + 1) + 0.79(\pm 0.57)\sigma + 6.12(\pm 0.14) \tag{20}$$

$$n = 58, \quad r = 0.900, \quad s = 0.264, \quad \pi_0 = 1.76(\pm 0.28)$$

For all practical purposes, the two equations are identical. Even the quality of the fit is the same. It seems reasonable to assume that chicken liver DHFR can be used as a model for tumor DHFR, the advantage being that the X-ray crystallographic structure for chicken DHFR[4] is known, while that for L1210 DHFR is not.

In Figure 24 the binding of (**18**); (X = CH_2OPh) to chicken DHFR is shown. The situation is different from *L. casei* DHFR. The outer phenyl ring falls into a hydrophobic cleft formed by Tyr-31 and Ile-60, where it (or smaller substituents) is completely desolvated. This makes sense in terms of the coefficients of 1 with π in equations (19) and (20). As with *L. casei* DHFR, Y does not play a role and π_Y is assigned the value of 0. Equations (19) and (20) lack the term in I of equation (15), showing an important difference in structure for equation (15). Also the term in σ does not occur with bacterial enzymes but does with vertebrate enzymes.

The coefficients with the σ terms in equations (19–21) are in good agreement (mean value 0.86). In an earlier study using DHFR from L5178Y tumor a ρ value of 0.90 was observed; the QSAR from

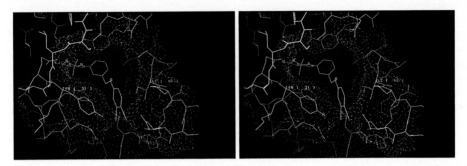

Figure 24 Chicken liver dihydrofolate reductase with model for binding of (**18**; X = CH$_2$OPh), displayed as in Figure 24

L5178Y cells gave a value of 0.88.[135] The high degree of consistency in these five examples encourages belief that when steric and hydrophobic factors can be accounted for, the Hammett equations will hold with enzymes or even in cell culture.

Equations (19–22) are based on 3-X substituents only; 4-X derivatives yield an equation of somewhat different form in which no σ term appears. At first it seemed rather strange that σ would hold only for *meta* substituents. There is no precedent from the field of physical organic chemistry for such an effect. Graphics analysis, however, shows that *para* substituents collide with Phe-49 in chicken liver DHFR. While this steric effect of *para* substituents can be accounted for by the use of steric constants,[130] it is not possible to establish a role for σ. As the σ effect is small, it may be that binding of the inhibitors is shifted by Phe-49 so that the electronic effect is lost.

An interesting aspect of Phe-49 is its apparent ability to accommodate rigid substituents. Groups of the type 4-C≡≡CR would appear to be unable to bind to chicken DHFR unless Phe-49 moves or unless the inhibitors can shift. Possibly both movements are involved since such groups do bind.

The results with the L1210 DHFR and triazines can be compared with L1210 cell culture. Equation (21) correlates 50% inhibition of normal L1210 cells and equation (22) correlates inhibition of L1210 cells highly resistant to methotrexate.[132] The dependence of activity on π and σ is much the same for equations (21) and (20); however, two additional indicator variables I_R and I_{OR} are required. I_R assumes the value of 1 for 3-alkyl groups and I_{OR} takes the value of 1 for 3-alkoxy groups. The alkyl groups are about four times as active, on the average, as expected and the alkoxy groups are half as active as expected. As yet there is no satisfactory explanation. These variables are not necessary for equations (19) and (20) and therefore suggest that either there is some difference in the structure of the DHFR *in vitro* and *in vivo* or some feature of the transport system is responsible.

$$\log(1/C) = 1.13(\pm 0.18)\pi - 1.20(\pm 0.21)\log(\beta 10^\pi + 1) + 0.66(\pm 0.23)I_R + 0.94(\pm 0.37)\sigma$$
$$- 0.32(\pm 0.17)I_{OR} + 6.72(\pm 0.13) \tag{21}$$
$$n = 61, \quad r = 0.890, \quad s = 0.241, \quad \pi_0 = 1.45(\pm 0.43)$$

$$\log(1/C) = 0.42(\pm 0.05)\pi - 0.15(\pm 0.05)MR + 4.83(\pm 0.11) \tag{22}$$
$$n = 62, \quad r = 0.941, \quad s = 0.220$$

20.3.11 SUMMARY

Molecular-modeling approaches to drug design based on the three-dimensional structures of drug receptors are still very new; research in this area began in the 1970s and has continued at a rapidly increasing rate as more receptor structures become available due to advances in X-ray crystallography. Most pharmaceutical companies and several universities now have research groups in macromolecular modeling as hardware and software have become better and more widely available; the number of publications in this area has increased dramatically in the last few years.

QSAR has not had an entirely logical development since its inception.[100] Various laboratories have simply tried the techniques (along with improvisations) in the attempt to correlate structural changes in a set of congeners with the changes they produce in a biological system—enzyme, cell or whole animal. The obvious difficulties with such an undertaking have discouraged many workers from even attempting so unlikely an enterprise. Even at the simplest level, the QSAR of purified enzymes, the problems are almost overwhelming. Except in a few instances, we have no idea of the shape of receptors, the nature of their surfaces or the degree of flexibility which may or may not exist. Lacking this information, 'fundamental' quantum chemical or thermodynamic approaches to SAR are largely ruled out. We are left with model-based (extrathermodynamic) hydrophobic, electronic and steric parameters, regression analysis and other empirical computerized tools to sort out what is occurring at the molecular level. The problem is even worse than it seems. While it is clear that the hydrophobic interaction is of the highest importance in biochemical processes we have no satisfactory definition of exactly what is meant by hydrophobic. We have only the operational definition of the partition coefficient ($\log P$) or π (or fragment constants).

At the receptor level not all of a ligand or one of its substituent parts may contact hydrophobic space. Hydrophobic space can gradually blend into polar space. How small a molecular unit of hydrophobicity can we define? A chlorine atom or a methyl group seem rather clear; however, a substituent such as SO_2NHMe or $COCH_2Cl$ presents serious difficulties. This has led some to attempt to define *atomic* hydrophobic constants[136] with which to explore smaller units of macromolecular surface. Even if we could precisely define the hydrophobic character of each atom in a complex fragment we would still face the problem of its orientation to an unknown surface. Due to the uncertainties in SAR our approach must be statistical in nature. One needs a redundancy in data points to establish, with some degree of certainty, the nature of the interaction of a particular atom or fragment. Moving from fragment constants to atomic constants increases the need for data beyond what will in general be possible for most laboratories to develop. Given the uncertainty in the structure of receptors and the uncertainty in our parameters, there are definite limits to the resolution of SAR problems.

The complexities increase when we attempt to formulate QSAR for cell culture studies. Although there is evidence that enzymes behave much the same in purified form as in the living cell, there are also established differences.[131, 132] Under these circumstances even if the stereoelectronic characteristics of the receptor are independent of the factors controlling the random walk of the ligands to the receptors, there will normally be two unrelated hydrophobically determined processes.[128] One hopes that the two processes are additive. In some instances the QSAR from intact cells may yield a better correlation than that with isolated enzyme (compare equation 22 with equation 20). This speaks for at least rough additivity. However, one still finds ambiguities. In comparing equations (22) and (20) it is seen that a break in correlation with π of X in equation (20) occurs at $\pi_0 = 1.76$, while nothing comparable to this occurs in equation (22). The explanation for this would appear to be that hydrophobic interactions in the random walk process overshadow those involved in receptor binding.

The problem of how to characterize, *via* QSAR, a receptor in a whole animal has not yet been studied. Here difficulties with multiple forms of receptors, metabolism and elimination are frightening to consider. Still, there are good QSARs for whole animal systems but exactly what they mean in terms of a receptor is not yet clear.

The situation with the electronic effect of substituents on biochemical SAR is not as bad as with hydrophobic effects. Here there is the foundation provided by the enormous success of the Hammett equation during the past 50 years. The Hammett equation provides excellent results with almost every kind of organic reaction in homogeneous solution. Moreover, it holds for reactions which are multiple step processes, which means that ρ can be additive. The difficulty with heterogeneous biochemical processes is that the role of hydrophobic and steric effects must be separated before the true value of ρ can be ascertained.

The Hammett equation is solvent dependent and this adds another complexity since we do not have any way of characterizing the solvent in a cell or a mouse. Nevertheless, there are now examples showing that the Hammett postulate does hold in biochemical systems. The electronic terms in equations (12) and (14) are in good agreement for the pyrazole inhibition of alcohol dehydrogenase *in vitro* and *in vivo*. A more impressive result, because of the greater variation in structure, is that from comparison of equation (21) with equations (19) and (20). It is most surprising that the coefficients with σ for such widely different systems are in good agreement. Equation (22) does not depend on σ and here we are left with the uncertainty that this might be due to a change in the conformation of DHFR in the resistant cell, or does the dependence on hydrophobicity in passive diffusion swamp out electronic effects?

That quantum chemistry can play a role in providing electronic parameters for QSAR has been known for some time.[137] The development of ever more powerful computers and better methods for calculations are good reasons for optimism in this area.

One of the most important benefits of QSAR is that it can be used to make comparisons of quite different systems. A reasonable hypothesis is that similar QSARs for different systems suggest similar mechanisms of action. Table 4 illustrates this for the electronic effect of substituents; there are many examples for hydrophobic effects.

Steric interactions between ligand and receptor have always been the ominous dark cloud hanging over QSAR. The surprising fact is that so much progress has been made, in many instances, without any allowance for steric effects.

Equations (1), (2), (7), (8) and (9) show that a kind of global steric parameter (MR) can be of use in rationalizing data even though it is extremely crude. MR is primarily a measure of volume and to a lesser extent polarizability of the loosely held electrons.[103] This may be most useful for inter-molecular steric effects. The parameter makes no allowance for how a substituent might orient itself with respect to the enzyme surface. One would expect more from the STERIMOL parameters of Verloop *et al.* (Chapter 18.4), but in the examples discussed above, this was not true.

In the example of equation (10) one sees how steric effects can be accounted for by indicator variables. While this approach provides no insight into mechanism, it is a means for separating steric effects.

Steric effects in equations (16) and (17) are negative in character and simply call attention to a possible steric problem without much information about its nature. In the examples of equations (16) and (17), MR is associated with only Y of the substituent X, but the negative MR term in equation (22) applies to all of X. If this is entirely a steric effect it would be hard to explain except in terms of a restriction in the receptor cavity not present in the purified DHFR, but present in DHFR of the resistant cells.

We have attempted to define three classes of protein surface: hydrophobic (red), polar (blue) and intermediate (yellow), knowing that there is a continuum of types. We encounter the same problem with the hydrophobicity of the surface as with the ligand. How small a unit of surface is it practical to define?

There have been many reports in which electrostatic surface potentials have been calculated for ligands and some for enzymes, but it is not yet clear how these can be put into numerical form for use in QSAR. Another question is, does the hydrophobic character of a surface parallel its electrostatic potential? We lack understanding about the hydrophobic hydration of the surface necessary for QSAR in terms of the potentials of the surface.

In mapping the surface of receptors using QSAR one must be alert to the behavior of vectors such as $\log P$, π, MR or E_s, that is, as larger and more heterogeneous substituents are included in the QSAR one may, for example, go beyond interaction with a hydrophobic surface into a polar area with or without steric features. Routine processing of data by multivariate techniques will not uncover such discontinuities. These will have to be deduced from a careful study of the residuals and the testing of *ad hoc* hypotheses. Developing a complex QSAR may still be as much of an art as a science.

Our discovery that in several instances a phenyl ring can rotate 180° to place a *meta* substituent into hydrophobic or polar space and that this would be reflected by a constant electronic effect of substituents, but a zero hydrophobic interaction of *meta* polar groups, was by no means anticipated. It seemed that polar groups, although not showing hydrophobic interaction, might interact *via* dispersion forces modeled by MR. So far, evidence for this is lacking in examples with chymotrypsin,[112] papain,[116] actinidin,[119, 120] ficin[121] and subtilisin.[138]

We expected to find many examples where MR would have a negative coefficient, signifying a detrimental steric effect. In fact, the number of QSAR with positive MR terms outweigh those with negative. The positive MR term could be interpreted as modeling London dispersion forces between ligand and enzyme. However, as Mulliken and Person pointed out, long ago, in the formation of complexes in solution such contributions usually cancel. The gain in intracomplex dispersion energy is approximately balanced by the loss of solvent donor and solvent acceptor dispersion energies.[139] From the evidence in hand it would seem that positive MR effects may often be associated with a buttressing or steadying effect of a polar part of the enzyme on the ligand. It is now quite clear that even small changes in the positioning of reacting organic compounds can have large effects on reaction rates.[111]

There has been speculation about what the coefficient with the π term in enzymic QSAR signifies.[68] Fersht has discussed this problem in terms of chymotrypsin.[140] In considering two sets of congeners with substituents assumed to be binding in the 'hydrophobic hole' Fersht noted a slope in one case for $\log(K_{cat}/K_m)$ of 2.2 and in another case of $\log K_i$ of 1.5 when the data were plotted against π. Assuming octanol to be the perfect model system with a coefficient of 1 for complete desolvation by an enzymic pocket, how are the figures of 1.5 and 2.2 accounted for? Fersht suggests that the cavity in chymotrypsin is hydrated and that it and the substrate are both desolvated on uniting; on partitioning into octanol only the solute surface is desolvated.

While this is a reasonable rationalization, results with other enzymes do not fit this picture. We have noted that in several instances where substituents appear to fall on a more or less flat hydrophobic region, a coefficient of about 0.5 for π or $\log P$ is found for papain,[116] actinidin[119, 120] and carbonic anhydrase.[44] In the case of alcohol dehydrogenase, where complete desolvation must occur, a slope of 1 is found.[127] When binding occurs in a hydrophobic cleft with papain a coefficient of 1 is found.[115] How to interpret the coefficient with π is somewhat of a mystery. If octanol is the perfect model and complete desolvation occurs, a coefficient between 1.5 and 2 might be expected. The difficulty is that in addition to partitioning effects the problem of steric effects is also present. The question of just how strongly hydrophobically hydrated a given surface area of protein is has received little attention. Also, hydrophobically determined partitioning may produce secondary conformational effects which could amplify or depress the coefficient with π. Like everything else connected with QSAR the situation is complex and unclear and much more work is needed before we can attach specific meaning to the terms of these equations.

The marriage of QSAR and molecular graphics dates only from 1982.[115] In spite of this short history, enough examples have been studied, with a variety of enzymes, to instill confidence in this powerful means of delineating the relative importance of the major factors involved in the union of ligands and receptors of macromolecules. While we cannot expect perfect answers (with r close to 1), we can confidently expect to obtain guidance in the study of enzymic processes in numerical terms which allow the comparison of results from different laboratories using different ligands and different macromolecules. In short we are provided with a statistically based numerical language which should carry us to higher levels of understanding.

Early on we believed that once we had the detailed three-dimensional structure of our receptor in hand, it would be easy to design an optimum drug; after all, much of classical medicinal chemistry and drug design is oriented towards defining a model of the receptor site based on the structures and activities of the compounds that interact with it. We have now had this opportunity with several different receptor structures and found that there are still formidable challenges to overcome, but receptor-based design and the variety of powerful modeling and QSAR methods that can be used with it undoubtedly provide the best possible chance for success in rational drug design.

ACKNOWLEDGEMENTS

We thank John A. Tainer, Michael G. Rossman and Paul A. Bartlett for providing slides for Figures 2, 6 and 11.

20.3.12 REFERENCES

1. C. R. Beddell, *Chem. Soc. Rev.*, 1984, **13**, 279.
2. W. G. J. Hol, *Angew. Chem., Int. Ed. Engl.*, 1986, **25**, 767.
3. F. C. Bernstein, T. F. Koetzle, G. J. B. Williams, E. F. Meyer, Jr., M. D. Brice, J. R. Rodgers, O. Kennard, T. Shimanouchi and M. Tasumi, *J. Mol. Biol.*, 1977, **112**, 535.
4. J. Kraut and D. A. Matthews, in 'Biological Macromolecules and Assemblies, Active Sites of Enzymes', ed. F. A. Jurnak and A. McPherson, Wiley, New York, 1987, vol. 3, p. 1.
5. K. K. Kannan, I. Vaara, B. Notstrand, S. Lövgren, A. Borell, K. Fridborg and M. Petef, in 'Drug Action at the Molecular Level', ed. G. C. K. Roberts, Macmillan, New York, 1977, p. 73.
6. H. Eklund, J.-P. Samama, L. Wallén, C.-I. Brändén, Å. Åkeson and T. A. Jones, *J. Mol. Biol.*, 1981, **146**, 561.
7. L. Sawyer, D. M. Shotton, J. W. Campbell, P. L. Wendell, H. Muirhead, H. C. Watson, R. Diamond and R. C. Ladner, *J. Mol. Biol.*, 1978, **118**, 137.
8. B. W. Dijkstra, K. H. Kalk, W. G. J. Hol and J. Drenth, *J. Mol. Biol.*, 1981, **147**, 97.
9. M. F. Perutz, G. Fermi, D. J. Abraham, C. Poyart and E. Bursaux, *J. Am. Chem. Soc.*, 1986, **108**, 1064.
10. D. C. Wiley, I. A. Wilson and J. J. Skehel, *Nature (London)*, 1981, **289**, 373.
11. J. N. Varghese, W. G. Laver and P. M. Colman, *Nature (London)*, 1983, **303**, 35.
12. T. J. Smith, M. J. Kremer, M. Luo, G. Vriend, E. Arnold, G. Kamer, M. G. Rossman, M. A. McKinlay, G. D. Diana and M. J. Otto, *Science (Washington, D.C.)*, 1986, **233**, 1286.
13. J. M. Hogle, M. Chow and D. J. Filman, *Science (Washington, D.C.)*, 1985, **229**, 1358.
14. M. Luo, G. Vriend, G. Kamer, I. Minor, E. Arnold, M. G. Rossman, U. Boege, D. G. Scraba, G. M. Duke and A. C. Palmenberg, *Science (Washington, D.C.)*, 1987, **235**, 182.
15. O. Herzberg and J. Moult, *Science (Washington, D.C.)*, 1987, **236**, 694.
16. J. Deisenhofer, O. Epp, K. Miki, R. Huber and H. Michel, *Nature (London)*, 1985, **318**, 618.
17. S. E. Ealick, T. J. Greenhough, Y. S. Babu, D. C. Carter, W. J. Cook, C. E. Bugg, S. A. Rule, J. Habach, J. R. Helliwell, J. D. Stoeckler, S. F. Chen and R. E. Parks, Jr., *Ann. N. Y. Acad. Sci.*, 1985, **451**, 311.
18. Y. S. Babu, J. S. Sack, T. J. Greenhough, C. E. Bugg, A. R. Means and W. J. Cook, *Nature (London)*, 1985, **315**, 37.
19. O. Herzberg and M. N. G. James, *Nature (London)*, 1985, **313**, 653.

20. L. W. Hardy, J. S. Finer-Moore, W. R. Montfort, M. O. Jones, D. V. Santi and R. M. Stroud, *Science (Washington, D.C.)*, 1987, **235**, 448.
21. A. McPherson, 'Preparation and Analysis of Protein Crystals', Wiley, New York, 1982.
22. J. A. McCammon and S. C. Harvey, 'Dynamics of Proteins and Nucleic Acids', Cambridge University Press, Cambridge, 1987.
23. J. G. Vinter, in 'Topics in Molecular Pharmacology: Molecular Graphics and Drug Design', ed. A. S. V. Burgen, G. C. K. Roberts and M. S. Tute, Elsevier, Amsterdam, 1986, vol. 3.
24. K. Wüthrich, 'NMR of Proteins and Nucleic Acids', Wiley, New York, 1986.
25. A. D. Kline, W. Braun and K. Wüthrich, *J. Mol. Biol.*, 1986, **189**, 377.
26. J. W. Pflugrath, G. Wiegand and R. Huber, *J. Mol. Biol.*, 1986, **189**, 383.
27. T. L. Blundell, B. L. Sibanda, M. J. E. Sternberg and J. M. Thornton, *Nature (London)*, 1987, **326**, 347.
28. T. A. Jones and S. Thirup, *EMBO J.*, 1986, **5**, 819.
29. D. B. Wetlaufer (ed.), 'The Protein Folding Problem', AAAS Selected Symposium 89, Westview Press, Boulder, CO, 1984.
30. F. E. Cohen, P. A. Kosen, I. D. Kuntz, L. B. Epstein, T. L. Ciardelli and K. A. Smith, *Science (Washington, D.C.)*, 1986, **234**, 349.
31. R. Langridge, T. E. Ferrin, I. D. Kuntz and M. L. Connolly, *Science (Washington, D.C.)*, 1981, **211**, 661.
32. M. L. Connolly, *Science (Washington, D.C.)*, 1983, **221**, 709.
33. F. M. Richards, *Annu. Rev. Biophys. Bioeng.*, 1977, **6**, 151.
34. P. A. Bash, N. Pattabiraman, C. Huang, T. E. Ferrin and R. Langridge, *Science (Washington, D.C.)*, 1983, **222**, 1325.
35. L. H. Pearl and A. Honegger, *J. Mol. Graphics*, 1983, **1**, 9.
36. M. L. Connolly, *J. Appl. Crystallogr.*, 1983, **16**, 548.
37. M. L. Connolly, *J. Am. Chem. Soc.*, 1985, **107**, 1118.
38. M. L. Connolly, *J. Mol. Graphics*, 1985, **3**, 19.
39. R. J. Feldmann, D. H. Bing, B. C. Furie and B. Furie, *Proc. Natl. Acad. Sci. USA*, 1978, **75**, 5409.
40. A. M. Lesk and K. D. Hardman, *Science (Washington, D.C.)*, 1982, **216**, 539.
41. J. S. Richardson, *Adv. Protein Chem.*, 1981, **34**, 167.
42. M. Carson and C. E. Bugg, *J. Mol. Graphics*, 1986, **4**, 121.
43. J. M. Burridge and S. J. P. Todd, *J. Mol. Graphics*, 1986, **4**, 220.
44. M. Recanatini, T. Klein, C. Yang, J. McClarin, R. Langridge and C. Hansch, *Mol. Pharmacol.*, 1986, **29**, 436.
45. P. K. Weiner, R. Langridge, J. M. Blaney, R. Schaefer and P. A. Kollman, *Proc. Natl. Acad. Sci. USA*, 1982, **79**, 3754.
46. U. C. Singh and P. A. Kollman, *J. Comput. Chem.*, 1984, **5**, 129.
47. E. D. Getzoff, J. A. Tainer, P. K. Weiner, P. A. Kollman, J. S. Richardson and D. C. Richardson, *Nature (London)*, 1983, **306**, 287.
48. D. Barry, unpublished results.
49. P. Kollman, in 'X-ray Crystallography and Drug Action', ed. A. S. Horn and C. J. De Ranter, Oxford University Press, Oxford, 1984, p. 63.
50. A. R. Fersht, *Trends Biochem. Sci. (Pers. Ed.)*, 1984, **9**, 145.
51. A. R. Fersht, J.-P. Shi, J. Knill-Jones, D. M. Lowe, A. J. Wilkinson, D. M. Blow, P. Brick, P. Carter, M. M. Y. Waye and G. Winter, *Nature (London)*, 1985, **314**, 235.
52. I. P. Street, C. R. Armstrong and S. G. Withers, *Biochemistry*, 1986, **25**, 6021.
53. P. A. Bartlett and C. K. Marlowe, *Science (Washington, D.C.)*, 1987, **235**, 569.
54. W. G. J. Hol and R. K. Wierenga, in 'X-ray Crystallography and Drug Action', ed. A. S. Horn and C. J. De Ranter, Oxford University Press, Oxford, 1984, p. 151.
55. S. K. Burley and G. A. Petsko, *Science (Washington, D. C.)*, 1985, **229**, 23.
56. T. E. Creighton, 'Proteins', Freeman, New York, 1984.
57. C. R. Cantor and P. R. Schimmel, 'Biophysical Chemistry', Freeman, New York, 1980.
58. G. E. Schulz and R. H. Schirmer, 'Principles of Protein Structure', Springer-Verlag, New York, 1979.
59. B. L. Bush, *Comput. Chem.*, 1984, **8**, 1.
60. J. M. Blaney, Ph.D. Dissertation, University of California, San Francisco, 1982.
61. J. M. Blaney, P. K. Weiner, A. Dearing, P. A. Kollman, E. C. Jorgensen, S. J. Oatley, J. M. Burridge and C. C. F. Blake, *J. Am. Chem. Soc.*, 1982, **104**, 6424.
62. E. F. Meyer, Jr., R. Radhakrishnan, G. M. Cole and L. G. Presta, *J. Mol. Biol.*, 1986, **189**, 533.
63. S. Naruto, I. Motoc, G. R. Marshall, S. B. Daniels, M. J. Sofia and J. A. Katzenellenbogen, *J. Am. Chem. Soc.*, 1985, **107**, 5262.
64. S. J. Wodak, J.-L. De Coen, S. J. Edelstein, H. Demarne and Y. Beuzard, *J. Biol. Chem.*, 1986, **261**, 14 717.
65. (a) I. D. Kuntz, J. M. Blaney, S. J. Oatley, R. Langridge and T. E. Ferrin, *J. Mol. Biol.*, 1982, **161**, 269; (b) R. L. Desjarlais, R. P. Sheridan, J. S. Dixon, I. D. Kuntz and R. Venkataraghavan, *J. Med. Chem.*, 1986, **29**, 2149.
66. (a) J. M. Blaney and G. M. Crippen, unpublished results; (b) M. Billeter, T. F. Havel and I. D. Kuntz, *Biopolymers*, 1987, **26**, 777.
67. P. A. Bash, U. C. Singh, F. K. Brown, R. Langridge and P. A. Kollman, *Science (Washington, D.C.)*, 1987, **235**, 574.
68. C. Hansch and T. E. Klein, *Acc. Chem. Res.*, 1986, **19**, 392.
69. C. R. Beddell, P. J. Goodford, F. E. Norrington, S. Wilkinson and R. Wootton, *Br. J. Pharmacol.*, 1976, **57**, 201.
70. C. R. Beddell, P. J. Goodford, G. Kneen, R. D. White, S. Wilkinson and R. Wootton, *Br. J. Pharmacol.*, 1984, **82**, 397.
71. P. J. Goodford, *J. Med. Chem.*, 1985, **28**, 849.
72. F. Takusagawa, B. M. Goldstein, S. Youngster, R. A. Jones and H. M. Berman, *J. Biol. Chem.*, 1984, **259**, 4714.
73. A. H.-J. Wang, G. Ughetto, G. J. Quigley and A. Rich, *Biochemistry*, 1987, **26**, 1152.
74. A. H.-J. Wang, G. Ughetto, G. J. Quigley, T. Hakoshima, G. A. van der Marel, J. H. van Boom and A. Rich, *Science (Washington, D.C.)*, 1984, **225**, 1115.
75. S. E. Sherman, D. Gibson, A. H.-J. Wang and S. J. Lippard, *Science (Washington, D.C.)*, 1985, **230**, 412.
76. M. L. Kopka, C. Yoon, D. Goodsell, P. Pjura and R. E. Dickerson, *J. Mol. Biol.*, 1985, **183**, 553.
77. D. B. McKay and T. A. Steitz, *Nature (London)*, 1981, **290**, 744.
78. C. O. Pabo and M. Lewis, *Nature (London)*, 1982, **298**, 443.

79. W. F. Anderson, D. H. Ohlendorf, Y. Takeda and B. W. Matthews, *Nature (London)*, 1981, **290**, 754.
80. R. W. Schevitz, Z. Otwinowski, A. Joachimiak, C. L. Lawson and P. B. Sigler, *Nature (London)*, 1985, **317**, 782.
81. J. A. McClarin, C. A. Frederick, B.-C. Wang, P. Greene, H. W. Boyer, J. Grable and J. M. Rosenberg, *Science (Washington, D.C.)*, 1986, **234**, 1526.
82. P. H. von Hippel and O. G. Berg, *Proc. Natl. Acad. Sci. USA*, 1986, **83**, 1608.
83. M. L. Kopka, C. Yoon, D. Goodsell, P. Pjura and R. E. Dickerson, *Proc. Natl. Acad. Sci. USA*, 1985, **82**, 1376.
84. T. P. Lybrand, S. C. Brown, S. Creighton, R. H. Shafer and P. A. Kollman, *J. Mol. Biol.*, 1986, **191**, 495.
85. F. F. Brown and P. J. Goodford, *Br. J. Pharmacol.*, 1977, **60**, 337.
86. M. Merrett, D. K. Stammers, R. D. White, R. Wootton and G. Kneen, *Biochem. J.*, 1986, **239**, 387.
87. D. A. Matthews, S. L. Smith, D. P. Baccanari, J. J. Burchall, S. J. Oatley and J. Kraut, *Biochemistry*, 1986, **25**, 4194.
88. L. F. Kuyper, B. Roth, D. P. Baccanari, R. Ferone, C. R. Beddell, J. N. Champness, D. K. Stammers, J. G. Dann, F. E. Norrington, D. J. Baker and P. J. Goodford, *J. Med. Chem.*, 1985, **28**, 303.
89. R. L. Magolda, W. C. Ripka, W. Galbraith, P. R. Johnson and M. S. Rudnick, in 'Prostaglandins, Leukotrienes, and Lipoxins', ed. J. M. Bailey, Plenum Press, New York, 1985, p. 669.
90. F. F. Davidson, E. A. Dennis, M. Powell and J. R. Glenney, Jr., *J. Biol. Chem.*, 1987, **262**, 1698.
91. B. W. Dijkstra, R. Renetseder, K. H. Kalk, W. G. J. Hol and J. Drenth, *J. Mol. Biol.*, 1983, **168**, 163.
92. S. Brunie, J. Bolin, D. Gewirth and P. B. Sigler, *J. Biol. Chem.*, 1985, **260**, 9742.
93. W. C. Ripka *et al.*, unpublished results.
94. J. G. N. de Jong, H. Amesz, A. J. Aarsman, H. B. M. Lenting and H. van den Bosch, *Eur. J. Biochem.*, 1987, **164**, 129.
95. D. E. Tronrud, H. M. Holden and B. W. Matthews, *Science (Washington, D.C.)*, 1987, **235**, 571.
96. J. M. Blaney, E. C. Jorgensen, M. L. Connolly, T. E. Ferrin, R. Langridge, S. J. Oatley, J. M. Burridge and C. C. F. Blake, *J. Med. Chem.*, 1982, **25**, 785.
97. J. D. McKinney, K. Chae, S. J. Oatley and C. C. F. Blake, *J. Med. Chem.*, 1985, **28**, 375.
98. U. Rickenbacker, J. D. McKinney, S. J. Oatley and C. C. F. Blake, *J. Med. Chem.*, 1986, **29**, 641.
99. L. G. Pedersen, T. A. Darden, S. J. Oatley and J. D. McKinney, *J. Med. Chem.*, 1986, **29**, 2451.
100. C. Hansch, P. P. Maloney, T. Fujita and R. M. Muir, *Nature (London)*, 1962, **194**, 178.
101. J. G. Topliss (ed.), 'Quantitative Structure–Activity Relationships in Drugs', Academic Press, New York, 1983.
102. The journal *Quantitative Structure–Activity Relationships*, published by Verlag Chemie since 1982, is devoted to the subject. In addition to research articles it publishes abstracts of all QSAR papers.
103. C. Hansch and A. J. Leo, 'Substituents Constants for Correlation Analysis in Chemistry and Biology', Wiley Interscience, New York, 1979.
104. C. Hansch, A. J. Leo and D. Nikaitani, *J. Org. Chem.*, 1972, **37**, 3090.
105. R. F. Rekker, 'The Hydrophobic Fragmental Constant', Elsevier, Amsterdam, 1977, p. 161.
106. C. Hansch and E. Coats, *J. Pharm. Sci.*, 1970, **59**, 731.
107. C. Hansch, C. Grieco, C. Silipo and A. Vittoria, *J. Med. Chem.*, 1977, **20**, 1420.
108. C. Grieco, C. Hansch, C. Silipo, R. N. Smith, A. Vittoria and K. Yamada, *Arch. Biochem. Biophys.*, 1979, **194**, 542.
109. C. Hansch and J. M. Blaney, in 'Drug Design: Fact or Fantasy?', ed. G. Jolles and K. R. H. Wooldridge, Academic Press, New York, 1984, p. 185.
110. N. R. C. Campbell, J. A. VanLoon, R. S. Sundaram, M. M. Ames, C. Hansch and R. Weinshilboum, *Mol. Pharmacol.*, 1987, **32**, 813.
111. F. M. Menger, *Acc. Chem. Res.*, 1985, **18**, 128.
112. L. Morgenstern, M. Recanatini, T. E. Klein, W. Steinmetz, C. Z. Yang, R. Langridge and C. Hansch, *J. Biol. Chem.*, 1987, **262**, 10767.
113. W. B. Schweizer and J. D. Dunitz, *Helv. Chim. Acta*, 1982, **65**, 1547.
114. C. D. Selassie, M. Chow and C. Hansch, *Chem.-Biol. Interact.*, 1988, **68**, 13.
115. R. N. Smith, C. Hansch, K. H. Kim, B. Omiya, G. Fukumura, C. D. Selassie, P. Y. C. Jow, J. M. Blaney and R. Langridge, *Arch. Biochem. Biophys.*, 1982, **215**, 319.
116. A. Carotti, R. N. Smith, S. Wong, C. Hansch, J. M. Blaney and R. Langridge, *Arch. Biochem. Biophys.*, 1984, **229**, 112.
117. G. R. Welch (ed.), 'The Fluctuating Enzyme', Wiley Interscience, New York, 1986.
118. I. G. Kamphuis, K. H. Kalk, M. B. A. Swarte and J. Drenth, *J. Mol. Biol.*, 1984, **179**, 233.
119. A. Carotti, C. Hansch, M. M. Mueller and J. M. Blaney, *J. Med. Chem.*, 1984, **27**, 1401.
120. A. Carotti, C. Hansch, M. M. Mueller and J. M. Blaney, *J. Med. Chem.*, 1985, **28**, 261.
121. A. Carotti, G. Casini and C. Hansch, *J. Med. Chem.*, 1984, **27**, 1427.
122. A. Carotti, C. Raguseo and C. Hansch, *Chem.-Biol. Interact.*, 1985, **52**, 279.
123. C. Hansch, J. McClarin, T. Klein and R. Langridge, *Mol. Pharmacol.*, 1985, **27**, 493.
124. C. Hansch, T. Klein, J. McClarin, R. Langridge and N. W. Cornell, *J. Med. Chem.*, 1986, **29**, 615.
125. C. Hansch, J. Schaeffer and R. Kerley, *J. Biol. Chem.*, 1972, **247**, 4703.
126. C. Hansch and J.-P. Björkroth, *J. Org. Chem.*, 1986, **51**, 5461.
127. N. W. Cornell, C. Hansch, K. H. Kim and K. Henegar, *Arch. Biochem. Biophys.*, 1983, **227**, 81.
128. C. Hansch, *Acc. Chem. Res.*, 1969, **2**, 232.
129. J. M. Blaney, C. Hansch, C. Silipo and A. Vittoria, *Chem. Rev.*, 1984, **84**, 333.
130. C. Hansch, B. A. Hathaway, Z.-R. Guo, C. D. Selassie, S. W. Dietrich, J . M. Blaney, R. Langridge, K. W. Volz and B. T. Kaufman, *J. Med. Chem.*, 1984, **27**, 129.
131. E. A. Coats, C. S. Genther, S. W. Dietrich, Z.-R. Guo and C. Hansch, *J. Med. Chem.*, 1981, **24**, 1422.
132. C. D. Selassie, C. D. Strong, C. Hansch, T. J. Delcamp, J. H. Freisheim and T. A. Khwaja, *Cancer Res.*, 1986, **46**, 744.
133. K. R. H. Wooldridge, *Eur. J. Med. Chem.*, 1980, **15**, 63.
134. S. W. Dietrich, R. N. Smith, S. Brendler and C. Hansch, *Arch. Biochem. Biophys.*, 1979, **194**, 612.
135. C. D. Selassie, C. Hansch, T. A. Khawaja, C. B. Dias and S. Pentecost, *J. Med. Chem.*, 1984, **27**, 347.
136. A. K. Ghose and G. M. Crippen, *J. Med. Chem.*, 1985, **28**, 333.
137. C. Hansch and T. Fujita, *J. Am. Chem. Soc.*, 1964, **86**, 1616.
138. A. Carotti, C. Raguseo and C. Hansch, *Quant. Struct.–Act. Relat.*, 1985, **4**, 145.
139. R. S. Mulliken and W. B. Person, *J. Am. Chem. Soc.*, 1969, **91**, 3409.
140. A. Fersht, 'Enzyme Structure and Mechanism', 2nd ed., Freeman, New York, 1985, p. 305.

21.1

The Extrathermodynamic Approach to Drug Design

Kyoto University, Japan

21.1.1 INTRODUCTION

21.1.1.1 Extrathermodynamic and Linear Free Energy Relationships

The term 'extrathermodynamic relationship' was first introduced by Leffler and Grunwald[1] for the analysis of organic reaction mechanisms. It means a relationship of quantities related to thermodynamic parameters such as free energies, enthalpies and entropies for various reactions. Among such quantities, simple relationships have often been found. They correlate thermodynamic and related parameters but the relationships themselves do not require the formal structure of thermodynamics; hence they are extrathermodynamic.

The Hammett–Taft type correlation equations[2, 3] for the analysis of substituent effects on various organic reactivities for series of compounds are examples of extrathermodynamic relationships. They represent relationships between logarithms of rate or equilibrium constants, *i.e.* free energy related quantities. They are formulated by using such free energy related parameters as σ, σ^* and E_s, empirically defined from certain standard reactions. Since the correlation equations are often linear with respect to at least one variable they are called linear free energy relationships.[4, 5]

The extrathermodynamic approach to drug design is directed mainly towards extending the Hammett–Taft type correlation analysis to the structure–activity relationships of drugs in biological systems and towards gaining insight into rational procedures for designing novel drugs. This was initiated by Hansch and co-workers in 1961–1964.[6–8] This chapter deals with the theory, procedure and application to drug design of the Hansch and related approaches.

21.1.1.2 The Nature of Drug Actions and the Extrathermodynamic Approach

Before entering into a detailed review of the Hansch approach, it is worth considering the conditions and requirements of drug actions to which the extrathermodynamic approach applies. In order to exhibit a certain biological activity, drug molecules, usually having a complex structure, must interact with a certain cellular component at the site of action or a particular receptor. Biological systems are composed of a number of heterogeneous phases, and the site at which drugs are administered is usually separated from the site of action. Thus the drug molecules must be transported through phase boundaries and undergo adsorption and desorption processes with proteins and membranes as well as partitioning between different phases before they reach the site of action. Moreover, the drug–receptor interaction at this site does not occur without perturbation by surrounding heterogeneous components such as water, serum protein, lipid particles, *etc.* Although the transport processes and the interaction with the site of action are essentially chemical and/or physicochemical in nature, they are far more complex than the homogeneous equilibria and rate processes of usual organic reactions. Thus, it would rarely be possible to rationalize the mechanism of drug action in terms of deterministic as well as microscopic mechanisms for individual steps of the transport and interaction processes. One of the advantages of the extrathermodynamic approach is that it does not require explicit identification of detailed microscopic mechanisms.

It is generally accepted that a drug initiates a chain of events which eventually leads to a specific biological effect but which does not involve the drug molecule after it triggers the mechanism through a drug–receptor interaction. For example, sucrose tastes sweet, but the role of sucrose molecules is to stimulate the taste buds, and they do not participate in the process of sensory conduction as such. The potency of the observable biological response is a direct reflection of the intensity of the chain of physiological events which, in turn, is determined by the degree of drug–receptor interaction. The degree of this interaction is controlled by the chemical and physicochemical properties of the partners and the drug concentration at the site of action, which, in turn, is governed by the drug concentration or dose at the site of administration as well as the physicochemical transport process.

The overall process, including the drug–receptor interaction and transport, is composed of a number of unit equilibrium and rate processes. For drug actions such as those observed using whole organisms, where the emergence of biological activity is rather slow and the amount of drug participating in the transport and interaction steps would be much lower than that at the site of administration, the rate of emergence of biological response can be formulated as equation (1), where C is the molar concentration or dose initially applied at the site of administration, which can be taken approximately as being unchanged. K is the overall constant for the consecutive chain of equilibria reaching toward the rate-determining step and k is the rate constant of the rate-determining step in the overall process. B is a proportionation factor connecting the intensity of drug–receptor interaction and the potency of observable biological activity. The rate-determining step is not necessarily the step for drug–receptor interaction. If one of the compartmental phases in the transport system traps a relatively high amount of drug molecules, the step of transport from this phase toward receptor could be rate determining, the rate constant of which is assumed as being equal to that of the drug–receptor interaction. The situations that can be expressed by equation (1) are thus based upon a steady state hypothesis. If the first step in the chain of processes is rate limiting, K need not be considered.

$$d(\text{response})/dt = BKkC \qquad (1)$$

For systems such as isolated tissues and organs and microorganisms in aqueous media where the drug is administered externally, transport occurs rather quickly and an equilibrium or pseudoequilibrium state is established for drug amounts between sites of administration and action. In such cases, the intensity of biological response could be unchanged after a short initial interval and formulated as equation (2), where C is the molar concentration or dose in the external phase. It is theoretically the value after the equilibrium has been established. Usually, the volume of the external phase is much greater than that of the biophases, and the drug amount in the biophases could be much lower than that in the external phase. Therefore, we can assume that C is the value initially applied as a first approximation.

$$\text{Response} = BKC \tag{2}$$

The potency of biological activity is often represented by an equipotent concentration or dose after a fixed time interval of testing such as I_{50}, LD_{50}, EC_{50} and MIC (minimum inhibitory concentration). Under conditions where these quantities are measured, the left-hand sides of equations (1) and (2) are constant throughout the series of compounds exhibiting a particular biological activity. Since B is also regarded as a constant for the given system, equations (1) and (2) can be rewritten in the form of equations (3) and (4), where C corresponds with the equipotent concentration or dose. If the mode of administration is fixed, the dose would be proportional to the concentration in the compartment where drug molecules were first applied. Since the logarithms of equilibrium and rate constants are free energy related, the $\log(1/C)$ (and also $\log C$) value is a free energy related biological parameter. If the potency is expressed as the ratio relative to that of a certain reference compound the situation is unchanged, *i.e.* the logarithm of relative potency is free energy related. For the extrathermodynamic approach to drug actions, therefore, $\log(1/C)$ or its counterparts should be used as the biological parameter regardless of whether the drug actions are exhibited under equilibrium or nonequilibrium conditions.

$$\log(1/C) = \log K + \log k + \text{constant} \tag{3}$$

$$\log(1/C) = \log K + \text{constant} \tag{4}$$

21.1.1.3 Extrathermodynamic Relationships among Biological Activities

As indicated in the above section, drug action is usually a chain or a network of complex unit processes. However, detailed identification of the individual processes is not required in the extrathermodynamic approach. Some examples showing this advantage will be presented in this section.

The first example is taken from a study of Portoghese.[9] He analyzed a possible relationship in analgesic activity between two series of compounds expressed as $\log ED_{50}$ ($\mu mol\,kg^{-1}$) measured by Eddy[10] using the hot plate method after subcutaneous application to mice. Figure 1 shows that the $\log ED_{50}$ value for the series of meperidine analogs (1; series 1) is linearly related with that for the series of reversed esters (2; series 2). In both series, R is varied from benzyl and cinnamyl to phenylbutyl. The correlation was formulated as equation (5).[9, 11]

$$\log ED_{50}(1) = 1.199(\pm 0.308)\log ED_{50}(2) + 1.357(\pm 0.260) \tag{5}$$

$$n = 5, \quad r = 0.990, \quad F(1,3) = 153.2, \quad s = 0.167$$

(1) R' = CO_2Et
(2) R' = OCOEt
(3) R' = OCOMe

In equation (5) and all subsequent correlation equations, n is the number of compounds or data points, r is the correlation coefficient, F is the value of the F ratio and s is the standard deviation.

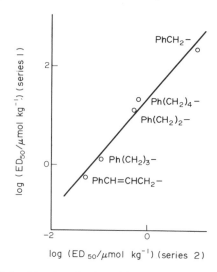

Figure 1 Extrathermodynamic relationship of analgesic activity between two series of *N*-substituted phenylpiperidine derivatives (reproduced from ref. 9 by permission of the American Chemical Society)

Figures in parentheses are the 95% confidence intervals of the regression coefficient and the constant term. Equation (5) indicates the existence of an extrathermodynamic relationship in the analgesic activity between two similar but different series of compounds. The fact that the slope is close to unity should be a consequence of almost identical modes of action including transport processes and the binding interaction with receptor. Since the lipophilic property does not differ much in corresponding pairs of compounds between series 1 and 2 (Figure 1), the effect of R substituents on the transport process is similar. Thus, the binding modes of two different analgesiophores (analgesic molecule other than the substituent R) should experience quite similar physicochemical environments on the receptor sites. Similar linear relationships were found in corresponding pairs of these compounds with such analogs as (3) having identical R substituents. In contrast, no relationship was found in the activity between compounds (1)–(3) and *N*-substituted morphine analogs (4). In the meperidine series (1), the replacement of *N*-methyl with *N*-cinnamyl enhanced the activity about 60 times, whereas in the morphine analogs it results in a loss of activity. This means that the binding mode of the analgesiophore in phenylpiperidine compounds (1)–(3) is different from that in the morphine series (4).

(4)

The second example comes from the studies of Timmermans and his group.[12,13] They found a linear relationship between the potencies of a series of phenyliminoimidazolines (5), structurally related to clonidine (5; X = 2,6-Cl$_2$), for decreasing cardiac frequency (bradycardia) and for decreasing mean arterial pressure (hypotension) following intravenous injection to normotensive rats as shown in Figure 2.[12] ED$_{25}$ is the dose (μmol kg^{-1}) required to induce a 25% decrease in cardiac frequency, and ED$_{30}$ is that required to invoke a 30% decrease in blood pressure. For 26 compounds where X is varied covering 2-, 2,6-di- and 2,4,6-tri-substitutions with various combinations of chloro, bromo, fluoro, methyl, methoxy and nitro substituents, equation (6) was formulated.[9,11]

$$\log(1/\text{ED}_{30})(\text{hypotension}) = 1.222(\pm 0.147)\log(1/\text{ED}_{25})(\text{bradycardia}) - 0.024(\pm 0.115) \qquad (6)$$

$$n = 26, \quad r = 0.962, \quad F(1,24) = 294.9, \quad s = 0.244$$

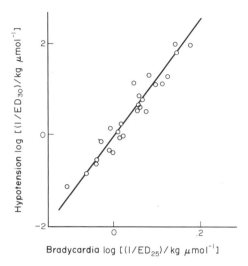

(5)

Figure 2 Extrathermodynamic relationship between bradycardic and hypotensive activities of clonidine analogs (5) (modified after ref. 12)

Equation (6) demonstrates an extrathermodynamic relationship between two different types of activity exhibited by a single series of analogs. The hypotensive activity of clonidine and its analogs is due to a stimulation of central α_2-adrenoceptors. Usually, their hypotensive activity is accompanied by bradycardic activity. Therefore, equation (6) with a slope close to unity suggests that bradycardic activity could also be initiated within the central nervous system by the 26 analogs included in the correlation.

Subsequently, Timmermans and co-workers noticed that the 3,4-dihydroxy analog of clonidine (5; X = 3,4-$(OH)_2$) was able to decrease heart rate at very low doses but was almost completely devoid of hypotensive activity after intravenous injection.[13] They thought that the very hydrophilic nature of this compound might make distribution into the brain difficult. They reinvestigated the log(dose)–response relationships for the bradycardic activity of a number of clonidine analogs covering a wider range of hydrophobicity and found that, whereas the more hydrophobic analogs show the regular sigmoidal curve, the less hydrophobic derivatives are characterized by a stepwise biphasic sigmoidal shape. They considered that one of the bisigmoidal phases observed at the lower concentration range corresponded to a peripheral mechanism (presynaptic cardic α_2-adrenoceptors) and the phase observed at the higher concentration range corresponded to the central bradycardia. In other words, the bradycardic activity of clonidine analogs consists of peripheral and central components. Although the peripheral mechanism is triggered by hydrophobic as well as hydrophilic compounds, potent central bradycardia and hypotension are limited to the hydrophobic analogs. The analogs included in equation (6) apparently belonged to the more hydrophobic group of compounds.

Of course, extrathermodynamic correlation analysis is not limited only to biological parameters. Free energy related physicochemical parameters usually used in the Hansch approach can be added as independent variables in the correlation equations. Thus, Timmermans and co-workers formulated equation (7) in their reinvestigation.[13] ED_{10} (bradycardia) was estimated from one of the sigmoidal log(dose)–response relationships assigned to the peripheral mechanism. Log P is the parameter for the hydrophobicity measured with 1-octanol/aqueous phosphate buffer (pH = 7.4). No log P term was required for the relationship with $\log(1/ED_{20})$ for the bradycardic activity attributed to the central mechanism as shown in equation (8).[13] Equations (7) and (8) clearly reinforce the hypothesis that the bradycardic activity of clonidine analogs following intravenous application can be brought about by peripheral cardic as well as by central α_2-adrenoceptors and the central mechanism requires a greater hydrophobicity than the peripheral bradycardia.

$$\log(1/ED_{20})(\text{hypotension}) = 1.039(\pm 0.26)\log(1/ED_{10})(\text{bradycardia}) + 0.566(\pm 0.19)\log P - 1.511 \tag{7}$$

$$n = 14, \quad r = 0.976, \quad F(2,11) = 111.2, \quad s = 0.199$$

$$\log(1/ED_{20})(\text{hypotension}) = 1.118(\pm 0.15)\log(1/ED_{20})(\text{bradycardia}) - 0.824 \tag{8}$$

$$n = 14, \quad r = 0.979, \quad F(1,12) = 283.01, \quad s = 0.177$$

The above two examples dealt with extrathermodynamic relationships between two free energy related biological activity indices. Without requiring knowledge of the detailed mechanisms by which drug molecules distribute and interact with receptors in such complex systems as the whole bodies of mice and rats, this approach offers not only procedures for detecting similar or dissimilar modes of drug–receptor interactions but also guidelines on how to proceed towards further clarification of the mode of drug action.

21.1.2 QUANTITATIVE STRUCTURE–ACTIVITY RELATIONSHIPS (QSAR) IN DRUG ACTIONS—THE HANSCH APPROACH

21.1.2.1 Early Trials of the Quantitative Approach

The extension of Hammett–Taft type correlation analyses to biological activities of drugs and agrochemicals had long been desired, since the correlations might be able to predict the activity of a given compound by virtue of its structure. The term 'structure' does not necessarily mean the structural formula, but rather the chemical and physicochemical properties inherent in that structure. Moreover, if the biological activity can be elucidated in chemical and physicochemical terms, invaluable information can be derived from the correlations for the mode of biological action on the molecular and/or submolecular levels.

One of the earliest examples comes from a study of Fukuto and Metcalf[14] who examined possible relationships between antiacetylcholinesterase activity of a series of substituted phenyl diethyl phosphates (6) and the Hammett constants of the ring substituents. Figure 3 illustrates the relationship for *meta-* and *para*-substituted derivatives, the ordinate being the scale of pI_{50}, the logarithm of the reciprocal of I_{50} concentration (mol dm^{-3}) against the enzyme preparation from the brain homogenate of houseflies. For the *para*-substituted derivatives, the enzyme inhibition is approximately linear with the σ^- constant. σ^- is a type of Hammett constant used when the critical reaction course involves phenoxide or free aniline formation and is strongly controlled by the electron-attracting through-resonance between the functional group and *para* substituents.[2] Equation (9) was formulated for the data on *para*-substituted derivatives.[11,14]

(6)

$$pI_{50} = 2.458(\pm 0.618)\sigma^- + 4.165(\pm 0.509) \tag{9}$$

$$n = 9, \quad r = 0.963, \quad F(1,7) = 88.3, \quad s = 0.367$$

The greater the electron withdrawal of the *para* substituents, the more potent the inhibitory activity. Acetylcholinesterase belongs to serine hydrolases and a nucleophilic attack of the hydroxy group of the side-chain of serine on the central phosphorus of the phosphate moiety initiates the enzyme–ligand interaction. The positive σ^- values and a slope close to that observed for the dissociation of substituted phenols $(\rho = 2.23)$[15] suggest that the mechanism of the enzyme inhibition is such that the phenoxide-leaving step is the rate-determining process leading to the phosphorylated enzyme. For the *meta*-substituted derivatives, the activity deviates from that expected for the *para* isomers upward or downward depending upon the substituents. Since the electronic effects of *meta* substituents should follow their σ value, the deviation could be due to effects other than electronic, and was recently shown to be attributable to effects that were mainly hydrophobic in nature.[16,17]

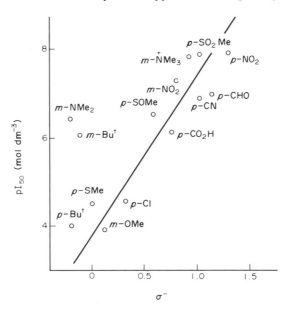

Figure 3 Relationship of antiacetylcholinesterase activity to Hammett constant for *para*- and *meta*-substituted phenyl diethyl phosphates (reproduced from ref. 14 by permission of the American Chemical Society)

However, at the time when the original study was done, no way was known of analyzing and separating the overlapping effects of substituents.

At about the turn of the century, long before the Hammett equation was publicized in 1933,[2] a physicochemical property such as the oil/water partition coefficient of certain series of compounds was recognized to relate in a quantitative way with their narcotic activity by Overton,[18] and Meyer.[19] Similar relationships were also noted in the narcosis of homologous series of compounds by Fühner,[20] who pointed out that the isonarcotic concentration decreases following an exponential progression as $1, 1/3, (1/3)^2, \ldots, (1/3)^n$ with an increase in carbon number. Moore[21] observed that the potency of insecticidal fumigants is related to their boiling point or inversely related to their vapor pressure.

These earlier observations were generalized by Ferguson[22] in 1939, recognizing that the equipotent or equitoxic concentration of series of compounds must be markedly influenced by their phase distribution relationships between biophase and exobiophase. His generalization was formulated in the form of equation (10), where C is the equipotent concentration, k and m are constants for a given system and A is a physicochemical constant related to phase distribution equilibria, such as aqueous solubility, partition coefficient ($P_{\text{water/oil}}$) and vapor pressure. Ferguson pointed out that equation (10) can be used to correlate a number of datasets of nonspecific toxicity of various series of organic compounds acting under equilibrium conditions. Equation (10) can be rewritten in the form of equation (11).

$$C = kA^m \tag{10}$$

$$\log(1/C) = m\log(1/A) + \text{constant} \tag{11}$$

An example shown in Figure 4 was taken from the work of Meyer and Hemmi.[23] The scale of the abscissa is converted to $\log P_{\text{oil/water}}$ in the usual sense, *i.e.* equal to $\log(1/P_{\text{water/oil}})$. It shows a clearly linear extrathermodynamic relationship between narcotic activity towards tadpoles and the partition coefficient measured with a system of oleyl alcohol/water. The greater the partition coefficient of the molecule, the lower is the required concentration in the exobiophase to exert equipotency.

21.1.2.2 Introduction to the Hansch Approach

Until the introduction of the Hansch approach in the early 1960s, there had been practically no examples of structure–activity relationships that followed the Meyer–Overton hypothesis or the Ferguson principle in terms of the extrathermodynamic style of equation (11). However, a large number of rather nonspecific inhibitory biological activities against enzyme preparations, nerve functions, red cells, chloroplasts, mitochondria and microbial growth, and toxicities against fishes

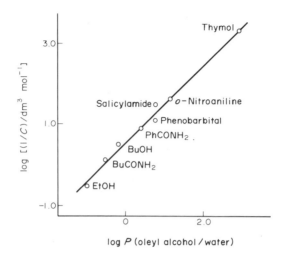

log *P* (oleyl alcohol / water)

Figure 4 Extrathermodynamic relationship between narcotic activity to tadpoles and hydrophobicity of various organic compounds (data from ref. 23)

and insects of series of homologs of alcohols, esters, anilines, phenols, ammonium ions, ureas and others, or sets of miscellaneous structurally unrelated compounds, have been shown to be generalized in the form of equation (12) by Hansch and co-workers.[24,25]

$$\log(1/C) = a \log P + \text{constant} \qquad (12)$$

In the Hansch approach, the parameter for the phase distribution relationship was standardized by log *P*, the free energy related parameter for the hydrophobicity of molecules, *P* being the partition coefficient measured in the 1-octanol/water system.[6-8,26,27] The term 'hydrophobicity' is used since the partitioning process into the organic solvent phase is driven mostly by the hydrophobic interaction of drugs with the organic solvent molecules. It has been used as a synonym of lipophilicity or liphophilic/hydrophilic balance in the Hansch approach. Some examples from a large compilation made by Hansch and Dunn[24] are listed in Table 1.

Table 1 Linear Hydrophobicity–Activity Relationships ($\log(1/C) = a \log P + b$)

Type of activity[a]	Compounds[a]	a[b]	b[b]	n	r	s	Equation number
−10 mV change in resting potential of lobster axon	ROH	0.872 (0.19)	−0.242 (0.16)	5	0.993	0.100	(13)
I_{50}, oxygen consumption, red blood cells	Misc.	0.911 (0.12)	0.114 (0.15)	14	0.977	0.218	(14)
I_{100}, frog heart beat	Misc.	0.903 (0.08)	0.130 (0.12)	33	0.974	0.195	(15)
Narcosis of tadpoles	Misc.	1.190 (0.25)	0.757 (0.35)	14	0.949	0.379	(16)
I_{50}, Hill reaction of chloroplasts	Carbamates	0.850 (0.32)	0.540 (0.98)	9	0.919	0.213	(17)
Colchicine-like mitosis, allium root tip	Misc.	0.964 (0.14)	0.600 (0.24)	22	0.955	0.390	(18)
Hemolysis of red blood cells of dogs	$RNH_2 \cdot HCl$	0.953 (0.32)	1.625 (0.46)	5	0.984	0.318	(19)
Denaturation, α-chymo-trypsinogen	ROH	0.434 (0.12)	−0.360 (0.11)	10	0.947	0.146	(20)
Rabbit liver P-450, conversion to P-420	Phenols	0.571 (0.08)	0.364 (0.19)	13	0.979	0.132	(21)
Denaturation of DNA of T4-phage	Amides	0.412 (0.06)	0.451 (0.06)	5	0.997	0.037	(22)
MIC, *C. albicans*	4-Hydroxybenzoates	0.690 (0.24)	0.616 (0.88)	7	0.956	0.251	(23)
MIC, *S. aureus*	Ethylenediamines	0.458 (0.11)	1.547 (0.60)	20	0.903	0.395	(24)
MED, hypnosis, mouse	Ureas	0.546 (0.09)	2.418 (0.12)	23	0.943	0.116	(25)
I_{25}, sheep liver esterase	ROH	0.749 (0.15)	3.694 (0.26)	19	0.931	0.322	(26)

[a] $I_n = n\%$ inhibitory concentration; MIC = minimum inhibitory concentration; MED = minimum effective dose; Misc. = set of miscellaneous compounds.
[b] Numbers in parentheses are 95% confidence intervals.

As mentioned before, the original Ferguson principle formulated as equation (10) or (11) was assumed to apply to systems at or near phase distribution equilibrium. Hansch and co-workers[24, 25] showed that equation (12) applies not only to systems at equilibrium but also to those considerably removed from equilibrium. For equilibrium cases, the $\log K$ value in equation (4) is linearly related to $\log P$. For nonequilibrium cases, both the $\log K$ and $\log k$ values in equation (3) are linearly related to $\log P$. In the latter situation, a linear free energy relationship is assumed to hold between logarithmic values of rate and equilibrium constants. There are a number of examples supporting this assumption in the Hammett–Taft type correlation analyses for physical organic systems.[4, 5]

It was well known, however, that the linear increase in biological activity does not continue infinitely with increasing $\log P$ value.[22] For certain sets of homologs or analogs covering a sufficiently wide range of hydrophobicity, the biological activity, having passed over a maximum, was often found to decrease with increasing hydrophobicity or the number of carbon atoms in homologs. This departure from linearity has sometimes been called the 'cut-off'.[22] Ferguson attempted to rationalize it by considering that the higher the carbon number in homologs the lower would be the solubility in water, so that concentrations high enough to induce a standard response could not be attained.[22] Whereas this might explain certain situations, the nonlinear relationship is probably due to the highly complex set of equilibria and rate processes through which the series of compounds reach the site of action. Thus, Hansch and co-workers considered that the transport of compounds could be a probabilistic 'random walk' process.[7, 8, 28] If the hydrophobicity is very low, the drug tends to remain in the first aqueous phase. On the other hand, if the hydrophobocity is very high, the drug is trapped in the first lipid phase in the transport process. There must be an optimum hydrophobicity for the transport process with the highest probability to reach the site of action in a given series of drugs acting on a given biological system. Since the departure from linearity is generally not precipitous, Hansch and co-workers proposed that the nonlinear dependence could be formulated by a second-order expression of hydrophobicity as equation (27), where $a \leq 0$. From a large number of examples compiled by Hansch and Clayton,[28] some representative correlations are shown in Table 2. Equation (27) corresponds with equations (3) and (4) where $(\log k + \log K)$ and $\log K$ are assumed to take the quadratic function of $\log P$.

$$\log(1/C) = a(\log P)^2 + b\log P + \text{constant} \tag{27}$$

Quite a few correlations with equation (12) are based on structurally unrelated miscellaneous compounds, as in the examples shown in Figure 4 and Table 1. These examples are believed to be due to implicitly nonspecific biological activities. No chemically as well as physicochemically specific mechanism other than hydrophobicity seems to decide their variations in potency. A number of other correlations with equation (12) are from data on homologs. Likewise, most of the correlations with equation (27) are based on homologs of such molecules as alkylguanidines, diacylureas, acetals,

Table 2 Parabolic Hydrophobicity–Activity Relationships ($\log(1/C) = a(\log P)^2 + b\log P + c$)

Type of activity[a]	Compounds[a]	a[b]	b[b]	c[b]	n	r	s	Equation number
Narcosis of tadpoles MKC (pH = 7.5),	ROH	−0.08 (0.07)	1.38 (0.34)	0.52 (0.34)	10	0.995	0.210	(28)
D. pneumoniae MHC, dove red	$RCHBrCO_2^-$	−0.20 (0.12)	1.21 (0.71)	2.88 (0.92)	8	0.893	0.596	(29)
blood cell I_{50}, rat liver	α-Monoglycerides	−0.24 (0.04)	1.98 (0.37)	0.17 (0.72)	8	0.989	0.107	(30)
mitochondria	Alkylguanidines	−0.10 (0.06)	0.28 (0.24)	2.77 (0.27)	11	0.982	0.239	(31)
MKC, *S. typhosa*	$R\overset{+}{N}Me_3$	−0.17 (0.10)	0.89 (0.24)	2.84 (0.53)	7	0.986	0.305	(32)
MIC, *S. aureus* Growth inhibition,	$PhCH_2\overset{+}{N}RMe_2$	−0.21 (0.04)	0.90 (0.14)	4.80 (0.14)	12	0.979	0.158	(33)
A. niger	5-Alkyl-8-hydroxy-quinolines	−0.13 (0.03)	1.20 (0.31)	−1.84 (0.64)	11	0.955	0.106	(34)
MED, hypnosis, mouse	N,N'-Diacylureas	−0.18 (0.09)	0.60 (0.22)	1.89 (0.13)	13	0.918	0.079	(35)
MIC, *M.tuberculosis*	Aminopyridines and anilines	−0.57 (0.14)	2.73 (0.70)	2.22 (0.75)	20	0.898	0.474	(36)

[a] Abbreviations are the same as those in Table 1, plus MKC = minimum killing concentration; MHC = minimum hemolytic concentration.
[b] Numbers in parentheses are 95% confidence intervals.

alkanoylates, α-bromoalkanoylates, barbiturates, aminopyridines, quaternary ammonium ions, monoglycerides, *etc.*, as in the examples in Table 2. This is not simply a chance occurrence. When elements other than methylene units are incorporated into a parent molecule, variations in the electronic structure become an important factor governing the potency. In general, in structure–activity relationships of a variety of drugs showing specific biological or pharmacological effects, the potency is not decided by $\log P$ and/or $(\log P)^2$ alone. For sets of compounds where various substituents with varying degrees of electronegativity are introduced, attempts to correlate the biological activity with only the Hammett–Taft type parameters also have no general utility, as reviewed by Hansen[29] as well as exemplified for the antiacetylcholinesterase activity of *meta*-substituted phenyl diethyl phosphates in the last section. Moreover, stereospecificity is sometimes very important for biologically active compounds to exhibit their activity, as postulated in the classic lock-and-key theory[30] of enzymic reactions as well as in the receptor theory.[31] Thus, in the interaction with site(s) of action, various types of steric effects are considered to arise in general.

Integrating all of the above situations, Hansch[32,33] proposed equation (37) as a more comprehensive generalized formulation for structure–activity relationships, where $a \leq 0$. For series of compounds where only substituents on a basal structure are varied, equation (37) can be converted into equation (38). π is a hydrophobicity parameter assigned to substituents. It is defined as the difference in $\log P$ values between substituted and unsubstituted derivatives.[26,27,34,35] The Taft steric parameter E_s is used in equations (37) and (38). The Taft E_s was originally defined as an intramolecular steric effect parameter for aliphatic substituents.[3] It was later extended by Kutter and Hansch[36] and shown to apply to the steric effect of aromatic substituents at various positions, even on intermolecular interactions in certain drug actions. Depending upon the mode of steric interactions, however, other steric parameters such as E_s',[37] E_s^c,[38] υ,[39] STERIMOL,[40] molecular volume,[41] molecular refractivity (MR)[42] and other shape parameters can be used in place of E_s.[43] Similarly, other Hammett–Taft type parameters[1,44] as σ^+, σ^0, σ^-, σ_I, σ_R and the Swain–Lupton $\mathcal{F}$ and $\mathcal{R}$,[45,46] as well as quantum chemical indices can often be used instead of σ as the electronic parameter.

$$\log(1/C) = a(\log P)^2 + b\log P + \rho\sigma + \delta E_s + \text{constant} \tag{37}$$

$$\log(1/C) = a\pi^2 + b\pi + \rho\sigma + \delta E_s + \text{constant} \tag{38}$$

In equations (37) and (38), a, b, ρ and δ are the coefficients showing the importance of each term. Depending upon the given sets of compounds and biological systems, these coefficients are not always significant. For example, in correlations formulated by equation (27) in Table 2, σ and E_s terms do not appear. As far as the sets of compounds included in these correlations are concerned, variations in stereoelectronic characteristics are not important in governing their potency. The separation of various effects and evaluation of coefficients are performed by multiple regression analysis using the least-squares method.[47] The level of significance of each coefficient is judged by such statistical procedures as the student t and F tests.[48]

Sometimes, hydrophobic and steric effects on the interaction with site(s) of action are specific to substituent positions.[49] For instance, *meta* but not *para* substituents exert a hydrophobic effect in the antiacetylcholinesterase activity of substituted phenyl diethyl phosphates, as shown in the last section. Using substituent parameters specific to substituent positions, regiospecific structural effects on the activity can be revealed. The electronic effect of aromatic substituents is not usually position specific, being decided by a common $\rho\sigma$ term for *meta* and *para* substituents. For *ortho* substituents, the σ value can be taken as that of the corresponding *para* substituent for ordinary electronic effects.[17,50] For the extra electronic and steric effects of *ortho* substituents on the side-chain functional group owing to their proximity, σ_I or $\mathcal{F}$ and E_s or other steric parameters can be used when such effects are significant.[17,50] Variables representing other structural effects can be further added to equations (37) and (38), such as for hydrogen-bond formation between substituents and receptor[51] and for specific potency increments assigned to certain structural units.[52] The latter are called indicator variables and are represented by zero or unity depending upon the absence or presence of the structural units. Sometimes, the proximity effects of *ortho* substituents are expressed by a single indicator variable as an approximation.

The Hansch approach, when it was initiated, was not as complete as described above. The earliest correlation was formulated for the plant growth activity of substituted phenoxyacetic acids (7) as shown in equation (39),[6,7] where C is the molar concentration causing a 10% elongation of avena coleoptile segments in 24 h. σ is the Hammett constant, which is taken to express the electronic effect of substituents directed toward the *ortho* position of the side-chain. In more recent years, the Hansch approach has been elaborated to utilize various steric parameters, to analyze regiospecific substituent effects as well as to include indicator variables and others. At every evolutionary step the versatility of the approach has been expanded remarkably. The number of successful analyses has been growing enormously, leading to the present state of development.

$$\log(1/C) = -1.98\pi^2 + 3.24\pi + 1.87\sigma + 4.16 \tag{39}$$

$$n = 21, \quad r = 0.881, \quad s = 0.484$$

$$X \text{—} \langle \text{benzene ring} \rangle \text{—OCH}_2\text{CO}_2\text{H}$$

(7)

21.1.2.3 Rationale of the Hansch Approach

As indicated in the last section, $\log P$ (where P = partition coefficient with the 1-octanol/water system) is used as the standard parameter for the hydrophobicity of drug molecules in the Hansch approach. The fact that $\log P$ (or π) as the hydrophobicity parameter generally works well in QSAR analyses, in spite of the various kinds of lipophilic phases involved in the process of drug action, is primarily based upon extrathermodynamic relationships among phase distribution properties of drugs such as partition coefficients and binding constants with proteins and tissues.

In fact, Collander[53] has shown that, if partition coefficients can be used in structure–activity correlations, linear relationships such as equation (40) should hold between pairs of solvent systems which are not too different from lipophilic/hydrophilic phases in biological systems (equation 40, where P_1 and P_2 are partition coefficients for a given drug in two different solvent systems). Leo and co-workers reexamined this relationship.[34, 54] Some examples of such correlations are shown in Table 3, where P_1 is defined as the partition coefficient for the 1-octanol/water system. While close correlations are observed between systems using similar solvents such as octanol, pentanol and oleyl alcohol, systems with solvents having different solvation patterns can be correlated only after categorization of solute classes according to their hydrogen-bonding characteristics (for a more complete discussion see ref. 51). Thus, lipophilic phases involved in drug actions to which $\log P_{\text{octanol}}$ applies could generally be regarded as being amphiprotic in nature, like alkanols.

$$\log P_2 = a \log P_1 + \text{constant} \tag{40}$$

Subsequently, Hansch and co-workers showed that equation (40) can be extended to other phase distribution properties, as shown in equations (50)–(53). In equations (51) and (52), C is the molar concentration of organic compounds required for one-to-one binding with protein.

Aqueous solubility (S) of organic liquids except for saturated aliphatic hydrocarbons[55]

$$\log(1/S) = 1.214(\pm0.05)\log P - 0.850(\pm0.11) \tag{50}$$

$$n = 140, \quad r = 0.955, \quad s = 0.344$$

Table 3 Extrathermodynamic Relationships between Partition Coefficients ($\log P_{\text{solv}} = a \log P_{\text{octanol}} + b$)

Solvent	Solutes	a^a	b^a	n	r	s	Equation number
Oleyl alcohol	Misc.	0.999 (0.06)	−0.575 (0.11)	37	0.985	0.225	(41)
Primary pentanols	Misc.	0.808 (0.07)	0.271 (0.09)	19	0.987	0.161	(42)
Primary butanols	Misc.	0.697 (0.02)	0.381 (0.03)	57	0.993	0.123	(43)
Methyl isobutyl ketone	Misc.	1.094 (0.07)	0.050 (0.11)	17	0.993	0.184	(44)
Isopentyl acetate	H-bond donors	1.027 (0.08)	0.072 (0.13)	22	0.986	0.209	(45)
Diethylether	H-bond donors	1.130 (0.04)	−0.170 (0.05)	71	0.988	0.186	(46)
Diethylether	H-bond acceptors	1.142 (0.13)	−1.070 (0.12)	32	0.957	0.326	(47)
Chloroform	H-bond donors	1.126 (0.12)	−1.343 (0.21)	28	0.969	0.308	(48)
Chloroform	H-bond acceptors	1.276 (0.14)	0.171 (0.17)	21	0.976	0.251	(49)

^{a} Numbers in parentheses are 95% confidence intervals

Binding with bovine serum albumin[56]

$$\log(1/C) = 0.751(\pm 0.07)\log P + 2.301(\pm 0.15) \tag{51}$$

$$n = 42, \quad r = 0.960, \quad s = 0.159$$

Binding with bovine hemoglobin[57]

$$\log(1/C) = 0.713(\pm 0.13)\log P + 1.512(\pm 0.33) \tag{52}$$

$$n = 17, \quad r = 0.950, \quad s = 0.160$$

Partition coefficient into red blood cell ghost[24]

$$\log P_{\text{red cell}} = 1.003(\pm 0.13)\log P - 0.883(\pm 0.39) \tag{53}$$

$$n = 5, \quad r = 0.998, \quad s = 0.082$$

Similar relationships were found for permeability of organic molecules through plant tissue as well as rabbit skin.[58] Thus it is apparent that the partition coefficient measured with 1-octanol/water systems can serve as a suitable model for almost all lipophilic interactions of drugs occurring in biological systems. No *a priori* reasoning existed for the selection of 1-octanol as the lipophilic organic phase, except that its higher boiling point made measurement of the partition coefficient easier when compared to the organic solvents used before, such as ether and chloroform. Other important properties are its amphiprotic nature enabling it to solubilize generally polar drug molecules and its ease of purification.

In the Hansch approach, the effect of the hydrophobicity of a series of drugs on biological activity is in general expressed by the parabolic function of $\log P$ or π as in equations (27), (37) and (38). Of course, the squared term is not always statistically significant. The parabolic function was originally introduced by intuition and subsequently supported empirically by a number of examples such as those listed in Table 2.[28] As mentioned before, Hansch and co-workers considered that the finding of site(s) of action by drug molecules could be a probabilistic random walk process. The probability A with which a drug molecule will reach the active site within a certain time was assumed to follow a normal Gaussian distribution with respect to the $\log P$ of compounds, as expressed in equation (54).[8] P_{opt} is the optimum P somewhere between zero and infinity. Taking the logarithm, equation (54) can be rewritten as equation (55).

$$A = a' \exp[-(\log P - \log P_{\text{opt}})^2 / b'] \tag{54}$$

$$\log A = a(\log P)^2 + b \log P + \text{constant} \tag{55}$$

In these equations, $a\,(\leq 0)$, b, a' and b' are constants. Thus, the parabolic relationship corresponds with an assumption that, all other things being equal, the overall rate constant or overall equilibrium constant for steps involved in the transport and drug–receptor interaction processes considered in formulating equations (3) and (4) is probabilistic and distributed normally according to the $\log P$ value in series of congeneric drugs as shown in equation (54). If specific electronic and steric effects participate in steps involved in the overall rate or equilibrium process, then parameters such as σ and E_s should be introduced leading to equations (37) and (38). In most cases, the drug–receptor interaction step is the step where such specific stereoelectronic effects are important, along with the hydrophobic interaction for binding that is related to $\log P$.

Hansch and co-workers attempted to rationalize the parabolic relationship with the use of a nonsteady state kinetic model.[59] They assumed a simple fluid lipid membrane as depicted in Figure 5, where k is the rate constant for passage from the aqueous to the lipid phase, and l is the rate constant for the reverse passage. Compartment 1 has a given concentration of solute, A_1^0, at zero time. The volume of each compartment and the surface area between compartments were taken to be the same for all. It was assumed that, in biological systems, we are considering a 'stirred' solution in each compartment. More compartments can be added to Figure 5 so that the drug must get across many membranes or undergo a number of adsorption–desorption steps with proteins and/or lipids before reaching the action site(s) in the final phase. For the sake of simplicity, the lipophilic nature of the lipid compartments was assumed to be the same so that single values for the rate constants k and l and $P(= k/l)$ can be used. In the final phase, the drug molecules were assumed to bind irreversibly with the action site(s) with a rate constant m.

The general set of differential equations to be analyzed is shown in equations (56)–(60). In these equations, A_i represents the concentration in the ith phase and A_n that in the final phase. Since A_1/A_n does not depend on A_1^0, an arbitrary initial concentration such as 10 and 100 can be employed for A_1^0. For a specific value of n, the partition coefficient ($P = k/l$) is varied over a certain

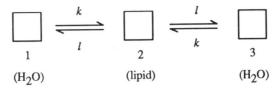

Figure 5 Three compartments for passive transport

range. The solution for the set of equations can be obtained by integrating over time t with the use of the Runge–Kutta approximation. Values of k and l were selected so that $k \times l = 1$, *i.e.* they assumed a reciprocal relation between hydrophobic and hydrophilic character. Figure 6 shows the plot of log (concentration) (log C) in the last compartment of the 20-barrier model when $t = 10$ (arbitrary time units) and $m = 1$ against log P. The solid line is a parabola fitted by the method of least squares to the calculated points. The fact that the points fit the line quite well was regarded as supporting the postulate for equation (54), at least for systems not at or near equilibrium. Hansch and Clayton further discussed other possible rationalizations of the nonlinear relationship.[28]

$$dA_1/dt = -kA_1 + lA_2 \tag{56}$$

$$dA_{2i}/dt = -2lA_{2i} + k(A_{2i-1} + A_{2i+1}) \tag{57}$$

$$dA_{2i+1}/dt = -2kA_{2i+1} + l(A_{2i} + A_{2i+2}) \tag{58}$$

$$dA_{n-1}/dt = -(l+m)A_{n-1} + kA_{n-2} \quad (n = \text{odd}) \tag{59a}$$

$$= -(k+m)A_{n-1} + lA_{n-2} \quad (n = \text{even}) \tag{59b}$$

$$dA_n/dt = mA_{n-1} \tag{60}$$

21.1.2.4 Other Models Related to the Hansch Approach—Combination with the Bilinear Model

Following publications that attempted to rationalize the nonlinear relationship in terms of the parabolic model, quite a few mathematical models for drug distribution and transport have been proposed. These models are discussed in Chapter 19.2, so this section will review in detail only a few models related to the Hansch approach.

McFarland[60] used a multicompartmental model similar to that shown in Figure 5. He defined the probability $p_{0,1}$ of drug molecules leaving the first aqueous phase (0) where they are applied and entering the first lipid phase (1) as being proportional to k and the probability $p_{1,0}$ of returning to the aqueous phase as being proportional to l. k and l are the rate constants shown in Figure 5. Since the

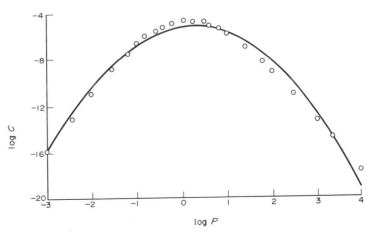

Figure 6 Relationship of drug concentration at $t = 10$ in the last compartment of 20-barrier model with log P (reproduced from ref. 59 by permission of the American Society for Pharmacology and Experimental Therapeutics)

sum of the two probabilities should be equal to unity, equations (61) and (62) can be derived.

$$p_{0,1} = k/(k + l) \tag{61}$$

$$p_{1,0} = l/(k + l) \tag{62}$$

Since $P = k/l$, equations (61) and (62) can be converted into equations (63) and (64).

$$p_{0,1} = P/(1 + P) \tag{63}$$

$$p_{1,0} = 1/(1 + P) \tag{64}$$

The probability $p_{0,n}$ of a certain drug molecule reaching the last compartment (n) is described by equation (65).

$$p_{0,n} = p_{0,1} \cdot p_{1,2} \cdots p_{n-1,n} = \frac{P^{n/2}}{(P + 1)^n} \tag{65}$$

If the last compartment is regarded as the receptor compartment, the concentration C_r at the receptor sites is proportional to $p_{0,n}$ and equation (66) can be derived for $\log C_r$, where a is a constant.

$$\log C_r = a \log P - 2a \log(P + 1) + \text{constant} \tag{66}$$

Equation (66) represents symmetrical linear ascending and descending sides and a 'parabolic' apical part. Although McFarland originally proposed that his model is supporting evidence for the parabolic relationship, it is not suited for regression analysis because the maximum $\log C_r$ always occurs at $\log P = 0$.

Kubinyi[61,62] reexamined the McFarland model, taking into consideration the difference in volume between lipophilic and aqueous phases in biological systems. If β is the ratio of the volumes (lipid/aqueous), equations (63) and (64) can be rewritten as equations (67) and (68).

$$p_{0,1} = \beta P/(\beta P + 1) \tag{67}$$

$$p_{1,0} = 1/(\beta P + 1) \tag{68}$$

The lipophilic nature of the receptor surface (P') may differ from that of the lipophilic compartments during transport (P). The volume ratio of the receptor phase to the last aqueous compartment (β') may not be equal to β. Thus, for the binding process from the last aqueous phase to the receptor site(s), the probability is expressed by equation (69).

$$p_{n,r} = \beta' P'/(\beta' P' + 1) \tag{69}$$

Since the 'volume' of the (lipophilic) receptor site(s) may well be much smaller than that of the last aqueous compartment, $\beta' P'$ could be negligible compared with unity, so ($\beta' P' + 1$) is nearly equal to unity. Therefore, equation (70) is given as the counterpart of equation (65).

$$p_{0,r} = \frac{\beta^{n/2} \cdot \beta' \cdot P^{n/2} \cdot P'}{(\beta P + 1)^n} \tag{70}$$

An extrathermodynamic relationship such as equation (40) could be valid between P and P'. Since β, β' and n are constants for a given system, equation (70) can be transformed into equation (71), where n' is a constant.

$$p_{0,r} = \frac{\text{constant} \cdot P^{n'}}{(\beta P + 1)^n} \tag{71}$$

From this equation, equation (72) can be derived for the relationship between biological activity and hydrophobicity of a set of drugs.

$$\log(1/C) = a \log P - b \log(\beta P + 1) + \text{constant} \tag{72}$$

Kubinyi[63] further derived equations very similar to equation (72) by reconsidering the Higuchi–Davis–Hyde model[64,65] where the applied dose of a drug is distributed in aqueous as well as various lipophilic phases under (quasi)equilibrium conditions in biological systems. Additional evidence for equation (72) was found not only from equilibrium simulations but also from drug transport simulations in open and closed multicompartment model systems using experimental k and l values for the drug transport (Figure 5).[63] In all cases, the resulting concentration–hydrophobicity relationships can be described by equation (72). In the course of these studies, Kubinyi disclosed that the product $k \times l$ is not unity but expressed by equation (73),[63] where a is a constant.

$$k \times l = \text{constant} \cdot P/(aP + 1)^2 \tag{73}$$

The $\log(\beta P + 1)$ term in equation (72) has interesting characteristics: for small P values, $(\beta P + 1)$ is approximately unity, so $\log(\beta P + 1)$ is zero; for large P values, $(\beta P + 1)$ is nearly equal to βP, therefore $\log(\beta P + 1)$ is nearly equal to $\log \beta P$, thus $\log(\beta P + 1)$ is linear with $\log P$. Biphasic functions with linear ascending and descending sides and a rounded apical part are represented by this model. The positive slope for the ascending side is expressed by a, the negative slope for the descending side is expressed by $(a - b)$, $-\log \beta$ is responsible for the distance of the $\log P_{opt}$ from the y axis. Four adjustable variables are required in general to describe such biphasic functions. While a, b, and 'constant' are coefficients of the linear terms and the intercept which can be calculated by linear multiple regression analysis, the coefficient β is that of the nonlinear term which must be estimated by an iteration method, either by a stepwise iteration 'by hand' or much more conveniently by a Taylor series iteration.[66] Because of the characteristic shape of the functions, Kubinyi named his model the bilinear model.

The bilinear model was originally derived from a rather intuitive concept such as the probability of drug molecules reaching the site(s) of action, following Hansch and co-workers and McFarland. The model has been rationalized as holding regardless of whether the drug actions occur under equilibrium or nonequilibrium conditions, theoretically. A number of relationships between hydrophobicity and activity for various series of drugs analyzed with the bilinear model have been compiled by Kubinyi.[62] Examples which were also analyzed by the parabolic model in Table 2 are listed in Table 4 for comparison. As stated by Kubinyi, the quality of the correlation generally seems better with the bilinear model than with the parabolic model. Kubinyi[63] showed that his bilinear model can also apply to various pharmacokinetic properties of drugs, which had been analyzed by Lien[67] using the parabolic model, such as buccal, gastrointestinal and colonic absorptions, transport through bladder, skin and the blood–brain barrier, and renal clearance. Thus, Kubinyi[68] proposed the use of equations (80) and (81) as counterparts of equations (37) and (38) for structure–activity relationships including electronic and steric effects of congeneric drugs exhibiting specific biological activities. The model is also applicable to nonlinear relationships of biological activity with electronic and steric effects, as will be shown in the following section.

$$\log(1/C) = a \log P - b \log(\beta P + 1) + \rho \sigma + \delta E_s + \text{constant} \tag{80}$$

$$\log(1/C) = a\pi - b \log(\beta \cdot 10^\pi + 1) + \rho \sigma + \delta E_s + \text{constant} \tag{81}$$

The bilinear model seems to be better rationalized theoretically than the parabolic model. It excludes discrepancy between linear and parabolic models. From the linear model represented by equation (12), the left-hand side of the 'parabola' in the parabolic model should be linear. Thus, the use of the bilinear model for hydrophobically nonlinear sets of drugs along with the linear model enables one to examine the slope of the ascending part of structure–hydrophobicity relationships according to a common standard.[25] One of the disadvantages of the bilinear model, however, is that it requires one more adjustable variable than the parabolic model. Secondly, the calculation should be made by iterative procedures which are more complex than the linear regression analysis for the parabolic model. Moreover, some care must be taken when using equations (80) and (81) in that, if not enough biological data of good quality are used, deviants on the right-hand side of the apex can markedly affect the slope of the ascending part. To obtain accurate data on more lipophilic and more water-insoluble compounds on the right-hand side is sometimes not easy. The variance of the biological data is often too large to allow a statistically significant decision as to whether the parabolic model or the bilinear model gives a better elucidation of the data. Nevertheless, when a nonlinear dependence is apparent in a structure–hydrophobicity relationship, it is worthwhile applying the bilinear model along with the parabolic model. At present, the parabolic model is most frequently used for QSAR correlation studies followed by the bilinear model. This seems to be due mainly to the simplicity of calculation of the parabolic model and the insufficient quality of the biological data.

21.1.2.5 Rules and Criteria in the Hansch Approach

In general, QSAR correlation equations are composed of various physicochemical parameter terms. The level of significance of each term must be examined statistically. At the same time, the

Table 4 Bilinear Hydrophobicity–Activity Relationships ($\log(1/C) = a\log P - b\log(\beta P + 1) + c$)

Type of activity	Compounds	a^a	b^a	c^a	$\log \beta$	n	r	s	Equation number
MKC (pH = 7.5), *D. Pneumoniae*	$RCHBrCO_2^-$	0.794 (0.62)	1.957 (1.22)	2.765 (0.91)	−3.599	8	0.934	0.530	(74)
MHC, dove red blood cell	α-Monoglycerides	0.831 (0.14)	1.650 (0.15)	1.516 (0.35)	−4.100	8	0.998	0.049	(75)
I_{50}, rat liver mitochondria	Alkylguanidines	0.825 (0.11)	0.925 (0.43)	3.498 (0.34)	0.594	10	0.997	0.103	(76)
MKC, *S. typhosa*	$R\overset{+}{N}Me_3$	1.002 (0.23)	1.787 (1.14)	2.512 (0.33)	−2.309	7	0.977	0.169	(77)
MIC, *S. aureus*	$Ph\overset{+}{N}RMe_2$	1.047 (0.19)	1.507 (0.19)	4.757 (0.12)	−1.438	12	0.993	0.100	(78)
Growth inhibition, *A. niger*	5-Alkyl-8-OH-quinolines	0.381 (0.11)	0.929 (0.24)	−0.604 (0.36)	−5.128	11	0.977	0.082	(79)

[a] Numbers in parentheses are 95% confidence intervals.

regression coefficient values assigned to each physicochemical parameter term should be reasonable in the light of our empirical knowledge of structure–reactivity relationships in physical organic systems as well as structure–activity relationships in related biological systems.

As shown in Table 1, which contains correlations according to the linear hydrophobicity model (equation 12), and in Table 4, which lists correlations with the bilinear model (equation 72), the slope of the ascending part of the $\log P$ term has a value approximately between 0.4 and 1.2. A slope close to unity for sets of miscellaneous drugs means that the drug molecules are almost completely dehydrated and engulfed in a hydrophobic pocket or cleft of lipophilic biomacromolecules in the critical step for their action, similar to partitioning into 1-octanol. For sets of homologs having a common polar functional group that may project into the aqueous phase, the engulfment is not necessarily required for the whole molecule but only for the hydrophobic part of the molecule. A slope close to 0.5 suggests that the dehydration of the drug molecules is about a half of that for the engulfment. There are two possible explanations for this. First, the drug molecules or their hydrophobic part could be held on a surface of the biomacromolecules so that they are about half dehydrated. Second, the hydrophobicity of a critical biolipid phase is lower than that of 1-octanol. If the hydrophobicity of this phase is closer to that of the water-saturated primary butanol, where the concentration of water is as high as 9.4 mol dm^{-3},[54,69] a slope significantly lower than unity could be observed even for complete engulfment, as easily anticipated from equation (43) in Table 3. Changes of target biomacromolecule conformation can sometimes have important consequences for the specific activity of drugs. For this effect, the drug may be involved in interactions much more sensitive than the usual hydrophobic bonding. Hansch and co-workers[56] observed a slope of 2.23 for the conformational perturbation of bovine serum albumin induced by miscellaneous organic compounds.

The above discussions indicate that the slope of the $\log P$ or π term should usually be between 0.4 and 1.2 in multiple regression equations. If some specific interactions such as the conformational perturbation of the target biomacromolecules are involved, the slope could be higher than 1.20 but, at the most, about 2.0.

A somewhat similar criterion applies to the regression coefficient of the electronic parameter. In Table 5, the ρ values of some physical organic systems are listed.[70] The magnitude of the ρ value is clearly associated with the position of the reaction site relative to the aromatic ring. For acid dissociation reactions (equations 82–86) the ρ value decreases progressively by a factor of about 1/2 as the distance of the site of electron pair migration increases by one bond.[70] For the rate-determining step of nucleophilic reactions (equations 87–89) a similar relationship is observed, where the sign of the ρ value is dependent on the direction of the electron pair migration. Although this type of relationship is not very sharp for complex multistep reactions and sometimes severely distorted by reaction conditions such as temperature and medium as well as by steric environment around the reaction center,[71,72] it still has important implications for the use of the magnitude of the ρ value in the elucidation of reaction mechanisms. One example is the acid dissociation reaction of substituted benzeneboronic acids.[70] In this case, $\rho = 2.18$ suggests that the reaction occurs according to equation (90) and not equation (91).

$$\text{ArB(OH)}_2 + 2\text{H}_2\text{O} \rightleftharpoons \text{ArB(OH)}_3^- + \text{H}_3\text{O}^+ \qquad (90)$$

$$\text{ArB(OH)}_2 + \text{H}_2\text{O} \rightleftharpoons \text{ArBO}_2\text{H}^- + \text{H}_3\text{O}^+ \qquad (91)$$

Table 5 Slopes of the Hammett Equations ($\log K = \rho\sigma + \text{constant}$)

Reaction system	i^a	ρ	Equation number
Ar—NH$_3^+$—H $\rightleftharpoons$ ArNH$_2$ + H$^+$ (25 °C)	2	2.94	(82)
Ar—O—H $\rightleftharpoons$ ArO$^-$ + H$^+$ (25 °C)	2	2.26	(83)
Ar—CO—O—H $\rightleftharpoons$ ArCO$_2^-$ + H$^+$ (25 °C)	3	1.00	(84)
Ar—CH$_2$—NH$_3^+$—H $\rightleftharpoons$ ArCH$_2$NH$_2$ + H$^+$ (25 °C)	3	1.05	(85)
Ar—CH$_2$—CO—O—H $\rightleftharpoons$ ArCH$_2$CO$_2^-$ + H$^+$ (25 °C)	4	0.56	(86)
Ar—NMe$_2$ + MeI $\xrightarrow[35\,°C]{\text{acetone}}$ ArN$^+$Me$_3$I$^-$	2	-2.39	(87)
Ar—CO$_2$Me + OH$^-$ $\xrightarrow[25\,°C]{60\,\%\ \text{acetone}}$ ArCO$_2^-$ + MeOH	2	2.38	(88)
Ar—CH$_2$—CO$_2$Et + OH$^-$ $\xrightarrow[25\,°C]{60\,\%\ \text{acetone}}$ ArCH$_2$CO$_2^-$ + EtOH	3	1.00	(89)

^{a} The electron pair migration occurs at the ith 'bond' counting from the aromatic system.

The empirical relationship shown above should hold in multiparameter QSAR correlation equations. For example, equation (92) was derived for substituent effects on the rate of transfer of substituted anilines catalyzed by carp vicera thiaminase originally studied by Mazrimas and co-workers.[73, 74]

$$\log k = 0.722(\pm 0.377)\pi - 1.842(\pm 0.484)\sigma^- + 0.557(\pm 0.404)E_s^{para} - 1.892(\pm 0.379) \tag{92}$$

$$n = 12, \quad r = 0.959, \quad s = 0.288$$

The slope of the π term is reasonable for a reaction occurring on the enzyme surface, where a regiospecific steric effect for the *para* substituents on the aniline ring expressed by E_s^{para} is also operative. Since the sign of the E_s term is positive, the size of the *para* substituents should be smaller for higher reactivity. The reaction occurs, under catalytic conditions, following equation (93) by a nucleophilic attack of the lone pair electrons of the nitrogen of substituted anilines (8) on the methylene bridge carbon atom of thiamine (9).

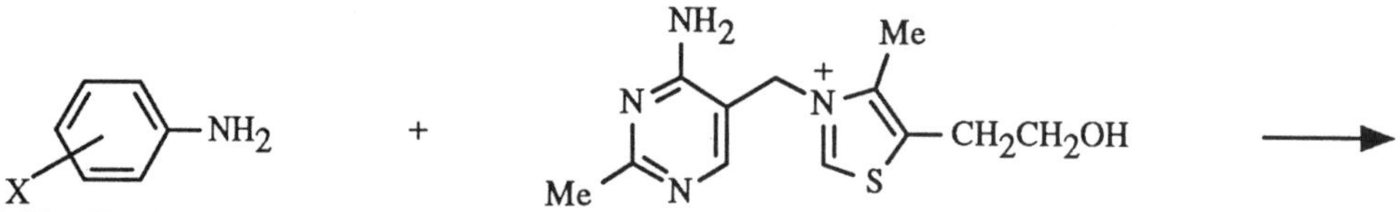

(8) X = H, 3-Me, 3-halo,
3-NO$_2$, 4-Me, 4-OMe,
4-OH, 4-halo, 4-Ac, 4-CN

(9)

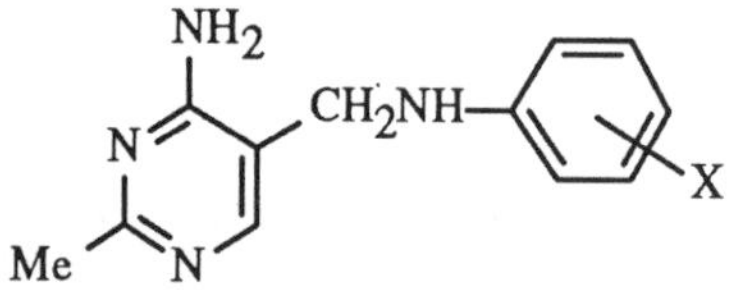

$$\tag{93}$$

Since the electron pair migration occurs at the second 'bond' place from the aniline aromatic ring, the ρ value should be close to that for equation (87) in Table 5, as indeed observed in equation (92). The substrate specificity for this enzymic reaction was originally rationalized only in terms of an electronic effect of aniline substituents, assuming a biphasic Hammett plot.[73] Equation (92) suggests that the biphasic mechanism is unlikely to occur and not only electronic but also steric and hydrophobic effects of substituent X participate in the reaction.

Another example is taken from the work of Unger and Hansch[75] on the analysis of adrenergic blocking activity of β-halo-β-arylethylamines (10). The analysis was originally done by Lien and Hansch[76] using the data of Graham and Karrar,[77] formulating equation (94). Subsequently, Cammarata[78] reanalyzed the same data giving equation (95).

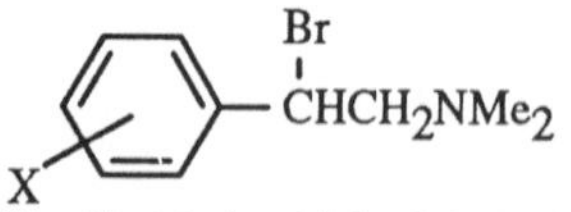

(10) X = H, 4-halo, 4-Me, 3-halo, 3-Me,
3,4-dihalo, 3,4-Me$_2$, 3,4-halomethyl

$$\log(1/ED_{50}) = 1.221\Sigma\pi - 1.587\Sigma\sigma + 7.888 \tag{94}$$

$$n = 22, \quad r = 0.918, \quad s = 0.238$$

$$\log(1/ED_{50}) = 0.747(\pm 0.123)\pi_m - 0.911(\pm 0.249)\sigma_m + 1.666(\pm 0.124)r_p + 5.769 \tag{95}$$

$$n = 22, \quad r = 0.961, \quad s = 0.168$$

Cammarata proposed that, from a statistical point of view, equation (95) could be selected as the best rationalization of the data. In equation (94) the σ and π terms summed for the disubstituted derivatives are significant, whereas in equation (95) electronic and hydrophobic parameters are considered only for *meta* substituents. r_p is the van der Waals radius for *para* substituents. Regiospecific hydrophobic and steric effects of substituents that can be expressed by π_m and r_p may be at work at the site of action of this series of compounds. Unger and Hansch[75] argued against the most disconcerting aspect of equation (95), where an electronic effect for just the *meta* substituents is indicated, because this is without precedent in the literature of physical organic chemistry. The proposed mechanism of action of this series of compounds is that the carbonium ion intermediate (**12**) is formed from the ethyleniminium ion (**11**) and reacts with a nucleophilic center Y^- at the α-adrenoceptor site, as shown in Scheme 1. Therefore, the carbonium ion intermediate (**12**) should be subject to a very strong electronic effect by *meta* as well as *para* substituents. The electronic effects of substituents on this type of carbonium ion would be better rationalized by σ^+ rather than the regular σ. Thus, Unger and Hansch finally derived equation (96).[75]

$$\log(1/ED_{50}) = 0.82(\pm 0.27)\Sigma\pi - 1.02(\pm 0.45)\Sigma\sigma^+ + 0.62(\pm 0.43)r_p + 7.06(\pm 0.55) \tag{96}$$

$$n = 22, \quad r = 0.964, \quad s = 0.164$$

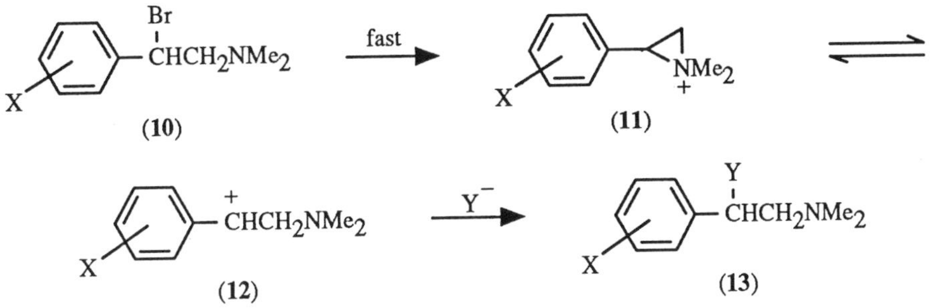

Scheme 1

The fact that the total electron-donating effect of substituents available to stabilize the carbonium ion (**12**) participates in the rate-determining step of the overall reaction is clearly suggested by equation (96). Although the size of the ρ value (1.02) seems a little bit lower than that expected from the distance of the reaction-center positive charge from the benzene ring, the reactions where the substituent electronic effect is correlated by σ^+ are those where the ρ value is most distorted by reaction conditions and substrate structures.[71] Nevertheless, the inclusion of the σ^+ term excludes the ammonium ion (**11**) as an important species unless the transition state is 'late' and essentially resembles the product (**13**). A further example was reported in work on herbicide design where a similar disconcerting correlation equation was derived with use of σ_p, but not σ_m, values and π values improperly assigned to substituents for a series of *meta*- and *para*-substituted phenylureas.[79]

An empirical rule seems to exist for the size of the regression coefficient of the E_s parameter term in correlation equations where the linear E_s term is significant. For some 10 QSAR examples of aromatic monoamine oxidase inhibitors, the regression coefficient was found to be at the most about 1.5.[74] For effects in physical organic systems, such as the steric effect of *ortho* substituents on the side-chain functional group, the coefficient of the E_s term is also lower than 1.5.[50] The value in equation (92) and also that in equation (96), which corresponds to -0.34 when r_p is corrected to E_s, are within this range, although the sign varies depending upon whether the steric 'bulkiness' is favorable or unfavorable to the drug action. Similar empirical rules could exist for other parameter terms, but they are not as apparent as those described above because an insufficient number of examples has been accumulated. Since the most important physicochemical parameters π and σ are usually significant in formulating QSAR multiparameter correlation equations, one can judge as to whether such correlations are acceptable or not by examining the coefficients associated with the π and σ terms.

Unger and Hansch proposed the criteria described below which must be considered before one identifies a 'best correlation equation' for a set of congeners.[75]

(a) Selection of independent variables. The widest possible number of independent variables must be examined. While σ, π and E_s have been most widely used, one must not overlook MR, various steric parameters, indicator variables and MO parameters. The parameters selected should be essentially independent of each other as an aid in rationalization of the mechanism of drug actions.

(b) Justification of the choice of independent variables. In the best correlation equations, each term must be validated by an appropriate statistical procedure. It is advantageous to examine regression analyses with all possible combinations of independent variables and then to use a forward selection procedure with sequential F tests to identify the 'best equation', generally that with the lowest standard deviation and all terms significant (usually over 95% level).

(c) Principle of parsimony. All things being equal, one should accept the simplest model.

(d) Number of independent variable terms. According to the suggestion of Topliss and Costello,[80] one should have at least five to six data points per variable in order to avoid chance correlations.

(e) Physical organic significance. The best correlation equation should be rationalized in terms of the principles of known physical organic and biomedicinal chemistry.

21.1.2.6 Other Modified Procedures

In the Hansch approach, the potency of a series of drugs is expressed by the logarithm of the reciprocal of the equieffective dose or concentration which is usually measured from dose–response relationships. In earlier stages of bioassay, however, especially those performed in industry, the potency is often recorded in the form of a rating score such as $-$, $\pm$, $+$, $2+$, etc. Moriguchi and co-workers developed an iterative procedure named the adaptive least-squares (ALS) method which applies to this situation.[81] The regular regression analysis is not suitable, because the rating score is an ordered category for the level of potency expressed by an integer that covers the ultimate potency values distributed in a certain range. Thus, the dependent variable might not necessarily be the fixed score, and the values of the score could be manipulated to adapt to the regression analysis with certain rules.

According to the ALS method, the discrimination of the ordered m groups ($m \geq 2$) of compounds is made using a single correlation equation such as equation (97), where L is the score, E_k is the kth independent variable ($k = 1 \sim p$) for the structural features and w_k is their weight coefficient. Initial rating scores a_j ($j = 1 \sim m$) for the members of class j are assumed to be expressed by equation (98), where n_i and n_j are the size of groups i and j, and n is that of the whole set. The discrimination points b_j ($j = 1 \sim m - 1$) between classes are defined as the midpoints between a_j and a_{j+1}.

$$L = \text{constant} + w_1 E_1 + w_2 E_2 + \cdots w_p E_p \tag{97}$$

$$a_j = \left[2 \left(2 \sum_{i=1}^{j-1} n_i + n_j \right) \Big/ n \right] - 2 \tag{98}$$

The procedure starts by introducing initial scores a_j [$= S_i^1$ ($i = 1 \sim n$)] in place of L in equation (97) to estimate w_i^1 values by the regular least-squares method. The superscript denotes the number of iteration times. The values of w_k^1 are the initial weight coefficients for the best fit of the initial score, now rewritten as S_i^1. All compounds are then classified on the basis of a calculated value for S_i^1, i.e. L_i^1, and the discrimination point as follows: if $L_i^1 \leq b_1$, then the ith compound belongs to class 1; if $b_1 < L_i^1 \leq b_2$, to class 2; and if $b_{m-1} < L_i^1$, then to class m. Classes are numbered in ascending order of activity. At iteration 2 and thereafter, the tentative score S_i^{t+1} ($t \geq 1$), is adapted as shown in equation (99). C_i^t is the correction term usually estimated according to equation (100), where $\delta_i = |L_i^t - b_k|$, b_k being the discrimination point (nearer to L_i^t) of the observed class for the ith compound. α and β are constants empirically selected as 0.45 and 0.10, respectively. The sign of C_i^t is chosen to correspond with that of $S_i^t - L_i^t$.

$$S_i^{t+1} = L_i^t \quad \text{(when correctly classified at iteration } t\text{)}$$

$$= L_i^t \pm C_i^t \quad \text{(when misclassified)} \tag{99}$$

$$C_i^t = 0.1/(\alpha + \delta_i)^2 + \beta \tag{100}$$

These adaptive iterations are generally performed 30 times and the 'best' discriminant function is selected where the number of misclassified compounds is lowest. For the quality of the discriminative correlation, a value corresponding to the mean square of errors and the Spearman rank correlation coefficient is evaluated at each ALS iteration time. Examples using this procedure for drug design practice will be shown later in this chapter (Sections 21.1.4.8 and 21.1.4.9).

Another approach which applies to cases where the potency is recorded as a rating score expressed by integers has been developed by Takahashi and co-workers.[82] Their approach, differing from the ALS procedure, uses the simplex technique in optimizing the weight coefficient E_k in equation (97) without adaptation of the independent variable. The simplex technique was first proposed by Ritter and co-workers[83] as a pattern classification procedure of two categorized groups. Takahashi and co-workers expanded this technique to classify multicategorized ordered groups such as the bioactive rating scores. The procedure which they named ORMUCS (ordered multicategorical classification using simplex technique) leads to a single discriminative correlation equation similar to that derived by the ALS procedure.

21.1.2.7 Procedure and Practice in the Formulation of Correlation Equations by the Hansch Approach

In publications dealing with QSAR analysis of certain series of drugs, final correlation equations are sometimes presented without detailed sequential procedures as to why and how terms of 'specific' parameters are selected or added to derive such equations. In this section, two examples are quoted from recent studies in order to illustrate these procedures.

The first example deals with the Ca^{2+} antagonistic activity of a set of compounds (14)[84] related to verapamil (15), a well-known Ca^{2+} antagonist. Compounds (14) were synthesized by Mitani and co-workers,[85] who hoped to make a new series of drugs by combining structural features of verapamil and the phenoxyalkylamine skeleton often found in α-blockers. Using the KCl-depolarized guinea-pig taenia coli, the pA_2 $(mol\,dm^{-3})$ values for 32 unsubstituted and mono- and di-substituted compounds were measured and are listed in Table 6.[84] The substituents of the monosubstituted compounds were selected in order to cover as wide a variety of physicochemical features (in terms of σ, π and B_5 at each of the substituent positions) as possible, the only limitation being their ease of synthesis by practical organic medicinal chemists. B_5 is one of the STERIMOL parameters representing the largest width of substituents from the bond axis connecting the α atom of substituents with the rest of molecule.[40,86] It is apparent that no single parameter is capable of rationalizing the variations in activity. A recent study by Mannhold and co-workers[87] of verapamil analogs shows that hydrophobic, electronic and/or steric parameters are significant in determining variations in various components of smooth muscle responses.

(14) (15)

First, Mitani and co-workers analyzed the pA_2 value of mono- and un-substituted compounds using single parameters.[84] The results indicated that the activity was correlated best by a quadratic equation of hydrophobic parameter π, as shown in equation (101) in Table 7 (see Figure 7), although the quality of correlation was not satisfactory. The π values of substituents used here were those estimated from the log P values of substituted anisoles.[35] In elaborating the correlation, it was noted that the activity of the *para*-substituted derivatives is lower than that of the corresponding *meta*-substituted isomers, irrespective of hydrophobicity and/or electron-withdrawing properties of the substituents. This may suggest that a steric effect specific to *para* substituents is unfavorable to the activity. Mitani and co-workers also noticed that electron-withdrawing substituents such as nitro and halogen at the *ortho* position lower the activity to an extent greater than at the *para* position. This is probably due to a proximity electron-withdrawing effect detrimental to the activity. Participation of other position-specific effects was clearly possible. For instance, hydrophilic substituents such as amino and hydroxymethyl at the *para* position are undoubtedly more unfavorable to the activity than hydrophobic substituents such as *n*-propyl and *t*-butyl. Various combinations of electronic, steric and hydrophobic parameters at each substituent position were therefore examined.

Since most *meta*-substituted derivatives show activity higher than the 'best' fit parabola in Figure 7, and the scatter in this subset seems inversely related to the electron-withdrawing properties of the substituents, Mitani and co-workers first examined a participation of 'ordinary' electronic effects of substituents at *ortho*, *meta* and *para* positions in addition to the quadratic correlation with π, as shown in Figure 8. $σ^0$ values, which are applicable to cases where the reaction center is insulated from conjugation with substituents, were used as the electronic parameters, instead of the regular σ, since $σ^0$ worked much better than σ. The $σ^0$ values of the corresponding *para* substituents were used for the $σ^0$ values of *ortho* substituents.[50] The coefficient (0.4) of the $σ^0$ term added to pA_2 on the ordinate was selected so that the plot for the *meta* derivatives was aligned as nicely as possible, as shown by the solid line. The plot for *ortho*- and *para*-substituted derivatives in Figure 8 generally deviates downward.

Assuming that the proximity electronic effect for the *ortho* substituents can be represented by the $\mathscr{F}$ constant of Swain–Lupton corrected by Hansch and co-workers[46] and considering the position-specific hydrophobic π and steric $\Delta B_5 [= B_5(X) - B_5(H)]$ terms for the *para* substituents, equation (105) was finally derived, as shown in Table 7 and Figure 9. The stepwise development

Table 6 Ca^{2+}-antagonistic Activity and Physicochemical Parameters of Verapamil Analogs (**14**)[84]

Substituents (X)		π	σ^0	$\mathscr{F}_{ortho}$	ΔB_5^{para}	π_{para}	pA_2 (mol dm^{-3})		
							Observed	Calculated (equation 106)	$\lvert\Delta\rvert$
1	H	0.00	0.00	0.00	0.00	0.00	8.19	8.50	0.31
2	2-Me	0.52	−0.12	−0.04	0.00	0.00	8.31	8.50	0.19
3	2-Pr^n	1.43	−0.13	−0.06	0.00	0.00	7.79	7.83	0.04
4	2-OMe	−0.08	−0.16	0.26	0.00	0.00	8.05	8.16	0.11
5	2-OEt	0.30	−0.14	0.22	0.00	0.00	8.10	8.18	0.08
6	2-F	0.21	0.17	0.43	0.00	0.00	7.95	7.78	0.17
7	2-Cl	0.77	0.27	0.41	0.00	0.00	7.69	7.56	0.13
8	2-NO$_2$	0.03	0.82	0.67	0.00	0.00	7.26	7.20	0.06
9	2-NH$_2$	−1.35	−0.38	0.02	0.00	0.00	8.15	7.89	0.26
10	3-Me	0.54	−0.07	0.00	0.00	0.00	8.48	8.41	0.07
11	3-Bu^t	1.85	−0.07	0.00	0.00	0.00	6.88	7.17	0.29
12	3-OMe	0.03	0.06	0.00	0.00	0.00	9.00	8.48	0.52
13	3-F	0.26	0.35	0.00	0.00	0.00	8.29	8.35	0.06
14	3-Cl	0.79	0.37	0.00	0.00	0.00	8.36	8.12	0.24
15	3-NO$_2$	−0.03	0.70	0.00	0.00	0.00	8.17	8.25	0.08
16	3-NH$_2$	−1.14	−0.14	0.00	0.00	0.00	8.13	8.04	0.09
17	3-CF$_3$	0.98	0.47	0.00	0.00	0.00	8.06	7.96	0.10
18	3-CH$_2$OH	−0.91	0.00	0.00	0.00	0.00	8.11	8.17	0.06
19	4-Me	0.52	−0.12	0.00	1.04	0.52	7.61	8.05	0.44
20	4-Pr^n	1.43	−0.13	0.00	2.49	1.43	6.88	6.90	0.02
21	4-Bu^t	1.82	−0.17	0.00	2.17	1.82	6.88	6.79	0.09
22	4-OMe	−0.08	−0.16	0.00	2.07	−0.08	7.64	7.27	0.37
23	4-F	0.21	0.17	0.00	0.35	0.21	8.02	8.31	0.29
24	4-Cl	0.77	0.27	0.00	0.80	0.77	7.73	8.04	0.31
25	4-NO$_2$	0.03	0.82	0.00	1.44	0.03	7.39	7.36	0.03
26	4-NH$_2$	−1.35	−0.38	0.00	0.97	−1.35	6.50	6.71	0.21
27	4-CN	−0.30	0.69	0.00	0.60	−0.30	7.64	7.72	0.08
28	4-CH$_2$OH	−1.00	0.05	0.00	1.70	−1.00	6.56	6.60	0.04
29	2,3-(OMe)$_2$	−0.05	−0.10	0.26	0.00	0.00	8.23	8.14	0.09
30	2,4-Me$_2$	1.04	−0.24	−0.04	1.04	0.52	7.97	7.83	0.14
31	2,5-Me$_2$	1.06	−0.19	−0.04	0.00	0.00	8.68	8.18	0.50
32	2,6-(OMe)$_2$	−0.16	−0.32	0.52	0.00	0.00	7.57	7.82	0.25

Table 7 Development of the Correlation Equations for the Ca^{2+}-antagonistic Activity of Verapamil Analogs (**14**)[84]

π^2	ΔB_5^{para}	π_{para}	$\mathscr{F}_{ortho}$	σ^0	Constant	n	r	s	F	Equation number
−0.34 (0.20)					8.04 (0.25)	28	0.56	0.51	11.74	(101)
−0.24 (0.17)	−0.43 (0.21)				8.17 (0.21)	28	0.77	0.40	17.90	(102)
−0.29 (0.15)	−0.51 (0.19)	0.39 (0.26)			8.22 (0.19)	28	0.84	0.35	19.26	(103)
−0.35 (0.11)	−0.58 (0.15)	0.43 (0.20)	−1.40 (0.64)		8.39 (0.16)	28	0.92	0.26	31.14	(104)
−0.40 (0.11)	−0.58 (0.14)	0.45 (0.18)	−1.28 (0.60)	−0.35 (0.32)	8.46 (0.16)	28	0.93	0.24	30.54	(105)
−0.39 (0.12)	−0.60 (0.14)	0.46 (0.19)	−1.50 (0.55)	−0.35 (0.30)	8.50 (0.16)	32	0.92	0.25	29.68	(106)

of equation (105) was justified statistically for 28 mono- and un-substituted derivatives. The addition of the σ^0 term next to the π^2 term was statistically insignificant, as was observed in the rather unsatisfactory overall correlation in Figure 8. The first-order π term was insignificant in each equation.

In equation (106) in Table 7 and Figure 9, four disubstituted compounds are also included, the quality of the correlation being satisfactory. Equations (105) and (106) are practically identical. The fact that the behavior of disubstituted compounds conforms to that of monosubstituted derivatives indicates that the effect of substituents is almost additive.

The hydrophobicity of substituents seems to exert a dual effect. One aspect is the effect on the transport process expressed by the π (or $\Sigma\pi$) value, independent of the substituent position, corresponding to the total hydrophobicity of the molecule. The other is the additional effect found only for *para* substituents, which are possibly participating in a position-specific hydrophobic

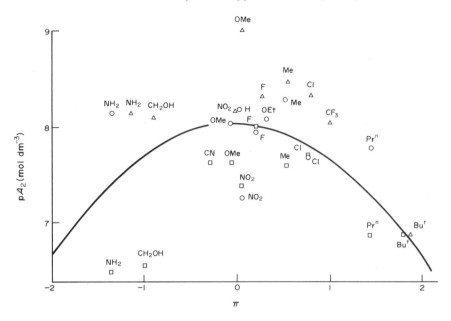

Figure 7 Relationship of Ca^{2+}-antagonistic activity with hydrophobicity of verapamil analogs (**14**). $\bigcirc$, $\triangle$ and $\square$ denote *ortho-*, *meta-* and *para-*substituents, respectively (reproduced from ref. 84 by permission of the Pharmaceutical Society of Japan)

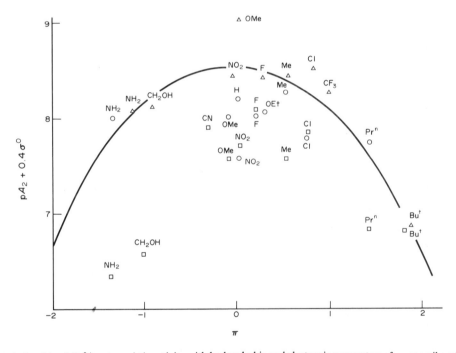

Figure 8 Relationship of Ca^{2+} antagonistic activity with hydrophobic and electronic parameters of verapamil analogs (**14**). $\bigcirc$, $\triangle$ and $\square$ denote *ortho-*, *meta-* and *para-*substituents, respectively (reproduced from ref. 84 by permission of the Pharmaceutical Society of Japan)

interaction at the receptor site. In Table 8 the intercorrelation between independent variables for 32 mono-, di- and un-substituted compounds is shown to be insignificant.

The above result shows that electron-donating substituents whose π values are close to zero are favorable to the activity. It also indicates that a smaller and yet hydrophobic *para* substituent is needed for high activity. Since these two requirements for *para* substituents are difficult to satisfy simultaneously, it was concluded that *para* substitution was best avoided. The most favorable substitution patterns expected from equation (106; Table 7) are those with smaller alkoxy groups at *ortho* and/or *meta* positions, as was experimentally observed for the *m*-methoxy derivative.

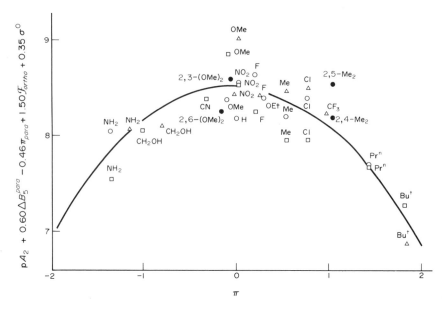

Figure 9 Relationship of Ca^{2+} antagonistic activity with hydrophobic, electronic and steric parameters of verapamil analogs (**14**). $\bigcirc$, $\triangle$ and $\square$ denote *ortho*-, *meta*- and *para*-substituents, respectively (reproduced from ref. 84 by permission of the Pharmaceutical Society of Japan)

Table 8 Correlation Coefficient (r) Matrix for Parameters Used to Derive Equation (106)[84]

	π^2	σ^0	$\mathscr{F}_{ortho}$	ΔB_5^{para}	π_{para}
π^2	1.00				
σ^0	0.40	1.00			
$\mathscr{F}_{ortho}$	0.34	0.16	1.00		
ΔB_5^{para}	0.31	0.10	0.29	1.00	
π_{para}	0.29	0.06	0.07	0.37	1.00

The second example is a QSAR study for phosphoramidothioate herbicides (**16**). An analog where the isopropylamino moiety is replaced by *s*-butylamino and X is 2-nitro-5-methyl has been developed in order to eradicate a wide range of annual grasses and annual broadleaved weeds.[88] Using the barnyard grass, *Echinochloa cruss-galli*, *L. Beauv.*, Yang and co-workers measured the herbicidal activity of 46 mono-, di-, tri- and un-substituted derivatives, listed in Table 9 in terms of pI_{50} ($mol\,dm^{-3}$), the I_{50} being the concentration in emulsified solution required for inhibiting 50% of the shoot elongation.[89]

$$X \overset{\displaystyle S}{\underset{\underset{\displaystyle NHPr^i}{|}}{-O-\overset{\|}{P}-OEt}}$$

(**16**)

Since earlier examples were known where the hydrophobicity of the compounds strongly affected their herbicidal activity on barnyard grass,[90] Yang and co-workers first looked for a possible correlation with π values. The π values used were derived for substituted phenyl dimethyl phosphates,[16] the structures of which are closely analogous to those of this series of compounds. Figure 10 shows plots of pI_{50} against $\Sigma\pi$, where the numbers refer to the compounds in Table 9. No overall direct relationship can be seen, but certain correlations are found for limited numbers of closely related compounds by a close inspection of the plots. For instance, for the set of mono-*ortho*-substituted derivatives (compounds 2–5 and 7 in Table 9; line a) and for the mono-*meta*- and *para*-substituted compounds (compounds 8 and 10–15 in Table 9; line b), activity seems to vary

Table 9 Physicochemical Parameters and pI_{50} Value of O-Aryl-O-Ethyl-N-Isopropylphosphoramidothioates (**16**)[89]

	Substituents (X)	$\Sigma\pi$	ΣE_s^{ortho}	E_s^{para}	π_{meta}	$\Sigma\sigma^0$	$\mathcal{F}_{ortho}$	pI_{50} Observed	pI_{50} Calculated (equation 109)	$\lvert\Delta\rvert$
1	H	0.00	0.00	0.00	0.00	0.00	0.00	4.03	4.15	0.12
2	2-NO$_2$	0.06	−1.01	0.00	0.00	0.82	0.67	5.64	5.03	0.61
3	2-Cl	0.53	−0.97	0.00	0.00	0.27	0.41	5.06	4.63	0.43
4	2-Br	0.76	−1.16	0.00	0.00	0.26	0.44	4.78	4.52	0.26
5	2-CN	−0.30	−0.51	0.00	0.00	0.69	0.51	4.61	4.61	0.00
6	2-OMe	−0.18	−0.55	0.00	0.00	−0.16	0.26	3.91	4.43	0.52
7	2-Me	0.47	−1.24	0.00	0.00	−0.12	−0.04	5.04	4.73	0.31
8	3-NO$_2$	−0.03	0.00	0.00	−0.03	0.70	0.00	4.20	4.31	0.11
9	4-NO$_2$	−0.04	0.00	−1.01	0.00	0.82	0.00	3.59[a]	5.03	
10	3-Cl	0.69	0.00	0.00	0.69	0.37	0.00	4.56	4.96	0.40
11	4-Cl	0.71	0.00	−0.97	0.00	0.27	0.00	4.41	4.46	0.05
12	4-Br	1.02	0.00	−1.16	0.00	0.26	0.00	3.87	4.17	0.30
13	3-Me	0.51	0.00	0.00	0.51	−0.07	0.00	4.78	4.72	0.06
14	4-Me	0.55	0.00	−1.24	0.00	−0.12	0.00	4.30	4.67	0.37
15	4-OMe	0.04	0.00	−0.55	0.00	−0.16	0.00	4.15	4.45	0.30
16	2-NO$_2$-5-Me	0.29	−1.01	0.00	0.51	0.75	0.67	6.09	5.73	0.36
17	2-NO$_2$-6-Me	0.41	−2.25	0.00	0.00	0.70	0.63	6.00	5.65	0.35
18	2-NO$_2$-4-Me	0.40	−1.01	−1.24	0.00	0.70	0.67	5.81	5.65	0.16
19	2-NO$_2$-6-Cl	0.41	−1.98	0.00	0.00	1.09	1.08	5.60	5.59	0.01
20	2-NO$_2$-4-Cl	0.53	−1.01	−0.97	0.00	1.09	0.67	5.49	5.51	0.02
21	2-NO$_2$-4-CF$_3$	0.84	−1.01	−2.40	0.00	1.35	0.67	5.95	6.16	0.21
22	2-NO$_2$-6-Br	0.55	−2.17	0.00	0.00	1.08	1.11	5.24	5.60	0.37
23	2,5-Me$_2$	0.92	−1.24	0.00	0.51	−0.19	−0.04	5.34	5.02	0.32
24	2,6-Cl$_2$	1.11	−1.94	0.00	0.00	0.54	0.82	4.83	4.60	0.23
25	2-Me-4-Cl	1.16	−1.24	−0.97	0.00	0.15	−0.04	4.85	4.57	0.28
26	2,4-Cl$_2$	1.23	−0.97	−0.97	0.00	0.54	0.41	4.74	4.38	0.36
27	3-Me-4-Cl	1.28	0.00	−0.97	0.51	0.20	0.00	4.78	4.36	0.42
28	2,6-Br$_2$	1.40	−2.32	0.00	0.00	0.52	0.88	4.09	4.28	0.19
29	2,4-Br$_2$	1.53	−1.16	−1.16	0.00	0.52	0.44	3.74	3.99	0.25
30	2-NO$_2$-4-Bu^t	1.62	−1.01	−2.78	0.00	0.65	0.67	4.87	4.74	0.13
31	3,4-Me$_2$	0.90	0.00	−1.24	0.51	−0.19	0.00	4.68	5.05	0.37
32	3-Me-4-NO$_2$	0.48	0.00	−1.01	0.51	0.75	0.00	3.58[a]	5.62	
33	2-Cl-4-NO$_2$	0.60	−0.97	−1.01	0.00	1.09	0.41	3.45[a]	5.45	
34	2-CHO	−0.21	−1.01	0.00	0.00	0.49	0.31	3.68[a]	4.90	
35	2-CO$_2$Me	0.13	−1.01	0.00	0.00	0.46	0.37	4.27[a]	4.91	
36	2-CO$_2$Et	0.59	−1.01	0.00	0.00	0.86	0.33	4.00[a]	4.78	
37	4-Bu^t	1.86	0.00	−2.78	0.00	−0.17	0.00	2.00[a]	3.22	
38	2,4-(NO$_2$)$_2$	−0.14	−1.01	−1.01	0.00	1.64	0.67	3.64[a]	5.89	
39	2,5-(NO$_2$)$_2$	−0.10	−1.01	0.00	−0.03	1.52	0.67	3.94[a]	5.18	
40	2,6-(NO$_2$)$_2$	−0.34	−2.02	0.00	0.00	1.64	1.34	4.09[a]	5.82	
41	2,4,6-Cl$_3$	1.81	−1.94	−0.97	0.00	0.81	0.82	3.62	3.76	0.14
42	2-NO$_2$-4,5-Me$_2$	0.63	−1.01	−1.24	0.51	0.63	0.67	5.99	6.24	0.25
43	2-NO$_2$-4,6-Cl$_2$	0.88	−1.98	−0.97	0.00	1.36	1.08	5.46	5.82	0.36
44	2,6-(NO$_2$)$_2$-4-Me	0.00	−2.02	−1.24	0.00	1.52	1.34	5.88[a]	6.65	
45	2,6-(NO$_2$)$_2$-4-Cl	0.13	−2.02	−0.97	0.00	1.91	1.34	5.34[a]	6.58	
46	2-Me-4,6-(NO$_2$)$_2$	0.33	−2.25	−1.01	0.00	1.52	0.63	3.92[a]	6.57	

[a] Not included in analyses.

parabolically with π. For each set of multiply substituted derivatives having a similar substitution pattern, activity tends to decrease with $\Sigma\pi$ [line (c) for 2-nitro-n-X compounds 16–20 and 22 (Table 9), line (e) for dimethyl, dihalo and halomethyl compounds 23–29 (Table 9), and line (d) for 2-nitro-n-X-n'-X' compounds 42–43 (Table 9)]. Moreover, the slopes of these lines (c–e) are similar to those for the supraoptimum part of the parabolic relations (a, b) for monosubstituted derivatives. Physicochemical properties other than the total hydrophobicity should also be important in determining the herbicidal activity. Thus, if these component physicochemical properties could be identified, they should reflect specificities to substituent positions, patterns or both, depending upon the subsets of compounds.

Besides the majority of compounds that belong to these sets, there are some irregularly behaving compounds the activities of which do not seem to fit the partial correlations for their closely related analogs. For example, the activities of 2,4-, 2,5- and 2,6-dinitro derivatives (compounds 38, 39 and 40

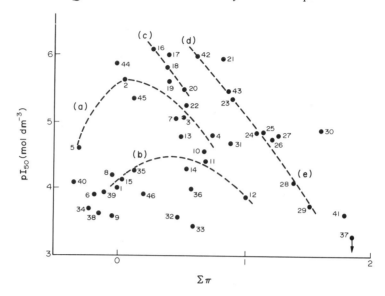

Figure 10 Relationship of herbicidal activity with hydrophobicity of aromatic substituents of phosphoramidothioates (**16**)
(reproduced from ref. 89 by permission of Academic Press)

in Table 9) are much lower than expected from the activity pattern of 2-nitro-*n*-X compounds (compounds 16–22 in Table 9). If they behaved regularly, their pI_{50} would be somewhere extrapolated parabolically from line (c) for the 2-nitro-*n*-X compounds. The activities of the 2-formyl, 2-methoxycarbonyl and 2-ethoxycarbonyl derivatives (compounds 34–36 in Table 9) are also lower than the parabola (line a) for the *ortho* derivatives. Another set of outliers includes those having a nitro group at the *para* position, such as the 3-methyl-4-nitro and 2-chloro-4-nitro derivatives (compounds 32 and 33 in Table 9). Their activities are much lower than those of the isomers having the nitro group at the *ortho* position (line c). Thus, Yang and co-workers first deleted these outliers in examining the other parameters to be combined with $\Sigma\pi$ for analysis, and hoped to rationalize their outlying behavior afterward.

In Figure 10, line (a) for mono-*ortho* substituted compounds is located in a higher activity region than line (b). Moreover, the activities of 2,6-disubstituted derivatives other than the dinitro compounds are higher than those expected for mono-*ortho* derivatives having the corresponding $\Sigma\pi$ value (compare lines (c–e) with line a). Thus, herbicidal activity seems to be enhanced by *ortho* substitutions, all other things being equal. Since both electron-withdrawing and -donating *ortho*-substituents seem to have similar effects within the partial correlations (compounds 5 and 7 (Table 9) on line (a) and compounds 23 and 24 (Table 9) on line (e) in Figure 10), the steric effect is more important than the electronic factor in rationalizing the activity enhancement of the *ortho* substituents. Preliminary examination was made to select the best steric parameter and its coefficient. The Taft–Kutter–Hansch E_s value was chosen. The coefficient δ (0.6) was taken so that the $[pI_{50} + \delta\Sigma E_s^{ortho}]$ values for mono- and di-*ortho*-substituted compounds lined up as neatly as possible graphically, as shown in Figure 11, where possible outliers have been deleted. The zero reference value of the E_s parameter was shifted to the unsubstituted compound. For 2,6-disubstituted compounds, the substituent E_s values were simply summed up.

In Figure 11, the plots for compounds having substituents at *meta* and *para* positions still generally deviate upward from parabolically lined-up values for *ortho* compounds. The effects of *meta* and *para* substituents are probably specific to their positions. Since the activities of the 2-nitro-4-trifluoromethyl and 2-nitro-4-*t*-butyl compounds with large but electronically opposite *para* substituents deviate similarly, steric bulk is also likely to be a very important factor for *para* substituents. The effect of *meta* substituents may be hydrophobic, since the plots for derivatives having hydrophobic *meta* substituents deviate upward, and the deviation of the plot for the *m*-nitro compound in which the substituent π value is close to zero is almost the same as that of the unsubstituted compound.

After these examinations, regression analyses were done using ΣE_s^{ortho}, E_s^{para} and π_{meta} along with $\Sigma\pi$ as parameters for the correlation, leading to equation (107). The coefficients of the ΣE_s^{para} and E_s^{para} terms are similar. In fact, they overlap partially if their 95% confidence intervals are taken into consideration. Thus the two terms were combined, giving equation (108). The difference between

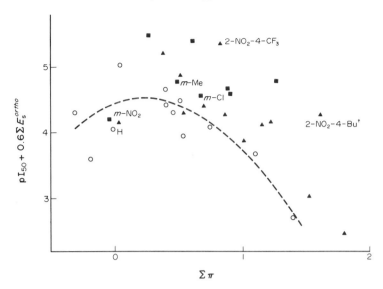

Figure 11 Relationship of herbicidal activity with hydrophobic and steric parameters of phosphoramidothioates (**16**). ○, derivatives with *ortho* and di-*ortho* substituents; ■, derivatives with *meta* substituents; ▲, derivatives with *para* substituents (reproduced from ref. 89 by permission of Academic Press)

observed and calculated pI_{50} values from equation (108) suggested participation of an electronic effect of the substituents. Thus, addition of the $\Sigma\sigma^0$ term yielded equation (109), although it is significant only at a level of 92% as examined by the F test. The addition of the first-order $\Sigma\pi$ term in equations (108) and (109) was not significant even at a level of 75%. In Table 9, the pI_{50} calculated from equation (109) is compared with the observed value.

$$pI_{50} = -0.795(\pm0.177)(\Sigma\pi)^2 + 1.578(\pm0.561)\pi_{meta} - 0.819(\pm0.181)\Sigma E_s^{ortho} - 0.629(\pm0.195)E_s^{para}$$
$$+ 4.134(\pm0.251) \tag{107}$$

$$n = 33, \quad r = 0.905, \quad s = 0.323$$

$$pI_{50} = -0.819(\pm0.182)(\Sigma\pi)^2 + 1.511(\pm0.578)\pi_{meta} - 0.735(\pm0.161)\Sigma E_s^{2,4,6} + 4.173(\pm0.257) \tag{108}$$

$$n = 33, \quad r = 0.893, \quad s = 0.336$$

$$pI_{50} = -0.760(\pm0.190)(\Sigma\pi)^2 + 1.544(\pm0.561)\pi_{meta} + 0.294(\pm0.348)\Sigma\sigma^0 - 0.631(\pm0.199)\Sigma E_s^{2,4,6}$$
$$+ 4.151(\pm0.251) \tag{109}$$

$$n = 33, \quad r = 0.904, \quad s = 0.325$$

In these analyses, the activity data of derivatives having two nitro groups at any position and having a nitro group at the *para* position were not included. In fact, the dinitro compounds are generally very insoluble. They may not be emulsified completely, so the value for their activity is somewhat inaccurate. Another reason for the low activity of dinitro and *p*-nitro compounds may be that the substituent electron-withdrawal is too high for the amidate side-chain to exist without suffering from possible base-catalytic degradation[91] metabolism *in vivo*. The electron-withdrawing effect of the *p*-, but not *o*- and *m*-nitro groups, could be greater than that expressed by σ^0, but perhaps corresponds with σ^-. Also excluded were the 2-formyl, 2-methoxycarbonyl and 2-ethoxy-carbonyl derivatives, as these substituents may be oxidized or hydrolyzed *in vivo*. The activity of the 4-*t*-butyl compound is probably too low to be estimated accurately.

Similar to equation (105) in Table 7 for verapamil analogs, equation (109) shows that the substituents affect the activity by hydrophobicity in two ways. One, represented by $\Sigma\pi$, is the effect of the hydrophobicity of the whole molecule, which governs the transport process to the target site of action. The other is a position-specific hydrophobic effect exhibited by *meta* substituents. Although this effect may be operative at the site of action, more compounds having a variety of *meta* substituents should be examined to draw a clearer picture of the *meta*-substituent effect. The negative ΣE_s term indicates that the bulkier the *ortho* and *para* substituents, the more potent the herbicidal activity. The steric bulk favorable to the activity may be any that is too big for detoxification enzymes.

The above two examples illustrate procedures for disentangling overlapping substituent effects, disclosing outliers and rationalizing their outlying behavior. Of course, the procedure used here can be done with the use of a display terminal. The stepwise graphical procedure is very useful for helping organic medicinal chemists to understand what they are doing physicochemically in substituent modifications at each position.

21.1.3 SELECTED EXAMPLES OF THE HANSCH APPROACH AND RELATED QSAR STUDIES

21.1.3.1 QSAR of Sulfa Drugs

The work published by Bell and Roblin on the sulfa drugs in 1942[92a] is perhaps the earliest example of QSAR studies, although they elucidated their quantitative result in a more or less qualitative way instead of formulating the correlation equations. As shown in Figure 12, they observed that a logarithmic plot of the bacteriostatic activities of some 40 sulfanilamides (**17**; substituted at the N^1-position with such groups as substituted phenyl, substituted heteroaryl and others) against their dissociation constants exhibits a biphasic relationship. Using their data, Silipo and Vittoria formulated equations (110) and (111) with the parabolic and bilinear models, respectively.[92b]

$$H_2N - \!\!\!\left\langle\!\!\!\!\bigcirc\!\!\!\!\right\rangle\!\!\! - SO_2NHR$$

$$(17)$$

$$\log(1/C) = 2.103(\pm 0.29)\mathrm{p}K_a - 0.155(\pm 0.02)\mathrm{p}K_A^2 - 1.351(\pm 0.96) \tag{110}$$

$$n = 39, \quad r = 0.939, \quad s = 0.321$$

$$\log(1/C) = 1.044(\pm 0.13)\mathrm{p}K_a - 1.640(\pm 0.18)\log(\beta \cdot 10^{\mathrm{p}K_a} + 1) + 0.275(\pm 0.65) \tag{111}$$

$$n = 39, \quad r = 0.956, \quad s = 0.275, \quad \log\beta = -5.96, \quad \mathrm{p}K_a(\mathrm{opt}) = 6.22$$

Equation (111), with the bilinear model, is slightly better than the parabolic equation (110). In these equations, C is the minimum inhibitory concentration (mol dm^{-3}) *in vitro* against proliferation of *E. coli* in a buffered (pH = 7) medium. They considered that the more negative the sulfonyl group of the sulfanilamide derivatives, the more closely they will resemble the *p*-aminobenzoate anion with which sulfanilamides compete for the dihydrofolate synthetic enzyme. Being affected by the negative charge on the adjacent amide nitrogen, the sulfonyl group of sulfanilamides in the ionized form is much more negative than in the nonionized form. Therefore, they postulated that the ionic form of

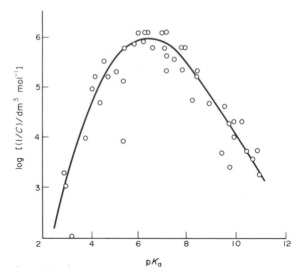

Figure 12 Relationship between bacteriostatic activity and acidity of N^1-substituted sulfanilamides (reproduced from ref. 92(a) by permission of the American Chemical Society)

any sulfanilamide should be much more active than the neutral form. As the electron-withdrawing power of the N^1-substituent increases, *i.e.* the pK_a value decreases, the relative amount of ionized form increases leading to higher activity. However, as the electron-withdrawing power of the N^1-substituent continues to increase, a greater part of the negative charge on the amide nitrogen could be withdrawn by the N^1-substituent and the sulfonyl group would therefore become less negative resulting in lower bacteriostatic activity. Up to a certain point, the decrease in activity with increasing electron-withdrawing power of the N^1-substituent is lower than the increase in activity due to increasing the proportion of highly active ions. A maximum would be expected, therefore, in the pK_a–activity relationship. The optimum pK_a value estimated from equations (110) and (111) is 6.2–6.8.

An alternative explanation for the 'parabolic' relationship was proposed by Brueckner[93] and Cowles.[94] They assumed that the negative ion which is responsible for the bacteriostatic activity penetrates to the site of action inside the cell with much more difficulty than the neutral form. It is not the extracellular ionic concentration which governs the potency but rather the intracellular ionic concentration. There should be an optimal dissociation constant, therefore, where the balance between the intrinsic activity and the penetration is most favorable to the overall bacteriostatic process.

Seydel and co-workers examined relationships of structure with not only antibacterial activity but also inhibition of folic acid synthesis in the cell-free system for sets of sulfanilamides. From their data for a set of N^1-phenylsulfanilamides (18) having various substituents on the N^1-benzene ring, equation (112) was derived for the minimum inhibitory concentration (mol dm^{-3}) against *E. coli* in a medium of pH $= 7.2$ (Sauton medium).[95] In the original correlation, highly ionized disubstituted derivatives having a pK_a lower than 7.0 were not used, although Seydel recognized the existence of an optimum pK_a value. The original data were reanalyzed[11] including disubstituted compounds to give the bilinear equation (112), indicating that the optimum pK_a value for this uniform set of sulfanilanilides is located at about 6.9. Although the regression coefficient of each term in equation (112) is about half that of the corresponding term in equation (111), the optimum pK_a values are very close, being slightly lower than the pH of the medium.

$$H_2N-\!\!\!\left\langle\!\!\!\bigcirc\!\!\!\right\rangle\!\!\!-SO_2NH-\!\!\!\left\langle\!\!\!\bigcirc\!\!\!\right\rangle\!\!\!\diagdown_X$$

(18) X = H, 2-OR, 2-alkyl, 2-halo, 3-alkyl, 3-OR, 3-halo, 3-NO$_2$, 3-CF$_3$,
4-alkyl, 4-OR, 4-halo, 4-NO$_2$, 4-CF$_3$, 4-CN, 4-Ac, 3,4-alkoxyhalo, *etc.*

$$\log(1/C) = 0.511(\pm 0.272)pK_a - 1.153(\pm 0.354)\log(\beta \cdot 10^{pK_a} + 1) + 2.439(\pm 1.753) \tag{112}$$

$$n = 37, \quad r = 0.923, \quad s = 0.188, \quad \log \beta = -7.0, \quad pK_a(\text{opt}) = 6.87$$

For the inhibition of folate biosynthesis of the same series of sulfanilanilides using a cell-free extract (pH $= 7.9$–8) prepared from *E. coli*, equation (113) was formulated,[96] where I_{50} (mol dm^{-3}) is the concentration of sulfanilanilides causing a 50% inhibition of folate synthesis. Addition of the π term for the participation of hydrophobicity in the inhibition process was not significant. Under conditions of the folate synthesis inhibition at pH $= 7.9$–8.0, seven compounds of $n = 14$ are highly ionized having a pK_a lower than 7.5 down to 5.7. With the cell-free system, permeability of the bacterial cell wall is not the critical factor. In fact, the slope of the descending part of equation (112) of $-0.64 (= 0.51 - 1.15)$, for compounds whose pK_a is much higher than 7.0 where the ionization is not important, is close to that of the pK_a term of equation (113). Seydel and co-workers[96] considered that these results support the hypothesis of Brueckner and Cowles rather than that proposed by Bell and Roblin. Without the permeation barrier, the activity of sulfanilamides should increase linearly with an increase in the electron-withdrawing power of the N^1-substituent.

$$pI_{50} = -0.427(\pm 0.066)pK_a + 8.232(\pm 0.509) \tag{113}$$

$$n = 14, \quad r = 0.971, \quad F = 200.85, \quad s = 0.120$$

Recently, Coats, Seydel and co-workers made quantitative analyses similar to the one described above for a series of dapsone analogs (19).[97] For the inhibition of cell-free extracts from dapsone-sensitive *Mycobacterium smegmatis* at pH $= 7.75$, equation (114) was derived, where ΔppmNH$_2$ is the chemical shift of the 4-amino protons of 4'-substituted derivatives relative to that of the unsubstituted compound. It was used because the Hammett σ constant was not known for some substituents. For substituents whose σ constants were known, ΔppmNH$_2$ correlated linearly with σ ($r = 0.969$). The coefficient associated with ΔppmNH$_2$ suggests that electron donation to the sulfonyl group results in increased activity. f_i is an indicator variable, being unity for the completely ionized carboxyl group of the 4'-CO$_2$H and 4'-NHCH$_2$CO$_2$H derivatives and zero for nonionized substituents. Although the original authors considered that the f_i term may indicate the existence of an additional binding point which attracts the negatively charged substituents, there is another explanation. The negatively charged substituents are more electron donating than the neutral ones with which the chemical shift was measured in DMSO. The ΔppmNH$_2$ value for the ionized groups may be overestimated so that its negative term in equation (114) should be corrected toward the positive direction.

$$H_2N-\!\!\!\!\overbrace{}-\!\!\!\!\overset{\displaystyle O}{\underset{\displaystyle O}{\overset{\|}{\underset{\|}{S}}}}-\!\!\!\!\overbrace{}-X$$

(19) X = NH$_2$, OMe, NO$_2$, H, OH, halo, Me, NHAc, NHMe, NHEt, NMe$_2$,
CO$_2$Me, CO$_2$H, CONHNH$_2$, NHCH$_2$CO$_2$Me, NHCH$_2$CO$_2$H

$$\text{pI}_{50}\ (\mu\text{mol dm}^{-3}) = -3.64(\pm 0.84)\Delta\text{ppmNH}_2 + 0.82(\pm 0.28)f_i - 1.06(\pm 0.13) \tag{114}$$

$$n = 17, \quad r = 0.95, \quad F = 59.0, \quad s = 0.18$$

They obtained a correlation equation almost identical with equation (114) for the folate synthesis inhibition in extracts from dapsone-resistant *M. smegmatis* with the same set of compounds. They considered that an increase in the amount of the folate synthetic enzyme by gene amplification in resistant cells is the most probable explanation for the high degree of resistance exhibited by the whole-cell cultures. If this mechanism of resistance is the case, it may be unrealistic to attempt to circumvent dapsone resistance by structural modification of the dapsone molecule. They also derived correlation equations very similar to equation (114) for the folate synthesis inhibition against preparations from dapsone-sensitive as well as dapsone-resistant strains of *M. lufu* and *M. leprae*. It is very likely that the enzymes isolated from resistant *M. smegmatis*, *M. lufu* and *M. leprae* cultures do not differ in receptor site from those of the corresponding sensitive strains and that the two cultivatable mycobacterial strains are reliable models for investigating the mechanism of resistance of *M. leprae*.

The addition of $\log P$, π or other hydrophobicity parameters to correlation equations for the inhibition of the cell-free enzyme systems, such as equation (114), did not improve the correlations. For the whole-cell culture systems, however, hydrophobicity of dapsone analogs was highly significant in governing the overall bacteriostatic activity. Thus equation (115) was formulated for the activity against the sensitive strain of *M. smegmatis*,[97] where C is the minimum inhibitory concentration (μmol dm^{-3}) for the whole-cell culture, pI$_{50}$ is the cell-free activity index used in equation (114) and $\log k'$ is the HPLC capacity factor linearly related to the π value ($r = 0.994$) of 4'-substituents for 15 out of 17 compounds. The slope of the pI$_{50}$ term is close to unity after possible transport factors across the cell wall are separated by the bilinear terms of hydrophobicity, indicating that the bacteriostatic activity of dapsone analogs is in fact due to the folate synthesis inhibition.

$$\log(1/C) = 0.729(\pm 0.384)\text{pI}_{50} + 0.933(\pm 0.728)\log k' - 1.438(\pm 0.675)\log(\beta \cdot k' + 1) + 1.532(\pm 2.332) \tag{115}$$

$$n = 17, \quad r = 0.899, \quad F = 9.22, \quad s = 0.329, \quad \log\beta = -0.35, \quad \log k'_{\text{opt}} = 0.621$$

They also reinvestigated the whole-cell and the cell-free activities of N^1-phenylsulfanilamides on *E. coli* using ΔppmNH$_2$ for the 4-amino protons as the electronic parameter.[97] In equation (116) for the bacteriostatic activity (C in μmol dm^{-3}), all compounds but one included were not highly ionized under conditions of the Sauton medium (pH = 7.2). In equation (117) for the cell-free system (pH = 7.75), about half the compounds were highly to moderately ionized, the conditions corresponding with those under which equation (114) was formulated.

$$\log(1/C) = 8.02(\pm 1.29)\Delta\text{ppmNH}_2 - 1.26(\pm 0.11) \tag{116}$$

$$n = 11, \quad r = 0.98, \quad F = 199.5, \quad s = 0.13$$

$$\text{pI}_{50}\ (\mu\text{mol dm}^{-3}) = 5.06(\pm 0.91)\Delta\text{ppmNH}_2 - 1.62(\pm 0.08) \tag{117}$$

$$n = 11, \quad r = 0.97, \quad F = 157.8, \quad s = 0.09$$

Seydel and co-workers stated that the structure–activity analyses provide evidence that the two classes of compounds, dapsone analogs and phenylsulfanilamides, interact with identical binding sites, *i.e.* dihydrofolate synthetase. The decisive factor is the generation of a negative charge at the sulfonyl group or its surroundings. The creation of such a charge is favored in the case of sulfones by electron-donating substituents and in the case of sulfanilamides by electron-withdrawing groups which facilitate ionization of the —SO_2NH— group. This is indicated by the change in sign of the regression coefficients of the $\Delta ppmNH_2$ term in equations (114) and (117). The most striking difference between these two classes of compounds seems to be between their characteristic cell-wall permeation behaviors. For sulfones, the permeability is bilinearly dependent on their hydrophobicity, as is apparent from the bilinear terms in equation (115). However, permeation is not the critical factor for sulfanilamides as far as they exist in a nonionized form in the medium. Their bacteriostatic activity is dependent only on their electronic structure, except for highly ionized analogs.

Although the QSAR studies of Seydel and co-workers clarified quite a few problems related to the mode of action of sulfa drugs, a couple of points still seem to require further elaboration. First, why is the transport of the neutral sulfones across the bacterial cell-wall governed by their hydrophobicity but not that of the nonionized sulfanilamides? The second point is the explanation of equations (113) and (117) for the folate synthesis inhibition of sulfanilamides. The pI_{50} value in these equations represents the activity of the equilibrated mixture of ionized and nonionized forms for appreciably ionized analogs. The fact that their 'apparent' activity index fits the correlation along with the activity of analogs which exist almost entirely as the neutral molecule within the single equation suggests that the parameters responsible for the intrinsic activity, such as K and k in equations (3) and (4), are identical for ionized and nonionized species. In general, physicochemical and pharmacokinetic properties are different between dissociable neutral molecule and the conjugate ion. The reason why two forms of sulfanilamides behave in the same manner at the target sites is still unclear. Moreover, although the relative amount of the ionized form increases with an increase in the electron-withdrawing power of the N^1-substituent, the negative charge on the sulfonyl group itself decreases continuously, as far as each of the ionic and neutral molecules is concerned. The increase in the proportion of negatively charged ionic species would outweigh the otherwise unfavorable electron-withdrawing effect on the activity. This is apparently paradoxical, since the activity is not determined by the relative abundance ratio of the two species but by the overall pI_{50}.

Recently, Hopfinger and co-workers tried to analyze the folic acid synthesis inhibition of dapsone analogs measured by Seydel using a molecular modeling method.[98] Since these molecules are quite flexible and possess multiple conformational energy minima, application of molecular shape analysis was not successful. However, the calculated free space intramolecular conformational entropy, S (cal mol^{-1} K^{-1}, 1 cal = 4.184 J) of these compounds was found to correlate with inhibition potency leading to equation (118).*

$$pI_{50} = -0.0075(\pm 0.0006)S + 6.30 \qquad (118)$$

$$n = 36, \quad r = 0.970, \quad F = 544.6, \quad s = 0.11$$

I_{50} (μmol dm^{-3}) was measured using extracts from *M. lufu* for 36 analogs having a greater variety of substituents than those used to derive equation (114). Inhibition potency increases as entropy decreases or the molecule is restricted conformationally. For this set of compounds, equation (119) was also derived, where MR_2 is the molecular refractivity of 3'-substituents. Equation (119) conforms with equation (114). Hopfinger and co-workers found that a correlation existed between S and the parameters used in equation (119) according to equation (120).

$$pI_{50} = -5.91\Delta ppmNH_2 + 0.507f_i + 0.047MR_2 + 5.06 \qquad (119)$$

$$n = 36, \quad r = 0.90, \quad F = 47.56, \quad s = 0.19$$

$$S = 776\Delta ppmNH_2 - 59.8f_i - 6.1MR_2 + 167.8 \qquad (120)$$

$$n = 36, \quad r = 0.91, \quad F = 52.2, \quad s = 0.24$$

Correlations given by equations (118) and (119) are two independent representations of common structure–activity data. Hopfinger and co-workers considered that the two QSARs mutually support one another, and made a successful prediction of activity of some untested compounds with equation (118). Whether such a multiple collinearity as shown by equation (120) has real physicochemical significance should be further substantiated.

* See the Note Added in Proof on page 556.

The QSAR procedures for series of ionizable drugs such as the sulfanilamides, whose degree of ionization varies depending upon the structure at biological pH, have led to quite a few controversies.[99-102] In cases where the ultimate active species, has been established, nonionized or ionized, the apparent total concentration required to exhibit a certain standard response should be corrected according to the dissociation constant and the pH of the medium in order to indicate the 'real' activity for either one of the species.[103-105] More common are cases where the active species are unknown. In some instances, *e.g.* for weakly acidic uncouplers,[106] both species are found to be equally active, similar to sulfanilamides. Martin demonstrated that the use of theoretically derived model-based nonlinear equations could resolve a considerable part of the problem.[107]

21.1.3.2 QSAR of Hapten–Antibody Interactions

Analyses of hapten–antibody interactions are important in understanding the nature of the immune response system. Hansch and co-workers attempted to apply QSAR analysis to this field using experimental data accumulated by Pauling, Pressman and co-workers.[108-110] The first example discussed here is for the effect of substituted benzoic acids as haptens on an antibody–antigen complex formation.[111] The antibodies were generated by injecting antigens (**20**; prepared by coupling diazotized *p*-aminobenzoic acid with serum proteins) into rabbits. Substituted benzoic acids inhibit complex formation to various extents depending upon the nature and position of substituents. Other proteins coupled with the diazotized *p*-aminobenzoic acid react with antibodies formed from the first antigen. From the potency of inhibition, the binding constant of haptens (benzoic acids) with antibody was evaluated in terms of K_{rel} relative to that of unsubstituted benzoic acid.

$$\text{protein} - \left[N = N - \langle \text{benzene} \rangle - CO_2^- \right]_x$$

(**20**)

Equation (121) was derived for the inhibition of the complex formation between antibody generated by the modified whole beef serum and ovalbumin antigen (**21**), where E_s is the Taft–Kutter–Hansch steric parameter. $\Sigma \pi$ is the summation of π values of substituents at various positions. The positive coefficient with E_s^{ortho} means that a large group in the *ortho* position of the hapten results in poor combination of hapten and antibody. The E_s^{meta} term with a small coefficient indicates that the size of *meta* substituents is not an important factor in the combination. The negative coefficient with the E_s^{para} term means that large groups in the *para* position prevent the complex formation. Since the antibody was generated by antigens where the *p*-diazotized benzoic acid is attached, the stereoelectronic complementarity between antibody and *p*-substituted benzoic acids could be highest among isomers. The larger the size of the *para* substituent of the hapten, the greater becomes the blockade of the hapten from displacing the antigen. The hydrophobicity of substituents does not seem to be very critical in the binding.

$$\text{ovalbumin} - \left[N = N - \langle \text{benzene} \rangle - CO_2^- \right]_x$$

(**21**)

$$\log K_{rel} = 0.864(\pm 0.07)E_s^{ortho} + 0.084(\pm 0.07)E_s^{meta} - 0.436(\pm 0.08)E_s^{para} + 0.166(\pm 0.07)\Sigma \pi$$
$$- 0.715(\pm 0.12) \tag{121}$$

$$n = 22, \quad r = 0.989, \quad s = 0.120$$

Similar analysis was made for the relative affinity of substituted benzenearsonic acids as haptens with their antibody, giving equation (122). The antibody was formed by coupling diazotized *p*-aminobenzenearsonic acid with whole sheep serum (**22**) and injecting the modified protein into rabbits.[111] In equation (122), the E_s^{meta} term is not significant at the 95% level. The overall similarity to equation (121) is apparent except for the rather small coefficient of the E_s^{ortho} term in equation (122). The carboxylate group is planar. The twisting conformational change of the carboxylate group would be detrimental for hapten binding. Of course, the smaller the *ortho* substituent, the lower the degree of the twist. This type of intramolecular steric effect seems to be included in the E_s^{ortho} term of equation (121) making the value of the coefficient more positive. The arsonate function is more symmetrical being tetrahedral so that *ortho* substituents cannot twist it into significantly different structures.

$$\text{protein} - \left[N=N - \underset{(22)}{\underset{\bigcirc}{}} - AsO_3H^- \right]_x$$

$$\log K_{rel} = 0.164(\pm 0.15)E_s^{ortho} - 0.154(\pm 0.20)E_s^{meta} - 0.360(\pm 0.15)E_s^{para} + 0.213(\pm 0.29) \qquad (122)$$

$$n = 16, \quad r = 0.850, \quad s = 0.220$$

With the use of different sets of steric parameters, Verloop and co-workers substantiated equation (121).[40] Including 14 additional compounds, where the E_s values of the substituents are not available but their STERIMOL parameters are known, equation (123) was formulated for the substituent effect of benzoic acid haptens. The STERIMOL B_1 was defined as the shortest width and B_4 as the 'largest' width of substituents, similar to B_5. The sign of corresponding terms in equations (121) and (123) is reversed according to the definitions of parameters used.

$$\log K_{rel} = -1.46 B_1^{ortho} + 0.50 B_4^{para} + 0.84 \qquad (123)$$

$$n = 36, \quad r = 0.953, \quad F = 164.3, \quad s = 0.302$$

When the E_s terms in equation (121) were replaced by the corresponding π terms, the correlation was much poorer. Thus, the binding site of haptens on the antibody is not hydrophobic. Pauling and Pressman[109] attempted to rationalize their experimental results by considering that the dispersion forces between hapten substituent and antibody are important. The dispersion forces of substituents could be modeled by the molar refractivity (MR) of substituents. In equations (121) and (122), however, MR was not capable of replacing E_s with improvement of the correlation. Nevertheless, the collinearity between E_s and MR of substituents at each position was rather high ($r^2 = 0.65-0.81$) for compounds included in these equations. In this situation, E_s might be accounting for some effect of dispersion forces (MR) in addition to modeling steric effects.

Hansch and co-workers[112] extended their analysis to other experimental data of Pressman and co-workers.[110] For the interaction of substituted succinanilides (23) as haptens with antibodies formed using whole beef serum treated with diazotized 3-aminopyridine (24), equation (124) was derived, where $-\Delta F_{rel}^0$ is the free energy change for complex formation relative to that of the unsubstituted compound. The positive coefficient with each of the MR (scaled by 1.0) terms indicates that, as Pauling and Pressman first suggested, dispersion forces are holding hapten and antibody together. Since collinearity between E_s and MR of substituents is still rather high ($r^2 = 0.87-0.88$) in this series of compounds, these conclusions seem to be somewhat compromised. The negative $\Sigma\pi$ term means that hydrophilic groups rather than hydrophobic groups enhance the binding, although the effect is not so important as that of MR. The $\Sigma\sigma$ term suggests that electron-withdrawing substituents potentiate the interaction.

$$X - \underset{(23)}{\underset{\bigcirc}{}} - NHCOCH_2CH_2CO_2^- \qquad\qquad \text{protein} - \left[N=N - \underset{(24)}{\underset{\bigcirc}{}} \right]_x$$

$$-\Delta F_{rel}^0 = 1.73(\pm 0.38)MR^{para} + 1.18(\pm 0.39)MR^{meta} - 0.24(\pm 0.22)\Sigma\pi + 0.52(\pm 0.52)\Sigma\sigma - 2.39(\pm 0.32) \qquad (124)$$

$$n = 20, \quad r = 0.946, \quad s = 0.277$$

For the binding of substituted pyridines as haptens with the same antibodies, equation (125) was formulated. Equation (125) differs from equation (124) in some respects, but, again, MR terms are the most important determinants of activity. Whereas the MR terms for *ortho* and *para* substituents are negative, the term for *meta* substituents is positive in this case. Substituents in the *meta* position promote binding, while *ortho* and *para* substituents inhibit it, possibly by intermolecular steric hindrance. Stereoelectronic complementarity similar to that observed in benzoic acid haptens as formulated by equation (121) is clearly seen in this example. The $\Sigma\pi$ term in equation (125) shows that hydrophobic forces play a significant but minor role in the hapten–antibody interaction.

$$-\Delta F^0_{rel} = 0.90(\pm 0.73)MR^{meta} - 1.65(\pm 0.90)MR^{para} - 1.00(\pm 0.83)MR^{ortho} + 0.60(\pm 0.38)\Sigma\pi$$
$$- 0.09(\pm 0.60) \tag{125}$$

$$n = 17, \quad r = 0.926, \quad s = 0.330$$

Hansch and Moser concluded that the above type of analyses clearly indicates that dispersion forces and/or steric effects but not typical hydrophobic forces are involved in the hapten–antibody interactions.[112] MR and π are nearly orthogonal and reasonably independent from each other for substituents at each of the positions in these analyses. The hydrophobicity parameter π is appropriate to model binding processes with dehydration of apolar parts of complementary molecular pairs. Since the binding of haptens with antibody is not hydrophobic, it must occur at sites constructed mostly of polar amino acid residues. Since low-molecular haptens are hydrated in the aqueous phase and also since the surface of a protein constructed of polar amino acids is no doubt saturated with water, it seems likely that a nonspecific bonding without dehydration occurs when MR is the significant parameter. According to a study by Franks,[113] two hydrated molecules or parts thereof appear to bind together by a process of mutual reinforcement of their 'clathrate' structures. Hansch and co-workers regarded this as a kind of 'freezing' together of two aqueous surfaces without significant dehydration. Thus, when the MR term is positive, the driving force for the binding should be linearly related to dispersion forces but the mechanism of the binding could be modeled mostly by the 'freezing together' of the hydrated organic compounds with a polar macromolecular surface.[42, 114] The degree of the 'freezing together' should increase with increasing MR value which is highly collinear with molecular volume. Hansch and Moser considered that the binding site of the antibody is predominantly polar in character. Antibodies recognize the surface of biomacromolecules primarily composed of polar amino acid residues.

21.1.3.3 QSAR of Antitumor Agents

The first QSAR study in this field was done by Khan and Ross for the aromatic aziridines (**25**) acting against Walker 256 tumors.[115] The result indicated that the lower the hydrophobicity, the more potent was the activity for a rather small number ($n = 8$) of compounds. In subsequent years, the interest in use of QSAR in cancer chemotherapy increased rapidly.

(**25**)

The nitrosoureas are an extremely active class of antitumor agents effective against solid tumors as well as leukemias. Following their preliminary studies suggesting an important role of hydrophobicity in determining potency as well as toxicity, Hansch, Montgomery and co-workers[116] tried to formulate QSAR correlation equations using a large number of analogs of nitrosourea (**26**; where X is either chloro or fluoro and R covers a variety of alkyl, cycloalkyl, thia- and dithia-cyclohexyl and oxacyclohexyl (monosaccharides) bearing various functional groups such as halogen, carboxy, alkoxy and hydroxy). For the action on intraperitoneally implanted L1210 leukemia in mice, equation (126) was derived for 90 compounds,[116] where C is the molar dose (mol kg^{-1}) of intraperitoneally applied drug producing a reduction of 1/1000 in the number of tumor cells. P_{opt} is the optimum P. The lowest $\log P$ value for 90 compounds was -1.48. Thus, equation (126) was considered to suggest that making more hydrophilic congeners should yield more active drugs. I_1 is an indicator variable assigned the value of unity for compounds where R is an α-substituted alkyl or cycloalkyl group. The negative sign shows a detrimental steric effect of branched substituents at the α position. I_2 is given a value of unity for compounds maintaining a nonoxidized sulfur in the thiacycloalkyl groups as R. This structural feature increases the potency about twice. Sulfur is easily oxidized *in vivo* yielding SO and SO$_2$. The oxidized molecules would have lower $\log P$ values, so I_2

$$N = O$$
$$XCH_2CH_2\overset{|}{N}CONHR$$

(26)

may account for a latent hydrophilic nature of these nitrosoureas. Congeners originally containing sulfonyl groups fit equation (126) well without the use of indicator variables. I_3 takes the value of unity when X is fluoro and the value of zero when X is chloro. The negative sign of this term shows that fluoro is inferior to chloro.

$$\log(1/C) = -0.13(\pm 0.07)\log P - 0.014(\pm 0.015)(\log P)^2 - 0.76(\pm 0.15)I_1 + 0.33(\pm 0.17)I_2$$
$$- 0.24(\pm 0.11)I_3 + 1.78(\pm 0.09) \tag{126}$$
$$n = 90, \quad r = 0.868, \quad s = 0.206, \quad \log P_{\text{opt}} = -4.4$$

With some additional compounds whose toxicity only was measured, equation (127) was derived, where C is the LD_{10} in mice. Since there is less variance in the LD_{10} value, the correlation is poorer in terms of r and s. For the toxicity, only one indicator variable was significant. The slope of the I_1 term is similar for equations (126) and (127), suggesting that the effect of α substitution in R is essentially equivalent in activity and toxicity. The difference in intercept between the two equations shows that about six times the dose is required to produce the LD_{10} effect as to cause the 1/1000 reduction in leukemia cells which is a measure of the therapeutic index.

$$\log(1/C) = -0.038(\pm 0.008)(\log P)^2 - 0.53(\pm 0.17)I_1 + 0.98(\pm 0.08) \tag{127}$$
$$n = 101, \quad r = 0.755, \quad s = 0.276, \quad \log P_{\text{opt}} = 0$$

The addition of a $\log P$ term to equation (127) was not significant. Thus, there appears to be a real difference in the $\log P_{\text{opt}}$ value for antitumor activity and toxicity. For better antileukemia drugs, one should search among more hydrophilic analogs than those included in equations (126) and (127). Introducing moieties which are able to ionize at the physiological pH is not recommended. The higher activity expected for highly hydrophilic ionized species may be offset by a loss of drug binding to serum protein and by the increased difficulty of ions in crossing biomembranes. It is preferable to make the R moiety an oligosaccharide. Introduction of a monosaccharide unit reduces $\log P$ by about 0.7 to 1.0. In fact, Suami and co-workers found that compounds where $R = \beta$-maltosyl, β-lactosyl and β-cellobiosyl increased the life-span four to five times acting on L1210 leukemia in mice for doses of 10 to 20 mg kg^{-1} day^{-1}.[117] Moreover, the maltosyl derivative was less toxic than the monosaccharides.

The second series of anticancer agents to be reviewed are the aniline mustards (27) that have been widely used in cancer chemotherapy. Hansch and his group examined the QSAR of this series of compounds extensively using published data.[118] Data for ED_{90} with rat Walker 256 carcinoma and those for LD_{50} from the work of Bardos and co-workers[119] on *para*- and un-substituted derivatives were formulated as equations (128) and (129), respectively. The indicator variable I_{Br} takes the value of zero in these equations when Y is chloro or iodo and the value of unity when Y is bromo. The positive slope with I_{Br} shows that bromo is the best group. π_{opt} applies to cases where Y is chloro or iodo technically. However, since the π value of aliphatic iodo differs from that of chloro, another factor must be associated with π to fortuitously allow the use of the single indicator variable for iodo and chloro. Log P for the parent compound (X = H, Y = Cl) is 2.90, so the maximum efficacy and toxicity for this series of compounds occur with rather lipophilic derivatives whose $\log P$ is about 2.0. There is practically no difference between equations (128) and (129) for antitumor activity and toxicity. No significant selectivity seems to be expected as far as the 'regular' mustards included in these correlations are concerned. The important and widely used agent chlorambucil (27; X = 4-$(CH_2)_3CO_2H$) is about 30 times more active [$\Delta\log(1/ED_{90}) = 1.5$] than expected by equation (128), where it is not included. Yet, its toxicity is well fitted by equation (129). The molecular features responsible for this selectivity are not clear, but QSAR can show that it is not due to lipophilic and/or electronic character of a unique sort. Some kind of ideal positioning of the CO_2H or CO_2^- group must be responsible for its selective action. One of the greatest assets of correlation equations is that they shed light on such unusual analogs as chlorambucil which can be then studied for possible new leads.

(27)

$$\log(1/ED_{90}) = -1.19(\pm 0.51)\sigma^- + 0.75(\pm 0.41)I_{Br} - 1.00(\pm 0.87)\pi - 0.53(\pm 0.55)\pi^2 + 3.84(\pm 0.33) \quad (128)$$

$$n = 14, \quad r = 0.940, \quad s = 0.291, \quad \pi_{opt} = -0.95$$

$$\log(1/LD_{50}) = -1.31(\pm 0.35)\sigma^- + 0.51(\pm 0.31)I_{Br} - 0.69(\pm 0.63)\pi - 0.35(\pm 0.39)\pi^2 + 3.87(\pm 0.24) \quad (129)$$

$$n = 18, \quad r = 0.932, \quad s = 0.272, \quad \pi_{opt} = -0.99$$

For antileukemia activity data for other sets of aniline mustards (27; Y = Cl) from the National Cancer Institute, Hansch and co-workers derived equations (130) and (131).[118] In these equations, C_{125} is the molar dose (mol kg^{-1}) producing a 25% increase ($T/C = 125$) in life-span acting against L1210 leukemia in mice. C_{180} is that required to elicit the T/C of 180 on P338 leukemia in mice. The indicator variable I_o takes the value of unity when *ortho* substituents are present on the aromatic ring to account for their proximity effects. The addition of the π^2 term in these equations did not improve the correlation. The $\log P_{opt}$ seems to be lower than the lowest $\log P (-0.70)$ of compounds included in these equations. Similar to equation (126), equations (130) and (131) suggest that the more hydrophilic derivatives should be more active on the 'liquid' tumor, leukemia. This is a remarkable contrast to the result derived from equation (128) for aniline mustards attacking the solid Walker tumor, where $\log P_{opt}$ is about 2.0. The $\log P_{opt}$ value seems to be dependent on tumor type.

$$\log(1/C_{125}) = -0.31(\pm 0.10)\pi - 0.96(\pm 0.54)\sigma^- + 0.86(\pm 0.37)I_o + 4.07(\pm 0.21) \quad (130)$$

$$n = 19, \quad r = 0.926, \quad s = 0.313$$

$$\log(1/C_{180}) = -0.34(\pm 0.11)\pi - 1.39(\pm 0.97)\sigma + 0.30(\pm 0.44)I_o + 4.13(\pm 0.21) \quad (131)$$

$$n = 16, \quad r = 0.914, \quad s = 0.311$$

To confirm the relationship of lipophilicity with tumor type, Hansch and co-workers tested a new set of *meta*- and *para*-substituted aniline mustards against the solid tumor, B-16 melanoma.[120] For 25% increase in life-span attacking melanoma in mice, equation (132) was derived.

$$\log(1/C_{125}) = -2.06(\pm 0.48)\sigma - 0.15(\pm 0.12)\pi - 0.13(\pm 0.07)\pi^2 + 4.13(\pm 0.20) \quad (132)$$

$$n = 19, \quad r = 0.936, \quad s = 0.303, \quad \pi_{opt} = -0.57$$

From the π_{opt} value, the $\log P_{opt}$ is calculated as about 2.3 which is in good agreement with the value from equation (128). Thus, for aniline mustards acting against solid tumors, $\log P_{opt}$ is definitely higher than that for the same type of compounds acting against leukemia. The parabolic dependence in equation (132) is much broader than that in equation (128). This can be attributed to differences in test methods. One analog (27; X = 4-CH$_2$CH(NH$_2$)CO$_2$H) was found to be 1000 times more potent than expected from equation (132). This might be due to an active transport into solid tumors specific to this type of amino acid mustard. The QSAR correlation was able to identify this unique type of analog in a similar fashion to the discovery of the specific selectivity of chlorambucil. Hansch and co-workers warned that excessive dependence on leukemia as a means for screening novel antitumor drugs tends to develop drugs which will not be ideally lipophilic for other types of tumors.[120]

Aniline mustards are referred to as alkylating agents and their active alkylating species is well established to be a cyclic aziridinium ion (28). The aziridinium ion is also the intermediate in hydrolysis. The ease of hydrolysis of aniline mustards may be related to the potency of the anticancer agent. Ross and co-workers made an extensive study of the hydrolysis of aniline mustards.[121] From their data on the percent hydrolysis of aniline mustards (27; Y = Br or Cl) in 50% aqueous acetone at 66 °C in 30 min, Hansch and co-workers formulated equation (133).[118]

$$\log(\% hyd) = -1.42(\pm 0.18)\sigma + 0.45(\pm 0.15)I_o + 0.70(\pm 0.11)I_{Br} + 1.21(\pm 0.06) \quad (133)$$

$$n = 42, \quad r = 0.952, \quad s = 0.157$$

(28)

In this equation, the same indicator variables as those in equations (128)–(131) are used. The positive slope of the I_o term shows that *ortho* substitution increases the rate of hydrolysis by twisting the nitrogen out of conjugation with the aromatic ring, thus making nitrogen a better nucleophile. The positive I_{Br} term indicates that the bromides are more rapidly hydrolyzed than the corresponding chlorides. The coefficients associated not only with these indicator variable terms but also with the σ term agree fairly well with those of the corresponding terms in equations (128)–(132). This is evidence for the fact that the critical process, where aromatic substituent effects are significant, is the aziridinium intermediate formation which is common between the hydrolysis and the anticancer action of aniline mustards. If hydrolysis and alkylation had occurred directly without passing through the cyclic intermediate, the σ term would no longer be significant. Unfortunately, the selection of aromatic substituents in the original studies make it difficult to decide about the relative merits of σ^- *vs.* σ for equations (128)–(133). However, the hydrolysis model is well justified in searching for more potent drugs. The size of the ρ value in equation (132) seems to conform to that expected for reactions where the site of electron pair migration is one bond away from the aromatic ring. That found in equations (128)–(131) and (133) is, however, a bit smaller. This may be due to the fact that a fraction of positive ρ for the step following the cyclic intermediate formation toward completion of the hydrolysis or alkylation is involved in the overall resultant ρ values. The above analyses can be regarded as an excellent example which satisfies the physical organic criteria of QSAR described in Section 21.1.2.5.

Another class of anticancer agents for which QSARs were examined extensively are the aromatic triazenes. Hansch and co-workers analyzed the antitumor activity of a number of substituted phenyltriazenes (**29**) against L1210 leukemia in mice to give equation (134),[122] where C_{140} is the dose (mol kg^{-1}) required for a T/C of 140, MR(2,6) is the sum of molar refractivity (scaled by 0.1) of substituents in the two *ortho* positions, and $E_s(R)$ is the Taft steric parameter for the R group at the side-chain terminal. The optimum log P value of this series of compounds seems to be higher than that observed for antileukemia activity of nitrosoureas and aniline mustards. Almost the same optimum log P values were found for imidazolyl- and pyrazolyl-triazenes (**30** and **31**). The negative MR(2,6) term suggests that both *ortho* positions should not be substituted for high activity. The positive $E_s(R)$ term, meaning that the smaller *N*-substituents are more favorable, limits one to dimethylamino substituents, since monomethylamino compounds are unstable.

$$\log(1/C_{140}) = 0.100(\pm 0.08)\log P - 0.042(\pm 0.02)(\log P)^2 - 0.312(\pm 0.11)\Sigma\sigma^+ - 0.178(\pm 0.08)MR(2,6)$$

$$+ 0.391(\pm 0.18)E_s(R) + 4.124(\pm 0.27) \tag{134}$$

$$n = 61, \quad r = 0.836, \quad s = 0.191, \quad \log P_{opt} = 1.18$$

(29)   (30)   (31)

The negative $\Sigma\sigma^+$ term may indicate that the introduction of more electron-releasing groups is favorable to the activity. However, according to the work of Kolar and Preussmann,[123] the rate of hydrolysis of this series of compounds is quite sensitive to the electronic effect of aromatic substituents. For the rate constant at 37 °C in aqueous medium of pH = 7.0 of 14 substituted phenyldimethyltriazenes (**29**; R = Me), equation (135) was derived.[122] Since the rate of hydrolysis is so enormously enhanced by electron-releasing groups, it is practically impossible to go beyond the 4-methoxy group (half-life = 12 min) in the use of electron-releasing functions.

$$\log(k_X/k_H) = -4.42(\pm 0.29)\sigma - 0.016(\pm 0.13) \tag{135}$$

$$n = 14, \quad r = 0.995, \quad s = 0.171$$

Since no obvious way to make more active compounds by modification of substituent effects seemed to exist, Hansch and co-workers examined the possibility of decreasing their toxicity so as to derive congeners with better therapeutic ratios. Thus, they determined the LD_{50} values for 11 phenyltriazenes (**29**; R = Me, Et or Bu) with a wide range of $\log P$ values (0.98 to 4.70) as well as a good spread in σ^+ values of X substituents (-0.78 to 0.66). Equation (136) was formulated on the LD_{50} for mice.[124] Substituting σ for σ^+ gives almost the same quality correlation. Addition of the $\log P$ term does not afford a significant reduction in the variance. Although the optimum $\log P$ is close to zero, significant changes in the toxicity are afforded only by large changes in $\log P$ because of the small coefficient of the $(\log P)^2$ term. Substituting $\log P$ of 6 in equation (136) with $\sigma^+ = 0$ yields $\log(1/LD_{50}) = 2.6$; that in equation (134) with $\sigma^+ = 0$ and $MR(2,6) = 0.2$ for the unsubstituted compound as the reference and $E_s(R) = -1.24$ (R = Me) yields $\log(1/C_{140}) = 2.7$. Thus, the therapeutic ratio is 1.04. Doing the same with $\log P = 1$ (optimum for potency) yields the ratio of 0.95. There is nothing to be gained therapeutically by modifying $\log P$. Furthermore, the σ^+ terms in equations (134) and (136) cancel each other as far as the therapeutic index is concerned. In addition, trying to increase potency by introducing substituents more electron-releasing than 4-methoxy will simply produce drugs too unstable.

$$\log(1/LD_{50}) = -0.024(\pm 0.013)(\log P)^2 - 0.264(\pm 0.16)\sigma^+ + 3.490(\pm 0.12) \tag{136}$$

$$n = 11, \quad r = 0.913, \quad s = 0.110$$

Another type of toxicity which must be considered for anticancer agents is carcinogenicity. Since the measurement of carcinogenic activity requires long-term testing as well as the high cost of using test animals, a simplified model system to detect possible carcinogenicity such as the Ames test has become widely used in recent years. The Ames test is ultimately a procedure to measure mutagenicity conducted with bacteria such as a strain of salmonella sensitive to mutation. About 90% of the chemicals causing mutations in this test are said to be carcinogenic in animals. Thus, Hansch and co-workers measured mutagenicity of phenyltriazenes (**29**) activated by simultaneous addition of rat liver microsomes.[125] This class of triazenes had been established as carcinogenic in animal tests. The QSAR for mutagenicity was derived as equation (137), where C is the molar concentration producing 30 mutations per 100 million bacteria. With σ instead of σ^+, the correlation was much poorer. The dependence of mutagenicity on the electronic effect in equation (137) is much greater than those in equations (134) and (136). If carcinogenicity parallels mutagenicity for triazenes, then less carcinogenic triazenes could be made by introducing more electron-withdrawing substituents. They also measured the mutagenicity along with carcinogenicity of aniline mustards (**27**; Y = Cl). For aniline mustards, however, no direct parallel was found between the two activities.[126] Hansch and co-workers concluded that the separation of toxic and antitumor activities of the triazenes is very difficult and recommended no further synthesis and testing of new derivatives of this series of compound.[124]

$$\log(1/C) = 1.09(\pm 0.17)\log P - 1.63(\pm 0.35)\sigma^+ + 5.58(\pm 0.95) \tag{137}$$

$$n = 17, \quad r = 0.974, \quad s = 0.315$$

A number of drugs have some unfavorable side effects. For drug design, one should consider minimizing the side effects as well as maximizing the principal target effect by structural optimization. It would be ideal to formulate QSAR correlations for both principal and unfavorable toxic side-effects. The optimal structure in terms of the principal activity as well as the minimum toxicity could be designed using these QSAR correlations rationally. Since the measurements of toxicity, carcinogenicity and mutagenicity are not easy, the QSAR correlation has often been formulated only for the principal effect.

The above studies performed by Hansch and co-workers on various sets of anticancer agents are good examples showing drug design principles derived from QSAR correlations not only for the principal activity but also for its difference between different types of targets as well as for unfavorable toxic side-effects. Moreover, their series of studies clearly indicated the versatility of QSAR procedures in cancer chemotherapy despite the great complexity of the problem. The biological activities were examined using such complex systems as the whole body of test animals. Furthermore, biological parameters such as C_{125} were determined from dose–response relationships where the dose was defined in a rather complicated way consisting of consecutive daily or weekly injections according to certain standards.

Denny and co-workers[127] made similar studies on antileukemic activity and toxicity of a very large number ($n = 509$ for activity and $n = 643$ for toxicity) of 9-anilinoacridines (**32**) having a variety of substituents not only on the acridine ring but also on the anilino benzene. One of the compounds in this class, *m*-AMSA (**32**; Y = 2-OMe-4-NHSO$_2$Me, X = X' = H), is a clinical agent. From the QSAR correlation equation with 13 parameter terms for the antileukemic activity in terms of log(1/C), where C is the molar dose required for T/C = 150, they showed that the most significant factor is the steric effect of substituents placed at various positions on the 9-anilinoacridine skeleton. Hydrophobic and electronic properties of substituents were also important. They demonstrated that the results are entirely consistent with a probable mechanism of action including the binding to double-stranded DNA by intercalation of the acridine chromophore between the base pairs and the positioning of the anilino group in the minor groove. The correlation for the toxicity was found to be somewhat similar to that for the antileukemic activity.

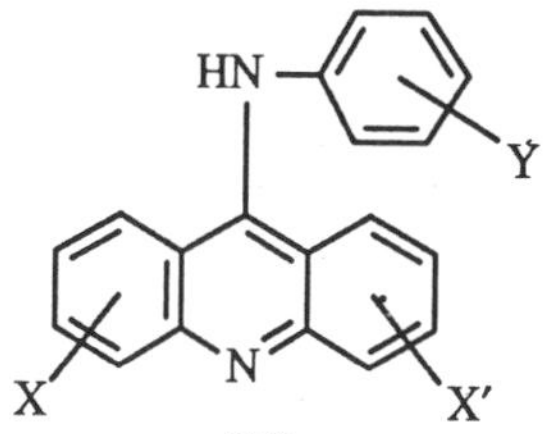

(32)

Yoshimoto and co-workers[128] analyzed the antileukemic activity of aziridinobenzoquinones (**33**) which resemble mitomycin C (**34**). Their results conform with those derived by Hansch and co-workers for the nitrosomethylurea antileukemic agents, in that the activity as well as the selectivity increases with an increase in the hydrophilic property of the molecule.

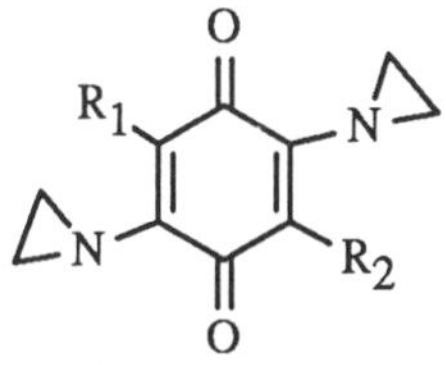

(33) R$_1$ = alkyl, alkoxy, hydroxyalkyl; R$_2$ = alkyl, acyl, aryl, carbamoyloxyalkyl

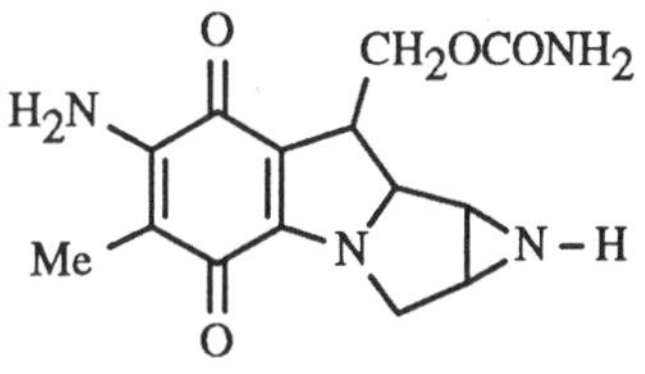

(34)

21.1.3.4 QSAR of Dihydrofolate Reductase Inhibitors

All living organisms use dihydrofolate reductase (DHFR) to synthesize heterocyclic bases that are components of DNA. However, the fact that the enzyme in different living organisms shows a significant variability is the basis for medicinal uses of inhibitors such as methotrexate (**35**) as anticancer, trimethoprim (**36**) as antibacterial and pyrimethamine (**37**) as antimalarial agents.

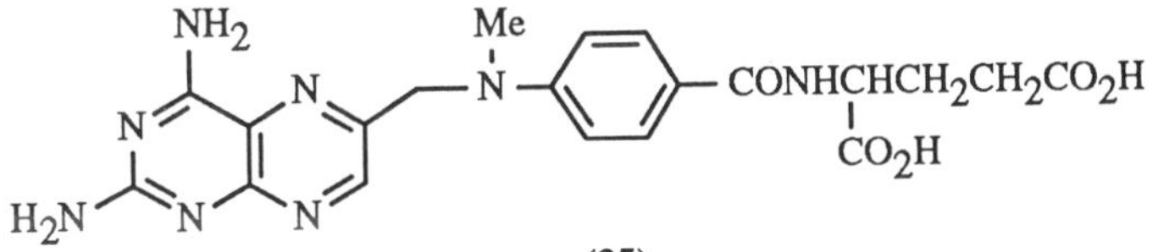

(35)

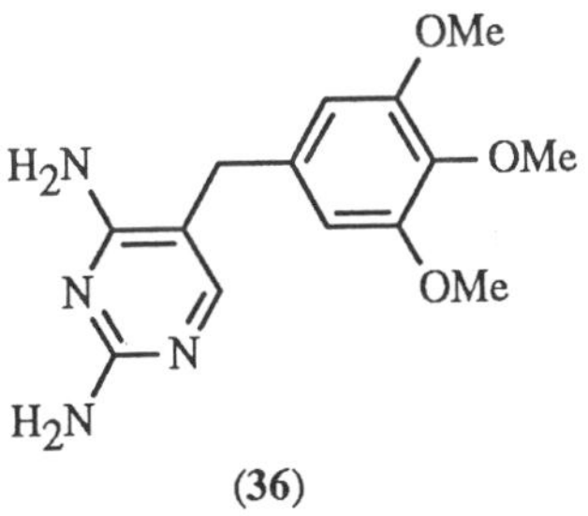

(36)

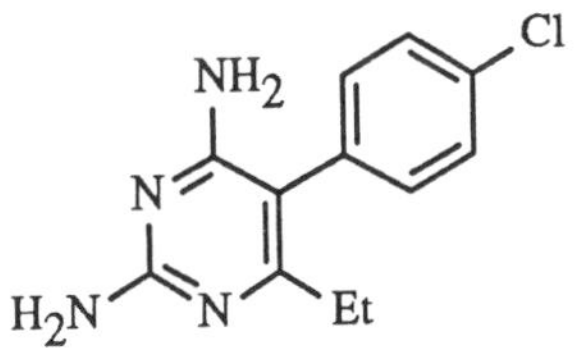

(37)

Following preliminary analyses for datasets consisting of a large number (n = 244) of triazines (**38**) published by Baker and his group, Hansch and co-workers have been investigating the QSAR of this and other groups of DHFR inhibitors extensively. Most of their work up to 1983 was summarized in their excellent review article.[129] More recently, they analyzed the inhibitory activities against purified DHFR preparations from tumor cells for triazines having not only such simple substituents as sulfamoyl, mesyl, acetyl, hydroxy, nitro, halogen, alkyl (C_1–C_8), alkoxy (C_1–C_{13}), alkenyl and alkynyl but also complex substituents such as —$CH_2OC_6H_4Y$ and —$OCH_2C_6H_4Y$ (Y = CN, alkoxy, SO_2NH_2, CH_2OH, NHCOMe, *etc.*) as X at *meta* and *para* positions on the benzene ring in (**38**).[130] The set of compounds was designed so as to cover a wide range in hydrophobicity in order to clearly elucidate its effect on the inhibitory activity. They also examined the QSAR for the growth inhibition of cultured tumor cells and compared the results with those derived from the enzyme inhibition.

(**38**)

For inhibition of purified DHFR from L1210 leukemia cells resistant to methotrexate, equations (138) and (139) were derived,[130] where K_1 ($mol\,dm^{-3}$) is the apparent inhibition constant. π' signifies that, for substituents of the types —$CH_2ZC_6H_4Y$ and —$ZCH_2C_6H_4Y$ (Z = O, S, Se), π_Y is set equal to zero. Since the substituents containing Y appear to have essentially the same effect on K_1 regardless of the hydrophobicity or size of Y, Y substituents are assumed not to contact the enzyme. For all alkoxy groups from methoxy to tetradecyloxy π' is set to zero, since they also show essentially the same effect. If they contact the enzyme, the gain in potency should be compensated in some way by a steric effect. I in equation (139) is an indicator variable that takes the value of unity for such bridged-type substituents as —$CH_2ZC_6H_4Y$ and —$ZCH_2C_6H_4Y$. Compounds containing these types of substituents at the *para* position seem to bind unusually tightly, probably because their location on DHFR is approximately the same as that taken by the *p*-aminobenzoyl moiety of folic acid. MR_2 for the *para* substituents accounts only for the binding of the α and β parts of substituents; for example, $MR_2(OCH_2Ph)$ is equal to $MR(OCH_2)$. As usual, MR values are scaled by 0.1. π_{opt} represents the value where the linearly increasing phase of the enzyme inhibition with increasing hydrophobicity of substituents turns to the descending linear phase. For 3-X substituted derivatives, the activity first rises with a slope close to unity. The slope of the second phase is practically equal to zero ($0.98 - 1.14 = -0.16$). This flat part of the bilinear curve probably means that portions of the larger substituents do not contact the enzyme. For 4-X substituted derivatives, the situation is similar except that the slope of the ascending phase is slightly lower than that for the 3-X isomers. The negative MR_2 term suggests the existence of a steric interaction of α and β parts of *para* substituents. As a result of graphic studies of DHFR from chicken liver, Hansch and co-workers assumed that the *para* substituents encounter an unfavorable steric interaction with Ile-60,[131] a residue which is probably conserved in L1210 DHFR.

For 3-X-triazines

$$pK_1 = 0.98(\pm0.14)\pi' - 1.14(\pm0.20)\log(\beta\cdot10^{\pi'} + 1) + 0.79(\pm0.57)\sigma + 6.12(\pm0.14) \tag{138}$$

$$n = 58, \quad r = 0.900, \quad s = 0.264, \quad \pi_{opt} = 1.76, \quad \log\beta = -0.979$$

For 4-X-triazines

$$pK_1 = 0.76(\pm0.15)\pi' - 0.83(\pm0.28)\log(\beta\cdot10^{\pi'} + 1) + 0.70(\pm0.25)I - 0.40(\pm0.26)MR_2$$
$$+ 6.47(\pm0.22) \tag{139}$$

$$n = 37, \quad r = 0.959, \quad s = 0.263, \quad \pi_{opt} = 2.11, \quad \log\beta = -1.10$$

For growth inhibition of cultured L1210 leukemia cells sensitive to methotrexate, equations (140) and (141) were formulated,[130] where I_{50} is the molar concentration required to inhibit the growth of cells by 50%. The overall π of substituents is used instead of truncated π'. This probably reflects the fact that the whole substituent must make hydrophobic contact in membrane penetration. In spite of the difference in the use of hydrophobicity parameters, the bilinear dependence of the activity is very similar for inhibitions of enzyme and cell growth. The slope of the ascending part of the cell growth inhibition is nearly 1.0, almost equivalent to that for the enzyme inhibition. It should be noted that the enzyme was obtained from the resistant leukemia cells. π_{opt} values for the sensitive cells around 1.6 are close to 1.8–2.1 for the purified enzyme. I_R is an indicator variable which takes the value of unity for alkyl groups and I_{OR} is that for alkoxy groups. In leukemia cells, the *m*- but not the *p*-alkyl-substituted compounds are about five times as active as expected on the basis of π and σ alone, while alkoxy substituents lower the activity by a factor of about 2.5 regardless of their positions. The I_{OR} term does not occur in equations (138) and (139) for the enzyme inhibition where the π' value is set to zero for alkoxy groups to account for the essentially constant activity of the alkoxy derivatives. Thus, the alkoxy groups seem to show some hydrophobic effect on the cell growth inhibition processes including membrane penetration. The negative sign with the I_{OR} term indicates that they still do not contact the receptor.

For 3-X-triazines

$$pI_{50} = 1.13(\pm 0.18)\pi - 1.20(\pm 0.21)\log(\beta \cdot 10^\pi + 1) + 0.66(\pm 0.23)I_R + 0.94(\pm 0.37)\sigma$$

$$- 0.32(\pm 0.17)I_{OR} + 6.72(\pm 0.13) \tag{140}$$

$$n = 61, \quad r = 0.890, \quad s = 0.241, \quad \pi_{opt} = 1.45, \quad \log\beta = -0.274$$

For 4-X-triazines

$$pI_{50} = 0.90(\pm 0.11)\pi - 0.97(\pm 0.14)\log(\beta \cdot 10^\pi + 1) - 0.45(\pm 0.20)MR_2 - 0.50(\pm 0.18)I_{OR}$$

$$+ 7.11(\pm 0.19) \tag{141}$$

$$n = 42, \quad r = 0.949, \quad s = 0.215, \quad \pi_{opt} = 1.83, \quad \log\beta = -0.706$$

Similar analyses for the growth inhibition of cultured L 1210 leukemia cells resistant to methotrexate gave equations (142) and (143).[130] Equations (142) and (143) are very different from those for the enzyme and susceptible cell inhibitions, especially in the hydrophobic terms. The π_{opt} value for the 3-X-triazines cannot be derived mathematically, since the dependence of activity on π is linear. However, it may be located at around six which is close to that for the 4-X-triazines. The value of six is far removed from that of about 1.8 found for isolated DHFR and susceptible cells. The difference of 4.2 log units is a remarkable aspect of the resistant cell QSAR, indicating that more lipophilic drugs are needed for methotrexate-resistant tumors. The sensitive cells interact with the inhibitor more like isolated enzyme, despite the fact that the enzyme is prepared from resistant cells.

For 3-X-triazines

$$pI_{50} = 0.42(\pm 0.05)\pi - 0.15(\pm 0.05)MR + 4.83(\pm 0.11) \tag{142}$$

$$n = 62, \quad r = 0.941, \quad s = 0.220$$

For 4-X-triazines

$$pI_{50} = 0.46(\pm 0.07)\pi - 0.47(\pm 0.31)\log(\beta \cdot 10^\pi + 1) + 4.29(\pm 0.13) \tag{143}$$

$$n = 41, \quad r = 0.935, \quad s = 0.330, \quad \pi_{opt} = 6.18, \quad \log\beta = -4.54$$

Quite similar QSAR correlations were found for the same type of triazines (**38**) acting on enzymes and cells of certain bacteria. For the inhibition of purified DHFR from *Lactobacillus casei* resistant to methotrexate by 3-X-triazines, equation (144) was formulated.[132] For the inhibition of *L. casei* cells sensitive and resistant to methotrexate, equations (145) and (146) were derived, respectively. In equations (144)–(146), the indicator variable I takes the value of unity for substituents such as $-CH_2ZC_6H_4Y$ but not for $-ZCH_2C_6H_4Y$, which differs from I in equation (139) where it takes care of both. MR_Y in equations (145) and (146) reflects a steric effect of Y substituents on the terminal benzene ring of the bridged-type substituents. Whereas Y in the isolated enzyme makes no contact with the enzyme, Y inhibits the binding in cells. In equation (145), the truncated π' was used for the sensitive cell growth inhibition although inhibitors must penetrate through the cell membranes. With the use of overall π, the correlation was slightly but significantly poorer ($r = 0.900$, $s = 0.435$, $\pi_{opt} = 3.48$).

$$pK_1 = 0.53(\pm 0.11)\pi' - 0.64(\pm 0.25)\log(\beta \cdot 10^{\pi'} + 1) + 1.49(\pm 0.25)I + 0.70(\pm 0.65)\sigma + 2.93(\pm 0.26) \tag{144}$$

$$n = 44, \quad r = 0.953, \quad s = 0.319, \quad \pi_{opt} = 4.31, \quad \log\beta = -3.66$$

$$pI_{50} = 0.80(\pm 0.15)\pi' - 1.06(\pm 0.27)\log(\beta \cdot 10^{\pi'} + 1) - 0.94(\pm 0.39)MR_Y + 0.80(\pm 0.56)I$$

$$+ 4.37(\pm 0.19) \tag{145}$$

$$n = 34, \quad r = 0.929, \quad s = 0.371, \quad \pi_{opt} = 2.94, \quad \log\beta = -2.45$$

$$pI_{50} = 0.45(\pm 0.05)\pi + 1.05(\pm 0.33)I - 0.48(\pm 0.24)MR_Y + 3.37(\pm 0.15) \tag{146}$$

$$n = 38, \quad r = 0.964, \quad s = 0.264, \quad \pi_{opt} > 5.9$$

In spite of these features different from those in correlations for L1210 enzymes and cells, there are certain similarities. Equation (146) for the resistant cells is extremely different from equations (144) and (145) for the isolated enzyme and sensitive cells that are similar to each other. Against the purified enzyme and sensitive cells, the inhibitory activity first increases with π' until the point where it is 3.0–4.5. Then, the slope becomes almost zero, indicating that portions of large substituents do not contact the enzyme. Against the resistant cells, however, the activity increases linearly with π at least up to 5.9 (π for the most hydrophobic substituent). Note that the enzyme comes from resistant cells. The situation is remarkably similar to that observed for L1210 enzyme and cells.

There are possible factors which could confer phenotype resistance on methotrexate-resistant cells, such as an increased cellular content of DHFR in excess of the amount inactivated by the inhibitors and alterations in the cell membrane. It is unlikely that the differences in QSAR for the sensitive and resistant cells are due to an altered form of DHFR in the resistant cells, since the QSAR for the sensitive cells is similar to that for the enzyme prepared from resistant cells. Hansch and co-workers thought that a difference in active transport in the membrane may constitute the single most important factor.[130] In mammalian cells and such bacteria as *L. casei*, there is an active transport system for folic acid which transports methotrexate as well as other DHFR inhibitors.[133,134] It has been reported that some triazine analogs utilize the same carrier as methotrexate and reduced folates in intact L1210 cells.[135] Thus, they proposed that the major difference in behavior between sensitive and resistant strains of tumor and *L. casei* cells to triazines exists in the mode of cell membrane transport. To examine this proposal, they tested the effect of triazines on methotrexate-sensitive and -resistant *Escherichia coli* systems.[136] The *E. coli* cells are not known to have the active transport system.

For the inhibition of purified DHFR from methotrexate-resistant *E. coli* by 3-X-triazines, equation (147) was formulated. Equation (147) is similar to equation (144) except for the size of the regression coefficients and intercept. Equation (147) can be compared with equations (148) and (149) for cultured cells sensitive and resistant to methotrexate, respectively. These equations seem to indicate that the inhibition of sensitive and resistant cells by the triazines occurs in much the same way if the cells lack the active transport system. Equations (148) and (149), without bilinear terms, are much simpler than equation (147) for the isolated enzyme. Being governed by the hydrophobicity alone, the passive diffusion of inhibitors to the DHFR inside the cells seems to be more rate limiting than interaction with the DHFR itself. They are similar to equations (142) and (143) for the growth inhibition of resistant L1210 cells and equation (146) for that of resistant *L. casei* in certain respects. These equations are much simpler than equations for the enzyme and susceptible cell inhibitions. The π_{opt} value is very or indeterminably high. The results appear to support the proposal that the greatest difference of methotrexate-resistant mutants from the sensitive cells is due to an impaired active transport system.

$$pK_1 = 1.16(\pm 0.25)\pi' - 1.10(\pm 0.29)\log(\beta \cdot 10^{\pi'} + 1) + 1.36(\pm 0.90)\sigma + 0.41(\pm 0.25)I + 5.80(\pm 0.28) \quad (147)$$

$$n = 31, \quad r = 0.930, \quad s = 0.280$$

$$pI_{50} = 0.51(\pm 0.06)\pi + 2.35(\pm 0.19) \qquad (148)$$

$$n = 18, \quad r = 0.973, \quad F_{1,16} = 281, \quad s = 0.261$$

$$pI_{50} = 0.54(\pm 0.07)\pi + 2.29(\pm 0.22) \qquad (149)$$

$$n = 18, \quad r = 0.969, \quad F_{1,16} = 245, \quad s = 0.299$$

Hansch and co-workers extended similar QSAR studies to benzylpyrimidines,[136–139] the family of compound (36) and quinazoline derivatives (39).[140,141] With benzylpyrimidines, however, sensitive and resistant *E. coli* cells interact in a quite different manner.[136] It is conceivable that a slight change in the DHFR could result in a different binding mode of pyrimidines by a conformational flexibility much greater than triazines. These types of comparative examinations for enzyme inhibitors acting on purified enzymes and on whole cells are believed to be fundamental to the study of the same inhibitors acting on the whole animals.

(39)

Hopfinger approached the QSAR of DHFR inhibitors from another direction by use of molecular shape analysis.[142, 143] In this approach, the analogs in a dataset are first examined by molecular mechanics to identify the most stable conformers. Then, a reference compound is selected with which the shape of all other congeners can be compared. The total common overlap volume of the reference compound and each of the congeners in a dataset is defined as V_0. A parameter S_0 ($S_0 = V_0^{2/3}$) was used as the independent variable in the correlation analysis of a dataset published by Baker and his group for triazines (38).[129] The original set of compounds ($n = 256$) was divided into seven groups according to conformational flexibility. As representatives from each group, 27 compounds were selected for the analysis. From the calculation using molecular mechanics, Hopfinger assumed that the most favorable conformation is such that the dihedral angle between benzene and triazine rings is 310°. Taking the most active 3,4-dichloro derivative (38; X = 3,4-Cl$_2$) as the reference compound, equation (150) was formulated for the inhibition of DHFR from either Walker 256 tumor or L1210 leukemia cells,[142] where I_{50} is the molar concentration of each compound for 50% *in vitro* inhibition. D_4 is a measure of the length of the *para* substituents beyond that of chloro. S_0 has the dimension of area, but is not a physical measure of common atomic surface areas. It is to be regarded as an alternative mathematical representation of V_0. Equation (150) means that the greater the hydrophobicity of aromatic substituents, the lower the activity. There are optimum values in the two shape descriptors, S_0 and D_4, for the activity. Hopfinger demonstrated that equation (150) was able to predict some highly active compounds.

$$pI_{50} = 1.384 S_0 - 0.0213(S_0)^2 - 0.434\Sigma\pi + 0.574 D_4 - 0.294(D_4)^2 - 15.66 \qquad (150)$$

$$n = 27, \quad r = 0.953, \quad s = 0.44$$

His group extended this molecular shape analysis to other DHFR inhibitors such as benzylpyrimidines related to trimethoprim (36)[144] and diaminoquinazolines (39).[145] In each of these analyses, they found that the QSAR correlation equation consists of terms of molecular shape descriptors and hydrophobicity similar to those appearing in equation (150). Hopfinger stated that his QSAR approach combined with molecular shape analysis was able to rationalize the activity of some compounds which were treated as outliers by Hansch and his group. In many. cases, however, he did not use the complete set of compounds in formulating his QSAR correlation equations, probably because of the cost in time and money of making calculations for the molecular shape analysis. If this situation is overcome and if the physical organic meaning of his molecular shape descriptors such as S_0 could be substantiated, his approach could be a powerful tool for drug design.

As shown in the examples outlined above, QSAR methodology is especially effective for the direct comparison of structural requirements for activity beween different compound classes showing the same type of activity as well as between the same type of compound series showing activities on different biological systems. Similarities and dissimilarities in molecular mechanisms of action between counterparts of comparison were clearly demonstrated and invaluable inferences for drug design principles were deduced from them.

21.1.3.5 Other QSAR Studies

Besides the above examples, an enormous number of extrathermodynamic QSAR studies have accumulated in the last two decades for various series of drugs. Almost every type of structure–activity relationship has been treated in terms of QSAR. Some important examples are: studies of muscarinic agents in order to elucidate the molecular mechanism of their binding with muscarinic receptor,[146, 147] studies of aldosterone analogs for binding with mineralocorticoid receptors in order to develop drugs for edematous diseases,[148] studies of the reaction mechanisms of a number of cysteine and serine hydrolases with various ligands in combination with X-ray crystallography and computer graphics,[149–152] studies of antimalarial drugs,[153, 154] studies of semisynthetic antibiotics,[155–157] studies of antihistaminic diphenhydramine analogs,[158, 159] studies of benzodiazepinooxazoles as minor tranquilizers,[160] and studies of various agrochemicals such as insecticides,[161–167] herbicides[168–172] and fungicides.[173–176] Environmental toxicology is also a field in which QSAR has been applied in the study of bioconcentration,[177, 178] toxicity of chemicals[179–182] and movement through soils.[183, 184] QSAR has provided insight into the structure–activity pattern of taste and olfactory compounds.[185–189] Drug metabolism and distribution[103, 105, 190, 191] and anesthesiology[51, 192] are also fields for QSAR application. Moreover, QSAR finds use in rationalizing the relative lethality of certain classes of drugs in forensic toxicology.[193, 194] Methodological studies of how to formulate the correlation equations for sets of ionizable drugs[99–101] and optically active compounds,[195–196] and series of compounds including drugs having hydrogen-bonding substituents besides those with nonhydrogen-bonding groups,[51] have expanded the versatility of the QSAR approach.

21.1.4 APPLICATIONS OF QSAR TO DRUG DESIGN PRACTICE

21.1.4.1 General Considerations

After formulation of a statistically significant as well as physicochemically meaningful correlation equation for a given set of compounds, the information contained in the equation can be used to design new compounds. According to the method of utilization of the information, examples could be classified into at least three categories.

(i) Extrapolation of certain parameters toward directions enhancing the potency. If the correlation is linear in terms of certain physicochemical parameters, structural modifications so as to extrapolate these parameters toward directions increasing the value of their terms should generate compounds of more potent activity. This corresponds to activity prediction beyond the substituent space span (SSS).[197] Since the linearity does not necessarily hold forever, the best policy is to attempt to gradually extend the extrapolation until the maximum potency is generated.

(ii) Identification of optimal structures with respect to certain parameters. If a parabolic (or bilinear) dependence of the activity on certain parameters is revealed, the optimum structure could exist within the SSS for the primary set of compounds. The structure can be optimized by being modified so that the value of the parabolic parameter terms is close to the maximum. Sometimes the best compounds can even be identified within the original set. After careful examinations around the structure of the 'best' compounds, one could make a decision to continue or discontinue the synthetic program.

(iii) Transposition of QSAR information to other series of compounds. In the above categories, the structural modifications are performed mainly on the basis of introduction or replacement of substituents. That is, the QSAR information derived from a set of compounds, $A—X_{1 \sim n}$, is utilized to design new structures, $A—X_m$, where A is the basal skeletal structure that is kept unchanged and X means variable substituents or substructures. There are examples where the information from $A—X_{1 \sim n}$ could be transposed to $A'—X_m$, A' being different from A. Although the structures of A and A' are sometimes similar to each other, being bioisosteric, these examples should be categorized separately.

In practice, there are examples utilizing various combinations of the above types so that the classification sometimes may not be explicit.

In recent years, successful applications of QSAR have been accumulating showing that designed compounds actually exhibit the predicted activity. This section will review these examples indicating performance of the QSAR procedure in drug design practice.

21.1.4.2 Design of Antiallergic Purinones

Such methylxanthines as caffeine (40; R = Me) and theophylline (40; R = H) are known to inhibit the antigen-induced release of histamine from human basophilic leukocytes.[198] Wooldridge and co-workers examined methylxanthines in the passive cutaneous anaphylactic (PCA) reaction mediated by reaginic antibodies in the rat and found that they were indeed inhibitors although the potency was low. They started a project to develop antiallergic drugs by molecular modification of methylxanthines.[199]

(40)

Methylxanthines possess a wide variety of pharmacological activity. After examination of available structural variants and screening of synthesized analogs, they found that 8-azatheophylline

(**41**; $R_1 = R_2 = Me$) was more potent than theophylline in the inhibition of rat PCA reaction, whereas the other pharmacological properties were much reduced. They prepared compounds having various alkyl groups at R_1 and R_2 positions in azaxanthine (**41**), and formulated equation (151) for 11 compounds, where A is the molar-corrected activity relative to that of disodium chromoglycate (DSCG) as the standard. Effects of only the R_2 substituent were statistically significant. The variations in R_1 did not affect the activity much. Each term was significant at a level higher than 99.5%. The high quality of the correlation of equation (151) encouraged them to believe that their bioassay system was sufficiently precise for the quantitative approach. The negative E_s term indicates that the bulky R_2 substituents are favorable for the higher activity. In fact, they found that a derivative (**41**; $R_1 = Me$, $R_2 = 4\text{-}NO_2C_6H_4CH_2$) showed an activity higher than that predicted by equation (151), being almost equivalent to that of DSCG.

(**41**)

$$\log A = -0.07\pi(R_2)^2 - 0.789E_s(R_2) + 1.37 \tag{151}$$

$$n = 11, \quad r = 0.942, \quad F = 31.4, \quad s = 0.153$$

Anticipating that bulky substituents at the R_2 or 3-positions of the azaxanthines and similar positions of related analogs are favorable to the activity, they continued further syntheses and found that 2-phenyl-8-azapurine-6-one (**42**; $X = H$) was four times more potent than DSCG. For the activity of the first synthesized benzene ring-substituted analogs (**42**; $X = H$, *o*-Me, *o*-Cl, *o*-OMe, *o*-OPr^i, *o*-OCH$_2$Ph, *m*-Me, *m*-OMe, *p*-Cl or *p*-OMe) equation (152) was derived, where Δv is the difference in the 1-NH stretching frequency in the benzene-ring-substituted 2-phenylazapurinones relative to that of the unsubstituted compound, and E_s is for the *o*-X substituents. Δv (cm^{-1}) varies from 0 (H, *m*-Me and *p*-Cl) to -128 (*o*-OPr^i), while E_s varies from 1.24 (H) to 0 (*o*-Me). The range of variations in Δv is about a hundredfold that of the E_s. Thus, the Δv term was justified at a level higher than 95% Equation (152) was taken to mean that such hydrogen-bond-accepting *ortho* substituents with high E_s values as alkoxy (OR, $E_s = 0.69$) are favorable to the activity. Hydrogen-bond formation with the *ortho* substituents reduces the NH stretching frequency. However, since the unsubstituted compound and those having nonhydrogen-bonding substituents at the *ortho* position were also active, they considered that the important feature required for high activity is not hydrogen bonding itself, but coplanarity of the azapurine ring with the substituted benzene. Thus, they synthesized 2-(*o*-alkoxyphenyl) derivatives and observed that the *o*-propoxy compound, zaprinast (**42**; $X = o\text{-}OPr$, M&B 22948) was about 40 times as active as DSCG.

(**42**)

$$\log A = -0.01\Delta v + 1.47E_s + 0.92 \tag{152}$$

$$n = 10, \quad r = 0.961, \quad F = 42.7, \quad s = 0.244$$

The effect of additional substituents was further investigated. For nine 2-methoxy-5-substituted phenyl derivatives (**43**; $X = H$, NO_2, NH_2, OH, OMe, Me, CF_3, Cl, Bu^t) equation (153) was formulated, where $\mathscr{R}$ is the Swain–Lupton resonance parameter. Substituent modifications designed to lower the π value as well as to increase the $\mathscr{R}$ value beyond the region of the original nine substituents were expected to yield more active compounds. As expected, quite a few such compounds (*e.g.* SO_2NH_2, SO_2Me and $CONH_2$ derivatives) were in fact highly active, being about

200 times more active than DSCG.

(43)

$$\log A = -0.74\pi + 0.87\mathcal{R} + 3.57 \tag{153}$$

$$n = 9, \quad r = 0.940, \quad F = 22.9, \quad s = 0.205$$

Some of these compounds as well as zaprinast were brought into developmental stages, but for some reason none of them have been clinically used. However, activities of the azaxanthines were enhanced about 200 times by structural modifications aided by QSAR. Moreover, structural information from the quantitative examination, such as the fact that coplanarity between the azapurinone and benzene rings is favorable to the activity and is facilitated as a result of intramolecular hydrogen-bonding, was skilfully utilized to design such compounds as (44) and (45), taking an extended planar aromatic system.[200] In fact, many of them showed outstanding anti-allergic activity.

(44)   (45)

21.1.4.3   Design of Antiallergic Pyranenamines

Cramer and co-workers used QSAR in designing another series of antiallergic drugs.[201] Their group found that a series of 3-[(arylamino)ethylidene]-5-acylpyrantriones (46; pyranenamines), sharing the γ-pyrone structure with, but otherwise differing considerably from, DSCG (47), are potent antiallergic compounds as evaluated by inhibition of the rat PCA reaction.[202]

(46)   (47)

Starting from pyranenamine (46; X = H) as the lead compound, various substituents X were introduced singly or multiply. 19 mono- and di-substituted derivatives were synthesized with halogen, trifluoromethyl, amino, hydroxy, dimethylamino and nitro groups. Their first policy for the introduction of substituents was mainly based upon the Topliss procedure.[203] In this procedure, the substituents which should be introduced next are selected according to an operational scheme. In this scheme, specific sequences of substituents to be introduced are proposed for the next selection at each sequential step according to the relative potencies of the preceding pair of compounds. For the first synthesized pyranenamines, the Topliss scheme did not work well. This was partly because of the assumption in the Topliss scheme that priority in substituent selection was placed on the more hydrophobic substituents. In the 19 pyranenamines, the hydrophilic 4-hydroxy derivative was most active.

Cramer and co-workers initially analyzed the structure–activity relationship graphically by plotting pI_{50} (mg kg^{-1} rat, i.v.) values against substituent parameters. The most promising graph was a 'three-dimensional' plot of the activity on the parameter space of π and σ, as shown in Figure 13. In this figure, the potency contour lines are drawn so that compounds showing similar potency are lined up as neatly as possible, although considerable scatter is present. It was apparent that substituents with low hydrophobicity and neutral electronic properties are favorable to the activity. They immediately predicted the activity of such compounds as the 2-hydroxy-5-acetyl-amino and 3-acetylamino-4-hydroxy derivatives to be higher than that of any compounds in the original set and the prediction was soon proved.

For the original 19 compounds, equation (154) was formulated at this stage. The quality of the correlation is not as good as one would like. In fact, the π term is not justified at the 95% level. They considered the correlation to be real, since the higher activity of compounds with hydrophilic substituents predicted from Figure 13 was proved. The F value shows that the correlation as a whole is significant at a level higher than 99%. The negative $(\Sigma\sigma)^2$ term would correspond with a biphasic Hammett relationship sometimes observed for multistep organic reactions when the rate-determining step changes with an increase in the σ value of the substituent in a series of aromatic derivatives.[15]

$$pI_{50} = -0.14(\pm 0.29)\Sigma\pi - 1.35(\pm 0.98)(\Sigma\sigma)^2 - 0.72 \tag{154}$$

$$n = 19, \quad r = 0.69, \quad F_{2,6} = 7.3, \quad s = 0.47$$

In accordance with the above information, they introduced hydrophilic polar substituents singly or in various combinations to synthesize 42 mono- and di-substituted derivatives. They observed results that were not expected from equation (154), such as that 3,5-di(acetylamino) substitution enhanced the activity about 100 times relative to the 4-hydroxy compound, and that the activity of 3-acylamino derivatives did not depend much on the chain length of the acyl group. Thus, the activity variations were thought to be decided not only by $\Sigma\sigma$ and $\Sigma\pi$ values but also by other structural features. For a total of 61 compounds, equation (155) was derived using rather complex descriptors.

$$pI_{50} = -0.30(\pm 0.12)\Sigma\pi - 1.50(\pm 0.67)(\Sigma\sigma)^2 + 2.00(\pm 2.00)\mathscr{F}\text{-}5 + 0.39(\pm 0.22)\,\#345\text{HBD}$$
$$- 0.63(\pm 0.33)\,\#\text{NHSO}_2 + 0.78(\pm 0.46)\text{MV} + 0.72(\pm 0.31)\text{CO}_2\text{-}4 - 0.75 \tag{155}$$

$$n = 61, \quad r = 0.88, \quad F_{7,53} = 25.1, \quad s = 0.40$$

$\mathscr{F}$-5 is the Swain–Lupton inductive-electronic parameter, corrected by Hansch and co-workers, for substituents at the 5-position,[46] $\#345$HBD is an indicator variable which takes a value of 0, 1, 2 or 3 according to the number of hydrogen-bonding substituents such as hydroxy, mercapto, amino, —NHCOR and —NHSO$_2$R (R = H, alkyl, NH$_2$ or aryl) located at the 3-, 4- and 5-positions, $\#$NHSO$_2$ is another indicator variable depending upon the number of substructures —NHSO$_2$—

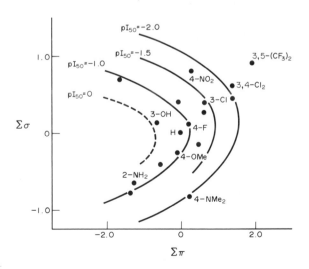

Figure 13 Dependence of antiallergic activity on the electronic and hydrophobic properties of substituents on the benzene ring of the primary set of pyranenamines (modified after ref. 201)

in these substituents, MV is the sum of the molar volume parameters defined by Moriguchi and co-workers[204] for the 3- and 5-substituents, and CO_2-4 is still another indicator variable for the presence of the —OCOR group (R = alkyl, aryl or NH_2) at the 4-position. After a variety of (sub)structural effects were separated, both $\Sigma\pi$ and $(\Sigma\sigma)^2$ terms are now significant over the 95% level in equation (155).

Although the lowest $\Sigma\pi$ value was lowered from -1.34 for the 2,6-dihydroxy compound included in the first 19 compounds to -1.94 for the 3,5-di(acetylamino) substitution in the second set, equation (155), without a significant negative $(\Sigma\pi)^2$ term, still indicates the possibility of a hydrophilicity optimum. Further efforts to lower the $\Sigma\pi$ value of substituents were carried out by preparing another set of compounds having such exotic groups as —NHC(=NH)NH_2, —CONH(barbiturate) and —NHCO(CHOH)$_2$H. The lowest $\Sigma\pi$ value was further lowered to -6.20 for 3,5-[NHCO(CHOH)$_2$H]$_2$ and -10.00 for 3,5-(NHCOCO$_2^-$)$_2$ in this final set. The regression equation best describing the QSAR of all 98 compounds was equation (156).

$$pI_{50} = -0.33(\pm0.11)\Sigma\pi - 0.034(\pm0.016)(\Sigma\pi)^2 + 4.3(\pm1.6)\mathscr{F}\text{-}5 + 1.3(\pm0.85)\mathscr{R}\text{-}5 - 1.7(\pm0.62)(\Sigma\sigma)^2$$
$$+ 0.73(\pm0.22)\#345HBD - 0.84(\pm0.34)\#HB(intra) - 0.69(\pm0.28)\#NHSO_2$$
$$+ 0.72(\pm0.35)CO_2\text{-}4 - 0.59 \tag{156}$$

$$n = 98, \quad r = 0.87, \quad F_{9,88} = 28.7, \quad s = 0.48$$

The negative $(\Sigma\pi)^2$ term is significant in equation (156), showing that a parabolic relationship exists between potency and hydrophilicity, the optimum being about -5.0. #HB(intra) is the number of intramolecular hydrogen-bonds between vicinally located amphiprotic substituents. $\mathscr{R}$-5 is the Swain–Lupton resonance-electronic parameter, suggesting a specific electronic effect of the substituent at the 5-position together with the $\mathscr{F}$-5 term. The MV term is not significant for the final set. The slope of the #345HBD term indicates that the potency increases about five times for every introduction of hydrogen-bonding substituents. However, if these substituents are introduced vicinally being capable of hydrogen bonding with each other, the potency decreases by about 1/7 as the slope of the #HB(intra) shows. The potency of the 3,5-[NHCO(CHOH)$_2$H]$_2$ derivative was the highest of all, with $pI_{50} = 3.0$. In this compound, two highly hydrophilic groups, being *meta*, do not form an internal hydrogen-bond, and give an estimated $\Sigma\pi$ value of -6.2 not far from the optimum. Moreover, the $(\Sigma\sigma)^2$ value (0.14) is close to zero. These features and others were considered to be combined most favorably in this compound. One of the practical aspects which should be mentioned in this example is that the pI_{50} values for about 2/3 of the compounds, where the potency was measured at only a single dose, were estimated by extrapolation. It was assumed that the log(dose)–response curve for the single-dose compounds is parallel with that where the potency was measured at more than one dose.

Unfortunately, the compound brought into clinical trials was not the 3,5-[NHCO(CHOH)$_2$H] derivative, which was the most potent in the intravenous PCA assay, but the 3-amino-4-hydroxy derivative with $pI_{50}(i.v.) = 0.1$. When orally administered, the rat PCA inhibition with the 3-amino-4-hydroxy derivative was observed to be higher than that with the 3,5-[NHCO(CHOH)$_2$H] derivative. Nevertheless, this example, showing that an approximately tenthousandfold potency enhancement from that of the primary set of compounds was achieved by 'enthusiastic' motivation in designing highly hydrophilic substituents, should be regarded as one of the successful applications of QSAR.

21.1.4.4 Design of Antidepressant 4-Anilinopyrimidines

Yoshimoto and co-workers found moderately potent antidepressant activity in some 4-anilino-pyrimidine derivatives.[205] These compounds (**48**) were originally synthesized as possible agricultural fungicides, so the substituents selected (in particular, those at the R_5, R_6 and X positions) were mostly rather hydrophobic, as R = H or Me; R_2 = H or NH_2; R_5 and R_6 = lower alkyl or halogen; R_5—R_6 = (CH$_2$)$_3$ or (CH$_2$)$_4$; X = H, alkyl, halogen, alkoxy or trifluoromethyl in the *ortho*, *meta* or *para* positions. They measured the activity of each anilinopyrimidine in terms of the percentage of inhibition (A) of ptosis caused by reserpine (2 mg kg^{-1}) in mice at the dose of 100 mg kg^{-1}, p.o. Among some 90 compounds synthesized first, about 30 having hydrophobic substituents such as bromo and butyl at the R_5 and R_6 positions were inactive. The substituent selection was biased, so variations in the value of physicochemical parameters such as σ and π of substituents at each position were not adequate in this set of compounds. Therefore these workers examined various combinations of indicator variables assigned to structural features along with the substituent parameters in QSAR analysis.[205]

(48)

For 59 active compounds for which the A (%) value ranged between 6 and 85, equation (157) was formulated, where the activity is expressed in logistic units. If we assume that the A value is proportional to the amount of each compound bound to a hypothetical receptor, the ratio $A/(100 - A)$ can be regarded as a binding equilibrium constant, which is physicochemically more meaningful than A alone for this kind of analysis. The seemingly large s value (0.810) corresponds to 0.352 if the activity index is transformed into the common logarithmic scale. I-1, I-4, I-5, I-6 and I-7 are the indicator variables that are statistically significant among the eight originally examined. I-1 refers to the substituted aniline moiety and I-4 and I-5 refer to the p-halogen and p-alkyl substitutions, respectively. I-6 is that for $R_5 =$ halogen and I-7 is for a structure bridged through $(CH_2)_3$ between R_5 and R_6. π_5 is the hydrophobicity parameter for the R_5 substituents taken to be that of the substituted aniline series. For the methylene bridging groups, half the $\Sigma \pi$ value was assigned to R_5 and half to R_6. σ_p is the Hammett constant of X substituents located at the *para* position. For *ortho* and *meta* substituents, the electronic effect term was not statistically significant.

$$\ln[A/(100 - A)] = -1.38(\pm 0.75)\pi_5 + 1.51(\pm 1.45)\sigma_p - 1.40(\pm 0.68)I\text{-}1 + 1.37(\pm 0.62)I\text{-}4 + 2.39(\pm 0.76)I\text{-}5$$
$$- 0.76(\pm 0.68)I\text{-}6 + 1.50(\pm 0.58)I\text{-}7 - 0.68(\pm 0.56) \quad (157)$$
$$n = 59, \quad r = 0.837, \quad s = 0.810$$

The I-1 term being negative means that substitution at the aniline moiety in general lowers the activity; the I-4 and I-5 terms being positive indicate that halogen and alkyl substitutions at the *para* position cancel the I-1 effect. The I-1, I-4, and I-5 terms suggest together that *meta* substitution is unfavorable for activity. Yoshimoto and co-workers obtained equations of a statistically similar correlation using indicator variables other than those used in equation (157). All these equations suggested that the electron-withdrawing groups at the *para* position of the aniline moiety (σ_p) and the $(CH_2)_3$ bridge between R_5 and R_6 (I-7) are favorable for activity. Thus, these workers designed and synthesized compounds (**49**; X = CN, SO$_2$Me and NO$_2$). These X substituents are strongly electron withdrawing but rather hydrophilic. Even though the π_5 term is for the R_5 substituents, its negative coefficient in equation (157) could be extended to the hydrophobicity of the whole molecule including the X substituents.

(49)

As predicted by equation (157) and its counterparts, these compounds had very strong antidepressant activity. Among them, the CN derivative (RS-2232) had the strongest activity (ED$_{50}$, 6.7 mg kg^{-1}, p.o.) in mice, about twice that of imipramine (ED$_{50}$, 11 mg kg^{-1}, p.o.). For compounds where the p-X substituent is either halogen or trifluoromethyl, the ED$_{50}$ (mg kg^{-1}) value was located between 30 and 70 when $R_5 = R_6 =$ Me, but between 20 and 50 when R_5—$R_6 = (CH_2)_3$. Thus, antidepressant activity was enhanced five to ten times by substituent modifications aided by QSAR. Unlike imipramine, RS-2232 does not cause undesirable side-effects such as anticholinergic and antihistaminic symptoms and was selected for preclinical trials.[206]

21.1.4.5 Design of Cerebral Vasodilating Benzyldiphenylmethylpiperazines

While searching for novel lead compounds for cerebral vasodilators, Ohtaka and co-workers noted that a clinically used coronary vasodilator trimetazidine (**50**) was reported to be distributed in the brain as well as the heart of dogs.[207, 208] Since a well-known cerebral vasodilator cinnarizine (**51**; X = Y = H) seemed much more hydrophobic than trimetazidine, they thought it pertinent to prepare a series of hydrophobic trimetazidine analogs for possible cerebral vasodilators. Among compounds synthesized by combining structural components of trimetazidine and cinnarizine, they found that 1-diphenylmethyl-4-(2,3,4-trimethoxybenzyl)piperazine (**52**; X = 2,3,4-(OMe)$_3$, Y = Z = H) showed considerable activity even with intraduodenal administration.[209] Thus, they started a project for selection of the best combination of X, Y and Z in structure (**52**).[210]

(50)   **(51)**

(52)

For 13 derivatives where Y and Z were fixed as H, and X was varied as H, 4-hydroxy, 4-methyl, 4-chloro, 4-fluoro, 4-acetylamino and 4-methoxycarbonyl, besides di- and tri-substitutions of methoxy at different positions, equation (158) was derived. For *ortho* substituents, the σ values of the corresponding *para* substituents were used in $\Sigma\sigma$. The potency was evaluated in terms of the maximum change of blood flow in vertebral arteries of dogs anesthetized with pentobarbital after intravenous administration of each compound (1 mg kg^{-1}). In equation (158), *A* is the molar-corrected potency relative to that of papaverine taken as unity. For the majority of compounds, the potency was measured at a single dose. Thus, the log *A* value was estimated assuming that the log(dose)–response curves are parallel within the series.

$$\log A = -1.08(\pm 0.37)\Sigma\sigma - 0.41 \tag{158}$$

$$n = 13, \quad r = 0.890, \quad F = 41.90, \quad s = 0.182$$

For 10 compounds where Y and Z are fixed as fluoro, equation (159) was formulated indicating that the effect of the benzyl substituent X is essentially identical with those included in equation (158) although the quality of the correlation is poorer. For nine compounds where X was 2,3,4-trimethoxy and Y and Z substituents were varied as combinations of H, fluoro, methyl, chloro and methoxy, equation (160) was derived. In equation (160), MR(L) stands for the value (scaled by 1/10) of the bulkier substituent (in terms of MR) of Y and Z. MR was shown to work slightly better than other steric parameters such as the STERIMOL B_4 and L values.

$$\log A = -0.81(\pm 0.50)\Sigma\sigma - 0.39 \tag{159}$$

$$n = 10, \quad r = 0.794, \quad F = 13.71, \quad s = 0.235$$

$$\log A = -1.43(\pm 0.70)MR(L) + 0.48 \tag{160}$$

$$n = 9, \quad r = 0.876, \quad F = 23.11, \quad s = 0.243$$

Equations (158) and (159) show that the electron-donating effect is the most important factor for substituents on the benzyl benzene ring, suggesting a role regulating the protonation equilibrium at the benzylic nitrogen of the piperazine skeleton. Equation (160) indicates that the smaller the substituent, the more favorable it is to activity at the diphenylmethyl side of the molecule. The aromatic fluoro substituent is slightly 'smaller' than hydrogen in terms of MR, being 0.09 *vs.* 0.10. Thus, it was originally anticipated that the highly electron-donating 4-dimethylamino group as X and the 'smallest' fluoro group as Y and Z substituents would be the optimum combination for activity. Since the 4-dimethylamino derivatives were acutely toxic, a compound with X = 2,3,4-$(OMe)_3$ ($\Sigma\sigma = -0.41$) and Y = Z = F (KB-2796) which showed the highest activity throughout the series was selected for further development. Although this compound showed activity only two to three times higher than that of papaverine as far as maximum blood flow is concerned, it exhibits a long-lasting effect with more than 20 times longer duration of action than papaverine at optimal doses.

In this example, the most active candidate designed using the QSAR information was dropped because of its toxicity. Thus, the compound actually selected was within the original set. Moreover, the span of the log *A* values was only 1.3. The selected compound (KB-2796) was in fact closely related to flunarizine (**51**; X = Y = F) which is also a well-known cerebral vasodilator. Nevertheless, Ohtaka and his group believe that the QSAR examinations reduced redundant syntheses and made their decision-making easier. KB-2796 is three to ten times as potent as flunarizine, depending upon the mode of administration.[209] The effect is likely to be due to a calcium antagonistic action on the vascular smooth muscle. It is now undergoing extensive clinical trials.

21.1.4.6 Design of a Novel Herbicidal *N*-Benzylacylamide, Bromobutide

Kirino and co-workers have recently developed[211] a novel broad spectrum herbicide, bromobutide (**53**), which is effective not only against annual but also against perennial weeds in paddy fields.

$$\text{—CMe}_2\text{NHCOCHBrCMe}_3$$

(**53**)

Kirino and co-workers previously found that fungicidal effects of *N*-substituted aminoacetonitriles (**54**) were dependent upon the steric features of *N*-substituents.[212] The 50% preventive dose, ED_{50} (mol) applied to a certain volume of soil, against 'yellows' of the Japanese radish, a soil-borne disease caused by *Fusarium oxysporum*, was analyzed as shown in equation (161). Equation (161) was formulated for 16 compounds where R is varied from *n*-alkyl to branched as well as cycloalkyls. E_s^c is the 'corrected' steric parameter emphasizing the effect of branching in addition to the steric bulk of R. The negative E_s^c term shows that the bulkier and the more branched the *N*-substituents, the higher is the activity.

$$\text{RNHCH}_2\text{CN}$$
(**54**)

$$pED_{50} = -0.606(\pm0.184)E_s^c + 1.518(\pm0.265) \tag{161}$$
$$n = 16, \quad r = 0.884, \quad s = 0.204$$

They also observed another example in which the steric bulk of substituents enhanced biological activity. For inhibitory activity against the shoot elongation of a barnyard grass, *Echinochloa crus-galli* var. *rumentaceus*, of a series of *N*-chloroacetyl-*N*-phenylglycine ethyl esters (**55**) with such various substituents as alkyl, alkoxy, halo, acyl, nitro, trifluoromethyl and phenyl at *ortho*, *meta* and *para* positions, they formulated equation (162).[213] They considered that the steric bulk of substituents has a role in protecting compounds against interaction with some detoxification enzyme and that the stability in soil or the plants is the critical factor determining these two types of agrochemical activity.

$$pI_{50} = -0.767(\pm0.158)E_s^{ortho} - 0.218(\pm0.196)E_s^{meta} + 3.990(\pm0.240) \tag{162}$$
$$n = 28, \quad r = 0.905, \quad s = 0.295$$

$$\text{(structure)} \quad \overset{\text{COCH}_2\text{Cl}}{\underset{\text{CH}_2\text{CO}_2\text{Et}}{\text{X}\!-\!\!\!\text{N}}}$$

$$(55)$$

They applied the above considerations to the structural optimization of a series of newly discovered herbicidal *N*-benzylacylamides (56), which have an amide moiety in common with the *N*-chloroacetyl-*N*-phenylglycine ethyl esters (55). Systematic modification of the structure was carried out to make the substituent R in (56) more and more bulky to increase protection against the possible hydrolytic detoxification mechanism.

$$\text{(structure)} \quad -\text{CMe}_2\text{NHCOR}$$

$$(56)$$

The analysis for the first set of 41 derivatives, in which R varies from simple normal and branched alkyls to such congested groups as Bu^{t}CHCl—, Bu^{t}CHBr—, Pr^{i}CMe$_2$— and CMe$_2$BrCHBr— gave equation (163).[211] I_{50} is the molar concentration required for 50% inhibition of shoot elongation of the seedlings of the bulrush, *Scirpus juncoides*, a perennial pest weed species. Compounds where R = CH$_2$Br and CH$_2$Cl are not included in equation (163), since their pI$_{50}$ values are about 1.5 units higher than those predicted by equation (163). This was attributed to higher reactivity of their halogen substituents, probably in an alkylation reaction. Groups having primary halogens, however, are not favorable as a structural feature, since the E_s^c value is made less negative and the π value is not modified greatly. The pI$_{50}$ values of compounds with haloalkyl groups where the halogen atom is not primary fit equation (163) quite well, as do those of compounds without haloalkyl groups. This suggests that the activity of this series of compounds is not due to their alkylating properties, even though they have α-haloacyl substituents. Substituents R having a double bond at the α,β-position and those having a β-halogen, which could give an α,β-double bond by an elimination reaction, were also deleted from equation (163). Since conjugation with the carbonyl function may perturb the reactivity of the amide moiety, the actual activities are about 0.5 log units lower than the values predicted by equation (163).

$$\text{pI}_{50} = -0.151(\pm0.093)\pi^2 + 0.983(\pm0.457)\pi - 0.350(\pm0.070)E_s^c + 2.877(\pm0.465) \qquad (163)$$

$$n = 41, \quad r = 0.933, \quad F = 83.00, \quad s = 0.267$$

Equation (163) indicates that although there is an optimum in hydrophobicity within the set of 41 compounds ($\pi_{opt} = 3.3$) the steric bulkiness of the R substituent in terms of E_s^c still needs to be augmented to obtain higher activity. Further structural modifications were made on the α-bromoacylamide structure (57). For 14 derivatives where R' is varied from a simple alkyl to such highly branched bulky groups as Pr^{i}CMe$_2$—, Bu^{i}CMe$_2$— and Et$_2$CMe—, equation (164) was derived to show that the optimum for the steric bulk of R' lies at about -3.1 in terms of E_s^c.[211] The E_s^c value for R' is used in equation (164), since the value for such highly congested groups as R'CHBr cannot be estimated with sufficient accuracy.

$$\text{(structure)} \quad -\text{CMe}_2\text{NHCOCHBrR'}$$

$$(57)$$

$$\text{pI}_{50} = -0.199(\pm0.152)(E_s^c)^2 - 1.227(\pm0.546)E_s^c + 4.393(\pm0.364) \qquad (164)$$

$$n = 14, \quad r = 0.940, \quad F = 41.67, \quad s = 0.242$$

Equation (164) indicates that the R substituent should not be made infinitely bulky. Beyond the optimum, the fit with its own receptor could be inhibited. In this respect, it should be noted that the steric bulk of the second largest subgroup R^2 of the R substituents [R = R^{1}R^{2}R^{3}C; $E_s^c(R^1) \geq E_s^c(R^2) \geq E_s^c(R^3)$] has a specific effect on the activity. When the R substituent is such that the R^2 subgroup is bulkier than ethyl, the activity is always considerably lower (by 0.7–1.4 log units) than that estimated by means of equation (163) using the normal $E_s^c(R)$ value. These derivatives also are not used in deriving equation (163). Even though the steric bulk is below the optimum, too great a congestion of the acyl moiety would be deleterious to their fit with their own receptor.

Although the details are not shown here, quantitative analyses of the effect of substituents in the benzylamine moiety confirmed that the α,α'-dimethyl substitution and lack of aromatic substitution are optimal for high activity.[214]

Using this information and taking into consideration a certain (but not too serious) collinearity between E_s^c and π values for the compounds included in equation (164), as well as their ease of preparation, bromobutide (57; R' = Bu^t), where $\pi(R) = 2.2$ and $E_s^c(R') = -2.5$, was selected as a promising novel paddy-field herbicide and taken forward to extensive field trials and toxicity tests. The activity of this compound, determined by the laboratory test against bulrush shoot growth, is about 100 times that of the chloroacetyl lead compound. The α,α'-dimethyl substitution also enhances the activity about tenfold.

The field trials showed that bromobutide in fact controls various weeds in paddy fields. Especially sensitive is the bulrush, which is killed at rates of 100–500 g ha^{-1} with complete safety to the rice plant.[215] The broad herbicidal spectrum has been shown to be due not only to its intrinsic activity, but also to its moderate degrees of persistency and its mobility in soil. It is now marketed in Japan and some Asian countries.

21.1.4.7 Design of a Novel Quinolone Carboxylic Acid Antibacterial Drug, Norfloxacin

Nalidixic acid (58) is the first member of an antibacterial drug family sharing a common γ-pyridone-β-carboxylic acid structure.[216] A number of efforts have been made to enhance the potency as well as to expand the antibacterial spectrum, by modification of its structure. Oxolinic acid (59) and pipemidic acid (60) are noteworthy examples among them. Oxolinic acid was characterized by its potent antibacterial activity and pipemidic acid was marked by its broader antibacterial spectrum, including its effect upon *Pseudomonas*.[217]

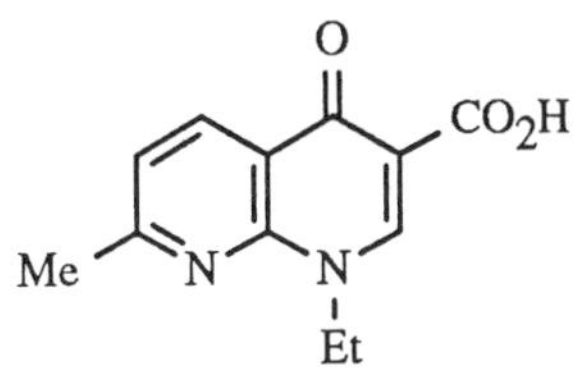

(58) 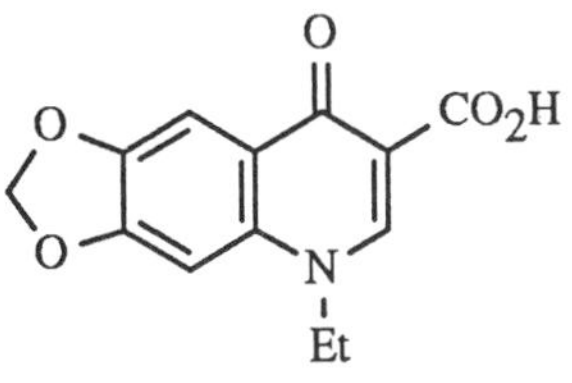   (59) 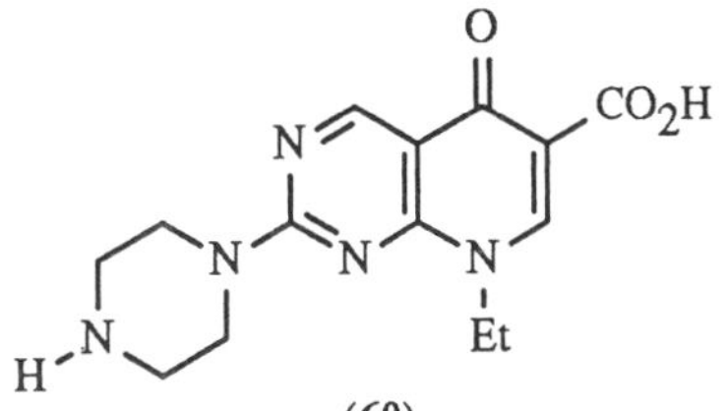   (60)

Under these circumstances, some 15 years ago, Koga and co-workers began their attempt to develop compounds having not only more potent activity and a broader spectrum, but also lower oral toxicity as well as a higher stability to metabolism than any other nalidixic acid analogs known at that time.[218] They selected 4-quinolone-3-carboxylic acid (61) as the reference compound, since its analogs can be prepared more easily than those having such polyazanaphthalene ring systems such as those in nalidixic and pipemidic acids.

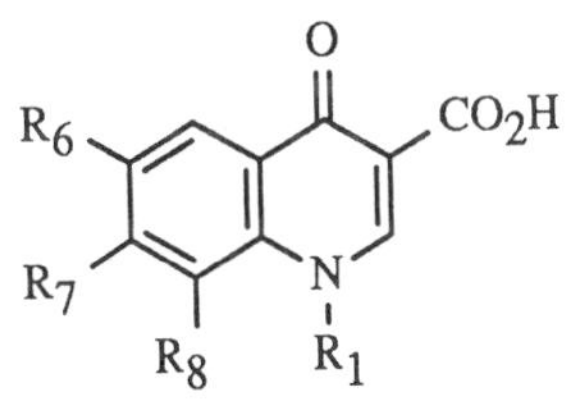

(61)

They synthesized analogs having various substituents inserted at different positions systematically. They determined the minimum inhibitory concentration (MIC in mol dm^{-3}) against *Escherichia coli* NIHJ JC-2 for each compound, since it was observed that the activity against this bacterium roughly parallels that against Gram-negative rod-shaped bacteria including *Pseudomonas*. The structure–activity relationship was first analyzed stepwise in terms of substituent effects at each position.[219]

For derivatives (**61**; $R^1 = Et$, $R^6 = H$, F, Cl, NO_2, Br, Me, OMe and I) equation (165) was derived. For the nitro compound, the E_s value estimated from the half-thickness in terms of the van der Waals dimension is used. Equation (165) shows that the effect of R^6 on the activity is mainly steric and the optimum steric dimension in terms of $E_s(= -0.66)$ is located in between those of fluoro and chloro.

$$\log(1/MIC) = -3.318(\pm 0.59)[E_s(6)]^2 - 4.371(\pm 0.85)[E_s(6)] + 3.924 \tag{165}$$

$$n = 8, \quad r = 0.989, \quad F = 112.29, \quad s = 0.108$$

For substituent effects at the 8-position, equation (166) was formulated for seven derivatives (**61**; $R^1 = Et$; $R^8 = H$, F, Cl, Me, OMe, Et and OEt). Equation (166) shows that the effect of R^8 is also steric. The optimum STERIMOL B_4 value is 1.83, which is close to that of chloro and methyl.

$$\log(1/MIC) = -1.016(\pm 0.46)[B_4(8)]^2 + 3.726(\pm 2.04)[B_4(8)] + 1.301 \tag{166}$$

$$n = 7, \quad r = 0.978, \quad F = 44.05, \quad s = 0.221$$

At the 2-position of the pipemidic acid (**60**) skeleton which corresponds to the 7-position of the quinolone carboxylic acid (**61**), such substituents as alkoxy, alkylthio, alkylamino and dialkylamino (including pyrrolidinyl and piperazinyl) had been known to favor activity, and the favorable effect had been considered to be due to their electron-donating properties.[220] The result for eight compounds (**61**; $R^1 = Et$; $R^7 = H$, NO_2, COMe, Cl, Me, OMe, NMe_2 and piperazinyl) was not in accordance with this earlier point of view. Although the introduction of substituents into the 7-position enhances the activity ten- to thirty-fold, the physicochemical factors responsible for the effect were not clear. In terms of the $\log(1/MIC)$ value, the activity varied only from 5 to 5.5 for compounds with substituents ranging from the electron-withdrawing nitro group to the electron-donating dimethylamino group.

Representing the effect of R^7 with an indicator variable $I(7)$, they combined the results for the three sets of 'mono'-substituted derivatives and derived equation (167). Equation (167) predicts that multiply-substituted derivatives having optimized substituents at each of the 6-, 7- and 8-positions ($E_s(6) = -0.65$, $I(7) = 1$ and $B_4(8) = 1.84$) show a $\log(1/MIC)$ value of 7.5, which corresponds to their being about 10 times more active than oxolinic acid (**59**).

$$\log(1/MIC) = -3.236(\pm 0.89)[E_s(6)]^2 - 4.210(\pm 1.26)[E_s(6)] + 1.358(\pm 0.40)I(7) - 1.024(\pm 0.32)[B_4(8)]^2$$

$$+ 3.770(\pm 1.43)[B_4(8)] + 1.251 \tag{167}$$

$$n = 21, \quad r = 0.978, \quad F = 67.50, \quad s = 0.205$$

In an attempt to prove the above prediction, as well as to elaborate the quantitative correlation, Koga and co-workers prepared a number of such multiply substituted derivatives. For the compounds where $R^1 = Et$ and $R^6 = F$, but R^7 is varied from such simple substituents as chloro, methyl and dimethylamino to such heterocyclic groups as pyrrolidinyl, piperazinyl and variously N'-substituted piperazinyl groups, equation (168) was formulated. $I(7N\text{-}CO)$ is an indicator variable for such 7-N-heterocyclic substituents as N'-acylpiperazinyl and 4-carbamoylpiperidinyl. The presence of this term indicates that the presence of carbonyl functions as a part of the 7-N-substituent lowers the activity. In equation (168), the 7-unsubstituted compound (**61**; $R^1 = Et$, $R^7 = H$) is not included, since the activity of this compound is about 0.8 log units lower than that expected from this equation. The effect of R^7 substituents, except for that of the carbonyl function, seems to be composed of two parts. One can be expressed by the $I(7)$ term in equation (167), and the other is determined by the hydrophobicity. The optimum π value is estimated as -1.38. The π value of the piperazinyl group is -1.74 (from experimentally measured log P values of the 7-piperazinyl and unsubstituted compounds) and that of the N'-methylpiperazinyl group is estimated as -1.24. These values are close to the optimum.

$$\log(1/MIC) = -0.244(\pm 0.05)[\pi(7)]^2 - 0.675(\pm 0.15)\pi(7) - 0.705(\pm 0.27)I(7N\text{-}CO) + 5.987 \tag{168}$$

$$n = 22, \quad r = 0.943, \quad F = 47.97, \quad s = 0.242$$

In the above analyses, the R^1 substituent was fixed as ethyl. In addition to ethyl, such R^1 substituents as vinyl, methoxy, 2-fluoroethyl and 2,2-difluoroethyl, which are sterically similar to ethyl, were known to be favorable for activity. Derivatives were prepared in which R^6 is fluoro, and R^7 is piperazinyl, but R^1 is varied as methyl, ethyl, vinyl, propyl, allyl, hydroxyethyl, benzyl and dimethylaminoethyl, and the results were analyzed for substituent effect to give equation (169), where L is the STERIMOL length parameter. There is an optimum length for the R^1 substituent at about 4.2 which coincides with that of ethyl ($L = 4.11$).

$$\log(1/MIC) = -0.492(\pm 0.18)[L(1)]^2 + 4.102(\pm 1.59)[L(1)] - 1.999 \qquad (169)$$

$$n = 8, \quad r = 0.955, \quad F = 25.78, \quad s = 0.126$$

At this point, it is possible to predict that the best substituents at each position are likely to be $R^1 = Et$, $R^6 = F$ or Cl, $R^7 = $ piperazinyl or N'-methylpiperazinyl and $R^8 = Cl$ or Me. By combining the results for some additional polysubstituted derivatives with those for compounds included in equations (167), (168) and (169) as well as with the value deleted from the formulation of equation (168), they finally derived equation (170) for 71 compounds.[219] This seems to confirm the prediction for the activities of polysubstituted compounds made by equation (167) formulated from activities of 'mono'-substituted compounds. Most physicochemical parameter terms appearing in equations (167), (168) and (169) are included in equation (170) as such, indicating that the substituent effects at various positions are almost additive in nature.

$$\log(1/MIC) = -0.362(\pm 0.25)[L(1)]^2 + 3.036(\pm 2.21)[L(1)] - 2.499(\pm 0.55)[E_s(6)]^2 - 3.345(\pm 0.73)[E_s(6)]$$

$$+ 0.986(\pm 0.24)I(7) - 0.734(\pm 0.27)I(7N\text{-}CO) - 1.023(\pm 0.23)[B_4(8)]^2 + 3.724(\pm 0.92)B_4(8)$$

$$- 0.205(\pm 0.05)[\Sigma\pi(6,7,8)]^2 - 0.485(\pm 0.10)\Sigma\pi(6,7,8) - 0.681(\pm 0.39)\Sigma\mathscr{F}(6,7,8) - 4.571 \qquad (170)$$

$$n = 71, \quad r = 0.964, \quad F = 64.07, \quad s = 0.274$$

In equation (170), a $\Sigma\pi(6,7,8)$ term is statistically significant. Since the collinearity between $\Sigma\pi(6,7,8)$ and $\log P$ is 0.92 for 71 compounds, the hydrophobicity of the whole molecule seems to play an important role, possibly in the transport process to the active site. By a joint effect of substituents, the negative $\Sigma\mathscr{F}(6,7,8)$ term becomes significant in equation (170) indicating that an inductively electron-donating effect on the 4-oxo function from the homocyclic moiety favors the activity through an electronic interaction with certain receptor sites.

With this information, and after considering the actual antibacterial effects against infections caused by various bacteria, as well as toxicity and the cost of synthesis, they selected norfloxacin (**62**) as the best compound ($\log(1/MIC) = 6.6$).[219] This compound shows much more potent antibacterial activity (sixteen- to five hundred-fold that of nalidixic acid depending upon the bacterial species) and a very broad antibacterial spectrum. It has been marketed recently. The most active compound included in equation (170) is a tricyclic 6,7,8-trisubstituted derivative (**63**; $X = CH_2$, $R = H$), $\log(1/MIC) = 7.2$, which seems to possess an ideally optimized structure in every respect suggested by the QSAR. It is interesting to note that ofloxacin (**63**; $X = O$, $R = Me$) developed and marketed more recently[221,] is very similar to the compound (**63**; $X = CH_2$, $R = H$).

(**62**)

(**63**)

Recently, 4-fluoro- and 2,4-difluoro-phenyl groups were found to be more favorable for R^1 by Chu[222] and Narita[223] and their respective co-workers. Although these R^1 substituents were not suggested by the original QSAR correlation study, Koga and co-workers were at least able to generate a useful product that could be marketed, and the role of QSAR analysis in this should not be disregarded. The QSAR correlation can be supplemented by additional results, which may provide more comprehensive information.

21.1.4.8 Design of Antihypertensive Quinazolines

6,7-Dimethoxy-2-(4-acyl-1-piperazinyl)-4-quinazolinamines (**64**; $n = 2$) are antihypertensive agents that block the post-synaptic α-adrenergic receptor site.[224] When Sekiya and co-workers started their project, prazosin (**64**; $n = 2$, $R = 2$-furyl) and E-643 (**64**; $n = 3$, $R = Pr$) were publicized members of this series of compounds. They first examined structures related to prazosin, such as tetrahydro-quinazoline derivatives (**65**), and found that these compounds are not always antihypertensive, but sometimes have various degrees of hypoglycemic activity.[225]

(**64**) R = H, Me, CH_2Ph, $COCH=CHPh$ (**65**) R = H, Me, CH_2Ph, $COCH=CHPh$

Among these tetrahydroquinazolines, the weakly hypoglycemic cinnamoyl derivative moderately reduced the blood pressure of spontaneously hypertensive (SH) rats. It did not, however, block α-adrenoceptors and the antihypertensive effect was less than that of prazosin. These workers tried to combine structural features of prazosin with those of the cinnamoyl tetrahydroquinazoline derivative to develop a new lead. As they expected, a compound (**66**; R_1 = Ph, R_2 = H) strongly blocked α-adrenoceptors and was also antihypertensive.[226] They proceeded further to optimize the lead structure of (**66**) by modifying R_1 and R_2, *e.g.* R_1 = substituted phenyl, thienyl and furyl, with alkyl, alkoxy, halo, trifluoromethyl and nitro; and R_2 = H and methyl.

(**66**)

The antihypertensive potency was measured using SH rats in terms of the blood-pressure reduction relative to the control after administration of certain doses orally. Since the dose–response relationship was not examined for most compounds, the activity was expressed on a four-grade rating according to potency. QSAR analysis was by the adaptive least-squares (ALS) method (Section 21.1.2.6).

For the first 29 compounds synthesized with structure (**66**), equation (171) was formulated,[227] where *n*(mis) is the number of compounds misclassified into the next class, and r_S is the Spearman rank correlation coefficient. RI is the retention index, a parameter of hydrophobicity defined by Baker and measured experimentally by HPLC.[228] ΔRI is the value relative to that of the unsubstituted cinnamoyl derivative (**66**; R_1 = Ph, R_2 = H). $\Sigma\sigma$ is the summation of the Hammett constants of substituents X_1 on the benzene ring (R_1). For thienyl and furyl derivatives, the replacement σ constant in which the hetero atom was regarded as a substituent[229] was used for the thia and oxa functions. I(2-OR) is an indicator variable which is used when an alkoxy group is located at the *ortho* position of the R_1 aromatic moiety. When the $\Sigma\pi$ value was used in place of ΔRI, a similar but slightly poorer discriminant function was derived.

$$A = -1.02\Delta RI - 2.36\Sigma\sigma - 1.84I(2\text{-}OR) + 0.77 \qquad (171)$$

$$n = 29 \ (4 \ \text{grades}), \quad n(\text{mis}) = 7, \quad r_s = 0.914$$

Equation (171) suggests that the less hydrophobic the molecule and the less electron-withdrawing the aromatic substituents, the higher the activity. With this suggestion, Sekiya and co-workers further modified structure (**66**), leading to more potent analogs. They decided that molecular modifications to reduce the hydrophobicity while holding the skeletal quinazoline in structure (**66**) would be most suitable. The introduction of such hydrophilic and electron-donating substituents as hydroxy and amino to the aromatic moiety may bring about further variables such as dissociation and hydrogen-bonding effects. They thought the above information could be extended beyond the original arylacryloyl compounds and replaced the whole arylacryloyl moiety with less hydrophobic acyl groups, such as *N*-substituted carbamoyl. The log *P* value of *N*-phenyl- and *N*-cyclopentyl-carbamoyl derivatives (**67**; R = Ph, cyclopentyl) was estimated as being lower than that of the cinnamoyl compound by about 0.6 and 1.0, respectively.

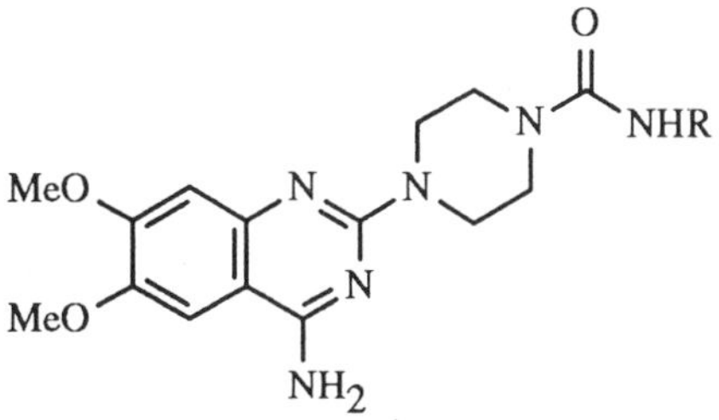

(**67**)

In fact, these carbamoyl derivatives were highly potent; the blood-pressure reduction compared to the control in SH rats was by more than 20% at the orally applied dose of 1 mg kg^{-1}.[227] In particular, the cyclopentylcarbamoyl derivative MY-5561 reduced blood pressure by 12% at a dose of 0.3 mg kg^{-1}. It strongly blocked α-adrenoceptors (pA_2 = 10.8) in rat thoracic aorta. These *in vivo* and *in vitro* activities are comparable to those of prazosin which gives 19% reduction at a dose of 0.3 mg kg^{-1} and has pA_2 = 10.3. The original cinnamoyl tetrahydroquinazoline compound (**65**; R = cinnamoyl) reduced blood pressure by 12% compared to the control with a dose of 25 mg kg^{-1}. Activity was enhanced in the cinnamoyl quinazoline (**66**; R_1 = Ph, R_2 = H); 14% reduction with a dose of 1 mg kg^{-1}. Thus, the structural modifications leading to MY-5561 backed up by QSAR analyses did increase antihypertensive activity about 80 times over that of the primary lead compound. MY-5561 has similar effects on renal and adrenal hypertensive rats, so it should be useful clinically.

21.1.4.9 Design of Anti-inflammatory and Analgesic Furoindolecarboxamides

Furo[3,2-*b*]indole compounds were first synthesized by Tanaka and his associates in 1979.[230] In later years, Kawashima and co-workers synthesized a number of substituted furoindole derivatives and screened them using various biological tests. They found that such amide derivatives as (**68**) had various degrees of analgesic and anti-inflammatory activities.[231] The R_1 substituent was H, methyl, fluoro, chloro, trifluoromethyl, methoxy or hydroxy. The R_2 substituent was confined to carbalkoxy groups since they had found that the carbalkoxy groups were not unfavorable to activities in related series where the amide side-chain was CONH(CH$_2$)$_3$N(CH$_2$)$_5$.[232]

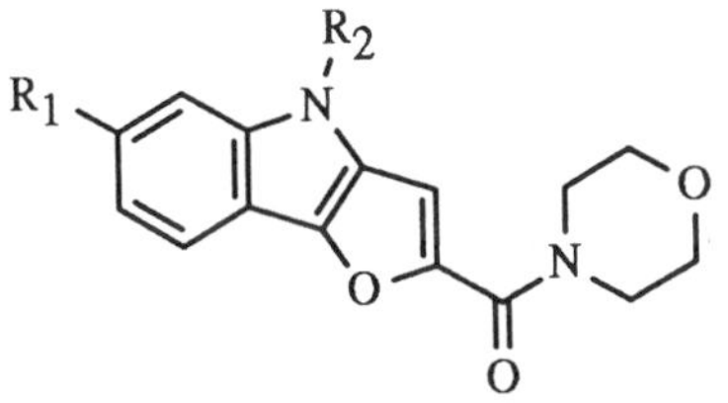

(**68**)

The analgesic activity was measured by the acetic acid writhing test in mice and the anti-inflammatory activity by the carrageenin edema test in rats each given 100 mg kg^{-1}, p.o. These activities were graded into three ordered classes according to the degree of inhibition. The QSAR analysis using the ALS method yielded equations (172) and (173).[233] In these equations, B_1 and L are the STERIMOL minimum width and length parameters. $\mathscr{F}$ is the inductive electronic parameter proposed by Swain and Unger[234] but not by Swain and Lupton. $N_c(R_2)$, the number of β-carbon atoms in the alkyl chain of carbalkoxy substituents R_2, accounts for the effect of branching of the ester moiety.

Analgesic activity

$$A = 1.96B_1(R_1) + 0.28L(R_2) + 1.12N_c(R_2) - 5.90 \tag{172}$$

$$n = 38 \text{ (3 grades)}, \quad n(\text{mis}) = 5, \quad r_s = 0.89$$

Anti-inflammatory activity

$$A = -1.78\mathscr{F}(R_1) + 3.06B_1(R_1) + 0.64N_c(R_2) - 4.29 \qquad (173)$$

$$n = 38 \text{ (3 grades)}, \quad n(\text{mis}) = 7, \quad r_S = 0.83$$

Equations (172) and (173) indicate that the wider the smallest width of R_1 substituents, the more potent are both activities. In other words, the more symmetrical and bulkier the R_1 substituents, the more favorable they are to the activity. Considering the availability of synthons, Kawashima and co-workers decided to make the R_1 substituent a trifluoromethyl group, although its electron-withdrawing character could be expected to interfere somewhat with the anti-inflammatory activity. The L and N_c terms suggest that the length and branching of R_2 substituents are advantageous to the activities. These workers expected that these structure–activity relationships could be transposed to analogs where R_2 was not restricted to carbalkoxy groups and synthesized derivatives where R_2 was modified into alkyl, alkenyl and acyl groups. In fact, derivatives having such branched acyl groups as COBu^i, COPr^i and COCH(Et)Bu and long-chain alkyl groups as butyl, pentyl and hexyl were highly potent in both activities. The COCH(Et)Bu derivative had particularly strong biological activities. The analgesic activity was 16 to 100 times as potent as that of tiaramide in the acetic acid writhing and other pharmacological tests. In the writhing test, the ED$_{50}$ value was 0.7 mg kg^{-1}. Antipyretic activity was about 3000 times as high as that of tiaramide in rats in the yeast-induced fever test, the ED$_{50}$ value being 0.06 mg kg^{-1}, p.o. The ED$_{50}$ value for the anti-inflammatory effect was of the order of 3 mg kg^{-1} in rats with the carrageenin edema assay, which is about 40 times more potent than tiaramide. Moreover, this derivative did not cause ulcers and did not kill mice at the oral dose of 100 mg kg^{-1}.

Although this compound and a few of its analogs seemed promising, they unfortunately caused mucosal irritation and Kawashima and co-workers gave up plans for their further development. Nevertheless, this was a successful application of lead optimization. The analgesic activity of the original compound (68; $R_1 = H$, $R_2 = CO_2Me$) was lower than 39% inhibition, and the anti-inflammatory activity was between 40% and 69% inhibition at a dose of 100 mg kg^{-1}, p.o. Thus, this structural modification enhanced the activities 30 to 150 times over that of the lead compound.

21.1.4.10 Design of a Novel Triazinone Herbicide, Metamitron

The development of the title compound is one of the earliest examples which skilfully utilized QSAR. Draber and co-workers noticed a report of the syntheses of 3-alkylthio-4-amino-6-aryl-1,2,4-triazin-5-one derivatives (69; $R_1 = SR$, $R_2 = $ aryl), published in 1964 by Dornow and his group,[235] and were interested in their novel structure. Anticipating that these derivatives might exhibit some biological activity, they followed the syntheses and in fact, found that a compound with $R_1 = SMe$ and $R_2 = Ph$ showed considerable herbicidal activity. They started intensive work on this new class of compounds, and soon synthesized metribuzine (69: $R_1 = SMe$, $R_2 = Bu^t$) which has been used since 1967 as a selective soy-bean herbicide.[236]

(69)

At about the same time, Trebst discovered that the mechanism of action of the triazinones was the inhibition of electron flow in photosystem II in plants.[237] Draber and co-workers, cooperating with Trebst, examined the QSAR of this series of compounds. For 13 compounds (69; $R_1 = SMe$, NHMe or OMe; $R_2 = $ simple lower straight chain or branched alkyl, cyclohexyl or Ph) equation (174) was formulated,[238] where I_{50} is the molar concentration required for 50% inhibition of the Hill reaction of chloroplasts isolated from spinach. The P value is that measured with a system of 1-octanol/pH = 7 buffer solution. In equation (174), derivatives (69; $R_1 = $ NHBu, NHPh, SEt and SPr^i) are not included since these compounds exhibited an activity considerably lower than that predicted by the correlation. They considered that this observation is probably due to the steric bulk of R_1 larger than that of methoxy, methylthio and methylamino. There might be a restriction in the bulk of R_1 substituents. Since equation (174) holds for imino, thio and oxo derivatives unless the bulk is larger than XMe (X = O, S or NH), differences in chemical properties, if any, were thought not to be important. Thus, they expected that simple small alkyl groups such as methyl may work as the R_1 substituent.

$$pI_{50} = 1.959 \log P - 0.486(\log P)^2 + 4.608 \qquad (174)$$

$$n = 13, \quad r = 0.953, \quad s = 0.259$$

Motivated by this QSAR insight, Draber and co-workers tried to synthesize compounds (**69**; R_1 = Me). The project was not easy, since entirely new synthetic methods were required to make these candidates with available synthons. After serious trials, they made (**69**; R_1 = Me, R_2 = Bu^t). As they expected, this compound showed activity in inhibiting the Hill reaction although it was rather low.[238] They continued to improve general synthetic procedures of this type of compounds and finally synthesized compound (**69**; R_1 = Me, R_2 = Ph), named metamitron. Metamitron, although its Hill-inhibitory activity is not extremely high (pI_{50} = 6.1), has been marketed as a highly selective sugar-beet herbicide since 1975. As far as the activity on the chloroplast level is concerned, an analog where R_1 was ethyl and R_2 was *m*-trifluoromethylphenyl was most potent (pI_{50} = 8.0).[238] This compound was unfortunately not selective.

21.1.4.11 Concluding Remarks

In this chapter only nine drug design examples are quoted from among a number of studies. For other examples, earlier reviews by Martin,[239] Unger,[240] Hansch[241] and Hopfinger[242] are recommended for an understanding of the scope of QSAR practice.

One may argue that compounds having higher activity predicted by QSAR could be synthesized sooner or later without QSAR solely by the intuition of organic chemists. However, the organic chemists themselves asserted that the QSAR approach was helpful in most of the examples in this chapter. Without the motivation provided by QSAR, projects for further syntheses would be difficult to undertake. Sometimes, the decision as to whether to continue or discontinue a project was much easier to make using QSAR.

In some examples, the selection of substituents in the primary set of compounds was biased so that the information derived from the QSAR was limited. As will be discussed in Chapter 21.2, each substituent in the set should be selected in such a manner that it conveys the maximum amount of information to the correlation equation. Substituents should be neither clustered nor linear in the substituent parameter space. Moreover, a sufficient number of compounds relative to the number of substituent parameters are required to avoid chance correlations.[80] Thus, substituent parameters such as σ, π and E_s should vary as widely and independently as possible in the primary set.[243, 244] Some examples do not necessarily meet these standards. There are a couple of reasons for this. First, practising organic chemists initially do not pay much attention to the behavior of substituents in substituent parameter space. Secondly, the synthetic difficulty and the unavailability of synthons sometimes severely restricts the variation of substituents. If the importance of the substituent selection in the primary test series is generally recognized, the efficiency of the QSAR application can be greatly improved. For multipositionally substituted compounds, the number of compounds in the primary set could be reduced by efficiently selecting substituents that are varied in various positions simultaneously using a specific statistical procedure called fractional factorial design.[245]

The examples of the application of QSAR to the the development of antiallergic agents (Sections 21.1.4.2 and 21.1.4.3), antidepressants (Section 21.1.4.4) and cerebral vasodilators (Section 21.1.4.5) can be categorized as utilizing QSAR information by extrapolation of parameters toward directions enhancing the activity. The acylamide herbicide development (Section 21.1.4.6) appears to use the extrapolation concept first. In this case, the best compound was identified from a parabolic function of certain parameter for the second set of compounds. Although the identified compound was included in the primary series, this example demonstrates both the extrapolation and identification concepts. The quinolone carboxylic acid example (Section 21.1.4.7) comprises the multiple identification of optimum substituents at various positions in a molecule and the assumption of additivity of their effects. For the antihypertensive (Section 21.1.4.8) and anti-inflammatory (Section 21.1.4.9) drugs and the triazinone herbicide (Section 21.1.4.10), the QSAR information first obtained was skilfully transposed to series of compounds structurally 'different' from that of the primary set, this being regarded as one of the procedures for lead evolution.

The compounds designed in three of the above examples are now marketed. The others reviewed here either await advanced trials or were for some reason dropped. The side effects, such as mucosal irritation observed in the furoindole anti-inflammatory agents (Section 21.1.4.9), and other toxicity that may be disclosed in later phases in drug development are beyond the scope of the QSAR analysis that identified the 'best' candidate at the earlier stages. Of course, the toxicities are often analyzable by QSAR if proper biological parameters can be selected (Section 21.1.3.3). However, the fact that compounds are brought into the preclinical development phase should be regarded as a measure of the success of QSAR. The recent accumulation of successful applications implies that QSAR is a routine procedure in drug design. In recent years, the use of molecular modeling and molecular graphics has been growing explosively in the field of computer-aided drug design. The classical QSAR and the newer computer-aided methods are believed to be complementary in the practice of drug design.

For further reading, there are a number of monographs [54, 246–253] and symposium proceedings[254–259] that can be recommended.

21.1.5 NOTE ADDED IN PROOF

Hopfinger and co-workers made a very serious correction for their work on QSAR of dapson analogs just recently.[260] The intramolecular entropy, S, was incorrectly calculated in their earlier work.[98] The revised QSAR, which replaces equation (118) on page 527, is as follows:

$$pI_{50} = 0.040(\pm 0.005)P_A + 0.469(\pm 0.052)\phi + 5.18(\pm 0.06)$$

$$n = 34, \quad r = 0.90, \quad F = 63.0, \quad s = 0.20$$

P_A is the sum of the thermodynamic probabilities for the molecule to take the postulated 'active' conformations and ϕ is the direction of the dipole moment of the substituted phenyl ring relative to the corresponding dipole for the most active analog. Two compounds were deleted from the original set to derive this correlation equation. From the uncorrected equation (118), where the single negative S term is sufficient to rationalize the variations in pI_{50}, they concluded that the inhibition potency of dapson analogs increases as the molecule is restricted conformationally. The P_A term in the corrected equation, representing the probability to take restricted conformations, seems to be in accord with their earlier statement. An additional conclusion is that the dipole moment of a substituted ring is important in the inhibition potency, suggesting some electrostatic interactions with the receptor. The discussions made in Section 21.1.3.1 relating to equation (118) should be read keeping the above correction in mind. Equation (120) is no longer valid.

21.1.6 REFERENCES

1. J. E. Leffler and E. Grunwald, 'Rates and Equilibria of Organic Reactions', Wiley, New York, 1963, p. 128.
2. L. P. Hammett, 'Physical Organic Chemistry', 2nd edn. McGraw-Hill, New York, 1970.
3. R. W. Taft, Jr., in 'Steric Effects in Organic Chemistry', ed. M. S. Newman, Wiley, New York, 1956, p. 556.
4. N. B. Chapman and J. Shorter (eds.), 'Advances in Linear Free Energy Relationships', Plenum Press, London, 1972.
5. N. B. Chapman and J. Shorter (eds.), 'Correlation Analysis in Chemistry', Plenum Press, London, 1978.
6. C. Hansch, P. P. Maloney, T. Fujita and R. M. Muir, *Nature (London)*, 1962, **194**, 178.
7. C. Hansch, R. M. Muir, T. Fujita, P. P. Maloney, F. Geiger and M. Streich, *J. Am. Chem. Soc.*, 1963, **85**, 2817.
8. C. Hansch and T. Fujita, *J. Am. Chem. Soc.*, 1964, **86**, 1616.
9. P. S. Portoghese, *J. Med. Chem.*, 1965, **8**, 609.
10. P. A. J. Janssen and N. B. Eddy, *J. Med. Pharm. Chem.*, 1960, **2**, 31.
11. T. Fujita, unpublished results of recalculations.
12. P. B. M. W. M. Timmermans and P. A. van Zwieten, *Arch. Int. Pharmacodyn. Ther.*, 1977, **228**, 237.
13. P. B. M. W. M. Timmermans, A. de Jonge and P. A. van Zwieten, *Quant. Struct.-Act. Relat.*, 1982, **1**, 8.
14. T. R. Fukuto and R. L. Metcalf, *J. Agric. Food Chem.*, 1956, **4**, 930.
15. O. Exner, in 'Advances in Linear Free Energy Relationships', ed. N. B. Chapman and J. Shorter, Plenum Press, London, 1972, p. 1.
16. K. Kamoshita and T. Fujita, unpublished results.
17. T. Fujita, *Anal. Chim. Acta*, 1981, **133**, 667.
18. E. Overton, *Z. Phys. Chem., Stoechiom. Verwandschaftsl.*, 1897, **22**, 189.
19. H. Meyer, *Arch. Exp. Pathol. Pharmakol.*, 1899, **42**, 109.
20. H. Fühner, *Arch. Exp. Pathol. Pharmakol.*, 1904, **51**, 1; 1904, **52**, 69.
21. W. Moore, *J. Agric. Res. (Washington, D.C.)*, 1916, **9**, 371; 1917, **10**, 365.
22. J. Ferguson, *Proc. R. Soc. London, Ser. B*, 1939, **127**, 387.
23. K. H. Meyer and H. Hemmi, *Biochem. Z.*, 1935, **277**, 39.
24. C. Hansch and W. J. Dunn, III, *J. Pharm. Sci.*, 1972, **61**, 1.
25. C. Hansch, D. Kim, A. Leo, E. Novellino, C. Silipo and A. Vittoria, *CRC Crit. Rev. Toxicol.*, 1989, **19**, 185.
26. T. Fujita, J. Iwasa and C. Hansch, *J. Am. Chem. Soc.*, 1964, **86**, 5175.
27. J. Iwasa, T. Fujita and C. Hansch, *J. Med. Chem.*, 1965, **8**, 150.

28. C. Hansch and J. M. Clayton, *J. Pharm. Sci.*, 1973, **62**, 1.
29. O. R. Hansen, *Acta. Chem. Scand.*, 1962, **16**, 1593.
30. E. Fischer, *Z. Physiol. Chem.*, 1898, **26**, 60.
31. A. J. Clark, 'The Mode of Action of Drugs on Cells', Williams and Wilkins, Baltimore, MD, 1937.
32. C. Hansch, *Acc. Chem. Res.*, 1969, **2**, 232.
33. C. Hansch, *Adv. Chem. Ser.*, 1972, **114**, 20.
34. A. Leo, C. Hansch and D. Elkins, *Chem. Rev.*, 1971, **71**, 525.
35. T. Fujita, *Prog. Phys. Org. Chem.*, 1983, **14**, 75.
36. E. Kutter and C. Hansch, *J. Med. Chem.*, 1969, **12**, 647.
37. J. A. MacPhee, A. Panaye and J.-E. Dubois, *Tetrahedron*, 1978, **34**, 3553; *J. Org. Chem.*, 1980, **45**, 1164.
38. C. K. Hancock, E. A. Meyers and B. J. Yager, *J. Am. Chem. Soc.*, 1961, **83**, 4211.
39. M. Charton, *Top. Curr. Chem.*, 1983, **114**, 57.
40. A. Verloop, W. Hoogenstraaten and J. Tipker, in 'Drug Design', ed. E. J. Ariëns, Academic Press, New York, 1976, vol. 7, p. 165.
41. A. Bondi, *J. Phys. Chem.*, 1964, **68**, 441.
42. C. Hansch and A. J. Leo, 'Substituent Constants for Correlation Analysis in Chemistry and Biology', Wiley, New York, 1979, p. 44.
43. T. Fujita and H. Iwamura, *Top. Curr. Chem.*, 1983, **114**, 119.
44. M. Charton, *Prog. Phys. Org. Chem.*, 1981, **13**, 119.
45. C. G. Swain and E. C. Lupton, *J. Am, Chem. Soc.*, 1968, **90**, 4328.
46. C. Hansch, A. Leo, S. H. Unger, K. H. Kim, D. Nikaitani and E. J. Lien, *J. Med. Chem.*, 1973, **16**, 1207.
47. C. Hansch, in 'Structure–Activity Relationships', ed. C. J. Cavallito, Pergamon Press, Oxford, 1973, vol. 1, p. 150.
48. C. A. Bennett and N. L. Franklin, 'Statistical Analysis in Chemistry and the Chemical Industry', Wiley, New York, 1963.
49. C. Hansch, E. W. Deutsch and R. N. Smith, *J. Am. Chem. Soc.*, 1965, **87**, 2738.
50. T. Fujita and T. Nishioka, *Prog. Phys. Org. Chem.*, 1976, **12**, 49.
51. T. Fujita, T. Nishioka and M. Nakajima, *J. Med. Chem.*, 1977, **20**, 1071.
52. M. Yoshimoto and C. Hansch, *J. Med. Chem.*, 1976, **19**, 71.
53. R. Collander, *Acta Chem. Scand.*, 1951, **5**, 774.
54. A. Leo and C. Hansch, *J. Org. Chem.*, 1971, **36**, 1539.
55. C. Hansch, J. E. Quinlan and G. L. Lawrence, *J. Org. Chem.*, 1968, **33**, 347.
56. F. Helmer, K. Kiehs and C. Hansch, *Biochemistry*, 1968, **7**, 2858.
57. K. Kiehs, C. Hansch and L. Moore, *Biochemistry*, 1966, **5**, 2602.
58. C. Hansch, *Farmaco, Ed. Sci.*, 1968, **23**, 293.
59. J. T. Penniston, L. Beckett, D. L. Bentley and C. Hansch, *Mol. Pharmacol.*, 1969, **5**, 331.
60. J. W. McFarland, *J. Med. Chem.*, 1970, **13**, 1192.
61. H. Kubinyi, *Arzneim.-Forsch.*, 1976, **26**, 1991.
62. H. Kubinyi, *J. Med. Chem.*, 1977, **20**, 625.
63. H. Kubinyi, *Arzneim.-Forsch.*, 1979, **29**, 1067.
64. T. Higuchi and S. S. Davis, *J. Pharm. Sci.*, 1970, **59**, 1376.
65. R. M. Hyde, *J. Med. Chem.*, 1975, **18**, 231.
66. H. Kubinyi and O. H. Kehrhahn, *Arzneim.-Forsch.*, 1978, **28**, 598.
67. E. J. Lien, in 'Drug Design', ed. E. J. Ariëns, Academic Press, New York, 1975, vol. 5, p. 81.
68. H. Kubinyi, *Farmaco, Ed. Sci.*, 1979, **34**, 248.
69. R. N. Smith, C. Hansch and M. Ames, *J. Pharm. Sci.*, 1975, **64**, 599.
70. P. R. Wells, 'Linear Free Energy Relationships', Academic Press, London, 1968, p. 25.
71. D. J. McLennan, *Tetrahedron*, 1978, **34**, 2331.
72. H. Tanida and T. Tsushima, *J. Am. Chem. Soc.*, 1970, **92**, 3397.
73. J. A. Mazrimas, P. -S. Song, L. L. Ingraham and R. D. Draper, *Arch. Biochem. Biophys.*, 1963, **100**, 409.
74. T. Fujita, *J. Med. Chem.*, 1973, **16**, 923.
75. S. H. Unger and C. Hansch, *J. Med. Chem.*, 1973, **16**, 745.
76. C. Hansch and E. J. Lien, *Biochem. Pharmacol.*, 1968, **17**, 709.
77. J. D. P. Graham and M. A. Karrar, *J. Med. Chem.*, 1963, **6**, 103.
78. A. Cammarata, *J. Med. Chem.*, 1972, **15**, 573.
79. T. Fujita, in 'QSAR and Strategies in the Design of Bioactive Compounds', ed. J. K. Seydel, VCH Verlagsgesellschaft, Weinheim, 1985, p. 207.
80. J. G. Topliss and R. J. Costello, *J. Med. Chem.*, 1972, **15**, 1068.
81. I. Moriguchi, K. Kamatsu and Y. Matsushita, *J. Med. Chem.*, 1980, **23**, 20.
82. Y. Takahashi, Y. Miyashita, H. Abe and S. Sasaki, *Bunseki Kagaku*, 1984, **33**, E487.
83. G. L. Ritter, S. R. Lowry, C. L. Wilkins and T. L. Isenhour, *Anal. Chem.*, 1975, **47**, 1951.
84. K. Mitani, T. Yoshida, T. Suzuki, E. Koshinaka, H. Kato, Y. Ito and T. Fujita, *Chem. Pharm. Bull.*, 1988, **36**, 776.
85. K. Mitani, T. Yoshida, S. Sakurai, K. Morikawa, Y. Iwanaga, E. Koshinaka, H. Kato and Y. Ito, *Chem. Pharm. Bull.*, 1988, **36**, 373.
86. A. Verloop, in 'Pesticide Chemistry, Human Welfare and Environment', ed. J. Miyamoto and P. C. Kearney, Pergamon Press, Oxford, 1983, p. 339.
87. R. Mannhold, R. Bayer, M. Ronsdorf and L. Martens, *Arzneim.-Forsch.*, 1987, **37**, 419.
88. M. Ueda, *Jpn. Pestic. Inf.*, 1975, **23**, 23.
89. H.-Z. Yang, Y.-J. Zhang, L.-X. Wang, H.-F. Tan, M.-R. Cheng, X.-D. Xing, R.-Y. Chen and T. Fujita, *Pestic. Biochem. Physiol.*, 1986, **26**, 275.
90. A. Fujinami, T. Satomi, A. Mine and T. Fujita, *Pestic. Biochem. Physiol.*, 1976, **6**, 287.
91. M. Eto, 'Organophosphorus Pesticides: Organic and Biological Chemistry', CRC Press, Cleveland, OH, 1974, p. 67.
92. (a) P. H. Bell and R. O. Roblin, *J. Am. Chem. Soc.*, 1942, **64**, 2905; (b) C. Silipo and A. Vittoria, *Farmaco, Ed. Sci.*, 1979, **34**, 858.
93. A. H. Brueckner, *Yale J. Biol. Med.*, 1943, **15**, 813.

94. P. B. Cowles, *Yale J. Biol. Med.*, 1942, **14**, 599.
95. J. K. Seydel, *J. Med. Chem.*, 1971, **14**, 724.
96. G. H. Miller, P. H. Doukas and J. K. Seydel, *J. Med. Chem.*, 1972, **15**, 700.
97. E. A. Coats, H.-P. Cordes, V. M. Kulkarni, M. Richter, K.-J. Schaper, M. Wiese and J. K. Seydel, *Quant. Struct.-Act. Relat.*, 1985, **4**, 99.
98. R. L. Lopez de Compadre, R. A. Pearlstein, A. J. Hopfinger and J. K. Seydel, *J. Med. Chem.*, 1987, **30**, 900.
99. T. Fujita, *J. Med. Chem.*, 1966, **9**, 797.
100. T. Fujita and C. Hansch, *J. Med. Chem.*, 1967, **10**, 991.
101. R. A. Scherrer and S. M. Howard, *J. Med. Chem.*, 1977, **20**, 53.
102. Y. C. Martin and J. J. Hackbarth, *J. Med. Chem.*, 1976, **19**, 1033.
103. Y. C. Martin and C. Hansch, *J. Med. Chem.*, 1971, **14**, 777.
104. T. Fujita, *J. Med. Chem.*, 1972, **15**, 1049.
105. T. Fujita, *Adv. Chem. Ser.*, 1973, **114**, 80.
106. H. Miyoshi, T. Nishioka and T. Fujita, *Biochim. Biophys. Acta*, 1987, **891**, 194, 293.
107. Y. C. Martin, in 'Physical Chemical Properties of Drugs', ed. S. H. Yalkowsky, A. A. Sinkula and S. C. Valvani, Dekker, New York, 1980, p. 49.
108. D. Pressman, S. M. Swingle, A. L. Grossberg and L. Pauling, *J. Am. Chem. Soc.*, 1944, **66**, 1731.
109. L. Pauling and D. Pressman, *J. Am. Chem. Soc.*, 1945, **67**, 1003.
110. A. Nisonoff and D. Pressman, *J. Am. Chem. Soc.*, 1957, **79**, 5565.
111. E. Kutter and C. Hansch, *Arch. Biochem. Biophys.*, 1969, **135**, 126.
112. C. Hansch and P. Moser, *Immunochemistry*, 1978, **15**, 535.
113. F. Franks, in 'Water: A Comprehensive Treatise', ed. F. Franks, Plenum Press, New York, 1974, vol. 4, chap. 1.
114. C. Hansch and D. F. Calef, *J. Org. Chem.*, 1976, **41**, 1240.
115. A. H. Khan and W. C. J. Ross, *Chem.-Biol. Interact.*, 1969, **1**, 27.
116. C. Hansch, A. Leo, C. Schmidt, P. Y. C. Jow and J. A. Montgomery, *J. Med. Chem.*, 1980, **23**, 1095.
117. T. Suami, T. Machinami and T. Hisamatsu, *J. Med. Chem.*, 1979, **22**, 247.
118. A. Panthananickal, C. Hansch, A. Leo and F. R. Quinn, *J. Med. Chem.*, 1978, **21**, 16.
119. T. J. Bardos, N. Datta-Gupta, P. Hebborn and D. J. Triggle, *J. Med. Chem.*, 1965, **8**, 167.
120. A. Panthananickal, C. Hansch and A. Leo, *J. Med. Chem.*, 1979, **22**, 1267.
121. W. C. J. Ross, *Ann. N.Y. Acad. Sci.*, 1958, **68**, 669.
122. G. J. Hatheway, C. Hansch, K. H. Kim, S. R. Milstein, C. L. Schmidt, R. N. Smith and F. R. Quinn, *J. Med. Chem.*, 1978, **21**, 563.
123. G. F. Kolar and R. Preussmann, *Z. Naturforsch., B*, 1971, **26**, 950.
124. C. Hansch, G. J. Hatheway, F. R. Quinn and N. Greenberg, *J. Med. Chem.*, 1978, **21**, 574.
125. B. H. Venger, C. Hansch, G. J. Hatheway and Y. U. Amrein, *J. Med. Chem.*, 1979, **22**, 473.
126. A. Leo, A. Panthananickal, C. Hansch, J. Theiss, M. Shimkin and A. W. Andrews, *J. Med. Chem.*, 1981, **24**, 859.
127. W. A. Denny, B. F. Cain, G. J. Atwell, C. Hansch, A. Panthananickal and A. Leo, *J. Med. Chem.*, 1982, **25**, 276.
128. M. Yoshimoto, H. Miyazawa, H. Nakao, K. Shinkai and M. Arakawa, *J. Med. Chem.*, 1979, **22**, 491.
129. J. M. Blaney, C. Hansch, C. Silipo and A. Vittoria, *Chem. Rev.*, 1984, **84**, 333.
130. C. D. Selassie, C. D. Strong, C. Hansch, T. J. Delcamp, J. H. Freisheim and T. A. Khwaja, *Cancer Res.*, 1986, **46**, 744.
131. C. Hansch, B. A. Hathaway, Z.-R. Guo, C. D. Selassie, S. W. Dietrich, J. M. Blaney, R. Langridge, K. W. Volz and B. T. Kaufman, *J. Med. Chem.*, 1984, **27**, 129.
132. E. A. Coats, C. S. Genther, S. W. Dietrich, Z.-R. Guo and C. Hansch, *J. Med. Chem.*, 1981, **24**, 1422.
133. D. Kessel, T. C. Hall and D. Roberts, *Cancer Res.*, 1968, **28**, 564.
134. G. B. Henderson, E. M. Zevely and F. M. Huennekens, *J. Bacteriol.*, 1979, **139**, 552.
135. J. R. Bertino and C. Lindquist, in 'Proceedings of the 8th Symposium of the Princess Takamatsu Cancer Research Fund—Advances in Cancer Chemotherapy, ed. S. K. Carter, A. Coldin, K. Kuretani, G. Mathé, Y. Sakurai, S. Tsukagoshi and H. Umezawa, Japan Scientific Societies Press, Tokyo, 1978, p. 155.
136. E. A. Coats, C. S. Genther, C. D. Selassie, C. D. Strong and C. Hansch, *J. Med. Chem.*, 1985, **28**, 1910.
137. R. L. Li, S. W. Dietrich and C. Hansch, *J. Med. Chem.*, 1981, **24**, 538.
138. R. L. Li, C. Hansch and B. T. Kaufman, *J. Med. Chem.*, 1982, **25**, 435.
139. R. L. Li, C. Hansch, D. Mattews, J. M. Blaney, R. Langridge, T. J. Delcamp, S. S. Susten and J. H. Freisheim, *Quant. Struct.-Act. Relat.*, 1982, **1**, 1.
140. J. Y. Fukunaga, C. Hansch and E. E. Steller, *J. Med. Chem.*, 1976, **19**, 605.
141. C. Hansch, J. Y. Fukunaga, P. Y. C. Jow and J. B. Hynes, *J. Med. Chem.*, 1977, **20**, 96.
142. A. J. Hopfinger, *J. Am. Chem. Soc.*, 1980, **102**, 7196.
143. A. J. Hopfinger, *Arch. Biochem. Biophys.*, 1981, **206**, 153.
144. A. J. Hopfinger, *J. Med. Chem.*, 1981, **24**, 818.
145. C. Battershell, D. Malhotra and A. J. Hopfinger, *J. Med. Chem.*, 1981, **24**, 812.
146. P. Pratesi, L. Villa, V. Ferri, C. de Micheli, E. Grana, C. Silipo and A. Vittoria, in 'Highlights in Receptor Chemistry', ed. C. Melchiorre and M. Giannella, Elsevier, Amsterdam, 1984, p. 225.
147. P. Pratesi, E. Grana, M. G. S. Barbone, M. I. la Rotonda, C. Silipo and A. Vittoria, *Farmaco, Ed. Sci.*, 1986, **41**, 335.
148. M. Yamakawa, K. Ezumi, M. Shiro, H. Nakai, S. Kamata, T. Matsui and N. Haga, *Mol. Pharmacol.*, 1987, **30**, 585.
149. R. N. Smith, C. Hansch, K. H. Kim, B. Omiya, G. Fukumura, C. D. Selassie, P. Y. C. Jow, J. M. Blaney and R. Langridge, *Arch. Biochem. Biophys.*, 1982, **215**, 319.
150. A. Carotti, C. Hansch, M. M. Mueller and J. M. Blaney, *J. Med. Chem.*, 1984, **27**, 1401; 1985, **28**, 261.
151. C. Hansch and T. E. Klein, *Acc. Chem. Res.*, 1986, **19**, 392.
152. C. Hansch and J. M. Blaney, in 'Drug Design: Fact or Fantasy?', ed. G. Jolles and K. R. H. Wooldridge, Academic Press, London, 1984, p. 185.
153. P. N. Craig and C. Hansch, *J. Med. Chem.*, 1973, **16**, 661.
154. K. H. Kim, C. Hansch, J. Y. Fukunaga, E. E. Steller, P. Y. C. Jow, P. N. Craig and J. Page, *J. Med. Chem.*, 1979, **22**, 366.
155. F. R. Quinn, J. S. Driscoll and C. Hansch, *J. Med. Chem.*, 1975, **18**, 332.
156. S. I. Fink, A. Leo, M. Yamakawa, C. Hansch and F. R. Quinn, *Farmaco., Ed. Sci.*, 1980, **35**, 965.

157. Y. C. Martin and K. R. Lynn, *J. Med. Chem.*, 1971, **14**, 1162.
158. E. Kutter and C. Hansch, *J. Med. Chem.*, 1969, **12**, 647.
159. A. F. Harms, W. Hespe, W. Th. Nauta, R. F. Rekker, H. Timmerman and J. de Vries, in 'Drug Design', ed. E. J. Ariëns, Academic Press, New York, 1975, vol. 6, p. 1.
160. M. Yoshimoto, T. Kamioka, T. Miyadera, S. Kobayashi, H. Takagi and R. Tachikawa, *Chem. Pharm. Bull.*, 1977, **25**, 1378.
161. K. Nishimura, K. Hirayama, T. Kobayashi, T. Fujita and G. Holan, *Pestic. Biochem. Physiol.*, 1986, **25**, 153.
162. M. Kiso, T. Fujita, M. Kurihara, M. Uchida, K. Tanaka and M. Nakajima, *Pestic. Biochem. Physiol.*, 1978, **8**, 33.
163. T. Nishioka, T. Fujita, K. Kamoshita and M. Nakajima, *Pestic. Biochem. Physiol.*, 1977, **7**, 107.
164. K. Nishimura, T. Kitahaba, N. Okajima and T. Fujita, *Pestic. Biochem. Physiol.*, 1985, **23**, 314.
165. Y. Nakagawa, T. Sotomatsu, K. Irie, K. Kitahara, H. Iwamura and T. Fujita, *Pestic. Biochem. Physiol.*, 1987, **27**, 143.
166. P. S. Magee, in 'Quantitative Structure–Activity Relationships of Drugs', ed. J. G. Topliss, Academic Press, New York, 1983, p. 393.
167. T. R. Fukuto, in 'Insecticide Biochemistry and Physiology', ed. C. F. Wilkison, Plenum Press, New York, 1976, p. 397.
168. H. Ohta, T. Jikihara, K. Wakabayashi and T. Fujita, *Pestic. Biochem. Physiol.*, 1980, **14**, 153.
169. E. Kakkis, V. C. Palmire, Jr., C. D. Strong, W. Bertsch, C. Hansch and U. Schirmer, *J. Agric. Food Chem.*, 1984, **32**, 133.
170. K. Mitsutake, H. Iwamura, R. Shimizu and T. Fujita, *J. Agric. Food Chem.*, 1986, **34**, 725.
171. I. Takemoto, R. Yoshida, S. Sumida and K. Kamoshita, *Pestic. Biochem. Physiol.*, 1985, **23**, 341.
172. C. Hansch, in 'Progress in Photosynthetic Research', ed. H. Metzner, Institute of Chemical Plant Physiology, University of Tübingen, 1969, vol. 3, p. 1685.
173. C. Takayama and A. Fujinami, *Pestic. Biochem. Physiol.*, 1979, **12**, 163.
174. C. Hansch and E. J. Lien, *J. Med. Chem.*, 1971, **14**, 653.
175. O. Kirino, S. Yamamoto and T. Kato, *Agric. Biol. Chem.*, 1980, **44**, 2149.
176. C. Takayama, O. Kirino, Y. Hisada and A. Fujinami, *Agric. Biol. Chem.*, 1987, **51**, 1547.
177. W. B. Neely, D. R. Branson and G. E. Blau, *Environ. Sci. Technol.*, 1974, **8**, 1113.
178. M. Uchida, S. Funayama and T. Sugimoto, *Nippon Noyaku Gakkaishi*, 1982, **7**, 181.
179. S. Bandiera, T. W. Sawyer, M. A. Campbell, T. Fujita and S. Safe, *Biochem. Pharmacol.*, 1983, **32**, 3803.
180. H. Geyer, P. Sheehan, D. Kotzias, D. Freitag and F. Korte, *Chemosphere*, 1982, **11**, 1121.
181. H. Levitan, *Proc. Natl. Acad. Sci. U.S.A.*, 1977, **74**, 2914.
182. H. Tanii and K. Hashimoto, *Arch. Toxicol.*, 1984, **55**, 47.
183. G. G. Briggs, *J. Agric. Food Chem.*, 1981, **29**, 1050.
184. M. Uchida and T. Kasai, *Nippon Noyaku Gakkaishi*, 1980, **5**, 553.
185. C. Hansch, *J. Med. Chem.*, 1970, **13**, 964.
186. H. Iwamura, *J. Med. Chem.*, 1980, **23**, 308.
187. H. Iwamura, *J. Med. Chem.*, 1981, **24**, 572.
188. R. J. Gardner, *Chem. Senses*, 1980, **5**, 185.
189. M. Asao, H. Iwamura, M. Akamatsu and T. Fujita, *J. Med. Chem.*, 1987, **30**, 1873.
190. C. Hansch, *Drug Metab. Rev.*, 1973, **1**, 1.
191. S. Nakagawa, K. Nishimura, N. Kurihara and T. Fujita, *Pestic. Biochem. Physiol.*, 1985, **24**, 182.
192. C. Hansch, A. Vittoria, C. Silipo and P. Y. C. Jow, *J. Med. Chem.*, 1975, **18**, 546.
193. A. C. Moffat and A. T. Sullivan, *J., Forensic Sci. Soc.*, 1981, **21**, 239.
194. L. A. King and A. C. Moffat, *Med. Sci. Law*, 1983, **23**, 193.
195. E. J. Lien, J. F. R. de Miranda and E. J. Ariëns, *Mol. Pharmacol.*, 1976, **12**, 598.
196. M. Yoshimoto and C. Hansch, *J. Org. Chem.*, 1976, **41**, 2269.
197. C. Hansch, in 'Biological Activity and Chemical Structure', ed. J. A. K. Buisman, Elsevier, Amsterdam, 1977, p. 47.
198. L. M. Lichtenstein and S. Margolis, *Science (Washington, D.C.)*, 1968, **161**, 902.
199. K. R. H. Wooldridge, in 'Drugs Affecting the Respiratory System', ed. D. L. Temple, American Chemical Society, Washington, DC, 1980, p. 117.
200. R. E. Ford, P. Knowles, E. Lunt, S. M. Marshall, A. J. Penrose, C. A. Ramsden, A. J. H. Summers, J. L. Walker and D. E. Wright, *J. Med. Chem.*, 1986, **29**, 538.
201. R. D. Cramer, III, K. M. Snader, C. R. Willis, L. W. Chakrin, J. Thomas and B. M. Sutton, *J. Med. Chem.*, 1979, **22**, 714.
202. K. M. Snader, L. W. Chakrin, R. D. Cramer, III, Y. M. Gelernt, C. K. Miao, D. H. Shah, J. W. Venslavsky, C. R. Willis and B. M. Sutton, *J. Med. Chem.*, 1979, **22**, 706.
203. J. G. Topliss, *J. Med. Chem.*, 1972, **15**, 1006.
204. I. Moriguchi, Y. Kanada and K. Komatsu, *Chem. Pharm. Bull.*, 1976, **24**, 1799.
205. H. Watanabe, S. Miyamoto, M. Yoshimoto, T. Kamioka, I. Nakayama, T. Kobayashi and T. Honda, *Chem. Pharm. Bull.*, 1987, **35**, 1452.
206. T. Kamioka, I. Nakayama, T. Karube, T. Yokoyama, N. Iwata, T. Honda and T. Kobayashi, *Jpn. J. Pharmacol.*, 1984, **36** (suppl.), 170.
207. S. Naito, S. Osumi, K. Sekishiro and M. Hirose, *Chem. Pharm. Bull.*, 1972, **20**, 682.
208. N. Toda, H. Usui, S. Osumi, M. Kanda and K. Kitao, *Arch. Int. Pharmacodyn. Ther.*, 1982, **260**, 230.
209. H. Ohtaka, T. Kanazawa, K. Ito and G. Tsukamoto, *Chem. Pharm. Bull.*, 1987, **35**, 3270.
210. H. Ohtaka and G. Tsukamoto, *Chem. Pharm. Bull.*, 1987, **35**, 4117.
211. O. Kirino, C. Takayama, H. Matsumoto and A. Mine, *Nippon Noyaku Gakkaishi*, 1983, **8**, 301.
212. O. Kirino, H. Ohshita, T. Oishi and T. Kato, *Agric. Biol. Chem.*, 1980, **44**, 31.
213. A. Fujinami, T. Satomi, A. Mine and T. Fujita, *Pestic. Biochem. Physiol.*, 1976, **6**, 287.
214. O. Kirino, K. Furuzawa, C. Takayama, H. Matsumoto and A. Mine, *Nippon Noyaku Gakkaishi*, 1983, **8**, 309.
215. O. Kirino, K. Furuzawa, H. Matsumoto, N. Hino and A. Mine, *Agric. Biol. Chem.*, 1981, **45**, 2669.
216. G. Y. Lesher, E. J. Froelich, M. D. Gruett, J. H. Bailey and R. P. Brundage, *J. Med. Pharm. Chem.*, 1962, **5**, 1063.
217. R. Albrecht, *Prog. Drug Res.*, 1977, **21**, 9.
218. H. Koga, A. Itoh, S. Murayama, S. Suzue and T. Irikura, *J. Med. Chem.*, 1980, **23**, 1358.
219. H. Koga, in 'Structure–Activity Relationships—Quantitative Approaches: Applications to Drug Design and Mode-of-Action Studies', ed. T. Fujita, Nankodo, Tokyo, 1982, p. 177.

220. S. Minami, in 'Drug Design', ed. S. Yamabe, Asakura Shoten, Tokyo, 1975, p. 145.
221. I. Hayakawa, T. Hiramitsu and Y. Tanaka, *Chem. Pharm. Bull.*, 1984, **32**, 4907.
222. D. T. W. Chu, P. B. Fernandes, A. K. Claiborne, E. Pihuleac, C. W. Nordeen, R. E. Maleczka and A. G. Pernet, *J. Med. Chem.*, 1985, **28**, 1558.
223. H. Narita, Y. Konishi, J. Nitta, H. Nagaki, Y. Kobayashi, Y. Watanabe, S. Minami and I. Saikawa, *Yakugaku Zasshi*, 1986, **106**, 795.
224. D. Cambridge, M. J. Davay and R. Massingham, *Med. J. Aust.*, 1977, **2** (Suppl. 1), 2.
225. T. Sekiya, H. Hiranuma, T. Kanayama and S. Hata, *Eur. J. Med. Chem.—Chim. Ther.*, 1980, **15**, 317.
226. T. Sekiya, H. Hiranuma, S. Hata, S. Mizogami, M. Hanazuka and S. Yamada, *J. Med. Chem.*, 1983, **26**, 411.
227. T. Sekiya, S. Imada, S. Hata and S. Yamada, *Chem. Pharm. Bull.*, 1983, **31**, 2779.
228. J. K. Baker, D. O. Rauls and R. F. Borne, *J. Med. Chem.*, 1979, **22**, 1301.
229. M. Charton, in 'Correlation Analysis in Chemistry', ed. N. B. Chapman and J. Shorter, Plenum Press, London, 1978, p. 175.
230. A. Tanaka, K. Yakushiji and S. Yoshina, *J. Heterocycl. Chem.*, 1977, **14**, 975.
231. Y. Nakashima, Y. Kawashima, M. Sato, S. Okuyama, F. Amanuma, K. Sota and T. Kameyama, *Chem. Pharm. Bull.*, 1985, **33**, 5250.
232. Y. Nakashima, Y. Kawashima, F. Amanuma, K. Sota, A. Tanaka and T. Kameyama, *Chem. Pharm. Bull.*, 1984, **32**, 4271.
233. Y. Kawashima, F. Amanuma, M. Sato, S. Okuyama, Y. Nakashima, K. Sota and I. Moriguchi, *J. Med. Chem.*, 1986, **29**, 2284.
234. C. G. Swain, S. H. Unger, N. R. Rosenquist and M. S. Swain, *J. Am. Chem. Soc.*, 1983, **105**, 492.
235. A. Dornow, H. Menzel and P. Marx, *Chem. Ber.*, 1964, **97**, 2173.
236. L. Eue, *Mededelingen Faculteit Landbouwwetenschappen Rijksuniversiteit Gent*, 1971, **36**, 1233.
237. W. Draber, K. H. Büchel, K. Dickore, A. Trebst and E. Pistorius, in 'Progress in Photosynthesis Research', ed. H. Metzner, Institute of Chemical Plant Physiology, University of Tübingen, 1969, vol. 3, p. 1789.
238. W. Draber, K. H. Büchel, H. Timmler and A. Trebst, *ACS Symp. Ser.*, 1974, **2**, 100.
239. Y. C. Martin, *J. Med. Chem.*, 1981, **24**, 229.
240. S. H. Unger, in 'Drug Design', ed. E. J. Ariëns, Academic Press, New York, 1980, vol. 9, p. 47.
241. C. Hansch, *Drug Dev. Res.*, 1981, **1**, 267.
242. A. J. Hopfinger, *J. Med. Chem.*, 1985, **28**, 1133.
243. P. N. Craig, *J. Med. Chem.*, 1971, **14**, 680, 1251.
244. C. Hansch, S. H. Unger and A. B. Forsythe, *J. Med. Chem.*, 1973, **16**, 1217.
245. S. Hellberg, M. Sjöström, B. Skagerberg, C. Wikström and S. Wold, *Acta Pharm. Jugosl.*, 1987, **37**, 53.
246. W. P. Purcell, G. E. Bass and J. M. Clayton, 'Strategy of Drug Design—A Molecular Guide to Biological Activity', Wiley, New York, 1973.
247. Y. C. Martin, 'Quantitative Drug Design', Dekker, New York, 1978.
248. J. G. Topliss (ed.), 'Quantitative Structure–Activity Relationships of Drugs', Academic Press, New York, 1983.
249. R. F. Rekker, 'The Hydrophobic Fragmental Constants', Elsevier, Amsterdam, 1977.
250. J. K. Seydel and K.-J. Schaper, 'Chemische Struktur und Biologische Aktivität von Wirkstoffen: Methoden der Quantitativen Struktur–Wirkung-Analyse', Verlag Chemie, Weinheim, 1979.
251. R. Franke, 'Theoretical Drug Design Methods', Elsevier, Amsterdam, 1984.
252. E. J. Lien, 'SAR, Side Effects and Drug Design', Dekker, New York, 1987.
253. T. Fujita (ed.), 'Structure–Activity Relationships—Quantitative Approaches: Significance in Drug Design and Mode-of-Action Studies', Nankodo, Tokyo, 1979.
254. F. Darvas (ed.), 'Chemical Structure–Biological Activity Relationships, Quantitative Approaches', Pergamon Press, Oxford, 1980.
255. J. K. Seydel (ed.), 'QSAR and Strategies in the Design of Bioactive Compounds', VCH Verlagsgesellschaft, Weinheim, 1985.
256. J. C. Dearden (ed.), 'Quantitative Approaches to Drug Design', Elsevier, Amsterdam, 1983.
257. M. Tichy (ed.), 'QSAR in Toxicology and Xenobiochemistry', Elsevier, Amsterdam, 1985.
258. D. Hadzi and B. Jerman-Blažič (eds.), 'QSAR in Drug Design and Toxicology', Elsevier, Amsterdam, 1987.
259. M. Kuchar (ed.), 'QSAR in Design of Bioactive Compounds', Prous, Barcelona, 1984.
260. R. L. Lopez de Compadre, R. A. Pearlstein, A. J. Hopfinger and J. K. Seydel, *J. Med. Chem.*, 1988, **31**, 2315.

21.2

The Design of Test Series and the Significance of QSAR Relationships

MICHAEL A. PLEISS and STEFAN H. UNGER

Syntex Research, Palo Alto, CA, USA

21.2.1 INTRODUCTION

The design of a new chemical entity targeted at a specific biological endpoint is a challenging and difficult task. Although the rewards can be many for the successful candidate, the costs involved can be staggering. Recent estimates in the United States have placed the costs from conception to final marketing of a new pharmaceutical at nearly $100 million.[1] Since most academic and industrial research programs work under both the restrictions of a finite budget and even more demanding time constraints, it becomes imperative that any new program utilize all available resources in the most efficient manner.

A typical scenario from an industrial setting sees the chemist paying a visit to the stockroom at the initiation of a new project to see what compounds the company has on hand in order to rapidly synthesize new congeners of an existing lead. More recently, this practice has been supplemented by a computer-aided search of the company's database in order to find existing in-house analogues that still exist in sufficient quantities and purity for screening. Many companies also perform blind screening of a large number of compounds, such as all compounds from Aldrich. It should be noted that blind screening is usually aimed at developing new leads, not optimization of an existing lead.

A common theme stemming from the above two examples is the *random* selection of compounds with no regard to the information to be gained and with little concern given to the overall experimental design of the test series. It is the intent of this chapter to provide the reader with a critical overview of the techniques utilized in the design and optimization of a test series, as well as a better understanding of why these techniques are so important for the successful outcome of any chemical optimization program.

21.2.2 THE DESIGN OF A TEST SERIES

21.2.2.1 The Need to Design a Test Series

'A day in the library will save you six weeks in the laboratory' is a common saying uttered more than once by most research advisors to their skeptical graduate students. The student begins to appreciate this advice only after he/she has spent the aforementioned six weeks in the laboratory 'rediscovering-the-wheel'. Unfortunately, this same principle holds in terms of the efforts exerted in the design of a test series. An appropriate corollary to this adage might be 'a day spent designing the compounds to be made will provide a gain in overall efficiency in terms of minimizing research dollars invested, maximizing information gained, as well as saving the time of many valuable research personnel'. The concept of good experimental design stated here is a lot older than QSAR itself. In fact, the notion that empirical models are heavily dependent on the experimental design is one that has been around for a long time in the statistical literature.[2] The need for sound experimental design should be as important to the investigator as is the route of synthesis and/or the best pharmacological method and screen. Unfortunately, it is generally brought into play *after* a project has been initiated and results have been generated, usually by one of the means described in the introduction.

Figure 1 depicts a common problem faced by the drug designer.[3–5] This graph clearly shows that a valid model cannot be formulated from the compounds prepared and tested. In fact, all the compounds convey essentially the same information within the limits of the assay. The researcher knows as much at this stage of the program as he/she did after the first compound was prepared and tested. Although all the compounds shown in Figure 1 can be considered 'active' for this particular assay, more real information would have been gained by having some 'inactive' or, especially, 'less active' compounds. The desire for 'inactive' test results and actually making compounds designed to be 'less active' might appear counter-intuitive to many medicinal chemists. In this case, however, intuition is wrong. Figure 2 demonstrates the impact that experimental design can have on the development of the model. This graph shows the results of a study utilizing the same test system as Figure 1, but with a carefully selected set of analogues. Obvious from this study is the *fewer* number of compounds employed (compared to Figure 1) related to the greater information gained by this set. In the example shown in Figure 1, the investigator is unable to fit any valid model to the data. The major difference between Figure 2 and Figure 1 is the fact that each point contributes to the empirical model developed. In addition, the data show enough spread in the variance to allow a model to be developed from the data. Table 1 provides a succinct overview of the components of a good experimental design and some of the related terms used to describe these components.

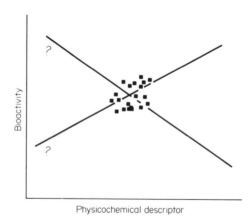

Figure 1 This simple 2-D plot demonstrates the problem when no unique variation among the compounds exists, *i.e.* all compounds convey the same information in this particular assay. The result is a non-valid, derivable statistical model. Two lines are drawn through the cluster of points; obviously, any other line would be equivalent because of the poor experimental design

Since most of the programs of current interest cannot be easily explained by the simple plots shown in Figures 1 and 2 (*i.e.* with a single physicochemical variable), there exist even more important, but less obvious, reasons for starting with a good, experimentally designed set of compounds. A well-designed test series must not only possess a normal distribution with maximum

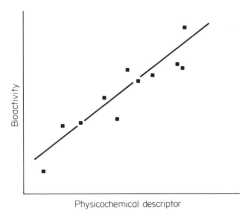

Figure 2 This 2-D plot shows the effect of good experimental design. Each compound in this case provides a unique amount of variation to the overall model. Notice that some of the compounds are not as active in this particular assay as the compounds shown in Figure 1; however, a valid statistical model is derivable from this data using *fewer* points than in Figure 1

Table 1 Components of a Good Experimental Design

Component	Related terms	Reason
Maximum variance	Range	Increases the range of validity of the information
Minimum covariance	Colinearity	Variables should be (ideally) totally independent of each other:
	Multicolinearity	both one-to-one (colinearity) and one-to-many (multicolinearity)
Normal distribution	Gaussian distribution	Most common statistical models will be applicable, *e.g.* multiple linear regression, unless they fail other tests
Number of derivatives		Must be sufficient to investigate all relevant variables
Synthetic feasibility		Need to convince synthetic chemist to actually make series

variance for one variable, but for all relevant variables *simultaneously*. Furthermore, each variable must vary independently (be orthogonal) to all others. Clearly, if substituents are chosen from the linear alkyl series over a large enough range in size, there will be a large variance. However, both lipophilicity and bulk will vary in a parallel manner (the two factors are 'colinear') and one cannot separate the effects of bulk and lipophilicity with *only* this choice of substituents.

Another problem is that the investigator usually hopes that the basic assumptions inherent in a particular statistical test are met by the data in his/her study. A parametric statistical test is most often selected for the analysis of the data. This class of statistics is based on the fact that the data follow a normal distribution, an assumption which can be easily violated with the type of data encountered in QSAR studies.[6] When this basic premise is assumed or ignored, severe problems can be encountered in any statistical model(s) developed, especially if the model is to be utilized in a predictive manner. Numerous univariate screens, as well as transformation methods have been described for normalizing distributions.[6] If these methods fail, then one is faced with using non-parametric or distribution free techniques to analyze the data. Rank transform regression, a non-parametric counterpart to regression analysis, has been applied successfully to QSAR.[6] Rank transform regression is also applicable in cases where no *quantitative* dependent variable can be determined.[6] This is a very common occurrence in the development of new herbicides and pesticides, where qualitative analyses provide the only data an analyst has to work with. Thus, non-parametric statistics opens up areas where many investigators had felt experimental and test series design did not apply, since only qualitative data could be generated. It should be pointed out that non-parametric techniques such as rank transform regression are also valid when applied to parametric data (*i.e.* data that follow a normal distribution).

In general, it is very hard to convince someone not to make a compound, especially if the overall synthesis is straightforward and the starting materials are readily available. On the other hand, the available physicochemical and *de novo* variables utilized in the development of the statistical model must be validated before the model can be used for predicting new analogues or aiding in the explanation of a mechanism. The actual statistical techniques used for this will be discussed in Section 21.2.3.1. However, non-statistical validation can be conducted by the preparation of bioisosteric analogues around the compound selected for advanced testing. This is a good way to test the current model—if the additional substituents selected are truly bioisosteric (*e.g.* the same cluster from a cluster analysis[7]), then similar activity should be displayed by the new moieties when compared to compounds already in the existing model. If the new substituents display widely differing bioactivity, then this is an indication that more ominous factors like pharmacokinetics and pharmacogenetics may be present and not accounted for in the present model. If this is the case then the existing model should be reevaluated.

There are some additional general cautions and remarks regarding the use of experimental design techniques which should be made at this point.[8] In practice, only a very small number of biological assays are sufficiently rapid or accurate enough to allow one to postpone synthetic target decisions until all previous results are in. Chemical intermediates must be ordered and programs must progress. Some chemists may be unwilling to follow statistically derived schemes without question, especially if: (a) some of the initial compounds are quite active; (b) synthesis of some of the more unusual analogues proves difficult; or (c) if intermediates are not delivered on time. Also, biologists often wait for groups of related compounds before testing in order to minimize interassay variation. Most of the literature examples are known already to have solutions, while in practice the QSAR have yet to be discovered and are often found to be complex. If leads turn out to be good ones, there is usually a large synthetic effort to both exploit the lead and protect any patent positions. Even if the original lead should fail, frequently other activities can be found and different compounds in the series might be optimal in these new bioassays. Structural diversity is therefore also useful in the longer run.

21.2.2.2 Selection of Compounds and Variables

21.2.2.2.1 *The compound as a probe*

It is useful to think of each compound synthesized as a probe which will report back to the investigator certain information about the 'environment' the probe detects. Unfortunately, the report is often in the form of a single number. We think of the number as being composed of several linearly additive components (contributions to the activity from steric, electronic, transport effects, *etc.*). There is no way to understand this single report in isolation, we have to see it in relation to many others. This is the fallacy of structure–activity relationships (compared to quantitative structure–activity relationships). With the 'one factor at a time' approach,[9] the factor you are thinking about at the time is usually thought to be responsible for the observed change. We have to use multivariate statistical models to decompose the total information of the set of compounds into its unique component parts.

Since each compound is a probe, each compound should be selected in such a manner that it conveys maximum information to the regression (*e.g.* Figure 2). Otherwise, we are wasting time and money and not gaining any additional information to show for the increased synthetic effort (*e.g.* Figure 1). The probes should be selected in such a manner as to sample the universe of physicochemical properties that is relevant to the biological system at hand. The following sections will review the methods available for selecting the probes in relation to the biological system(s) being studied, as well as discuss the advantages and disadvantages of each technique.

21.2.2.2.2 *Nonstatistical methods for selecting compounds*

(i) *Intuition*, etc.

Many methods have been developed over the years for compound selection. These range in ease of use and complexity, but most people tend to follow the path of least resistance. In terms of experimental design and test series selection this generally translates into selecting the most synthetically feasible analogues to make. Chemical intuition is utilized in this case because it is felt that the greater the degree of resemblance (*i.e.* substructural features in common) between two compounds, the more chance of obtaining an active compound. This approach might lead to compounds with the desired activity, but, without a statistical model, one never knows the maximum possible activity available from the series. This simplistic approach generally results, in the long run, in more work for the chemist, since more compounds will be necessary to focus in on the target and to generate a sufficient amount of unique variation (*cf.* Figure 1), *i.e.* information. Furthermore, removing undesirable side-effects is difficult without parallel statistical models. Attempting to optimize a complicated pharmacological profile by intuition is virtually hopeless.

Martin[10] has reported that experience at Abbott has shown that adding experimental design strategies to the intuition of experienced medicinal chemists *decreases* the average number of analogues required to investigate one physical property from 11 to 4, or a 2.75-fold increase in the information gained per compound synthesized.

(ii) Craig plots

Over the years, various schemes have been developed to increase the efficiency and effectiveness of series optimization on the one hand, while still providing the chemist with a simple to use (*i.e.* non-computerized) procedure for compound selection on the other. The problem of interdependence among the physicochemical variables, or colinearity, was one of the earliest problems to be noted as a way of increasing the efficiency of a test series design. This problem was first addressed by Craig,[11] who introduced the concept of 2-D plots of variables as a way for the investigator to easily select substituents from the four quadrants of the 2-D plot. This would increase the chemist's chances of dealing with compounds that provide unique information, and thus provide a more efficient overall experimental design.

Craig and Hansch[12] have demonstrated the usefulness of this technique utilizing a large series of phenanthrene aminomethanols designed as potential antimalarials. In the initial series of 60 compounds a plot of π *versus* σ values (*i.e.* Craig plot) for substituents on the aromatic ring resulted in nearly a straight line. Several substituents from other quadrants of a π *versus* σ plot were suggested and made, and the high degree of colinearity was broken. This study eventually produced one of the largest regression equations known in the QSAR field.[13]

Although this method can easily be extended to 3-D plots without the aid of computers,[14] the problem of new compound design by rational means encompasses more complexity than a simple 2-D or 3-D plot can portray. In other words, all variables under consideration for a study must be examined as a unit, and not two or three at a time.

One of the more important benefits of examining all variables as a unit is an appreciation for the omnipresent problem of colinearity. The Craig approach[11] was the foundation for many of the more advanced techniques to be discussed in the following sections.

(iii) Topliss operational schemes

In 1972, Topliss proposed a simple operational scheme for analogue selection[15,16] that allowed the concurrent study of hydrophobic, electronic and steric effects on aromatic rings. This was the first nonmathematical technique that considered all the important physicochemical parameters *simultaneously*. It was designed to maximize the chances of synthesizing the most potent compound(s) in the series as early as possible and is a simple decision tree approach. The only assumption made is that the starting compound is the unsubstituted phenyl compound and that its bioactivity has been measured. The procedure continues by testing the effect of a hydrophobic substituent (*p*-chloro) on the system, and classifies the potency as either greater than, equal to, or less than the activity of its predecessor, the parent phenyl compound. The results of this initial comparison determine how one proceeds (*i.e.* the next analogue to synthesize) at the following step in the scheme, based on a judgement as to which properties (hydrophobic, electronic or steric) most often determine the bioactivity. The process is repeated, governed by the results of the previous choice and the new bioactivity until a stopping point is reached along one branch of the decision tree. This approach is attractive when new analogue synthesis is difficult and/or slow and test results are relatively rapid. This operational scheme takes into consideration the fact that it is possible to exceed the optimum value of a physicochemical variable (*i.e.* π, σ and/or E_s), which would result in a compound being less active than its predecessor. The next compound suggested would have this possibility in mind. It also allows the basic principles of QSAR to be applied without the use of statistical procedures and computers, a very attractive feature to many chemists.

Topliss has also proposed a similar scheme for aliphatic side-chain substitution.[15,16] This type of situation is seen when groups adjacent to a carbonyl (—COR), amino (—NHR), or amide function (—CONHR, —NHCOR), where R is the variable substituent in question, are varied. This scheme covers all those moieties that are not directly substituted on an aromatic ring, so many other substituents could also be envisioned. The operation of this scheme is similar to the aromatic procedure just described. A methyl substitution serves as the parent compound. Instead of examining the effects of a chloro group, the first substituent prepared is an isopropyl. The same logic has been applied in the aliphatic side-chain problem, namely the most likely effect one would first expect would be from an increase in hydrophobicity. As was the case with the aromatic scheme, the bioactivity at this step is compared to the preceding point, and ranked as either being less than, equal to, or greater than the previous compound. The next analogue to be synthesized is determined, based on the outcome of this comparison, and is guided by the physicochemical properties. Like the aromatic case, this procedure continues, based on a comparison of the bioactivities at each step, until a stopping point is reached along one branch of the decision tree. The aliphatic side chain scheme differs from the aromatic system mainly in the physicochemical variables which have been considered in the development of these operational schemes. This will ultimately be reflected in the substituents considered. The aliphatic tree is based on π, σ^* (the aliphatic sigma value from Taft[17]), and E_s^c (the corrected E_s value according to Hancock and co-workers[18]). The inclusion of cycloalkyl groups in the aliphatic tree marks the main difference between the two schemes. The presence of these groups allows one to probe the effect of hydrogen bonding on the system while imposing a minimal change in steric influences at the same time. This is not the case in the aromatic operational scheme, where special effects like hydrogen bonding are not considered.

The operational schemes by Topliss[15,16] offer the investigator an easy flow chart to navigate on the road to the most active compound. In addition, the biodata can be qualitative in nature since we are only concerned with the difference in ranking among the two bioactivities being compared. Thus, biodata reported as $+ + + +$, $+ + +$, $+ +$, *etc.* would be just as useful as highly refined IC_{50} values in these operational schemes.

One drawback is that the bioassay utilized must be reproducible from day-to-day, since the potency of the new analogue under consideration is based on the bioactivity of all the previous prepared analogues. In other words, the scheme does not properly account for experimental error. However, if one is dealing with a simple *in vitro* screen, it might be possible to rerun the previous prepared analogues, along with the new compound, so this potential source of error could be circumvented.

These schemes have been employed in the preparation of a host of pharmaceuticals,[14,19–57] herbicides,[58,59] insecticides,[60] as well as serving as an impetus for many of the techniques to be discussed. An interesting sideline regarding the above reports[14,19–60] is the large number of investigators that have misapplied this rather straightforward technique. The majority of the misuses deal with the fact that the operational schemes were applied at the end of the synthetic phase to 'check to make sure the right compounds had been prepared'. A line from ref. 29 sums up another typical violation: 'Approximate choices of substituents in the benzyl group were made by intuitive, rational (Topliss tree), and practical considerations, *e.g.* which aromatic aldehydes were conveniently accessible'.

Several papers[14,35–38] have been published reporting 'false predictions' after proper application of these operational schemes. This can be attributed to several factors, including a series designed to optimize potency and not necessarily spanned substituent space,[8] thus a degree of colinearity may result among the physicochemical variables. In addition, the schemes are designed around the main physicochemical variables (π, σ and E_s), and would tend to 'fail' if other descriptors accounted for the major variance in the bioactivity. In other words, in order to have confidence in the optimal compound selected, the investigator should perform a follow-up study with additional rationally selected analogues (to break any colinearities and also consider other variables) and conduct a rigorous regression analysis of the entire set. Although the Topliss schemes have some limitations, their popularity as a design tool will probably remain due to the ease of planning and carrying out a new project.

(iv) Topliss manual method

In 1977, Topliss proposed another strategy for the selection of substituents for synthesis which is referred to as the manual method.[61] Unlike the operational schemes[15,16] just discussed, in which a single compound selection is guided by a decision tree, the manual method involves a batchwise analysis of a selected, small group of five aromatic substituents: H (unsubstituted), 4-Cl, 3,4-Cl$_2$, 4-Me and 4-OMe. Like the operational schemes, these substituents were designed to be easy to synthesize. This initial set of compounds is prepared and ordered according to potency. The potency order in the group is then compared to the tabulated potency order calculated for various parameter combinations relating to hydrophobic, electronic and steric effects and the appropriate parameter dependency is established. New substituent selection for potentially more active analogues is guided by this parameter dependency.

The manual method has been employed in the experimental design of test series for numerous pharmaceuticals.[14,31,62-68] Cramer and co-workers[14] found this technique, like the previous operational scheme, to be of little or no value in the optimization of their potent antiallergic pyranenamines. These workers have found that strict adherence to the manual method would have produced nothing but derivatives less active than the unsubstituted starting compound. The most active compound from this work[14] was 1000 times more potent than any of the original series members. Cramer has attributed this to experimental error which was probably wrong for one or more of the initial potency measurements which would be excessive compared to the physico-chemical parameter relationships that actually exist.

Martin and Panas[69] have rigorously evaluated several of the more popular techniques utilized in QSAR for experimental design, including the Topliss manual method.[61] They have applied factor and cluster analyses to the initial and secondary sets suggested by Topliss (Tables II and III in ref. 61) and have found that the sets contain a high degree of multicolinearity and do not adequately span substituent space.[8] Thus, the conclusions drawn for the operational schemes discussed above, also hold true for this approach. Although the expected potency order for some of the parameter dependencies is quite similar—which is attributable in part to a very limited set of initial compounds—the procedure was designed to quickly reach the region of optimal substitution, and not to be used for precise model definition. This technique,[61] like the operational schemes,[15,16] is equally suited for either qualitative or quantitative bioactivity data. In fact, the similarity in parameter dependencies is probably assisted by some of the error inherent in the bioactivity.

(v) Fibonacci search technique

In 1974 Bustard[70] proposed the Fibonacci search technique as a way to discover the optimum bioactivity in a series using the minimum number of compounds. The technique is based on the series of numbers 1, 2, 3, 5, 8, 13, 21, . . . After the second, each number remaining is the sum of the two preceding numbers. This technique is well known to mathematicians. It was Fibonacci, an Italian mathematician, who discovered that this series forms the basis for an efficient search for the optimum value of some property. Bustard proposed that this technique could be utilized to find the optimal length of an aliphatic side-chain. The first two analogues synthesized are those correspond-ing to the two Fibonacci numbers below the uppermost limit (*i.e.* if the uppermost limit is 21 methylene units, then the first analogues prepared would contain eight and 13 methylene units). After these initial compounds are tested, the search space is reduced by excluding all compounds between the least active and its nearest boundary. The remaining compounds are renumbered, starting with the one with the value with the lowest Fibonacci number. The process outlined above is repeated until the final comparison is in a single step. In an ideal situation, the optimum should be found in the minimum number of steps. A major problem with this procedure, unlike the previous techniques discussed above, is that the investigator needs a good understanding of the relevant physicochemical properties operating on the system before the synthesis begins.

A modification of the Fibonacci search technique has been proposed by Santora and Auyang,[71] in which several physical properties could be considered at one time, unlike the Bustard approach which only operated on a single property (a situation that is less likely to occur with real biodata). They suggested that one consider as the search variables the sum of π and σ or ΔpK_a, rather than any one of these properties alone. Although some improvement might be seen in certain cases, this modification would not be useful in the majority of cases studied in QSAR.

Zeelen[72] has applied this technique to the design of a series of testosterone esters and found it an effective way to rapidly locate the necessary ester length for optimal activity. The technique seems very well suited for this type of optimization process.

Deming[73] has pointed out that the Fibonacci search is applicable only to response surfaces[74] that increase smoothly toward the optimum from both sides—the search will fail on response surfaces on which there is superimposed noise arising from experimental error or lack of a normal distribution of the data. When these problems are serious, the optimum value probably will not be found. In addition the technique is based on the assumption[73] that the difference in activity found is biologically significant.

It appears that the Fibonacci search technique is most useful in a retrospective analysis of a series where one physicochemical property dominates. The technique may be more useful in determining the optimum of a series when one is faced with a small number of data points and/or a parabolic equation cannot be generated. Like the previous techniques discussed in this section, this procedure is equally well suited for either qualitative or quantitative biodata.

(vi) Simplex methods

Simplex methods are used to find the optimum value of a response surface or function[74,75] (the actual analytical form of the function does not need to be known) and have been utilized extensively in chemistry.[76] The technique was originally described by Spendley and co-workers,[77] proposed for optimization in drug design by Darvas[78,79] and modified for this application by Gilliom and co-workers.[80] A simplex is a geometric figure defined by a number of points equal to the number of dimensions, n, of the hyperspace plus one ($n + 1$). In a two independent variable design the simplex is a triangle, with three independent variables the simplex is a tetrahedron, *etc.* This nonstatistical technique can therefore be classified both as a manual as well as a computationally based procedure, since the two independent variable design can be done graphically. The original application in drug design by Darvas[78] was presented as a manual, two independent variable design, utilizing π and σ. Algorithms and computer programs[81] exist for higher order optimizations.

The method is best understood by examining the two independent variable design. The simplex formed in this case will be a triangle of any three analogues on the response surface. Three analogues are selected that reside in a plane (if 2-D, or hyperspace if a higher dimension), based on the two independent variables in which bioactivity is expected to depend. Darvas[79] has suggested that analogues be selected using either cluster analysis[7] or nonlinear mapping.[82] The simplex would start with one analogue from each cluster. All analogues of interest (*i.e.* all synthetically feasible substituents) could also be plotted on the independent variable plane (or hyperspace) and three analogues selected (generally the unsubstituted lead and two analogues) for initiation of the simplex. The analogue of this trio that has the poorest response in the hyperface of the other two analogues is reflected to form a new simplex (*i.e.* triangle). This reflection will occur through the midpoint of the line joining the two most potent analogues of the original triangle. The new simplex contains the two most potent analogues of the original simplex and the new analogue determined by the reflection. The new analogue may be the most potent analogue, result in no change, or be the worst analogue tested to date. This process of reflection and simplex formation is continued until an analogue is located that demonstrates maximal activity (*i.e.* when the last simplex falls on an existing simplex, or no more synthetically feasible analogues are available in the direction of maximal bioactivity). Besides reflection, the simplex is also capable of undergoing expansion and contraction.[83] If a design with more than two independent variables is desired, an analogous procedure to the one outlined above will be needed, which can be easily evaluated by computer.[81] Additional details can be obtained by consulting one of the primary references listed above,[78–80] or any of the many texts on QSAR methodology.[83–85]

This method has only been applied to QSAR by a few groups,[78–80] and in all instances only as retrospective analyses. The original application of Darvas[78] included studies on natriuretic sulfonamides, phenoxyacetic acids with auxin activity, bactericidal benzyl alcohols, and diethyl phenyl phosphates which inactivated acetylcholinesterase. A follow-up study by Darvas[79] also contained a wide range of biotypes. On average, the optimum analogue was found after six analogues entered the simplex (from sets containing an average of 17 compounds). The paper by Gilliom and co-workers[80] also concentrated on retrospective analyses, mainly those dealing with the pressor effects and toxicities of biologically active indole, 1-methylindole and benzo[*b*]thiophene derivatives. These workers have also demonstrated the predictive power of this procedure by the *minimization* of the pressor effect of the above derivatives which resulted in a compound with a very weak pressor effect. Note that *optimization* does not always imply that *maximal* biological response is desired.

Simplex methods are best suited for series in which synthesis is slow and testing is rapid, analogous to the Topliss operational schemes.[15,16] An advantage that this technique has over the Topliss procedures is a greater selection of substituents from which to choose. This technique, like other optimization methods, does not guarantee that the investigator will find the global maximum rather than just a local maximum.[79] The method is applicable to both quantitative and qualitative biodata. The technique is also robust in terms of experimental error, *only* if the investigator repeats the testing of compounds.[8] Simplex methodology, like the Fibonacci search technique, can locate optima (*i.e.* parabolic relationships) that cannot be demonstrated by standard statistical analysis. Simplex methods are dependent on the proper choice of independent variables; an incorrect choice of parameters or unexplained variance in the data matrix that cannot be quantified can cause the optimization to fail. This may be attributed to the problem of obtaining a 'false optimum', an example of this is seen with the natriuretic sulfonamides studied by Darvas[78] (*i.e.* 3-NO_2,4-CF_3). Finally, the resulting set of analogues may be (multi)colinear if the optimization has simply climbed a hill from base to top.

(vii) Austel's methods for test series design

Austel has introduced two methods for series design in QSAR. The first is based on set theory,[86,87] in which intersections among substructural features/physicochemical properties and biological activities are investigated. The second manual method is based on factorial experimental design,[88-92] a technique which minimizes the number of compounds needed to investigate the effect of several parameters on bioactivity. This method is capable of handling both physicochemical and topological variables. Several reviews[87,93-95] have been written describing these techniques and their application to QSAR. The following sections will provide a succinct overview of these procedures, as well as applications from the literature.

(a) Set theory in drug design. Structure–activity relationships consider simultaneously both a biological response plus substructural features and/or physicochemical properties. Thus, one is interested in knowing the common intersection between the desired biological response and the various topological/physicochemical variables. Austel[86,87] has proposed the use of set theory for lead optimization, where one uses a series of intersecting circles (Venn diagrams) to depict the various 'universes' of biological activity. Thus, within a set of all compounds with a given bioactivity are defined specific subsets, each defining a certain pharmacological requirement necessary to meet the goals of the optimization program. These subsets can include selectivity for a given receptor, duration of action, oral activity, *etc.* The object of this procedure is to characterize (*i.e.* assign membership of the compounds under study to one or more of the subsets) and locate the common intersection of these subsets in terms of substructural and/or physicochemical information. This technique has been utilized in the development of clenbuterol, an orally active analogue of isoproterenol,[87] as well as a new class of cardiotonic compounds with a 2-phenylbenzimidazole nucleus.[87,96] Set theory has also been applied in the evaluation of benzo[*de*]isoquinoline-1,3-diones and related compounds screened for antitumor activity against P388 lymphocytic leukemia and L1210 lymphoid leukemia.[97] A compound (NSC 308847) was selected for preclinical toxicology studies as a result of this analysis.[97]

(b) Factorial designs. Factorial design of analogues is based on 2^n factorial experimental design, a well-established mathematical technique,[9,98,99] and combines the ease of application of the Topliss manual method[61] with the need for a unique spread in variance, orthogonal variables and synthetic practicability.[88-92] Factorial experimental design has several advantages over the Topliss manual method, which is designed around a limited number of aromatic substituents with known physicochemical values. In addition, the Topliss manual method also imposes restrictions on the site of attachment and the requirements regarding substructure or the makeup of physicochemical variables. In 2^n factorial experiments (*i.e.* test series), where n refers to the number of varied substituent positions times the number of variables, the descriptors are usually expressed as a nominal variable (*i.e.* yes/no or $+/-$, see ref. 6 for more explanation) and the test series is designed so that every combination occurs once in the test series. It is easy to see that this type of design can require a large number of compounds, especially if more than one position is varied on the molecule.

An extension of factorial design, fractional factorial design, can be used to reduce the number of compounds required.[9,98,99] This modification allows the investigator to span all the main factors using a minimum number of analogues. The resulting test series is well defined in terms of spanned substituent space,[8] resulting in a fairly orthogonal set of variables. These procedures have been applied to design a series of analgesics,[88] antiinflammatory agents,[88] cardiotonics,[91] analgesic/antipyretics,[100] as well as in the design of a test series for a Free–Wilson analysis.[92]

(viii) Parameter focusing and related techniques

Parameter focusing is a simple plotting technique introduced by Magee.[101] Sets of physicochemical parameters are used as coordinates for 2-D plots in conjunction with an indication of the bioactivity of each compound as being either active or inactive. The objective is to locate a cluster of active compounds within a confined region of the plot, which in turn can be attributed to the physicochemical parameters under investigation. It is assumed that for any relevant parameter there is an optimum region for the active compounds.

This technique is similar to that proposed by Darvas[78] for sequential simplex design. Related plots were utilized by Unger and co-workers[102] in the design and synthesis of potent tuberin analogues as well as by Cramer and co-workers[14,33] in their study and optimization of the antiallergic pyranenamines. Magee[101] has applied this technique on many different series of herbicides with good results. Parameter focusing is good with either quantitative or qualitative data and only requires that the biodata be ranked according to activity classification. Thus, this type of technique is well suited for most of the biodata generated by agricultural-related research centers.[6] This technique is also well suited for small sample size (seven or more compounds are needed for the 2-D focusing, or a minimum of three compounds to define an activity cluster). It is an application of the principle of bioisosterism so that greatly improved activity would only occur by good fortune.

McFarland and Gans[103] have extended Magee's parameter-focusing technique.[101] Their extension, which they refer to as cluster significance analysis (CSA; see Chapter 22.2), attempts to evaluate the statistical significance of an apparent cluster from the parameter-focusing technique. Two advantages of CSA are apparent over the original parameter-focusing technique. The first benefit removes any chance association that might be present in the determination of a clustered group of actives. The second advantage removes the need to only examine 2-D plots of physicochemical variables, since they have computerized their algorithms.

CSA operates by first determining the tightness of a cluster. First, all variables are standardized to unity,[6,104] then for a reputed active cluster the mean squared distance (MSD) is calculated. MSD is the squared distance between each pair of points in the cluster which is summed and then divided by the number of pairs in the cluster. MSD is thus an index which puts greater weights on the outliers. The MSD values for all possible combinations of clusters of equal size to the active cluster under consideration are calculated and compared to the original active cluster and a probability factor is computed from all those calculated MSDs that have MSDs equal to or less than the MSD of the active group. Using this calculated probability, a significance can be assigned to the relationship of the physicochemical variables and activity cluster and it can be determined if this cluster would have arisen by chance alone.

The CSA methodology[103] has been used to reevaluate a data set of Martin and co-workers[105] dealing with aminotetralin and aminoindan monoamine oxidase inhibitors. This is the data set used by Martin to introduce the multivariate technique of linear discriminant analysis (LDA) to QSAR. Both LDA and CSA arrive at the same selection of primary variables, but differed on the choice of secondary terms.

CSA has also been used[103] to study a larger set of compounds ($n = 32$ with 15 'actives') to discover any relationships that might exist between physicochemical descriptors and mutagenicity among some aminoacridine derivatives. There are 565, 722, 720 combinations of 32 items taken 15 at a time, which translates into the number of MSDs needed to calculate the necessary significance probabilities. This is not a manageable number, so the authors have proposed a modified method in which the probability factors are estimated from a large random sample of all the possible combinations. Again, CSA gave the same results as the original work.

Magee[106,107] has also developed a program called RANGEX, which is useful for selecting bioisosteric substituents within a given upper and lower limit of a known analogue from a data bank of substituent constants. Its biggest use is in finding novel substituents that are bioisosteric with a given lead. It has been used extensively in the herbicide and pesticide fields. Analogous APL programs have been developed and utilized at Syntex by Unger.[8]

Although all the techniques covered in this section can be classified as manual or nonstatistical-based techniques, the information gained from each in terms of experimental design is quite varied. All of the techniques discussed have been applied to practical problems in drug design. In applying these techniques, one must be constantly aware that they were designed to assist in getting the investigator in the region of the optimal analogue in a rapid, and hopefully synthetically feasible, manner. Once this is reached, one can add additional compounds that will break any multicolinearities, as well as increase spanned substituent space.[8] When this is achieved the investigator can examine the data for potential statistical models. Table 2 provides an overview of the various nonstatistical techniques for experimental design as an aid to the user trying to apply these techniques to a potential QSAR problem.

21.2.2.2.3 *Statistical methods for selecting compounds*

All of the methods which are discussed in this chapter assume that the variables that are proposed are 'relevant' to the problem at hand. Clearly, if we include a variable and try to optimize compound selection based on that variable (which will depend upon other variables chosen), and if that variable is not really important, we have optimized based on the wrong criteria. For example, if we make a selection so as to maximize some electronic effect then the substituents selected will give more weight to electron-donating and -accepting properties. If we are also optimizing lipophilicity, then we will probably try to choose compounds that are hydrophobic or hydrophilic, but also strongly electron-donating and -withdrawing. Therefore, we will probably try to select, for example, alkyl sulfones. If

Table 2 Comparison of Non-statistical Techniques used for Experimental Design in QSAR

Criteria[a]	Techniques[b]									
	1	*2*	*3*	*4*	*5*	*6*	*7*	*8*	*9*	*10*
1. Easy to use	Y	Y	Y	Y	Y	P	Y	P	Y	P
2. All possible substituents available	Y	Y	N	N	N	Y	Y	Y	Y	Y
3. Qualitative biodata useable	Y	Y	Y	Y	Y	Y	Y	Y	Y	Y
4. Quantitative biodata useable	Y	Y	Y	Y	Y	Y	Y	Y	Y	Y
5. Handling of inactives	Y	Y	N	P	N	Y	Y	Y	Y	Y
6. Robust in terms of experimental error	N	N	N	P	P	P	P	P	P	P
7. Maximizes variance—spanned substituent space	N	P	N	P	N	P	N	Y	N	N
8. Minimizes covariance	N	P	N	N	N	N	N	P	N	P
9. Handling of synthetic feasibility	Y	Y	P	P	P	Y	N	P	Y	Y
10. All types of substitutions allowed	Y	Y	N	N	Y	Y	Y	Y	Y	Y
11. Handling of multiple substituents	P	N	N	N	N	P	P	Y	P	P
12. Handles all relevant physicochemical variables	P	P	N	N	N	P	P	P	P	P
13. Requires reproducible bioassay from day-to-day	N	Y	Y	Y	Y	Y	Y	Y	Y	Y
14. Useful for determining optimum	N	N	P	P	Y	Y	P	P	Y	Y
15. Robust in terms of choice of variables	N	N	N	N	N	P	P	P	N	P
16. Multivariate (*i.e.* more than two parameters allowed)	N	P	P	P	N	Y	Y	Y	N	Y
17. Robust in terms of distribution of data	N	N	N	N	N	N	N	N	N	N
18. Significance testing	N	N	N	N	N	N	N	N	N	Y
19. Quantitative correlation	N	N	N	N	N	N	N	N	N	N
20. Qualitative interpretation	Y	Y	Y	Y	Y	Y	Y	Y	Y	Y
21. Useful for prediction	P	P	Y	Y	Y	Y	P	Y	P	P

[a] The following scale was used in judging the techniques, based on the expected results with a moderately skilled user operating a 'canned' computer program: Y = yes; N = no; P = possibly. [b] The techniques are as follows (lead reference(s) are listed in parentheses): 1. Intuition; 2. Craig plots (11.12); 3. Topliss operational schemes (15, 16); 4. Topliss manual method (61); 5. Fibonacci search techniques (70, 71); 6. Simplex methods (78–80); 7. Set theory (86, 87); 8. Factorial designs (88–92); 9. Parameter focusing (101); 10. Cluster significance analysis (103).

electronic effects are found not to be important at all, the choice of alkyl sulfones would be irrelevant and synthesis time could have been better spent on other types of substituents. The simple answer is to 'not include irrelevant variables'. Unfortunately, this is usually impossible since we do *not* know the answer (*i.e.* the relevant variables) at the outset of a program. These are only known at the end of the QSAR study.

The only solution to this dilemma is to make an educated guess based on related series. Methods based on principal components analytical techniques have the advantage of giving better overall variables; however, there is no guarantee that only 'relevant' variables will be found in the data matrix.

(i) Maximum variance methods: cluster methods

The work of Hansch and co-workers[7] provided the first application of multivariate statistics to the field of QSAR. These workers have introduced cluster analysis as a means of rationally designing a series of analogues from a larger, finite population of relevant variables.[108] This procedure can be considered an outgrowth of the work of Craig,[11] discussed in the previous section, which brought the notion of colinearity to light in relation to compound selection and QSAR equation development. These workers have greatly extended the Craig plot notion to an *n*-dimensional variable problem (instead of the 2-D Craig approach). Cluster analysis by itself will not guarantee a lack of multicolinearity. If applied properly, the technique will provide the user with a good spread of variance in the variables. It is *essential*[7,8,109] that the investigator always check for the presence of multicolinearity by examining the determinant of the variance–covariance matrix, $\det(X'X)\sigma^2$ (or simply $\det(X'X)$, since σ^2, the variance of the predicted equation, is a constant). Other techniques, such as the method of Farrar and Glauber,[110,111] may be used to check multicolinearity directly (an APL algorithm for this procedure is given and discussed in ref. 8). This is required whether the test series was designed from a published cluster,[7,108] or generated using one of the many canned statistical packages or published algorithms.[8]

Cluster analysis is used to group substituents with relevant physical properties into groups of similar substituents based on their calculated distances in multidimensional space. Normally the Euclidean (city block) distance is used. The grouping algorithm can use one of several techniques, typically the centroid (grand mean) of each existing substituent cluster replaces the individual substituents and is used to measure the distance to the as yet unclustered substituents.

Distance measurement is utilized in one of two principle classification techniques depending on the type of question being asked. The first method is known as the k-nearest-neighbor (kNN) approach which finds the distance between an unknown and the two closest members (when $k = 3$) of the data set.[112-114] The second method considers the distance between clusters to be the distance between the centroid of the clusters.[113-115] This method generally utilizes a hierarchical clustering algorithm, which is popular in many of the statistical packages and published algorithms like CLU,[8] CLUSANL,[7] BMDP2M,[116] CLUSTAN,[117] CLUSTER,[118-120] packages with their basis in biometrics,[121] as well as others (this is not meant to be a comprehensive list of all the many commercial packages and algorithms that exist). In order to design an optimal test series, the investigator would select one analogue from each cluster (usually from the median of the cluster) and then examine this subset for possible multicolinearities and synthetic feasibility. If nonorthogonal variables result, the suspect substituent is replaced with another member of the same cluster, the new subset is rechecked for uniqueness, and this cycle is repeated until orthogonality is optimized, along with the caveat regarding synthetic feasibility.

Unger[122,123] has demonstrated the utility of cluster analysis with a 'retrospective' design of a test series from a known QSAR regression of 104 phenanthrene aminomethanols which exhibited high antimalarial activity in the mouse.[12] Random subsets consisting of a substituent from each of ten clusters was selected from the larger data set.[12] Each subset was subjected to a regression analysis and the correct equation (*i.e.* same as reported in ref. 12) was generated with approximately the same statistical significance as the original equation from the total data matrix. This example demonstrates the power a good experimental design can have on the outcome of a project—10 well-chosen substituents generated the same information as 104!

Unger and co-workers[102] have designed and synthesized a series of tuberin analogues (*para*-substituted N-(β-styryl)formamides) as potential antibacterial agents that demonstrate the usefulness of cluster analysis as more than just a technique for lead optimization.[8] Equation (1) shows the QSAR that was generated during the course of this work.

$$pA = -0.3(\pm 1.4) + 2.9(\pm 1.1)\log P - 0.5(\pm 0.2)(\log P)^2 - 2.6(\pm 0.9)\mathcal{F}^2 - 1.7(\pm 0.5)\mathcal{R}^2 + 0.4(\pm 0.3)D_{ha} \quad (1)$$

$$n = 19, s = 0.161, r = 0.952, F_{5,13} = 25$$

In this equation, $pA = \log(1/MIC_{molar})$, $\log P$ was determined experimentally,[124] $\mathcal{F}$ and $\mathcal{R}$ are the *corrected* Swain and Lupton electronic component constants[125,126] and $D_{ha} = 1$ if the substituent was a hydrogen bond acceptor. Equation (1) is somewhat unique in that there is no linear term for either $\mathcal{F}$ or $\mathcal{R}$, which suggested that the optimum electronic effect occurred with electronically neutral substituents and, thus, the aromatic moiety may not be necessary for activity. The hypothesis was shown to be true when the phenyl moiety was replaced with isolipophilic and isoelectronic linear or cyclic alkyl groups, like cyclohexyl and n-pentyl, both of which possessed high activity. This rather remarkable finding can be attributed to a large extent to the experimental design phase of this project which utilized cluster analysis using the $\det(X'X)$ criteria.[7] Starting with a well-designed, initial test series allowed these workers to develop a model that could go from lead optimization to lead generation. Unusual substituents were suggested because of physicochemical reasons that might not have occurred to the chemist and that possibly represented a unique system to the patent attorneys.[8] Another interesting outcome of this study is the fact that the 'lead generation phase' was conducted *without* extrapolation outside spanned substituent space—the final outcome produced novel and more potent compounds that fell within the bounds of the initial vector space!

Coats and co-workers[127] have applied cluster analysis in the experimental design of a follow-up study to their initial work[128] on 7-substituted 4-hydroxyquinoline-3-carboxylic acids as inhibitors of cellular respiration. In the initial study, 15 compounds were designed to minimize covariance between the physicochemical parameters π, MR and σ_p.* The primary objective of this study was to assess the transport properties of these analogues. The compounds were evaluated for their ability to inhibit the respiration of Erlich ascites cells as a whole cell model (studying transport properties, *etc.*) and for their ability to inhibit mitochondrial malic acid dehydrogenase (m-MDH) as an intracellular target enzyme model. The correlation analyses conducted indicated that membrane transport was controlled by lipophilicity (equation was linear in $+\pi$), while enzyme inhibition was controlled by a

---

* The experimental design phase for this study (ref. 128) was conducted prior to publication of the original cluster analysis procedure described in ref. 7. The authors have, in fact, accomplished the same goal with the procedures described in ref. 128.

strong polar interaction between the 7-substituent and the enzyme-binding site (equation was linear in $+$ MR).[128] The follow-up study[127] was conducted to investigate the effect of changing the size of the bridge[129] between the aromatic ring and the 7 position of the quinoline nucleus. The substituents on the aromatic ring were selected from published clusters[7,108] based upon π, MR, $\mathscr{F}$ and $\mathscr{R}$. Increasing bridge size ($-CH_2O-$, $-CH_2CH_2O-$ and $-OCH_2CH_2O-$) had a minimal effect on activity. Another goal of these studies was the development of selective inhibitors of malate dehydrogenase. In addition to m-MDH, the compounds were also evaluated as inhibitors of cytoplasmic malate dehydrogenase (s-MDH) and of skeletal muscle lactate dehydrogenase (LDH-M_4). Correlation analyses of the data with the three enzymes allowed a comparison of the binding sites in quantitative terms, while examination of the data on inhibition of ascites tumor cellular respiration afforded an indication of membrane transport. As was found in the earlier study, the whole cell model (*i.e.* membrane transport model) was a linear function of lipophilicity. QSAR on the three enzymatic systems indicated that the mitochondrial enzyme (m-MDH) was quite different from the two cytoplasmic enzymes, s-MDH and LDH-M_4. These results[127,128,130] were the first to clearly show the distinction between membrane transport and receptor binding and help to emphasize the importance of a well-designed test series on the outcome of the model. They also demonstrated the importance of optimizing both *in vitro* (*i.e.* enzyme) and *in* (or *ex*) *vivo* (*i.e.* whole cell or animal) results *concurrently*. Unfortunately, most workers first maximize *in vitro* results (independent of any *in vivo* testing), which will generally put them outside the acceptable 'lipophilicity window'[83,131,132] in which the compound chosen for advanced study will show *in vivo* activity.

Dunn and co-workers[133,134] have also utilized published clusters[7,108] in their experimental design of substituted 1-phenyl-3,3-dimethyltriazenes as potential antitumor compounds. This study is somewhat analogous to the work of Coats and co-workers,[128] in that the study was initiated before the cluster analysis procedure[7] was published with the preparation and testing of seven analogues.[134] Unlike the Coats study, these initial congeners were not well chosen and correlation analysis on this set was not very successful due to multicolinearities among the variables. Examination of the published clusters[7] indicated that two of the initial substituents were from cluster 1, while the remaining five all came from cluster 3. An additional seven analogues were chosen from the published clusters and combined with the original seven so as to minimize the multicolinearity problem. The combined set could be fitted by one regression equation in which the Hammett σ constant accounted for 85% of the explained variance. The most potent compound was the 4-Pr^n analogue which was one of the substituents added from the published clusters. Dunn and co-workers[133,134] have also investigated the mechanism of action of these compounds. They have utilized the results of Kolar and Preussmann,[135] who determined the half-life towards hydrolysis for several of the same compounds studied by Dunn. Dunn and co-workers have generated a regression equation utilizing these data and have shown that hydrolysis, like the antitumor activity, is also dependent upon σ. This is a rather remarkable finding and suggests that the hydrolysis of the triazene to the reputed diazonium cation may be the species responsible for the cytotoxic response. Thus, this study demonstrates that cluster analysis allowed for a meaningful model to be developed that was not possible with the initial, randomly selected set in addition to helping to find unexplored areas of substituent space which contained the substituent that resulted in maximum potency. Cluster analysis was also useful in helping to narrow down the potential mechanism of action of these compounds.

Unlike the examples discussed to date, in which the object of clustering was substituent selection,[7,108] Horváth and co-workers[136] utilized cluster analysis of variables to indicate their similarity by the magnitude of the correlation coefficients. Examination of a tree diagram can clearly illustrate how the clusters are formed at each step by using an amalgamation rule based on the maximum similarity over all pairings of the variables between the two clusters. Thus, both clustering techniques (substituents and variables) should be conducted during the experimental design phase of a project in order to maximize unique variation.

Cluster analysis has been used extensively in the design and optimization of many pharmaceuticals[102,127,133,134, 136 – 156] as well as herbicides and pesticides.[157,158] The original procedure[7] has been modified and extended by several groups.[159 – 164] Although the examples discussed above each point out a different strength of the technique, the procedure is not without its shortcomings. The most common criticism heard is the lack of synthetic feasibility from the published[7,108] or internally generated clusters,[69] although the method was intended to find the 'best' selection from a larger list chosen by the chemist who, in turn, chose from the master list of substituents. Borth and co-workers[164] along with Schaper[165,166] have added a difficulty of synthesis measure to the original procedure[7] (see section below on D-optimal design for further discussion on this topic). Martin and

Panas[69] have also suggested a second shortcoming dealing with too little variation existing among electronic properties, especially when only monosubstituted substituents are considered. Regardless, cluster analysis has been linked with a renewed interest in experimental design in QSAR and is indirectly responsible for many of the techniques to be discussed in the next section.

(ii) Maximum variance methods: multidimensional separation methods

As mentioned in the previous section, the application of cluster analysis to QSAR[7,108] rapidly changed the way investigators thought about potential (and existing) projects. The following sections will detail these newer applications and their impact on experimental design in QSAR.

(a) D-optimal designs. As mentioned above, clustering substituents does not guarantee lack of (multi)colinearity, only that a reasonable variance will be achieved in the design.[7] Both maximum variance and minimum multicolinearity (called covariance) can be combined in a single matrix, the variance–covariance matrix $(X'X)$. The determinant of this matrix gives a single number which is maximum for substituents showing maximum variance and minimum covariance. The use of $\det(X'X)$ is well known in the statistical literature as the 'D-optimal' design.[167–170]

The reason that D-optimal designs are useful for multiple regression QSAR studies can be seen in equation (2), the matrix formulation of the linear least squares problem. The solution for the regression coefficients, shown in equation (3), contains the variance–covariance matrix. The errors in fitting b are shown in equation (4) and are seen to be inversely related to this matrix. Thus maximizing the variance–covariance matrix minimizes the errors in fitting the model. Other criteria, such as maximizing the trace of $(X'X)$, are also possible. It should be clear that the $\det(X'X)$ depends upon the model (selected variables, powers of variables, *etc.*) and since one does not usually know the model when in the design stage, one must make an educated guess.

$$Y = Xb + e \tag{2}$$

$$b = (X'X)^{-1} X'Y \tag{3}$$

$$V(b) = (X'X)^{-1} s^2 \tag{4}$$

The $\det(X'X)$ can not only be used to evaluate a given design, it can be used to select a design given a relevant choice of variables. Unger[8] in collaboration with statistician Dr. J. D. Johnson of Syntex Biostatistics have published three methods for deriving D-optimal models from sparsely populated substituent space. The STEPUP method allows one to start with some already synthesized analogues and add the best possible n additional analogues. The STEPDOWN method allows one to start with a list of all possible (or synthetically feasible) analogues and select the best subset containing the desired number of substituents. The third method allows the SWAPping of substituents in the model for those not included and allows for correction of the model. APL language programs have been published.[8]

The STEPDOWN procedure was applied to a list of 171 aromatic substituent constants using π^2, π, $\mathscr{F}$, $\mathscr{R}$, MR, HA and HD and were unstandardized. The value of $\det(X'X)$ for the 20 'best' selected substituents was 100 times larger than the largest value obtained in 100 randomly selected sets.[8] Herrmann[171] has described an information theoretic approach to optimal selection of test series in drug design and has achieved similarly good designs which are improved over those using simple maximum variance.

Borth and co-workers[164] have extended the D-optimal method described above to include a difficulty of synthesis measure in the mathematical equation. Unger[8] incorporated this information by allowing the chemist to choose all synthetically feasible substituents from the complete list of candidates and then using the relevant variables to select the desired subset. It is felt that this is the more reasonable way to go, since the difficulty of synthesis will vary greatly depending on the parent ring system the chemist is attempting to modify.

(b) Wellcome distance mapping approach. Wootton and co-workers[172] at Wellcome have published a modification of an idea originally suggested by Kennard and Stone.[173] This is a technique for selecting a well-spread set of derivatives including those from 'messy design' situations (*i.e.* a situation in which all the combinations of variables are not physically realizable). The Wellcome algorithm operates such that new analogues are selected that are at least a minimum distance from each point already included in the design and also as close to the center of gravity of the included points as possible. The program functions in an interactive mode and the user specifies the minimum distance constraint and the initial starting point. The program uses a data bank[174] of 35 position-dependent π, $\mathscr{F}$ and $\mathscr{R}$ as well as position-independent MR, aromatic substituents. The original $\mathscr{F}$

and $\mathscr{R}$ values from Swain and Lupton[175] have been scaled by application of a positional weighing scheme.[176] Unger[8] has pointed out that in the construction of this data bank,[174] the authors used the original $\mathscr{F}$ and $\mathscr{R}$ values from Swain and Lupton, and not the *corrected* values from Hansch and co-workers.[125,126] This resulted in an incorrect value for $\mathscr{F}$ being used. Although this approach is limited to aromatic substitution, it differs from the previous techniques discussed in that it allows for multiple substituents to be present.

Several potential problems (in addition to the wrong substituent values for $\mathscr{F}$) are inherent to this procedure. Like cluster analysis,[7] this technique will generally result in a design with sufficient variance in the data; however, it too may be multicolinear to some extent and it is *essential* that this be checked utilizing one of the techniques[7,8,109-111] described above in the section above on cluster methods. Another problem deals with the lack of a precise selection for the minimum distance constraint utilized. All of the data are scaled from 0.0–1.0 to ensure that each variable is given a proper weighing and a distance constraint of 0.3 is normally used. The choice of distance constraint will also affect the number of final points generated. Since the program is interactive, it stops at each step and presents the user with all possible substituents that meet the required minimum distance constraint. The user responds with the substituent he/she feels is most synthetically feasible. The choice of the initial analogue will affect the choice of the remaining analogues. The choice is left up to the investigator, but hydrogen is normally used. Problems could arise if other analogues (instead of hydrogen) are utilized first, or if the user chooses other than the preferred analogue at each step. The major concern would be that potentially a very different test series could arise. The user of this procedure must decide which series is then 'best' (*i.e.* most unique), both from a statistical and from a synthetic standpoint. The authors[172] have only suggested that a simple correlation matrix be constructed to test for the uniqueness. The previous section on D-optimal design suggested much better ways to test the proposed series for multicolinearity. It should be pointed out that the Wellcome distance mapping approach is also well suited for starting from a small series of existing compounds, not just a single, initial point. However, if the starting point is an existing series, problems can arise if multicolinearity is present to begin with,[177,178] especially if the investigator does not take the time to screen the resulting new series for potential multicolinearities.[69,179]

As outlined above, the distances in substituent space from the initial analogue to all other theoretical analogues are computed and the second analogue for synthesis is proposed as the one closest in substituent space to the initial analogue, yet greater than the preset minimum distance constraint from it. Any analogues closer than this are rejected since they reside in the same 'cluster' and would provide redundant information to the test series. This procedure is repeated, rejecting analogues which reside too close in substituent space to selected analogues, and choosing further analogues according to their distance from the center of gravity in substituent space of the selected analogues. This process is continued until no analogues are left to choose from. A problem arises if the selected analogue is synthetically unfeasible or otherwise undesirable. The program automatically outputs all analogues greater than the minimum distance constraint from those previously chosen. Any of these 'hits' may be selected, but Wootton and co-workers[172] advise sticking with the initial choice if at all possible to avoid potential multicolinearities.

The Wellcome group has utilized their distance mapping technique[172] to design a well-spread set (H (fixed, initial), NH_2, NMe_2, NO_2, Br, SO_2Me, Bu^n, Pr^iO, CO_2Et, F, NHCOMe and *n*-pentylO) of 12 *para* substituents. Cognizant of the problems discussed above, Unger[8] has 'reevaluated' the Wellcome approach by first correcting the problem with their $\mathscr{F}$ values pointed out in the introductory remarks to this section (the entire, corrected data matrix has been published in ref. 8, Table X). Using this corrected, standardized data matrix, along with an APL implementation of the Wellcome algorithm (DISTPLAN)[8]* resulted in a different series: H (fixed, initial), Et, OEt, CO_2Me, Cl, CN, OH, SO_2NH_2, NH_2, NMe_2, Bu^nO and Bu^n. Problems were encountered in finding a suitable minimum distance that would yield the necessary 12 substituents. Application of the D-optimal, STEPDOWN algorithm[8] yielded another design: H (fixed, initial), Bu^n, Bu^t, CF_3, *n*-pentylO, OPh, NH_2, NMe_2, NO_2, SO_2Me, SO_2NH_2 and F. The two test series proposed by Unger appeared to be fairly equivalent from a statistical standpoint (equivalent eigenvalues, values of determinant and no significant multicolinearity[110,111]).

Unger[8] has uncovered greater differences in designs when the full (95 substituents) data matrix is utilized. In this case, the differences between STEPDOWN (D-optimal) and DISTPLAN (Wellcome distance mapping) are more pronounced.

---

* The Wellcome group have implemented their algorithm in FOCAL on a PDP8-e computer and will send a copy on request. This apparent incompatibility in both hardware/software has caused some concern in the literature (see refs. 69 and 123). The authors in ref. 181 make mention that they have implemented the algorithm, but give no further details.

Wootton[178] has proposed a modified version of the initial[172] procedure, which is concerned not only with an insufficient spread in the variance, but also addresses the multicolinearity problem by maximizing the determinant of the correlation matrix[179] (as in D-optimal design). The problem with choice of the minimum distance still is present, as is the problem with the $\mathscr{F}$ parameter values in the Wellcome data bank.

This technique has been applied to several interesting design situations. The Wellcome group has used this method to design a series of arylamidinoureas as potential antimalarial agents[123,180] and a series of analogues of the insecticide methoxychlor.[123,177] More recently, Doweyko and co-workers[181] utilized the Wellcome technique to prepare a series of substituted 2-[(phenylmethyl)sulfonyl]pyridine 1-oxides as a new class of preemergent herbicides. Several problems are apparent from this study, including: (a) improper parameter value generation (the authors have combined data from the Wellcome data bank,[174] as well as from Hansch and Leo[108] and this will result in incorrect $\mathscr{F}$ values being used, as mentioned previously); (b) a simple correlation matrix was utilized to assess multicolinearity, which is similar to the approach taken in the original Wellcome study;[172] (c) minimum distance criteria of 0.14 to 0.35 were applied, depending on the structural subclass (an insufficient spread in variance will result when distances of less than 0.3 are utilized[178]); (d) although the resultant regression equations explained between 79–93% of the variation, the equations were not used for the prediction of new analogues, nor were any cross-validation studies conducted.

Although this method provides an interesting approach to experimental design, it appears that the multicolinearity problem will prevent it from ever having the popularity of cluster analysis.[7] When applied in a statistically sound fashion, it appears to be just as good in series design as some of the more complex constrained optimization procedures that have been developed and utilized.[8]

(c) PCMM technique. PCMM, a technique introduced by Franke and co-workers[84,179] in 1980, is a combination of Principal Component analysis and Multidimensional Mapping, and thus is an extension of the Wellcome distance mapping approach (see above). It was the first technique developed that identified the multicolinearity problem neglected by the original Wellcome approach.[172] Wootton[178] proposed his modification in 1983 and addressed the multicolinearity problem by taking the proposed test series from the original procedure,[172] calculating the determinant of the correlation matrix, and repeating this process with different substituents until the determinant criterion was maximized.

The PCCM approach takes the opposite approach from the Wootton modification and starts with the fact that a hyperplane can be fitted to selected substituents if multicolinearity is present (the higher the multicolinearity, the better the fit of the hyperplane). Multicolinearity could be removed by simply removing those substituents closest to the hyperplane. In order to accomplish this, Franke and co-workers have divided the total substituent space into two sets. Set 1 contains those substituents closest to and set 2 the substituents distant from the hyperplane judged by Euclidean distances computed with the aid of principal component analysis between the substituents and the hyperplane. These two subsets of substituent space each serve as the data bank of substituents for the Wellcome distance mapping approach[172] and this procedure is applied so a higher percentage of potential substituents come from set 2, the more distant cluster (Franke and co-workers[179] suggest selecting 30% from set 1 and 70% from set 2). This is done in an iterative fashion in order to check the new positioning of the hyperplane after each substituent selection.

PCMM[181] has been applied to the same examples studied by the Wellcome group[172] and by Hansch and co-workers,[7] using the substituent space spanned by the variables π, $\mathscr{F}$, $\mathscr{R}$ and MR. PCMM appears to maximize variance as well as cluster analysis,[7] and slightly less than the Wellcome distance mapping approach. This approach appears to minimize collinearity, determined from the determinant of the correlation matrix, better than either cluster analysis or the Wellcome distance mapping approach. However, the synthetic feasibility of many of the analogues selected is questionable.

(d) TMIC method. When less than 50 substituents are being considered for experimental design a modification of the PCMM technique, called the TMIC method has been proposed by Franke and co-workers[84,182] and is based on Two-dimensional spectral Mapping of Intra-class Correlation matrices. The TMIC method allows one to present results in a very simple and instructive way in the form of a two-dimensional map. The test series is generated by simple inspection ('rational intuition') by the chemist, unlike the PCMM technique which is somewhat removed from the end-user, and, as such, often results in analogues with questionable synthetic feasibility. The same spectral map can result in several different test series, so synthetic feasibility can be easily taken into account.

Franke and co-workers[182] have applied TMIC to the same examples discussed by the Wellcome group.[172] It performed as well as the PCMM method on these examples, with the added feature of simple compound selection from the spectral map by the chemist to make it a better technique (than PCMM) in terms of synthetic feasibility. Franke and co-workers[183] have also applied TMIC in the design of a test series of 2-anilinopyrimidines as potential fungicidal agents. They were able to demonstrate that a training set of 13 substituents selected using this method provided the same information and arrived at the same model as a much larger series of analogues previously prepared.

A related approach has been described by Ordukhanyan and co-workers[184] in which factor analysis was utilized as the data reduction technique on a set of 90 substituents in which π, σ_m, σ_p and MR were used as the physicochemical parameters. Four eigenvalues were found and the parameter values for the substituent population were replaced by the weights of the general factors. The substituents were grouped according to the sign of the weight factor for the second through fourth factors, or into one of eight possible classes. This means of classification can be thought of as a hierarchical procedure similar to cluster analysis;[7] however, distances are not measured in the usual Euclidean sense, but by the angle between the vectors. A second classification was also conducted using the varimax factors. These results have been plotted in the form of projection plots of the factor weights on a plane formed by the various factors to allow simple selection of substituents by visual selection. Thus, this technique is similar to the TMIC procedure in its ultimate goal. The authors have suggested several test series for use from this analysis, but have not compared it to existing techniques.

(e) Wooldridge rational substituent set. In 1980, Wooldridge[185,186] proposed a rational substituent set consisting of 15 congeners based on a set of criteria proposed for an ideal substituent set that were very similar to those published elsewhere.[187] Parameters included π, $\mathscr{F}$, $\mathscr{R}$, E_s and MR. The necessary values were taken from the compilation of Hansch and co-workers,[125,126] so the corrected $\mathscr{F}$ values were used. E_s for this work was found in the work of Taft,[17] calculated from van der Waals radii according to the procedure of Charton,[188] or estimated from models.

The actual criterion used for the series design or the total number of substituents considered has not been disclosed. The substituents have not been weighted by position, like the Wellcome distance mapping technique.[172] A simple correlation matrix has been provided which shows a substantial colinearity between π and MR. The other parameters appear to be rather orthogonal, based on this simple correlation matrix; however, no determination of multicolinearity[8,110,111] has been provided. Wooldridge[185,186] also has not subjected his data matrix to any comparable procedures (*i.e.* any of the other design methods discussed in this chapter). Thus, this series should be used with caution.

A few applications utilizing this rational test series have appeared in the literature. Wooldridge initially applied it to a set of triazenes prepared as potential antibacterials.[189] *De facto* QSAR regression analysis on the set of 52 compounds resulted in a parabolic relationship in π with antibacterial activity *in vitro*. The optimal π value (5.25) resulted in excellent *in vitro* results, but poor response when screened *in vivo*. This is a fairly typical occurrence; Wooldridge and co-workers attributed this to binding to serum albumin, which in reality is true, but also *predictable*—the optimum *in vitro* responder was not a member of the *in vivo* 'spanned *pharmacokinetic* space', which can be thought of in the same vein as 'spanned substituent space'.[8]

Hansch and Silipo[190] have tested some of these same triazenes against a purified tumor dihydrofolate reductase. Their correlation results[190,191] indicated that the *meta* congeners were correlated by a parabolic relationship in π while the *para* congeners by a linear term in MR. This relationship also held for the combined set.[191]

Wooldridge and co-workers[185,186] attempted to apply a similar approach to their antibacterial activity (*i.e.* design based on π and MR). In doing so, they have synthesized the additional compounds missing from the original study in order to have the substituents necessary for their rational set. However, one of the points made above regarded a colinearity problem between π and MR ($r^2 = 0.49$), thus the properties these workers are trying to optimize are not even orthogonal!

This approach has also been utilized by Ramsden and co-workers[192] in the design and synthesis of a series of antiallergic compounds related to disodium cromoglycate. The simple correlation matrix showed a high degree of colinearity between MR and E_s ($r^2 = 0.63$) and between π^2 and MR ($r^2 = 0.57$).[192] These types of simple colinearities make it difficult to uniquely define the desired action of the physicochemical parameters, and make the use of this design set questionable.

(f) Wold multivariate approach. Wold[193] has remarked that the field of statistics has reached the level of sophistication that *both* the independent and dependent variables in QSAR can be optimized as the multivariate processes that they actually represent. This statement was directed at the pharmacological systems that produce the dependent variables subjected to QSAR analysis and the fact that the pharmacologists are still attempting to measure a single number that determines the bioactivity. Wold has used the area of process control for analogy—40 years ago only a single variable (like pH or reaction temperature) was considered sufficient to optimize the yield of a given

reaction. It did not take long to realize that one variable provided insufficient information—today, hundreds of different probes are used to contribute to this process optimization. It stands to reason that a pharmacological system should be no different than the process control system just mentioned. In fact, it is much more complex, and, thus, more important that bioactivity be considered as a multivariate process. Wold has provided a final analogy that should answer any questions as to why such an analysis is necessary. A pharmacologist, in designing and performing an experiment, would never consider the results meaningful if the experiment were only conducted on one rat; variability in biopharmaceutics and pharmacokinetics from rat to rat would cause only the most lax scientist to put meaning into such a result. Likewise, one should not put his/her faith into a single number for an explanation of the bioactivity.

This concept is not new, in fact, it has been expounded by many other workers in the field, including Mager;[194,195] however, Wold and co-workers[196–204] were the first to apply the necessary statistics to the above ideas in a logical and reproducible manner. This multivariate approach to QSAR developed by Wold and co-workers[197] is dependent on three major areas. The first is defined as multivariate characterization (MVC), in which the 'variables' under question, whether they be biological activity, chemical structure, steric effects or other influences should be quantified with as many relevant measurements as possible in order to provide an extractable set of latent variables, or 'principal properties' to describe the actual dimensionality of the system. The second is referred to as multivariate design (MVD) where the investigator changes all variables simultaneously in a system (rather than 'one variable at a time', OVAT, or 'change one substituent at a time', COST, as is the normal practice followed in most studies) by using techniques such as factorial designs (see Section 21.2.2.2.2) or simplex methods (also in Section 21.2.2.2.2). Thus, a training set is constructed using the 'principal properties' from the first step. These 'principal properties' are used in the measurement of the bioactivity response(s). The third area deals with multivariate analysis (MVA) of the accumulated data using PLS (partial least-squares modeling in latent variables, or projections to latent structures), a technique developed by Herman Wold,[205] and applied to chemical applications by his son, Svante Wold.[200,206,207] This technique is useful for the situation where both the bioactivity and the structure descriptors (independent variables) are multivariate. PLS can be thought of as the application of principal components (PC) analysis to both the X and Y blocks (matrices) followed by a check for the maximum linear relationship between the PCs of the two blocks. Multiple regression is equivalent to the case where the Y block consists of a single variable. A more detailed description of PLS can be found in refs. 200, 205 and 207. Wold[197] refers to these three areas as the integrated multivariate approach, or MV³.

MV³ multivariate design analysis has been applied to several areas of interest to QSAR, most notably peptides,[196,197,199,202] anesthetic activity and toxicity of halogenated ethyl methyl ethers,[201] mutagenicity of chlorinated aliphatic hydrocarbons[200] and calcium channel antagonists.[203] The peptide work has probably produced the most remarkable results to date.

Table 3 provides an overview of the various statistical techniques for experimental design as an aid to the user trying to apply these techniques to a potential QSAR problem.

21.2.3 THE SIGNIFICANCE OF QSAR RELATIONSHIPS

21.2.3.1 Basic Statistical Concepts

There are many excellent texts[9,115,168,208–214] and even some specialized QSAR treatments[83–85] covering basic statistical concepts, including experimental design[9,98,99] and optimization methodology.[74,215] We include here a very brief summary for ease of reference.

In the early years of QSAR, the concern over 'statistical validity' was usually overlooked in favor of testing the applicability of the expanded Hammett–Taft approach to physical organic chemistry as proposed by Hansch and co-workers.[216] It soon became apparent that good data analysis was as important to a good QSAR as meaningful physical organic substituents and reliable biodata. Shortly thereafter several 'guidelines for a good QSAR equation' began to appear in the literature.[191,217–219] Hansch[220] has published a succinct, but extremely clear, introduction to regression analysis and probability testing. The paper by Martin and Panas[69] includes guidelines for series design. Martin[83] has provided a good introduction to the fundamental statistics necessary for QSAR.

The following statistical and probability measures should accompany any regression analysis, regardless if it is an equation used in the development of the model, or if it is actually the 'best'

Table 3 Comparison of Statistical Techniques used for Experimental Design in QSAR

Criteria[a]	Techniques[b]						
	1	*2*	*3*	*4*	*5*	*6*	*7*
1. Easy to use	Y	Y	Y	Y	Y	Y	P
2. All possible substituents available	Y	Y	N	N	N	N	Y
3. Qualitative biodata useable	N	N	N	N	N	N	N
4. Quantitative biodata useable	Y	Y	Y	Y	Y	Y	Y
5. Handling of inactives	N	N	N	N	N	N	P
6. Robust in terms of experimental error	Y	Y	Y	Y	Y	Y	Y
7. Maximizes variance—spanned substituent space	Y	Y	Y	Y	Y	Y	Y
8. Minimizes covariance	P	Y	P	Y	Y	N	P
9. Handling of synthetic feasibility	Y	Y	P	P	P	N	P
10. All types of substitutions allowed	Y	Y	N	N	Y	N	Y
11. Handling of multiple substituents	P	P	Y	Y	P	N	P
12. Handles all relevant physicochemical variables	P	P	N	N	P	N	P
13. Requires reproducible bioassay from day-to-day	N	N	N	N	N	N	N
14. Useful for determining optimum	P	P	P	P	P	N	P
15. Robust in terms of choice of variables	P	P	N	N	P	N	P
16. Multivariate (*i.e.* more than two parameters allowed)	Y	Y	Y	Y	Y	Y	Y
17. Robust in terms of distribution of data	Y	Y	N	Y	Y	N	Y
18. Significance testing	N	Y	N	N	N	N	P
19. Quantitative correlation	Y	Y	Y	Y	Y	Y	Y
20. Qualitative interpretation	Y	Y	Y	Y	Y	Y	Y
21. Useful for prediction	P	P	P	P	P	N	P

[a] The following scale was used in judging the techniques, based on the expected results with a moderately skilled user operating a 'canned' computer program: Y = yes; N = no; P = possibly. [b] The techniques are as follows (lead reference(s) are listed in parentheses): 1. Cluster methods (7, 8, 108); 2. D-optimal designs (8, 164, 171); 3. Wellcome distance mapping (172, 178); 4. PCMM technique (84, 179); 5. TMIC method (84, 182); 6. Wooldridge rational substituent set (185, 186); 7. Wold multivariate approach (196 204).

equation. First, the number of compounds or data points, n, used to develop the equation is important. This number, in conjunction with the number of variables in the equation, gives a quick indication of the degrees of freedom available, or an initial check on the stability of the model. It is important that the compounds *excluded* from the model, and the reason they were not utilized, be indicated. Excluding one or two points for no apparent reason in order to improve the overall statistics is not valid on any grounds, and probably is an indication that the model is ill-conditioned (*i.e.* poor design that is weighted on a few points more heavily than others). Second, and probably the most important statistic[69,83]—but more often than not just reported because it is required by the journal—is the standard deviation, s, which indicates the overall quality or suitability of the correlation and is used as a measure of dispersion or scatter of the observation from the mean. Martin[69,83] provides an interesting way of determining the necessary variation of a physicochemical property needed to effectively cause a suitable range in bioactivity which involves the manipulation of the equation for r^2, the fraction of the variance in the data which is explained by the correlation equation. In doing so, it becomes apparent that for a 'useful equation, *i.e.* one with at least an r^2 of 0.80, the standard deviation for the parameter in question must be approximately 0.75. Martin[69] has pointed out that hydrophobic and steric substituents usually meet this criterion; however, electronic variables are consistently very low, indicating that it may be difficult designing a series with useful variation in electronic properties. In order to determine the overall goodness-of-fit or significance of the equation, the correlation coefficient, r, is determined from the sum of squares due to regression and the sum of squares due to error (or unexplained variance). As more and more variance is accounted for by the sum of squares due to regression (*i.e.* as the sum of squares due to error approaches zero), the value for r^2, the square of the correlation coefficient (called the 'determinant of correlation'), approaches one. In this manner, r^2 can be envisioned as the fraction of total variance in the data which is explained by the regression model. It should be noted that a high r^2 value does not imply that a good model has been developed as this requires hypothesis testing of the model.

Hypothesis testing of the model is usually accomplished by means of the F test, which tests the likelihood that the equation developed did not arise by chance alone and is therefore significant. Two variations of this test are used. The first, a sequential F test, is used in the stepwise development of a model; it is a test to see if the sum of squares due to regression has increased significantly over the equation with one less variable added. The second variation is a partial F test, a variant of the sequential test, which is calculated for each coefficient of the variables in the model. Each term must be significant. The F test measures the 'internal' statistical likelihood of the significance of the equation. It is not an 'external' measure, as are the various jack-knife and leave-N-out procedures, which will be discussed in Section 21.2.3.2.

In addition, each variable in the model has associated with it a 95% confidence interval, which is based on a generalized t test with s. Thus the estimation of the expected bioactivity of a new analogue is directly proportional to the size of s.

An alternative to the F test is the generalized Student's t value, in which the statistical significance of the calculated regression coefficient, b, is assessed. In the case of simple regression, the t value is simply the square root of the F test value. Regardless if either a F test or Student's t test is conducted the calculated result should be reported with the tabulated theoretical values, usually at the 95% confidence level.

Several areas of statistics/probability theory have continued to afflict the QSAR practitioner. Included at the top of the list is the problem of cause and effect in regression analysis. The fact that a straight line (in the 2-D case) can be fitted to the data does not necessarily suggest a cause and effect. Martin[10] demonstrates a good example of this with a QSAR analysis of the pargyline data at Abbott, where the 'only significant correlation was a positive one between pI_{50} and the date of synthesis'. In this example the results were a little obvious; however, in many others this may not be the case. A similar nemesis is the problem of colinearity. One of the major reasons for attempting an experimental design in the first place is to avoid this problem. In actuality, this problem is twofold. The first problem is the proper detection of the presence of multicolinearity and what to do about it if one has no choice. The second problem was the subject of the previous sections; namely, how to plan your design to avoid the multicolinearity dilemma in the first place. Of the two, the first problem is by far the more abused. The major difficulty lies in the distinction between *simple* and *multiple* colinearities. A simple colinearity, or correlation, determines the interdependence between two variables. The closer this measure is to 1, the more related are the two, or the higher the colinearity. The two variables are said to be orthogonal, or unique, as this measure approaches zero. These values are obtained from the correlation matrix. However, this only considers variables two at a time, not in the multivariate environment that they usually exist in. In other words, a variable might show no significant pairwise (*i.e.* simple) colinearities with all the other variables in the model; however, it might be highly interrelated to two or more variables (*i.e.* multiple colinearity). Several methods have been proposed to determine whether multiple correlations exist,[8,110,111,221,222] including factor analysis of the correlation matrix.[69,83,123,223]

Another problem that has just started receiving attention,[6] is the proper choice of statistical analysis. Most investigators, by default, choose parametric regression analysis for this means, without paying attention to the inherent assumptions of the technique during the experimental design phase of the project. This is usually done because of the wide availability of statistical packages containing such techniques. The major assumption with any parametric analysis is the need for a normal distribution[6]—if the analogues to be synthesized have been chosen utilizing one of the techniques just discussed along with the proposed test series being checked for multi-colinearities, there should not be any problems in the application of a parametric statistical technique. A good example of such problems can be seen with a QSAR analysis published by Lien and co-workers.[224] The authors in this study developed an equation to explain the affinity of 128 quaternary ammonium compounds towards cholinergic receptors, using multiple regression analysis, without first screening the data to see if the assumptions of this parametric technique had been met. Although the authors felt they had developed an excellent equation with good statistics to back it up, the validity of the equation was questioned when compounds suggested by these workers were made by Lambrecht *et al.*[225] and found to be many orders of magnitude less active than predicted. In fact, one of the compounds proposed by Lien and co-workers to be a potent cholinergic agonist actually turned out to be an antagonist when prepared by Lambrecht *et al.* Reanalysis of the data[226] showed that the data did not follow a normal distribution. It was also shown that rank transform regression,[6,226] a nonparametric, distribution-free equivalent to parametric regression analysis, was the proper statistical technique for the analysis of such data, as can be seen from the results of equation (5) and Tables 4 and 5, where the compounds proposed by the original authors,[224] along with the actual data,[225] demonstrates the validity of equation (5) over the original equation developed by Lien and co-workers. In equation (5) R (variable name) indicates the appropriate rank of the variable. The variables are as described in the paper by Lien and co-workers.[224] The results are reported for the rank transform regression equation: this means that the standard deviation, s, must be transformed before comparison can be made to the original parametric equation.

$$R(\log K) = 0.46(\pm 0.05) \, R(n_{OH}) + 1.26(\pm 0.12) \, R(\log \mu_R) - 0.01(\pm 0.0) \, (R(\log \mu_R)^2) - 27.49 \quad (5)$$

$$n = 128, r = 0.943, s = 12.49, F_{3,124} = 332$$

$$r^2 \text{ of } R(n_{OH}) \textit{ versus } R(\log \mu_R) = 0.12$$
$$r^2 \text{ of } R(n_{OH}) \textit{ versus } R(\log \mu_R)^2 = 0.11$$

Table 4 Reexamination of the Results from Lien and Co-workers[224]

Variable[a]	Skewness[b]	Kurtosis[b]
n_{OH}	2.24	3.05
μ_R	−0.57	−1.00
$\log \mu_R$	−1.00	0.23
$(\pi\text{-N}\!\!\equiv\!\!{}^+)^2$	1.78	1.74

[a] See ref. 224 for definitions. [b] See ref. 6 for detailed description. Values of approximately zero for these two measures indicate that the data are normally distributed.

Table 5 Problems with Prediction using Non-normally Distributed Data

Compound Predicted by Lien and Co-workers Based on Results from Ref. 224

Predicted affinity from ref. 224: 11.75
Predicted affinity from equation (5)[a]: 6.86
Compound synthesized, tested and reported in ref. 225: observed $\log K = 5.50 \pm 0.12$

[a] Ref. 226 using methodology described in ref. 6.

21.2.3.2 The Importance of Predictions

There is only one method of testing hypotheses that is absolutely guaranteed to give an accurate picture of the QSAR and that is by making predictions, synthesizing the new compounds and having them tested in the *same* biological system* upon which the QSAR model was based. All other methods of 'validating' statistical models are subject to some type of bias or weakness,[227] including split-halves and cross-validation,[6,227-230] as well as bootstrapping and jack-knife[231-233] techniques.

The ultimate test for a QSAR equation is the generation of a novel entity that makes it into the clinic or, hopefully, the market. QSAR as a formal technique was 25 years old in 1987, which makes it of sufficient vintage to have some 'success stories' under its belt. Hansch,[234,235] Hopfinger[236,237]

---

* It is important to realize that the *same* biological system implies that the assay will be done in a similar time frame as the previous determinations (this will assist in keeping experimental errors relatively constant). A similar time frame refers to such things as using the same batch or preparation of enzyme, same supplier and batch of animals, same technician, same lab, etc. It is always best to include standards or rerun knowns. Although this all seems very obvious, papers still appear in the literature where these guidelines are not followed.

and Martin[10] have documented many of the achievements from the USA and Fujita[238] has outlined the success stories from Japan, several of which have actually made it to the marketplace. The majority of these accomplishments utilized one or more of the experimental design techniques discussed in this chapter.

21.2.4 THE FUTURE

Craig plots[11,12] represented one of the first attempts at planning substituents. Since the approach was pencil and paper based, it was limited to examining two variables at a time. With the advent of 3-D computer graphics, it now becomes possible to easily view four (three axes plus color) dimensions simultaneously. Cramer and co-workers[239] and Unger and Pleiss[240] have Evans and Sutherland PS-300 real time color graphics display programs for examining multiple variables simultaneously. Also, an Apple Macintosh program[241] has recently become available for similar manipulations. Such techniques allow the choice of maximum variance and minimum covariance subsets using visual cues. In the future, it would also be possible to pick individual substituents, display a goodness-of-pick parameter, tie the substituents in with a known intermediate parent structure, perform an automatic search of a commercial chemical intermediate database and generate purchase orders for the available compounds.

21.2.5 CONCLUSION

Many methods have been devised to help in the planning of test series. Each method has certain advantages and disadvantages. It should be clear from the above discussion that it is always better to plan a series using any favorite method than to simply make the synthetically expedient members of a series. Planning has several benefits, including the following. (1) Planned series are easier to analyze by statistical QSAR. (2) Planned series more often than not contain 'unusual' substituents which extend the range of validity of the statistical QSAR. This assures a more global QSAR solution and therefore synthesis of the most active (selective, *etc.*) analogues possible. The patent position is also broadened and strengthened. (3) Synthesis of analogues with unusual substituents also extends the possibility of discovering 'unexpected' uses for the series at a later time. (4) Planned series can reduce the synthetic effort required to thoroughly understand the SAR of a lead compound.

The only real drawback to planning a series is that it takes extra effort on the part of the chemist to carry out the plan and to request replacement substituents if a particular target substituent cannot be made. If the overall benefits of planning are explained to the chemist, perhaps some of the resistance will disappear.

Most of the discussion in this chapter has been centered around aromatic or aliphatic substituents on a fixed parent molecule. Many of these same methods can also be applied to more diverse structures if the structures are represented by variables that are consistent, global properties calculated using any of the computational schemes such as CLOGP3[242,243] for lipophilicity, quantum mechanical techniques for charge or orbital characteristics, molecular mechanics for conformational properties, and programs like STERIMOL[244,245] for steric contributions.

ACKNOWLEDGEMENTS

This is contribution number 744 from the Syntex Institute of Organic Chemistry. We would like to thank Dr. Richard D. Cramer, III (Evans Sutherland/Tripos) for a copy of ref. 239 prior to publication and Dr. Christine E. Brotherton-Pleiss (Syntex Research, IBOC) for her assistance in the preparation of this chapter.

21.2.6 REFERENCES

1. E. J. Lien, 'SAR: Side Effects and Drug Design', Dekker, New York, 1987.
2. R. A. Fisher, 'The Design of Experiments', Oliver and Boyd, Edinburgh, 1935.
3. W. S. Cleveland, 'The Elements of Graphing Data', Wadsworth, Monterey, CA, 1985.
4. E. R. Tufte, 'The Visual Display of Quantitative Information', Graphics Press, Cheshire, CT, 1984.
5. B. S. Everitt, 'Graphical Techniques for Multivariate Data', Heinemann, London, 1978.

6. M. A. Pleiss, in 'QSAR Design of Bioactive Compounds', ed. M. Kuchar, Prous, Barcelona, 1984, pp. 403–424 (*Chem. Abstr.*, 1985, **103**, 101 362v).
7. C. Hansch, S. H. Unger and A. B. Forsythe, *J. Med. Chem.*, 1973, **16**, 1217.
8. S. H. Unger, in 'Drug Design', ed. E. J. Ariëns, Academic Press, New York, 1980, vol. 9, pp. 75–85.
9. G. E. P. Box, W. G. Hunter and J. S. Hunter, 'Statistics for Experimenters', Wiley, New York, 1978.
10. Y. C. Martin, *J. Med. Chem.*, 1981, **24**, 229.
11. P. N. Craig, *J. Med. Chem.*, 1971, **14**, 680.
12. P. N. Craig and C. H. Hansch, *J. Med. Chem.*, 1973, **16**, 661.
13. K. H. Kim, C. Hansch, J. Y. Fukunaga, E. E. Steller, P. Y. C. Jow, P. N. Craig and J. Page, *J. Med. Chem.*, 1979, **22**, 366.
14. R. D. Cramer, III, K. M. Snader, C. R. Willis, L. W. Chakrin, J. Thomas and B. M. Sutton, *J. Med. Chem.*, 1979, **22**, 714.
15. J. G. Topliss, *J. Med. Chem.*, 1972, **15**, 1006.
16. J. G. Topliss and Y. C. Martin, in 'Drug Design', ed. E. J. Ariëns, Academic Press, New York, 1975, vol. 5, pp. 1–21.
17. R. W. Taft, Jr., in 'Steric Effects in Organic Chemistry', ed. M. S. Newman, Wiley, New York, 1956, p. 598.
18. C. K. Hancock, E. A. Meyers and B. J. Yager, *J. Am. Chem. Soc.*, 1961, **83**, 4211.
19. J. E. Tompkins, *J. Med. Chem.*, 1986, **29**, 855.
20. D. S. Shewach, J.-W. Chern, K. E. Pillote, L. B. Townsend and P. E. Daddona, *Cancer Res.*, 1986, **46**, 519.
21. C. A. Lipinski, J. L. LaMattina and L. A. Hohnke, *J. Med. Chem.*, 1985, **28**, 1628.
22. A.-M. M. E. Omar, N. H. Eshba and H. M. Salama, *Arch. Pharm.* (*Weinheim, Ger.*), 1984, **317**, 701 (*Chem. Abstr.*, 1984, **101**, 171 160k).
23. L. L. Setescak, F. W. Dekow, J. M. Kitzen and L. L. Martin, *J. Med. Chem.*, 1984, **27**, 401.
24. R. Imhof, E. Kyburz and J. J. Daly, *J. Med. Chem.*, 1984, **27**, 165.
25. I. N. Gracheva, I. R. Kovel'man, A. I. Tochilkin, I. V. Verevkina, D. I. Ioffina and V. Z. Gorkin, *Khim.-Farm. Zh.*, 1983, **17**, 1055 (*Chem. Abstr.*, 1983, **100**, 138 914r).
26. A. Ilczuk, *Acta Pol. Pharm.*, 1982, **39**, 337 (*Chem. Abstr.*, 1983, **99**, 98 808m).
27. M. A. El-Dawy, A.-M. M. E. Omar, A. M. Ismail and A. A. B. Hazzaa, *J. Pharm. Sci.*, 1983, **72**, 45.
28. F. El-Fehail Ali, P. A. Dandridge, J. G. Gleason, R. D. Krell, C. H. Kruce, P. G. Lavanchy and K. M. Snader, *J. Med. Chem.*, 1982, **25**, 947.
29. N. Finch, T. R. Campbell, C. W. Gemenden and H. J. Povalski, *J. Med. Chem.*, 1980, **23**, 1405.
30. M. Prost, V. Van Cromphaut, W. Verstraeten, M. Dirks, C. Tornay, M. Colot and M. de Clavière, *Eur. J. Med. Chem.-Chim. Ther.*, 1980, **15**, 215.
31. S. Takemura, H. Terauchi, Y. Miki, K. Nakano, Y. Inamori, K. Miyazeki and H. Nishimura, *Yakugaku Zasshi*, 1979, **99**, 779 (*Chem. Abstr.*, 1980, **92**, 41 510j).
32. V. Hagen, E. Morgenstern, E. Goeres, R. Franke, W. Sauer and G. Heine, *Pharmazie*, 1980, **35**, 183.
33. R. D. Cramer, III, K. M. Snader, C. R. Willis, L. W. Chakrin, J. Thomas and B. M. Sutton, *ACS Symp. Ser.*, 1980, **118**, 159.
34. A. A. Ordukhanyan, M. A. Landau, E. A. Rudzit, Sh. L. Mndzhoyan and Yu. Z. Ter-Zakharyan, *Khim.-Farm. Zh.*, 1980, **14**, 65 (*Chem. Abstr.*, 1980, **92**, 191 871s).
35. I. W. Mathison and R. J. Pennington, *J. Med. Chem.*, 1980, **23**, 206.
36. C. H. Cashin, J. Fairhurst, D. C. Horwell, I. A. Pullar, S. Sutton, G. H. Timms, E. Wildsmith and F. Wright, *Eur. J. Med. Chem.-Chim. Ther.*, 1978, **13**, 495.
37. K. H. Baggaley, M. Heald, R. M. Hindley, B. Morgan, J. L. Tee and J. Green, *J. Med. Chem.*, 1975, **18**, 833.
38. C. B. Chapleo, M. Myers, P. L. Myers, J. F. Saville, A. C. B. Smith, M. R. Stillings, I. F. Tulloch, D. S. Walter and A. P. Welbourn, *J. Med. Chem.*, 1986, **29**, 2273.
39. W. J. Ross, R. G. Harrison, M. R. J. Jolley, M. C. Neville, A. Todd, J. P. Verge, W. Dawson and W. J. F. Sweatman, *J. Med. Chem.*, 1979, **22**, 412.
40. J. Wolinski and W. Maniecka, *Acta Pol. Pharm.*, 1978, **35**, 621 (*Chem. Abstr.*, 1979, **91**, 91 536x).
41. P. C. Ruenitz and C. M. Mokler, *J. Med. Chem.*, 1979, **22**, 1142.
42. H. Horstmann, S. Schütz, R. Gönnert and P. Andrews, *Eur. J. Med. Chem.-Chim. Ther.*, 1978, **13**, 475.
43. N. Finch, T. R. Campbell, C. W. Gemenden, M. J. Antonaccio and H. J. Povalski, *J. Med. Chem.*, 1978, **21**, 1269.
44. H. Horstmann, E. Moeller, E. Wehinger and K. Meng, *ACS Symp. Ser.*, 1978, **83**, 125–139.
45. H. H. Ong, V. B. Anderson and J. C. Wilker, *J. Med. Chem.*, 1978, **21**, 758.
46. G. May, D. Peteri and K. Hummel, *Arzneim.-Forsch.*, 1978, **28**, 732.
47. J. A. Waters, *J. Med. Chem.*, 1978, **21**, 628.
48. T. Aono, Y. Araki, M. Imanishi and S. Noguchi, *Chem. Pharm. Bull.*, 1978, **26**, 1153.
49. A. J. van der Broek, A. I. A. Broess, M. J. van der Heuvel, H. P. de Jongh, J. Leemhuis, K. H. Schönemann, J. Smits, J. de Visser, N. P. van Vliet and F. J. Zeelen, *Steroids*, 1977, **30**, 481.
50. K. Rehse, T. Lang and N. Rietbrock, *Arch. Pharm.* (*Weinheim, Ger.*), 1977, **310**, 979 (*Chem. Abstr.*, 1978, **88**, 146 122e).
51. P. Bouvier, D. Branceni, M. Prouteau, E. Prudhommeaux and C. Viel, *Eur. J. Med. Chem.-Chim. Ther.*, 1976, **11**, 271.
52. P. H. Morgan and I. W. Mathison, *J. Pharm. Sci.*, 1976, **65**, 635.
53. R. I. Mrongovius, *Eur. J. Med. Chem.-Chim. Ther.*, 1975, **10**, 474.
54. J. K. Sugden and M. R. T. Saberi, *Eur. J. Med. Chem.-Chim. Ther.*, 1979, **14**, 189.
55. I. W. Mathison and P. H. Morgan, *J. Med. Chem.*, 1974, **17**, 1136.
56. H. Nagatomi, K. Ando, M. Kawasaki, B. Yasui, Y. Miki and S. Takemura, *Chem. Pharm. Bull.*, 1979, **27**, 1021.
57. H. Brunner, R. Kroiss, M. Schmidt and H. Schönenberger, *Eur. J. Med. Chem.-Chim. Ther.*, 1986, **21**, 333.
58. J. Jung, C. Rentzea and W. Rademacher, *J. Plant Growth Regul.*, 1986, **4**, 181.
59. G. Smith, C. H. L. Kennard, A. H. White and B. W. Skelton, *J. Agric. Food Chem.*, 1981, **29**, 1046.
60. C. C. Dary and L. K. Cutkomp, *Pestic. Sci.*, 1984, **15**, 443.
61. J. G. Topliss, *J. Med. Chem.*, 1977, **20**, 463.
62. B. Blank, A. J. Krog, G. Weiner and R. G. Pendleton, *J. Med. Chem.*, 1980, **23**, 837.
63. J. R. Dimmock, K. Shyam, N. W. Hamon, B. M. Logan, S. K. Raghavan, D. J. Harwood and P. J. Smith, *J. Pharm. Sci.*, 1983, **72**, 887.
64. J. R. Dimmock, S. K. Raghavan, B. M. Logan and G. E. Bigam, *Eur. J. Med. Chem.-Chim. Ther.*, 1983, **18**, 248.
65. J. M. Yeung, L. A. Corleto and E. E. Knaus, *J. Med. Chem.*, 1982, **25**, 720.

66. I. Virsis, D. A. Silaraya, B. Grinberga, A. Prikulis, I. Gerbashevska and M. Orenis, *Khim.-Farm. Zh.*, 1981, **15**, 23 (*Chem. Abstr.*, 1982, **96**, 118 054f).
67. A. B. Sahasrabudhe, B. V. Bapat and S. N. Kulkarni, *Indian J. Chem., Sect. B*, 1981, **20**, 495 (*Chem. Abstr.*, 1981, **95**, 168 907m).
68. G. B. Bennett, R. G. Babington, M. A. Deacon, P. L. Eden, S. P. Kerestan, G. H. Leslie, E. A. Ryan, R. B. Mason and H. E. Minor, *J. Med. Chem.*, 1981, **24**, 490.
69. Y. C. Martin and H. N. Panas, *J. Med. Chem.*, 1979, **22**, 784.
70. T. M. Bustard, *J. Med. Chem.*, 1974, **17**, 777.
71. N. J. Santora and K. Auyang, *J. Med. Chem.*, 1975, **18**, 959.
72. F. J. Zeelen, *Abh. Akad. Wiss. DDR, Abt. Math., Naturwiss., Tech.*, 1978, 333 (*Chem. Abstr.*, 1979, **91**, 68 774g).
73. S. N. Deming, *J. Med. Chem.*, 1976, **19**, 977.
74. A. I. Khuri and J. A. Cornell, 'Response Surfaces: Designs and Analyses', Dekker, New York, 1987.
75. S. N. Deming and S. L. Morgan, *Anal. Chem.*, 1973, **45**, 278A.
76. S. N. Deming and S. L. Morgan, *Anal. Chim. Acta*, 1983, **150**, 183.
77. W. Spendley, G. R. Hext and F. R. Himsworth, *Technometrics*, 1962, **4**, 441.
78. F. Darvas, *J. Med. Chem.*, 1974, **17**, 799.
79. F. Darvas, L. Kovacs and A. Eory, *Abh. Akad. Wiss. DDR, Abt. Math., Naturwiss., Tech.*, 1978, 311 (*Chem. Abstr.*, 1979, **91**, 32 476e).
80. R. D. Gilliom, W. P. Purcell and T. R. Bosin, *Eur. J. Med. Chem.-Chim. Ther.*, 1977, **12**, 187.
81. D. M. Olsson, *J. Qual. Technol.*, 1974, **6**, 53.
82. G.L. Kirschner and B. R. Kowalski, in 'Drug Design', ed. E. J. Ariëns, Academic Press, New York, 1979, vol. 8, pp. 73–131.
83. Y. C. Martin, 'Quantitative Drug Design: A Critical Introduction', Dekker, New York, 1978.
84. R. Franke, 'Theoretical Drug Design Methods', Akademie-Verlag, Berlin, 1984 (this is the revised English translation of 'Optimierungsmethoden in der Wirkstofforschung—Quantitative Struktur-Wirkungs-Analyse', published in 1980).
85. J. K. Seydel and K.-J. Schaper, 'Chemische Struktur und Biologische Aktivitat von Wirkstoffen', Verlag Chemie, Weinheim, 1979.
86. V. Austel and E. Kutter, *Arzneim.-Forsch.*, 1981, **31**, 130.
87. V. Austel and E. Kutter, in 'Drug Design', ed. E. J. Ariëns. Academic Press, New York, 1980, vol. 10, pp. 1–69.
88. V. Austel, *Eur. J. Med. Chem.-Chim. Ther.*, 1982, **17**, 9.
89. V. Austel, *Eur. J. Med. Chem.-Chim. Ther.*, 1982, **17**, 339.
90. V. Austel, *Quant. Struct.-Act. Relat. Pharmacol., Chem. Biol.*, 1983, **2**, 59.
91. V. Austel, in 'Quantitative Approaches to Drug Design', Pharmacochemistry Library, vol. 6, ed. J. C. Dearden, Elsevier, New York, 1983, pp. 223–229 (*Chem. Abstr.*, 1983, **99**, 187 093s).
92. V. Austel, in 'QSAR Strategies in the Design of Bioactive Compounds, Proceedings of the European Symposium on Quantitative Structure–Activity Relationships, 5th', ed. J. K. Seydel, VCH, Weinheim, Germany, 1985, pp. 247–250 (*Chem. Abstr.*, 1986, **104**, 101 943f).
93. V. Austel, in 'Fortschritte der Arzneimittelforschung, Gesamtkongress aus Deutsche Pharmazeutische Wissenschaft, 1st', ed. H. Oelschlaeger, Wissenschafliche Verlagsgesellschaft, Stuttgart, Germany, 1983, pp. 69–76 (*Chem. Abstr.*, 1985, **102**, 214 420k).
94. V. Austel, *Top. Curr. Chem.*, 1983, **114**, 7 (*Chem. Abstr.*, 1983, **99**, 151 593r).
95. V. Austel, in 'X-Ray Crystallography in Drug Action, Course for the International School of Crystallography, 9th', ed. A. S. Horn and C. J. De Ranter, Oxford University Press, Oxford, 1984, pp. 441–460 (*Chem. Abstr.*, 1985, **102**, 14g).
96. E. Kutter and V. Austel, *Arzneim.-Forsch.*, 1981, **31**, 135.
97. K. D. Paull, M. Nasr and V. L. Narayanan, *Arzneim.-Forsch.*, 1984, **34**, 1243.
98. W. G. Cochran and G. M. Cox, 'Experimental Designs', 2nd edn., Wiley, New York, 1957.
99. R. G. Petersen, 'Design and Analysis of Experiments', Dekker, New York, 1985.
100. V. Gomez-Parra, F. Sanchez and C. Yague, *Arch. Pharm. (Weinheim, Ger.)*, 1986, **319**, 552 (*Chem. Abstr.*, 1986, **105**, 115 026s).
101. P. S. Magee, in 'Pesticide Chemistry: Human Welfare and Environment, Proceedings of the International Congress on Pesticide Chemistry, 5th, Kyoto', ed. J. Miyamoto and P. C. Kearney, Pergamon Press, Oxford, 1983, vol. 1, pp. 251–260 (*Chem. Abstr.*, 1983, **99**, 34 405v).
102. I. T. Harrison, W. Kurz, I. J. Massey and S. H. Unger, *J. Med. Chem.*, 1978, **21**, 588.
103. J. W. McFarland and D. J. Gans, *J. Med. Chem.*, 1986, **29**, 505.
104. A. M. Stoddard, *Biometrics*, 1979, **35**, 765.
105. Y. C. Martin, J. B. Holland, C. H. Jarboe and N. Plotnikoff, *J. Med. Chem.*, 1974, **17**, 409.
106. P. S. Magee, in 'Computer-assisted Drug Design', ed. E. C. Olson and R. E. Christoffersen, *ACS Symp. Ser.*, 1979, **112**, 319.
107. P. S. Magee, *CHEMTECH*, 1981, **11**, 378.
108. C. Hansch and A. J. Leo, 'Substituent Constants for Correlation Analysis in Chemistry and Biology', Wiley, New York, 1979, chap. VI, pp. 48–63.
109. S. H. Unger and C. Hansch, in 'Proceedings of the 167th Meeting of the American Chemical Society', American Chemical Society, Washington, D. C., 1974, Abstract No. CHLT004.
110. D. E. Farrar and R. R. Glauber, *Rev. Econ. Stat.*, 1969, **49**, 92.
111. Y. Haitovsky, *Rev. Econ. Stat.*, 1969, **51**, 486.
112. J. A. Hartigan and M. A. Wong, *Appl. Stat.*, 1979, **28**, 100.
113. J. A. Hartigan, 'Clustering Algorithms', Wiley, New York, 1975.
114. G. W. Milligan, *Psychometrika*, 1980, **45**, 325.
115. P. E. Green, 'Analyzing Multivariate Data', The Dryden Press, Hinsdale, IL, 1978.
116. BMDP Statistical Software, ed. W. J. Dixon, University of California Press, Berkeley, CA, USA, 1983, pp. 456–463.
117. CLUSTAN, 16 Kingsburgh Road, Edinburgh EH12 6DZ, Scotland.
118. SAS Institute Inc., Box 8000, Cary, NC 27511, USA.
119. SPSS Inc., 444 North Michigan Avenue, Chicago, IL, USA.

120. STATGRAPHICS, STSC, Inc., 2115 East Jefferson Street, Rockville, MD 20852, USA.
121. P. H. A. Sneath and R. R. Sokal, 'Numerical Taxonomy; The Principles and Practice of Numerical Classification', Freeman, San Francisco, 1973.
122. S. H. Unger, unpublished observation.
123. Y. C. Martin, in 'Drug Design', ed. E. J. Ariëns, Academic Press, New York, 1979, vol. 8, pp. 1–72.
124. S. H. Unger, J. R. Cook and J. S. Hollenberg, *J. Pharm. Sci.*, 1978, **67**, 1364.
125. C. Hansch, A. J. Leo, S. H. Unger, K. H. Kim, D. Nikaitani and E. J. Lien, *J. Med. Chem.*, 1973, **16**, 1207.
126. C. Hansch, S. D. Rockwell, P. Y. C. Jow, A. J. Leo and E. E. Steller, *J. Med. Chem.*, 1977, **20**, 304.
127. E. A. Coats, K. J. Shah, S. R. Milstein, C. S. Genther, D. M. Nene, J. Roesener, J. Schmidt, M. A. Pleiss, E. Wagner and J. K. Baker, *J. Med. Chem.*, 1982, **25**, 57.
128. K. J. Shah and E. A. Coats, *J. Med. Chem.*, 1977, **20**, 1001.
129. B. R. Baker, 'Design of Active-Site-Directed Irreversible Enzyme Inhibitors', Wiley, New York, 1967.
130. E. A. Coats, in 'Physical Chemical Properties of Drugs', ed. S. H. Yalkowsky, A. A. Sinkula and S. C. Valvani, Dekker, New York, 1980, pp. 111–139.
131. C. Hansch and J. M. Clayton, *J. Pharm. Sci.*, 1973, **62**, 1.
132. C. Hansch, *Drug Dev. Res.*, 1981, **1**, 267.
133. W. J. Dunn, III and M. J. Greenberg, *J. Pharm. Sci.*, 1977, **66**, 1416.
134. W. J. Dunn, III, M. J. Greenberg and S. S. Callejas, *J. Med. Chem.*, 1976, **19**, 1299.
135. G. F. Kolar and R. Preussmann, *Z. Naturforsch., Teil B*, 1971, **26**, 950.
136. B.-K. Chen, C. Horváth and J. R. Bertino, *J. Med. Chem.*, 1979, **22**, 483.
137. K. Awano and S. Suzue, *Chem. Pharm. Bull.*, 1986, **34**, 2833.
138. R. Lisciani, S. Lembo, S. Cozzolino, M. I. La Rotonda, C. Silipo and A. Vittoria, *Farmaco, Ed. Sci.*, 1986, **41**, 89 (*Chem. Abstr.*, 1986, **104**, 179 735j).
139. P. Dallet, J.-P. Dubost, J.-C Colleter, E. Audry and M.-H. Creuzet, *Eur. J. Med. Chem.-Chim. Ther.*, 1985, **20**, 551.
140. J. Bompart, G. Pastor, L. Giral and R. Alvart, *Ann. Pharm. Fr.*, 1984, **42**, 537 (*Chem. Abstr.*, 1985, **102**, 215 069q).
141. M. Calas, A. Barbier, L. Giral, B. Balmayer and E. Despaux, *Eur. J. Med. Chem.-Chim. Ther.*, 1982, **17**, 497.
142. P. Pratesi, L. Villa, V. Ferri, C. De Micheli, E. Grana, M. G. S. Barbone, C. Silipo and A. Vittoria, *Farmaco, Ed. Sci.*, 1981, **36**, 749 (*Chem. Abstr.*, 1981, **95**, 180 674r).
143. V. E. Golender and A. B. Rozenblit, *Zh. Vses. Khim. Ova*, 1980, **25**, 28 (*Chem. Abstr.*, 1980, **92**, 140 306d).
144. M. Calas, C. Pages, G. Pastor, L. Giral and E. Despaux, *Eur. J. Med. Chem.-Chim. Ther.*, 1979, **14**, 529.
145. J. A. Kiritsy, D. K. Yung and D. E. Mahony, *J. Med. Chem.*, 1978, **21**, 1301.
146. E. A. Coats, S. R. Milstein, M. A. Pleiss and J. A. Roesener, *J. Med. Chem.*, 1978, **21**, 804.
147. G. L. Tong, M. Cory, W. W. Lee, D. W. Henry and G. Zbinden, *J. Med. Chem.*, 1978, **21**, 732.
148. B. Blank, N. W. DiTullio, L. Deviney, J. T. Roberts and H. L. Saunders, *J. Med. Chem.*, 1977, **20**, 577.
149. T. R. Herrin, J. M. Pauvlik, E. V. Schuber and A. O. Geiszler, *J. Med. Chem.*, 1975, **18**, 1216.
150. E. A. Coats, S. R. Milstein, G. Holbein, J. McDonald, R. Reed and H. G. Petering, *J. Med. Chem.*, 1976, **19**, 131.
151. P. Pratesi, L. Villa, V. Ferri, C. De Micheli, E. Grana, C. Grieco, C. Silipo and A. Vittoria, *Farmaco, Ed. Sci.*, 1979, **34**, 571 (*Chem. Abstr.*, 1979, **91**, 140 446d).
152. K. K. Shatemirova, G. A. Davydova, I. V. Verevkina and A. I. Tochilkin, *Vopr. Med. Khim.*, 1977, **23**, 609 (*Chem. Abstr.*, 1977, **87**, 196 269h).
153. C. Grieco, C. Silipo and A. Vittoria, *Farmaco, Ed. Sci.*, 1976, **31**, 917 (*Chem. Abstr.*, 1977, **86**, 65 346n).
154. J. P. Tollenaere, H. Moereels and M. Protiva, *Eur. J. Med. Chem.-Chim. Ther.*, 1976, **11**, 293.
155. C. Grieco, C. Silipo and A. Vittoria, *Farmaco, Ed. Sci.*, 1976, **31**, 824 (*Chem. Abstr.*, 1977, **86**, 25 696v).
156. F. Giordano, C. Grieco, C. Silipo and A. Vittoria, *Boll. Soc. Ital. Biol. Sper.*, 1975, **51**, 1069 (*Chem. Abstr.*, 1976, **84**, 174 411j).
157. E. L. Plummer, *ACS Symp. Ser.*, 1984, **255**, 297.
158. E. L. Plummer and D. S. Pincus, *J. Agric. Food Chem.*, 1981, **29**, 1118.
159. P. Willett, *J. Chem. Inf. Comput. Sci.*, 1985, **25**, 78.
160. D. Bawden, *Anal. Chim. Acta*, 1984, **158**, 363.
161. P. Willett, *J. Chem. Inf. Comput. Sci.*, 1984, **24**, 29.
162. G. W. Adamson and D. Bawden, *J. Chem. Inf. Comput. Sci.*, 1981, **21**, 204.
163. D. Coomans and D. L. Massart, *Anal. Chim. Acta*, 1981, **133**, 225.
164. D. M. Borth, R. J. McKay and J. R. Elliott, *Technometrics*, 1985, **27**, 25.
165. K.-J. Schaper, in 'Quantitative Approaches to Drug Design', Pharmacochemistry Library, vol. 6, ed. J. C. Dearden, Elsevier, New York, 1983, pp. 235–236.
166. K.-J. Schaper, *Quant. Struct.-Act. Relat. Pharmacol., Chem. Biol.*, 1983, **2**, 111.
167. M. J. Box and N. R. Draper, *Technometrics*, 1971, **13**, 731.
168. N. R. Draper and H. Smith, 'Applied Regression Analysis', 2nd edn., Wiley, New York, 1981.
169. D. W. Gaylor and J. A. Merrill, *Technometrics*, 1968, **10**, 73.
170. T. J. Mitchell, *Technometrics*, 1974, **16**, 203.
171. E. C. Herrmann, in 'Quantitative Approaches to Drug Design', Pharmacochemistry Library, vol. 6, ed. J. C. Dearden, Elsevier, New York, 1983, pp. 231–232.
172. R. Wootton, R. Cranfield, G. C. Sheppey and P. J. Goodford, *J. Med. Chem.*, 1975, **18**, 607.
173. R. W. Kennard and L. A. Stone, *Technometrics*, 1969, **11**, 137.
174. F. E. Norrington, R. M. Hyde, S. G. Williams and R. Wootton, *J. Med. Chem.*, 1975, **18**, 604.
175. C. G. Swain and E. C. Lupton, Jr., *J. Am. Chem. Soc.*, 1968, **90**, 4328.
176. S. G. Williams and F. E. Norrington, *J. Am. Chem. Soc.*, 1976, **98**, 508.
177. P. J. Goodford, A. T. Hudson, G. C. Sheppey, R. Wootton, M. H. Black, G. J. Sutherland and J. C. Wickham, *J. Med. Chem.*, 1976, **19**, 1239.
178. R. Wootton, *J. Med. Chem.*, 1983, **26**, 275.
179. W. J. Streich, S. Dove and R. Franke, *J. Med. Chem.*, 1980, **23**, 1452.
180. R. Cranfield, P. J. Goodford, F. E. Norrington, W. H. G. Richards, G. C. Sheppey and S. G. Williams, *Br. J. Pharmacol.*, 1974, **52**, 87.

181. A. M. Doweyko, A. R. Bell, J. A. Minatelli and D. I. Relyea, *J. Med. Chem.*, 1983, **26**, 475.
182. S. Dove, W. J. Streich and R. Franke, *J. Med. Chem.*, 1980, **23**, 1456.
183. G. Krause, M. Klepel and R. Franke, in 'Quantitative Approaches to Drug Design', Pharmacochemistry Library, vol. 6, ed. J. C. Dearden, Elsevier, New York, 1983, pp. 233, 234.
184. A. A. Ordukhanyan, A. S. Kabankin and M. A. Landau, *Khim.-Farm. Zh.*, 1979, **13**, 21 (*Chem. Abstr.*, 1979, **90**, 202 962j). An English translation of this article appears in *Pharm. Chem. J.* (*Engl. Transl.*), 1979, **13**, 245–251.
185. K. R. H. Wooldridge, *Eur. J. Med. Chem.-Chim. Ther.*, 1980, **15**, 63.
186. K. R. H. Wooldridge, *Chem. Ind.*, 1980, 478.
187. S. H. Unger and C. Hansch, *J. Med. Chem.*, 1973, **16**, 745.
188. M. Charton, *J. Am. Chem. Soc.*, 1969, **91**, 615.
189. R. J. A. Walsh, K. R. H. Wooldridge, D. Jackson and J. Gilmour, *Eur. J. Med. Chem.-Chim. Ther.*, 1977, **12**, 495.
190. C. Hansch and C. Silipo, *J. Med. Chem.*, 1974, **17**, 661.
191. S. W. Dietrich, R. N. Smith, S. Brendler and C. Hansch, *Arch. Biochem. Biophys.*, 1979, **194**, 612.
192. R. E. Ford, P. Knowles, E. Lunt, S. M. Marshall, A. J. Penrose, C. A. Ramsden, A. J. H. Summers, J. L. Walker and D. E. Wright, *J. Med. Chem.*, 1986, **29**, 538.
193. S. Wold, in 'Drug Design: Fact or Fantasy?', ed. G. Jolles and K. R. H. Wooldridge, Academic Press, New York, 1984, pp. 253–254.
194. P. P. Mager, in 'Drug Design', ed. E. J. Ariëns, Academic Press, New York, 1980, vol. 9, pp. 187–236.
195. P. P. Mager, 'Multidimensional Pharmacochemistry: Design of Safer Drugs', Academic Press, New York, 1984.
196. S. Hellberg, M. Sjöström, B. Skagerberg, C. Wikström and S. Wold, *Acta Pharm. Jugosl.*, 1987, **37**, 53 (*Chem. Abstr.*, 1987, **107**, 17 234e).
197. S. Wold, M. Sjöström, R. Carlson, T. Lundstedt, S. Hellberg, B. Skagerberg, C. Wikström and J. Öhman, *Anal. Chim. Acta*, 1986, **191**, 17.
198. S. Wold and W. J. Dunn, III, *J. Chem. Inf. Comput. Sci.*, 1983, **23**, 6.
199. S. Hellberg, M. Sjöström, B. Skagerberg and S. Wold, *J. Med. Chem.*, 1987, **30**, 1126.
200. W. J. Dunn, III, S. Wold, U. Edlund, S. Hellberg and J. Gasteiger, *Quant. Struct.-Act. Relat. Pharmacol., Chem. Biol.*, 1984, **3**, 131.
201. S. Hellberg, S. Wold, W. J. Dunn, III, J. Gasteiger and M. G. Hutchings, *Quant. Struct.-Act. Relat. Pharmacol., Chem. Biol.*, 1985, **4**, 1.
202. S. Hellberg, M. Sjöström and S. Wold, *Acta Chem. Scand. Ser. B*, 1986, **40**, 135.
203. P. Berntsson and S. Wold, *Quant. Struct.-Act. Relat. Pharmacol., Chem. Biol.*, 1986, **5**, 45.
204. S. Wold, W. J. Dunn, III and S. Hellberg, in 'Drug Design: Fact or Fantasy?', ed. G. Jolles and K. R. H. Wooldridge, Academic Press, New York, 1984, pp. 95–117.
205. H. Wold, in 'Systems under Indirect Observation', ed. K. G. Jöreskog and H. Wold, North Holland, Amsterdam, 1982, part II, pp. 1–54.
206. D. Johnels, U. Edlund, H. Grahn, S. Hellberg, M. Sjöström, S. Wold, S. Clementi and W. J. Dunn, III, *J. Chem. Soc., Perkin Trans. 2*, 1983, 863.
207. S. Wold, A. Ruhe, H. Wold and W. J. Dunn, III, *SIAM J. Sci. Stat. Comput.*, 1984, **5**, 735.
208. D. G. Kleinbaum and L. L. Kupper, 'Applied Regression Analysis and Other Multivariable Methods', Duxbury Press, North Scituate, MA, 1978.
209. A. A. Afifi and S. P. Azen, 'Statistical Analysis—A Computer Oriented Approach', Academic Press, 2nd edn., New York, 1979.
210. C. Daniel and F. S. Wood, 'Fitting Equations to Data', 2nd edn., Wiley, New York, 1980.
211. R. R. Sokal and F. J. Rohlf, 'Biometry: The Principles and Practice of Statistics in Biological Research', 2nd edn., W. H. Freeman, San Francisco, 1981.
212. T. H. Wonnacott and R. J. Wonnacott, 'Regression: A Second Course in Statistics', Wiley, New York, 1981.
213. W. J. Dixon and F. J. Massey, Jr., 'Introduction to Statistical Analysis', 3rd edn., McGraw-Hill, New York, 1969.
214. R. H. Myers, 'Classical and Modern Regression With Applications', Duxbury Press, Boston, 1986.
215. B. S. Everitt, 'Introduction to Optimization Methods', Chapman and Hall, New York, 1987.
216. T. Fujita, J. Iwasa and C. Hansch, *J. Am. Chem. Soc.*, 1964, **86**, 5175.
217. P. N. Craig, C. H. Hansch, J. W. McFarland, Y. C. Martin, W. P. Purcell and R. Zahradník, *J. Med. Chem.*, 1971, **14**, 447.
218. A. Cammarata, R. C. Allen, J. K. Seydel and E. Wempe, *J. Pharm. Sci.*, 1970, **59**, 1496.
219. W. P. Purcell, *Eur. J. Med. Chem.-Chim. Ther.*, 1975, **10**, 335.
220. C. Hansch, in 'International Encyclopedia of Pharmacology and Therapeutics: Section 5, Structure–Activity Relationships', ed. C. J. Cavallito, Pergamon Press, New York, vol. I, 1973, pp. 150–158.
221. R. F. Gunst and R. L. Mason, *Biometrics*, 1977, **33**, 249.
222. R. L. Mason, R. F. Gunst and J. T. Webster, *Commun. Stat.*, 1975, **4**, 277.
223. R. Franke, *Farmaco, Ed. Sci.*, 1979, **34**, 545 (*Chem. Abstr.*, 1979, **91**, 82 767w).
224. E. J. Lien, E. J. Ariëns and A. J. Beld, *Eur. J. Pharmacol.*, 1976, **35**, 245.
225. G. Lambrecht, U. Moser and E. Mutschler, *Eur. J. Med. Chem.-Chim. Ther.*, 1980, **15**, 305.
226. M. A. Pleiss, unpublished observations.
227. R. D. Snee, *Technometrics*, 1977, **19**, 415.
228. B. Efron, *J. Am. Stat. Assoc.*, 1983, **78**, 316.
229. H. T. Eastment and W. J. Krzanowski, *Technometrics*, 1982, **24**, 73.
230. J. C. Baskerville and J. H. Toogood, *Technometrics*, 1982, **24**, 9.
231. D. A. Freedman and S. C. Peters, *J. Am. Stat. Assoc.*, 1984, **79**, 97.
232. B. Efron and G. Gong, *Am. Stat.*, 1983, **37**, 36.
233. S. W. Dietrich, N. D. Dreyer, C. Hansch and D. L. Bentley, *J. Med. Chem.*, 1980, **23**, 1201.
234. C. Hansch, *J. Med. Chem.*, 1976, **19**, 1.
235. C. Hansch, in 'Pharmacochemical Library', ed. J. A. Keverling Buisman, Elsevier, New York, 1977, pp. 47–61 (*Chem. Abstr.*, 1978, **89**, 52 958a).
236. A. J. Hopfinger, *Pharm. Int.*, 1984, **5**, 224 (*Chem. Abstr.*, 1985, **102**, 16 919s).
237. A. J. Hopfinger, *J. Med. Chem.*, 1985, **28**, 1133.

238. T. Fujita, in 'Drug Design: Fact or Fantasy?', ed. G. Jolles and K. R. H. Wooldridge, Academic Press, New York, 1984, pp. 19–33.
239. R. D. Cramer, III, J. D. Bunce, D. E. Patterson and I. E. Frank, *Quant. Struct.–Act. Relat. Pharmacol., Chem. Biol.*, 1988, **7**, 18.
240. S. H. Unger and M. A. Pleiss, unpublished result, (program MDD—Multivariate Data Display).
241. MacSpin, *J. Am. Chem. Soc.*, 1987, **109**, 2863.
242. A. J. Leo, in 'QSAR Strategies in the Design of Bioactive Compounds, Proceedings of the European Symposium of Quantitative Structure-Activity Relationships, 5th', ed. J. K. Seydel, VCH, Weinheim, Germany, 1985, pp. 294–298 (*Chem. Abstr.*, 1986, **104**, 148 214j).
243. CLOGP3 is one of the components of the Daylight MedChem Software package. For more information, contact: Dr. David Weininger, Daylight Chemical Information Systems, Inc., 3951 Claremont Street, Irvine, CA 92714, USA.
244. A. Verloop, W. Hoogenstraaten and J. Tipker, in 'Drug Design', ed. E. J. Ariëns, Academic Press, New York, 1976, vol 7, pp. 165–207.
245. A. Verloop, in 'QSAR Strategies in the Design of Bioactive Compounds, Proceedings of the European Symposium of Quantitative Structure-Activity Relationships, 5th', ed. J. K. Seydel, VCH, Weinheim, Germany, 1985, pp. 98–104 (*Chem. Abstr.*, 1986, **104**, 14 330k).

21.3

The Free–Wilson Method and
its Relationship to the
Extrathermodynamic Approach

HUGO KUBINYI
BASF, Ludwigshafen, FRG

21.3.1 INTRODUCTION

All research in natural sciences aims to derive general conclusions from observations and planned experiments, to discover and recognize the laws of nature and to describe them in a quantitative manner. In this way astronomy, physics and even chemistry have developed step by step into disciplines based on a solid mathematical foundation. It is therefore not surprising that relationships between chemical structures and biological activities were also sought at a very early stage in drug research. From their investigations on the biological action of alkaloids, Crum Brown and Frazer concluded in 1868 that the 'physiological' activity Φ of these compounds has to be a function of their chemical structure C (equation 1).[1] Of course they were unable to calculate biological activities by using this equation, and even today it is impossible to apply equation (1) in a direct manner, but the slight modification to give equation (2), where ΔC is a certain change of the chemical structure and $\Delta \Phi$ is the resulting change in biological activity, leads straight to the concept of Free–Wilson analysis.[2]

$$\Phi = f(C) \tag{1}$$

$$\Delta\Phi = f(\Delta C) \tag{2}$$

All medicinal chemists, whether they are acquainted with the principles of the Free–Wilson approach or not, follow this strategy: each change of chemical structure of a lead compound results in a definite change of biological activity and, as a first approximation, this change is independent of all other modifications in the molecule. Following these guidelines, the drug designer tries to optimize the lead structure step by step. Formulated in mathematical terms, the biological activity BA should be—within a series of closely related compounds—a function of the activity contributions a_i of all different substituents X_i ($X_i = 1$ when the group X_i is present, otherwise $X_i = 0$) and of the biological activity μ of the 'naked' parent structure of the series (equation 3). Equation (3) does not tell us whether the group contributions to biological activity are additive or not, but it assumes that each a_i has a constant value, irrespective of all other chemical changes in the molecule.

$$\text{BA} = f(a_i X_i, \mu) \tag{3}$$

In drug research, the following situation is very common: the medicinal chemist prepares a series of chemically related compounds and after some time the pharmacologist provides the biological data. What is the next step? Is there an easy way to start a structure–activity analysis? Of course, one can try to apply the principles of the extrathermodynamic approach (Chapter 21.1) and perform a Hansch analysis.[3] However, most often, especially in the first stages of lead structure optimization, many different substituent positions of the parent molecule are varied and only a few different substituents in each position are tested, making Hansch analysis very difficult or even impossible. For such cases Free and Wilson, in 1964, proposed a mathematical model (equation 4)[2] based on an additivity concept of biological activity contributions within congeneric series. In equation (4), a_i are the activity contributions of the substituents X_i, referring to the overall mean of biological activity values, μ. If μ is not defined as the overall mean but as the biological activity value of the unsubstituted parent molecule of the series (a definition which corresponds to a later version of Free–Wilson analysis[4]) then the a_i values can be defined (equation 5) in an identical manner to Hammett σ values (equation (6); $k = $ rate constant, $\sigma_X = $ electronic parameter of substituent X, $\rho = $ reaction constant)[5] or Hansch π values (equation (7); $P = $ partition coefficient, $\pi_X = $ lipophilicity parameter of substituent X).[6]

$$\text{BA} = \Sigma a_i + \mu \tag{4}$$

$$\text{BA}_{\text{RX}} - \text{BA}_{\text{RH}} = a_X \tag{5}$$

$$\log k_{\text{RX}} - \log k_{\text{RH}} = \rho\sigma_X \tag{6}$$

$$\log P_{\text{RX}} - \log P_{\text{RH}} = \pi_X \tag{7}$$

In equation (5), a_X is the activity contribution of the group X, with reference to hydrogen in the same position of the same compound. Although it was not formulated by Free and Wilson in their original paper, nowadays biological activity values BA are always on a logarithmic scale, making the analogy to the definitions of σ and π perfect.

Indeed, the Free–Wilson approach is the only real structure–activity model because fragment contributions are derived from biological activity values by linear multiple regression analysis, while, in the extrathermodynamic approach, properties and not structural elements are correlated with the biological data. On the other hand, the applicability and usefulness of Free–Wilson analysis, in comparison to the extrathermodynamic approach, is more limited. While Hansch analysis leads to general conclusions and, in favorable cases, to a better understanding of drug action at the molecular level, Free–Wilson analysis gives only a quantification of the effects of chemical changes on the biological activity.

21.3.2 HISTORY

While equation (1) may be seen as the first formulation of a Free–Wilson-type equation, it was not applied until nearly 90 years later. Although many relationships between physicochemical properties, such as partition coefficients, solubility, vapor pressure, *etc.*, and unspecific biological actions, such as toxic, narcotic, bactericidal, fungicidal and hemolytic activities, have been recognized since the end of the 19th century, it was not until 1956 that Bruice, Kharasch and Winzler first formulated a real structure–activity relationship for a more specific type of biological activity. For the thyromimetic activities of 47 analogues of (1) they formulated equation (8),[7] where $\Sigma f = f_X + f_Y + f_{OR'}$.

(1)

$$\log(\% \text{ thyroxine-like activity}) = k \cdot \Sigma f + c \qquad (8)$$

Without giving mathematical details, they stated that all the substituent constants f were entirely empirical and were derived in a similar way to Hammett σ constants (equation 6). Close relationships were found between experimental activity values and the sums of f values. Bruice *et al.* stressed the empirical nature of this correlation but emphasized that it was a useful summary of experimental results and a rationale for further research in this field. On reflection, it is very remarkable that log(activity) values were used for the analysis and that the authors were able to interpret their group contributions f in physicochemical terms: they recognized that the thyromimetic activity is related to the electron-attracting or -releasing character of the groups X, Y and OR' by a comparison of the different f values with the electronic properties of these substituents.

Fried and Borman followed a similar approach to derive enhancement factors (biological activity values on a linear scale) for different substituents and different biological tests in a series of mineralo- and glucocorticoidal steroids.[8] As a result, they noted that the introduction of one or more structural elements into the parent molecule changed the biological activity of the total molecule in a characteristic manner, largely independent of the presence of other activity-modifying groups, and that this was true not only in a qualitative sense, but also to an unusual extent quantitatively.

A few years later, and shortly before the important contributions of Hansch and Fujita[3] and Free and Wilson[2] were published, Zahradnik[9–12] tried to apply the extrathermodynamic approach to structure–activity relationships in a very strict manner. In comparing many different unspecific biological activities he derived equation (9), where τ_i is the biological activity value of the ith member of a congeneric series, τ_{Et} is the corresponding quantity of a reference compound (in this case the ethyl derivative) and α and β are constants. The value of β, which was called the constant of biological activity of a substituent group R, is a quantity describing the properties of the group, and its value depends neither on the nature of the rest of the molecule nor on the biological system, the latter being characterized by α. In comparing equation (9) and the Hammett equation (6), β corresponds to σ and α corresponds to ρ.

$$\log \tau_i - \log \tau_{Et} = \alpha\beta \qquad (9)$$

A number of examples have been presented to support equation (9), but today we know that its applicability is limited to unspecific biological activities and only to examples where the biological effect is a linear function of lipophilicity. Due to the fact that the lipophilicity of organic compounds is an additive property (equation 7),[6] equation (9) holds true in such cases. However, for other biological activities, where specific interactions between a drug and its biological counterpart are responsible for the effect, equation (9) is too restrictive.

In 1964, Hansch and Fujita[3] published their paper 'ρ–σ–π Analysis, A Method for the Correlation of Biological Activity and Chemical Structure' and in the same year Free and Wilson[2] published 'A Mathematical Contribution to Structure Activity Studies'. Both papers mark the beginning of tremendous progress in quantitative structure–activity analyses in two different directions: (i) the extrathermodynamic approach, where physicochemical properties are correlated with biological activity values (Hansch analysis); and (ii) the *de novo* approach (Free–Wilson analysis), where empirical parameters are derived from activity values to describe the dataset in a quantitative manner.

By using a very simple example (Table 1; equations 10 and 11) Free and Wilson explained their model, where μ is the overall average of biological activities and the other terms are the individual activity contributions. Equation (11), however, has two shortcomings, which were not recognized by Free and Wilson: firstly, linear activity values were used instead of logarithmic values and, secondly, the different symbols a and b for positions R_1 and R_2 are somewhat misleading. Another problem of equation (11) had already been noticed and solved, but not in the most satisfactory way. The application of equation (11) to the compounds of Table 1 gives four equations with five unknowns. However, since μ is the overall average of biological activity values, equations (12) and (13) lead to a simplification. Substitution of either a[H] or a[Me] in equation (11) by equation (12) and of b[NMe$_2$] or b[NEt$_2$] in equation (11) by equation (13) gives four equations with three unknowns which are readily solvable by linear multiple regression analysis. Two more realistic examples were presented to explain the model further.

$$\text{response} = \text{average} + \text{effect of } R_1 + \text{effect of } R_2 \tag{10}$$

$$LD_{50} = \mu + a[H] + a[Me] + b[NMe_2] + b[NEt_2] \tag{11}$$

$$a[H] + a[Me] = 0 \tag{12}$$

$$b[NMe_2] + b[NEt_2] = 0 \tag{13}$$

A much better solution for the last problem was given by Fujita and Ban several years later:[4] instead of taking the overall average for μ, they selected a reference compound for each set, usually the unsubstituted parent molecule of the series, and defined μ as the calculated biological activity value of the reference compound. This modification not only leads to logical simplifications but also to a much easier calculation procedure.

At approximately the same time as Free and Wilson's work, Bocek and Kopecky tested an additive model (equation 14), a multiplicative model (equation 15) and a combined model (equation 16) to correlate toxicity data of disubstituted benzenes.[13, 14] In comparing equations

Table 1 LD_{50} Values of Analgesics (2)[2]

(2)

	R_1		
R_2	*H*	*Me*	*Average*
NMe$_2$	2.13	1.64	1.885
NEt$_2$	1.28	0.85	1.065
Average	1.705	1.245	1.475

$\mu = 1.475$
$a[H] = 1.705 - 1.475 = 0.230$
$a[Me] = 1.245 - 1.475 = -0.230$
$b[NMe_2] = 1.885 - 1.475 = 0.410$
$b[NEt_2] = 1.065 - 1.475 = -0.410$

(14)–(16), where a, b, d and e are activity contributions of the substituents X and Y, they found that only equation (16), which corresponds to a Free–Wilson model with additional interaction terms $e_X e_Y$, was appropriate, giving a statistically significant correlation between observed and calculated values.

$$\log[\mathrm{LD}_{50}]_{HH} - \log[\mathrm{LD}_{50}]_{XY} = a_X + a_Y \tag{14}$$

$$\log[\mathrm{LD}_{50}]_{HH} - \log[\mathrm{LD}_{50}]_{XY} = d_X \cdot d_Y \tag{15}$$

$$\log[\mathrm{LD}_{50}]_{HH} - \log[\mathrm{LD}_{50}]_{XY} = b_X + b_Y + e_X \cdot e_Y \tag{16}$$

When Singer and Purcell[15] looked for relationships between the extrathermodynamic approach and the Free–Wilson method, they found that, for theoretical reasons, the linear Hansch model and Free–Wilson analysis correspond to each other and that the Bocek–Kopecky model (equation 16) corresponds to the parabolic Hansch model because both of them formally include interaction terms. Although these interrelationships have been questioned since then,[16, 17] there is now sufficient evidence that this view is indeed correct.[18, 19]

Indicator variables have long been used in Hansch analyses to combine sets of compounds where a continuous parameter cannot describe the differences in biological activity.[20, 21] The use of such dummy parameters has been rationalized as the extrathermodynamic approach assisted by the Free–Wilson method[22] and logically both approaches have been combined into a mixed model,[23] either in a linear form (equation 17) or with additional nonlinear terms (*e.g.* π^2, equation 18). In equations (17) and (18), C is a molar concentration producing a definite biological effect, Σa_i is a Free–Wilson part and $\Sigma k_j \Phi_j$ is a linear Hansch part, which can be substituted further by a nonlinear Hansch parameter, *e.g.* π^2. Equations (17) and (18) combine the advantages of both approaches and are suitable for cases where for one position of the lead structure there is sufficient structural variation to derive a meaningful Hansch correlation, while for another position the structural variation is too narrow and only Free–Wilson-type parameters can be used.

$$\log(1/C) = \Sigma a_i + \Sigma k_j \Phi_j + c \tag{17}$$

$$\log(1/C) = k\pi^2 + \Sigma a_i + \Sigma k_j \Phi_j + c \tag{18}$$

In reviewing the historical development of Free–Wilson analysis, it is important to notice that many problems were treated and discussed controversially over a long period of time. Such problems were, for example, the use of linear or logarithmic values for biological activities, the use of the classical Free–Wilson model or the Fujita–Ban modification, the choice of the reference compound, the inclusion or elimination of single point determinations, the relationship to the extrathermodynamic approach and so on. Nowadays, however, all these problems have been overcome.

21.3.3 PRACTICAL AND THEORETICAL ASPECTS OF FREE–WILSON ANALYSIS

As this work is intended to be a desk reference for the medicinal chemist and the biologist, theory and mathematical details will not be dealt with in depth. All aspects are illustrated with typical examples; references to theoretical work are given wherever necessary.

21.3.3.1 Design of Test Series

The most important prerequisite for the successful application of quantitative structure–activity relationships in drug design is that of congenericity within a series of compounds.[24] While in chemistry the term congeneric is not easy to define, when applied to Free–Wilson analysis it simply means that all compounds must have the same parent skeleton and the substituents should not chemically influence each other in a nonadditive manner, *e.g.* by hydrogen bonds, conjugation, mesomeric or push–pull effects, or by steric hindrance.

In drug research, the cost of the synthesis and testing of a single compound is usually much higher than the statistical treatment of data derived from these compounds. Thus it is important to ensure that the design of the test series is optimal, because in Free–Wilson analysis a number of problems may result from ill-conditioned sets of compounds.[25]

The minimum number of compounds needed for Free–Wilson analysis is one for each parameter of the regression equation, *i.e* one for the reference compound and one for each substituent differing from the corresponding substituent in the reference compound. Thus, at least two different positions of structural variation are needed for a statistically meaningful Free–Wilson analysis. If there are j different positions of structural variation with n_i different substituents in each position, the minimum number of analogues is given by equation (19),[26] while the number of theoretically possible analogues, provided there is no symmetry in the parent structure, is given by equation (20). For a series with three positions of substitution and five different groups in each position, the minimum number of analogues is $4 + 4 + 4 + 1 = 13$ and the maximum number is $5 \times 5 \times 5 = 125$. The actual number of analogues selected for synthesis and testing will always be a compromise between the ease of synthesis and the statistical requirements. The higher the number of analogues included in the analysis, then the higher the number of degrees of freedom and, therefore, the higher the reliability of the results. On the other hand, the effort and cost increase while the number of predictable biological activities decreases. All recommendations on the required number of degrees of freedom and on the number of analogues per parameter are completely arbitrary. They depend on the design of the series, whether all substituents are evenly distributed or not, on the quality of the biological data and on the degree to which the additivity principle applies to the biological activities under investigation. Usually each substituent should occur at least two to four times in the series and they all should be present in comparable frequencies. Two particular problems, often observed in retrospective analyses, should be avoided in prospective planning: firstly, if a substituent occurs only once in a series, a single point determination without any statistical significance results. The value of the group contribution for this substituent includes the total experimental error of a single compound. Secondly, if two substituents always occur together, no individual group contributions for each substituent can be derived.[27, 28] They have to be treated as a single substituent in either one of both positions with no substituent in the other position,[29] otherwise regression analysis will fail because of linear dependence in the substituent matrix.[27, 30] Even more complex linear dependences can arise in unbalanced groups of compounds, making the application of Free–Wilson analysis impossible or leading to unreliable results (Section 21.3.3.8).[29]

$$N_{min} = \sum_{j}(n_i - 1) + 1 \tag{19}$$

$$N_{max} = n_1 \cdot n_2 \cdot \ldots \cdot n_{j-1} \cdot n_j \tag{20}$$

Austel proposed a manual design of test series for Free–Wilson analyses, based on 2^n-factorial schemes[31] to avoid such shortcomings and to arrive at a minimum number of compounds with optimal selection of substituents. In his example with three, four and five different substituents in three positions of the molecule (**3**; Table 2), he was able to cover each substituent in each position

Table 2 Manual Design of Test Series;[31]
Test Set of 10 Compounds Selected by an
Extended Factorial Design

(3)

No.	R_1	R_2	R_3
1	H	SMe	OMe
2	Me	OMe	OH
3	Cl	NH_2	OMe
4	H	Me	OH
5	Me	H	OMe
6	Cl	SMe	H
7	Me	OMe	NH_2
8	H	NH_2	H
9	Cl	Me	NH_2
10	H	H	H

two to four times with the minimum number of analogues, $N = 2 + 4 + 3 + 1 = 10$. Of course, a larger series of compounds has to be considered in this case in order to perform a statistically meaningful Free–Wilson analysis. The disadvantage of such a factorial design is the neglect of synthetic accessibility of the selected compounds.

A quantitative procedure to extract an optimal set out of all possible analogues, based on the maximization of the determinant of the substituent matrix, has been proposed.[32] With this procedure it is also possible to select different groups of compounds which are easy to synthesize and to compare the information content of the resulting substituent matrices with that of the optimal set.

21.3.3.2 Biological Activity Values

The correct choice of the biological activity parameter is of the utmost importance in quantitative structure–activity analyses, a fact often neglected in Free–Wilson analysis because different scales for the activity were used in early times, *e.g.* linear values of doses producing a definite biological effect;[2, 33, 34] linear values of relative biological activities referring to a reference compound;[2, 35] linear values of percent biological effect;[36] linear values of percent biological effect divided by the dose;[37] linear values of percent relative biological effect referring to a reference compound;[38] numbers of classes with different biological activity;[39] logarithms of doses producing a definite effect;[4, 25, 40] logarithms of relative biological activities referring to a reference compound;[16, 35, 41] and logarithms of percent biological effect.[7] Most often, $mg\,kg^{-1}$ doses were used instead of molar doses; the use of linear and logarithmic values in Free–Wilson analysis has been reviewed.[42]

In the first comparisons of linear and logarithmic biological activity values, a better fit of the data was obtained in the logarithmic scale.[35, 37] Of course, the sums of the squared deviations calculated from the antilog values of the logarithmic regressions are larger than those from the linear regressions (Table 3), but this is an inappropriate comparison. Thus, the statement that statistics is unable to decide which of the two scales should be used in Free–Wilson analyses[35, 37] is wrong: the higher correlation coefficients (standard deviations s cannot be compared due to the different scales) of the logarithmic regressions demonstrate that the Free–Wilson additivity concept applies better in this scale than in the linear one. A large number of successful Free–Wilson analyses now confirms the validity of the additivity concept in the logarithmic scale.

There are some more arguments for the logarithmic scale: due to the close practical and theoretical relationship to the extrathermodynamic approach,[15, 18, 19] only logarithmic values of biological activities should be used in Free–Wilson analyses because only these values are linear free energy related parameters. The experimental errors of biological data follow, as a first approximation, a normal distribution in the logarithmic scale and not in the linear scale and, last but not least, no negative values of calculated biological activities (which do occur in the linear scale[33, 35, 37]) can result.

Thus negative logarithms of molar doses C producing a definite biological effect, such as $\log(1/C)$, $\log(1/LD_{50})$, $\log(1/ED_{20})$, $\log(1/K_i)$, pI_{50}, *etc.*, are the appropriate activity parameters for Free–Wilson analysis; the inverse doses are taken to obtain higher values for the more active analogues.

Linear values may give satisfactory correlations in such cases where the variance of the biological activity values is small and therefore linear and logarithmic values are closely intercorrelated.

Relative biological activities referring to a standard should be used only when this reference compound can be included in the analysis, otherwise the uncertainty of the biological activity value of this one compound is added to all other activity values, thereby increasing the standard error of

Table 3 Comparison of Free–Wilson Analyses with Linear and Logarithmic Biological Activity Values

Compounds and type of biological activity	Linear values[a] $BA = \Sigma a_i + \mu$		Logarithmic values[a] $\log BA = \Sigma a_i + \mu$		
	r	$\Sigma\Delta^2$	r	$\Sigma\Delta^{2b}$	Ref.
Sympathomimetics ($n = 30$), uptake 1 inhibition	0.83	881 244	0.96	1 988 149	35
Sympathomimetics ($n = 12$), uptake 2 inhibition	0.93	1 068 801	0.98	1 385 783	35
Carbamates ($n = 55$), antitumor activity	0.80	527.6	0.93	920.8	37

[a] r = correlation coefficient, $\Sigma\Delta^2$ = sum of squared deviations between observed and calculated biological activity values. [b] Calculated from the antilogarithms of the results of the logarithmic regression.

the data. In cases where biological effects are measured using fixed doses, a probit or logit transformation should be performed to estimate the doses needed to produce identical biological effects. The direct use of percent biological effect values is inappropriate due to the sigmoidal shape of the biological response against log(dose) curves.

All biological activity values should be as accurate as possible. They should cover a range much larger than the standard deviations of these values, which should be known. The standard deviation of the resulting regression equation can never be smaller than the experimental error of the observed data. If this occurs, it is an indication of overprediction, probably resulting from too many parameters in the equation. All compounds included in the analysis should act by the same mechanism, a demand whose fulfillment is not known, especially in the case of complex enzyme–inhibitor or drug–receptor interactions, where closely related analogues may bind in very different manners.

Although the biological activity values need not be normally distributed (only the residuals resulting from the regression must be), a grouping of values should be avoided. The larger the activity differences between two or more different classes of compounds are, the lower the reliability of the parameters predicting the activity differences within the classes will be. In such cases, each group should be analyzed separately. They may be combined in one equation when the different correlations correspond to each other, with major differences only in the intercepts μ.

21.3.3.3 The Classical Free–Wilson Model

Comprehensive reviews on the Free–Wilson model have been published.[26–28,42–46] In addition, the Free–Wilson method is treated together with other methods in reviews on quantitative structure–activity relationships.[47–64] The names '*de novo* approach' and 'mathematical model' are synonyms for the Free–Wilson model, used because the activity contributions are derived from biological data only by mathematical means, without describing properties by physicochemical parameters as in Hansch analysis.

The basic concept of Free–Wilson analysis is the additivity principle of biological activity values.[2] It assumes that, within a congeneric series of compounds sharing a common parent structure, the substituents make additive and constant contributions to the biological activity irrespective of all other structural changes in the molecule. Although such a strict definition does not appear very reliable, it has proven its utility for the exploration of quantitative structure–activity relationships in a large number of cases, not only in Free–Wilson analysis but also in the extrathermodynamic approach.

For every compound of a congeneric series the biological activity BA_i can be expressed as the sum of the biological activity contributions a_{jk} of the substituents (indicated by k) in each position j, and the activity contribution of the common parent structure, μ (equation 21). In equation (21), X_{jk} has a value of one when the substituent X_k is present in the position j, otherwise its value is zero. If the BA values are inverse molar doses, such as $1/C$, $1/K_i$, *etc.*, then activity-enhancing substituents have positive a_{jk} values, while activity-lowering substituents have negative values (for the use of logarithmic biological activity values see Section 21.3.3.2).

$$\log BA_i = \sum_j a_{jk} X_{jk} + \mu \tag{21}$$

In the original Free–Wilson model, symmetry conditions are assumed for each position of substitution.[2] They define that for each position j the sums of the group contributions a_{jk} of all members of the series equal zero (equation 22). From equations (21) and (22) it follows that μ must be the average of biological activity values of all molecules (equation 23).

$$\sum_i a_{jk} X_{jk} = 0 \qquad \text{(for each value of } j\text{)} \tag{22}$$

$$\mu = (1/n) \cdot \sum_i \log BA_i \tag{23}$$

The calculation procedure can best be explained by an example. Structures and biological activity values of a group of antiadrenergic N,N-dimethyl-α-bromophenethylamines (4) are given in Table 4, together with a structural matrix where the presence or absence of each X and Y substituent is indicated by one or zero. Table 4 corresponds to a set of equations (24), *e.g.* equations (25) and (26) for compounds 1 and 17; in equations (25) and (26) a_{11}, a_{13}, a_{22} and a_{24} are the activity

Table 4 Antiadrenergic Activities of *N,N*-Dimethyl-α-bromophenethylamines (**4**);[65] Structures, Structural Matrix and Log(1/C) Values

No.	X	Y	Meta substituents (X)						Para substituents (Y)						Log(1/C)[a]
			H	F	Cl	Br	I	Me	H	F	Cl	Br	I	Me	
1	H	F	1							1					8.16
2	H	Cl	1								1				8.68
3	H	Br	1									1			8.89
4	H	I	1										1		9.25
5	H	Me	1											1	9.30
6	F	H		1					1						7.52
7	Cl	H			1				1						8.16
8	Br	H				1			1						8.30
9	I	H					1		1						8.40
10	Me	H						1	1						8.46
11	Cl	F			1					1					8.19
12	Br	F				1				1					8.57
13	Me	F						1		1					8.82
14	Cl	Cl			1						1				8.89
15	Br	Cl				1					1				8.92
16	Me	Cl						1			1				8.96
17	Cl	Br			1							1			9.00
18	Br	Br				1						1			9.35
19	Me	Br						1				1			9.22
20	Me	Me						1						1	9.30
21	Br	Me				1								1	9.52
22	H	H	1						1						7.46
Sums			6	1	4	5	1	5	6	4	4	4	1	3	191.32

[a] $C = ED_{50}$ in mol kg^{-1}.

contributions of *m*-hydrogen, *m*-chloro, *p*-fluoro and *p*-bromo.

$$\log(1/C) = a_{1k}[X_k] + a_{2k}[Y_k] + \mu \tag{24}$$

$$\log(1/C) = a_{11} + a_{22} + \mu \tag{25}$$

$$\log(1/C) = a_{13} + a_{24} + \mu \tag{26}$$

Thus, the structural matrix of Table 4 corresponds to a set of 22 equations with 12 variables (the a_{jk} values) and a constant term μ, which should be readily solvable by linear multiple regression analysis. However, this set of equations contains linear dependences (singularities), making a unique solution impossible. In all equations the sum of X_{jk} values is one for each position of substitution (equations 27 and 28), because there is always one and only one substituent in each position. By applying the symmetry equations (22) (sometimes called restriction equations) to substitute one variable at each position by the other variables, the linear dependences can be removed. In the case of Table 4, the symmetry equations are given by equations (29) and (30) and, after arbitrarily selecting one substituent, *e.g.* hydrogen, for each position, equations (31) and (32) are derived from equations (29) and (30). A new matrix (Table 5) results from the elimination of the *m*-hydrogen and *p*-hydrogen columns by using these equations. Table 5 represents the classical Free–Wilson data matrix which no longer contains linear dependences and thus can be solved by linear multiple regression analysis.

$$[m\text{-H}] + [m\text{-F}] + [m\text{-Cl}] + [m\text{-Br}] + [m\text{-I}] + [m\text{-Me}] = 1 \tag{27}$$

$$[p\text{-H}] + [p\text{-F}] + [p\text{-Cl}] + [p\text{-Br}] + [p\text{-I}] + [p\text{-Me}] = 1 \tag{28}$$

$$6a_{11} + a_{12} + 4a_{13} + 5a_{14} + a_{15} + 5a_{16} = 0 \tag{29}$$

Table 5 Antiadrenergic Activities of *N,N*-Dimethyl-α-bromophenethylamines (Table 4); Data Matrix Used for Regression Analysis[29]

No.	Meta substituents (X)					Para substituents (Y)					Log (1/C)
	F	Cl	Br	I	Me	F	Cl	Br	I	Me	
1	−1/6	−2/3	−5/6	−1/6	−5/6	1					8.16
2	−1/6	−2/3	−5/6	−1/6	−5/6		1				8.68
3	−1/6	−2/3	−5/6	−1/6	−5/6			1			8.89
4	−1/6	−2/3	−5/6	−1/6	−5/6				1		9.25
5	−1/6	−2/3	−5/6	−1/6	−5/6					1	9.30
6	1					−2/3	−2/3	−2/3	−1/6	−1/2	7.52
7		1				−2/3	−2/3	−2/3	−1/6	−1/2	8.16
8			1			−2/3	−2/3	−2/3	−1/6	−1/2	8.30
9				1		−2/3	−2/3	−2/3	−1/6	−1/2	8.40
10			·		1	−2/3	−2/3	−2/3	−1/6	−1/2	8.46
11		1				1					8.19
12			1			1					8.57
13				1		1					8.82
14		1					1				8.89
15			1				1				8.92
16				1			1				8.96
17		1						1			9.00
18			1					1			9.35
19				1				1			9.22
20				1					1		9.30
21			1							1	9.52
22	−1/6	−2/3	−5/6	−1/6	−5/6	−2/3	−2/3	−2/3	−1/6	−1/2	7.46
Sums	0	0	0	0	0	0	0	0	0	0	191.32

$$6a_{21} + 4a_{22} + 4a_{23} + 4a_{24} + a_{25} + 3a_{26} = 0 \tag{30}$$

$$a_{11} = -(1/6) \cdot a_{12} - (2/3) \cdot a_{13} - (5/6) \cdot a_{14} - (1/6) \cdot a_{15} - (5/6) \cdot a_{16} \tag{31}$$

$$a_{21} = -(2/3) \cdot a_{22} - (2/3) \cdot a_{23} - (2/3) \cdot a_{24} - (1/6) \cdot a_{25} - (1/2) \cdot a_{26} \tag{32}$$

Equation (33) is the result of the regression analysis. 95% confidence intervals are given for each regression coefficient. The correlation coefficient $r = 0.969$, the standard deviation $s = 0.194$ and the F value indicate the good fit of the additive Free–Wilson model. The activity contributions of *m*-hydrogen and *p*-hydrogen can be calculated from equations (31) and (32) to be $a_{11} = -0.252$ and $a_{21} = -0.623$. Equation (33) is only one possible result, depending on the selection of the substituents removed from the original data matrix by the symmetry equations. If, for example, the *m*-fluoro and *p*-iodo columns are removed from Table 4 by using equations (34) and (35), equation (36) results.[29] The difference between equations (33) and (36) is the appearance of the [*m*-H] and [*p*-H] terms and the lack of the [*m*-F] and [*p*-I] terms in equation (36), all other terms being identical. Of course the statistical parameters of both equations are identical, too. The terms lacking in equation (36) can be derived from equations (34) and (35) to be $a_{12} = -0.553$ and $a_{25} = 0.806$.

$$\begin{aligned}
\log(1/C) = &-0.553(\pm 0.45)[m\text{-F}] - 0.045(\pm 0.20)[m\text{-Cl}] + 0.182(\pm 0.17)[m\text{-Br}] + 0.327(\pm 0.45)[m\text{-I}] \\
&+ 0.202(\pm 0.17)[m\text{-Me}] - 0.283(\pm 0.20)[p\text{-F}] + 0.144(\pm 0.20)[p\text{-Cl}] + 0.397(\pm 0.20)[p\text{-Br}] \\
&+ 0.806(\pm 0.45)[p\text{-I}] + 0.633(\pm 0.24)[p\text{-Me}] + 8.696(\pm 0.09)
\end{aligned} \tag{33}$$

$$n = 22, \quad r = 0.969, \quad s = 0.194, \quad F = 16.99$$

$$a_{12} = -6a_{11} - 4a_{13} - 5a_{14} - a_{15} - 5a_{16} \tag{34}$$

$$a_{25} = -6a_{21} - 4a_{22} - 4a_{23} - 4a_{24} - 3a_{26} \tag{35}$$

$$\begin{aligned}
\log(1/C) = &-0.252(\pm 0.16)[m\text{-H}] - 0.045(\pm 0.20)[m\text{-Cl}] + 0.182(\pm 0.17)[m\text{-Br}] + 0.327(\pm 0.45)[m\text{-I}] \\
&+ 0.202(\pm 0.17)[m\text{-Me}] - 0.623(\pm 0.17)[p\text{-H}] - 0.283(\pm 0.20)[p\text{-F}] + 0.144(\pm 0.20)[p\text{-Cl}] \\
&+ 0.397(\pm 0.20)[p\text{-Br}] + 0.633(\pm 0.24)[p\text{-Me}] + 8.696(\pm 0.09)
\end{aligned} \tag{36}$$

$$n = 22, \quad r = 0.969, \quad s = 0.194, \quad F = 16.99$$

The theory of linear equations and the problem of linear dependences as applied to Free–Wilson analysis have been discussed in the literature.[29, 30, 66]

From either equation, biological activities can be calculated by using equation (24), *e.g.* equation (37) for compound 1 or equation (38) for compound 17. All calculated biological activity values and the differences between observed and calculated values are given in Table 6. For compounds 4, 6 and 9 the residuals are zero because the calculated $\log(1/C)$ values result from single point determinations; the activity contributions for [m-F], [m-I] and [p-I] include the experimental error of one activity value. Some problems of single point determinations are discussed in Section 21.3.3.7.

$$\log(1/C) = a_{11} + a_{22} + \mu = -0.252 - 0.283 + 8.696 = 8.161 \tag{37}$$

$$\log(1/C) = a_{13} + a_{24} + \mu = -0.045 + 0.397 + 8.696 = 9.048 \tag{38}$$

The biological activity values of compounds with other combinations of substituents, *e.g.* m-iodo and p-iodo, can be predicted; a $\log(1/C)$ value of 9.83, higher than all other $\log(1/C)$ values of Table 4, is predicted for this compound (equation 39). As in the extrathermodynamic approach, predictions far outside the range of observed biological activities are not very reliable. However, the risk of totally wrong predictions is much smaller in Free–Wilson analysis, because only activity values for new combinations of substituents already included in the series can be predicted.

$$\log(1/C) = a_{15} + a_{25} + \mu = 0.327 + 0.806 + 8.696 = 9.829 \tag{39}$$

The greatest disadvantage of Free–Wilson analysis is the definition of the symmetry equations. For all positions of substitution these equations must be derived and the corresponding matrix must be built up. After each exchange, addition or elimination of compounds to or from the data set, new symmetry equations and a new matrix for regression analysis must be calculated. Due to the definition of the constant term μ as the average of the activity values, the value of μ changes to a certain extent after each addition or deletion of a compound with high or low activity. Consequently, the activity contributions of all substituents are changed, making the comparison of different Free–Wilson analyses in the same data set very difficult.

Table 6 Antiadrenergic Activities of *N,N*-Dimethyl-α-bromophenethylamines (Table 4); Comparison of Observed and Calculated Biological Activity Values[18, 29]

No.	X	Y	Log(1/C) values Observed	Log(1/C) values Calculated	Residuals $y_{obs} - y_{calc}$
1	H	F	8.16	8.161	−0.001
2	H	Cl	8.68	8.589	0.091
3	H	Br	8.89	8.841	0.049
4	H	I	9.25	9.250	0.000
5	H	Me	9.30	9.077	0.223
6	F	H	7.52	7.520	0.000
7	Cl	H	8.16	8.028	0.132
8	Br	H	8.30	8.255	0.045
9	I	H	8.40	8.400	0.000
10	Me	H	8.46	8.275	0.185
11	Cl	F	8.19	8.368	−0.178
12	Br	F	8.57	8.595	−0.025
13	Me	F	8.82	8.615	0.205
14	Cl	Cl	8.89	8.796	0.094
15	Br	Cl	8.92	9.023	−0.103
16	Me	Cl	8.96	9.043	−0.083
17	Cl	Br	9.00	9.048	−0.048
18	Br	Br	9.35	9.275	0.075
19	Me	Br	9.22	9.295	−0.075
20	Me	Me	9.30	9.531	−0.231
21	Br	Me	9.52	9.511	0.009
22	H	H	7.46	7.821	−0.361

21.3.3.4 The Fujita–Ban Modification

Following the concepts of the Hammett equation (equation 6), the Zahradnik approach (equation 9) and the Bocek–Kopecky model (equations 14–16), Cammarata formulated a modified

Free–Wilson model (equation 40).[16, 41]

$$\log BA_i - \log BA_H = \sum_j b_{jk} X_{jk} \qquad (40)$$

In equation (40), BA_H is the observed biological activity value of the unsubstituted parent compound of the series, with hydrogen atoms in all positions of structural variation. As a consequence, the group contributions b_{jH} of the hydrogen substituents are assigned a value of zero. In this way the activity contributions b_{jk} of all other substituents become defined in an equivalent manner to linear free energy related parameters. However, this procedure is inadequate from a statistical point of view,[18, 29, 44, 66] because $\log BA_H$ is an observed activity value including the same experimental error as all other $\log BA_i$ values. Thus, the standard error of all $\log BA_i$ values is additionally increased.

Fujita and Ban recognized and considered this important fact and reformulated equation (40) to equation (41),[4] where the constant term μ is now defined as the calculated biological activity value of the unsubstituted parent compound of the series. All activity contributions b_{jk} refer to hydrogen in the same position of substitution.

$$\log BA_i = \sum_j b_{jk} X_{jk} + \mu \qquad (41)$$

A much simpler calculation procedure, compared to the classical model, results from equation (41) which, again, can best be explained by the example given in Table 4. Due to the fact that all activity contributions b_{jk} refer to $b_{jH} = 0$, the hydrogen columns are removed from the original matrix for each position of substitution. The resulting data matrix (Table 7) no longer contains any linear dependences, thus no symmetry equations are needed. The set of equations presented by this matrix can be solved directly with linear multiple regression analysis (equation 42).

$$\begin{aligned}
\log(1/C) = &-0.301(\pm 0.50)[m\text{-}F] + 0.207(\pm 0.29)[m\text{-}Cl] + 0.434(\pm 0.27)[m\text{-}Br] + 0.579(\pm 0.50)[m\text{-}I] \\
&+ 0.454(\pm 0.27)[m\text{-}Me] + 0.340(\pm 0.30)[p\text{-}F] + 0.768(\pm 0.30)[p\text{-}Cl] + 1.020(\pm 0.30)[p\text{-}Br] \\
&+ 1.429(\pm 0.50)[p\text{-}I] + 1.256(\pm 0.33)[p\text{-}Me] + 7.821(\pm 0.27) \qquad (42)
\end{aligned}$$

$$n = 22, \quad r = 0.969, \quad s = 0.194, \quad F = 16.99$$

The value of the constant term μ (the calculated biological activity value of the unsubstituted parent compound) is identical with the value predicted by the classical Free–Wilson model (see

Table 7 Antiadrenergic Activities of *N,N*-Dimethyl-α-bromophenethylamines (Table 4); Fujita–Ban Matrix[18, 29]

No.	Meta substituents (X)					Para substituents (Y)					Log (1/C)
	F	Cl	Br	I	Me	F	Cl	Br	I	Me	
1						1					8.16
2							1				8.68
3								1			8.89
4									1		9.25
5										1	9.30
6	1										7.52
7		1									8.16
8			1								8.30
9				1							8.40
10					1						8.46
11		1				1					8.19
12			1			1					8.57
13				1		1					8.82
14		1					1				8.89
15			1				1				8.92
16					1		1				8.96
17		1						1			9.00
18			1					1			9.35
19					1			1			9.22
20					1					1	9.30
21			1							1	9.52
22											7.46

Table 6, compound 22). Also the statistical parameters r, s and F and all calculated biological activity values are identical, indicating the close relationships between both versions.

Fujita and Ban's conjecture that the biological activity of the unsubstituted compound should be known in order to apply their version[4] is not true. Any other compound can be taken as the reference compound,[27] even a substituent combination not included in the original series,[29] *e.g.* *m*-fluoro and *p*-iodo for the data set of Table 4. Now these two columns are removed instead of the two hydrogen columns, leading to the definition that $b_{m\text{-}F} = 0$, $b_{p\text{-}I} = 0$ and $\mu =$ calculated biological activity of a 3-fluoro-4-iodo analogue (equation 43). Again the statistical parameters r, s and F and all calculated biological activity values are identical with those from equations (33), (36) and (42).[29]

$$
\begin{aligned}
\log(1/C) = {} & 0.301(\pm 0.50)[m\text{-}H] + 0.508(\pm 0.51)[m\text{-}Cl] + 0.735(\pm 0.50)[m\text{-}Br] + 0.880(\pm 0.60)[m\text{-}I] \\
& + 0.755(\pm 0.50)[m\text{-}Me] - 1.429(\pm 0.50)[p\text{-}H] - 1.089(\pm 0.50)[p\text{-}F] - 0.661(\pm 0.50)[p\text{-}Cl] \\
& -0.409(\pm 0.50)[p\text{-}Br] - 0.173(\pm 0.52)[p\text{-}Me] + 8.949(\pm 0.66)
\end{aligned}
\tag{43}
$$

$$
n = 22, \quad r = 0.969, \quad s = 0.194, \quad F = 16.99
$$

Table 8 demonstrates that the results from different Fujita–Ban analyses (equations 42 and 43) are linear transformations of the results from the classical Free–Wilson model (equations 33 and 36), differing only in the value of the constant term μ, to which all activity contributions refer.[29, 44, 66] The linear relationships between the results of Free–Wilson and Fujita–Ban analyses are presented by equations (44) and (45), where a_{jZ} are the activity contributions of the reference substituents Z in position j and μ_{FB} and μ_{FW} are the constant terms from both analyses.

$$
b_{jk} = a_{jk} - a_{jZ}
\tag{44}
$$

$$
\mu_{FB} = \mu_{FW} + \sum_{j} a_{jZ}
\tag{45}
$$

The only disadvantage of the Fujita–Ban version comes from the differences in the confidence intervals of the regression coefficients (Table 9). Due to the fact that for each position of substitution one b_k value is arbitrarily forced to zero, the uncertainty of this value is distributed to all other b_k values. As long as the reference substituents are well represented in the series (equation 42), the differences between the classical Free–Wilson model and the Fujita–Ban version are small. However, they become much larger when less well-represented substituents are chosen as reference groups (equation 43; Table 9).

In symmetric molecules, where two or more positions of substitution are equivalent (or considered to be equivalent), the Fujita–Ban matrix is derived from the structures in a similar manner as before. One substituent for each position of substitution is chosen as the reference substituent and for all other substituents a number indicates how often the group occurs in equivalent positions. An

Table 8 Antiadrenergic Activities of *N*,*N*-Dimethyl-α-bromophenethylamines (Table 4); Comparison of Activity Contributions from Different Free–Wilson Analyses[29]

Substituent	Classical Free–Wilson model a_{jk}	Fujita–Ban modification Equation (42) b_{jk}	$b_{jk} - a_{jk}$	Equation (43) b_{jk}	$b_{jk} - a_{jk}$
m-H	−0.252	0[a]	0.252	0.301	0.553
m-F	−0.553	−0.301	0.252	0[a]	0.553
m-Cl	−0.045	0.207	0.252	0.508	0.553
m-Br	0.182	0.434	0.252	0.735	0.553
m-I	0.327	0.579	0.252	0.880	0.553
m-Me	0.202	0.454	0.252	0.755	0.553
p-H	−0.623	0[a]	0.623	−1.429	−0.806
p-F	−0.283	0.340	0.623	−1.089	−0.806
p-Cl	0.144	0.768	0.624[b]	−0.661	−0.805[b]
p-Br	0.397	1.020	0.623	−0.409	−0.806
p-I	0.806	1.429	0.623	0[a]	−0.806
p-Me	0.633	1.256	0.623	−0.173	−0.806
μ	8.696	7.821	0.875[c]	8.949	−0.253[d]

[a] By definition. [b] Deviation due to rounding errors. [c] $\mu_{FW} - \mu_{FB} = 8.696 - 7.821 = 0.252 + 0.623 = 0.875$. [d] $\mu_{FW} - \mu_{FB} = 8.696 - 8.949 = 0.553 - 0.806 = -0.253$.

Table 9 Antiadrenergic Activities of *N*,*N*-Dimethyl-α-bromophenethylamines (Table 4); Comparison of 95% Confidence Intervals[29]

Substituent	Classical Free–Wilson model	Fujita–Ban modification	
		Equation (42)	Equation (43)
m-H	$\pm 0.16^{a}$	0^{c}	± 0.50
m-F	$\pm 0.45^{b}$	± 0.50	0^{c}
m-Cl	± 0.20	± 0.29	± 0.51
m-Br	± 0.17	± 0.27	± 0.50
m-I	± 0.45	± 0.50	± 0.60
m-Me	± 0.17	± 0.27	± 0.50
p-H	$\pm 0.17^{a}$	0^{c}	± 0.50
p-F	± 0.20	± 0.30	± 0.50
p-Cl	± 0.20	± 0.30	± 0.50
p-Br	± 0.20	± 0.30	± 0.50
p-I	$\pm 0.45^{b}$	± 0.50	0^{c}
p-Me	± 0.24	± 0.33	± 0.52
μ	± 0.09	± 0.27	± 0.66

[a] From equation (36) only. [b] From equation (33) only. [c] By definition.

example is given in Table 10.[28, 36] Another example of this procedure is equation (46), where the antibacterial activities of chlorinated phenols against *S. aureus* are described by the single parameter [Cl].[23] Corresponding to the number of chlorine atoms in the molecule, [Cl] takes values of zero to five; structural isomers have identical [Cl] values. Equation (46) can be regarded as the simplest form of Free–Wilson analysis.

$$\log(1/C) = 0.503(\pm 0.13)[\text{Cl}] + 2.578 \tag{46}$$

$$n = 9, \quad r = 0.960, \quad s = 0.256, \quad F = 83.06$$

In homologous series, the number of additional methylene groups of each homolog, with reference to the lowest member of the series, may be used as a Free–Wilson parameter, provided the additivity of group contributions holds for all compounds.

Nowadays, the Fujita–Ban modification (equation 41) is used nearly exclusively for Free–Wilson analysis because of its ease of calculation. The addition, elimination or exchange of compounds causes no problems and the different results can be compared directly. The values of the activity contributions can be easily interpreted; they are related to linear free energy parameters and directly comparable to the activity contributions derived from extrathermodynamic structure–activity relationships (see Section 21.3.5).

21.3.3.5 The Bocek–Kopecky Model

At the same time as Free and Wilson published their model, Bocek and Kopecky found, working independently, that the toxicity of a series of *para*-disubstituted benzenes could best be described by equation (47), which corresponds to a Free–Wilson model with additional interaction terms.[13] Neither an additive model (equation 14) nor a multiplicative model (equation 15) gave appropriate results. In equation (47), b_X and b_Y are Fujita–Ban-type group contributions of the substituents X and Y, referring to hydrogen in the same position, while e_X and e_Y are also empirical parameters, accounting for the nonadditivity of the group contributions b_X and b_Y. Using this model they obtained a correlation coefficient $r = 0.981$ (b_i and e_i values given in Table 11).

$$\log[\text{LD}_{50}]_{\text{HH}} - \log[\text{LD}_{50}]_{\text{XY}} = b_X + b_Y + e_X \cdot e_Y \tag{47}$$

Later on they derived a corresponding relationship for the *meta*-disubstituted benzenes[14] and were able to combine both data sets into one equation, using different e_X and e_Y values for the *meta* and *para* positions (Table 11).

In an attempt to reexamine these data, Cammarata and Bustard committed some serious errors.[17] Firstly, they used recalculated data instead of observed values which were unavailable to them; secondly, they took all possible combinations of substituents instead of the combinations actually tested and included in the original calculation; and thirdly, from a comparison of only two different

Table 10 Structures of Hypoglycemic Piperidinosulfamylsemicarbazides (**5**);[28, 36] Fujita–Ban
Matrix, Based on $R_1 = R_2 = H$ and $R_3 = $ —CH$_2$—

(**5**)

R_1	R_2	R_3	Fujita–Ban matrix R_1 and R_2					R_3
			Me	*Et*	*OMe*	—(CH$_2$)$_4$—	—(CH$_2$)$_5$—	—(CH$_2$)$_2$—
H	H	—CH$_2$—						
H	H	—(CH$_2$)$_2$—						1
H	Me	—CH$_2$—	1					
H	Me	—(CH$_2$)$_2$—	1					1
Me	Me	—CH$_2$—	2					
Me	Me	—(CH$_2$)$_2$—	2					1
Me	Et	—(CH$_2$)$_2$—	1	1				1
Et	Et	—(CH$_2$)$_2$—		2				1
—(CH$_2$)$_4$—		—CH$_2$—				1		
—(CH$_2$)$_4$—		—(CH$_2$)$_2$—				1		1
—(CH$_2$)$_5$—		—(CH$_2$)$_2$—					1	1
H	OMe	—(CH$_2$)$_2$—			1			1

Table 11 Toxicity of Disubstituted Benzenes; Results of Bocek and Kopecky[13, 14]

Compounds and parameters[a]	Parameter values					
	NO_2	*Cl*	*OH*	*Me*	*H*	NH_2
Para-*disubstituted benzenes*[13]						
b_i	0.565	0.328	0.318	0.217	0.005	−0.026
e_i	0.59	−0.07	0.53	0.04	−0.04	−0.87
Meta-*disubstituted benzenes*[14]						
b_i	0.601	0.248	0.260	0.149	0.004	0.015
e_i	0.90	−0.49	−0.13	−0.22	−0.06	−0.41
***All compounds*[14]**						
b_i	0.516	0.295	0.294	0.191	0.014	0.015
e_i^{meta}	0.84	−0.45	−0.04	−0.13	0.00	−0.37
e_i^{para}	0.68	−0.04	0.57	0.07	0.00	−0.83

[a] For explanation of symbols see text.

equations, with and without an interaction term, they concluded that nonadditive models are not superior, which is an unjustified generalization.

While there is no doubt that the interaction model of Bocek and Kopecky (equation 47) is indeed appropriate in cases of nonadditivity of group contributions,[15, 18, 19] two problems obstruct its practical application: firstly, the number of degrees of freedom sharply decreases by the additional interaction terms and, secondly, nonlinear regression analysis, with all its problems of parameter estimation and iteration, is necessary to obtain the best values for the interaction terms. Consequently, there are only a few examples in the literature where the ideas of Bocek and Kopecky have been followed to describe nonadditive structure–activity relationships with corresponding models.[18, 67–69]

A descriptive example, where an interaction term can be used to account for nonlinear effects in Free–Wilson analysis, is given in Table 12. If the hypnotic activity of substituted ureas is described by normal Free–Wilson analysis, the resulting equation is not significant at the 95% probability level (equation 48; $F = 2.93$). However, when a Bocek–Kopecky-type interaction term $N_1 \cdot N_2$ is added, the highly significant equation (49) results;[18] N_1 and N_2 were arbitrarily chosen to consider nonlinear effects depending on the chain lengths of the acyl groups R_1 and R_2. A comparison of the statistical parameters shows the much better fit of the data by equation (49).

$$\log(1/C) = -0.003(\pm0.22)[\text{prop}] + 0.149(\pm0.22)[\text{but}] + 0.186(\pm0.23)[\text{pent}]$$
$$+ 0.242(\pm0.29)[\text{hex}] + 0.511(\pm0.47)[\text{hept}] + 2.039(\pm0.34) \tag{48}$$
$$n = 13, \quad r = 0.822, \quad s = 0.135, \quad F = 2.93$$

$$\log(1/C) = 0.430(\pm0.17)[\text{prop}] + 0.910(\pm0.28)[\text{but}] + 1.249(\pm0.39)[\text{pent}] + 1.487(\pm0.45)[\text{hex}]$$
$$+ 1.813(\pm0.49)[\text{hept}] - 0.085(\pm0.03)N_1 \cdot N_2 + 1.930(\pm0.13) \tag{49}$$
$$n = 13, \quad r = 0.982, \quad s = 0.049, \quad F = 26.59$$

Equation (49) is a simplification of the Bocek–Kopecky model (equation 47). In this example, as well as in many other cases, it would have been impossible to apply the Bocek–Kopecky model in its extended form because of the large number of possible interaction terms.

Interaction terms can also be used to describe other nonadditive effects, *e.g.* resulting from intramolecular hydrogen bridges or from unfavorable steric interactions between adjacent groups. For the hydroxylation of substituted phenethylamines by dopamine β-hydroxylase, Fujita and Ban[4] derived equation (50). The correlation was improved when an interaction term [OH:OMe] was added to account for nonadditive effects resulting from the simultaneous occurrence of a *m*-methoxy group and a *p*-hydroxy group in the molecule (equation 51). If only one of these substituents is present, the contributions to biological activity are described by the corresponding group contribution. When both groups occur together, the interaction term corrects the sum of otherwise independent group contributions.

$$\log A = 0.054(\pm0.41)[p\text{-OH}] - 1.345(\pm0.46)[p\text{-OMe}] - 0.012(\pm0.34)[m\text{-OH}] - 0.376(\pm0.42)[m\text{-OMe}]$$
$$-0.305(\pm0.37)[\alpha\text{-Me}] - 0.555(\pm0.49)[N\text{-Me}] + 0.080(\pm0.51)[m'\text{-OMe}] + 1.906(\pm0.40) \tag{50}$$
$$n = 16, \quad r = 0.962, \quad s = 0.238, \quad F = 14.01$$

$$\log A = 0.147(\pm0.36)[p\text{-OH}] - 1.334(\pm0.39)[p\text{-OMe}] + 0.054(\pm0.29)[m\text{-OH}] - 0.180(\pm0.41)[m\text{-OMe}]$$
$$-0.343(\pm0.31)[\alpha\text{-Me}] - 0.649(\pm0.42)[N\text{-Me}] + 0.026(\pm0.43)[m'\text{-OMe}]$$
$$- 0.326(\pm0.34)[\text{OH:OMe}] + 1.875(\pm0.33) \tag{51}$$
$$n = 16, \quad r = 0.978, \quad s = 0.194, \quad F = 19.06$$

The significance of interaction terms should be tested statistically (see Section 21.3.3.6) and there should be reasonable explanations for such interaction terms. While the interaction term in equation (49) can be explained by a nonlinear lipophilicity–activity relationship (see Section 21.3.5.5), the negative interaction of *m*-methoxy and *p*-hydroxy (equation 51) may be explained either by a hindrance of the binding of the hydroxyl group to the enzyme, by the steric bulk of the methoxy group or by a weakening of this interaction due to intramolecular hydrogen bonding.

Table 12 Hypnotic Activities of N,N'-Diacylureas $R_1NHCONHR_2$; Structures and Matrix Used for Fujita–Ban Analysis (equation 48) and Bocek–Kopecky Analysis (equation 49)[18]

Substituents[a]		Matrix for Fujita–Ban analysis[a]					Interaction term,	
R_1	R_2	Prop	But	Pent	Hex	Hept	$N_1 \cdot N_2$[b]	Log (1/C)
Acet	Prop	1					6	1.84
Prop	Prop	2					9	2.06
Acet	But		1				8	2.16
Prop	But	1	1				12	2.23
Acet	Pent			1			10	2.27
But	But		2				16	2.40
Prop	Pent	1		1			15	2.35
Acet	Hex				1		12	2.46
But	Pent		1	1			20	2.38
Prop	Hex	1			1		18	2.25
Acet	Hept					1	14	2.55
Pent	Pent			2			25	2.32
But	Hex		1		1		24	2.28

[a] Acet, prop, but, pent, hex, hept = acetyl, propanoyl, butanoyl, pentanoyl, hexanoyl and heptanoyl. [b] N_1 and N_2 = number of carbon atoms of substituents R_1 and R_2.

21.3.3.6 Regression Analysis and Statistical Parameters

When the linear dependences in the original structural matrix (*e.g.* Table 4) are removed either by the classical Free–Wilson procedure (*e.g.* Table 5) or by the Fujita–Ban procedure (*e.g.* Table 7) the resulting data matrices can be solved by standard programs for linear multiple regression analysis. Any program package that includes the calculation of the most important statistical parameters (see below), *e.g.* the BMDP programs,[70] can be used. A comfortable program for Hansch and Free–Wilson analysis, including graphics and several other important features, was described by Esaki.[71]

The calculation procedure for the classical Free–Wilson model has been demonstrated step by step.[27, 28, 43, 44, 46] A computer program for regression analysis has been published[28] and another program, including the generation of the symmetry equations and the transformation of the structural matrix, has been described.[72] Also, for the Fujita–Ban modification, the calculation procedure has been explained step by step.[73] A simplified algorithm can be derived for the Fujita–Ban modification[73] and corresponding computer programs are mentioned in the literature.[74–76] A calculation procedure for an approximative solution to the Free–Wilson model has been proposed.[44, 77, 78] However, standard programs for linear multiple regression analysis are readily available nowadays, thus these algorithms should not be used any longer.

While regression analysis[79–81] need not be explained in this context, the meaning and valuation of the statistical parameters in Free–Wilson analysis must be discussed. As in the extrathermodynamic approach, the statistical parameters to be presented with the results should include 95% confidence intervals of the regression coefficients (not standard errors, which are lower by a factor of approximately 2–3), the correlation coefficient r, the standard deviation s and the F value.[82]

The correlation coefficient r is a relative measure for the fit of the data. It is calculated from the residual variance (*i.e.* the variance not explained by the regression), $\Sigma(y_{obs} - y_{calc})^2$, and the overall variance of the data, $\Sigma(y_{obs} - y_{mean})^2$, by equation (52). The smaller the unexplained variance is, the higher the correlation coefficient r will be, approaching a value of unity. However, the correlation coefficient r also depends on the overall variance of the data; the higher this value is, the higher the correlation coefficient will be. Thus, the inclusion of single point determinations (see Section 21.3.3.7) increases the value of r when these additional points are far from y_{mean}, without a better fit of all other y values. The use of adjusted r values was recommended to account for this problem (equation (53); n = number of compounds, k = number of variables).[83]

$$r = \sqrt{\left(1 - \frac{\Sigma(y_{obs} - y_{calc})^2}{\Sigma(y_{obs} - y_{mean})^2}\right)} \tag{52}$$

$$r_{adj} = \sqrt{\left(1 - \frac{(n-1)(1-r^2)}{n-k}\right)} \tag{53}$$

In contrast to the correlation coefficient r, the standard deviation s is an absolute measure for the fit of the data, depending only on the unexplained variance and the number of degrees of freedom, $n - k - 1$ (equation 54). The addition or deletion of a single point determination does not change its value because the number of degrees of freedom remains constant.

The significance of a regression equation is tested by its F value (equation 55). In equation (55), r^2 corresponds to the explained variance, $1 - r^2$ corresponds to the unexplained variance and their ratio is corrected by the number of degrees of freedom, $n - k - 1$, and by the number of variables, k. The higher the explained variance is, the higher the F value of the equation will be. On the other hand, variables not contributing to the explained variance diminish the F value of the equation. Although the F values of Free–Wilson equations cannot be as high as they are in the extrathermodynamic approach because of the usually much larger number of variables, each Free–Wilson equation should be checked for its overall significance by a comparison of its F value with the tabulated $F_{95\%, k, n-k-1}$ value.

$$s = \sqrt{\left(\frac{\Sigma(y_{obs} - y_{calc})^2}{n-k-1}\right)} \tag{54}$$

$$F_{k, n-k-1} = \frac{r^2(n-k-1)}{(1-r^2)k} \tag{55}$$

$$F_{k_2-k_1, n-k_2-1} = \frac{(r_2^2 - r_1^2)(n - k_2 - 1)}{(1 - r_2^2)(k_2 - k_1)} \tag{56}$$

In linear multiple regression analysis, the inclusion of additional variables is checked by a partial *F* test (usually called sequential *F* test if $k_2 - k_1 = 1$) (equation (56); the numbers 1 and 2 indicate two regressions with the smaller and the larger number of variables). In Free–Wilson analysis the number of variables is determined by the structural matrix, thus a partial *F* test cannot be performed. Only the significance of interaction terms should be checked by a sequential *F* test. For the example given in Table 12 (equations 48 and 49) the inclusion of the interaction term $N_1 \cdot N_2$ is justified by the *F* test ($F_{1,6} = 47.56$; $F_{95\%, 1,6} = 5.99$). On the other hand, the interaction term [OH:OMe] in equation (51) is not significant at the 95% probability level ($F_{1,7} = 5.03$; $F_{95\%, 1,7} = 5.59$; the *F* values given in ref. 4 are wrong).

Confidence intervals can be calculated for each regression coefficient b_i from the diagonal elements c_{ii} of the inverted matrix, the standard deviation *s* and the Student *t* value by equation (57). In linear multiple regression analysis each regression coefficient should be justified at the 95% significance level by a confidence interval not including the value of zero, *i.e.* the absolute value of the confidence interval must not be larger than the absolute value of the regression coefficient. While this also applies to Free–Wilson analysis, parameters which are not significant are not removed from the regression equation. The fact that the activity contribution of a group is not significantly different from the activity contribution of the reference group is also a valuable piece of information. In the Fujita–Ban version the values of the confidence intervals depend on the choice of the reference groups, which further complicates a correct interpretation of these confidence intervals (compare Table 9).[29] The use of standard errors of the regression coefficients (equation 58) instead of 95% confidence intervals is not good practice and should be avoided.

$$95\% \text{ confidence interval of } b_i = s \cdot t_{95\%, n-k-1} \cdot \sqrt{c_{ii}} \tag{57}$$

$$\text{standard error of } b_i = s \cdot \sqrt{c_{ii}} \tag{58}$$

For interaction terms, the significance of these additional parameters can be derived directly from the confidence intervals. The 95% confidence interval of the $N_1 \cdot N_2$ term of equation (49) proves the significance of this parameter, while for the [OH:OMe] term in equation (51) the 95% confidence interval includes the value of zero, which, in both cases, corresponds to the results of the sequential *F* test.

The confidence interval of a predicted *y* value can be estimated by equation (59).[26]

$$y = y_{\text{calc}} \pm s \cdot t_{95\%, n-k-1} \cdot \sqrt{\left(1 + \frac{1}{n} + \frac{\Sigma(x_i - x_{\text{mean}})^2}{\Sigma(y_{\text{obs}} - y_{\text{mean}})^2}\right)} \tag{59}$$

21.3.3.7 Single Point Determinations

Usually single point determinations do not cause problems. In the example of Table 4, three compounds lead to single point determinations (compounds 4, 6 and 9). Thus, the resulting activity contributions of the substituents occurring only once in the data matrix (*m*-F, *m*-I and *p*-I in Table 4) are merely the differences between the observed biological activities of the compounds and the overall average μ (in Free–Wilson analysis) or the calculated biological activity of the reference compound (in the Fujita–Ban version). Consequently, such single point based activity contributions have no statistical significance; nevertheless they are useful as first approximations for the real values of the activity contributions. If compounds 4, 6 and 9 and the corresponding substituent columns are removed from Table 7, equation (60) results instead of equation (42). All regression coefficients and all confidence intervals are identical in both equations, the only difference being that equation (60) no longer contains the [*m*-F], [*m*-I] and [*p*-I] terms. Of course all biological activity values calculated from equation (60) are identical with those from equation (42).

$$\log(1/C) = 0.207(\pm 0.29)[m\text{-Cl}] + 0.434(\pm 0.27)[m\text{-Br}] + 0.454(\pm 0.27)[m\text{-Me}] + 0.340(\pm 0.30)[p\text{-F}]$$
$$+ 0.768(\pm 0.30)[p\text{-Cl}] + 1.020(\pm 0.30)[p\text{-Br}] + 1.256(\pm 0.33)[p\text{-Me}] + 7.821(\pm 0.27) \tag{60}$$
$$n = 19, \quad r = 0.958, \quad s = 0.194, \quad F = 17.33$$

The correlation coefficient *r* is somewhat smaller in equation (60) because the unexplained variance (the sum of squared deviations between calculated and observed $\log(1/C)$ values) is identical but the overall variance is smaller. The standard deviation *s* does not change its value because the number of degrees of freedom remains constant. In equation (42) there are 22 compounds and 10 variables, while in equation (60) there are 19 compounds and 7 variables, both leading to 11 degrees of freedom. The F values are nearly identical because the smaller variance of $\log(1/C)$ values and the smaller number of variables have opposite effects on the value of F.

Normally there is no reason to eliminate single point based substituents from a Free–Wilson analysis. However, if the number of single point determinations is large, as compared to the number of well-represented substituents, the statistical reliability of the result decreases. An example is given in Table 13. When the biological activity values of 23 hallucinogenic phenylalkylamines (6) are described by 15 variables, the r, s and F values indicate an excellent fit of the data and high statistical significance. However, when eight compounds, leading to eight single point determinations of group contributions, are removed from the data set, much worse r and F values result. This demonstrates that the apparently better fit resulted only from the larger variance of the y values in the larger group, the unexplained variance and the standard deviation s being identical in both groups.

Table 14 presents an example[41] where the number of degrees of freedom is zero, because 11 compounds (7) are described by 10 variables and the constant term μ. While it is easy to recognize that the values for [7-NO$_2$], [5-OH], [7-NH$_2$], [9-NH$_2$], [7-Br], [9-NO$_2$] and [9-NMe$_2$] are single point determinations (*e.g.* equation 61), based on compounds 1, 5, 6, 7, 9, 10 and 11, the other dependences are not seen at a first glance. However, equations (62)–(64) reveal that the remaining activity contributions are also based on single points. Of course the correlation coefficient $r = 1.000$ of such an analysis is meaningless; the standard deviation s and the F value cannot be calculated.

$$[\text{7-NO}_2] = \log k_1 - \log k_8 = 2.874 - 1.975 = 0.899 \tag{61}$$

$$[\text{6}'\text{-Me}] = \log k_3 - \log k_2 = 2.604 - 2.714 = -0.110 \tag{62}$$

$$[\text{7-Cl}] = \log k_3 - \log k_4 = 2.604 - 2.434 = 0.170 \tag{63}$$

$$[\text{6-OH}] = \log k_2 - (\log k_3 - \log k_4) - \log k_8 = 2.714 - 2.604 + 2.434 - 1.975 = 0.569 \tag{64}$$

While this is an extreme example, many published Free–Wilson equations suffer from an exceedingly large number of variables, including many single point determinations. The results may be useful to sort substituents according to their activity contributions, but one should be aware of the fact that the statistical parameters of such equations are without any significance.

Table 13 Hallucinogenic Properties of Phenylalkylamines (6);[84] Activity Contributions and Statistical Parameters Derived from Two Different Free–Wilson Analyses

(6)

R	All compounds	Without single point values
2-OMe	0.392 ($\pm$0.30)	0.392 ($\pm$0.30)
3-OMe	-0.211 ($\pm$0.30)	-0.211 ($\pm$0.30)
4-OMe	0.081 ($\pm$0.34)	0.081 ($\pm$0.34)
5-OMe	0.216 ($\pm$0.35)	0.216 ($\pm$0.35)
6-OMe	0.410 ($\pm$0.46)	0.410 ($\pm$0.46)
3,4—OCH$_2$O—	0.165 ($\pm$0.44)	0.165 ($\pm$0.44)
4,5—OCH$_2$O—	0.284 ($\pm$0.48)	0.284 ($\pm$0.48)
2,3—OCH$_2$O—	0.050 ($\pm$0.53)	—
4-OEt	0.264 ($\pm$0.55)	—
4-Br	1.754 ($\pm$0.55)	—
4-Me	0.934 ($\pm$0.55)	—
4-Et	1.054 ($\pm$0.55)	—
4-Pr^n	0.984 ($\pm$0.55)	—
4-Bu^n	0.674 ($\pm$0.55)	—
4-(CH$_2$)$_4$Me	0.134 ($\pm$0.55)	—
μ	0.349 ($\pm$0.47)	0.349 ($\pm$0.47)
Number of compounds (n)	23	15
Correlation coefficient (r)	0.985	0.896
Standard deviation (s)	0.182	0.182
F value (F)	15.28[a]	4.08

[a] The published[84] $F = 12.28$ is wrong.

Table 14 Antibacterial Activity of Substituted Tetracyclines (7); Structural Matrix and Inhibition Constants k against *Escherichia coli*[41]

(7)

No.	Structural matrix					Log k
	R_5	R_6	$R_{6'}$	R_7	R_9	
1	H	H	H	NO_2	H	2.874
2	H	OH	H	Cl	H	2.714
3	H	OH	Me	Cl	H	2.604
4	H	OH	Me	H	H	2.434
5	OH	OH	Me	H	H	2.400
6	H	H	H	NH_2	H	2.259
7	H	H	H	H	NH_2	2.161
8	H	H	H	H	H	1.975
9	H	H	H	Br	H	1.714
10	H	H	H	H	NO_2	1.647
11	H	H	H	H	NMe_2	1.374

As mentioned before, there can be no clear recommendation on the number of degrees of freedom necessary for a Free–Wilson analysis. The statistical significance of the result will increase with an increase in the number of degrees of freedom, but will also require a greater synthetic and biological effort. A reasonable compromise will always be the best strategy.

21.3.3.8 Complex Linear Dependences

While the linear dependences for each position of substitution within a structural matrix can easily be removed either by the classical Free–Wilson procedure or by the Fujita–Ban modification, complex linear dependences may arise in unbalanced groups of compounds.[27–30]

Hudson *et al.*[25] were unable to find a unique solution for a data set including 23 compounds and 20 variables. They called the data set 'ill-conditioned' because two different calculations, both using the classical Free–Wilson model with different substitutions from the symmetry equations, gave completely different results. On reexamining this problem, Craig[27] recognized three linear dependences in their structural matrix; substituents B and M always occurred together (compounds 3 and 4), and G and T (compound 5) and Q and Z (compound 6) also only occurred together. Thus, no unique solution can be obtained using standard programs for linear multiple regression analysis.

Purcell *et al.*[28] gave a similar example of a linear dependence (Table 15). After transformation of the structural matrix, where the substituents A_1 and B_1 always occur together, by using the symmetry equations, a Free–Wilson matrix with 15 equations and five unknowns results. However, due to the linear dependence ($A_1 + B_1 = 1$ for all compounds), no solution or only a meaningless solution is obtained for the group contributions a_1 and b_1. While compounds 1, 2 and 3 must be eliminated in the classical Free–Wilson model to get a unique solution, this need not be done in the Fujita–Ban modification. If, for example, A_2, B_2 and C_2 are used as reference substituents, A_1 and B_1 can be combined to a pseudosubstituent D;[29] the Fujita–Ban matrix for this modification is given in Table 15. The value of d is the activity contribution of the A_1B_1 substitution, based on the A_2B_2 substitution.

Even more complex linear dependences can arise in unbalanced groups of compounds.[29] In Table 16 the substituents A_1–A_3 always occur together with B_1–B_3, and the substituents

Table 15 Linear Dependence in a Structural Matrix;[28] Compounds, Structural Matrix and Fujita–Ban Matrix[29,a]

Compound	Structural matrix									Fujita–Ban matrix				
	A_1	A_2	A_3	B_1	B_2	B_3	C_1	C_2	C_3	D	C_1	A_3	B_3	C_3
$A_1B_1C_1$	1		1				1			1	1			
$A_1B_1C_2$	1		1					1		1				
$A_1B_1C_3$	1		1						1	1				1
$A_2B_2C_1$		1			1		1				1			
$A_2B_2C_2$		1			1			1						
$A_2B_2C_3$		1			1				1					1
$A_3B_3C_1$			1			1	1			1	1	1		
$A_3B_3C_2$			1			1		1			1	1		
$A_3B_3C_3$			1			1			1		1	1	1	
$A_2B_3C_1$		1				1	1			1		1		
$A_2B_3C_2$		1				1		1				1		
$A_2B_3C_3$		1				1			1			1	1	
$A_3B_2C_1$			1		1		1			1	1			
$A_3B_2C_2$			1		1			1			1			
$A_3B_2C_3$			1		1				1		1			1

[a] Reference substituents A_2, B_2 and C_2; $D = A_1 + B_1$.

A_4–A_6 always occur together with B_4–B_6, which corresponds to two linear dependences (equations 65 and 66).

$$A_1 + A_2 + A_3 = B_1 + B_2 + B_3 \tag{65}$$

$$A_4 + A_5 + A_6 = B_4 + B_5 + B_6 \tag{66}$$

Due to these linear dependences, the classical Free–Wilson model and the Fujita–Ban modification can only be applied to subsets, *i.e.* to the first nine compounds or to the last nine compounds separately. If, for example, A_1 and B_4 are taken as reference substituents in the Fujita–Ban version, the linear dependence is given by equation (67). Addition of a compound A_1B_4 or A_3B_5 to the series eliminates the linear dependence, but then a new problem arises: Free–Wilson analysis or Fujita–Ban analysis forces a solution where all group contributions are influenced by the exper-

Table 16 Complex Linear Dependencies in a Structural Matrix;[29] Structures and Fujita–Ban Matrix[a]

Compound	Fujita–Ban matrix									
	A_2	A_3	A_4	A_5	A_6	B_1	B_2	B_3	B_5	B_6
A_1B_1						1				
A_1B_2							1			
A_1B_3								1		
A_2B_1	1					1				
A_2B_2	1						1			
A_2B_3	1							1		
A_3B_1		1				1				
A_3B_2		1					1			
A_3B_3		1						1		
A_4B_4			1							
A_4B_5			1						1	
A_4B_6			1							1
A_5B_4				1						
A_5B_5				1					1	
A_5B_6				1						1
A_6B_4					1					
A_6B_5					1				1	
A_6B_6					1					1

[a] Reference substituents A_1 and B_4.

imental error of the activity value of the added compound. This example again demonstrates the importance of proper test-series design for reliable and meaningful results.

$$A_4 + A_5 + A_6 + B_1 + B_2 + B_3 = 1 \tag{67}$$

A closer inspection of Table 2 (Section 21.3.3.1) shows that such complex linear dependences are also relevant in practical examples. Although the series was designed by a factorial scheme[31] to avoid an unbalanced grouping of substituents, a linear dependence arises from R_2 = methoxy and methyl and R_3 = hydroxy and amino (equation 68). If a compound of this set contains either R_3 = methoxy or R_3 = methyl, it also contains R_2 = hydroxy or R_3 = amino and *vice versa* (compounds 2, 4, 7 and 9); thus, it is impossible to separate the activity contributions of these substituents. The linear dependence can only be removed by a replacement of substituents in the original set, *e.g.* R_3 = amino instead of R_3 = methoxy in compound 1.

$$[\text{2-OMe}] + [\text{2-Me}] = [\text{3-OH}] + [\text{3-NH}_2] \tag{68}$$

An easy way to detect hidden linear dependences in Fujita–Ban matrices is to look at the matrix determinant.[30] If the value of the determinant is zero or near zero, no unique solution of the system of equations can be obtained. For the example of Table 2 the value of the determinant is <0.001 (deviations due to rounding errors; the determinant was calculated from the diagonal elements of the triangularized Fujita–Ban matrix). After replacement of R_3 in compound 1 this value increases to 1.00. Usually the determinant is much larger (*e.g.* 138.56 for the matrix given in Table 7; 48.91 for Table 13); its value neither depends on the choice of reference substituents nor on the inclusion or deletion of compounds leading to single point determinations.[85]

21.3.4 APPLICATIONS OF FREE–WILSON ANALYSIS

This chapter reviews applications of Free–Wilson analysis in order to demonstrate the utility of this approach in the quantitative description of structure–activity relationships. Practical examples illustrate the derivation of Hansch equations for definite positions of substitution (Section 21.3.4.1), additivity and nonadditivity in Free–Wilson analysis (Section 21.3.4.2), some applications of Free–Wilson analysis to rigid molecules (Sections 21.3.4.3 and 21.3.4.4), predictions of biological activities for other parent structures (Section 21.3.4.4), the comparison of activity contributions for different biological activities (multivariate Free–Wilson analysis,[53, 86] Section 21.3.4.5) and the application of Free–Wilson analysis to optically active compounds (Sections 21.3.4.4 and 21.3.4.6) and peptides (Section 21.3.4.7).

It should be noted here that many published Free–Wilson analyses, including examples reviewed in this chapter, are numerically wrong for two reasons: firstly, typing errors, either in the manuscripts or in the publications, are common for such large data sets and, secondly, different computer programs may give different numerical results, especially when single precision arithmetic is used in the calculations.

The application of Free–Wilson analysis in drug design and in other fields is reviewed in Section 21.3.4.8 of this chapter.

21.3.4.1 Antimalarials

The very first application of Free–Wilson analysis to a large data set was the quantitative description of the antimalarial activities of a series of 69 2-phenylquinolinylmethanols[87] (**8**; Table 17), followed by a corresponding analysis of 43 phenanthreneaminoalkylmethanols.[88] In the case of the phenylquinolinylmethanols, Hansch-type correlations were derived for positions 4′, 7 and 8 (equations 69–71), indicating the importance of the lipophilic and electronic properties of the aromatic substituents.[87] Following the strategy of looking at different positions of a lead structure in a different manner, Craig and Hansch[88] derived equation (72) for 102 phenanthreneaminoalkylmethanols, which confirmed the conclusions drawn from the Free–Wilson analysis of the phenylquinolinylmethanols. In a later investigation, including many more compounds, equation (73) was derived for 122 phenylquinolinylmethanols[51, 89] and equation (74) was derived for 646 analogues[89] with different parent structures (for an explanation of the parameter symbols see the original references).

Table 17 Antimalarial Activity of 2-Phenylquinolinylmethanols (8); Substituent Contributions from Free–Wilson Analysis[87]

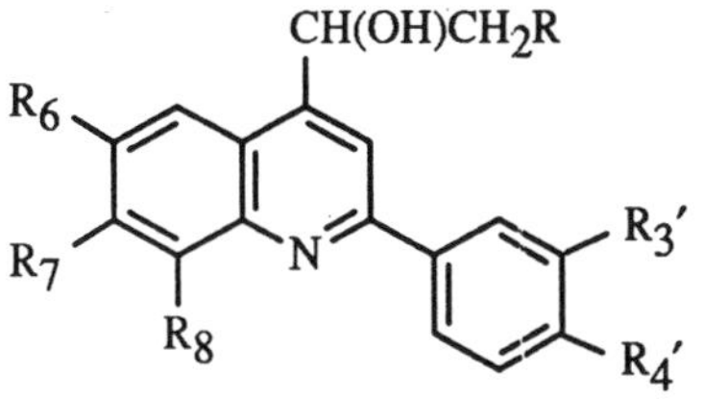

$$CH(OH)CH_2R$$

(8)

Group	Substituent	a_i	Group	Substituent	a_i
R	6-Methyl-2-piperidyl	0.397	$R_{4'}$	I	0.445
	2-Piperidyl	0.323		CF_3	0.145
	CH_2NBu_2	0.262		Cl	0.123
	$CH_2N(Me)Pr^i$	−0.036		F	0.044
	$CH_2N(C_6H_{13})_2$	−0.077		H	−0.150
	$CH_2N(isopentyl)_2$	−0.174		OMe	−0.203
	CH_2NHCH_2Ph	−0.193		Me	−0.240
	$CH_2N(C_8H_{17})_2$	−0.284	R_6	CF_3	0.435
	$CH_2N(CH_2CH_2)_2NMe$	−0.431		Cl	0.143
	CH_2NEt_2	−0.432		OMe	−0.064
	$CH_2(piperidyl)$	−0.451		H	−0.239
	$CH_2NH(cyclopropyl)$	−0.573	R_7	CF_3	0.927
	$CH_2N(CH_2CH_2OEt)_2$	−0.574		Cl	0.487
	$CH_2N(C_7H_{15})_2$	−0.697		F	0.209
	$CH_2NH(1-adamantyl)$	−0.906		H	−0.168
	$CH_2N(CH_2CH_2)_2O$	−0.973		OMe	−0.489
$R_{3'}$	OMe	0.562	R_8	CF_3	0.774
	Cl	0.361		Cl	0.321
	H	−0.055		Me	−0.078
	CF_3	−0.221		H	−0.311
μ		3.39			
Number of Compounds (n)		69			
Correlation coefficient (r)		0.905			
Standard deviation (s)		0.359			
F value (F)		4.55			

Position 4'

$$a_i = 0.220(\pm 0.35)\pi + 0.626(\pm 0.94)\sigma_{meta} - 0.232(\pm 0.23) \qquad (69)$$

$$n = 7, \quad r = 0.895, \quad s = 0.133, \quad F = 8.06$$

Position 7

$$a_i = 1.811(\pm 0.65)\sigma_{para} - 0.010(\pm 0.19) \qquad (70)$$

$$n = 5, \quad r = 0.981, \quad s = 0.123, \quad F = 77.7$$

Position 8

$$a_i = 0.959(\pm 0.78)\pi - 0.395(\pm 0.57) \qquad (71)$$

$$n = 4, \quad r = 0.966, \quad s = 0.151, \quad F = 27.9$$

$$\log(1/C) = 0.396(\pm 0.134)\pi_y + 0.270(\pm 0.105)\pi_x + 0.654(\pm 0.280)\sigma_x + 0.878(\pm 0.269)\sigma_y$$
$$+ 0.137(\pm 0.087)\pi_{sum} - 0.015(\pm 0.009)(\pi_{sum})^2 + 2.335(\pm 0.194) \qquad (72)$$

$$\text{optimal } \pi_{sum} = 4.44 \ (3.25-5.21)$$

$$n = 102, \quad r = 0.913, \quad s = 0.258$$

$$\log(1/C) = 0.621(\pm 0.15)\Sigma\sigma + 0.365(\pm 0.16)\Sigma\pi + 0.365(\pm 0.16)\log P - 0.043(\pm 0.017)(\log P)^2$$

$$- 0.400(\pm 0.19)\text{c-side} + 0.476(\pm 0.19)\text{MR-4}' + 0.341(\pm 0.25)\text{Me-6,8} + 0.248(\pm 0.15)\text{2-Pip}$$

$$+ 2.060(\pm 0.15) \tag{73}$$

$$\log P_o = 4.23\ (3.7\text{–}4.7)$$

$$n = 122, \quad r = 0.834, \quad s = 0.274$$

$$\log(1/C) = 0.576(\pm 0.09)\Sigma\sigma + 0.168(\pm 0.05)\Sigma\pi + 0.105(\pm 0.05)\log P - 0.167(\pm 0.07)\log(\beta P + 1)$$

$$- 0.169(\pm 0.10)\text{c-side} + 0.319(\pm 0.136)\text{CNR}_2 - 0.139(\pm 0.06)\text{AB} - 0.795(\pm 0.06) < 3\text{-cures}$$

$$+ 0.278(\pm 0.11)\text{MR-4}'\text{-Q} + 0.252(\pm 0.18)\text{Me-6,8-Q} + 0.084(\pm 0.10)\text{2-Pip} + 0.151(\pm 0.19)\text{NBrPy}$$

$$- 0.683(\pm 0.22)\text{Q2P378} + 0.267(\pm 0.11)\text{Py} + 2.726(\pm 0.15) \tag{74}$$

$$\log\beta = -3.959 \qquad \log P_o = 4.19$$

$$n = 646, \quad r = 0.898, \quad s = 0.309$$

More examples for the translation of Free–Wilson analyses into meaningful extrathermodynamic relationships will be given in Section 21.3.5.

21.3.4.2 Acetylcholine Analogues

Acetylcholine analogues, where the acetoxy group has been replaced by simple alkyl or alkoxy groups, are agonists or partial agonists, while analogues, where these groups or the acetoxy group are further substituted by aromatic or cycloaliphatic rings, are antagonists of acetylcholine. A group of analogues XCH_2CH_2Y with all theoretically possible combinations of 16 different X substituents with eight different Y groups ($n = 16 \times 8 = 128$) was prepared and tested for their affinity to the postganglionic acetylcholine receptor.[90]

The large number of compounds compared to the number of variables, the wide range of $\log K$ values (from 3.7 to 9.8 covering more than six decades) and the very low standard errors of these data make the set ideally suited for a Free–Wilson analysis. If X and Y groups are considered as individual substituents, the group contributions of Table 18 result.[85] Two different analyses, one including all compounds, the other including only the antagonists, give nearly identical group contributions. The excellent fit of the data and the narrow confidence limits indicate a perfect verification of the Free–Wilson additivity concept. However, if the group contributions of phenyl or cyclohexyl within the X groups are compared for different X substituents (Table 19), there are significant differences, depending on the nature of the other substituents in X. The first phenyl or cyclohexyl group in X contributes approximately 1.3–1.7 log units to the affinity. If the second group introduced into X is identical with the first, the same group contributions result in aliphatic X residues. However, in carboxylic acid esters the activity contribution of this second group is much larger (≈ 2.1–2.5) and it is even larger (≈ 3.1–3.2) when the first and second group are different. Accordingly, the exchange of phenyl by cyclohexyl does not change the value of the group contributions significantly if there is only one ring in X. The exchange of one phenyl group in the diphenyl analogues by cyclohexyl increases the activity significantly (≈ 0.9–1.1), while the exchange of the phenyl group in a cyclohexyl,phenyl analogue by cyclohexyl decreases the biological activity. Possible explanations for these deviations have been discussed.[90] The conclusion for Free–Wilson analysis, as well as for pattern recognition methods, is that the molecules and partial structures should not be dissected into excessively small pieces (compare Section 21.3.6.2).

21.3.4.3 Corticosteroids

Steroids are relatively rigid molecules, with substituents far apart and a specific mode of action caused by interaction with definite receptors. Thus, Free–Wilson analysis should be very suitable for describing the biological activity values of these compounds.

Justice[91] derived the group contributions presented in Table 20 for the glucocorticoid activity of 44 pregn-4-ene-3,20-diones (9). Although the value of the correlation coefficient r indicates some deviations from additivity of group contributions, the standard deviation s is acceptable if one considers that a highly specific biological activity is investigated in a complex animal model after subcutaneous application of the compounds. (The original values, presented in Table 20, cannot be

Table 18 Affinity of Acetylcholine Analogues XCH_2CH_2Y to the Postganglionic Receptor;[90] Free–Wilson Group Contributions of X and Y and Statistical Parameters[85]

Substituent		Group contributions	
		All compounds	Antagonists
X	EtO	-2.479 (± 0.16)	-2.446 (± 0.25)
	Pr^n	-2.175 (± 0.16)	-2.152 (± 0.19)
	$PhCH_2CO_2$	-1.228 (± 0.16)	-1.399 (± 0.15)
	$PhCH_2CH_2O$	-1.177 (± 0.16)	-1.348 (± 0.15)
	$PhCH_2CH_2CH_2$	-0.909 (± 0.16)	-1.080 (± 0.15)
	$C_6H_{11}CH_2CO_2$	-1.035 (± 0.16)	-1.205 (± 0.15)
	$C_6H_{11}CH_2CH_2O$	-0.819 (± 0.16)	-0.990 (± 0.15)
	$C_6H_{11}CH_2CH_2CH_2$	-0.683 (± 0.16)	-0.854 (± 0.15)
	Ph_2CHCO_2	0.872 (± 0.16)	0.701 (± 0.15)
	Ph_2CHCH_2O	-0.070 (± 0.16)	-0.240 (± 0.15)
	$Ph_2CHCH_2CH_2$	0.374 (± 0.16)	0.203 (± 0.15)
	$Ph(C_6H_{11})CHCO_2$	2.035 (± 0.16)	1.864 (± 0.15)
	$Ph_2C(OH)CO_2$	2.047 (± 0.16)	1.876 (± 0.15)
	$(C_6H_{11})_2CHCO_2$	1.467 (± 0.16)	1.296 (± 0.15)
	$(C_6H_{11})_2CHCH_2O$	0.806 (± 0.16)	0.636 (± 0.15)
	$Ph(C_6H_{11})C(OH)CO_2$	2.975 (± 0.16)	2.804 (± 0.15)
Y	$\overset{+}{N}Me_3$	-0.261 (± 0.11)	-0.260 (± 0.11)
	$\overset{+}{N}(Et)Me_2$	0.081 (± 0.11)	0.127 (± 0.11)
	$\overset{+}{N}(Me)Et_2$	0.139 (± 0.11)	0.154 (± 0.10)
	$\overset{+}{N}Et_3$	0.128 (± 0.11)	0.103 (± 0.10)
	$\overset{+}{N}(Me)CH_2(CH_2)_2CH_2$	-0.009 (± 0.11)	0.000 (± 0.11)
	$\overset{+}{N}(Et)CH_2(CH_2)_2CH_2$	0.093 (± 0.11)	0.096 (± 0.10)
	$\overset{+}{N}(Me)CH_2(CH_2)_3CH_2$	-0.012 (± 0.11)	-0.037 (± 0.10)
	$\overset{+}{N}(Et)CH_2(CH_2)_3CH_2$	-0.158 (± 0.11)	-0.183 (± 0.10)
μ		6.499 (± 0.04)	6.670 (± 0.04)
Number of compounds (n)		128	120
Correlation coefficient (r)		0.991	0.991
Standard deviation (s)		0.231	0.218
F value (F)		257.94	238.09

Table 19 Affinity of Acetylcholine Analogues XCH_2CH_2Y to the Postganglionic Receptor;[90] Nonadditivity of the Group Contributions of Ph and C_6H_{11} in X (see Table 18)

Group X	Group contributions, a_i			$a_j - a_i$
	$R = H$	$R = Ph$	$R = C_6H_{11}$	
$RCH_2CH_2CH_2$	-2.175	-0.909		1.266
RCH_2CH_2O	-2.479	-1.177		1.302
$PhCH(R)CH_2CH_2$	-0.909	0.374		1.283
$PhCH(R)CH_2O$	-1.177	-0.070		1.107
$PhCH(R)CO_2$	-1.228	0.872		2.100
$C_6H_{11}CH(R)CO_2$	-1.035	2.035		3.070
$RCH_2CH_2CH_2$	-2.175		-0.683	1.492
RCH_2CH_2O	-2.479		-0.819	1.660
$C_6H_{11}CH(R)CH_2O$	-0.819		0.806	1.625
$C_6H_{11}CH(R)CO_2$	-1.035		1.467	2.502
$PhCH(R)CO_2$	-1.228		2.035	3.263
$RCH_2CH_2CH_2$		-0.909	-0.683	0.226
RCH_2CH_2O		-1.177	-0.819	0.358
RCH_2CO_2		-1.228	-1.035	0.193
$PhCH(R)CO_2$		0.872	2.035	1.163
$PhCH(R)(OH)CO_2$		2.047	2.975	0.928
$C_6H_{11}CH(R)CO_2$		2.035	1.467	-0.568

Table 20 Glucocorticoid Activity of Pregn-4-ene-3,20-diones (9);
Fujita–Ban Group Contributions[91]

(9)

Group	a_i	Group	a_i
1,2 Double bond	0.564	11-CO	1.920
2α-Me	0.611	11β-OH	2.010
		16α-Me	0.702
6,7 Double bond and 6-Me	1.129	16α-OH	0.043
6,7 Double bond	−0.108	17-MeCO$_2$	0.162
6α-Me	0.782	17-OH	0.536
6α-F	0.793	21-MeCO$_2$	0.630
9α-F	0.652	21-OH	0.754
9α-Cl	0.312	21-F	0.022
μ	−3.388		
Number of compounds (n)	44		
Correlation coefficient (r)	0.868		
Standard deviation (s)	0.455		
F value (F)	4.69		

reproduced by using the substituent matrix given in ref. 91. Only after correcting the structure of compound 3 are regression coefficients and statistical parameters very similar to the published values obtained.)

The specificity of the interaction of these compounds with their receptor can be derived from a comparison of group contributions of equal substituents in different positions. While a hydroxyl group increases the activity in the 17- or 21-position, no such effect is observed for the 16α-position. The highest activity increase is observed after introduction of an 11β-hydroxyl group or an 11-keto group. The effect of a 6-methyl group depends significantly on the presence or absence of a 6,7 double bond. While fluorine increases the biological activity in the 6α- or 9α-position, it has no effect in the 21-position, *etc.*

A more comprehensive discussion of these structure–activity relationships has been given by Justice.[91] Quantitative structure–activity relationships of steroids have been reviewed by Zeelen.[92]

21.3.4.4 Benzomorphan Analgesics

Katz *et al.*[93] analyzed a series of benzomorphans with ED$_{50}$ values determined *in vivo* by the hot-plate method in mice using the Fujita–Ban modification of the Free–Wilson model. 99 benzo-morphans with 36 different substituents included 70 (−) enantiomers or racemic mixtures with substituents occurring at least twice in the structural matrix (compounds 1–70), 16 (−) enantiomers or racemates with substituents occurring only once (single point determinations, compounds 71–86) and 13 biologically less active (+) enantiomers (compounds 87–99).

The results of three different Free–Wilson analyses, one including all compounds, the second without the (+) isomers, and the third additionally excluding all single points, are given in Table 21. In the first analysis, two extra parameters [+] and [−] account for the activity differences between (+) enantiomers, (−) enantiomers and the racemates. The value 0.173 for the (−) enantiomers is not too far from the theoretical value 0.30 (if one enantiomer is biologically inactive and the other is responsible for the activity of the racemate). The difference to the value −0.968 for the (+) enantiomers shows that these isomers are approximately 15 times less active than the (−) enantiomers. In the other two analyses no extra parameter was used to differentiate between the racemates and the (−) enantiomers. The statistical parameters of all three analyses are nearly

Table 21 Analgesic Activity of Benzomorphans (**10**); Activity Contributions and Statistical Parameters[93]

(**10**)

Group	Substituent	All compounds	Compounds 1–86	Compounds 1–70
R_1	$OCOAr(3-C_5H_4N)$	1.49	1.694	1.694
	OCOMe	0.869	0.923	0.923
	OH	0.803	0.984	0.984
	OCOEt	0.612	0.641	—
	$OCOPr^n$	0.552	0.581	—
	$OCOBu^n$	0.367	0.729	—
	OMe	0.239	0.349	0.349
	H	0	0	0
	NO_2	−0.108	0.036	—
	F	−0.454	−0.463	—
	Cl	−0.704	−0.713	—
R_2	$(CH_2)_2Ar(2-C_4H_3S)$	2.54	2.44	—
	$(CH_2)_2Ph$	1.58	1.62	1.62
	$CH_2\overline{CHCH_2C}{=}CH_2$	1.53	2.14	—
	$(CH_2)_2COPh$	1.11	1.10	1.10
	Me	0.835	0.924	0.924
	$CH_2\overline{CHCH_2CH_2}$	0.133	0.022	0.022
	H	0	0	0
	$CH_2CH{=}CMe_2$	−0.414	−0.242	−0.242
R_3	Me	0.166	0.204	0.204
	Et	0.103	0.035	0.035
	H	0	0	0
	OCOMe	−0.037	−0.078	−0.078
	OH	−0.729	−0.809	−0.809
R_4	Pr^n	0.586	0.507	—
	Et	0.518	0.712	0.712
	Me	0.433	0.486	0.486
	H	0	0	0
	OH	−0.318	−0.412	−0.412
	CH_2OH	−0.552	−0.703	—
R_5	Ph	0.552	0.336	0.336
	OCOMe	0.385	0.411	—
	Et	0.290	0.239	0.239
	Pr^n	0.236	0.160	0.160
	OCOEt	0.175	0.201	—
	Me	0.114	0.110	0.110
	Bu^n	0.112	−0.043	—
	H	0	0	0
	$(CH_2)_4Me$	−0.098	−0.253	—
	OH	−0.505	−0.479	—
	$(CH_2)_5Me$	−0.528	−0.683	—
(+)	Enantiomer	−0.968	—	—
(−)	Enantiomer	0.173	—	—
(±)	Enantiomer	0	—	—
μ		0.420	0.305	0.305
Number of compounds (n)		99	86	70
Correlation coefficient (r)		0.893	0.909	0.879
Standard deviation (s)		0.466	0.457	0.457
F value (F)		6.23	6.45	8.35

identical and the biological activity values calculated from the group contributions are in good agreement with the observed data.

The substituent constants obtained from this analysis were also used to estimate the activities of six morphinans (**11**). The good correlation between the observed and calculated activities (equation 75)[93] demonstrates that the group contributions obtained from one parent structure can be extended to another series of compounds with a different but chemically related skeleton. The value of the intercept of equation (75) indicates that the activity contribution of a cyclohexyl ring is much higher than the activity contributions of two ethyl groups in the same position. Whether the deviation of the slope (0.769 instead of 1) is fortuitous or not cannot be determined from such a small number of compounds. At the 95% probability level the difference to unity is not significant.

(11)

$$\log(1/C_{obs}) = 0.769(\pm 0.35)\log(1/C_{calc}) + 4.052(\pm 1.02) \tag{75}$$

$$n = 6, \quad r = 0.950, \quad s = 0.254, \quad F = 37.12$$

21.3.4.5 Thrombin, Plasmin and Trypsin Inhibitors

In drug research, different types of biological activities are often simultaneously investigated to find compounds with higher selectivity, fewer side effects or lower toxicity. Of course, Free–Wilson analysis can be used to compare the activity contributions from different test models in order to derive an optimal pattern of substitution. Mager[53, 59, 86, 94, 95] proposed the name multivariate Free–Wilson analysis for this procedure. However, no special multivariate method is needed to derive the individual activity contributions; all values are calculated by ordinary Free–Wilson analysis. The PLS (partial least squares) method, developed by H. and S. Wold,[96–100] may be suitable for such cases, but its applicability to Free–Wilson analysis has never been tested.

Lukovits[101–103] recommended a comparison of activity contributions from different biological models to find correlations between related models. However, the same information can be derived directly and much more easily from the original data. Therapeutic ratios, often taken as measures of selectivity, should not be used in quantitative structure–activity analyses because ratios include the experimental error of two values, thus increasing the standard error of the data.

Labes and Hagen[104] analyzed the inhibition of three different serine proteases (the blood-coagulating enzyme thrombin, the fibrinolytic enzyme plasmin and the pancreatic enzyme trypsin) by a series of 84 substituted benzamidines (**12**) and obtained the activity contributions presented in Table 22 (these values cannot be reproduced by using the structures given in the original reference; only after correcting the structures of compounds 2–4 and 12 can regression coefficients and r values similar to those of Table 22 be obtained).

Significant selectivity differences can be seen when the group contributions for the inhibition of thrombin and plasmin are compared. While the combination of a —CH$_2$CH(Br)CO$_2$— bridge with a phenyl group, bearing hydrogen, *p*-methyl or *p*-chloro, or an α-naphthyl group, may lead to a hundredfold higher inhibition of thrombin, the combination of a —CH$_2$CH(Br)CO— or a —OCH$_2$CO— bridge with polar substituents such as —CH$_2$OH or —C$_6$H$_4$-*p*-OH leads to ten- to thirty-fold higher activity against plasmin. Analogous selectivity differences have been observed for trypsin. However, it should be noted that some of these conclusions are based on single points. In addition, deviations may occur due to nonlinear structure–activity relationships. As in all other cases, results and predictions can only be used to generate new hypotheses which have to be verified or disproved by the synthesis and testing of new analogues.

21.3.4.6 Optically Active Phenethylamines

In one of the very first papers on the application of Free–Wilson analysis, Ban and Fujita investigated the norepinephrine uptake inhibition by optically active sympathomimetic amines.[35]

Table 22 Inhibition of Thrombin, Plasmin and Trypsin by Substituted Benzamidines (12); Activity Contributions and Statistical Parameters[104]

(12)

Group		Thrombin	Plasmin	Trypsin
X	$-CH_2-$	0	0	0
	$-CO-$	−1.02	−0.93	−0.75
	$-CO_2-$	−0.92	−0.18	−0.58
	$-CONH-$	−0.57	−0.51	0.04
	$-CH_2CO-$	0.21	−0.03	0.03
	$-CH_2CO_2-$	0.84	0.39	0.21
	$-CH_2NH-$	0.20	−0.08	0.38
	$-CH_2S-$	−0.03	0.06	0.11
	$-CH_2SO_2-$	−0.51	−0.49	−0.43
	$-(CH_2)_2CO-$	0.43	0.64	0.60
	$-CH_2CH(Br)CO-$	0.98	1.63	1.16
	$-CH=CHCO-$	0.10	0.12	−0.04
	$-(CH_2)_2CO_2-$	0.56	0.36	0.09
	$-CH_2CH(Br)CO_2-$	1.15	0.00	0.40
	$-CH=CHCO_2-$	0.06	−0.25	0.41
	$-CH(Cl)CH(Ph)CO-$	0.23	−0.18	−0.27
	$-OCH_2-$	−0.14	−0.12	−0.13
	$-OCH_2CO-$	−0.26	0.99	0.46
	$-OCH_2CO_2-$	0.47	0.83	0.77
	$-NHCO-$	−0.37	−0.24	0.21
	$-SO_2O-$	−0.85	−1.25	−1.13
	$-SO_2NH-$	−1.16	−1.21	−0.76
Y	Me	0	0	0
	CH_2Cl	0.35	0.13	0.17
	CH_2OH	−0.04	0.28	0.21
	CH_2OEt	0.56	0.25	0.59
	Et	0.01	−0.06	0.19
	Pr^n	0.23	−0.13	0.29
	Ph	0.72	0.03	0.43
	$C_6H_4\text{-}p\text{-}Cl$	0.77	0.08	0.99
	$C_6H_4\text{-}p\text{-}Me$	0.79	0.02	0.70
	$C_6H_4\text{-}p\text{-}OH$	0.21	0.54	0.22
	$C_6H_4\text{-}p\text{-}OMe$	0.43	0.21	0.91
	$C_6H_4\text{-}p\text{-}NH_2$	0.25	0.15	0.55
	$C_6H_4\text{-}p\text{-}NO_2$	0.54	0.28	0.69
	$C_6H_4\text{-}p\text{-}Ph$	0.47	0.44	1.14
	$C_6H_4\text{-}p\text{-}OPh$	1.01	0.43	1.31
	CH_2Ph	0.58	0.34	0.60
	$\alpha\text{-}C_{10}H_7$	1.13	0.27	1.09
μ		0.55	0.50	1.20
Number of compounds (n)		83	82	84
Correlation coefficient (r)		0.90	0.96	0.91
Standard deviation (s)		0.39	0.24	0.35

They considered substituents leading to (R) and (S) configurations as different substituents and assigned values of 0.5 to both substituents in racemates; for diastereomeric mixtures they assumed a 50:50 mixture of both racemates.

The classical Free–Wilson model was used to calculate the activity contributions of the individual substituents. However, the transformation of the structural matrix to the matrix for regression analysis is even more complex when values different from one or zero are in the columns to be eliminated. For this reason and because the original paper contains numerical errors, all values were recalculated using the Fujita–Ban modification (Table 23). As in all other cases, the generation of the matrix for regression analysis is very simple: after selecting the unsubstituted parent compound phenethylamine as the reference structure, all hydrogen columns are removed from the structural

Table 23 Norepinephrine Uptake Inhibiting Activity of Sympathomimetic Amines (13);[35] Compounds, Fujita–Ban Matrix and Biological Activity Values

(13)

Compound	X		Y		R_1	R_2	R_3	R_4	R_5	
	OH	OMe	OH	OMe	OH	OH	Me	Me	Me	Log BA
(−)-Metaraminol			1		1			1		3.158
Dopamine	1		1							2.813
(±)-α-Methyldopamine	1		1				0.5	0.5		2.785
(+)-Amphetamine								1		2.785
(±)-Hydroxyamphetamine	1						0.5	0.5		2.785
(−)-Nordefrin	1		1		1			1		2.740
(−)-Norepinephrine	1		1		1					2.610
(±)-Nordefrin	1		1		0.5	0.5	0.5	0.5		2.408
Tyramine	1									2.389
(±)-Amphetamine							0.5	0.5		2.380
Metatyramine			1							2.332
(+)-Methylamphetamine								1	1	2.217
(±)-Norepinephrine	1		1		0.5	0.5				2.215
N-Methyldopamine	1		1						1	2.161
(−)-Epinephrine	1		1		1				1	2.041
Mephentermine							1	1	1	2.041
Phenethylamine										2.000
(±)-Octopamine	1				0.5	0.5				1.929
(+)-Norepinephrine	1		1			1				1.898
(±)-Epinephrine	1		1		0.5	0.5			1	1.892
(±)-Phenylpropanolamine					0.5	0.5	0.5	0.5		1.740
(−)-Ephedrine					1			1	1	1.699
(−)-Amphetamine							1			1.477
(±)-Phenylethanolamine[a]					0.5	0.5				1.362
(−)-Phenylephrine			1		1				1	1.301
p-Methoxyphenethylamine		1								1.041
(±)-Oxedrine[a]	1				0.5	0.5			1	0.954
(±)-Metanephrine	1			1	0.5	0.5			1	0.415
(±)-Normetanephrine[a]	1			1	0.5	0.5				−0.260
3,4-Dimethoxyphenethylamine		1		1						−0.260

[a] The assignment (+) in ref. 35 is wrong, see ref. 105.

matrix (Table I of ref. 35). In contrast to the classical Free–Wilson model, it makes no difference whether values of 0, 0.5 or 1 are in these columns because all hydrogen group contributions are zero.

The results (Table 24) show not only an excellent fit of the data, but also a high degree of specificity and stereoselectivity of the interaction of these compounds with their receptor. While hydroxyl groups in the *meta* and *para* positions of the aromatic ring increase biological activity, a side-chain hydroxyl group decreases biological activity. The negative effect of the aliphatic hydroxyl group is higher in the (S) position (R_2 = OH) than in the (R) position (R_1 = OH). Aromatic methoxy groups reduce biological activity, indicating that the hydrogen donor properties of the aromatic hydroxyl groups are of the utmost importance for interaction with the receptor. The highest stereoselectivity can be seen for the α-methyl group: while a methyl group in the (R) position (R_3 = Me) leads to biologically less active compounds, a methyl group in the (S) position (R_4 = Me) increases biological activity by a factor of 5–6.

A comparison of this example with that of Section 21.3.4.4 shows two different ways to account for optical isomers: in the benzomorphan example, extra parameters were used for the (+) and (−) enantiomers, while in this example each substituent is considered separately in the (R) and (S) configurations. The shortcoming of the first method is neglect of the well-known fact that the activity differences between optical isomers are not constant. Normally, activity differences are higher the

Table 24 Norepinephrine Uptake Inhibiting Activity of Sympathomimetic Amines (13);[35] Activity Contributions and Statistical Parameters (Recalculated Values[85])

Position	Group	Activity contribution ($\pm$ confidence limits)
X	H	0
	OH	0.355 (± 0.28)
	OMe	-0.879 (± 0.53)
Y	H	0
	OH	0.417 (± 0.27)
	OMe	-1.410 (± 0.42)
R_1	H	0
	OH	-0.379 (± 0.32)
R_2	H	0
	OH	-0.785 (± 0.45)
R_3	H	0
	Me	-0.288 (± 0.41)
R_4	H	0
	Me	0.763 (± 0.32)
R_5	H	0
	Me	-0.467 (± 0.24)
μ		1.975 (± 0.29)
Number of compounds (n)		30
Correlation coefficient (r)		0.963
Standard deviation (s)		0.276
F value (F)		28.78

higher the biological activity of the active enantiomer (Pfeiffer's rule[106]). The different description of (*R*) and (*S*) substituents accounts for this phenomenon, but it is applicable only when at least the relative configurations within the series are known.

21.3.4.7 Peptides

Peptides nowadays play an important role as leads in the design of new drugs. The major goals of structural variation are an increase of activity and selectivity as well as an increase of metabolic stability. Every peptide chemist, whether acquainted with the Free–Wilson model or not, implicitly follows the additivity concept of Free–Wilson analysis, assuming that each structural change within the peptide changes the biological activity in a constant manner, irrespective of some other chemical variations in the molecule.

The first proof for this assumption was given by Greven and de Wied[107] in an investigation of the behavioral potencies of 34 ACTH analogues in pole-jumping experiments in rats. For a larger group of compounds the activity contributions of Table 25 were derived.[108] In analogy to normal Free–Wilson analysis, the activity contributions of side chains of amino acids are estimated, based on the contribution of a reference amino acid which is set to zero. If amino acids are deleted from the lead structure, these deletions are considered by extra parameters. This is, of course, an empirical procedure, not consistent with the demand for a constant backbone of all compounds.

The results of two different analyses are presented in Table 25. The first includes all variables, while the second is derived by stepwise regression analysis,[81] including only variables with regression coefficients significantly different from zero. The graphic display of such results is illustrated by Figure 1. Substituents above the zero level increase biological activity; some other changes are without significant influence on biological activity, while some deletions diminish biological activity.

A few other Free–Wilson analyses of biological activities of peptides have been published.[109–116]

Table 25 Behavioral Potencies of a Series of ACTH-(4–16) Analogues in Pole-Jumping Experiments; Activity Contributions ($\pm$ Standard Errors) and Statistical Parameters of Two Different Free–Wilson Analyses[108]

Position	Group	Activity contributions[a]			
		All variables		Stepwise regression	
4	Met	0		0	
	Deletion	−0.46	($\pm$0.50)	−0.52	($\pm$0.26)
	Met(O)	1.42	($\pm$0.27)	1.33	($\pm$0.21)
	Met(O$_2$)	1.20	($\pm$0.23)	1.20	($\pm$0.18)
	D-Met	0.68	($\pm$0.37)	0.59	($\pm$0.30)
	β-Ala	0.92	($\pm$0.37)	0.79	($\pm$0.30)
	Aib	0.32	($\pm$0.31)	—	
	Val	0.18	($\pm$0.37)	—	
5	Glu	0		0	
	Deletion	0.08	($\pm$0.60)	—	
	Ala	−0.01	($\pm$0.39)	—	
	D-Glu	0.33	($\pm$0.32)	—	
6	His	0		0	
	Deletion	−0.69	($\pm$0.41)	−0.65	($\pm$0.31)
	Ala	0.04	($\pm$0.38)	—	
	D-His	0.64	($\pm$0.32)	0.51	($\pm$0.25)
7	Phe	0		0	
	Deletion	−1.31	($\pm$0.42)	−1.19	($\pm$0.28)
	Tyr	0.19	($\pm$0.37)	—	
	Ala	0.34	($\pm$0.42)	—	
8	Arg	0		0	
	Deletion	0.27	($\pm$0.60)	—	
	Lys	0.23	($\pm$0.39)	—	
	D-Lys	2.04	($\pm$0.39)	1.92	($\pm$0.14)
9	Trp	0		0	
	Deletion	0.18	($\pm$0.57)	—	
	Phe	0.03	($\pm$0.24)	—	
10	Gly	0		0	
	Deletion	−0.21	($\pm$0.36)	—	
11–16	Deletion	0		0	
	LPVGLL[b]	1.33	($\pm$0.34)	1.37	($\pm$0.21)
	D-LPVGLL[b]	2.89	($\pm$0.42)	3.05	($\pm$0.22)
μ		−0.24		−0.14	
Number of compounds (n)		52		52	
Number of variables (k)		24		11	
Correlation coefficient (r)		0.986		0.984	
Standard deviation (s)		0.464		0.406	
F value (F)		40		112	

[a] $\pm$ Standard error.
[b] LPVGLL = ACTH-(11–16)-amide = Lys-Pro-Val-Gly-Lys-Lys-NH$_2$.

21.3.4.8 Review of Applications

Applications of the Free–Wilson model (also including the Fujita–Ban modification and Bocek–Kopecky analyses) are demonstrated in Tables 26–29. The ideas of Free–Wilson analysis had been used long before this model was explicitly formulated to derive the electronic Hammett parameter σ (equation 6, Section 21.3.1)[5] and the same concept of additivity of group contributions was followed to derive lipophilicity contributions π (equation 7, Section 21.3.1)[6] and fragment contributions f from partition coefficients P.[117–120] Neither application is reviewed here (see Chapter 18.7). Although many examples do not meet the statistical criteria discussed in Section 21.3.3.6, and although many of them contain numerical errors, most published applications of Free–Wilson analysis in drug design (Table 26), in enzyme and receptor binding (Table 27), and in agrochemistry (Table 28) are covered. A few examples where the Free–Wilson model was used in physical chemistry are listed in Table 29.

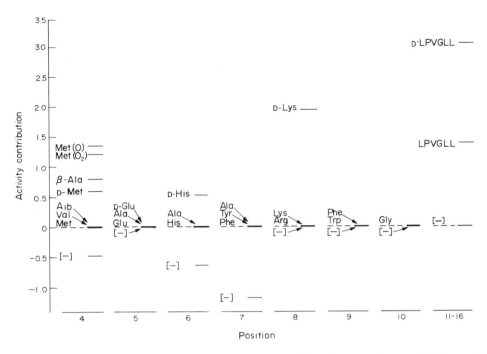

Figure 1 Behavioral potencies of ACTH-(4–16) analogues in pole-jumping experiments. Activity contributions ([−] = deletion) from a stepwise regression analysis (Table 25; reproduced from ref. 108 by permission of Sigma Chemie Publishing Foundation)

21.3.5 RELATIONSHIPS BETWEEN FREE–WILSON ANALYSIS AND THE EXTRATHERMODYNAMIC APPROACH

21.3.5.1 The Equivalence of the Activity Contributions

In the extrathermodynamic approach (Volume 4, Chapter 21.1), logarithmic biological activity values are correlated with physicochemical properties. Most often this approach is called Hansch analysis, because Hansch and Fujita[3] were the first to treat more specific structure–activity relationships rather than simple lipophilicity dependences by combining different physicochemical properties Φ_j into one equation (equation 76). If all physicochemical parameters Φ_j are additive constitutive properties, such as π, MR, σ or E_s, and as long as there are only linear terms in equation (76), activity contributions corresponding to Free–Wilson group contributions can be calculated from a Hansch equation by equation (77).

$$\log(1/C) = k_1\Phi_1 + k_2\Phi_2 + \ldots + k_n\Phi_n + c = \Sigma k_j\Phi_j + c \qquad (76)$$

$$a_i = \sum_j k_j\Phi_{ij} \qquad (77)$$

In equation (77), Φ_{ij} are the physicochemical properties Φ_j of the substituent X_i. If these properties have a value of zero for hydrogen, which is the case for π and σ values, but not for MR or E_s values, c in equation (77) is, like μ in the Fujita–Ban modification of Free–Wilson analysis, the theoretical biological activity value of the unsubstituted parent compound of the series.

These interrelationships between Free–Wilson analysis and the Hansch model were first recognized and theoretically proven by Singer and Purcell,[15] who also stated that this relationship between the models does not apply to parabolic Hansch equations, where instead the Bocek–Kopecky model is theoretically related (Section 21.3.5.5). Although this view has been questioned by Cammarata,[16, 17] it was confirmed theoretically and by practical examples[18, 19] in later investigations.

The close interrelationship between linear Hansch equations and the Free–Wilson model can best be illustrated by the example of Table 4. Some physicochemical parameters relevant to this structure–activity relationship are listed in Table 30 (the reasons for the use of σ^+ values instead of σ values were discussed by Unger and Hansch[234]). Of the many Hansch equations derived for this

Table 26 Applications of Free–Wilson Analysis in Medicinal Chemistry; *In Vitro* and *In Vivo* Models

Biological activity	Ref.	Biological activity	Ref.
Anabolic activity	121	Glucocorticoid activity	8, 91
Analgesic activity	2, 53, 93–95, 122	Hallucinogenic activity	84
Androgenic activity	121	Hypnotic activity	18
Anthelmintic activity	123	Hypoglycemic activity	28, 36, 47
Antiadrenergic activity	16, 18, 29, 73, 76, 78, 101–103, 124	Inhibition of oxytocin contraction	109, 113–115
Antiallergic activity	125, 126	Learning and behavior	107, 108
Antibacterial activity	2, 23, 26, 28, 41, 45, 46, 52, 73, 78, 127–134	Local anesthetic activity	163
Anticholinergic activity	135	Mineralocorticoid activity	8
Anticonvulsant activity	136, 137	Muscle relaxant activity	136
Antifungal activity	18, 23, 30, 138	Neuroleptic activity	164
Antihypertensive activity	139, 140	Ovicidal activity	165
Antiinflammatory activity	38, 75, 141–143	Pharmacokinetic properties	166–169
Antilipemic activity	74	Progestational activity	170
Antimalarial activity	25, 51, 87, 88	Psychostimulant activity	171
Antimicrobial activity	144–146	Radioprotective activity	69, 172–175
Antimitotic activity	147, 148	Saluretic activity	161
Antimycoplasmal activity	39	Sedative activity	136
Antitumor activity	37, 149–155	Spasmolytic activity	61, 62, 176–178
Antiulcer activity	67	Thyromimetic activity	7, 18
Antiviral activity	156–159	Toxicity	2, 13, 14, 17, 53, 68, 74, 142, 165, 172, 175, 179–187
Bradykinin potentiation	111	Tranquilizing activity	180, 188
Diuretic activity	160, 161	Tuberculostatic activity	189
Estrogenic activity	162	Vermicidal activity	165

Table 27 Applications of Free–Wilson Analysis in Drug–Receptor Binding, Enzyme Inhibition and Related Test Models

Biological activity	Ref.	Biological activity	Ref.
Acetylcholinesterase inhibition	175, 190, 191	Muscarine receptor binding	164
Adrenergic α-receptor binding	164	Norepinephrine uptake inhibition	35, 101, 102, 206
Albumin binding	192	Opiate receptor binding	110, 112
Benzodiazepine receptor binding	193–195	Phenylethanolamine *N*-methyltransferase inhibition	102
Butyrylcholinesterase inhibition	18, 33, 34, 40, 175, 196	Phenylethanolamine *N*-methyltransferase substrate properties	4, 101, 102
Carboxylesterase inhibition	175	Plasmin inhibition	104
Dihydrofolate reductase inhibition	197, 198	Prostaglandin synthetase inhibition	207
Dopamine β-hydroxylase substrate properties	4, 102	Renin inhibition	116
Dopamine receptor binding	164, 199	Ribosomal erythromycin binding	208
Inhibition of brain oxidative metabolism	94	Serotonin receptor binding	164
Microsomal oxidation	200	Serotonin uptake inhibition	209
Mitochondrial uncoupling activity	201, 202	Thrombin inhibition	104
Monoamine oxidase inhibition	203–205	Trypsin inhibition	104

Table 28 Applications of Free–Wilson Analysis in Agrochemistry

Biological activity	Ref.	Biological activity	Ref.
Antifungal activity	210–212	Inhibition of cell division	147, 217
Herbicidal activity	213–215	Insecticidal activity	218–220
Hill reaction inhibition	45, 216	Plant growth stimulation	144, 145, 221

Table 29 Applications of Free–Wilson Analysis in Physical Chemistry

Property	Ref.	Property	Ref.
Cyclodextrin complexation	222	NMR spectral data	230
Electronic absorption spectra	223	Partition coefficients	231
Fastness properties of dyes	224–227	Partitioning of ion pairs	232
Fluorescence intensity	228	Permeability through synthetic membrane	233
Molecule geometry	229	R_m values	166

data set[16, 18, 234, 235] only a few are cited here. Equations (78) and (79)[234] describe the biological activity values by the lipophilicity and the electronic properties of the aromatic substituents. While in equation (78) the *meta* and *para* positions are assumed to be equivalent, equation (79), where different regression coefficients are derived for the *meta* and *para* substituents, gives a better fit of the data. One of the possible explanations for these differences might be a steric hindrance of the *meta* substituents. Consequently, equation (80) was derived,[18] which gives a comparably good fit with only three parameters.

$$\log(1/C) = 1.15(\pm 0.19)\pi - 1.47(\pm 0.38)\sigma^+ + 7.82 \tag{78}$$

$$n = 22, \quad r = 0.944, \quad s = 0.197$$

$$\log(1/C) = 0.83(\pm 0.27)\pi_{meta} + 1.33(\pm 0.20)\pi_{para} - 0.92(\pm 0.50)\sigma^+_{meta} - 1.89(\pm 0.57)\sigma^+_{para} + 7.80 \tag{79}$$

$$n = 22, \quad r = 0.966, \quad s = 0.164$$

$$\log(1/C) = 1.26(\pm 0.19)\pi - 1.46(\pm 0.34)\sigma^+ + 0.21(\pm 0.17)E_s^{meta} + 7.62 \tag{80}$$

$$n = 22, \quad r = 0.959, \quad s = 0.173$$

Equations (81)–(86) result for the activity contributions a_i if equation (77) is applied to equations (78)–(80). The E_s^{meta} values in equation (85) are normalized to $E_{s, H} = 0$ in order to obtain comparable a_i values from all different Hansch equations.

$$a_{i, meta} = 1.15\pi_{meta} - 1.47\sigma^+_{meta} \tag{81}$$

$$a_{i, para} = 1.15\pi_{para} - 1.47\sigma^+_{para} \tag{82}$$

$$a_{i, meta} = 0.83\pi_{meta} - 0.92\sigma^+_{meta} \tag{83}$$

$$a_{i, para} = 1.33\pi_{para} - 1.89\sigma^+_{para} \tag{84}$$

$$a_{i, meta} = 1.26\pi_{meta} - 1.46\sigma^+_{meta} + 0.21(E_s^{meta} - E_{s,H}) \tag{85}$$

$$a_{i, para} = 1.26\pi_{para} - 1.46\sigma^+_{para} \tag{86}$$

Table 30 Physicochemical Properties of Aromatic Substituents[18, 234]

Substituent	π_{meta}	π_{para}	σ^+_{meta}	σ^+_{para}	E_s
H	0	0	0	0	1.24
F	0.13	0.15	0.35	−0.07	0.78
Cl	0.76	0.70	0.40	0.11	0.27
Br	0.94	1.02	0.41	0.15	0.08
I	1.15	1.26	0.36	0.14	−0.16
Me	0.51	0.52	−0.07	−0.31	0

Craig[88] argued that the Free–Wilson model and the extrathermodynamic approach give different quantitative results, but this statement is valid only if the group contributions derived from Free–Wilson analysis and from Hansch analysis are not transformed to $a_H = 0$ and/or in the case of nonadditivity of the group contributions. Provided that the biological data are accurate and reliable and that both models give good correlations between observed and calculated $\log(1/C)$ values, which are basic conditions for a comparison of both models, the group contributions from Free–Wilson analysis and from a linear Hansch model are closely correlated (Tables 31 and 32). If the Fujita–Ban modification of Free–Wilson analysis is used and if all physicochemical properties Φ_j are normalized to $\Phi_H = 0$, the activity contributions from both models are also numerically equivalent.[18]

Another important fact can be seen by a comparison of the statistical parameters given in Table 31. The correlation coefficients from a linear Hansch analysis can never be larger than the correlation coefficient of a corresponding Free–Wilson analysis. If equation (77) is understood as the physicochemical meaning of Free–Wilson activity contributions, linear Hansch analysis can only give approximations to the best set of additive parameters derived by the Free–Wilson analysis. Each group contribution a_i may be interpreted as the weighted sum of a large number of different physicochemical properties.

Thus, the value of the correlation coefficient of a Free–Wilson analysis marks the upper limit of the correlation coefficient that can be obtained by a Hansch analysis, including only linear terms and additive physicochemical parameters.[18] In other words, when Free–Wilson analysis of a series gives a good description of the data, there may be a good chance to derive a linear Hansch equation of

Table 31 Antiadrenergic Activities of *N,N*-Dimethyl-α-bromophenethylamines (Table 4); Activity Contributions from Free–Wilson Analysis (Table 8, equation 42) and Hansch Analysis (equations 78–80) and Statistical Parameters[18]

| Substituent | Activity contributions a_i, calculated from | | | |
| | *Free–Wilson analysis* | *Hansch analysis* | | |
		Equation (78)	*Equation (79)*	*Equation (80)*
m-H	0	0	0	0
m-F	−0.30	−0.37	−0.21	−0.44
m-Cl	0.21	0.29	0.26	0.17
m-Br	0.43	0.48	0.40	0.34
m-I	0.58	0.79	0.62	0.63
m-Me	0.45	0.69	0.49	0.48
p-H	0	0	0	0
p-F	0.34	0.28	0.33	0.29
p-Cl	0.77	0.64	0.72	0.72
p-Br	1.02	0.95	1.07	1.07
p-I	1.43	1.24	1.41	1.38
p-Me	1.26	1.05	1.28	1.11
μ/c	7.82	7.82	7.80	7.88[a]
Number of compounds (*n*)	22	22	22	22
Correlation coefficient (*r*)	0.969	0.944	0.966	0.959
Standard deviation (*s*)	0.194	0.197	0.164	0.173

[a] Calculated from $\mu = c + 0.21\ E_s^{m\text{-H}}$

Table 32 Antiadrenergic Activities of *N,N*-Dimethyl-α-bromophenethylamines; Correlation between Activity Contributions from Free–Wilson Analysis and Hansch Analysis (Table 31)

| Activity contributions from | Correlation coefficients (*r*) | | |
| | *Hansch analysis* | | |
	Equation (78)	*Equation (79)*	*Equation (80)*
Free–Wilson analysis	0.965	0.997	0.992
Hansch analysis, equation (78)		0.966	0.979
Hansch analysis, equation (79)			0.990

corresponding quality. If Free–Wilson analysis fails, nonlinear terms have to be considered in both models (Section 21.3.5.5).

21.3.5.2 The Derivation and Improvement of Hansch Equations

The derivation of a Hansch equation is relatively easy in series where only one position of substitution is varied. The compounds can be sorted according to their biological activity values and the relationship to one or two physicochemical parameters may be seen at a glance or from a Craig plot.[236] In series where two or more positions of a lead structure are varied simultaneously, this procedure can only be applied when the different positions of the molecule contribute to the biological activity in the same manner. In all other cases, especially in the case of complex drug–receptor or enzyme–inhibitor interactions, the direct search for a Hansch equation is more of a trial and error procedure than a rational derivation.

Cammarata and Yau[41] and Fujita and Ban[4] were the first to demonstrate that the analysis of Free–Wilson group contributions by the linear multiple regression model may lead to Hansch equations which allow predictions of biological activity values for compounds with substituents outside the original Free–Wilson matrix. There are now many examples in the literature where Free–Wilson analysis has helped to get a first impression as to which physicochemical properties might be responsible for the activity contributions in the different positions of the lead structure (compare equations (69)–(71), Section 21.3.4.1, for example). Obviously, it is much easier to derive Hansch equations for each position independently after prior separation of the activity contributions by a Free–Wilson analysis. Some early applications of this approach have been reviewed.[23]

Free–Wilson analysis may also be used to improve the fit of a Hansch equation by comparing the activity contributions from both models. Again, the strategy can best be explained by an example. Equation (87) was derived[237] for the compounds of Table 33 (compounds 19 and 25 were excluded). For all compounds, equation (88) results where the σ term is no longer significant at the 95% probability level and the correlation coefficient r drops to 0.88. On the other hand, a Free–Wilson analysis of the same data set indicates nearly perfect additivity of the group contributions (Table 24). Equivalence of the *ortho* and *ortho'* positions and likewise of the *meta* and *meta'* positions was assumed for this Free–Wilson analysis. A coefficient of two was given to *ortho,ortho'*- and *meta,meta'*-disubstituted compounds (Table 33).

$$\log(1/C) = 0.691(\pm0.14)\log P + 0.428(\pm0.51)\sigma + 1.213 \tag{87}$$

$$n = 26, \quad r = 0.911, \quad s = 0.216$$

$$\log(1/C) = 0.665(\pm0.15)\log P + 0.500(\pm0.57)\sigma + 1.235 \tag{88}$$

$$n = 28, \quad r = 0.879, \quad s = 0.241$$

A comparison of the Free–Wilson group contributions with those derived from Hansch analysis (equations 89 and 90; Table 34) shows significant differences for some substituents, especially for *ortho* substituents and for the *p*-methyl group. These differences indicate that the electronic properties of the substituents do not contribute to the biological activity and that, instead, a steric hindrance caused by the *ortho* substituents may be responsible for their lower activity contributions.

$$a_{i,\mathrm{R}} = 0.665\pi + 0.500\sigma \tag{89}$$

$$a_{i,\mathrm{X;Y}} = 0.665\pi \tag{90}$$

Consequently, equation (91) was derived[18] where $\log P$ in combination with a steric parameter for the *ortho* substituents and without the electronic parameter σ now gives a much better description of the data, including all compounds. Addition of the electronic parameter σ to this equation does not lead to a further improvement of the fit.

$$\log(1/C) = 0.741(\pm0.11)\log P + 0.214(\pm0.08)E_{\mathrm{s}}^{ortho} + 0.846 \tag{91}$$

$$n = 28, \quad r = 0.942, \quad s = 0.170$$

Of course, one can never be sure whether a certain Hansch equation is indeed the correct explanation of the real structure–activity relationship by merely comparing the statistical parameters of different equations. However, the simplest model that fits the data and that is compatible with the (often hypothetical) biological mechanism of action of the compounds should be taken.[234]

Table 33 Antifungal Activity of Phenyl Ethers (**14**); Structures, Fujita–Ban Matrix and Biological Activity Values[18, 237]

(14)

No.	R	X	Y	2-Me	3-Me	4-Me	2-Cl	4-Cl	X OH	Y OH	Log(1/C) (observed)
1	2-Me	OH	OH	1					1	1	2.26
2	2-Cl	OH	OH				1		1	1	2.31
3	4-Cl	OH	OH					1	1	1	2.31
4	2,6-Cl$_2$	OH	OH				2		1	1	2.37
5	2,4-Cl$_2$	OH	OH				1	1	1	1	2.61
6	2-Me-4-Cl	OH	OH	1				1	1	1	2.33
7	3-Me-4-Cl	OH	OH		1			1	1	1	2.90
8	2-Me-6-Cl	OH	OH	1			1		1	1	2.33
9	2,6-Me$_2$-4-Cl	OH	OH	2				1	1	1	2.76
10	3,5-Me$_2$-4-Cl	OH	OH		2			1	1	1	3.24
11	2,6-Cl$_2$-4-Me	OH	OH			1	2		1	1	3.10
12	2-Me	OH	H	1					1		2.46
13	2-Cl	OH	H				1		1		2.84
14	4-Cl	OH	H					1	1		2.81
15	2,6-Cl$_2$	OH	H				2		1		3.04
16	2,4-Cl$_2$	OH	H				1	1	1		3.35
17	2-Me-4-Cl	OH	H	1				1	1		3.30
18	3-Me-4-Cl	OH	H		1			1	1		3.30
19	2-Me-6-Cl	OH	H	1			1		1		2.70
20	2,6-Me$_2$-4-Cl	OH	H	2				1	1		3.51
21	3,5-Me$_2$-4-Cl	OH	H		2			1	1		3.68
22	2,6-Cl$_2$-4-Me	OH	H			1	2		1		3.47
23	2-Me	H	OH	1						1	2.79
24	4-Cl	H	OH					1		1	3.07
25	2-Me-6-Cl	H	OH	1			1			1	2.78
26	2,6-Me$_2$-4-Cl	H	OH	2				1		1	3.51
27	3,5-Me$_2$-4-Cl	H	OH		2			1		1	3.93
28	2,6-Cl$_2$-4-Me	H	OH			1	2			1	3.67

Header structure: Fujita–Ban matrix (R): 2-Me, 3-Me, 4-Me, 2-Cl, 4-Cl; then X OH, Y OH.

Table 34 Antifungal Activity of Phenyl Ethers (Table 33); Comparison of Activity Contributions and Statistical Parameters from Free–Wilson Analysis and from Hansch Analysis[18]

Substituent	Physicochemical parameters[18] π	σ[a]	E_s[b]	a_i values, calculated from Free–Wilson analysis	Hansch analysis Equation (88)	Equation (91)
o-H	0	0	1.24	0	0	0
o-Me	0.68	−0.14	0	0.20	0.38	0.24
o-Cl	0.59	0.21	0.27	0.27	0.50	0.23
m-H	0	0	—	0	0	0
m-Me	0.51	−0.07	—	0.41	0.30	0.38
p-H	0	0	—	0	0	0
p-Me	0.52	−0.17	—	0.59	0.26	0.39
p-Cl	0.70	0.23	—	0.52	0.58	0.52
X = OH	−0.96	—	—	−0.65	−0.64	−0.71
Y = OH	−0.76	—	—	−0.54	−0.51	−0.56
Number of compounds (*n*)				28	28	28
Correlation coefficient (*r*)				0.967	0.879	0.942
Standard deviation (*s*)				0.145	0.241	0.170

[a] Only for aromatic substituents. [b] Only for *ortho* substituents.

Under these circumstances, equation (91) is to be preferred to equation (88). Of course, the activity contributions calculated from equation (91) (equations 92 and 93) are much closer to the Free–Wilson group contributions than those calculated from equation (88; Table 34).

$$a_{i,\,ortho} = 0.741\pi + 0.214(E_s^{ortho} - E_{s,H}) \tag{92}$$

$$a_{i,\,meta;\,para;\,X;\,Y} = 0.741\pi \tag{93}$$

21.3.5.3 The Mixed Approach

Due to the close theoretical relationships and the numerical equivalence of activity contributions derived from Free–Wilson analysis and from linear Hansch analysis (Section 21.3.5.1), both models can be combined to a mixed approach (equation 94),[23] which includes Free–Wilson group contributions a_i to describe one group of substituents and physicochemical parameters Φ_j to describe another group of substituents. The mixed approach is of special value for series of compounds where there is sufficient structural variation in one position of substitution to derive a Hansch equation, but only little structural variation in another. As Hansch formulated in an analogous context (characterizing the use of different types of indicator variables in the extra-thermodynamic approach, see Section 21.3.5.4), equation (94) incorporates the best aspects of the two quantitative approaches used in structure–activity studies.[22]

$$\log(1/C) = \Sigma a_i + \Sigma k_j \Phi_j + c \tag{94}$$

Equations (95) and (96) were derived[238] for the binding of two different groups of amides to papain (Table 35). The regression coefficients of both equations are comparable, indicating that the activity contributions are identical in both subsets. Thus, equations (95) and (96) can be combined, following equation (94), to equation (97), which contains an additional Free–Wilson parameter I ($I = 1$ for mesylamides).[238] The regression coefficient of I is the activity contribution of a mesyl group, based on the benzoyl group as a reference substituent. The correlation coefficient of equation (97) is much higher than the correlation coefficients of equations (95) and (96) because the range of $\log(1/C)$ values is now much larger. However, the standard deviations of

Table 35 Papain–Ligand Interactions; Structures, Biological Activity Values and Parameters[238]

(15)

X	R	$\log(1/K_m)$	MR	σ	I
4-OH	SO_2Me	2.05	0.28	−0.37	1
4-OMe	SO_2Me	2.13	0.79	−0.27	1
4-Me	SO_2Me	2.08	0.56	−0.17	1
3-Me	SO_2Me	2.23	0.56	−0.07	1
H	SO_2Me	1.79	0.10	0.00	1
4-F	SO_2Me	1.95	0.09	0.06	1
3-OMe	SO_2Me	2.29	0.79	0.12	1
4-CHO	SO_2Me	2.33	0.69	0.42	1
4-Cl	SO_2Me	2.38	0.60	0.23	1
3-F	SO_2Me	1.98	0.09	0.34	1
4-COMe	SO_2Me	2.57	1.12	0.50	1
3-NO_2	SO_2Me	2.53	0.74	0.71	1
4-NO_2	SO_2Me	2.71	0.74	0.78	1
4-NH_2	COPh	3.58	0.54	−0.66	0
4-Me	COPh	4.02	0.56	−0.17	0
H	COPh	3.77	0.10	0.00	0
4-Cl	COPh	4.00	0.60	0.23	0
4-F	COPh	3.69	0.09	0.06	0
3-NO_2	COPh	4.74	0.74	0.71	0
4-NO_2	COPh	4.85	0.74	0.78	0

equations (95)–(97) demonstrate that the quality of fit is comparable for the subsets and for the whole series.

Mesylamides (compounds 1–13 of Table 35)

$$\log(1/K_m) = 0.53(\pm 0.23)MR + 0.37(\pm 0.20)\sigma + 1.88(\pm 0.13) \tag{95}$$

$$n = 13, \quad r = 0.935, \quad s = 0.105, \quad F = 34.51$$

Benzamides (compounds 14–20 of Table 35)

$$\log(1/K_m) = 0.77(\pm 0.67)MR + 0.73(\pm 0.37)\sigma + 3.62(\pm 0.34) \tag{96}$$

$$n = 7, \quad r = 0.971, \quad s = 0.148, \quad F = 32.85$$

All compounds

$$\log(1/K_m) = 0.57(\pm 0.26)MR + 0.56(\pm 0.19)\sigma - 1.92(\pm 0.15)I + 3.74(\pm 0.17) \tag{97}$$

$$n = 20, \quad r = 0.990, \quad s = 0.148, \quad F = 272.03$$

Another example for the combination of the Hansch model with the Free–Wilson model is given in equation (98),[23] where the substituents X of structure (16) are described by the physicochemical parameters π, σ and E'_s, while the substituents Y are described by the Free–Wilson parameters [I] for iodine and [Me] for methyl in this position. Both activity contributions are based on bromine as a reference substituent.

(16)

$$\log A = 1.699(\pm 0.34)\pi_X - 2.059(\pm 0.70)\sigma_X + 1.713(\pm 0.39)E'_s + 0.234(\pm 0.20)[I]$$
$$- 0.532(\pm 0.26)[Me] - 1.792 \tag{98}$$

$$n = 25, \quad r = 0.943, \quad s = 0.349, \quad F = 30.37$$

The combination of nonlinear parameters with Free–Wilson variables is discussed in Section 21.3.5.5.

21.3.5.4 Indicator Variables

Indicator variables (sometimes called dummy variables) are frequently used in linear multiple regression analysis[80, 81] to account for specific features that cannot be described by continuous variables. They take the value of one or zero, depending on the presence or absence of the feature to be described. Bearing this definition in mind, Free–Wilson analysis can be interpreted as a regression analysis approach using only indicator variables[22] and the mixed approach is the combination of the Hansch model with special indicator variables.

Since 1970, indicator variables have been used in the extrathermodynamic approach on an empirical basis[20, 21] to obtain activity contributions of specific structural moieties that cannot be described by physicochemical parameters. Most often, separate equations are derived for different subsets and, after proving the equivalence of these equations, they are combined with the help of one or more indicator variables.

Martin and Lynn[21] derived equations (99) and (100) for the antibacterial properties of N-alkyl-substituted lincomycins (17; Table 36). Generally the *trans* isomers show higher activity than the corresponding *cis* isomers. Thus, a better description of the data can be achieved by including an indicator variable T accounting for these differences (equations 101 and 102; $T = 1$ for *trans* isomers). The regression coefficients of equations (101) and (102) are comparable. Only the constant term is different in both equations, indicating a negative activity contribution of the N-ethyl group, compared to the N-methyl group. Consequently, both groups of compounds can be described by one equation if a further indicator variable is added (equation 103; $Et = 1$ for N-ethyl analogues). Both indicator variables can be interpreted as Free–Wilson-type parameters, T describing the

Table 36 *In Vitro* Activity of Lincomycins (**17**) against *Sarcina lutea*; Structures, Matrix for Regression Analysis and Biological Activity Values[21]

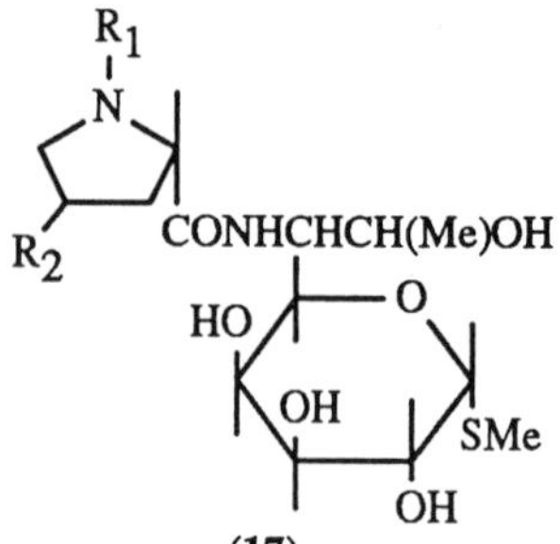

(17)

R_1	R_2		Matrix for regression analysis			$Log(1/C)$ (observed)
			π	T	Et	
Me	H		0.00	0	0	−0.164
Me	Et	(trans)	1.00	1	0	0.882
Me	Pr	(trans)	1.50	1	0	1.390
Me	Bu	(trans)	2.00	1	0	1.697
Me	C_5H_{11}	(trans)	2.50	1	0	1.892
Me	C_6H_{13}	(trans)	3.00	1	0	1.903
Me	C_7H_{15}	(trans)	3.50	1	0	1.510
Me	C_8H_{17}	(trans)	4.00	1	0	1.321
Me	Pr	(cis)	1.50	0	0	1.089
Me	Bu	(cis)	2.00	0	0	1.489
Me	C_5H_{11}	(cis)	2.50	0	0	1.616
Me	C_6H_{13}	(cis)	3.00	0	0	1.669
Me	C_7H_{15}	(cis)	3.50	0	0	1.288
Me	C_8H_{17}	(cis)	4.00	0	0	1.099
Et	H		0.50	0	1	−0.277
Et	Pr	(trans)	2.00	1	1	1.375
Et	Bu	(trans)	2.50	1	1	1.440
Et	C_5H_{11}	(trans)	3.00	1	1	1.824
Et	C_6H_{13}	(trans)	3.50	1	1	1.635
Et	C_8H_{17}	(trans)	4.50	1	1	0.910
Et	Pr	(cis)	2.00	0	1	1.074
Et	Bu	(cis)	2.50	0	1	1.206
Et	C_5H_{11}	(cis)	3.00	0	1	1.461
Et	C_6H_{13}	(cis)	3.50	0	1	1.413
Et	C_8H_{17}	(cis)	4.50	0	1	0.910

activity contribution of a *trans* configuration, compared to the *cis* configuration, and *Et* describing the activity contribution of an *N*-ethyl group, based on the activity contribution of an *N*-methyl group.

N-Methyl derivatives

$$\log act = -0.274\pi^2 + 1.458\pi - 0.238 \tag{99}$$

$$n = 14, \quad r = 0.961, \quad \bar{s} = 0.155, \quad F = 68.27$$

N-Ethyl derivatives

$$\log act = -0.284\pi^2 + 1.448\pi - 0.308 \tag{100}$$

$$n = 11, \quad r = 0.963, \quad s = 0.167, \quad F = 51.76$$

N-Methyl derivatives

$$\log act = -0.262\pi^2 + 1.402\pi + 0.196T - 0.282 \tag{101}$$

$$n = 14, \quad r = 0.980, \quad s = 0.117, \quad F = 83.51$$

N-Ethyl derivatives

$$\log act = -0.298\pi^2 + 1.528\pi + 0.228T - 0.527 \tag{102}$$

$$n = 11, \quad r = 0.984, \quad s = 0.115, \quad F = 75.62$$

All compounds

$$\log \text{act} = -0.242(\pm 0.04)\pi^2 + 1.388(\pm 0.22)\pi + 0.232(\pm 0.14)T - 0.203(\pm 0.14)Et - 0.413(\pm 0.26) \tag{103}$$

$$n = 25, \quad r = 0.960, \quad s = 0.162, \quad F = 60.21$$

On a more empirical basis, Hansch[20] combined equations (104) (*meta* isomers) and (105) (*para* isomers) to equation (106) for the inhibition of cholinesterase by diethyl phenyl phosphates. X is a dummy parameter that accounts for the basic stereoelectronic differences in the inhibitory mechanism betweeen *meta* and *para* isomers ($X = 1$ for *meta* isomers).

Meta-substituted compounds

$$-\log I_{50} = -1.090(\pm 0.59)E_s + 1.576(\pm 1.40)\sigma + 4.499 \tag{104}$$

$$n = 5, \quad r = 0.993, \quad s = 0.248$$

Para-substituted compounds

$$-\log I_{50} = 2.490(\pm 0.44)\sigma^- + 4.184(\pm 0.37) \tag{105}$$

$$n = 8, \quad r = 0.985, \quad s = 0.254$$

All compounds

$$-\log I_{50} = -0.966(\pm 0.36)E_s + 2.287(\pm 0.44)\sigma^- - 1.201(\pm 0.96)X + 5.519 \tag{106}$$

$$n = 13, \quad r = 0.980, \quad s = 0.313$$

Indicator variables are recommended in Hansch analysis whenever special structural features cannot be adequately parameterized by continuous variables. The use of indicator variables was a major breakthrough in the quantitative description of specific enzyme–inhibitor and drug–receptor interactions, as was elegantly demonstrated by the work of Hansch and his group. Most of their examples are listed, together with some other applications, in Table 37 (some early applications of indicator variables in Hansch analyses have been reviewed[23]). A clean classification of the indicator variables used in these analyses is impossible. Many of them are Free–Wilson-type variables, but some others, accounting for intramolecular hydrogen bonding, hydrogen donor and acceptor properties, *ortho* effects, differences between positional isomers, *cis–trans* isomerism, different parent structures, different test models, *etc.*, are merely empirical parameters. Despite the enormous value of such variables, a warning must be given: their misuse may lead to a meaningless combination of pattern recognition with the extrathermodynamic approach.

Table 37   The Use of Indicator Variables in Hansch Analyses

Biological activity	Ref.	Biological activity	Ref.
Adenosine receptor binding	239	Chymotrypsin inhibition	250–254
Antiadrenergic activity	124	Complement inhibition	22, 255, 256
Antiallergic activity	125, 240	Dihydrofolate reductase inhibition	249, 250, 256–261
Antibacterial activity	21	Guanine deaminase inhibition	256, 262
Antifungal activity	241	Inhibition of various enzymes	256
Antiinflammatory activity	75	Papain–ligand interactions	238
Antimalarial activity	89	Thyromimetic activity	18, 23, 263
Antitumor activity	242–247	Toxicity	53
Carbonic anhydrase inhibition	248, 249	Trypsin inhibition	264
Cholinesterase inhibition	20		

21.3.5.5   Nonlinear Equations

In their comparison of Free–Wilson analysis with the extrathermodynamic approach, Singer and Purcell[15] found that nonlinear Hansch equations are also related to Free–Wilson analysis. The nonlinear terms can be seen as interaction terms of the type used in the Bocek–Kopecky modification (Section 21.3.3.5). The validity of this view can be confirmed by the example given in Table 12, where the biological activity values can be described, using nonlinear Hansch analysis, by equation (107).[265] Because of the linear relationship between lipophilicity and the number of carbon atoms (equation 108), equation (107) can be reformulated into equation (109) and, after splitting N into N_1 and N_2 (the number of carbon atoms of the substituents R_1 and R_2) into equation (110).[18]

Equation (110) corresponds to the Bocek–Kopecky model (equation (49); Section 21.3.3.5) with additive parameters $a_i = -0.044N_i^2 + 0.760N_i$ and an interaction term $e_1e_2 = -0.088N_1 \cdot N_2$, which is numerically equivalent to the interaction term of equation (49). Thus, nonlinear terms can be added to a Free–Wilson equation to detect nonlinear dependence of biological activity values on certain physicochemical properties.

$$\log(1/C) = -0.177(\pm 0.09)(\log P)^2 + 0.599(\pm 0.22)\log P + 1.893 \tag{107}$$

$$n = 13, \quad r = 0.918, \quad s = 0.079$$

$$\log P = 0.50N - 2.60 \tag{108}$$

$$\log(1/C) = -0.044(\pm 0.02)N^2 + 0.760(\pm 0.33)N - 0.862 \tag{109}$$

$$n = 13, \quad r = 0.918, \quad s = 0.079$$

$$\log(1/C) = -0.044(N_1^2 + N_2^2) + 0.760(N_1 + N_2) - 0.088N_1 \cdot N_2 - 0.862 \tag{110}$$

The data of Table 38 can be correlated by a linear as well as by a nonlinear Hansch equation (equations 111 and 112)[237] and, correspondingly, these data can be described by a linear Free–Wilson equation (113) and by a Bocek–Kopecky-type equation (114),[23] where π^2 is added as a nonlinear parameter. The inclusion of this term leads to a significant improvement of the fit, as can be seen by a comparison of the statistical parameters and by a sequential F test ($F_{1,9} = 9.68$).

$$\log(1/C) = 0.619(\pm 0.15)\log P + 1.715 \tag{111}$$

$$n = 18, \quad r = 0.914, \quad s = 0.275, \quad F = 80.79$$

$$\log(1/C) = -0.190(\pm 0.09)(\log P)^2 + 1.859(\pm 0.56)\log P + 0.627(\pm 0.39)\sigma - 0.092 \tag{112}$$

$$n = 18, \quad r = 0.975, \quad s = 0.160, \quad F = 90.35$$

$$\log(1/C) = 0.822(\pm 0.19)[p\text{-Cl}] + 0.526(\pm 0.19)[o\text{-Me}] + 1.051(\pm 0.38)[o\text{-Pr}^i] + 0.757(\pm 0.33)[o\text{-Bu}^t]$$
$$+ 1.426(\pm 0.38)[o\text{-C}_6\text{H}_{11}] + 1.536(\pm 0.38)[o\text{-Ph}] + 0.470(\pm 0.19)[m\text{-Me}] + 2.363(\pm 0.26) \tag{113}$$

$$n = 18, \quad r = 0.977, \quad s = 0.183, \quad F = 29.93$$

$$\log(1/C) = -0.124(\pm 0.09)\pi^2 + 1.235(\pm 0.33)[p\text{-Cl}] + 0.674(\pm 0.18)[o\text{-Me}] + 1.541(\pm 0.45)[o\text{-Pr}^i]$$
$$+ 1.726(\pm 0.75)[o\text{-Bu}^t] + 2.522(\pm 0.84)[o\text{-C}_6\text{H}_{11}] + 2.263(\pm 0.60)[o\text{-Ph}]$$
$$+ 0.618(\pm 0.18)[m\text{-Me}] + 2.187(\pm 0.23) \tag{114}$$

$$n = 18, \quad r = 0.989, \quad s = 0.134, \quad F = 50.10$$

The mixed approach (Section 21.3.5.3) can be extended to a nonlinear form (equations 115 and 116),[23, 266] where a_i are additive Free–Wilson group contributions, Φ_j are different physicochemical properties and π^2 (or a comparable term) is a parameter accounting for the nonlinear dependence of biological activity values.

$$\log(1/C) = k\pi^2 + \Sigma a_i + c \tag{115}$$

$$\log(1/C) = k\pi^2 + \Sigma a_i + \Sigma k_j \Phi_j + c \tag{116}$$

Many of the equations cited in Table 37 are nonlinear equations where either the parabolic Hansch model[3, 267] (equation 117; $\Phi = \log P$, π, MR, E_s, etc.) or the bilinear model[268–271] (equations 118 and 119; $\Phi = \pi$, MR or other additive property) have been combined with Free–Wilson-type variables or with other indicator variables. Such combinations are not only justified by the theoretical relationships between the Free–Wilson model, the Bocek–Kopecky model and the linear and nonlinear Hansch model, but they are also the most powerful tool for the quantitative description of complex structure–activity relationships. A few examples (equations 120–127; for data and explanations of the symbols see the original references) are given here to illustrate this statement and to demonstrate the proper use of indicator variables in nonlinear Hansch equations (compare also equations 73, 74 and 99–103).

$$\log(1/C) = a\Phi^2 + b\Phi + c \tag{117}$$

$$\log(1/C) = a\log P - b\log(\beta P + 1) + c \tag{118}$$

$$\log(1/C) = a\Phi - b\log(\beta \cdot 10^\Phi + 1) + c \tag{119}$$

Table 38 Growth Inhibitory Activity of Substituted Phenols (**18**)
against *Aspergillus niger*; Structures, Physicochemical Parameters and
Biological Activity Values[23, 237]

(**18**)

R	Log P	π	σ	Log (1/C)
H	1.46	0.00	0.00	2.35
4-Cl	2.39	0.93	0.23	3.35
2-Me	1.96	0.56	−0.14	2.70
2-Me-4-Cl	2.89	1.49	0.09	3.70
3-Me	2.02	0.56	−0.07	2.68
3-Me-4-Cl	2.95	1.49	0.16	3.70
2,6-Me$_2$	2.46	1.12	−0.28	3.35
2,6-Me$_2$-4-Cl	3.39	2.05	−0.05	4.40
3,5-Me$_2$	2.58	1.12	−0.14	3.26
3,5-Me$_2$-4-Cl	3.51	2.05	0.09	4.22
2-Pr^i	2.76	1.53	−0.23	3.35
2-Pr^i-4-Cl	3.69	2.46	0.00	4.30
2-Bu^t-5-Me	3.70	2.54	−0.59	3.70
2-Bu^t-4-Cl-5-Me	4.63	3.47	−0.36	4.30
2-Cyclohexyl	3.97	2.51	−0.23	4.00
2-Cyclohexyl-4-Cl	4.90	3.44	0.00	4.40
2-Ph	3.59	1.96	0.00	4.10
2-Ph-4-Cl	4.52	2.89	0.23	4.52

Hydrolysis of acylamino acid esters by chymotrypsin[253]

$$\log(k_{cat}/K_m) = 0.76(\pm 0.14)MR_1 + 3.19(\pm 0.35)MR_2 + 0.56(\pm 0.13)MR_3 + 1.30(\pm 0.26)\sigma_3^* - 2.27(\pm 0.28)I$$
$$- 0.32(\pm 0.08)(MR_2)^2 - 0.067(\pm 0.02)MR_1 \cdot MR_2 \cdot MR_3 - 3.21(\pm 0.61) \quad (120)$$
$$n = 77, \quad r = 0.988, \quad s = 0.369$$

Inhibition of dihydrofolate reductase from mouse and rat tumors by triazenes[257, 261]

$$\log(1/C) = 0.680(\pm 0.12)\pi\text{-}3 - 0.118(\pm 0.03)(\pi\text{-}3)^2 + 0.230(\pm 0.07)MR\text{-}4 - 0.024(\pm 0.009)(MR\text{-}4)^2$$
$$+ 0.238(\pm 0.12)I\text{-}1 - 2.530(\pm 0.27)I\text{-}2 - 1.991(\pm 0.29)I\text{-}3 + 0.877(\pm 0.23)I\text{-}4$$
$$+ 0.686(\pm 0.14)I\text{-}5 + 0.704(\pm 0.16)I\text{-}6 + 6.489(\pm 0.16) \quad (121)$$
$$n = 244, \quad r = 0.923, \quad s = 0.377$$

Inhibition of dihydrofolate reductase from rat liver by 2,4-diaminoquinazolines[258, 261]

$$\log(1/C) = 0.810(\pm 0.12)MR\text{-}6 - 0.0635(\pm 0.017)(MR\text{-}6)^2 + 0.775(\pm 0.12)\pi\text{-}5 - 0.734(\pm 0.49)I\text{-}1$$
$$- 2.145(\pm 0.38)I\text{-}2 - 0.544(\pm 0.21)I\text{-}3 - 1.395(\pm 0.41)I\text{-}4 + 0.776(\pm 0.37)I\text{-}6$$
$$- 0.197(\pm 0.12)(MR\text{-}6 \cdot I\text{-}1) + 4.924(\pm 0.23) \quad (122)$$
$$n = 101, \quad r = 0.961, \quad s = 0.441$$

Inhibition of guanine deaminase by 9-phenylguanines[256]

$$\log(1/C) = 1:176(\pm 0.25)MR\text{-}3 + 0.403(\pm 0.11)\pi\text{-}4 - 3.417(\pm 0.44)I\text{-}1 + 1.608(\pm 0.29)I\text{-}3 - 0.127(\pm 0.05)(MR\text{-}3)^2$$
$$- 0.618(\pm 0.25)I\text{-}2 + 0.994(\pm 0.43)E_s\text{-}2 + 3.659(\pm 0.50) \quad (123)$$
$$n = 92, \quad r = 0.941, \quad s = 0.366$$

Inhibition of trypsin by benzamidines[264]

$$\log(1/K_i) = -0.59(\pm 0.49)MR\text{-}4 + 0.88(\pm 0.52)\log(\beta \cdot 10^{MR\text{-}4} + 1) + 0.23(\pm 0.07)\pi\text{-}3' - 0.74(\pm 0.20)\sigma$$
$$+ 0.20(\pm 0.30)I\text{-}M + 0.65(\pm 0.22)I\text{-}1 + 0.43(\pm 0.19)I\text{-}2 + 0.51(\pm 0.15)I\text{-}3 + 1.38(\pm 0.28) \quad (124)$$
$$n = 104, \quad r = 0.924, \quad s = 0.222$$

Growth inhibition of methotrexate-sensitive murine tumor cells by 3-substituted triazines[247]

$$\log(1/C) = 1.40(\pm 0.23)\pi - 1.65(\pm 0.26)\log(\beta \cdot 10^\pi + 1) + 0.88(\pm 0.57)\sigma + 0.52(\pm 0.20)I$$
$$- 0.25(\pm 0.24)OR + 0.63(\pm 0.33)DO + 7.94(\pm 0.21) \tag{125}$$
$$n = 64, \quad r = 0.904, \quad s = 0.298$$

Growth inhibition of methotrexate-resistant murine tumor cells by 3-substituted triazines[247]

$$\log(1/C) = 0.63(\pm 0.20)\pi - 0.26(\pm 0.25)\log(\beta \cdot 10^\pi + 1) - 0.17(\pm 0.07)MR$$
$$- 0.33(\pm 0.24)OR + 5.11(\pm 0.19) \tag{126}$$
$$n = 61, \quad r = 0.878, \quad s = 0.335$$

Antitumor activity of 9-anilinoacridines[246]

$$\log(1/D_{50}) = 0.63(\pm 0.27)\Sigma\pi - 0.75(\pm 0.23)\log(\beta \cdot 10^{\Sigma\pi} + 1) - 1.01(\pm 0.09)\Sigma\sigma - 1.21(\pm 0.36)R_{BS}$$
$$- 0.26(\pm 0.16)MR\text{-}2 + 4.95(\pm 0.75)MR\text{-}3 - 5.13(\pm 0.86)\log(\beta \cdot 10^{MR\text{-}3} + 1)$$
$$- 0.67(\pm 0.12)I_{3,6} - 1.67(\pm 0.20)E_s\text{-}3' - 1.57(\pm 0.21)(E_s\text{-}3')^2 + 0.58(\pm 0.13)I\text{-}NO_2$$
$$+ 0.87(\pm 0.31)I_{DAT} + 0.52(\pm 0.17)I_{BS} + 9.24(\pm 1.33) \tag{127}$$
$$n = 509, \quad r = 0.893, \quad s = 0.305$$

21.3.6 MODELS RELATED TO FREE–WILSON ANALYSIS AND TO PATTERN RECOGNITION

21.3.6.1 The Reduced Free–Wilson Model

The number of variables in Free–Wilson analysis is determined by the number of different substituents in the positions of structural variation. In favorable cases this number of variables can be reduced without a significant loss of information, for example by assuming that the activity contributions of substituents in different positions of a molecule or of all methylene groups within a homologous series are identical (compare Section 21.3.3.4); or by assigning only one variable to different substituents, which are chemically related and for which identical activity contributions are presumed; or by considering only activity contributions significantly different from zero (compare Table 25, Section 21.3.4.7).

An example where these strategies have been followed is given in Table 39. After considering thousands of regression equations, the following variables were found to be most significant:[197] *I*-1, *I*-2, *I*-7, *I*-8, *I*-9, *I*-10, *I*-13, *I*-15, *I*-20, *I*-4 · *I*-8 and *I*-8 · *I*-17. Of the 2047 possible linear combinations of these variables, equation (128) was the best equation using only seven variables.[197]

$$\log(1/C) = 0.365(\pm 0.12)I\text{-}1 + 1.013(\pm 0.12)I\text{-}8 - 0.784(\pm 0.19)I\text{-}9 + 0.419(\pm 0.20)I\text{-}13$$
$$- 0.220(\pm 0.09)I\text{-}15 + 0.513(\pm 0.18)I\text{-}20 + 0.674(\pm 0.23)I\text{-}4 \cdot I\text{-}8 + 7.174(\pm 0.07) \tag{128}$$
$$n = 105, \quad r = 0.903, \quad s = 0.229$$

Mager[53, 187] introduced the name 'reduced Free–Wilson model' for a procedure where non-significant variables are eliminated from a normal Free–Wilson analysis. This can be done by simply eliminating all variables not significant at the 95% probability level, by a backward elimination procedure or by a stepwise regression analysis.[81] Darvas[272] used the name BEL-FREE (= backward elimination of variables from Free–Wilson analysis) for an approach where the variables which provide the widest confidence intervals for a compound to be predicted are eliminated from the regression equation.

The reduced Free–Wilson model can be seen as a transition from normal Free–Wilson analysis, where the number of variables is predetermined, to the pattern recognition approach[55, 122, 273–277] (Volume 4, Chapter 22.3), where a large number of parameters, encoding different structural features, are tested for their significance to separate groups with different properties.

Some examples where the reduced Free–Wilson model or a related approach have been used are listed in Table 40. The value of this strategy should not be overemphasized. The advantage of the higher number of degrees of freedom is counterbalanced by an increase of ambiguity of the analysis, caused by the more or less arbitrary elimination of variables. As mentioned before, the significance of group contributions in the Fujita–Ban modification depends not only on the quality of the data

Table 39 Inhibition of Dihydrofolate Reductase by 2,4-Diaminopyrimidines (19); Indicator Variables Studied in the Regression Analysis[197]

(19)

Variable	Structural moiety	Variable	Structural moiety
I-1	$X = —CH_2—$	I-15	$4'-SO_2F$
I-2	$Y = 2-Cl$	I-16	$3'-SO_2F$
I-3	$Y = 3-Cl$	I-17	SO_2F plus other substituent on the same ring
I-4	$Y = 3-Me$	I-18	I-12 + I-13
I-5	Only substituent 3-position	I-19	L-1210/FR8 enzyme
I-6	Only substituent 4-position	I-20	L-1210/0 enzyme
I-7	Substituent in 4 plus other substituent in 2, 3 or 6	I-21	I-8 + I-9
I-8	Bridge 4-NHCONH	I-22	I-10 + I-11
I-9	Bridge 4-NHCO	I-23	$Z = 2'-Cl$
I-10	Bridge $4-(CH_2)_{1-2}-NHCONH$ or $3-CH_2NHCONH$	I-24	$Z = 3'-Me$
I-11	Bridge $4-(CH_2)_{1-2}-NHCO$ or $3-CH_2NHCO$	I-25	$Z = 4'-Me$
I-12	Bridge $4-(CH_2)_{1-2}-NHSO_2$ or $3-CH_2NHSO_2$	I-26	$3'-$ or $5'-SO_2F$
I-13	Bridge $4-NHSO_2$	I-27	Liver enzyme
I-14	Bridge $4-CH_2CH_2$ or $3-CH_2CH_2$	I-28	$Z = 2'-OMe$

Cross-product terms studied

I-4·I-21	I-15·I-10	I-16·I-11
I-4·I-8	I-15·I-11	I-16·I-22
I-8·I-17	I-15·I-21	I-6·I-22·I-16
I-15·I-5	I-15·I-22	I-6·I-22·I-17
I-15·I-6	I-16·I-10	I-4·I-8·I-17

Table 40 Applications of the Reduced Free–Wilson Model and of Related Approaches

Application	Ref.	Application	Ref.
Antiadrenergic activity	76	Benzodiazepine–receptor binding	193
Antibacterial activity	23, 73, 129	Dihydrofolate reductase inhibition	197
Anticonvulsant activity	272	Learning and behavior	108
Antifungal activity	272	Toxicity	53, 185–187
Antiinflammatory activity	272		

but also on the choice of the reference substituents (compare Table 9, Section 21.3.3.4, and Section 21.3.3.6).

21.3.6.2 The Hyperstructure Concept

Some different QSAR approaches follow the concept of a hyperstructure that is biologically active and assign activity contributions to the parent skeleton and to each fragment of this hyperstructure, as in Free–Wilson analysis.

In an analysis of the biological activity of symmetrical and unsymmetrical disulfides XSSX′, Schaad *et al.*[30] looked at the molecules in two different ways. Firstly, they took —SS— as the parent structure and X and X′ as individual substituents, as in a normal Free–Wilson analysis, and secondly they dissected the molecule into $X_1X_2SSX_3X_4$ and derived activity contributions for X_1–X_4. Both analyses gave comparable results.

This concept of a hyperstructure that can be dissected into small fragments is realized in a more rigorous manner in the DARC/PELCO approach[278-282] (DARC = description, acquisition, retrieval and computer-aided design; PELCO = perturbation of environments which are limited, concentric and ordered). The hyperstructure is made up from the parent skeleton and from all fragments of linear and branched substituents at this frame, ordered in concentric spheres.

An example (Table 41) illustrates the close relationship to the Free–Wilson model. Values of one and zero are given to A_i and B_i, depending on the presence or absence of methylene or methyl in the corresponding positions of the hyperstructure (**20**). For C_i the numbers of carbon atoms exceeding the chain —A_1B_1 are taken. Nine compounds were used to derive the activity contributions of all fragments (equation 129; PC′ = phenol coefficient)[283] and to calculate the biological activity values of six compounds not included in the DARC/PELCO analysis. However, this result is trivial because all biological activity values are linearly related to lipophilicity (equation 130).[284]

$$\log PC' = 0.392(\pm 0.21)A_1 + 0.268(\pm 0.21)A_2 + 0.158(\pm 0.21)A_3 + 0.446(\pm 0.15)B_1$$
$$+ 0.533(\pm 0.04)C_i - 2.05(\pm 0.16) \tag{129}$$

$$n = 9, \quad r = 1.000, \quad s = 0.050, \quad F = 655.1$$

$$\log PC' = 1.024(\pm 0.06)\log P - 1.536(\pm 0.07) \tag{130}$$

$$n = 15, \quad r = 0.996, \quad s = 0.090, \quad F = 1592.9$$

The greatest disadvantage of the DARC/PELCO model arises from the large number of variables needed.[46] The results from DARC/PELCO analyses have been compared with the corresponding results from Hansch analyses and from normal Free–Wilson analyses.[127, 128, 283, 285] No real advantages over these methods were established. On the other hand, similar approaches have been followed for the derivation of the fragment contributions of lipophilicity[117-120] from partition

Table 41 Antibacterial Activity of Aliphatic Alcohols (**20**); DARC/PELCO Analysis[283]

$$\text{HO} \overset{A_1B_1C_1}{\underset{A_3}{\overset{|}{—}}} \!\!\! A_2$$

(**20**)

| Alcohol | DARC/PELCO matrix | | | | | Log PC′ | |
	A_1	A_2	A_3	B_1	C_1	Observed	Calculated
Methanol	0	0	0	0	0	−2.05	−2.05
Ethanol	1	0	0	0	0	−1.70	−1.66
n-Propanol	1	0	0	1	0	−1.19	−1.21
n-Butanol	1	0	0	1	1	−0.67	−0.68
n-Octanol	1	0	0	1	5	1.46	1.45
Isopropanol	1	1	0	0	0	−1.39	−1.39
t-Butanol	1	1	1	0	0	−1.19	−1.23
t-Hexanol	1	1	1	1	1	−0.31	−0.25
t-Pentanol	1	1	1	1	0	−0.77	−0.78
n-Pentanol	1	0	0	1	2	−0.13	−0.15
n-Hexanol	1	0	0	1	3	0.40	0.38
n-Heptanol	1	0	0	1	4	0.92	0.91
s-Butanol	1	1	0	1	0	−0.92	−0.94
s-Pentanol	1	1	0	1	1	−0.44	−0.41
s-Hexanol	1	1	0	1	2	0.04	0.12

Table 42 QSAR Methods Based on Hyperstructure Concepts

Method	Ref.	Method	Ref.
Minimal steric difference (MSD)	286	Topological pharmacophores	44, 291–295
Minimal topological difference (MTD)	286–288	LOGANA	292, 293, 295
Steric maps (SIBIS method)	289, 290	LOCON	292, 294, 295

coefficients *P*. Thus, for certain groups of compounds and for large data sets the DARC/PELCO approach may be appropriate; in all other cases normal Free–Wilson analysis should be used (compare Section 21.3.4.2).

Other approaches based on a more or less hypothetical hyperstructure and on logical concepts comparable to Free–Wilson analysis are summarized in Table 42. Some of these methods use logical operators, such as *and*, *or* and *not*, instead of linear multiple regression analysis to derive information on the pharmacophore topology from the biological data.

21.3.7 SCOPE AND LIMITATIONS OF FREE–WILSON ANALYSIS

The additivity of group contributions to biological activity is the basic concept of Free–Wilson analysis. This additivity is to be expected for drugs that interact with a receptor or an enzyme by noncovalent forces as long as there are no major differences in the binding mode of all molecules within the series.

The binding constants of 200 drugs and enzyme inhibitors have been used by Andrews *et al.*[296] to calculate AVERAGE ΔG (average energy resulting from all group energies) values of functional groups (equation 131; DOF = degrees of freedom; see also Volume 4, Chapter 18.8). From equation (131), average binding energies can be calculated by summing up the intrinsic binding energies of all functional groups and carbon atoms and then subtracting two entropy-related terms (14 kcal mol^{-1} for the loss of overall rotational and translational entropy and 0.7 kcal mol^{-1} for each degree of conformational freedom; 1 kcal = 4.184 kJ). Although the result is only a number corresponding to the average fit of a ligand to its binding site, its comparison with the observed binding constant gives a measure for the goodness of fit to the binding site. Large differences between experimental and calculated values were observed: biotin, estradiol and diazepam bind much better to their receptors than the average, while methotrexate, ouabain, thyroxine, buprenorphine and ketanserin bind less well than the average. Equation (131) corresponds to the additivity concept of Free–Wilson analysis. However, the differences between experimental and average binding energies clearly indicate the limitations of this concept. Minor structural changes of a lead compound may not alter the binding mode, but larger variations, especially those influencing the conformational flexibility of the molecule, may change the overall fit of the ligand. In such a case the additivity concept will fail.

$$\text{AVERAGE } \Delta G = 0.7n_{C(sp^2)} + 0.8n_{C(sp^3)} + 11.5n_{N^+} + 1.2n_N + 8.2n_{CO_2^-} + 10.0n_{\text{phosphate}} + 2.5n_{OH}$$

$$+ 3.4n_{C=O} + 1.1n_{O,S} + 1.3n_{\text{hal}} - 0.7n_{\text{DOF}} - 14 \tag{131}$$

Equation (131) is an independent confirmation for the use of logarithmic biological activity values in Free–Wilson analysis, because binding constants *K* and binding energies ΔG are related by equation (132).

$$\Delta G = -RT\ln K \tag{132}$$

Especially in the early stages of lead structure optimization, the Free–Wilson approach can be of great help. Only from the structures and the biological data can the relevant information be extracted, provided that the series has been designed in the right manner, with structural variations in at least two different positions. In all cases where the additivity concept applies, activity contributions for each structural variation are the result of the analysis. In all other cases nonlinear models or specific interaction terms have to be considered. As the structural variation within a congeneric series proceeds, the identity or nonidentity of group contributions of identical substituents in different positions of the parent structure indicate specificity or nonspecificity of the interaction of the drugs with the biological system. In addition, the values of the activity contributions can be correlated with physicochemical properties and from these correlations Hansch equations or mixed equations, including physicochemical terms and Free–Wilson variables, can be derived for the original activity values.

Several structural features, such as *cis–trans* isomerism, optical enantiomers, intramolecular hydrogen bonds, *ortho* effects, *etc.*, that cannot be adequately treated by physicochemical parameters can be described by Free–Wilson variables. The most important progress in QSAR in recent years has come from the combination of the extrathermodynamic approach with such Free–Wilson-type indicator variables to describe complex structure–activity relationships of large data sets.

The limitations of Free–Wilson analysis come from the impossibility of predicting biological activity values for compounds with structural features that are not included in the analysis. Only from a correlation of the activity contributions with other properties or from an intuitive analysis of the effects responsible for activity changes caused by definite groups will predictions outside the original substituent matrix be possible. Another limitation arises in groups where one position of substitution has been varied extensively as compared to all other positions. In such a case, these group contributions are all based on single data points and are thus without any statistical significance. Again the combination of Hansch analysis and Free–Wilson analysis can solve the problem.

21.3.8 SUMMARY

Free–Wilson analysis is a simple but valuable tool of QSAR. Its concept of the additivity of group contributions to biological activity is logical and easy to understand. All calculations can be done by standard programs of linear multiple regression analysis. The results indicate whether the additivity concept is fulfilled or not and the values of the group contributions give a first impression as to which properties of the substituents might be responsible for the biological activity. The results can also be used as input for other QSAR methods, *e.g.* Hansch analysis. Because of the theoretical relationship between the extrathermodynamic approach and the Free–Wilson model and their differences in practical applicability, both models can be combined to a mixed approach, which is the most powerful method of classical QSAR today.

21.3.9 REFERENCES

1. A. Crum Brown and T. R. Fraser, *Trans. R. Soc. Edinburgh*, 1868–69, **25** (I), 151.
2. S. M. Free, Jr. and J. W. Wilson, *J. Med. Chem.*, 1964, **7**, 395.
3. C. Hansch and T. Fujita, *J. Am. Chem. Soc.*, 1964, **86**, 1616.
4. T. Fujita and T. Ban, *J. Med. Chem.*, 1971, **14**, 148.
5. N. B. Chapman and J. Shorter (eds.), 'Advances in Linear Free Energy Relationships', Plenum Press, London, 1972.
6. A. Leo, C. Hansch and D. Elkins, *Chem. Rev.*, 1971, **71**, 525.
7. T. C. Bruice, N. Kharasch and R. J. Winzler, *Arch. Biochem. Biophys.*, 1956, **62**, 305.
8. J. Fried and A. Borman, in 'Vitamins and Hormones. Advances in Research and Applications', ed. R. S. Harris, G. F. Marrian and K. V. Thimann, Academic Press, New York, 1958, vol. 16, p. 303.
9. R. Zahradnik and M. Chvapil, *Experientia*, 1960, **16**, 511.
10. R. Zahradnik, *Experientia*, 1962, **18**, 534.
11. R. Zahradnik, *Arch. Int. Pharmacodyn. Ther.*, 1962, **135**, 311.
12. M. Chvapil, R. Zahradnik and B. Čmuchalová, *Arch. Int. Pharmacodyn. Ther.*, 1962, **135**, 330.
13. K. Boček, J. Kopecký, M. Krivucová and D. Vlachová, *Experientia*, 1964, **20**, 667.
14. J. Kopecký, K. Boček and D. Vlachová, *Nature (London)*, 1965, **207**, 981.
15. J. A. Singer and W. P. Purcell, *J. Med. Chem.*, 1967, **10**, 1000.
16. A. Cammarata, *J. Med. Chem.*, 1972, **15**, 573.
17. A. Cammarata and T. M. Bustard, *J. Med. Chem.*, 1974, **17**, 981.
18. H. Kubinyi and O.-H. Kehrhahn, *J. Med. Chem.*, 1976, **19**, 578.
19. L. J. Schaad, B. A. Hess, Jr., W. P. Purcell, A. Cammarata, R. Franke and H. Kubinyi, *J. Med. Chem.*, 1981, **24**, 900.
20. C. Hansch, *J. Org. Chem.*, 1970, **35**, 620.
21. Y. C. Martin and K. R. Lynn, *J. Med. Chem.*, 1971, **14**, 1162.
22. C. Hansch and M. Yoshimoto, *J. Med. Chem.*, 1974, **17**, 1160.
23. H. Kubinyi, *J. Med. Chem.*, 1976, **19**, 587.
24. R. F. Rekker, *Pharmacochem. Libr.*, 1985, **8**, 3.
25. D. R. Hudson, G. E. Bass and W. P. Purcell, *J. Med. Chem.*, 1970, **13**, 1184.
26. Y. C. Martin, 'Quantitative Drug Design. A Critical Introduction', Dekker, New York, 1978.
27. P. N. Craig, *Adv. Chem. Ser.*, 1972, **114**, 115.
28. W. P. Purcell, G. E. Bass and J. M. Clayton, 'Strategy of Drug Design: A Guide to Biological Activity', Wiley, New York, 1973.
29. H. Kubinyi and O.-H. Kehrhahn, *J. Med. Chem.*, 1976, **19**, 1040.
30. L. J. Schaad, R. H. Werner, L. Dillon, L. Field and C. E. Tate, *J. Med. Chem.*, 1975, **18**, 344.
31. V. Austel, in 'QSAR and Strategies in the Design of Bioactive Compounds. Proceedings of the Fifth European Symposium on Quantitative Structure Activity Relationships', ed. J. K. Seydel, VCH, Weinheim, 1985, p. 247.
32. C. Cativiela, J. Elguero, D. Mathieu, E. Melendez and R. Phan Tan Luu, *Eur. J. Med. Chem.—Chim. Ther.*, 1983, **18**, 359.
33. W. P. Purcell, *Biochim. Biophys. Acta*, 1965, **105**, 201.
34. J. G. Beasley and W. P. Purcell, *Biochim. Biophys. Acta*, 1969, **178**, 175.
35. T. Ban and T. Fujita, *J. Med. Chem.*, 1969, **12**, 353.
36. W. R. Smithfield and W. P. Purcell, *J. Pharm. Sci.*, 1967, **56**, 577.
37. W. P. Purcell and J. M. Clayton, *J. Med. Chem.*, 1968, **11**, 199.
38. R. T. Buckler, *J. Med. Chem.*, 1972, **15**, 578.
39. C. E. Berkoff, P. N. Craig, B. P. Gordon and C. Pellerano, *Arzneim.-Forsch.*, 1973, **23**, 830.
40. J. M. Clayton and W. P. Purcell, *J. Med. Chem.*, 1969, **12**, 1087.
41. A. Cammarata and S. J. Yau, *J. Med. Chem.*, 1970, **13**, 93.
42. E. Gaebler, R. Franke and P. Oehme, *Pharmazie*, 1976, **31**, 1.

43. R. Franke, 'Optimierungsmethoden in der Wirkstoff-Forschung. Quantitative Struktur–Wirkungs-Analyse', Akademie-Verlag, Berlin, 1980.
44. R. Franke, 'Theoretical Drug Design Methods', Elsevier, Amsterdam, 1984.
45. C. Grieco, C. Silipo and A. Vittoria, *Farmaco, Ed. Sci.*, 1976, **31**, 607.
46. J. K. Seydel and K.-J. Schaper, 'Chemische Struktur und Biologische Aktivität von Wirkstoffen. Methoden der Quantitativen Struktur–Wirkung-Analyse', Verlag Chemie, Weinheim, 1979.
47. A. Cammarata and K. S. Rogers, 'Advances in Linear Free Energy Relationships', ed. N. B. Chapman and J. Shorter, Plenum Press, London, 1972, p. 401.
48. K. C. Chu, in 'Burger's Medicinal Chemistry', ed. M. E. Wolff, Wiley, New York, 1980, vol. 1, p. 393.
49. P. N. Craig, in 'Chemical Information Systems', ed. J. E. Ash and E. Hyde, Horwood, Chichester, 1975, p. 259.
50. A. C. Glasser, *Methods Find. Exp. Clin. Pharmacol.*, 1984, **6**, 563.
51. C. Hansch, *Drug Dev. Res.*, 1981, **1**, 267.
52. P. J. Lewi, in 'Drug Design', ed. E. J. Ariëns, Academic Press, New York, 1976, vol. 7, p. 209.
53. P. P. Mager, *Med. Res. Rev.*, 1983, **3**, 435.
54. R. Osman, H. Weinstein and J. P. Green, *ACS Symp. Ser.*, 1979, **112**, 21.
55. A. J. Stuper, W. E. Brugger and P. C. Jurs, 'Computer Assisted Studies of Chemical Structure and Biological Function', Wiley, New York, 1979.
56. H. Bruns, *Chem.-Ztg.*, 1972, **96**, 417.
57. T. Esaki, *Kagaku to Yakugaku no Kyoshitsu*, 1975, **48**, 55 (*Chem. Abstr.*, 1976, **84**, 98 956y).
58. R. Franke and P. Oehme, *Pharmazie*, 1973, **28**, 489.
59. H. Mager, *Sci. Pharm.*, 1977, **45**, 71.
60. A. Sabljic and N. Trinajstic, *Kem. Ind.*, 1979, **28**, 467 (*Chem. Abstr.*, 1980, **92**, 103 849n).
61. O.-E. Schultz, *Pharm. Ztg.*, 1975, **120**, 1449.
62. O.-E. Schultz, *Pharm. Ztg.*, 1976, **121**, 73.
63. I. Schwartz, *Stud. Cercet. Chim.*, 1973, **21**, 721 (*Chem. Abstr.*, 1973, **79**, 73 404a).
64. Z. Zhang and S. Wang, *Yiyao Gongye*, 1983, 30 (*Chem. Abstr.*, 1984, **100**, 114 331x).
65. J. D. P. Graham and M. A. Karrar, *J. Med. Chem.*, 1963, **6**, 103.
66. L. J. Schaad and B. A. Hess, Jr., *J. Med. Chem.*, 1977, **20**, 619.
67. J. Elguero and A. Fruchier, *Afinidad*, 1982, **39**, 548 (*Chem. Abstr.*, 1983, **98**, 154 901e).
68. N. S. Antonov, M. I. Gevenyan and L. T. Tseirova, *Khim.–Farm. Zh.*, 1982, **16**, 325.
69. V. K. Mukhomorov, *Khim.–Farm. Zh.*, 1982, **16**, 1086.
70. L. Bures and T. Havranek, *Cesk. Farm.*, 1985, **34**, 126 (*Chem. Abstr.*, 1985, **103**, 64 386h).
71. T. Esaki, *Anal. Chim. Acta*, 1981, **133**, 657.
72. W. Meiske, E. Gäbler and R. Franke, *Pharmazie*, 1976, **31**, 740.
73. H. Kubinyi, *Arzneim.-Forsch.*, 1977, **27**, 750.
74. V. Gombar, V. K. Kapoor and H. Singh, *Arzneim.-Forsch.*, 1982, **32**, 7.
75. V. Gombar, V. K. Kapoor and H. Singh, *Arzneim.-Forsch.*, 1983, **33**, 1226.
76. V. Gombar, *Arzneim.-Forsch.*, 1986, **36**, 1014.
77. T. Roesner, R. Kuehne and R. Franke, *Abh. Akad. Wiss. DDR, Abt. Math., Naturwiss., Tech.*, 1978, 317 (*Chem. Abstr.*, 1979, **91**, 14 619z).
78. T. Roesner, R. Franke and R. Kuehne, *Pharmazie*, 1978, **33**, 226.
79. G. W. Snedecor and W. G. Cochran, 'Statistical Methods', The Iowa State University Press, Ames, IA, 1973.
80. C. Daniel and F. S. Wood, 'Fitting Equations to Data', Wiley, New York, 1980.
81. N. R. Draper and H. Smith, 'Applied Regression Analysis', Wiley, New York, 1981.
82. P. N. Craig, C. H. Hansch, J. W. McFarland, Y. C. Martin, W. P. Purcell and R. Zahradnik, *J. Med. Chem.*, 1971, **14**, 447.
83. R. F. Rekker and H. M. de Kort, personal communication, 1979.
84. M. C. Bindal, P. Singh and S. P. Gupta, *Arzneim.-Forsch.*, 1982, **32**, 719.
85. H. Kubinyi, unpublished results.
86. P. P. Mager, 'Drug Design', ed. E. J. Ariëns, Academic Press, New York, 1980, Vol. 10, p. 343.
87. P. N. Craig, *J. Med. Chem.*, 1972, **15**, 144.
88. P. N. Craig and C. H. Hansch, *J. Med. Chem.*, 1973, **16**, 661.
89. K. H. Kim, C. Hansch, J. Y. Fukunaga, E. E. Steller, P. Y. C. Jow, P. N. Craig and J. Page, *J. Med. Chem.*, 1979, **22**, 366.
90. F. B. Abramson, R. B. Barlow, M. G. Mustafa and R. P. Stephenson, *Br. J. Pharmacol.*, 1969, **37**, 207.
91. J. B. Justice, Jr., *J. Med. Chem.*, 1978, **21**, 465.
92. F. J. Zeelen, *Quant. Struct.–Act. Relat.*, 1986, **5**, 131.
93. R. Katz, S. F. Osborne and F. Ionescu, *J. Med. Chem.*, 1977, **20**, 1413.
94. P. P. Mager and A. Seese, *Pharmazie*, 1981, **36**, 427.
95. P. P. Mager and A. Seese, *Pharm. Unserer Zeit*, 1981, **10**, 97.
96. H. Wold, in 'Systems under Indirect Observation', ed. K. G. Joereskog and H. Wold, North-Holland, Amsterdam, 1982, part 2, p. 1.
97. S. Wold, H. Martens and H. Wold, in 'Matrix Pencils. Lecture Notes in Mathematics', ed. B. Kågström and A. Ruhe, Springer, Heidelberg, 1983, p. 286.
98. B. Nordén, U. Edlund, D. Johnels and S. Wold, *Quant. Struct.–Act. Relat.*, 1983, **2**, 73.
99. W. J. Dunn, III, S. Wold, U. Edlund, S. Hellberg and J. Gasteiger, *Quant. Struct.–Act. Relat.*, 1984, **3**, 131.
100. S. Hellberg, S. Wold, W. J. Dunn, III, J. Gasteiger and M. G. Hutchings, *Quant. Struct.–Act. Relat.*, 1985, **4**, 1.
101. I. Lukovits, *Mol. Pharmacol.*, 1982, **22**, 725.
102. I. Lukovits, in 'QSAR in Design of Bioactive Compounds', ed. M. Kuchar, Prous, Barcelona, 1984, p. 359.
103. I. Lukovits, *Acta Pharm. Jugosl.*, 1986, **36**, 219.
104. D. Labes and V. Hagen, *Pharmazie*, 1979, **34**, 554.
105. A. S. V. Burgen and L. L. Iversen, *Br. J. Pharmacol.*, 1965, **25**, 34.
106. P. A. Lehmann, J. F. Rodrigues de Miranda and E. J. Ariëns, *Prog. Drug Res.*, 1976, **20**, 101.
107. H. M. Greven and D. de Wied, in 'Frontiers of Hormone Research', ed. F. J. H. Tilders, D. F. Swaab and T. B. van Wimersma Greidanus, Karger, Basel, 1977, vol. 4, p. 140.
108. J. Kelder and H. M. Greven, *Recl. Trav. Chim. Pays-Bas*, 1979, **98**, 168.

109. V. Pliška, *Experientia*, 1978, **34**, 1190.
110. V. V. Ezhov, P. F. Potashnikov and G. A. Sokol'skii, *Khim.–Farm. Zh.*, 1980, **14**, 52 (*Chem. Abstr.*, 1981, **94**, 25 418w).
111. K. J. Schaper, *Eur. J. Med. Chem.—Chim. Ther.*, 1980, **15**, 449.
112. D. Maysinger, M. Movrin and M. Ljubic, *Acta Pharm. Jugosl.*, 1982, **32**, 177 (*Chem. Abstr.*, 1983, **98**, 47 102y).
113. V. Pliška, in 'Perspectives in Peptide Chemistry', ed. A. Eberle, R. Geiger and T. Wieland, Karger, Basel, 1981, p. 221.
114. V. Pliška, *J. Steroid Biochem.*, 1984, **20**, 1512.
115. V. Pliška and J. Heiniger, *Pharmacochem. Libr.*, 1987, **10**, 263.
116. D. Nisato, J. Wagnon, G. Callet, D. Mettefeu, J. L. Assens, C. Plouzane, B. Tonnerre and J. L.Fauchere, *Pharmacochem. Libr.*, 1987, **10**, 277.
117. R. F. Rekker, 'The Hydrophobic Fragmental Constant. Its Derivation and Application. A Means of Characterizing Membrane Systems', Elsevier, Amsterdam, 1977.
118. R. F. Rekker and H. M. de Kort, *Eur. J. Med. Chem.—Chim. Ther.*, 1979, **14**, 479.
119. A. Leo, P. Y. C. Jow, C. Silipo and C. Hansch, *J. Med. Chem.*, 1975, **18**, 865.
120. C. Hansch and A. Leo, 'Substituent Constants for Correlation Analysis in Chemistry and Biology', Wiley-Interscience, New York, 1979.
121. I. Urbankova, *Cesk. Farm.*, 1984, **33**, 331 (*Chem. Abstr.*, 1985, **102**, 40 063d).
122. H. Wijnne, 'Biological Activity and Chemical Structure, Proceedings of IUPAC IUPHAR Symposium', ed. J. A. Keverling Buisman, Elsevier, Amsterdam, 1977, p. 211.
123. C. Pellerano, L. Savini, C. E. Berkoff, J. Thomas and P. Actor, *Farmaco, Ed. Sci.*, 1975, **30**, 965.
124. P. A. Borea, A. Bonora, V. Bertolasi and G. Gilli, *Arzneim.-Forsch.*, 1980, **30**, 1613.
125. P. A. Borea, *Arzneim.-Forsch.*, 1982, **32**, 325.
126. P. A. Borea, *Boll.—Soc. Ital. Biol. Sper.*, 1981, **57**, 633 (*Chem. Abstr.*, 1981, **95**, 125 878k).
127. B. Duperray, M. Chastrette, M. Cohen Makabeh and H. Pacheco, *Eur. J. Med. Chem.—Chim. Ther.*, 1976, **11**, 323.
128. L. H. Hall and L. B. Kier, *Eur. J. Med. Chem.—Chim. Ther.*, 1978, **13**, 89.
129. Y. C. Martin, P. H. Jones, T. J. Perun, W. E. Grundy, S. Bell, R. R. Bower and N. L. Shipkowitz, *J. Med. Chem.*, 1972, **15**, 635.
130. D. Maysinger, M. Birus and M. Movrin, *Pharm. Acta Helv.*, 1981, **56**, 151.
131. Y. Miyashita, Y. Takahashi, Y. Yotsui, H. Abe and S. Sasaki, *CODATA Bull.*, 1981, **41**, 37 (*Chem. Abstr.*, 1981, **95**, 180 655k).
132. A.-M. Noel-Artis, G. Berge, P. Fulcrand and J. Castel, *Eur. J. Med. Chem.—Chim. Ther.*, 1985, **20**, 25.
133. H. Sun, L. Xu and M. Shen, *Huadong Huagong Xueyuan Xuebao*, 1982, 309 (*Chem. Abstr.*, 1983, **98**, 86 082m).
134. B. L. Tóth-Martinez, Z. Dinya and F. Hernádi, in 'Advances in Pharmacological Research and Practice, Proceedings of the Congress of the Hungarian Pharmacological Society, 3rd, Budapest, 1979', ed. J. Knoll, Pergamon Press, Oxford, 1980, vol. 3, p. 339.
135. B. Tinland, *Farmaco, Ed. Sci.*, 1975, **30**, 935.
136. P. A. Borea, G. Gilli and V. Bertolasi, *Farmaco, Ed. Sci.*, 1979, **34**, 1073.
137. J. Lapszewicz, J. Lange, S. Rump and K. Walczyna, *Acta Pharm. Suec.*, 1977, **14** (Suppl.), 48.
138. S. L. Galdino, I. R. Pitta and C. Luu-Duc, *Farmaco, Ed. Sci.*, 1986, **41**, 59.
139. V. M. Kulkarni, *Curr. Sci.*, 1977, **46**, 801.
140. B. Tinland, C. Decoret and J. Badin, *Pharmacol. Res. Commun.*, 1972, **4**, 195.
141. J. Badin and B. Tinland, *Res. Commun. Chem. Pathol. Pharmacol.*, 1973, **6**, 1099.
142. E. Mizuta, N. Suzuki, Y. Miyake, M. Nishikawa and T. Fujita, *Chem. Pharm. Bull.*, 1975, **23**, 5.
143. B. Tinland and J. Badin, *Farmaco, Ed. Sci.*, 1974, **29**, 886.
144. J. Halgas, V. Sutoris, P. Foltinova and V. Sekerka, *Chem. Zvesti*, 1983, **37**, 799 (*Chem. Abstr.*, 1984, **100**, 103 234s).
145. J. Halgas, V. Sutoris, V. Sekerka, P. Foltinova and E. Solcaniova, *Chem. Zvesti*, 1983, **37**, 663 (*Chem. Abstr.*, 1984, **100**, 156 529y).
146. H. Sun, Z. Chen, G. Xu and L. Xu, *Yao Hsueh Hsueh Pao*, 1982, **17**, 107 (*Chem. Abstr.*, 1982, **96**, 192 936m).
147. Lj. Butula and D. Maysinger, *Pharm. Acta Helv.*, 1981, **56**, 273.
148. R. D. Pop, I. Schwartz, M. Coman, A. Muresan and I. Simiti, *Rev. Roum. Biochim.*, 1979, **16**, 135 (*Chem. Abstr.*, 1980, **92**, 174 204j).
149. A. Chiriac, O. Dragomir, F. Motoc and I. Motoc, *Univ. Timisoara, [Prepr.], Ser. Chim.*, 1979, 1 (*Chem. Abstr.*, 1980, **93**, 197 549k).
150. A. U. De and A. K. Ghose, *Indian J. Chem., Sect. B*, 1978, **16**, 513.
151. A. U. De and D. Pal, *J. Indian Chem. Soc.*, 1976, **53**, 1049.
152. D. Maysinger, M. Birus and M. Movrin, *Acta Pharm. Jugosl.*, 1979, **29**, 15 (*Chem. Abstr.*, 1979, **91**, 32 658r).
153. R. F. Rekker, in 'Developments in Pharmacology', ed. D. N. Reinhoudt, T. A. Connors, H. M. Pinedo and K. W. van de Poll, Nijhoff, The Hague, 1983, vol. 3, p. 23.
154. I. Simiti, I. Schwartz and M. Coman, *Rev. Roum. Biochim.*, 1974, **11**, 139 (*Chem. Abstr.*, 1975, **82**, 132 793p).
155. B. Tinland, *Farmaco, Ed. Sci.*, 1976, **31**, 888.
156. R. Franke, D. Labes, M. Tonew, W. Zschiesche and L. Heinisch, *Acta Biol. Med. Ger.*, 1975, **34**, 491.
157. H. J. Michel, R. Franke and H. Willitzer, *Abh. Akad. Wiss. DDR, Abt. Math., Naturwiss., Tech.*, 1978, 89 (*Chem. Abstr.*, 1979, **91**, 82 981m).
158. J. Thomas, C. E. Berkoff, W. B. Flagg, J. J. Gallo, R. F. Haff, C. A. Pinto, C. Pellerano and L. Sävini, *J. Med. Chem.*, 1975, **18**, 245.
159. B. Tinland, *Res. Commun. Chem. Pathol. Pharmacol.*, 1974, **8**, 571.
160. E. Mizuta, K. Nishikawa, K. Omura and Y. Oka, *Chem. Pharm. Bull.*, 1976, **24**, 2078.
161. J. Reiter, L. Toldy, I. Schäfer, E. Szondy, J. Borsy and I. Lukovits, *Eur. J. Med. Chem.—Chim. Ther.*, 1980, **15**, 41.
162. Z.-M. Tang, J.-J. Wu, X.-Q. Mao, M.-Y. Chen and Y.-M. Li, *Yao Hsueh Hsueh Pao*, 1980, **15**, 410 (*Chem. Abstr.*, 1981, **95**, 906m).
163. B. Tinland, *Farmaco, Ed. Sci.*, 1973, **28**, 831.
164. J. Kelder, Th. de Boer, J. S. de Graaf and J. H. Wieringa, in 'QSAR and Strategies in the Design of Bioactive Compounds. Proceedings of the Fifth European Symposium on Quantitative Structure Activity Relationships', ed. J. K. Seydel, VCH, Weinheim, 1985, p. 162.

165. A. Meister, M. Tschaepe and E. Schroetter, *Pharmazie*, 1977, **32**, 174.
166. P. Fernandez Gomez and J. L. Vila Jato, *Arch. Farmacol. Toxicol.*, 1984, **10**, 199 (*Chem. Abstr.*, 1985, **102**, 197 525d).
167. J. Kvetina, M. Laznicek, M. Kvetinova and K. Waisser, in 'Biopharmaceutics and Pharmacokinetics, European Congress, 2nd', ed. J. M. Aiache and J. Hirtz, Lavoisier, Paris, 1984, vol. 2, p. 451.
168. M. Laznicek, K. Waisser, J. Kvetina and P. Beno, *Pharmacochem. Libr.*, 1985, **8**, 249.
169. K. Waisser, M. Laznicek and J. Kvetina, *Cesk. Farm.*, 1985, **34**, 359 (*Chem. Abstr.*, 1986, **104**, 45 347a).
170. F J. Zeelen, in 'Biological Activity and Chemical Structure, Proceedings of the IUPAC IUPHAR Symposium', ed. J. A. Keverling Buisman, Elsevier, Amsterdam, 1977, p. 147.
171. F. Darvas, Z. Budai, L. Petöcz and I. Kosóczky, *Res. Commun. Chem. Pathol. Pharmacol.*, 1975, **12**, 243.
172. G. Grassy, A. Terol, A. Belly, Y. Robbe, J.-P. Chapat, R. Granger, M. Fatome and L. Andrieu, *Eur. J. Med. Chem.—Chim. Ther.*, 1975, **10**, 14.
173. B. Hu, D. Zhong, R. Huang, M. Li, S. Li, X. Song, W. Tang, C. Zhang, Y. Song *et al.*, *Zhongguo Yixue Kexueyuan Xuebao*, 1985, **7**, 6 (*Chem. Abstr.*, 1985, **103**, 192 400v).
174. V. M. Kulkarni, *Indian J. Chem., Sect. B*, 1976, **14**, 190.
175. K. Waisser, M. Machacek and M. Celadnik, in 'QSAR in Design of Bioactive Compounds', ed. M. Kuchar, Prous, Barcelona, 1984, p. 425.
176. R. Baumes, H. C. N. Tien Duc, J. Elguero and A. Fruchier, *An. Quim., Ser. C*, 1983, **79**, 128 (*Chem. Abstr.*, 1984, **101**, 203 886v).
177. A. Boucherle, H. Cousse, G. Mouzin, L. Dussourd d'Hinterland and J. F. Queffelec, *Boll. Chim. Farm.*, 1976, **115**, 89 (*Chem. Abstr.*, 1976, **85**, 206b).
178. H. Cousse, G. Mouzin and L. Dussourd d'Hinterland, *Chim. Ther.*, 1973, **8**, 466.
179. E. S. Balynina, L. A. Timofievskaya and M. R. Zel'tser, *Gig. Tr. Prof. Zabol.*, 1982, 35 (*Chem. Abstr.*, 1982, **96**, 212 001h).
180. F. Darvas, A. Lopata, Z. Budai and L. Petöcz, *Pharmacochem. Libr.*, 1985, **8**, 199.
181. E. O. Dillingham, R. W. Mast, G. E. Bass and J. Autian, *J. Pharm. Sci.*, 1973, **62**, 22.
182. V. V. Ezhov, B. I. Dan'shin, P. F. Potashnikov and G. A. Sokol'skii, *Zh. Vses. Khim. Otva.*, 1978, **23**, 224 (*Chem. Abstr.*, 1978, **89**, 36 467j).
183. L. H. Hall, L. B. Kier and G. Phipps, *Environ. Toxicol. Chem.*, 1984, **3**, 355.
184. M. L. Krasovitskaya and N. E. Ainbinder, *Deposited Doc.*, VINITI 2749-83, 1983, 1 (*Chem. Abstr.*, 1984, **101**, 85 078h).
185. P. P. Mager, A. Seese, H. Hikino, T. Ohta, M. Ogura, Y. Ohizumi, C. Konno and T. Takemoto, *Pharmazie*, 1981, **36**, 717.
186. P. P. Mager, A. Seese and K. Takeya, *Pharmazie*, 1981, **36**, 381.
187. P. P. Mager, H. Mager and A. Barth, *Sci. Pharm.*, 1979, **47**, 265.
188. F. Darvas, A. Lopata, Z. Budai and L. Petöcz, in 'QSAR and Strategies in the Design of Bioactive Compounds. Proceedings of the Fifth European Symposium on Quantitative Structure Activity Relationships', ed. J. K. Seydel, VCH, Weinheim, 1985, p. 324.
189. K. Waisser, O. Leifertova and J. Vanzura, *Cesk. Farm.* 1986, **35**, 55 (*Chem. Abstr.*, 1986, **105**, 3361r).
190. V. Deljac, D. Maysinger, M. Maksimovic, L. Radovic and Z. Binenfeld, *Naucno-Teh. Pregl.*, 1982, **32**, 35 (*Chem. Abstr.*, 1983, **98**, 12 483w).
191. M. Maksimovic, D. Maysinger, V. Deljac and Z. Binenfeld, *Acta Pharm. Jugosl.*, 1981, **31**, 159 (*Chem. Abstr.*, 1982, **96**, 99 047n).
192. D. Maysinger, M. Birus and M. Movrin, *Acta Pharm. Jugosl.*, 1980, **30**, 9 (*Chem. Abstr.*, 1980, **93**, 142 658q).
193. P. A. Borea, *Arzneim.-Forsch.*, 1983, **33**, 1086.
194. P. A. Borea and V. Ferretti, *Biochem. Pharmacol.*, 1986, **35**, 2836.
195. H. G. Schauzu and P. P. Mager, *Pharmazie*, 1983, **38**, 490.
196. W. P. Purcell and J. M. Clayton, in 'Molecular Orbital Studies in Chemical Pharmacology', ed. L. B. Kier, Springer, New York, 1970, p. 145.
197. C. Hansch, C. Silipo and E. E. Steller, *J. Pharm. Sci.*, 1975, **64**, 1186.
198. G. Naray-Szabo, *THEOCHEM* 1986, **31**, 197 (*Chem. Abstr.*, 1986, **105**, 90 807g).
199. H. G. Schauzu and P. P. Mager, *Pharmazie*, 1983, **38**, 562.
200. H. Singh, V. Gombar and D. V. S. Jain, *Proc.—Indian Acad. Sci., Sect. A, Chem. Sci.*, 1980, **89A**, 77 (*Chem. Abstr.*, 1980, **93**, 106 805b).
201. B. Tinland, *Farmaco, Ed. Sci.*, 1975, **30**, 423.
202. B. Tinland, *Farmaco, Ed. Sci.*, 1976, **31**, 233.
203. P. Fulcrand, G. Berge, A.-M. Noel, P. Chevallet, J. Castel and H. Orzalesi, *Eur. J. Med. Chem.—Chim. Ther.*, 1978, **13**, 177.
204. H. Orzalesi, J. Castel, P. Fulcrand, G. Bergé, A.-M. Noël and P. Chevallet, *C. R. Hebd. Seances Acad. Sci., Ser. C*, 1974, **279**, 709.
205. A. Prikulis, B. Grinberga, I. Katlaps and V. Grinshtein, *Latv. PSR Zinat. Akad. Vestis, Kim. Ser.*, 1982, 181 (*Chem. Abstr.*, 1982, **97**, 51 649d).
206. V. Gombar, *Proc.—Indian Acad. Sci., Sect. A, Chem. Sci.*, 1982, **91**, 255 (*Chem. Abstr.*, 1982, **97**, 155 963t).
207. R. J. Gryglewski, Z. Ryznerski, M. Gorczyca and J. Krupińska, *Adv. Prostaglandin Thromboxane Res.*, 1976, **1**, 117.
208. D. V. S. Jain and V. Gombar, *Int. J. Quantum Chem.*, 1981, **20**, 419.
209. A. J. Bigler, K. P. Boegesoe, A. Toft and V. Hansen, *Eur. J. Med. Chem.—Chim. Ther.*, 1977, **12**, 289.
210. B. Bordas, M. Kovacs, M. Tuske, F. Darvas and G. Matolcsy, *Abh. Akad. Wiss. DDR, Abt. Math., Naturwiss., Tech.*, 1979, 333 (*Chem. Abstr.*, 1980, **92**, 141 627w).
211. O. Kirino, C. Takayama, A. Fujinami, K. Yanagi and M. Minobe, *Nippon Noyaku Gakkaishi*, 1984, **9**, 351 (*Chem. Abstr.*, 1985, **102**, 19 470z).
212. A. Lopata, F. Darvas, K. Valkó, G. Mikite, E. Jakucs and A. Kis-Tamás, *Pestic. Sci.*, 1983, **14**, 513.
213. S. P. Gupta and P. Singh, *Indian J. Chem., Sect. B*, 1978, **16**, 411.
214. W. P. Purcell, M. Martin and R. Carbo, *Afinidad*, 1976, **33**, 159 (*Chem. Abstr.*, 1976, **85**, 42 008u).
215. D. Schoenfelder and R. Franke, *Abh. Akad. Wiss. DDR, Abt. Math., Naturwiss., Tech.*, 1978, 303 (*Chem. Abstr.*, 1979, **91**, 19 651p).
216. L. K. Gibbons, E. F. Koldenhoven, A. A. Nethery, R. E. Montgomery and W. P. Purcell, *J. Agric. Food Chem.*, 1976, **24**, 203.

217. D. Maysinger and M. Movrin, *Arzneim.-Forsch.*, 1980, **30**, 1839.
218. B. Bordás, F. Darvas, M. Tüske and A. Lopata, in 'Advances in Pharmacological Research and Practice, Proceedings of the Congress of the Hungarian Pharmacological Society, 3rd, Budapest, 1979', ed. J. Knoll, Pergamon Press, Oxford, 1980, vol. 3, p. 331.
219. Z. Dinya, T. Timár, S. Hosztafi, A. Fodor, P. Deák, A. Somogyi and M. Berényi, in 'QSAR and Strategies in the Design of Bioactive Compounds. Proceedings of the Fifth European Symposium on Quantitative Structure Activity Relationships', ed. J. K. Seydel, VCH, Weinheim, 1985, p. 403.
220. O. Kirino and J. E. Casida, *J. Agric. Food Chem.*, 1985, **33**, 1208.
221. E. P. Serebryakov, N. A. Epstein, N. P. Yasinskaya and A. B. Kaplun, *Phytochemistry*, 1984, **23**, 1855.
222. A. Lopata, *Magy. Kem. Lapja*, 1986, **41**, 41 (*Chem. Abstr.*, 1986, **105**, 197 039k).
223. C. Cativiela, J. I. G. Laureiro and J. Elguero, *Gazz. Chim. Ital.*, 1986, **116**, 119.
224. R. Carpignano, P. Savarino, E. Barni, G. Di Modica and S. S. Papa, *J. Soc. Dyers Colour.*, 1985, **101**, 270 (*Chem. Abstr.*, 1986, **104**, 7192p).
225. R. Carpignano, P. Savarino, G. Di Modica and G. Scavia, *Tinctoria*, 1984, **81**, 97 (*Chem. Abstr.*, 1984, **101**, 172 983t).
226. R. Carpignano, E. Barni, G. Di Modica, R. Grecu and G. Bottaccio, *Dyes Pigm.*, 1983, **4**, 195 (*Chem. Abstr.*, 1983, **99**, 55 053u).
227. R. Grecu, M. Pieroni and R. Carpignano, *Dyes Pigm.*, 1981, **2**, 305 (*Chem. Abstr.*, 1981, **95**, 221 286m).
228. K. Yamamoto, H. Sunada, N. Yonezawa and A. Otsuka, *Bunseki Kagaku*, 1979, **28**, 205 (*Chem. Abstr.*, 1979, **91**, 29 856y).
229. J.-P. Schmit and G. G. Rousseau, *J. Steroid Biochem.*, 1978, **9**, 921.
230. C. Cativiela, J. I. Garcia and J. Elguero, *An. Quim.*, *Ser. C*, 1987, **83**, 278.
231. S. Inoue, A. Ogino, M. Kise, M. Kitano, S. Tsuchiya and T. Fujita, *Chem. Pharm. Bull.*, 1974, **22**, 2064.
232. H. K. Lee, Y. W. Chien, T. K. Lin and H. J. Lambert, *J. Pharm. Sci.*, 1978, **67**, 847.
233. M. R. Gasco, M. Trotta, M. E. Carlotti and R. Carpignano, *Int. J. Pharm.*, 1984, **18**, 235.
234. S. H. Unger and C. Hansch, *J. Med. Chem.*, 1973, **16**, 745.
235. C. Hansch and E. J. Lien, *Biochem. Pharmacol.*, 1968, **17**, 709.
236. P. N. Craig, *J. Med. Chem.*, 1971, **14**, 680.
237. C. Hansch and E. J. Lien, *J. Med. Chem.*, 1971, **14**, 653.
238. C. Hansch and D. F. Calef, *J. Org. Chem.*, 1976, **41**, 1240.
239. H. W. Hamilton, D. F. Ortwine, D. F. Worth, E. W. Badger, J. A. Bristol, R. F. Bruns, S. J. Haleen and R. P. Steffen, *J. Med. Chem.*, 1985, **28**, 1071.
240. R. E. Ford, P. Knowles, E. Lunt, S. M. Marshall, A. J. Penrose, C. A. Ramsden, A. J. H. Summers, J. L. Walker and D. E. Wright, *J. Med. Chem.*, 1986, **29**, 538.
241. W. Dittmar, E. Druckrey and H. Urbach, *J. Med. Chem.*, 1974, **17**, 753.
242. A. Panthananickal, C. Hansch, A. Leo and F. R. Quinn, *J. Med. Chem.*, 1978, **21**, 16.
243. C. Hansch, A. Leo, C. Schmidt, P. Y. C. Jow and J. A. Montgomery, *J. Med. Chem.*, 1980, **23**, 1095.
244. S. I. Fink, A. Leo, M. Yamakawa, C. Hansch and F. R. Quinn, *Farmaco, Ed. Sci.*, 1980, **35**, 965.
245. T. A. Khwaja, S. Pentecost, C. D. Selassie, Z. Guo and C. Hansch, *J. Med. Chem.*, 1982, **25**, 153.
246. W. A. Denny, B. F. Cain, G. J. Atwell, C. Hansch, A. Panthananickal and A. Leo, *J. Med. Chem.*, 1982, **25**, 276.
247. C. D. Selassie, C. Hansch, T. A. Khwaja, C. B. Dias and S. Pentecost, *J. Med. Chem.*, 1984, **27**, 347.
248. C. Hansch, J. McClarin, T. Klein and R. Langridge, *Mol. Pharmacol.*, 1985, **27**, 493.
249. C. Hansch and T. E. Klein, *Acc. Chem. Res.*, 1986, **19**, 392.
250. M. Yoshimoto and C. Hansch, *J. Med. Chem.*, 1976, **19**, 71.
251. M. Yoshimoto and C. Hansch, *J. Org. Chem.*, 1976, **41**, 2269.
252. C. Hansch, C. Grieco, C. Silipo and A. Vittoria, *J. Med. Chem.*, 1977, **20**, 1420.
253. C. Grieco, C. Hansch, C. Silipo, R. N. Smith, A. Vittoria and K. Yamada, *Arch. Biochem. Biophys.*, 1979, **194**, 542.
254. C. Silipo, C. Hansch, C. Grieco and A. Vittoria, *Arch. Biochem. Biophys.*, 1979, **194**, 552.
255. C. Hansch, M. Yoshimoto and M. H. Doll, *J. Med. Chem.*, 1976, **19**, 1089.
256. C. Silipo and C. Hansch, *J. Med. Chem.*, 1976, **19**, 62.
257. C. Silipo and C. Hansch, *J. Am. Chem. Soc.*, 1975, **97**, 6849.
258. J. Y. Fukunaga, C. Hansch and E. E. Steller, *J. Med. Chem.*, 1976, **19**, 605.
259. C. Hansch, B. A. Hathaway, Z. Guo, C. D. Selassie, S. W. Dietrich, J. M. Blaney, R. Langridge, K. W. Volz and B. T. Kaufman, *J. Med. Chem.*, 1984, **27**, 129.
260. B. A. Hathaway, Z. Guo, C. Hansch, T. J. Delcamp, S. S. Susten and J. H. Freisheim, *J. Med. Chem.*, 1984, **27**, 144.
261. J. M. Blaney, C. Hansch, C. Silipo and A. Vittoria, *Chem. Rev.*, 1984, **84**, 333.
262. C. Silipo and C. Hansch, *Mol. Pharmacol.*, 1974, **10**, 954.
263. S. W. Dietrich, M. B. Bolger, P. A. Kollman and E. C. Jorgensen, *J. Med. Chem.*, 1977, **20**, 863.
264. M. Recanatini, T. Klein, C.-Z. Yang, J. McClarin, R. Langridge and C. Hansch, *Mol. Pharmacol.*, 1986, **29**, 436.
265. C. Hansch, A. R. Steward, S. M. Anderson and D. Bentley, *J. Med. Chem.*, 1968, **11**, 1.
266. H. Kubinyi, *Prog. Drug Res.*, 1979, **23**, 97.
267. C. Hansch and J. M. Clayton, *J. Pharm. Sci.*, 1973, **62**, 1.
268. H. Kubinyi, *Arzneim.-Forsch.*, 1976, **26**, 1991.
269. H. Kubinyi, *J. Med. Chem.*, 1977, **20**, 625.
270. H. Kubinyi, *Farmaco, Ed. Sci.*, 1979, **34**, 248.
271. H. Kubinyi, in 'QSAR in Design of Bioactive Compounds', ed. M. Kuchar, Prous, Barcelona, 1984, p. 321.
272. F. Darvas, J. Roehricht, Z. Budai and B. Bordas, in 'Advances in Pharmacological Research and Practice, Proceedings of the Congress of the Hungarian Pharmacological Society, 3rd, Budapest, 1979', ed. J. Knoll, Pergamon Press, 1980, vol. 3, p. 25.
273. B. R. Kowalski and C. F. Bender, *J. Am. Chem. Soc.*, 1972, **94**, 5632.
274. A. Cammarata and G. K. Menon, *J. Med. Chem.*, 1976, **19**, 739.
275. G. K. Menon and A. Cammarata, *J. Pharm. Sci.*, 1977, **66**, 304.
276. G. L. Kirschner and B. R. Kowalski, in 'Drug Design', ed. E. J. Ariëns, Academic Press, New York, 1979, vol. 8, p. 73.
277. P. C. Jurs, J. T. Chou and M. Yuan, *ACS Symp. Ser.*, 1979, **112**, 103.
278. J.-E. Dubois, D. Laurent and A. Aranda, *J. Chim. Phys. Phys.—Chim. Biol.*, 1973, **70**, 1608.

279. J.-E. Dubois, D. Laurent and A. Aranda, *J. Chim. Phys. Phys.—Chim. Biol.*, 1973, **70**, 1616.
280. A. Aranda, *C. R. Hebd. Seances Acad. Sci., Ser. C*, 1973, **276**, 1301.
281. J.-E. Dubois, D. Laurent, P. Bost, S. Chambaud and C. Mercier, *Eur. J. Med. Chem.—Chim. Ther.*, 1976, **11**, 225.
282. J.-E. Dubois, C. Mercier and A. Panaye, *Acta Pharm. Jugosl.*, 1986, **36**, 135.
283. B. Duperray, M. Chastrette, M. Cohen Makabeh and H. Pacheco, *Eur. J. Med. Chem.—Chim. Ther.*, 1976, **11**, 433.
284. E. J. Lien, C. Hansch and S. M. Anderson, *J. Med. Chem.*, 1968, **11**, 430.
285. C. Mercier and J.-E. Dubois, *Eur. J. Med. Chem.—Chim. Ther.*, 1979, **14**, 415.
286. A. T. Balaban, A. Chiriac, I. Motoc and Z. Simon, 'Steric Fit in QSAR. Lecture Notes in Chemistry', Springer, Berlin, 1980, vol. 15.
287. Z. Simon, S. Holban and I. Motoc, *Rev. Roum. Biochim.*, 1979, **16**, 141 (*Chem. Abstr.*, 1980, **92**, 71 845q).
288. Z. Simon, D. Ciubotariu and A. T. Balaban, in 'QSAR and Strategies in the Design of Bioactive Compounds. Proceedings of the Fifth European Symposium on Quantitative Structure Activity Relationships', ed. J. K. Seydel, VCH, Weinheim, 1985, p. 370.
289. I. Motoc, *Quant. Struct.-Act. Relat.*, 1984, **3**, 43.
290. I. Motoc, *Quant. Struct.-Act. Relat.*, 1984, **3**, 47.
291. S. Huebel, T. Roesner and R. Franke, *Pharmazie*, 1980, **35**, 424.
292. W. J. Streich and R. Franke, *Quant. Struct.-Act. Relat.*, 1985, **4**, 13.
293. R. Franke and W. J. Streich, *Quant. Struct.-Act. Relat.*, 1985, **4**, 51.
294. R. Franke and W. J. Streich, *Quant. Struct.-Act. Relat.*, 1985, **4**, 63.
295. R. Franke, S. Huebel and W. J. Streich, *EHP, Environ. Health Perspect.*, 1985, **61**, 239.
296. P. R. Andrews, D. J. Craik and J. L. Martin, *J. Med. Chem.*, 1984, **27**, 1648.

22.1

Substructural Analysis and Compound Selection

PAUL N. CRAIG

National Library of Medicine, Bethesda, MD, USA

22.1.1   CONCEPT OF CHEMICAL STRUCTURE

22.1.1.1   Development of Knowledge about Molecular Structure

Although some ancient Greeks first conceived of the existence of invisibly small building blocks, which they named atoms, from which all matter was constructed, it was not until the 18th century that the beginnings of modern atomic structure were laid down. Dalton, Lavoisier and Avogadro helped achieve a better understanding of structure and in 1859 Kekulé first proposed the modern organic chemical structural formulas for aliphatic compounds. In 1865 he proposed the benzene ring structure for aromatic compounds. In a remarkably short period of time, by use of a combination of intuition and deductive reasoning, the structures of many thousands of organic compounds were determined by a growing band of chemists in many countries.[1]

22.1.1.2   Use of Substructures in Organic Chemistry

The enormous variety of organic compounds led to the need to classify them in order to simplify the problem of learning about each and every compound. The stepwise progression of aliphatic chains from one carbon atom to the frequently encountered 16- and 18-carbon linear chains, and the gradual changes in physical and chemical properties of such simple derivatives as alcohols, halides, amines, acids, *etc.* as the chain length increased, were readily noted. Some groupings were seen to impart definitive chemical properties, such as amino and acid groups, and the recognition that such groups provided predictable chemical reactivity to molecules led to the designation of these as 'functional' groups. Since these groups are partial representations of complete molecules, they have become known as 'substructures'.

Such groups may vary in size from one atom (*e.g.* fluorine) to large groups such as thiosemicarbazone, or ring systems (*e.g.* phenanthrene). Combined with the representation of organic structures by two-dimensional structures, which can usually serve to convey a reasonable approximation of the actual three-dimensional structure, these tools have enabled chemistry to progress to the point where more than 8 000 000 structures, mostly organic, are currently known.[2] The understanding of mechanisms whereby the chemical transformations occur is a 20th century development, which usually requires knowledge of the three-dimensional structures actually involved in the reaction. Knowledge of these mechanisms is desirable in attempts to model chemical and biological activities. Even in the absence of such knowledge, however, it is often possible to model chemical and biological properties, although the utility of such models is usually more limited than when complete knowledge of the mechanism of activity is known.

22.1.2   PRINCIPLES OF SUBSTRUCTURE SEARCHING

22.1.2.1   The Problems of Specificity and Genericity

Substructures are abstract man-made concepts, of varying size, whose definition is arbitrary, and the utility of a given substructure is the sole measure of its validity.

The terms 'specific' and 'generic' have special meaning in the chemical information field. They refer to the results of retrieval of information from the mass of literature and data available. If an inquiry results in exactly the desired information, it is considered to be 100% specific.

The problem is exemplified by consideration of a series of carboxylic acids. Substructure searching at a generic level for carboxylic acids would classify all compounds containing the group 'CO_2H' together. More specific searching would classify them into aliphatic, aromatic and heterocyclic carboxylic acids. Further specific classification would divide the aliphatic acids into unsubstituted straight chain acids, alicyclic acids, halogenated straight chain acids, *etc.* Rather than set up a rigid classification scheme it is helpful to provide the flexibility to generate the classification specificity as desired. This is illustrated in Section 22.1.5.2, where a set of acids is sorted by a variety of substructures.

22.1.2.2 The Problem of 'False Drops'

This term comes from the information retrieval field, where it aptly describes the problem of retrieving information of only the desired level of specificity. False drops may arise from faulty or inconsistent manual encoding at input, but more often is inherent in the generic nature of the particular code and search methods used. The retrieval of all organic structures which contain a carbonyl group ($C{=}O$) illustrates the subjective nature of this problem. When applied to a search for aliphatic aldehydes in a database, the search for $C{=}O$ will also retrieve all ketones, esters, acids, amides, *etc.*, including complex structures such as semicarbazones. If one is working with a small collection of chemicals, containing mostly aldehydes, one may be content to search for all carbonyl groups, and manually weed out the undesired nonaldehydes. But if one is working with a large number of diverse compounds, the use of a more specific target than $C{=}O$, such as $RC(H)({=}O)$, where C is attached to an aliphatic carbon, will ensure the retrieval of only the desired aliphatic aldehydes. The important factors are the number and complexity of the chemicals under consideration, together with the requirements which are to be made of the system. Thus, the requirements of Chemical Abstracts to handle some 8 000 000 compounds are different from those involving sets of 10 000 or 100 000 compounds. Sets of a few dozen to several thousand compounds are used in QSAR studies or toxicity model building and the exact specificity of assignment of substructures is often critical.

22.1.3 USE OF SUBSTRUCTURES IN THE CHEMICAL INFORMATION FIELD

22.1.3.1 The Beilstein Classification System

The first consistent effort to classify the mushrooming numbers of organic chemicals was made by Friederich Beilstein, who developed a hierarchical system which first considered aliphatic hydrocarbons, followed by their simple derivatives such as halides, alcohols, amines, in order of increasing complexity. Saturated hydrocarbons were followed by alkenes, aromatic ring compounds and heterocycles, in a rigorously defined sequence, based on molecular formula.[3] Effective use of Beilstein requires an in-depth knowledge of organic chemistry, and it remains a valuable source of summary data for organic compounds.[4]

22.1.3.2 Development of Substructure Coding

22.1.3.2.1 The CBCC system

The first major attempt to encode chemical structures and their related biological activities was made by the Chemical–Biological Coordination Center (CBCC) of the National Research Council in 1946. The need for efficient retrieval of structures became obvious during the World War II crash research projects, which were coordinated by the Office of Scientific Research and Development. Thousands of chemicals were prepared and studied for various biological properties, including antimalarial, insecticidal and rodenticidal activities. It was realized that mechanical aids were needed, and the success with which the Hollerith punched cards handled enormous amounts of data at the US Patent Office was put to use by CBCC. Building on a chemical code developed by Frear at Penn State in 1942, they developed a comprehensive chemical-coding system which utilized the capabilities of the IBM punched card equipment then available.[5] These code terms ('keys') were assigned manually, and experienced chemists were required for this task. The assignment of such codes inevitably becomes subjective in nature, and different encoders may not always exactly duplicate each others' encoding for highly complicated structures. Thus the use of such manually assigned codes carries two deficits: they are expensive in terms of man-power required, and they are subject to human fallibility, which can lead to false drops or incomplete retrieval.

The CBCC codes were designed to allow a choice of the desired level of specificity; this is shown in Table 1, where the codes for the amides are listed. At the time of encoding, a fourth digit is assigned, which designates the number of occurrences of the coded feature in the molecule. Thus, succinamide (a diamide) would be assigned code '65L2.' This design permits one to retrieve all amides by searching for codes beginning with 65; or all tertiary amides by searching for codes with 650, 651 or 652.

Table 1 CBCC Codes for Amides

Class	Structure	Code
Amides, tertiary	HC(=O)NRR'	650
	R = heterocyclic, R' = heterocyclic, aromatic carbocyclic, alicyclic, or aliphatic	651
	R = aromatic carbocyclic, R' = aromatic carbocyclic, alicyclic, or aliphatic	652
Amides, secondary	RC(=O)NHR' or RC(OH)=NR	
	R and R' are heterocyclic	65A
	R and R' are aromatic carbocyclic	65B
	R and R' are aliphatic or alicyclic (R may be H)	65C
	R is heterocyclic	
	and R' is aromatic carbocyclic	65D
	and R' is alicyclic or aliphatic	65E
	R is aromatic carbocyclic	
	and R' is heterocyclic	65F
	and R' is alicyclic or aliphatic	65G
	R is H, alicyclic or aliphatic	
	and R' is heterocyclic	65H
	and R' is aromatic carbocyclic	65I
Amides, primary	RC(=O)NH$_2$ or RC(OH)=NH	
	R is heterocyclic	65J
	R is aromatic carbocyclic	65K
	R is H, alicyclic or aliphatic	65L

The need for rapid retrieval and classification of chemical structures was especially felt by the mushrooming pharmaceutical industry in the 1950s, and several simplified coding schemes were described for use with 'keysort' (manual) or IBM cards (see, for example, the system developed at Merrell).[6]

22.1.3.2.2 *The Documentation Ring code*

CBCC-type codes were designed to retrieve specific functional groups, but did not permit easy generic retrieval of related groups or similar ring systems. Such capability is desired by medicinal chemists; this requirement was met by the Steidle code,[7] which was the basis for the Documentation Ring code, used originally by a group of German pharmaceutical companies and in 1964, by the Ringdoc System. Using IBM punched cards initially, it was soon computerized, and the Ring code served the needs of medicinal chemists by permitting generic searching of the chemical and patent literature. Genericity was obtained by using combinations of punches, often in the same column, to encode related structures. This 'overpunching' leads to false drops, but does allow the encoding of a greater variety of information than is provided by simple use of the 960 holes in the 80-column by 12-row IBM card. An example of the Steidle coding system is shown in Table 2. Much redundancy is built into the coding scheme; this gives it the ability to retrieve generic substances with considerable flexibility of choice. As for CBCC codes, these keys are assigned manually, which can lead to inconsistencies.

Table 2 Some Ring Codes for 2-Phenyl-4-aminobenzoic Acid

Column/punch	Description	Column/punch	Description
19/0	R—NH$_2$	70/y	Isolated (unfused) aromatic ring
19/x	19/0 Group is attached to a double bond	70/3	2 aromatic rings
23/2	—C(=Y)(OH)	74/8	3 ring substituents oriented 1,2,4
23/7	Above Y = O	75/0	Ring–ring link
23/x	23/2 Group is attached to a double bond	76/3	Ring–C(=X)Z
69/y	1 ring	78/5	C=X conjugated with phenyl ring
69/x	2 rings		

The Ring code does not allow specific searching for complex functional groups, and the number of false drops becomes increasingly burdensome as the size of the data collection increases.

22.1.3.2.3 *The Smith Kline and French code*

By 1962, the punched card sorters were being replaced by computers, which both speeded the retrieval and freed the designers from the limitations imposed by the 960 holes in an IBM punched card. At Smith Kline and French (SK&F) the CBCC code was expanded by about 50%, and the best features of the CBCC and Ring codes were combined, so the user could search either specifically or generically, as required.[8] However, the code terms were still assigned manually. This 'extended' SK&F code was used in the first study which used substructural features to estimate biological activity.[9]

22.1.3.3 Efficient Storage of Chemical Structures

22.1.3.3.1 *Connection tables*

The substructure codes so far discussed use arbitrary codes to describe groups of atoms of various sizes and complexity. In 1965 several groups reported similar approaches to the handling of chemical structures by the use of matrices which describe each atom and its bonds to other atoms (Table 3).[10] The matrices permit accurate representation and efficient storage of large numbers of structures, and further developments have used this 'connection table' approach to permit substructure searching and to generate graphical structure displays for the more than 8 000 000 compounds in the CAS ONLINE database.[11]

Table 3 A Sample Connection Table for Acetic Acid[a]

	C1	C2	O3	O4	
C1	X	1	0	0	Numbering of atoms for CH$_3$C(=O)OH:
C2	1	X	2	1	C^1—C^2(=O^3)—O^4 (hydrogens are omitted)
O3	0	2	X	0	
O4	0	1	0	X	

[a] 0 = no bond; 1 = single bond; 2 = double bond; X = identical atom

22.1.3.3.2 *Augmented fragments*

The use of simple pairs of atoms and the bond types which connect them results in several thousands of such combinations. Although they have been used as substructures, they are so generic that they lead to many false drops. To overcome this problem, workers at Ciba (Basel) introduced the 'augmented fragment' (AF), which increased the specificity of retrieval.[12] An AF includes for each atom a description of every other atom to which it is bonded, as well as the types of bonds involved (Table 4). The use of AFs greatly reduced the number of false drops in substructure searching.

Table 4 Augmented Connectivity Fragments for Acetic Acid

For C^2 ;	C—$\underline{C}$(=O)O	[from C^1—$\underline{C}^2$(=O^3)—O^4]
For C^1 ;	$\underline{C}$—C(=)—	[from $\underline{C}^1$—C^2]
For O^3 ;	$\underline{O}$=C(—)—	[from $\underline{O}^3$=C^2]
For O^4 ;	$\underline{O}$—C(=)—	[from $\underline{O}^4$—C^2]

An example of another type of AF is given by Hodes *et al.*, who define 'triplet' ganglia as sets of three attached nonhydrogen atoms, together with their pendant bonds.[61]

22.1.3.3.3 The CIDS chemical code

During the late 1960s the US Army developed a substructure code expressly designed for use with a computer, as a part of an information system.[13] The assignment of structural search keys, made by computer algorithms from connection tables, includes keys which describe both the cyclic or acyclic nature and the functional groups present in the molecule. This computer assignment of code terms simultaneously reduced the costs of encoding and ensured consistency of assignment of codes.

There are 147 keys which define the cyclic systems and 348 keys which define functional groups. This code, augmented with the log of the partition coefficient and the molecular weight, was used in the first study which estimated LD_{50} values for a diverse set of chemicals from substructural features[14] (see Section 22.1.7.1.1).

22.1.4 LINEAR NOTATION SYSTEMS

22.1.4.1 Wiswesser Line Notation System (WLN)

In 1950 Wiswesser developed and reported on a workable linear notation system which could encode complex chemical structures by a linear string of simple symbols. The WLN underwent several modifications which improved its efficiency, and was widely used because of the relative simplicity of most notations, and the ease of both manual and machine use of the notations.[15] The use of printed permuted indexes of WLNs provided facile identification of many substructures; this approach is used in indexing chemicals by the Institute for Scientific Information.[16]

An illustration of how such a permuted index can provide a good degree of substructure searching is shown in Table 5.

Table 5 Substructures Identified by Permuted WLN Index Entries

(Example is 4-chloro-2-acetoxybenzoic acid)	
WLN (unpermuted) = QVR BOV1 DG	
Permutation generates the following entries (in alphabetical sequence)	
$\underline{G}$*QVR BOV1 D	(chlorine substituent)
$\underline{OV}$1 DG*QVR B	(acetoxy group)
$\underline{QVR}$ BOV1 DG	(acid group)
$\underline{V}$1 DG*QVR BO	(aceto group)
$\underline{1}$ DG*QVR BOV	(methyl group—part of aceto group)

* marks beginning of normal WLN

Almost at the same time that the WLN system was developed, a competing linear notation system was under development by Dyson in England. This system was more complex than the WLN system; together with the Hayward notation system, developed later at the US Patent Office in the 1960s, these systems never achieved large-scale usage. The WLN system was accepted widely, and is still in frequent use, due to its relative simplicity and utility, with or without access to a computer.

The WLN was used by ICI chemists as the basis for a substructure retrieval system called CROSSBOW.[51] It was able to handle about 90% of the searches desired by ICI chemists. However, it was found necessary to use the WLN to generate connection tables (see Section 22.1.3.3.1) and use these to perform more complex substructure searches which were unable to be handled by direct search of the WLN itself.

22.1.4.2 The SMILES Notation System

22.1.4.2.1 *SMILES*

Despite the popularity of the WLNs, encoding required careful training to ensure that the correct form of the WLN was assigned. In the early 1980s, Weininger developed the *S*implified *M*olecular *I*nput *L*ine *E*ntry *S*ystem (SMILES) notation, which is now used in the automated estimation of partition coefficients at Pomona College.[17] The SMILES notations proved to be very easy to learn and use. The code can be assigned in many different forms, which, unlike the WLN, are equally valid, because the computer converts them by algorithms into a canonical form for computer manipulation. The system, developed by Weininger to handle SMILES notations, permits substructure searching with precise definition of specific targets, or with various levels of generic targets, as desired (see Section 22.1.5.2).

22.1.4.2.2 *GENIE and SMARTS*

Substructure searching with SMILES is carried out by a language called GENIE (*GEN*eralized *I*somorph *E*numeration). The target language used in GENIE is SMARTS (*SM*iles *AR*bitrary *T*arget *S*pecification), which is an extension of SMILES. It differs from SMILES in that fragments may be defined by SMARTS, whereas only complete molecules are usually described by SMILES notations. An illustration of their use is given in Section 22.1.5.2.

22.1.5 SUBSTRUCTURE SEARCH SYSTEMS

22.1.5.1 Existing Systems

22.1.5.1.1 *The NIH–EPA Chemical Information System*

The NIH–EPA Chemical Information System (CIS) was developed to permit full-structure or substructure searching, as desired. The system utilizes connection tables and permits searching for substructures which are imbedded into ring systems,[18] as well as simple searches. Searches can be carried out in several ways, using the *S*tructure *A*nd *N*omenclature *S*earch *S*ystem (SANSS). Searches are initially run using either CIDS codes (see Section 22.1.3.3.3) or atom-centered fragments and ring probes. The results from the intersection of these relatively generic screens are then subjected to atom-by-atom and bond-by-bond exact match substructure searching. Two-dimensional structure diagrams are displayed for each compound. About 250 000 chemicals are contained in the CIS files; all of these are linked to chemical, physical or biological data in some of the more than 20 types of files in the system.

22.1.5.1.2 *Chemical Abstracts Service (CAS) ONLINE system*

The CAS ONLINE system provides similar capabilities for all 8 000 000 compounds in the CAS online computer files. Based on connection tables, searching is carried out by either exact matching or by intersection of results from substructure searches. The use of substructures is a great improvement over the use of nomenclature for searching the Chemical Abstracts files. The system has been designed to provide access to these files by use of desktop terminals or microcomputers.[11]

22.1.5.1.3 *The MACCS system*

The MACCS system, offered by Molecular Design Ltd. (MDL), offers similar capabilities, and is available for use on microcomputers as well as larger computers for use with custom files.[19] The system is based on connection tables. A light pen or mouse is required for its use, and the system is user friendly, with multiple on-screen menus. The CHEMBASE system (MDL) allows chemical structures to be drawn readily on the screen by use of stored substructure fragments.

22.1.5.1.4   The Mechanical Chemical Code (MCC)

This cipher, developed by Lefkovitz et al.[20] for Eastman Kodak, combines many features of notation systems and connection tables and served the information needs at Kodak quite well for many years. Algorithms were written to link the MCC with CAS connection tables. It was used by Tinker in his modeling of mutagenesis data (see Section 22.1.7.1.3).

22.1.5.2   Use of Substructure Searching to Achieve Specificity

A series of organic carboxylic acids will be sorted into categories of varying specificity, to illustrate the use of the substructure search system, GENIE, to achieve specificity in searching. Table 6 shows the characters used in SMILES and Table 7 describes the logical search operators used in GENIE searches. Table 8 lists the SMILES notations for some acids and Table 9 shows the definitions of the targets which sort these acids into the groups listed in Table 10.

Vectors are predefined atom/bond combinations; their use both simplifies and speeds searching of large databases. In combination with the logical search operators, and using special characters provided for generic atoms and bonds, the retrieval can be made as specific as desired.

The utility of negative expressions in targets is shown by the following uses of vectors and targets: by re-defining the vector '$CX' as 'C[!C]' (aliphatic carbon attached by a single bond to any atom except aliphatic carbon), one may write a target 'O=C[O;H1][$CX]' which is equivalent to target 2 in Table 9 when used to sort the acids in Table 8. The target 'O=C[O;H1][C;!$CX]' is equivalent

Table 6   Symbols Used in SMILES Notations[a]

C = aliphatic carbon	c = aromatic carbon
O = oxygen	o = aromatic oxygen
N = nitrogen	n = aromatic nitrogen
S = sulfur	s = aromatic sulfur
F, Cl, Br, I = halogen	P = phosphorus

[a] Symbols for all other elements are contained in brackets: [Sn], [Ba], *etc.* as are unusual valence states for the common elements

Table 7   Logical Search Operators[a]

Symbol	Expression	Meaning
Exclamation	!e1	NOT e1
Ampersand	e1&e2	e1 AND e2
Comma	e1,e2	e1 OR e2
Semicolon	e1;e2	e1 AND e2

[a] These are listed in descending order of precedence. Thus, e1&e2,e3 is (e1&e2),e3 and e1;e2,e3 is e1;(e2,e3)

Table 8   SMILES for Acids

No.	SMILES	Acid
1	CC(=O)O	Acetic acid
2	ClCC(=O)O	Chloroacetic acid
3	ClCC1C(=O)O	Cyclopropanecarboxylic acid
4	c1ccccc1C(=O)O	Benzoic acid
5	n1c(C(=O)O)cccc1	2-Pyridinecarboxylic acid
6	n1cc(C(=O)O)ccc1	3-Pyridinecarboxylic acid
7	n1ccc(C(=O)O)cc1	4-Pyridinecarboxylic acid
8	c12c(C(=O)O)cccc1cccc2	1-Naphthoic acid
9	O=C(O)C(=O)O	Oxalic acid
10	n1c(C(=O)O)c[nH]c1	Imidazole-4-carboxylic acid

Table 9 GENIE Targets

No.	Target	Description
1	O=C([O;H1])[C;R0;!$CX;!$CDX][a]	RCH$_2$CO$_2$H
2	O=C([O;H1])[$CX]	Halogenated aliphatic-CO$_2$H
3	O=C([O;H1])[C;R]	Alicyclic-CO$_2$H
4	O=C([O;H1])C	Generic nonarom-CO$_2$H
5	O=C([O;H1])c1ccccc1	Phenyl-CO$_2$H, fused or not
6	O=C([O;H1])c[c;R1][c;R1][c;R1][c;R1][c;R1]	Unfused phenyl-CO$_2$H
7	O=C([O;H1])c[c;R2]	Arom-CO$_2$H *ortho* to fusion
8	O=C([O;H1])[c,C;R][N,n,O,o,S,s;R]	Hetero-2-CO$_2$H
9	O=C([O;H1])[c,C;R][c,C;R][N,n,O,o,S,s;R]	Hetero-3-CO$_2$H
10	O=C([O;H1])[c,C;R][c,C;R][c,C;R][N,n,O,o,S,s;R]	Hetero-4-CO$_2$H
11	O=C([O;H1])c	CO$_2$H on arom-type carbon atom
12	O=C([O;H1])[$CDX]	X = CCO$_2$H, where X = O,N,S

[a] $CX is a vector for any carbon–halogen bond. $CDX is a vector for a carbon doubly bonded to O, N, S. R # = atom in brackets must be in # number of rings; R0 = not in a ring

Table 10 Acids Classified by Table 9 Targets

Acid no. (Table 8)		Targets assigned (see Table 9)
1	Acetic acid	1, 4
2	Chloroacetic acid	2, 4
3	Cyclopropanecarboxylic acid	3, 4
4	Benzoic acid	5, 6, 11
5	2-Pyridinecarboxylic acid	8, 11
6	3-Pyridinecarboxylic acid	9, 11
7	4-Pyridinecarboxylic acid	10, 11
8	1-Naphthoic acid	5, 7, 11
9	Oxalic acid	4, 12
10	Imidazole-4-carboxylic acid	8, 9, 11

to target 1 of Table 9; it stipulates that the carboxyl group must be attached to an aliphatic carbon, which, in turn, must *not* be attached to any element other than aliphatic carbon and hydrogen (note the effect of doubling the exclamation marks in vector and target definitions above). This flexibility is of great help when substructural keys are used to develop models for estimating physical or biological properties of compounds.

22.1.6 USE OF SUBSTRUCTURES IN QSAR

22.1.6.1 Free–Wilson Methodology—The Additivity Model

The original concept of Bruice, Kharasch and Winzler was a test of the assumption that each substructural feature which differed in a series of closely related analogs could be assigned a constant coefficient which reflects a consistent contribution or detraction from the overall biological activity of the molecule. They showed that this concept of additivity of the group contributions was valid for a series of thyroxine-like compounds.[21] Free and Wilson used the same concept and tested the methodology on several types of biological activity.[22] Although the method was used successfully from time to time, it was not as flexible nor as simple to use as was the multiple parameter approach, and the use of a successful additivity study was limited to the particular series under study.

In 1976 Kubinyi re-examined the methodology and developed an approach which combined the additivity and multiple parameter models (see Chapter 21.3 for a detailed review by Kubinyi).

22.1.6.2 Use of 'Indicator' Variables in Multiple Parameter Studies

Hansch and collaborators introduced the concept of assigning dichotomous 'dummy' or 'indicator' variables (given values of '1' or '0' for the presence or absence of the feature) to series of chemicals with similar biological properties, but which vary by the presence or absence of one structural feature (such as a ring atom). These series were then combined into a single class for QSAR studies.[23] In only two dimensions, the use of such indicator variables may be seen as bringing two parallel lines together, with the coefficient for the indicator variable (which results from solution of the regression equation) representing the distance between the parallel lines. A much more complex (multidimensional) example is illustrated by the inclusion of some 60 different types of antimalarials into one master equation which well expresses the QSAR of 646 compounds.[24]

22.1.6.3   Pattern Recognition Studies with Substructures

The first attempt to use pattern recognition methods in QSAR work was made by Chu *et al.* in 1974 in an effort to improve the selection of chemicals for testing as antitumor agents.[25] Several augmented connectivity fragments were used, together with some ring codes, and three different pattern recognition methods were tried with varying degrees of success. Further studies at NCI have used the Hodes heuristical method (see Section 22.1.8.1).

22.1.6.4   Molecular Connectivity Indices

The molecular connectivity index (MCI) concept was proposed by Randic in 1974. Kier and Hall expanded on this concept, and performed a large number of studies to test the utility of the MCI.[26] Since they are abstract numerical values, it is difficult for one to visualize the relationship of these indexes to substructures. However, they are rigorously defined in terms of graph theory, and represent topological parameters which reflect the degree of branching, as well as taking into consideration the types of atoms and bonds in a molecule. The use of several types of MCIs has been shown to provide parameters which, like the partition coefficient, can be used to obtain regression equations, which often well reflect the relationship between structure and either physical or biological activity for a series of chemicals.[27] The use of MCI parameters, along with other parameters such as substructural features, partition coefficient, *etc.*, has often proved to be beneficial. The MCI parameters alone can be used to help estimate physical parameters, such as dissociation constants, boiling points, partition coefficients, *etc.* One advantage of the MCIs is that they can be readily obtained for any well-defined molecule of less than 50 nonhydrogen atoms.

22.1.6.5   Substructural Analysis

In 1974, Cramer *et al.* published the first study in which an intentionally diverse set of compounds was analyzed by their substructural features (obtained by use of the SK&F code) in an attempt to improve the selection of chemicals for study in the rat paw edema test, as possible antiinflammatory agents.[9] The frequency of occurrence of each substructure in active compounds relative to its occurrence in inactive compounds was used to provide fragment weights. These were then added to give an overall weight for the molecule, which was compared to the average weights for active and inactive compounds in order to predict the activity of the untested compound. The results were of marginal value for the intended purpose, but they did show a statistically significant confirmation that the weights (coefficients) for the substructures could be considered as additive properties.

22.1.6.6   Use of Substructures to Estimate Partition Coefficients

Rekker has shown that partition coefficients can be calculated as an additive property of substructural features of molecules.[28] Leo, Hansch and colleagues have also studied several methods to achieve computer estimates of partition coefficients; they achieved the best results when the fragments were defined in terms of 'isolating carbons' and 'fragments.' An isolating carbon atom is defined as any carbon atom which is not doubly or triply bonded to a noncarbon atom. Isolating carbons may be bonded by aromatic, single or double bonds. When the isolating carbons, together with the attached hydrogen atoms, are identified (using the SMILES and GENIE programs—see Section 22.1.4.2), the remaining atoms in the molecule are considered to be fragments. To obtain the best possible accuracy, the introduction of several correction factors (indicator variables) was found to be necessary. (see Chapter 18.7 for an in-depth discussion by Leo of the CLOG P programs).

22.1.6.7 The Use of Different Types of Parameters in QSAR—The Collinearity Problem

A major contribution of Hansch and collaborators was the combination of the partition coefficient, the Hammett sigma constant, steric parameters, molar refractivities, *etc.* into a 'multiple parameter' analysis of QSAR. These parameters can be considered as 'quasi-thermodynamic' properties and have been shown to be additive properties (see Chapter 21.1).

The use of different types of parameters introduced the statistical problems associated with the collinearity of these factors. The difficulty of assignment of cause and effect relationships when one is attempting correlations involving parameters with various degrees of collinearity was illustrated in early studies of the QSAR of antimalarial agents.[29] This problem is critical when one is studying the mechanisms of activity; however, for modeling purposes with many different parameters, these problems can be minimized by controlling the sequence in which parameters are introduced into the stepwise regression[30] (for an example, see Section 22.1.7.1.1).

22.1.7 MODELING TOXIC EFFECTS OF CHEMICALS

22.1.7.1 Mammalian and Bacterial Systems

22.1.7.1.1 First LD_{50} model

The addition of indicator variables to partition and polar parameters has been well established, when using closely related series of compounds, and where the same mechanism of biological action may be assumed to hold.[23] The LD_{50} models were the first attempt to correlate widely diversified structures having different types of mechanisms of activity. This differentiates these models from the usual QSAR equations. They belong to the class of mathematical models which are widely used in engineering, economics, *etc.* and are developed by statistical methods. Assessment of their utility and validity is also judged by statistical measures.

The initial study using substructures to estimate biological activity[9] encouraged the application of this method to the field of toxicology. A study[14] was based on 549 compounds for which rat oral LD_{50} values were reported in the 1974 Toxic Substances List.[55] Although Cramer *et al.*[9] had used only substructural features, for the LD_{50} study these were augmented by inclusion of partition coefficients (log P, n-octanol/water) and molecular weights. The method combined features of the multiple parameter and additivity QSAR methods; in addition, as many diverse types of chemical structures as possible were used simultaneously, following up on the work of Cramer *et al.*[9] This also meant that many different mechanisms of lethality were being considered together under the broad classification of lethality as an endpoint.

The use of partition coefficient data, together with the removal of outliers by clustering techniques, reduced the number of compounds to 525 for consideration in the final model. Then 100 compounds were randomly withheld, and the model was developed with the remaining 425 compounds. The 421 substructures used were assigned by computer from the US Army CIDS codes.[13] Only 134 fragment codes were assigned seven or more times for the 425 compounds; these 134 codes were used in the regression. A stepwise regression method was used,[30] and molecular weight (M) was entered last into the equation, after the other parameters had been studied (the high correlation between the M and many of the other parameters, including log P, resulted in the loss of many parameters if the M were entered earlier into the stepwise regression). Log(log $1/C + 1$) was used to obtain a more normal (Gaussian) distribution of the biological data. (This was not found necessary in the models developed later from larger datasets.) The resulting equation contained 25 fragment codes, the partition coefficient (log P) and the molecular weight (M). It 'explained' 49% of the variance in the data ($r^2 = 0.493$, $r = 0.702$). When the modeling process was tested by estimation of the 100 compounds which had been randomly withheld from the model-building step, a standard deviation of $\pm 0.48 \log 1/C$ units was obtained for the estimates, when compared with the experimental values.

22.1.7.1.2 LD_{50} models from larger datasets

The first model illustrated the potential utility of this approach to toxicity estimation, and was followed by development of LD_{50} models from 1000 and 2000 compounds. Noticeable improvement resulted when the model was developed from 1000 compounds, with a further slight improvement when 2000 compounds were used.[31] These larger models were developed with the CROSSBOW system structure codes, which were more readily available than the CIDS codes, but were less detailed.

Reproducibility of laboratory LD_{50} data is very poor, as shown by a study performed by the EEC.[32] This is the main reason LD_{50} model statistics improve asymptotically when larger numbers of compounds are studied, as shown in Table 11.

Table 11 Regression Statistics for LD_{50} Models

No. of compounds studied	425	1000	1500	2000
r^2	0.493	0.562	0.523	0.524
r	0.702	0.750	0.723	0.724
SE of estimate	(0.48)[a]	0.598	0.620	0.623
No. of variables	29	88	77	103

[a] This is the standard deviation of residuals of a test subset; the equation was developed using $\log(\log 1/C + 1)$, so an SE is not comparable to the SE for the other models, for which $\log 1/C$ was used.

The advantage of using the models developed from as many compounds as possible is that the inclusion of a greater variety of chemicals permits the estimation of more diverse compounds than will be the case if models derived from fewer compounds are used. The accuracy of the estimates will not be improved, however, due mainly to the inherent limitation of the poor reproducibility of the LD_{50} test itself.

22.1.7.1.3 *Mutagenesis models*

(i) *Classical QSAR studies*

Kier *et al.* used molecular connectivity indices to correlate the mutagenicity of 15 nitrosamines.[33] Hansch *et al.* correlated the Ames test mutagenicity of a series of 15 organoplatinum compounds to an electron withdrawal parameter.[34] Both of these studies provide guidance for those working within these series of chemicals.

(ii) *ADAPT models*

Mutagenicity has been modeled by Jurs *et al.* for 105 diversified mutagens and nonmutagens, using their 'ADAPT' system.[35] They employed a combination of substructural features with parameters (such as moments of inertia, ratio of surface area to volume, *etc.*) which are related to the overall size and shape of molecules. These were then related to the mutagenicity results by several pattern recognition techniques and by discriminant analysis. The results were not useful for extrapolating to untested compounds, giving almost the same estimates as a random choice. This is probably due to the lack of a sufficient number of diverse compounds. (It is best to have at least 200–300 diverse compounds as a minimum for modeling.)

When the ADAPT programs were applied by Nesnow *et al.* to a homogeneous data set of 21 aliphatic nitrosamines,[36] it was found that four descriptors were sufficient to accurately model the mutagenic properties of these closely related substances. Two of the parameters are geometric in nature, based on principal moments (and not highly intercorrelated); another is the path 1 molecular connectivity index and the fourth is the σ charge on the nitrogen to which the nitroso group is attached, as calculated by the del-Re method. It is noteworthy that these same descriptors were found by Jurs and Rose[48] to be important in their study of the carcinogenic activity of nitrosamines. The classification results of Nesnow *et al.* are shown in Table 12.

(iii) *CASE models*

Klopman *et al.* have developed an additivity approach (called the CASE method) for correlation studies. An example is shown for aromatic amines as mutagens.[37] The method uses substructure units as large as eight to ten or more attached atoms, in an attempt to find larger size structural

Table 12 Classification Results for Model of Mutagenic Aliphatic
N-Nitrosamines by the ADAPT Method

	% correct classification		
Routine	Active	Inactive	Overall
Linear learning machine	100 (16/16)	100(5/5)	100
Bayes linear classifier	81.3(13/16)	80(4/5)	81
Bayes quadratic classifier	93.8(15/16)	80(4/0)	90.5
Iterative least squares	100 (16/16)	100(5/5)	100
Linear discriminant analysis	100 (16/16)	80(4/5)	95.2

features which may be correlated to the biological activity in a series of related chemicals. In applying their method to an analysis of the mutagenic properties of a series of aromatic amines, the classification table given in Table 13 summarizes their results for the Ames test data on tester strain TA100. They estimated 86% of the 107 amines to within ± 1 activity category; 16% were incorrectly estimated as false negatives or positives.

Table 13 Summary of Calculated *Versus* Experimental Results for TA100

	Calculated					
Experimental	−	+	+ +	+ + +	+ + + +	Total
−	39	1	0	2	1	43
+	6	0	0	2	0	8
+ +	5	0	0	2	3	10
+ + +	0	0	0	5	6	11
+ + + +	2	0	0	5	28	35
Total	52	1	0	16	38	107

(iv) Substructure models

Utilizing the concepts first described in Section 22.1.7.1.1 for the LD_{50} models, but using a discriminant analysis equation (because the endpoint was dichotomous), Enslein *et al.* developed two models which estimate the probability that a chemical will be mutagenic in an Ames test.[38] The models differ in that one of them was based on Ames mutagenesis tests for 700 compounds, run with tester strains TA100 and TA1535, which detect base pair mutations. The other model used these tests and tests run with strains which detect frame shift mutagens for a total of 805 compounds. In both models, a heterogeneous set of chemicals was studied. Test results were obtained with and without S9 enzyme activation, by closely controlled protocols, as evaluated by the 'Genetox' program criteria. These data have been assembled and evaluated by the Environmental Mutagen Information Center.[39] On checking for consistency, the models have about 5% of indeterminates, and of the 95% of the chemicals for which the models give a classification, about 95% are correctly classified. These results are summarized in the classification tables (Tables 14 and 15).

Table 14 Mutagenesis Model Using All Ames Tester Strains

	Number	Correctly classified		Incorrectly classified
Total compounds	805			
Indeterminates	53			
Classifiable	752			
Positive compounds	509	480		29
Negative compounds	243	236		7
Indeterminates:		53/805 = 6.6%		
False positives:		7/752 = 0.9%	or	7/243 = 2.9%
False negatives:		29/752 = 3.9%	or	29/509 = 5.7%
Correctly classified:		716/752 = 95.2%		

Table 15 Mutagenesis Model Using Base Pair Substitution Strains

	Number	Correctly classified	Incorrectly classified
Total compounds	700		
Indeterminates	37		
Classifiable	663		
Positive compounds	469	442	27
Negative compounds	194	181	13
Indeterminates		$37/700 = 5.3\%$	
False positives		$27/663 = 4.1\%$ or $27/469 = 5.8\%$	
False negatives		$13/663 = 2.0\%$ or $13/194 = 6.7\%$	
Correctly classified		$623/670 = 92.9\%$	

(v) Use of the Hodes method for mutagenesis data

Tinker has reported a modification of the Hodes method (see Section 22.1.8.1) by which he studied the mutagenesis data (Ames test) for more than 1000 diverse chemicals.[40] In extending the Hodes method, Tinker quantified the mutagenicity data into five activity categories and developed a discriminant analysis model to use for the estimation of untested structures. The 'mechanical chemical code' of Lefkovitz was used to provide ring codes, and CAS connection tables were used to provide the augmented fragments used in the study. The results obtained by using the model to estimate the Ames test values for 34 compounds as they were being tested showed that 88% (30/34) were estimated to within ± 1 category, and 76% were correctly estimated to the correct category.

(vi) Effectiveness of these mutagenic models

These various models differ in the databases to which they have been applied, as well as being significantly different in the statistical methods used for their development. Therefore it is interesting to note that all of them have about the same degree of self-consistency, from 88 to 95%. This suggests that the major limitation is the accuracy and reproducibility of the biological test itself. The models differ in their utility for estimation of as yet untested compounds because such utility is determined primarily by the scope of chemicals used in their derivation.

22.1.7.1.4 Carcinogenesis studies

(i) Quantum chemistry models

In 1955 the Pullmans correlated electronic indices for a series of polycyclic aromatic hydrocarbons (PAH) with carcinogenic properties.[41] Nagata *et al.* considered the frontier electron density of π electron systems to be the determining factor in carcinogenesis of the PAH.[42] Hansch and Fujita correlated the carcinogenic activities of the butter yellow series, using σ and π values.[43]

The many studies which succeeded these attempts have been reviewed and summarized by Loew *et al.*, who propose that these methods may be used to aid in the selection of compounds for study as carcinogens.[44]

(ii) ADAPT models

Jurs *et al.* have used a combination of parameters which pertain to molecular size and shape, molecular connectivity indices and indicator variables to model the carcinogenic activity of polycyclic aromatic hydrocarbons,[45] aromatic amines,[46,47] nitrosamines[48] and a heterogeneous set of chemicals.[49] The statistical methods used by Jurs differ somewhat from those employed by Enslein *et al.* The classification results which compare the use of several methods for analysis of the nitrosamines data are given in Table 16. Table 17 shows the results obtained from analysis of the heterogeneous data set.[49]

(iii) Substructure models

Enslein and Craig used substructure fragments and molecular weight to develop a model from 343 compounds for which the International Agency for Research on Cancer (IARC) had published an

Table 16 Carcinogenic *N*-Nitroso Compounds—ADAPT Classification for 22 Descriptors and 150 Compounds

Analysis method used	Total number wrong	% correct +	% correct −	Total
Iterative least squares	8	97.3	86.8	94.7
Adaptive least squares	4	99.1	92.1	97.3
Bayes linear function	17	92.0	79.0	88.7
Bayes quadratic function	11	95.5	84.2	92.7
KNN-1st nearest neighbor	28	90.2	55.3	81.3
KNN-3 nearest neighbors	31	92.0	79.0	88.7
Sequential simplex optimization	23	94.6	55.3	84.7
Linear discriminant function	17	92.0	79.0	88.7
Linear learning machine	4	99.1	92.1	97.3

Table 17 Statistical Analyses of Carcinogenic Models Derived from 209 Compounds Using 26 Descriptors

Method	No. correctly classified +	No. correctly classified −	% correctly classified +	% correctly classified −	Total
Bayes (linear)	100	65	76.9	82.3	79.0
Bayes (quadratic)	115	75	88.5	94.9	90.9
Learning machine	108	50	83.1	63.3	75.6
Iterative least squares	121	65	93.1	82.3	89.0
K nearest neighbor	105	49	80.8	62.0	73.7

evaluation of the results of carcinogenesis testing, as reported in volumes 1–17 (1972–1978) of the IARC Monographs.[50] The IARC classifications of 'non' or 'indefinite' carcinogens were combined, and these 120 compounds were compared with 223 compounds which IARC considered as definite animal or human carcinogens. Substructure descriptors were based on the WLN notations and were those developed by the 'CROSSBOW' programs, originally written by ICI chemists in England.[51] Molar refractivities were also calculated for these compounds, but did not enter into the final equation when allowed to compete with the substructure keys. The resulting equation (model), derived from more than 36 classes of chemicals, contained 76 substructure functions and the molecular weight.[52] The model correctly classified about 90% of the carcinogens and about 80% of the non and indefinite carcinogens used in its derivation. It also had a 3.6% false negative classification. The classification table is given in Table 18.

Table 18 Classification Table for 343 Carcinogens

IARC classification	Classification by discriminant equation Non or indefinite No.	Non or indefinite %	Indeterminate No.	Indeterminate %	Definite No.	Definite %
Indeterminate: 0.4–0.6						
Non or indefinite	96	80.0	10	8.3	14	11.7
Definite	11	4.9	9	4.0	203	91.0
Indeterminate: 0.3–0.7						
Non or indefinite	93	77.5	14	11.7	13	10.8
Definite	8	3.6	21	9.4	194	87.0

In the next three years following this publication, IARC reversed their classification for three of the chemicals used in this model; this serves to illustrate the difficulty of making simple discriminations about carcinogenicity, and suggests that any such model probably cannot be expected to exceed 90–95% consistency.

Enslein *et al.* have used 335 NCI–NTP evaluations of carcinogenicity[53] to develop a new carcinogenesis model, employing a different set of substructure keys (MOLSTAC), which was based on the enhanced SK&F Chemical Code (see Section 22.1.3.2.3) but has been greatly expanded. MOLSTAC includes specific code terms for about 390 ring systems and 150 types of functional groups, in contrast with CROSSBOW, which has 40 functional groups and 10 specific ring systems (other ring systems are encoded generically in CROSSBOW). MOLSTAC has included keys to identify electron-donating and electron-withdrawing groups on aromatic rings. This model, based on 335 NCI–NTP studies, also uses molecular connectivity indices. It correctly classifies 94.9% of the compounds used in its derivation.[54] The classification table is given in Table 19. In the development of this model, eight chemicals were identified as outliers by the use of clustering and other statistical methods. These data were not used in the final model.

Table 19   Discriminant Analysis Classification for NCI–NTP Model

'Actual' class	Model classification		
	Negative	Indeterminate	Positive
Negative:	217	4	7
Positive:	3	3	101
% Indeterminate:	$7/335 = 2.1\%$		
% False positives:	$7/328 = 2.1\%$		
% False negatives:	$3/328 = 0.9\%$		
Overall accuracy:	$318/335 = 94.9\%$		

22.1.7.1.5 *Draize skin irritation models*

A study of rabbit skin irritation bioassay data in the Registry of Toxic Effects of Chemical Substances (RTECS)[55] showed the necessity of going to the original literature in order to 'harmonize' data obtained by different protocols. Thus 774 chemicals were identified as having skin irritation data with similar test criteria, which were combined into one database.[56] Substructure models (using MOLSTAC keys and MCIs) were prepared for ring compounds and for nonring compounds, since the two classes showed different distributions of irritation severity. Nonring compounds were further differentiated by locating or estimating pK_a data; as anticipated, the strong acids or bases were more irritating. It was found necessary to develop two discriminant equations to handle the nonring compounds: the first discriminated between the severe irritants and the combination of negative-plus mild-plus moderate irritants; the second equation discriminated between the mild and moderates (Tables 20 and 21).

Table 20   Rabbit Skin Irritation (Nonring Models) Three-Way Discriminant Analysis

Actual class	Discriminant classification					
	Neg/mild/moderate		Severe-pKa		Severe-no pKa	
	No.	%	No.	%	No.	%
Neg/mild/moderate	192	92.3	1	0.5	15	7.2
Severe-pKa	3	8.8	31	91.2	0	0
Severe-no pKa	3	16.7	0	0	15	83.3

Table 21   Rabbit Skin Irritation (Nonring Models) Negatives *versus* Mild/Moderates

Actual class	Discriminant classification		
	Negative	Indeterminate	Mild/moderate
Negative	20	2	3
Mild/moderate	11	10	167
False positives:	$3/201 = 1.5\%$ or $3/25 = 12.0\%$		
False negatives:	$11/201 = 5.5\%$ or $11/188 = 5.9\%$		
Indeterminates:	$12/213 = 5.6\%$		
$187/201 = 93\%$ of the classifiable compounds are correctly classified			

22.1.7.1.6 *Draize eye irritation model*

The Draize eye irritation model (using MOLSTAC keys and MCIs) was developed for about 1100 compounds, after first 'harmonizing' the experimental results from the two major sources of data.[57] As for the skin irritation data, it was found useful to separate the ring compounds from nonring compounds, and to develop two discriminant equations for each set. The eye irritation data models had a higher proportion of indeterminates than most other toxicity models. This is believed to be due to the noise in the data, *i.e.* the difficulty of reproducing the test results. The results are summarized in Tables 22–25.

Table 22 Rabbit Eye Irritation (Nonring Models) Negative *versus* All Others

	Discriminant equation classification		
Actual class	*Negative*	*Indeterminate*	*Other*
Negative	55	18	16
Other	35	70	394
Indeterminates:	$88/588 = 15\%$		
False other:	$16/500 = 3.2\%$	or	$16/71 = 22.5\%$
False negatives:	$35/500 = 7.0\%$	or	$35/429 = 8.2\%$
Overall accuracy:	$449/500 = 89.9\%$		
$F = 25.3$ with $(12; 575)$ DF; $p \ll 0.001$			

Table 23 Rabbit Eye Irritation (Nonring Models) Severe *versus* All Others

	Discriminant equation classification		
Actual class	*Other*	*Indeterminate*	*Severe*
Other	247	107	11
Severe	29	76	123
Indeterminates:	$183/593 = 30.9\%$		
False severes:	$11/410 = 2.7\%$	or	$11/258 = 4.3\%$
False others:	$29/410 = 7.1\%$	or	$29/152 = 19.1\%$
Overall accuracy:	$370/410 = 90.2\%$		
$F = 15.2$ with $(28; 564)$ DF; $p \ll 0.001$			

Table 24 Rabbit Eye Irritation (Ring Models) Negative/Mild *versus* Moderate/Severe

	Discriminant equation classification		
Actual class	*Neg/mild*	*Indeterminate*	*Mod/severe*
Neg/mild	125	85	11
Mod/severe	17	74	230
Indeterminates:	$159/542 = 29.3\%$		
False mod/severe:	$11/383 = 2.9$	or	$11/136 = 8.1\%$
False neg/mild:	$17/383 = 4,4\%$	or	$17/247 = 6.9\%$
Overall accuracy:	$355/383 = 92.7\%$		
$F = 8.63$ with $(43; 501)$ DF; $p \ll 0.001$			

Table 25 Rabbit Eye Irritation (Ring Models) Severe *versus* All Others

	Discriminant equation classification		
Actual class	*All other*	*Indeterminate*	*Severe*
Other	246	84	27
Severe	23	48	115
Indeterminates:	$132/543 = 24.3\%$		
False severes:	$27/411 = 6.6\%$	or	$27/273 = 9.9\%$
False others:	$23/411 = 5.6\%$	or	$23/138 = 16.7\%$
Overall Accuracy:	$361/411 = 87.8\%$		
$F = 10$ with $(36; 506)$ DF; $p \ll 0.001$			

22.1.7.1.7   *Teratogenesis model*

A substructural model was developed from compounds for which the results of teratogenesis studies were reported in the two leading compilations of such data.[58] The results for 670 compounds were scored and submitted to a panel of teratogenesis experts for review. From this review a database of 430 compounds resulted, and a discriminant analysis equation was obtained, using CROSSBOW substructure keys and the molecular weight as parameters. The equation had poorer statistics than those for the mutagenesis or carcinogenesis models, and classified 22% of the compounds as indeterminates rather than assigning them as teratogens and nonteratogens.

Enslein has since withdrawn this model from the HDI toxicological repertoire, based on discussions with leading teratologists, who are no longer performing the same type of teratology studies that were used in the model.[59] The older test often was carried out at doses which were toxic to the mother, and thus provided many false positives. The newer approach is to find compounds which produce teratogenic effects at doses which are apparently safe for the mother.

22.1.8   SUBSTRUCTURES IN COMPOUND SELECTION

22.1.8.1   Hodes Heuristic Method

22.1.8.1.1   *Development of concept*

Faced with the task of selecting some 15 000 test compounds per year for an antitumor screening program, Hodes *et al.* developed a 'statistical-heuristic' method to automate the selection from several hundred-thousand candidate chemicals available to the National Cancer Institute (NCI) for testing.[60] The structural features used were a combination of ring fragments and augmented fragments (see Section 22.1.3.3.2). The initial study used the same training set (136 compounds) previously studied by Chu *et al.*[25] The method used a modification of the substructural approach of Cramer *et al.* and was based on the relative frequency of occurrence of a substructure in 'active' and 'inactive' compounds. The method does not employ matrix inversion in its operation, and thus is readily applied to very large sets of chemicals.

22.1.8.1.2   *Extension to large datasets*

These initial results were encouraging, and the method was further developed using more than 15 000 compounds in the training set, with the deliberate withholding of well-known active 'lead' compounds.[61] These were withheld to aid in the selection of novel structures, rather than to unduly bias the selection based upon existing structures.

In 1986 Hodes reported that inclusion of octanol/water partition coefficient data with the substructural features, in a two-component approach, resulted in a significant improvement over the previous method.[62] The training set is now almost 100 000 compounds, and consists of all chemicals which have given definitive test results (negative and positive) in the NCI *in vivo* prescreen. The entire Pomona College Medicinal Chemistry Project database of 4013 experimental octanol/water log *P* values (version of July, 1983) was divided into four ranges as shown in Table 26.

Table 26 Distribution of Pomona College Database Into Four Ranges

Log P range	No. of compounds	Log P range	No. of Compounds
−5 to −1	270	1 to 3	1990
−1 to 1	1094	3 to 8	659

Each subset was taken as the active set to produce four sets of fragments by Hodes 'heuristic' method. These fragment weights signified the likelihood that a compound with a given fragment would belong to that particular range of log *P* values. Then the training set of 92 963 compounds was run through each of the log *P* models, using the newly created weighting factors. Thus each fragment now had different weights, depending on which log *P* range was in use. A training set (20% of the entire 92 963 set) was run through the heuristic model, and each compound was evaluated by summing the fragment weights for its assigned $\log P$ range. Table 27 shows how this inclusion of $\log P$ information improved the selection of both the highly active compounds (group A) and the moderately active compounds (group C) in the 20% subset studied.

Table 27 Comparison of Original and Log P Modified Models[a]

| | Cumulative percent actives | | | |
| | Original | | Log P modification | |
Percentile	A[b]	C	A	C
99	27	10	37	11
98	39	16	48	20
95	71	30	76	33
90	88	44	86	47
80	92	59	97	63
70	97	70	99	73

[a] Original model (see ref. 54); $\log P$ model (see ref. 55).
[b] A = highly active in P388 screen, C = moderately active.

This method is in current use at NCI in the selection of compounds for the *in vivo* prescreen.

22.1.8.2 Need to Conserve Resources in Toxicity Testing

22.1.8.2.1 Selection of chemicals for carcinogenicity testing

The average cost of a lifetime carcinogenesis bioassay in two species exceeds $1 000 000 by the time all costs, from preparatory studies to histological assays, are included. From start to finish it requires about 3.5–4.5 years to complete one study. Thus a major commitment of man-power, animals, supplies and facilities is required when a new bioassay is contemplated. The National Toxicology Program (NTP)[63] commissioned a study by the National Academy of Sciences–National Research Council (NAS–NRC) to consider the best ways to select candidate compounds for testing, since a major part of the NTP budget is tied up in the lifetime carcinogenesis bioassay area. The NRC report[64] considered the possibility of using modeling techniques in the compound selection process to be premature at the time of the report (1983).

22.1.8.2.2 Protection of animals

There has been increasing opposition to the use of living animals in research and testing activities from animal rights groups as well as some scientists. The Federal Government has increased its own efforts to ensure proper and reasonable treatment of animals, as summarized in a recent Office of Technology Assessment (OTA) report. This pressure from concerned citizens has encouraged development of alternatives to animal testing, and has motivated the granting mechanisms of various government agencies to fund further research in these areas.

The OTA report gave encouragement to further investigation of the LD_{50} substructure model as a possible replacement for many LD_{50} tests.[65]

22.1.8.3 The ECETOC Study

A six-man task force was formed by the European Chemical Industry Ecology and Toxicology Centre (ECETOC) to study structure–activity relationships in toxicology and ecotoxicology. In their report, a review of the available methods for QSAR study was made, together with an evaluation of 18 studies which discuss the various methods.[66] As examples of the substructural approach, they reviewed the LD_{50} and teratology models of Enslein *et al.*; the aromatic amines model of Jurs *et al.* served as an example of pattern recognition; and one of Klopman's papers on carcinogens was also reviewed.

22.1.9 LIMITATIONS AND UTILITY OF TOXICITY MODELS FOR PREDICTION

22.1.9.1 Limitations Based on Confidence Limits of Estimates

Of the existing models, only the LD_{50} model will give an estimate of a numerical value for a toxicological endpoint; the statistical parameters for its estimates are shown in Table 11. Although it would be better if the standard error of estimate were lower (SE $= \pm 0.62 \log 1/C$; or from four- to one quarter-fold times the estimate), the large variability attached to experimentally determined LD_{50} values[32] shows that it is impossible to expect better results than this, either from the model or from an experiment. A well-performed experiment should give a value with a better standard error of estimate than would be expected from the model (statistical steps employed in all model development necessarily result in some diffusion of the average accuracy for all endpoints).

However, this fact must be weighed against the financial costs and the required accuracy must be carefully considered.

The models based on discriminant analysis are best evaluated in terms of percent of accurate estimates, as shown by new test results. Since it is often impractical to await such results (the number of new carcinogenesis tests, for example, is very low due to the extreme costs involved), it has been customary to withhold some of the data from the model development process, and then see how well they are predicted. Unfortunately, this is really self-defeating, as it has been shown that this practice is but a sampling test of how well the compounds and parameters are distributed among the total population, and gives no guidance as to the extrapolatability of the model to other chemicals.[67]

22.1.9.2 Utility Based on Financial Constraints

The costs of toxicity testing in animals varies greatly, depending on the test involved. Acute toxicity testing (single dose), such as the LD_{50}, requires several days, and from 10 to 50 animals; costs might range from $300 to $1000. At the other extreme, carcinogenicity testing requires from two to three years of chronic dosing and many hundreds of animals are involved; costs can range from $800 000 to more than $1 000 000 over a four-year period.

In contrast to these costs, an estimate for a single toxicity endpoint, such as an LD_{50} or carcinogenesis estimate, costs about $300 if carried out on a custom basis.[59]

If the requirement for testing is a legal one, such as for regulatory purposes, there is no substitute at present for the actual animal test. Thus the development of new drugs, pesticides, food additives or any new product with the potential for human exposure, must necessarily include considerable costs for toxicity testing.

However, the developer of a new product is often faced with the dilemma of selecting a chemical for future development from a group of potential candidates. Rather than expend the considerable time and money required to test all of the candidates, the use of toxicity models, based on substructural fragments and other parameters, should be considered as a method for reducing the number of compounds for which animal tests will be required. This must be done carefully, with consideration of the similarity between the candidate chemical and the chemicals which were used in the model development as a guide. It should be obvious that it is impossible for such models to provide meaningful estimates for chemicals which contain structural features not represented in the original databases from which the models were developed. Thus the user of such models must have complete access to the structures and test results for all chemicals which were in the database, so that he may be assured of the similarities and differences between his compound and those in the database from which the model was developed. One should not use these models unless such knowledge is available and considered at every use of the model.

22.1.10 REFERENCES

1. L. F. Fieser and M. Fieser, 'Organic Chemistry', 1st edn., Heath, Boston, 1944, chap. 1.
2. Chemical Abstracts Service, *CAS Online News*, Feb. 1987.
3. 'Beilstein's Handbuch der Organischen Chemie', Springer-Verlag, Berlin (some 325 volumes in this series). (For an English introduction to the use of Beilstein see: E. H. Huntress, 'A Brief Introduction to the Use of Beilstein's Handbuch der Organischen Chemie', 2nd edn., Wiley, New York, 1938).
4. Beilstein is converting to a computerized online database.
5. Chemical–Biological Coordination Center, National Research Council, 'A Method of Coding Chemicals For Correlation and Classification', Washington, DC, 1950.

6. K. W. Wheeler, E. R. Andrews, F. Fallon, G. L. Krueger, F. P. Palopoli and E. L. Schumann, *Am. Doc.*, 1958, **9**, 198.
7. W. Steidle, *Pharm. Ind.*, 1957, **19**, 88; The Ringdoc System is a product of Derwent Publications Ltd., London, England.
8. P. N. Craig and H. M. Ebert, *J. Chem. Doc.*, 1969, **9**, 141.
9. R. D. Cramer, III, G. Redl and C. E. Berkoff, *J. Med. Chem.* 1974, **17**, 533.
10. *J. Chem. Doc.*, 1965, issues 1 and 2, **5** contain a series of such papers resulting from a symposium. These developments are analyzed in depth by M. F. Lynch, J. M. Harrison, W. G. Town and J. E. Ash, 'Computer Handling of Chemical Structure Information', MacDonald, London, 1971, p. 67.
11. F. B. Winer, *Drug Inf. J.*, 1983, **17**, 277.
12. W. Graf, H. K. Kaindl, H. Kniess, B. Schmidt and R. Warszawski, *J. Chem. Inf. Comput. Sci.*, 1979, **19**, 51.
13. M. Milne, D. Lefkovitz, H. Hill and R. Powers, *J. Chem. Doc.*, 1972, **12**, 183.
14. K. Enslein and P. N. Craig, *J. Environ. Pathol. Toxicol.*, 1978, **2**, 115.
15. E. G. Smith and P. A. Baker, 'The Wiswesser Line-Formula Chemical Notation (WLN)', 3rd edn., Chemical Information Management, Cherry Hill, NJ, 1975.
16. The 'Chemical Substructure Index' is a permuted WLN index to the 'Current Abstracts of Chemistry and Index Chemicus' (published by the Institute for Scientific Information, Philadelphia, PA).
17. Medicinal Chemistry Project, Department of Chemistry, Pomona College, Claremont, CA.
18. R. J. Feldmann, G. W. A. Milne, S. R. Heller, A. Fein, J. A. Miller and B. Koch, *J. Chem. Inf. Comput. Sci.*, 1977, **17**, 157.
19. W. T. Wipke, J. G. Nourse and T. Moock, in 'Computer Handling of Generic Chemical Structures', ed. J. M. Barnard, Gower, Aldershot, 1984, p. 167.
20. D. Lefkovitz and A. R. Gennaro, *J. Chem. Doc.*, 1970, **10**, 86.
21. T. C. Bruice, N. Kharasch and R. J. Winzler, *Arch. Biochem. Biophys.*, 1956, **62**, 305.
22. S. M. Free, Jr. and J. W. Wilson, *J. Med. Chem.*, 1964, **7**, 395.
23. C. Silipo and C. Hansch, *J. Am. Chem. Soc.*, 1975, **97**, 6849.
24. K. H. Kim, C. Hansch, J. Y. Fukunaga, E. E. Steller, P. Y. C. Jow, P. N. Craig and J. Page, *J. Med. Chem.*, 1979, **22**, 366.
25. K. C. Chu, R. J. Feldmann, M. B. Shapiro, G. F. Hazard, Jr. and R. I. Geran, *J. Med. Chem.*, 1975, **18**, 539.
26. L. B. Kier and L. H. Hall, 'Molecular Connectivity in Chemistry and Drug Research', Academic Press, New York, 1976.
27. L. B. Kier and L. H. Hall, 'Molecular Connectivity in Structure-Activity Analysis', Research Studies Press, Letchworth, UK, 1986.
28. R. F. Rekker, 'The Hydrophobic Fragmental Constant', Elsevier, Amsterdam, 1977, Pharmacochemistry Library, vol. 1.
29. P. N. Craig, *J. Med. Chem.*, 1971, **14**, 680.
30. D. W. Marquardt and R. D. Snee, *Am. Stat.*, 1975, **29**, 3.
31. K. Enslein, T. Lander, M. Tomb and P. N. Craig, 'A Predictive Model for Estimating Rat Oral LD_{50} Values,' Princeton Scientific Publishers, Princeton, NJ, 1983.
32. W. Lingk, in 'Quality Assurance of Toxicological Data', ed. W. J. Hunter and C. Morris, Report EVR-7270EN, EEC, Luxembourg, 1982.
33. L. B. Kier, R. J. Simons and L. H. Hall, *J. Pharm. Sci.*, 1978, **67**, 725.
34. C. Hansch, B. H. Venger and A. Panthananickal, *J. Med. Chem.*, 1980, **23**, 459.
35. T. R. Stouch and P. C. Jurs, *EHP, Environ. Health Perspect.*, 1985, **61**, 329.
36. S. Nesnow, R. Langenbach and M. J. Mass, *EHP, Environ. Health Perspect.*, 1985, **61**, 345.
37. G. Klopman, M. R. Frierson and H. S. Rosenkranz, *Environ. Mutagen.*, 1985, **7**, 624.
38. K. Enslein, B. W. Blake, M. E. Tomb and H. Borgstedt, *In Vitro Toxicol.*, 1986/87, **1**, 33.
39. Environmental Mutagen Information Center, Oak Ridge National Laboratory, Oak Ridge, TN, USA.
40. J. F. Tinker, *J. Comput. Chem.*, 1981, **2**, 231.
41. A. Pullman and B. Pullman, *Adv. Cancer Res.*, 1955, **3**, 117.
42. C. Nagata, K. Fukui, T. Yonezawa and Y. Tagashira, *Cancer Res.* 1955, **15**, 233.
43. C. Hansch and T. Fujita, *J. Am. Chem. Soc.*, 1964, **86**, 1616.
44. G. H. Loew, M. Poulsen, E. Kirkjian, J. Ferrell, B. S. Sudhindra and M. Rebagliati, *EHP, Environ. Health Perspect.*, 1985, **61**, 69.
45. M. Yuan and P. C. Jurs, *Toxicol. Appl. Pharmacol.*, 1980, **52**, 294.
46. K. Yuta and P. C. Jurs, *J. Med. Chem.*, 1981, **24**, 241.
47. K. Yuta and P. C. Jurs, *Yakugaku Zasshi*, 1984, **104** (5), 496 (*Chem. Abstr.*, 1984, **101**, 85 399).
48. S. L. Rose and P. C. Jurs, *J. Med. Chem.*, 1982, **25**, 769.
49. P. C. Jurs, J. T. Chou and M. Yuan, *J. Med. Chem.*, 1979, **22**, 476.
50. 'IARC Monographs on the Evaluation of the Carcinogenic Risk of Chemicals to Humans', IARC, Lyon, 1972; 41 volumes have been published up to 1985.
51. D. L. Eakin and E. Hyde, in 'Computer Representation and Manipulation of Chemical Information', ed. W. T. Wipke, S. R. Heller, R. J. Feldmann and E. Hyde, Wiley, New York, 1974, pp. 1–30.
52. K. Enslein and P. N. Craig, *J. Toxicol. Environ. Health*, 1982, **10**, 521.
53. NCI–NTP Technical Report Series. Begun as NCI Carcinogenesis Bioassay Reports, the series now numbers more than 300; National Toxicology Program, Research Triangle Park, NC, USA.
54. K. Enslein, H. H. Borgstedt, M. E. Tomb, B. W. Blake and J. B. Hart, *Toxicol. Ind. Health*, 1987, **3**, 267.
55. 'Registry of Toxic Effects of Chemical Substances'. Published by the National Institute for Occupational Safety and Health, from 1972 up to the present. US Government Printing Office, Washington, DC, USA.
56. 'HDI Toxicol. Newsletter,' 1984, **3**, 1.
57. 'HDI Toxicol. Newsletter,' 1987, **6**, 1.
58. K. Enslein, T. R. Lander and J. R. Strange, *Teratogen. Carcinog. Mutagen.*, 1983, **3**, 289.
59. K. Enslein, private communication, 1987.
60. L. Hodes, G. F. Hazard, R. I. Geran and S. Richman, *J. Med. Chem.*, 1977, **20**, 469.
61. L. Hodes, *ACS Symp. Ser.*, 1979, **112**, 583.
62. L. Hodes, *J. Med. Chem.*, 1986, **29**, 2207.
63. The National Toxicology Program was formed from segments of the National Cancer Institute, the National Institute for Environmental Health Sciences, the Food and Drug Administration and the National Institute for Occupational Safety and Health in 1979, to coordinate major toxicological testing by the government.

64. 'Strategies to Determine Needs and Priorities for Toxicity Testing', National Academy Press, Washington, DC, vol. 1.
65. US Congress, Office of Technology Assessment, 'Alternatives to Animal Use in Research, Testing and Education', US Government Printing Office, Washington, DC, OTA-BA-273, 1986, p. 182.
66. 'Structure–Activity Relationships in Toxicology and Ecotoxicology: An Assessment,' ECETOC Monograph No. 8, European Chemical Industry Ecology and Toxicology Centre, Brussels, 1986.
67. S. M. Snapinn and J. D. Knoke, *Technometrics*, 1985, **27**, 199.

22.2

Linear Discriminant Analysis and Cluster Significance Analysis

JAMES W. McFARLAND

and

DANIEL J. GANS
Pfizer Inc., Groton, CT, USA

22.2.1 INTRODUCTION

Despite rapid advances in the QSAR field over the past 25 years, the relationship of biological activity to chemical structure is still frequently described in qualitative terms. Thus, one commonly finds in the literature statements such as: 'activity is found in those members of the series in which a halogen is located at the *ortho* position'. This is not necessarily a rejection of QSAR methods; it may be merely that quantitative biological data or measured physical properties of the compounds are not available.

Not much can be done if both of these elements are missing. However, statistical techniques exist that can treat situations in which there are only *qualitative* biological data but for which *quantitative* physical data are available. The present chapter will discuss two such methods: linear discriminant analysis (LDA) and cluster significance analysis (CSA). For each procedure we will treat situations in which the biological activity is expressed as a dichotomy, *e.g.* 'active' or 'inactive'. Both methods attempt to achieve the same end: to determine in particular drug series those physical parameters which influence biological activity, and on that basis predict the activity class of new members.

LDA has a long tradition in statistics, and has been employed with increasing frequency in medicinal chemistry since its introduction there in 1974.[1] CSA on the other hand is a recent innovation.[2] These two methods are somewhat similar, but sometimes one will be more useful than the other.

Discriminant analysis includes more than LDA; there are also other methods such as quadratic discriminant analysis. Because there has been little application of the latter in medicinal chemistry, we will emphasize LDA in this chapter. LDA is capable of dealing with classifications involving more than two groups, and we will introduce the mathematics in enough detail to make this plain, but our major interest will be in dichotomies. This accords with the fact that the majority of medicinal chemistry applications to date are of this type.

CSA is not be confused with another approach used in QSAR, which is often called simply 'cluster analysis'.[3] This latter technique is quite different; it is a method to obtain clusters of related objects from previously undifferentiated data. CSA, on the other hand, is a method to analyze the statistical significance of apparent clusters. It is limited to analyzing two-group classifications as will be seen.

22.2.2 LINEAR DISCRIMINANT ANALYSIS

22.2.2.1 A Graphical Overview

LDA is a mathematical technique which handles numerical data directly. However, it is often convenient to display the data graphically in order to develop an intuitive sense of a particular case. In LDA membership in the activity classes can be considered the dependent variable; in a graph class membership is indicated by a distinguishing symbol. The independent variables will comprise the axes of the graph, and thereby create a 'parameter space' in which the dependent variable will reside. The illustrations that follow are two-dimensional. This, however, is for the convenience of the printed page. One or three or more dimensions are equally valid, and are treated similarly.

The goal of the analyst is to find independent variables which allow the maximum separation between the members of one group and those of the other. The independent variables so discovered are thus identified as parameters which probably have an influence on biological activity. The region of the 'parameter space' in which the actives are found defines the most likely positions for any new actives.

Figure 1 presents an idealized situation. The objects of the data set are divided into groups by some biological response (*e.g.* 'active' or 'inactive'); these objects are then plotted in a graph. The axes X and Y in this case may represent $\log P$ and pK_a, for example. It is clear from the figure that there is a sharp separation of the members of one group from those of the other. In LDA a (mathematically linear) boundary separating the two classes is then sought. This boundary then predicts the classification of new objects from knowledge of their independent variables. In the example, objects falling above and to the left of the boundary are likely to be 'active' ($\triangle$) while those falling below and to the right will tend to be 'inactive' ($\bigcirc$).

The statistical significance of the group separation can be determined by methods associated with LDA. The principle is to locate the centroids of each group, and from knowledge of the variability in each group determine whether or not the centroids are in fact separated by more than chance would allow. The farther apart the centroids and the smaller the variability in each group, the greater the likelihood that the chosen parameters truly separate the groups, and that the membership of new objects can be reasonably predicted.

Figure 2 represents a less idealized situation. Here the separation of the two groups is still evident from the distance between the centroids, but the boundary no longer classifies the membership of the

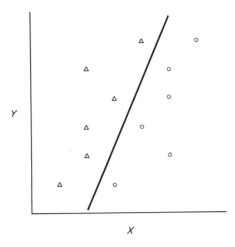

Figure 1   Idealized data for linear discriminant analysis. The plot of the two data types ($\triangle$ and $\bigcirc$) is against two arbitrary parameters, X and Y, which allow the data types to be accurately separated in the parameter space

two groups perfectly. One member from each group would be misclassified. Nevertheless, classification is apt to be more successful than in the situation represented in Figure 3 where the two classes are intermixed completely.

In Figure 2 the centroids are closer together than those in Figure 1, but statistical analysis shows the separation to be still significant. In Figure 3 the centroids of the two classes coincide exactly, and

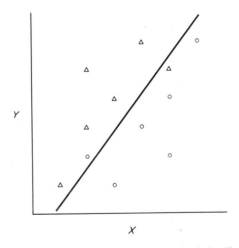

Figure 2   A variation of the data in Figure 1; some misclassification occurs

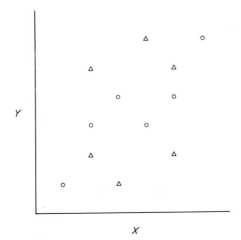

Figure 3   Another variation of the data in Figure 1: the two data types are no longer separated

there is no statistical meaning to be derived from this situation. A more formal description of LDA follows.

22.2.2.2 The Mathematical Apparatus

LDA is a statistical method for classifying observations in multidimensional space into one of several groups. To the medicinal chemist the observations represent compounds in a congeneric series. The location coordinates of the compounds in k-dimensional space ($k \geq 1$) are the values of k associated physicochemical parameters, and the r groups ($r \geq 2$) are defined by the biological responses of the compounds. In the examples we will discuss $r = 2$, with the groups comprising 'actives' and 'inactives', but LDA applies for larger r as well, and initially in our discussion we will not restrict r to be 2.

The development below is similar to that of Anderson.[4] Another useful reference is Kendall.[5] The three major statistical computer packages available in the US, BMDP, SAS and SPSS-X, all include programs or procedures which can perform LDA. BMDP has the program BMDP7M;[6] SPSS-X provides DISCRIMINANT;[7] SAS offers DISCRIM.[8] The cited manuals for these packages also can furnish helpful explanations. The terminology presented below is in fairly common use, but nomenclature is not fully standardized for LDA, and the reader is cautioned that usage in different references may vary.

For completeness we use vector and matrix notation in the development which follows. Both vectors and matrices are represented by bold letters, lower case for the former and upper case for the latter. The prime (') indicates transpose; unprimed vectors are assumed in column form. The theoretically minded reader can find information on handling vectors and matrices, and the definitions of statistical vectors and matrices used below, in many textbooks on multivariate statistics, such as for example Morrison.[9] The reader unfamiliar with vector–matrix notation is emphatically assured, however, that its use is simply to add some precision to the development, and that if one confines oneself to the prose below one will not miss anything essential.

22.2.2.2.1 The framework

Suppose we have in hand a series whose parameter values are known and whose biological responses (groups) have been determined. Let there be n_i such compounds in the ith group, $i = 1, \ldots, r$, and let $N = n_1 + \cdots + n_r$ be the total number of compounds. For now we assume we are dealing with a fixed set of k parameters, all of which we wish to use. Thus each of the N tested compounds is represented by a point x in k-dimensional parameter space. The N compounds are also known as the 'training set'. The problem is this: given a new compound not in the training set, whose parameter values are known and are represented by the point x, how can we use the information in the training set to arrive at a rule for predicting the response group into which the new compound will fall?

22.2.2.2.2 Assumptions

In employing LDA it is assumed — *assumption (1)*: that the typical observation of the ith response group can be described by the *multivariate normal* probability density function

$$p_i(x) = (2\pi)^{-\frac{k}{2}} |\Sigma|^{-\frac{1}{2}} \exp\left\{-\tfrac{1}{2}(x - \mu_i)'\Sigma^{-1}(x - \mu_i)\right\} \tag{1}$$

This density is a generalization of the familiar normal or Gaussian probability density function of one dimension.

Here μ_i is the centroid of the probability distribution in k-dimensional space. Contours of constant probability density are hypersurfaces in this space. These hypersurfaces form concentric ellipsoids, whose size and orientation are described by the covariance matrix Σ. Note that in equation (1) μ_i can vary with i, but Σ cannot. This means that while the groups can have different centroids—*assumption (2)*: all groups have the same covariance matrix Σ. Thus all r groups are assumed to be represented by 'probability clouds' of a specific ellipsoidal shape with the same size and orientation. The only difference allowed among these 'clouds' is that they may be displaced one from another in parallel fashion.

Assumptions 1 and 2 are clearly restrictive. Still, useful analyses have been made with LDA. Further discussion about this occurs below.

22.2.2.2.3 The classification functions

We will also suppose that, before the location x of the new compound is known, it is possible to specify probabilities q_i, $i = 1, \ldots, r$, that it will fall into group i. The q_i are called prior probabilities because they are determined prior to the observation x. More will be said about them later.

Given now the location x for the new compound, Bayes' theorem gives posterior probabilities

$$\pi_i(x) = \frac{q_i p_i(x)}{\sum_{j=1}^{r} q_j p_j(x)} \tag{2}$$

Here the π_i combine the information in the q_i and in the location of x to give group probabilities specific for (or 'after' determining) the location x.

Now x, though it is arbitrary, can be regarded as fixed; we are really interested in how $\pi_i(x)$ varies with i, for that tells us which response group (or groups) is (are) the best candidate(s) for the compound's membership. To this end we will feel free to write 'const.' for any expression which is constant for, *i.e.* does not depend on, i. Thus we may write

$$\pi_i(x) = \text{const. } q_i p_i(x) \tag{3}$$

Owing to the form of equation (1), it is easier to work with logarithms; thus

$$\ln \pi_i(x) = \text{const.} + \ln q_i + \ln p_i(x) \tag{4}$$

$$= \text{const.} + \ln q_i + \text{const.} - \tfrac{1}{2}(x - \mu_i)' \Sigma^{-1}(x - \mu_i) \tag{5}$$

$$= \text{const.} + \ln q_i - \tfrac{1}{2}(x - \mu_i)' \Sigma^{-1}(x - \mu_i) \tag{6}$$

Some matrix algebra reduces this to

$$\ln \pi_i(x) = \text{const.} + \ln q_i + x' \Sigma^{-1} \mu_i - \tfrac{1}{2}\mu_i' \Sigma^{-1} \mu_i \tag{7}$$

(allowing again one constant independent of i to be absorbed into another).

Equation (7) gives π_i in the logarithmic scale. If one transformed back from logarithms to the raw scale, the constant term in equation (7) would become a multiplicative constant, still independent of i. Accordingly, discarding it would not affect the *relative* values of the posterior probabilities, which are the quantities of interest.

Now the μ_i and Σ are usually unknown. The best we can do is replace them with their estimates $\bar{x}_i$ and S from the training set (S is the pooled within-groups estimate). Doing so, and discarding the constant in equation (7) as discussed, gives

$$f_i(x) = \ln q_i + x' S^{-1} \bar{x}_i - \tfrac{1}{2} \bar{x}_i' S^{-1} \bar{x}_i \tag{8}$$

as an estimate of the logarithm of the relative posterior probability for the group i at point x.

The $f_i(x)$, $i = 1, \ldots, r$, are often called the *classification functions*. Because the logarithm increases with its argument, the larger $f_i(x)$ is, the greater the estimated likelihood that the point x 'belongs' in the group i. Thus, the best choice of predicted group for the point x is the group i for which $f_i(x)$ is largest. This establishes the classification rule.

In addition to the intuitive appeal of this derivation, it is also true[10] that, except for the approximation in introducing sample quantities as estimates in equation (8), the rule minimizes the overall probability of misclassification error for the new compound.

Interest often ends with classification into the best group using the above rule. However, the posterior probabilities themselves contain information potentially useful to the analyst, because they give a quantitative measure of the likelihood of each type of response for the compound. Because $\exp f_i(x)$ gives merely the relative posterior probability for group i, the actual probability must be recovered from

$$\hat{\pi}_i(x) = \frac{\exp f_i(x)}{\sum\limits_{j=1}^{r} \exp f_j(x)} \tag{9}$$

(where the caret denotes an estimate using sample quantities). However, it should not often be necessary to use equation (9), because all three package programs mentioned above can provide direct classification of new (*i.e.* non-training) observations and posterior probabilities for them in the same run as the original classification analysis using the training set. (But it should nonetheless be noted that, in distinction to our usage, SAS DISCRIM employs the term 'linear discriminant function' in referring to $f_i(x)$.)

22.2.2.2.4 *The discriminant function*

The above is a self-contained derivation and is sufficient for the problem. Some additional insight may be gained, however, by further pursuit. Let us now suppose $r = 2$. The above classification rule is to choose group 1 if

$$f_1(x) - f_2(x) > 0 \tag{10}$$

and group 2 if the inequality is reversed. (In the boundary case where the left member equals 0, $\hat{\pi}_1(x) = \hat{\pi}_2(x) = 1/2$ and we may choose either group.)

Now from equation (8), with some algebraic manipulation, it follows that

$$f_1(x) - f_2(x) = \ln(q_1/q_2) + x'S^{-1}(\bar{x}_1 - \bar{x}_2) - \tfrac{1}{2}(\bar{x}_1 + \bar{x}_2)'S^{-1}(\bar{x}_1 - \bar{x}_2) \tag{11}$$

If we put

$$D(x) = x'S^{-1}(\bar{x}_1 - \bar{x}_2) - \tfrac{1}{2}(\bar{x}_1 + \bar{x}_2)'S^{-1}(\bar{x}_1 - \bar{x}_2) \tag{12}$$

then our rule is to choose group 1 if

$$D(x) > \ln(q_2/q_1) \tag{13}$$

$D(x)$ is often called the *linear discriminant function*.

Equation (12) shows that $D(x)$ is a linear function of the argument x. This means that the separating boundary between the two regions of classification [given by equality instead of the inequality in equation (13)] is mathematically a linear subspace of dimension one smaller than that of the whole parameter space. Thus for $k = 1$, 2 or 3, the boundary is a point, line or plane respectively. In the general case it is termed simply a hyperplane. Whatever the dimension k, the boundary separates the whole space into two similar unbounded half-spaces.

The position of the boundary, however, depends on $\ln(q_2/q_1)$. If q_1 (and thus q_2) is allowed to vary, but the training set remains unchanged, the boundary hyperplane will move along a fixed direction, sweeping out parallel images of itself. For sufficiently extreme values of $\ln(q_2/q_1)$, the centroids of both samples will actually be found on the same side of the boundary.

22.2.2.2.5 *Choice of prior probabilities*

The prior probabilities q_i are thus important. Each q_i should reflect the chance that a new congener, with nothing known or assumed about its parameter values, will fall into the ith biological response category. In the absence of compelling reasons to do otherwise, it is probably best to let the q_i be proportional to the size of each response group in the training set, *i.e.* to put

$$q_i = \frac{n_i}{N} \tag{14}$$

The reader is warned, however, that many computer programs, including the three mentioned above, will by default set

$$q_i = \frac{1}{r} \tag{15}$$

unless specifically instructed otherwise.

22.2.2.2.6 *Selection of parameters*

We have assumed thus far that the analyst knows from the outset which parameters are relevant to the analysis. This is often not really the case, of course.

BMDP7M and SPSS-X DISCRIMINANT provide methods for determining which parameters to include in an LDA. SAS STEPDISC[11] also does this. These methods work by adding or deleting parameters one at a time according to one of several criteria.

One measure of the separating ability of a set of parameters is Wilks' lambda test. Formally, it stands outside of LDA, but it is based on the same assumptions 1 and 2 above. Wilks' lambda provides a test of the hypothesis that in the space made up from the specified parameters, all r group centroids coincide. A low significance probability (p-value) for this test tends to imply that at least some of the centroids differ, suggesting that LDA with this set of parameters may be successful. When $r = 2$, Wilks' test is somewhat similar to CSA (discussed below), for both yield p-values for differences between groups, using a set of parameters all together. BMDP7M, DISCRIMINANT and STEPDISC all give Wilks' test.*

Given a set of parameters, it is possible directly to test the extra separating power contributed by an additional parameter. One way to do this involves an analysis-of-covariance method;[12] this is a valid method within the framework of the LDA assumptions. The existence of such an associated 'partial' test is an advantage of LDA.

22.2.2.2.7 *Importance of the LDA assumptions*

The assumptions 1 and 2 are restrictive and it is problematic in any given case whether they will be met. Lachenbruch[13] has summarized available information on the effects on LDA of violation of these assumptions. Violation of multivariate normality (assumption 1) can increase greatly the overall misclassification probability. Violation of identical covariance matrices (assumption 2) may not have much effect if 'the matrices are not too different', but LDA 'can be considerably affected' if they differ greatly.

Despite the intuitive appeal and apparent utility of the LDA method, these considerations ought perhaps to sound a cautionary note to the potential user. Some may also be a bit uneasy over the need to provide prior probabilities.

If assumption 1 holds but 2 does not, the quadratic discriminant function[14] is appropriate; however, it does not seem often to have been used in the QSAR literature. Programs which can perform discriminant analysis using separate covariance matrices for each group are not necessarily performing quadratic discrimination.[15]

We observe that the 'linearity' of LDA (in reality the linearity of equation 12) actually stems from the distributional assumptions 1 and 2. It is not a separate assumption.

22.2.2.2.8 *Misclassification probabilities*

Lachenbruch[16] has reviewed available knowledge on estimation of misclassification rates. Most programs provide a table giving the classification results for the training set based on the LDA generated from that set. Error rates estimated in this way tend to be too low; the 'bias . . . can be bad for very small samples', although 'for large samples [this method] is quite satisfactory'. This bias can virtually be removed by classifying each compound in the training set from a separate LDA generated from the training set but without that compound. Results of this so-called 'jack-knifed' classification are given automatically by BMDP7M, in addition to the ordinary classification table.

---

* For $r = 2$, Mahalanobis' D^2 or Hotelling's T^2 provide tests equivalent to Wilks'. SPSS-X DISCRIMINANT can furnish an assortment of Wilks' lambda values, a situation which might be a bit confusing. For a particular set of parameters, what is wanted (as with the other two programs) is the lambda test for that set. In DISCRIMINANT, this is found at the head of the output for the step in question (if 'stepping'); it appears with an 'equivalent F' and a significance probability is given. If step output is not obtained ('direct-entry method'), an essentially equivalent test is available under the heading 'Canonical Discriminant Functions'. The row labelled 'After Function 0' must be used; a chi-squared appears instead of an 'equivalent F' and a significance probability is again provided.

22.2.2.3 Applications

22.2.2.3.1 *Bitter/sweet aldoximes*

A good example of the use of LDA comes from the work of Kier[17] in which the two response classes are related to the taste receptor. Taking data from Acton and Stone,[18] Kier was able to separate bitter-tasting aldoximes (R—CH=N—OH) from the sweet-tasting ones by using molecular connectivity indices calculated from the structures. A typical sweet-tasting aldoxime would be cyclohex-1-enecarboxaldehyde oxime, while a typical bitter-tasting one would be bicyclo[3.2.2]nona-6,8-diene-6-carboxaldehyde oxime. Figure 4 is a plot of the data used. The axes are $^1\chi$ and $^4\chi_p$. These parameters are described in greater detail elsewhere,[19] but they represent essentially the size of the molecule in terms of the number of bonds, and the size of the molecule as well as the substituents, respectively.

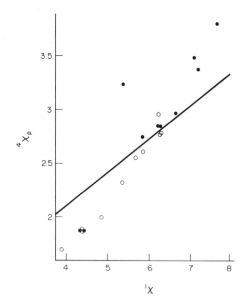

Figure 4   Sweet and bitter aldoximes separated by the molecular connectivity indices $^1\chi$ and $^4\chi_p$: sweet = ○; bitter = ●
(reproduced from ref. 17 with permission of the American Pharmaceutical Association)

As can be seen from Figure 4, there is a good separation between the bitter and sweet molecules. As obtained by Kier, the discriminant function is

$$1.21\ ^1\chi\ -\ 3.88\ ^4\chi_p \tag{16}$$

with boundary value -3.27, *i.e.* when the value of the expression (equation 16) exceeds -3.27 the related molecule is classified as sweet, and when less than -3.27 as bitter. The function seems to classify the two types of responses fairly accurately. Of the 20 compounds considered only one sweet and two bitter aldoximes were misclassified, but these are not 'jack-knifed' error rate estimates (see above). When equation (16) was applied to nine new (non-training) compounds from the Acton–Stone set, seven were correctly predicted as to the response category, one was predicted incorrectly, and one was on the boundary.

From these findings, Kier was able to make several significant deductions concerning the structure–taste relationships: (1) the larger, more branched substituents on the aldoximes increase the tendency for bitterness; (2) smaller substituents, with four or fewer first row atoms, increase the tendency towards sweetness; and (3) unsaturation or heteroatoms in the structure have no significant influence on the taste category to which the compounds belong.

22.2.2.3.2 *Table of LDA applications*

LDA has been applied to a number of medicinal chemistry problems. Most of those in the literature involve specific complicating factors, and as such cannot be summarized with full justice in the space allotted here. For the interested reader we have provided a selection of relevant applications in Table 1.

Table 1   Selected Applications of Linear Discriminant Analysis

Class of compounds	Biological activity	Ref.
Aminotetralins	Monamine oxidase inhibition	1
Isothiosemicarbazones	Virostatic	20
Triazenes	Antitumor	21
Pyrimido[5,4-*d*]pyrimidines	Antiviral	22
Naphthoquinones	Antitumor	23
o-Toluenesulfonylureas	Hypo- and hyper-glycemic	24
Pyrimidines	Antibacterial	25
Amines	5 therapeutic categories	26
Quinazolines	Thymidylate synthetase inhibition	27
Steroids	5 therapeutic categories	28
Benzoguanamines	Antiulcer	29
Phenylacetic acids	Antiinflammatory	29
Aminouracils	Antiinflammatory	29
N-Aryl-4-quinolinamines	Analgesic	30
2-Anilinopyrimidines	Fungicidal	31
Quinolines	Antiviral	32
Nitrobenzenes	Musk odor	33
Various types	Antineoplastic	34

22.2.3   CLUSTER SIGNIFICANCE ANALYSIS

22.2.3.1   A Graphical Overview

CSA is a general statistical method, but it was inspired by a procedure for using graphics as a means to determine those physical parameters that influence biological activity. That procedure is called 'parameter focusing', a concept developed by Magee.[35] It is concerned with a congeneric series of compounds for which there are numeric physical data, but for which the biological results fall only into one of two classes, for example 'active' and 'inactive'.

'Parameter focusing' works as follows. A graph such as Figure 5 is constructed as described in Section 22.2.2.1. When the 'active' compounds are clustered in a relatively confined region of the graph this group is said to be 'focused', and the graph's parameters (or at least one of them) are thus identified as likely predictors of biological activity.

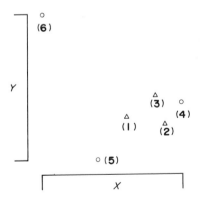

Figure 5   A two-dimensional plot of the six members of a hypothetical series of biologically active compounds: the active members (△); the inactives (○) (reprinted from ref. 2 with permission of the American Chemical Society)

Why is this so? Our expectation would be that if the physical parameters of interest are *not* related to the biological activity then the 'actives' would be scattered randomly about the graph. Therefore, when the 'actives' do cluster we must infer that the parameters in fact influence activity; 'focused' clusters signal non-random events. The trick is to judge when a group is 'focused'. To put it another way: how can we know that what appears to be a 'focused' cluster did not arise merely by chance?

Some experts[36] imply that 'experience' is a good guide to recognizing significant patterns in graphically displayed data, but in our view this is begging the question. What is needed is an objective method for gauging the reality of an apparent cluster. However, prior to the introduction of CSA there was not, to our knowledge, a standard mathematical technique that addressed this specific problem. The present section of this chapter will discuss the CSA solution. Its principles are simple, but it is computation-intensive, and thus requires a high-speed computer.

We begin, however, with an example so elementary that elaborate calculation is not needed. In Figure 5 there are six compounds: three actives and three inactives. The actives appear to be 'focused', but this tendency will be shown to fall short of statistical significance at the usual 0.05 probability level.

The active set is three in number. Under the assumption (the 'null hypothesis') that activity is unrelated to position in parameter space, any set of three of the compounds may be viewed as equally likely to have formed an 'active set' as any other. There are 20 combinations in which these six compounds can be taken three at a time. Compounds (1)–(3) comprise the active group actually obtained. Only one set of three is more tightly clustered: compounds (2)–(4). The groups composed of compounds (1), (2) and (4), and (1), (3) and (4) are close in size to the active group, but are nevertheless larger. All other groups include compounds (5) and/or (6) and are therefore much more loosely clustered. Hence, including the active group, exactly two sets of size three, (1)–(3) and (2)–(4), are at least as tightly clustered as the actual active set. The chance, then, that confinement as close or closer than that observed would have arisen randomly may be put at 2/20 or 0.10.

Cases this simple, of course, rarely occur; commonly more compounds are involved and many distance calculations are needed. For these, computer-based algorithms are needed.

22.2.3.2 The Mathematical Apparatus

22.2.3.2.1 *Calculating the probability*

Still considering the hypothetical case in Figure 5, let us see how we may put the above thoughts on a general basis. What is needed is a suitable definition of the tightness of a cluster. To this end, for any candidate 'active' group the mean squared distance (MSD) among its three compounds may be calculated by taking the squared distance between each pair of points in the group and then dividing the sum of the squared distances by the number of pairs

$$\text{Total squared distance} = (x_1 - x_2)^2 + (y_1 - y_2)^2 + (x_1 - x_3)^2 + (y_1 - y_3)^2 + (x_2 - x_3)^2$$
$$+ (y_2 - y_3)^2 \tag{17}$$

$$\text{MSD} = (\text{total squared distance})/3 \tag{18}$$

This MSD is an index of tightness. Other definitions are conceivable—for instance the mean of the ordinary distances. The MSD, however, seems to us preferable to the latter because it puts greater weight on outliers. The MSD also lends itself to considerable savings in computation as shown in the appendix of our original paper.[2]

As stated, if X and Y play no role in determining activity, the observed active cluster in Figure 5 would be a chance aggregation, and all other possible clusters of the same size would be as likely to have arisen as the observed one. We have seen that there are 20 combinations of size three. Once the MSD for the active cluster has been computed, the MSDs for all combinations are calculated in the same way, and are compared to that of the active group. The number of groups (including the active one itself) which have MSDs equal to or less than the MSD of the active group is designated as A. The probability (p) that a cluster at least as tight as the one observed would have arisen by chance alone then is given by

$$p = A/20 \tag{19}$$

or in the actual case

$$p = 2/20 = 0.10 \tag{20}$$

This significance probability or p-value thus indicates the significance of the evidence that there is a relationship of activity to X and/or Y. It has the same interpretation as, for example, the p-value of an F-test in multiple regression, because it gives the probability that results *at least as suggestive* of relationship as those actually obtained would have occurred by chance alone. As always, the lower the p-value the less tenable the chance explanation.

Dimensions other than two are easily treated; one has only to modify the definition of the squared distance appropriately. Thus, in the one-dimensional case, the terms containing y in equation (17) are dropped. For the three-dimensional situation, corresponding terms for Z (a third parameter) are added to the equation. Higher dimensions are treated in the obvious way. Series containing greater numbers of compounds are more burdensome in terms of calculations, but the method is the same.

We do, however, recommend that, prior to calculating MSDs, the parameters be 'autoscaled', *i.e.* each parameter's values have their mean subtracted followed by division by their standard deviation. This will tend to equalize the influence of the various parameters in the distance calculations, adjusting for any differences in units or range.

Helpful details on the computations involved are given in the appendix of the original paper.[2] However, rather than repeat that information, we give actual FORTRAN computer programs in Section 22.2.3.4. There are two: CSA1 and CSA2. The former is used in situations where it is practical to make a complete enumeration of all clusters containing the same number of compounds as the active group. CSA2 is for cases where this would be too time consuming; it uses random sampling of the set of all possible clusters. A case using CSA1 is given in the first example below (Section 22.2.3.3.1). The second example requires the use of CSA2 for its various solutions (see Section 22.2.3.3.2).

22.2.3.2.2 *Predicting the class of new members*

While the above discussion gives details on how the significance of an 'active' cluster is determined, predictions of activity are made by observing the location of new compounds, either real or projected, on the graph. Thus, a judgment must be made whether a new point is close enough to the 'active' region to be considered a member of that class. This will be easier to do in some situations than in others. But in all cases it can be done with more confidence once the genuineness of the active clustering has been established; this is the utility of CSA within the graphical context.

Thus unlike LDA, CSA does not provide an objective rule for classifying new compounds. On the other hand, CSA does not require the specialized assumptions of LDA, and hence will be more broadly applicable. In fact CSA would apply to any situation in which it is desired to test whether an apparent spatial localization of a distinguished subset, with respect to the region occupied by the full set of observations, is real.

22.2.3.3 Applications

22.2.3.3.1 *Lasalocid derivatives*

Westley *et al.*[37] prepared derivatives of the ionophore lasalocid and evaluated their antibacterial activities. To better understand the structure–activity relationships in this series they also measured the pK_a's and partition coefficients (P) of selected compounds. (However, in the ensuing calculations we will use log P.) The relevant data are presented in Table 2. In commenting later on this group of compounds, Westley[38] considered separately the effects that $pK_{a'}$ and P had on antibacterial activity. He asserted 'that any deviation in the acidity of the carboxyl group in lasalocid has a detrimental effect on the antibacterial activity of the antibiotic'. This is a perfectly reasonable suggestion, but CSA now offers the possibility of testing it for statistical significance.

Figure 6 displays in one dimension the relationship between antibacterial activity and $pK_{a'}$. The more active members are clustered in the middle, but there are also some less active ones in this region. Therefore, the same degree of acidity as lasalocid does not guarantee good activity. Application of CSA to this set of data shows that such a clustering could have arisen under pure chance with a fairly high probability ($p = 0.143$). Therefore, a connection between $pK_{a'}$ and activity is not confirmed by the present data. To be sure it also has not been eliminated, but a proposal to prepare derivatives that are substantially more or less acidic than lasalocid should not be discouraged.

In assessing the influence of P on activity, Westley[38] concluded that there was at least some relationship: that those derivatives with P approximately the same or up to twice as large as that of lasalocid were likely to be the more active ones. The data for this relationship are shown in Figure 7. Here, the application of CSA supports his conjecture ($p = 0.002$).

We should also consider the possibility that there is a joint influence of $pK_{a'}$ and log P upon activity. In this case, however, there are measurements of log P and $pK_{a'}$ common to only nine of the compounds. The graphical representation of these data in two dimensions is given in Figure 8. The indicated probability ($p = 0.008$) would be considered significant normally, but in this case we note

Table 2 The Apparent $pK_{a'}$, Logarithm of the Partition Coefficient (log P), and Relative *in vitro* Antibacterial Activity of Lasalocid and Some of its Derivatives[a]

Compound	$pK_{a'}$	log P	in vitro activity[b]
Lasalocid	4.4	2.83	1
Bromolasalocid	3.9	3.25	1
Chlorolasalocid	4.0	3.12	1
Iodolasalocid	3.9	3.19	1
Nitrolasalocid	2.4	2.46	0
Aminolasalocid	6.0	0.28	0
N-Acetylaminolasalocid	4.3	1.04	0
Lasalocid acetate	4.15	0.95	0
Bromolasalocid acetate	4.25	1.49	0
Diazolasalocid	6.6[c]		0
Benzylideneaminolasalocid		0.86	0
Lasalocid methyl ether		2.54	0
Lasalocid pentanoate		2.08	0
Lasalocid octanoate		2.21	0
Lasalocid decanoate		2.75	0
Lasalocid bromobenzoate		3.48	0

[a] Adapted from data presented in the literature (see ref. 37). Compounds for which there is neither $pK_{a'}$ nor log P data are omitted. [b] Activity = 1 where antibacterial potency is greater than 50% of that of lasalocid (active), and activity = 0 where it is less (inactive). There is a reasonably large gap between the potencies of the compounds so distinguished. [c] This $pK_{a'}$ value taken from ref. 38.

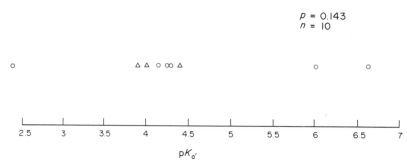

Figure 6 Lasalocid derivatives: antibacterial activity as a function of $pK_{a'}$; active compounds (△) and inactives (○). The triangle at $pK_{a'} = 3.9$ represents two data points (reproduced from ref. 39 with permission of the American Chemical Society)

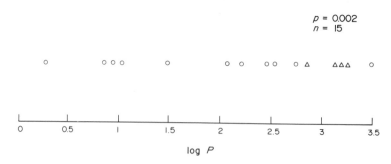

Figure 7 Lasalocid derivatives: antibacterial activity as a function of log P (reproduced from ref. 39 with permission of the American Chemical Society)

that this is a *higher* probability than that given by log P separately, *i.e.* two parameters operating together give a weaker result than one alone. In our previous work[2,39] we suggested that in situations of this kind, as a rough rule of thumb, where one parameter alone (but not the other) results in a lower probability than the two together, the parameter that results in the lower probability could be considered as contributory and the other possibly spurious. Only those parameters that together result in a probability lower than either one separately might be said to have a significant *joint* effect on activity. This guiding rule can extend to higher numbers of interacting parameters.

Further details of the CSA approach to this problem are presented elsewhere.[39]

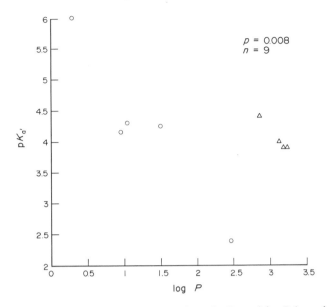

Figure 8 Lasalocid derivatives: antibacterial activity as a function of $pK_{a'}$ and $\log P$ (reproduced from ref. 39 with permission of the American Chemical Society)

22.2.3.3.2 *Mutagenicity among aminoacridines*

The relationship between frameshift mutagenicity and the DNA-binding affinity of some amino-acridine derivatives was reported in 1981 by Ferguson and Baguley.[40] Their data are displayed in Figure 9. The general structure of the compounds involved, some 32 of them, is given by structure (**7**), whilst structure (**8**) is of a specific compound, the 4-carboxamide analog, which we identify as an outlier. A filled triangle designates this compound in Figures 10 and 11. We argue that this amide differs from the other compounds because the substituent group is located favorably and is suitably constituted to interact with DNA by hydrogen bonding. None of the others is so endowed.

Figure 9 differs from the others in this series in that only the horizontal dimension represents a physical property, the DNA binding constant K, with the vertical dimension presenting the maximum reversion frequency which is the biological response. It is obvious that there are two

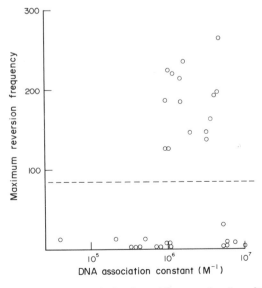

Figure 9 Maximum reversion frequencies (Ames test) of aminoacridines as a function of their DNA association constants. This graph differs from the others in that it consists of a physical property in one dimension and a measure of biological activity in the other (reproduced from ref. 40 with permission of Elsevier Scientific Publishing Company)

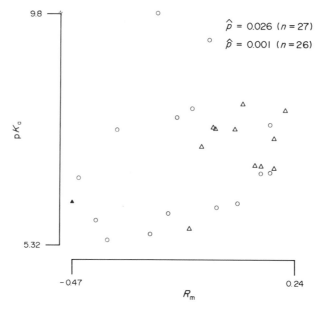

(7)   (8) = ▲

distinct groups of mutagens: the weak ones along the horizontal axis and the strong ones in the cloud of points above the dotted line. It is also clear from these data that the strong mutagens are not likely to be correlated with the DNA binding constant by regression techniques. If the points are visualized as projected onto the horizontal axis, then there is an obvious 'focused' cluster of actives in one-dimensional space (log K space).

The inactives extend on either side, and the two classes mingle only to a minor extent in the lower boundary region. CSA (using the CSA2 FORTRAN program) shows this clustering to be very highly significant ($p = 0.00003 \pm 0.00002$). As stated, the CSA2 algorithm involves random sampling; the expression for p gives 95% confidence limits incorporating the sampling uncertainty.

While this correlation is perfectly satisfactory for predicting the mutagenicity of other congeneric aminoacridines, we wanted to ask a new question: how is mutagenicity related to the more commonly measured or calculated physical constants used in QSAR? The answer to this would give us a better idea of the component forces associated with aminoacridine DNA binding. For these compounds R_m hydrophobicity constants and pK_a's have been reported by Ferguson and Denny.[41] When the biological data are plotted using these new parameters, further interesting clusters are observed.

Figure 10 shows the data plotted in the dimensions of R_m and pK_a. Note the previously mentioned outlier (filled triangle) at the far left. There is another a little closer in, but because we could discern no objective reason to exclude it, it was included in all probability calculations. The first probability value at the top of the graph is the p-value (estimated from sampling) with the first outlier included. The lower value is the result without it. As can be seen there is a great improvement: $p = 0.026 \pm 0.001$ *versus* $p = 0.0013 \pm 0.0003$.

Figure 10 A two-dimensional plot of active (△) and inactive (○) aminoacridine derivatives as frameshift mutagens: the lipophilicity parameter, R_m, *versus* the acidity parameter, pK_a. The filled triangle (▲) represents the 4-$CONH_2$ analog which is excluded as an outlier to give the lower $\hat{p}$ value (0.001) (reproduced from ref. 2 with permission of the American Chemical Society)

We also considered the influence of group dipole moment (μ) on the mutagenicity. Figure 11 shows the data plotted in the dimensions of pK_a (again) and μ. As before, with the outlier included the p-value is much higher than when it is excluded. Here, with the outlier excluded, the evidence of association is even stronger than before: $p = 0.0004 \pm 0.0001$.

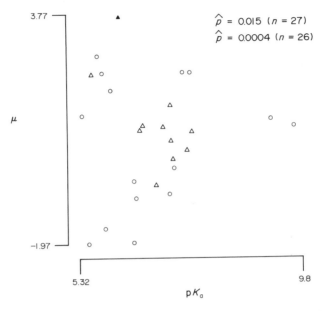

Figure 11 Similar to Figure 10: the acidity parameter, pK_a, *versus* the group dipole moment, μ (reproduced from ref. 2 with permission of the American Chemical Society)

Figure 12 is an attempt to consider simultaneously all three dimensions. For the stereo figure the weak mutagens are represented by the open spheres, the strong mutagens by filled ones. The carboxamide outlier is the filled sphere in the lower left corner. When it is excluded the three dimensions together give the best result: $p = 0.00014 \pm 0.00005$.

What these results suggest is that each of the three parameters appears to play a role in determining mutagenicity, and likely represents an important binding force for the aminoacridine to the DNA. One of the compounds, the carboxamide, behaves differently from all the others, but from its structure we deduced that this is because particular hydrogen bonding interactions play a role as well. Thus, CSA has helped us extract the maximum information from the qualitative biological data, and has given insights into structure–mutagenicity relationships among aminoacridines that have escaped notice previously.

Additional details to the CSA approach to this problem are found in the original literature.[2]

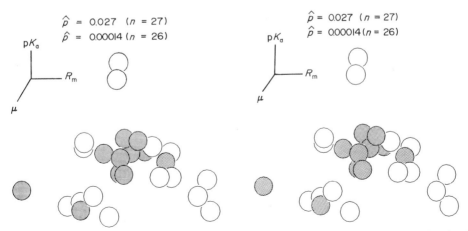

Figure 12 A stereoscopic view combining the data of Figures 10 and 11. Mutagenic aminoacridines (◉); inactives (○). The outlier (4-CONH₂ analog) is in the lower left corner (reproduced from ref. 42 with permission of Elsevier Scientific Publishing Company)

22.2.3.4 The Computer Algorithms

In this section we list the two computer programs, CSA1 and CSA2. The programs have identical initial and final segments, but differ in the middle portions. We present the common initial segment in Section 22.2.3.4.1 and the final in Section 22.2.3.4.4. The middle portion of CSA1 is given in Section 22.2.3.4.2 while that of CSA2 appears in Section 22.2.3.4.3. Thus to implement either program, the appropriate middle section should be appended to the end of the common initial segment, and the common final segment then should be appended to the result.

These programs are written in ANSI FORTRAN 77 and run as is on VAX/750 or 780 computers. On other equipment some modification may be needed; the comments at the beginning and in the body of the listings will be helpful in this case.

If more parameters are available in the input data file than are actually to be used in the analysis, these programs have some limited capability for using·a reduced number: if t are available, the first k can be used, where $k = 1, 2, \ldots$, or t. Choosing any other subset, however, will require the reordering of parameters in the data file itself; the programs do not facilitate such a choice.

22.2.3.4.1 The initial segment for programs CSA1 and CSA2

```
C -- INPUT FILE INFORMATION:

C --    1 RECORD PER COMPOUND FORMATTED AS FOLLOWS --

C --        COLS 1- 20:  ARBITRARY INFORMATION IDENTIFYING COMPOUND; ANY
C --                     CHARACTERS CAN BE PLACED HERE, INCLUDING
C --                     TWENTY BLANKS

C --        REMAINING COLUMNS CONTAIN NUMERIC DATA IN 10-COLUMN FIELDS
C --        (COLS 21-30, 31-40, ETC.).  DATA SHOULD BE RIGHT JUSTIFIED
C --        IN FIELD, WITH ANY MINUS SIGN IMMEDIATELY PRECEDING NUMBER.
C --        (IT IS OPTIONAL WHETHER TO PUNCH DECIMAL FOR INTEGER-VALUED
C --        NUMBER; IF NOT PUNCHED, INTEGER IS ASSUMED)

C --        FIRST FIELD:  NUMBER INDICATING ACTIVITY OR ACTIVITY CATEGORY
C --        (COLS 21-30)    (A HIGHER NUMBER MUST INDICATE GREATER
C --                        ACTIVITY; NUMBERS NEED NOT BE INTEGERS)

C --                       NOTE: FOR CONVENIENCE ANY NUMBER OF POSSIBLE
C --                       ACTIVITY VALUES IS ALLOWED HERE.  AT EXECU-
C --                       TION PROGRAM ASKS FOR 'MINIMUM VALUE FOR
C --                       RESPONSE'; THEN DICHOTOMIZES COMPOUNDS AS
C --                       'RESPONDING' OR 'NON-RESPONDING' USING
C --                       FURNISHED VALUE AS CUTOFF

C --        NEXT FIELDS:  VALUE FOR FIRST PARAMETER AND, OPTIONALLY,
C --        (COLS 31-40,   VALUE FOR SECOND PARAMETER, THIRD, ETC., NOT
C --        41-50, ETC.)  TO EXCEED TEN PARAMETERS

C --    IF THE NUMBER OF PARAMETERS SPECIFIED IN RESPONSE TO QUERY AT
C --    EXECUTION IS LESS THAN NUMBER ON RECORDS, ONLY THE INITIAL
C --    PARAMETERS THROUGH THE NUMBER SPECIFIED WILL BE USED

C --    NOTE THE PROGRAM READS INPUT FILE ON CHANNEL 4 (AT STATEMENT 1);
C --    WRITES OUTPUT FILE ON CHANNEL 21.  ON VAX SYSTEM NO STATE-
C --    MENTS ARE NEEDED DEFINING OR OPENING FILES; ON OTHER SYSTEMS,
C --    SUCH STATEMENTS MAY BE REQUIRED

C -- FORMAT STATEMENTS 801 & 804 USE FREE FORMAT (VAX RULES) FOR
C --    ACCEPTING INFORMATION INTERACTIVELY.  IF SYSTEM EMPLOYED
C --    DOES NOT SUPPORT FREE FORMAT, THESE STATEMENTS WILL NEED
C --    MODIFICATION.  (NOTE THAT IN THE CSA2 PROGRAM, STATEMENT 801
C --    IS USED IN INPUTTING TWO DIFFERENT VARIABLES -- 'K' & 'NSS')

C -- NOTE: IF GET ERROR MESSAGE THAT A MAXIMUM LIMIT HAS BEEN EXCEEDED,
C --    LOCATE STATEMENT RESPONSIBLE FOR MESSAGE.  IT WILL HAVE A COMMENT
C --    OF THE FORM '! CONST = . . . .' AT ITS END.  CHANGE DIMENSIONS
C --    INDICATED IN COMMENT (SEE COMMON & DIMENSION STATEMENTS AT START
C --    OF PROGRAM AND COMMON STATEMENT IN SUBPROGRAM 'ASD').  ALSO
C --    CHANGE TO NEW DIMENSION VALUE THE INTEGER CONSTANT IN THE
C --    STATEMENT RESPONSIBLE FOR MESSAGE & IN THE STATEMENT FOLLOWING IT

        COMMON /LABEL/ I(20),X(10,100),K,LITN,Q(100),CONST4
        DIMENSION ID(100,5),Y(100),IY(100)

C -- ELICIT INFORMATION ON DATA INTERACTIVELY

2       TYPE 800
        ACCEPT 801,K
        IF(K.GT.10)TYPE 802          ! CONST = 1ST DIM OF X(.,.)
```

```
        IF(K.GT.10)GO TO 2
        TYPE 803
        ACCEPT 804,YLIM


C -- READ DATA FROM INPUT FILE, SET I(.) FOR 'RESPONDING' COMPOUNDS


        JJ=1
        INDRES=1

1       READ(4,810,END=9)(ID(JJ,LZ),LZ=1,5),Y(JJ),(X(II,JJ),II=1,K)

        IF(Y(JJ).LT.YLIM)GO TO 5

          IF(INDRES.GT.20)TYPE 806    ! CONST = DIM OF I(.)
          IF(INDRES.GT.20)STOP
          I(INDRES)=JJ
          INDRES=INDRES+1

5       JJ=JJ+1

          IF(JJ.GT.100)TYPE 805       ! CONST = DIM OF Q(.), Y(.), IY(.),
          IF(JJ.GT.100)GO TO 9        ! 1ST DIM OF ID(.,.), & 2ND DIM OF
                                      ! X(.,.)
        GO TO 1

9       N=JJ-1
        LITN=INDRES-1

        TYPE 815,N,LITN

        IF(LITN.LE.1.OR.LITN.GE.N)TYPE 807
        IF(LITN.LE.1.OR.LITN.GE.N)STOP
```

22.2.3.4.2   *The middle section for CSA1*

CSA1 implements the procedure for small data sets. This program computes an exact significance probability (*p*-value) by calculating MSDs for all possible subsets containing the same number of compounds as the 'active' group actually observed.

```
C -- WRITE INPUT DATA, RELATED INFO TO OUTPUT FILE


        WRITE(21,830)

        DO 30 JJ=1,N
30      IY(JJ)=' '

        DO 31 M=1,LITN
31      IY(I(M))='*'

        DO 32 JJ=1,N
32      WRITE(21,831)(ID(JJ,LZ),LZ=1,5),Y(JJ),IY(JJ),(X(II,JJ),II=1,K)

        WRITE(21,832)YLIM
        WRITE(21,815)N,LITN
        WRITE(21,833)


C -- SET CONSTANTS


        CONST1=FLOAT(N)
        CONST2=FLOAT(N-1)
        CONST3=2./FLOAT(LITN-1)
        CONST4=CONST3/FLOAT(LITN)


C -- STANDARDIZE X(II,.)


        DO 10 II=1,K

        XBAR=0.
        DO 11 JJ=1,N
11      XBAR=XBAR+X(II,JJ)
        XBAR=XBAR/CONST1

        SUMSQ=0.
        DO 12 JJ=1,N
12      SUMSQ=SUMSQ+(X(II,JJ)-XBAR)**2
        S=SQRT(SUMSQ/CONST2)

        DO 13 JJ=1,N
13      X(II,JJ)=(X(II,JJ)-XBAR)/S

10      CONTINUE
```

```
C -- COMPUTE Q(JJ)

        DO 20 JJ=1,N

        Q(JJ)=0.

          DO 21 II=1,K
21        Q(JJ)=Q(JJ)+X(II,JJ)**2
20        Q(JJ)=CONST3*Q(JJ)


C -- CALCULATE AVG SQ DIST FOR OBSERVED SUBSET (RESPONDING COMPOUNDS)

        ASDOBS=ASD(DUMMY)


C -- INITIALIZE COUNTERS, I(.)

        NSUBS=0
        NLE=0

        DO 50 M=1,LITN
50      I(M)=M


C -- CALCULATE AVG SQ DIST FOR PRESENT SUBSET, UPDATE COUNTERS

75      NSUBS=NSUBS+1
        IF(ASD(DUMMY).LE.ASDOBS)NLE=NLE+1


C -- STEP TO NEXT SUBSET

        DO 100 J=LITN,1,-1

        IF(I(J).EQ.N-LITN+J)GO TO 100

        I(J)=I(J)+1
        IF(J.EQ.LITN)GO TO 75

        DO 110 M=J+1,LITN
110     I(M)=I(J)+M-J
        GO TO 75

100     CONTINUE


C -- ALL SUBSETS INVESTIGATED; COMPUTE FINAL DATA, OUTPUT

        P=FLOAT(NLE)/FLOAT(NSUBS)
        TYPE 820,NSUBS,NLE,P
        WRITE(21,820)NSUBS,NLE,P
800     FORMAT(/1X,'ENTER NUMBER OF PARAMETERS: ',$)
801     FORMAT(I)
802     FORMAT(/1X,'TOO MANY -- 10 MAX')
803     FORMAT(1X,'ENTER MINIMUM VALUE FOR RESPONSE (MUST USE',
     &   ' DECIMAL POINT): ',$)
804     FORMAT(F)
805     FORMAT(/1X,'*** 100 COMPOUNDS MAX -- USING FIRST 100 ***')
806     FORMAT(/1X,'*** TOO MANY RESPONSES -- 20 MAX -- FATAL ERROR',
     &   ' ***')
807     FORMAT(/1X,'*** NUMBER OF RESPONSES = 0, 1, OR NUMBER COMP',
     &   'OUNDS -- CANNOT'/1X,'COMPUTE P-VALUE -- FATAL ERROR ***')

810     FORMAT(5A4,11F10.0)

815     FORMAT(/1X,'NUMBER OF COMPOUNDS = ',I5
     &   /1X,'NUMBER OF RESPONSES = ',I5)
820     FORMAT(/1X,'TOTAL SUBSETS = ',28X,I10/1X,'SUBSETS AT LEAST AS',
     &   ' TIGHTLY CLUSTERED = ',4X,I10/1X,'P-VALUE = ',34X,F10.6/)

830     FORMAT(1X,'CLUSTER SIGNIFICANCE ANALYSIS (CSA) -- TOTAL ENUMER',
     &   'ATION'//1X,'COMPOUND ID',11X,'ACTIVITY',5X,'RAW VALUES FOR PA',
     &   'RAMETERS USED IN THE ANALYSIS . . . .'/)
```

22.2.3.4.3 *The middle section for CSA2*

For large data sets an exact computation of the significance probability can be too time-consuming even with modern computers. CSA2 estimates the significance probability by taking a random sample from the collection of all possible subsets of the appropriate size.

For its success this method depends on the random number generator used. All generators are not alike; we advise that you read with care all the annotations pertaining to this aspect of the program. If you are not familiar with the statistical and programming aspects of your random number generator (including its initialization), we suggest that you obtain the help of a local computer scientist in implementing CSA2.

```
C -- ELICIT SAMPLE SIZE INTERACTIVELY

        TYPE 808
        ACCEPT 801,NSS


C -- OBTAIN INITIAL SEED FOR RANDOM NUMBER GENERATOR INTERACTIVELY

        TYPE 809
        ACCEPT 801,ISEED


C -- WRITE INPUT DATA, RELATED INFO TO OUTPUT FILE

        WRITE(21,830)

        DO 30 JJ=1,N
30      IY(JJ)=' '

        DO 31 M=1,LITN
31      IY(I(M))='*'

        DO 32 JJ=1,N
32      WRITE(21,831)(ID(JJ,LZ),LZ=1,5),Y(JJ),IY(JJ),(X(II,JJ),II=1,K)

        WRITE(21,832)YLIM
        WRITE(21,815)N,LITN
        WRITE(21,833)


C -- SET CONSTANTS

        CONST1=FLOAT(N)
        CONST2=FLOAT(N-1)
        CONST3=2./FLOAT(LITN-1)
        CONST4=CONST3/FLOAT(LITN)
        CONST5=FLOAT(LITN)


C -- STANDARDIZE X(II,.)

        DO 10 II=1,K

        XBAR=0.
        DO 11 JJ=1,N
11      XBAR=XBAR+X(II,JJ)
        XBAR=XBAR/CONST1

        SUMSQ=0.
        DO 12 JJ=1,N
12      SUMSQ=SUMSQ+(X(II,JJ)-XBAR)**2
        S=SQRT(SUMSQ/CONST2)

        DO 13 JJ=1,N
13      X(II,JJ)=(X(II,JJ)-XBAR)/S

10      CONTINUE


C -- COMPUTE Q(JJ)

        DO 20 JJ=1,N
        Q(JJ)=0.

        DO 21 II=1,K
21      Q(JJ)=Q(JJ)+X(II,JJ)**2

20      Q(JJ)=CONST3*Q(JJ)


C -- CALCULATE AVG SQ DIST FOR OBSERVED SUBSET (RESPONDING COMPOUNDS)

        ASDOBS=ASD(DUMMY)


C -- INITIALIZE COUNTERS

        NSUBS=0
        NLE=0
```

```
C -- CHOOSE A RANDOM SUBSET

50        M=1
          F1=CONST5
          F2=CONST1

          DO 60 JJ=1,N

          IF(M.GT.LITN)GO TO 75          ! IF UNCERTAINTY IN SELECTION ENDED,
          IF(M.EQ.LITN-N+JJ)GO TO 55     ! AVOID FURTHER TESTS USING RAN

C --      STATEMENT BELOW CALLS 'RAN', THE VAX SYSTEM RANDOM NUMBER
C --      GENERATOR, WHICH, ON SUCCESSIVE CALLS, RETURNS PSEUDO-RANDOM
C --      NUMBERS INDEPENDENTLY AND UNIFORMLY DISTRIBUTED ON THE
C --      INTERVAL FROM 0 TO 1.  ON OTHER THAN VAX SYSTEMS, CARE
C --      MUST BE TAKEN THAT THE GENERATING ROUTINE USED IN PLACE OF
C --      'RAN' HAS THESE STATISTICAL PROPERTIES

          IF(RAN(ISEED).GT.F1/F2)GO TO 60

55           I(M)=JJ
             M=M+1
             F1=F1-1.

60        F2=F2-1.


C -- CALCULATE AVG SQ DIST FOR RANDOM SUBSET, UPDATE COUNTERS

75        NSUBS=NSUBS+1
          IF(ASD(DUMMY).LE.ASDOBS)NLE=NLE+1


C -- CHECK WHETHER SAMPLE SIZED REACHED

          IF(NSUBS.LT.NSS)GO TO 50


C -- ALL SUBSETS INVESTIGATED; COMPUTE FINAL DATA, OUTPUT

          P=FLOAT(NLE)/FLOAT(NSUBS)
          PLORMI=1.96*SQRT(P*(1.-P)/FLOAT(NSUBS))
          TYPE 820,NSUBS,NLE,P,PLORMI
          WRITE(21,820)NSUBS,NLE,P,PLORMI
800       FORMAT(/1X,'ENTER NUMBER OF PARAMETERS: ',$)
801       FORMAT(I)
802       FORMAT(/1X,'TOO MANY -- 10 MAX')
803       FORMAT(1X,'ENTER MINIMUM VALUE FOR RESPONSE (MUST USE',
     &    ' DECIMAL POINT): ',$)
804       FORMAT(F)
805       FORMAT(/1X,'*** 100 COMPOUNDS MAX -- USING FIRST 100 ***')
806       FORMAT(/1X,'*** TOO MANY RESPONSES -- 20 MAX -- FATAL ERROR',
     &    ' ***')
807       FORMAT(/1X,'*** NUMBER OF RESPONSES = 0, 1, OR NUMBER COMP',
     &    'OUNDS -- CANNOT'/1X,'COMPUTE P-VALUE -- FATAL ERROR ***')
808       FORMAT(/1X,'ENTER NUMBER OF SUBSETS TO BE SAMPLED: ',$)
809       FORMAT(1X,'ENTER ODD INTEGER AS RANDOM SEED (9 DIGITS MAX): ',$)

810       FORMAT(5A4,11F10.0)

815       FORMAT(/1X,'NUMBER OF COMPOUNDS = ',I5
     &    /1X,'NUMBER OF RESPONSES = ',I5)
820       FORMAT(/1X,'SUBSETS SAMPLED = ',30X,I10/1X,'SUBSETS AT LEAST',
     &    ' AS TIGHTLY CLUSTERED = ',8X,I10//1X,'ESTIMATED P-VALUE = ',
     &    28X,F10.6/1X,'UNCERTAINTY IN ESTIMATE (95% CONFIDENCE) = ',5X,
     &    F10.6/)

830       FORMAT(1X,'CLUSTER SIGNIFICANCE ANALYSIS (CSA) -- RANDOM SAMP',
     &    'LING'//1X,'COMPOUND ID',11X,'ACTIVITY',5X,'RAW VALUES FOR PA',
     &    'RAMETERS USED IN THE ANALYSIS . . . .'/)
```

22.2.3.4.4   *The final segment for programs CSA1 and CSA2*

```
831       FORMAT(1X,5A4,F10.4,A1,10F10.4)
832       FORMAT(/1X,'* INDICATES RESPONSE (ACTIVITY AT LEAST = ',F11.4,
     &    ')')
833       FORMAT(/1X,130('-'))

          END


          FUNCTION ASD(DUMMY)


C -- COMPUTES AVG SQ DIST FOR ARBITRARY SUBSET OF SIZE LITN CONSISTING
C --    OF COMPOUNDS I(1), . . . , I(LITN)

C -- NOTE: 'DUMMY' IS A DUMMY ARGUMENT AND IS NOT USED
```

```
      COMMON /LABEL/ I(20),X(10,100),K,LITN,Q(100),CONST4

      SUM1=0.
      DO 1 M=1,LITN
1     SUM1=SUM1+Q(I(M))

      SUM2=0.
      DO 2 II=1,K

       T=0.
       DO 3 M=1,LITN
3      T=T+X(II,I(M))

2     SUM2=SUM2+T**2

      ASD=SUM1-CONST4*SUM2
      RETURN

      END
```

22.2.3.4.5 Test set for CSA1 and CSA2

Once you have CSA1 and/or CSA2 running on your system, it is a good idea to verify them by running them on the hypothetical data provided in Table 3. CSA1 should yield precisely the same output as furnished in Table 4.

The format of the output from CSA2 is given in Table 5. However, for a given run the last three lines of output will not match exactly those of the table, because the random-sampling scheme is subject to sampling variability. Nonetheless, the estimated p-value, within its presented uncertainty, should be consistent with the correct exact value of 0.7 on approximately 95% of runs.

Table 3 Sample Input Data for CSA1 and CSA2[b]

		Parameter values[a]			
Compound ID	Activity value	# 1	# 2	# 3	# 4
H	4.4	1.8	8.9	7.3	−12.6
CH3	2.2	0.7	8.1	2.7	−9.1
CH2F	3.3	4.9	−5.3	5.1	16.4
CH2Cl	5.5	3.6	11.6	6.2	−2.3
CH2Br	1.1	2.1	1.7	3.8	−4.1

[a] These should be raw values. The programs autoscale them automatically. [b] Note: Only the five data records should actually be in the input file (the heading and other text lines should be omitted). See comments in the initial program segment listing (Section 22.2.3.4.1) for required spacing and format of fields within data records.

Table 4 Output for Sample Data—CSA1

CLUSTER SIGNIFICANCE ANALYSIS (CSA)—TOTAL ENUMERATION

COMPOUND ID	ACTIVITY	RAW VALUES FOR PARAMETERS USED . . .[a]			
H	4.4000*	1.8000	8.9000	7.3000	−12.6000
CH3	2.2000	0.7000	8.1000	2.7000	−9.1000
CH2F	3.3000*	4.9000	−5.3000	5.1000	16.4000
CH2Cl	5.5000*	3.6000	11.6000	6.2000	−2.3000
CH2Br	1.1000	2.1000	1.7000	3.8000	−4.1000

* INDICATES RESPONSE (ACTIVITY AT LEAST = 3.0000)

NUMBER OF COMPOUNDS = 5
NUMBER OF RESPONSES = 3

TOTAL SUBSETS =	10
SUBSETS AT LEAST AS TIGHTLY CLUSTERED =	7
P-VALUE =	0.700000

[a] Note: if not all parameters used in run (see comments in initial program segment listing), only values for those used will be displayed here.

Because there are only a small number of possible subsets with this hypothetical data, CSA2 will do a great deal of repeated sampling in this example. Repetition of the same subset in the sample is allowed in general,[2] although in practice one will have far less of it than with this small data set.

Table 5   Output for Sample Data—CSA2

CLUSTER SIGNIFICANCE ANALYSIS (CSA)—RANDOM SAMPLING

COMPOUND ID	ACTIVITY	RAW VALUES FOR PARAMETERS USED . . .[a]			
H	4.4000*	1.8000	8.9000	7.3000	−12.6000
CH3	2.2000	0.7000	8.1000	2.7000	−9.1000
CH2F	3.3000*	4.9000	−5.3000	5.1000	16.4000
CH2Cl	5.5000*	3.6000	11.6000	6.2000	−2.3000
CH2Br	1.1000	2.1000	1.7000	3.8000	−4.1000

* INDICATES RESPONSE (ACTIVITY AT LEAST = 3.0000)

NUMBER OF COMPOUNDS = 5
NUMBER OF RESPONSES = 3

SUBSETS SAMPLED =	50000
SUBSETS AT LEAST AS TIGHTLY CLUSTERED =	34997
ESTIMATED P-VALUE =	0.699940
UNCERTAINTY IN ESTIMATE (95% CONFIDENCE) =	0.004017

[a] Note at bottom of Table 4 applies here as well.

22.2.4   ADDENDUM

Applications of CSA1 to antihypertensive prazosin analogs and to anticoccidal acridinediones have recently been described.[43]

22.2.5   REFERENCES

1. Y. C. Martin, J. B. Holland, C. H. Jarboe and N. Plotnikoff, *J. Med. Chem.*, 1974, **17**, 409.
2. J. W. McFarland and D. J. Gans, *J. Med. Chem.*, 1986, **29**, 505.
3. C. Hansch, S. H. Unger and A. B. Forsythe, *J. Med. Chem.*, 1973, **16**, 1217.
4. T. W. Anderson, 'An Introduction to Multivariate Statistical Analysis', Wiley, New York, 1958, pp. 126–153.
5. M. G. Kendall, 'Multivariate Analysis', Griffin, London, 1975, pp. 145–169.
6. W. J. Dixon (ed.), 'BMDP Statistical Software', 1985 Printing, University of California Press, Berkeley, pp. 519–537.
7. SPSS Inc., 'SPSS-X User's Guide', 2nd edn., SPSS Inc., Chicago, 1986, pp. 688–712.
8. SAS Institute Inc., 'SAS User's Guide: Statistics, Version 5 Edition', SAS Institute Inc., Cary, NC, 1985, pp. 317–333.
9. D. F. Morrison, 'Multivariate Statistical Methods', McGraw-Hill, New York, 1967.
10. P. A. Lachenbruch, 'Discriminant Analysis' in 'Encyclopedia of Statistical Sciences', ed. S. Kotz and N. L. Johnson, Wiley, New York, vol. 2, 1982, pp. 389–397, specifically p. 390.
11. See ref. 8, pp. 749–762.
12. See ref. 6, p. 520.
13. P. A. Lachenbruch, ref. 10, p. 393.
14. See ref. 10; p. 390.
15. See ref. 7, p. 706.
16. P. A. Lachenbruch, ref. 10, p. 392.
17. L. B. Kier, *J. Pharm. Sci.*, 1980, **69**, 416.
18. E. M. Acton and H. Stone, *Science (Washington, D.C.)*, 1976, **193**, 584.
19. L. B. Kier and L. H. Hall, 'Molecular Connectivity in Chemistry and Drug Research', Academic Press, New York, 1976.
20. R. Franke and W. Meisske, *Acta Biol. Med. Ger.*, 1976, **35**, 73.
21. W. J. Dunn, III and M. J. Greenberg, *J. Pharm. Sci.*, 1977, **66**, 1416.
22. M. Tonew, W. Laass, E. Tonew, R. Franke, H. Goldner and W. Zschiesche, *Acta Virol. (Engl. Ed.)*, 1978, **22**, 287.
23. G. Prakash and E. M. Hodnett, *J. Med. Chem.*, 1978, **21**, 369.
24. S. Dove, R. Franke, O.L. Mndshojan, W. A. Schkuljev and L. W. Chashakjan, *J. Med. Chem.*, 1979, **22**, 90.
25. C. C. Smith, C. S. Genther and E. A. Coats, *Eur. J. Med. Chem.-Chim. Ther.*, 1979, **14**, 271.
26. D. R. Henry and J. H. Block, *J. Med. Chem.*, 1979, **22**, 465.
27. B.-K. Chen, C. Horvath and J. R. Bertino, *J. Med. Chem.*, 1979, **22**, 483.
28. D. R. Henry and J. H. Block, *Eur. J. Med. Chem.-Chim. Ther.*, 1980, **15**, 133.
29. A. Ogino, S. Matsumura and T. Fujita, *J. Med. Chem.*, 1980, **23**, 437.
30. P. Broto, G. Moreau and C. Vandycke, *Eur. J. Med. Chem.-Chim. Ther.*, 1984, **19**, 79.

31. G. Krause, M. Klepel and R. Franke, in 'QSAR and Strategies in the Design of Bioactive Compounds', ed. J. K. Seydel, VCH, Weinheim, 1985, p. 416.
32. V. K. Gombar, *Arzneim.-Forsch.*, 1985, **35**, 1633.
33. M. Chastrette, D. Zakarya and A. Elmouaffek, *Eur. J. Med. Chem.-Chim. Ther.*, 1986, **21**, 505.
34. I. K. Pajeva, Z. C. Lateva and G. V. Dimitrov, in 'QSAR in Drug Design and Toxicology', ed. D. Hadži and B. Jerman-Blažič, Elsevier, Amsterdam, 1987, p. 49.
35. P. S. Magee, in 'IUPAC Pesticide Chemistry: Human Welfare and the Environment', ed. J. Miyamoto and P. C. Kearney, Pergamon Press, Oxford, 1983, p. 251.
36. J. M. Chambers, W. S. Cleveland, B. Kleiner and P. A. Tukey, 'Graphical Methods for Data Analysis', Duxbury Press, Boston, 1983, p. 317.
37. J. W. Westley, E. P. Oliveto, J. Berger, R. H. Evans, Jr., R. Glass, A. Stempel, T. Voldemar and T. Williams, *J. Med. Chem.*, 1973, **16**, 397.
38. J. W. Westley, in 'Polyether Antibiotics', ed. J. W. Westley, Dekker, New York, 1983, vol. 2, p. 65.
39. J. W. McFarland and D. J. Gans, *J. Med. Chem.*, 1987, **30**, 46.
40. L. R. Ferguson and B. C. Baguley, *Mutat. Res.*, 1981, **82**, 31.
41. L. R. Ferguson and W. A. Denny, *J. Med. Chem.*, 1980, **23**, 269.
42. J. W. McFarland and D. J. Gans, in 'QSAR in Drug Design and Toxicology', ed. D. Hadži and B. Jerman-Blažič, Elsevier, Amsterdam, 1987, p. 25.
43. J. W. McFarland and D. J. Gans, in 'QSAR: Quantitative Structure–Activity Relationships in Drug Design', ed. J. L. Fauchère, Liss, New York, 1989, p. 199.

22.3
Pattern Recognition Techniques in Drug Design

WILLIAM J. DUNN, III
University of Illinois, Chicago, IL, USA

and

SVANTE WOLD
Umeå University, Sweden

22.3.1 WHAT IS PATTERN RECOGNITION?

Quantitative structure–activity relationships (QSAR) are models which relate the variation in measured biological response for a series of compounds to the variation in chemical structure within the series. The models can give a quantitative prediction of response, as in the traditional Hansch approach,[1] or they can give qualitative predictions of biological activity.[2] In some case it may be possible to obtain both. Methods of data analysis which derive models that give qualitative results will be referred to as pattern recognition or classification methods and those which give models capable of both qualitative and quantitative predictions will be referred to as generalized pattern recognition.[2] One property of the models used in QSAR is that they are local models.[3] This is to say that they are valid for limited changes in chemical structure and biological activity.

Classification methods can be divided into two rather broad categories. The first is unsupervised classification, often also called cluster analysis or pattern cognition. Here no prior knowledge of categories or classes of compounds in a data set is available and the objective is to find natural groups, classes and clusters, in the data. In the case of supervised classification, prior knowledge of the classes of some compounds (the training sets) is used to train the method with this retrospective information and use this to recognize the class of new compounds. This chapter will deal only with supervised classification methods, often also called pattern recognition or discriminant analysis.

In recent years applied mathematics has provided researchers with many new tools for data analysis. Methods of classification have their beginnings in the work of the great statistician Fisher,[4] who developed data analytic methods for discriminating one group of objects from another. Fisher's interests were in classifying biological individuals, such as plants, into classes according to species and strains, and this was the main motivation for him laying the foundations for discriminant analysis as we know it today. The first applications of classification methods in drug design[5] involved discriminant analysis and the objective of these applications was to classify chemical compounds according to the pharmacological response they elicit when they interact with a biological system. Since this initial paper by Martin *et al.*,[5] many applications of classification techniques to chemical structure–biological response data have been published and the use of pattern recognition techniques in QSAR studies has become a standard practice in drug design research.

22.3.2 THE BASIS OF PATTERN RECOGNITION AS USED IN QSAR AND DRUG DESIGN

The primary objective of QSAR is to derive models which predict the biological activity for unknown or untested compounds. In addition, if the predictions are highly significant from a statistical point of view, they should provide some understanding of how a change in structure within the series can effect a change in the biological activity. In order for the prediction objective to be fully satisfied, two conditions must be fulfilled by the model and the data. The first is that the model should reproduce the biological activity as well as possible for the compounds on which they are based. Second, the predictions should be done with as few parameters as possible compared to the number of degrees of freedom in the data. It is well known that empirical models can exactly reproduce a given data set when the number of adjustable parameters equals or exceeds the number of data elements.[6] Therefore, the second condition is most important if the primary objective of the use of QSAR is to be met.

There are a number of pattern recognition methods available for application to drug design problems. These techniques have different requirements regarding the two conditions discussed in the former section. The application of pattern recognition in drug design is based on the same assumptions about the relationship between chemical structure and biological activity as the more traditional methods of QSAR and this is the analogy principle of Hammett.[7] This states, in effect, that a similar change in structure will effect a similar change in (biological) activity. Thus one assumes that the change of substituent in a studied compound, say from methyl to chloro, will cause a change in biological activity which can be modeled in terms of the same 'effects' as in an ensemble of standard chemical reactions. The change from methyl to chloro induces a certain change in pK_a of *para*-substituted benzoic acids, change in distribution coefficient in 1-octanol/water of *para*-substituted benzene, *etc.* These changes can be described numerically in different physicochemical variables corresponding to pK_a, $\log P$, *etc.* In this way, QSAR and pattern recognition are linear free energy based approaches to structure–biological activity problems.

Pattern recognition methods are also subject to the same criteria regarding statistical degrees of freedom and significance as other modeling techniques. In order to obtain classification rules with good predictability these criteria must be satisfied. One commonly held misconception about the use of pattern recognition in QSAR is that a large number of compounds is essential. As long as the above-mentioned criteria are met even for small numbers of compounds, pattern recognition methods can give highly reliable predictions.

Another misconception about pattern recognition is that it is mathematically complex and difficult to understand. This is far from the truth. In fact, pattern recognition is a very important part of chemistry and medicinal chemistry. Chemists are trained to recognize patterns and associate them with some aspect of chemical structure. For example, associating the strong absorption of a compound in the IR at 1680 cm^{-1} with the presence of a carbonyl group is pattern recognition, as is associating the catechol amine functionality with adrenergic receptor activity. Computational pattern recognition is simply a quantitative extension of these conceptual models. As long as straightforward rules about statistics and chemistry are obeyed, pattern recognition techniques can be used to provide answers to classification problems.

Figure 1 illustrates the data table for a pattern recognition problem. Here the compounds are described by a data table of biological activity data and chemical structure descriptors. The biological activity data may be categorical or continuous and may consist of one or several types of activity. The chemical descriptors for the compounds are features or variables which are either measured experimentally or obtained from theoretical calculations. All of the variables for a compound represent a pattern, a data vector.

From prior knowledge about the activities of the compounds, those with similar activities, *e.g.* agonist or antagonist, can be placed in classes. The objective of the use of pattern recognition methods is then to derive models which will classify the training compounds correctly and predict the class assignment of similar, unknown compounds. The derivation of models or classification functions is termed training and the data of compounds of known classes constitute the training set. If generalized pattern recognition results are desired, also the level of agonist or antagonist activity can be predicted.

Two types of training methods are sometimes discussed[8] and these are parametric and nonparametric.[9,10] In theory, parametric methods are used when each class or training set is known to be modeled by a certain distribution which is characterized by a set of parameters, the values of some of which are unknown. With parametric methods the training set data are used for estimating the unknown parameters and these then define discriminant functions.

With nonparametric methods no explicit assumptions are made about the statistical distribution of the descriptors or variables and the associated classification rules or functions. The classification rules are assumed to be approximated by some functions, and coefficients in these are adjusted to give optimum classification results. The linear learning machine[10] and *k*-nearest neighbor[9] methods are considered to be nonparametric and Fisher's discriminant function is considered to be a parametric method.[10]

In practice, however, the distinction between parametric and nonparametric methods is of less importance, and there is little difference in their actual use.

Of greater importance is the fact that in QSAR studies, it cannot always be assumed that all of the known and unknown compounds in the study belong to one of the given classes. Therefore one must anticipate that there may be new, unexpected classes represented in the data set and the method of pattern recognition should be able to deal with this classification result. In most pattern recognition studies of biologically active compounds, the mechanism of action of the classes of compounds may or may not be known in advance but some common sense should be used in order that the training sets are not arbitrarily defined. Few would think of putting all known carcinogens in one class, for

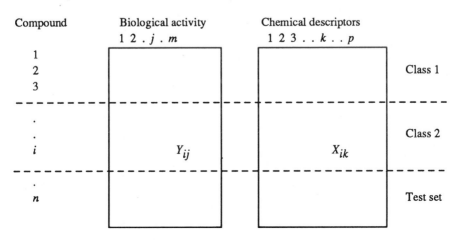

Figure 1   Data tables for a two-class pattern recognition problem

example, and it is always a good practice to involve an expert in drug action in a pattern recognition study.

In order to illustrate how pattern recognition methods work, a geometric interpretation can be used. A coordinate system with axes in the units of the variables can be constructed as in Figure 2. In most cases the number of variables will be greater than three for each compound but for illustrative convenience a three-dimensional coordinate system will be used here. It is standard mathematical notation to use upper case bold letters to denote data in block or matrix form. That notation will be used here. The elements of data matrices will be indicated as lower case letters.

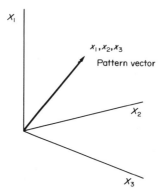

Figure 2 Three-dimensional representation of a pattern vector with coordinates x_1, x_2, x_3

The coordinate system defines the feature or pattern or data space. In this space the compounds are represented by points. The basis of all methods of pattern recognition is that compounds which are similar should be near each other and cluster in pattern space. The assumption has been made that chemical structure and biological activity are functionally related, which further assumes that the compounds of a specified pharmacological class must be pharmacologically similar, *i.e.* that they exert their effect by a common mechanism of action. If the training sets have been defined in accordance with this, for multiclass problems, clusters of compounds will form in different regions of pattern space. In this example, two classes of compounds are illustrated.

22.3.3 INFORMATION FROM PATTERN RECOGNITION

Pattern recognition analyses can give different levels of information depending on the objective of the analysis.[11,12] These have been termed *four levels of pattern recognition* and the formulation of classification results in terms of these four levels of information is important in designing a classification study.

At the first and lowest level of classification (level I), it is of interest to predict the assignment of an unknown or untested substance into one of a few given classes. This is a limited objective and rarely a sufficient one for QSAR studies since one can rarely be certain that the test compounds belong to one of the training set classes. At a higher level (level II) the classification result that the test compound(s) may belong to a new or undiscovered class is allowed. This is equivalent to allowing the test compound(s) to be an 'outlier' from the defined classes. At levels I and II only qualitative results are possible.

In many QSAR applications, not only the activity class of compounds is available, but quantitative biological activity data are also available. Pattern recognition at level III (1) involves the prediction of the qualitative class assignment with the possibility for outliers; and (2) predicts a single biological response variable. Level IV involves step (1) above followed by (2) the prediction of *several* biological response variables. At this level, the problem is multivariate. The most general formulation for classification problems is at level IV and is the natural one for QSAR problems due to the broad pharmacological profile considered in drug design and development problems. In the area of toxicity and risk assessment,[11] a single activity parameter contains only limited toxicity information about the human exposure hazards of a compound. Hence, QSAR studies should be done on batteries of tests,[13] and level IV is the natural level at which classification studies should be done.

22.3.4 A GEOMETRIC BASIS OF METHODS OF PATTERN RECOGNITION

In order to classify compounds as being members of the different classes in Figure 1, it is necessary to establish the regions of pattern space occupied by the training compounds. The methods of pattern recognition differ as to how this is done, and can thus be divided into different types. The first are the hyperplane methods, *e.g.* discriminant analysis[14] and the linear learning machine,[8] which derive a discriminant line, plane or hyperplane that separates one class from the other. This is illustrated in Figure 3 for the two classes which lie on different sides of a discriminant line.

To illustrate how the discriminant function is calculated consider a two-class problem in which compounds in the groups are described by two variables X_1 and X_2 as shown in Figure 4. The variables for the two groups are Y_1 and Y_2, which may be categorical, *i.e.* 2 for active and 1 for inactive compounds, or continuous, in which the two classes are weakly active and strongly active. The two groups are clustered in the figure with some overlap of the two classes. Discriminant analysis, and the so-called hyperplane methods, calculate a discriminant function which is a linear combination of X_1 and X_2 as shown in equation (1).

$$Y = v_1 X_1 + v_2 X_2 \tag{1}$$

The discriminant function passes through the origin as shown in Figure 5. It follows that an infinite number of discriminant functions can be calculated. The orthogonal projection of the two classes onto the discriminant function shows that some separation occurs but it follows that other functions are possible and some will be better than others. The optimal one is the one that gives minimal overlap of the training sets.

A separation criterion can be defined as the ratio of the between group variance, SS_b, and within group variance, SS_w, as in equation (2). The larger this ratio the better the class separation. It is obtained by finding the values of v_1 and v_2 in equation (1) which make U in equation (2) a maximum. In this way this technique has properties similar to regression methods.

$$U = SS_b / SS_w \tag{2}$$

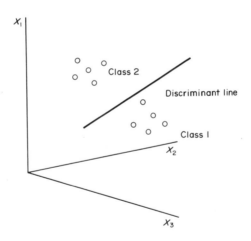

Figure 3 Three-dimensional representation of two classes separated by a discriminant line

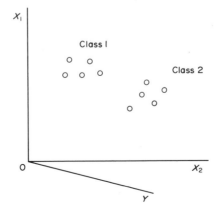

Figure 4 Two-dimensional representation of two classes with the projection on to the $0Y$ line

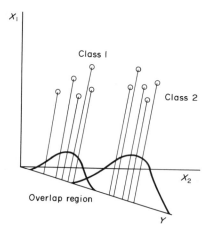

Figure 5 Projection of the two classes on to the line $0Y$ showing the two classes separated with a region of class overlap

Orthogonal to this linear combination, a line can be constructed, as shown in Figure 6, which will separate the two classes. The representation of test compounds in the pattern space results in the unknowns being on one side of the plane or the other. It must be known in advance, however, that the unknowns which are to be classified are members of these well-defined training sets, as unknowns which are not members of the training sets will also be classified as members of one class or the other. This assumption can rarely be met in drug design and QSAR work, however.

Another shortcoming of the hyperplane methods is that they select variables for their discriminant functions which *differentiate* between classes. These may not be the variables or descriptors which contain information about class membership *per se*, and can lead to ambiguities if, in a later interpretive step, a physicochemical basis of class membership is to be deduced.

The second type of pattern recognition is the distance-based methods. An example is the k-nearest neighbor method[9] in which classification is based on the proximity of a compound to its k-nearest neighbors, in which k is a small, odd number such as 1, 3, 5 or 7. The distance is often Euclidian so that it is easy to program and calculate classification results with this method. The k-nearest neighbor method is illustrated in Figure 7 for the two-class problem.

The third class of pattern recognition methods is called class-modeling or projection methods.[15] These methods are based on modeling with principal components[15] or projections to latent structures (PLS).[16,17] The most widely used class-modeling method is the SIMCA[15] method, which can give both qualitative and quantitative classification results. The modeling methods approximate the regular variation of the data within a class by a point, a line or a plane. Around the model, a tolerance interval is constructed from the statistical scatter of the data. The result is that the volumes occupied by the classes are defined mathematically so that classification of test compounds is based on where in pattern space they are located. An advantage of projection methods is that the models, *i.e.* points, lines and planes, can be defined to satisfy different criteria depending on the objective of the data analysis.

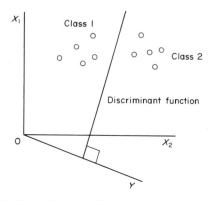

Figure 6 The two classes in Figure 5 separated by the discriminant function orthogonal to line $0Y$

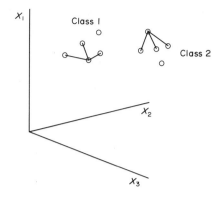

Figure 7 Two classes with a member of each class with its three nearest neighbors

The projection of the data for two classes from a p-dimensional space onto a lower dimensional (hyper)plane is illustrated geometrically and in matrix notation in Figure 8. This is equivalent to deriving disjoint principal components models (equation 3) for the two classes with the position in hyperspace of each compound approximated by the class mean, x, and their principal component scores, ts.

$$x_{i,k} = x + \sum_{a}^{A} t_{i,a} b_{a,k} + e_{i,k} \tag{3}$$

The bs in equation (3) are from the direction matrix, P, in Figure 8, and give the weight of the descriptor variables in the principal components model. The number of principal components required to approximate the data structure for each class corresponds to whether the classes are approximated by a point, $A = 0$, a line, $A = 1$, a plane, $A = 2$ or a hyperplane, $A = 3$. The residuals, $e_{i,k}$, are used to construct tolerance intervals around each class and these define volumes within which members of each class have the highest probability of being observed. Classification of test set compounds is based on where they are situated relative to the training sets 'volumes'.

The PLS method operates at level IV and therefore provides qualitative and quantitative estimates of biological activity. The X- and Y-data in Figure 1 are approximated by principal component-like models as in equations (4) and (5). The compound scores, or latent variables, t and u, are calculated so as to: (a) explain the maximum variance in the X- and Y-data; and also (b) extract those latent variables which are optimally correlated.

$$x_{i,k} = \sum_{a}^{A} t_{i,a} b_{a,k} + e_{i,k} \tag{4}$$

$$y_{i,j} = \sum_{a}^{A} u_{i,a} q_{a,j} + e_{i,j} \tag{5}$$

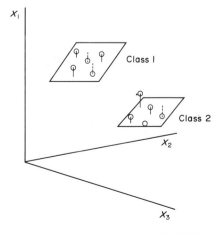

Figure 8 Two classes projected on to their SIMCA class models

This leads to the inner relation (equation 6), which can be used to predict the Y-data from the X-data. Level IV pattern recognition with PLS regression is shown in Figure 9.

$$u = b*t \rightarrow y = t*b*u \tag{6}$$

With the SIMCA method, the principal components are calculated by the *N*onlinear *I*terative *PA*rtial *L*east *S*quares, or NIPALS,[17] algorithm. PLS regression uses a variation of NIPALS.

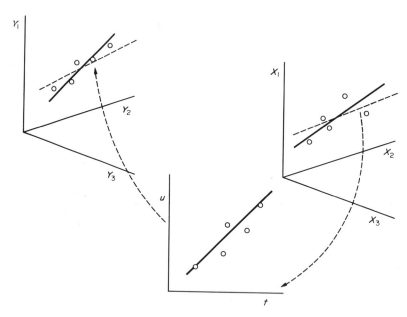

Figure 9 Representation of PLS regression through the inner relation, $u = b*t$. The solid lines in X- and Y-space are the principal components models and the dashed lines are the PLS models. These are slightly skewed or weighted to account for the correlation between the two data blocks

22.3.4.1 Steps in Pattern Recognition Analyses

Pattern recognition studies are usually done in steps. In the first phase of the study, the objectives of the analysis and the level of information required are determined. After this the training sets are established and the chemical variables are included to describe the compounds. Preprocessing (see Section 22.3.4.3) of the variables is often carried out and preliminary classification models are derived to classify the training set compounds. A variable selection process is sometimes used to determine variable significance and unimportant variables are deleted. From the reduced variable set new refined classification models are derived. The refined models are validated by applying them to data on compounds reserved as a test set.

22.3.4.2 The Nature of the Variables used in a Pattern Recognition Study

Published reports of the application of pattern recognition to biological data are extremely varied with regard to the approach to molecular description and this reflects mainly our general lack of understanding about chemical structure as it is relevant to biological activity. First, and most important, the variables must be relevant to the biological activity problem. Variables obtained experimentally from systems which model the biological system best satisfy this requirement. Such variables are the logarithm of the partition coefficient, $(\log P)$,[18-21] molar refractivity,[20,22] solubility, and pK_a or pK_b. Substituent constants such as Hammett's σ,[23,24] Hansch and Fujita's π,[20] Taft's E_s,[25] and Verloöp's B and L STERIMOL[26] parameters have been used with much success in QSAR and pattern recognition.

Other variables, such as the type and count of substructure fragments[27] and molecular connectivity indices[28] calculated from two-dimensional formulae are frequently used in QSAR and pattern recognition analyses. It is rather easy to derive large numbers of these variables by computer. When used with inappropriate methods of data analysis, such as the hyperplane methods, the noncritical use of these large sets of variables can lead to overfitting and models with poor predictive capability.

In principle, any parameter which contains information about the chemical structure of the compounds under analysis can be used. Early pattern recognition studies of biologically active compounds based on mass spectral data[29,30] have been published but it should be realized that no one single variable will contain sufficient information required for compound description and that the problem is generally a multivariable one.

In order for a model to be useful in drug design, not only must it be able to predict the activity of a compound with a given structure but it should be possible for the chemist to construct a compound from a given set of parameters that will place it in a range of activity estimated by the model. This is not possible using substructure and connectivity based descriptors. Substructure results cannot be extrapolated to new types of substructures, and given a value for a connectivity index a compound cannot be designed from this parameter. In view of these severe limitations, the use of such parameters is not recommended.

It has been recognized for some time that there can be considerable redundancy or collinearity in linear free energy related substituent constant and equilibrium constant data.[31] It has also been shown that variables, such as the $\log P$ for a compound, can be interpreted in more fundamental terms,[21] the major factor being the surface area of the nonpolar part of the hydrated solute. Such observations lead one to conclude that alternative approaches to molecular description may be desirable. One such approach has been the 'principal property' approach.[32] A basic assumption of this approach is that by starting with a large number of variables that are measured on systems which model biological systems, these can then be reduced to a basis set of abstract 'principal properties' by principal components analysis. These principal properties can be used as independent variables in the traditional sense to construct QSAR.

One approach to studying the QSAR of peptides and proteins has been the use of molecular modeling and graphical display of the peptides and macromolecules. A limitation of this approach is that it is qualitative. A more quantitative approach to molecular description would offer many advantages.

The difficulty with describing the change in structure within a series of peptides, in which one or more amino acid residues are changed, is construction of a data matrix to describe this change. This is illustrated for the pentapeptide, $H_3N^+CHR^1CONHCHR^2CONHCHR^3CONHCHR^4CONHCHR^5CO_2^-$. For peptides which contain only naturally occurring amino acids, the variation in primary structure will result from the change in properties that are due to the substituents on the α carbons in the backbone. Using a substituent effect approach, each R group could be described by a group of substituent constants. In theory this could be done except that few substituent constants are available for the R groups, and, if available, they are not generally applicable to use in the environment of a peptide.

To circumvent this, Hellberg *et al.*[32] have measured or taken from the literature a number of properties of the natural amino acids and their simple derivatives. These are given in Table 1.

Principal components analysis of this property matrix resulted in three significant components.[32] The principal component scores that result for each amino acid are a set of three abstract variables, or principal properties, for each acid. These can be used as variables to describe an amino acid residue in a peptide, thus providing an avenue to the problem of description. The principal properties, z_1, z_2 and z_3 are given in Table 2. Their use will be illustrated in Section 22.3.5.6.6.

22.3.4.3 Data Preprocessing, Scaling and Weighting

Variables commonly used in pattern recognition are often based on different metrics. This is especially true of the linear free energy based parameters when substituent constants such as Hammett's σ,[23,24] Hansch and Fujita's π,[20] Taft's E_s,[25] and Verloop's B and L STERIMOL[26] parameters are used together in the same data set. Some type of preprocessing of the variables is required for this, to prevent variables with large variation from dominating the analysis over other variables with small but significant variation. One type of such preprocessing is called autoscaling[33] and the scaled variable, $x_{i,k}$, for compound i, results from subtracting the mean, x_k, of a column from the variable and dividing by the standard deviation, s_k, as shown in equation (7).

$$x_{i,k} = (x_{i,k} - x_k)/s_k \tag{7}$$

The scaled variable has zero mean and unit variance. This preprocessing is also referred to as regularization.

Table 1   Variables Used to Characterize the Amino Acids

Variable number	Property
1	Molecular weight
2	pK_{CO_2H} (CO_2H on C_α)
3	pK_{NH_2} (NH_2 on C_α)
4	pI, pH at the isoelectric point
5	Substituent van der Waals volume
6	^{1}H NMR shift for C_α—H (cation)
7	^{1}H NMR shift for C_α—H (dipolar)
8	^{1}H NMR shift for C_α—H (anion)
9	^{13}C NMR for C=O carbon
10	^{13}C NMR for C_α—H
11	^{13}C NMR for C=O in tetrapeptide
12	^{13}C NMR for C_α—H in tetrapeptide
13	R_f for 1-N-(4-nitrobenzofurazono) amino acids in ethyl acetate/pyridine/water
14	Slope of plot $1/(R_f - 1)$ *versus* mol % water in paper chromatography
15	dG of transfer of amino acids from organic solvent to water
16	Hydration potential or free energy of transfer from vapor phase to water
17	R_f, salt chromatography
18	log P, partition coefficient for amino acids in octanol/water
19	log D, distribution coefficient at pH 7.1 for acetylamide derivatives of amino acids in octanol/water
20	d$G = RT$ ln$f$; f = fraction buried/accessible amino acids in 22 proteins
21–29	HPLC retention times of amino acids for nine combinations of three pH and three eluent mixtures

Table 2   Principal Properties of the 20 Amino Acids

Amino acid		z_1	z_2	z_3
Ala	(A)	0.07	−1.73	0.09
Val	(V)	−2.69	−2.53	−1.29
Leu	(L)	−4.19	−1.03	−0.98
Ile	(I)	−4.44	−1.68	−1.03
Pro	(P)	−1.22	0.88	2.23
Phe	(F)	−4.92	1.30	0.45
Trp	(W)	−4.75	3.65	0.85
Met	(M)	−2.49	−0.27	−0.41
Lys	(K)	2.84	1.41	−3.14
Arg	(R)	2.88	2.52	−3.44
His	(H)	2.41	1.74	1.11
Gly	(G)	2.23	−5.36	0.30
Thr	(T)	0.92	−2.09	−.1.40
Ser	(S)	1.96	−1.63	0.57
Tyr	(Y)	−1.39	2.32	0.01
Cys	(C)	0.71	−0.97	4.13
Asn	(N)	3.22	1.45	0.84
Gln	(Q)	2.18	0.53	−1.14
Asp	(D)	3.64	1.13	2.36
Glu	(E)	3.08	0.39	−0.07

Sometimes variables are given weights in accordance with their significance in discrimination or classification. This variable weighting can be arbitrary and it should be done with caution, mainly because it often leads to overfitting and overestimation of the class separation. One frequently used weighting procedure of this type is called Fisher weighting.[29, 34] The 'Fisher ratio', R_k, for a variable k which occurs in two training sets, is calculated by equation (8) where X_1 is the mean of the variable in class 1 and X_2 is the variable mean in class 2. s_1 and s_2 are the standard deviations of these variables in the two classes.

$$R_k = (X_1 - X_2)^2/(s_1^2 + s_2^2) \tag{8}$$

This ratio is then used to weight the variables and tends to give variables large weight which have very different means and little variation. This type of weighting is used mainly with hyperplane methods and should be used only when the total number of variables is small compared with the number of compounds in the training set.

22.3.4.4 Data Structure in Pattern Recognition Problems

One of the key assumptions of pattern recognition is that the training sets will form well-defined clusters in pattern space. This follows from ones experience if they have attempted to model structure–biological activity data using traditional methods of QSAR since only data on active compounds are analyzed. Therefore, it must be expected that data for classes of compounds will behave similarly. Two types of data structure have been recognized and observed and these are termed symmetric and asymmetric data[35] structure.

22.3.4.4.1 *Well-defined classes: symmetric data structure*

The fact that compounds which are chemically and pharmacologically similar should cluster in pattern space is an extension of the Hammett[7] analogy principle as discussed earlier. In the ideal case, all of the classes in a pattern recognition study will be subject to and satisfy this condition. Examples of such problems which can be expected to behave in this way are classes of compounds which interact at the same receptor but exert different or opposite effects. Receptor agonists and antagonists, as well as enzyme substrates and inhibitors, can be expected to form well-defined classes in the same pattern space.[36]

Symmetric data structure is illustrated in Figure 10. When symmetric data structure is observed all of the methods of classification which will operate at level I can give reliable classification results if the assumptions about independence of variables and conditions regarding degrees of freedom and adjustable parameters in the analysis are met.

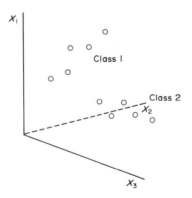

Figure 10 Symmetric data structure for two well-defined classes

22.3.4.4.2 *The problem of active versus inactive classification: asymmetric data structure*

In a number of structure–biological activity investigations, the classification problem is often formulated as active *versus* inactive. Examples are toxic *versus* nontoxic, carcinogenic *versus* noncarcinogenic, etc. This has been termed an asymmetric classification problem because one of the classes is usually not a well-defined and homogeneous class.[35] This type of data structure is shown in Figure 11. Attempts to model the active and inactive classes in such cases often leads to failure and this is due to the fact that the active class is often homogeneous, while the inactive class is not.

Asymmetric data structure, also known as embedded structure, can be understood in terms of control theory.[37] For a class of active compounds, the structure–activity system, which consists of the active compound and the biological system, will respond well for small changes in structure. The result will be small and regular changes in biological activity. However, if the structural change is too drastic or if it exceeds the requirements for the receptor or enzyme in the biological system, a discontinuity in the structure–activity relationship will occur and inactivity will result. This corresponds to moving away from the active class in any of the p dimensions of descriptor space. One would not expect the inactive class of compounds of form a homogeneous class that can be modeled but to be scattered randomly in descriptor space around the active class.

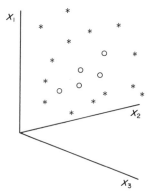

Figure 11   Asymmetric data structure with a well-defined class (open circles) and a nonstructured class (starred objects). This is sometimes referred to as embedded structure

In this sense, all nonactive compounds are to be considered 'outliers' of the active class(es). This leads to the conclusion that a compound which is classified as being 'nonactive' is nonactive by the same mechanism as the active class. This does not discount the result that the compound could be active, *e.g.* toxic, by a different mechanism. It would, in this case, represent a new class of active compounds. This problem arises not from the way the data are analyzed but from the way the problem is formulated. The problem is the same with biological model systems. Lack of activity in a battery of cell tests does not necessarily imply a lack of activity *in vivo*.

22.3.5   APPLICATION OF METHODS OF PATTERN RECOGNITION IN QSAR AND DRUG DESIGN

The use of pattern recognition techniques in analyzing structure–biological activity data was the result of applications of these techniques to chemical data in the late 1960s and early 1970s. These resulted mainly from the efforts of Isenhour and his students[38] and Kowalski and Bender.[33,39] The early applications in QSAR used the linear learning machine[8] following the lead of these workers and selected applications are reviewed here due to their historical significance.

22.3.5.1   The Linear Learning Machine

The linear learning machine has been discussed by Nilsson[8] and has been applied to biological structure–activity data to classify psychotropic drugs as sedatives or tranquilizers.[34] The classification procedure, termed an adaptive binary classifier,[27] involves representing the structure of a compound by binary and numeric substructural fragment descriptors. Binary variables, indicating the presence or absence of substructural fragments and the counts of number of such fragments in a compound are used to represent each compound as a vector, $X = (x_1, x_2, x_3, \ldots, x_n)$ where $(n - 1)$ of the elements are the binary and substructural descriptors and x_n is an extra component whose value is optimized and added to each vector during the training phase. If a linear decision surface or hyperplane can be constructed from the variables, the classes are said to be linearly separable.[8] In this study such a surface was calculated from initially 69 variables which would classify approximately 90% of the 219 (140 tranquilizers and 79 sedatives) compounds correctly.

Using this technique a correction feedback[34] procedure was designed to select the variables which give an acceptable correct classification rate. This procedure works by leaving out variables one at a time and checking the classification accuracy after the variable has been deleted. If the accuracy is not affected, the variable is left out and the procedure continued. This strategy selects variables which distinguish one class from another so that the classification function may contain little information about the role of structure in determining the sedative or tranquilizing properties of the compounds.

The majority of the subsequent applications of the linear learning machine to structure–biological activity pattern recognition problems involve classification of compounds as toxic *versus* nontoxic, carcinogenic *versus* noncarcinogenic. This is the type of problem which can lead to asymmetric data structure, as discussed earlier, and this technique is now known not to perform well with this type of data structure, especially when applied to compounds described by large numbers of binary and substructural descriptors.[27] The high probability of obtaining chance correlations has recently been discussed with regard to the analysis of multivariate chemical data by pattern recognition.[40]

The linear learning machine has also been criticized on other grounds.[41] The linear discriminant function found is the first encountered that will separate the training sets. This may not be the only one and may not be the best that satisfies the prediction objective of QSAR. Also, the training sets cannot be assumed to be linearly separable and usually are not. If the latter is the case, the algorithm will not terminate.

22.3.5.2 Linear Discriminant Analysis

Linear discriminant analysis and the linear learning machine are quite similar in that both methods search for surfaces which separate classes. The difference, however, has been discussed.[10] For two groups of compounds belonging to classes A and B, with mean vectors m_A and m_B and common covariance matrix, V, the Fisher discriminant function can be written as in equation (9).

$$D(y) = ((m_A - m_B) - \tfrac{1}{2}(m_A + m_{Bj}))'V^{-1}(m_A - m_B) \tag{9}$$

The m_A, m_B and V terms are not known but if it can be assumed that they can be estimated from the sample means, y_A, y_B, and the pooled covariance matrix S, then the discriminant function can be calculated. The function is linear and nonlinear data are often observed in QSAR. Most published applications of this approach should test both linear and nonlinear discriminant functions.

One of the more significant papers, in the sense that it focused the attention of researchers on the use of classification methods in QSAR, was published by Kowalski and Bender[33] in 1974. In this work, purines and pyrimidines from the data base of the National Cancer Institute were classified as active or inactive against Adenocarcinoma 755 (CA 755). The data used to describe the compounds were counts of substructural fragments present in each compound. Twenty features were used in all and the training sets contained 200 compounds; 87 of these were active and 113 were considered inactive. The ratio of compounds to features was 10, which is apparently acceptable for such a study, provided that the compounds are independent of each other, which in cases with sets of diverse compounds often is doubtful.[2]

Using linear discriminant analysis trained by least squares, an overall correct classification rate of 90% was obtained and using a feedback error correction procedure a 91.5% classification rate was obtained. The k-nearest neighbor method was slightly better with a 93.5% rate. This paper, retrospectively, was harshly criticized by Mathews,[41] based on the personal observation that any trained pharmacologist could, from an examination of the structures of the 200 compounds, develop a decision scheme which would do as well. Other applications of linear discriminant analysis involved the classification of monoamine oxidase inhibitors into four groups[42] and triazenes as active or nonactive against Sarcoma-180 in mice.[43]

One advantage of linear discriminant analysis is that the discriminant functions are given in terms of the variables used in the analysis and interpretation of the functions is straightforward.

The method has two serious limitations. The main one is that it can only operate at level I as discussed earlier. This assumes that the compounds to be classified are to be in one of the training sets. This assumption can rarely be made in QSAR even if the problem is rigidly formulated as, for example, agonist *versus* antagonist. The other limitation is that the data structure for the problem may be asymmetric, in which case the classes will not be linearly or nonlinearly separable.

Furthermore, linear discriminant analysis is a least squares method in which the degrees of freedom are determined by the number of compounds in the training sets. If the number of descriptors approaches the number of compounds the chance of spurious correlations becomes so great that the results are meaningless.

22.3.5.3 PLS Discriminant Analysis

A regularized form of discriminant analyses can be developed using PLS modeling with binary y-variables. If the data structure for a classification problem can be expected to be symmetric, binary variables can be assigned to the training set data to indicate class assignment. In some cases biological data are reported qualitatively as $+$, $++$, $+++$ and $++++$ to indicate general levels of activity for a group of compounds. In an exploratory analysis of such data it may be of interest to attempt to classify the compounds into four groups. This is a level III pattern recognition problem and the PLS method can be used to illustrate this for the data in Table 3. These are data for classes of β-adrenergic agonists and antagonists (1) reported by Mukherjee *et al.*[44] These data have been analyzed previously[36,45] and the reader is referred to these reports for the specific structures. In Table 3 compounds **1–15** are agonists and **16–32** are antagonists. The remaining compounds are the test set. Variables 1–9 are physicochemical variables indicating the change in chemical structure within the series. These will be referred to as the *x*-data. The biological, or *y*-data, are variables 10–13. Variable 10 is the agonist or antagonist activity, variable 11 is the negative log of the binding constant to the β-adrenergic receptor, variable 12 is the intrinsic activity for the agonists and variable 13 is an indicator variable with a value of 1 for agonists and 2 for antagonists.

$$X-\underset{Y}{\underset{|}{\bigcirc}}-\underset{\overset{|}{R}}{CH}-\underset{\overset{|}{R^1}}{CH}-NHR^2$$

(1)

Applying PLS, with *y* the indicator variables and *x* the physicochemical variables, showed that three PLS components were significant. This indicates that there is information in the physicochemical data related to agonist and antagonist activity. A plot of the first two PLS components

Table 3   Structure–Activity Data for β-Adrenergic Agonists and Antagonists (**1**)

Compound	pK_a	f_ϕ	f_{R2}	f_{R1}	σ^*_{R1}	E_{sR1}	σ_p	B_p	L_p	Act	pK_b	IA	D
Agonists													
1	8.93	1.14	0.70	0.19	0.49	1.24	−0.37	2.74	1.93	4.39	4.55	0.87	1.00
2	8.93	1.14	1.23	0.19	0.49	1.24	−0.37	2.74	1.93	4.42	4.74	0.71	1.00
3	9.29	1.14	0.19	0.70	0.00	0.00	−0.37	2.74	1.93	5.00	5.07	0.75	1.00
4	9.90	1.14	0.19	1.64	−0.19	−0.47	−0.37	2.74	1.93	5.85	5.77	1.00	1.00
5	9.90	1.14	1.23	1.64	−0.19	−0.47	−0.37	2.74	1.93	4.35	4.62	0.72	1.00
6	9.93	1.14	1.23	2.35	−0.20	−0.51	−0.37	2.74	1.93	4.51	4.41	0.64	1.00
7	9.19	1.14	0.19	2.83	−0.13	−0.93	−0.37	2.74	1.93	6.33	6.17	1.10	1.00
8	9.19	1.14	0.19	2.56	−0.13	−0.93	−0.37	2.74	1.93	6.37	6.17	1.10	1.00
9	10.03	1.14	0.19	2.42	−0.08	−0.38	−0.37	2.74	1.93	4.68	4.33	0.25	1.00
10	10.29	1.14	0.19	3.36	−0.13	−0.93	−0.37	2.74	1.93	5.04	4.62	0.25	1.00
11	9.29	1.14	0.19	2.43	−0.30	−1.60	−0.37	2.74	1.93	7.10	7.22	1.20	1.00
12	10.22	1.14	0.19	2.95	−0.08	−0.38	−0.37	2.74	1.93	5.04	4.64	0.17	1.00
13	9.94	−0.07	0.19	1.64	−0.19	−0.47	−0.37	2.74	1.93	6.00	5.62	0.28	1.00
14	9.77	−0.07	0.19	1.64	−0.19	−0.47	−0.37	2.74	1.93	5.48	6.19	0.24	1.00
15	9.29	−0.07	0.19	3.80	−0.30	−1.60	−0.37	2.74	1.93	7.10	7.85	0.27	1.00
Antagonists													
16	8.93	2.66	0.19	0.19	0.49	1.24	0.00	2.00	1.00	3.51	4.08	—	2.00
17	9.29	0.55	0.19	0.70	0.00	0.00	0.00	2.00	1.00	3.66	4.19	—	2.00
18	9.29	1.36	0.19	0.70	0.00	0.00	0.00	2.00	1.00	3.87	4.28	—	2.00
19	9.61	1.36	0.19	1.23	−0.10	0.07	0.00	2.00	1.00	4.29	4.66	—	2.00
20	9.90	2.04	0.19	1.64	−0.19	−0.47	0.23	3.52	1.80	5.89	5.38	—	2.00
21	9.90	1.36	0.19	1.64	−0.19	−0.47	−0.37	2.74	1.93	4.96	4.82	—	2.00
22	8.93	1.36	0.70	0.19	0.49	1.24	0.00	2.00	1.00	4.52	4.46	—	2.00
23	9.90	3.34	0.19	1.64	−0.19	−0.47	0.23	3.52	1.80	6.40	6.24	—	2.00
24	9.90	0.55	0.19	1.64	−0.19	−0.47	0.03	4.06	3.08	5.80	5.89	—	2.00
25	8.46	1.90	0.02	0.19	0.49	1.24	0.00	2.00	1.00	3.85	4.29	—	2.00
26	9.29	1.90	0.70	0.70	0.00	0.00	0.00	2.00	1.00	4.07	5.04	—	2.00
27	9.90	−0.94	1.23	1.64	−0.19	−0.47	0.28	5.50	3.48	5.35	4.85	—	2.00
28	9.03	1.36	0.70	2.42	−0.08	−0.38	−0.37	2.74	1.93	5.74	5.06	—	2.00
29	8.16	1.36	0.70	2.77	−0.13	−0.93	0.37	2.74	1.93	6.62	5.85	—	2.00
30	9.29	1.36	0.70	3.90	−0.13	−0.93	−0.37	2.74	1.93	6.89	6.74	—	2.00
31	8.16	1.04	0.70	2.77	−0.13	−0.93	−0.37	2.74	1.93	7.22	7.12	—	2.00
32	10.26	1.96	0.70	2.24	−0.30	−1.60	0.00	2.00	1.00	5.64	5.11	—	2.00
Test set													
33	8.93	1.14	0.19	0.19	0.49	1.24	−0.37	2.74	1.93	4.04	1.85	—	0.00
34	8.93	1.36	0.19	0.19	0.49	1.24	−0.37	2.74	1.93	1.50	1.85	—	0.00
35	8.93	1.36	0.19	0.19	0.49	1.24	−0.37	2.74	1.93	1.50	1.85	—	0.00
36	8.93	1.14	0.19	0.19	0.49	1.24	−0.37	2.74	1.93	1.50	1.85	—	0.00
37	9.80	1.90	0.19	0.19	0.49	1.24	0.00	2.00	1.00	1.50	1.85	—	0.00

from the x-data for the agonists and antagonists is shown in Figure 12. Compounds **1–15** form a cluster in the same region of the plot while the antagonists, **16–32**, are largely in the same region. Four of the antagonists overlap into the agonist class. One observation can be made about the change in structure as it affects agonist or antagonist activity. The agonists form a very tight cluster while the antagonists have much more structural variation as noted from the scatter in the plot for this group. This illustrates the highly specific structural requirements for receptor stimulation and the local nature of any model which might describe it.

PLS discriminant analysis is stable also in the case in which the number of x-variables approaches the number of compounds. In this case the ratio is $9:32$ or approximately 3. Using linear discriminant analysis on such data would result in a considerable risk of chance correlations.

Another advantage of PLS regression is that it can be used to select variables which discriminate between the training sets. PLS regression models the systematic variation in the x-data as well as in the y-data. Therefore, one can define modeling power, or MPOW_{pls}, as in equation (10), in which the $s_{k,x}$ is the standard deviation of the kth variable before it is modeled by PLS discriminant analysis and s_k is the standard deviation after modeling.

$$\text{MPOW}_{\text{pls}} = 1 - s_k/s_{k,x} \tag{10}$$

In this particular example variables 4, 5, 6 and 7 had MPOW_{pls} of 0.37, 0.53, 0.48 and 0.72, respectively and thus contain class discrimination information. These are the hydrophobicity of the amino substituent, Taft σ^* of the amino substituent, E_s or the size of the amino substituent, and the Hammett σ for the *para* substituent.

One drawback of PLS discriminant analysis is that it is not well described in the statistical literature and thus its advantages and limitations are not well understood. There is a considerable literature on the use of PLS in the multivariate calibration problem and the interested reader is referred to these.[46] The above example is one of the first applications of PLS discriminant analysis in the QSAR literature.

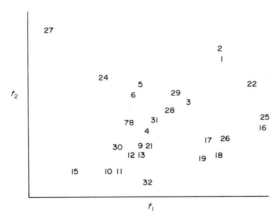

Figure 12 Principal components score plot for the first two components from the X-block in Table 3. Agonists are compounds **1–15** and antagonists are compounds **16–32**

22.3.5.4 The k-Nearest Neighbor (kNN) Method

The k-nearest neighbor method[9] of pattern recognition is one of the most easily applied and widely used classification techniques. It is based on the premise that an unknown pattern vector will also be most similar biologically to the pattern vectors to which it is nearest. The distance is usually Euclidian, but if the data are binary the Hamming distance is given. The Euclidian distance, D, between vectors x_1 and x_2, is calculated by equation (11) and is graphically illustrated in Figure 7 for $k = 3$. kNN was used extensively in earlier QSAR applications.[29,33]

$$D = \left[\sum_k (x_1 - x_2)^2 \right]^{1/2} \tag{11}$$

The earlier discussed example of linear discriminant analysis of Kowalski and Bender[39] in the classification of compounds active or inactive against Adenocarcinoma 755 compared linear discriminant analysis with the k-nearest neighbor results and found that it performed as well or slightly better. The classification by Ting *et al.*[29] of sedatives and tranquilizers from their mass spectra also used the k-nearest neighbor method with Fisher weighting of the variables. This report has been criticized, mainly due to the problem it addressed.[47]

One of the major advantages of the technique is its simplicity. It works at level II in our pattern recognition scheme as it will accommodate the result that a compound can be an outlier. It will also work with asymmetric data structure and no training is required. If the classification at level II is required and a small data set is to be analyzed, k-nearest neighbors is a good method, provided that the training compounds are representative and well distributed over the class domains. Its major disadvantage is that it can be computationally demanding, especially if the number of compounds becomes large. It can also require considerable computer memory, since no data reduction can be carried out and all vectors must be stored.

22.3.5.5 ALLOC

ALLOC is a new method of supervised pattern recognition which is under development by Coomans *et al.*[48,49] It differs from the other techniques in that it uses the cumulative potential of an object in its position with the training compounds to make a class assignment. The cumulative potential is calculated from a continuous Gaussian function, even though others should work. The test compound is classified into the class which gives rise to the largest cumulative potential. The boundary between two classes is calculated and is determined as the points at which the potentials are equal between the two learning sets.

When compared with other techniques, ALLOC is nonparametric. It makes no assumptions about the distribution of compounds in the training sets as does linear discriminant analysis. Nor does it make any assumptions about the shape and homogeneity of the classes. It will work with asymmetric data and in some ways is similar to k-nearest neighbor. It has been compared with a number of different methods using Fisher's iris data and gives very good results at level I. It has been applied to medical diagnostic data on hyper- and hypo-thyroid patients.[49] No applications to QSAR have been reported due to the fact that it is a rather new technique.

22.3.5.6 SIMCA Pattern Recognition

The SIMCA method of pattern recognition[15] is an acronym for a number of phrases related to similarity, chemistry and analogy. It is a method of pattern recognition that embodies several routines for classification, depending on the level of pattern recognition required by an analysis. Of the methods of pattern recognition available for QSAR studies, this is the method of choice. The alternative methods, such as linear discriminant analysis, are optimal for classification based on differences in means of classes. In QSAR studies and higher level pattern recognition the objective is to explain variation about means. As a result SIMCA does not operate at level I, as outlier detection is automatic. It is based on classification by the analogy principle of Hammett[7] and defines an 'analogy region' in pattern space. If a compound is not inside this region for a class, it is an outlier.

At level II, the data for a pattern recognition problem are the X- or Y-data in Figure 1. X- or Y-data are used since classification can be based on the analysis of chemical descriptor or pharmaeological response data at level II. Level III and IV classification of a compound can be obtained by correlating the position in physicochemical pattern space of its class with a level of biological response using principal components regression or, the more general method, PLS regression. These are illustrated with the data in Table 3. A detailed discussion of the SIMCA method is presented in the following sections (Sections 22.3.5.6.1 to 22.3.5.6.6).

22.3.5.6.1 *Training set assignment*

With all pattern recognition methods, the selection of compounds for the training set must be done with some prior knowledge of the pharmacological profile of the substances in the analysis. This is based on testing or response data obtained from the literature or from a data base so in most cases it is rather straightforward. In some problems this is not the case, especially in the analysis of compounds that have been evaluated under poorly defined biological conditions such as the Ames tests for mutagenicity. In such *in vivo* screens effects such as metabolism and specific transport, which might occur for some compounds and not others, are difficult to anticipate and placing compounds in a class with those which are nonspecific can lead to misleading results. In such cases, some form of exploratory data analysis should be used. PLS discriminant analysis is one type of technique which can be used. Another is the use of principal component plots of the proposed training set data.

The data in Table 3 for the agonists and antagonists can be used to illustrate this procedure. The data for compounds **1–32** are subjected to a principal components analysis (see equation 3 and Figure 8). The principal components are extracted along the first two axes of greatest variation and should explain a large amount of the variance in the data. If the chemical descriptor data are sufficient for class separation (this can be tested with PLS discriminant analysis), the compounds will usually be observed in different parts of the principal components plots. If not, the descriptor data should be evaluated or preliminary class assignments of the training compounds are not correct. A principal components plot of the data for compounds **1–32** in Table 3 is given in Figure 13. This shows that the classes are well separated in pattern space.

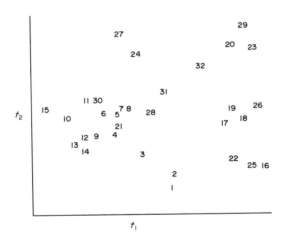

Figure 13 PLS discriminant plot of the first two PLS component scores or latent variables from the *X*- and *Y*-blocks in Table 3. The compound identities are the same as in Figure 12

22.3.5.6.2 *Modeling the training set data*

There are two aspects of modeling the training set data. The first is determining the optimum dimensionality of the class models and the other is determining the relevance of the chemical variables to the particular problem. Once the training classes have been selected, it is necessary that each be approximated by a principal components model. In most problems, more variables are input into the analysis than are necessary for class assignment so it is desirable to delete those that contain 'noise', while at the same time finding the number of principal components for the class models. Variable selection can be done in two ways: (a) using the principal components modeling power, $MPOW_{pc}$; or (b) using $MPOW_{pc}$ and $MPOW_{pls}$. The first criterion selects variables which model the classes, while the second finds variables which contain both class-modeling and class discrimination information.

$MPOW_{pc}$ is calculated as in equation (10) but is derived only from principal components modeling of the *X*-data for a class. If level II classification is desired, this approach to variable selection will suffice while the second approach should be used for higher order classification problems.

The other aspect of optimizing the class models is determining model dimensionality. This can be understood by referring to the equation for a principal components or PLS model in Figure 9. If the compounds within a class are indeed similar, as assumed, there will be collinearity in their variables so that the dimensionality of the data describing them will be less than the number of variables. In order to find the number of components which can reproduce the data to a predetermined degree, criteria regarding the residuals in equation (3) must be used.

This is a rather controversial aspect of the use of principal components and PLS modeling and is discussed by Bergman and Gittins.[10] To understand this consider the composition of the error term in equation (3). A set of data will not satisfy the relation in equation (12) for two reasons. The first is that $x_{i,k}$ may contain observed or experimental error and the second is that this function may not be the correct one for estimating y. Therefore, the residual (equation 13) can be composed of experimental error, E, model error, F, or both. Some workers prefer to reduce the residuals to experimental error, which ignores the possibility that the data may contain model error. A more realistic approach is to use a procedure which will give the best prediction of the class assignment.

$$x_{i,k} = x_k + \sum_a^A t_{i,a} b_{a,k} \tag{12}$$

$$e_{i,k} = E + F \tag{13}$$

This can be done using a cross-validation technique[50] (not to be confused with model validation or classification validation). This procedure works as follows. Assume a model with no principal components, *i.e.* equation (3) with $A = 0$, and we wish to compare to this the model with $A = 1$. Begin by deleting a fraction, say one-fifth, of the elements of the data for the training set. Derive a model with $A = 1$ from the remaining elements and use this model to predict *only* the deleted elements. Store the sum of the squares of the prediction errors (PRESS), restore the data matrices and delete another fraction of the elements. Calculate a model from the remaining elements, calculate the prediction errors of the deleted elements, add their squares to PRESS, and restore the matrix. Continue this cycle until each element has been deleted once and only once. Compare the PRESS.es from the model with $A = 1$ with that of $A = 0$ when corrected for differences in degrees of freedom. If the ratio is less than 1, the new dimension in the model is retained and the process continued until a stopping point is reached.

The result is a model optimized for prediction of the training set data. From experience with applying cross-validation to physicochemical data for QSAR studies, this approach extracts the systematic variation in the data and typically explains 50–75% of the variance in the data with 3–5 principal or PLS components. This usually does not appear to be a satisfying result to those who deal exclusively with multiple regression and use the correlation coefficient, r^2, as the sole criterion for model development, but it has given good results in classification studies.

22.3.5.6.3 *Level II pattern recognition analysis of β-adrenergic agonists and antagonists*

The agonists and antagonists (**1**) were modeled by separate principal components models with three components determined by cross-validation. For the agonists only descriptors 1–6 were variable. These had high modeling powers. For the antagonists descriptors 1–9 were variable and had high modeling powers. In this case the classes were modeled in a different pattern space. The class models explained 86% and 79% of the variance in the data for the two classes. Fitting the data for the training and test sets to these class models resulted in all of the agonists being correctly identified. The distance of each compound to the class models is given in Table 4. Five of the antagonists compounds **17**, **22**, **24**, **26** and **32**, were classified as agonists and compound **29** was outside both classes. This is an 81% correct classification rate. The test compounds were classified as agonists except for compound **37**, which was classified as an antagonist.

These results can be displayed graphically by plotting the distances in Table 4 as shown in Figure 14. On the *x*-axis is the distance to class 1 and on the *y*-axis the distance to class 2. The class standard deviations are also given and it can be seen that the agonists are close to class 1 and far from class 2, while most of the antagonists are close to class 2. There are some which are inside both classes but nearer to class 2.

22.3.5.6.4 *Level III pattern recognition analysis of β-adrenergic agonists and antagonists*

To illustrate a level III pattern recognition analysis PLS regression can be used to classify the agonists and antagonists (**1**) and then estimate their level of activity. Variable 10 is the dependent variable and two PLS components are found to be significant for the agonists while one PLS component was significant for the antagonists. A plot of the observed and predicted activities for each of the classes and the fit of the test set compounds to the two PLS models is also given.

Table 4 Distances of β-Adrenergic Agents to Class Models

Compound	Distance to Class 1	Distance to Class 2	Compound	Distance to Class 1	Distance to Class 2
1	0.48	1.05	20	1.23	0.71
2	0.43	1.38	21	0.64	0.71
3	0.56	0.68	22	0.40	0.69
4	0.50	0.72	23	2.46	1.01
5	0.82	0.94	24	0.51	0.84
6	0.86	0.90	25	1.20	0.46
7	0.46	0.74	26	0.88	0.48
8	0.41	0.71	27	2.06	1.47
9	0.44	0.76	28	0.68	0.62
10	0.59	0.92	29	1.54	0.99
11	0.62	0.80	30	1.09	0.78
12	0.55	0.86	31	1.42	0.94
13	1.00	0.90	32	1.38	0.96
14	0.98	0.87	33	0.91	0.94
15	0.74	1.00	34	0.89	0.92
16	1.55	0.39	35	0.89	0.92
17	0.71	0.54	36	0.91	0.94
18	0.65	0.35	37	1.22	0.52
19	0.65	0.46			

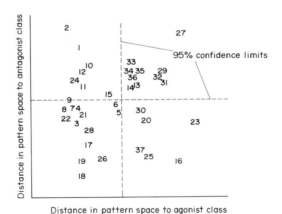

Figure 14 Plot of the distances of the training and test set compounds from Table 3 showing the fit of each compound to the two classes

The level II results for the test set compounds predicts that they are agonists with the exception of compound 37 which is outside both classes. The level III results suggest correctly that they are expected to be very weak agonists. These results are summarized in Figure 15.

22.3.5.6.5 *Level IV pattern recognition analysis of β-adrenergic agonists and antagonists*

As discussed earlier, this is the level at which pattern recognition studies in QSAR should be done. This is especially true for the analysis of toxicity or carcinogenicity data in which the responses of suspect compounds are commonly obtained in several biological systems. Appropriate data are available for a level IV analysis of the β-adrenergic agents (**1**) and the results of such an analysis are discussed below.

For the agonist class three biological responses are given in Table 3 (variable 10, receptor stimulant activity; variable 11, receptor binding affinity; and variable 12, intrinsic activity). For the antagonists, the first two variables above are appropriate for a PLS analysis as dependent variables.

The PLS predictions of antagonist receptor binding activity for the training and test sets were also quite good. The test compounds are all correctly predicted to have low binding affinity.

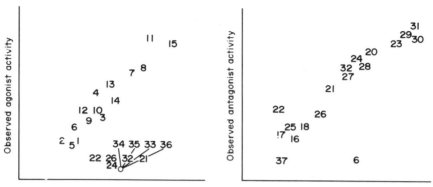

Figure 15 PLS predicted activities for the agonists and antagonists in Table 3. The training set objects are predicted to be very weak agonists. Compound **6** is predicted incorrectly to be similar to the antagonists

22.3.5.6.6 *PLS regression of peptide structure–activity data*

One of the more interesting developments in QSAR in the past few years has been the attempts to model structure–activity data for peptides. One obstacle to developing QSAR for peptides has been developing suitable molecular descriptors for amino acid units in the peptide chain. This has appeared to be at least partially overcome with the development of what have been termed principal property scales for the amino acids. This was discussed earlier (Section 22.3.4.2) and the principal properties for the 20 natural amino acids are given in Table 2.

A number of peptide QSAR have recently been published using this approach.[32] To illustrate the utility of the principal properties in QSAR of peptides, the data of Kettner *et al.*[51] on the inhibition of the proteases, trypsin and thrombin by a series of chloromethyl tripeptides, A_1—A_2—Arg—CH_2Cl, will be used. A_1 and A_2 are the amino acid residues. The compounds are irreversible inhibitors in which the alkylation step is of similar magnitude and the difference in activity is the result of differences in affinity for the active site. The activity data are given in Table 5.

Using the principal properties in Table 2 as the *X*-data and the inhibition data in Table 5 as the *Y*-data, a PLS regression was carried out. Two significant PLS components were found by cross-validation, which explained 75% of the variation in the inhibition data. Using a two component model, the activity of each dipeptide was predicted and these are plotted in Figure 16. This and other recently published examples of the use of amino acid principal properties as descriptors show the simplicity and power of their use and suggest that they could be useful in other aspects of QSAR research.

22.3.6 THE IMPORTANCE OF COMPOUND SELECTION FOR PATTERN RECOGNITION STUDIES

In most of the cases cited above the pattern recognition studies were carried out on so-called 'historical' data. This means that the data were obtained from the literature or the compounds selected for study were not selected as the result of any experimental design. This does not distract from the significance of the pattern recognition studies themselves, but limits the validity of the classification models and discriminant functions. The point is that for the prediction objectives of pattern recognition to be met, the classification models must be based on training sets designed considering the multivariate nature of the problem.

With regard to designing peptides for QSAR studies, it has been suggested by Rudinger[52] to change one amino acid position at a time. While this may appear to be natural to a naive reader, it is in fact a very inefficient and COSTly (Change One Substituent at a Time) way to design series. This rationale leads to series with little if any information about structure–activity relationships. This is especially true for studying peptide QSAR since so many peptides are possible for a given problem.

Instead, factorial or fractional factorial designs should be used. This leads to data with much higher information content.[6, 32, 53] When the design variables are numerous—larger than three or four—the design problem is simplified by using a fractional factorial 2^{q-r} design. Here $q = m*j$ with $m =$ the number of varied positions, and $j =$ the number of variables for each amino acid. The reduction factor (r) is chosen so that 2^{q-r} is larger than q. Consider, for example, the case when four

Table 5 Inhibition Data for Peptides (A$_1$—A$_2$—Arg—CH$_2$Cl) against the Enzymes Trypsin and Thrombin, $k = \log(10 + k_{app}/[\text{I}] \times 10^4)$

Compound	A$_1$-A$_2$	Trypsin	Thrombin
1	Pro-Phe	1.20	1.00
2	acGly-Gly	1.99	1.03
3	Ala-Phe	1.66	1.00
4	Gly-Val	1.71	1.07
5	Pro-Gly	1.90	1.05
6	Phe-Ala	2.50	1.25
7	Ile-Pro	2.66	1.71
8	Val-Val	1.63	1.08
9	Glu-Gly	2.80	1.09
10	Ile-Leu	1.73	1.18
11	Val-Pro	2.64	1.31

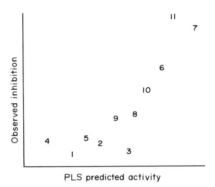

Figure 16 Observed and predicted inhibition of thrombin by the tripeptides in Table 5. The compound number corresponds to that in Table 5

positions in a peptide are varied, and each of these positions is characterized with the three scales z_1 to z_3. In this case with $r = 8$, a set of $2^{(4*3)-8} = 16$ peptides can be designed. This should allow for deviations from additivity that can result from intramolecular interactions, *etc.*

Standard rules[6] for the design of the series can be applied as follows. (a) A full factorial design is constructed for four columns with A, B, C and D. The first column, A, consists of alternate minus and plus signs. The second column, B, consists of minus–minus, plus–plus signs, the third, C, consists of four minus signs followed by four plus signs, and the fourth column consists of eight minus signs followed by eight plus signs. These columns are assigned to the z_1 values for the four varied positions. (b) Three additional columns are constructed from A, B, C and D by multiplying the signs of three of these together and then assigning them to z_2 for positions 2, 3 and 4. Thus z_2 for position 2 is ABC, for position 3 it is BCD, *etc.* (c) The last column for each position is for z_3, the sign of which is obtained by AB for position 1, BC for position 2, AC for position 3 and BD for position 4.

Using these rules the design matrix shown in Table 6 can be constructed. This matrix contains the signs of the z-values for use in selecting the amino acid residues to be used in each position. The first peptide designed would be the one with z_1 and z_2 minus and z_3 plus. From the principal properties in Table 2 there are no amino acids with this combination of signs but three have the combination of minus, minus, minus. These are Ile, Leu and Met. Ile has the greater negative values of z_1 and z_2 so it can be used for positions 1, 2, 3 and 4. Table 6 was constructed using these rules and represents a basis set of tetrapeptides or peptides in which four positions are to be varied in a structure–activity study. This basis set spans the physicochemical property space represented by the z-data and contains maximum structure–activity information.

22.3.7 DISCUSSION OF THE PHILOSOPHY OF USE AND SELECTION OF A METHOD FOR A GIVEN PROBLEM

The strategy for QSAR development in a given problem depends on different but coupled parts, as discussed above. The crucial quality governing the choice of methodology is the complexity of the

Table 6   Fractional Factorial Design Matrix for a Tetrapeptide of a Peptide with Variation in Four Positions

Number	Position 1 z_1, z_2, z_3			Position 2			Position 3			Position 4		
	A	ABC	AB	B	BCD	BC	C	ACD	AC	D	ABC	BD
1	−	−	+	−	−	+	−	−	+	−	−	+
2	+	+	−	−	−	−	−	−	+	−	+	+
3	−	+	−	+	+	+	−	−	+	−	+	−
4	+	−	+	+	+	−	−	+	−	−	−	−
5	−	+	+	−	+	−	+	+	−	−	−	+
6	+	−	−	−	+	−	+	−	+	−	+	+
7	−	−	+	+	−	+	+	+	−	−	+	−
8	+	+	+	+	−	+	+	−	+	−	−	−
9	−	−	+	−	+	+	−	+	+	+	+	−
10	+	+	−	−	+	+	−	−	−	+	−	−
11	−	+	−	+	−	−	−	+	+	+	−	+
12	+	−	+	+	−	−	−	−	−	+	+	+
13	−	+	+	−	−	−	+	−	−	+	+	−
14	+	−	−	−	−	−	+	+	+	+	−	−
15	−	−	−	+	+	+	+	−	−	+	−	+
16	+	+	+	+	+	+	+	+	+	+	+	+

relationship to be modeled. This complexity is a function both of the structural complexity of the molecular system to be modeled, and the anticipated range of structural variation. If both are moderate or large, the total complexity is large, which demands a highly multivariate structural description, may be a design in latent variables (principal properties), and certainly multivariate modeling with SIMCA and PLS. At the other limit, a narrow structural variation in a set of simple molecules may be described by one or two variables, *e.g.* lipophilicity and charge separation, which can be used as design variables for training set selection, and as descriptor variables in a simple regression (*e.g.* Hansch model) or discriminant analysis model.

We must realize that it is mainly the *range of structural variation* that determines the QSAR complexity. A narrow variation in a large molecule, say the change at a single amino acid position in an enzyme, may be described by just two or three variables—*e.g.* the z-scales discussed above—and hence treated as a simple QSAR. On the other hand, a total variation in halogenation pattern and hydrocarbon backbone in a set of aliphatic halogenated hydrocarbons with maximum five carbons, may need 20 variables to capture the more important chemical factors, and hence multivariate design and modeling.

However, to get a good picture of a system involving large molecules, the variation of many structural factors is usually needed. Therefore, in practice, the size of the investigated system is correlated to the complexity of the QSAR.

We are now in the position to structure the choice of pattern recognition methods as a function of the problem complexity. Following the conceptual division above with four levels of pattern recognition, the two highest also providing quantitative relationships, we include multiple regression and Free–Wilson type additive modeling as simple level III methods, that can be combined with linear discriminant analysis or kNN analysis for classification. We note also that PC modeling includes kNN as a special case, and PLS modeling includes multiple regression as a special case. This makes PCA, SIMCA and PLS suitable for use over the whole range of complexity, albeit with a varying degree of model complexity from single one component models in two or three variables for simple systems, to sets of multiple component hierarchical models in hundreds or even thousands of variables for systems of very high complexity.

Simple systems. The structural variation can be described by at most six or seven fairly uncorrelated descriptors, typically including total lipophilicity and electronic effects. Fractional factorial designs (plus center points) in these variables can be used for constructing the training set. The number of biological activity variables is small, at most two or three. Multiple regression (Hansch modeling), or Free–Wilson additive modeling may be used for quantitative modeling, and linear discriminant analysis (quadratic for the asymmetric case), kNN or ALLOC for the classification. As said above, PCA, SIMCA and PLS can also be used if so is desired.

Systems of moderate complexity. The structural variation demands a larger number of descriptors than five or six or seven variables, which then have a tendency to be collinear. With principal components analysis principal properties may be derived that can be used for designing test series (*e.g.* fractional factorials plus center points), and SIMCA and PLS can be used for relating the structural variation to the qualitative and quantitative measures of biological activity (*Y*). The latter (*Y*) should be multivariate to allow, among other things, the checking of pharmacological similarity inside classes of compounds. If *k*NN is used for classification, local Mahalanobis metrics should be used because of collinearities, *not* Euclidean metrics.

Stepwise selection of variables together with multiple regression or linear discriminant analysis is *not* recommended due to the resulting high risk of spurious correlations.

Systems of large complexity. Pattern recognition, as used in QSAR, is a multivariable approach. The philosophical outcome of this has been discussed generally by Mager.[54] Hundreds to thousands of structural descriptors are needed together with large numbers of activity variables. To accomplish a design of a training set, and to get a reasonable chance to interpret results, the descriptor variables must be 'chunked' to principal properties (latent variables) by the use of principal components and PLS, possibly in hierarchical form. Traditional pattern recognition methods are not useful in this domain, only projection methods in modified form are useful. Here much research remains to be done, however, and applications in this area are still rare and far from routine.

22.3.8 REFERENCES

1. C. Hansch, *Acc. Chem. Res.*, 1969, **2**, 232.
2. S. Wold and W. J. Dunn, III, *J. Chem. Inf. Comput. Sci.*, 1983, **23**, 6.
3. M. Sjöström and S. Wold, *Acta Chem. Scand., Ser. B*, 1981, **35**, 537.
4. R. A. Fisher, 'Statistical Methods for Research Workers', 2nd edn., Oliver and Boyd, Edinburgh, 1928.
5. Y. C. Martin, J. B. Holland, C. H. Jarboe and N. Plotnikoff, *J. Med. Chem.*, 1974, **17**, 409.
6. G. E. P. Box, W. G. Hunter and J. S. Hunter, 'Statistics for Experimenters', Wiley, New York, 1978, chap. 16.
7. L. P. Hammett, 'Physical Organic Chemistry', McGraw-Hill, New York, 1940, chap. 7.
8. N. J. Nilsson, 'Learning Machines', McGraw-Hill, New York, 1965.
9. K. Varmuza, 'Pattern Recognition in Chemistry', Springer-Verlag, Berlin, 1980.
10. S. E. Bergman and J. C. Gittins, 'Statistical Methods for Pharmaceutical Research Planning', Dekker, New York, 1985, chap. 1.
11. S. Wold, W. J. Dunn, III and S. Hellberg, *EHP Environ. Health Perspec.*, 1985, **61**, 257.
12. C. Albano, W. J. Dunn, III, U. Edlund, E. Johansson, B. Norden, M. Sjöström and S. Wold, *Anal. Chim. Acta*, 1978, **103**, 429.
13. J. McCann and B. N. Ames, *Proc. Nat. Acad. Sci. USA*, 1976, **73**, 950.
14. P. A. Lachenbruch, 'Discriminant Analysis', Hafner Press, New York, 1975.
15. S. Wold, *Pattern Recognition*, 1976, **8**, 127.
16. S. Wold, A. Ruhe, H. Wold and W. J. Dunn, III, *SIAM J. Sci. Stat. Comput.*, 1984, **5**, 735.
17. H. Wold, in 'Perspectives in Probability and Statistics, Papers in Honor of M. S. Bartlett', ed. J. Gani, Academic Press, New York, 1975.
18. Y. C. Martin, 'Quantitative Drug Design, A Critical Introduction', Dekker, New York, 1978.
19. J. K. Seydel and K.-J. Schaper, 'Chemische Struktur und Biologische Aktivatät von Wirkstoffen Methoden der Quantitativen Struktur-Wirkung-Analyse' Verlag Chemie, Weinheim, 1979.
20. C. Hansch and A. J. Leo, 'Substituent Constants for Correlation Analysis in Chemistry and Biology', Wiley, New York, 1979.
21. W. J. Dunn, III, M. G. Koehler and S. Grigoras, *J. Med. Chem.*, 1987, **30**, 1121.
22. W. J. Dunn, III and E. M. Hoonett, *Eur. J. Med. Chem.—Chim. Ther.*, 1977, **12**, 113.
23. L. P. Hammett, 'Physical Organic Chemistry', 2nd edn., McGraw-Hill, New York, 1970.
24. O. Exner, in 'Correlation Analysis in Chemistry, Recent Advances', ed. N. B. Chapman and J. Shorter, Plenum Press, New York, 1978, p. 439.
25. R. W. Taft, Jr., in 'Steric Effects in Organic Chemistry', ed. M. S. Newman, Wiley, New York, 1956.
26. A. Verloop, W. Hoogenstraaten and J. Tipker, in 'Drug Design', ed. E. J. Ariëns, Academic Press, New York, 1976, vol. 7, p. 165.
27. A. J. Stuper, W. E. Brügger and P. C. Jurs, 'Computer Assisted Studies of Chemical Structure and Biological Function', Wiley, New York, 1979.
28. L. B. Kier and L. H. Hall, 'Molecular Connectivity in Structure–Activity Analysis', Research Studies Press, Letchworth, UK, 1986.
29. K. H. Ting, R. C. T. Lee, G. W. A. Milne, M. Shapiro and A. M. Guarino, *Science (Washington, D.C.)*, 1973, **180**, 417.
30. K. C. Chu, R. J. Feldman, M. B. Shapiro, G. F. Hazard, Jr. and R. I. Geran, *J. Med. Chem.*, 1975, **18**, 539.
31. C. Hansch, S. H. Unger and A. B. Forsythe, *J. Med. Chem.*, 1973, **16**, 1217.
32. S. Hellberg, M. Sjöström, B. Skagerberg and S. Wold, *J. Med. Chem.*, 1987, **30**, 1126.
33. B. R. Kowalski and C. F. Bender, *J. Am. Chem. Soc.*, 1972, **94**, 5632.
34. A. J. Stuper and P. C. Jurs, *J. Am. Chem. Soc.*, 1975, **97**, 182.
35. W. J. Dunn, III and S. Wold, *J. Med. Chem.*, 1980, **23**, 595.
36. W. J. Dunn, III, S. Wold and Y. C. Martin, *J. Med. Chem.*, 1978, **21**, 922.
37. O. I. Elgerd, 'Control Systems Theory', McGraw-Hill, New York, 1967.
38. P. C. Jurs, B. R. Kowalski and T. L. Isenhour, *Anal. Chem.*, 1969, **41**, 21.
39. B. R. Kowalski and C. F. Bender, *J. Am. Chem. Soc.*, 1974, **96**, 916.
40. P. C. Jurs, *Science (Washington, D.C.)*, 1986, **232**, 1219.

41. R. J. Mathews, *J. Am. Chem. Soc.*, 1975, **97**, 935.
42. Y. C. Martin, J. B. Holland, C. H. Jarboe and N. Plotnikoff, *J. Med. Chem.*, 1974, **17**, 409.
43. W. J. Dunn, III and M. J. Greenberg, *J. Pharm. Sci.*, 1977, **66**, 1416.
44. C. Mukherjee, M. G. Caron, D. Mulliken and R. J. Lefkowitz, *Mol. Pharmacol.*, 1976, **12**, 16.
45. S. Woid, W. J. Dunn, III and S. Hellberg, in 'Drug Design: Fact or Fantasy', ed. G. Jolles and K. R. H. Wooldridge, Academic Press, London, 1984, p. 95.
46. S. Wold, K. Esbensen and P. Geladi, *Chemo. Intel. Lab. Sys.*, 1987, **2**, 37.
47. C. L. Perrin, *Science (Washington, D.C.)*, 1974, **183**, 551.
48. D. Coomans, D. L. Massart, I. Broeckaert and A. Tassin, *Anal. Chim. Acta*, 1981, **133**, 215.
49. D. Coomans, M. Derde, D. L. Massart and I. Broeckaert, *Anal. Chim. Acta*, 1981, **133**, 241.
50. S. Wold, *Technometrics*, 1978, **20**, 397; H. T. Eastment and W. J. Krzanowski, *Technometrics*, 1982, **24**, 73.
51. C. Kettner, S. Springhorn, E. Shaw, W. Müller and H. Fritz, *Hoppe-Seyler's Z. Physiol. Chem.*, 1978, **359**, 1183.
52. J. Rudinger, in 'Drug Design', ed. E. J. Ariëns, Academic Press, New York, 1971, vol. 2, p. 319.
53. V. Austel, *Eur. J. Med. Chem.—Chim. Ther.*, 1982, **17**, 339.
54. P. P. Mager, 'Multidimensional Pharmacochemistry: Design of Safer Drugs', Academic Press, New York, 1984.

22.4

The Distance Geometry Approach to Modeling Receptor Sites

ARUP K. GHOSE

Nucleic Acid Research Institute, Costa Mesa, CA, USA

and

GORDON M. CRIPPEN

University of Michigan, Ann Arbor, MI, USA

22.4.1 INTRODUCTION

22.4.1.1 Historical Background

Until recently, drug research has consisted of a trial and error approach. On average it is necessary to synthesize and biologically evaluate 10 000–15 000 compounds to get an effective drug. Medicinal chemists never considered the unsuccessful trials as total failure, since biological activity was not simply 'yes or no', but rather it had some scale. Using intuition, they tried to interpret the effect structural changes made on the biological activity. The interpretation was then used to design the next possible candidate. However, since it is extremely difficult to define the structural changes in a systematic way, and the biological effect on animals is a consequence of a very complex process, qualitative interpretations most often proved to be incorrect. Chemists for their own interests studied many properties of the compounds and established a definite relationship with the chemical structure. Which of these properties should be considered in order to understand the change in biological activity, and what should be the systematic approach to study the relationship? The answer to these questions was first suggested by Hansch *et al.*[1] They introduced the octanol–water partition coefficient and a few other physical properties to correlate with biological activity, and used statistics to study the nature of the relationship. Almost at the same time a group of scientists tried to give a qualitative interpretation of the biological activity in terms of comparable relative positions of two or more atoms in molecules having diverse structure but similar biological activity.[2] In fact both structural and physicochemical properties are equally important to biological activity. In order to take both these properties into account using reasonable amounts of computer time, the distance geometry approach was developed.[3]

22.4.1.2 Some Difficulties in Conventional QSAR

To get the essence of the distance geometry approach to receptor modeling, let us consider some of the important changes in the conventional (Hansch-type) QSAR.[4] The initial tendency in the QSAR studies was to correlate the biological activity with the overall physicochemical properties of the drug molecules. The exploratory step was to select the appropriate physicochemical property and appropriate function. Structural specificity for the biological activity is a long-known phenomenon, and often the structural change is not appropriately reflected in the overall physicochemical property. QSAR workers, therefore, started correlating the physicochemical property of some parts of the molecule with the biological activity, particularly different physicochemical properties for different parts. This change was very fruitful since correlation of different physicochemical properties for different parts of the molecule may be interpreted as due to the difference in the environment around the ligand molecule at the active site. Such an interpretation is especially true if the biological activity results primarily from the affinity of the drug molecule for the receptor. The inherent presupposition here, however, is that we know the superimposition of the ligand molecules; in other words, how the ligands bind at the receptor site. Unfortunately not only is such knowledge extremely difficult to gather, but even for small relatively rigid molecules the assumption may be treacherous. Let us take a simple hypothetical example to illustrate the problem. Suppose one has measured the binding of 3-substituted phenols with a biological receptor. In order to explain the biological activity, QSAR chemists often assume that the difference in activity, at least partly, is caused by the interaction of the 3-substituent with the receptor. If, for example, a hydrophobic group increases the activity and a hydrophilic one decreases it, one assumes that the receptor atoms surrounding the 3-substituent of the ligand are hydrophobic in nature. This process is known as receptor mapping. In other words, we assume that the 3-substituent always goes to the same region, even if it experiences a repulsive interaction. The aromatic ring here could easily flip over (Figure 1) to avoid such interaction, if it is not forbidden for some other reason such as a steric factor. Another

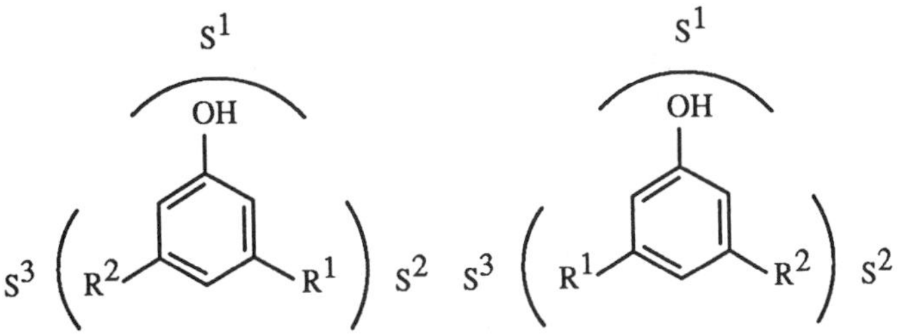

Figure 1 The two possible ways of interaction of the 3,5-disubstituted phenols. The actual approach will be determined by the interaction energy with the active site. S^is represent the hypothetical receptor surface interacting with the ligand atoms or groups

problem arises in the case of 3,5-disubstituted derivatives. If we want to correlate the biological activity with one substituent, then the problem is to decide the appropriate substituent. If we want to correlate the biological activity with both the substituents separately, then the problem is to decide which one is to be considered as the 3-substituent and which one as the 5-substituent. Obviously the biological receptor does not know the rule by which we number these positions. Relative to the phenolic group the two substituents have the same positions. The receptor has its own selection process which is governed by the interaction energy. Change in orientation or conformation has been observed experimentally where the ligand–receptor binding has been studied at the molecular level.[5] Such considerations were rarely made in Hansch[1]- or Free–Wilson[6]-type QSAR studies. The situation becomes worse if the ligand molecules are conformationally very flexible. An *a priori* assumption of superimposition becomes so difficult if the ligands belong to diverse classes, that a unified interpretation totally fails. Distance geometry receptor modeling tries to overcome these situations by examining the actual three-dimensional structure of the ligand molecule. Apart from these advantages, it is conceptually very similar to the conventional Hansch approach to receptor mapping. Incidentally, whatever is done in distance geometry analysis can also be done using three-dimensional atomic coordinates, generally at the cost of substantially increased computation time.

22.4.2 MOLECULAR STRUCTURE AND DISTANCE GEOMETRY

22.4.2.1 Rigid Molecules

The three-dimensional structure of a molecule can be represented in various ways. The concept of a chemical bond is so deeply embedded in the chemist's mind that quite often molecules are represented in terms of bond distance, bond angle and torsion angle. The concept of crystal symmetry led the crystallographer to prefer fractional coordinates. However, atomic Cartesian coordinates are by far the most popular choice since they can be used directly to calculate the distance between any two atoms. In distance geometry the structure is represented by a distance matrix. One may dispute the efficiency of such representation on the ground that the geometry of a molecule can be represented by $3n-6$ ($3n-5$ for a linear molecule) coordinates, where n is the number of atoms, while there are $n(n-1)/2$ interatomic distances. For a molecule containing four or more atoms the number of interatomic distances is much higher. It is true that given these $3n-6$ independent distances, one can evaluate the other distances using geometric and trigonometric theories, but the distance matrix representation has advantages in structural comparison of two or more molecules.

22.4.2.2 Flexible Molecules

Most organic molecules are flexible. Flexibility may come from bond stretching, bond bending, and, most important, rotation around single bonds. For acyclic molecules one can easily express the flexibility by giving the range of bond distance, bond angle and torsion angle. Although all structures satisfying these ranges are not energetically equivalent, such a representation is a useful summary of the various possibilities. The situation, however, is much more complicated for cyclic molecules. Due to the constraints of ring closure, many of these variables become dependent on others.[7] Here flexibility is represented by the range of values of the independent variables, although many combinations of values of the independent variables satisfying these ranges may not give acceptable values for the dependent variables. Distance geometry representation of the flexibility of a molecule goes even one step further. Flexibility of a molecule is expressed by a distance range matrix showing the upper and lower bounds on the distance between each atom pair. Due to the constraints demanded by a molecular structure and interdependency of distances, many sets of distances satisfying this distance range matrix are unacceptable. However, distance geometry[8] allows one to decide not only whether the distances are acceptable in three-dimensional space, but calculates the closest set of these distances that is acceptable in a three-dimensional space. This problem will be dealt with in greater detail in the next section. The ease of handling this otherwise extremely difficult problem has encouraged the development of distance geometry.

22.4.3 DIMENSIONALITY OF SPACE, INTERDEPENDENCY OF ATOMIC DISTANCES AND DISTANCE GEOMETRY

22.4.3.1 Necessary Coordinates in Three-Dimensional Space

Suppose we have a molecule containing n atoms, and we want to know the number of coordinates necessary to represent the internal structure of the molecule. Since we are interested only in the internal structure and not in its position relative to a fixed frame of reference, the frame of reference can be defined relative to some atoms of the molecule. We can assume that the origin is at the nucleus of the first atom, the x axis is toward the second atom and the xy plane is defined by the plane of the first three atoms. Here the coordinates of the first atom are always fixed at (0,0,0), and the y and z coordinates of the second atom, and the z coordinate of the third atom are all fixed at zero. In other words, for the first atom we need no coordinates, for the second one coordinate, for the third two coordinates and for each of the fourth and higher atoms we need three coordinates. Altogether the number of independent variables is

$$N_c = [0 + 1 + 2 + 3(n - 3)] = 3n - 6 \qquad (1)$$

The number is independent of the coordinate system used to represent the molecule. The structure can also be represented in terms of bond length, bond angle and torsion angle. In that case for the first atom we need nothing, for the second atom the bond length, for the third atom one bond length and one bond angle and for each of the fourth and higher atoms, one bond length, one bond angle and one torsion angle. However, in both these cases we considered our well-known three-dimensional space.

22.4.3.2 Necessary Coordinates in Higher Dimensions

The number of coordinates will be different if the dimensionality changes. Let us consider a hypothetical space of dimension m. In order to define the space we need m suitable atoms. Here the number of coordinates is shown by equation (2)

$$N_c = [0 + 1 + 2 + \ldots + (m - 1) + m(n - m)] = mn - m(m + 1)/2 \qquad (2)$$

where n is the number of atoms in the molecule. On the other hand, the number of interatomic distances in the molecule is given by equation (3).

$$N_d = n(n - 1)/2 \qquad (3)$$

The condition that all these distances will be independent can be obtained by equating equation (2) with (3) to give equation (4).

$$mn - m(m + 1)/2 = n(n - 1)/2 \quad \text{or} \quad (n - m)(n - m - 1) = 0 \qquad (4)$$

Equation (4) clearly shows that the dimension of space should be at least $n - 1$, in order that all the distances be independent. Even under such conditions, the distances are subject to some inequality constraints. In order to illustrate the problem, let us consider a one-dimensional space having three points. Equation (2) shows that the number of independent variables here is two. For example, we can give the 1–2 and 2–3 distances independently, but due to colinearity the 1–3 distance will be the sum (or difference) of the 1–2 and 2–3 distances, and we cannot independently change it. The

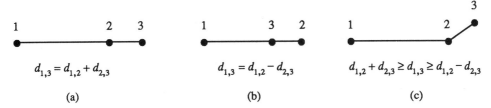

$$d_{1,3} = d_{1,2} + d_{2,3}$$

(a)

$$d_{1,3} = d_{1,2} - d_{2,3}$$

(b)

$$d_{1,2} + d_{2,3} \geq d_{1,3} \geq d_{1,2} - d_{2,3}$$

(c)

Figure 2 Position of atom 3 in a two-dimensional space for fixed 1–2 and 2–3 distances. ds represent the distance of the indicated points. (a) and (b) represent two limiting positions, (c) represents one of the intermediate positions

situation will be different if the space is two-dimensional. Here the number of independent variables becomes three. Three points can form different angles to attain different values of the 1–3 distance. Even here the 1–3 distance should be greater than or equal to the difference between the 1–2 and 2–3 distances and less than or equal to their sum, due to the triangle inequality (Figure 2). In actual molecules the problem is even more complex due to a large number of dependent variables.[9] However, if the independently assigned distances can be brought back to values in three dimensions in a physically realistic way, the enormous number of low energy conformations of a molecule can be easily considered in any process that needs a thorough search of the conformational space. The computer program and mathematics for obtaining the Cartesian coordinates from a set of consistent or inconsistent distance information was first suggested by Crippen.[10] The coordinates obtained by this approach give distances that correspond most closely to the independently chosen distances. This process is called *embedding*.

22.4.4 EMBEDDING POINTS IN THREE-DIMENSIONAL SPACE FROM DISTANCE INFORMATION IN HIGHER DIMENSIONS

Here we want to summarize the algorithm for obtaining three-dimensional coordinates from interatomic distance ranges. For a detailed explanation of these steps, see ref. 8. (i) Note the absolute handedness of the chiral centers; (ii) consider all combinations of three points and decrease the upper distance bound if the triangle inequality (the sum of two sides of a triangle is greater than or equal to the third side) is violated; (iii) similarly, increase the lower distance bounds for 'inverse triangle inequality' violations (as in Figure 2c); (iv) for each pair of atoms randomly select a trial distance between the new upper and lower bounds (random selection is not necessarily the only or best way to select the trial distances); (v) find the three-dimensional coordinates that best agree with the trial distances;[11] and (vi) refine the three-dimensional coordinates by minimizing a penalty function arising from the violation of the distance bounds or chirality constraints.

22.4.5 DISTANCE GEOMETRY AND MOLECULAR SUPERIMPOSITION

22.4.5.1 Rigid Molecules

Let us first consider the superimposition of two rigid molecules by the conventional approach. First, we have to define the mode of superimposition (in other words, which atom of the second molecule will be placed on which atom of the first). Second, we have to roughly match the two structures by placing three widely spaced atoms of the first molecule upon the corresponding atoms of the other by a rigid translation and rotation of the first molecule. Third, we have to improve the superimposition by an iterative technique.[12] If one is interested in all possible superimpositions, then all combinations should be considered. If we define molecular superimposition as the superimposition of one or more atoms of molecule 1 on the same number of atoms of molecule 2, the total number of possibilities is given by equation (5)

$$N = \sum_{i=1}^{n_2} [{}^{n_1}C_{n_2+1-i}] \times [{}^{n_2}P_{n_2+1-i}] \qquad (5)$$

where n_1 and n_2 are the number of atoms in the two molecules, and $n_1 \geq n_2$; ${}^{n_1}C_{n_2+1-i}$ stands for combination of n_1 taking $n_2 + 1 - i$ at a time $(n_1!/(n_1 - n_2 - 1 + i)!(n_2 + 1 - i)!)$ and ${}^{n_2}P_{n_2+1-i}$ stands for permutation of n_2 taking $n_2 + 1 - i$ at a time $(n_2!/(n_2 + 1 - i)!)$. We have to take a permutation in the second case, since if we pick up the same atom sets from the two molecules, a change in the order in one will lead to different superimposition. Even for medium-sized molecules this number becomes astronomically large. Since each superimposition entails calculating the translation and rotation matrix and applying them on each atom of the molecule, searching all possibilities is a very slow process. This technique is practical only for a limited number of superimpositions. Distance geometry allows a very fast way of checking the superimposition of one molecule over another. If atoms $n_1, n_2, \ldots, n_p$ of molecule n are superimposed on atoms $m_1, m_2, \ldots, m_p$ of molecule m respectively, a necessary condition for superimposition is

$$|d_{n_i, n_j} - d_{m_i, m_j}| \leq \delta \tag{6}$$

where d represents the distance between the indicated points, and δ represents the distance error acceptable for a superimposition. For rigid molecules, the distance geometry binding modes are similar to those obtained from three-dimensional coordinate manipulation, except that the distances do not change under reflection, while the coordinate system changes its handedness. In other words, the distance check alone permits superimposition of enantiomers. However, this problem can be solved by comparing the signed volume of the tetrahedron formed by four noncoplanar points.[9] The actual number of superimpositions may be somewhat less than the number given by equation (5), since some superimpositions involving a small number of atoms may be a subset of a superimposition with more atoms. Kuhl *et al.*[13] considered those possibilities and devised a very efficient algorithm for determining the various atom sets ('clique' in graph theoretical terms) of superimposing molecules.

22.4.5.2 Flexible Molecules

If the molecule is flexible the problem becomes worse. Earlier we had two combinatorial problems (choosing a set of atoms out of all atoms in each of the two molecules). Here we are adding one more combinatorial problem to the already lengthy procedure (comparing each conformation of the second molecule with all conformations of the first molecule). In distance geometry the problem can be solved in two different ways.

(i) Evaluate the upper and lower bounds on the distances between each atom pair in the energetically allowed conformations. Suppose the subscript u represents the molecule having higher upper bound and l for the molecule having lower upper bound for a particular interatomic distance. The contact condition is that the two distance ranges should have some overlapping region. If not, their ranges should not be separated by more than a preassigned small value δ, mathematically

$$d_l^{max} + \delta \geq d_u^{min} \tag{7}$$

Two important distance range cases are illustrated in Figure 3. Any superimposition allowed that way may be geometrically incorrect in three-dimensional space due to the interdependency of the distances. The feasibility of such a superimposition in three dimensions can be ultimately checked by *embedding* the common distance range of the superimposed atoms and the essential part of the molecular tree, that is the atoms connecting the superimposed atoms. This is faster than the ensemble distance geometry where one takes all atoms.[14]

(ii) Alternatively, one can form one distance range matrix to represent a small energetically allowed region of conformational space and use equation (7) as the contact condition without carrying out the embedding step. The distance range matrix here very closely satisfies the requirement of three-dimensional space and therefore guarantees an approximate superimposition.[9]

22.4.6 MODELING THE GEOMETRY OF THE ACTIVE SITE CAVITY BY DISTANCE GEOMETRY

22.4.6.1 The Basic Idea

Distance geometry modeling of the structure of the active site cavity is based on a very simple idea. Suppose we have two flexible ligand molecules m and n, and the atoms m_i and m_j of molecule m and atoms n_i and n_j of molecule n occupy the same respective regions of the active site. The distance between the *i*th and *j*th atoms in the two molecules must be very close in their active conformations (the conformations in which they bind with the receptor). Since in the distance geometry represen-

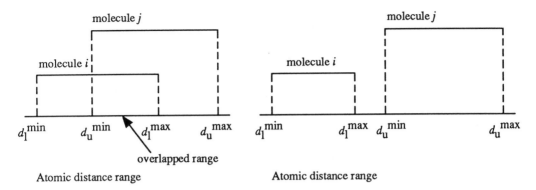

Figure 3 Left: interatomic distance having common distance range, right: having no common range. d_l represents the distance having lower upper limit, and d_u having higher upper limit

tation of the flexible molecules atomic distances have ranges, the active conformations should be represented by a common distance range. If we have several molecules, such comparisons will gradually decrease the range, thereby contracting the possible conformational region. Ultimately, embedding these distances will give the three-dimensional structure of the site pockets accommodating the ligand atoms.

22.4.6.2 Formulation of Hypothesis about the Mode of Superimposition of Ligands

The reasoning above has one flaw: we have not stated how to determine 'equivalent' ligand atoms at the receptor site. If one has the three-dimensional structure of the receptor, docking the molecule at the active site may give a possible answer. The approach giving the best binding energy gives the type of superimposition we are interested in. In the absence of the explicit structure of the receptor, which is the real life of a medicinal chemist, the answer to this question ought to be based on some assumptions, although it can be iteratively improved. Most of the approaches are based on a common ground, that is the structural similarity among ligands having similar biological activity. However, there are two difficulties. Although one can superimpose a molecule on another in an infinite number of ways, what should be the basis of preferring one of these superimpositions over another? For the present problem it should not be decided on the basis of the chemical or physical properties of the ligand alone. We can determine the binding energy of exactly the same set of ligands with two different receptors, and our method should not suggest the same superimposition (and hence the same active site geometry) for the two receptors. The other problem is including the binding energies of the ligands with the receptor to decide their initial approximate superimpositions. The three methods: (i) guided search; (ii) exhaustive search and (iii) partial search, illustrate the simplification of this otherwise complex problem.

22.4.6.2.1 Guided search

The guided search works when the medicinal chemist suggests a superimposition and wants to check its geometrical feasibility only. If it is feasible, one accepts it as the hypothesis and proceeds to the next step, until the model fails to fit and predict the biological data. At this point a modified hypothesis is used to create a new model. The steps in the guided search are very simple:

(i) Generate the ligand molecules using standard bond lengths and bond angles. There are standard interactive graphics programs for this, or one can connect fragments of crystallographic structures.
(ii) Generate all conformations by incrementing all combinations of the dihedral angles at a suitable interval. Use the energetically or sterically allowed conformations to generate the upper and lower distance bound between each atom pair in the molecule. If the conformational energy is the basis of deciding the allowed conformations, then one should use conformations having energy less than a specified amount compared to the global minimum. Sterically allowed conformations include those in which the van der Waals surfaces of atoms separated by rotatable bonds do not overlap.
(iii) Using the idea of bioisosterism[15] determine a set of atoms from each ligand molecule as a possible trial for superimposition. In the single distance range representation, determine the

common distance ranges for all superimposed atoms in the set of ligand molecules under study. If all molecules have common distance ranges for all superimposed atoms, use these to determine the site point coordinates by embedding.[16] Addition of the atoms connecting these superimposed atoms, as is done in ensemble distance geometry,[14] is sometimes helpful, since that may put more constraints on the geometry. Ensembling is helpful only if the connection between the superimposed atoms differs either in nature or in conformational features.

In the single distance representation, existence of common distance ranges for the superimposed atoms does not guarantee the superimposition in the three-dimensional space; even simple embedding using only the common distance ranges of the superimposed atoms does not guarantee this. Superimposition is properly verified by including the common distance ranges of the superimposed atoms, combined with the distances between the atoms connecting the superimposed atoms, as in the ensemble approach.

In order to simplify the situation conceptually, at the cost of computation, one can represent the energetically allowed regions of the conformational space by more than one distance range matrix. Each distance range matrix is generated from a small low energy region of the conformational space. The presence of at least one distance range matrix satisfying equation (6) warrants the approximate superimposition. Embedding the common distance ranges from these matrices will give the coordinates of the site point.

The guided search has the advantages that it is computationally much faster, and allows a hypothesis to be tested. The disadvantages are: (i) the model is biased by the intuition of the scientist; and (ii) if the structures of the ligand molecules are very diverse, it is extremely difficult to formulate any good hypothesis.

22.4.6.2.2 Exhaustive search

The exhaustive search is very similar to the above approach, except that all possible superimpositions on a selected molecule should be tried. This is the most desirable approach, but in most cases it is computationally too lengthy. In terms of atom to atom superimposition the number of superimpositions of even a medium-sized molecule over a similar one is enormous. The best superimposition can be estimated from 'property matching' of the atoms.[17] One such function is

$$F = \sum_{i=1}^{n_1} [|\phi_{1,i}| - |\phi_{1,i} - \phi_{2,i}|] \tag{8}$$

where n_1 represents the number of atoms in the first molecule (the molecule with which we want to compare), $\phi_{1,i}$ represents a physicochemical property of the ith atom (see next section for a detailed discussion on atomic physicochemical properties) of the first molecule, and $\phi_{2,i}$ represents the same property of the atom of the second molecule (the molecule we want to compare) superimposed on the ith atom of the first. If no atom is superimposed, the value of the property should be considered zero. The physical idea behind this function is that any deviation from the first molecule will put a penalty on the second molecule and the maximum value can be attained only by a molecule identical to the first one. If the interaction is a quadratic function of the property, and we assume that the first molecule has attained the maximum interaction at each atom, this function is a rough measure of its binding energy. This conclusion is true if this property is solely or largely responsible for the binding. Unfortunately, there is no guarantee that the currently available best bound ligand is the ultimate one, and also it is difficult to tell in advance which physicochemical property is largely responsible for the binding. Due to these problems one can think of other artificial functions for this purpose, such as

$$F = \sum_{i=1}^{n_1} \frac{\phi_{1,i}/|\phi_{1,i}|}{\phi_{2,i}/|\phi_{2,i}|} |\phi_{2,i}| \tag{9}$$

The idea behind this function is that if the sign of the property of the atoms in two molecules matches, the function gets a score equal to the absolute value of the property of the atom in the second molecule. If the interaction of the ligand with the receptor is a linear function of the property, this function will be proportional to its binding energy. The best values of the function in equation (8 or 9) can be correlated with the binding energy of the ligands to decide the best hypothesis. However, the use of these functions is still under development, and it would be unwise to comment on their success or relative merit at this stage. See ref. 17 for the current status of this approach.

22.4.6.2.3　*Partial search*

The enormity of the number of superimpositions of even medium-sized molecules hinders the total search over all possible superimpositions. A 'partial search' is helpful in such a situation. Here a limited number of 'important' atoms are used for the initial superimposition. The superimposition of the rest of the atoms are considered only as a consequence of the superimposition of the 'important' atoms. Although it is difficult to select the important atoms in the binding process, heteroatoms, carbonyl or carboxyl carbon and hydrogen attached to heteroatoms are the most likely candidates. This approach, no doubt, is a better way to tackle the problem rather than removing the 'unimportant' atoms from the molecule as was done in an earlier approach.[3]

22.4.6.3　Site Point, Site Cavity and Site Pocket

The coordinates determined by the above procedure represent the approximate relative locations of the nuclei of the ligand atoms at the receptor site. Although we call them 'site points', they are by no means site atoms. The minimum distance from the site point to the van der Waals surface of the site is the size of the atom occupying it. In other words, if we expand each site pocket to a sphere having a radius equal to the van der Waals radius of the atom occupying it, we can get an approximate shape of the 'site cavity' accommodating the incoming ligand. At this point it is difficult to tell whether a ligand atom remains in the solvent during binding with the receptor, or in a region surrounded by site atoms and so is a part of the receptor cavity. However, if a point is located in the solvent, it will be reflected by its small contribution towards binding energy. This feature will be considered in the next section. Due to the approximate nature of the location of the site points and the tolerance of the site in accommodating a ligand atom, each site point is given a small radius. Such a sphere is termed a 'site pocket'. A ligand atom nucleus occupying any part of a site pocket is assumed to experience equal interaction with the receptor. The boundary formed by the surface of the sphere is not a physical barrier, but signifies only the change in the nature or intensity of the interaction.

22.4.7　ATOMIC PHYSICOCHEMICAL PROPERTIES FOR EVALUATING THE GOODNESS OF MOLECULAR SUPERIMPOSITION AND MODELING INTERMOLECULAR INTERACTION

Until now we have considered a molecule as a collection of points characterized by the position of the nuclei. The recognition of the ligand by the receptor is done not only by the relative positions of the ligand atoms but also by the 'nature' of the atoms. Certain physicochemical properties which are important in the molecular recognition process by the receptor can be used to express their nature. The most important properties among these are formal charge density, electrostatic potential at the van der Waals surface, atomic refractivity and hydrophobicity. Any standard molecular orbital program[18] can be used to calculate the first two properties. However, assigning atomic refractivity and hydrophobicity is a difficult problem. Hydrophobicity has the additional disadvantage of being a poorly understood property.[19] The medicinal chemists use water–octanol partition coefficients (P) as a measure of hydrophobicity. Both molar refractivity and octanol–water partition coefficient are partly additive and partly constitutive. Various empirical methods are available to calculate these molecular properties (see Chapter 18.7). In these approaches, the atoms or molecular fragments are given some standard values, and the change in their values due to intramolecular interaction is covered by some correction factor. The approach taken by the authors was to use atomic values in order to relate the contribution of each atom in its site pocket to the total binding. To get such values, a large number of atom types were defined to cover the difference due to the structural environment. Finally it was assumed that simple addition of these values will give the molecular values. In this approach the intramolecular interaction was kept hidden in the atom classification. The atomic hydrophobicity[20] and refractivity[21] currently being used by the authors are shown in Table 1.

At the beginning of the model-building process, one is not sure about the relative importance of the various properties in the recognition process by the receptor. If one is interested in using more than one property to decide the goodness of molecular superimposition using expression (8) or (9), it is advisable to scale the properties, in order to give them equal weight.[17]

Table 1 Classification of Atoms, Their Contributions to Molar Refractivity and Hydrophobicity

Type	Description[a]	Atomic refractivity[b]	Partition Coefficient[c]
	C in		
1	CH_3R, CH_4	2.3000	−0.6037
2	CH_2R_2	2.3071	−0.4295
3	CHR_3	2.4926	−0.3426
4	CR_4	2.3000	−0.1155
5	CH_3X	3.4006	−1.0578
6	CH_2RX	3.2624	−0.8188
7	CH_2X_2	3.6770	−0.1540
8	CHR_2X	3.0137	−0.5995
9	$CHRX_2$	3.2250	0.0095
10	CHX_3	3.2401	0.5134
11	CR_3X	2.6140	−0.4807
12	CR_2X_2	3.1488	0.2853
13	CRX_3	2.3010	0.5335
14	CX_4	3.3559	1.1114
15	$=CH_2$	3.5071	−0.1654
16	$=CHR$	4.4814	−0.1033
17	$=CR_2$	3.7781	−0.2330
18	$=CHX$	3.6211	−0.0649
19	$=CRX$	4.4310	−0.7814
20	$=CX_2$	3.2000	0.1734
21	$\equiv CH$	3.4161	0.0859
22	$\equiv CR$, $R=C=R$	4.3043	0.1335
23	$\equiv CX$	3.4905	—
24	R--CH--R	3.4127	−0.0220
25	R--CR--R	4.3725	0.1596
26	R--CX--R	3.8182	−0.0064
27	R--CH--X	2.5001	0.0245
28	R--CR--X	2.5000	0.1114
29	R--CX--X	2.7967	−0.2378
30	X--CH--X	2.5000	0.2921
31	X--CR--X	—	0.8471
32	X--CX--X	2.5000	0.3002
33	R--CH$\cdot\cdot$X	3.4372	0.0183
34	R--CR$\cdot\cdot$X	3.4494	−0.2625
35	R--CX$\cdot\cdot$X	3.1048	−0.2959
36	Al—CH=X	3.8251	−0.1243
37	Ar—CH=X	4.5401	0.3310
38	Al—C(=X)—Al	3.7529	0.5353
39	Ar—C(=X)—R	4.1288	−0.2182
40	R—C(=X)—X, R—C$\equiv$X, X=C=X	2.7938	0.0278
41	X—C(=X)—X	2.4165	0.3514
42	X--CH$\cdot\cdot$X	3.0606	−0.3040
43	X--CR$\cdot\cdot$X	2.5001	−0.0102
44	X--CX$\cdot\cdot$X	2.5001	−0.1746
45	Unused	—	—
	H attached to[d]		
46	$C_{sp^3}^0$, having no X attached to next C	1.1461	0.4234
47	$C_{sp^3}^1$, $C_{sp^2}^0$	0.8000	0.3610
48	$C_{sp^3}^2$, $C_{sp^2}^1$, C_{sp}^0	0.8006	0.1183
49	$C_{sp^3}^3$, $C_{sp^2}^2$, $C_{sp^2}^3$, C_{sp}^3	0.8001	−0.1573
50	Heteroatom	0.8000	−0.2106
51	α-C	1.0026	0.1869
52	$C_{sp^3}^0$, having one X attached to next carbon	1.1461	0.3546
53	$C_{sp^3}^0$, having two X attached to next carbon	1.1461	0.2676
54	$C_{sp^3}^0$, having three X attached to next carbon	1.1461	0.3528
55	$C_{sp^3}^0$, having four or more X attached to next carbon	1.1461	—
	O in		
56	Alcohol	1.4430	−0.0876
57	Phenol, enol, carboxyl OH	1.4090	0.1665
58	=O	1.6506	−0.2473
59	Al—O—Al	1.2000	0.0380
60	Al—O—Ar, Ar_2O	1.8434	0.1938
	R$\cdot\cdot$O$\cdot\cdot$R, R—O—C=X		
61[e]	--O	1.6001	1.4968
62–65	Unused	—	—

Table 1 (*Contd.*)

Type	Description[a]	Atomic refractivity[b]	Partition Coefficient[c]
	N in		
66	$Al-NH_2$	2.5001	−0.2577
67	Al_2NH	2.5001	0.0266
68	Al_3N	2.5377	0.1680
69	$Ar-NH_2, X-NH_2$	3.6195	−0.0362
70	$Ar-NH-Al$	2.9832	0.0274
71	$Ar-NAl_2$	3.9733	0.5799
72	$RCO-N{<}, {>}N-X{=}X$	3.0059	−0.2736
73	$Ar_2NH, Ar_3N, Ar_2N-Al, R\cdots N\cdots R^f$	2.6295	0.3323
74	$R{\equiv}N, R{=}N-$	3.1464	0.1849
75	$R--N--R^g, R--N--X$	4.5123	−0.0545
76	$Ar-NO_2, R--N(--R)--O^h, RO-NO_2$	4.7725	−2.6143
77	$Al-NO_2$	3.0389	−2.6455
78	$Ar-N{=}X, X-N{=}X$	3.6838	0.5466
79–80	Unused	—	—
	F attached to		
81	$C^1_{sp^3}$	0.8060	0.4093
82	$C^2_{sp^3}$	0.8000	0.1590
83	$C^3_{sp^3}$	1.3484	0.1890
84	$C^1_{sp^2}$	0.8000	0.5035
85	$C^{2-4}_{sp^2}, C^1_{sp}, C^4_{sp}, X$	1.6440	0.2550
	Cl attached to		
86	$C^1_{sp^3}$	5.3647	0.9282
87	$C^2_{sp^3}$	5.6484	0.4659
88	$C^3_{sp^3}$	5.6858	0.4381
89	$C^1_{sp^2}$	5.0000	0.9036
90	$C^{2-4}_{sp^2}, C^1_{sp}, C^4_{sp}, X$	5.9312	0.5302
	Br attached to		
91	$C^1_{sp^3}$	8.3379	1.0474
92	$C^2_{sp^3}$	8.5393	0.5809
93	$C^3_{sp^3}$	8.8635	0.5407
94	$C^1_{sp^2}$	8.0866	1.1743
95	$C^{2-4}_{sp^2}, C^1_{sp}, C^4_{sp}, X$	9.0569	0.8656
	I attached to		
96	$C^1_{sp^3}$	13.7535	1.4378
97	$C^2_{sp^3}$	13.6306	—
98	$C^3_{sp^3}$	13.4586	—
99	$C^1_{sp^2}$	12.8876	1.7028
100	$C^{2-4}_{sp^2}, C^1_{sp}, C^4_{sp}, X$	13.5530	0.8654
101–105	Unused halogens		
	S in		
106	$R-SH$	7.7751	0.7412
107	$R_2S, RS-SR$	7.3151	0.7598
108	$R{=}S$	9.2916	0.2968
109	$R-SO-R$	5.3957	−0.2515
110	$R-SO_2-R$	5.4662	0.0425

[a] R represents any group linked through carbon; X represents any heteroatom (O, N, S and halogens); Al and Ar represent aliphatic and aromatic groups respectively; $=$ represents double bond; $\equiv$ represents triple bond; $--$ represents aromatic bonds as in benzene or delocalized bonds such as the N—O bond in nitro group; $\cdots$ represents aromatic *single* bonds such as the C—N bond in pyrrole. [b] Atomic refractivity of only one atom. [c] 850 compounds were used to evaluate these atomic partition coefficient values, and they were tested by predicting the partition coefficient of 125 compounds.[20b] [d] The subscript represents hybridization and the superscript its formal oxidation number. [e] As in nitro, $=$N oxides. [f] Pyrrole-type structure. [g] Pyridine-type structure. [h] Pyridine *N*-oxide type structure.

22.4.8   ALTERNATE BINDING MODES: EVALUATION AND JUSTIFICATION

By the term alternate binding mode, we mean which atom of the ligand goes to which site pocket. Due to rigid rotation and translation and conformational flexibility a ligand molecule can occupy a site cavity in an almost infinite number of ways. Evaluation of the geometrically feasible binding modes of the ligand at the receptor site is an important part of the present approach. Since we do not know the functional form of the interaction, we cannot guarantee that the binding mode suggested from the guided search or exhaustive search, using the simple function as in equations (8) or (9), will be the energetically optimal binding mode once such an interaction function has been suggested. In the next step, therefore, the important alternate geometrically feasible binding modes are evaluated. In order to keep the number of alternate binding modes to a minimum, we recommend keeping the rigid rotation and translation to a minimum and concentrating on the binding modes arising mainly from the low energy conformations. It is also advisable to neglect any binding mode in which a large part of the ligand goes beyond the site cavity. The alternate binding modes have several effects on the calculation. They expose various obvious faults in the original binding hypothesis formulated on the basis of either 'guided search' or 'exhaustive search'. One bad effect is that, in general, it puts more constraints on the problem, thereby making the curve fitting more difficult. However, in one example the effect of alternate binding modes was tested and found to increase the predictive power of the model.[22]

22.4.9 MODELING THE INTERACTION BETWEEN THE LIGAND AND THE RECEPTOR

If we knew the structure of the active receptor site, the interaction of the ligand with the receptor could be modeled in terms of the van der Waals, electrostatic and hydrogen bonding interactions. The explanation of the binding constant also requires the entropic factors arising from the flexibility of the ligand and the receptor and the structuring of the biophase around them before and after binding.[23] Since the structure of the receptor is not known in most cases, in the distance geometry based 3-D QSAR, the problem is solved by modeling the interaction as a function of one or more physicochemical parameters of the ligands. Here also we recommend the three properties: octanol–water partition coefficient, atomic refractivity and formal charge density from any standard molecular orbital approach. The basic idea here is very simple. When an atom of the ligand enters a site pocket, it interacts with the site and the interaction energy is a function of one or more physicochemical properties of the atom, where the nature of the function is determined by the characteristics of the active site. Summing up the interactions in the various parts of the active site will give the binding energy. Assuming a linear function for the interaction, the binding energy of a particular binding mode is given by

$$E_{calc} = -CE_c + \sum_{i=1}^{n_s} \sum_{j=1}^{n_p} \left[C_{i',j} \sum_{k=1}^{n_o} P_j(t_k) \right] \tag{10}$$

where E_c is the energy of the conformation under consideration; Cs are coefficients, characteristic of site type and the physicochemical property (determined by some optimization technique as discussed in the next section); i' is the type of the site i; n_s represents the number of site pockets; n_p represents the number of physicochemical properties to be correlated with the site; n_o represents the number of atoms occupying the site pocket; and $P_j(t_k)$ represents the jth physicochemical parameter of the occupied atom of type t_k. The advantage of using physicochemical parameters to model the interaction of the ligand with the receptor is well known. Each physicochemical property in use represents a particular type of force active in the ligand receptor interaction. A good correlation indicates the nature of the receptor atoms surrounding the hypothetical pocket in our model. If the interaction between the ligand and the receptor is represented in terms of some abstract types, as was done in the early studies of distance geometry,[3] predicting the binding energy of a ligand containing a foreign atom is not possible. Sometimes physicochemically similar atoms are assigned very different interactions, which makes the physical interpretation of the model difficult. These are also the factors responsible for the popularity of Hansch-type QSAR over the Free–Wilson approach. Atomic physicochemical parameters were used for modeling the interaction, rather than atom group parameters. A large substituent occupies a large space, and because of the differences in the nature of the site atoms surrounding such a substituent, it may experience different types of interactions in different regions. Such differences may be modeled only if the physicochemical properties of the subfragments are used. The conformation of a group can also affect the interaction with the receptor. Such changes are difficult to study using overall physicochemical properties.

22.4.10 MATHEMATICAL LIMITATION ON THE NUMBER AND TYPES OF THE SITE POCKETS

In order to differentiate the nature of interaction at the various regions of the site cavity, the site pockets should be given different types. However, it is not obvious how many site pockets are of major importance in the binding process and how many different types should be used to differentiate their nature of interaction. There are at least some physical or mathematical limitations on these numbers: (i) a site pocket occupied by only one molecule should not be included in the model unless it has the same type as another pocket which does interact with more than one molecule; (ii) a site pocket occupied by only one type of atom in different molecules should not be used for the study; however, all such pockets combined can be studied as a single parameter; (iii) the classification of the site types should not be such that a particular type will be occupied by the same number and types of atoms; and (iv) the number of adjustable parameters either should be much lower than the number of compounds used in the data set, or the optimized parameters should be tested by predicting the biological activity of a respectable number of compounds not included in the model-building process.

22.4.11 EVALUATION OF THE RECEPTOR DEPENDENT PARAMETERS: THE OPTIMIZATION PROCEDURE

The theory behind the evaluation of the receptor dependent parameters, the Cs of equation (10), is conceptually very simple, although the mathematical description may be somewhat cumbersome to the practicing medicinal chemist. In conventional QSAR, there is only one expression for the calculated binding energy. They follow the lock and key interpretation for the drug–receptor interaction. Even if a part of the ligand experiences repulsion, no rigid rotation or translation or conformational change is allowed to avoid the interaction. The problem in such a case is straightforward. Assign the values of the Cs of equation (10) so that the calculated binding energy has minimum deviation from the observed value. Mathematically

$$\text{minimize } F = \sum_{i=1}^{m} (E_{\text{calc}} - E_{\text{obsd}})^2 \tag{11}$$

where m represents the number of molecules. This optimization is the well-known least squares solution.

In the present approach we have more than one geometrically feasible binding mode. Different modes represent different expressions for the calculated binding energy (equation 10). A relatively straightforward situation is where we know or can hypothesize the binding mode. In this case, we have to satisfy some additional conditions to make the model physically realistic. The calculated binding energy of the hypothesized mode should be higher (energetically more favorable) than that of the rest of the binding modes. Mathematically, this is shown by equation (12)

$$\text{minimize } F = \sum_{i=1}^{m} (E_{\text{calc}}(\text{HM}) - E_{\text{obsd}})^2 \tag{12}$$

under the constraints of equation (13)

$$E_{\text{calc}}(\text{HM}) > E_{\text{calc}}(\text{RM}) \tag{13}$$

where HM stands for the hypothesized mode and RM represents the rest of the geometrically feasible binding modes. This constrained solution is equivalent to the standard quadratic programming problem. Standard computer programs[24, 25] are available to solve the problem.

The situation may not be that simple in many cases. If we knew the parameters, C, we could determine the calculated binding energy as well as the optimal binding mode by evaluating the binding energy of the various geometrically allowed binding modes. However, the Cs can be evaluated only on the basis of some hypothesized binding modes. This is an interlocking problem. Furthermore, the hypothesized binding modes may not be mathematically consistent. Therefore a revised algorithm is used for the optimization. The present form of this algorithm is slightly different from the one reported earlier[26] and is presented in Figure 4.

The advantage of this algorithm is that it permits keeping the hypothesized binding mode until it is found to be mathematically inconsistent. In that situation it changes the mode to the

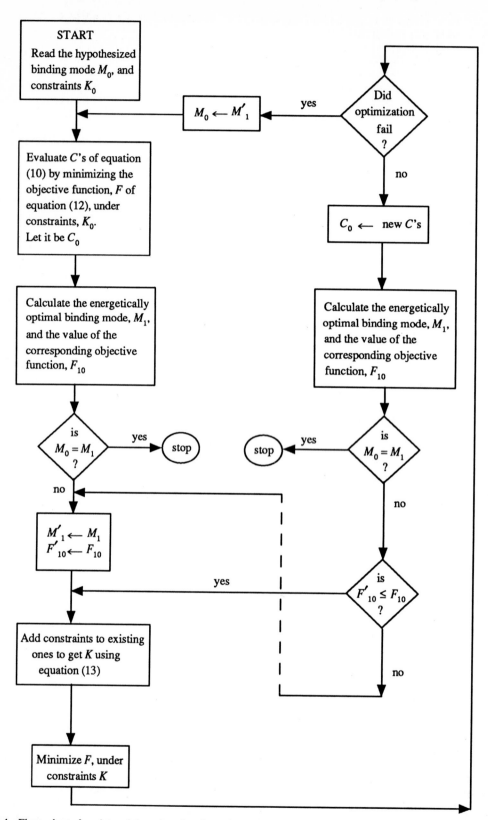

Figure 4 Flow chart for determining the site dependent parameters. For minimizing the objective function F (see equation 12) standard quadratic programming is used. M_0 represents the starting binding mode which determines the objective function F. Initially the optimization procedure is not aware of the alternate geometrically feasible binding modes. The unconstrained or not sufficiently constrained solution often finds a different geometrically binding mode to be energetically optimal

one found to have the best fit in the ongoing optimization. In a one-dimensional problem the binding mode will be mathematically inconsistent if, for example, the two constraints shown in equations (14) and (15) apply.

$$C > k \tag{14}$$

$$C < k \text{ (where } k \text{ is a constant)} \tag{15}$$

22.4.12 DIFFICULTY OF JUDGING DISTANCE GEOMETRY RESULTS BY STANDARD STATISTICS

Compared to conventional QSAR, distance geometry based three-dimensional QSAR usually requires more parameters. Statistically minded readers should be cautious in estimating the validity of the approach from the number of adjustable parameters alone. In conventional QSAR, each molecule represents one equation. Some function of the observed biological activity represents the dependent variable (b_i) of equation (16). The dependent variable is assumed to be the linear function of a few properties (x_{ij}) of the ligands.

$$\sum_{j=1}^{m} a_j x_{ij} = b_i \quad i = 1, \ldots, n \tag{16}$$

If any two sets of properties are linearly independent, an exact solution of the problem can be obtained only when $m = n$. While such a solution is good for estimating the existing data, it often does not have any predictive power. If m is less than n, in general no exact solution can be obtained. However, using the principle of least squares, it is possible to get a set of values for a_j that can give a good estimate of the dependent variables. Statistics indicates that if a good estimate of the dependent variables are obtained in terms of a relatively small number of parameters (a_j), then the predictive power of equation (16) will be high.

Although distance geometry based receptor modeling in its final presentation expresses the binding energy in terms of an equation like (16), it has one basic difference with conventional QSAR. It tries to partly account for the dynamic behavior of the ligand–receptor interaction. (The reader should not confuse this aspect with molecular dynamics calculations). It assumes that the ligand molecule is conformationally flexible and has free rotation or translation relative to the hypothesized active site (see Section 22.4.8). Such factors may give a large number of possible binding modes. The binding energy corresponding to these possible binding modes can be calculated using equation (16). The hypothetical mode can be the actual mode only if it yields maximum binding energy (see equation (13) and Section 22.4.11 for details). These are *inequalities*, and they mathematically closely resemble equations. If the geometric interpretation of an equation is a curve, then an inequality represents one side of that curve. Alternatively, we can think that an equation excludes from the solution both sides of a curve in the parameter space, whereas an inequality excludes only one side. If we fix the number of adjustable parameters, it is well known in curve fitting that increasing the number of equations always poses difficulties in getting a good fit. Our inequalities work in the same way. In other words, distance geometry receptor modeling often needs more adjustable parameters than conventional QSAR does, as a consequence of these extra inequality constraints. The presence of inequalities also demanded a different mathematical technique to solve our problem (quadratic programming instead of simple least squares). The different mathematical technique, in turn, makes the statistical evaluation of the method difficult, if not impossible.

22.4.13 APPLICATION OF THE METHOD

In its various levels of improvement, the method was applied to different data sets. In the initial stages it was applied to construct a model which could explain the biological activity of one class of compound. Initially, the inhibition data of 68 quinazolines of *S. faecium* dihydrofolate reductase inhibitors were fitted with 11 site points.[3] Later it was found that removal of two site points increased the goodness of fit. However, decreasing the number beyond nine worsened the fit.[27] In the next phase it was applied to generate a unified model for different classes of compounds where they are known to interact at the same receptor site. These models were also applied to predict the biological activity of a large number of compounds not included in the model-building process. A model was constructed from the inhibition of rat liver dihydrofolate reductase by 33 triazines and 15 quinazolines. The model successfully predicted the biological activity of 91 compounds, mostly quinazolines.[28] Guided by the success of the above model, we generated another unified model, using 62 compounds of five different classes of rat liver dihydrofolate inhibitors. The classes of compounds included triazines, quinazolines, pyrroloquinazolines, pyrimidines and pyridopyrimidines. The model successfully predicted the biological activity of 33 compounds of five different classes, one molecule of which was a member of a new class not included in the original data set.[29] Finally, one very convenient way of using physicochemical properties in the distance geometry approach was suggested. This approach gave a very clear picture of the complementary receptor site geometry of *E. coli* dihydrofolate reductase.[22] The model was constructed from 25 pyrimidines and 14 triazines. The general structures of these compounds are shown in Figure 5. It consisted of 19 site pockets of nine different types. The binding properties of these site pockets are illustrated in Figure 6. In order to check the effect of the alternate geometrically feasible binding modes, the study was made in two different ways: one in which the alternate binding modes were kept and in the other where they were removed. As expected, the data set was better fitted in the absence of the alternate binding modes due to the absence of inequality constraints. However, when the parameters obtained from these two different studies were allowed to predict the biological activity of five triazines and five pyrimidines, the study using alternate binding modes was found to be superior. The statistics of the two studies are given in Table 2. The constructed active site model was compared with the X-ray crystal structure of methotrexate bound (Figure 7) to *E. coli* dihydrofolate reductase. The comparison showed three interesting features: (i) all the site pockets are sterically accessible, *i.e.* they do not overlap with the receptor site atoms; (ii) three site points portrayed as hydrogen bonding are surrounded by groups capable of hydrogen bonding and (iii) the pockets showing high correlation with the hydrophobic property of the ligand are surrounded by various hydrophobic groups.[22] The model gave a few interesting explanations of the biological activity of the various ligands. Unlike benzylpyrimidine, the benzyltriazine was a weakly bound ligand. It showed that the rotation of the benzyl group relative to the heterocyclic ring was totally different in the two systems. The steric hindrance between the two methyl groups and the phenyl ring in the triazine keeps the phenyl ring perpendicular to the heterocyclic ring, while in the pyrimidine the phenyl ring comes very close to the plane of the heterocyclic ring. This, in turn, led us to suggest the synthesis of some demethylated benzyltriazines as a possible inhibitor of *E. coli* dihydrofolate reductase. Unlike most triazines, 3-phenylpropyltriazine is found to be a strongly bound ligand. The model suggested that the phenyl ring in this compound goes to a strongly hydrophobic site pocket which should be in a remote position from the heterocyclic ring. It also suggested the synthesis of some benzylpyrimidines in which one hydrogen of the methylene group has been replaced by —CH$_2$CH$_2$Ph. With some modification of the original method, Linschoten *et al.*[30] applied the distance geometry method for mapping the turkey erythrocyte β-receptor. They described the binding data using nine site points. Sheridan *et al.*[14] using their ensemble distance geometry, evaluated the nicotinic pharmacophore geometry.

Given the interaction coefficients (*C*s of the interaction function), geometry of the site cavity and the structure of a molecule, one can easily calculate its binding energy from the geometrically feasible binding modes. The actual binding energy will correspond to the binding mode yielding highest binding energy. The interaction coefficient and the site geometry can also be used to predict structures expected to have higher binding energy.[31]

A largely modified approach[17] of this method was used to model the binding site cavity of parainfluenza virus from the virus rating of 28 ribonucleosides. The model predicted various interesting aspects of the ligand–receptor interaction and the structural requirement for better antiviral activity. Some of these findings confirmed earlier conclusions from extensive structure–activity studies among these compounds. It supported the binding conformation of ribavirin, a broad antiviral agent; however, it showed that the binding conformation is not the minimum energy conformation. The difference resulted from the orientation of the amide group. It

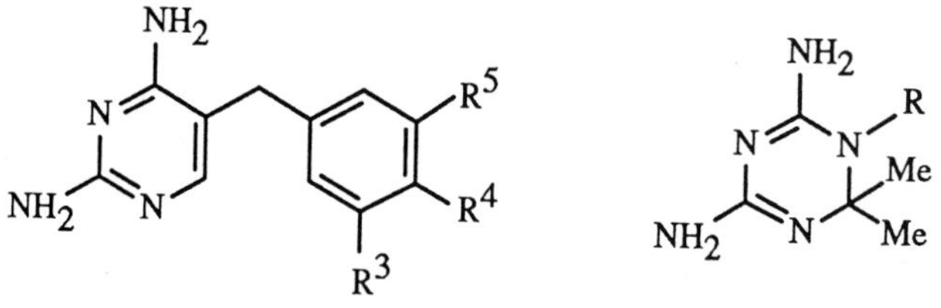

Figure 5 General structure of the pyrimidines and triazines used to construct the *E. coli* dihydrofolate reductase active site model

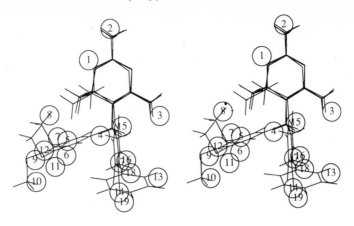

Figure 6 Two triazines and a pyrimidine derivative superimposed over the active site model

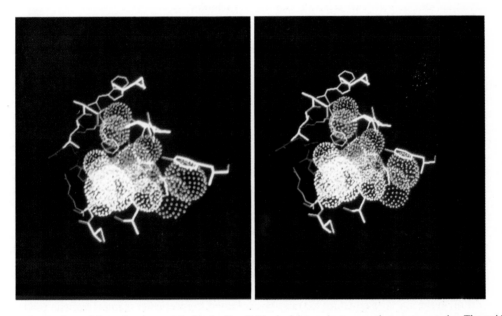

Figure 7 Comparison of the active site model with the *E. coli* dihydrofolate reductase–methotrexate complex. The residues within 5 Å of methotrexate are shown in the picture for clarity. The color coding in the picture is as follows: methotrexate: red; oxygens in the receptor: magenta; nitrogens in the receptor: green; carbons: white; hydrogen-bonding site pockets: yellow; hydrophobic site pockets: cyan; dispersive or partly dispersive partly hydrophobic site pockets: white

Table 2 Statistics of *E. coli* Dihydrofolate Reductase Site Model

Study	Purpose	No. of compounds	No. of variables	No. of constraints	Correlation coefficients	Standard deviation	Max. error
Ia	Data fitting	39	13	11	0.893	0.530	1.29
Ib	Prediction	10	0	0	0.941	0.893	2.10
IIa	Data fitting	39	13	0	0.961	0.326	0.96
IIb	Prediction	10	0	0	0.779	2.627	7.86

suggested that five-membered heterocyclic nucleosides will better fit the binding site for this virus. The nucleosides containing five- and six-membered fused heterocyclic rings experience steric repulsion at the six-membered ring. It also gave the rationale behind the failure of some weakly active nucleosides having properly oriented amide groups and five-membered heterocyclic rings.

22.4.14 POSSIBLE ABUSE OF THE METHOD

Like most scientific techniques, one can abuse these computational methods. The distance geometry receptor modeling may also be abused if the philosophy of this type of calculation is not properly understood. The biological activity of a different class of compounds may come from the binding of the ligand either in the same region or in different regions. If they interact at different regions, construction of a unified model may be nonsense. Getting the receptor site geometry from the common distance range of different ligands may be misleading if the range is not appreciably smaller than the flexible ligands. This type of situation may arise if the ligand molecules are relatively flexible and do not differ conformationally from one another. Here embedding can give an infinite number of solutions. Calculation of the conformational energy of the ligands for each possible site geometry may help. For example, some solutions may give low energy conformations for the strongly bound ligands only. The danger here is that if a strongly bound ligand does not bind in its global minimum energy conformation, such an approach will yield an incorrect solution. By the term biological activity we mean the affinity of the ligand for the purified receptor. Any other biological activity includes contributions from several other factors, depending on the particular type of activity. If the biological activity is not the affinity of the ligand for the purified receptor, proper care should be taken to separate the other variables.

22.4.15 COMPARISON WITH TWO DIFFERENT EXTREMES OF THE QUANTITATIVE DRUG DESIGN APPROACHES

Quantitative drug design has two extreme approaches. In the Hansch-type approach a purely empirical correlation is made between the biological activity and the physicochemical properties of the overall ligand or some parts of it. This approach is computationally very fast, and the correlation of different physicochemical properties with different parts of the ligand molecule gives some idea about the nature of interaction around different parts of the ligand with the receptor. However, due to drastic simplification of the otherwise very complex biological process, its success is limited even after twenty years of extensive use by a large number of workers in this field.

The other extreme involves the identification and purification of the biological receptor, crystallization of the ligand–receptor complex and solution of the structure of the complex using X-ray crystallography. This part involves the experimental identification of the active site. Once it is done, one can remove the ligand molecule, insert other molecules in a comparable orientation, and minimize the energy of the complex with respect to the geometry of the ligand atoms and the site atoms close to the active site. The interaction thus calculated corresponds to the internal energy or enthalpy in many cases. The stability of the ligand–receptor complex, however, is related to the free energy change. The entropic factor may be separated either experimentally by the effect of temperature on the stability constant, or may be evaluated theoretically using a realistic model of the biophase and its structuring around the ligand and the active site before and after binding. Unfortunately, such a direct solution is an extremely difficult and slow process. Also the structure of the biophase is not well understood. The structure of a ligand–receptor complex has been solved in a very few cases. From the medicinal chemist's point of view, this direct approach can seldom be employed due to the lack of input data.

The distance geometry approach to receptor modeling is intermediate between these two extremes. It takes the three-dimensional structure of the ligand molecules in terms of energetically allowed conformations. In the absence of the explicit structure of the receptor, it tries to hypothesize it in terms of site pockets. It also tries to figure out the nature of the site atoms surrounding a site pocket from the nature of the physicochemical property of the ligand that correlates the binding energy of the ligand with the receptor. Being an indirect approach, like most other similar approaches,[32] it has some steps that may seem to be subjective. However, current research is directed towards increasing its objectivity.

22.4.16 SUMMARY

In order to rescue the reader from the various complexities of distance geometry receptor modeling, we want to collect the most important steps of the approach in a straightforward way.

(i) Generate an approximate three-dimensional structure of the ligand.

(ii) Generate the low energy conformations using constant valence structure conformational analysis.

(iii) Evaluate the minimum and maximum distances between the ligand atoms in its energetically allowed conformations.

(iv) Make a hypothesis regarding the binding mode of the ligand at the receptor site. A computerized search can be used to make a good hypothesis at this stage.

(v) Evaluate the common distance range of the superimposed atoms in the various ligands.

(vi) Use the common distance range of the superimposed atoms and the distance range of the atoms connecting the superimposed atoms in 'embedding' the atoms in three-dimensional space.

(vii) Evaluate the alternate binding modes due to rigid rotations and translations and conformational flexibility. In order to keep the number to a manageable level, it is advisable to keep the rigid rotation and translation to a minimum and to take only the low energy conformations of the ligand molecule.

(viii) Classify the site pockets into different types to differentiate the nature or intensity of interaction. One should keep in mind that the number of site pocket types should be kept to a minimum in order to get statistically reliable parameters.

(ix) Evaluate the empirical parameters using a constrained least squares technique (quadratic programming).

(x) If the parameters fit the binding data of the training set well, use the model to predict the binding energy of some ligands not included in the data set.

(xi) If the predictive power of the model is acceptable, use it to predict structures having good or better biological activity.

22.4.17 REFERENCES

1. C. Hansch and T. Fujita, *J. Am. Chem. Soc.*, 1964, **86**, 1616.
2. L. B. Kier, 'Molecular Orbital Theory in Drug Research', Academic Press, New York, 1971.
3. G. M. Crippen, *J. Med. Chem.*, 1979, **22**, 988.
4. Y. C. Martin, *J. Med. Chem.*, 1981, **24**, 229.
5. J. T. Bolin, D. J. Filman, D. A. Matthews, R. C. Hamlin and J. Kraut, *J. Biol. Chem.*, 1982, **257**, 13650.
6. S. M. Free, Jr. and J. W. Wilson, *J. Med. Chem.*, 1964, **7**, 395.
7. J. B. Hendrickson, *J. Am. Chem. Soc.*, 1961, **83**, 4537.
8. G. M. Crippen, in 'Distance Geometry and Conformational Calculations: Chemometric Research Studies', ed. D. Bawden, Wiley, Chichester, 1981, vol. 1.
9. A. K. Ghose and G. M. Crippen, *J. Comput. Chem.*, 1985, **6**, 350.
10. G. M. Crippen, *J. Comput. Phys.*, 1977, **24**, 96.
11. G. M. Crippen and T. F. Havel, *Acta Crystallogr., Sect. A*, 1978, **34**, 282.
12. C. E. KenKnight, *Acta Crystallogr., Sect. A*, 1984, **40**, 708.
13. F. S. Kuhl, G. M. Crippen and D. K. Friesen, *J. Comput. Chem.*, 1984, **5**, 24.
14. R. P. Sheridan, R. Nilakantan, J. S. Dixon and R. Venkataraghavan, *J. Med. Chem.*, 1986, **29**, 899.
15. C. W. Thornber, *Chem. Soc. Rev.*, 1979, **8**, 563.
16. G. M. Crippen, *J. Comput. Phys.*, 1978, **26**, 449.
17. A. K. Ghose, G. M. Crippen, G. R. Revankar, P. A. McKernan, D. F. Smee and R. K. Robins, *J. Med. Chem.*, 1989, **32**, 746.
18. W. J. Hehre, L. Radom, P. v. R. Schleyer and J. A. Pople, 'Ab initio Molecular Orbital Theory', Wiley, New York, 1986.
19. A. Ben-Naim, 'Hydrophobic Interactions', Plenum Press, New York, 1980.
20. (a) A. K. Ghose and G. M. Crippen, *J. Comput. Chem.*, 1986, **7**, 565; (b) A. K. Ghose, A. Pritchett and G. M. Crippen, *J. Comput. Chem.*, 1988, **9**, 80.
21. (a) A. K. Ghose and G. M. Crippen, *J. Chem. Inf. Comput. Sci.*, 1987, **27**, 21; (b) V. N. Viswanadhan, A. K. Ghose, G. R. Revankar and R. K. Robins, *J. Chem. Inf. Comput. Sci.*, 1989, **29**, in press.
22. A. K. Ghose and G. M. Crippen, *J. Med. Chem.*, 1985, **28**, 333.
23. J. M. Blaney, P. K. Weiner, A. Dearing, P. A. Kollman, E. C. Jorgensen, S. J. Oatley, J. M. Burridge and C. C. F. Blake, *J. Am. Chem. Soc.*, 1982, **104**, 6424.
24. J. L. Kuester and J. H. Mize, 'Optimization Techniques with Fortran', McGraw-Hill, New York, 1973, p. 106.
25. A. Ravindran, *Commun. ACM.*, 1972, **15**, 818; 1974, **17**, 157.
26. A. K. Ghose and G. M. Crippen, in 'Quantitative Approaches to Drug Design', ed. J. C. Dearden, Elsevier, Amsterdam, 1983, p. 99.
27. A. K. Ghose and G. M. Crippen, *J. Med. Chem.*, 1982, **25**, 892.
28. A. K. Ghose and G. M. Crippen, *J. Med. Chem.*, 1983, **26**, 996.
29. A. K. Ghose and G. M. Crippen, *J. Med. Chem.*, 1984, **27**, 901.
30. M. R. Linschoten, T. Bultsma, A. P. Ijzerman and H. Timmerman, *J. Med. Chem.*, 1986, **29**, 278.
31. G. M. Crippen, *Quant. Struct.–Act. Relat.*, 1983, **2**, 95.
32. A. J. Hopfinger, *J. Med. Chem.*, 1985, **28**, 1133.

Subject Index